EXPERIMENTAL DESIGN: PROCEDURES FOR THE BEHAVIORAL SCIENCES

SECOND EDITION

EXPERIMENTAL DESIGN: PROCEDURES FOR THE BEHAVIORAL SCIENCES

SECOND EDITION

ROGER E. KIRK BAYLOR UNIVERSITY

 BROOKS/COLE PUBLISHING COMPANY

Brooks/Cole Publishing Company
A Division of Wadsworth, Inc.

Printed in the United States of America

10 9 8 7 6 5 4 3   ———   93 92 91 90 89 88 87 86 85 84

Library of Congress Cataloging in Publication Data

Kirk, Roger E.
    Experimental design.

    Includes bibliographical references and
index.
    1. Experimental design.   2. Psychometrics.
3. Psychological research.   4. Educational
research.   I. Title.   [DNLM: 1. Research
design.   2. Behavioral sciences—Methods.
BF 76.5 K59e]
BF39.K55   1982   300'.724   82-9532
ISBN 0-8185-0286-X   AACR2

Subject Editor: C. Deborah Laughton
Manuscript Editor: Margaret E. Hill
Production Coordinator: Joan Marsh
Interior Design: Deborah S. Schneider
Cover Design: Katherine Minerva
Illustrations: Julia Gecha
Typesetting: Interactive Composition Corporation
Production: Unicorn Production Services, Inc.

PREFACE

The second edition of *Experimental Design*, like the first, was written to serve both as a text and as a reference book for students and researchers in the behavioral sciences and education. In-depth coverage is provided for recent developments in statistics, such as new multiple comparison procedures, the circularity assumptions associated with block designs, the partition of interactions into interpretable contrast-contrast interactions, and the analysis of factorial designs with unequal sample sizes and missing observations. The most significant change in the second edition is the inclusion of two kinds of computational algorithms for sums of squares: the traditional sum of squares approach and the general linear model approach. The latter approach is discussed in Chapter 5 and in optional starred sections throughout the remainder of the book. This material can be omitted without loss of continuity by those who are interested only in the traditional approach to analysis of variance.

Review exercises that give readers an opportunity to test their understanding of key concepts and procedures are provided at the end of each chapter. Answers to the starred exercises are contained in Appendix F. Most of the research examples were taken from behavioral science and education journals; however, the data sets have been reduced to a manageable size.

The organization of the second edition parallels that of the first edition. A minor change is the division of Chapter 7 into three chapters. New to the second edition are Chapters 5 and 10 entitled, respectively, General Linear Model Approach to ANOVA and Hierarchical Designs. The extensive treatment of hierarchical designs reflects their increasing importance in educational research.

It is assumed that the reader has had a course in algebra and basic statistical inference. A review of statistical inference is given in Chapter 1 for those whose skills are rusty. Appendix D reviews the elementary matrix algebra that is used in the optional starred sections on the general linear model. The text provides material for a two-semester graduate course. Chapters 1 through 9 provide sufficient material for a one-semester graduate course on experimental design.

It is a pleasure to express my appreciation to James Carlson, University of Ottawa, Raymond Collier, University of Minnesota, D. Gene Davenport, St. Louis University, Silas Halperin, Syracuse University, Richard J. Harris, University of New Mexico, John D. Hundleby, University of Guelph, Donald L. Meyer, University of Pittsburgh, Robert Newman, University of Southern California, B. Kent Parker, West Virginia University, Robert E. Prytula, Middle Tennessee State University, Larry Toothaker, University of Oklahoma, Allan Wolach, Illinois Institute of Technology, and Leroy Wolins, Iowa State University, for their revision suggestions or for their thoughtful comments on the revised edition. The production staff at Brooks/Cole did an outstanding job.

I am indebted to the Literary Executor of the late Sir Ronald A. Fisher, F.R.S., to Frank Yates, F.R.S., and to Oliver and Boyd Ltd., Edinburgh, for permission to reprint Tables E.2, E.4, E.6, and E.12 from their book *Statistical Tables for Biological, Agricultural and Medical Research*.

I am also indebted to E. S. Pearson and H. O. Hartley, editors of *Biometrika Tables for Statisticians*, Vol. 1, and to the *Biometrika* trustees for permission to reprint Tables E.5, E.7, E.10, E.14, and E.19; to D. B. Duncan and the editor of *Biometrics* for permission to reprint Table E.8; to C. W. Dunnett and the editor of the *Journal of the American Statistical Association* for permission to reprint Table E.9 (one-tailed values); to C. W. Dunnett and the editor of *Biometrics* for permission to reprint Table E.9 (two-tailed values); to McGraw-Hill Book Co. for permission to reprint Table E.11, to T. L. Bratcher and the editor of the *Journal of Quality Technology* for permission to reprint Table E.15; to O. J. Dunn and the editor of the *Journal of the American Statistical Association* for permission to reprint Table E.16; to P. A. Games and the editor of the *Journal of the American Statistical Association* for permission to reprint Table E.17; and to M. R. Stoline and the editor of the *Journal of the American Statistical Association* for permission to reprint Table E.18.

Portions of this book were written while on several sabbaticals provided by Baylor University. I am grateful to the administration of Baylor University and in particular to Abner V. McCall, Herbert H. Reynolds, John S. Belew, and William G. Toland for providing an environment that encourages creative and scholarly activity.

Roger E. Kirk

CONTENTS

3   MULTIPLE COMPARISON TESTS

4   COMPLETELY RANDOMIZED DESIGN

5   GENERAL LINEAR MODEL APPROACH TO ANOVA

9 COMPLETELY RANDOMIZED FACTORIAL DESIGN WITH THREE OR MORE TREATMENTS AND RANDOMIZED BLOCK FACTORIAL DESIGN

10 HIERARCHICAL DESIGNS

11· SPLIT-PLOT FACTORIAL DESIGN: DESIGN WITH GROUP-TREATMENT CONFOUNDING

12 CONFOUNDED FACTORIAL DESIGNS: DESIGNS WITH GROUP-INTERACTION CONFOUNDING

13 FRACTIONAL FACTORIAL DESIGNS: DESIGNS WITH TREATMENT-INTERACTION CONFOUNDING

14 ANALYSIS OF COVARIANCE

1 INTRODUCTION TO BASIC CONCEPTS IN EXPERIMENTAL DESIGN

1.1   INTRODUCTION

The term *experimental design* refers to a plan for assigning experimental conditions to subjects and the statistical analysis associated with the plan. The design of an experiment to investigate a scientific or research hypothesis involves a number of interrelated activities:

1. Formulation of statistical hypotheses that are germane to the scientific hypothesis. A statistical hypothesis is a statement about one or more parameters of a population. Statistical hypotheses are rarely identical to scientific hypotheses, but are testable formulations of scientific hypotheses.

2. Determination of the experimental conditions (independent variables) to be employed and the extraneous conditions (nuisance variables) that must be controlled.

3. Specification of the number of experimental units (subjects) required and the

1

population from which they are to be sampled.*

4.   Specification of the procedure for assigning the experimental conditions to the subjects.

5.   Determination of the measurement to be recorded (dependent variable) for each subject and the statistical analyses that will be performed.

In short, an experimental design indicates the way in which an experiment is to be performed.

SUBJECT MATTER AND GENERAL ORGANIZATION OF THIS BOOK

The concepts and procedures involved in carrying out the five preceding interrelated activities constitute the subject matter of this book. But experimental design is only one of the many facets of scientific research. The most important one—identifying relevant research questions—is outside the scope of this book. The reader should remember the truism that a carefully conceived and executed design is of no value if the scientific hypothesis that led to the experiment is without merit.

Chapters 1 through 3 provide an overview of experimental designs, a review of basic statistical concepts, and a presentation of statistical tools that are used throughout the remainder of the book. A detailed examination of the designs begins in Chapter 4. This examination includes a description of the design, conditions, and assumptions under which the design is appropriate; computational examples; and advantages and disadvantages of the design.

Two kinds of computational algorithms are provided for each design. The first, referred to as the *traditional sums of squares approach*, uses scalar algebra and is suitable for pocket calculators. The second, called the *general linear model approach*, uses matrix algebra and requires a computer or calculator. The latter approach is new to this edition of *Experimental Design* and is discussed in optional (starred) sections. This approach is particularly useful with complex designs having unequal sample sizes.

---

*   An experimental unit is that entity that is allocated an experimental condition independently of other such entities. An experimental unit may contain several observational units; for example, in an educational experiment the experimental unit is often the classroom, but the individual students are the observational units. The terms experimental unit and subject will be used interchangeably.

1.2   FORMULATION OF PLANS FOR COLLECTION AND ANALYSIS OF DATA

ACCEPTABLE RESEARCH HYPOTHESES

Some questions cannot currently be subjected to scientific investigation. For example, the questions "Can three or more angels sit on the head of a pin?" and "Does life exist in more than one galaxy in the universe?" cannot be answered because no procedures now exist for observing either angels or other galaxies. Scientists confine their research hypotheses to questions for which procedures can be devised that offer the possibility of arriving at an answer. This does not mean that the question concerning the existence of life on other galaxies can never be investigated. Indeed, with continuing advances in space science, it is likely that this question eventually will be answered.

An *experiment* involves the manipulation of one or more variables by an experimenter in order to determine the effect of this manipulation on another variable. Questions that provide the impetus for experimental research should be statable in the logical form of the general implication. That is, a question should be reducible to the form, *if A, then B*. For example, *if* albino rats are subjected to microwave radiation, *then* their food consumption will decrease. This research hypothesis can be investigated because procedures are available both for manipulating radiation level and for measuring food consumption of rats.

Much research departs from this pattern in that nature rather than the experimenter manipulates the variables. It would be unethical, for example, to study the effects of prenatal malnutrition on IQ by deliberately providing pregnant women with inadequate diets. Instead, the question is investigated by locating children whose mothers were malnourished during pregnancy and comparing their IQs with those of children whose mothers were not malnourished. Such research is referred to as *observational studies, correlational studies*, or *investigations*. Observational studies pose special problems for the experimenter who wants to make causal inferences.*

DISTINCTION BETWEEN DEPENDENT AND INDEPENDENT VARIABLES

In the radiation example cited earlier, the presence or absence of radiation is designated as the *independent variable*—the variable that is under the control of the experimenter. The terms independent variable and treatment will be used interchangeably. The *dependent variable* is the amount of food consumed by the rats. The dependent variable reflects any effects associated with manipulation of the independent variable.

---

*   These problems are examined in detail by Cook and Campbell (1979).

SELECTION OF DEPENDENT VARIABLE

The choice of an appropriate dependent variable may be based on theoretical considerations, although in many investigations the choice is determined by practical considerations. In this example, other dependent variables that could be measured include the following:

1. Activity level of rat in an activity cage
2. Body temperature of rat
3. Emotionality of rat as evidenced by amount of defecation and urination
4. Problem-solving ability
5. Weight of rat in grams
6. Speed of running in a straight-alley maze
7. Visual discrimination capacity
8. Frequency of mating behavior.

Several independent variables can be employed in an experiment, but the designs described in this book are limited to the assessment of one dependent variable at a time. If it is necessary to evaluate two or more dependent variables simultaneously, a multivariate analysis of variance design can be used.* The selection of the most fruitful variables to measure should be determined by a consideration of the sensitivity, reliability, distribution, and practicality of the possible dependent variables. From previous experience, an experimenter may know that one dependent variable is more sensitive than another to the effects of a treatment or that one dependent variable is more reliable, that is, gives more consistent results, than another variable. Because behavioral research generally involves a sizable investment in time and material resources, the dependent variable should be reliable and maximally sensitive to the phenomenon under investigation. Choosing a dependent variable that possesses these two characteristics may minimize the amount of time and effort required to investigate a research hypothesis.

Another important consideration in selecting a dependent variable is whether the observations within each treatment level (or combination of treatment levels in the case of multitreatment experiments) would be approximately normally distributed. The assumption of normality, discussed in Chapter 2, is required for the experimental designs described in Chapters 4 through 14. In some cases it may be possible to *transform* nonnormally distributed observations so that the associated population distributions are normal. This procedure is described in Chapter 2. If theoretical considerations do not dictate the selection of a dependent variable and if several alternative

---

*   For a discussion of these designs, see Bock (1975), Cooley and Lohnes (1971), Finn (1974), Harris (1975), Tatsuoka (1971), and Timm (1975).

variables are equally sensitive and reliable, in addition to being normally distributed, an experimenter should select the variable that is most easily measured.

SELECTION OF INDEPENDENT VARIABLE

In the radiation example cited earlier, the independent variable was the presence or absence of radiation. Such a treatment is said to have two treatment levels. If the experimenter is interested in the nature of the relationship between radiation dosages and food consumption, three or more levels of radiation can be employed. The levels could consist of 0 microwatts, 20,000 microwatts, 40,000 microwatts, and 60,000 microwatts of radiation. This particular treatment is an example of a *quantitative* independent variable in which different treatment levels constitute different amounts of the independent variable.

In general, when the independent variable is quantitative in character, there is little interest in the exact values of the treatment levels used in the experiment. In the radiation example, the research hypothesis could also be investigated using three other levels of radiation, say, 25,000, 50,000, and 75,000 microwatts in addition to the zero-microwatt control level. The treatment levels should be chosen so as to cover a sufficiently wide range to detect effects of the independent variable if real effects exist. In addition, the number and spacing of the levels should be sufficient to define the shape of the function relating the independent and dependent variables. This is necessary if an experimenter is interested in performing a trend analysis as described in Chapter 4.

Selection of appropriate levels of the independent variable may be based on results of previous experiments or on theoretical considerations. In some research areas, it may be helpful to carry out a small pilot experiment to select treatment levels prior to the main experiment.

Under the conditions described in Chapters 2 and 4, the levels of a quantitative independent variable may be selected randomly from a population of treatment levels. If this procedure is followed, an experimenter can extrapolate from the results of the experiment to treatment levels that are not included in the experiment. If the treatment levels are not randomly sampled, the results of an experiment are applicable only to the specific levels included in the experiment.

Often a different type of independent variable is used. For example, if the treatment levels consisted of unmodulated radiation, amplitude-modulated radiation, and pulse-modulated radiation, the treatment is designated as a *qualitative* independent variable. The different treatment levels represent different *kinds* rather than different *amounts* of the independent variable. The distinction between quantitative and qualitative treatments is important in connection with trend analysis. The particular levels of a qualitative independent variable employed in an experiment are generally of specific interest to an experimenter. And the levels chosen are usually dictated by the research hypothesis.

CONTROL OF NUISANCE VARIABLES

In addition to independent and dependent variables, all experiments include one or more *nuisance variables*. Nuisance variables are undesired sources of variation in an experiment that may affect the dependent variable. As the name implies, the effects of nuisance variables are of no interest per se. In the radiation example, potential nuisance variables include sex of the rats, variation in weight of the rats prior to the experiment, presence of infectious diseases in one or more cages where the rats are housed, temperature variation among the cages, and differences in previous feeding experiences of the rats. Unless controlled, nuisance variables can bias the outcome of an experiment. For example, if rats in the radiated groups suffer from some undetected disease, differences among the groups would reflect the effects of the disease in addition to radiation effects—if the latter effects exist.

Four approaches can be used to control nuisance variables. One approach is to hold the nuisance variable constant for all subjects. For example, use only male rats of the same weight. Although an experimenter may attempt to hold all nuisance variables constant, the probability is high that some variable will escape attention. A second approach, one that is used in conjunction with the first, is to assign subjects randomly to the experimental conditions. Then known as well as unsuspected sources of variation or bias are distributed over the entire experiment and thus do not affect just one or a limited number of treatment levels. In this case, an experimenter increases the magnitude of random variation among observations in order to minimize systematic bias effects, that is, the effects of nuisance variables that bias all observations in one or more treatment levels in the same manner. Random variation can be taken into account in evaluating the outcome of an experiment, whereas it is difficult or impossible to account for systematic nuisance effects. A third approach to controlling nuisance variables is to include the variable as one of the factors in the experimental design. This approach is illustrated in Section 1.3.

The three preceding approaches for controlling nuisance variables illustrate the application of *experimental control* as opposed to the fourth approach which is *statistical control*. In some experiments it may be possible—through the use of regression procedures (see Chapter 14)—to remove the effects of a nuisance variable statistically. This use of statistical control is referred to as the *analysis of covariance*.

CLASSIFICATION OF INDEPENDENT AND NUISANCE VARIABLES

All independent and nuisance variables in behavioral research can be classified in one of three general categories: organismic, environmental, and task variables. In the radiation example, the independent variable of radiation can be classified as an environmental variable. The nuisance variables listed earlier as sex, weight, prior experience, and infectious diseases are examples of organismic variables. The other nuisance variable of temperature variation among the cages is an example of an environmental variable. This radiation experiment does not include a task variable. A task variable could be introduced into the experiment by requiring the rats to perform easy, medium, and difficult visual discriminations before gaining access to food. The effect of the visual discrimination on food consumption represents an additional independent variable that can be classified as a task variable. In the design of experiments, this classification scheme may help an experimenter determine nuisance variables that should be controlled.

EFFICIENCY AND EXPERIMENTAL DESIGN

A goal of research is to draw valid conclusions about the effects of an independent variable and to do this as efficiently as possible. Generally several experimental designs can be used in testing a statistical hypothesis. However, alternative designs that are equally valid for testing a hypothesis are rarely equally efficient. The efficiency of alternative research procedures can be defined in a number of ways. For example, efficiency can be defined in terms of time required to collect data, cost of data collection, ratio of information obtained to cost, and so on. A commonly used index for assessing the relative efficiency of two experimental designs is the ratio of their experimental error variances. *Experimental error variance* is that portion of the variance in the dependent variable that is attributed to extraneous sources. The main sources of experimental error are inherent variability in the behavior of subjects and lack of uniformity in the conduct of the experiment.

The relative efficiency index just described does not take into account the cost of data collection. An index that, while difficult to estimate, does provide a more complete assessment of the relative efficiency of two designs is

$$\text{Relative efficiency} = \frac{\left(\dfrac{n_2 C_2}{\hat{\sigma}_1^2}\right)\left(\dfrac{df_1 + 1}{df_1 + 3}\right)}{\left(\dfrac{n_1 C_1}{\hat{\sigma}_2^2}\right)\left(\dfrac{df_2 + 1}{df_2 + 3}\right)},$$

where $\hat{\sigma}^2$ is the estimate of error variance per observation, n is the number of subjects in each treatment level, C is the cost of collecting data per subject, df is the error variance degrees of freedom, and the subscripts 1 and 2 designate the two experimental designs (Federer, 1955, 13). If the ratio is less than one, the second design is more efficient than the first. The converse is true if the ratio is greater than one. The formula calls attention to four factors that affect the efficiency of experimental designs. Unfortunately an experimental design that is advantageous with respect to one factor may not be advantageous with respect to the others. For example, if a design has the desirable attribute of a small error variance, it may have a high cost per subject or a small number of degrees of freedom for error variance, or it may require a large number of subjects. The problem facing an investigator is to select an experimental design that represents the best compromise obtainable within the constraints of the research situation.

DETERMINATION OF SAMPLE SIZE

Replication, the observation of two or more experimental units under identical experimental conditions, is a basic technique in experimental design. As we will see, it accomplishes two important objectives: it enables an experimenter to obtain an estimate of experimental error or error variation and it permits a more precise estimate of treatment effects.

Unfortunately, specifying the number of subjects required for an experiment is often one of the more puzzling problems in experimental design. This need not be so. Five factors must be considered in choosing a sample size:

1.   Minimum treatment effects one is interested in detecting

2.   Number of treatment levels

3.   Population error variance

4.   Probability of making a type I error

5.   Power that is desired, where power is defined as the probability of rejecting a false null hypothesis.*

In general, the third factor—population error variance—is unknown. It may be possible to make a reasonable estimate of the population error variance on the basis of previous experiments or a pilot study, in which case procedures in Section 4.3 can be used to determine the sample size. If it is not possible to estimate the population error variance, the sample size can still be estimated by expressing the minimum treatment effects of interest as a multiple of the unknown error variance. This sounds a bit complicated, but as we will see in Section 4.3 it is really quite simple.

The required sample size should be routinely estimated before beginning an experiment. If these preliminary calculations indicate that the power of the experimental design is inadequate, the experimenter may choose not to conduct the experiment or may modify it so as to increase its power. The two most common procedures for increasing power are to increase the size of the sample and to employ an experimental design that provides a more precise estimate of treatment effects and smaller error effects. The latter alternative is explored in the following overview of types of experimental designs.

1.3   OVERVIEW OF TYPES OF EXPERIMENTAL DESIGNS

An almost bewildering array of experimental designs is available. It is not surprising then that some experimenters approach the selection of an appropriate design with misgivings. The problem is magnified because there is no universally accepted nomen-

---

*   Section 1.6, Review of Statistical Inference, provides an elementary discussion of these factors.

clature for analysis of variance, ANOVA, designs (some of them having as many as five different names) and most nomenclatures were not developed to explicate the structural similarities among different designs. It turns out, fortunately, that all complex designs can be constructed from and understood in terms of three relatively simple ANOVA designs, called *building block* designs.* These designs are the completely randomized design (CR), randomized block design (RB), and Latin square design (LS). They provide the organizational structure for the design nomenclature and classification system that is outlined in Table 1.3-1.** The letter p in the table denotes the number of levels of a treatment. If a design includes a second or third treatment, the number of their levels is denoted by q and r, respectively.

The category *systematic designs* in the outline is of historical interest only. According to Leonard and Clark (1939), agricultural field research employing systematic designs on a practical scale dates back to 1834. Prior to the work of Fisher in the 1920s and 1930s, as well as that of Neyman and Pearson on the theory of statistical inference, investigators used systematic schemes rather than randomization procedures for assigning treatment levels to plots of land or other suitable experimental units—hence the designation systematic designs for these early field experiments. Impetus for this early experimental research came from a need to improve agricultural techniques. Today the nomenclature of experimental design is replete with terms from agriculture. Systematic designs in which the randomization principle is not followed do not provide a valid estimate of error variance and hence are not subject to powerful tools of statistical analysis such as analysis of variance.

Modern principles of experimental design, particularly the principle of random assignment of treatment levels to experimental units, received general acceptance as a result of Fisher's work (1922, 1923, 1935). Experimental designs using the randomization principle are called *randomized designs*. Randomized designs can be subdivided into several distinct categories based on (1) whether the experimental units are subdivided into homogeneous blocks or groups prior to assignment of treatment levels and (2) the number of restrictions placed on the random assignment of treatment levels to the experimental units. Table 1.3-1 contains three pseudocategories: factorial experiments, hierarchical experiments, and randomized designs with one or more covariates. The pseudocategories are so designated because the designs there do not represent distinct kinds of experimental designs but, instead, consist of a combination of two or more building block designs. A brief overview of some of the simpler designs follows.

COMPLETELY RANDOMIZED DESIGN

The simplest experimental design from the standpoint of assignment of treatment levels to experimental units and statistical analysis is the completely randomized design. The design is denoted by the letters CR-p, where p refers to the number of

---

*   This point is developed more fully in Kirk (1972).

**   More complete classification systems have been proposed by Cox (1943); Doxtator, Tolman, Cormany, Bush, and Jensen (1942); and Federer (1955, 6–12). Miniature nomenclatures for a limited number of ANOVA designs have been presented by Lindquist (1953) and Winer (1971).

TABLE 1.3-1   Outline of Experimental Designs Described in This Book.

Experimental Design	Abbreviated Designation
I. Systematic Designs	
II. Randomized Designs With One Treatment	
A. Experimental units not subdivided on any basis other than randomization prior to assignment of treatment levels; no restrictions on random assignment other than the option of assigning each level to the same number of subjects	
1. Completely randomized design	CR-p
B. Experimental units subdivided on some nonrandom basis and/or one or more restrictions on random assignment	
1. Randomized block design	RB-p
2. Latin square design	LS-p
3. Generalized randomized block design	GRB-p
4. Graeco-Latin square design	GLS-p
5. Hyper-Graeco-Latin square design	HGLS-p
*6. Balanced incomplete block design	BIB-p
*7. Youden square design	YBIB-p
*8. Partially balanced incomplete block design	PBIB-p
III. Randomized Designs With Two or More Treatments	
A. Factorial experiments: Designs in which all treatments are crossed	
1. Designs without confounding	
a. Completely randomized factorial design	CRF-pq
b. Randomized block factorial design	RBF-pq
2. Design with group-treatment confounding	
a. Split-plot factorial design	SPF-$p \cdot q$
3. Designs with group-interaction confounding	
a. Randomized block completely confounded factorial design	RBCF-p^k
b. Randomized block partially confounded factorial design	RBPF-p^k
c. Latin square confounded factorial design	LSCF-p^k
4. Designs with treatment-interaction confounding	
a. Completely randomized fractional factorial design	CRFF-p^k
b. Randomized block fractional factorial design	RBFF-p^k
c. Latin square fractional factorial design	LSFF-p^k
d. Graeco-Latin square fractional factorial design	GLSFF-p^k
B. Hierarchical experiments: Designs in which one or more treatments are nested	
1. Designs with complete nesting	
a. Completely randomized hierarchical design	CRH-$pq(A)$
b. Randomized block hierarchical design	RBH-$pq(A)$
2. Designs with partial nesting	
a. Completely randomized partial hierarchical design	CRPH-$pq(A)r$
b. Randomized block partial hierarchical design	RBPH-$pq(A)r$
IV. Randomized Designs With One or More Covariates	
A. A covariate can be used with all of the designs described above. When this is done, the letters AC are added to the nomenclature, as in the following examples.	
1. Completely randomized analysis of covariance design	CRAC-p
2. Randomized block analysis of covariance design	RBAC-p
3. Latin square analysis of covariance design	LSAC-p
4. Completely randomized factorial analysis of covariance design	CRFAC-pq
5. Split-plot factorial analysis of covariance design	SPFAC-$p \cdot q$

* Not included in this edition of *Experimental Design*.

treatment levels. The main features of the design will be illustrated by means of an experiment to help cigarette smokers break the habit. The independent variable is type of therapy; the dependent variable is the number of cigarettes smoked per day six months after therapy. For notational convenience, the therapies are called treatment A. The levels of treatment A corresponding to the specific therapies are denoted by the lowercase letter a and a subscript—a_1 is behavior therapy, a_2 is hypnosis, and a_3 is drug therapy. A particular but unspecified level of treatment A is designated by a_j, where j ranges over values $j = 1, \ldots, p$. In our example, p is equal to three. The number of cigarettes smoked per day six months after therapy by subject i in treatment level j is denoted by Y_{ij}. The hypothesis that we want to test is

$$H_0: \quad \mu_1 = \mu_2 = \mu_3$$

where the μ_j's denote the mean cigarette consumption for the respective populations.

Assume that 45 smokers who want to stop are available to participate in the experiment. The random numbers table in Appendix E.2 is used to randomly assign treatment levels to the subjects with the restriction that 15 subjects receive each therapy. The phrase "randomly assigned" is important. Randomization is used to distribute the idiosyncratic characteristics of the subjects over the three treatment levels so that they will not selectively bias the outcome of the experiment. To put it another way, randomization provides a basis for obtaining unbiased estimates of the treatment effects. As we will see in Chapter 2, randomization also provides a basis for obtaining an unbiased estimate of the random error variation in the experiment and it helps to ensure that the error effects are statistically independent.

Thus far we have identified the hypothesis that we want to test, $\mu_1 = \mu_2 = \mu_3$, and described the manner in which the three treatment levels are assigned to the subjects. In the following paragraphs, we will discuss the composite nature of a score, describe the experimental design model equation for a type CR-p design, and explicate the meaning of the terms treatment effect and error effect.

A block diagram of the type CR-3 design for our smoking experiment is shown in Figure 1.3-1. Consider the cigarette consumption of subject 3 in treatment level a_2. Six months after therapy this subject is smoking five cigarettes a day ($Y_{32} = 5$). What factors have affected the value of Y_{32}? One factor is the efficacy of the therapy received—hypnosis in this case. Other factors are the subject's cigarette consumption prior to therapy, level of motivation to stop smoking, and the weather the day Y_{32} was recorded, to mention only a few. In short, Y_{32} is a composite that reflects (1) the effects of treatment level a_2, (2) the effects unique to subject 3 in a_2, (3) the effects attributable to chance fluctuations in subject 3's behavior, and (4) the effects attributable to environmental conditions. Our conjectures about Y_{32} or any other score can be expressed more formally by an *experimental design model* equation. For this completely randomized design, we assume that each of the 45 scores is the sum of three terms in the model equation

(1.3-1) $$Y_{ij} = \mu + \alpha_j + \epsilon_{i(j)} \qquad (i = 1, \ldots, n; j = 1, \ldots, p)$$

where μ = the grand population mean (μ is a constant for the 45 scores and reflects their general elevation)

α_j = the treatment effect for level j and is equal to $\mu_j - \mu$, the deviation of the grand mean from the jth population mean (α_j is a constant for the 15 scores in population j and reflects the elevation or depression of these scores resulting from using therapy j)

$\epsilon_{i(j)}$ = the error effect associated with Y_{ij} and is equal to $Y_{ij} - \mu - \alpha_j$.

The meaning of the term *treatment effect* seems fairly clear, but that for *error effect* remains elusive. Consider Figure 1.3-1 again. There is considerable variation among the Y_{ij}'s of subjects who receive the same type of therapy. This variation among the Y_{ij}'s must be due to differences among the error effects, $\epsilon_{i(j)}$, since the other parameters in the model equation $Y_{ij} = \mu + \alpha_j + \epsilon_{i(j)}$ are constants for subjects receiving the same therapy. What are the sources of this variation? We listed some of the sources earlier: effects unique to a subject, chance fluctuations in a subject's behavior, and effects attributable to extraneous environmental conditions—in short, all effects not due to a treatment. These then constitute error effect. To put it another way, error effect in this experimental design model equation is that portion of a score that remains after the treatment effect and grand mean have been subtracted from it. An experimenter attempts, by using an appropriate design and experimental controls, to minimize the size of error effects. Designs described in subsequent sections permit an experimenter to accomplish this by isolating sources of variation that would ordinarily be included in the error effect.

RANDOMIZED BLOCK DESIGN

In behavioral and educational research, the experimental units are often people whose aptitudes and experiences differ markedly. Individual differences are inevitable, but it is often possible to isolate or partition out a portion of these effects so that they do not appear in estimates of treatment and error effects. One ANOVA design for accomplishing this is a randomized block design, denoted by the letters RB-p.

A type RB-p design isolates the effects of a nuisance variable by means of a

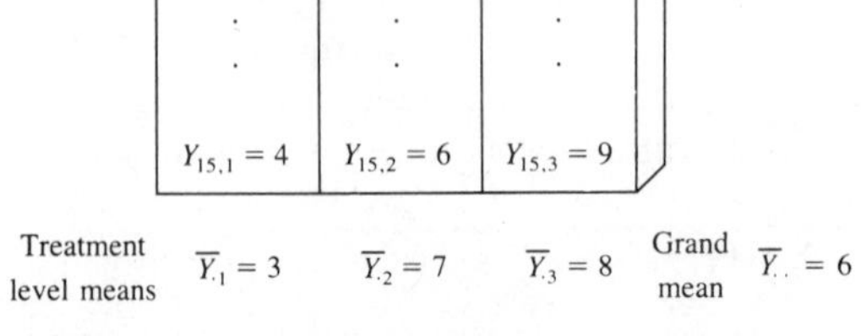

FIGURE 1.3-1 Layout for type CR-3 design with 45 subjects. The cigarette consumption six months after therapy for subject i in treatment level j is denoted by Y_{ij}. The mean consumption of the subjects who received treatment level j is denoted by $\overline{Y}_{.j}$; the mean consumption for all three treatment levels (grand mean) is denoted by $\overline{Y}_{..}$.

	a_1	a_2	a_3
	$Y_{11} = 2$	$Y_{12} = 8$	$Y_{13} = 3$
	$Y_{21} = 0$	$Y_{22} = 10$	$Y_{23} = 9$
	$Y_{31} = 3$	$Y_{32} = 5$	$Y_{33} = 8$
	.	.	.
	.	.	.
	.	.	.
	$Y_{15,1} = 4$	$Y_{15,2} = 6$	$Y_{15,3} = 9$

| Treatment level means | $\overline{Y}_{.1} = 3$ | $\overline{Y}_{.2} = 7$ | $\overline{Y}_{.3} = 8$ | Grand mean $\overline{Y}_{..} = 6$ |

blocking procedure whereby subjects who are relatively homogeneous with respect to the nuisance variable are assigned to the same block. Let's reconsider the smoking experiment cited earlier. It is reasonable to assume that difficulty in breaking the smoking habit is related to the number of cigarettes that a person smokes. The design of the experiment can be improved by isolating this variable. Suppose that instead of assigning the treatment levels randomly to the 45 subjects, we formed 15 blocks each containing three subjects whose cigarette consumption was similar prior to the experiment. The subjects in block one each smoked less than a half pack a day; the subjects in block two, one-half to three-fourths pack a day; and so on. After all of the blocks have been formed, the three types of therapy are assigned randomly to the subjects within each block. A block diagram of this type RB-3 design is shown in Figure 1.3-2. With this design we can test two hypotheses:

H_0: $\mu_{\cdot 1} = \mu_{\cdot 2} = \mu_{\cdot 3}$ (Treatment level population means are equal.)

H_0: $\mu_{1 \cdot} = \mu_{2 \cdot} = \cdots = \mu_{15 \cdot}$. (Block population means are equal.)

The experimental design model equation for this type RB-3 design is

$$Y_{ij} = \mu + \alpha_j + \pi_i + \epsilon_{ij} \qquad (i = 1, \ldots, n; j = 1, \ldots, p)$$

where μ = the grand population mean

α_j = the treatment effect for level j and is equal to $\mu_{\cdot j} - \mu$, the deviation of the grand mean from the jth population mean

π_i = the block effect for block i and is equal to $\mu_{i \cdot} - \mu$, the deviation of the grand mean from the ith population block mean (π_i reflects the elevation or depression of Y_{ij} due to smoking a certain number of cigarettes per day prior to therapy)

ϵ_{ij} = the error effect associated with Y_{ij} and is equal to $Y_{ij} - \mu - \alpha_j - \pi_i$.*

The error effect for this experimental design model equation is that portion of a score that remains after the treatment effect, *block effect*, and grand mean have been subtracted from it. The sum of the squared error effects for this design $\Sigma\Sigma \, \epsilon_{ij}^2 = \Sigma\Sigma \, (Y_{ij} - \mu - \alpha_j - \pi_i)^2$ will be smaller than the sum for the completely randomized design $\Sigma\Sigma \, \epsilon_{ij}^2 = \Sigma\Sigma(Y_{ij} - \mu - \alpha_j)^2$ if π_i is greater than zero. Blocking with respect to the nuisance variable, number of cigarettes smoked per day, has enabled us to isolate this variable and remove it from the error effect. The result is a more efficient and powerful experimental design.

LATIN SQUARE DESIGN

The last building block design to be described is a Latin square design. The name derives from an ancient puzzle that is concerned with the number of different ways

---

* For now we will ignore the possibility that cigarette consumption interacts with type of therapy. Section 6.3 describes an experimental design model equation that includes a block-treatment interaction term $(\pi\alpha)_{ij}$.

Latin letters can be arranged in a square matrix so that each letter appears once and only once in each row and column. The design, designated by the letters LS-p, enables an experimenter to isolate the effects of *two* nuisance variables. The levels of one nuisance variable are assigned to the rows of the square; the levels of the other are assigned to columns. The levels of the treatment are assigned to the cells of the square.

Let's return again to the cigarette smoking experiment. With a Latin square design we can isolate the effects of cigarette consumption and the effects of a second nuisance variable, say, length of time that a person has smoked. This advantage comes at a price. The randomization procedure for a Latin square design is more complex than that for a randomized block design. Also, the number of rows and columns of a Latin square must each equal the number of treatment levels—which is three in our example. We can assign three levels of cigarette consumption to the rows of the square and three levels of the other nuisance variable, duration of the smoking habit, to the columns. A block diagram of the design is shown in Figure 1.3-3. This design lets us test three hypotheses:

H_0: $\mu_{1..} = \mu_{2..} = \mu_{3..}$ (Treatment level population means are equal.)

H_0: $\mu_{.1.} = \mu_{.2.} = \mu_{.3.}$ (Row population means are equal.)

H_0: $\mu_{..1} = \mu_{..2} = \mu_{..3}$ (Column population means are equal.)

The experimental design model equation for this type LS-3 design is

$$Y_{ijkl} = \mu + \alpha_j + \beta_k + \gamma_l + \epsilon_{pooled} \qquad (i = 1, \ldots, n; j = 1, \ldots, p;$$
$$k = 1, \ldots, p; l = 1, \ldots, p)$$

FIGURE 1.3-2   Layout for a type RB-3 design. Subjects who are relatively homogeneous with respect to cigarette consumption prior to the experiment are assigned to the same block. The cigarette consumption six months after therapy for a subject in block i and treatment level j is denoted by Y_{ij}. The mean consumption of subjects in treatment level j is denoted by $\overline{Y}_{.j}$; that for subjects in block i is denoted by $\overline{Y}_{i.}$.

	a_1	a_2	a_3	Block means
Block 1	$Y_{11} = 0$	$Y_{12} = 0$	$Y_{13} = 3$	$\overline{Y}_{1.} = 1$
Block 2	$Y_{21} = 2$	$Y_{22} = 1$	$Y_{23} = 3$	$\overline{Y}_{2.} = 2$
Block 3	$Y_{31} = 3$	$Y_{32} = 5$	$Y_{33} = 4$	$\overline{Y}_{3.} = 4$
.	.	.	.	.
.	.	.	.	.
.	.	.	.	.
Block 15	$Y_{15.1} = 8$	$Y_{15.2} = 14$	$Y_{15.3} = 14$	$\overline{Y}_{15.} = 12$
Treatment level means	$\overline{Y}_{.1} = 3$	$\overline{Y}_{.2} = 7$	$\overline{Y}_{.3} = 8$	Grand mean $\overline{Y}_{..} = 6$

where μ = the grand population mean

α_j = the treatment effect for level j and is equal to $\mu_{j..} - \mu$

β_k = the row effect for level k and is equal to $\mu_{.k.} - \mu$

γ_l = the column effect for level l and is equal to $\mu_{..l} - \mu$

ϵ_{pooled} = a pooled error effect. (The nature of this error effect is discussed in Chapter 7.)

The sum of the squared error effects for this design, $\Sigma\Sigma\epsilon^2_{pooled} = \Sigma\Sigma(Y_{ijkl} - \mu - \alpha_j - \beta_k - \gamma_l)^2$, will ordinarily be smaller than the sum for the randomized block design, $\Sigma\Sigma\epsilon^2_{ij} = \Sigma\Sigma(Y_{ij} - \mu - \alpha_j - \pi_i)^2$. Thus, by isolating the effects of two nuisance variables, we have obtained a more efficient and powerful experimental design. The discussion in this section has focused on the relative efficiency of the three building block designs. In subsequent chapters we will see that increased efficiency is usually accompanied by more restrictive statistical assumptions and greater complexity in the randomization and analysis of data.

FIGURE 1.3-3   Layout for a type LS-3 design. The cigarette consumption six months after therapy for subject i in treatment level a_j, row b_k, and column c_l is denoted by Y_{ijkl}. Five subjects are assigned to each cell of the 3×3 Latin square.

	c_1 < 1 year	c_2 1–3 years	c_3 > 3 years	Row means
b_1 < 1 pack/day	a_1 $Y_{1111} = 0$ $Y_{2111} = 2$ · · · · · $Y_{5111} = 1$	a_2 $Y_{1212} = 1$ $Y_{2212} = 3$ · · · · · $Y_{5212} = 0$	a_3 $Y_{1313} = 1$ $Y_{2313} = 3$ · · · · · $Y_{5313} = 2$	$\overline{Y}_{.1.} = 2$
b_2 1–3 packs/day	a_2 $Y_{1221} = 1$ $Y_{2221} = 3$ · · · · · $Y_{5221} = 6$	a_3 $Y_{1322} = 3$ $Y_{2322} = 0$ · · · · · $Y_{5322} = 9$	a_1 $Y_{1123} = 4$ $Y_{2123} = 9$ · · · · · $Y_{5123} = 10$	$\overline{Y}_{.2.} = 7$
b_3 > 3 packs/day	a_3 $Y_{1331} = 0$ $Y_{2331} = 5$ · · · · · $Y_{5331} = 5$	a_1 $Y_{1132} = 4$ $Y_{2132} = 8$ · · · · · $Y_{5132} = 6$	a_2 $Y_{1233} = 12$ $Y_{2233} = 6$ · · · · · $Y_{5233} = 9$	$\overline{Y}_{.3.} = 9$
Column means	$\overline{Y}_{..1} = 4$	$\overline{Y}_{..2} = 6$	$\overline{Y}_{..3} = 8$	Grand mean $\overline{Y}_{...} = 6$
Treatment level means	a_1 $\overline{Y}_{1..} = 3$	a_2 $\overline{Y}_{2..} = 7$	a_3 $\overline{Y}_{3..} = 8$	

The completely randomized, randomized block, and Latin square designs provide the organizational framework for the classification system and nomenclature used throughout this book. All of the ANOVA designs to be described represent an extension of one or a combination of these designs. The next section describes a factorial design that is constructed from two completely randomized designs.

COMPLETELY RANDOMIZED FACTORIAL DESIGN

The simplest factorial design, from the standpoint of data analysis and assignment of treatment levels to subjects, is based on a completely randomized building block design and hence is called a completely randomized factorial design. Factorial designs differ from those covered previously in that two or more treatments can be evaluated simultaneously.* A two-treatment completely randomized factorial design is denoted by the letters CRF-pq, where p and q stand for the number of levels of treatments A and B, respectively.

Consider an experiment to evaluate the effects of two treatments A and B on speed of reading. Suppose that treatment A consists of two levels of room illumination: a_1 is 5 foot-candles and a_2 is 30 foot-candles. Treatment B consists of three levels of type size: b_1 is 6-point type, b_2 is 12-point type, and b_3 is 18-point type. The designation for this design is a type CRF-23 design, where 2 refers to the two levels of treatment A and 3 refers to the three levels of treatment B. The design contains six treatment combinations: a_1b_1, a_1b_2, . . . , a_2b_3. Assume that 30 sixth-graders are available to participate in the experiment. The six treatment combinations are randomly assigned to the children with the restriction that five receive each combination. A block diagram of this type CRF-23 design is shown in Figure 1.3-4.

FIGURE 1.3-4 Layout for a type CRF-23 design with 30 subjects. The reading speed of subject i in treatment combination jk is denoted by Y_{ijk}.

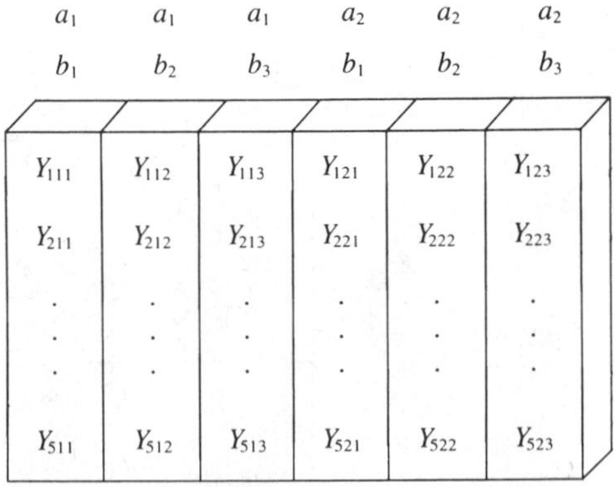

a_1 b_1	a_1 b_2	a_1 b_3	a_2 b_1	a_2 b_2	a_2 b_3
Y_{111}	Y_{112}	Y_{113}	Y_{121}	Y_{122}	Y_{123}
Y_{211}	Y_{212}	Y_{213}	Y_{221}	Y_{222}	Y_{223}
.	.	.	.	.	.
.	.	.	.	.	.
.	.	.	.	.	.
Y_{511}	Y_{512}	Y_{513}	Y_{521}	Y_{522}	Y_{523}

---

*   The distinction between a treatment and a nuisance variable that is included as one of the factors in, say, a randomized block design is in the mind of the experimenter. A nuisance variable is included in a design for the purpose of improving the efficiency and power of the design; a treatment is included because it is related to the scientific hypothesis that an experimenter wants to test. This distinction has important implications for the statistical analysis, as we will see.

Each of the 30 scores in Figure 1.3-4 is assumed to be the sum of five terms in the model equation

$$Y_{ijk} = \mu + \alpha_j + \beta_k + (\alpha\beta)_{jk} + \epsilon_{i(jk)} \qquad (i = 1, \ldots, n; j = 1, \ldots, p; \\ k = 1, \ldots, q)$$

where
μ = the grand mean

α_j = the treatment effect for level a_j and is equal to $\mu_{j\cdot} - \mu$

β_k = the treatment effect for level b_k and is equal to $\mu_{\cdot k} - \mu$

$(\alpha\beta)_{jk}$ = the interaction effect for levels a_j and b_k; it represents the joint effects of the two levels

$\epsilon_{i(jk)}$ = the error effect for subject i in treatment combination jk and is equal to $Y_{ijk} - \mu - \alpha_j - \beta_k - (\alpha\beta)_{jk}$.

With this design we can test three hypotheses:

H_0: $\mu_{1\cdot} = \mu_{2\cdot}$ (Treatment A population means are equal.)

H_0: $\mu_{\cdot 1} = \mu_{\cdot 2} = \mu_{\cdot 3}$ (Treatment B population means are equal.)

H_0: $\mu_{jk} - \mu_{j'k} - \mu_{jk'} + \mu_{j'k'} = 0$ for all j, j', k, and k', $j \neq j'$ and $k \neq k'$. (All AB interaction effects equal zero.)*

The last hypothesis is unique to factorial designs. It states that the joint effect (interaction) of treatments A and B is equal to zero for all combinations of the two treatments. Two treatments are said to interact if differences in performances under the levels of one treatment are different at two or more levels of the other treatment. Figure 1.3-5 illustrates two possible outcomes of our reading experiment: part (a) illustrates the presence of an interaction, part (b) illustrates the absence of an interaction. We will have more to say about interactions in Chapter 8.

Earlier we observed that a completely randomized design is the building block for a type CRF-pq design. The type CRF-23 design diagrammed in Figure 1.3-4, for

FIGURE 1.3-5   Two possible outcomes of the reading experiment. Part (a) illustrates an interaction between treatments A and B; part (b) illustrates the absence of an interaction. Non-parallelism of the lines signifies interaction.

(a)

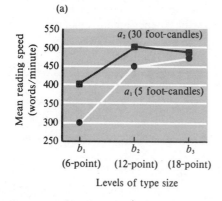

(b)

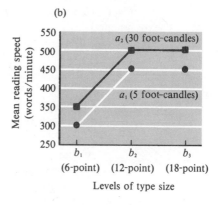

*   The notation j and j', for example, denotes two different values of the subscript.

example, is constructed from the treatment levels of a type CR-2 design, a_1 and a_2, and a type CR-3 design, b_1, b_2, and b_3. The factorial design contains all combinations of the two sets of treatment levels: a_1b_1, a_1b_2, and so on. The randomization procedures for the two types of designs are the same. In the factorial design, the treatment combinations are assigned randomly to the experimental units; in the CR-p design, the treatment levels are assigned randomly. At first glance, the experimental design model equations for the two types of designs appear to be different:

$$\text{CR-}p \qquad Y_{ij} = \mu + \alpha_j + \epsilon_{i(j)}$$
$$\text{CRF-}pq \qquad Y_{ijk} = \mu + \alpha_j + \beta_k + (\alpha\beta)_{jk} + \epsilon_{i(jk)}.$$

If we let τ_{jk} denote the effects of the treatments and interaction in the type CRF-pq design, we see that its model equation has the same form as that for the type CR-p design,

$$\text{CRF-}pq \qquad Y_{ijk} = \mu + \tau_{jk} + \epsilon_{i(jk)}.$$

We have now briefly described four of the 28 types of designs listed in Table 1.3-1. The description has highlighted ways in which the designs differ: (1) randomization, (2) experimental design model equation, (3) number of treatments, (4) inclusion of a nuisance variable as a factor in the design, and (5) sensitivity and power. In the chapters that follow, we will discover other ways in which designs differ: (1) use of crossed, nested, and partially nested treatments, (2) presence or absence of confounding, and (3) use of a covariate. There is a wide variety of designs available, and it is important to clearly identify them in describing research. Although one often sees such statements as ". . . a two-treatment factorial design was used . . . ," it should be evident that a more precise description is required. This description could refer to any of the ten factorial designs listed in Table 1.3-1.

QUESTIONS TO CONSIDER IN SELECTING AN APPROPRIATE DESIGN

On what basis should an experimenter decide which of the wide array of designs to use? Selection of the *best* experimental design for a particular research problem requires (1) a knowledge of the research area and (2) a knowledge of different experimental designs. To arrive at the best experimental design, an experimenter must consider the following questions:

1.   What kinds of data are required to test the statistical hypotheses?
    a. How many treatment levels should be used?
    b. Should the treatment levels used in the experiment be selected on an a priori basis or by random sampling from a population of treatment levels?
    c. Should a factorial experiment be used so that interaction effects can be evaluated?
    d. Are all treatments and all treatment levels of equal interest to the experimenter? Experimental designs may be used that sacrifice power in evaluating some treatments in order to gain power in evaluating other treatments.

2.   Is the proposed sample of subjects large enough to provide adequate precision in testing the statistical hypotheses?

    a. Do the available subjects represent a random sample from the population of interest to the experimenter?

    b. Can the subjects be stratified into homogeneous blocks?

    c. Does the nature of the experiment permit each subject to be observed under more than one treatment level?

    d. Will the treatment(s) produce physical or psychological injury to the subjects? The use of potentially injurious treatments precludes the employment of human subjects.

3.   Is the power of the proposed experimental design adequate to test the statistical hypotheses?

    a. What size treatment effects does the experimenter consider to be of practical interest?

    b. What are the consequences of committing type I and type II errors?

4.   Does the proposed experimental design provide maximum efficiency in testing the statistical hypotheses?

    a. Would efficiency be improved more by using a design employing blocks of homogeneous subjects or by using random assignment of a large number of subjects to the treatment levels?

    b. Can efficiency be increased more by the use of a larger sample size or by controlling additional nuisance variables during the conduct of the experiment?

    c. Can efficiency be increased by the measurement of one or more characteristics related to the dependent variable in order to use regression techniques?

    d. Can efficiency be increased more by the use of a complex experimental design that requires considerable time to plan and analyze or by using a simple design but a large number of subjects? If subjects are plentiful and time required to obtain the data is sufficient, a simple design using a large number of subjects may be more efficient than a complex design that involves costly planning and statistical analysis.

It should be apparent that the question "What is the best experimental design to use?" is not easily answered. Statistical as well as nonstatistical factors must be considered.

ROLE OF EXPERIMENTER AND STATISTICIAN

It is the conviction of the author that the selection of the best experimental design for a particular research problem can be most expeditiously accomplished when the roles of experimenter and statistician are performed by the same person. This is essentially the same position taken by Finney (1960, 3), who states, "to write of the 'experimenter' and the 'statistician' as though they are separate persons is often convenient; the one is concerned with undertaking a piece of research comprehensively and accurately yet with reasonable economy of time and materials, the other is to provide technical advice and assistance on quantitative aspects both in planning and in interpretation . . . the statistician can produce good designs only if he understands something of the particular field of research, and the experimenter will receive better help if he knows the general principles of design and statistical analysis. Indeed, the two roles can be combined when an experimenter with a little mathematical knowledge is prepared to learn enough of the theory of design to be able to design his own experiments."

CRITERIA FOR EVALUATING AN EXPERIMENTAL DESIGN

Many different sets of criteria could be given for evaluating an experimental design. The following questions appear to be basic and should be considered in evaluating any experimental design.

1. Does the design permit an experimenter to calculate a valid estimate of the experimental effects and error effects?

2. Does the data-collection procedure produce reliable results?

3. Does the design provide maximum efficiency within the constraints imposed by the experimental situation?

4. Does the design possess sufficient power to permit an adequate test of the statistical hypotheses?

5. Does the experimental procedure conform to accepted practices and procedures used in the research area? Other things being equal, an experimenter should use procedures that offer an opportunity for comparison of the findings with the results of other investigations.

1.4   THREATS TO VALID INFERENCE-MAKING

Two goals of research are to draw valid conclusions concerning the effects of an independent variable and to make valid generalizations to populations and settings of interest. Cook and Campbell (1979), drawing on the earlier work of Campbell and Stanley (1966), have identified four categories of threats to these goals:*

1. *Statistical conclusion validity* is concerned with threats to valid inference-

---

*   The lists of categories and threats to valid inference-making are Cook and Campbell's; responsibility for the interpretation of items in their lists is mine.

making that result from random error and the illadvised selection of statistical procedures.

2.   *Internal validity* is concerned with correctly concluding that an independent variable is, in fact, responsible for variation in the dependent variable.

3.   *Construct validity of causes or effects* is concerned with the possibility that operations that are meant to represent the manipulation of a particular independent variable or the measurement of a particular dependent variable can be construed in terms of other variables.

4.   *External validity* is concerned with the generalizability of research findings to and across populations of subjects and settings.

This book is concerned with statistical conclusion validity and to a lesser degree with internal validity. For this reason, only Cook and Campbell's (1979, Ch. 2) threats to valid inference-making in these two categories are given. The reader is encouraged to consult the original source. It is must reading for those who do research in field settings.*

THREATS TO STATISTICAL CONCLUSION VALIDITY

1.   *Low statistical power.*   One may fail to reject a false null hypothesis because of inadequate sample size, failure to control or isolate irrelevant sources of variation, and choice of an inefficient test statistic.

2.   *Violated assumptions of statistical tests.*   Test statistics require the tenability of certain assumptions; if these assumptions are not met, incorrect inferences may result.

3.   *Fishing and the error rate problem.*   With certain test statistics, the probability of drawing one or more erroneous conclusions increases as a function of the number of tests performed. This threat to valid inference-making is discussed in detail in Chapter 3.

4.   *The reliability of measures.*   The use of a dependent variable having low reliability may inflate the estimate of error variance and lead to a type II error.

5.   *The reliability of treatment implementation.*   Failure to standardize the administration of treatment levels may inflate the estimate or error variance and lead to a type II error.

6.   *Random irrelevancies in the experimental setting.*   Variation in environment (physical, social, and so on) in which a treatment level is administered may inflate the estimate of error variance and lead to a type II error.

7.   *Random heterogeneity of respondents.*   Idiosyncratic characteristics of experimental units may inflate the estimate of error variance and lead to a type II error.

---

*   Other recommended books include Barber (1976); Huck and Sandler (1979); Rosenthal and Rosnow (1969); and Webb, Campbell, Schwartz, and Sechrest (1966).

THREATS TO INTERNAL VALIDITY

1. *History*.   Events other than the administration of a treament level that occur between the time the treatment level is assigned to experimental units and the dependent variable measurement is made may affect the dependent variable.

2. *Maturation*.   Processes not related to the administration of a treatment level that occur within subjects simply as a function of the passage of time (growing older, stronger, larger, more experienced, and so on) may affect the dependent variable.

3. *Testing*.   Repeated testing of experimental units may result in familiarity with the testing situation or recall of information that can affect the dependent variable.

4. *Instrumentation*.   Changes in the calibration of measuring instruments, shifts in the criteria used by observers and scorers, and shifts in the metric in different ranges of a test can affect the measurement of the dependent variable.

5. *Statistical regression*.   When the measurement of the dependent variable is not perfectly reliable, there is a tendency for extreme scores to regress or move toward the mean. Statistical regression operates to (a) increase the scores of experimental units originally found to score low on a test, (b) decrease the scores of experimental units originally found to score high on a test, and (c) not affect the scores of experimental units found to score at the mean of the test. The amount of statistical regression is inversely related to the reliability of the test.

6. *Selection*.   Differences among the dependent variable means may reflect prior differences among the experimental units assigned the various levels of the independent variable.

7. *Mortality*.   The loss of experimental units may differentially affect the measure of the dependent variable across treatment levels.

8. *Interactions with selection*.   Some of the foregoing threats to internal validity may interact with selection to produce effects that are confounded with or indistinguishable from treatment effects. Among these are selection-maturation effects, selection-history effects, and selection-motivation effects. For example, selection-maturation effects occur when experimental units with different maturation schedules are assigned to different treatment levels.

9. *Ambiguity about the direction of causal influence*.   In some types of research, for example correlational studies, it may be difficult to determine whether X is responsible for the change in Y or vice versa. This ambiguity is not present, for example, when X is known to occur before Y.

10. *Diffusion or imitation of treatments*.   Sometimes the independent variable involves information that is selectively presented to experimental units in the various treatment levels. If the units in different levels can communicate with

one another, differences among the treatment levels may be compromised.

11. *Compensatory equalization of treatments.* When experimental units in some treatment levels receive goods or services generally believed to be desirable, cognizant persons or agencies may, in the spirit of fairness, provide compensatory goods or services to units not scheduled to receive them. For example, a teacher may provide special tutoring to students in a control group who are falling behind those in a special educational program.

12. *Compensatory rivalry by respondents receiving less desirable treatments.* When experimental units in some treatment levels receive goods or services generally believed to be desirable and this becomes known to units in treatment levels not receiving those goods and services, social competition may motivate the latter units to attempt to reverse or reduce the anticipated effects of the desirable treatment levels. Saretsky (1972) named this the "John Henry" effect in honor of the steel driver who, upon learning that his output was being compared to that of a steam drill, worked so hard that he outperformed the drill and died of overexertion.

13. *Resentful demoralization of respondents receiving less desirable treatments.* If experimental units learn that the treatment level to which they have been assigned received less desirable goods or services, they may experience feelings of resentment and demoralization. Their response may be to perform at an abnormally low level, thereby increasing the magnitude of the difference between their performance and that of units assigned to the desirable treatment level.

Threats to statistical conclusion validity and internal validity are present in every experiment. The experimenter's task is awesome—to try to rule out all rival or alternative explanations for observed differences in the dependent variable save the independent variable. This is a deductive process requiring statistical sophistication, subject-matter expertise, insight, and the capacity for self-criticism. Cook and Campbell's lists of threats to valid inference-making provide the experimenter with an excellent starting point. And the random assignment of treatment levels to experimental units simplifies the deductive process somewhat since this helps to rule out such threats as maturation, selection, and selection-maturation. However, random assignment does not rule out the internal validity threats of history, testing, instrumentation, statistical regression, mortality, diffusion or imitation of treatments, compensatory equalization of treatments, compensatory rivalry, and resentful demoralization. Random assignment is of limited help when it comes to threats to statistical conclusion validity. As mentioned earlier, it does lend credence to the assumption that error effects in some experimental designs are independent. The chain of logic between an experimenter's scientific hypothesis and conclusions regarding the truth or falsity of the hypothesis contains both deductive and inductive links and many opportunities for error. We will have more to say about the logic involved in testing scientific hypotheses when we review statistical inference in Section 1.6.

1.5   ETHICAL TREATMENT OF RESEARCH SUBJECTS

In recent years we have witnessed a renewed resolve to protect the rights and interests of human and animal subjects. Codes of ethics for research with human subjects have been adopted by a number of professional societies. Of particular interest are those of the American Psychological Association (1973), American Sociological Association (1971), and the American Anthropological Association (1971). These codes specify what is required and what is forbidden. In addition, they point out the ideal practices of the profession as well as ethical pitfalls. The seventies also saw the passage of laws to govern the conduct of research with human subjects. One law, which is enforced by the U.S. Department of Health, Education, and Welfare, requires that all research funded by HEW involving human subjects must be reviewed by an institutional review board (Weinberger, 1974, 1975). As a result, most institutions where research is conducted have human subject committees that screen all research proposals. These committees can disapprove research proposals or require additional safeguards for the welfare of subjects.

In addition to codes of ethics of professional societies, legal statutes, and peer review, perhaps the most important regulatory force within society is the individual experimenter's ethical code. Experimenters should be familiar with the codes of ethics and statutes relevant to their research areas and incorporate them into their personal code of ethics.

Space does not permit an extensive examination of ethical issues here. For this the reader should consult the references above and the thorough and balanced treatment by Diener and Crandall (1978). However, we can not leave the subject without listing some general guidelines.

1. An experimenter should be knowledgeable about issues of ethics and values, take these into account in making research decisions, and accept responsibility for decisions and actions that have been taken. The experimenter is also responsible for the ethical behavior of collaborators, assistants, and employees who have parallel obligations.

2. Subjects should be informed of all aspects of research that might be expected to influence their willingness to participate. Failure to make full disclosure places an added responsibility on the experimenter to protect the welfare and dignity of the subject. Subjects should understand that they have the right to decline to participate in an experiment and withdraw at any time; strong pressure should not be used to gain cooperation.

3. Research subjects should be protected from physical and mental discomfort, harm, and danger. If the risk of such consequences exists, an experimenter must inform the subject of this. If harm does befall a subject, the experimenter has an obligation to remove or correct the consequences.

4. Special care should be taken to protect the rights and interests of the less

powerful subjects such as children, minorities, patients, the poor, and prisoners.

5. Research deception should never be used without a prior careful ethical analysis. When the methodological requirements of a study demand concealment or deception, the experimenter should take steps to ensure the subject's understanding of the reasons for this action and afterwards restore the quality of the relationship that existed. Where scientific or other compelling reasons require that this information be withheld, the experimenter acquires a special responsibility to assure that there are no damaging consequences for the subject.

6. Private information about participants may be collected only with their consent. All such research information is confidential. Publication should be in a form that protects the subject's identity unless the subject agrees otherwise.

7. After data are collected, the experimenter must provide the subjects with information regarding the nature of the study and relevant findings.

8. Results of research should be reported accurately and honestly, without omissions that might seriously affect their interpretation.

A number of guides for research with animal subjects have been published. Those engaged in such research should be familiar with the *Guide for the Care and Use of Laboratory Animals* (1974).

*1.6 REVIEW OF STATISTICAL INFERENCE

SCIENTIFIC AND STATISTICAL HYPOTHESES

People are by nature inquisitive. We ask questions, develop hunches, and sometimes put our hunches to the test. Over the years a formalized procedure for testing hunches has evolved—the scientific method. It involves (1) observing nature, (2) asking questions, (3) formulating hypotheses, (4) conducting experiments, and (5) developing theories and laws. Let's examine in detail the third characteristic, formulating hypotheses.

A *scientific hypothesis* is a testable supposition that is tentatively adopted to account for certain facts and to guide in the investigation of others. It is a statement about nature that requires verification. Some examples of scientific hypotheses are the following: The child-rearing practices of parents affect the personalities of their

---

*   This section provides an elementary review of statistical inference. The reader who has a good background in statistical inference can skip this section.

offspring. College students involved in student government have higher than average IQs. Cigarette smoking is associated with high blood pressure. Children who feel insecure engage in overt aggression more frequently than children who feel secure. These hypotheses have three characteristics in common with all scientific hypotheses. (1) They are intelligent, informed guesses about phenomena of interest. (2) They can be stated in the *if–then* form of an implication, for example, "*if* John smokes, *then* he will show signs of high blood pressure." (3) Their truth or falsity can be determined by observation or experimentation.

Statistical inference is a form of reasoning whereby rational decisions about states of nature can be made on the basis of incomplete information. Rational decisions often can be made without resorting to statistical inference, as when a scientific hypothesis concerns some limited phenomenon that is directly observable, for example, "this rat is running." The truth or falsity of this hypothesis can be determined by observing the rat. Many scientific hypotheses, on the other hand, refer to phenomena that cannot be directly observed, that is, to populations with elements that are so numerous that it is impossible or impractical to view them all, for example, "all rats run under condition X." Likewise it is impossible to observe all parents rearing their children, all student government leaders, all smokers, or all insecure children. If a scientific hypothesis cannot be evaluated directly by observing all members of a population it may be possible to evaluate the hypothesis indirectly by statistical inference. Classical statistical inference encompasses two complementary topics: hypothesis testing and interval estimation. We will consider hypothesis testing first.

The initial step in evaluating a scientific hypothesis is to express it in the form of a statistical hypothesis. A *statistical hypothesis* is a statement about one or more parameters of a population distribution that requires verification. For example, $\mu > 115$ is a statistical hypothesis; it states that the population mean is greater than 115. Another statistical hypothesis can be formulated that states that the mean is equal to or less than 115, that is, $\mu \leq 115$. These hypotheses, $\mu \leq 115$ and $\mu > 115$, are mutually exclusive and exhaustive; if one is true, the other must be false. They are examples, respectively, of the *null hypothesis* H_0 and the *alternative hypothesis* H_1. The null hypothesis is the one whose tenability is actually tested. If on the basis of this test the null hypothesis is rejected, only the alternative hypothesis remains tenable. According to convention, the alternative hypothesis is formulated so that it corresponds to the experimenter's scientific hunch.* The process of choosing between H_0 and H_1 is called *hypothesis testing*.

THE ROLE OF LOGIC IN EVALUATING A SCIENTIFIC HYPOTHESIS

We turn now to the experimenter's ultimate objective—evaluation of the scientific hypothesis. This evaluation involves a chain of deductive and inductive logic that begins and ends with the scientific hypothesis. The chain is diagrammed in Figure

---

*   The merits of this convention have been extensively debated by Binder (1963); Edwards (1965); Grant (1962); Wilson and Miller (1964); and Wilson, Miller, and Lower (1967).

1.6-1. First, by means of deductive logic the scientific hypothesis and its negation are expressed as two mutually exclusive and exhaustive statistical hypotheses that make predictions concerning a population parameter. These predictions, denoted by H_0 and H_1, are made about the population mean, median, variance, and so on. If, as is usually the case, all the elements in the population cannot be observed, a random sample is obtained from the population. The sample provides an estimate of the unknown population parameter.

The process of deciding whether to reject the null hypothesis is called a *statistical test*. The decision is based on (1) a test statistic computed from a random sample from the population, (2) hypothesis testing conventions, and (3) a decision rule. These are described in subsequent paragraphs. The outcome of the statistical test is the basis for the final link in the chain shown in Figure 1.6-1, an inductive inference concerning the probable truth or falsity of the scientific hypothesis.

Logic therefore plays a key role in hypothesis testing. It is the basis for arriving at both the statistical hypothesis that is tested and the final decision regarding the scientific hypothesis. If errors occur in the deductive or inductive links in the chain of logic, the statistical hypothesis subjected to test may have little or no bearing on the original scientific hypothesis, and/or the inference concerning the scientific hypothesis may be incorrect. Consider the scientific hypothesis that cigarette smoking is associated with high blood pressure. If this hypothesis is true, a measure of central tendency such as mean blood pressure should be higher for the population of smokers than for nonsmokers. The statistical hypotheses are

$$H_0: \quad \mu_1 - \mu_2 \leq 0$$
$$H_1: \quad \mu_1 - \mu_2 > 0$$

where μ_1 and μ_2 designate unknown population means for smokers and nonsmokers, respectively. The null hypothesis H_0 states in effect that the mean blood pressure of

FIGURE 1.6-1   The evaluation of a scientific hypothesis using deductive and inductive logic.

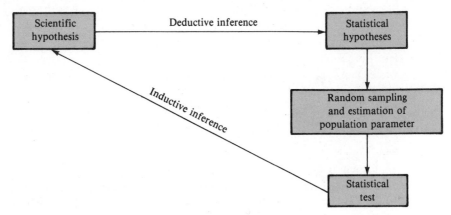

smokers is equal to or less than that of nonsmokers; H_1 states that the mean blood pressure of smokers is greater than that of nonsmokers. These hypotheses follow logically from the original scientific hypothesis. Suppose the experimenter formulated the statistical hypotheses in terms of population variances, for example,

$$H_0: \quad \sigma_1^2 - \sigma_2^2 \le 0$$

$$H_1: \quad \sigma_1^2 - \sigma_2^2 > 0$$

where σ_1^2 and σ_2^2 denote the population variances of smokers and nonsmokers, respectively. A statistical test of this null hypothesis, which states in effect that the variance of blood pressure for the population of smokers is equal to or less than the variance for nonsmokers, would have little bearing on the original scientific hypothesis. However, it would be relevant if the experimenter were interested in determining whether the two populations differed in dispersion.

The reader should not infer that for any scientific hypothesis there is only one pertinent null hypothesis. A null hypothesis stating that the correlation between cigarette smoking and blood pressure is zero bears more directly on the scientific hypothesis than the one involving population means. If cigarette smoking is associated with high blood pressure, there should be a positive correlation, $\rho > 0$, between cigarette consumption and blood pressure. The statistical hypotheses are

$$H_0: \quad \rho \le 0$$

$$H_1: \quad \rho > 0$$

where ρ denotes the population correlation coefficient for cigarette consumption and blood pressure. So we see that both creativity and deductive skill are required to formulate pertinent statistical hypotheses.

SAMPLING DISTRIBUTIONS, THE CENTRAL LIMIT THEOREM, AND TEST STATISTICS

At this point in our review we need to back up and examine several concepts that play a key role in statistical inference: sampling distributions, the central limit theorem, and test statistics. Inferential statistics is concerned with reasoning from a sample to the population—from the particular to the general. Such reasoning is based on a knowledge of the sample-to-sample variability of a statistic, that is, on its sampling behavior. Before data have been collected we can speak of a sample statistic such as $\overline{Y}$ in terms of probability. Its value is yet to be determined and will depend on which score values happen to be randomly selected from the population. Thus, at this stage of an investigation a sample statistic is a random variable, since it is computed from score values obtained by random sampling. Like any random variable, a sample statistic has a probability distribution that gives the probability associated with each value of the statistic over all possible samples of the same size that could be drawn from the population. The distribution of a statistic is called a *sampling distribution* to distinguish it from a probability distribution for, say, a score value. The sampling distribution of the mean is particularly important in the discussion that follows.

The characteristics of the sampling distribution of the mean are succinctly stated in the *central limit theorem,* one of the most important theorems in statistics. In one form, the theorem states that if random samples are selected from a population with mean μ and finite standard deviation σ, as the sample size n increases the distribution of $\overline{Y}$ approaches a normal distribution with mean μ and standard deviation $\sigma/\sqrt{n}$. Probably the most significant point is that regardless of the shape of the sampled population, the means of sufficiently large samples will be nearly normally distributed. Just how large is sufficiently large? This depends on the shape of the sampled population; the more a population departs from the normal form, the larger n must be. For most populations encountered in the behavioral sciences and education, a sample size of 100 is sufficient to produce a nearly normal sampling distribution of $\overline{Y}$. The tendency for the sampling distributions of statistics to approach the normal distribution as n increases helps to explain why the normal distribution is so important in statistics.

It is useful to distinguish sample statistics from *test statistics*. The former are used to describe characteristics of samples or estimate population parameters; the latter are used to test hypotheses about population parameters. An example of a test statistic is z

$$z = \frac{\overline{Y} - \mu_0}{\sigma/\sqrt{n}}$$

where $\overline{Y}$ is the mean of a sample, μ_0 is the hypothesized value of the population mean according to the null hypothesis, σ is the population standard deviation, and n is the size of the sample used to compute $\overline{Y}$. This test statistic is used to test a hypothesis concerning a population mean μ. Like all statistics, it has a sampling distribution. If the null hypothesis is true and Y is approximately normally distributed or n is large, its sampling distribution is the standard normal distribution with mean equal to zero and standard deviation (standard error) equal to one. In the following section we will see how all of this fits together in testing a hypothesis about μ.

HYPOTHESIS TESTING USING A ONE-SAMPLE z TEST STATISTIC

Let's assume we are interested in testing the scientific hypothesis "student government leaders have higher IQs than the average college student." The corresponding statistical hypothesis is $\mu > \mu_0$, where μ denotes the unknown population mean of student leaders and μ_0 denotes the population mean of college students. Assume also that μ_0 and the population standard deviation, σ, of college students are known to equal 115 and 15, respectively. The first step is to state the null and alternative hypotheses,

$$H_0: \quad \mu \leq 115$$
$$H_1: \quad \mu > 115$$

where μ_0 has been replaced by 115, the known population mean of college students. As written, the null hypothesis is inexact because it states a whole range of possible values for the population mean. However, one exact value is specified—$\mu = 115$—and that is the value actually tested. If the null hypothesis $\mu = 115$ can be rejected, then the hypothesis $\mu < 115$ is automatically rejected. Obviously, if $\mu = 115$ is considered improbable because the sample statistic exceeds 115, then any $\mu < 115$ would be considered even less probable.

A relatively small number of test statistics are used to evaluate hypotheses about population parameters. As we will see in Section 2.1, the principal ones are denoted by z, t, χ^2, and F. A test statistic is called a z *statistic* if its sampling distribution is the standard normal distribution; a test statistic is called a t *statistic* if its sampling distribution is a t distribution, and so forth. The choice of a test statistic is determined by (1) the hypothesis to be tested, (2) the information about the population that is known, and (3) the assumptions about the population that appear to be tenable. Which test statistic should be used to test the hypothesis H_0: $\mu \leq 115$? Since the hypothesis concerns the mean of a single population, the population standard deviation is known, and the population is assumed to be approximately normal, the appropriate test statistic is

$$z = \frac{\overline{Y} - \mu_0}{\sigma/\sqrt{n}}.$$

Having specified the statistical hypotheses and test statistic, the next step in the hypothesis testing process is to specify a sample size. In doing this we must keep in mind that to use the standard normal distribution Appendix Table E.3 the sampling distribution of our z test statistic must be approximately normal. The sampling distribution of z will be normal if the sampling distribution of $\overline{Y}$ is normal, since z is a linear transformation of $\overline{Y}$; that is, $z = a + b\overline{Y}$ where $a = -\mu_0/(\sigma/\sqrt{n})$ and $b = 1/(\sigma/\sqrt{n})$. The sampling distribution of $\overline{Y}$ will be normal if the population distribution of Y is normal or if the sample is fairly large, say, $n > 100$. This follows from the central limit theorem discussed earlier. In the case of student government leaders' IQs the population distribution is probably nearly normal, since college students' IQs are approximately normally distributed, but to be on the safe side we will rely on a large sample and obtain an n of 100. Later we will describe a more rational way to determine the value of n. For practical reasons (limitations on the availability of research funds and time, and the difficulty of securing subjects) we do not want to specify a larger n than is needed.

We can now specify the sampling distributions of the random variables $\overline{Y}$ and z given the information (1) $\mu_0 = 115$, (2) $\sigma = 15$, and (3) $n = 100$; and the assumptions (1)* $\mu = 115$ and (2) the population distribution of Y is approximately normal. The sampling distribution of $\overline{Y}$ will be normal with mean equal to 115 and standard error equal to $\sigma/\sqrt{n} = 15/\sqrt{100} = 1.5$. The sampling distribution of z will also be normal with mean equal to zero and standard error equal to one because z is simply a linear transformation of $\overline{Y}$. We have now specified the sampling distribution of the z test statistic.

When decisions are based on incomplete information there is always the risk

---

*   This "assumption" is the null hypothesis to be tested and hopefully rejected.

of making an error. For example, we might decide that $\mu > 115$ when in fact $\mu \leq 115$. We need to specify an acceptable risk of making this kind of error, that is, rejecting H_0 when it is true. We will touch on this subject here and return to it later. Considering the sample-to-sample variability of random variables, we would not expect the mean of a single random sample $\overline{Y}$ to exactly equal the predicted value μ_0, although $\mu = \mu_0$. We would probably be willing to attribute a small discrepancy between $\overline{Y}$ and μ_0 to chance, but if the discrepancy is large enough, we would be inclined to believe that μ_0 is incorrect and the null hypothesis should be rejected. According to hypothesis-testing convention, a large discrepancy is one that would occur by chance five times or fewer in 100 if the null hypothesis is true. To state it another way, given that H_0 is true, a discrepancy between $\overline{Y}$ and μ_0 with probability less than or equal to .05 is considered to be sufficient evidence for believing that $\mu \neq \mu_0$. Therefore, such a discrepancy leads to rejection of the null hypothesis.

A probability of .05 is by convention the largest risk an experimenter is willing to take of rejecting a true null hypothesis—declaring, for example, that $\mu > 115$ when in fact $\mu \leq 115$. Such a probability, called a *level of significance,* is designated by the Greek letter α. For $\alpha = .05$, and H_1: $\mu > 115$, the region for rejecting H_0, called the *critical region,* is shown in Figure 1.6-2. The location and size

FIGURE 1.6-2 Sampling distribution of $\overline{Y}$ and z given that H_0 is true. The critical region, which corresponds in this example to the upper .05 portion of the sampling distribution, defines values of $\overline{Y}$ and z that are improbable if the null hypothesis H_0: $\mu \leq 115$ is true. Hence, if the z test statistic falls in the critical region, the null hypothesis is rejected. The value of z that cuts off the upper .05 portion can be found in the standard normal distribution Table E.3 and is 1.645. It can be shown that the corresponding $\overline{Y}$ value is given by

$$\mu_0 + z\sigma_{\overline{Y}} = 115 + 1.645(1.5) = 117.47.$$

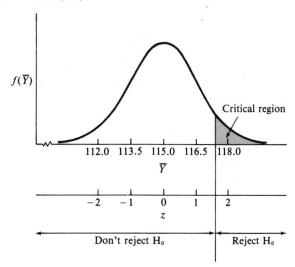

of the critical region are determined, respectively by H_1 and α.

   A decision to adopt the .05, .01, or any other level of significance is generally based on hypothesis-testing conventions, which have evolved over the past 50 years. Unfortunately these conventions do not always lead to decisions that are optimal for an experimenter's purposes. We will return to the problem of selecting a level of significance later.

   The final step in testing H_0: $\mu \leq 115$ is to obtain a sample from a population of interest, compute the test statistic, and make a decision. The *decision rule* is "Reject H_0 if the test statistic falls in the critical region; otherwise don't reject H_0." The procedures just described for testing H_0: $\mu \leq 115$ can be succinctly summarized as follows.

Step 1.   State the statistical hypotheses:

$$H_0: \quad \mu \leq 115$$
$$H_1: \quad \mu > 115$$

Step 2.   Specify the test statistic:

$$z = \frac{\overline{Y} - \mu_0}{\sigma/\sqrt{n}}$$

Step 3.   Specify the sample size:

$$n = 100$$

and the sampling distribution:

   Standard normal distribution because
   σ^2 is known and the population distribution
   of Y is approximately normal in form.

Step 4.   Specify the level of significance:

$$\alpha = .05$$

Step 5.   Obtain a random sample of size n, compute z, and make a decision. *Decision rule:*   Reject H_0 if z falls in the upper 5% of the sampling distribution of z; otherwise, don't reject H_0.

   We will now illustrate the use of

$$z = \frac{\overline{Y} - \mu_0}{\sigma/\sqrt{n}}.$$

Let's assume that a random sample of 100 student government leaders has been obtained from the population of leaders at the Big Ten universities and that the mean IQ of this sample is 117. The number 117 is called a *point estimate* of μ; it is the best guess we can make concerning the unknown value of μ. How improbable is the sample mean of 117 if the population mean is 115? Would it occur five or fewer times in 100 by chance? Stated another way, does the sample statistic $\overline{Y} = 117$ fall in the critical region, which for our example is the upper 5% of the sampling distribution? To answer this question we will first transform the random variable $\overline{Y}$ into a z random variable with sampling distribution (if H_0 is true) identical to the standard normal distribution listed in Appendix Table E.3. If z falls in the upper 5% of the sampling distribution of z, we know that $\overline{Y}$ also falls in the upper 5% of the sampling distribution of $\overline{Y}$.* The linear transformation of $\overline{Y}$ into z is a convenience that enables us to use the standard normal distribution table. The z statistic for our example is

$$z = \frac{\overline{Y} - \mu_0}{\sigma/\sqrt{n}} = \frac{117 - 115}{15/\sqrt{100}} = \frac{2}{1.5} = 1.33.$$

According to the standard normal distribution Table E.3, the value of z that cuts off the upper .05 region of the sampling distribution is 1.645, that is, $z_{.05} = 1.645$. The subscript .05 denotes the proportion of the sampling distribution that falls above the z value. Since $z = 1.33$ is less than $z_{.05} = 1.645$, the observed z falls short of the upper .05 critical region. This is illustrated in Figure 1.6-3. According to our decision rule, we fail to reject H_0 and therefore conclude that our sample data do not indicate that student government leaders at the Big Ten universities have higher IQs than the average college student. Two points need to be emphasized. First, we have not proven that the null hypothesis is true—only that the evidence does not warrant its rejection. Second, our conclusion has been restricted to the population from which we sampled, namely, the Big Ten universities.

In the following sections we will discuss a number of concepts that round out the review of hypothesis testing.

FIGURE 1.6-3   Sampling distribution of z under the null hypothesis. Since $z = 1.33$ falls short of the critical region, H_0 is not rejected.

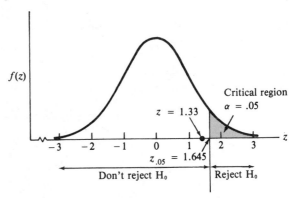

ONE-TAILED AND TWO-TAILED TESTS

A statistical test for which the critical region is in either the upper or lower tail of the sampling distribution is called a *one-tailed test*. If the critical region is in both the upper and lower tails of the sampling distribution, the statistical test is called a *two-tailed test*. A one-tailed test is used whenever the experimenter makes a *directional* prediction concerning the phenomenon of interest, for example, that student government leaders have higher IQs than the average college student.

Often we do not have sufficient information to make a directional prediction about a population parameter; we simply believe the parameter is not equal to the value specified by the null hypothesis. This situation calls for a two-tailed, or *nondirectional,* test. The statistical hypotheses for a two-tailed test have the following form:

$$H_0: \quad \mu = \mu_0$$
$$H_1: \quad \mu \neq \mu_0.$$

For a two-tailed test the region for rejecting the null hypothesis lies in both the upper and lower tails of the sampling distribution. The two-tailed critical region is shown in Figure 1.6-4. A one-tailed, or directional, hypothesis is called for when the experimenter's original hunch is expressed in such terms as "more than," "less than," "increased," or "decreased." Such a hunch indicates that the experimenter has quite a bit of knowledge about the research area. The knowledge could come from previous research, a pilot study, or perhaps theory. If the experimenter is interested only in determining whether something happened or whether there is a difference without specifying the direction of the difference, a two-tailed, or nondirectional, hypothesis

FIGURE 1.6-4   Critical regions for two-tailed test; $H_0: \quad \mu = \mu_0$; $H_1: \quad \mu \neq \mu_0$; $\alpha = .025 + .025 = .05$.

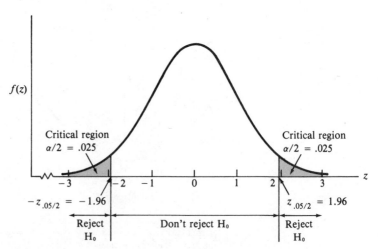

should be used. Generally, significance tests in the behavioral sciences are non-directional because most experimenters lack the information necessary to make directional predictions.

How does the choice of a one- or two-tailed test affect the probability of rejecting a false null hypothesis? We will answer this question by means of an illustration. Assume that $\alpha = .05$ and the following hypotheses have been advanced.

$$H_0: \quad \mu = \mu_0$$
$$H_1: \quad \mu \neq \mu_0$$

If the z statistic falls in either the upper or lower .025 region of the sampling distribution, the result is said to be significant at the .05 level of significance because $.025 + .025 = .05$. The values of z that cut off the upper and lower $.05/2 = .025$ region are $-z_{.05/2} = -1.96$ and $z_{.05/2} = 1.96$. An observed z test statistic is significant at the .05 level if its value is greater than 1.96 or less than -1.96; or, more simply, if its absolute value $|z|$ is greater than 1.96. Now consider the hypotheses

$$H_0: \quad \mu \leq \mu_0$$
$$H_1: \quad \mu > \mu_0;$$

again $\alpha = .05$. If the z statistic falls in the upper tail of the sampling distribution, the result is said to be significant at the .05 level of significance. The critical regions for the two cases are shown in Figures 1.6-3 and 1.6-4. From an inspection of these figures it should be apparent that the size of the difference $\overline{Y} - \mu_0$ necessary to reach the critical region for a two-tailed test is larger than that required for a one-tailed test. Consequently an experimenter is less likely to reject a false null hypothesis with a two-tailed test than with a one-tailed test. However, if the directional hunch is incorrect, the rejection region will be in the wrong tail and the experimenter will most certainly fail to reject the null hypothesis even though it is false. An experimenter is rewarded for making a correct directional prediction and penalized for making an incorrect directional prediction. In the absence of sufficient information for using a one-tailed test, the experimenter should play it safe and use a two-tailed test.

EXACT VERSUS INEXACT HYPOTHESES

A statistical hypothesis can be either exact or inexact. The hypothesis $H_0: \quad \mu = 100$ is an exact hypothesis. The hypothesis $H_0: \quad \mu \leq 100$ is inexact, because instead of specifying a single value for μ it specifies a range of values—those equal to or less than 100.

If the null hypothesis is exact, the alternative hypothesis can be either exact or inexact; for example,

$$H_0: \quad \mu = 100 \quad (\text{exact } H_0) \qquad\qquad H_0: \quad \mu = 100 \quad (\text{exact } H_0)$$

$$\text{or}$$

$$H_1: \quad \mu = 115 \quad (\text{exact } H_1) \qquad\qquad H_1: \quad \mu \neq 100 \quad (\text{inexact } H_1).$$

If the null hypothesis is inexact, the alternative hypothesis must also be inexact; for example,

$$H_0: \quad \mu \leq 100 \quad \text{(inexact } H_0) \qquad H_0: \quad \mu \geq 100 \quad \text{(inexact } H_0)$$
$$\text{or}$$
$$H_1: \quad \mu > 100 \quad \text{(inexact } H_1) \qquad H_1: \quad \mu < 100 \quad \text{(inexact } H_1).$$

In the behavioral sciences inexact alternative hypotheses are the rule rather than the exception because it is rarely possible to make a precise numerical prediction about the experimental outcome.

As we have seen from the preceding examples, null hypotheses can be either exact or inexact, but the hypothesis testing procedure is the same in both cases. If the null hypothesis is inexact, say $\mu \geq \mu_0$, it still embodies at least one exact statement, namely $\mu = \mu_0$, and this is the hypothesis that is actually tested.

TYPE I AND TYPE II ERRORS

When the null hypothesis is tested, an experimenter's decision will be either correct or incorrect. An incorrect decision can be made in two ways. The experimenter can reject the null hypothesis when it is true; this is called a *type I error*. Or the experimenter can fail to reject the null hypothesis when it is false; this is called a *type II error*. Likewise, a correct decision can be made in two ways. If the null hypothesis is true and the experimenter does not reject it, a *correct acceptance* has been made; if the null hypothesis is false and the experimenter rejects it, a *correct rejection* has been made. The two kinds of correct decisions and the two kinds of errors are summarized in Table 1.6-1.

The probability of making a type I error is determined by the experimenter when the level of significance α is specified. If $\alpha = .05$, the probability of making a type I error is .05. The level of significance also determines the probability of a correct acceptance of a true null hypothesis, since this probability is equal to $1 - \alpha$.

TABLE 1.6-1   Decision Outcomes Categorized

		True Situation	
		H_0 *True*	H_0 *False*
Experimenter's Decision	*Fail to Reject* H_0	Correct acceptance Probability $= 1 - \alpha$	Type II error Probability $= \beta$
	Reject H_0	Type I error Probability $= \alpha$	Correct rejection Probability $= 1 - \beta$

The probability, symbolized by β, of making a type II error and the probability of making a correct rejection, which is equal to $1 - \beta$, are determined by a number of variables: (1) the level of significance, (2) the size of the sample, (3) the size of the population standard deviation, (4) the magnitude of the difference between μ and μ_0, and (5) whether a one- or two-tailed test is used. The probability of making a correct rejection, $1 - \beta$, is called the *power* of a satistical test. To compute β and power it is necessary either to know μ or to specify a value of μ that would be of interest. The latter approach is usually necessary, since in any practical situation μ is unknown.

Perhaps an example will help clarify the meaning of α, $1 - \alpha$, β, and $1 - \beta$. Earlier we tested the hypothesis that student government leaders have higher IQs than the average college student, and we failed to reject the null hypothesis. Suppose for purposes of exposition that the sample mean equaled 117.47 instead of 117. In this case, the z statistic is

$$z = \frac{\overline{Y} - \mu_0}{\sigma/\sqrt{n}} = \frac{117.47 - 115}{15/\sqrt{100}} = 1.647.$$

Also assume that the population mean of student government leaders' IQs is equal to 118. This information is illustrated in Figure 1.6-5. Two sampling distributions are shown in this figure, one associated with the null hypothesis and the other with the true IQ of student government leaders. The region corresponding to a type II error (labeled β) is determined from

$$z = \frac{\overline{Y} - \mu}{\sigma/\sqrt{n}} = \frac{117.47 - 118}{15/\sqrt{100}} = \frac{-.53}{1.50} = -.35$$

FIGURE 1.6-5 Regions corresponding to probabilities of making a type I error (α) and a type II error (β). The size and location of the region corresponding to a type I error are determined by α and H_1, respectively. If for a given n, σ, H_0, and true H_1 α is made smaller, the probability of making a type II error is increased.

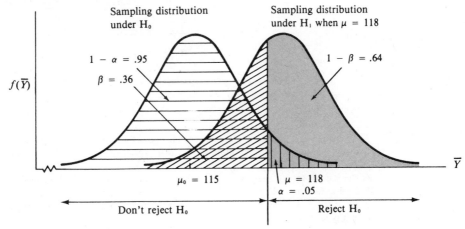

where μ, the true population mean, is substituted for μ_0 in the z formula. According to the standard normal distribution Table E.3, the area below a z of -0.35 is .36. Thus, if the true parameter is 118, the probability of making a type II error (β) is equal to .36 and the probability of making a correct rejection (power) is equal to $1 - \beta = 1 - .36 = .64$. A power of .64 is considerably less than the minimum usually considered acceptable, which is .80.* Table 1.6-2 summarizes the probabilities associated with the possible decision outcomes when $\mu = 115$ and when $\mu = 118$.

For the example we have described, the probability of making a correct decision is larger when the null hypothesis is true (Probability $= .95$) than when the null hypothesis is false (Probability $= .64$). It is also apparent that the probability of making a type I error (.05) is much smaller than the probability of making a type II error (.36). In most research situations the experimenter follows the convention of setting $\alpha = .05$ or $\alpha = .01$. This convention of choosing a small numerical value for α is based on the notion that a type I error is very bad and should be avoided. Unfortunately, as the probability of a type I error is made smaller and smaller, the probability of a type II error increases and vice versa. This can be seen from an examination of Figure 1.6-5; if the vertical line cutting off the upper α region is moved to the right or to the left, the region designated β is made, respectively, larger or smaller. In our example the decision rule, reject H_0 if z falls in the upper .05 region of the sampling distribution, is weighted in favor of deciding that the population mean is less than or equal to 115 rather than greater than 115 because the associated probabilities are .95 and .64, respectively.

TABLE 1.6-2 . Probabilities Associated with the Decision Process

		True Situation	
		$\mu = 115$	$\mu = 118$
Experimenter's Decision	$\mu \leq 115$	Correct aceptance $1 - \alpha = .95$	Type II error $\beta = .36$
	$\mu > 115$	Type I error $\alpha = .05$	Correct rejection $1 - \beta = .64$

---

*   The selection of .80 as the minimum acceptable power is a convenient rule of thumb and reflects the view that type I errors are more serious than type II errors. Consider the value of the ratio p(type II error)$/p$(type I error) when the conventional .05 level of significance is adopted and power is equal to .80. Under these conditions the ratio is $.20/.05 = 4$. The probability of a type II error is four times larger than that for a type I error. An experimenter who adopts $\alpha = .05$ and $1 - \beta = .80$ is saying in effect that a type I error is considered to be four times more serious than a type II error.

DETERMINING THE *n* REQUIRED TO ACHIEVE
AN ACCEPTABLE $1-\beta$ AND $\mu-\mu_0$

As Figure 1.6-5 shows, power can be increased by adopting a large numerical value for α, but this increases the probability of making a type I error and may therefore be undesirable. Often the simplest way to increase the power of a statistical test is to increase the sample size. Until now we haven't said very much about specifying n except that it should be large enough—but not too large. There is a more rational approach to specifying sample size. The factors we have been discussing (α, $1 - \beta$, σ, n, and $\mu - \mu_0$) are interrelated; if we specify any four of them, the fifth is determined. The appropriate size of n can be estimated, therefore, if we know or specify acceptable values for α, $1 - \beta$, σ, and $\mu - \mu_0$. The student-government leader experiment, where $\alpha = .05$, $\sigma = 15$, and $\mu_0 = 115$, will be used to illustrate the procedure.

Suppose we want the power of the experiment to be .80. That leaves one factor, $\mu - \mu_0$, unspecified. There is no way of knowing $\mu - \mu_0$ without measuring all the student government leaders in the population. However, we can specify the minimum IQ difference between μ and μ_0 that we would be interested in finding—if in fact $\mu \neq \mu_0$. Suppose this difference is 3 IQ points; then $\mu - \mu_0 = 118 - 115 = 3$. By specifying that $\mu - \mu_0 = 3$, we are saying that any difference less than 3 points is too small to be of practical interest. The formula for determining the sample size* is

$$n = \frac{(z_\alpha - z_\beta)^2}{(\mu - \mu_0)^2/\sigma^2}$$

where z_α is the value of z that cuts off the α region of the sampling distribution of μ_0 and z_β is the value that cuts off the β region of the sampling distribution of μ. These regions are shown in Figure 1.6-5. If the alternative hypothesis is nondirectional, z_α is replaced by $z_{\alpha/2}$. For our example, $z_{.05} = 1.645$ and $-z_{.20} = -0.84$. Substituting in the formula gives

$$n = \frac{[1.645 - (-0.84)]^2}{(118 - 115)^2/(15)^2} = 154.4$$

which when rounded to the next larger integer value is 155. Thus, the minimum sample size required to detect a 3-point IQ difference with $\alpha = .05$ and power equal to .80 is 155.

This sample size gives the experimenter a fighting chance of detecting a difference considered worth finding. We can be more concrete. Suppose we repeated the experiment 100 times using random samples of 155 student government leaders. If the null hypothesis is true, we would expect to reject it five times (a type I error) and to fail to reject it 95 times (a correct acceptance). If, however, the null hypothesis

---

*   The derivation of the formula is given in Kirk (1978, 235–236).

is false and the true difference between μ and μ_0 is 3 IQ points, we would expect to reject the null hypothesis 80 times (a correct rejection) and to fail to reject it 20 times (a type II error).

Suppose we consider type I and type II errors to be equally serious. If we set both errors at .05, the required sample size is

$$n = \frac{[1.645 - (-1.645)]^2}{(118 - 115)^2/(15)^2} \simeq 271.$$

To increase the power from .80 to .95, other things being equal, we have to increase the sample size from 155 to 271—a 75% increase. Research always involves a series of tradeoffs, as this example illustrates for power and sample size. Another tradeoff, as we will see, involves n and $\mu - \mu_0$; the larger $\mu - \mu_0$ is, the smaller is the n required to reject the null hypothesis.

In many experiments the dependent variable is a new untried measure or one with which the experimenter has had little experience. For example, one may have developed a new test of assertiveness for which there are no norms or a new apparatus for measuring complex reaction time. In such cases, it is difficult if not impossible to specify the minimum difference $\mu - \mu_0$ that would be worth detecting. Fortunately there is an alternative procedure for estimating n that does not require either an estimate of $\mu - \mu_0$ or a knowledge of σ. If the difference we want to detect is divided by the population standard deviation $(\mu - \mu_0)/\sigma$, the resulting score is similar to a standard score. Cohen (1969, 18–25) calls this relative measure an *effect size* and denotes it by the symbol d. It expresses the magnitude of the difference $\mu - \mu_0$ in standard deviation units. The value of d for our student-government leader experiment is

$$d = \frac{\mu - \mu_0}{\sigma} = \frac{118 - 115}{15} = \frac{3}{15} = 0.2.$$

Hence the difference we wanted to detect in that experiment was 0.2 of a standard deviation. Cohen refers to a d value of 0.2 as a *small-effect size*. A medium-effect size is one for which $d = 0.5$, and a large-effect size is one for which $d = 0.8$. Using Cohen's rule of thumb, a medium-effect size for the IQ data is one for which $\mu - \mu_0 = 122.5 - 115 = 7.5$, since

$$d = \frac{7.5}{15} = 0.5.$$

Similarly a large-effect size corresponds to $\mu - \mu_0 = 127 - 115 = 12$, since

$$d = \frac{12}{15} = 0.8.$$

The specification of the minimum difference between μ and μ_0 that one wants to detect in terms of effect size simplifies the formula for estimating n. If d is substituted for $(\mu - \mu_0)/\sigma$, the formula becomes

$$n = \frac{(z_\alpha - z_\beta)^2}{(\mu - \mu_0)^2/\sigma^2} = \frac{(z_\alpha - z_\beta)^2}{d^2}.$$

To compute n, all we have to specify is (1) d, (2) the probability of a type I error, and (3) power. Using Cohen's rule of thumb concerning the interpretation of $d = 0.2, 0.5$, and 0.8, the n's necessary to detect small-, medium-, and large-effect sizes for the student-government leader experiment are, respectively,

$$n = \frac{[1.645 - (-0.84)]^2}{(0.2)^2} \simeq 155$$

$$n = \frac{[1.645 - (-0.84)]^2}{(0.5)^2} \simeq 25$$

$$n = \frac{[1.645 - (-0.84)]^2}{(0.8)^2} \simeq 10.$$

As in the earlier computation of n, the probability of a type I error is .05 ($z_{.05} = 1.645$) and the power is equal to 0.80 ($-z_{.20} = -0.84$).

It is obvious that one's sample size can be too small, resulting in insufficient power. But n can also be too large, resulting in wasted time and resources. An experimenter can avoid these problems by using the formulas just described to make a rational choice of sample size. This procedure has two other less obvious benefits: it focuses attention on the interrelationships among n, α, $1 - \beta$, σ, and $\mu - \mu_0$; and it forces the experimenter to think about the size of the difference $\mu - \mu_0$ that would be worth detecting.

Critics of hypothesis testing procedures have observed that the population mean is rarely equal to the exact value specified by the null hypothesis, and hence by obtaining a large enough sample virtually any H_0 can be rejected. For this reason it is important to distinguish between statistical significance, which leads to the decision that $\mu \neq \mu_0$, and practical significance, which means that the difference $\mu - \mu_0$ is large enough to be useful in the real world. By estimating the n required to reject a difference of interest, an experimenter increases the chances of detecting practical differences. One of the challenges of research is to design an experiment that (1) has adequate power for detecting a meaningful or practical difference between μ and μ_0, (2) uses minimum resources, n, (3) provides adequate protection against making a type I error, and (4) minimizes the effects of extraneous variables.

MORE ABOUT TYPE I AND TYPE II ERRORS

In many research situations the cost of committing a type I error can be large relative to that of a type II error. It is a serious matter to commit a type I error by deciding that a new medication is more effective than conventional therapies in halting the production of cancer cells and therefore can be used in place of other medical procedures. On the other hand, falsely deciding that the new medication is not more effective, a type II error, would result in withholding the medication from the public and lead to further research. The further research would eventually demonstrate the effectiveness of the new medication. In this example a type I error is more costly and should be avoided more than a type II error. The use of the .01, or even the .001, level of significance seems warranted. However, in other research situations that do not involve life and death, a type I error may be less costly than a type II error. For example, an experimenter who makes a type II error may discontinue a promising line of research whereas a type I error would lead to further exploration into a blind alley. Faced with these two alternatives many experimenters would set the level of significance equal to .05 or even .20, preferring to make a type I error rather than a type II error.

It is apparent that the loss functions associated with type I and type II errors must be known before a rational choice of α can be made. Unfortunately, experimenters in the behavioral sciences are generally unable to specify the losses associated with the two kinds of errors, and therein lies the problem. The problem is resolved by using the conventional but arbitrary .05 or .01 level of significance.

I hope that this discussion has dispelled the magical aura that surrounds the .05 and .01 levels of significance—their use in hypothesis testing is simply a convention. A test of significance yields the probability of committing an error in rejecting the null hypothesis. It embodies no information concerning the importance or usefullness of the result obtained. It is just one bit of information used in making a decision concerning a scientific hypothesis.

INTERVAL ESTIMATION

We turn now to interval estimation, the other side of classical statistical inference. Although we can never know the value of a parameter except by measuring all of the elements in the population, a random sample can be used to specify a segment or interval on the number line such that the parameter has a high probability of lying on the segment. The segment is called a *confidence interval*. Interval estimation procedures are used less frequently in the behavioral sciences and education than hypothesis testing procedures even though they provide more information regarding the outcome of an experiment. Experimenters have been slow to adopt the recommendations of statisticians that research reports should include confidence interval information.

In many research situations an experimenter has a choice between testing a null hypothesis or computing a confidence interval. Consider the student-government leader experiment described earlier. The hypotheses of interest were

$$H_0: \quad \mu \leq 115$$
$$H_1: \quad \mu > 115$$

and the level of significance was $\alpha = .05$. A one-sided 95% confidence interval corresponding to this one-tailed hypothesis is given by

$$\overline{Y} - \frac{z_{.05}\sigma}{\sqrt{n}} \leq \mu^*$$

---

*   See Kirk (1978, 285–287) for the derivation of the confidence interval inequality.

$$117 - \frac{1.645(15)}{\sqrt{100}} \leq \mu$$

$$114.53 \leq \mu.$$

We can be fairly confident that the population mean is equal to or greater than 114.53. The degree of our confidence is represented by the *confidence coefficient* $100(1 - .05)\% = 95\%$, where .05 is the level of significance. This confidence interval indicates values of the parameter μ that are consistent with the observed sample statistic. It also contains a range of values of μ_0 for which the null hypothesis is nonrejectable at the .05 level of significance. To put it another way, this confidence interval can be used to test all one-tailed hypotheses of interest, not just H_0: $\mu \leq 115$. For example, we know that H_0: $\mu \leq 113$ and H_0: $\mu \leq 112$ would be rejected, but not H_0: $\mu \leq 115$ or H_0: $\mu \leq 116$.

A two-sided confidence interval can be constructed that is analogous to a two-tailed hypothesis. It's given by

$$\overline{Y} - \frac{z_{\alpha/2}\sigma}{\sqrt{n}} \leq \mu \leq \overline{Y} + \frac{z_{\alpha/2}\sigma}{\sqrt{n}}.$$

A two-sided $100(1 - .05)\% = 95\%$ confidence interval for the student-government leader experiment is

$$117 - \frac{1.96(15)}{\sqrt{100}} \leq \mu \leq 117 + \frac{1.96(15)}{\sqrt{100}}$$

$$114.06 \leq \mu \leq 119.94.$$

We can be 95% confident that the population mean is between 114.06 and 119.94. We could increase our confidence that the interval included the unknown parameter by replacing $z_{.05/2}$ with $z_{.01/2}$; the resulting interval

$$117 - \frac{2.576(15)}{\sqrt{100}} \leq \mu \leq 117 + \frac{2.576(15)}{\sqrt{100}}$$

$$113.14 \leq \mu \leq 120.86$$

is a $100(1 - .01)\% = 99\%$ confidence interval. Note that as our confidence that we have captured μ increases, so does the size of the interval.

This concludes the review of statistical inference. After reading this section, one should be able to test a hypothesis about μ and construct a confidence interval for μ. In addition, the following concepts should be familiar.

scientific hypothesis	point estimate
statistical inference	decision rule
null and alternative hypotheses	one- and two-tailed tests
sampling distribution	exact and inexact statistical hypotheses
central limit theorem	type I and II errors
test statistic	power

level of significance   effect size
critical region   confidence coefficient

1.7   REVIEW EXERCISES*

1.   [1.1] For each of the following, identify the experimental unit (EU) and the observational unit (OU).

   †**a)** Fraternities at a large state university were randomly sampled and members asked to complete several scales of the California Psychological Inventory.

   b) Cars at a roadblock were stopped at random and the occupants searched for illegal drugs.

   †**c)** Twenty students in an introductory psychology class were selected by random sampling and asked to participate in an experiment.

   d) Time to run a straight-alley maze was recorded for each of five randomly sampled rats from ten cages.

   e) Telephone numbers obtained by random sampling from a directory were called and the respondents asked their political preference.

2.   [1.2] Which of the following are acceptable research hypotheses?

   a) Right-handed people tend to be taller than left-handed people.

   b) Behavior therapy is more effective than hypnosis in helping smokers kick the habit.

   c) Most clairvoyant people are able to communicate with beings from outer space.

   d) Rats are likely to fixate an incorrect response if it is followed by an intense noxious stimulus.

3.   [1.2] For each of the following experiments, identify the (i) independent variable, (ii) dependent variable, and (iii) possible nuisance variables.

   †**a)** Televised scenes portraying physical, cartoon, and verbal violence were shown to 20 preschool children. The facial expressions of the children were videotaped and then classified by judges.

   †**b)** Power spectral analyses of changes in cortical EEG were made during a five-hour period of morphine administration in ten female Sprague-Dawley rats.

   c) The effects of four amounts of flurazepam on hypnotic suggestability in males and females was investigated.

   d) The keypecking rate of 20 female Silver King pigeons on fixed ratio reinforcement schedules of FR10, FR50, and FR100 was recorded.

†**4.**   [1.2] For the independent variables in Exercise 3, indicate which are quantitative and which are qualitative.

---

*   The daggers before exercise numbers or part numbers denote exercises or portions thereof for which answers are given in Appendix F. The number in brackets following the exercise number denotes the section that is reviewed in the exercise.

5.   [1.2] What are the two most common procedures for increasing power?

6.   [1.2] What is a commonly used index of the relative efficiency of two experimental designs?

7.   [1.2] Define the following terms:

   a) Nuisance variable   **b)** Experimental error variance   **c)** Replication.

†**8.**   [1.3] What is the major problem with using systematic designs in which the randomization principle is not followed?

9.   [1.3] For each of the following experiments or investigations indicate (i) type of experimental design, (ii) null hypothesis (exclude nuisance variables), and (iii) experimental design model equation.

   †**a)** Forty-five executives were assigned to one of nine categories on the basis of years of experience (b_1 is less than 3 years, b_2 is 3 to 6 years, b_3 is more than 6 years) and educational attainment (c_1 is less than 3 years of college, c_2 is a college graduate, c_3 has some graduate work). Five executives were in each category. The independent variable was type of training used to increase speed in composing complex business letters (a_1 is preparing outline of letter prior to dictating it, a_2 is making a list of the major points to be covered prior to dictating letter, a_3 is silently dictating letter prior to dictating it). Treatment level a_1 was paired with b_1 and c_1; a_2 was paired with b_1 and c_2, and so on. The dependent variable was the average time taken to dictate five letters following two weeks of practice with the assigned practice procedure.

   †**b)** The effects of isolation at 90, 120, 150, and 180 days of age on subsequent combative behavior of Mongolian gerbils (*Meriones unguiculatus*) was investigated. Eighty gerbils were randomly assigned to one of the four isolation conditions with twenty in each condition. The number of combative encounters was recorded when the gerbils were two years old. It was hypothesized that the number of combative encounters would increase with earlier isolation.

   c) Dreams of a random sample of 50 English-Canadian and 50 French-Canadian students were analyzed in terms of the proportion of achievement-oriented elements such as achievement imagery, success, and failure. The students were matched on the basis of reported frequency of dreaming. It was hypothesized that the French-Canadian students' dreams would contain a higher proportion of achievement-oriented elements.

   d) Pictures of human faces posing six distinct emotions were obtained. The faces and their mirror reversals were split down the midline, and left-side and right-side composites were constructed. This resulted in 12 pictures. Six hundred introductory psychology students were randomly assigned to one of 12 groups with 50 in each group. Each student rated one of the 12 pictures on a seven-point scale in terms of the intensity of the emotion expressed. It was hypothesized that left-side composites would express the most intense emotion.

   e) Ninety Sprague-Dawley rats performed a simple operant barpress response and were given either partial reinforcement, partial reinforcement followed by continuous reinforcement, or partial reinforcement preceded by continuous reinforcement. The dependent variable was rate of responding following the training condition. The experimental conditions were randomly assigned to the rats; there were 30 rats in each condition.

10.   [1.3] Construct block diagrams similar to those in Figure 1.3-1 through 1.3-4 for the following designs.

   †**a)** Type CR-3 design, $n = 10$

   †**b)** Type CRF-32 design, $n = 3$

c) Type RB-4 design, $n = 6$

d) Type CRF-222 design, $n = 2$

e) Type LS-3 design, $n = 3$

11.   [1.3] The following data on running time (in seconds) in a straight-alley maze have been obtained in a type CRF-33 design.

| | a_1 | a_1 | a_1 | a_2 | a_2 | a_2 | a_3 | a_3 | a_3 |
	b_1	b_2	b_3	b_1	b_2	b_3	b_1	b_2	b_3
$\overline{Y}_{.jk} =$	9	8	5	7	7	5	6	5	5

Magnitude of reinforcement   *Hours of deprivation*

a_1 = small $b_1 = 10$

a_2 = medium $b_2 = 15$

a_3 = large $b_3 = 20$

a) Graph the interaction.

b) Give a verbal description of the interaction.

†12.   [1.4] For the experiment in Exercise 9(a), list the threats to internal validity.

13.   [1.4] For the experiment in Exercise 9(b), list the threats to internal validity.

14.   [1.4] For the experiment in Exercise 9(d), list the threats to internal validity.

†15.   [1.6] When is it necessary to use the techniques of statistical inference in evaluating a scientific hypothesis?

16.   [1.6] According to convention,

a) Which statistical hypothesis corresponds to the experimenter's scientific hunch?

b) Which is the hypothesis that is actually tested?

†17.   [1.6] Assume that an experimenter has a hunch that insecure children engage in overt aggression more frequently than children who feel secure. Let μ_1 and μ_2 denote the mean daily number of aggressive acts, respectively, of insecure and secure children. State H_0 and H_1 for the experiment.

18.   [1.6] State H_0 and H_1 for the scientific hunch in Exercise 2(a). Let μ_1 and μ_2 denote the population mean height, respectively, for right- and left-handed people.

19.   [1.6] State H_0 and H_1 for the scientific hunch in Exercise 2(b). Let μ_1 and μ_2 denote the mean number of cigarettes smoked by the hypothetical populations of smokers who receive behavior therapy and hypnosis, respectively.

20.   [1.6] State H_0 and H_1 for the scientific hunch in Exercise 2(d). Let μ_1 and μ_2 denote the mean number of incorrect responses in the hypothetical populations of rats who receive an intense noxious stimulus and no noxious stimulus, respectively.

21.   [1.6] Which of the following are examples of statistical hypotheses?

†a) H_0: $\mu = 100$ **b)** H_1: $\rho > 0$ **†c)** H_1: $\sigma^2 > 0$

d) H_0: $S^2 \leq 50$ **†e)** H_0: $\overline{Y}_1 - \overline{Y}_2 = 0$ **f)** H_1: $\mu_1 - \mu_2 < 5$

g) H_0: $\overline{Y} \leq 0$ **h)** H_0: $S_1^2 - S_2^2 = 0$

†**22.**   [1.6] Distinguish a sampling distribution from a sample (frequency) distribution.

23.   [1.6] How is the dispersion of the sampling distribution of $\overline{Y}$ related to σ and n?

†**24.**   [1.6] For the past several years the mean arithmetic-achievement score for a population of ninth-grade students has been 45 with a standard deviation of 15. After participating in an experimental teaching program a random sample of 100 students had a mean score of 50.

   a) List the five steps you would follow in testing the hypothesis that the new program is superior to the old program, and supply the required information. Let $\alpha = .05$.

   b) State the decision rule.

25.   [1.6] For the data in Exercise 24, draw the sampling distribution associated with the null hypothesis and indicate the regions that lead to rejection and nonrejection of the null hypothesis.

26.   [1.6] Assume that the Pd (Psychopathic deviate) scale of the Minnesota Multiphasic Personality Inventory has been given to a random sample of 30 men classified as habitual criminals. The experimenter wants to test the hypothesis that habitual criminals have higher Pd scores than noncriminals. The latter population is known to be normally distributed, with mean and standard deviation equal to 50 and 10, respectively.

   a) List the five steps you would follow in testing the scientific hypothesis, let $\alpha = .05$.

   b) State the decision rule.

27.   [1.6] For the data in Exercise 26, draw the sampling distribution associated with the null hypothesis and indicate the regions that lead to rejection and nonrejection of the null hypothesis.

†**28.**   [1.6] Why should an experimenter always specify H_0, H_1, α, and n before collecting data?

29.   [1.6] For each of the following statistical hypotheses, sketch the sampling distribution, designate the critical region(s), and indicate the size. In each case z is the test statistic.

   †**a)** H_0: $\mu = 60$ **b)** H_0: $\mu \leq 100$ **c)** H_0: $\mu \geq 25$
       H_1: $\mu \neq 60$ H_1: $\mu > 100$ H_1: $\mu < 25$
       $\alpha = .01$ $\alpha = .05$ $\alpha = .005$

†**30.**   [1.6] Under what condition is a one-tailed test less powerful than a two-tailed test?

31.   [1.6] Which of the hypotheses in Exercise 29 are exact and which are inexact?

32.   [1.6] Indicate the type of error or correct decision for each of the following.

   †**a)** A true null hypothesis was rejected.

   †**b)** The null hypothesis is false and the experimenter rejected it.

   c) The experimenter failed to reject a false null hypothesis.

   d) The experimenter did not reject a true null hypothesis.

   e) A false null hypothesis was rejected.

   f) The experimenter rejected the null hypothesis when he or she should have failed to reject it.

33.   [1.6] List the ways in which an experimenter can increase the power of an experimental methodology. What are their relative merits?

34.   [1.6] What are the advantages of confidence interval procedures over hypothesis testing procedures?

†35.   [1.6] How is the size of a confidence interval related to the following?
       a) Size of population standard deviation   **b)** Sample size   **c)** Confidence coefficient

2 FUNDAMENTAL ASSUMPTIONS IN ANALYSIS OF VARIANCE

2.1 BASIC SAMPLING DISTRIBUTIONS IN ANALYSIS OF VARIANCE

One of the distinguishing characteristics of the scientific method is the formulating and testing of hypotheses concerning population parameters. Tests of statistical hypotheses require the a priori formulation of decision rules as well as a knowledge of the sampling distributions of test statistics. The formulation of decision rules and the distribution of $z = (\overline{Y} - \mu_0)/(\sigma/\sqrt{n})$ were discussed in Section 1.6. Here we will describe several other important sampling distributions.

The most important sampling distributions in the behavioral sciences are the binomial, normal, t, chi-square, and F distributions. The first three distributions are primarily used in drawing inferences about the central tendency of populations. The last two distributions, chi-square and F, are useful in drawing inferences about variability or variance as well as central tendency. Because the analysis of variability is such an important aspect of experimental design, we will review the essential features of the chi-square and F distributions and examine the interrelations between these distributions and the t distribution.

The basic assumptions associated with the analysis of variance are developed in this chapter. This material assumes a knowledge of elementary rules of summation and expectations of random variables. These rules are reviewed in Appendices A and B.

CHI-SQUARE DISTRIBUTION

In 1876 F. R. Helmert derived the chi-square distribution, but it was Karl Pearson who in 1900 first used it as a means of testing hypotheses. For purposes of illustration, let us assume that there is a population of scores Y that is normally distributed. The mean of the distribution of Y is $E(Y) = \mu$, where $E(Y)$ refers to the expectation or expected value of the random variable Y. The expected value of a continuous random variable is simply the long-run average of the variable over an indefinite number of samplings. The variance of the distribution of Y is $E[Y - E(Y)]^2 = \sigma^2$ (see Appendix B). Suppose that a random sample of size one is repeatedly drawn from this normally distributed population. Each observation can be expressed in standardized form as

$$z = \frac{Y - \mu}{\sigma}.$$

The square of this random variable, that is,

$$z^2 = \frac{(Y - \mu)^2}{\sigma^2}$$

is called chi-square, $\chi^2_{(1)}$. The distribution of the random variable $\chi^2_{(1)}$, where Y is normally distributed with mean μ and variance σ^2, is a chi-square distribution with one degree of freedom. What does the distribution of $\chi^2_{(1)}$ look like? Because $\chi^2_{(1)}$ is a squared quantity, it can range over only nonnegative real numbers from zero to positive infinity, whereas Y and z can range over all real numbers. The distribution of $\chi^2_{(1)}$ is very positively skewed because approximately 68% of the sampling distribution lies between zero and one. The remaining 32% of the distribution lies between one and positive infinity.

If two random samples of independent observations Y_1 and Y_2 are drawn from the normally distributed population, they can be expressed in standardized form as

$$z_1^2 = \frac{(Y_1 - \mu)^2}{\sigma^2} \quad \text{and} \quad z_2^2 = \frac{(Y_2 - \mu)^2}{\sigma^2}.$$

Two observations are said to be independent if the probability of one observation occurring is unaffected by the occurrence of the other; that is, if $p(Y_1|Y_2) = p(Y_1)$. The sum $z_1^2 + z_2^2 = \chi^2_{(2)}$ over repeated independent samplings has a chi-square distribution with two degrees of freedom. The sampling distribution of $\chi^2_{(2)}$ is somewhat less positively skewed than that for $\chi^2_{(1)}$. For n independent observations drawn at random from a normal population with mean equal to μ and population variance equal to σ^2,

$$\chi^2_{(n)} = \frac{\sum_{i=1}^{n}(Y_i - \mu)^2}{\sigma^2} = \sum_{i=1}^{n} z_i^2$$

with n degrees of freedom. The form of a chi-square distribution depends on the number of degrees of freedom, ν. The chi-square distributions for several different degrees of freedom are shown in Figure 2.1-1. The chi-square distribution is positively skewed for small degrees of freedom, but as ν increases the distribution approaches the normal form with mean and variance, respectively,

$$E(\chi^2_{(\nu)}) = \nu \qquad \text{and} \qquad V(\chi^2_{(\nu)}) = 2\nu.$$

For $\nu > 2$, the mode is equal to $\nu - 2$. The distribution of χ^2 depends only on degrees of freedom.

It follows from the definition of χ^2 that the sum of two *independent* χ^2's is itself distributed as χ^2 with degrees of freedom equal to $\nu_1 + \nu_2$, where ν_i designates the respective degrees of freedom. That is,

$$\chi^2_{(\nu_1)} + \chi^2_{(\nu_2)} = \chi^2_{(\nu_1 + \nu_2)}.$$

In research situations, the value of a sample mean $\overline{Y}$ is known or can be determined, but the value of the population mean is unknown. A practical question arises concerning the distribution of $\sum_{i=1}^{n}(Y_i - \overline{Y})^2/\sigma^2$: What is the form of the distribution if the sample mean is substituted for the population mean? We will now show that for random samples of n observations from a normally distributed population, $\sum_{i=1}^{n}(Y_i - \overline{Y})^2/\sigma^2$ has a chi-square distribution but with degrees of freedom equal to $n - 1$ instead of n.

We begin with the following identity:

$$(Y_i - \mu) = (Y_i - \overline{Y}) + (\overline{Y} - \mu).$$

Squaring both sides, we have

$$(Y_i - \mu)^2 = [(Y_i - \overline{Y}) + (\overline{Y} - \mu)]^2$$
$$= (Y_i - \overline{Y})^2 + 2(Y_i - \overline{Y})(\overline{Y} - \mu) + (\overline{Y} - \mu)^2.$$

FIGURE 2.1-1   Forms of the χ^2 distribution for different degrees of freedom.

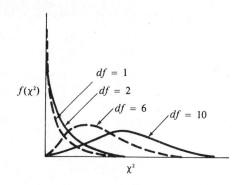

Summing over all of the squared deviations from μ in the sample gives

$$\sum_{i=1}^{n}(Y_i - \mu)^2 = \sum_{i=1}^{n}[(Y_i - \overline{Y})^2 + 2(Y_i - \overline{Y})(\overline{Y} - \mu) + (\overline{Y} - \mu)^2]$$

(2.1-1)
$$= \sum_{i=1}^{n}(Y_i - \overline{Y})^2 + n(\overline{Y} - \mu)^2.$$

This follows because the sum over the n observations of the term $(\overline{Y} - \mu)^2$ on the right is a constant. According to Rule A.1 in Appendix A, the sum of a constant is equal to n times the constant. In addition, the sum of the product $2(Y_i - \overline{Y})(\overline{Y} - \mu)$ equals zero because, by Rule A.3, the sum of a variable times a constant is equal to the constant times the sum of the variable. Thus,

$$\sum_{i=1}^{n}[2(Y_i - \overline{Y})(\overline{Y} - \mu)] = 2(\overline{Y} - \mu)\sum_{i=1}^{n}(Y_i - \overline{Y}).$$

But the sum of the deviations of observations from their mean $\sum_{i=1}^{n}(Y_i - \overline{Y})$ is equal to zero. Dividing both sides of equation (2.1-1) by the population variance σ^2 gives

(2.1-2)
$$\frac{\sum_{i=1}^{n}(Y_i - \mu)^2}{\sigma^2} = \frac{\sum_{i=1}^{n}(Y_i - \overline{Y})^2}{\sigma^2} + \frac{n(\overline{Y} - \mu)^2}{\sigma^2}.$$

The term on the left is based on n randomly sampled observations from a normal population and is thus distributed as $\chi^2_{(n)}$ with $\nu = n$.

The last term on the right of equation (2.1-2) can be rewritten as

$$\frac{n(\overline{Y} - \mu)^2}{\sigma^2} = \frac{(\overline{Y} - \mu)^2}{\sigma^2/n} = \chi^2_{(1)}.$$

In this form it is apparent that it is also a chi-square random variable with $\nu = 1$. This follows because the distribution of $\overline{Y}$ for a normal population is normal with mean equal to μ and variance equal to σ^2/n. We can rewrite equation (2.1-2) as

$$\chi^2_{(n)} = \frac{\sum_{i=1}^{n}(Y_i - \overline{Y})^2}{\sigma^2} + \chi^2_{(1)}.$$

If the two terms on the right of equation (2.1-2) are independent, it follows from the addition property of chi-square that $\sum_{i=1}^{n}(Y_i - \overline{Y})^2/\sigma^2$ must have a chi-square distribution with degrees of freedom equal to $n - 1$. Therefore

(2.1-3)
$$\chi^2_{(n)} = \chi^2_{(n-1)} + \chi^2_{(1)}.$$

An easily followed proof of the independence of the terms on the right of equation (2.1-2) is given by Lindquist (1953, 31–35).

We saw from the foregoing that the term $\sum_{i=1}^{n}(Y_i - \overline{Y})^2/\sigma^2$, based on the sample mean, must have a chi-square distribution with degrees of freedom equal to

$n - 1$. To understand the relationship between the chi-square and the F distributions, it is helpful to express $\sum_{i=1}^{n}(Y_i - \overline{Y})^2/\sigma^2$ as

$$\frac{\sum_{i=1}^{n}(Y_i - \overline{Y})^2}{\sigma^2} = \frac{(n - 1)\hat{\sigma}^2}{\sigma^2}$$

where $\hat{\sigma}^2$ is an unbiased estimator of the population variance. This equivalence follows because

$$\hat{\sigma}^2 = \frac{\sum_{i=1}^{n}(Y_i - \overline{Y})^2}{n - 1}$$

and

$$\sum_{i=1}^{n}(Y_i - \overline{Y})^2 = (n - 1)\hat{\sigma}^2.$$

Thus, the ratio of $(n - 1)\hat{\sigma}^2$ to σ^2 is a random variable that is distributed as chi-square with $n - 1$ degrees of freedom. The hypothesis

$$H_0: \quad \sigma_1^2 = \sigma_0^2$$

that concerns a single population variance σ_1^2 can be tested by means of the chi-square distribution. The test statistic is $(n - 1)\hat{\sigma}_1^2/\sigma_0^2$, where $\hat{\sigma}_1^2$ is an estimator of σ_1^2 computed from a sample and σ_0^2 is the value of the population variance specified by the null hypothesis. The probability of obtaining a χ^2 as large as that observed in an experiment if the null hypothesis is true can be determined from the chi-square table in Appendix E.

F DISTRIBUTION

The experimental designs described in this book all involve tests concerning two population variances rather than a single variance. The sampling distribution that is used for testing a hypothesis about two population variances is the F distribution. The chi-square random variable was described in some detail in previous paragraphs because an F random variable can be defined as the ratio of two independent chi-square variables, each divided by its degrees of freedom. That is,

$$F = \frac{\chi^2_{(\nu_1)}/\nu_1}{\chi^2_{(\nu_2)}/\nu_2}.$$

The distribution of this ratio was determined by R. A. Fisher (1924) and given the name F in his honor by G. W. Snedecor (1934).

Let us imagine that there are two populations of scores Y_1 and Y_2, each having normal distributions. Assume that both populations have the same variance σ^2 but not necessarily the same means. Suppose that two independent random samples of size n_1

and n_2 are drawn from the two populations. Unbiased estimates of σ^2 based on $\hat{\sigma}_1^2$ and $\hat{\sigma}_2^2$ are computed for the samples. We know, from the discussion of the chi-square random variable, that

$$\frac{(n-1)\hat{\sigma}^2}{\sigma^2} = \chi_{(n-1)}^2.$$

This can be written as

$$\hat{\sigma}_1^2 = \frac{\sigma^2 \chi_{(\nu_1)}^2}{\nu_1}$$

for the first population and as

$$\hat{\sigma}_2^2 = \frac{\sigma^2 \chi_{(\nu_2)}^2}{\nu_2}$$

for the second population.
The ratio

$$\frac{\hat{\sigma}_1^2}{\hat{\sigma}_2^2} = \frac{\sigma^2 \chi_{(\nu_1)}^2/\nu_1}{\sigma^2 \chi_{(\nu_2)}^2/\nu_2} = \frac{\chi_{(\nu_1)}^2/\nu_1}{\chi_{(\nu_2)}^2/\nu_2}$$

for each possible pair of samples from the two populations is the random variable F. This is true, of course, only if both population variances are equal to σ^2. It follows that an F ratio can be used to test the following kinds of hypotheses:

$$H_0: \quad \sigma_1^2 = \sigma_2^2$$
$$H_0: \quad \sigma_1^2 \leq \sigma_2^2$$
$$H_0: \quad \sigma_1^2 \geq \sigma_2^2.$$

The most common use of an F ratio is in testing hypotheses regarding the equality of three or more population means. The rationale underlying the use of an F ratio in testing hypotheses of the form

$$H_0: \quad \mu_1 = \mu_2 = \mu_3 = \cdots = \mu_p$$

is discussed in Section 2.4.

The distribution of F depends on only two parameters, ν_1 and ν_2. The distribution of F for $\nu_1 = 9$ and $\nu_2 = 15$ is shown in Figure 2.1-2. Table E.5 in the appendix provides values of F that cut off the upper α portion of an F distribution. In analysis

FIGURE 2.1-2   Form of F distribution for $\nu_1 = 9$ and $\nu_2 = 15$.

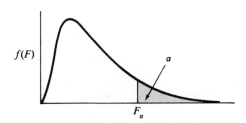

of variance, only the upper percentage points of the F distribution are required because in practice F is taken as the ratio of $\hat{\sigma}^2_{\text{larger}}$ to $\hat{\sigma}^2_{\text{smaller}}$. If the lower percentage points of the F distribution are needed, they can be readily computed from

$$F_{1-\alpha;\nu_1,\nu_2} = \frac{1}{F_{\alpha;\nu_2,\nu_1}}.$$

That is, the value of F in the lower portion of the F distribution can be found by computing the reciprocal of the corresponding value in the upper portion, with degrees of freedom for numerator and denominator reversed.

As Figure 2.1-2 shows, the distribution of F for ν_1 and ν_2 is nonsymmetrical. It approaches normality for very large values of ν_1 and ν_2. Because an F variable is the ratio of two independent chi-square variables divided by their respective degrees of freedom, it can only range over nonnegative real numbers from zero to positive infinity. The mean and variance of an F distribution are, respectively,

$$E(F) = \frac{\nu_2}{\nu_2 - 2} \qquad \text{for } \nu_2 > 2$$

$$V(F) = \frac{2\nu_2^2(\nu_1 + \nu_2 - 2)}{\nu_1(\nu_2 - 2)^2\,(\nu_2 - 4)} \qquad \text{for } \nu_2 > 4.$$

It can be shown that the mean is always greater than unity, whereas the mode is always less than unity. Hence the F distribution is positively skewed.

RELATIONSHIP BETWEEN THE t DISTRIBUTION AND THE χ^2 AND F DISTRIBUTIONS

The exact distribution of the t statistic, or t ratio as it is sometimes called, was derived by W. S. Gosset (1908), who wrote under the pseudonym of Student. Gosset is credited with starting a trend toward the development of exact statistical tests, that is, tests independent of knowledge or assumptions concerning population parameters. This development marked the beginning of the modern era in mathematical statistics. Consider a random sample of scores Y from a normally distributed population with mean μ and variance σ^2. The mean of this random sample can be expressed in standardized form as

$$z = \frac{\overline{Y} - \mu}{\sigma\sqrt{n}}.$$

We know that z is normally distributed with mean equal to zero and variance equal to unity. The one-sample t statistic can be defined as

$$t = \frac{z}{\sqrt{\dfrac{\chi^2_{(n-1)}}{(n-1)}}}.$$

Thus, a t statistic is simply the ratio of a standardized normal variable z to the square root of a chi-square variable divided by its degrees of freedom.

The one-sample t statistic can be expressed in a form more familiar to the reader by replacing z with $(\overline{Y} - \mu)/(\sigma/\sqrt{n})$ and $\chi^2_{(n-1)}$ with $(n - 1)\hat{\sigma}^2/\sigma^2$.

$$t = \frac{z}{\sqrt{\dfrac{\chi^2_{(n-1)}}{(n - 1)}}} = \frac{\dfrac{\overline{Y} - \mu}{\sigma/\sqrt{n}}}{\sqrt{\dfrac{(n - 1)\hat{\sigma}^2/\sigma^2}{n - 1}}} = \frac{\dfrac{\overline{Y} - \mu}{\sigma/\sqrt{n}}}{\hat{\sigma}/\sigma} = \frac{\overline{Y} - \mu}{\hat{\sigma}/\sqrt{n}}$$

Note that the computation of t does not require a knowledge of the population parameter σ^2.

The reader may have surmised from the foregoing that t is related to F. What is the relationship between t and F? To answer this question, let us consider the square of the random variable t with ν equal to $n - 1$ degrees of freedom,

$$t^2 = \frac{z^2}{\chi^2_{(n-1)}/(n - 1)} = \frac{\chi^2_{(1)}/1}{\chi^2_{(n-1)}/(n - 1)}.$$

The numerator of this ratio is a chi-square variable, as defined earlier, with one degree of freedom; and the denominator is also a chi-square variable divided by its degrees of freedom, where $\nu = n - 1$. It can also be shown that the numerator and denominator are independent because $\hat{\sigma}^2$ is independent of $\overline{Y}$ for a normal population. It follows from this that t^2 is identical to an F ratio, with 1 and ν_2 degrees of freedom. That is $t^2_{(\nu)} = F_{(1,\nu_2)}$, where $\nu = \nu_2$.

The distribution of t is symmetrical with mean and variance, respectively,

$$E(t) = 0$$
$$V(t) = \nu/(\nu - 2) \qquad \text{for } \nu > 2.$$

It ranges over values from negative to positive infinity. The distribution of t depends on only one parameter ν. Percentage points of Student's t distribution are given in Table E.4. For large values of n, the t distribution approaches the normal distribution. In this connection, note that the normal curve is the *parent* distribution for the t, F, and χ^2 distributions. For $n > 30$, the t distribution may, for most practical purposes, be regarded as normally distributed.

SUMMARY OF BASIC ASSUMPTIONS ASSOCIATED WITH THE F DISTRIBUTION

The F distribution is the theoretical model against which an F statistic is evaluated. The alternative statistical hypothesis, together with the level of significance discussed in Section 1.6, specifies the region of the F distribution for rejection of the null hypothesis. Hypothesis-testing procedures using the F distribution are based on certain assumptions that are necessary for the mathematical justification of the procedure. The following paragraphs briefly summarize those assumptions that have already been described in connection with the F and χ^2 distributions. We will see later that even when certain assumptions associated with the F distribution are violated, it may still provide an adequate approximation for purposes of hypothesis testing.

Because F is a random variable formed from the ratio of two independent chi-square variables, each divided by its degrees of freedom, it follows that the assumptions of χ^2 are also assumptions of F. A basic assumption associated with the derivation of the chi-square distribution is that the population is normally distributed. It is further assumed that random sampling from this normally distributed population is employed. It follows from the definition of a random sample that the selection of one observation is independent of the selection of another observation. The letters *NID* are often used as an abbreviation for the assumptions that variables are normally and independently distributed.

In the discussion of the random variable F, it was also assumed that the numerator and denominator of the ratio $\hat{\sigma}_1^2/\hat{\sigma}_2^2$ were independent and that the null hypothesis was true, that is, that the population variances were equal.

In summary, hypothesis testing based on the F distribution as the theoretical model involves the following assumptions.

1. Observations are drawn from normally distributed populations.

2. Observations are random samples from the populations.

3. The numerator and denominator of the F ratio are estimates of the same population variance, σ_ϵ^2.

4. The numerator and denominator of the F ratio are independent.

Two sets of assumptions are involved in using analysis of variance to test hypotheses: those associated with the derivation of the F sampling distribution and those associated with the experimental design model for a particular design. The former apply to all designs; those associated with the experimental design model vary from one design to another. We will see in Sections 2.4 and 2.6 that there is considerable overlap between the two sets of assumptions.

2.2 PARTITION OF TOTAL SUM OF SQUARES

The experimental design model equations for several of the simpler experimental designs were described in Section 1.3. We saw that each score in an experiment is a composite that represents all sources of variability that have affected the score. For example, a score may reflect (1) the effects of a treatment level, (2) effects unique to a subject who received the treatment level, (3) effects attributable to chance fluctuations in the subject's behavior, and (4) effects attributable to other uncontrolled variables such as environmental conditions.

The experimental design model equation for a design purports to represent all the sources of variability that affect individual scores. The total variance for *np* scores in an experiment is given by $\sum_{i=1}^{n}\sum_{j=1}^{p}(Y_{ij} - \overline{Y}..)^2/np$. The problem, then, is to partition the total variance into its component parts according to the appropriate model. We will see that the *breakdown* of the total variance is different for each experimental design. Actually the partitioning will be done with the total sum of squares given by $\sum_{i=1}^{n}\sum_{j=1}^{p}(Y_{ij} - \overline{Y}..)^2$ rather than with the total variance.

It will be shown in this section that the total sum of squares, designated by the letters *SSTO*, for a completely randomized design can be partitioned into two components: sum of squares between groups, *SSBG*, and sum of squares within groups, *SSWG*.

PARTITION OF *SSTO* INTO *SSBG* AND *SSWG* FOR TYPE CR-*p* DESIGN

We saw in Section 1.3 that the experimental design model equation for a completely randomized design is

(2.2-1) $$Y_{ij} = \mu + \alpha_j + \epsilon_{i(j)} \qquad (i = 1, \ldots, n; \quad j = 1, \ldots, p).$$

The values of the parameters μ, α_j, and $\epsilon_{i(j)}$ in equation (2.2-1) are unknown, but they can be estimated from sample data as follows.

Statistic	Parameter estimated	Interpretation of parameter
$\overline{Y}..$	μ	General elevation of scores.
$\overline{Y}._j - \overline{Y}..$	α_j	Elevation or depression of scores attributable to the *j*th treatment level.
$Y_{ij} - \overline{Y}._j$	$\epsilon_{i(j)}$	Error effect unique to subject *i* in treatment level *j*.

The meaning of the terms in the first column can be determined from Table 2.2-1.

TABLE 2.2-1   Layout for Type CR-*p* Design (Entry is Y_{ij}; $i = 1, \ldots, n$; $j = 1, \ldots, p$)

a_1	a_2		a_j		a_p	
Y_{11}	Y_{12}	$\cdots$	Y_{1j}	$\cdots$	Y_{1p}	
Y_{21}	Y_{22}	$\cdots$	Y_{2j}	$\cdots$	Y_{2p}	
$\vdots$	$\vdots$		$\vdots$		$\vdots$	
Y_{i1}	Y_{i2}	$\cdots$	Y_{ij}	$\cdots$	Y_{ip}	
$\vdots$	$\vdots$		$\vdots$		$\vdots$	
Y_{n1}	Y_{n2}	$\cdots$	Y_{nj}	$\cdots$	Y_{np}	
$\dfrac{\sum_{i=1}^{n} Y_{i1}}{n} = \overline{Y}._1$	$\dfrac{\sum_{i=1}^{n} Y_{i2}}{n} = \overline{Y}._2$	$\cdots$	$\dfrac{\sum_{i=1}^{n} Y_{ij}}{n} = \overline{Y}._j$	$\cdots$	$\dfrac{\sum_{i=1}^{n} Y_{ip}}{n} = \overline{Y}._p$	$\dfrac{\sum_{j=1}^{p}\sum_{i=1}^{n} Y_{ij}}{np} = \overline{Y}..$

We will now show that

$$SSTO = SSBG + SSWG.$$

The starting point for this derivation is the experimental design model equation. We begin by replacing the parameters in equation (2.2-1) with statistics and rearranging terms as follows.

$$Y_{ij} - \mu = \alpha_j \qquad + \epsilon_{i(j)}$$
$$Y_{ij} - \overline{Y}_{..} = (\overline{Y}_{.j} - \overline{Y}_{..}) + (Y_{ij} - \overline{Y}_{.j})$$

Next we square both sides of the equation.

$$(Y_{ij} - \overline{Y}_{..})^2 = [(\overline{Y}_{.j} - \overline{Y}_{..}) + (Y_{ij} - \overline{Y}_{.j})]^2$$

This equation is for a single score, but we can treat each of the np scores in the same way and sum the resulting equations to obtain

$$\sum_{j=1}^{p} \sum_{i=1}^{n} (Y_{ij} - \overline{Y}_{..})^2 = \sum_{j=1}^{p} \sum_{i=1}^{n} [(\overline{Y}_{.j} - \overline{Y}_{..}) + (Y_{ij} - \overline{Y}_{.j})]^2$$

$$= \sum_{j=1}^{p} \sum_{i=1}^{n} [(\overline{Y}_{.j} - \overline{Y}_{..})^2 + 2(\overline{Y}_{.j} - \overline{Y}_{..})(Y_{ij} - \overline{Y}_{.j}) + (Y_{ij} - \overline{Y}_{.j})^2].$$

According to Rule A.5 in Appendix A, if addition or subtraction is the only operation to be performed before summation, the summation can be distributed. Thus,

$$\sum_{j=1}^{p}\sum_{i=1}^{n}(Y_{ij} - \overline{Y}_{..})^2 = \sum_{j=1}^{p}\sum_{i=1}^{n}(\overline{Y}_{.j} - \overline{Y}_{..})^2 + \sum_{j=1}^{p}\sum_{i=1}^{n}2(\overline{Y}_{.j} - \overline{Y}_{..})(Y_{ij} - \overline{Y}_{.j}) + \sum_{j=1}^{p}\sum_{i=1}^{n}(Y_{ij} - \overline{Y}_{.j})^2$$

(2.2-2)
$$= n\sum_{j=1}^{p}(\overline{Y}_{.j} - \overline{Y}_{..})^2 + 2\sum_{j=1}^{p}(\overline{Y}_{.j} - \overline{Y}_{..})\sum_{i=1}^{n}(Y_{ij} - \overline{Y}_{.j}) + \sum_{j=1}^{p}\sum_{i=1}^{n}(Y_{ij} - \overline{Y}_{.j})^2.$$

The specific rules used in arriving at the first two terms on the right of (2.2-2) are as follows. In the first term on the right,

$$\sum_{j=1}^{p} \sum_{i=1}^{n}(\overline{Y}_{.j} - \overline{Y}_{..})^2 = n\sum_{j=1}^{p}(\overline{Y}_{.j} - \overline{Y}_{..})^2. \qquad \text{(Rules A.1, A.2)}$$

Rules A.1 and A.2 apply because $(\overline{Y}_{.j} - \overline{Y}_{..})$ is a variable with respect to summing over j but a constant with respect to summing over i; note that j is the only subscript. For the middle term

$$\sum_{j=1}^{p} \sum_{i=1}^{n}2(\overline{Y}_{.j} - \overline{Y}_{..})(Y_{ij} - \overline{Y}_{.j}) = 2\sum_{j=1}^{p}(\overline{Y}_{.j} - \overline{Y}_{..})\sum_{i=1}^{n}(Y_{ij} - \overline{Y}_{.j}). \qquad \text{(Rules A.3, A.6)}$$

Rules A.3 and A.6 apply because 2 is a constant and $(\overline{Y}_{.j} - \overline{Y}_{..})$ involves only the outside index of summation, j. On reflection it should be apparent that

$$\sum_{i=1}^{n}(Y_{ij} - \overline{Y}_{.j}) = 0$$

for each of the $j = 1, \ldots, p$ treatment levels. Hence the entire middle term on the right side of (2.2-2) is equal to zero. This leads finally to the desired partition of the total sum of squares:

(2.2-3)
$$\sum_{j=1}^{p}\sum_{i=1}^{n}(Y_{ij} - \overline{Y}_{..})^2 = n\sum_{j=1}^{p}(\overline{Y}_{.j} - \overline{Y}_{..})^2 + \sum_{j=1}^{p}\sum_{i=1}^{n}(Y_{ij} - \overline{Y}_{.j})^2$$
$$SSTO \quad = \quad SSBG \quad + \quad SSWG.$$

This partition of the total sum of squares is appropriate for a completely randomized design. A different partition is required for other designs. In each case, however, the starting point for the derivation is the experimental design model equation.

Now that we have partitioned *SSTO* into *SSBG* and *SSWG*, we need to show what this has to do with testing the hypothesis that $\mu_1 = \mu_2 = \cdots = \mu_p$ or the equivalent hypothesis that $\alpha_1 = \alpha_2 = \cdots = \alpha_p = 0$. Before doing this we will develop more convenient formulas for computing *SSTO*, *SSBG*, and *SSWG* and discuss the concepts of degrees of freedom and expected values of mean squares.

COMPUTATIONAL FORMULA FOR *SSTO*

The partition of the total sum of squares in equation (2.2-3) does not provide formulas that are convenient for computational purposes. We will now derive a more convenient formula for computing *SSTO*.

$$\sum_{j=1}^{p}\sum_{i=1}^{n}(Y_{ij} - \overline{Y}_{..})^2 = \sum_{j=1}^{p}\sum_{i=1}^{n}(Y_{ij}^2 - 2\overline{Y}_{..}Y_{ij} + \overline{Y}_{..}^2)$$

$$= \sum_{j=1}^{p}\sum_{i=1}^{n}Y_{ij}^2 - 2\overline{Y}_{..}\sum_{j=1}^{p}\sum_{i=1}^{n}Y_{ij} + np\overline{Y}_{..}^2 \quad \text{(Rules A.1, A.2, A.3, A.5)}$$

$$= \sum_{j=1}^{p}\sum_{i=1}^{n}Y_{ij}^2 - 2\frac{\sum_{j=1}^{p}\sum_{i=1}^{n}Y_{ij}}{np}\sum_{j=1}^{p}\sum_{i=1}^{n}Y_{ij} + np\frac{\left(\sum_{j=1}^{p}\sum_{i=1}^{n}Y_{ij}\right)^2}{n^2p^2}$$

$$\text{because} \quad \overline{Y}_{..} = \frac{\sum_{j=1}^{p}\sum_{i=1}^{n}Y_{ij}}{np}$$

$$\sum_{j=1}^{p}\sum_{i=1}^{n}(Y_{ij} - \overline{Y}_{..})^2 = \sum_{j=1}^{p}\sum_{i=1}^{n}Y_{ij}^2 - 2\frac{\left(\sum_{j=1}^{p}\sum_{i=1}^{n}Y_{ij}\right)^2}{np} + \frac{\left(\sum_{j=1}^{p}\sum_{i=1}^{n}Y_{ij}\right)^2}{np}$$

(2.2-4)
$$= \sum_{j=1}^{p}\sum_{i=1}^{n}Y_{ij}^2 - \frac{\left(\sum_{j=1}^{p}\sum_{i=1}^{n}Y_{ij}\right)^2}{np} \quad \text{or} \quad \sum_{j=1}^{p}\sum_{i=1}^{n_j}Y_{ij}^2 - \frac{\left(\sum_{j=1}^{p}\sum_{i=1}^{n_j}Y_{ij}\right)^2}{N}$$

where n_j denotes the number of scores in treatment level j and $N = n_1 + n_2 + \cdots + n_p$. The second formula on the right must be used when the n_j's are not equal. Writing out the preceding computational formula consumes both time and space. We will find it convenient in subsequent chapters to use abbreviated symbols for the two terms in this formula. The abbreviated symbols are

$$\sum_{j=1}^{p} \sum_{i=1}^{n} Y_{ij}^{2} = [AS] \quad \text{and} \quad \frac{\left(\sum_{j=1}^{p} \sum_{i=1}^{n} Y_{ij} \right)^{2}}{np} = [Y].$$

The total sum of squares can be written, using these symbols, as

$$SSTO = [AS] - [Y].$$

COMPUTATIONAL FORMULA FOR *SSBG*

The computational formula for *SSBG* can be derived as follows.

$$n \sum_{j=1}^{p} (\overline{Y}_{j} - \overline{Y}_{..})^{2} = n \sum_{j=1}^{p} (\overline{Y}_{j}^{2} - 2\overline{Y}_{..}\overline{Y}_{j} + \overline{Y}_{..}^{2})$$

$$= n \sum_{j=1}^{p} \overline{Y}_{j}^{2} - 2n\overline{Y}_{..} \sum_{j=1}^{p} \overline{Y}_{j} + np\overline{Y}_{..}^{2} \quad \text{(Rules A.1, A.2, A.3, A.5)}$$

$$= n \sum_{j=1}^{p} \frac{\left(\sum_{i=1}^{n} Y_{ij} \right)^{2}}{n^{2}} - 2n \frac{\sum_{j=1}^{p} \sum_{i=1}^{n} Y_{ij}}{np} \frac{\sum_{j=1}^{p} \sum_{i=1}^{n} Y_{ij}}{n} + np \frac{\left(\sum_{j=1}^{p} \sum_{i=1}^{n} Y_{ij} \right)^{2}}{n^{2} p^{2}}$$

$$\text{because} \quad \overline{Y}_{..} = \frac{\sum_{j=1}^{p} \sum_{i=1}^{n} Y_{ij}}{np} \quad \text{and} \quad \overline{Y}_{j} = \frac{\sum_{i=1}^{n} Y_{ij}}{n}$$

$$= \sum_{j=1}^{p} \frac{\left(\sum_{i=1}^{n} Y_{ij} \right)^{2}}{n} - 2 \frac{\left(\sum_{j=1}^{p} \sum_{i=1}^{n} Y_{ij} \right)^{2}}{np} + \frac{\left(\sum_{j=1}^{p} \sum_{i=1}^{n} Y_{ij} \right)^{2}}{np}$$

$$= \sum_{j=1}^{p} \frac{\left(\sum_{i=1}^{n} Y_{ij} \right)^{2}}{n} - \frac{\left(\sum_{j=1}^{p} \sum_{i=1}^{n} Y_{ij} \right)^{2}}{np} \quad \text{or} \quad \sum_{j=1}^{p} \frac{\left(\sum_{i=1}^{n_j} Y_{ij}^{2} \right)^{2}}{n_j} - \frac{\left(\sum_{j=1}^{p} \sum_{i=1}^{n_j} Y_{ij} \right)^{2}}{N}$$

where $N = n_1 + n_2 + \cdots + n_p$. The second formula on the right must be used when the n_j's are not equal. The abbreviated symbol for $\sum_{j=1}^{p} (\sum_{i=1}^{n} Y_{ij})^{2}/n$ is $[A]$. The between-groups sum of squares can be written, using abbreviated symbols, as

$$SSBG = [A] - [Y].$$

COMPUTATIONAL FORMULA FOR *SSWG*

The computational formula for *SSWG* can be derived as follows.

$$\sum_{j=1}^{p} \sum_{i=1}^{n} (Y_{ij} - \overline{Y}_{\cdot j})^2 = \sum_{j=1}^{p} \sum_{i=1}^{n} (Y_{ij}^2 - 2\overline{Y}_{\cdot j}Y_{ij} + \overline{Y}_{\cdot j}^2)$$

$$= \sum_{j=1}^{p} \sum_{i=1}^{n} Y_{ij}^2 - 2\sum_{j=1}^{p} \overline{Y}_{\cdot j} \sum_{i=1}^{n} Y_{ij} + n\sum_{j=1}^{p} \overline{Y}_{\cdot j}^2$$

(Rules A.1, A.2, A.3, A.5, A.6)

$$= \sum_{j=1}^{p} \sum_{i=1}^{n} Y_{ij}^2 - 2\sum_{j=1}^{p} \frac{\left(\sum_{i=1}^{n} Y_{ij}\right)}{n} \sum_{i=1}^{n} Y_{ij} + n\sum_{j=1}^{p} \frac{\left(\sum_{i=1}^{n} Y_{ij}\right)^2}{n^2}$$

because $\overline{Y}_{\cdot j} = \dfrac{\sum_{i=1}^{n} Y_{ij}}{n}$

$$= \sum_{j=1}^{p} \sum_{i=1}^{n} Y_{ij}^2 - 2\sum_{j=1}^{p} \frac{\left(\sum_{i=1}^{n} Y_{ij}\right)^2}{n} + \sum_{j=1}^{p} \frac{\left(\sum_{i=1}^{n} Y_{ij}\right)^2}{n}$$

$$= \sum_{j=1}^{p} \sum_{i=1}^{n} Y_{ij}^2 - \sum_{j=1}^{p} \frac{\left(\sum_{i=1}^{n} Y_{ij}\right)^2}{n} \quad \text{or} \quad \sum_{j=1}^{p} \sum_{i=1}^{n} Y_{ij}^2 - \sum_{j=1}^{p} \frac{\left(\sum_{i=1}^{n_j} Y_{ij}\right)^2}{n_j}$$

The second formula on the right must be used when the n_j's are not equal. The within-groups sum of squares can be written, using abbreviated symbols, as

$$SSWG = [AS] - [A].$$

Note that only three terms are needed in computing *SSTO*, *SSBG*, and *SSWG*; these are [*AS*], [*A*], and [*Y*].

DEGREES OF FREEDOM

The term *degrees of freedom* refers to the number of observations with values that can be assigned arbitrarily. We will now determine the degrees of freedom associated with $SSBG$, $SSWG$, and $SSTO$.

Consider $SSBG = n\sum_{j=1}^{p}(\overline{Y}_j - \overline{Y}_{..})^2$. If we have $p = 3$ sample means each based on n observations, they are related to the grand mean by the equation

$$\frac{\overline{Y}_1 + \overline{Y}_2 + \overline{Y}_3}{3} = \overline{Y}_{..}$$

If $\overline{Y}_{..} = 4$ and we arbitrarily specify $\overline{Y}_1 = 3$ and $\overline{Y}_2 = 7$, then $\overline{Y}_3$ must equal 2 since $(3 + 7 + 2)/3 = 4$. Alternatively, if we specify $\overline{Y}_1 = 5$ and $\overline{Y}_2 = 1$, then $\overline{Y}_3$ must equal 6, since $(5 + 1 + 6)/3 = 4$. Given the value of the grand mean, we are free to assign any values to two of the three means, but having done so the third is determined. Hence the number of degrees of freedom for $SSBG$ is $p - 1$, one less than the number of treatment means.

The degrees of freedom associated with $SSWG$ is $p(n - 1)$. To see why this is true, consider $SSWG = \sum_{j=1}^{p} \sum_{i=1}^{n}(Y_{ij} - \overline{Y}_j)^2$ and let $p = 3$ and $n = 8$. For each of the $j = 1, \ldots, 3$ treatment levels we can compute $n = 8$ deviations.

$$\overbrace{(Y_{1j} - \overline{Y}_j)^2}^{\text{1st}} + \overbrace{(Y_{2j} - \overline{Y}_j)^2}^{\text{2nd}} + \cdots + \overbrace{(Y_{8j} - \overline{Y}_j)^2}^{\text{8th}}$$

Each score in the jth treatment level is related to the jth mean by

$$\frac{Y_{1j} + Y_{2j} + \cdots + Y_{8j}}{8} = \overline{Y}_j.$$

Seven of the scores can take any value, but the eighth is determined, since the sum of the scores divided by eight must equal $\overline{Y}_j$. Hence there are $n - 1 = 8 - 1 = 7$ degrees of freedom associated with the jth treatment level. If $n_1 = n_2 = \cdots = n_p$, each of the p treatment levels has $n - 1$ degrees of freedom. Thus there are $p(n - 1)$ degrees of freedom associated with $SSWG$. If the n_j's are not equal, the degrees of freedom for $SSWG$ is $(n_1 - 1) + (n_2 - 1) + \cdots + (n_p - 1) = N - p$, where N is the total number of scores.

The same line of reasoning can be used to show that when $n_1 = n_2 = \cdots = n_p$, the total sum of squares has $np - 1$ degrees of freedom. This follows since each of the $np = (8)(3) = 24$ scores is related to the grand mean by

$$\frac{Y_{11} + Y_{21} + \cdots + Y_{83}}{24} = \overline{Y}_{..}$$

Hence, $np - 1 = 23$ of the scores can take any value, but the twenty-fourth must be assigned so that the mean of the scores equals $\overline{Y}_{..}$. If the n_j's are not equal, the number of degrees of freedom is $n_1 + n_2 + \cdots + n_p - 1 = N - 1$.

2.3   EXPECTATION OF MEAN SQUARES

In Section 2.2 we saw that the total sum of squares for a completely randomized design can be partitioned into between- and within-groups sums of squares. A careful examination of these terms provides an intuitive understanding of the sources of variation that they represent. The within-groups sum of squares is composed of variation due to differences among individual subjects who receive the same treatment level. Some variation among the scores of subjects in the same treatment level can always be expected because of individual differences. Since subjects are randomly assigned to the treatment levels, this source represents chance variation. Differences among the scores of subjects who are assigned to different treatment levels reflect chance variation and, in addition, the systematic effects of the particular treatment levels.

   In summary, the within-groups sum of squares is an estimate of chance variation, whereas the between-groups sum of squares is an estimate of chance variation plus the effects of treatment levels if such effects are present. A more systematic exposition of these points requires an understanding of the concept of the expected value of a random variable. Before pursuing this further, two models—the fixed-effects and random-effects models—will be described. ·

FIXED-EFFECTS EXPERIMENTAL DESIGN MODEL

The partition of the total sum of squares in Section 2.2 involved no special assumptions concerning populations or sampling procedures. However, certain assumptions are required if one is to draw inferences concerning population parameters from sample data. General assumptions associated with an F random variable were described in Section 2.1. In addition to these assumptions, special assumptions concerning the experimental design model equation are necessary. The model equation for a completely randomized design is

$$Y_{ij} = \mu + \alpha_j + \epsilon_{i(j)} \quad (i = 1, \ldots, n; j = 1, \ldots, p).$$

We will now describe the assumptions for a *fixed-effects model*. This model is appropriate for experiments in which all treatment levels about which inferences are to be drawn are included in the experiment. If the experiment were replicated, the same treatment levels would be included in the replication. Under these conditions, conclusions drawn from the experiment apply only to the actual p treatment levels in the experiment. If the p treatment levels included in the experiment are a random sample from a much larger population of P levels, the model is a *random-effects model*. Special assumptions appropriate for this model are described later. For the fixed-effects model, $\alpha_j = \mu_j - \mu$ is a constant fixed effect for all observations within population j, but it may vary for each of the $j = 1, \ldots, p$ treatment populations. It follows that

$$\sum_{j=1}^{p} \alpha_j = 0.$$

The expected value of the constant α_j is

$$E(\alpha_j) = \alpha_j \qquad \text{(Rule B.3)}$$

and

$$E\left(\sum_{j=1}^{p}\alpha_j\right) = E(\alpha_1 + \alpha_2 + \alpha_3 + \cdots + \alpha_p) = 0. \qquad \text{(Rule B.3)}$$

Also,

$$E\left(\sum_{j=1}^{p}\alpha_j\right)^2 = E(\alpha_1 + \alpha_2 + \cdots + \alpha_p)^2$$

$$= E[\alpha_1(\alpha_1 + \alpha_2 + \cdots + \alpha_p) + \alpha_2(\alpha_1 + \alpha_2 + \cdots + \alpha_p)$$
$$+ \cdots + \alpha_p(\alpha_1 + \alpha_2 + \cdots + \alpha_p)]$$

but

$$(\alpha_1 + \alpha_2 + \cdots + \alpha_p) = 0.$$

Thus,
$$E\left(\sum_{j=1}^{p}\alpha_j\right)^2 = 0$$

and

$$E\left(\sum_{j=1}^{p}\alpha_j^2\right) = E(\alpha_1^2 + \alpha_2^2 + \cdots + \alpha_p^2)$$

$$= \alpha_1^2 + \alpha_2^2 + \cdots + \alpha_p^2 = \sum_{j=1}^{p}\alpha_j^2. \qquad \text{(Rule B.3)}$$

Because both μ and α_j are constants for all observations within population j, the only source of variation among these observations is that due to the error effect, $\epsilon_{i(j)} = Y_{ij} - \mu - \alpha_j$. It is assumed that the distribution of $\epsilon_{i(j)}$ for each treatment population is normal, with mean equal to zero and variance equal to σ_ϵ^2. It is also assumed that the $\epsilon_{i(j)}$'s are independent, both within each treatment level and across all treatment levels. If subjects are randomly sampled and randomly assigned to treatment levels, the value of $\epsilon_{i(j)}$ for any observation can be assumed to be independent of the values of $\epsilon_{i(j)}$'s for other observations.

In summary, it is assumed that $\epsilon_{i(j)}$ is a random variable that is normally and independently distributed (*NID*), with mean equal to zero and variance equal to σ_ϵ^2. The expected value of $\epsilon_{i(j)}$, according to the assumptions that have been made, is

$$E(\epsilon_{i(j)}) = \mu_\epsilon = 0 \qquad \text{(Rule B.1)}$$

$$E\left(\sum_{i=1}^{n}\epsilon_{i(j)}\right) = n\mu_\epsilon = 0 \qquad \text{(Rule B.8)}$$

$$E\left(\sum_{i=1}^{n}\epsilon_{i(j)}^2\right) = n(\mu_\epsilon^2 + \sigma_\epsilon^2) = n\sigma_\epsilon^2 \qquad \text{(Rule B.10)}$$

$$E\left(\sum_{i=1}^{n}\epsilon_{i(j)}\right)^2 = n(n\mu_\epsilon^2 + \sigma_\epsilon^2) = n\sigma_\epsilon^2 \qquad \text{(Rule B.11)}$$

$$E\left(\sum_{j=1}^{p}\sum_{i=1}^{n}\epsilon_{i(j)}^2\right) = np(\mu_\epsilon^2 + \sigma_\epsilon^2) = np\sigma_\epsilon^2 \qquad \text{(Rule B.10)}$$

$$E\left(\sum_{j=1}^{p}\sum_{i=1}^{n}\epsilon_{i(j)}\right)^2 = np(np\mu_\epsilon^2 + \sigma_\epsilon^2) = np\sigma_\epsilon^2. \qquad \text{(Rule B.11)}$$

The expected value of the grand mean μ is

$$E(\mu) = \mu. \qquad \text{(Rule B.3)}$$

Finally, it is assumed that a score Y_{ij} is an estimate of the linear combination of the parameters specified in the experimental design model equation.

EXPECTATION OF BETWEEN-GROUPS MEAN SQUARE

The computational formula for a between-groups sum of squares was given in Section 2.2 as

(2.3-1)
$$SSBG = \sum_{j=1}^{p} \frac{\left(\sum_{i=1}^{n}Y_{ij}\right)^2}{n} - \frac{\left(\sum_{j=1}^{p}\sum_{i=1}^{n}Y_{ij}\right)^2}{np}.$$

At the beginning of this section, we expressed the idea that this sum of squares is an estimate of chance variation, plus treatment effects if the latter are present in an experiment. A more formal exposition of this idea for a fixed-effects experimental design model involves the use of the concept of an expected value.

It will be shown that the expected value of the between-groups mean square is

$$E(MSBG) = \sigma_\epsilon^2 + \frac{n\sum_{j=1}^{p}\alpha_j^2}{p-1}.$$

This final result is obtained by first determining the expected values of the two terms in (2.3-1) separately.

The model equation for a completely randomized design was given earlier as

$$Y_{ij} = \mu + \alpha_j + \epsilon_{i(j)} \quad (i = 1, \ldots, n; j = 1, \ldots, p).$$

The terms $(\mu + \alpha_j + \epsilon_{i(j)})$ can be substituted for Y_{ij} in $\sum_{j=1}^{p} (\sum_{i=1}^{n} Y_{ij})^2/n$ as follows:

$$E\left[\sum_{j=1}^{p} \frac{\left(\sum_{i=1}^{n} Y_{ij}\right)^2}{n}\right]$$

$$= E\left\{\sum_{j=1}^{p} \frac{\left[\sum_{i=1}^{n}(\mu + \alpha_j + \epsilon_{i(j)})\right]^2}{n}\right\}$$

$$= \frac{1}{n}\sum_{j=1}^{p} E\left[\sum_{i=1}^{n}(\mu + \alpha_j + \epsilon_{i(j)})\right]^2 \qquad \text{(Rules B.3, B.6)}$$

$$= \frac{1}{n}\sum_{j=1}^{p} E\left[n\mu + n\alpha_j + \sum_{i=1}^{n}\epsilon_{i(j)}\right]^2 \qquad \text{(Rules A.1, A.2, A.5)}$$

$$= \frac{1}{n}\sum_{j=1}^{p} E\left[n^2\mu^2 + n^2\alpha_j^2 + \left(\sum_{i=1}^{n}\epsilon_{i(j)}\right)^2 + 2n^2\mu\alpha_j + 2n\mu\sum_{i=1}^{n}\epsilon_{i(j)} + 2n\alpha_j\sum_{i=1}^{n}\epsilon_{i(j)}\right]$$

$$= \frac{1}{n}\sum_{j=1}^{p}(n^2\mu^2 + n^2\alpha_j^2 + n\sigma_\epsilon^2 + 2n^2\mu\alpha_j) \qquad \text{(Rules B.3, B.4, B.5, B.6, B.8,}$$

$$\text{B.11)}$$

because $\quad E\left(\sum_{i=1}^{n}\epsilon_{i(j)}\right)^2 = n(n\mu_\epsilon^2 + \sigma_\epsilon^2) = n\sigma_\epsilon^2 \quad$ and $\quad E\left(\sum_{i=1}^{n}\epsilon_{i(j)}\right) = n\mu_\epsilon = 0$

$$= np\mu^2 + n\sum_{j=1}^{p}\alpha_j^2 + p\sigma_\epsilon^2 + 2n\mu\sum_{j=1}^{p}\alpha_j. \qquad \text{(Rules A.1, A.3, A.5)}$$

But $\sum_{j=1}^{p}\alpha_j = 0$. Thus,

(2.3-2) $$E\left[\sum_{j=1}^{p} \frac{\left(\sum_{i=1}^{n} Y_{ij}\right)^2}{n}\right] = np\mu^2 + n\sum_{j=1}^{p}\alpha_j^2 + p\sigma_\epsilon^2.$$

If $(\mu + \alpha_j + \epsilon_{i(j)})$ is substituted for Y_{ij} in $(\sum_{j=1}^{p} \sum_{i=1}^{n} Y_{ij})^2/np$, following the preceding procedure,

$$E\left[\frac{\left(\sum_{j=1}^{p} \sum_{i=1}^{n} Y_{ij}\right)^2}{np}\right] = E\left\{\frac{\left[\sum_{j=1}^{p} \sum_{i=1}^{n}(\mu + \alpha_j + \epsilon_{i(j)})\right]^2}{np}\right\}$$

$$= \frac{1}{np} E\left[\left(np\mu + n\sum_{j=1}^{p}\alpha_j + \sum_{j=1}^{p}\sum_{i=1}^{n}\epsilon_{i(j)}\right)^2\right] \quad \text{(Rules A.1, A.2, A.3, A.5)}$$

$$= \frac{1}{np} E\left[n^2p^2\mu^2 + n^2\left(\sum_{j=1}^{p}\alpha_j\right)^2 + \left(\sum_{j=1}^{p}\sum_{i=1}^{n}\epsilon_{i(j)}\right)^2 + 2n^2p\mu\sum_{j=1}^{p}\alpha_j\right.$$

$$\left. + 2np\mu\sum_{j=1}^{p}\sum_{i=1}^{n}\epsilon_{i(j)} + 2n\sum_{j=1}^{p}\alpha_j\sum_{j=1}^{p}\sum_{i=1}^{n}\epsilon_{i(j)}\right]$$

$$= \frac{1}{np}(n^2p^2\mu^2 + np\sigma_\epsilon^2) \quad \text{(Rules B.3, B.4, B.5, B.6, B.8, B.11)}$$

because $\sum_{j=1}^{p}\alpha_j = 0,$

$$E\left(\sum_{j=1}^{p}\sum_{i=1}^{n}\epsilon_{i(j)}\right)^2 = np(np\mu_\epsilon^2 + \sigma_\epsilon^2)$$

$$= np\sigma_\epsilon^2, \quad \text{and}$$

$$E\left(\sum_{j=1}^{p}\sum_{i=1}^{n}\epsilon_{i(j)}\right) = np\mu_\epsilon = 0.$$

Hence,

(2.3-3)
$$E\left[\frac{\left(\sum_{j=1}^{p}\sum_{i=1}^{n}Y_{ij}\right)^2}{np}\right] = np\mu^2 + \sigma_\epsilon^2.$$

The expected value of the between-groups sum of squares is

$$E\left[\sum_{j=1}^{p}\frac{\left(\sum_{i=1}^{n}Y_{ij}\right)^2}{n} - \frac{\left(\sum_{j=1}^{p}\sum_{i=1}^{n}Y_{ij}\right)^2}{np}\right] = \left(np\mu^2 + n\sum_{j=1}^{p}\alpha_j^2 + p\sigma_\epsilon^2\right) - (np\mu^2 + \sigma_\epsilon^2)$$

$$= (p-1)\sigma_\epsilon^2 + n\sum_{j=1}^{p}\alpha_j^2.$$

A mean square (*MS*) is obtained by dividing a sum of squares by its degrees of freedom. For an experiment with p treatment levels, the between-groups degrees of freedom is equal to $p - 1$. The expected value of *MSBG* is

$$E(MSBG) = \frac{(p - 1)\sigma_\epsilon^2 + n \sum_{j=1}^{p} \alpha_j^2}{p - 1} = \sigma_\epsilon^2 + \frac{n \sum_{j=1}^{p} \alpha_j^2}{p - 1}.$$

If the null hypothesis $\alpha_j = 0$ for all j is true, then

$$E(MSBG) = \sigma_\epsilon^2$$

that is, *MSBG* is an estimator of the population error variance.

EXPECTATION OF WITHIN-GROUPS MEAN SQUARE

The computational formula for the within-groups sum of squares was given in Section 2.3 as

(2.3-4)
$$SSWG = \sum_{j=1}^{p} \sum_{i=1}^{n} Y_{ij}^2 - \sum_{j=1}^{p} \frac{\left(\sum_{i=1}^{n} Y_{ij} \right)^2}{n}.$$

We have seen that

$$E \left[\sum_{j=1}^{p} \frac{\left(\sum_{i=1}^{n} Y_{ij} \right)^2}{n} \right] = np\mu^2 + n \sum_{j=1}^{p} \alpha_j^2 + p\sigma_\epsilon^2.$$

The expected value of the first term in (2.3-4) can be determined by substituting $(\mu + \alpha_j + \epsilon_{i(j)})$ for Y_{ij} as follows.

$$E \left(\sum_{j=1}^{p} \sum_{i=1}^{n} Y_{ij}^2 \right) = E \left[\sum_{j=1}^{p} \sum_{i=1}^{n} (\mu + \alpha_j + \epsilon_{i(j)})^2 \right]$$

$$= E \left[\sum_{j=1}^{p} \sum_{i=1}^{n} (\mu^2 + \alpha_j^2 + \epsilon_{i(j)}^2 + 2\mu\alpha_j + 2\mu\epsilon_{i(j)} + 2\alpha_j\epsilon_{i(j)}) \right]$$

$$= E \left[np\mu^2 + n \sum_{j=1}^{p} \alpha_j^2 + \sum_{j=1}^{p} \sum_{i=1}^{n} \epsilon_{i(j)}^2 + 2n\mu \sum_{j=1}^{p} \alpha_j + 2\mu \sum_{j=1}^{p} \sum_{i=1}^{n} \epsilon_{i(j)} \right.$$

$$\left. + 2 \sum_{j=1}^{p} \alpha_j \sum_{i=1}^{n} \epsilon_{i(j)} \right] \quad \text{(Rules A.1, A.2, A.3, A.5, A.6)}$$

$$= np\mu^2 + n \sum_{j=1}^{p} \alpha_j^2 + np\sigma_\epsilon^2 \quad \text{(Rules B.3, B.4, B.5, B.8, B.10)}$$

because

$$E\left(\sum_{j=1}^{p}\sum_{i=1}^{n}\epsilon_{i(j)}^2\right) = np\,(\mu_\epsilon^2 + \sigma_\epsilon^2) = np\sigma_\epsilon^2$$

$$\sum_{j=1}^{p}\alpha_j = 0$$

$$E\left(\sum_{j=1}^{p}\sum_{i=1}^{n}\epsilon_{i(j)}\right) = np\mu_\epsilon = 0$$

$$E\left(\sum_{i=1}^{n}\epsilon_{i(j)}\right) = n\mu_\epsilon = 0.$$

Hence, the expected value of the within-groups sum of squares is

$$E\left[\sum_{j=1}^{p}\sum_{i=1}^{n}Y_{ij}^2 - \sum_{j=1}^{p}\frac{\left(\sum_{i=1}^{n}Y_{ij}\right)^2}{n}\right] = \left(np\mu^2 + n\sum_{j=1}^{p}\alpha_j^2 + np\sigma_\epsilon^2\right)$$

$$- \left(np\mu^2 + n\sum_{j=1}^{p}\alpha_j^2 + p\sigma_\epsilon^2\right)$$

$$= p\,(n-1)\sigma_\epsilon^2.$$

The degrees of freedom for the within-groups sum of squares is $p\,(n-1)$. It follows then that the expected value of $MSWG$ is

$$E(MSWG) = \frac{p\,(n-1)\sigma_\epsilon^2}{p\,(n-1)} = \sigma_\epsilon^2$$

that is, $MSWG$ is an estimator of the population error variance.

EXPECTATION OF TOTAL MEAN SQUARE

The computational formula for a total sum of squares was given in Section 2.3 as

$$SSTO = \sum_{j=1}^{p}\sum_{i=1}^{n}Y_{ij}^2 - \frac{\left(\sum_{j=1}^{p}\sum_{i=1}^{n}Y_{ij}\right)^2}{np}.$$

We have already obtained the expected values of the two terms in the formula. Replacing the terms in the $SSTO$ formula by their expected values, we obtain

$$E(SSTO) = \left(np\mu^2 + n\sum_{j=1}^{p}\alpha_j^2 + np\sigma_\epsilon^2\right) - (np\mu^2 + \sigma_\epsilon^2)$$

$$= (pn-1)\sigma_\epsilon^2 + n\sum_{j=1}^{p}\alpha_j^2.$$

The degrees of freedom for the total sum of squares is $pn - 1$. Hence, the expected value of $MSTO$ is

$$E(MSTO) = \frac{(pn - 1)\sigma_\epsilon^2 + n\sum_{j=1}^{p} \alpha_j^2}{pn - 1} = \sigma_\epsilon^2 + \frac{n\sum_{j=1}^{p} \alpha_j^2}{pn - 1}.$$

If one word could characterize the procedure for determining the expected values of $MSBG$, $MSWG$, and $MSTO$ for the fixed-effects model, it would be tedious. An alternative simpler procedure that works for some experimental designs is described in Section 8.9. Before moving on to the random-effects model, we will summarize the most important points in the preceding discussion.

The fixed-effects model applies when the experiment contains all of the treatment levels about which an experimenter wants to draw conclusions. The assumptions we have made in deriving the expected values for the fixed-effects model for a completely randomized design are as follows.

1. The model equation $Y_{ij} = \mu + \alpha_j + \epsilon_{i(j)}$ reflects the sum of all the sources of variation that affect Y_{ij}.

2. The experiment contains all of the treatment levels, a_j's, of interest.

3. The error effect, $\epsilon_{i(j)}$, (a) is independent of all other $\epsilon_{i(j)}$'s, and (b) is normally distributed within each treatment population, with (c) mean equal to zero and (d) variance equal to σ_ϵ^2.

And finally, when the null hypothesis is true, the expected values of $MSBG$ and $MSWG$ are

$$E(MSBG) = \sigma_\epsilon^2 \quad \text{and} \quad E(MSWG) = \sigma_\epsilon^2;$$

when the null hypothesis is false, the expected values are

$$E(MSBG) = \sigma_\epsilon^2 + \frac{n\sum_{j=1}^{p} \alpha_j^2}{p - 1} \quad \text{and} \quad E(MSWG) = \sigma_\epsilon^2.$$

EXPECTATIONS OF MEAN SQUARES FOR RANDOM-EFFECTS EXPERIMENTAL DESIGN MODEL

Sometimes an experimenter wants to draw conclusions about more treatment levels than can be included in the experiment. This requires obtaining a random sample of p treatment levels from the population of P levels. The results of the experiment can then be generalized to the P levels in the population. The random-effects experimental design model for a type CR-p design is

$$Y_{ij} = \mu + \alpha_j + \epsilon_{i(j)} \quad (i = 1, \ldots, n; j = 1, \ldots, p)$$

where (1) α_j is a random variable (random treatment effect) that is normally and independently distributed with mean equal to zero and variance equal to σ_α^2 and, (2)

Y_{ij}, μ, and $\epsilon_{i(j)}$ are as defined for the fixed-effects model. The expected value of α_j, according to the assumptions that have been made, is

$$E(\alpha_j) = \mu_\alpha = 0 \qquad \text{(Rule B.1)}$$

$$E\left(\sum_{j=1}^{p} \alpha_j\right) = p\mu_\alpha = 0 \qquad \text{(Rule B.8)}$$

$$E\left(\sum_{j=1}^{p} \alpha_j^2\right) = p(\mu_\alpha^2 + \sigma_\alpha^2) = p\sigma_\alpha^2 \qquad \text{(Rule B.10)}$$

$$E\left(\sum_{j=1}^{p} \alpha_j\right)^2 = p(p\mu_\alpha^2 + \sigma_\alpha^2) = p\sigma_\alpha^2. \qquad \text{(Rule B.11)}$$

The expectations of the μ and $\epsilon_{i(j)}$ terms are identical to those for the fixed-effects model. Following the procedures employed for the fixed-effects model, it can be shown that

$$E\left(\sum_{j=1}^{p} \sum_{i=1}^{n} Y_{ij}^2\right) = np\mu^2 + np\sigma_\alpha^2 + np\sigma_\epsilon^2$$

$$E\left[\sum_{j=1}^{p} \frac{\left(\sum_{i=1}^{n} Y_{ij}\right)^2}{n}\right] = np\mu^2 + np\sigma_\alpha^2 + p\sigma_\epsilon^2$$

$$E\left[\frac{\left(\sum_{j=1}^{p} \sum_{i=1}^{n} Y_{ij}\right)^2}{np}\right] = np\mu^2 + n\sigma_\alpha^2 + \sigma_\epsilon^2$$

$$E(SSBG) = (np\mu^2 + np\sigma_\alpha^2 + p\sigma_\epsilon^2) - (np\mu^2 + n\sigma_\alpha^2 + \sigma_\epsilon^2)$$
$$= (p - 1)\sigma_\epsilon^2 + n(p - 1)\sigma_\alpha^2$$

$$E(SSWG) = (np\mu^2 + np\sigma_\alpha^2 + np\sigma_\epsilon^2) - (np\mu^2 + np\sigma_\alpha^2 + p\sigma_\epsilon^2)$$
$$= p(n - 1)\sigma_\epsilon^2.$$

Dividing the sums of squares by their degrees of freedom gives

$$E(MSBG) = \frac{(p - 1)\sigma_\epsilon^2 + n(p - 1)\sigma_\alpha^2}{p - 1} = \sigma_\epsilon^2 + n\sigma_\alpha^2$$

and

$$E(MSWG) = \frac{p(n - 1)\sigma_\epsilon^2}{p(n - 1)} = \sigma_\epsilon^2.$$

When the null hypothesis is true, $\sigma_\alpha^2 = 0$ and $E(MSBG) = E(MSWG) = \sigma_\epsilon^2$.

The expected values of the mean squares for the fixed- and random-effects models are very similar for a completely randomized design. This will not be true for more complex experimental designs. A knowledge of expected values is particularly important for complex designs because this information determines which mean squares should be used in testing various null hypotheses. This point is discussed at length in Section 8.9.

2.4   THE *F* RATIO IN ANALYSIS OF VARIANCE

We will now examine the rationale for using the ratio $F = MSBG/MSWG$ to test hypotheses of the form

Fixed-Effects Model

H_0: $\mu_1 = \mu_2 = \cdots = \mu_p$

where p is the number of treatment levels in the experiment

H_1: $\mu_j \neq \mu_{j'}$ for some j and j'

or

H_0: $\alpha_1 = \alpha_2 = \cdots = \alpha_p = 0$

H_1: $\alpha_j \neq 0$ for some j

Random-Effects Model

H_0: $\mu_1 = \mu_2 = \cdots = \mu_P$

where P is the number of treatment levels in the population

H_1: $\mu_j \neq \mu_{j'}$ for some j and j'

or

H_0: $\sigma_\alpha^2 = 0$

H_1: $\sigma_\alpha^2 \neq 0$.

It may seem paradoxical to use the ratio of two variances to test a hypothesis about means, but this makes sense considering the nature of the *F* random variable and the expected values of *MSBG* and *MSWG*. In Section 2.1 the *F* random variable was defined as the ratio of two independent estimators of a population variance. We have just seen that when the null hypothesis is true, $E(MSBG) = \sigma_\epsilon^2$ and $E(MSWG) = \sigma_\epsilon^2$, that is, *MSBG* and *MSWG* are both estimators of the population error variance. Furthermore, it can be shown that when *Y* is normally distributed, *MSBG* and *MSWG* are statistically independent.* Under these conditions the ratio

$$F = \frac{MSBG}{MSWG}$$

---

*   This follows because *MSWG* depends only on the variances of the samples while *MSBG* depends only on the means of the samples. Since the mean and variance of a sample from a normal population are independent, it follows that *MSBG* and *MSWG* are also independent.

over repeated random samples is distributed as F with $\nu_1 = p - 1$ and $\nu_2 = p(n - 1)$ degrees of freedom. Hence, the sampling distribution of F can be used to determine the probability of an F ratio as large or larger than that obtained, given that the null hypothesis is true. As we have seen, when the null hypothesis is false,

$$E(MSBG) > E(MSWG)$$

and this produces, on the average, a larger value of F. How large should an obtained F be before the prudent experimenter decides that the null hypothesis is probably false? According to convention, an F ratio that falls in the upper 5% of the sampling distribution of F is considered to be sufficient evidence for rejecting the null hypothesis. Values of F that cut off the upper .05 region of the sampling distribution of F for various degrees of freedom are given in Appendix Table E.5.

An F ratio, as just defined, always provides a one-tailed test of the null hypothesis. Ratios less than $\nu_2/(\nu_2 - 2)$ have no meaning with respect to the null hypothesis. Such ratios may occur (1) as a result of chance because both the numerator and denominator are subject to sampling error, (2) because of failure to randomize properly, or (3) because one or more assumptions associated with the derivation of the F sampling distribution or the experimental design model are not tenable.

2.5   EFFECTS OF FAILURE TO MEET ASSUMPTIONS IN ANALYSIS OF VARIANCE

As we have seen, analysis of variance involves two sets of assumptions: those associated with the F random variable and those associated with the experimental design model. The former assumptions apply in testing hypotheses for any ANOVA design; the latter assumptions vary from one design to the next. The two sets of assumptions for a fixed-effects type CR-p design overlap to some degree as can be seen from the following lists.

F Assumptions

A1.   Observations are drawn from normally distributed populations.

A2.   Observations are random samples from the populations.

A3.   The numerator and denominator are estimates of the same population variance, σ_ϵ^2.

A4.   The numerator and denominator of the F ratio are independent.

Model Assumptions

B1.   The model equation $Y_{ij} = \mu + \alpha_j + \epsilon_{i(j)}$ reflects the sum of all the sources of variation that affect Y_{ij}.

B2.　　The experiment contains all of the treatment levels, a_j's, of interest.

B3.　　The error effect, $\epsilon_{i(j)}$, (a) is independent of all other $\epsilon_{i(j)}$'s, and (b) is normally distributed within each treatment population, with (c) mean equal to zero and (d) variance equal to σ_ϵ^2.

We will now briefly discuss these assumptions and the consequences of violating them. For an in-depth discussion, the reader is referred to Glass, Peckham, and Sanders (1972). At the outset it should be observed that for real data some of the assumptions will always be violated. For example, the underlying populations from which samples are drawn are never exactly normally distributed with equal variances. The important question then is not whether the assumptions are violated, but rather whether violations have serious effects on the significance level and the power of the F test. Cochran and Cox (1957, 91) have observed that a test performed at the .05 level, for example, may actually be made at the .04 or .09 level. Also, a loss of power results when assumptions are not fulfilled because it is often possible to construct a more powerful test than that using the F ratio if the correct model can be specified. Fortunately, as we will see, the F test is robust with respect to violation of a number of assumptions. That is, the test is not very sensitive to departures from some of its assumptions.

ASSUMPTION OF NORMALITY (A1, B3b) AND INDEPENDENCE OF *MSBG* AND *MSWG* (A4)

The classic studies by Pearson (1931) and Norton as cited by Lindquist (1953) indicate that the F test is quite robust with respect to violation of the normality assumption. Skewed populations have very little effect on either the level of significance or the power of the F test for the fixed-effects model. Platykurtic (flat) and leptokurtic (peaked) populations also have little effect on significance level but can have an appreciable effect on power when the sample n_j's are small. In general, an experimenter need not be concerned about moderate departures from normality provided that the populations are homogeneous in form, for example, all positively skewed and slightly leptokurtic. It is usually possible by examining sample frequency distributions of Y_{ij} or $\hat{\epsilon}$, in the case of complex designs, to detect cases in which an ANOVA will lead to gross errors in interpreting the outcome of an experiment.

Further support for the robustness of the F test with respect to nonnormality comes from a different kind of investigation. Lunney (1970) studied the effect of using a dichotomous dependent variable, the ultimate in nonnormality, on the significance level of the F test. The actual significance levels were found to be quite close to the nominal significance levels when the sample n_j's were equal. This result does not hold for unequal n_j's. Similar results were obtained by Hsu and Feldt (1969) who investigated four dependent scale lengths: 2, 3, 4, and 5 points. They reported that the five-point scale with sample n_j's as small as 11 gave excellent control of significance level in the presence of moderate heterogeneity of variance, platykurtosis, and skewness. Interestingly, the three-point scale was superior to the four-point scale. With larger n_j's, even the two-point scale resulted in adequate control of the significance level.

The assumption (B3b) that the $\epsilon_{i(j)}$'s within each population are normally distributed is equivalent to the assumption that the Y_{ij}'s in each of the populations are normally distributed (A1). This follows since according to the linear model equation $Y_{ij} = \mu + \alpha_j + \epsilon_{i(j)}$ the $\epsilon_{i(j)}$'s are the only source of variation within a particular treatment population.

It can be shown that the numerator and denominator of the F ratio are independent (assumption A4) if the populations are normally distributed or approximately so (Lindquist, 1953, 31–35).

It is sometimes possible to transform nonnormally distributed scores so that the underlying population if subjected to the transformation would approach normality. Procedures for selecting an appropriate transformation are described in Section 2.6. Considering the robustness of the F test to nonnormality when the n_j's are equal, the use of a transformation for this purpose will rarely be advantageous (Games and Lucas, 1966).

ASSUMPTION OF RANDOM SAMPLING (A2) AND INDEPENDENCE OF ERROR EFFECTS (B3a)

The validity of an experiment depends on random sampling and/or random assignment of treatment levels to the experimental units. Random assignment is used to distribute the idiosyncratic characteristics of subjects over the treatment levels so that they will not selectively bias the outcome of the experiment.

Random sampling and assignment ensure that the error effects $\epsilon_{i(j)}$'s are independent. The errors for subjects i and i' are independent if $p(\epsilon_{i(j)}|\epsilon_{i'(j)}) = p(\epsilon_{i(j)})$. That is, the errors are independent if knowing subject i''s error tells us nothing about subject i's error. For a type CR-p design with or without equal n_j's, nonindependence of errors seriously affects both the level of significance and the power of the F test. Some experimental designs permit nonindependence of scores but still require that the errors be independent; we will return to this topic in Section 6.4

ASSUMPTION THAT NUMERATOR AND DENOMINATOR ESTIMATE THE SAME VARIANCE (A3)

This assumption corresponds to the null hypothesis. It is advanced in the hope that it can be rejected.

ASSUMPTION THAT $Y_{ij} = \mu + \alpha_j + \epsilon_{i(j)}$ (B1)

Assumption B1 states that a score is the sum of three components: the overall elevation of scores, μ; the elevation or depression of scores attributable to the jth treatment level, α_j; and all effects not attributable to the jth treatment level, $\epsilon_{i(j)}$. It is possible for treatment and error effects to combine in some other manner, say, multiplicatively. In such cases, one of the transformations described in Section 2.6 may produce an additive model.

A type CR-p design contains one treatment with p levels randomly assigned to experimental units. If, for example, an experimenter wants to observe the experimental units under more than one treatment level, a different ANOVA design must be used. The choice of an incorrect design can seriously affect the level of significance and power of the F test.

ASSUMPTION THAT THE EXPERIMENT CONTAINS ALL TREATMENT LEVELS OF INTEREST (B2)

This assumption identifies the model $Y_{ij} = \mu + \alpha_j + \epsilon_{i(j)}$ as a fixed-effects model and distinguishes it from a random-effects model in which the p treatment levels are a random sample from the population of P levels. As we have seen, the expectation of $MSBG$ is different for the fixed- and random-effects models as is the nature of the null hypothesis tested.

ASSUMPTION THAT $E(\epsilon_{i(j)}) = 0$ FOR EACH TREATMENT POPULATION (B3c)

Assumption B3c states that the mean of the error effects within each treatment population is equal to zero. This follows from the way an error effect is defined.

$$\epsilon_{i(j)} = Y_{ij} - \mu - \alpha_j$$
$$= Y_{ij} - \mu_j$$

and

$$E(Y_{ij} - \mu_j) = E(Y_{ij}) - \mu_j$$
$$= \mu_j - \mu_j = 0.$$

ASSUMPTION OF HOMOGENEITY OF VARIANCE (B3d)

Assumption B3d states that the variances of the p populations are homogeneous. It can be written as

$$H_0: \quad \sigma_1^2 = \sigma_2^2 = \cdots = \sigma_p^2$$
$$H_1: \quad \sigma_j^2 \neq \sigma_{j'}^2 \quad \text{for some } j \text{ and } j'.$$

The F test is robust with respect to moderate violations of this assumption provided that the number of observations in the samples is equal (Box, 1954a,b; Cochran, 1947; Norton as cited by Lindquist, 1953). When the variances are quite heterogeneous, Rogan and Keselman (1977) found that the actual significance level may be appreciably larger than the nominal level. For samples of unequal size, moderate violation of the homogeneity assumption can have a marked effect on the test of significance.

According to Box (1953, 1954a), the nature of the bias for this latter case may be positive or negative. The actual significance level will exceed the nominal level when the smaller samples are drawn from the more heterogeneous populations; the actual level will be less than the nominal level when the smaller samples are drawn from the more homogeneous populations.

A number of statistics are available for testing the hypothesis

$$H_0: \quad \sigma_1^2 = \sigma_2^2 = \cdots = \sigma_p^2.$$

When sample n_j's are equal, there is little reason to test the homogeneity of variance assumption prior to performing an analysis of variance. Occasionally an experimenter's research hypothesis involves population variances, in which case one of the statistics to be described may be of interest.

Tests of homogeneity of variance can be classified as either robust or not robust to nonnormality. Two easy-to-use tests that fall in the latter category are those of Hartley (1940, 1950) and Cochran (1941).* Both tests require equal or approximately equal sample n's. Hartley's test statistic is

$$F_{max} = \frac{\text{largest of } p \text{ variances}}{\text{smallest of } p \text{ variances}} = \frac{\hat{\sigma}_{j\,largest}^2}{\hat{\sigma}_{j\,smallest}^2}$$

with degrees of freedom equal to p and $n - 1$, where

$$\hat{\sigma}_j^2 = \left[\sum_{i=1}^{n} Y_{ij}^2 - \frac{\left(\sum_{i=1}^{n} Y_{ij} \right)^2}{n} \right] \Big/ (n - 1)$$

p is the number of variances, and n is the number of observations within each treatment level. Critical values for the distribution of F_{max} are given in Table E.10. The hypothesis of homogeneity of variance is rejected if F_{max} is greater than the tabled value for $F_{max,\alpha}$. If the n's for the treatment levels differ only slightly, the largest of the n's can be used for purposes of determining the degrees of freedom for this test. This procedure leads to a slight positive bias in the test, that is, in rejecting the hypothesis of homogeneity more frequently than it should be rejected.

Cochran's (1941) test statistic is

$$C = \frac{\hat{\sigma}_{j\,largest}^2}{\sum_{j=1}^{p} \hat{\sigma}_j^2}$$

where $\sigma_{j\,largest}^2$ is the largest of the p treatment variances and $\sum_{j=1}^{p} \hat{\sigma}_j^2$ is the sum of all of the variances. The degrees of freedom for this test are equal to p and $n - 1$ as defined for the F_{max} test. Critical values for the distribution of C are given in Table E.11.

A test proposed by Box (1953) and modified by Scheffé (1959, 83) called the

---

*   The well-known test by Bartlett (1937) is also not robust to nonnormality. It is not discussed here because it is more complex to perform than the tests by Hartley and Cochran.

Box-Scheffé test has been found to be quite robust to nonnormality (Martin and Games, 1977).* It has the added advantage that it can be used when the sample n's are unequal. The first step in performing the test is to randomly divide the observations in each of the p samples into v_j subsamples. According to Games, Keselman, and Clinch (1979) the optimum subsample size, n_{hj}, is the nearest integer value of $(n_j)^{1/2}$. If, for example, n_j is equal to eight, the subsample size should be $(8)^{1/2} \cong 3$. This results in $v_j = 3$ subsamples of size $n_{hj} = 3, 3,$ and 2. The computational procedures for the Box-Scheffé test are illustrated in Table 2.5-1. The procedure involves performing an analysis of variance on the logarithms of the subsample variances. According to part (iii) of the table, the null hypothesis

$$H_0: \quad \log_e \sigma_1^2 = \log_e \sigma_2^2 = \log_e \sigma_3^2 = \log_e \sigma_4^2$$

remains tenable. There is no reason for believing that the four populations from which the samples were obtained have heterogeneous variances.

2.6 TRANSFORMATIONS

A transformation is any systematic alteration in a set of scores whereby certain characteristics of the set are changed and other characteristics remain unchanged. Three reasons for performing transformations in analysis of variance are

1. To achieve homogeneity of error variance.
2. To achieve normality of error effects.
3. To obtain additivity of effects.

Because the F distribution is relatively unaffected by heterogeneity of variance and lack of normality, the first two reasons for performing a transformation are less compelling than the third.** In this context, additivity of effects implies that the effects do not interact.

Obtaining additivity of effects is particularly important in designs, such as a randomized block design, where a *residual* mean square (*MSRES*) is used as an estimate of experimental error. For example, if block and treatment effects are not additive, the expected value of the residual mean square is

$$E(MSRES) = \sigma_\epsilon^2 + \sigma_{\pi\alpha}^2$$

instead of

$$E(MSRES) = \sigma_\epsilon^2$$

where $\sigma_{\pi\alpha}^2$ refers to the interaction of blocks and treatment levels. Interaction is said

---

*   For a discussion of other robust tests the reader is referred to Games, Keselman, and Clinch (1979) and O'Brien (1979).

**   Budescu and Appelbaum (1981), for example, report that the use of variance stabilizing transformations has little effect on the significance level and power of the F test.

TABLE 2.5-1 Computational Procedures for Box-Scheffé Test for Homogeneity of Variance

(i) Data and notation [Y_{hij} denotes a score for experimental unit i in subsample h and treatment level j; $h = 1, \ldots, v_j$ subsamples; $i = 1, \ldots, n_j$ experimental units; $j = 1, \ldots, p$ treatment levels (a_j)]:

$$\hat{\sigma}^2_{hj} = \left[\sum_{i=1}^{n_{hj}} Y^2_{hij} - \frac{\left(\sum_{i=1}^{n_{hj}} Y_{hij}\right)^2}{n_{hj}} \right] \Big/ \nu_{hj}$$

$$\nu_{hj} = n_{hj} - 1$$

$$\nu_{\cdot j} = \sum_{h=1}^{v_j} \nu_{hj}$$

$$\nu = \sum_{h=1}^{v_j} \sum_{j=1}^{p} \nu_{hj}$$

$$T_{hj} = \log_e \hat{\sigma}^2_{hj}$$

$$\overline{T}_{\cdot j} = \sum_{h=1}^{v_j} \nu_{hj} T_{hj} / \nu_{\cdot j}$$

$$\overline{T}_{\cdot \cdot} = \sum_{j=1}^{p} \nu_{\cdot j} \overline{T}_{\cdot j} / \nu$$

Y_{hi1}	$\hat{\sigma}^2_{h1}$	ν_{h1}	T_{h1}	Y_{hi2}	$\hat{\sigma}^2_{h2}$	ν_{h2}	T_{h2}	Y_{hi3}	$\hat{\sigma}^2_{h3}$	ν_{h3}	T_{h3}	Y_{hi4}	$\hat{\sigma}^2_{h4}$	ν_{h4}	T_{h4}
3				4				6				8			
3	.3333	2	−1.0986	5	.3333	2	−1.0986	5	.3333	2	−1.0986	9	.3333	2	−1.0986
2				4				6				8			
2				3				8				11			
2	.3333	2	−1.0986	3	.3333	2	−1.0986	7	.3333	2	−1.0986	7	4.3333	2	1.4663
1				2				7				10			
6	4.5000	1	1.5041	4	.5000	1	−.6931	5	.5000	1	−.6931	10	.5000	1	−.6931
3				3				6				9			

$\nu_{\cdot 1} = 5$ $\overline{T}_{\cdot 1} = -.5781$ $\nu_{\cdot 2} = 5$ $\overline{T}_{\cdot 2} = -1.0175$ $\nu_{\cdot 3} = 5$ $\overline{T}_{\cdot 3} = -1.0175$ $\nu_{\cdot 4} = 5$ $\overline{T}_{\cdot 4} = .0085$

$\overline{T}_{\cdot \cdot} = -.6512$

(ii) Computational symbols:

$$\nu \overline{T}^2_{\cdot \cdot} = [Y] = 20(-.6512)^2 = 8.4812$$

$$\sum_{h=1}^{v_j} \sum_{j=1}^{p} \nu_{hj} T^2_{hj} = [AS] = 2(-1.0986)^2 + 2(-1.0986)^2 + \cdots + 1(-.6931)^2 = 24.9005$$

$$\sum_{j=1}^{p} \nu_{\cdot j} \overline{T}^2_{\cdot j} = [A] = 5(-.5781)^2 + 5(-1.0175)^2 + \cdots + 5(.0085)^2 = 12.0244$$

(iii) Computational formulas and F test:

$$SSBG = [A] - [Y] = 3.5432$$

$$SSWG = [AS] - [A] = 12.8761$$

TABLE 2.5-1 (continued)

$$MSBG = SSBG/(p - 1) = 3.5432/(4 - 1) = 1.1811$$

$$MSWG = SSWG/\sum_{j=1}^{p} (v_j - 1) = 12.8761/(2 + 2 + 2 + 2) = 1.6095$$

$$F = \frac{MSBG}{MSWG} = 0.73$$

$$F_{.05;3,8} = 4.07$$

to be present when the dependent variable behaves differently for different blocks of subjects. The expected value of the treatment mean square for a fixed-effects treatment is

$$E(MSA) = \sigma_\epsilon^2 + \frac{n\sum_{j=1}^{p} \alpha_j^2}{(p - 1)}.$$

If the null hypothesis is true, then, according to Section 2.4, the numerator and denominator of the ratio

$$F = \frac{MSA}{MSRES}$$

should provide independent estimates of the same population error variance, σ_ϵ^2. It is apparent from an examination of the expected values of the two mean squares that this can occur only if $\sigma_{\pi\sigma}^2 = 0$. We will return to this point in Chapter 6.

Sometimes nonadditivity of effects results from the particular choice of a scale of measurement for the dependent variable, and additivity may be achieved by transforming or changing the scale of measurement.

Fortunately a transformation that accomplishes any one of the three objectives listed will often accomplish the other two objectives. In general, a transformation can be used whenever the means of the treatment levels and the variances of the error effects are proportional and whenever the error-effects distribution shapes are homogeneous. It is not always possible to find an appropriate transformation for a set of data. For example, if any of the following conditions are present, no transformation exists that will make the data more suitable for analysis of variance:

1. Means of treatment levels are approximately equal but variances of the error effects are heterogeneous.
2. Means of treatment levels vary independently of error effect variances.
3. Error effect variances are homogeneous but treatment level distributions are heterogeneous in form.

If no transformation is appropriate and if the departures from the required assumptions are gross, an experimenter may be able to use a nonparametric ANOVA (see Marascuilo and McSweeney, 1977) or a different criterion measure. The choice of a dependent variable in the behavioral sciences is often arbitrary; a different choice may fulfill the requirements of additivity, normality, and homogeneity.

A number of procedures exist for determining which transformation is appropriate for a set of data. Several methods are described by Box, Hunter, and Hunter (1978, 231–241, 334–336); Olds, Mattson, and Odeh (1956); and Tukey (1949). One procedure is to follow general rules concerning situations in which a given transformation is often successful. This approach will be emphasized in describing each of the types of transformations. Alternative procedures for selecting a transformation are described later.

SQUARE-ROOT TRANSFORMATION

For certain types of data, treatment level means and variances tend to be proportional, as in a Poisson distribution where $\mu = \sigma^2$. This kind of distribution often results when the dependent variable is a frequency count of events having a small probability of occurrence, for example, number of errors at each choice point in a relatively simple multiple T maze. The data can often be normalized for this type of situation by taking the square root of each of the scores. A transformed score Y' is given by

$$Y' = \sqrt{Y}.$$

If any Y is less than 10, a more appropriate transformation is given either by

$$Y' = \sqrt{Y + .5} \quad \text{or} \quad Y' = \sqrt{Y} + \sqrt{Y + 1}.$$

The latter transformation has been recommended by Freeman and Tukey (1950). Tables of $\sqrt{Y} + \sqrt{Y + 1}$ are reproduced in Mosteller and Bush (1954). The effects of performing a square-root transformation are shown for the data in Table 2.6-1. An examination of the means and variances of the transformed scores shows

TABLE 2.6-1   Original and Transformed Scores for a Type CR-3 Design*

| | Original Scores | | | Transformed Scores $Y' = \sqrt{Y + .5}$ | |
a_1	a_2	a_3	a_1	a_2	a_3
3	6	12	1.87	2.55	3.54
0	4	6	.71	2.12	2.55
4	2	6	2.12	1.58	2.55
2	4	10	1.58	2.12	3.24
2	7	6	1.58	2.74	2.55
$\bar{Y} = 2.2$	4.6	8.0	1.57	2.22	2.89
$\hat{\sigma}^2_{\epsilon_j} = 2.2$	3.8	8.0	.28	.20	.22

* Note that $\hat{\sigma}^2_j = \hat{\sigma}^2_{\epsilon_j}$ for a type CR-p design.

that they are no longer proportional; additionally, the variances are more homogeneous. These transformed scores are more suitable than the original scores for an analysis of variance.

LOGARITHMIC TRANSFORMATION

If treatment means and standard deviations tend to be proportional, a logarithmic transformation may be appropriate. A transformed score Y' is given by

$$Y' = \log_{10} Y \qquad \text{or} \qquad Y' = \log_{10}(Y + 1).$$

The latter formula is used when some scores are zero or very small. Logarithmic transformations have been found to be useful when the dependent variable is some measure of reaction time and the data are positively skewed.

RECIPROCAL TRANSFORMATION

If the squares of treatment means are proportional to standard deviations, a reciprocal transformation may be appropriate. A transformed score Y' is given by

$$Y' = \frac{1}{Y} \qquad \text{or} \qquad Y' = \frac{1}{Y + 1}.$$

The latter formula should be used if any scores are equal to zero. A reciprocal transformation may be useful when the dependent variable is reaction time.

ANGULAR OR INVERSE SINE TRANSFORMATION

The angular transformation is given by

$$Y' = 2 \arcsin \sqrt{Y}$$

where Y is a proportion. It is not necessary to solve for Y' in the preceding formula; values of Y from .001 to .999 are given in Table E.13. The transformed values in Table E.13 are in radians. Bartlett (1947) suggests that $1/2n$ or $1/4n$ be substituted for $Y = $ zero and $1 - 1/2n$ or $1 - 1/4n$ be substituted for $Y = 1$, where n is the number of observations on which each proportion is based. An angular transformation may be useful when means and variances are proportional and the distribution has a binomial form. This condition may occur when the number of trials is fixed and Y is the probability of a correct response that varies from one treatment level to another.

SELECTING A TRANSFORMATION

We have described situations where particular transformations have been found to be useful. An alternative approach to selecting a transformation is to apply each of the transformations to the largest and smallest score in each treatment level. The range within each treatment level is then determined and the ratio of the largest to the smallest range is computed. The transformation that produces the smallest ratio is selected as the most appropriate one. This procedure is illustrated in Table 2.6-2 for the data in Table 2.6-1. On the basis of this procedure, a square-root formation would be selected for these data.

Once an appropriate transformation is selected and the data analyzed on the new scale, all inferences regarding treatment effects must be made with respect to the new scale. In most behavioral research situations, inferences based on log Y's or $\sqrt{Y}$'s, for example, are just as meaningful as inferences based on untransformed scores.

If additivity of treatment effects is the principal concern of an experimenter, the appropriateness of a particular transformation can be determined by the test of nonadditivity described in Section 6.3. This test provides a means of determining if effects are additive for the untransformed scores and for any transformations that may be tried. A mathematically sophisticated exposition of general issues involved in the use of transformations is given by Box and Cox (1964).

TABLE 2.6-2   Transformations Applied to Largest and Smallest Scores in Table 2.6-1

	\multicolumn{3}{c}{*Treatment Levels*}	$\dfrac{\text{Range}_{\text{largest}}}{\text{Range}_{\text{smallest}}}$
	a_1	a_2	a_3	
Largest score (L)	4	7	12	
Smallest score (S)	0	2	6	
Range =	4	5	6	$6/4 = 1.50$
$\sqrt{L + .5}$	2.12	2.74	3.54	
$\sqrt{S + .5}$	.71	1.58	2.55	
Range =	1.41	1.16	.99	$1.41/.99 = 1.42$
$\log(L + 1)$	.6990	.9031	1.1139	
$\log(S + 1)$	.0000	.4771	.8451	
Range =	.6990	.4260	.2688	$.6990/.2688 = 2.60$
$1/(L + 1)$	.200	.125	.077	
$1/(S + 1)$	1.000	.333	.143	
Range =	.800	.208	.066	$.800/.066 = 12.12$

2.7 REVIEW EXERCISES

1. [2.1] Whose name is most closely associated with the derivation of each of the following sampling distributions?

   a) Chi-square

   b) F

   c) t

2. [2.1] For each of the following chi-square random variables, determine the mean and variance.

   †a) $\chi^2_{(20)}$ c) $\chi^2_{(6)}$

   †b) $\chi^2_{(16)}$ d) $\chi^2_{(12)}$

†3. [2.1] Use the summation rules in Appendix A to prove that $\sum_{i=1}^{n}(Y_i - \overline{Y}) = 0$.

4. [2.1] List the assumptions associated with the use of the chi-square sampling distribution.

5. [2.1] Compute the value of F that cuts off the lower $1 - \alpha$ region of the sampling distribution for the following.

   †a) $F_{1-.05;3,20}$ c) $F_{1-.05;3,10}$

   †b) $F_{1-.01;5,15}$ d) $F_{1-.01;10,30}$

6. [2.1] For each of the following F random variables, determine the mean and variance.

   †a) $F_{6,30}$ c) $F_{5,20}$

   †b) $F_{10,50}$ d) $F_{3,20}$

7. [2.1] For each of the following t random variables, determine the mean and variance.

   †a) t_{20} c) t_{100}

   †b) t_{60} d) t_{30}

8. [2.2] Explain why the experimental design model equation is the starting point for partitioning the total sum of squares.

9. [2.2] The partition of $SSTO$ into $SSBG$ and $SSWG$ and the derivation of convenient computational formulas require the use of six summation rules (A.1, A.2, . . . , A.6). These rules are described in Appendix A. For each of the following, where $i = 1, \ldots, n$ and $j = 1, \ldots, p$, indicate the summation rule(s) that applies.

   †a) $\displaystyle\sum_{j=1}^{p} \overline{Y}_{\cdot j} = \sum_{j=1}^{p} \overline{Y}_{\cdot j}$

   †b) $\displaystyle\sum_{j=1}^{p} \sum_{i=1}^{n} (\overline{Y}_{\cdot j} - \overline{Y}_{\cdot\cdot})^2 = n \sum_{j=1}^{p} (\overline{Y}_{\cdot j} - \overline{Y}_{\cdot\cdot})^2$

   †c) $\displaystyle\sum_{j=1}^{p} \sum_{i=1}^{n} 2(\overline{Y}_{\cdot j} - \overline{Y}_{\cdot\cdot})(Y_{ij} - \overline{Y}_{\cdot j}) = 2 \sum_{j=1}^{p} (\overline{Y}_{\cdot j} - \overline{Y}_{\cdot\cdot}) \sum_{i=1}^{n} (Y_{ij} - \overline{Y}_{\cdot j})$

†d) $\sum_{i=1}^{n} \overline{Y}_{.j} = n\overline{Y}_{.j}$

e) $\sum_{j=1}^{p} \sum_{i=1}^{n} (Y_{ij} - \overline{Y}_{.j})^2 = \sum_{j=1}^{p} \sum_{i=1}^{n} (Y_{ij} - \overline{Y}_{.j})^2$

f) $\sum_{j=1}^{p} \sum_{i=1}^{n} (Y_{ij}^2 - 2\overline{Y}_{..}Y_{ij} + \overline{Y}_{..}^2) = \sum_{j=1}^{p} \sum_{i=1}^{n} Y_{ij}^2 - 2\overline{Y}_{..} \sum_{j=1}^{p} \sum_{i=1}^{n} Y_{ij} + np\overline{Y}_{..}^2$

g) $\sum_{j=1}^{p} \sum_{i=1}^{n} 2\overline{Y}_{.j}Y_{ij} = 2\sum_{j=1}^{p} \overline{Y}_{.j} \sum_{i=1}^{n} Y_{ij}$

h) $\sum_{j=1}^{p} \sum_{i=1}^{n} \overline{Y}_{.j}^2 = n\sum_{j=1}^{p} \overline{Y}_{.j}^2$

10.   [2.2] Determine the degrees of freedom for *SSTO*, *SSBG*, and *SSWG* for the following type CR-*p* designs.

†a) CR-3 with $n = 10$

b) CR-5 with $n = 6$

†c) CR-4 with $n_1 = 3$, $n_2 = 4$, $n_3 = 4$, $n_4 = 5$

d) CR-3 with $n_1 = 6$, $n_2 = 6$, $n_3 = 5$

†e) CR-6 with $n = 8$

f) CR-3 with $n_1 = 7$, $n_2 = 10$, $n_3 = 8$

g) CR-2 with $n = 20$

11.   [2.3] For each of the following experiments or investigations, indicate whether a fixed-effects or random-effects model is appropriate.

†a) Random samples of white and black employed mothers were queried concerning the number and nature of the physical symptoms that would lead them to not send a child to school.

†b) Twenty male Wistar rats were subjected to 22 hours of food deprivation. Tetrahydrocannabinol was injected intraperiteally at five randomly selected times following the deprivation period. Activity level was recorded following each injection.

c) The effects of completing 1, 2, 3, or 4 courses in the social sciences on liberalizing political attitudes were investigated using a random sample of 200 college students.

d) The amount of social interaction initiated by 4–6 year old children classified as mildly, moderately, severely, and nonhandicapped was investigated.

e) Eight randomly selected dosages of testosterone propionate were administered to 80 immature male domestic ducks of the Rouen breed. The dependent variable was the number of different patterns of sexual behavior that were exhibited following administration of the drug.

12.   [2.3] Assume a fixed-effects model for a type CR-*p* design. Give the expected value of each of the following and indicate the relevant rule(s) (B.1, B.2, . . . , B.10) from Appendix B.

†a) $E(n^2\mu^2)$

†f) $E\left(2n\alpha_j \sum_{i=1}^{n} \epsilon_{i(j)}\right)$

†b) $E(n^2\alpha_j^2)$

g) $E\left[n^2\left(\sum_{j=1}^{p} \alpha_j\right)^2\right]$

†c) $E\left(\sum_{i=1}^{n} \epsilon_{i(j)}\right)^2$ †h) $E\left(\sum_{j=1}^{p} \sum_{i=1}^{n} \epsilon_{i(j)}\right)^2$

d) $E\left(\sum_{i=1}^{n} \epsilon_{i(j)}\right)$ i) $E(n^2 p^2 \mu^2)$

e) $E\left(2n\mu \sum_{i=1}^{n} \epsilon_{i(j)}\right)$ j) $E(2n^2 \mu \alpha_j)$

13. [2.3] Do Exercise 12 assuming a random-effects model.

14. [2.3] Derive the expected values of $MSBG$ and $MSWG$ in a type CR-p design assuming a random-effects model.

†15. [2.3] Compare the scope of the null hypotheses for the fixed- and random-effects models for a type CR-p design.

†16. [2.4] Explain why H_0: $\sigma_\alpha^2 = 0$ is a test of the hypothesis that $\alpha_1 = \alpha_2 = \cdots = \alpha_P = 0$.

17. [2.4] Explain the rationale for using the F statistic, which is the ratio of two independent variances, to test the hypothesis of equality of means.

†18. [2.5] Qualify the statement, "The F test is quite robust with respect to violation of the normality assumption."

19. [2.5] Discuss the statement, "The F test is not appropriate for dichotomous data since such data depart radically from the normal distribution."

20. [2.5] What purposes do random sampling and random assignment serve?

†21. [2.5] Qualify the statement, "The F test is robust with respect to moderate violation of the homogeneity of variance assumption."

22. [2.5] Classify the tests of homogeneity in terms of those which are (1) robust or not robust to nonnormality and (2) appropriate for equal or unequal n_j's.

†23. [2.5] Three programs for improving the reading skills of sixth graders were evaluated. Twenty-seven boys were randomly assigned to one of three groups; each group used one of the programs. Treatment level a_1 was the Delacato program, a_2 was the Stanford program, and a_3 was a control condition. The Iowa Test of Reading Skills was administered at the beginning and end of the school year. The dependent variable was the amount of increase from the pretest to the posttest. The following data were obtained.

a_1	a_2	a_3
20	15	12
18	20	15
18	13	18
23	12	20
22	16	18
17	17	17
15	21	10
13	15	24
21	13	16

Test the hypothesis that the population variances are homogeneous using **(a)** the F_{max} test, **(b)** Cochran's C test, and **(c)** the Box-Scheffé test. For the latter test, assume that each of the three samples have been randomly divided into three subsamples. The first three observations in each sample represent one subsample; the next three, the second subsample; and the last three, the third subsample. Let α equal .05. **(d)** What is your decision regarding the null hypothesis?

24.   [2.5] The experiment in Exercise 23 was repeated with a random sample of 27 boys in the fifth grade. The following data were obtained.

a_1	a_2	a_3
17	10	16
13	4	7
16	9	6
13	5	18
9	15	19
10	5	2
15	12	11
11	16	20
15	9	5

Test the hypothesis that the population variances are homogeneous; follow the instructions in Exercise 23.

†25.   [2.6] Determine if a transformation would be useful for the data in Exercise 24, and, if so, which one by using the first approach described in the text.

†26.   [2.6] Determine if a transformation would be appropriate for the data in Exercise 24, and, if so, which one by using the procedure shown in Table 2.6-2.

27.   [2.6] The legibility of four versions of a dial was investigated. Twenty pilots and navigators with over 2000 hours of flying time were randomly assigned to one of four groups; each group read one version of the dials. Pictures of the dials were projected on a screen for 600 milliseconds using a slide projector. Each subject read 100 settings on a dial. The dependent variable was the number of reading errors. The following data were obtained.

Dial

a_1	a_2	a_3	a_4
5	3	15	10
11	6	8	6
7	3	10	6
7	5	6	12
5	3	13	6

a) Use the first approach described in the text to determine which, if any, transformation is appropriate for these data.

b) Use the procedure illustrated in Table 2.6-2 to determine which, if any, transformation is appropriate for these data.

†**28.** [2.6] Give rules of thumb for deciding which transformation is appropriate for a set of data.

29. [2.6] The effects of lesions in the parafascicular nucleus of 25 male Norway rats (Rattus norvegicus) on running time in a straight-alley maze was investigated. The rats were randomly assigned to one of five groups, subject to the restriction that each group contained five rats. The extent of the lesions differed for each group. The following data are running times in seconds for the five groups.

a_1	a_2	a_3	a_4	a_5
28	7	6	177	184
22	11	9	151	146
54	30	26	110	131
19	6	7	117	110
32	11	7	135	134

a) Use the first approach described in the text to determine which, if any, transformation is appropriate for these data.

b) Use the procedure in Table 2.6-2 to determine which, if any, transformation is appropriate for these data.

†**30.** [2.6] If data are not suitable for ANOVA and an appropriate transformation can not be found, what recourses does an experimenter have?

3 MULTIPLE COMPARISON TESTS

3.1 INTRODUCTION TO MULTIPLE COMPARISON TESTS

The most common use of analysis of variance is in testing the hypothesis that $p \geq 3$ population means are equal. Other important uses are described in subsequent chapters. If the overall hypothesis of equality of means is rejected, an experimenter is still faced with the problem of deciding which of the means are not equal. Thus an overall F test is often merely the first step in analyzing a set of data. A significant F ratio indicates that something has happened in an experiment that has a small probability of happening by chance. The purpose of this chapter is to describe a variety of procedures for pinpointing what has happened. Specifically we will examine a number of test statistics for deciding which population means in an experiment are not equal. But first, several important concepts need to be defined.

CONTRAST DEFINED

A *contrast* or *comparison* among means is a difference among the means, with appropriate algebraic signs. We will use the symbols ψ_i and $\hat{\psi}_i$ to denote, respectively, the ith contrast among population means and a sample estimate of the ith contrast. For example, $\hat{\psi}_i = \overline{Y}_j - \overline{Y}_{j'}$ is the ith contrast between sample means for treatment levels j and j'. If an experiment contains p equal to three treatment levels, contrasts involving two and three means may be of interest:

(3.1-1)
$$\begin{array}{ll} \hat{\psi}_1 = \overline{Y}_1 - \overline{Y}_2 & \hat{\psi}_4 = (\overline{Y}_1 + \overline{Y}_2)/2 - \overline{Y}_3 \\ \hat{\psi}_2 = \overline{Y}_1 - \overline{Y}_3 & \hat{\psi}_5 = (\overline{Y}_1 + \overline{Y}_3)/2 - \overline{Y}_2 \\ \hat{\psi}_3 = \overline{Y}_2 - \overline{Y}_3 & \hat{\psi}_6 = (\overline{Y}_2 + \overline{Y}_3)/2 - \overline{Y}_1. \end{array}$$

The contrasts on the right involve the average of two means versus a third mean.

More formally, a contrast or comparison among sample means is a linear combination of means with known coefficients, c_j's, such that (1) at least one coefficient is not equal to zero and (2) the coefficients sum to zero. That is,

$$\hat{\psi}_i = c_1\overline{Y}_1 + c_2\overline{Y}_2 + \cdots + c_p\overline{Y}_p$$

where $c_j \neq 0$ for some j and $\Sigma_{j=1}^p c_j = 0$. If two of the coefficients are equal to 1 and -1 and all of the other coefficients are equal to zero, the contrast is called a *pairwise comparison;* otherwise it is a *nonpairwise comparison.* The contrasts in (3.1-1) can be expressed as linear combinations of the $\overline{Y}_j$'s by the appropriate choice of coefficients.

$$\begin{array}{llll} \hat{\psi}_i = c_1\overline{Y}_1 & + c_2\overline{Y}_2 & + c_3\overline{Y}_3 \\ \hline \hat{\psi}_1 = (1)\overline{Y}_1 & + (-1)\overline{Y}_2 & + (0)\overline{Y}_3 & = \overline{Y}_1 - \overline{Y}_2 \\ \hat{\psi}_2 = (1)\overline{Y}_1 & + (0)\overline{Y}_2 & + (-1)\overline{Y}_3 & = \overline{Y}_1 - \overline{Y}_3 \\ \hat{\psi}_3 = (0)\overline{Y}_1 & + (1)\overline{Y}_2 & + (-1)\overline{Y}_3 & = \overline{Y}_2 - \overline{Y}_3 \\ \hat{\psi}_4 = (½)\overline{Y}_1 & + (½)\overline{Y}_2 & + (-1)\overline{Y}_3 & = (\overline{Y}_1 + \overline{Y}_2)/2 - \overline{Y}_3 \\ \hat{\psi}_5 = (½)\overline{Y}_1 & + (-1)\overline{Y}_2 & + (½)\overline{Y}_3 & = (\overline{Y}_1 + \overline{Y}_3)/2 - \overline{Y}_2 \\ \hat{\psi}_6 = (-1)\overline{Y}_1 & + (½)\overline{Y}_2 & + (½)\overline{Y}_3 & = (\overline{Y}_2 + \overline{Y}_3)/2 - \overline{Y}_1 \end{array}$$

Note that for each contrast, $c_j \neq 0$ for some j and $\Sigma_{j=1}^p c_j = 0$. For convenience, coefficients of contrasts are often chosen so that the sum of their absolute values is equal to two. That is,

$$\sum_{j=1}^p |c_j| = 2$$

where $|c_j|$ indicates that the sign of c_j is always taken to be plus. All six contrasts above satisfy this property. For example, the sum of the absolute value of the coefficients for $\hat{\psi}_1$ and $\hat{\psi}_4$ are, respectively,

$$\begin{array}{l} |1| + |-1| + |0| = 1 + 1 + 0 = 2 \\ |½| + |½| + |-1| = ½ + ½ + 1 = 2. \end{array}$$

The number of pairwise comparisons that can be defined for p means is $p(p - 1)/2$. Contrasts $\hat{\psi}_1$, $\hat{\psi}_2$, and $\hat{\psi}_3$ exhaust the $3(3 - 1)/2 = 3$ pairwise comparisons among the three means. The situation is quite different when we consider the number of nonpairwise comparisons that can be defined for $p \geq 3$ means. The number is infinite. Three nonpairwise comparisons for p equal to three means were given earlier: $\hat{\psi}_4$, $\hat{\psi}_5$, and $\hat{\psi}_6$. Other examples are

$$\hat{\psi}_7 = (1/3)\overline{Y}_1 + (2/3)\overline{Y}_2 + (-1)\overline{Y}_3$$
$$\hat{\psi}_8 = (1/4)\overline{Y}_1 + (3/4)\overline{Y}_2 + (-1)\overline{Y}_3$$
$$\hat{\psi}_9 = (1/5)\overline{Y}_1 + (4/5)\overline{Y}_2 + (-1)\overline{Y}_3.$$

This particular pattern of coefficients can be extended indefinitely.

ORTHOGONAL CONTRASTS

As we have seen there is an infinite number of contrasts that can be formulated for $p \geq 3$ means. However, most of these contrasts can be expressed as linear combinations of other contrasts and as such involve redundant information. For example, $\hat{\psi}_3 = \overline{Y}_2 - \overline{Y}_3$ defined earlier is equal to $\hat{\psi}_2 - \hat{\psi}_1$

$$\hat{\psi}_3 = \overbrace{(\overline{Y}_1 - \overline{Y}_3)}^{\hat{\psi}_2} - \overbrace{(\overline{Y}_1 - \overline{Y}_2)}^{\hat{\psi}_1} = \overline{Y}_2 - \overline{Y}_3$$

and $\hat{\psi}_4 = (\overline{Y}_1 + \overline{Y}_2)/2 - \overline{Y}_3$ is equal to $(1/2)\hat{\psi}_2 + (1/2)\hat{\psi}_3$

$$\hat{\psi}_4 = \overbrace{(\overline{Y}_1 - \overline{Y}_3)/2}^{(1/2)\hat{\psi}_2} + \overbrace{(\overline{Y}_2 - \overline{Y}_3)/2}^{(1/2)\hat{\psi}_3} = (\overline{Y}_1 + \overline{Y}_2)/2 - \overline{Y}_3.$$

Sometimes an experimenter is interested in contrasts that are mutually nonredundant and uncorrelated. Such contrasts are called *orthogonal contrasts*. A simple rule exists for determining whether contrasts are orthogonal. Let $\hat{\psi}_i$ and $\hat{\psi}_{i'}$ denote two contrasts and c_{ij} and $c_{i'j}$ ($j = 1, \ldots, p$) their respective coefficients. The contrasts are orthogonal if

$$\sum_{j=1}^{p} c_{ij} c_{i'j} = 0$$

for the equal n case, or

$$\sum_{j=1}^{p} \frac{c_{ij} c_{i'j}}{n_j} = 0$$

for the unequal n case.* Consider the contrasts $\hat{\psi}_1 = (1)\overline{Y}_1 + (-1)\overline{Y}_2$ and

---

* If the means are $NID(\mu_j, \sigma_\epsilon^2/n_j)$, orthogonality of contrasts is equivalent to statistical independence of the contrasts. The correlation between contrasts i and i' is given by

$$\rho_{ii'} = \left(\sum_{j=1}^{p} c_{ij} c_{i'j}/n_j\right) \bigg/ \sqrt{\left(\sum_{j=1}^{p} c_{ij}^2/n_j\right)\left(\sum_{j=1}^{p} c_{i'j}^2/n_j\right)}.$$

$\hat{\psi}_2 = (1)\overline{Y}_1 + (-1)\overline{Y}_3$, and assume that the n_j's are equal. These contrasts are not orthogonal since

$$
\begin{array}{llcccc}
c_{1j} & \text{for } \hat{\psi}_1 = c_{11}c_{12}c_{13} = & 1 & -1 & 0 \\
c_{2j} & \text{for } \hat{\psi}_2 = c_{21}c_{22}c_{23} = & 1 & 0 & -1 \\
\end{array}
$$

$$\sum_{j=1}^{p} c_{1j}c_{2j} = 1 + 0 + 0 = 1.$$

However, contrasts $\hat{\psi}_1 = (1)\overline{Y}_1 + (-1)\overline{Y}_2$ and $\hat{\psi}_4 = (\frac{1}{2})\overline{Y}_1 + (\frac{1}{2})\overline{Y}_2 + (-1)\overline{Y}_3$ are orthogonal.

$$
\begin{array}{llccc}
c_{1j} & \text{for } \hat{\psi}_1 = c_{11}c_{12}c_{13} = & 1 & -1 & 0 \\
c_{4j} & \text{for } \hat{\psi}_4 = c_{41}c_{42}c_{43} = & \frac{1}{2} & \frac{1}{2} & -1 \\
\end{array}
$$

$$\sum_{j=1}^{p} c_{1j}c_{4j} = \frac{1}{2} \quad -\frac{1}{2} \quad 0 = 0$$

The latter two contrasts, $\hat{\psi}_1$ and $\hat{\psi}_4$, exhaust one of the possible sets of orthogonal contrasts among three means. Two other sets of orthogonal contrasts are

$$\hat{\psi}_2 = (1)\overline{Y}_1 + (-1)\overline{Y}_3 \quad \text{and} \quad \hat{\psi}_5 = (\frac{1}{2})\overline{Y}_1 + (\frac{1}{2})\overline{Y}_3 + (-1)\overline{Y}_2$$

since

$$
\begin{array}{llccc}
c_{2j} & \text{for } \hat{\psi}_2 = c_{21}c_{22}c_{23} = & 1 & 0 & -1 \\
c_{5j} & \text{for } \hat{\psi}_5 = c_{51}c_{52}c_{53} = & \frac{1}{2} & -1 & \frac{1}{2} \\
\end{array}
$$

$$\sum_{j=1}^{p} c_{2j}c_{5j} = \frac{1}{2} \quad 0 \quad -\frac{1}{2} = 0$$

and

$$\hat{\psi}_3 = (1)\overline{Y}_2 + (-1)\overline{Y}_3 \quad \text{and} \quad \hat{\psi}_6 = (\frac{1}{2})\overline{Y}_2 + (\frac{1}{2})\overline{Y}_3 + (-1)\overline{Y}_1$$

since

$$
\begin{array}{llccc}
c_{3j} & \text{for } \hat{\psi}_3 = c_{31}c_{32}c_{33} = & 0 & 1 & -1 \\
c_{6j} & \text{for } \hat{\psi}_6 = c_{61}c_{62}c_{63} = & -1 & \frac{1}{2} & \frac{1}{2} \\
\end{array}
$$

$$\sum_{j=1}^{p} c_{3j}c_{6j} = 0 \quad \frac{1}{2} \quad -\frac{1}{2} = 0.$$

For $p \geq 4$ means, an infinite number of sets of orthogonal contrasts exists. Table 3.1-1 gives seven sets involving four means.

A general principle can be stated. If an experiment contains p treatment levels, the number of orthogonal contrasts in any set is equal to $p - 1$. Furthermore, an orthogonal set provides a basis for constructing all other contrasts. That is, all other contrasts can be expressed as linear combinations of those in an orthogonal set. For example, consider Set 1 in Table 3.1-1. The six contrasts in Sets 2 and 3 can be expressed as linear combinations of the three orthogonal contrasts in Set 1 as follows:

Set 2 $\begin{cases} \hat{\psi}_4 = (\hat{\psi}_1 - \hat{\psi}_2)/2 + \hat{\psi}_3 & = \overline{Y}_{\cdot 1} - \overline{Y}_{\cdot 3} \\ \hat{\psi}_5 = (-\hat{\psi}_1 + \hat{\psi}_2)/2 + \hat{\psi}_3 = \overline{Y}_{\cdot 2} - \overline{Y}_{\cdot 4} \\ \hat{\psi}_6 = (\hat{\psi}_1 + \hat{\psi}_2)/2 & = (\overline{Y}_{\cdot 1} + \overline{Y}_{\cdot 3})/2 - (\overline{Y}_{\cdot 2} + \overline{Y}_{\cdot 4})/2 \end{cases}$

Set 3 $\begin{cases} \hat{\psi}_7 = (\hat{\psi}_1 + \hat{\psi}_2)/2 + \hat{\psi}_3 & = \overline{Y}_{\cdot 1} - \overline{Y}_{\cdot 4} \\ \hat{\psi}_8 = (-\hat{\psi}_1 - \hat{\psi}_2)/2 + \hat{\psi}_3 = \overline{Y}_{\cdot 2} - \overline{Y}_{\cdot 3} \\ \hat{\psi}_9 = (\hat{\psi}_1 - \hat{\psi}_2)/2 & = (\overline{Y}_{\cdot 1} + \overline{Y}_{\cdot 4})/2 - (\overline{Y}_{\cdot 2} + \overline{Y}_{\cdot 3})/2. \end{cases}$

TABLE 3.1-1   Sets of Orthogonal Contrasts Among Four Means (Numbers in the table are the coefficients of the contrasts)

Set	\multicolumn{4}{c}{$c_1\overline{Y}_{\cdot 1} + c_2\overline{Y}_{\cdot 2} + c_3\overline{Y}_{\cdot 3} + c_4\overline{Y}_{\cdot 4}$}				Contrast
1	1	−1	0	0	$\hat{\psi}_1 = \overline{Y}_{\cdot 1} - \overline{Y}_{\cdot 2}$
	0	0	1	−1	$\hat{\psi}_2 = \overline{Y}_{\cdot 3} - \overline{Y}_{\cdot 4}$
	1/2	1/2	−1/2	−1/2	$\hat{\psi}_3 = (\overline{Y}_{\cdot 1} + \overline{Y}_{\cdot 2})/2 - (\overline{Y}_{\cdot 3} + \overline{Y}_{\cdot 4})/2$
2	1	0	−1	0	$\hat{\psi}_4 = \overline{Y}_{\cdot 1} - \overline{Y}_{\cdot 3}$
	0	1	0	−1	$\hat{\psi}_5 = \overline{Y}_{\cdot 2} - \overline{Y}_{\cdot 4}$
	1/2	−1/2	1/2	−1/2	$\hat{\psi}_6 = (\overline{Y}_{\cdot 1} + \overline{Y}_{\cdot 3})/2 - (\overline{Y}_{\cdot 2} + \overline{Y}_{\cdot 4})/2$
3	1	0	0	−1	$\hat{\psi}_7 = \overline{Y}_{\cdot 1} - \overline{Y}_{\cdot 4}$
	0	1	−1	0	$\hat{\psi}_8 = \overline{Y}_{\cdot 2} - \overline{Y}_{\cdot 3}$
	1/2	−1/2	−1/2	1/2	$\hat{\psi}_9 = (\overline{Y}_{\cdot 1} + \overline{Y}_{\cdot 4})/2 - (\overline{Y}_{\cdot 2} + \overline{Y}_{\cdot 3})/2$
4	1	−1	0	0	$\hat{\psi}_{10} = \overline{Y}_{\cdot 1} - \overline{Y}_{\cdot 2}$
	1/2	1/2	−1	0	$\hat{\psi}_{11} = (\overline{Y}_{\cdot 1} + \overline{Y}_{\cdot 2})/2 - \overline{Y}_{\cdot 3}$
	1/3	1/3	1/3	−1	$\hat{\psi}_{12} = (\overline{Y}_{\cdot 1} + \overline{Y}_{\cdot 2} + \overline{Y}_{\cdot 3})/3 - \overline{Y}_{\cdot 4}$
5	1/2	1/2	−1/2	−1/2	$\hat{\psi}_{13} = (\overline{Y}_{\cdot 1} + \overline{Y}_{\cdot 2})/2 - (\overline{Y}_{\cdot 3} + \overline{Y}_{\cdot 4})/2$
	1/2	−1/2	−1/2	1/2	$\hat{\psi}_{14} = (\overline{Y}_{\cdot 1} + \overline{Y}_{\cdot 4})/2 - (\overline{Y}_{\cdot 2} + \overline{Y}_{\cdot 3})/2$
	−1/2	1/2	−1/2	1/2	$\hat{\psi}_{15} = (\overline{Y}_{\cdot 2} + \overline{Y}_{\cdot 4})/2 - (\overline{Y}_{\cdot 1} + \overline{Y}_{\cdot 3})/2$
6	1/2	1/2	−1/2	−1/2	$\hat{\psi}_{16} = (\overline{Y}_{\cdot 1} + \overline{Y}_{\cdot 2})/2 - (\overline{Y}_{\cdot 3} + \overline{Y}_{\cdot 4})/2$
	2/3	−2/3	1/3	−1/3	$\hat{\psi}_{17} = (2\overline{Y}_{\cdot 1} + \overline{Y}_{\cdot 3})/3 - (2\overline{Y}_{\cdot 2} + \overline{Y}_{\cdot 4})/3$
	−1/3	1/3	2/3	−2/3	$\hat{\psi}_{18} = (\overline{Y}_{\cdot 2} + 2\overline{Y}_{\cdot 3})/3 - (\overline{Y}_{\cdot 1} + 2\overline{Y}_{\cdot 4})/3$
7	1/2	1/2	−1/2	−1/2	$\hat{\psi}_{19} = (\overline{Y}_{\cdot 1} + \overline{Y}_{\cdot 2})/2 - (\overline{Y}_{\cdot 3} + \overline{Y}_{\cdot 4})/2$
	3/4	−3/4	1/4	−1/4	$\hat{\psi}_{20} = (3\overline{Y}_{\cdot 1} + \overline{Y}_{\cdot 3})/4 - (3\overline{Y}_{\cdot 2} + \overline{Y}_{\cdot 4})/4$
	−1/4	1/4	3/4	−3/4	$\hat{\psi}_{21} = (\overline{Y}_{\cdot 2} + 3\overline{Y}_{\cdot 3})/4 - (\overline{Y}_{\cdot 1} + 3\overline{Y}_{\cdot 4})/4$

As we have seen, there are always $p - 1$ nonredundant questions that can be answered from the data in an experiment. However, an experimenter may not be interested in all of the $p - 1$ questions. For example, an experimenter may want to test the hypotheses that $\mu_1 - \mu_2 = 0$ and $\mu_3 - \mu_4 = 0$ but not that $(\mu_1 + \mu_2)/2 - (\mu_3 + \mu_4)/2 = 0$. The latter hypothesis may have no meaning in terms of the objectives of the experiment. Also, not all interesting questions involve orthogonal contrasts. In an experiment with four treatment levels, each of the six pairwise comparisons among means may be associated with a question that the experimenter seeks to answer.

In summary, the analysis of variance provides an overall test of the hypothesis that $\mu_1 = \mu_2 = \cdots = \mu_p$ or $\alpha_j = 0$ for all j. This test is equivalent to a simultaneous test of the hypothesis that all possible contrasts among means are equal to zero. The degrees of freedom for $MSBG$ in a completely randomized design is $p - 1$, which is also the number of orthogonal contrasts that can be constructed from p means. If an overall F test is significant, an experimenter can be certain that some set of orthogonal contrasts contains at least one significant contrast among the means. The contrast or contrasts that are significant may or may not be ones that are of interest to the experimenter. An F test in ANOVA is an overall test that indicates whether something has happened. It remains for an experimenter to carry out follow-up tests to determine what has happened. The following sections describe procedures for carrying out tests of (1) planned orthogonal contrasts, (2) planned nonorthogonal contrasts, and (3) unplanned nonorthogonal contrasts.

3.2   A PRIORI ORTHOGONAL CONTRASTS

In planning an experiment, one often has a specific set of hypotheses that the experiment is designed to test. Tests involving these hypotheses are referred to as *a priori* or *planned* tests. This situation may be contrasted with another in which an investigator believes that a treatment affects the dependent variable and the experiment is designed to accept or reject this notion. If an overall F test indicates that at least one contrast is not equal to zero, interest turns to determining which contrast or contrasts among means is significant. Tests that are used for *data snooping*—that is, for evaluating a subset of all possible contrasts following a significant overall test—are referred to as *a posteriori, unplanned,* or *post hoc* tests.

A PRIORI ORTHOGONAL TESTS USING A t STATISTIC

Hypotheses of the form

$$H_0: \quad \psi_1 = 0$$
$$H_0: \quad \psi_2 = 0$$
$$\cdots$$

$$H_0: \quad \psi_i \quad = 0$$
$$\cdots$$
$$H_0: \quad \psi_{p-1} = 0$$

where $\psi_i = c_1 \mu_1 + c_2 \mu_2 + \cdots + c_p \mu_p$ and the $p - 1$ contrasts are mutually orthogonal and a priori, can be tested using a t statistic. It is not necessary to perform an overall test of significance prior to testing planned orthogonal contrasts. An overall test using, say, an F statistic simply answers the question, "Did anything happen in the experiment?" If a specific set of hypotheses for orthogonal contrasts has been advanced, an experimenter is not interested in answering this general question. Rather, one is interested in answering a limited number—$p - 1$ or fewer—specific questions from the data. In such cases, it is recommended that each hypothesis be evaluated at α level of significance. The rationale for this recommendation is discussed later.

A t statistic for testing the hypothesis $H_0: \quad \psi_i = 0$ is given by

$$(3.2\text{-}1) \qquad t = \frac{\hat{\psi}_i}{\hat{\sigma}_{\psi_i}} = \frac{\displaystyle\sum_{j=1}^{p} c_j \overline{Y}_{.j}}{\sqrt{MS_{error} \displaystyle\sum_{j=1}^{p} \frac{c_j^2}{n_j}}} = \frac{c_1 \overline{Y}_{.1} + c_2 \overline{Y}_{.2} + \cdots + c_p \overline{Y}_{.p}}{\sqrt{MS_{error}\left(\dfrac{c_1^2}{n_1} + \dfrac{c_2^2}{n_2} + \cdots + \dfrac{c_p^2}{n_p} \right)}}$$

where $\hat{\sigma}_{\psi_i}$ is the standard error of the ith contrast and MS_{error} is a pooled estimator of the population error variance. For a completely randomized design,

$$MS_{error} = MSWG = \left[\sum_{j=1}^{p} \sum_{i=1}^{n} Y_{ij}^2 - \sum_{j=1}^{p} \frac{\left(\displaystyle\sum_{i=1}^{n} Y_{ij} \right)^2}{n} \right] \Bigg/ p(n-1)$$

with $p(n - 1)$ degrees of freedom. If the sample sizes are not equal, the formula for unequal n_j's in Section 2.2 is used to compute $MSWG$. Under the assumptions that

1.   The observations are drawn from normally distributed populations or the sample n_j's are fairly large,
2.   The observations are random samples from the populations,
3.   The null hypothesis is true, and
4.   The variances of the $j = 1, \ldots, p$ populations are equal to σ_ϵ^2,

the t statistic (3.2-1) is distributed as Student's t with $p(n - 1)$ degrees of freedom. The critical values of t that cut off the upper α and $\alpha/2$ regions of Student's t distribution for ν degrees of freedom are given in Appendix Table E.4 and denoted by $t_{\alpha,\nu}$ and $t_{\alpha/2,\nu}$, respectively. The null hypothesis for a two-tailed test is rejected if the absolute value of t, $|t|$, exceeds $t_{\alpha/2,\nu}$. For $H_0: \quad \mu_j - \mu_{j'} \leq 0$, t must exceed $t_{\alpha,\nu}$; for $H_0: \quad \mu_j - \mu_{j'} \geq 0$, $-t$ must be less than $-t_{\alpha,\nu}$.

Multiple t statistics use the same error mean square in the denominator. As a result, the tests of significance are not statistically independent even though the contrasts are statistically independent. Research by Norton and Bulgren as cited by Games (1971) indicates that when the degrees of freedom for MS_{error} are moderately large, say 40, multiple t tests can, for all practical purposes, be regarded as independent.

COMPUTATIONAL EXAMPLE OF A PRIORI ORTHOGONAL TESTS USING A t STATISTIC

The use of multiple t statistics to test hypotheses about a priori orthogonal contrasts will be illustrated for an experiment in which five qualitative treatment levels have been randomly assigned to 50 subjects. Ten subjects receive each treatment level. Assume that the five treatment means are

$$\overline{Y}_1 = 36.7, \quad \overline{Y}_2 = 48.7, \quad \overline{Y}_3 = 43.4, \quad \overline{Y}_4 = 47.2, \quad \text{and} \quad \overline{Y}_5 = 40.3.$$

Also assume that the treatment populations are approximately normally distributed and that the variances are homogeneous. The layout for this experiment corresponds to that for a completely randomized design; hence, $MSWG$ is the appropriate estimator of the common population error variance. The estimate is 28.8 with degrees of freedom equal to $p(n - 1) = 5(10 - 1) = 45$. The experiment has been designed to test the hypotheses in Table 3.2-1. The .05 level of significance is adopted for each test. The reader can easily verify from the coefficients in Table 3.2-1 that the $p - 1 = 4$ contrasts are mutually orthogonal. The t statistics are

$$t = \frac{\hat{\psi}_1}{\hat{\sigma}_{\psi_1}} = \frac{(1)48.7 + (-1)43.4}{\sqrt{28.8\left[\frac{(1)^2}{10} + \frac{(-1)^2}{10}\right]}} = \frac{5.300}{2.400} = 2.21$$

$$t = \frac{\hat{\psi}_2}{\hat{\sigma}_{\psi_2}} = \frac{(1)47.2 + (-1)40.3}{\sqrt{28.8\left[\frac{(1)^2}{10} + \frac{(-1)^2}{10}\right]}} = \frac{6.900}{2.400} = 2.88$$

$$t = \frac{\hat{\psi}_3}{\hat{\sigma}_{\psi_3}} = \frac{(\frac{1}{2})48.7 + (\frac{1}{2})43.4 + (-\frac{1}{2})47.2 + (-\frac{1}{2})40.3}{\sqrt{28.8\left[\frac{(\frac{1}{2})^2}{10} + \frac{(\frac{1}{2})^2}{10} + \frac{(-\frac{1}{2})^2}{10} + \frac{(-\frac{1}{2})^2}{10}\right]}} = \frac{2.300}{1.697} = 1.36$$

$$t = \frac{\hat{\psi}_4}{\hat{\sigma}_{\psi_4}} = \frac{(1)36.7 + (-\frac{1}{4})48.7 + (-\frac{1}{4})43.4 + (-\frac{1}{4})47.2 + (-\frac{1}{4})40.3}{\sqrt{28.8\left[\frac{(1)^2}{10} + \frac{(-\frac{1}{4})^2}{10} + \frac{(-\frac{1}{4})^2}{10} + \frac{(-\frac{1}{4})^2}{10} + \frac{(-\frac{1}{4})^2}{10}\right]}}$$

$$= \frac{-8.200}{1.897} = -4.32.$$

The critical value required to reject the null hypothesis is, according to Appendix Table E.4, $t_{.05/2,45} = 2.02$. Thus, the null hypothesis can be rejected for contrasts $\psi_1, \psi_2,$ and ψ_4.

TABLE 3.2-1   Statistical Hypotheses and Associated Orthogonal Coefficients

Contrast	Coefficients of Contrast					Hypotheses
	c_1	c_2	c_3	c_4	c_5	
ψ_1	0	1	-1	0	0	H_0: $\mu_2 - \mu_3 = 0$ H_1: $\mu_2 - \mu_3 \neq 0$
ψ_2	0	0	0	1	-1	H_0: $\mu_4 - \mu_5 = 0$ H_1: $\mu_4 - \mu_5 \neq 0$
ψ_3	0	$\frac{1}{2}$	$\frac{1}{2}$	$-\frac{1}{2}$	$-\frac{1}{2}$	H_0: $(\mu_2 + \mu_3)/2 - (\mu_4 + \mu_5)/2 = 0$ H_1: $(\mu_2 + \mu_3)/2 - (\mu_4 + \mu_5)/2 \neq 0$
ψ_4	1	$-\frac{1}{4}$	$-\frac{1}{4}$	$-\frac{1}{4}$	$-\frac{1}{4}$	H_0: $\mu_1 - (\mu_2 + \mu_3 + \mu_4 + \mu_5)/4 = 0$ H_1: $\mu_1 - (\mu_2 + \mu_3 + \mu_4 + \mu_5)/4 \neq 0$

A PRIORI ORTHOGONAL TESTS USING AN F STATISTIC

We saw in Section 2.1 that t^2 is identical to F with 1 and ν_2 degrees of freedom. It is sometimes convenient to perform a priori orthogonal tests using an F statistic rather than a t statistic. The formula for an F statistic is

$$F = \frac{\hat{\psi}_i^2}{\hat{\sigma}_{\hat{\psi}_i}^2} = \frac{\left(\sum_{j=1}^{p} c_j \overline{Y}_{\cdot j}\right)^2}{MS_{\text{error}} \sum_{j=1}^{p} \frac{c_j^2}{n_j}} = \frac{(c_1 \overline{Y}_{\cdot 1} + c_2 \overline{Y}_{\cdot 2} + \cdots + c_p \overline{Y}_{\cdot p})^2}{MS_{\text{error}} \left(\frac{c_1^2}{n_1} + \frac{c_2^2}{n_2} + \cdots + \frac{c_p^2}{n_p}\right)}$$

with 1 and ν_2 degrees of freedom. If $MSWG$ is used as an estimator of the common population error variance, ν_2 is equal to $p(n - 1)$ when the n's are equal and $N - p$ when they are unequal.

CONFIDENCE INTERVALS FOR A PRIORI ORTHOGONAL CONTRASTS

Emphasis in this book is placed on significance tests as opposed to confidence intervals. This emphasis is in line with contemporary practice in the behavioral sciences and education. Many mathematical statisticians, however, prefer confidence interval procedures, and there is much merit in their position. Most of the hypothesis testing procedures described in this chapter can also be used to establish $100(1 - \alpha)\%$ confidence intervals for population contrasts.

A $100(1 - \alpha)\%$ confidence interval for an a priori orthogonal contrast is given by

$$\hat{\psi}_i - \hat{\psi}(c) \leq \psi_i \leq \hat{\psi}_i + \hat{\psi}(c)$$

where $\quad \hat{\psi}(c) = t_{\alpha/2,\nu} \sqrt{MS_{\text{error}} \sum_{j=1}^{p} \frac{c_j^2}{n_j}}$

$$\hat{\psi}_i = \sum_{j=1}^{p} c_j \overline{Y}_j$$

$$\psi_i = \sum_{j=1}^{p} c_j \mu_j.$$

To illustrate, a 95% confidence interval for contrast ψ_1 in Table 3.2-1, which involves the difference between μ_2 and μ_3, is given by

$$(48.7 - 43.4) - 4.8 \leq \psi_1 \leq (48.7 - 43.4) + 4.8$$
$$5.3 - 4.8 \leq \psi_1 \leq 5.3 + 4.8$$
$$0.5 \leq \psi_1 \leq 10.1$$

where $\quad \hat{\psi}(c) = 2.02 \sqrt{28.8 \left(\frac{(1)^2}{10} + \frac{(-1)^2}{10} \right)}$

$$= 2.02(2.4) = 4.8.$$

The two boundaries

$$(\overline{Y}_2 - \overline{Y}_3) - \hat{\psi}(c) = 0.5 \quad \text{and} \quad (\overline{Y}_2 - \overline{Y}_3) + \hat{\psi}(c) = 10.1$$

of the confidence interval are called the $100(1 - .05)\% = 95\%$ confidence limits. This confidence interval indicates values of $\mu_2 - \mu_3$ that are consistent with the observed sample means. After the samples have been obtained and point estimates of μ_2 and μ_3 have been substituted in the confidence interval, it is incorrect to say that the probability is .95 that the difference between the population means lies between 0.5 and 10.1. Once the interval has been computed, the difference $\mu_2 - \mu_3$ either is or is not in the interval. This point can be clarified by realizing that there are many 95% confidence intervals over all possible random samples. Some of these confidence intervals will include the true difference and others will not. If one confidence interval is sampled at random, the probability is .95 that it will include the true difference.

If, as in the present example, a confidence interval does not include zero, the hypothesis that the contrast equals zero can also be rejected. Confidence interval procedures permit an experimenter to reach the same kind of decision as tests of significance. In addition, confidence interval procedures permit an experimenter to consider simultaneously all possible null hypotheses, not just that the contrast equals zero. If the hypothesized value of the contrast lies outside the $100(1 - \alpha)\%$ confidence interval, the null hypothesis can be rejected at α level of significance. The size of the confidence interval also provides information concerning the error variation associated with an estimate and, hence, the strength of an inference. The preference of many mathematical statisticians for confidence interval procedures over significance tests is understandable since both procedures involve the same assumptions, but confidence interval procedures provide an experimenter with more information.

ROBUST PROCEDURES FOR A PRIORI ORTHOGONAL CONTRASTS

Significance tests and confidence intervals using a t statistic involve the assumptions that the populations are approximately normally distributed and the variances are homogeneous. The t statistic like the F statistic is robust with respect to violation of both assumptions provided that the number of observations in the samples is equal. However, Boik (1975) and Kohr and Games (1977) have shown that when the sample sizes or the absolute values of the contrast coefficients, $|c_j|$, are unequal (for example, ½, ½, 1), the t statistic is not robust to heterogeneity of variance. In the discussion that follows we will consider some test statistics that are robust under these conditions. Pairwise comparisons will be considered first.

When the population variances are unequal, the pooled estimator, MS_{error}, in the t statistic can be replaced by the individual variance estimators for populations j and j' as follows

(3.2-2)
$$t' = \frac{\hat{\psi}_i}{\hat{\sigma}_{\psi_i}} = \frac{c_j \overline{Y}_{.j} + c_{j'} \overline{Y}_{.j'}}{\sqrt{\dfrac{\hat{\sigma}_j^2}{n_j} + \dfrac{\hat{\sigma}_{j'}^2}{n_{j'}}}} .$$

The earliest attempts to determine the sampling distribution of t' were made by Behrens (1929) and enlarged upon by Fisher (1935). No exact solution for this problem exists. Three approximate solutions have been proposed: (1) Cochran and Cox (1957, 100); (2) Dixon and Massey (1957, 123), Satterthwaite (1946), and Smith (1936); and (3) Welch (1947).* In general, there is close agreement among these approximate solutions; accordingly only those of Cochran and Cox and Welch will be described.

Cochran and Cox's procedure uses the t' statistic defined in (3.2-2). The critical value of t' is given by

$$t'_{\alpha/2, \nu} = \frac{(\hat{\sigma}_j^2/n_j)t_{\alpha/2, \nu_j} + (\hat{\sigma}_{j'}^2/n_{j'})t_{\alpha/2, \nu_{j'}}}{(\hat{\sigma}_j^2/n_j) + (\hat{\sigma}_{j'}^2/n_{j'})}$$

where $t_{\alpha/2, \nu_j}$ and $t_{\alpha/2, \nu_{j'}}$ are the tabled t values at $\alpha/2$ level of significance for $\nu_j = n_j - 1$ and $\nu_{j'} = n_{j'} - 1$ degrees of freedom, respectively. The critical value for t' will always be between the ordinary t values for ν_j and $\nu_{j'}$ degrees of freedom. For a one-tailed test, values of t_{α, ν_j} and $t_{\alpha, \nu_{j'}}$ are used. If $n_j = n_{j'}$, then $t' = t$ and the conventional t value with $n_j - 1$ degrees of freedom can be used. The t' test is conservative because the critical value for t' tends to be slightly too large.

Welch's (1947, 1949) procedure for pairwise contrasts also uses the t' statistic defined in (3.2-2). Tables of the distribution of t' have been prepared by Aspin

---

* A different approach to the problem has been suggested by Johnson (1978). He used the Cornish-Fisher expansion to derive a corrected form of the t statistic. Properties of the data are used to adjust t so that the resulting statistic is approximately distributed as Student's t distribution.

(1949). An approximation to the critical value of t' can be obtained from Student's t distribution with degrees of freedom equal to

$$\nu' = \frac{\left(\dfrac{\hat{\sigma}_j^2}{n_j} + \dfrac{\hat{\sigma}_{j'}^2}{n_{j'}}\right)^2}{\dfrac{\hat{\sigma}_j^4}{n_j^2(n_j - 1)} + \dfrac{\hat{\sigma}_{j'}^4}{n_{j'}^2(n_{j'} - 1)}}.$$

This procedure provides a relatively powerful test that is robust under all conditions that have been investigated (Scheffé, 1970; Wang, 1971).

For nonpairwise contrasts, Welch's (1947) t' test statistic is

$$t' = \frac{\hat{\psi}_i}{\hat{\sigma}_{\psi_i}} = \frac{c_1 \overline{Y}_{\cdot 1} + c_2 \overline{Y}_{\cdot 2} + \cdots + c_p \overline{Y}_{\cdot p}}{\sqrt{\dfrac{c_1^2 \hat{\sigma}_1^2}{n_1} + \dfrac{c_2^2 \hat{\sigma}_2^2}{n_2} + \cdots + \dfrac{c_p^2 \hat{\sigma}_p^2}{n_p}}}$$

with degrees of freedom equal to

$$\nu' = \frac{v^2}{w}$$

where

$$v = \frac{c_1^2 \hat{\sigma}_1^2}{n_1} + \frac{c_2^2 \hat{\sigma}_2^2}{n_2} + \cdots + \frac{c_p^2 \sigma_p^2}{n_p}$$

$$w = \frac{c_1^4 \hat{\sigma}_1^4}{n_1^2(n_1 - 1)} + \frac{c_2^4 \hat{\sigma}_2^4}{n_2^2(n_2 - 1)} + \cdots + \frac{c_p^4 \hat{\sigma}_p^4}{n_p^2(n_p - 1)}.$$

According to Kohr and Games (1977), this procedure provides reasonable protection against type I errors when the variances are heterogeneous and the sample sizes or the absolute values of the coefficients of a contrast are unequal. Under these conditions, Welch's procedures appear to be good choices for evaluating planned orthogonal contrasts.

Earlier it was recommended that each hypothesis for planned orthogonal contrasts be evaluated at α level of significance. In the sections that follow we will examine the rationale for this recommendation and describe procedures for evaluating planned nonorthogonal and unplanned nonorthogonal contrasts.

3.3   CONCEPTUAL UNIT FOR ERROR RATE

If an experimenter tests C independent contrasts, each at α level of significance, the probability of making one or more type I errors is

$$\text{Prob. of one or more type I errors} = 1 - (1 - \alpha)^C$$

which is approximately equal to $C\alpha$ for small values of α.

The rationale underlying $1 - (1 - \alpha)^C$ is as follows. We know that if a contrast is tested at α level of significance, the probability of not making a type I error is $1 - \alpha$. If C independent contrasts are each tested at α level of significance, the probability of not making a type I error for the first, and the second, . . . , and the Cth null hypothesis is, according to the multiplication rule for independent events,

$$\overbrace{(1 - \alpha)(1 - \alpha) \cdots (1 - \alpha)}^{C} = (1 - \alpha)^C.$$

This is the probability of retaining all C null hypotheses when they are true. The probability of not retaining all null hypotheses when they are true is

Prob. of one or more type I errors $= 1 -$ (Prob. of retaining all null hypotheses when they are true)

$$= 1 - (1 - \alpha)^C.$$

As the number of independent tests increases, so does the probability of obtaining spuriously significant results. For example, if α is equal to .05 and an experimenter tests three, five, or ten independent contrasts, the probability of one or more type I errors is, respectively,

$$1 - (1 - .05)^3 = .14$$
$$1 - (1 - .05)^5 = .23$$
$$1 - (1 - .05)^{10} = .40.$$

Hence, if enough t statistics are computed, each at α level of significance, an experimenter will probably reject one or more null hypotheses even though they are all true. This problem can be minimized by restricting the use of multiple t tests to a priori orthogonal contrasts. The corresponding probability of making one or more type I errors for nonindependent contrasts is difficult to compute, although it can be shown to be less than or equal to $1 - (1 - \alpha)^C$.* The basic question can be raised, "Should the probability of committing a type I error be set at α for each test or should the probability of one or more errors be set at α or less for some larger conceptual unit such as the collection of tests?" This question, which has been debated extensively in the literature, does not have a simple answer. Answers have varied, depending on the nature of the contrasts that are of interest to the experimenter. As we will see, the most relevant considerations appear to be whether the contrasts are orthogonal and whether they are planned in advance.

The meaning of the term significance level is unambiguous for experiments with two treatment levels, but not so when the experiment contains three or more

---

*   Harter (1957) and Pearson and Hartley (1942, 1943) describe procedures for computing this probability.

treatment levels. The ambiguity arises because a significance level or *error rate* (*ER*) can be defined for a number of different conceptual units: the individual contrast, family of contrasts, and experiment. These conceptual units are described next.

ERROR RATE PER CONTRAST

The error rate per contrast, α_{PC}, is the probability that a contrast will be falsely declared significant. We can think of this in more concrete terms. Suppose that hypotheses for many many contrasts are tested and somehow we are able to count the number of erroneous conclusions, then

$$ER \text{ per contrast } (\alpha_{PC}) = \frac{\text{Number of contrasts falsely declared significant}}{\text{Number of contrasts}} .$$

When a t statistic is used to test a priori orthogonal contrasts each at α level of significance, the conceptual unit for error rate is the individual contrast. Controlling the error rate per contrast, however, allows the probability of making one or more type I errors for the set of contrasts to increase as the number of tests increases.

ERROR RATE PER EXPERIMENT AND EXPERIMENTWISE

An alternative strategy to controlling error rate per contrast is to adopt the experiment as the conceptual unit for error rate. Consider performing many experiments in which hypotheses for $C = 5$ contrasts are tested in each experiment and again assume that we are able to count the number of erroneous conclusions. Two error rates can be defined for this situation:

$$ER \text{ per experiment } (\alpha_{PE}) = \frac{\text{Number of contrasts falsely declared significant}}{\text{Number of experiments}}$$

and

$$ER \text{ experimentwise } (\alpha_{EW}) = \frac{\text{Number of experiments with at least one contrast falsely declared significant}}{\text{Number of experiments}} .$$

The first error rate is the long-run average number of erroneous statements made per experiment. The error rate experimentwise is the probability that one or more erroneous statements will be made in an experiment and is less conservative than the error rate per experiment. The experimentwise error rate is a probability, whereas the error rate per experiment is not a probability, but, rather, the expected number of errors per experiment. The use of the experimentwise error rate is based on the premise that it is as serious to make one erroneous statement in an experiment as it is to make, say, five such statements. The two error rates are numerically almost identical for small values of α. As a result, Miller (1966, 10) observes that a choice between them is essentially a matter of taste. The relationship between error rate experimentwise and error rate per contrast for C orthogonal contrasts is

$$\alpha_{EW} = 1 - (1 - \alpha_{PC})^C.$$

For nonorthogonal contrasts, the relationship is

$$\alpha_{EW} \leq 1 - (1 - \alpha_{PC})^C.$$

The error rate experimentwise cannot exceed the error rate per experiment, that is, $\alpha_{EW} \leq \alpha_{PE}$.

When an F statistic is used to test the overall null hypothesis

$$H_0: \quad \mu_1 = \mu_2 = \cdots = \mu_p$$

in a single-treatment analysis of variance design at α level of significance, the experiment is the conceptual unit for error rate. Suppose that the overall null hypothesis is rejected. At this point, the interest usually shifts to determining which pairwise comparisons, if any, among means are significant. It is generally recommended that the error rate for the collection of a posteriori nonorthogonal follow-up tests equal that for the overall F test. As we will see, a variety of data snooping statistics have been developed for this purpose.

ERROR RATE PER FAMILY AND FAMILYWISE

In multitreatment ANOVA designs another conceptual unit for error rate can be adopted—the family. A family of contrasts consists of all contrasts of interest that are associated with a particular treatment or interaction. A factorial experiment with two treatments, for example, contains the three families of contrasts: one associated with treatment A, a second associated with treatment B, and a third associated with the AB interaction. Two new error rates can be defined for such experiments:

$$ER \text{ per family } (\alpha_{PF}) = \frac{\text{Number of contrasts falsely declared significant}}{\text{Number of families}}$$

$$ER \text{ familywise } (\alpha_{FW}) = \frac{\text{Number of families with at least one contrast falsely declared significant}}{\text{Number of families}}.$$

In multitreatment ANOVA designs it is customary to use an F statistic to test each of the overall null hypotheses at α level of significance. In other words, contemporary practice favors the family rather than the experiment as the conceptual unit for error rate. This is reasonable considering the special nature of the treatments and interactions in factorial and hierarchical designs—they are planned in advance and mutually orthogonal.* Consequently, instead of designing an experiment with, say,

---

*   The orthogonality of sources of variations in ANOVA is discussed by Box, Hunter, and Hunter (1978, Ch. 6).

two treatments and one interaction, one could choose to design three separate experiments to test the three overall null hypotheses. Then one would have three separate experiments instead of one experiment with three a priori orthogonal families. But in either case, an F statistic would be used to test each of the three overall null hypotheses at α level of significance. The approach of combining the three families into one experiment has the advantage of providing more degrees of freedom for estimating the experimental error.*

Three conceptual units for error rate have been described: the contrast, family, and experiment. They are all identical for an experiment involving one contrast. The error rates become more disparate as the number of families and contrasts within families increases.

WHAT IS THE CORRECT CONCEPTUAL UNIT FOR ERROR RATE

The relative merits of making the contrast or some larger unit, such as the family or experiment, the conceptual unit for error rate has been extensively debated: Duncan (1955), McHugh and Ellis (1955), Ryan (1959, 1960, 1962), and Wilson (1962). The answer to the question, "What is the correct conceptual unit for error rate?" depends, as we have seen, on the nature of the contrasts. The following discussion summarizes recommendations for three categories of contrasts: (1) a priori and orthogonal, (2) a priori and nonorthogonal, and (3) a posteriori and nonorthogonal.

The interpretation of significance level for an experiment involving only one contrast is unambiguous. However, the situation becomes more complicated when an experiment involves several contrasts. If orthogonal contrasts have been planned in advance, contemporary practice favors the contrast as the conceptual unit for error rate. We noted in Section 3.1 that testing planned orthogonal contrasts is equivalent to partitioning the data so that each test involves nonredundant, independent pieces of information. The same can be said for treatments and interactions in multitreatment ANOVA designs. In this case, contemporary practice favors the family of contrasts associated with a treatment or interaction as the conceptual unit for error rate.

Nonorthogonal contrasts involve redundant information—the outcome of one test is not independent of those for other tests. Here contemporary practice favors adopting a larger conceptual unit for error rate, either the family or, in the case of single-treatment experiments, the experiment. Mathematical statisticians have developed a wide variety of test statistics for controlling error rate at or less than α for various collections of tests. For example, test statistics have been developed for controlling error rate at or less than α for (1) the comparison of a control-group mean with $p - 1$ experimental-group means, (2) any set of C contrasts among means, (3) all $p(p - 1)/2$ pairwise comparisons among means, and (4) all possible contrasts among means. Test statistics in the first two categories are well suited to evaluating

---

* As noted in Section 3.2, when the same error term is used in a series of tests, the tests are not statistically independent even though the sources being tested are independent.

a priori nonorthogonal contrasts; those in the latter two categories are more often used for a posteriori nonorthogonal contrasts, although exceptions occur. Hard and fast rules are difficult to formulate because, as we will see, the various test statistics differ markedly in power. The problem facing an experimenter is to select the test statistic that provides the desired kind of protection and the maximum power. In general, test statistics that were designed for testing a select, limited number of contrasts are more powerful than those designed to test all pairwise comparisons or all possible contrasts. Hence, when possible, it is to the experimenter's advantage to specify a select, limited number of contrasts in advance. The strategy of using more powerful test statistics for planned tests and less powerful statistics for data snooping is discussed in Section 4.4.

3.4   A PRIORI NONORTHOGONAL CONTRASTS

A researcher, in planning an experiment, often has a specific set of C hypotheses that the experiment is designed to test. Often the associated contrasts are not orthogonal as in comparing a control-group mean with $p - 1$ experimental group means, comparing each mean with every other mean, or making, say, $C = p + 2$ tests among p means. If the contrasts are planned in advance and relatively limited in number, one of the test statistics in this section can be used. If the contrasts are a posteriori or if the number is relatively large, one of the test statistics in Section 3.5 should be considered. Often several test statistics will provide acceptable protection against making one or more type I errors. In such cases, the experimenter is encouraged to compute the critical difference necessary to reject the null hypotheses for the alternative test procedures and use the one that gives the smallest critical difference (Miller, 1966, 18). As we will see, this is one of the more useful techniques for choosing among alternative test procedures.

DUNN'S MULTIPLE COMPARISON PROCEDURE

Dunn's procedure is appropriate for testing hypotheses concerning C planned contrasts among means. The originator of the procedure is unknown. Dunn (1961) has examined the properties of the procedure in detail and has prepared tables that facilitate its use. Consequently it is referred to as *Dunn's multiple comparison procedure.* Some writers refer to it as the *Bonferroni t procedure,* since it is based on the Bonferroni or Boole inequality.

Dunn (1961) used the Bonferroni inequality in showing that the error rate experimentwise* cannot exceed the sum of C per contrast error rates, that is,

---

*   The term experimentwise is used here because the following computational example involves a type CR-p design. If a multitreatment design were used, we would control the error rate familywise.

$$\alpha_{EW} \le \sum_{i=1}^{C} \alpha_{PC_i}$$

where α_{PC_i} is the per contrast error rate for the ith contrast. Thus, if each of C contrasts is tested at α/C level of significance, the error rate experimentwise cannot exceed α. The procedure basically consists of splitting up α among a set of C planned contrasts. For example, if we want to test C equal to 2 contrasts and we want the error rate experimentwise to be less than or equal to .05, each contrast should be tested at $.05/2 = .025$ level of significance. Then, $\alpha_{EW} \le .025 + .025 = .05$. It is not necessary to perform an overall test of significance prior to testing the planned contrasts.

Dunn developed the procedure using Student's t distribution and the t statistic described in Section 3.2, but it is applicable to other test statistics. The t statistic will be denoted by tD when it is used with Dunn's procedure. The tD statistic is

$$tD = \frac{\hat{\psi}_i}{\hat{\sigma}_{\psi_i}} = \frac{\sum_{j=1}^{p} c_j \overline{Y}_{\cdot j}}{\sqrt{MS_{error} \sum_{j=1}^{p} \frac{c_j^2}{n_j}}} = \frac{c_1 \overline{Y}_{\cdot 1} + c_2 \overline{Y}_{\cdot 2} + \cdots + c_p \overline{Y}_{\cdot p}}{\sqrt{MS_{error} \left(\frac{c_1^2}{n_1} + \frac{c_2^2}{n_2} + \cdots + \frac{c_p^2}{n_p} \right)}}.$$

We will denote the critical value for this tD statistic by $tD_{\alpha/2;C,\nu}$. The ith null hypothesis H$_0$: $\psi_i = 0$ is rejected if $|tD| \ge tD_{\alpha/2;C,\nu}$, where $tD_{\alpha/2;C,\nu}$ is the two-tailed critical value of tD in Appendix Table E.16*, C is the number of planned contrasts among p means, and ν is the degrees of freedom for MS_{error}. The critical values in Table E.16 were obtained from Student's t distribution; it can be shown for a two-tailed test that

$$tD_{\alpha/2;C,\nu} = t_{(\alpha/2)/C,\nu}.$$

For example, $tD_{.05/2;5,20} = t_{(.05/2)/5,20} = 2.845$, where $t_{(.05/2)/5,20}$ cuts off the upper $(.05/2)/5 = .005$ portion of Student's t distribution.

Let's assume that an experimenter is interested in testing the following hypotheses involving a priori nonorthogonal contrasts:

H$_0$: $\psi_1 = \mu_1 - \mu_2 = 0$ H$_0$: $\psi_3 = \mu_4 - \mu_5 = 0$

H$_0$: $\psi_2 = \mu_1 - \mu_3 = 0$ H$_0$: $\psi_4 = (\mu_1 + \mu_2 + \mu_3)/3 - (\mu_4 + \mu_5)/2 = 0$.

Let the experimentwise error rate equal .01. Assume that the sample means are $\overline{Y}_1 = 36.7$, $\overline{Y}_2 = 48.7$, $\overline{Y}_3 = 43.4$, $\overline{Y}_4 = 47.2$, $\overline{Y}_4 = 47.2$, and $\overline{Y}_5 = 40.3$ and that $MSWG = 28.8$ with $p(n - 1) = 5(10 - 1) = 45$ degrees of freedom. These same data were used to illustrate the t statistic in Section 3.2. The values of the test statistics are

$$tD = \frac{\hat{\psi}_1}{\hat{\sigma}_{\psi_1}} = \frac{(1)36.7 + (-1)48.7}{\sqrt{28.8 \left[\frac{(1)^2}{10} + \frac{(-1)^2}{10} \right]}} = \frac{-12.000}{2.400} = -5.00$$

---

*   More complete tables are provided by Dayton and Schafer (1973).

$$tD = \frac{\hat{\psi}_2}{\hat{\sigma}_{\psi_2}} = \frac{(1)36.7 + (-1)43.4}{\sqrt{28.8\left[\dfrac{(1)^2}{10} + \dfrac{(-1)^2}{10}\right]}} = \frac{-6.700}{2.400} = -2.79$$

$$tD = \frac{\hat{\psi}_3}{\hat{\sigma}_{\psi_3}} = \frac{(1)47.2 + (-1)40.3}{\sqrt{28.8\left[\dfrac{(1)^2}{10} + \dfrac{(-1)^2}{10}\right]}} = \frac{6.900}{2.400} = 2.88$$

$$tD = \frac{\hat{\psi}_4}{\hat{\sigma}_{\psi_4}} = \frac{(1/3)36.7 + (1/3)48.7 + (1/3)43.4 + (-1/2)47.2 + (-1/2)40.3}{\sqrt{28.8\left[\dfrac{(1/3)^2}{10} + \dfrac{(1/3)^2}{10} + \dfrac{(1/3)^2}{10} + \dfrac{(-1/2)^2}{10} + \dfrac{(-1/2)^2}{10}\right]}}$$

$$= \frac{-0.817}{1.549} = -0.53.$$

The critical value of tD in Table E.16 is $tD_{.01/2;4,45} \cong 3.21$; hence, only the null hypothesis for ψ_1 can be rejected.

    If an experimenter had advanced one-tailed hypotheses, the required value of $tD_{.01;4,45}$ could not be obtained from Table E.16. An approximate value of $tD_{\alpha;C,\nu}$ that cuts off the upper α/C proportion of Student's t distribution for ν degrees of freedom can be determined from the standard normal distribution by

$$tD_{\alpha;C,\nu} \cong z_{\alpha/C} + \frac{z_{\alpha/C}^3 + z_{\alpha/C}}{4(\nu - 2)}$$

where $z_{\alpha/C}$ is obtained from Appendix Table E.3 (Peiser, 1943). For our example,

$$tD_{.01;4,45} = t_{.01/4,45} \cong 2.81 + \frac{(2.81)^3 + 2.81}{4(45 - 2)} = 2.96$$

where 2.81 is the value of z that cuts off the upper $.01/4 = .0025$ proportion of the standard normal distribution.

    If Dunn's multiple comparison procedure is used to test hypotheses concerning all pairwise comparisons among means and the sample n_j's are equal, it is computationally more convenient to compute the *critical difference* $\hat{\psi}(D)$ that a comparison must exceed in order to be declared significant. This difference is given by

$$\hat{\psi}(D) = tD_{\alpha/2;C,\nu}\sqrt{2MS_{error}/n}.$$

Suppose that an experimenter wants to evaluate all pairwise comparisons among the five means given earlier. One can compute $5(5 - 1)/2 = 10$ such comparisons. The values of the comparisons are given in Table 3.4-1, where the means have been ordered from the smallest to the largest. The critical difference that a pairwise comparison must exceed is

$$\hat{\psi}(D) = tD_{.01/2;10,45}\sqrt{2(28.8)/10}$$
$$= 3.520(2.40) = 8.45.$$

TABLE 3.4-1 Absolute Value of Differences Among Means ($MS_{error} = 28.8$, $p = 5$, and $n = 10$)

	$\overline{Y}_{.1}$	$\overline{Y}_{.5}$	$\overline{Y}_{.3}$	$\overline{Y}_{.4}$	$\overline{Y}_{.2}$
$\overline{Y}_{.1} = 36.7$	—	3.6	6.7	10.5*	12.0*
$\overline{Y}_{.5} = 40.3$		—	3.1	6.9	8.4
$\overline{Y}_{.3} = 43.4$			—	3.8	5.3
$\overline{Y}_{.4} = 47.2$				—	1.5
$\overline{Y}_{.2} = 48.7$					—

*$p < .01$

Two differences, the ones that are starred in Table 3.4-1, exceed 8.45. Thus, the null hypotheses for these differences can be rejected.

In both of the preceding examples, the level of significance α was divided evenly among the contrasts by using Dunn's table (Table E.16). This procedure of dividing α evenly among C contrasts is appropriate if an experimenter considers the consequences of making a type I error to be equally serious for all contrasts. If this is not true, α can be allocated unequally among the C contrasts in a manner that reflects the experimenter's a priori concern for type I and II errors. Let's assume that the .05 level of significance is adopted for a collection of C equal to five a priori contrasts. The use of Dunn's table in Appendix E amounts to testing each of the five contrasts at α_i, where $\alpha_i = \alpha/C = .05/5 = .01$. Suppose that the consequences of making a type I error are not equally serious for all contrasts. If this is true, the experimenter can allocate α_i unequally among the contrasts in a manner that reflects concern about type I errors, so long as the sum of α_i for $i = 1, \ldots, C$ contrasts is equal to α, the significance level selected for the collection of tests. For example, the five values of α_i could be $\alpha_1 = .02$, $\alpha_2 = .01$, $\alpha_3 = .01$, $\alpha_4 = .005$, and $\alpha_5 = .005$. The error rate for the collection of the five tests is equal to $.02 + .01 + .01 + .005 + .005 = .05$, which is the same value that would have been obtained if α were divided equally among the five tests.

Dunn's procedure can also be used to establish C simultaneous $100(1 - \alpha)\%$ confidence intervals for a collection of population contrasts. The degree of one's confidence that all C contrasts are simultaneously in their respective confidence intervals is greater than or equal to $1 - \Sigma_{i=1}^{C} \alpha_{PC_i}$. The confidence interval is given by

$$\hat{\psi}_i - \hat{\psi}(D) \le \psi_i \le \hat{\psi}_i + \hat{\psi}(D)$$

where $\hat{\psi}(D) = tD_{\alpha/2;C,\nu}\sqrt{MS_{error} \sum_{j=1}^{p} \frac{c_j^2}{n_j}}$

$$\hat{\psi}_i = \sum_{j=1}^{p} c_j \overline{Y}_{.j}$$

$$\psi_i = \sum_{j=1}^{p} c_j \mu_j.$$

ŠIDÁK'S MODIFICATION OF DUNN'S PROCEDURE

Significance levels and confidence coefficients for Dunn's procedure are approximate since they are based on the additive Bonferroni inequality. For C nonorthogonal contrasts, we saw that

$$\alpha_{EW} \leq \sum_{i=1}^{C} \alpha_{PC_i}.$$

It is evident from this additive inequality that Dunn's procedure provides an upper bound to the error rate experimentwise; that is, the error rate experimentwise cannot exceed $\sum_{i=1}^{C} \alpha_{PC_i}$. For small values of α_{EW}, the approximation is excellent. However, an even better approximation to the upper bound for α_{EW} is provided by a multiplicative inequality proved by Šidák (1967). According to Šidák, the experimentwise error rate for nonorthogonal contrasts is always less than or equal to $1 - (1 - \alpha)^C$.* The following relations can be shown to hold for nonorthogonal contrasts

$$\alpha_{EW} \leq 1 - (1 - \alpha_{PC})^C < \sum_{i=1}^{C} \alpha_{PC_i}.$$

It follows that instead of testing each contrast at α_{EW}/C level of significance to control α_{EW} as in Dunn's procedure, each contrast can be tested at $1 - (1 - \alpha_{EW})^{1/C}$ level of significance. Differences between the critical values for the two procedures are negligible for $\alpha_{EW} < .01$. Suppose an experimenter plans to test five nonorthogonal contrasts and wants to control the probability of making one or more errors at or less than .05. If $\alpha_{EW} = .05$ is split evenly among the five contrasts, use of the additive and multiplicative inequalities would result in testing each contrast at, respectively,

Additive inequality $\qquad \dfrac{\alpha_{EW}}{C} = \dfrac{.05}{5} = .01$

Multiplicative inequality $\qquad 1 - (1 - \alpha_{EW})^{1/C} = 1 - (1 - .05)^{1/5} = .010206.$

Use of .010206 instead of .01 requires a slightly smaller critical value and critical difference. Hence the multiplicative inequality leads to a more powerful test and a narrower confidence interval than the Bonferroni additive inequality. Note, however, that it is relatively easy to allocate α unequally among C contrasts using the additive inequality; this is not the case for the multiplicative inequality. In subsequent discussions we will refer to the procedure based on the additive inequality as Dunn's procedure and that based on the multiplicative inequality as the Dunn-Šidák procedure.

Games (1977) developed a table of critical values for the t statistic based on the multiplicative inequality $\alpha_{EW} \leq 1 - (1 - \alpha_{PC})^C$; these values are given in Appendix Table E.17. The t statistic will be denoted by tDS when it is used with the Dunn-Šidák procedure. The tDS statistic is

---

* Dunn (1958) proved this earlier for $p = 2$, 3, or any p with a special set of variances and covariances (variance-covariance matrix).

$$tDS = \frac{\hat{\psi}_i}{\hat{\sigma}_{\psi_i}} = \frac{\sum_{j=1}^{p} c_j \overline{Y}_{.j}}{\sqrt{MS_{\text{error}} \sum_{j=1}^{p} \frac{c_j^2}{n_j}}}.$$

We will denote the value that t must exceed for C contrasts by $tDS_{\alpha/2;C,\nu}$. The critical difference that a contrast must exceed is denoted by $\hat{\psi}(DS)$ and is equal to

$$\hat{\psi}(DS) = tDS_{\alpha/2;C,\nu} \sqrt{MS_{\text{error}} \sum_{j=1}^{p} \frac{c_j^2}{n_j}}.$$

If an experimenter tests five nonorthogonal contrasts at $\alpha_{EW} = .05$ and there are 60 degrees of freedom for experimental error, the critical values for Dunn's and Šidák's procedures are, respectively,

$$tD_{\alpha/2;5,60} = 2.660$$
$$tDS_{\alpha/2;5,60} = 2.653.$$

Both procedures control the probability of making one or more type I errors at or less than α; however, the Dunn-Šidák procedure is always slightly more powerful.

Because these procedures are not restricted to orthogonal contrasts but are applicable to any number of planned contrasts, the reader may wonder why they have not replaced the multiple t statistic described in Section 3.2. The answer is to be found in a comparison of the length of the confidence interval for the procedures. For $C \geq 2$, a confidence interval based on the Dunn or Dunn-Šidák procedures is always longer than the corresponding interval based on the multiple t statistic procedure. The advantage of being able to make all planned contrasts, not just those that are orthogonal, is gained at the expense of an increase in the probability of making type II errors.

Dunn (1961) compared her procedure with a posteriori procedures developed by Tukey (1953) and Scheffé (1953). Games (1977) made a similar comparison for the Dunn-Šidák and Scheffé procedures. The Tukey and Scheffé procedures, which are described in Section 3.5, were developed for exploring interesting contrasts suggested by an inspection of the data. Dunn and Games have shown that when there are many means in an experiment, and the number of contrasts that an experimenter wants to evaluate is relatively small, the Dunn and Dunn-Šidák procedures produce shorter confidence intervals than either of the a posteriori procedures.

On the other hand, if p is small and C is relatively large, the a posteriori procedures lead to shorter confidence intervals. This results, in part, because the length of the confidence interval for the Dunn and Dunn-Šidák procedures depends on C, the number of contrasts among p means, whereas with both a posteriori procedures the length depends on p, the number of means. For planned contrasts, an experimenter can always determine in advance which multiple comparison procedure will lead to the smallest critical difference for a contrast or shortest confidence interval.

DUNNETT'S TEST FOR CONTRASTS INVOLVING A CONTROL MEAN

The object of many experiments is to compare $p - 1$ treatment means with a control-group mean. Dunnett (1955) has developed a multiple comparison procedure for this purpose, that is, for testing $p - 1$ a priori null hypotheses of the form

$$H_0: \quad \psi_1 = \mu_1 - \mu_p = 0$$
$$H_0: \quad \psi_2 = \mu_2 - \mu_p = 0$$
$$\cdots$$
$$H_0: \quad \psi_{p-1} = \mu_{p-1} - \mu_p = 0.$$

More specifically, Dunnett's procedure is applicable to a special set of $p - 1$ a priori nonorthogonal comparisons—those in which the correlation between any two contrasts is 0.5. This occurs when $p - 1$ means are compared with a control-group mean. To illustrate, consider the following contrasts where $n = 10$ and $\overline{Y}_4$ is the control-group mean.

	$\overline{Y}_{.1}$	$\overline{Y}_{.2}$	$\overline{Y}_{.3}$	$\overline{Y}_{.4}$
$\hat{\psi}_1 = 1$	0	0	$-1 = \overline{Y}_{.1} - \overline{Y}_{.4}$
$\hat{\psi}_2 = 0$	1	0	$-1 = \overline{Y}_{.2} - \overline{Y}_{.4}$
$\hat{\psi}_3 = 0$	0	1	$-1 = \overline{Y}_{.3} - \overline{Y}_{.4}$

The correlation between the ith and the i'th contrasts is given by

$$\rho_{ii'} = \left(\sum_{j=1}^{p} c_{ij} c_{i'j} / n_j \right) \bigg/ \sqrt{\left(\sum_{j=1}^{p} c_{ij}^2 / n_j \right) \left(\sum_{j=1}^{p} c_{i'j}^2 / n_j \right)}.$$

The correlations for the ith and the i'th contrasts are

$$\rho_{12} = (1/10)/\sqrt{(2/10)(2/10)} = 0.5$$
$$\rho_{13} = (1/10)/\sqrt{(2/10)(2/10)} = 0.5$$
$$\rho_{23} = (1/10)/\sqrt{(2/10)(2/10)} = 0.5.$$

For such contrasts, Dunnett's procedure controls the probability of falsely rejecting one or more null hypotheses at α_{EW}. It is not necessary to perform an overall test of significance prior to testing the planned comparisons.

The test statistic for Dunnett's procedure is the t statistic described in Section 3.2. We will denote this statistic by tD' when it is used with Dunnett's procedure.

$$tD' = \frac{\hat{\psi}_i}{\hat{\sigma}_{\psi_i}} = \frac{c_j \overline{Y}_{.j} + c_{j'} \overline{Y}_{.j'}}{\sqrt{MS_{\text{error}} \left(\dfrac{c_j^2}{n_j} + \dfrac{c_{j'}^2}{n_{j'}} \right)}}$$

If the sample n's are equal, the statistic simplifies to

$$tD' = \frac{\hat{\psi}_i}{\hat{\sigma}_{\psi_i}} = \frac{c_j \overline{Y}_j + c_{j'} \overline{Y}_{j'}}{\sqrt{\dfrac{2MS_{error}}{n}}}.$$

The two-tailed critical value for this t statistic is denoted by $tD'_{\alpha/2;p,\nu}$, where p is the number of means, including the control mean, and ν is the degrees of freedom for MS_{error}. The one-tailed critical value is denoted by $tD'_{\alpha,p,\nu}$. The critical values are given in Appendix Table E.9.

Dunnett's multiple comparison procedure will be illustrated for the data in Table 3.4-1; let $\overline{Y}_1$ be the control mean. Four pairwise comparisons involving the control mean can be made among these five means. The critical difference $\hat{\psi}(D')$ that a comparison must exceed for a two-tailed test at $\alpha_{EW} = .01$ is given by

$$\hat{\psi}(D') = tD'_{.01;5,45}\sqrt{2MS_{error}/n}$$
$$= 3.167\sqrt{2(28.8)/10}$$
$$= 7.60.$$

From an examination of Table 3.4-1, it is apparent that the absolute value of two comparisons involving $\overline{Y}_1$ exceeds 7.60; these are $|\overline{Y}_1 - \overline{Y}_2| = 12$ and $|\overline{Y}_1 - \overline{Y}_4| = 10.5$. Hence the null hypotheses for these comparisons can be rejected.

Dunnett's procedure can also be used to establish simultaneous $100(1 - \alpha)\%$ confidence intervals for $p - 1$ comparisons involving a control mean. A confidence interval is given by

$$\hat{\psi}_i - \hat{\psi}(D') \leq \psi_i \leq \hat{\psi}_i + \hat{\psi}(D')$$

where $\hat{\psi}(D') = tD'_{\alpha/2;p,\nu}\sqrt{2MS_{error}/n}$
$\hat{\psi}_i = c_j \overline{Y}_j + c_{j'} \overline{Y}_{j'}$
$\psi_i = c_j \mu_j + c_{j'} \mu_{j'}.$

Dunnett's procedure and the Dunn and Dunn-Šidák procedures can be used to evaluate a priori nonorthogonal contrasts. However, Dunnett's procedure is restricted to $p - 1$ comparisons where the correlation between each pair of contrasts equals 0.5. For this application, Dunnett's procedure is exact whereas the Dunn and Dunn-Šidák procedures only provide an upper bound to the experimentwise error rate. Insight concerning the relative efficiency of the three multiple comparison procedures can be obtained from

$$\text{Relative efficiency} = \frac{(\text{Critical difference for more efficient test})^2}{(\text{Critical difference for less efficient test})^2} \times 100.$$

Suppose an experimenter plans to compare four treatment means with a control mean. Let $MS_{error} = 28.8$, $n = 10$, $\nu = 45$, and $\alpha_{EW} = .01$. The critical differences for the Dunn, Dunn-Šidák, and Dunnett procedures are, respectively,

$$\hat{\psi}(D) = tD_{.01/2;4,45}\sqrt{2MS_{error}/n} = 3.20667(2.40) = 7.696$$
$$\hat{\psi}(DS) = tDS_{.01/2;4,45}\sqrt{2MS_{error}/n} = 3.20167(2.40) = 7.684$$
$$\hat{\psi}(D') = tD'_{.01/2;5,45}\sqrt{2MS_{error}/n} = 3.16667(2.40) = 7.600.$$

The efficiency of the Dunn and Dunn-Šidák procedures relative to Dunnett's procedure is, respectively,

$$\text{Relative efficiency} = \frac{[\hat{\psi}(D')]^2}{[\hat{\psi}(D)]^2} \times 100 = \frac{(7.600)^2}{(7.696)^2} \times 100 = 97.5\%$$

$$\text{Relative efficiency} = \frac{[\hat{\psi}(D')]^2}{[\hat{\psi}(DS)]^2} \times 100 = \frac{(7.600)^2}{(7.684)^2} \times 100 = 97.8\%.$$

For comparing $p - 1$ treatment means with a control mean, Dunnett's procedure is the method of choice.

A measure of the relative length of the confidence interval for any two procedures is given by

$$\text{Relative length} = \frac{\text{Length of shorter confidence interval}}{\text{Length of longer confidence interval}} \times 100.$$

According to the following computations, the length of the confidence interval for Dunnett's procedure is approximately 99% as long as that for the Dunn and Dunn-Šidák procedures.

$$\text{Relative length} = \frac{\hat{\psi}(D')}{\hat{\psi}(D)} \times 100 = \frac{(7.600)}{(7.696)} \times 100 = 98.8\%$$

$$\text{Relative length} = \frac{\hat{\psi}(D')}{\hat{\psi}(DS)} \times 100 = \frac{(7.600)}{(7.684)} \times 100 = 98.9\%$$

Dunnett (1964) has described modifications of his procedure that can be used when the variance of the control group is not equal to the variance of the $p - 1$ treatment groups.

3.5   A POSTERIORI NONORTHOGONAL CONTRASTS

Many experiments are designed to determine if any treatment effects are present. If a test of significance leads to rejection of the overall null hypothesis, attention is directed to exploring the data in order to find the source of the effects. A number of procedures has been developed for data snooping. Most are appropriate for evaluating all pairwise comparisons among means; one, Scheffé's S method, can be used to evaluate all contrasts among means. Unless indicated otherwise, all of the procedures assume that

1. The observations are drawn from normally distributed populations or the sample n_j's are fairly large.
2. The observations are random samples from the populations.
3. The null hypothesis is true.
4. The variances of the $j = 1, \ldots, p$ populations are equal to σ_ϵ^2.

As we will see, the procedures differ markedly in the protection they offer against making type I and II errors.

LEAST SIGNIFICANT DIFFERENCE TEST

The first a posteriori test described here is also one of the oldest. In 1935, Fisher (1949, 56–58) described a multiple comparison procedure called the *least significant difference* (*LSD*) test. This test consists of first performing a test of the overall null hypothesis that $\mu_1 = \mu_2 = \cdots = \mu_p$ by means of an F statistic. If the overall null hypothesis is rejected, multiple t statistics are used to evaluate all pairwise comparisons among means. If the overall F statistic is not significant, no further tests are performed. The error rate experimentwise is equal to α for the overall F test. However, if subsequent tests are performed, the conceptual unit for error rate is the individual comparison. Thus the *LSD* test is not consistent with respect to the error rate protection at the two stages of the test. This procedure has been widely used in research but is not generally recommended by statisticians—for an exception, see Carmer and Swanson (1973).

If the F statistic is significant, the least significant difference between two means, $\hat{\psi}(LSD)$, is

$$\hat{\psi}(LSD) = t_{\alpha/2,\nu} \sqrt{\frac{2MS_{\text{error}}}{n}}$$

where $t_{\alpha/2,\nu}$ is the upper $\alpha/2$ percentage point from Student's t distribution (Appendix Table E.4) and ν is the degrees of freedom associated with MS_{error}, the denominator of the F statistic. If the absolute value of a comparison $|\hat{\psi}| = |\overline{Y}_{.j} - \overline{Y}_{.j'}|$ exceeds $\hat{\psi}(LSD)$, the comparison is declared significant. This procedure is convenient if the n's are equal because $\hat{\psi}(LSD)$ need only be computed once for any set of comparisons. If the n's are not equal, multiple comparisons among means can be made using the formula (3.2-1) for the t statistic in Section 3.2.

The use of the *LSD* test can lead to an anomalous situation in which the overall F statistic is significant, but none of the pairwise comparisons is significant. This situation can arise because the overall F test is equivalent to a simultaneous test of the hypothesis that all possible contrasts among means are equal to zero. The contrast that is significant, however, may involve some linear combination of means such as $\mu_1 - (\mu_2 + \mu_3)/2$ rather than $\mu_1 - \mu_2$.

TUKEY'S *HSD* TEST

One of the more widely used a posteriori procedures for evaluating all pairwise comparisons among means was developed by Tukey (1953). This test, which is called the *HSD* (*honestly significant difference*) test or *WSD* (*wholly significant difference*) test, has been the subject of numerous investigations.* The test sets the experimentwise error rate at α for the collection of all pairwise comparisons. The basic assumptions of normality, homogeneity of variance, and so on, discussed at the beginning of Section 3.5 are also required. In addition, the n's in each treatment level must be equal. Alternative procedures for the unequal n and unequal variance cases are described later.

Tukey's *HSD* test is based on the sampling distribution of the studentized range statistic which, like the t distribution, was derived by William Sealey Gossett. The studentized range statistic q is

$$q = \frac{\overline{Y}_{\text{largest}} - \overline{Y}_{\text{smallest}}}{\sqrt{\dfrac{MS_{\text{error}}}{n}}}$$

where $\overline{Y}_{\text{largest}}$ and $\overline{Y}_{\text{smallest}}$ are the largest and smallest of p sample means, MS_{error} is an estimator of the unknown common population error variance, and $\sqrt{MS_{\text{error}}/n}$ is the standard error of the range of means, denoted by $\sigma_{\overline{Y}}$. The sampling distribution of q depends on the number of sample means used in computing the range, $\overline{Y}_{\text{largest}} - \overline{Y}_{\text{smallest}}$. It is reasonable to expect that on the average the size of the range for, say, three independent samples is larger than that for two samples and increases as the number of samples increases. This increase in the size of the range, $\overline{Y}_{\text{largest}} - \overline{Y}_{\text{smallest}}$, as the number of means increases is reflected in the studentized range distribution. Selected percentage points for the distribution of q are given in Appendix Table E.7. To enter the table, two values are required: the degrees of freedom for MS_{error} and p, the number of means on which the range, $\overline{Y}_{\text{largest}} - \overline{Y}_{\text{smallest}}$, is based. For a completely randomized design, an estimator of the population error variance is $MSWG$ with $p(n-1)$ degrees of freedom.

The critical difference, $\hat{\psi}(HSD)$, that a pairwise comparison must exceed to be declared significant is, according to Tukey's procedure,

$$\hat{\psi}(HSD) = q_{\alpha;p,\nu}\sqrt{\frac{MS_{\text{error}}}{n}}$$

where $q_{\alpha;p,\nu}$ is obtained from Appendix Table E.7 for α level of significance, p means, and ν degrees of freedom associated with MS_{error}. We will illustrate the procedure for the data in Table 3.5-1; assume that MS_{error} is equal to 28.8 with $p(n-1) = 5(10-1) = 45$ degrees of freedom. The .01 level of significance will be used. For these data, $\hat{\psi}(HSD)$ corresponding to the .01 level of significance for a two-tailed test is equal to

---

*   For a summary of research since 1953 and bibliographies see Keselman and Rogan (1977) and Miller (1977).

TABLE 3.5-1 Absolute Value of Differences Among Means ($MS_{error} = 28.8$, $p = 5$, and $n = 10$)	$\overline{Y}_{.1}$	$\overline{Y}_{.5}$	$\overline{Y}_{.3}$	$\overline{Y}_{.4}$	$\overline{Y}_{.2}$
$\overline{Y}_{.1} = 36.7$	—	3.6	6.7	10.5*	12.0*
$\overline{Y}_{.5} = 40.3$		—	3.1	6.9	8.4*
$\overline{Y}_{.3} = 43.4$			—	3.8	5.3
$\overline{Y}_{.4} = 47.2$				—	1.5
$\overline{Y}_{.2} = 48.7$					—

*$p < .01$

$$\hat{\psi}(HSD) = q_{.01;5,45}\sqrt{\frac{MS_{error}}{n}} = 4.893\sqrt{\frac{28.8}{10}} = 8.30.$$

A test of the overall null hypothesis that $\mu_1 = \mu_2 = \cdots = \mu_p$ is provided by a comparison of the largest pairwise difference between means, $\hat{\psi} = \overline{Y}_{largest} - \overline{Y}_{smallest}$, with the critical difference $\hat{\psi}(HSD)$. This test procedure, which utilizes a range statistic, is an alternative to the overall F test. For most sets of data, the range and F tests lead to the same decision concerning the overall null hypothesis. However, the F test is generally more powerful. According to Table 3.5-1, the difference between the largest and smallest means is equal to 12.0. Because this difference exceeds $\hat{\psi}(HSD) = 8.30$, the overall null hypothesis is rejected. An examination of Table 3.5-1 indicates that three pairwise comparisons—those starred—exceed the critical difference and hence are declared significant at the .01 level. It should be noted that the values of $q_{\alpha;p,\nu}$ in Appendix Table E.7 are appropriate for testing two-tailed hypotheses; this is true for all of the a posteriori procedures described in Section 3.5.

It is instructive to compare the critical difference for Tukey's HSD test with those for the LSD, Dunn, and Dunn-Šidák tests. The critical differences are

$$\hat{\psi}(HSD) = q_{.01;5,45}\sqrt{\frac{MS_{error}}{n}} = 4.893(1.697) = 8.30$$

$$\hat{\psi}(LSD) = t_{.01/2,45}\sqrt{\frac{2MS_{error}}{n}} = 2.689(2.400) = 6.45$$

$$\hat{\psi}(D) = tD_{.01/2;10,45}\sqrt{\frac{2MS_{error}}{n}} = 3.520(2.400) = 8.45$$

$$\hat{\psi}(DS) = tDS_{.01/2;10,45}\sqrt{\frac{2MS_{error}}{n}} = 3.519(2.400) = 8.45.$$

As expected, the LSD test is the most powerful because at the second stage of the testing procedure it does not control the error rate at α for the collection of tests. The least sensitive procedures in this example are those of Dunn and Dunn-Šidák. They become more powerful relative to Tukey's HSD test as the number of comparisons among the p means is reduced. For example, if an experimenter had planned to make only eight instead of all ten pairwise comparisons among means, the critical difference for the Dunn-Šidák procedure would have been only

$$\hat{\psi}(DS) = tDS_{.01/2;8,45}\sqrt{\frac{2MS_{\text{error}}}{n}} = 3.443(2.400) = 8.26$$

which is less than that for Tukey's procedure.

Tukey's procedure can be used to establish $100(1 - \alpha)\%$ simultaneous confidence intervals for all pairwise population contrasts. The confidence interval is given by

$$\hat{\psi}_i - \hat{\psi}(HSD) \leq \psi_i \leq \hat{\psi}_i + \hat{\psi}(HSD)$$

where $\hat{\psi}(HSD) = q_{\alpha;p,\nu}\sqrt{\dfrac{MS_{\text{error}}}{n}}$

$\quad\quad\quad \hat{\psi}_i = \overline{Y}_{.j} - \overline{Y}_{.j'}$

$\quad\quad\quad \psi_i = \mu_j - \mu_{j'}.$

The test statistic for Tukey's procedure is

$$q = \frac{\hat{\psi}_i}{\hat{\sigma}_{\overline{Y}}} = \frac{\overline{Y}_{.j} - \overline{Y}_{.j'}}{\sqrt{\dfrac{MS_{\text{error}}}{n}}}.$$

If $|q| \geq q_{\alpha;p,\nu}$, the null hypothesis, H_0: $\mu_j - \mu_{j'} = 0$, is rejected. Tukey's procedure can also be used with the conventional t statistic (3.2-1) described in Section 3.2.

$$t = \frac{\hat{\psi}_i}{\hat{\sigma}_{\psi_i}} = \frac{\overline{Y}_{.j} - \overline{Y}_{.j'}}{\sqrt{\dfrac{2MS_{\text{error}}}{n}}}$$

The critical value for this t statistic is $q_{\alpha;p,\nu}/\sqrt{2}$, since $t = q/\sqrt{2}$.

Tukey's procedure can be extended to test nonpairwise contrasts. However, this is not recommended since for this application it is less powerful than Scheffé's procedure, which is described later.

Earlier we noted that Tukey's procedure assumes equal n's. This is necessary in order for the sample means to have the same variance σ^2/n. The procedure also assumes that the p population variances are homogeneous. In the following sections we will describe procedures for evaluating pairwise comparisons that do not require these assumptions.

SPJØTVOLL AND STOLINE'S MODIFICATION OF THE HSD TEST

Over the years a variety of a posteriori procedures have been suggested for evaluating pairwise comparisons when the sample n's are unequal: Dunn (1974), Gabriel (1978b), Hochberg (1974), Kramer (1956), Šidák (1967), Scheffé (1953), Spjøtvoll and Stoline (1973), and Tukey (1953). Ury (1976) compared the procedures of Dunn-Šidák (Šidák, 1967), Hochberg (1974), Scheffé (1953), and Spjøtvoll and Stoline (1973). He concluded that when the n's are similar the preferred procedure was that due to Spjøtvoll and Stoline. When the n's are quite different or a very high level of significance is adopted ($\alpha < .01$), one of the other procedures may be preferred (Ury, 1976; Stoline, 1978). Probably the most common situation involving unequal n's in the behavioral sciences and education is one in which the n's are similar.

The Spjøtvoll-Stoline test, which is a generalization of Tukey's test, is appropriate for unequal n's and is referred to as the T' test. The basic assumptions of normality, homogeneity of variance, and so on, discussed in Section 3.5 are required for this test. Of course the test should be preceded by a significant test of the overall null hypothesis. The T' test is based on the studentized augmented range distribution.* The critical difference $\hat{\psi}(T')$ that a pairwise comparison must exceed to be declared significant is

$$\hat{\psi}(T') = q'_{\alpha;p,\nu}\sqrt{\frac{MS_{\text{error}}}{n_{\min}}}$$

where $q'_{\alpha;p,\nu}$ is obtained from Appendix Table E.18 for α level of significance, p means, and ν degrees of freedom associated with MS_{error}; and $n_{\min}$ is the minimum of n_j and $n_{j'}$, the sample sizes used to compute $\overline{Y}_j$ and $\overline{Y}_{j'}$, respectively. When p exceeds eight treatment means, the value of $q'_{\alpha;p,\nu}$ can be obtained from the studentized range distribution (Appendix Table E.7).

Simultaneous $100(1 - \alpha)\%$ confidence intervals for the Spjøtvoll-Stoline procedure are given by

$$\hat{\psi}_i - \hat{\psi}(T') \leq \psi_i \leq \hat{\psi}_i + \hat{\psi}(T')$$

where $\quad \hat{\psi}(T') = q'_{\alpha;p,\nu}\sqrt{\frac{MS_{\text{error}}}{n_{\min}}}$

$$\hat{\psi}_i = \overline{Y}_j - \overline{Y}_{j'}$$

$$\psi_i = \mu_j - \mu_{j'}.$$

The test statistic for the Spjøtvoll-Stoline procedure is

$$q'T' = \frac{\hat{\psi}_i}{\hat{\sigma}_{\overline{Y}}} = \frac{\overline{Y}_j - \overline{Y}_{j'}}{\sqrt{\dfrac{MS_{\text{error}}}{n_{\min}}}}.$$

If $|q'T'| \geq q'_{\alpha;p,\nu}$, the null hypothesis, H_0: $\mu_j - \mu_{j'} = 0$, is rejected.

TUKEY-KRAMER MODIFICATION OF THE *HSD* TEST

Tukey (1953) and Kramer (1956) independently proposed a modification of the *HSD* test for the case in which the sample n's are unequal but the basic assumptions of normality, homogeneity of variance, and so on are tenable. We will refer to the test as the Tukey-Kramer procedure. For this test, the critical difference, $\hat{\psi}(TK)$, that a pairwise comparison must exceed to be declared significant is

---

*   See Scheffé (1959, 78) for a description of the studentized augmented range.

$$\hat{\psi}(TK) = q_{\alpha;p,\nu}\sqrt{MS_{\text{error}}\left(\frac{1}{n_j} + \frac{1}{n_{j'}}\right)\bigg/2}$$

where $q_{\alpha;p,\nu}$ is obtained from the studentized range distribution (Appendix Table E.7) for α level of significance, p means, and ν degrees of freedom associated with MS_{error}.

The Tukey-Kramer procedure is thought by some to be liberal, that is, to not control the experimentwise error rate at or less than α. Research by Dunnett (1980) and unpublished research cited by Dunnett (1980), however, suggests that the procedure is conservative at least for experiments involving moderate to severe imbalance among sample n's. Furthermore the size of the critical difference, $\hat{\psi}(TK)$, is always less than that for the Spjøtvoll-Stoline procedure. Pending further research, the Spjøtvoll-Stoline procedure is recommended when sample n's are approximately equal; the Tukey-Kramer procedure is recommended when there is a moderate or large imbalance among the sample n's. We turn next to two procedures that are appropriate for experiments with unequal population variances and equal or unequal sample sizes.

ROBUST PROCEDURES FOR A POSTERIORI NONORTHOGONAL PAIRWISE COMPARISONS

The problem of evaluating pairwise comparisons when the variances are unequal or both the variances and the n's are unequal has also received considerable attention in recent years.* The most promising a posteriori procedures are those of Games and Howell (1976) and Tamhane (1977, 1979). Both procedures use the Behrens-Fisher statistic described in Section 3.2 to estimate the standard error of a contrast, σ_{ψ_i}, and the Welch procedure for determining the degrees of freedom for the standard error of the contrast. However, the Games-Howell procedure uses the studentized range distribution while the Tamhane procedure uses Student's t distribution and the Šidák multiplicative inequality. Since the procedures are used for data snooping, they should be preceded by a significant test of the overall null hypothesis.

The critical difference $\hat{\psi}(GH)$ that a pairwise comparison must exceed in order to reject the hypothesis H_0: $\mu_j - \mu_{j'} = 0$ for the Games and Howell procedure is

$$\hat{\psi}(GH) = q_{\alpha;p,\nu'}\sqrt{\left(\frac{\hat{\sigma}_j^2}{n_j} + \frac{\hat{\sigma}_{j'}^2}{n_{j'}}\right)\bigg/2}$$

where $q_{\alpha;p,\nu'}$ is obtained from the studentized range distribution in Appendix Table E.7 for α level of significance, p means, and degrees of freedom equal to

$$\nu' = \frac{\left(\dfrac{\hat{\sigma}_j^2}{n_j} + \dfrac{\hat{\sigma}_{j'}^2}{n_{j'}}\right)^2}{\dfrac{\hat{\sigma}_j^4}{n_j^2(n_j - 1)} + \dfrac{\hat{\sigma}_{j'}^4}{n_{j'}^2(n_{j'} - 1)}}.$$

---

*   See, for example, Brown and Forsythe (1974); Dalal (1978); Games and Howell (1976); Hochberg (1976); Keselman and Rogan (1977); Keselman, Games, and Rogan (1979); Miller (1966, 43); Spjøtvoll (1972); Tamhane (1977, 1979); and Ury and Wiggins (1971).

The critical difference, $\hat{\psi}(T2)$, for the Tamhane procedure is

$$\hat{\psi}(T2) = tDS_{\alpha/2;C,\nu'} \sqrt{\frac{\hat{\sigma}_j^2}{n_j} + \frac{\hat{\sigma}_{j'}^2}{n_{j'}}}$$

where $tDS_{\alpha/2;C,\nu'}$ is obtained from the t distribution using the Šidák multiplicative inequality (Appendix Table E.17) for $C = p(p - 1)/2$ pairwise comparisons and degrees of freedom, ν', as defined for the Games-Howell procedure. If at least one of the following conditions holds, ν' is set equal to $n_j + n_{j'} - 2$ (Tamhane, 1979; Ury and Wiggins, 1971).

1. $9/10 \le n_j/n_{j'} \le 10/9$

2. $9/10 \le (\hat{\sigma}_j^2/n_j)/(\hat{\sigma}_{j'}^2/n_{j'}) \le 10/9$

3. $4/5 \le n_j/n_{j'} \le 5/4$ and $1/2 \le (\hat{\sigma}_j^2/n_j)/(\hat{\sigma}_{j'}^2/n_{j'}) \le 2$

4. $2/3 \le n_j/n_{j'} \le 3/2$ and $3/4 \le (\hat{\sigma}_j^2/n_j)/(\hat{\sigma}_{j'}^2/n_{j'}) \le 4/3$

The procedures of Tamhane and Games and Howell appear to be excellent choices for evaluating pairwise comparisons when the variances or the variances and sample n's are unequal. Monte Carlo sampling studies indicate that even when the unknown variances are homogeneous, use of the procedures does not lead to a substantial loss of power (Keselman, Games, and Rogan, 1979). The Games-Howell procedure is slightly more powerful than the Tamhane procedure (Tamhane, 1979). However, the Tamhane procedure controls the experimentwise error rate at α or less while the Games-Howell procedure can become slightly liberal (Keselman, Games, and Rogan, 1979; Tamhane, 1979). A choice between them depends on whether one prefers slightly greater power or better control of the experimentwise error rate.

SCHEFFÉ'S S TEST

The Scheffé S procedure (1953) is one of the most flexible, conservative, and robust data snooping procedures available. If the overall F statistic is significant, Scheffé's procedure can be used to evaluate all a posteriori contrasts among means, not just the pairwise comparisons. In addition, it can be used with unequal n's. The error rate experimentwise is equal to α for the infinite number of possible contrasts among $p \ge 3$ means. Since an experimenter always evaluates a subset of the possible contrasts, Scheffé's procedure tends to be conservative. It is much less powerful than Tukey's HSD procedure for evaluating pairwise comparisons, for example, and consequently is recommended only when complex contrasts are of interest. Scheffé's procedure uses the F sampling distribution and, like ANOVA, is robust with respect to nonnormality and heterogeneity of variance.

The critical difference $\hat{\psi}(S)$ that a contrast must exceed to be declared significant is given by

$$\hat{\psi}(S) = \sqrt{(p - 1)F_{\alpha;\nu_1,\nu_2}} \sqrt{MS_{\text{error}} \sum_{j=1}^{p} \frac{c_j^2}{n_j}}$$

where p is the number of means in the experiment (family) and $F_{\alpha;\nu_1,\nu_2}$ is obtained from Appendix Table E.5 for ν_1 equal to $p - 1$ degrees of freedom and ν_2 equal to the degrees of freedom associated with MS_{error}. For purposes of comparison, the value of $\hat{\psi}(S)$ will be computed for a pairwise comparison using the data in Table 3.5-1. We will assume that the overall null hypothesis $\mu_1 = \mu_2 = \cdots = \mu_p$ has been rejected using an F statistic. The value of the critical difference is

$$\hat{\psi}(S) = \sqrt{(5-1)3.77} \sqrt{28.8 \left[\frac{(1)^2}{10} + \frac{(-1)^2}{10} \right]} = 3.883(2.400) = 9.32.$$

The critical difference for Tukey's procedure is only $\hat{\psi}(HSD) = 8.30$. Hence, if one wanted to evaluate only pairwise comparisons, Scheffé's procedure would be a poor choice. The advantage of Scheffé's procedure, the ability to evaluate all possible contrasts, comes at a price—low power.

Scheffé's procedure can be used to establish $100(1 - \alpha)\%$ simultaneous confidence intervals for all population contrasts. The confidence interval is given by

$$\hat{\psi}_i - \hat{\psi}(S) \leq \psi_i \leq \hat{\psi}_i + \hat{\psi}(S)$$

where $\hat{\psi}(S) = \sqrt{(p-1)F_{\alpha;\nu_1,\nu_2}} \sqrt{MS_{error} \sum_{j=1}^{p} \frac{c_j^2}{n_j}}$

$\hat{\psi}_i = \sum_{j=1}^{p} c_j \overline{Y}_{.j}$

$\psi_i = \sum_{j=1}^{p} c_j \mu_j.$

The test statistic for Scheffé's procedure is

$$F = \frac{\hat{\psi}_i^2}{\hat{\sigma}_{\psi_i}^2} = \frac{\left(\sum_{j=1}^{p} c_j \overline{Y}_{.j} \right)^2}{MS_{error} \sum_{j=1}^{p} \frac{c_j^2}{n_j}}.$$

To be significant, F must exceed F', where $F' = (p-1)F_{\alpha;\nu_1,\nu_2}$.

BROWN-FORSYTHE *BF* PROCEDURE

A procedure, which is a modification of Scheffé's S test, for evaluating all a posteriori contrasts among means following a significant F test of the overall null hypothesis has been described by Brown and Forsythe (1974). Their procedure is even more robust with respect to heterogeneity of variance than Scheffé's. This is accomplished by using the Behrens-Fisher statistic described in Section 3.2 to estimate the standard error of a contrast σ_{ψ_i} and the Welch (1947) procedure for determining the degrees of freedom for the standard error of the contrast. Like Scheffé's procedure, it can be used with unequal sample n's. The experimentwise error rate is less than or equal to α.

The critical difference $\hat{\psi}(BF)$ that a contrast must exceed in order to reject the null hypothesis for the population contrast is given by

$$\hat{\psi}(BF) = \sqrt{(p-1)F_{\alpha;\nu_1,\nu_2'}}\ \sqrt{\frac{c_1^2\hat{\sigma}_1^2}{n_1} + \frac{c_2^2\hat{\sigma}_2^2}{n_2} + \cdots + \frac{c_p^2\hat{\sigma}_p^2}{n_p}}$$

where $F_{\alpha;\nu_1,\nu_2'}$ is obtained from Appendix Table E.5 for ν_1 equal to $p-1$ degrees of freedom and ν_2' equal to

$$\nu_2' = \frac{\left(\sum_{j=1}^{p} \frac{c_j^2\,\hat{\sigma}_j^2}{n_j}\right)^2}{\sum_{j=1}^{p} \frac{c_j^4\,\hat{\sigma}_j^4}{n_j^2\,(n_j-1)}}$$

degrees of freedom.

If an experimenter is interested in evaluating only pairwise comparisons and the population variances are believed to be heterogeneous, the procedures of Games and Howell or Tamhane described earlier should be used since for this case they are more powerful.

NEWMAN-KEULS TEST

A different approach to evaluating a posteriori pairwise comparisons stems from the work of Student (1927), Newman (1939), and Keuls (1952). The Newman-Keuls procedure is based on a stepwise or layer approach to significance testing. Sample means are ordered from the smallest to the largest, as in Table 3.5-2. The largest difference, which involves means that are $r = p$ steps apart, is tested first at α level of significance; if significant, means that are $r = p - 1$ steps apart are tested at α level of significance and so on. The Newman-Keuls procedure provides an r-mean significance level equal to α for each group of r ordered means; that is, the probability of falsely rejecting the hypothesis that all means in an ordered group are equal is α. It follows that the concept of error rate applies neither on an experimentwise nor on a per comparison basis—the actual error rate falls somewhere between the two. The

TABLE 3.5-2 Absolute Value of Differences Among Means ($MS_{error} = 28.8$, $p = 5$, and $n = 10$)

	$\overline{Y}_{.1}$	$\overline{Y}_{.5}$	$\overline{Y}_{.3}$	$\overline{Y}_{.4}$	$\overline{Y}_{.2}$
$\overline{Y}_{.1} = 36.7$	–	3.6	6.7	10.5*	12.0*
$\overline{Y}_{.5} = 40.3$		–	3.1	6.9	8.4*
$\overline{Y}_{.3} = 43.4$			–	3.8	5.3
$\overline{Y}_{.4} = 47.2$				–	1.5
$\overline{Y}_{.2} = 48.7$					–

*$p < .01$

Newman-Keuls procedure, like Tukey's procedure, requires equal sample n's. It differs from all of the tests described previously in that it cannot be used to construct confidence intervals. The stepwise or multistage character of the test has no counterpart in confidence intervals.

The critical difference $\hat{\psi}(W_r)$ that two means separated by r steps must exceed to be declared significant is, according to the Newman-Keuls procedure,

$$\hat{\psi}(W_r) = q_{\alpha;r,\nu} \sqrt{\frac{MS_{error}}{n}}$$

where $q_{\alpha;r,\nu}$ is obtained from Appendix Table E.7 for α level of significance, means separated by r steps, and ν degrees of freedom associated with MS_{error}. The subscript r designates the number of steps separating ordered means. Consider the following means from Table 3.5-2.

$\overline{Y}_{\cdot 1}$	$\overline{Y}_{\cdot 5}$	$\overline{Y}_{\cdot 3}$	$\overline{Y}_{\cdot 4}$	$\overline{Y}_{\cdot 2}$
36.7	40.3	43.4	47.2	48.7

Mean one is defined as being five steps away from mean two, four steps away from mean four, and so on. Two values are required to enter the studentized range table—the degrees of freedom for MS_{error} and r. For $p(n-1) = 45$ degrees of freedom, the respective values of $q_{.01;r,45}$ from Appendix Table E.7 are

$$q_{.01;2,45} = 3.800, \quad q_{.01;3,45} = 4.340, \quad q_{.01;4,45} = 4.663, \quad q_{.01;5,45} = 4.893.$$

The critical differences, $\hat{\psi}(W_r)$, are the products of

$$q_{.01;r,45} \sqrt{\frac{MS_{error}}{n}}$$

where $MS_{error} = 28.8$
$n = 10$.

$$\hat{\psi}(W_2) = 3.800 \sqrt{\frac{28.8}{10}} = 6.45$$

$$\hat{\psi}(W_3) = 4.340 \sqrt{\frac{28.8}{10}} = 7.37$$

$$\hat{\psi}(W_4) = 4.663 \sqrt{\frac{28.8}{10}} = 7.91$$

$$\hat{\psi}(W_5) = 4.893 \sqrt{\frac{28.8}{10}} = 8.30$$

The critical difference is the difference that two means r steps apart must exceed in order to reject the null hypothesis H_0: $\mu_j - \mu_{j'} = 0$.

There is a prescribed sequence in which tests on pairwise comparisons must be performed. As noted earlier, the means must first be arranged in order of increasing size as in Table 3.5-2. The first test is made in row one for $\overline{Y}_1$ versus $\overline{Y}_2$. Because these two means are separated by five steps, the critical difference that this comparison must exceed is 8.30. The difference in Table 3.5-2 exceeds this critical difference. This first test is a range test of the overall null hypothesis. If it had been insignificant, no other comparisons would have been tested. The next comparison tested is $\overline{Y}_1$ versus $\overline{Y}_4$ in row one. The critical difference for this comparison is 7.91. This comparison is also significant. This procedure is continued in row one until a nonsignificant comparison is found. In this example, the comparison $\overline{Y}_1$ versus $\overline{Y}_3$ does not exceed 7.37, so no further tests are made in this row or in the following rows to the left of the column headed by $\overline{Y}_4$. The next test is the comparison of $\overline{Y}_5$ versus $\overline{Y}_2$ in row two. The critical difference for this test is 7.91. This process is repeated for each successive row until a nonsignificant comparison is found or until the column at which tests were stopped in the preceding row is reached. In this example, the Newman-Keuls procedure leads to three significant pairwise comparisons—those starred in Table 3.5-2. This is the same number that was declared significant by Tukey's *HSD* procedure but one more than was obtained by Scheffé's procedure.

It should be noted that the Newman-Keuls and Tukey procedures require the same critical difference, 8.30, for the first comparison that is tested. The Tukey procedure uses this critical difference for all of the remaining tests while the Newman-Keuls procedure reduces the size of the critical difference, depending on the number of steps separating the ordered means. As a result, the Newman-Keuls test is more powerful than Tukey's test. Remember, however, that the Newman-Keuls procedure does not control the experimentwise error rate at α.

Frequently a test of the overall null hypothesis $\mu_1 = \mu_2 = \cdots = \mu_p$ is performed with an F statistic in ANOVA rather than with a range statistic. If the F statistic is significant, Shaffer (1979) recommends using the critical difference $\hat{\psi}(W_{r-1})$ instead of $\hat{\psi}(W_r)$ to evaluate the largest pairwise comparison at the first step of the testing procedure. The testing procedure for all subsequent steps is unchanged. She has shown that the modified procedure leads to greater power at the first step without affecting control of the type I error rate. This makes *dissonances,* in which the overall null hypothesis is rejected by an F test without rejecting any one of the proper subsets of comparisons, less likely.

DUNCAN'S NEW MULTIPLE RANGE TEST

Duncan (1955) has developed a procedure for evaluating a posteriori pairwise comparisons that shares many of the features of the Newman-Keuls procedure including a stepwise approach to significance testing, prescribed sequence for performing tests, and absence of confidence interval procedures. The r-mean significance level for this test is equal to $1 - (1 - \alpha)^{r-1}$. For five ordered means in a group and $\alpha = .01$, the probability of falsely rejecting the hypothesis that all means in the ordered group are equal is $1 - (1 - .01)^{5-1} = .0394$. The corresponding r-mean significance level for the Newman-Keuls procedure is .01 and remains at .01 for each group of ordered means. For Duncan's procedure, the r-mean significance level decreases as r increases, for example

$$2\text{-means} \quad 1 - (1 - .01)^{2-1} = .01$$
$$3\text{-means} \quad 1 - (1 - .01)^{3-1} = .0199$$
$$4\text{-means} \quad 1 - (1 - .01)^{4-1} = .0297$$
$$5\text{-means} \quad 1 - (1 - .01)^{5-1} = .0394.$$

Duncan (1955) has argued that if $p > 2$, an experiment is more likely to contain significant contrasts than if the experiment contains only two means. Thus, as p increases, the test should become more powerful. He reasons further that p means can be used to form $p - 1$ orthogonal contrasts, each of which can be tested at α level of significance. We saw in Section 3.3 that for $p - 1$ orthogonal contrasts, the error rate experimentwise is $1 - (1 - \alpha)^{p-1}$. Duncan's procedure provides the same r-mean significance level as that tolerated for $r - 1$ orthogonal contrasts.

The critical difference $\hat{\psi}(W_r)$ that two means separated by r steps must exceed to be declared significant is, according to Duncan's procedure,

$$\hat{\psi}(W_r) = q_{\alpha;r,\nu} \sqrt{\frac{MS_{\text{error}}}{n}}$$

where $q_{\alpha;r,\nu}$ is obtained from Duncan's table in the Appendix (Table E.8) for α level of significance, means separated by r steps, and ν degrees of freedom associated with MS_{error}. For the data in Table 3.5-2, where $MS_{\text{error}} = 28.8$ and $n = 10$, the critical differences are

$$\hat{\psi}(W_2) = q_{.01;2,45} \sqrt{\frac{28.8}{10}} = 3.800(1.697) = 6.45$$

$$\hat{\psi}(W_3) = q_{.01;3,45} \sqrt{\frac{28.8}{10}} = 3.967(1.697) = 6.73$$

$$\hat{\psi}(W_4) = q_{.01;4,45} \sqrt{\frac{28.8}{10}} = 4.077(1.697) = 6.92$$

$$\hat{\psi}(W_5) = q_{.01;5,45} \sqrt{\frac{28.8}{10}} = 4.153(1.697) = 7.05.$$

The sequence in which the tests are carried out is identical to that for the Newman-Keuls procedure. An examination of Table 3.5-2 reveals that the null hypotheses for four of the pairwise comparisons can be rejected: $\mu_1 - \mu_2$, $\mu_1 - \mu_4$, $\mu_5 - \mu_2$, and $\mu_5 - \mu_4$. This is one more significant test than was obtained using the Newman-Keuls and Tukey *HSD* procedures.

An extension of Duncan's new multiple range test for the case of unequal sample n's is described by Kramer (1956). Other applications of the test are discussed by Duncan (1957). Scheffé (1959, 718) has criticized the justification originally advanced for Duncan's test.

3.6 COMPARISON OF MULTIPLE COMPARISON PROCEDURES

A variety of multiple comparison procedures has been described in this chapter. The procedures and their salient characteristics are summarized in Table 3.6-1. Most of the procedures have been illustrated using the same set of data involving five means. An indication of the relative power of the procedures can be obtained by comparing the size of their critical difference. For purposes of this comparison, we will assume that the multiple t procedure and Dunnett's procedure are used to evaluate $p - 1$ a priori contrasts and the other procedures are used to evaluate all ten pairwise comparisons among the five means. This comparison is not fair to the Dunn and Dunn-Šidák procedures because they were designed for evaluating C a priori contrasts, nor is it fair to the Scheffé procedure which is not recommended for evaluating pairwise contrasts. Nevertheless an examination of the critical differences in Table 3.6-2 is instructive.

The relative merits of various multiple comparison procedures has engendered much debate among statisticians in recent years. Each of the procedures in Table 3.6-1 has been recommended by one or more statisticians (Carmer and Swanson, 1973; Dunnett, 1980; Einot and Gabriel, 1975; Gabriel, 1978a; Games and Howell, 1976; Keselman, Games, and Rogan, 1979; Keselman and Rogan, 1977, 1978; Ramsey, 1978; Tamhane, 1979; Thomas, 1974; Ury, 1976). Several statisticians (Games, 1971; Hopkins and Anderson, 1973; Hopkins and Chadbourn, 1967) have developed helpful guides in the form of flowcharts for selecting a multiple comparison procedure. Here too one finds little agreement. The problem facing an experimenter is to select the test statistic that provides the desired kind of protection against type I errors and at the same time provides maximum power. The characteristics of the most frequently recommended procedures have been described in some detail along with pertinent references so that an experimenter can make informed choices.

3.7 REVIEW EXERCISES

1.　　[3.1] In order for $\psi_i = c_1\mu_1 + c_2\mu_2 + \cdots + c_p\mu_p$ to be a contrast, what conditions must the coefficients satisfy?

†2.　　[3.1] Distinguish between a pairwise comparison and a contrast.

3.　　[3.1] List the coefficients for the following contrasts.

　　†a) μ_1 versus μ_2　　†b) μ_1 versus mean of μ_2 and μ_3

　　c) μ_1 versus mean of μ_2, μ_3, and μ_4　　†d) Mean of μ_1 and μ_2 versus mean of μ_3 and μ_4

　　e) Mean of μ_1 and μ_2 versus mean of μ_3, μ_4, and μ_5

　　†f) μ_1 versus the weighted mean of μ_2 and μ_3, where μ_2 is weighted twice as much as μ_3

TABLE 3.6-1   Summary of Characteristics of Multiple Comparison Procedures

	Pairwise Comparisons Only	p − 1 Pairwise Comparisons	Complex Comparisons	C Complex Comparisons	Equal n's Only	Un-equal n's	Homo-geneous Variances	Hetero-geneous Variances
	Test is appropriate for							
A Priori Orthogonal Contrasts								
Multiple t			x			x	x	
Multiple F			x			x	x	
Multiple t with Behrens-Fisher and Welch Procedures			x			x		x
A Priori Nonorthogonal Contrasts								
Dunn D				x		x	x	
Dunn-Šidák DS		x		x		x	x	
Dunnett D'						x	x	x*
A Posteriori Nonorthogonal Contrasts								
Fisher LSD	x					x	x	
Tukey HSD	x				x		x	
Spjøtvoll-Stoline T'	x					x	x	
Tukey-Kramer TK	x					x	x	
Games-Howell GH	x					x		x
Tamhane T2	x					x		x
Scheffé S			x			x	x	
Brown-Forsythe BF			x					
Newman-Keuls	x				x		x	
Duncan	x				x		x	x

*With modification

TABLE 3.6-2   Comparison of Critical Difference for Several Multiple Comparison Procedures

Test	Critical Difference			
	A Priori Orthogonal Contrasts			
Multiple t	6.45 for $p - 1 = 4$ contrasts			
	A Priori Nonorthogonal Contrasts			
Dunnett Dunn-Šidák Dunn	7.60 for $p - 1 = 4$ contrasts with a control group mean 8.45 for $C = 10$ pairwise comparisons 8.45 for $C = 10$ pairwise comparisons			
	A Posteriori Nonorthogonal Contrasts *Number of steps separating means*			
	2	3	4	5
Fisher *LSD*	6.45	same	same	same
Duncan	6.45	6.73	6.92	7.05
Newman-Keuls	6.45	7.37	7.91	8.30
Tukey *HSD*	8.30	same	same	same
Scheffé *S*	9.32	same	same	same

for 10 pairwise comparisons

g) The weighted mean of μ_1 and μ_2 versus the weighted mean of μ_3 and μ_4, where μ_1 and μ_3 are weighted twice as much as μ_2 and μ_4

4. [3.1] Which of the following meet the requirements for a contrast?

†**a)** $\mu_1 - \mu_2$ †**b)** $2\mu_1 - \mu_2 - \mu_3$ †**c)** $\mu_1 - (1/3)\mu_2 - (1/3)\mu_3$

d) $(1^1/_2)\mu_1 - \mu_2 - (1^1/_2)\mu_3$ **e)** $\mu_1 - (1/4)\mu_2 - (1/4)\mu_3 - (1/4)\mu_4$

†**f)** $(3/4)\mu_1 - (1/4)\mu_2 - (1/4)\mu_3 - (1/4)\mu_4$

g) $(1/2)\mu_1 + (1/2)\mu_2 - (1/3)\mu_3 - (1/3)\mu_4 - (1/3)\mu_5$

5. [3.1] Which contrasts in Exercise 4 satisfy $|c_1| + |c_2| + \cdots + |c_p| = 2$?

6. [3.1] Indicate the number of pairwise comparisons that can be constructed for the following designs.

†**a)** Type CR-3 design   **b)** Type CR-4 design

†**c)** Type CR-5 design   **d)** Type CR-6 design

7. [3.1] Which of the following sets of contrasts are orthogonal? Assume that the n's are equal.

†a) $\mu_1 - \mu_2$ †b) $\mu_1 - \mu_2$ c) $\mu_1 - \mu_2$

     $\mu_1 - \mu_3$ $\mu_3 - \mu_4$ $(1/2)\mu_1 + (1/2)\mu_2 - (1/2)\mu_3 - (1/2)\mu_4$

†d) $(3/4)\mu_1 - (3/4)\mu_2 + (1/4)\mu_3 - (1/4)\mu_4$

     $(1/2)\mu_1 + (1/2)\mu_2 - (1/2)\mu_3 - (1/2)\mu_4$

e) $(1/2)\mu_1 + (1/2)\mu_2 - (1/2)\mu_3 - (1/2)\mu_4$

     $(2/3)\mu_1 - (2/3)\mu_2 + (1/3)\mu_2 - (1/3)\mu_4$

8.   [3.1] Construct four sets of orthogonal contrasts among five means.

†9.   [3.2] The religious dogmatism of members of four church denominations in a large midwestern city was investigated. A random sample of 30 members from each denomination took a paper and pencil test of dogmatism. The sample means were $\overline{Y}_1 = 64$, $\overline{Y}_2 = 73$, $\overline{Y}_3 = 61$, and $\overline{Y}_4 = 49$; $MSWG = 120$; $\nu_2 = 4(30 - 1) = 116$. The experimenter advanced the following a priori null hypotheses: $\mu_1 - \mu_2 = 0$, $\mu_3 - \mu_4 = 0$, $(\mu_1 + \mu_2)/2 - (\mu_3 + \mu_4)/2 = 0$.

a) Use a t statistic to test these null hypotheses; let $\alpha = .01$.

b) Construct $100(1 - .01)\%$ confidence intervals for these a priori contrasts.

c) [3.4] Compute the correlations among the contrasts.

d) Assume that the sample variances are $\hat{\sigma}_1^2 = 62$, $\hat{\sigma}_2^2 = 73$, $\hat{\sigma}_3^2 = 80$, and $\hat{\sigma}_4^2 = 265$. Use Welch's t' statistic to test the null hypotheses.

10.   [3.2] The effectiveness of three approaches to drug education in junior high school was investigated. The approaches were providing objective scientific information about the physiological and psychological effects of drug usage, a_1; examining the psychology of drug use, a_2; and a control condition in which the chemical nature of various drugs was examined, a_3. Sixty-three students who did not use drugs were randomly assigned to one of three groups with 21 in each group. The treatment levels were randomly assigned to the groups. At the conclusion of the education program the students evaluated its effectiveness; a high score signified effectiveness. The sample means were $\overline{Y}_1 = 25.8$, $\overline{Y}_2 = 26.7$, and $\overline{Y}_3 = 22.1$; $MSWG = 16.4$; and $\nu_2 = 3(21 - 1) = 60$. The experimenter advanced the following a priori null hypotheses: $\mu_1 - \mu_2 = 0$ and $(\mu_1 + \mu_2)/2 - \mu_3 = 0$.

a) Use a t statistic to test these null hypotheses; let $\alpha = .05$.

b) Construct $100(1 - .05)\%$ confidence intervals for these a priori contrasts.

c) [3.4] Compute the correlation between the contrasts.

d) Assume that the sample variances are $\hat{\sigma}_1^2 = 10.6$, $\hat{\sigma}_2^2 = 9.2$, and $\hat{\sigma}_3^2 = 29.4$. Use Welch's t' statistic to test the null hypotheses.

†11.   [3.3] For the experiment in Exercise 9, what is the error rate

a) Per contrast   **b)** Experimentwise?

12.   [3.3] For the experiment in Exercise 10, what is the error rate

a) Per contrast   **b)** Experimentwise?

†13.   [3.3] Suppose that 1000 experiments involving a type CR-4 design have been performed and in each experiment hypotheses for all possible pairwise comparisons have been tested. Assume that 50 type I errors are committed and these occur in 35 of the 1000 experiments. Compute the following:

a) Error rate per contrast   **b)** Error rate per experiment   **c)** Error rate experimentwise.

14. [3.3] Suppose that 1000 experiments involving a type CR-5 design have been performed and in each experiment hypotheses for all possible pairwise comparisons have been tested. Assume that 90 type I errors are committed and these occur in 70 of the 1000 experiments. Compute the following:

a) Error rate per contrast   b) Error rate per experiment   c) Error rate experimentwise.

15. [3.3] Compute α_{EW} for the following:

†a) Three a priori hypotheses involving orthogonal contrasts are each tested at $\alpha_{PC} = .01$.

†b) Four a priori hypotheses involving nonorthogonal contrasts are each tested at $\alpha_{PC} = .05$.

c) Four a priori hypotheses involving orthogonal contrasts are each tested at $\alpha_{PC} = .05$.

d) Five a priori hypotheses involving nonorthogonal contrasts are each tested at $\alpha_{PC} = .01$.

16. [3.3] For each of the following, indicate the recommended conceptual unit for error rate.

a) A priori orthogonal contrasts   b) A priori nonorthogonal contrasts

c) A posteriori nonorthogonal contrasts

†**17.** [3.4] The effects of four dosages of ethylene glycol on the reaction time of 20 chimpanzees were investigated. The animals were randomly assigned to one of four groups with five in each group. Those assigned to group a_1, the control group, received a placebo; those assigned to group a_2 received 2 cc of the drug; those assigned to group a_3 received 4 cc; and those assigned to group a_4 received 6 cc. The sample means were $\overline{Y}_1 = 0.28$ sec., $\overline{Y}_2 = 0.29$ sec., $\overline{Y}_3 = 0.31$ sec., and $\overline{Y}_4 = 0.39$ sec.; $MSWG = 0.002$ and $\nu_2 = 4(5-1) = 16$. The experimenter advanced a priori hypotheses concerning all pairwise comparisons among means.

a) Use Dunn's procedure to test these hypotheses by comparing $\hat{\psi}_i$ with $\hat{\psi}(D)$. Construct a table like Table 3.4-1. Let $\alpha_{EW} = .05$.

b) Construct $1 - (1 - .05)\%$ confidence intervals for these a priori contrasts.

c) Compute the correlations among the contrasts; assume that $c_{1j} = 1\ -1\ 0\ 0$, $c_{2j} = 1\ 0\ -1\ 0$, et cetera.

d) Compare the relative efficiency of Dunn's procedure with that of the Dunn-Šidák procedure.

e) Suppose that the experimenter is only interested in the $p - 1 = 3$ contrasts involving the control group. For this case, compare the relative efficiency of the Dunn and Dunn-Šidák procedures with Dunnett's procedure.

18. [3.4] The effects of information regarding a rape victim's past sexual behavior on perceived culpability were investigated. One hundred twenty-four college students were randomly assigned to one of four groups with 31 in each group. The students read specially written newspaper stories describing testimony at a trial. One newspaper account of the trial, condition a_1, indicated that the victim had an inactive sexual history. According to other accounts the victim refused to discuss her past sexual experience, a_2; the judge prohibited testimony regarding past sexual history, a_3; and no mention of past sexual experience came up, a_4. The students rated the culpability of the victim on a ten-point scale; the higher the rating the more culpable the victim. The sample means were $\overline{Y}_1 = 4.2$, $\overline{Y}_2 = 7.1$, $\overline{Y}_3 = 3.3$, and $\overline{Y}_4 = 4.5$; $MSWG = 14.08$ and $\nu_2 = 4(31 - 1) = 120$. The experimenter advanced a priori hypotheses concerning all pairwise comparisons among means.

a) Use Dunn's procedure to test these hypotheses by comparing $\hat{\psi}_i$ with $\hat{\psi}(D)$. Construct a table like Table 3.4-1. Let $\alpha_{EW} = .05$.

b) Construct $100 (1 - .05)\%$ confidence intervals for these a priori contrasts.

c) Compute the correlations among the contrasts; assume that $c_{1j} = 1\ -1\ 0\ 0$, $c_{2j} = 1\ 0\ -1\ 0$, et cetera.

d) Compare the relative efficiency of Dunn's procedure with that of the Dunn-Šidák procedure.

e) Suppose that the experimenter is only interested in the $p - 1 = 3$ contrasts involving treatment level a_4. For this case, compare the relative efficiency of the Dunn and Dunn-Šidák procedures with Dunnett's procedure.

†**19.** [3.5] Exercise 17 described an experiment to evaluate the effects of four dosages of ethylene glycol on the reaction time of chimpanzees.

a) Use Tukey's procedure to test the overall null hypothesis $\mu_1 = \mu_2 = \mu_3 = \mu_4$. If this hypothesis is rejected, proceed to test all pairwise comparisons. Construct a table like Table 3.5-1. Let $\alpha_{EW} = .05$.

b) Construct $1 - (1 - .05)\%$ confidence intervals for all pairwise comparisons.

c) If you did Exercise 17(d), compare the relative efficiency of the Tukey and Dunn-Šidák procedures.

d) Use the Newman-Keuls test to evaluate all pairwise comparisons.

e) Use Duncan's new multiple range test to evaluate all pairwise comparisons.

20. [3.5] Exercise 18 described an experiment to evaluate the effects of information regarding a rape victim's past sexual behavior on perceived culpability.

a) Use Tukey's procedure to test the overall null hypothesis $\mu_1 = \mu_2 = \mu_3 = \mu_4$. If this hypothesis is rejected, proceed to test all pairwise comparisons. Construct a table like Table 3.5-1. Let $\alpha_{EW} = .05$.

b) Construct $1 - (1 - .05)\%$ confidence intervals for all pairwise comparisons.

c) If you did Exercise 18(d), compare the relative efficiency of the Tukey and Dunn-Šidák procedures.

d) Use the Newman-Keuls test to evaluate all pairwise comparisons.

e) Use Duncan's new multiple range test to evaluate all pairwise comparisons.

†**21.** [3.5] Exercise 10 described an experiment to evaluate the effectiveness of three approaches to drug education in junior high school. Assume that the overall null hypothesis was rejected at the .05 level of significance.

a) Use Scheffé's procedure to test the following null hypotheses:

$$\mu_1 - \mu_3 = 0, \qquad \mu_2 - \mu_3 = 0, \qquad \text{and} \qquad (\mu_1 + \mu_2)/2 - \mu_3 = 0.$$

Let $\alpha_{EW} = .05$.

b) Suppose that the sample variances for this problem are $\hat{\sigma}_1^2 = 4.1$, $\hat{\sigma}_2^2 = 13.3$, and $\hat{\sigma}_3^2 = 31.8$. Use the Brown-Forsythe procedure to test the given null hypotheses.

22. [3.5] The effects of simulator training involving synergistic 6-degree-of-freedom platform motion on the acquisition of basic approach and landing skills of 63 undergraduate pilot trainees were investigated. The trainees were randomly divided into three groups. Those in group a_1 received ten sorties with platform motion in the Advanced Simulator for Pilot Training; those in group a_2 also received ten sorties but without motion. Trainees in group a_3, the control group, received the standard syllabus of preflight and flightline instructions. The dependent variable was instructor-pilot ratings of trainee performance in a T-37 aircraft. The sample means were

$\overline{Y}_{.1} = 16.2$, $\overline{Y}_{.2} = 15.1$, and $\overline{Y}_{.3} = 11.4$; $MSWG = 39.94$; and $\nu_2 = 3(21 - 1) = 60$. Assume that the overall null hypothesis was rejected at the .05 level.

a) Use Scheffé's procedure to test the following null hypotheses:

$$\mu_1 - \mu_2 = 0 \qquad \text{and} \qquad (\mu_1 + \mu_2)/2 - \mu_3 = 0.$$

Let $\alpha_{EW} = .05$.

b) Suppose that the sample variances for this problem are $\hat{\sigma}_1^2 = 28.12$, $\hat{\sigma}_2^2 = 31.63$, and $\hat{\sigma}_3^2 = 60.07$. Use the Brown-Forsythe procedure to test the given null hypotheses.

4 COMPLETELY RANDOMIZED DESIGN

4.1 DESCRIPTION OF DESIGN

Chapters 1, 2, and 3 introduced some of the basic concepts and statistical tools used in experimental design. In the remaining eleven chapters those designs that appear to have the greatest potential usefulness to researchers in the behavioral sciences and education will be examined in detail.

One of the simplest experimental designs from the standpoint of data analysis and assignment of subjects to treatment levels is the completely randomized design. For convenience, this design is designated as a *type CR-p design,* where p stands for the number of treatment levels. The type CR-p design is appropriate for experiments that meet, in addition to the general assumptions of analysis of variance discussed in Chapter 2, the following two conditions.

1.   One treatment with $p \geq 2$ treatment levels. The levels of the treatment can differ either quantitatively or qualitatively. When the experiment contains only two treatment levels, the F test is analogous to a t test for independent samples.

2.   Random assignment of treatment levels to experimental units, with each

experimental unit designated to receive only one level. If the treatment levels are of equal interest, it may be advantageous to assign each level to the same number of experimental units, although this is not necessary. Actually flexibility in this respect is one of the advantages of the design.

It is apparent from the foregoing that the completely randomized design is applicable to a broad range of experimental situations. The design is one of three basic designs that may be used by itself or in combination to form more complex experimental designs. The other two designs that may also be used by themselves or as *building blocks* for complex designs are the randomized block design and the Latin square design. The use of these designs as building blocks in constructing other designs is described in Chapters 8 through 14.

EXPERIMENTAL DESIGN MODEL FOR A TYPE CR-*p* DESIGN

The experimental design model for a completely randomized design was discussed in detail in Sections 1.3, 2.2, and 2.3. Here we will summarize the key features of the fixed-effects model. We assume that a score Y_{ij} for subject i in treatment level j is a composite that reflects the sum of the following terms:

(4.1-1) $$Y_{ij} = \mu + \alpha_j + \epsilon_{i(j)} \qquad (i = 1, \ldots, n; j = 1, \ldots, p)$$

where μ = the overall population mean (μ is a constant for all scores and reflects their general elevation)

α_j = the effect of treatment level j and is subject to the restriction $\Sigma_{j=1}^{p} \alpha_j = 0$

$\epsilon_{i(j)}$ = the experimental error that is $NID(0, \sigma_\epsilon^2)$.

The assumptions associated with this model are discussed in Sections 2.3 and 2.5. This fixed-effects model, which is also called model I, is the most commonly used model for a type CR-*p* design. For this model, it is assumed that the experiment includes all treatment levels of interest. The case in which the treatment levels have been randomly sampled from a population of levels, the random-effects model, is dicussed in Section 4.7.

4.2   EXPLORATORY DATA ANALYSIS

The emphasis in this book is on *confirmatory data analysis*—using samples to tell us something about the populations from which they came and assessing the precision of our inferences concerning the populations. But every confirmatory data analysis should be preceded by an *exploratory data analysis*—looking at data to see what they seem to say. Eyeballing data in this computer age may seem a bit old-fashioned to some, but it is, nevertheless, an important first step in any confirmatory data analysis. Such an exploration may uncover, for example, suspected data recording errors, assumptions that appear untenable, and unsuspected promising lines of investigation. Several exploratory techniques will be presented in the following discussion. For an in-depth coverage, the reader should refer to Tukey (1977).

Assume that we are interested in the effects of sleep deprivation on hand-steadiness. We have conducted an experiment in which 32 subjects were randomly divided into four groups of eight subjects each. Four levels of sleep deprivation were then randomly assigned to the four groups. The four levels of sleep deprivation, which will be referred to as *treatment levels,* are 12, 24, 36, and 48 hours. The letters and subscripts a_1, a_2, a_3, and a_4 will be used to designate the four treatment levels, respectively. The original research hypothesis that led to this experiment was based on the idea that differences in hand-steadiness would exist among subjects assigned to the four treatment levels. A hypothetical set of data for this experiment are shown in Table 4.2-1(i). The frequency distribution in part (ii) and the descriptive statistics in part (iii) support the original research hypothesis—at least there are large differences among the sample means. Furthermore, there is no reason to doubt the tenability of the normality and homogeneity of variance assumptions since the sample distributions appear to be reasonably symmetrical and the dispersions fairly homogeneous. Procedures for testing the hypothesis of homogeneity of variance are described in Section 2.5.

It is often more helpful to examine error effects or residuals, $\hat{\epsilon}_{i(j)}$,

$$\hat{\epsilon}_{i(j)} = Y_{ij} - \overline{Y}_{\cdot j}$$

rather than observations, Y_{ij}, in determining the aptness of a particular experimental design model. In Figure 4.2-1(a) the information in Table 4.2-1(ii) has been displayed

---

TABLE 4.2-1   Summary of Sleep-Deprivation Data

(i) *Data:*

	Treatment Levels		
a_1	a_2	a_3	a_4
3	4	7	7
6	5	8	8
3	4	7	9
3	3	6	8
1	2	5	10
2	3	6	10
2	4	5	9
2	3	6	11

TABLE 4.2-1 (continued)

(ii) *Frequency distribution*:

Y_{ij}	a_1	a_2	a_3	a_4
	f	f	f	f
11				1
10				11
9				11
8			1	11
7			11	1
6	1		111	
5		1	11	
4		111		
3	111	111		
2	111	1		
1	1			

(iii) *Descriptive statistics*:

	a_1	a_2	a_3	a_4
$\overline{Y}_{\cdot j}$	2.75	3.50	6.25	9.00
$\hat{\sigma}_j$	1.49	0.93	1.04	1.31

$$\overline{Y}_{\cdot j} = \sum_{i=1}^{n} Y_{ij}/n \qquad \hat{\sigma}_j = \sqrt{\left[\sum_{i=1}^{n} Y_{ij}^2 - \frac{\left(\sum_{i=1}^{n} Y_{ij}\right)^2}{n}\right]/(n-1)}$$

in the form of a frequency distribution of residuals. The residual plots for each treatment level show about the same extent of scatter. Again, there is no reason to doubt the tenability of the normality and homogeneity of variance assumptions. Figure 4.2-1(b) displays a different kind of information. Here, residuals have been plotted against the order in which the steadiness measurements were collected. An interesting phenomenon appears in treatment level a_4. The size of the residuals increases as a function of the order in which the measurements were collected. One would certainly want to review the data collection procedures for this treatment level.

FIGURE 4.2-1   (a) Frequency distributions of residuals, $\hat{\epsilon}_{i(j)} = Y_{ij} - \overline{Y}_{.j}$. (b) Plot of residuals versus order in which measurements within each treatment level were obtained.

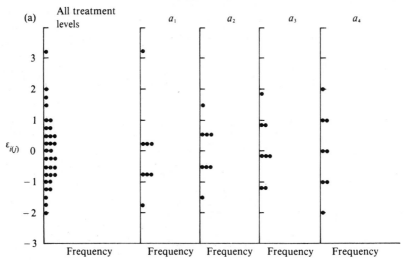

Frequency distribution of residuals

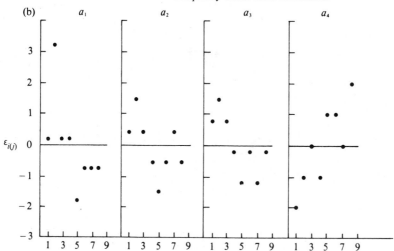

Time order of measurements within each treatment level

OUTLIERS

Occasionally one encounters data having one or more observations that appear to deviate markedly from other observations in the sample. Such observations are called *outliers*. In a residual plot, they are points that lie perhaps four or more standard deviations from zero. To detect such points, it is helpful to plot $\hat{\epsilon}_{i(j)}/\sqrt{MS_{\text{error}}}$ in place of $\hat{\epsilon}_{i(j)}$. Then the ordinate of the graph represents standard deviation units around a mean of zero. Tests for outliers have been described by Grubbs (1969) and Barnett and Lewis (1978).

Outliers call for detective work on the part of an experimenter since they may be merely extreme manifestation of the random variability inherent in data or the result of deviations from prescribed experimental procedure, recording errors, equipment malfunctions, and so on. In the former instance, they should be retained and processed in the same manner as the other observations. If some physical explanation for the outlier can be found, an experimenter may (1) replace the observation with new data, (2) correct the observation if records permit, or (3) reject the observation and *winsorize*. In the simplest case of winsorization—one high or one low observation—the high and low observations are replaced, respectively, by the next highest and next lowest observations. With, say, two high observations, the two highest and two lowest are replaced, respectively, by the third highest and third lowest observations, and so on. The resulting data are then processed as if they were the original data but with modified degrees of freedom. This approach is described by Dixon and Tukey (1968). The most deviant residual in Figure 4.2-1(a) is 3.25 in treatment level a_1. It lies

$$\hat{\varepsilon}_{i(j)}/\sqrt{MS_{\text{error}}}{}^* = 3.25/\sqrt{1.464} = 2.7$$

standard deviations above the mean of zero and is well within the expected range of variability. We will assume that a careful review of the experimental procedures finds no reason for rejecting any of the data. Having performed an exploratory data analysis, we are now ready for the next step which is a confirmatory data analysis.

4.3 COMPUTATIONAL EXAMPLE FOR A TYPE CR-*p* DESIGN

The null hypothesis we want to test for the hand-steadiness data in Table 4.2-1, assuming a fixed-effects model (model I), is

$$H_0: \quad \mu_1 = \mu_2 = \mu_3 = \mu_4 \qquad\qquad H_0: \quad \alpha_j = 0 \text{ for all } j$$

versus   or

$$H_1: \quad \mu_j \neq \mu_{j'} \text{ for some } j \text{ and } j' \qquad H_1: \quad \alpha_j \neq 0 \text{ for some } j.$$

The level of significance adopted is $\alpha = .05$. Procedures for computing the sums of squares used in testing the null hypothesis are illustrated in Table 4.3-1. The *AS* Summary Table is so designated because variation among the 32 scores reflects the effects of the treatment (*A*) and subjects (*S*). The computational scheme in parts (ii) and (iii) of the table uses the abbreviated symbols, [*AS*], [*A*], and [*Y*], which were introduced in Section 2.2. This abbreviated notation is particularly useful in presenting

---

*   The computation of MS_{error} is illustrated in Table 4.3-2.

TABLE 4.3-1   Computational Procedures for a Type CR-4 Design

(i) Data and notation [Y_{ij} denotes a score for experimental unit i in treatment level j; $i = 1, \ldots, n$ experimental units (s_i); $j = 1, \ldots, p$ levels of treatment $A(a_j)$]:

AS Summary Table
Entry is Y_{ij}

	a_1	a_2	a_3	a_4
	3	4	7	7
	6	5	8	8
	3	4	7	9
	3	3	6	8
	1	2	5	10
	2	3	6	10
	2	4	5	9
	2	3	6	11
$\sum\limits_{i=1}^{n} Y_{ij} =$	22	28	50	72

(ii) Computational symbols:

$$\sum_{i=1}^{n} \sum_{j=1}^{p} Y_{ij} = 3 + 6 + 3 + \cdots + 11 = 172.0$$

$$\frac{\left(\sum\limits_{i=1}^{n} \sum\limits_{j=1}^{p} Y_{ij} \right)^2}{np} = [Y] = \frac{(172.0)^2}{(8)(4)} = 924.5$$

$$\sum_{i=1}^{n} \sum_{j=1}^{p} Y_{ij}^2 = [AS] = (3)^2 + (6)^2 + (3)^2 + \cdots + (11)^2 = 1160.0$$

$$\sum_{j=1}^{p} \frac{\left(\sum\limits_{i=1}^{n} Y_{ij} \right)^2}{n} = [A] = \frac{(22)^2}{8} + \frac{(28)^2}{8} + \cdots + \frac{(72)^2}{8} = 1119.0$$

(iii) Computational formulas:

$SSTO = [AS] - [Y] = 1160.0 - 924.5 = 235.5$

$SSBG = [A] - [Y] = 1119.0 - 924.5 = 194.5$

$SSWG = [AS] - [A] = 1160.0 - 1119.0 = 41.0$

the complex designs described in the latter part of the book. An ANOVA table summarizing the results of the analysis is shown in Table 4.3-2. The mean square (*MS*) in each row is obtained by dividing the sum of squares (*SS*) by the degrees of freedom (*df*) in its row. Recall from Section 2.3 that a *MS* is an estimator of a population variance and is given by

$$MS = \frac{SS}{df} = \hat{\sigma}^2.$$

The *F* statistic is obtained by dividing the mean square in row 1 by the mean square in row 2. This is indicated symbolically by [½]. According to Appendix Table E.5, the value of *F* that cuts off the upper .05 region of the sampling distribution for 3 and 28 degrees of freedom is 2.95. Since the obtained *F* exceeds the table value, $F > F_{.05;3,28}$, the null hypothesis is rejected. It is customary to include in an ANOVA table the approximate probability value, *p value*, associated with the *F* statistic. A decision to reject or not reject the null hypothesis should be based on the experimenter's preselected level of significance, .05 in our example. The inclusion of the *p* value permits readers to, in effect, set their own level of significance.

In reporting the results of an experiment, a descriptive summary of the data—means, standard deviations, and perhaps a graph—should always precede the reporting of significance tests. For the hand-steadiness experiment, the descriptive statistics in Table 4.2-1(iii) provide an adequate summary of the data. The results of the significance test can be presented either by means of a table (see Table 4.3-2) or in the text. Using the latter form, one might say "We can infer from the analysis of variance that the hand-steadiness population means differ, $F(3,28) = 44.28$, $p < .01$." If the experimental design is complex and requires reporting numerous *F* statistics, the *Publication Manual of the American Psychological Association* (1974) recommends a tabular presentation.

TABLE 4.3-2   ANOVA Table for Type CR-4 Design

Source	SS	df	MS	F	E(MS) Model I
1. Between groups (Sleep-deprivation levels)	194.500	$p - 1 = 3$	64.833	[½] 44.28*	$\sigma_\epsilon^2 + n\sum_{j=1}^{p} \alpha_j^2/(p-1)$
2. Within groups	41.000	$p(n-1) = 28$	1.464		σ_ϵ^2
3. Total	235.500	$np - 1 = 31$			

*$p < .01$

Having rejected the overall null hypothesis, the next step in the analysis is to decide which population means differ. Multiple comparison procedures for this purpose are described in Section 4.4. Before turning to this topic, we will describe procedures for computing the power of the F test and selecting a sample size.

POWER OF THE ANALYSIS OF VARIANCE F TEST

The power, denoted by $1 - \beta$, of an F test for the fixed-effects model is the probability of rejecting the null hypothesis given that it is false; that is

$$1 - \beta = \text{Prob}\{\text{Reject } H_0 \mid H_0 \text{ is false}\}$$
$$= \text{Prob}\{F > F_{\alpha; \nu_1, \nu_2} \mid H_0 \text{ is false}\}.$$

A knowledge of power is useful for assessing the discriminating ability of a test and for determining the sample size to employ.*

If the null hypothesis is true, $F = MSBG/MSWG$ is distributed as a *central* F random variable (Appendix Table E.5). If the null hypothesis is false, $F = MSBG/MSWG$ is distributed as a *noncentral* F random variable. The latter distribution depends on three parameters: ν_1, ν_2, and a noncentrality parameter δ, where

$$\delta = \sqrt{\sum_{j=1}^{p} n\alpha_j^2 / \sigma_\epsilon^2}.$$

The value of δ is determined by the sizes of the squared-treatment effects relative to σ_ϵ^2. To select an appropriate member of the family of noncentral F distributions required in calculating power, it is necessary to know or specify $\sum_{j=1}^{p} n\alpha_j^2 / \sigma_\epsilon^2$. Special charts based on a procedure by Tang (1938) that simplify the calculation of power have been prepared. These charts are reproduced in Appendix Table E.14. The parameter ϕ is entered in the charts and is given by

$$\phi = \frac{\sqrt{\sum_{j=1}^{p} \alpha_j^2 / p}}{\sigma_\epsilon / \sqrt{n}}.$$

To use the charts, the following are specified:

1. α, probability of rejecting the null hypothesis when it is true.
2. $\sum_{j=1}^{p} \alpha_j^2$, sum of squared treatment effects.
3. n, size of the jth sample.
4. σ_ϵ^2, error variance.
5. ν_1 and ν_2, degrees of freedom for treatment and error effects, respectively.

---

*   The reader should find the exchanges among Brewer (1972, 1974); Cohen (1973); Dayton, Schafer, and Rogers (1973); and Meyer (1974a, 1974b) helpful in clarifying practical issues surrounding the calculation and interpretation of power.

In practice, the sizes of population treatment effects, α_j^2's, are unknown. However, one can (1) estimate the treatment effects from sample data or a pilot study or (2) specify the minimum effects that one would be interested in finding if in fact the null hypothesis is false.

The calculation of power will be illustrated for the data summarized in Table 4.3-2. Sample estimates of the parameters required in computing ϕ are obtained from Table 4.3-1. An unbiased estimator of the sum of squared treatment effects is given by

$$\sum_{j=1}^{p} \hat{\alpha}_j^2 = \frac{(p-1)}{n} (MSBG - MSWG) = \frac{4-1}{8} (64.833 - 1.464) = 23.76.$$

The rationale behind the formula for estimating $\sum_{j=1}^{p} \alpha_j^2$ is probably not apparent to the reader. This formula is based on the expected values of the mean squares. In section 2.3 we saw that the expected values of *MSBG* and *MSWG* are

$$E(MSBG) = \sigma_\epsilon^2 + \frac{n \sum_{j=1}^{p} \alpha_j^2}{p-1} \quad \text{and} \quad E(MSWG) = \sigma_\epsilon^2.$$

Hence,

$$\frac{p-1}{n} (MSBG - MSWG) = \frac{(p-1)}{n} \left(\hat{\sigma}_\epsilon^2 + \frac{n \sum_{j=1}^{p} \hat{\alpha}_j^2}{p-1} - \hat{\sigma}_\epsilon^2 \right)$$

$$= \frac{p-1}{n} \left(\frac{n \sum_{j=1}^{p} \hat{\alpha}_j^2}{p-1} \right) = \sum_{j=1}^{p} \hat{\alpha}_j^2.$$

An unbiased estimator of σ_ϵ^2 is given by

$$\hat{\sigma}_\epsilon^2 = MSWG = 1.464.$$

For these data,

$$\hat{\phi} = \frac{\sqrt{23.76/4}}{\sqrt{1.464}/\sqrt{8}} = 5.7.$$

According to Appendix Table E.14, a power of approximately .99 is associated with $\phi = 2.60$, with $\nu_1 = 3$ and $\nu_2 = 28$. Unfortunately for purposes of this illustration, the data in Table 4.3-1 yield a value of $\hat{\phi}$ outside the range covered by Appendix Table E.14. An experimenter can conclude, however, that the probability of rejecting a false null hypothesis is greater than .99.

If the necessary information is available, power calculations should be made before rather than after an experiment is performed. It may be found, for example, that the contemplated sample size is so small that it gives less than a 20% chance of detecting treatment effects considered of practical interest. On the other hand, an experimenter may find that the contemplated sample size is wastefully large.

DETERMINATION OF SAMPLE SIZE

Calculation of the power of an F test in ANOVA provides a basis for determining the number of subjects that should be included in an experiment.* In the example just described we saw that the power exceeded .99. It is evident that for this experiment we could have used fewer subjects and still achieved an acceptable power. Many experimenters prefer a power between .80 and $1 - \alpha$.

Suppose we had adopted a power of .95 for the hand-steadiness experiment. The number of subjects needed to achieve this power can be estimated using the formula

$$\hat{\phi} = \frac{\sqrt{\sum_{j=1}^{p} \hat{\alpha}_j^2 / p}}{\hat{\sigma}_\epsilon / \sqrt{n}}$$

and trial and error. The number is $np \cong (3)(4) = 12$. This is a rough approximation since the use of three subjects in each treatment level gives a power in excess of .99 while the use of two subjects gives a power below .95. The computation of $\hat{\phi}$ for the two cases is as follows.

$$\hat{\phi} = \frac{\sqrt{23.76/4}}{\sqrt{1.464}/\sqrt{3}} = 3.5$$

$$\nu_1 = p - 1 = 3 \quad \text{and} \quad \nu_2 = p(n - 1) = 4(3 - 1) = 8$$

$$\hat{\phi} = \frac{\sqrt{23.76/4}}{\sqrt{1.464}/\sqrt{2}} = 2.8$$

$$\nu_1 = p - 1 = 3 \quad \text{and} \quad \nu_2 = p(n - 1) = 4(2 - 1) = 4$$

If an accurate estimate of σ_ϵ is not available from previous research or from a pilot study, the procedure just described for calculating n is not satisfactory. An alternative procedure that does not require a knowledge of σ_ϵ can be used. According to this procedure, differences among treatment effects, and hence means, of interest to an experimenter are specified in terms of units of the unknown σ_ϵ. For example, an experimenter can determine the size of n necessary to achieve a given power if the largest difference among means is equal to $C\sigma_\epsilon$, where C is any number greater than zero. Many possible choices of α_j's yield a maximum difference of $C\sigma_\epsilon$. It can be shown that $\sum_{j=1}^{p} \alpha_j^2$ is minimal when two of the treatment effects, say α_j and $\alpha_{j'}$, are equal to $-C\sigma_\epsilon/2$ and $C\sigma_\epsilon/2$ and all other $p - 2$ effects are equal to zero. For this

---

*   For an in-depth discussion of procedures for estimating sample size, see Cohen (1969).

case,

$$\sum_{j=1}^{p} \alpha_j^2 = \left[-\frac{C\sigma_\epsilon}{2}\right]^2 + \left[\frac{C\sigma_\epsilon}{2}\right]^2 + 0 + \cdots + 0 = \frac{2C^2\sigma_\epsilon^2}{4} = \frac{C^2\sigma_\epsilon^2}{2}.$$

Because power increases with an increase in $\sum_{j=1}^{p} \alpha_j^2$, it follows that a choice of values for α_j other than these will always lead to greater power. Hence, if the sample size necessary to achieve a given power is computed for the preceding treatment effects, one can be certain that any other combination of effects for which the maximum difference is $C\sigma_\epsilon$ will yield a power greater than that specified. The formula for computing n when the size of treatment effects is expressed in units of σ_ϵ is as follows:

$$\phi = \frac{\sqrt{\sum_{j=1}^{p} \alpha_j^2/p}}{\sigma_\epsilon/\sqrt{n}} = \frac{\sqrt{[C^2\sigma_\epsilon^2/2]/p}}{\sigma_\epsilon/\sqrt{n}} = \sqrt{n}\sqrt{C^2/2p}.$$

Let us assume that an experiment contains four treatment levels and that the experimenter is interested in detecting differences among means such that the largest difference is equal to $1\sigma_\epsilon$. If C, α, $1 - \beta$, and ν_1 are specified as 1, .05, .80, and 3, respectively, the value of n can be estimated by trial and error from the formula

$$\phi = \sqrt{n}\sqrt{C^2/2p} = \sqrt{n}\sqrt{(1)^2/2(4)} = \sqrt{n}\,(.3536)$$

and Appendix Table E.14. For $n = 20$, Table E.14 can be entered to determine if a power of .80 is associated with $\phi = \sqrt{20}(.3536) = 1.58$, $\nu_1 = 3$, $\nu_2 = p(n - 1) = 4(20 - 1) = 76$, and $\alpha = .05$. According to Table E.14, the power associated with $n = 20$ falls just short of $1 - \beta = .80$. If $n = 25$, the values of ϕ and ν_2 are $\phi = \sqrt{25}(.3536) = 1.77$ and $\nu_2 = 96$. According to Table E.14, the probability of rejecting a false null hypothesis for $n = 25$ is just beyond the .80 level.

Tables prepared by Bratcher, Moran, and Zimmer (1970) further simplify the estimation of sample size. Their tables, which are abridged in Appendix Table E.15, are entered with C, α, $1 - \beta$, and p. Earlier we determined by trial and error that more than 20 but fewer than 25 subjects per treatment level would be required in a type CR-4 design to reject a false null hypothesis where $C = 1$, $\alpha = .05$, and $1 - \beta = .80$. The required n can be determined directly from Table E.15 and is equal to 23.

The preceding procedures for estimating n do not require a priori information concerning the size of $\sum_{j=1}^{p} \alpha_j^2$ and σ_ϵ. This is an advantage over the procedure described earlier. Considering the simplicity of the latter approaches, one should always estimate the sample size required to achieve a given power prior to doing an experiment. The only information that must be specified is α, $1 - \beta$, p, and the largest difference between treatment means of interest expressed as a multiple, C, of the unknown σ_ϵ. If preliminary calculations indicate that the required sample size is too large, an experimenter may choose not to conduct the experiment or modify it so as to reduce the required number of subjects. The modification may involve selecting a less stringent level of significance, settling for lower power, redefining the maximum difference between means that is of interest, or redesigning the experiment so as to obtain a more precise estimate of treatment effects and a smaller error term.

4.4   PROCEDURES FOR TESTING DIFFERENCES AMONG MEANS

A PRIORI TESTS

If all orthogonal contrasts among means of interest to the experimenter have been specified prior to collection of the data, multiple t statistics can be used to test the $p - 1$ null hypotheses. As discussed in Section 3.2, it is not necessary to perform an overall test of significance prior to evaluating a priori orthogonal contrasts.

Procedures for evaluating a priori orthogonal contrasts were discussed in Section 3.2. Here we will illustrate the procedures for the hand-steadiness data in Table 4.3-1. Recall that if an experiment contains p equal to four treatment levels, one can test, at most, hypotheses for $p - 1$ orthogonal contrasts. Suppose that an experimenter is interested in testing the following null hypotheses:

$$H_0: \quad \mu_1 - \mu_2 = 0$$

$$H_0: \quad \mu_3 - \mu_4 = 0$$

$$H_0: \quad (\mu_1 + \mu_2)/2 - (\mu_3 + \mu_4)/2 = 0.$$

The sample means are $\overline{Y}_1 = 2.75$, $\overline{Y}_2 = 3.50$, $\overline{Y}_3 = 6.25$, and $\overline{Y}_4 = 9.00$; the error term, $MSWG$, is obtained from Table 4.3-2 and is equal to 1.464. The t statistics are

$$t = \frac{\hat{\psi}_i}{\hat{\sigma}_{\psi_i}} = \frac{c_1 \overline{Y}_1 + c_2 \overline{Y}_2 + \cdots + c_p \overline{Y}_p}{\sqrt{MSWG\left[\dfrac{c_1^2}{n_1} + \dfrac{c_2^2}{n_2} + \cdots + \dfrac{c_p^2}{n_p}\right]}}$$

$$t = \frac{\hat{\psi}_1}{\hat{\sigma}_{\psi_1}} = \frac{(1)2.75 + (-1)3.50}{\sqrt{1.464\left[\dfrac{(1)^2}{8} + \dfrac{(-1)^2}{8}\right]}} = \frac{-0.750}{0.605} = -1.24$$

$$t = \frac{\hat{\psi}_2}{\hat{\sigma}_{\psi_2}} = \frac{(1)6.25 + (-1)9.00}{\sqrt{1.464\left[\dfrac{(1)^2}{8} + \dfrac{(-1)^2}{8}\right]}} = \frac{-2.750}{0.605} = -4.55$$

$$t = \frac{\hat{\psi}_3}{\hat{\sigma}_{\psi_3}} = \frac{(\tfrac{1}{2})2.75 + (\tfrac{1}{2})3.50 + (-\tfrac{1}{2})6.25 + (-\tfrac{1}{2})9.00}{\sqrt{1.464\left[\dfrac{(\tfrac{1}{2})^2}{8} + \dfrac{(\tfrac{1}{2})^2}{8} + \dfrac{(-\tfrac{1}{2})^2}{8} + \dfrac{(-\tfrac{1}{2})^2}{8}\right]}} = \frac{-4.500}{0.428} = -10.51.$$

The critical value of $t_{.05/2,28}$ is 2.048. Since $|t| > t_{.05/2,28}$ for the second and third contrasts, we can conclude that the population contrasts are not equal to zero. The sign of the t statistics has no significance in this example since the null hypotheses are nondirectional (two-tailed). If directional hypotheses had been advanced, the sign of the t's would have determined whether they were in the correct tail of the sampling distribution.

The Dunn-Šidák test, described in Section 3.4, is appropriate for testing a relatively small number of hypotheses involving a priori nonorthogonal contrasts. The experimentwise error rate for this test is equal to or less than α. The error rate for the multiple t test described earlier is equal to α for each contrast. A computational example for the Dunn-Šidák test will not be given since its test statistic is identical to that for the multiple t statistic. The two procedures differ in that the Dunn-Šidák statistic is referred to Appendix Table E.17 whereas the t statistic is referred to Appendix Table E.4.

A POSTERIORI TESTS

If an experimenter is interested in data snooping, that is, in making tests suggested by an inspection of the data, an a posteriori test statistic should be used. A significant overall test for differences among the means is required before proceeding with a posteriori tests.

We will illustrate two procedures for evaluating a posteriori contrasts. The first, developed by Tukey (1953), is useful for evaluating all comparisons among pairs of means. The second procedure, due to Scheffé (1959), can be used to evaluate all possible contrasts among the means, not just those involving two means. The two procedures were introduced in Section 3.5. Tukey's test statistic is given by

$$q = \frac{\hat{\psi}_i}{\hat{\sigma}_{\overline{Y}}} = \frac{c_j \overline{Y}_{.j} + c_{j'} \overline{Y}_{.j'}}{\sqrt{\dfrac{MSWG}{n}}}.$$

To reject the null hypothesis H_0: $\mu_j - \mu_{j'} = 0$ for populations j and j', $|q|$ must exceed $q_{\alpha;p,p(n-1)}$, the critical value obtained from Appendix Table E.7. Alternatively, a null hypothesis involving population means j and j' is rejected if $|\overline{Y}_j - \overline{Y}_{j'}|$ exceeds the critical difference $\hat{\psi}(HSD)$, where

$$\hat{\psi}(HSD) = q_{\alpha;p,\nu} \sqrt{\frac{MSWG}{n}}.$$

The critical difference for the hand-steadiness data is

$$\hat{\psi}(HSD) = q_{.05;4,28} \sqrt{\frac{1.464}{8}} = 3.86(0.428) = 1.65.$$

It can be seen from Table 4.4-1 that five comparisons exceed this critical difference.

Using Tukey's test statistic, all six pairwise comparisons among means can be evaluated. To evaluate nonpairwise contrasts, Scheffé's procedure should be used. Scheffé's test statistic is

$$F = \frac{\hat{\psi}_i^2}{\hat{\sigma}_{\psi_i}^2} = \frac{\left(\displaystyle\sum_{j=1}^{p} c_j \overline{Y}_j \right)^2}{MSWG \displaystyle\sum_{j=1}^{p} \frac{c_j^2}{n_j}}.$$

	$\overline{Y}_{.1}$	$\overline{Y}_{.2}$	$\overline{Y}_{.3}$	$\overline{Y}_{.4}$
TABLE 4.4-1 Absolute Value of Differences Among Means				
$\overline{Y}_{.1} = 2.75$	$-$	0.75	3.50*	6.25*
$\overline{Y}_{.2} = 3.50$		$-$	2.75*	5.50*
$\overline{Y}_{.3} = 6.25$			$-$	2.75*
$\overline{Y}_{.4} = 9.00$				

*$p > .05$

The null hypothesis H_0: $\sum_{j=1}^{p} c_j \mu_j = 0$ is rejected if $F > F'$, where F' is equal to $(p - 1) F_{\alpha;\nu_1,\nu_2}$ and $F_{\alpha;\nu_1,\nu_2}$ is obtained from Appendix Table E.5. To illustrate, suppose that for the hand-steadiness data we want to test several null hypotheses involving pairwise comparisons and the complex contrast $\psi_1 = \mu_1 - (\mu_2 + \mu_3 + \mu_4)/3$. The test statistic for evaluating the complex contrast is

$$F = \frac{\hat{\psi}_1^2}{\hat{\sigma}_{\psi_1}^2} = \frac{[(1)2.75 + (-\frac{1}{3})3.50 + (-\frac{1}{3})6.25 + (-\frac{1}{3})9.00]^2}{1.464\left[\frac{(1)^2}{8} + \frac{(-\frac{1}{3})^2}{8} + \frac{(-\frac{1}{3})^2}{8} + \frac{(-\frac{1}{3})^2}{8}\right]}$$

$$= \frac{(-3.50)^2}{0.244} = 50.20.$$

To reject the null hypothesis, F must exceed $F' = (4 - 1)2.95 = 8.85$. Since $F > F'$, we can conclude that μ_1 is not equal to the mean of μ_2, μ_3, and μ_4. We can also evaluate $\hat{\psi}_1$ by computing the critical difference $\hat{\psi}(S)$ that $\hat{\psi}_1$ must exceed to be declared significant. The critical difference is

$$\hat{\psi}(S) = \sqrt{(p - 1) F_{\alpha;\nu_1,\nu_2}} \sqrt{MSWG \sum_{j=1}^{p} \frac{c_j^2}{n_j}}$$

$$= 8.85(0.244) = 2.16.$$

USE OF A PRIORI AND A POSTERIORI TESTS IN THE SAME EXPERIMENT

In an experiment, $p - 1$ a priori orthogonal contrasts can be evaluated. An experimenter may be interested in only one or two of these contrasts. Once the contrasts of interest have been evaluated, the remainder of the data can be explored using a posteriori procedures. The combined use of a priori and a posteriori tests in the same experiment will be illustrated for the data in Table 4.3-1. Let us assume that one a priori comparison, $\hat{\psi}_1 = \overline{Y}_3 - \overline{Y}_4$, is of particular interest. The t statistic for this comparison is equal to

$$t = \frac{(1)6.25 + (-1)9.00}{\sqrt{2(1.464)/8}} = -4.55.$$

This comparison, which is significant beyond the .01 level, accounts for one of the $p - 1 = 3$ degrees of freedom for $SSBG$. The remaining $p - 1 - 1 = 2$ degrees of freedom can be used in making an overall test of significance. This test is performed with the sum of squares for $\hat{\psi}_1$ removed from $SSBG$ and is given by

$$F = \frac{(SSBG - SS\hat{\psi}_1)/(p - 1 - 1)}{MSWG} = \frac{(194.50 - 30.25)/(4 - 1 - 1)}{1.464}$$

$$= \frac{82.125}{1.464} = 56.10$$

where $SS\hat{\psi}_1 = \dfrac{[c_3 \overline{Y}_3 + c_4 \overline{Y}_4]^2}{\dfrac{c_3^2}{n_3} + \dfrac{c_4^2}{n_4}} = \dfrac{[(1)6.25 + (-1)9.00]^2}{\dfrac{(1)^2}{8} + \dfrac{(-1)^2}{8}}$

$$= 30.25.$$

In this example, $F > F_{.05;2,28}$. Thus, an experimenter can explore all possible remaining contrasts using a posteriori tests. If the overall test is insignificant, no a posteriori tests are made.

Only one planned contrast was involved in this example. If two planned contrasts had been involved, there would have been $p - 1 - 2 = 1$ degree of freedom left for the overall test. The F ratio for this case would have the form

$$F = \frac{(SSBG - SS\hat{\psi}_1 - SS\hat{\psi}_2)/(p - 1 - 2)}{MSWG}$$

with ν_1 equal to 1 and ν_2 equal to 28. This procedure enables an experimenter to use powerful a priori tests for important contrasts and, in addition, explore all possible remaining comparisons if the overall F test is significant.

The multiple-comparison procedures illustrated in this section are appropriate for a fixed-effects model. If a random-effects model is appropriate, an experimenter's principal interest is in testing the hypothesis that the population variance σ_α^2 is equal to zero. Because treatment levels for a random-effects model are randomly sampled from a population of levels, there is little interest in determining which contrasts among the means are not equal to zero.

4.5 TESTS FOR TRENDS IN THE DATA

QUANTITATIVE VERSUS QUALITATIVE TREATMENT LEVELS

Differences among treatment levels may be either quantitative or qualitative. If the levels represent different amounts of a single common variable, the treatment is referred to as a quantitative treatment. The example involving four levels of sleep deprivation described in Section 4.2 is such a variable. The four treatment levels can obviously be ordered; that is, described in terms of more or less of the variable. If the levels cannot be meaningfully ordered along a single continuum, the treatment is designated as qualitative treatment.

PURPOSES OF TREND ANALYSIS

An experimenter's principal interest in designs involving a qualitative treatment is in determining whether there are differences among the p means. If the treatment levels differ quantitatively and if the sizes of the intervals separating the treatment levels can be specified, the experimenter can turn to more penetrating questions. For example, the experimenter may want to know the nature of the relationship between the independent variable (treatment levels) and the dependent variable (observed performance under each level). If the treatment levels represent a quantitative variable, the following types of questions may be posed.

Question 1.   Is there a trend in the population means? This question can be rephrased as: are the population means for the dependent variable influenced by changes in the independent variable?

Question 2.   Is the trend of the dependent variable population means linear or nonlinear?

Question 3.   If the trend of the dependent variable population means is nonlinear, what higher-degree equation is required to provide a satisfactory fit for the means?

Question 4.   Are the patterns of the predicted and observed means the same? That is, does a particular equation based on the data provide a satisfactory *fit* for the means?

Question 5.   Is the trend of the population means for one treatment the same for different levels of a second treatment?

This last question is not applicable to a completely randomized design, for this design has only one treatment variable. We will return to this question in Chapter 8, which describes factorial experiments.

TEST FOR PRESENCE OF TREND

An answer to the first question, "Is there a trend in the population means?" is provided by $F = MSBG/MSWG$. If $F > F_{\alpha; \nu_1, \nu_2}$, it is concluded that the population means for the dependent variable are influenced by changes in the independent variable. If there is no trend, the population means will have the appearance shown by the squares in Figure 4.5-1. A trend is illustrated by the circles. A trend is indicated whenever dependent variable means are not equal at two or more levels of the independent variable.

FIGURE 4.5-1 Squares illustrate lack of trend; circles illustrate presence of trend.

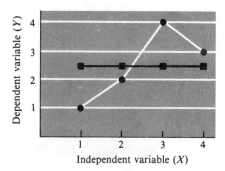

INTRODUCTION TO TEST FOR LINEARITY OF TREND

If the F test indicates a trend in the population, the experimenter may want to determine the nature of this trend. In the broadest terms, a trend is either linear or nonlinear. Procedures for determining the nature of the trend are illustrated in the following sections. The question "What is the simplest equation that provides a satisfactory description of the trend?" can be answered most easily through the use of orthogonal polynomials. A given set of data can often be described by many different equations. A polynomial equation containing one less trend component than there are treatment levels, $p - 1$ terms, will always provide a fit for the data. The question "Is a polynomial equation the most appropriate way to fit a set of data?" is outside the scope of this book. An introduction to curve-fitting procedures can be found in Neter and Wasserman (1974). Orthogonal polynomials are introduced here because they provide a convenient way of determining whether a trend is linear or nonlinear.

USE OF ORTHOGONAL POLYNOMIALS IN FITTING A TREND

A polynomial is an algebraic expression containing more than one term. For example, a polynomial of degree $p - 1$ is

$$\beta_0 + \beta_1 X + \beta_2 X^2 + \beta_3 X^3 + \beta_4 X^4 + \cdots + \beta_{p-1} X^{p-1}$$

where
$\beta_0 = \text{constant}$

$\beta_1 X = \text{linear component}$

$\beta_2 X^2 = \text{quadratic component}$

$\beta_3 X^3 = \text{cubic component}$

$\beta_4 X^4 = \text{quartic component}$

$\beta_{p-1} X^{p-1} = p - 1\text{th component } (p - 1 \text{ is a positive integer}).$

An equation of the form $Y = \beta_0 + \beta_1 X$ is a first-degree or linear equation, $Y = \beta_0 + \beta_1 X + \beta_2 X^2$ is a second-degree or quadratic equation, and $Y = \beta_0 + \beta_1 X + \beta_2 X^2 + \beta_3 X^3$ is a third-degree or cubic equation. These equations, when graphed, have the general form shown in Figure 4.5-2. Note that at least two data points are

FIGURE 4.5-2   Illustration of linear (a), quadratic (b), and cubic (c) relations between independent and dependent variables.

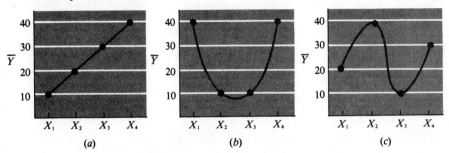

required to define a straight line and, hence, to estimate the Y intercept, β_0, and the slope of the line, β_1. Similarly, at least three data points are required to estimate the parameters, β_0, β_1, and β_2 of a quadratic equation, and so on.

Suppose that we want to fit a polynomial equation to p population means, μ_j. The equation is

(4.5-1)
$$\mu_j = \beta_0 + \beta_1 X_j + \beta_2 X_j^2 + \beta_3 X_j^3 + \cdots + \beta_{p-1} X_j^{p-1}.$$

A disadvantage of this particular equation is that the various terms (trend components) are not associated with an orthogonal decomposition of the trend. As we will see, instead of finding values of $\beta_0, \beta_1, \ldots, \beta_{p-1}$ in (4.5-1) it is more useful to find values of $\beta_0', \beta_1', \cdots, \beta_{p-1}'$ in the equation

(4.5-2)
$$\mu_j = \beta_0' + \beta_0' c_{1j} + \beta_2' c_{2j} + \beta_3' c_{3j} + \cdots + \beta_{p-1}' c_{p-1,j}$$

where
$$c_{1j} = a_1 + X_j$$
$$c_{2j} = a_2 + b_2 X_j + X_j^2$$
$$c_{3j} = a_3 + b_3 X_j + c_3 X_j^2 + X_j^3$$
$$\cdots$$
$$c_{p-1,j} = a_{p-1} + b_{p-1} X_j + c_{p-1} X_j^2 + d_{p-1} X_j^3 + \cdots + X_j^{p-1}.$$

If the c_{ij}'s, called *orthogonal polynomial coefficients*, are properly chosen, each of the $\beta_i' c_{ij}$'s in (4.5-2) represents an orthogonal component—linear, quadratic, and so on—of the trend. The derivation of orthogonal polynomial coefficients is illustrated in Appendix C. When the levels of the independent variable are separated by equal intervals and the sample n's are equal, the orthogonal polynomial coefficients for p equal to 3 through 10 can be obtained from Appendix Table E.12. However, if either of these conditions is not satisfied, the coefficients must be derived following the somewhat tedious procedure in Appendix C.

The orthogonal polynomial coefficients in Table E.12 were derived so that each set represents one and only one trend or form of relationship. It can be shown that $SSBG$ can be partitioned into $p - 1$ sums of squares reflecting orthogonal contrasts among means. A sum of squares for the ith contrast among means, for example, is given by

$$SS\hat{\psi}_i = n\left[\sum_{j=1}^{p}(c_{ij}\overline{Y}_{.j})\right]^2 / \sum_{j=1}^{p}c_{ij}^2 \quad \text{or} \quad \left[\sum_{j=1}^{p}\left(c_{ij}\sum_{i=1}^{n}Y_{ij}\right)\right]^2 / n\sum_{j=1}^{p}c_{ij}^2.$$

And $SS\hat{\psi}_1 + SS\hat{\psi}_2 + \cdots + SS\hat{\psi}_{p-1} = SSBG$, where $\hat{\psi}_1, \hat{\psi}_2, \cdots, \hat{\psi}_{p-1}$ are orthogonal.

Similarly, $SSBG$ can be partitioned into $p - 1$ sums of squares reflecting orthogonal trend contrasts. If p is equal to 4, the best prediction equation is of no higher degree than $p - 1 = 3$. The meaning of these statements can be made clearer by using an example. Consider an experiment with p equal to four treatment levels. The coefficients for the linear, quadratic, and cubic trend contrasts are shown in Table 4.5-1. The $\overline{Y}_{.j}$ values in Figure 4.5-2(a) are 10, 20, 30, and 40. If these values are multiplied by the c_{1j}, c_{2j}, and c_{3j} coefficients in the table, the following trend contrasts are obtained:

$$\hat{\psi}_{\text{lin}} = \sum_{j=1}^{p}(c_{1j}\overline{Y}_{.j}) = -3(10) - 1(20) + 1(30) + 3(40) = 100$$

$$\hat{\psi}_{\text{quad}} = \sum_{j=1}^{p}(c_{2j}\overline{Y}_{.j}) = +1(10) - 1(20) - 1(30) + 1(40) = 0$$

$$\hat{\psi}_{\text{cubic}} = \sum_{j=1}^{p}(c_{3j}\overline{Y}_{.j}) = -1(10) + 3(20) - 3(30) + 1(40) = 0.$$

Since the relationship in Figure 4.5-2(a) is clearly linear, only the linear coefficients are useful in describing the trend. The products $\Sigma_{j=1}^{p}(c_{2j}\overline{Y}_{.j})$ and $\Sigma_{j=1}^{p}(c_{3j}\overline{Y}_{.j})$ sum to zero, which indicates that neither the quadratic nor the cubic coefficients are useful in describing this trend. The $\overline{Y}_{.j}$ values in Figure 4.5-2(b) are 40, 10, 10, and 40. Since this trend is quadratic, it may be anticipated that $\hat{\psi}_{\text{lin}} = \Sigma_{j=1}^{p}(c_{1j}\overline{Y}_{.j})$ and $\hat{\psi}_{\text{cubic}} = \Sigma_{j=1}^{p}(c_{3j}\overline{Y}_{.j})$ will equal zero but $\hat{\psi}_{\text{quad}} = \Sigma_{j=1}^{p}(c_{2j}\overline{Y}_{.j})$ will be greater than zero. This can be demonstrated as follows:

$$\hat{\psi}_{\text{lin}} = \sum_{j=1}^{p}(c_{1j}\overline{Y}_{.j}) = -3(40) - 1(10) + 1(10) + 3(40) = 0$$

$$\hat{\psi}_{\text{quad}} = \sum_{j=1}^{p}(c_{2j}\overline{Y}_{.j}) = +1(40) - 1(10) - 1(10) + 1(40) = 60$$

$$\hat{\psi}_{\text{cubic}} = \sum_{j=1}^{p}(c_{3j}\overline{Y}_{.j}) = -1(40) + 3(10) - 3(10) + 1(40) = 0.$$

Using the same procedure, it can be shown for the data in Figure 4.5-2(c) that $\hat{\psi}_{\text{lin}}$ and $\hat{\psi}_{\text{quad}} = 0$ but $\hat{\psi}_{\text{cubic}} > 0$. It is apparent from these illustrations that each set of polynomial coefficients has been derived so as to describe one and only one trend or form of relationship. It can also be shown that the $p - 1 = 3$ sets of trend coefficients are mutually orthogonal. By definition, trend coefficients are mutually

TABLE 4.5-1 Coefficients for Linear, Quadratic, and Cubic Trend Contrasts	*Polynomial Trend*	*Coefficients of Trend*			
		X_1	X_2	X_3	X_4
	Linear (c_{1j})	-3	-1	$+1$	$+3$
	Quadratic (c_{2j})	$+1$	-1	-1	$+1$
	Cubic (c_{3j})	-1	$+3$	-3	$+1$

orthogonal if the pairwise product of their coefficients sum to zero; that is, if $\sum_{j=1}^{p} c_{1j}c_{2j}$, $\sum_{j=1}^{p} c_{1j}c_{3j}$, and $\sum_{j=1}^{p} c_{2j}c_{3j} = 0$. For example, the product of c_{1j} and $c_{2j} = (-3)(+1) + (-1)(-1) + (+1)(-1) + (+3)(+1) = 0$.

COMPUTATIONAL PROCEDURE FOR TESTING LINEARITY OF TREND

Let us turn to computational procedures for answering the second question listed earlier that can be answered by trend analysis. The basic problem confronting an experimenter is in deciding how many trend components are required to adequately describe the relationship between X and Y. As a first approximation, an experimenter can determine if a linear equation provides a satisfactory description of the trend. If it does not, quadratic, cubic, or higher-degree equations can be employed. As we will see, analysis of variance can be used to determine whether or not any particular trend component accounts for a portion of the variance of Y and thus contributes to the overall goodness of fit. By employing only those trend components that are found to be significant, an experimenter can write a polynomial equation that represents the simplest possible description of the relationship between the dependent variable (Y) and the independent variable (X).

The use of orthogonal polynomial coefficients in determining whether the linear component of a polynomial equation makes a contribution to describing the trend is illustrated in Table 4.5-2. The null hypothesis can be stated as

$$\text{H}_0\text{:}\quad \beta_1' = 0 \quad \text{or} \quad \text{H}_0\text{:}\quad \psi_{\text{lin}} = 0$$

where β_1' is the coefficient of the linear trend and ψ_{lin} is the linear contrast. The linear population contrast for this example is

$$\psi_{\text{lin}} = (-3)\mu_1 + (-1)\mu_2 + (1)\mu_3 + (3)\mu_4.$$

The values of $\sum_{i=1}^{n} Y_{ij} = 22, 28, 50,$ and 72 in part (i) of the table are hand-steadiness sums for the four levels of sleep deprivation. The original data appear in Table 4.3-1. Since $F > F_{.05;1,28}$, it is concluded that the linear component of the trend is significant. The test procedure described here is appropriate for planned orthogonal trend tests. It is not necessary to test the overall null hypothesis prior to testing planned orthogonal contrasts. The conceptual unit for error rate is the individual trend contrast. If an experimenter performs $p - 1 = 3$ orthogonal trend tests each at $\alpha = .05$, the error rate experimentwise is equal to

$$1 - (1 - .05)^3 = .14.$$

TEST FOR DEPARTURE FROM LINEARITY

If an experimenter has not advanced specific hypotheses with respect to, say, the quadratic and cubic trend components, an overall test for these higher-order components can be performed. This test is referred to as a test for departure from linearity. Computational procedures for this test are also illustrated in Table 4.5-2. Because

---

TABLE 4.5-2 Computational Procedures for Testing Significance of Linear Contrast and Departure from Linearity

---

(i) Treatment sums and coefficients for linear contrast [$\sum_{j=1}^{n} Y_{ij}$ obtained from Table 4.3-1; linear orthogonal polynomial coefficients, c_{1j}, obtained from Appendix Table E.12]:

	a_1	a_2	a_3	a_4
$\sum_{i=1}^{n} Y_{ij} =$	22	28	50	72
$c_{1j} =$	-3	-1	1	3

$$\hat{\psi}_{\text{lin}} = \sum_{j=1}^{p} \left(c_{1j} \sum_{i=1}^{n} Y_{ij} \right) = (-3)22 + (-1)28 + (1)50 + (3)72 = 172$$

(ii) Computation of sum of squares ($SS\hat{\psi}_{\text{dep from lin}}$ is discussed in the following section of the text):

$$SS\hat{\psi}_{\text{lin}} = \frac{\hat{\psi}_{\text{lin}}^2}{n\sum_{j=1}^{p} c_{1j}^2} = \frac{(172)^2}{8[(-3)^2 + (-1)^2 + (1)^2 + (3)^2]} = 184.9$$

$$df = 1$$

$$SS\hat{\psi}_{\text{dep from lin}} = SSBG - SS\hat{\psi}_{\text{lin}} = 194.5 - 184.9 = 9.6$$

$$df = p - 2 = 2$$

(iii) Computation of F ratio:

$$F = \frac{MS\hat{\psi}_{\text{lin}}}{MSWG} = \frac{SS\hat{\psi}_{\text{lin}}/1}{SSWG/p(n-1)} = \frac{184.9/1}{41.0/28} = 126.30$$

$$df = 1, p(n-1)$$

$$F_{.05;1,28} = 4.20$$

$$F = \frac{MS\hat{\psi}_{\text{dep from lin}}}{MSWG} = \frac{SS\hat{\psi}_{\text{dep from lin}}/2}{SSWG/p(n-1)} = \frac{9.6/2}{41.0/28} = 3.28$$

$$df = p - 2; p(n-1)$$

$$F_{.05;2,28} = 3.34$$

$F < F_{.05;2,28}$, it is concluded that the trend does not depart from linearity. Variability of the means around the best-fitting straight line is assumed to represent nothing more than error variability. If a test for departure from linearity is significant, Scheffé's method or the Dunn-Šidák procedure described in Chapter 3 can be used to determine which higher-order trend components are significant. This procedure enables an experimenter to assign the same error rate experimentwise to the collection of tests as that allotted to the F test for the overall null hypothesis.

An experimenter may have reason to believe that the linear, quadratic, cubic, and so on trend components all contribute to the overall trend. In this case each trend contrast for which a priori hypotheses have been advanced can be tested without first performing an overall test. Computational procedures for testing the quadratic and cubic contrasts are described in the following section. It should be noted that one or more tests of higher-order trend contrasts may be significant, although the test for departure from linearity is insignificant. This apparent inconsistency can occur if, for example, the individual trend hypothesis is employed as the conceptual unit for error rate.

COMPUTATIONAL PROCEDURE FOR TESTING QUADRATIC AND CUBIC TREND CONTRASTS

A third question that can be answered by trend analysis is, "If the trend of dependent variable population means is nonlinear, what higher-degree equation is required to provide a satisfactory fit for the means?" Let us assume that a priori hypotheses with respect to the presence of linear, quadratic, and cubic trend components have been advanced. The quadratic and cubic null hypotheses can be stated as

$$H_0: \quad \beta_2' = 0 \qquad \text{or} \qquad H_0: \quad \psi_{\text{quad}} = 0$$

and

$$H_0: \quad \beta_3' = 0 \qquad \text{or} \qquad H_0: \quad \psi_{\text{cubic}} = 0$$

where β_2' and β_3' are the coefficients of the quadratic and cubic trends, respectively, and ψ_{quad} and ψ_{cubic} are the quadratic and cubic contrasts. The *SSBG* in the sleep-deprivation example has $p - 1 = 3$ degrees of freedom. It will be shown that *SSBG* can be partitioned into linear, quadratic, and cubic trend components with one degree of freedom per component.

A general principle is that *SSBG* can be partitioned into as many trend components as there are degrees of freedom for treatment levels. Procedures for testing the significance of the quadratic and cubic contrasts are illustrated in Table 4.5-3. The quadratic and cubic orthogonal coefficients are obtained from Appendix Table E.12. It is apparent from Table 4.5-3 that $F > F_{.05;1,28}$ for the quadratic contrast but not for the cubic contrast. Thus the trend of the data involves some curvature. An inspection of Figure 4.5-3 lends support to this conclusion.

The sum of the linear, quadratic, and cubic sum of squares is equal to

TABLE 4.5-3 Computational Procedures for Testing Significance of Quadratic and Cubic Contrasts

(i) Treatment sums and coefficients for quadratic and cubic contrasts [$\sum_{i=1}^{n} Y_{ij}$ obtained from Table 4.3-1; quadratic and cubic orthogonal polynomial coefficients, c_{2j} and c_{3j}, obtained from Appendix Table E.12]:

	a_1	a_2	a_3	a_4	
$\sum_{i=1}^{n} Y_{ij} =$	22	28	50	72	
$c_{2j} =$	1	-1	-1	1	
$c_{3j} =$	-1	3	-3	1	
$\hat{\psi}_{quad} = \sum_{j=1}^{p} (c_{2j} Y_{ij}) =$	22	-28	-50	$72 =$	16
$\hat{\psi}_{cubic} = \sum_{j=1}^{p} (c_{3j} Y_{ij}) =$	-22	84	-150	$72 =$	-16

(ii) Computation of sum of squares:

$$SS\hat{\psi}_{quad} = \frac{\hat{\psi}_{quad}^2}{n\sum_{j=1}^{p} c_{2j}^2} = \frac{(16)^2}{8[(1)^2 + (-1)^2 + (-1)^2 + (1)^2]} = 8.0$$

$$df = 1$$

$$SS\hat{\psi}_{cubic} = \frac{\hat{\psi}_{cubic}^2}{n\sum_{j=1}^{p} c_{3j}^2} = \frac{(-16)^2}{8[(-1)^2 + (3)^2 + (-3)^2 + (1)^2]} = 1.6$$

$$df = 1$$

(iii) Computation of F ratio:

$$F = \frac{MS\hat{\psi}_{quad}}{MSWG} = \frac{SS\hat{\psi}_{quad}/1}{SSWG/p(n-1)} = \frac{8.0/1}{41.0/28} = 5.46$$

$$F = \frac{MS\hat{\psi}_{cubic}}{MSWG} = \frac{SS\hat{\psi}_{cubic}/1}{SSWG/p(n-1)} = \frac{1.6/1}{41.0/28} = 1.09$$

$$F_{.05;1,28} = 4.20$$

$184.9 + 8.0 + 1.6 = 194.5$. This sum is equal to *SSBG*. Thus, *SSBG* has been partitioned into three orthogonal trend components. Of the 194.5 units of variation due to differences in sleep deprivation, the linear component accounts for 184.9, or 95.1% of the variation. The quadratic and cubic components account for 4.1% and 0.8%, respectively. A summary of the trend analyses that have been performed is shown in Table 4.5-4.

TABLE 4.5-4   ANOVA Table Summarizing Trend Tests

	Source	SS	df	MS	F
1	Between groups (Sleep-deprivation levels)	194.500	$p - 1 = 3$	64.833	$[1/2]$ 44.28**
1a	Linear trend	184.900	1	184.900	$[1a/2]$ 126.30**
1b	Dep from lin trend	9.600	$p - 2 = 2$	4.800	$[1b/2]$ 3.28
1c	Quadratic trend	8.000	1	8.000	$[1c/2]$ 5.46*
1d	Cubic trend	1.600	1	1.600	$[1d/2]$ 1.09
2	Within groups	41.000	$p(n - 1) = 28$	1.464	
3	Total	235.500	$np - 1 = 31$		

$*p < .05.$
$**p < .01.$

FIGURE 4.5-3   Trend of the sample means is designated by $\overline{Y}_{.j \text{ observed}}$. The predicted means are computed in Table 4.5-5 and discussed later.

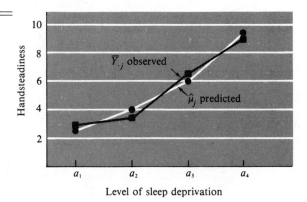

In experiments involving many treatment levels, it is unlikely that tests of trend components beyond the cubic or quartic degree will add materially to the experimenter's understanding of the data. The usual practice, then, is to test trend components for which there are a priori hypotheses and then pool the remaining components in a test for higher-order trends by $F = MS\ HIGHER\text{-}ORDER\ TRENDS/MSWG$. The *MS HIGHER-ORDER TRENDS* is given by

$$MS\ HIGHER\text{-}ORDER\ TRENDS = \frac{SSBG\ -\ SS\ LOWER\text{-}ORDER\ TRENDS}{(p\ -\ 1)\ -\ df\ LOWER\text{-}ORDER\ TRENDS}.$$

The *SS LOWER-ORDER TRENDS* is the sum of the *SS* for each trend component not included as one of the higher-order components. The *df LOWER-ORDER TRENDS* is the sum of the *df* for each of the lower-order sum of squares.

DESCRIBING THE TREND BY A POLYNOMIAL EQUATION

Assume that a priori hypotheses concerning the linear, quadratic, and cubic trend components have been advanced and that the analyses indicate that hand-steadiness is not a simple linear function of amount of sleep deprivation but is a curvilinear function. It is concluded that a polynomial equation of the form $\hat{\mu}_j = \hat{\beta}_0 + \hat{\beta}_1 X + \hat{\beta}_2 X^2$ is adequate to describe the relationship between the independent and dependent variables. This equation includes only those components that were shown to be significant and thus represents the simplest (lowest possible degree) description of the trend. It might be of interest to an experimenter to compare the predicted values, $\hat{\mu}_j$, with the obtained values, $\overline{Y}_j$, or to display the data in graphic form for only the significant components of the trend. Computation of the predicted values for each level of the independent variable based on the significant components of the trend is easily accomplished by means of orthogonal polynomials. Computational procedures for this purpose are illustrated in the following.

We noted earlier that instead of solving for the $\hat{\beta}_i$'s in

$$\hat{\mu}_j = \hat{\beta}_0 + \hat{\beta}_1 X_j + \hat{\beta}_2 X_j^2 + \cdots + \hat{\beta}_{p-1} X_j^{p-1}$$

it is more convenient to solve for the $\hat{\beta}_i'$'s in

$$\hat{\mu}_j = \hat{\beta}_0' + \hat{\beta}_1' c_{1j} + \hat{\beta}_2' c_{2j} + \cdots + \hat{\beta}_{p-1}' c_{p-1,j}$$

using orthogonal polynomial coefficients. The procedures for computing $\hat{\beta}_0'$, $\hat{\beta}_1'$, and $\hat{\beta}_2'$ are illustrated in Table 4.5-5. The predicted means based on a second-degree polynomial equation are compared with the observed sample means in Figure 4.5-3. The fit appears quite good. An exact fit would be obtained if a third-degree polynomial equation were used. This is of little significance because any set of p means can always be fitted exactly by a polynomial equation of, at most, degree $p - 1$. The best description of p means is the *simplest* equation that most nearly fits the points. We saw earlier that for the hand-steadiness data this is a second-degree polynomial equation. A more general approach to solving for regression coefficients, $\hat{\beta}_i$'s, is presented in Chapter 5.

TESTING GOODNESS OF FIT

A fourth question that can be answered by trend-analysis procedures is whether the patterns of the predicted and observed means are the same. For example, if a second-degree equation has been fitted to the sleep-deprivation data, we can determine if the equation provides a satisfactory fit. Predicted values for the treatment levels were computed in Table 4.5-5 and plotted in Figure 4.5-3. An F statistic can be used to determine if the predicted values provide a satisfactory fit for the pattern of the obtained values. Computational procedures for an F test for departure from patterns are shown in Table 4.5-6. Because $F < F_{.05;1,28}$, it can be concluded that the second-degree equation provides a satisfactory fit for the data. The *F DEP. FROM LIN* test

TABLE 4.5-5 Computational Procedures for Polynomial Equation

(i) Terms obtained from Tables 4.5-2 and 4.5-3:

$$\hat{\beta}_0' = \frac{\sum_{i=1}^{n} \sum_{j=1}^{p} Y_{ij}}{np} = \frac{172}{(8)(4)} = 5.375$$

$$\hat{\beta}_1' = \frac{\hat{\psi}_{\text{lin}}}{n \sum_{j=1}^{p} c_{1j}^2} = \frac{172}{8[(-3)^2 + (-1)^2 + (1)^2 + (3)^2]} = 1.075$$

$$\hat{\beta}_2' = \frac{\hat{\psi}_{\text{quad}}}{n \sum_{j=1}^{p} c_{2j}^2} = \frac{16}{8[(1)^2 + (-1)^2 + (-1)^2 + (1)^2]} = 0.500$$

$$\hat{\beta}_3' = \frac{\hat{\psi}_{\text{cubic}}}{n \sum_{j=1}^{p} c_{3j}^2} = \frac{-16}{8[(-1)^2 + (3)^2 + (-3)^2 + (1)^2]} = -0.100$$

(ii) Polynomial equations (The values of c_{1j} and c_{2j} are given in Table 4.5-1):

$$\hat{\mu}_j = \hat{\beta}_0' \quad + \hat{\beta}_1' c_{1j} \quad + \hat{\beta}_2' c_{2j}$$
$$\hat{\mu}_1 = 5.375 + 1.075(-3) + 0.500(1) = 2.65$$
$$\hat{\mu}_2 = 5.375 + 1.075(-1) + 0.500(-1) = 3.80$$
$$\hat{\mu}_3 = 5.375 + 1.075(1) + 0.500(-1) = 5.95$$
$$\hat{\mu}_4 = 5.375 + 1.075(3) + 0.500(1) = 9.10$$

TABLE 4.5-6 Computational Procedures for Determining if Patterns of Predicted, $\hat{\mu}_j$, and Observed, $\overline{Y}_{\cdot j}$, Means are Different

(i) Data:

a_j	n_j	$\overline{Y}_{\cdot j}$	$\hat{\mu}_j$	$(\overline{Y}_{\cdot j} - \hat{\mu}_j)^2$	$n_j(\overline{Y}_{\cdot j} - \hat{\mu}_j)^2$
a_1	8	2.75	2.65	.01	.08
a_2	8	3.50	3.80	.09	.72
a_3	8	6.25	5.95	.09	.72
a_4	8	9.00	9.10	.01	.08

$$\sum_{j=1}^{p} = 1.60$$

TABLE 4.5-6 (continued)

(ii) Computation of F ratio:

$$SS\ DEP.\ FROM\ PATTERN = \sum_{j=1}^{p} n_j(\overline{Y}_j - \hat{\mu}_j)^2 = 1.600$$

$$MS\ DEP.\ FROM\ PATTERN = \frac{SS\ DEP.\ FROM\ PATTERN}{p - k}$$

where k is the number of constants (e.g., $\hat{\beta}_0'$, $\hat{\beta}_1'$, and so on) in equation

$$= \frac{1.600}{4 - 3} = 1.600$$

$$F = \frac{MS\ DEP.\ FROM\ PATTERN}{MSWG} = \frac{1.600}{1.464} = 1.09$$

$$F_{.05;1,28} = 4.20$$

described earlier is a special case of the more general *F DEP. FROM PATTERN* test. When a set of data has been fitted by a linear equation, the two ratios yield identical *F* values.

A POSTERIORI TREND ANALYSIS PROCEDURES

The trend tests described previously fall in the category of a priori or planned tests. If the experimenter wishes to engage in trend snooping, a test statistic that controls the error rate at α for the collection of trend tests should be used. Either the Dunn procedure, which divides the critical region among the trend tests, or Scheffé's method can be used for this purpose. If the Dunn procedure is used, the observed value of an F statistic is compared with $F_{\alpha/C;\nu_1,\nu_2}$, where C is the number of trend comparisons in the family. Before engaging in trend snooping, it must be ascertained that the F statistic for differences between groups is significant. That is, there should be evidence for believing that there is a trend in the data.

4.6   MEASURES OF ASSOCIATION

In the previous section, procedures for determining if there is a trend in data were described. If a trend is present, an experimenter may want to determine the *strength* of the trend. Several measures of strength of association between independent and dependent variables are described in the following. If both independent (X) and dependent (Y) variables are quantitative, a descriptive index of linear correlation is given by

$$r = \sqrt{\frac{SS\hat{\psi}_{\text{lin}}}{SSTOT}}.$$

For the sleep-deprivation data summarized in Table 4.5-4, the $SS\hat{\psi}_{lin} = 184.9$ and $SSTOT = 235.5$. The linear correlation between sleep deprivation and hand-steadiness for the sample data is

$$r = \sqrt{\frac{184.9}{235.5}} = .89.$$

A measure that describes the curvilinear correlation for a sample is given by

$$\eta = \sqrt{\frac{SSBG}{SSTOT}}.$$

It was concluded, on the basis of a priori testing procedures, that the trend of the sleep-deprivation data is curvilinear. Under these conditions, η provides a better index of relationship than r. The value of η for the data in Table 4.5-4 is

$$\eta = \sqrt{\frac{194.5}{235.5}} = .91.$$

Two useful estimators of strength of association between X and Y for a population are $\hat{\rho}_I$, intraclass correlation, and $\hat{\omega}^2$, omega squared. Hays (1981, 382) points out that $\hat{\rho}_I$ and $\hat{\omega}^2$ are identical in general meaning. They provide estimates of total strength of association between X and Y when the independent variable is quantitative or qualitative in character. Both $\hat{\omega}^2$ and $\hat{\rho}_I$ indicate the proportion of variance in Y accounted for by specifying the X treatment-level classification. $\hat{\rho}_I$ applies to the random-effects model and $\hat{\omega}^2$ to the fixed-effects model. The random-effects model is described in Section 4.7.

The $\hat{\rho}_I$ statistic is given by

$$\hat{\rho}_I = \frac{MSBG - MSWG}{MSBG + (n - 1)MSWG}.$$

It is maximal when scores within each treatment level are identical and differ only between treatment levels (Haggard, 1958).

The $\hat{\omega}^2$ statistic, which has been described by Hays (1981, 290–291,349), is given by

$$\hat{\omega}^2 = \frac{SSBG - (p - 1)MSWG}{SSTO + MSWG}.$$

For the hand-steadiness data in Table 4.3-2,

$$\hat{\omega}^2 = \frac{194.5 - (4 - 1)1.464}{235.5 + 1.464} = 0.80.$$

We can conclude that the independent variable of sleep deprivation accounts for 80% of the variance in the hand-steadiness scores. Not only is the association statistically significant, as is evident from the significant $F = MSBG/MSWG$ statistic, but it is also quite strong.

The $\hat{\omega}^2$ statistic can also be used to assess the proportion of variance in Y accounted for by the ith contrast. The formula is

$$\hat{\omega}^2_{Y|\psi_i} = \frac{SS\hat{\psi}_i - MSWG}{SSTO + MSWG}.$$

For example, consider the sums of squares for the two significant trend contrasts in Table 4.5-4. The proportion of variance accounted for by the linear and quadratic contrasts are, respectively,

$$\hat{\omega}^2_{Y|\psi_{lin}} = \frac{184.5 - 1.464}{235.5 + 1.464} = 0.772$$

$$\hat{\omega}^2_{Y|\psi_{quad}} = \frac{8.0 - 1.464}{235.5 + 1.464} = 0.028.$$

A significant F statistic for treatment effects in a completely randomized design indicates that there is an association between X and Y. The statistics $\hat{\omega}^2$ and $\hat{\rho}_I$ indicate the strength of the association. This is an important bit of information in evaluating the outcome of an experiment because a *trivial* association between X and Y may achieve statistical significance if the sample is sufficiently large. An alternative method of interpreting the importance of sources of variation is in terms of variance components. This approach is discussed by Scheffé (1959). It cannot be emphasized too much that an F statistic only provides information concerning the likelihood that population effects are or are not equal to zero. An F statistic does not tell us whether the effects are large or small. This latter information is provided by an examination of variance components.

A word of caution is appropriate here. In interpreting measures of strength of association for the fixed-effects model it must be remembered that the treatment levels are selected a priori rather than by random sampling. The presence of a truncated range or extreme values of a quantitative independent variable or the use of a multi-dimensional qualitative independent variable can markedly affect the value of $\hat{\omega}^2$. Note also that, although $\hat{\omega}^2$ and $\hat{\rho}_I$ are computed as the ratio of unbiased estimators, they are biased estimators of the corresponding parameters. In general the ratio of two unbiased estimators is not itself an unbiased estimator. For a discussion of the merits of these and other measures of strength of association the reader is referred to Dwyer (1974), Glass and Hakstian (1969), Halderson and Glasnapp (1972), Levin (1967), and Vaughan and Corballis (1969).

4.7   RANDOM-EFFECTS MODEL

The experimental design model equation for a completely randomized design was given in Section 4.1 as

$$Y_{ij} = \mu + \alpha_j + \epsilon_{i(j)} \qquad (i = 1, \ldots, n; j = 1, \ldots, p).$$

There we assumed that the levels of the treatment represented fixed effects. If instead the p treatment levels are drawn at random from a population of P levels, where P is large relative to p, the treatment may be regarded as a random variable. The treatment effects, α_j's, for this case are assumed to be $NID(0, \sigma_\alpha^2)$. As before, μ is a constant and $\epsilon_{i(j)}$ is $NID(0, \sigma_\epsilon^2)$. This model is designated as a random-effects model or model II.

A comparison of the expected values of the mean squares for the two models is given in Table 4.7-1. The derivation of $E(MS)$ is discussed in Section 2.3. If p is small relative to P, the coefficient $(1 - p/P)$ of σ_α^2 in model II approaches one and the $E(MSBG)$ becomes $\sigma_\epsilon^2 + n\sigma_\alpha^2$. For both models, a test of the null hypothesis H_0: $\alpha_j = 0$ for all j (model I) or H_0: $\alpha_\alpha^2 = 0$ (model II) is given by

$$F = \frac{MSBG}{MSWG} = \frac{\text{Error effects} + \text{Treatment effects}}{\text{Error effects}}$$

where, under H_0, $MSBG$ and $MSWG$ are two independent estimators of the error variation. This ratio illustrates a basic principle. The expected value of the numerator of an F statistic should always contain one more term than the expected value of the denominator. The $MSBG$ is an unbiased estimator of σ_ϵ^2, the error variance, plus a term that will equal zero only if there are no treatment effects. If the null hypothesis is true, $E(MSBG)$ is equal to σ_ϵ^2. If any treatment effect exists, then $E(MSBG)$ is greater than σ_ϵ^2. An F test can be regarded as a procedure for deciding, on the basis of sample data, which of the model equations

$$Y_{ij} = \mu + \epsilon_{i(j)}$$
$$Y_{ij} = \mu + \alpha_j + \epsilon_{i(j)} \qquad (i = 1, \ldots, n; j = 1, \ldots, p)$$

underlies observations in the population.* If the null hypothesis is rejected, the second equation is adopted; if not, the first equation remains tenable.

As we have seen, the fixed- and random-effects models are identical except for the assumptions regarding the nature of the treatment effects. This difference is important because it determines the nature of the conclusions that can be drawn from an experiment. For the fixed-effects model, conclusions are restricted to the p treatment populations represented in the experiment. For the random-effects model, conclusions apply to the P treatment populations that were sampled.

Power, $1 - \beta$, for the random effects model can be estimated, following

---

TABLE 4.7-1   Comparison of $E(MS)$ for Models I and II

Source	Model I $E(MS)$	Model II $E(MS)$
$MSBG$	$\sigma_\epsilon^2 + n\Sigma\alpha_j^2/(p - 1)$	$\sigma_\epsilon^2 + n(1 - p/P)\sigma_\alpha^2$
$MSWG$	σ_ϵ^2	σ_ϵ^2

---

*This view is explored in detail in Chapter 5.

Scheffé (1959), from

$$F = \frac{F_{\alpha;(p-1),p(n-1)}}{1 + n[\hat{\rho}_I/(1 - \hat{\rho}_I)]}$$

where $F_{\alpha;(p-1),p(n-1)}$ is obtained from Appendix Table E.5 for α level of significance. Ordinarily this F statistic will be less than one. Hence to estimate power, compute $1/F$ and determine its significance level for $\nu_1 = p(n-1)$ and $\nu_2 = p - 1$ degrees of freedom. The significance level determined in this way is denoted by $\hat{\beta}$; power is estimated from $1 - \hat{\beta}$.

4.8 ADVANTAGES AND DISADVANTAGES OF TYPE CR-*p* DESIGN

The major advantages of the completely randomized design are as follows.

1. There is simplicity in the layout of the design.

2. Statistical analysis and interpretation of results are straightforward.

3. It does not require the use of equal sample sizes for each treatment level.

4. It allows for the maximum number of degrees of freedom for the error sum of squares.

5. It does not require a subject to participate under more than one treatment level or the use of subjects who have been matched on an appropriate variable.

The major disadvantages of the design are as follows.

1. The effects of differences among subjects are *controlled* by random assignment of the subjects to treatment levels. For this to be effective, subjects should be relatively homogeneous or a large number of subjects should be used.

2. When many treatment levels are included in the experiment, the required sample size may be prohibitive.

4.9 REVIEW EXERCISES

1. Two approaches to learning problem solving strategies—more specifically, generating alternative solutions—were investigated. Thirty sixth-graders were randomly assigned to one of the two approaches and a control condition. Treatment level a_1, referred to as the training condition, involved participating in five sessions per week during three consecutive weeks. Students

assigned to this condition observed a videotape introduction for ten minutes, practiced the skill for 15 minutes, observed peer models via videotape for 15 minutes, and observed a videotaped review for ten minutes. Treatment level a_2, a film-and-discussion condition, was conducted concurrently with the training condition and for the same amount of time. Films related to generating alternative solutions were shown followed by group discussions. The students in the control condition, treatment level a_3, did not receive any form of training. At the conclusion of the experiment, five problem situations were presented and the students instructed to write down as many solutions to each one as they could. The dependent variable was the number of solutions proposed, summed across the five problems. The following data were obtained. (Experiment suggested by Poitras-Martin, Danielle, and Steve, Gerald L. Psychological education: A skills-oriented approach. *Journal of Counseling Psychology*, 1977, *24*, 153–157.)

a_1	a_2	a_3
11	11	7
12	14	18
19	10	16
13	9	11
17	12	9
15	13	10
17	10	13
14	8	14
13	14	12
16	11	12

†**a)** [4.2] Perform an exploratory data analysis on these data (see Table 4.2-1) and interpret the analysis.

†**b)** [4.3] Test the hypothesis H_0: $\alpha_j = 0$ for all j; let $\alpha = .05$. Construct an ANOVA table summarizing the results of the analysis.

†**c)** [4.3] Calculate the power of the test in Exercise 1(b).

†**d)** [4.3] Determine by trial and error the number of subjects required to achieve a power of approximately .80.

†**e)** [4.4] Use Tukey's statistic to determine which population means differ.

†**f)** [4.6] Compute $\hat{\omega}^2$ for these data.

g) Prepare a "results and discussion section" appropriate for the *Journal of Counseling Psychology*.

2. The effect of instructions-to-learn on performance on a delayed recall test was investigated. Twenty male and female college undergraduate volunteers were randomly assigned to two instructional conditions: the subjects assigned to treatment level a_1 were informed of a subsequent recall test prior to presentation of a word list and were told to use any kind of rehearsal that they felt would aid their recall; the subjects in treatment level a_2 were not informed of a subsequent recall test. Thirty concrete nouns were shown to the subjects. Each noun was presented for one second with a nine-second interstimulus interval. As each noun was shown, the subjects were required to write it down. Twenty-four hours later the subjects were given a ten-minute written recall test. The dependent variable was the number of nouns recalled. The following data were obtained. (Experiment suggested by McDaniel, Mark A., and Masson, Michael E. Long-term retention: When incidental semantic processing fails. *Journal of Experimental Psychology: Human Learning and Memory*, 1977, *3*, 270–281.)

a_1	a_2
3	15
6	8
12	10
9	7
8	5
17	4
15	9
11	11
14	9
11	12

†a) [4.2] Perform an exploratory data analysis on these data (see Table 4.2-1) and interpret the analysis.

†b) [4.3] Test the hypothesis H_0: $\mu_1 = \mu_2$; let $\alpha = .05$. Construct an ANOVA table summarizing the results of the analysis.

†c) [4.4] Test the null hypothesis in Exercise 2(b) using a t statistic. Compare t^2 with the F obtained in Exercise 2(b).

3. The effect of written instructions designed to maximize subject attention to hypnotic facilitative information was investigated. The subjects were 36 hypnotically naive male and female college students who scored in the low and moderate range on the Harvard Group Scale of Hypnotic Susceptibility. The subjects were randomly assigned to one of four groups with nine subjects in each group. Subjects in the programmed active information group, treatment level a_1, read a booklet concerning hypnosis. Interspersed throughout the booklet were incomplete sentences designed to test the subject's knowledge of the material. Answers were provided on the following page of the booklet. Subjects in the active information group, treatment level a_2, read a booklet covering the same information but the booklet did not contain the self-testing feature. Subjects in the passive information group, treatment level a_3, read a booklet covering the historical development of hypnosis. This booklet did not contain information about how to experience hypnosis. Subjects in the control group, treatment level a_4, were given several magazines and told to browse through them in a relaxed manner. Following this phase of the experiment, subjects took the Stanford Hypnotic Susceptibility Scale, Form C. The dependent variable was the subject's score on this scale. The following data were obtained. (Experiment suggested by Diamond, Michael Jay, Steadman, Clarence, Harada, David, and Rosenthal, Joseph. The use of direct instructions to modify hypnotic performance: The effects of programmed learning procedures. *Journal of Abnormal Psychology*, 1975, *84*, 109–113.)

a_1	a_2	a_3	a_4
4	10	4	4
7	6	6	2
5	3	5	5
6	4	2	7
10	7	10	5
11	8	9	1
9	5	7	3
7	9	6	6
8	7	7	4

a) [4.2] Perform an exploratory data analysis on these data (see Table 4.2-1) and interpret the analysis.

b) [4.3] Test the hypothesis H_0: $\alpha_j = 0$ for all j; let $\alpha = .05$. Construct an ANOVA table summarizing the results of the analysis.

c) [4.3] Compute the power of the test in Exercise 3(b).

d) [4.3] Determine by trial and error the number of subjects required to obtain a power of approximately .84.

e) [4.4] Use Tukey's statistic to determine which population means differ. Construct a table like Table 4.4-1.

f) [4.6] Compute $\hat{\omega}^2$ for these data.

g) Prepare a "results and discussion section" for the *Journal of Abnormal Psychology*.

4.   An experiment was designed to evaluate the effects of different levels of training on children's ability to acquire the concept of equilateral triangle. Fifty three-year-old children were recruited from day-care facilities and randomly assigned to one of five groups with ten children in each group. Each group contained an equal number of boys and girls. Children in treatment level a_1 (visual condition) were shown 36 blocks, one at a time, and instructed to look at them but not to touch them. Children in treatment level a_2 (visual plus motor condition) looked at the blocks and were permitted to play with them. They were also asked to perform specific tactile-kinesthetic exercises, such as tracing the perimeter of the blocks with their index finger. Children in treatment level a_3 (visual plus verbal condition) looked at the blocks and were told to notice differences in their shape, color, size, and thickness. Children in treatment level a_4 (visual plus motor plus verbal condition) used a combination of visual, motor, and verbal means of stimulus predifferentiation. Children in treatment level a_5 (control condition) engaged in unrelated play activity. All training was done individually. The day after training the children were shown a "target" block for five seconds and then asked to identify the block in a group of seven blocks. This was repeated six times using different target blocks. The dependent variable was the number of target blocks correctly identified. The following data were obtained. (Experiment suggested by Nelson, Gordon K. Concomitant effects of visual, motor, and verbal experiences in young children's concept development. *Journal of Educational Psychology*, 1976, *68*, 466–473.)

a_1	a_2	a_3	a_4	a_5
0	2	2	2	1
1	3	3	4	0
3	4	4	5	2
1	2	4	3	1
1	1	2	2	1
2	1	1	1	2
2	2	2	3	1
1	2	3	3	0
1	3	2	2	1
2	4	2	4	3

a) [4.2] Perform an exploratory data analysis on these data (see Table 4.2-1) and interpret the analysis.

b) [4.3] Test the hypothesis H_0: $\alpha_j = 0$ for all j; let $\alpha = .05$. Construct an ANOVA table summarizing the results of the analysis.

c) [4.3] Compute the power of the test in Exercise 4(b).

d) [4.3] Determine by trial and error the number of subjects required to obtain a power of approximately .80.

e) [4.4] Use Tukey's statistic to determine which population means differ. Construct a table like Table 4.4-1.

f) [4.6] Compute $\hat{\omega}^2$ for these data.

g) Prepare a "results and discussion section" for the *Journal of Educational Psychology*.

5. [4.3] For the following designs, estimate the number of subjects required to achieve a power of .80, where the largest difference among means is equal to $C\sigma_\epsilon$.

†a) Type CR-3 design; let $\alpha = .05$ and $C = 0.8$

†b) Type CR-4 design; let $\alpha = .01$ and $C = 1.2$

c) Type CR-4 design; let $\alpha = .05$ and $C = 1.0$

d) Type CR-5 design; let $\alpha = .05$ and $C = 1.4$

†6. Section 4.1 described an experiment concerning the effects of sleep deprivation on hand-steadiness. Assume that a second sleep-deprivation experiment was performed in which the dependent variable was simple reaction time to the onset of a light. The following data (in hundredths of a second) have been obtained.

a_1 (12 hrs.)	a_2 (24 hrs.)	a_3 (36 hrs.)	a_4 (48 hrs.)
20	21	25	26
20	20	23	27
17	21	22	24
19	22	23	27
20	20	21	25
19	20	22	28
21	23	22	26
19	19	23	27

a) [4.2] Perform an exploratory data analysis on these data (see Table 4.2-1) and interpret the analysis.

b) [4.5] Test the hypothesis that there is no trend in the population; let $\alpha = .05$.

c) [4.5] Test the hypothesis that the nonlinear components of the trend equal zero; let $\alpha = .05$.

d) [4.5] (i) Test the null hypothesis for each trend component at $\alpha = .05$. (ii) Write the simplest polynomial equations for $\hat{\mu}_1$, $\hat{\mu}_2$, $\hat{\mu}_3$, and $\hat{\mu}_4$ neccessary to adequately describe the trend.

e) [4.5] What percentage of the total trend in the sample data is accounted for by the linear and quadratic components of the trend?

f) [4.6] Compute $\hat{\omega}^2_{Y|\psi_{\text{lin}}}$ and $\hat{\omega}^2_{Y|\psi_{\text{quad}}}$.

7. The effects of viewing "mugshots" on accuracy of eyewitness identification was investigated. Twenty-four subjects observed a videotape of six men whom they were later asked to identify in a recognition test. The subjects were randomly assigned to one of four groups. Subjects in group a_4 searched through a sequence of 75 mugshots to identify the suspects, those in group

a_3 searched through 50 mugshots, and those in group a_2 searched through 25 mugshots. Subjects in a_1 spent an equivalent amount of time looking for articles about crime in *Time* magazine. Following this, the subjects were shown pictures that included the suspects and asked to identify them. The dependent variable is the number of suspects identified. The following data were obtained.

a_1	a_2	a_3	a_4
5	4	3	0
6	3	0	1
3	6	1	0
4	3	2	2
5	5	2	1
4	4	1	2

a) [4.2] Perform an exploratory data analysis on these data (see Table 4.2-1) and interpret the analysis.

b) [4.5] Test the hypothesis that there is no trend in the population; let $\alpha = .05$.

c) [4.5] Test the overall hypothesis that the nonlinear components of the trend equal zero; let $\alpha = .05$.

d) [4.5] Assume that a priori hypotheses about the $p - 1$ trend components have been advanced. Test the null hypotheses for the linear, quadratic, and cubic trend components; let $\alpha = .05$ for each test.

e) [4.5] Write the simplest polynomial equations for $\hat{\mu}_1$, $\hat{\mu}_2$, $\hat{\mu}_3$, and $\hat{\mu}_4$ necessary to adequately describe the trend.

f) [4.5] What percentage of the total trend in the sample data is accounted for by the linear and cubic trend components?

g) [4.6] Compute $\hat{\omega}^2_{Y|\psi_{\text{lin}}}$ and $\hat{\omega}^2_{Y|\psi_{\text{cubic}}}$.

8. [4.6] How does r differ from η?

†9. [4.7] How do model I and model II differ for a type CR-p design?

5 GENERAL LINEAR MODEL APPROACH TO ANOVA [*]

5.1 INTRODUCTION TO THE GENERAL LINEAR MODEL

This chapter introduces the general linear model and matrix operations for computing sums of squares and estimating parameters in a type CR-p design. The traditional approach to the analysis of ANOVA designs described in Chapters 2 and 4 evolved during the heyday of mechanical desk calculators. It continues to be the preferred approach when computations are performed with modern desk and pocket calculators. In contrast, a digital computer and efficient software packages for manipulating matrices are required for the successful implementation of the general linear model

---

[*] The reader who is interested only in the traditional approach to analysis of variance can, without loss of continuity, omit this chapter and similarly marked sections. A familiarity with elementary matrix algebra is assumed in this chapter. Appendix D provides a survey of the basic concepts.

approach.* The growing popularity of this approach in the last decade parallels the increasing accessibility of high-speed computers. Before describing the general linear model we will examine some similarities and differences between analysis of variance and regression analysis.

COMPARISON OF ANALYSIS OF VARIANCE AND REGRESSION ANALYSIS

Most discussions of analysis of variance and regression analysis have tended to emphasize differences between the two approaches to the study of the relation between a dependent variable and one or more independent variables. As we will see, the techniques have much in common, and, in fact, both the experimental design model and the regression model are subsumed under the general linear model.

In the typical regression situation an experimenter is interested in the magnitude of the relation between one or more quantitative independent variables, X_1, $X_2, \ldots, X_{h-1}$, and a quantitative dependent variable, Y. Stated another way, the experimenter wants to learn the extent to which variation in the independent variables is associated with variation in the dependent variable. The value of each independent variable is selected in advance so that Y_i is observed for a combination of $h - 1$ fixed X values. In addition, the experimenter must also specify the nature of the relation between the independent variables and the dependent variable. When there is only one independent variable, the relation can be represented as a line in two-dimensional space where the fixed values of X are plotted on the horizontal axis and Y on the vertical axis. For two independent variables, the relation can be represented as a response surface in three-dimensional space, and so on. Regression analysis can also be used with qualitative independent variables, such as sex and type of therapy, in which case the question concerning the nature of the relation becomes meaningless.

In the analysis of variance situation, the experimenter is also interested in the extent to which variation in one or more quantitative or qualitative independent variables is associated with variation in a quantitative dependent variable. And as in regression analysis, the value of each independent variable is selected in advance. However, in analysis of variance, even though the independent variables are quantitative, no assumption is made regarding the nature of the relation between the independent variables and the dependent variable. This is a key difference between the two situations. Analysis of variance ignores the magnitude of differences among the levels of each independent variable and the nature of the relation between X_1, $X_2, \ldots, X_{h-1}$ and Y; regression analysis uses this information. When the independent variable is qualitative, such information does not exist, and there is no fundamental difference between the two situations. As we will see, analysis of variance and

---

* Relatively expensive pocket calculators are available that will perform operations on matrices, including inversion of a matrix. Computer software packages for performing matrix operations are widely available. An interactive matrix program written in Fortran 77 is described in the *Instructor's Manual for Experimental Design.*

regression analysis in which the independent variable is treated as if it is qualitative lead to equivalent results.*

GENERAL LINEAR MODEL

In this section we will express the regression and experimental design models in matrix notation. We will see that the matrix representation, which is called the *general linear model,* is the same for both models. The reader who is not familiar with elementary matrix algebra will find a brief introduction to those aspects employed in this book in Appendix D. More extensive presentations are given by Graybill (1969), Green and Carroll (1976), Hohn (1964), Searle (1966), and Timm (1975, Ch. 1).

A *linear model* consists of two parts: a model equation and associated assumptions. The model equation relates the observable random variable to underlying parameters and random variables in a linear manner. The assumptions specify the nature of the random components and any restrictions that the parameters must satisfy. Consider the general linear regression model

(5.1-1)
$$Y_i = \beta_0 + \beta_1 X_{i1} + \beta_2 X_{i2} + \cdots + \beta_{h-1} X_{i,h-1} + \epsilon_i \qquad (i = 1, \ldots, N)$$

where

Y_i = the value of the dependent variable for experimental unit i

$\beta_0, \beta_1, \ldots, \beta_{h-1}$ = unknown parameters of the model

$X_{i1}, X_{i2}, \ldots, X_{i,h-1}$ = $j = 1, \ldots, h - 1$ values of the independent variables measured without error for the ith observation

ϵ_i = a random error term with mean equal to 0, $E(\epsilon_i) = 0$; variance equal to σ_ϵ^2, $V(\epsilon_i) = \sigma_\epsilon^2$; and ϵ_i and $\epsilon_{i'}$ are uncorrelated so that the covariances equal zero, $COV(\epsilon_i, \epsilon_{i'}) = 0$, for all i and i' where $i \neq i'$.**

Several features of the model deserve further comment.

1. The observed value of Y_i is the sum of two components: the constant predictor term $\beta_0 + \beta_1 X_{i1} + \beta_2 X_{i2} + \cdots + \beta_{h-1} X_{i,h-1}$ and the random error term ϵ_i. The error term reflects that portion of Y_i not accounted for by the independent

---

* For purposes of exposition it is convenient to distinguish between analysis of variance and regression analysis in which the independent variables are treated qualitatively. Some writers prefer a more parsimonious classification scheme. For example, according to Scheffé (1959, 192–193), the analysis of variance is a group of statistical methods in which all independent variables are treated qualitatively; in regression analysis, all independent variables are quantitative and treated quantitatively.

** Rules of expectation, variance, and covariance are given in Appendix B.

variables. Since the errror term is a random variable, Y_i is also a random variable.

2. The expected value of the error term equals zero, $E(\epsilon_i) = 0$; it follows that

$$E(Y_i) = E(\beta_0 + \beta_1 X_{i1} + \beta_2 X_{i2} + \cdots + \beta_{h-1} X_{i,h-1} + \epsilon_i)$$
$$= \beta_0 + \beta_1 X_{i1} + \beta_2 X_{i2} + \cdots + \beta_{h-1} X_{i,h-1} + E(\epsilon_i)$$

(5.1-2)
$$= \beta_0 + \beta_1 X_{i1} + \beta_2 X_{i2} + \cdots + \beta_{h-1} X_{i,h-1}.$$

3. If the independent variables are quantitative, the unknown parameters can be interpreted as follows. The parameter β_0 is the Y intercept. If the scope of the model includes $X_{i1} = 0$, $X_{i2} = 0$, . . . , $X_{i,h-1} = 0$; β_0 gives the mean response for those values of the independent variables. The β_1, . . . , β_{h-1} are partial regression coefficients, or weights applied to the X_{ij}'s in order to optimally predict Y_i. The parameter β_1, for example, indicates the change in the mean response per unit increase in X_1 when X_2, X_3, . . . , X_{h-1} are held constant. The interpretation of the parameters when the independent variables are qualitative is presented in Sections 5.4 and 5.7.

4. When there is only one independent variable, the model is a *simple regression model;* when there are two or more independent variables, it is a *multiple regression model.*

5. The error term, ϵ_i, is assumed to have constant variance σ_ϵ^2, that is, $V(\epsilon_i) = \sigma_\epsilon^2$. It follows then that the variance of Y_i is

$$V(Y_i) = V(\beta_0 + \beta_1 X_{i1} + \beta_2 X_{i2} + \cdots + \beta_{h-1} X_{i,h-1} + \epsilon_i)$$
$$= V(\epsilon_i)$$

(5.1-3)
$$= \sigma_\epsilon^2.$$

6. The error terms are assumed to be uncorrelated: the value of ϵ_i is not related to the value of $\epsilon_{i'}$ for all i and i'. Since the ϵ_i's are uncorrelated, the Y_i's are also uncorrelated.

Model (5.1-1) is called a linear model because it is *linear in the parameters*—all of the parameters appear in the first power; none is multiplied or divided by other parameters, raised to powers, transformed to logarithms, appear as exponents, and so on. The model is also *linear in the independent variables* because the variables appear only in the first power. The general linear regression model can also represent models that are not linear in the independent variable. For example, the curvilinear regression model

$$Y_i = \beta_0 + \beta_1 X_i + \beta_2 X_i^2 + \epsilon_i$$

with one independent variable, X_i, a first degree term, $\beta_1 X_i$, and a second degree term, $\beta_2 X_i^2$, can be written in the form of model equation (5.1-1) by letting $X_i = X_{i1}$ and $X_i^2 = X_{i2}$. This gives

$$Y_i = \beta_0 + \beta_1 X_{i1} + \beta_2 X_{i2} + \epsilon_i \qquad (i = 1, \ldots, N).$$

Model equation (5.1-1) represents the following system of equations.

$$\begin{aligned}
Y_1 &= \beta_0 + \beta_1 X_{11} + \beta_2 X_{12} + \cdots + \beta_{h-1} X_{1,h-1} + \epsilon_1 \\
Y_2 &= \beta_0 + \beta_1 X_{21} + \beta_2 X_{22} + \cdots + \beta_{h-1} X_{2,h-1} + \epsilon_2 \\
&\ \cdot \qquad \cdot \qquad \cdot \qquad \cdot \qquad \qquad \cdot \qquad\quad \cdot \\
Y_N &= \beta_0 + \beta_1 X_{N1} + \beta_2 X_{N2} + \cdots + \beta_{h-1} X_{N,h-1} + \epsilon_N
\end{aligned}$$

(5.1-4)

This system can be written in matrix form as

$$
\begin{bmatrix} Y_1 \\ Y_2 \\ \cdot \\ \cdot \\ \cdot \\ Y_N \end{bmatrix}
=
\begin{bmatrix}
1 & X_{11} & X_{12} & \cdot & X_{1,h-1} \\
1 & X_{21} & X_{22} & \cdot & X_{2,h-1} \\
 & & & \cdot & \\
 & & & \cdot & \\
 & & & \cdot & \\
1 & X_{N1} & X_{N2} & \cdot & X_{N,h-1}
\end{bmatrix}
\begin{bmatrix} \beta_0 \\ \beta_1 \\ \cdot \\ \cdot \\ \cdot \\ \beta_{h-1} \end{bmatrix}
+
\begin{bmatrix} \epsilon_1 \\ \epsilon_2 \\ \cdot \\ \cdot \\ \cdot \\ \epsilon_N \end{bmatrix}
$$

or

(5.1-5)
$$
\underset{N \times 1}{\mathbf{y}} = \underset{N \times h}{\mathbf{X}} \quad \underset{h \times 1}{\boldsymbol{\beta}} + \underset{N \times 1}{\boldsymbol{\epsilon}}
$$

where $\mathbf{y}$ = an $N \times 1$ vector of observations

$\mathbf{X}$ = an $N \times h$ matrix of constants

$\boldsymbol{\beta}$ = a $h \times 1$ vector of parameters

$\boldsymbol{\epsilon}$ = an $N \times 1$ vector of random errors with $E(\boldsymbol{\epsilon}) = \mathbf{0}$, that is,

$$
\begin{bmatrix} E(\epsilon_1) \\ E(\epsilon_2) \\ \cdot \\ \cdot \\ \cdot \\ E(\epsilon_N) \end{bmatrix}
=
\begin{bmatrix} 0 \\ 0 \\ \cdot \\ \cdot \\ \cdot \\ 0 \end{bmatrix}
$$

and $V(\boldsymbol{\epsilon}) = \sigma_\epsilon^2 \mathbf{I}$, that is,

$$
V(\boldsymbol{\epsilon}) = \sigma_\epsilon^2
\begin{bmatrix}
1 & 0 & \cdots & 0 \\
0 & 1 & \cdots & 0 \\
\cdot & \cdot & & \cdot \\
\cdot & \cdot & & \cdot \\
\cdot & \cdot & & \cdot \\
0 & 0 & \cdots & 1
\end{bmatrix}
=
\begin{bmatrix}
\sigma_\epsilon^2 & 0 & \cdots & 0 \\
0 & \sigma_\epsilon^2 & \cdots & 0 \\
\cdot & \cdot & & \cdot \\
\cdot & \cdot & & \cdot \\
\cdot & \cdot & & \cdot \\
0 & 0 & \cdots & \sigma_\epsilon^2
\end{bmatrix}.
$$

Furthermore, it follows from (5.1-2) and (5.1-3) that

$$ E(\mathbf{y}) = E(\mathbf{X}\boldsymbol{\beta} + \boldsymbol{\epsilon}) = \mathbf{X}\boldsymbol{\beta} \quad \text{and} \quad V(\mathbf{y}) = V(\mathbf{X}\boldsymbol{\beta} + \boldsymbol{\epsilon}) = \sigma_\epsilon^2 \mathbf{I}. $$

The reader who is familiar with matrix algebra will see the equivalence of (5.1-4) and (5.1-5). Those who are encountering matrix algebra for the first time should read Appendix D, particularly the rules for matrix multiplication (Appendix

D.2-4). To obtain Y_1 in (5.1-4), for example, we multiply row one of $\mathbf{X}$ by $\boldsymbol{\beta}$ and add the first element of $\boldsymbol{\epsilon}$ to the product as follows.

$$
\begin{bmatrix} Y_1 \\ \\ \\ \\ \\ \\ \end{bmatrix} = \begin{bmatrix} 1 & X_{11} & X_{12} & \cdots & X_{1,h-1} \\ \\ \\ \\ \\ \\ \end{bmatrix} \begin{bmatrix} \beta_0 \\ \beta_1 \\ \cdot \\ \cdot \\ \cdot \\ \beta_{h-1} \end{bmatrix} + \begin{bmatrix} \epsilon_1 \\ \\ \\ \\ \\ \\ \end{bmatrix}
$$

$$Y_1 = \beta_0 + \beta_1 X_{11} + \beta_2 X_{12} + \cdots + \beta_{h-1} X_{1,h-1} + \epsilon_1$$

Equation (5.1-5) is the *general linear model* equation, which is used in a variety of univariate applications. In the following paragraphs we will write the experimental design model equation in ANOVA as (5.1-5).

Consider the experimental design model for a completely randomized design.

(5.1-6) $$Y_{ij} = \mu + \alpha_j + \epsilon_{i(j)} \qquad (i = 1, \ldots, n; j = 1, \ldots, p)$$

where Y_{ij} = the value of the dependent variable for experimental unit i in treatment level j

μ = a fixed unknown parameter representing the population mean

α_j = a fixed unknown parameter representing the effect of treatment level j

$\epsilon_{i(j)}$ = a random error term with mean equal to 0, $E(\epsilon_{i(j)}) = 0$; variance equal to σ_ϵ^2, $V(\epsilon) = \sigma_\epsilon^2$; and $\epsilon_{i(j)}$ and $\epsilon_{i'(j)}$ are uncorrelated so that $COV(\epsilon_{i(j)}, \epsilon_{i'(j)}) = 0$ for all i and i' where $i \neq i'$.

If there are two subjects in each of the $j = 1, \ldots, p$ treatment levels, model (5.1-6) represents the following system of equations.

$$
\begin{aligned}
Y_{11} &= \mu + \alpha_1 + & \cdot & & + \epsilon_{1(1)} \\
Y_{21} &= \mu + \alpha_1 + & \cdot & & + \epsilon_{2(1)} \\
Y_{12} &= \mu + & \alpha_2 + \cdot & & + \epsilon_{1(2)} \\
Y_{22} &= \mu + & \alpha_2 + \cdot & & + \epsilon_{2(2)} \\
& \cdot \quad \cdot \quad \cdot \quad \cdot \quad \cdot \quad \cdot \quad \cdot \quad \cdot \quad \cdot \quad \cdot \quad \cdot \\
Y_{1p} &= \mu + & \cdot + \alpha_p & & + \epsilon_{1(p)} \\
Y_{2p} &= \mu + & \cdot + \alpha_p & & + \epsilon_{2(p)}
\end{aligned}
$$

This system can be rewritten using an *indicator variable* that takes on the integer values 0 and 1. The indicator variable is used to indicate whether a parameter is or is not included in an equation. If a parameter appears in an equation, it is multiplied by 1, otherwise it is multiplied by 0. Using the indicator variable, (5.1-6) becomes

(5.1-7)

$$Y_{11} = \mu 1 + \alpha_1 1 + \alpha_2 0 + \cdot + \alpha_p 0 + \epsilon_{1(1)}$$
$$Y_{21} = \mu 1 + \alpha_1 1 + \alpha_2 0 + \cdot + \alpha_p 0 + \epsilon_{2(1)}$$
$$Y_{12} = \mu 1 + \alpha_1 0 + \alpha_2 1 + \cdot + \alpha_p 0 + \epsilon_{1(2)}$$
$$Y_{22} = \mu 1 + \alpha_1 0 + \alpha_2 1 + \cdot + \alpha_p 0 + \epsilon_{2(2)}$$

$$\cdot \quad \cdot \quad \cdot \quad \cdot \quad \cdot \quad \cdot \quad \cdot \quad \cdot \quad \cdot \quad \cdot \quad \cdot \quad \cdot \quad \cdot$$

$$Y_{1P} = \mu 1 + \alpha_1 0 + \alpha_2 0 + \cdot \cdot + \alpha_p 1 + \epsilon_{1(p)}$$
$$Y_{2P} = \mu 1 + \alpha_1 0 + \alpha_2 0 + \cdot \cdot + \alpha_p 1 + \epsilon_{2(p)}.$$

The equations in this system, like those in the regression model, are linear in the parameters. Furthermore, they are linear in the indicator variables. We can write (5.1-7) in matrix form as

$$
\begin{bmatrix} Y_{11} \\ Y_{21} \\ Y_{12} \\ Y_{22} \\ \vdots \\ Y_{1p} \\ Y_{2p} \end{bmatrix}
=
\begin{bmatrix} 1 & 1 & 0 & \cdot & 0 \\ 1 & 1 & 0 & \cdot & 0 \\ 1 & 0 & 1 & \cdot & 0 \\ 1 & 0 & 1 & \cdot & 0 \\ \cdot & \cdot & \cdot & \cdot & \cdot \\ 1 & 0 & 0 & \cdot & 1 \\ 1 & 0 & 0 & \cdot & 1 \end{bmatrix}
\begin{bmatrix} \mu \\ \alpha_1 \\ \alpha_2 \\ \vdots \\ \alpha_p \end{bmatrix}
+
\begin{bmatrix} \epsilon_{1(1)} \\ \epsilon_{2(1)} \\ \epsilon_{1(2)} \\ \epsilon_{2(2)} \\ \vdots \\ \epsilon_{1(p)} \\ \epsilon_{2(p)} \end{bmatrix}
$$

or

(5.1-8)

$$
\begin{array}{cccccc}
\mathbf{y} & = & \mathbf{X} & \boldsymbol{\beta} & + & \boldsymbol{\epsilon} \\
N \times 1 & & N \times h & h \times 1 & & N \times 1
\end{array}
$$

where

$\mathbf{y}$ = an $N \times 1$ vector of observations

$\mathbf{X}$ = an $N \times h$ matrix of constants $(0,1)$

$\boldsymbol{\beta}$ = a $h \times 1$ vector of parameters $(\mu, \alpha_1, \ldots, \alpha_p)$

$\boldsymbol{\epsilon}$ = an $N \times 1$ vector of random errors with $E(\boldsymbol{\epsilon}) = \mathbf{0}$ and $V(\boldsymbol{\epsilon}) = \sigma_\epsilon^2 \mathbf{I}$.

Note that $\mathbf{X}$ is a matrix of indicator variable values. Such a matrix is called a *structural matrix*. Note also that the form of the general linear model equations (5.1-5) and (5.1-8) is identical for the regression and experimental design models. Thus, both models are subsumed under the general linear model. In the following sections we will see how to solve for the vector of unknowns, $\boldsymbol{\beta}$, in the general linear model. Six solutions will be illustrated: three involve different coding schemes for the independent variable in a qualitative regression model (Sections 5.4 and 5.7), two solutions involve a less than full rank experimental design model (Section 5.9), and one involves a full rank experimental design model (Section 5.10).

5.2   ESTIMATING PARAMETERS IN THE GENERAL LINEAR REGRESSION MODEL

In Chapters 2 and 4 we learned how to estimate and test hypotheses about the parameters of an experimental design model. In this section we will see how this is done for a regression model.

There are several well-recognized methods for estimating the parameters of the general linear regression model. The most frequently used method and the one we will outline here is called the *method of least squares*. It is concerned with minimizing a function of the error in predicting Y_i from $\hat{\beta}_0 + \hat{\beta}_1 X_{i1} + \hat{\beta}_2 X_{i2} + \cdots + \hat{\beta}_{h-1} X_{i,h-1}$, where the $\hat{\beta}$'s are parameter estimators. An error $\hat{\epsilon}_i$, also called a *residual*, is the difference between the value of the ith score Y_i and the value predicted for that score $\hat{Y}_i$, that is, $\hat{\epsilon}_i = Y_i - \hat{Y}_i$. The objective of the method of least squares is to find estimators of the parameters $\beta_0, \beta_1, \ldots, \beta_{h-1}$ that minimize the sum of the squared errors

$$\sum_{i=1}^{N} \hat{\epsilon}_i^2 = \sum_{i=1}^{N} (Y_i - \hat{Y}_i)^2$$

(5.2-1)
$$= \sum_{i=1}^{N} [Y_i - (\hat{\beta}_0 + \hat{\beta}_1 X_{i1} + \hat{\beta}_2 X_{i2} + \cdots + \hat{\beta}_{h-1} X_{i,h-1})]^2.$$

In matrix notation, (5.2-1) is written

$$\hat{\epsilon}'\hat{\epsilon} = (\mathbf{y} - \hat{\mathbf{y}})'(\mathbf{y} - \hat{\mathbf{y}})$$

(5.2-2)
$$= (\mathbf{y} - \mathbf{X}\hat{\beta})'(\mathbf{y} - \mathbf{X}\hat{\beta}).$$

The method of finding numerical values for $\hat{\beta}$ that minimizes $\hat{\epsilon}'\hat{\epsilon}$ uses the differential calculus and hence is beyond the scope of this text.* We will simply report here in matrix form the least squares normal equations.** They are

(5.2-3)
$$\mathbf{X}'\mathbf{X}\hat{\beta} = \mathbf{X}'\mathbf{y}.$$

To obtain the vector of parameter estimates, $\hat{\beta}$, we premultiply both sides of (5.2-3) by the inverse of $\mathbf{X}'\mathbf{X}$, assuming that the inverse exists (see Appendix D.2-6).

$$(\mathbf{X}'\mathbf{X})^{-1}\mathbf{X}'\mathbf{X}\hat{\beta} = (\mathbf{X}'\mathbf{X})^{-1}\mathbf{X}'\mathbf{y}$$

Since $(\mathbf{X}'\mathbf{X})^{-1}\mathbf{X}'\mathbf{X} = \mathbf{I}$, the identity matrix, we have

(5.2-4)
$$\hat{\beta} = (\mathbf{X}'\mathbf{X})^{-1}\mathbf{X}'\mathbf{y}.$$

This matrix equation gives the set of parameter estimates $\hat{\beta}_0, \hat{\beta}_1, \ldots, \hat{\beta}_{h-1}$ that minimizes $\hat{\epsilon}'\hat{\epsilon}$ or $\sum_{i=1}^{N} (Y_i - \hat{Y}_i)^2$. Two important properties of the least squares esti-

---

*   Searle (1966, 203–207) describes the method in detail.

**   Normal equations are a set of equations derived by the method of least squares to obtain estimates of the parameters $\beta_0, \beta_1, \ldots, \beta_{h-1}$.

mators are succinctly stated in the Gauss-Markov theorem. For model (5.1-5) the least squares estimators for $\beta_0, \beta_1, \ldots, \beta_{h-1}$ are unbiased and have minimum variance among all unbiased linear estimators. Thus,

$$E(\hat{\beta}_0) = \beta_0, \, E(\hat{\beta}_1) = \beta_1, \, \ldots, \, E(\hat{\beta}_{h-1}) = \beta_{h-1}$$

and the variances of the sampling distributions of $\hat{\beta}_0, \hat{\beta}_1, \ldots, \hat{\beta}_{h-1}$ are smaller than those of any other unbiased estimators that are linear functions of the observations $Y_1, Y_2, \ldots, Y_N$. In deriving the least squares estimators it is not necessary to make any assumptions regarding the functional form of the distribution of the error term ϵ_i. Later when we test hypotheses about the parameters it will be necessary to assume that the ϵ_i's (and hence the Y_i's) are normally distributed. Then the assumption in models (5.1-1), (5.1-5), (5.1-6), and (5.1-8) that the error terms are uncorrelated so that $COV(\epsilon_i, \epsilon_{i'}) = 0$ for $i \neq i'$ becomes one of independence of the error terms.

PARTITION OF THE TOTAL SUM OF SQUARES

In the previous section, a procedure for obtaining estimates of the parameters of the general linear regression model was described. The question then arises as to whether one or more of the $\beta_1, \beta_2, \ldots, \beta_{h-1}$ parameters differs from zero. In this and subsequent sections we will lay the foundation for using an F statistic to test the null hypothesis that the parameters $\beta_1, \beta_2, \ldots, \beta_{h-1}$ are equal to zero against the alternative that $\beta_j \neq 0$ for some j, where $j = 1, \ldots, h - 1$.

We begin by partitioning the total sum of squares $SSTO$ for a regression model. The partition is similar to that for a type CR-p design model in Section 2.2-1. There we started with the identity

$$Y_{ij} - \overline{Y}.. = (Y_{ij} - \overline{Y}._j) + (\overline{Y}._j - \overline{Y}..)$$

and after squaring both sides and summing over the pn scores obtained

$$\sum_{j=1}^{p}\sum_{i=1}^{n} (Y_{ij} - \overline{Y}..)^2 = \sum_{j=1}^{p}\sum_{i=1}^{n}(Y_{ij} - \overline{Y}._j)^2 + n \sum_{j=1}^{p} (\overline{Y}._j - \overline{Y}..)^2$$

$$SSTO \quad = \quad SSWG \quad + \quad SSBG.$$

We saw that $SSTO$ reflects the total dispersion of scores around the overall mean and is equal to $SSWG$, the error variation or dispersion of scores around their respective treatment means, plus $SSBG$, a measure of the dispersion of treatment means around the overall mean.

A similar partition for the regression model is obtained from

(5.2-5)
$$Y_i - \overline{Y}. = (Y_i - \hat{Y}_i) + (\hat{Y}_i - \overline{Y}.)$$

where $Y_i - \hat{Y}_i$ = a prediction error, the deviation of a score from the regression line or regression surface

$\hat{Y}_i - \overline{Y}.$ = the deviation of a fitted regression value from the overall mean.

These deviations for a simple regression model are illustrated in Figure 5.2-1. To obtain sums of squares we square both sides of (5.2-5) and sum over the N scores as follows.

$$\sum_{i=1}^{N} (Y_i - \overline{Y}.)^2 = \sum_{i=1}^{N} [(Y_i - \hat{Y}_i) + (\hat{Y}_i - \overline{Y}.)]^2$$

$$= \sum_{i=1}^{N} [(Y_i - \hat{Y}_i)^2 + 2(\hat{Y}_i - \overline{Y}.)(Y_i - \hat{Y}_i) + (\hat{Y}_i - \overline{Y}.)^2]$$

$$= \sum_{i=1}^{N} (Y_i - \hat{Y}_i)^2 + 2 \sum_{i=1}^{N} (\hat{Y}_i - \overline{Y}.)(Y_i - \hat{Y}_i) + \sum_{i=1}^{N} (\hat{Y}_i - \overline{Y}.)^2$$

It can be shown that the middle term on the right is equal to zero. Hence

(5.2-6)
$$\sum_{i=1}^{N} (Y_i - \overline{Y}.)^2 = \sum_{i=1}^{N} (Y_i - \hat{Y}_i)^2 + \sum_{i=1}^{N} (\hat{Y}_i - \overline{Y}.)^2$$

$$SSTO \quad = \quad SSE \quad + \quad SSR$$

where SSE denotes the sum of squares due to error and SSR, the sum of squares due to regression. These formulas are not the most convenient ones to use in computing sums of squares. Before presenting more usable formulas, we will discuss the degrees of freedom for the sums of squares.

FIGURE 5.2-1 Graphic illustration of the deviations represented by $(Y_i - \overline{Y}.)$, $(Y_i - \hat{Y}_i)$, and $(\hat{Y}_i - \overline{Y}.)$. For 2, 3, 4, . . . independent variables, the line denoted by $\hat{Y}_i$ is replaced by surfaces in 3, 4, 5, . . . dimensional space and deviations are taken from the surface.

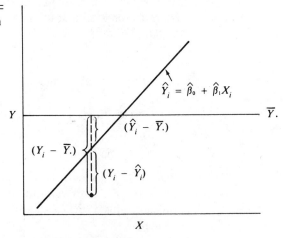

DEGREES OF FREEDOM, MEAN SQUARES, AND COMPUTATIONAL FORMULAS

We can think of degrees of freedom as the number of observations whose values are free to vary or as the number of independent observations for a source of variation minus the number of independent parameters estimated in computing the variation. Following the first definition, the total sum of squares has $N - 1$ degrees of freedom because values of $N - 1$ of the observations are free to vary, but the Nth one is determined since $\Sigma(Y_i - \overline{Y})$ must equal zero. According to the second definition, when we compute the total sum of squares $\Sigma(Y_i - \overline{Y})^2$, we replace the unknown population mean μ by $\overline{Y}$, which is a function of the sample Y's. Hence we lose one degree of freedom, leaving $N - 1$. Following the same line of reasoning, $SSE = \Sigma(Y_i - \hat{Y}_i)^2$ has $N - h$ degrees of freedom because the function $\hat{Y}_i$ involves estimating h parameters, $\beta_0, \beta_1, \ldots, \beta_{h-1}$. The sum of squares for regression $SSR = \Sigma(\hat{Y}_i - \overline{Y})^2$ has $h - 1$ degrees of freedom; $\hat{Y}_i$ involves h independent parameters, but the deviations $\hat{Y}_i - \overline{Y}$ are subject to the constraint that $\Sigma(\hat{Y}_i - \overline{Y}) = 0$.

A mean square is obtained by dividing a sum of squares by its degrees of freedom. The mean squares for the regression model are

$$MSTO = SSTO/(N - 1)$$
$$MSR = SSR/(h - 1)$$
$$MSE = SSE/(N - h).$$

In regression analysis we are usually interested in testing the hypothesis

$$H_0: \quad \beta_1 = \beta_2 = \cdots = \beta_{h-1} = 0$$

versus $\beta_j \neq 0$ for some j, where $j = 1, \ldots, h - 1$. The null hypothesis can be written in matrix form as

(5.2-7)
$$H_0: \quad \underset{(h-1)\times h}{\mathbf{C}'} \; \underset{h\times 1}{\boldsymbol{\beta}} = \underset{(h-1)\times 1}{\mathbf{0}}$$

where

$$\underset{(h-1)\times h}{\mathbf{C}'} = \begin{bmatrix} 0 & 1 & 0 & \cdot & 0 \\ 0 & 0 & 1 & \cdot & 0 \\ 0 & 0 & 0 & \cdot & 0 \\ \cdot & \cdot & \cdot & \cdot & \cdot \\ 0 & 0 & 0 & \cdot & 1 \end{bmatrix} \quad \underset{h\times 1}{\boldsymbol{\beta}} = \begin{bmatrix} \beta_0 \\ \beta_1 \\ \beta_2 \\ \vdots \\ \beta_{h-1} \end{bmatrix} \quad \underset{(h-1)\times 1}{\mathbf{0}} = \begin{bmatrix} 0 \\ 0 \\ 0 \\ \vdots \\ 0 \end{bmatrix} .$$

This hypothesis is referred to as a partial joint hypothesis, since it concerns $\beta_1, \beta_2, \ldots, \beta_{h-1}$ but not β_0. The total regression hypothesis is

(5.2-8)
$$H_0: \quad \underset{h\times h}{\mathbf{C}'} \; \underset{h\times 1}{\boldsymbol{\beta}} = \underset{h\times 1}{\mathbf{0}}$$

where

$$\underset{h\times h}{\mathbf{C}'} = \begin{bmatrix} 1 & 0 & 0 & \cdot & 0 \\ 0 & 1 & 0 & \cdot & 0 \\ 0 & 0 & 1 & \cdot & 0 \\ \cdot & \cdot & \cdot & \cdot & \cdot \\ 0 & 0 & 0 & \cdot & 1 \end{bmatrix} \quad \underset{h\times 1}{\boldsymbol{\beta}} = \begin{bmatrix} \beta_0 \\ \beta_1 \\ \beta_2 \\ \vdots \\ \beta_{h-1} \end{bmatrix} \quad \underset{h\times 1}{\mathbf{0}} = \begin{bmatrix} 0 \\ 0 \\ 0 \\ \vdots \\ 0 \end{bmatrix} .$$

In the discussion that follows, our interest will be in testing (5.2-7) rather than (5.2-8); the inclusion of the hypothesis that β_0, the Y intercept, is equal to zero has little relevance. We observed earlier that the partition of the total sum of squares in (5.2-6) does not provide formulas that are convenient for computational purposes. The desired formulas in matrix notation are as follows.

$$MSTO = SSTO/(N - 1) = \left[\mathbf{y}'\mathbf{y} - \left(\sum_{i=1}^{N} Y_i \right)^2 \Big/ N \right] \Big/ (N - 1)*$$

$$MSR = SSR/(h - 1) = \{(\mathbf{C}'\hat{\boldsymbol{\beta}})'[\mathbf{C}'(\mathbf{X}'\mathbf{X})^{-1}\mathbf{C}]^{-1}(\mathbf{C}'\hat{\boldsymbol{\beta}})\}/(h - 1)$$

$$MSE = SSE/(N - h) = (\mathbf{y}'\mathbf{y} - \hat{\boldsymbol{\beta}}'\mathbf{X}'\mathbf{y})/(N - h),**$$

where $\mathbf{y}$ = an $N \times 1$ vector of observations

$$\sum_{i=1}^{N} Y_i = \text{the sum of all observations in } \mathbf{y}$$

N = the number of observations in $\mathbf{y}$

$\mathbf{C}'$ = a coefficient matrix that determines which parameters of $\boldsymbol{\beta}$ are included in the null hypothesis

$\hat{\boldsymbol{\beta}}$ = an $h \times 1$ vector of parameter estimates

$\mathbf{X}$ = an $N \times h$ matrix of constants.

For testing (5.2-7), the formula for MSR simplifies to

$$MSR = \left[\hat{\boldsymbol{\beta}}'\mathbf{X}'\mathbf{y} - \left(\sum_{i=1}^{N} Y_i \right)^2 \Big/ N \right] \Big/ (h - 1).$$

EXPECTATION OF MEAN SQUARES

It can be shown that the expectations of the mean squares are

$$E(MSR) = \sigma_{\epsilon}^2 + \{(\mathbf{C}'\boldsymbol{\beta})'[\mathbf{C}'(\mathbf{X}'\mathbf{X})^{-1}\mathbf{C}]^{-1}(\mathbf{C}'\boldsymbol{\beta})\}/(h - 1)$$

and $E(MSE) = \sigma_{\epsilon}^2 .$

---

*   The term $(\sum_{i=1}^{N} Y_i)^2/N$ can be written as $\mathbf{y}'\mathbf{J}\mathbf{y}/N$, where $\mathbf{J}$ is an $N \times N$ matrix with every element unity; for simplicity, scalar notation is used.

**   From (5.2-2),

$$\hat{\boldsymbol{\epsilon}}'\hat{\boldsymbol{\epsilon}} = (\mathbf{y} - \mathbf{X}\hat{\boldsymbol{\beta}})'(\mathbf{y} - \mathbf{X}\hat{\boldsymbol{\beta}})$$
$$= (\mathbf{y}' - \hat{\boldsymbol{\beta}}'\mathbf{X}')(\mathbf{y} - \mathbf{X}\hat{\boldsymbol{\beta}})$$
$$= \mathbf{y}'\mathbf{y} - \mathbf{y}'\mathbf{X}\hat{\boldsymbol{\beta}} - \hat{\boldsymbol{\beta}}'\mathbf{X}'\mathbf{y} + \hat{\boldsymbol{\beta}}'\mathbf{X}'\mathbf{X}\hat{\boldsymbol{\beta}}.$$

It follows from (5.2-3) that

$$\mathbf{y}'\mathbf{X} = \boldsymbol{\beta}'\mathbf{X}'\mathbf{X}.$$

Thus
$$\hat{\boldsymbol{\epsilon}}'\hat{\boldsymbol{\epsilon}} = \mathbf{y}'\mathbf{y} - \hat{\boldsymbol{\beta}}'\mathbf{X}'\mathbf{X}\hat{\boldsymbol{\beta}} - \hat{\boldsymbol{\beta}}'\mathbf{X}'\mathbf{y} + \hat{\boldsymbol{\beta}}'\mathbf{X}'\mathbf{X}\hat{\boldsymbol{\beta}}$$
$$= \mathbf{y}'\mathbf{y} - \hat{\boldsymbol{\beta}}'\mathbf{X}'\mathbf{y}.$$

If all of the $j = 1, \ldots, h - 1$ parameters are equal to zero ($\beta_1 = \beta_2 = \cdots = \beta_{h-1} = 0$), $E(MSR)$ is equal to σ_ϵ^2. If, however, $\beta_j \neq 0$ for some j ($j = 1, \ldots, h - 1$), $E(MSR)$ is greater than σ_ϵ^2. This suggests an approach to testing the null hypothesis

$$H_0: \quad \beta_1 = \beta_2 = \cdots = \beta_{h-1} = 0.$$

If MSR is substantially greater than MSE, we would suspect that the null hypothesis is false, in which case the alternative hypothesis

$$H_1: \quad \beta_j \neq 0 \quad \text{for some } j, \text{ where } j = 1, \ldots, h - 1$$

is tenable. On the other hand, if MSR and MSE are of the same order of magnitude, this would suggest that MSR estimates only σ_ϵ^2 and the null hypothesis should not be rejected. As you have probably anticipated, the statistic for testing the null hypothesis is

$$F = \frac{MSR}{MSE}$$

with $h - 1$ and $N - h$ degrees of freedom. We can justify the use of F here by recourse to Cochran's theorem. For our purposes, the theorem can be stated as follows.

If N observations are obtained from the same normal population with mean μ and variance σ_ϵ^2, and $SSTO$ with $N - 1$ degrees of freedom is partitioned into m sums of squares $SS_1, SS_2, \ldots, SS_m$, with degrees of freedom $df_1, df_2, \ldots, df_m$; then the m terms SS_j/σ_ϵ^2 are independent χ^2 variables with df_j degrees of freedom if and only if $df_1 + df_2 + \cdots + df_m = N - 1$.

Earlier we partitioned $SSTO$ into SSR and SSE, and the $N - 1$ degrees of freedom for the total sum of squares into $h - 1$ and $N - h$. For model (5.1-1) we assumed that $E(Y_i) = \beta_0 + \beta_1 X_{i1} + \beta_2 X_{i2} + \cdots + \beta_{h-1} X_{i,h-1}$ and $V(Y_i) = \sigma_\epsilon^2$. According to Cochran's theorem, if the N observations are obtained from a normal population and the null hypothesis $\beta_1 = \beta_2 = \cdots = \beta_{h-1} = 0$ is true, in which case $E(Y_i) = \beta_0 = \mu$ for all i, then SSR/σ_ϵ^2 and SSE/σ_ϵ^2 are independent χ^2 variables with $h - 1$ and $N - h$ degrees of freedom. Further, the F statistic

$$F = \frac{MSR}{MSE} = \frac{\dfrac{SSR/\sigma_\epsilon^2}{h - 1}}{\dfrac{SSE/\sigma_\epsilon^2}{N - h}} = \frac{\chi^2_{(\nu_1)}/\nu_1}{\chi^2_{(\nu_2)}/\nu_2}$$

is the ratio of two independent chi-square variables, each divided by its degrees of freedom. Recall from Section 2.1 that this is the definition of an F random variable. Thus, if the null hypothesis is true, $F = MSR/MSE$ is distributed as the F distribution with $h - 1$ and $N - h$ degrees of freedom.

Our discussion of the regression model is motivated by an interest in analysis of variance. Accordingly, we have emphasized those topics that are relevant to this interest and omitted a number of traditional topics that are not. In the following discussion, we will further restrict the coverage to qualitative regression models. Recall from Section 5.1 that these models are indistinguishable from experimental design models. In the next section, three schemes are described for coding the independent variables of a qualitative regression model so as to obtain a test of the hypothesis that $\alpha_1 = \alpha_2 = \cdots = \alpha_p = 0$ for a completely randomized design.

5.3 REGRESSION MODEL APPROACH TO ANOVA

A qualitative regression model equation with $h - 1$ independent variables $(X_{i1}, X_{i2}, \ldots, X_{i,h-1})$

$$Y_i = \beta_0 + \beta_1 X_{i1} + \beta_2 X_{i2} + \cdots + \beta_{h-1} X_{i,h-1} + \epsilon_i \quad (i = 1, \ldots, N)$$

can be formulated so that a test of the null hypothesis H_0: $\beta_1 = \beta_2 = \cdots = \beta_{h-1} = 0$ provides an indirect test of the hypothesis H_0: $\alpha_1 = \alpha_2 = \cdots = \alpha_p = 0$ for a type CR-p design. And functions of $\hat{\beta}_0, \hat{\beta}_1, \hat{\beta}_2, \ldots, \hat{\beta}_{h-1}$ can be used to estimate the ANOVA parameters $\mu, \alpha_1, \alpha_2, \ldots, \alpha_p$. To accomplish this it is necessary to establish a correspondence between $p - 1$ of the treatment levels of a type CR-p design and the $X_{i1}, X_{i2}, \ldots, X_{i,h-1}$ independent variables of a regression model.

Consider the hand-steadiness data in Table 4.3-1. The experiment has p equal to four treatment levels. The regression model equation for this experiment can be written as

$$Y_i = \beta_0 X_0 + \beta_1 X_{i1} + \beta_2 X_{i2} + \beta_3 X_{i3} + \epsilon_i \quad (i = 1, \ldots, N)$$

where X_0 is always equal to one and the $h - 1 = 3$ qualitative independent variables X_{i1}, X_{i2}, X_{i3} are coded as follows.

$$X_{i1} = \begin{cases} 1, & \text{if an observation appears in } a_1 \\ 0, & \text{otherwise} \end{cases}$$

$$X_{i2} = \begin{cases} 1, & \text{if an observation appears in } a_2 \\ 0, & \text{otherwise} \end{cases}$$

$$X_{i3} = \begin{cases} 1, & \text{if an observation appears in } a_3 \\ 0, & \text{otherwise} \end{cases}$$

For example, the observation Y_2 in treatment level a_1 is coded

$$Y_2 = \beta_0 + \beta_1 1 + \beta_2 0 + \beta_3 0 + \epsilon_2 = \beta_0 + \beta_1 + \epsilon_2.$$

Observation Y_9 in treatment level a_2 is coded

$$Y_9 = \beta_0 + \beta_1 0 + \beta_2 1 + \beta_3 0 + \epsilon_9 = \beta_0 + \beta_2 + \epsilon_9$$

and Y_{27} in level a_4 is coded

$$Y_{27} = \beta_0 + \beta_1 0 + \beta_2 0 + \beta_3 0 + \epsilon_{27} = \beta_0 + \epsilon_{27}.$$

The **X** matrix for the sleep-deprivation experiment consists of four columns: the first column always contains ones, the second through the fourth columns contain the coded values for X_{i1}, X_{i2}, and X_{i3}. The pattern of ones and zeros for the **X** matrix is as follows.

$$\begin{array}{c} \mathbf{X} \\ {\scriptstyle N \times h} \end{array}$$

$$
\begin{array}{cccc}
\mathbf{x}_0 & \mathbf{x}_1 & \mathbf{x}_2 & \mathbf{x}_3 \\
\end{array}
$$

$$
a_1 \left\{
\begin{bmatrix}
1 & 1 & 0 & 0 \\
1 & 1 & 0 & 0 \\
1 & 1 & 0 & 0 \\
\cdot & \cdot & \cdot & \cdot \\
1 & 1 & 0 & 0 \\
\hline
1 & 0 & 1 & 0 \\
1 & 0 & 1 & 0 \\
1 & 0 & 1 & 0 \\
\cdot & \cdot & \cdot & \cdot \\
1 & 0 & 1 & 0 \\
\hline
1 & 0 & 0 & 1 \\
1 & 0 & 0 & 1 \\
1 & 0 & 0 & 1 \\
\cdot & \cdot & \cdot & \cdot \\
1 & 0 & 0 & 1 \\
\hline
1 & 0 & 0 & 0 \\
1 & 0 & 0 & 0 \\
1 & 0 & 0 & 0 \\
\cdot & \cdot & \cdot & \cdot \\
1 & 0 & 0 & 0 \\
\end{bmatrix}
\right.
$$

(a_1, a_2, a_3, a_4 label the four blocks of rows.)

On reflection it should be apparent that only $h - 1 = p - 1 = 3$ independent variables are needed to code observations for $p = 4$ treatment levels. Consider the sleep-deprivation experiment again. If $X_{i1} = 0$, $X_{i2} = 0$, and $X_{i3} = 0$, we know that the observation is in treatment level a_4. Including a fourth independent variable $X_{i4} = 1$ would provide no information not already provided by the first $p - 1$ independent variables.* The required number of independent variables $(h - 1)$ in the regression model is always equal to the number of degrees of freedom associated with the treatment sum of squares in the type CR-p design. The complete **X** matrix for the hand-steadiness data in Table 4.3-1 is given in Table 5.4-1.

---

*   Furthermore, including a redundant variable in **X** would make $(\mathbf{X'X})$ singular, and the inverse $(\mathbf{X'X})^{-1}$ would not exist (see Appendix D.2-6).

ALTERNATIVE CODING SCHEMES FOR THE X MATRIX

The use of the coding scheme 0, 1 for the indicator variable is often referred to as *dummy coding*. A second scheme called *effect coding* can be used instead. It is as follows.

$$X_{i1} = \begin{cases} 1, \text{ if observation appears in } a_1 \\ -1, \text{ if observation appears in } a_p \\ 0, \text{ otherwise} \end{cases}$$

$$X_{i2} = \begin{cases} 1, \text{ if observation appears in } a_2 \\ -1, \text{ if observation appears in } a_p \\ 0, \text{ otherwise} \end{cases}$$

. . . .

$$X_{i,h-1} = \begin{cases} 1, \text{ if observation appears in } a_{p-1} \\ -1, \text{ if observation appears in } a_p \\ 0, \text{ otherwise} \end{cases}$$

This coding scheme looks a bit more complicated, but the resulting $\mathbf{X}$ matrix is like that for dummy coding except for the last treatment level a_p, where the independent variables $X_{i1}, X_{i2}, \ldots , X_{i,h-1}$ all receive ones instead of zeros. To illustrate the $\mathbf{X}$ matrix for this coding scheme, let p equal 4 and the number of subjects in each treatment level equal 2. The $\mathbf{X}$ matrix using effect coding is as follows.

$$
\begin{array}{c}
\quad\quad\quad \mathbf{x}_0 \quad \mathbf{x}_1 \quad \mathbf{x}_2 \quad \mathbf{x}_3 \\
a_1 \left\{ \begin{array}{c} \\ \\ \end{array} \right. \\
a_2 \left\{ \begin{array}{c} \\ \\ \end{array} \right. \\
a_3 \left\{ \begin{array}{c} \\ \\ \end{array} \right. \\
a_4 \left\{ \begin{array}{c} \\ \\ \end{array} \right.
\end{array}
\begin{bmatrix}
1 & 1 & 0 & 0 \\
1 & 1 & 0 & 0 \\
1 & 0 & 1 & 0 \\
1 & 0 & 1 & 0 \\
1 & 0 & 0 & 1 \\
1 & 0 & 0 & 1 \\
1 & -1 & -1 & -1 \\
1 & -1 & -1 & -1
\end{bmatrix}
$$

A third coding scheme called *orthogonal coding* is sometimes used. The distinguishing feature of orthogonal coding is that the columns of the $\mathbf{X}$ matrix denoted by $\mathbf{x}_0, \mathbf{x}_1, \mathbf{x}_2, \ldots , \mathbf{x}_{h-1}$ are mutually orthogonal, that is, $\Sigma_{i=1}^{N} X_{ij}X_{ij'} = 0$ for all j and $j', j \neq j'$ (see Appendix D.1-8).

In Section 4.5 we performed a trend analysis using orthogonal polynomial coefficients. A regression model could be used to perform the analysis by coding the independent variables with the linear, quadratic, and cubic orthogonal coefficients as follows.

$$
\begin{array}{cccc}
\mathbf{X_0} & \mathbf{X_1} & \mathbf{X_2} & \mathbf{X_3}
\end{array}
$$

$$
\begin{array}{l}
a_1 \left\{ \begin{array}{l} \\ \\ \end{array} \right. \\
a_2 \left\{ \begin{array}{l} \\ \\ \end{array} \right. \\
a_3 \left\{ \begin{array}{l} \\ \\ \end{array} \right. \\
a_4 \left\{ \begin{array}{l} \\ \\ \end{array} \right.
\end{array}
\begin{bmatrix}
1 & -3 & 1 & -1 \\
1 & -3 & 1 & -1 \\
1 & -1 & -1 & 3 \\
1 & -1 & -1 & 3 \\
1 & 1 & -1 & -3 \\
1 & 1 & -1 & -3 \\
1 & 3 & 1 & 1 \\
1 & 3 & 1 & 1
\end{bmatrix}
$$

As we will see later, the resulting estimates of β_1, β_2, and β_3 reflect the contribution to SSR of the linear, quadratic, and cubic trend components, respectively. Alternatively, orthogonal coding can be used to partition SSR into $p - 1$ a priori orthogonal contrasts among means as was done in Section 4.4. Suppose we are interested in the following orthogonal contrasts: $\psi_1 = \mu_1 - \mu_2$, $\psi_2 = \mu_3 - \mu_4$, $\psi_3 = (\mu_1 + \mu_2)/2 - (\mu_3 + \mu_4)/2$. The **X** matrix for these contrasts is as follows. The estimates of β_1, β_2, and β_3 reflect the contribution to SSR of $\hat{\psi}_1$, $\hat{\psi}_2$, and $\hat{\psi}_3$, respectively.

$$
\begin{array}{cccc}
\mathbf{X_0} & \mathbf{X_1} & \mathbf{X_2} & \mathbf{X_3}
\end{array}
$$

$$
\begin{array}{l}
a_1 \left\{ \begin{array}{l} \\ \\ \end{array} \right. \\
a_2 \left\{ \begin{array}{l} \\ \\ \end{array} \right. \\
a_3 \left\{ \begin{array}{l} \\ \\ \end{array} \right. \\
a_4 \left\{ \begin{array}{l} \\ \\ \end{array} \right.
\end{array}
\begin{bmatrix}
1 & 1 & 0 & 1/2 \\
1 & 1 & 0 & 1/2 \\
1 & -1 & 0 & 1/2 \\
1 & -1 & 0 & 1/2 \\
1 & 0 & 1 & -1/2 \\
1 & 0 & 1 & -1/2 \\
1 & 0 & -1 & -1/2 \\
1 & 0 & -1 & -1/2
\end{bmatrix}
$$

In the following sections we will illustrate the computational procedures for the qualitative regression model using the three types of coding.

5.4 ANOVA VIA THE REGRESSION MODEL APPROACH WITH DUMMY CODING

A qualitative regression model with dummy coding can be used to analyze data for a type CR-p design. Our example will use the data in Table 4.3-1 where the traditional sum of squares approach to ANOVA is illustrated. The scheme for coding the independent variables of the regression model is described in Section 5.3. The statistical hypothesis that we will test at $\alpha = .05$ is

$$H_0: \quad \beta_1 = \beta_2 = \beta_3 = 0$$
$$H_1: \quad \beta_j \neq 0 \quad \text{for some } j \; (j = 1, 2, 3).$$

In matrix notation the null hypothesis is

$$H_0: \underset{\substack{\mathbf{C}' \\ (h-1)\times h}}{\begin{bmatrix} 0 & 1 & 0 & 0 \\ 0 & 0 & 1 & 0 \\ 0 & 0 & 0 & 1 \end{bmatrix}} \underset{\substack{\boldsymbol{\beta} \\ h\times 1}}{\begin{bmatrix} \beta_0 \\ \beta_1 \\ \beta_2 \\ \beta_3 \end{bmatrix}} = \underset{\substack{\mathbf{0} \\ (h-1)\times 1}}{\begin{bmatrix} 0 \\ 0 \\ 0 \end{bmatrix}}.$$

The only restriction on $\mathbf{C}'$ is that its rows be linearly independent (see Appendix D.1-5). Procedures for computing the terms required to test the null hypothesis are illustrated in Table 5.4-1; the results are summarized in Table 5.4-2.

TABLE 5.4-1   Computational Procedures for a Type CR-4 Design Using a Regression Model with Dummy Coding

(i)   Data and basic matrices ($N = 32$, $h = 4$):

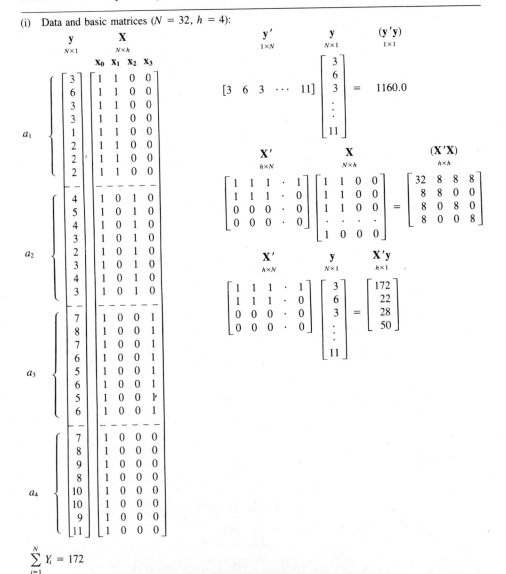

$$\sum_{i=1}^{N} Y_i = 172$$

TABLE 5.4-1 (continued)

$$
\underset{h \times h}{(\mathbf{X'X})^{-1}} \qquad \underset{h \times 1}{(\mathbf{X'y})} \qquad \underset{h \times 1}{\hat{\boldsymbol{\beta}}}
$$

$$
\begin{bmatrix}
\tfrac{1}{8} & -\tfrac{1}{8} & -\tfrac{1}{8} & -\tfrac{1}{8} \\
-\tfrac{1}{8} & \tfrac{2}{8} & \tfrac{1}{8} & \tfrac{1}{8} \\
-\tfrac{1}{8} & \tfrac{1}{8} & \tfrac{2}{8} & \tfrac{1}{8} \\
-\tfrac{1}{8} & \tfrac{1}{8} & \tfrac{1}{8} & \tfrac{2}{8}
\end{bmatrix}
\begin{bmatrix}
172 \\ 22 \\ 28 \\ 50
\end{bmatrix}
=
\begin{bmatrix}
9.00 \\ -6.25 \\ -5.50 \\ -2.75
\end{bmatrix}
$$

$$
\underset{1 \times h}{\hat{\boldsymbol{\beta}}'} \qquad\qquad \underset{h \times 1}{(\mathbf{X'y})} \qquad \underset{1 \times 1}{(\hat{\boldsymbol{\beta}}'\mathbf{X'y})}
$$

$$
\begin{bmatrix} 9.00 & \cdots & -2.75 \end{bmatrix}
\begin{bmatrix}
172 \\ 22 \\ 28 \\ 50
\end{bmatrix}
= \quad 1119.0
$$

(ii) Computation of sum of squares:

$$
SSTO = \mathbf{y'y} - \frac{(\sum Y_i)^2}{N} = 1160.0 - \frac{(172)^2}{32} = 1160.0 - 924.5 = 235.5
$$

$$
SSR = \hat{\boldsymbol{\beta}}'\mathbf{X'y} - \frac{(\sum Y_i)^2}{N} = 1119.0 - \frac{(172)^2}{32} = 1119.0 - 924.5 = 194.5
$$

$$
SSE = \mathbf{y'y} - \hat{\boldsymbol{\beta}}'\mathbf{X'y} = 1160.0 - 1119.0 = 41.0
$$

(iii) Computation of mean squares:

$$
MSR = \frac{SSR}{(h-1)} = \frac{194.5}{4-1} = 64.833
$$

$$
MSE = \frac{SSE}{(N-h)} = \frac{41.0}{32-4} = 1.464
$$

TABLE 5.4-2 Analysis of Variance Table for Sleep-Deprivation Data

Source	SS	df	MS	F	
1 Regression	194.5	$h - 1 = 3$	64.833	[1/2]	44.28*
2 Error	41.0	$N - h = 28$	1.464		
3 Total	235.5	$N - 1 = 31$			

*$p < .01$

Since $F > F_{.05;3,28}$, the null hypothesis is rejected. A comparison of Table 5.4-2 with Table 4.3-2 reveals that $SSR = SSBG = 194.5$, $SSE = SSWG = 41.0$, the F statistic is equal to 44.28 for both analyses, and in both cases the null hypothesis is rejected. One begins to suspect that a test of the null hypothesis for the regression model has some bearing on the tenability of the null hypothesis

$$H_0: \quad \alpha_1 = \alpha_2 = \alpha_3 = \alpha_4 = 0$$

for the experimental design model. We will examine this relationship next.

CORRESPONDENCE BETWEEN PARAMETERS OF THE REGRESSION AND EXPERIMENTAL DESIGN MODELS

The expected values of Y for the experimental design and regression models for the sleep-deprivation experiment are

Experimental design model $\quad E(Y_{ij}) = \mu + \alpha_j = \mu_j$

Regression model $\quad E(Y_i) = \beta_0 + \beta_1 X_{i1} + \beta_2 X_{i2} + \beta_3 X_{i3}$

since $E(\epsilon_{i(j)}) = 0$ (Section 2.3) and $E(\epsilon_i) = 0$ (Section 5.1). Consider the ith observation in treatment level a_4; the respective expectations are

$$E(Y_{i4}) = \mu + \alpha_4$$
$$= \mu_4$$
$$E(Y_i) = \beta_0 + \beta_1 0 + \beta_2 0 + \beta_3 0$$
$$= \beta_0.$$

Equating these two expectations for treatment level a_4, we find the following correspondence between the parameters:

(5.4-1a)
$$E(Y_{i4}) = \beta_0 = \mu + \alpha_4$$
$$= \mu_4.$$

The correspondence for treatment levels a_1 through a_3 is

(5.4-1b)
$$E(Y_{i1}) = \beta_0 + \beta_1 = \mu + \alpha_1 = \mu_1$$
$$\beta_1 = \alpha_1 - \alpha_4 = \mu_1 - \mu_4$$

(5.4-1c)
$$E(Y_{i2}) = \beta_0 + \beta_2 = \mu + \alpha_2 = \mu_2$$
$$\beta_2 = \alpha_2 - \alpha_4 = \mu_2 - \mu_4$$

(5.4-1d)
$$E(Y_{i3}) = \beta_0 + \beta_3 = \mu + \alpha_3 = \mu_3$$
$$\beta_3 = \alpha_3 - \alpha_4 = \mu_3 - \mu_4.$$

For the general case, β_0 is equal to μ_p; β_j is equal to $\mu_j - \mu_p$ for $j = 1, \ldots, h - 1$. If the null hypothesis, $\beta_1 = \beta_2 = \beta_3 = 0$, is true, it follows that $\beta_0 = \mu + \alpha_4 = \mu$ and that $\alpha_1 = \alpha_2 = \alpha_3 = \alpha_4 = 0$. This is why the statistic $F = MSR/MSE$ that is used to test $\beta_1 = \beta_2 = \beta_3 = 0$ for the regression model also provides a test of $\alpha_1 = \alpha_2 = \alpha_3 = \alpha_4 = 0$, which is the hypothesis tested by $F = MSBG/MSWG$ in a completely randomized design.

We can obtain estimates of population means, μ_j's, by substituting the values of $\hat{\boldsymbol{\beta}}$ from Table 5.4-1 into equations (5.4-1a, b, c, and d).

$$\bar{Y}_{.1} = \hat{\beta}_0 + \hat{\beta}_1 = 9.00 + (-6.25) = 2.75$$
$$\bar{Y}_{.2} = \hat{\beta}_0 + \hat{\beta}_2 = 9.00 + (-5.50) = 3.50$$
$$\bar{Y}_{.3} = \hat{\beta}_0 + \hat{\beta}_3 = 9.00 + (-2.75) = 6.25$$
$$\bar{Y}_{.4} = \hat{\beta}_0 \qquad\quad = 9.00$$

These means are identical to those given in Table 4.2-1 for the hand-steadiness data. It is a simple matter to construct a $p \times h$ coefficient matrix $\mathbf{W}$ such that the product of $\mathbf{W}$ and $\hat{\boldsymbol{\beta}}$ is equal to the vector of treatment level means $\bar{\mathbf{y}}$ for a type CR-p design.

(5.4-2)
$$\underset{p\times h}{\mathbf{W}} \ \underset{h\times 1}{\hat{\boldsymbol{\beta}}} = \underset{p\times 1}{\bar{\mathbf{y}}}$$

where $\bar{\mathbf{y}}' = [\bar{Y}_{.1}\ \bar{Y}_{.2}\ \bar{Y}_{.3} \cdots \bar{Y}_{.p}]$. Based on the pattern of relationships in (5.4-1), the $\mathbf{W}$ matrix for the general case has the following form.

(5.4-3)
$$\underset{p\times h}{\mathbf{W}} =
\begin{bmatrix}
1 & 1 & 0 & 0 & 0 & \cdot & 0 \\
1 & 0 & 1 & 0 & 0 & \cdot & 0 \\
1 & 0 & 0 & 1 & 0 & \cdot & 0 \\
1 & 0 & 0 & 0 & 1 & \cdot & 0 \\
\cdot & \cdot & \cdot & \cdot & \cdot & \cdot & \cdot \\
1 & 0 & 0 & 0 & 0 & \cdot & 1 \\
1 & 0 & 0 & 0 & 0 & \cdot & 0
\end{bmatrix}
\begin{matrix}
\text{1st row} \\ \text{2nd row} \\ \text{3rd row} \\ \text{4th row} \\ \cdot \\ p - \text{1th row} \\ p\text{th row}
\end{matrix}$$

For the data in Table 5.4-1, the vector of treatment level means is obtained from

(5.4-4)
$$\underset{p\times h}{\mathbf{W}} \qquad\qquad \underset{h\times 1}{\hat{\boldsymbol{\beta}}} \qquad \underset{p\times 1}{\bar{\mathbf{y}}}$$
$$\begin{bmatrix}
1 & 1 & 0 & 0 \\
1 & 0 & 1 & 0 \\
1 & 0 & 0 & 1 \\
1 & 0 & 0 & 0
\end{bmatrix}
\begin{bmatrix}
9.00 \\ -6.25 \\ -5.50 \\ -2.75
\end{bmatrix} =
\begin{bmatrix}
2.75 \\ 3.50 \\ 6.25 \\ 9.00
\end{bmatrix} .$$

Note that the pattern of ones and zeros in $\mathbf{W}$ follows that for the $\mathbf{X}$ matrix in Table 5.4-1. As we will see, when $\mathbf{X}$ is a matrix of indicator variables, the solution to $\hat{\boldsymbol{\beta}} = (\mathbf{X}'\mathbf{X})^{-1}\mathbf{X}'\mathbf{y}$ depends on the particular coding scheme used—dummy coding, effect coding, and so on. However, by choosing the rows of $\mathbf{W}$ from the appropriate structural matrix, the product $\mathbf{W}\hat{\boldsymbol{\beta}}$ has the same value regardless of the coding scheme.

5.5   ALTERNATIVE CONCEPTION OF THE TEST OF $\beta_1 = \beta_2 = \cdots = \beta_{h-1} = 0$

We will digress briefly and examine an alternative way of thinking about a test of, say, H_0: $\beta_1 = \beta_2 = \beta_3 = 0$ versus H_1: $\beta_j \neq 0$ for some $j = 1, 2, 3$. The model we have used up until now to describe Y_i is called a *full* or *unrestricted model* and is

$$\text{Full model} \quad Y_i = \beta_0 + \beta_1 X_{i1} + \beta_2 X_{i2} + \beta_3 X_{i3} + \epsilon_i.$$

For the time being we will denote the error sum of squares for this model by $SSE(F)$. We know from Section 5.2 that this error sum of squares is given by

$$SSE(F) = \sum_{i=1}^{N} (Y_i - \hat{Y}_i)^2 \quad \text{where } \hat{Y}_i = \hat{\beta}_0 + \hat{\beta}_1 X_{i1} + \hat{\beta}_2 X_{i2} + \hat{\beta}_3 X_{i3}.$$

According to the null hypothesis, $\beta_1 = \beta_2 = \beta_3 = 0$; this leads to a second model for describing Y_i called the *reduced model*

$$\text{Reduced model} \quad Y_i = \beta_0 + \epsilon_i.$$

The error sum of squares for this model, $SSE(R)$, is given by

$$SSE(R) = \sum_{i=1}^{N} (Y_i - \hat{Y}_i)^2, \quad \text{where } \hat{Y}_i = \hat{\beta}_0.$$

Our task is to choose between these two models. Suppose the null hypothesis is false. This means that one or more of the terms $\beta_1 X_{i1}, \beta_2 X_{i2}, \beta_3 X_{i3}$ is required in the model in order to minimize the sum of the squared discrepancies $\Sigma (Y_i - \hat{Y}_i)^2$. For this case, the full model provides a better fit to the Y_i's than the reduced model, and $SSE(F)$ is smaller than $SSE(R)$. Thus, when the null hypothesis is false we expect the difference $SSE(R) - SSE(F)$ to be positive and large. On the other hand, suppose that the difference $SSE(R) - SSE(F)$ is very small. This tells us that the sum of the squared discrepancies $\Sigma (Y_i - \hat{Y}_i)^2$ is about the same for the full model as it is for the reduced model. Furthermore, the terms $\beta_1 X_{i1}, \beta_2 X_{i2}$, and $\beta_3 X_{i3}$ are not of very much help in reducing the squared discrepancies and they can be left out of the model. Therefore a very small difference, $SSE(R) - SSE(F)$, provides evidence in favor of H_0: $\beta_1 = \beta_2 = \beta_3 = 0$.

It should be apparent from the foregoing that the magnitude of the difference $SSE(R) - SSE(F)$ is relevant to a test of the null hypothesis. The actual test is carried out using a function of $SSE(R) - SSE(F)$, namely

$$F = \frac{\dfrac{SSE(R) - SSE(F)}{df(R) - df(F)}}{\dfrac{SSE(F)}{df(F)}}$$

where $df(R)$ and $df(F)$ are the degrees of freedom associated with the reduced and full models, respectively.

Let us consider $SSE(F)$ and $SSE(R)$ more closely. The error sum of squares for the full model is simply SSE. For the reduced model, $\hat{Y}_i = \hat{\beta}_0 = \overline{Y}.$ and

$$SSE(R) = \sum_{i=1}^{N} (Y_i - \hat{Y}_i)^2 = \sum_{i=1}^{N} (Y_i - \overline{Y}.)^2.$$

But note that the formula on the right is identical to that for $SSTO$ given in (5.2-6). Thus, $SSE(R) - SSE(F) = SSTO - SSE$. Furthermore, according to (5.2-6), $SSTO - SSE = SSR$, which is the sum of squares due to regression. It follows that the F test statistic can be written as

$$F = \frac{\dfrac{SSE(R) - SSE(F)}{(N-1) - (N-h)}}{\dfrac{SSE(F)}{N-h}} = \frac{\dfrac{SSTO - SSE}{(N-1) - (N-h)}}{\dfrac{SSE}{N-h}} = \frac{\dfrac{SSR}{h-1}}{\dfrac{SSE}{N-h}} = \frac{MSR}{MSE}.*$$

If $F \geq F_{\alpha;h-1,N-h}$, the null hypothesis is rejected and we conclude that the full model provides a better fit to the Y_i's than the reduced model. This suggests that one or more of the parameters β_1, β_2, and β_3 are not equal to zero. If $F < F_{\alpha;h-1,N-h}$, the null hypothesis is not rejected.

ALTERNATIVE NOTATION FOR ERROR AND REGRESSION SUMS OF SQUARES

Sometimes it is useful to consider models that are somewhere between $Y_i = \beta_0 + \epsilon_i$ and $Y_i = \beta_0 + \beta_1 X_{i1} + \cdots + \beta_{h-1} X_{i,h-1} + \epsilon_i$, that is, models for which some but not all $\beta_1, \ldots, \beta_{h-1}$ may be equal to zero. This is particularly true for regression models in which the independent variables are treated quantitatively. A comparison, for example, of the error sum of squares for model equation

(5.5-1)
$$Y_i = \beta_0 + \beta_1 X_{i1} + \beta_2 X_{i2} + \epsilon_i$$

---

*   Some writers express this ratio in terms of the *coefficient of multiple determination*, R^2, which represents that proportion of the total variability among the Y scores that is accounted for by the independent variables $X_1, X_2, \ldots, X_{h-1}$. One formula for computing R^2 is

$$R^2 = \frac{SSTO - SSE}{SSTO} = \frac{SSR}{SSTO}.$$

It follows that $SSR = SSTO(R^2)$ and $SSE = SSTO(1 - R^2)$. Thus,

$$F = \frac{\dfrac{SSR}{h-1}}{\dfrac{SSE}{N-h}} = \frac{\dfrac{SSTO(R^2)}{h-1}}{\dfrac{SSTO(1-R^2)}{N-h}} = \frac{\dfrac{R^2}{h-1}}{\dfrac{1-R^2}{N-h}}.$$

versus that for

(5.5-2) $$Y_i = \beta_0 + \beta_1 X_{i1} + \beta_2 X_{i2} + \beta_3 X_{i3} + \epsilon_i$$

would indicate the contribution of X_3 to minimizing the sum of the squared discrepancies $(Y_i - \hat{Y}_i)^2$ over and above that due to X_1 and X_2.

We could denote the error sum of squares for the full model (5.5-2) by $SSE(F)$ and that for the reduced model (5.5-1) by $SSE(R)$. However, it would be useful to have a more explicit notation. One scheme for denoting error sums of squares and mean squares lists the independent variables that are included in the model in parentheses following SSE and MSE. For example, the error sums of squares for the full and reduced models are denoted by $SSE(X_1\ X_2\ X_3)$ and $SSE(X_1\ X_2)$, respectively. The regression sum of squares reflecting the contribution of X_3 over and above that due to including X_1 and X_2 in the model is denoted by

$$SSR(X_3 \mid X_1\ X_2) = SSE(X_1\ X_2) - SSE(X_1\ X_2\ X_3).$$

To determine if X_3 can be dropped from the full model, we would use the test statistic

$$F = \frac{\dfrac{SSE(R) - SSE(F)}{df(R) - df(F)}}{\dfrac{SSE(F)}{df(F)}} = \frac{\dfrac{SSE(X_1\ X_2) - SSE(X_1\ X_2\ X_3)}{(N-3)-(N-4)}}{\dfrac{SSE(X_1\ X_2\ X_3)}{N-4}} = \frac{MSR(X_3 \mid X_1\ X_2)}{MSE(X_1\ X_2\ X_3)}$$

with $4 - 3 = 1$ and $N - 4$ degrees of freedom.

The computation of $SSE(X_1\ X_2\ X_3)$ is identical to that for SSE as described previously; $SSE(X_1\ X_2)$ is computed by eliminating $\mathbf{x}_3$ in the $\mathbf{X}$ matrix and performing the computations using the two remaining independent variables $(\mathbf{x}_1, \mathbf{x}_2)$ and $\mathbf{x}_0$. For purposes of illustration we will perform this analysis on the data in Table 5.4-1. The error sum of squares for the reduced model is given by $SSE(X_1\ X_2) = \mathbf{y'y} - \hat{\boldsymbol{\beta}}'\mathbf{X'y} = 1160.00 - 1088.75 = 71.25$, where

$$\underset{3\times 3}{(\mathbf{X'X})^{-1}} \quad \underset{3\times 1}{(\mathbf{X'y})} \quad \underset{3\times 1}{\hat{\boldsymbol{\beta}}}$$

$$\begin{bmatrix} 1/16 & -1/16 & -1/16 \\ -1/16 & 3/16 & 1/16 \\ -1/16 & 1/16 & 3/16 \end{bmatrix} \begin{bmatrix} 172 \\ 22 \\ 28 \end{bmatrix} = \begin{bmatrix} 7.625 \\ -4.875 \\ -4.125 \end{bmatrix}$$

and

$$\underset{1\times 3}{\hat{\boldsymbol{\beta}}'} \qquad \underset{3\times 1}{(\mathbf{X'y})} \quad \underset{1\times 1}{(\boldsymbol{\beta}'\mathbf{X'y})}$$

$$[7.625 \ -4.875 \ -4.125] \begin{bmatrix} 172 \\ 22 \\ 28 \end{bmatrix} = 1088.75.$$

The value of $SSE(X_1\ X_2\ X_3)$, which is equivalent to SSE, is obtained from Table

5.4-1. The test statistic is

$$F = \frac{\dfrac{SSE(X_1\ X_2) - SSE(X_1\ X_2\ X_3)}{(32-3)-(32-4)}}{\dfrac{SSE(X_1\ X_2\ X_3)}{(32-4)}} = \frac{\dfrac{71.25-41.00}{1}}{\dfrac{41.00}{28}} = \frac{30.25}{1.464} = 20.66$$

which exceeds $F_{.05;1,28} = 4.20$. On the basis of the correspondence between the parameters of the regression and ANOVA models shown in (5.4-1), we can conclude that the contribution to the regression sum of squares of the difference $\alpha_3 - \alpha_4$ over and above that of $\alpha_1 - \alpha_4$ and $\alpha_2 - \alpha_4$ is not due to chance. Such hypotheses are ordinarily not of interest in analysis of variance.

A second scheme for denoting sums of squares is called *reduction notation*; it is ordinarily used with experimental design models.* In reduction notation, following Searle (1971a), the sum of squares due to regression SSR is denoted by $R(\quad)$, where the parentheses contain the parameters that have been fitted in the model. For example, the reductions in sums of squares for models (5.5-1) and (5.5-2) are denoted by $R(\beta_0\ \beta_1\ \beta_2)$ and $R(\beta_0\ \beta_1\ \beta_2\ \beta_3)$, respectively. The additional reduction in sum of squares due to fitting model (5.5-2) over and above that due to fitting model (5.5-1) is denoted by

$$R(\beta_3 \mid \beta_0\ \beta_1\ \beta_2) = R(\beta_0\ \beta_1\ \beta_2\ \beta_3) - R(\beta_0\ \beta_1\ \beta_2).$$

For purposes of comparison, this same reduction in sum of squares using the first notation is written as

$$SSR(X_3 \mid X_1\ X_2) = SSE(X_1\ X_2) - SSE(X_1\ X_2\ X_3).$$

5.6 TESTING DIFFERENCES AMONG MEANS USING THE REGRESSION MODEL APPROACH

We turn now to procedures for estimating population means and testing hypotheses about differences among means. We saw in Section 5.4 that with dummy coding

$$\beta_0 + \beta_1 = \mu_1$$
$$\beta_0 + \beta_2 = \mu_2$$
$$\beta_0 + \beta_3 = \mu_3$$
$$\beta_0 \qquad = \mu_4.$$

---

*   The use of this scheme and the one just described can be misleading. This point is explored by Speed and Hocking (1976) and by Speed, Hocking, and Hackney (1978).

Furthermore, these equations can be written as $\mathbf{W}\,\boldsymbol{\beta} = \boldsymbol{\mu}$, where

$$\underset{p \times h}{\mathbf{W}} = \begin{bmatrix} 1 & 1 & 0 & 0 \\ 1 & 0 & 1 & 0 \\ 1 & 0 & 0 & 1 \\ 1 & 0 & 0 & 0 \end{bmatrix}, \qquad \underset{h \times 1}{\boldsymbol{\beta}} = \begin{bmatrix} \beta_0 \\ \beta_1 \\ \beta_2 \\ \beta_3 \end{bmatrix}, \text{ and } \qquad \underset{p \times 1}{\boldsymbol{\mu}} = \begin{bmatrix} \mu_1 \\ \mu_2 \\ \mu_3 \\ \mu_4 \end{bmatrix}.$$

An unbiased estimator for $\boldsymbol{\mu}$ is given by

$$\mathbf{W}\,\hat{\boldsymbol{\beta}} = \bar{\mathbf{y}}$$

where $\hat{\boldsymbol{\beta}}$ is the least squares solution for $\boldsymbol{\beta}$ and $\bar{\mathbf{y}}' = [\bar{Y}_1\ \bar{Y}_2\ \bar{Y}_3\ \bar{Y}_4]$.

In Section 3.1 we saw that the ith contrast among population means is a linear combination

(5.6-1) $$\psi_i = c_1 \mu_1 + c_2 \mu_2 + \cdots + c_p \mu_p$$

such that (1) at least one coefficient is not equal to zero and (2) the coefficients sum to zero. Using vector notation, (5.6-1) can be written as

(5.6-2) $$\psi_i = \underset{1 \times p}{[c_1\ c_2\ \cdots\ c_p]} \underset{p \times 1}{\begin{bmatrix} \mu_1 \\ \mu_2 \\ \vdots \\ \mu_p \end{bmatrix}}.$$

An unbiased estimator for the ith population contrast is given by

(5.6-3) $$\hat{\psi}_i = \underset{1 \times p}{\mathbf{c}'}\ \underset{p \times 1}{\bar{\mathbf{y}}}.$$

We now have all the ingredients for testing hypotheses of the form

$$H_0:\quad c_1 \mu_1 + c_2 \mu_2 + \cdots + c_p \mu_p = 0$$

except a test statistic. The test statistics presented in Chapter 3 can be expressed either as $\hat{\psi}_i / \hat{\sigma}_{\psi_i}$, the ratio of a contrast to the standard error of the contrast, or as $\hat{\psi}_i / \hat{\sigma}_{\bar{Y}}$, the ratio of a contrast to the standard error of a mean.

Dunn's statistic Dunn-Šidák statistic Dunnett's statistic t statistic	$\dfrac{\hat{\psi}_i}{\hat{\sigma}_{\psi_i}}$
Duncan's statistic Newman-Keuls statistic Tukey's statistic	$\dfrac{\hat{\psi}_i}{\hat{\sigma}_{\bar{Y}}}$
F statistic Scheffé's statistic	$\dfrac{\hat{\psi}_i^2}{\hat{\sigma}_{\psi_i}^2}$

To test a null hypothesis regarding the ith population contrast, we need to compute $\hat{\psi}_i$ and either $\hat{\sigma}_{\psi_i}$ or $\hat{\sigma}_{\bar{Y}}$, depending on which test statistic is used. The computation of $\hat{\sigma}_{\psi_i}$ and $\hat{\sigma}_{\bar{Y}}$ is described next.

An unbiased estimator for $\sigma_{\psi_i}^2$, the population variance error of the ith contrast, is given by

$$\hat{\sigma}_{\psi_i}^2 = MSE\left[\frac{c_1^2}{n_1} + \frac{c_2^2}{n_2} + \cdots + \frac{c_p^2}{n_p}\right] = MSE\left\{ \underset{1\times p}{\mathbf{c}'} \left[\underset{p\times h}{\mathbf{W}} \underset{h\times h}{(\mathbf{X}'\mathbf{X})^{-1}} \underset{h\times p}{\mathbf{W}'} \right] \underset{p\times 1}{\mathbf{c}} \right\}$$

where $\mathbf{c}'$ is a vector of coefficients that defines the ith contrast, $\mathbf{W}$ is given by (5.4-3) and $\mathbf{X}$ is a structural matrix.* An unbiased estimator for $\sigma_{\bar{Y}}^2$ is given by

$$\hat{\sigma}_{\bar{Y}}^2 = \frac{MSE}{n}.$$

The data in Table 5.4-1 will be used to illustrate the computation of $\hat{\psi}_i/\hat{\sigma}_{\psi_i}$ and $\hat{\psi}_i/\hat{\sigma}_{\bar{Y}}$ for the following null hypotheses.

H_0: $\mu_1 - \mu_2 = 0$ where $\mathbf{c}_1' = [1 \quad -1 \quad 0 \quad 0]$

H_0: $\mu_3 - \mu_4 = 0$ where $\mathbf{c}_2' = [0 \quad 0 \quad 1 \quad -1]$

H_0: $(\mu_1 + \mu_2)/2 - (\mu_3 + \mu_4)/2 = 0$ where $\mathbf{c}_3' = [\frac{1}{2} \quad \frac{1}{2} \quad -\frac{1}{2} \quad -\frac{1}{2}]$

These hypotheses were chosen because they were used in Section 4.4 to illustrate the computation of the t and q test statistics. The computational procedures are illustrated in Table 5.6-1. The various test statistics are given in part (iii) of the table. A comparison of the values of $\hat{\psi}_i/\hat{\sigma}_{\psi_i}$ and $\hat{\psi}_i/\hat{\sigma}_{\bar{Y}}$ with those for the t and q statistics in Section 4.4 reveals that they are identical.

In Sections 5.4 and 5.6 we have analyzed the data for a type CR-p design using the qualitative regression model with dummy coding. As we have seen, the results are the same as those obtained using the conventional ANOVA approach in Chapter 4. Although the results are the same, the amount of computational labor is much greater for the regression approach. This difference is unimportant if one has access to a digital computer and efficient software packages for manipulating matrices. In the remainder of the chapter we illustrate two alternative coding schemes for the regression model and two experimental design models.

---

* Hypotheses about $\beta_0, \beta_1, \ldots, \beta_{h-1}$ can be tested by using linear functions of $\hat{\beta}_0, \hat{\beta}_1, \ldots, \hat{\beta}_{h-1}$. Let $\hat{L}_i$ denote a linear function of the $\hat{\beta}_j$'s. The variance error of the linear function, $\hat{\sigma}_{L_i}^2$, is given by $MSE[\mathbf{c}_i'(\mathbf{X}'\mathbf{X})^{-1}\mathbf{c}_i]$, where $\mathbf{c}_i'$ is a vector of coefficients that defines the function. The statistic $\hat{L}_i/\hat{\sigma}_{L_i}$ can be evaluated in the same way as $\hat{\psi}_i/\hat{\sigma}_{\psi_i}$; it differs from $\hat{\psi}_i/\hat{\sigma}_{\psi_i}$ in that the coefficients of the vector $\mathbf{c}_i'$ need not sum to zero. For example, to test H_0: $\beta_1 = 0$ for the data in Table 5.4-1, $\mathbf{c}' = [0 \ 1 \ 0 \ 0]$, $\hat{L} = \mathbf{c}'\hat{\boldsymbol{\beta}} = -6.25$, $\hat{\sigma}_L = \sqrt{1.464(2/8)} = 0.605$, and $\hat{L}/\hat{\sigma}_L = -10.33$. If the test statistic is t, the critical value at $\alpha = .05$ for $N - h$ degrees of freedom is 2.048. Since $|10.33| > 2.048$, the null hypothesis is rejected.

TABLE 5.6-1 Computational Procedures for Testing Differences Among Means (Dummy coding)

(i) Computation of means, $\bar{\mathbf{y}}$, and contrasts, $\hat{\psi}_i$ ($h = 4$, $n = 8$, $p = 4$):*

$$
\underset{p \times h}{\mathbf{W}}
\begin{bmatrix}
1 & 1 & 0 & 0 \\
1 & 0 & 1 & 0 \\
1 & 0 & 0 & 1 \\
1 & 0 & 0 & 0
\end{bmatrix}
\qquad
\underset{h \times 1}{\hat{\boldsymbol{\beta}}}
\begin{bmatrix}
9.00 \\
-6.25 \\
-5.50 \\
-2.75
\end{bmatrix}
=
\underset{p \times 1}{\bar{\mathbf{y}}}
\begin{bmatrix}
2.75 \\
3.50 \\
6.25 \\
9.00
\end{bmatrix}
$$

$$
\underset{1 \times p}{\mathbf{c}_1'} \qquad \underset{p \times 1}{\bar{\mathbf{y}}} \qquad \hat{\psi}_1
$$

$$
\begin{bmatrix} 1 & -1 & 0 & 0 \end{bmatrix}
\begin{bmatrix}
2.75 \\
3.50 \\
6.25 \\
9.00
\end{bmatrix}
= -0.75
$$

$$
\underset{1 \times p}{\mathbf{c}_2'} \qquad \underset{p \times 1}{\bar{\mathbf{y}}} \qquad \hat{\psi}_2
$$

$$
\begin{bmatrix} 0 & 0 & 1 & -1 \end{bmatrix}
\begin{bmatrix}
2.75 \\
3.50 \\
6.25 \\
9.00
\end{bmatrix}
= -2.75
$$

$$
\underset{1 \times p}{\mathbf{c}_3'} \qquad \underset{p \times 1}{\bar{\mathbf{y}}} \qquad \hat{\psi}_3
$$

$$
\begin{bmatrix} \tfrac{1}{2} & \tfrac{1}{2} & -\tfrac{1}{2} & -\tfrac{1}{2} \end{bmatrix}
\begin{bmatrix}
2.75 \\
3.50 \\
6.25 \\
9.00
\end{bmatrix}
= -4.50
$$

(ii) Computation of $\hat{\sigma}_{\bar{Y}}^2$ and $\hat{\sigma}_{\hat{\psi}_i}^2$:**

$$
\hat{\sigma}_{\bar{Y}}^2 = \frac{MSE}{n} = \frac{1.464}{8} = 0.183
$$

$$
\hat{\sigma}_{\hat{\psi}_1}^2 = 1.464
\left\{
\underset{1 \times p}{\mathbf{c}'}
\begin{array}{c} \overbrace{}^{MSE} \end{array}
\right.
\begin{bmatrix} 1 & -1 & 0 & 0 \end{bmatrix}
\underset{p \times h}{\mathbf{W}}
\begin{bmatrix}
1 & 1 & 0 & 0 \\
1 & 0 & 1 & 0 \\
1 & 0 & 0 & 1 \\
1 & 0 & 0 & 0
\end{bmatrix}
\underset{h \times h}{(\mathbf{X}'\mathbf{X})^{-1}}
\begin{bmatrix}
\tfrac{1}{8} & -\tfrac{1}{8} & -\tfrac{1}{8} & -\tfrac{1}{8} \\
-\tfrac{1}{8} & \tfrac{2}{8} & \tfrac{1}{8} & \tfrac{1}{8} \\
-\tfrac{1}{8} & \tfrac{1}{8} & \tfrac{2}{8} & \tfrac{1}{8} \\
-\tfrac{1}{8} & \tfrac{1}{8} & \tfrac{1}{8} & \tfrac{2}{8}
\end{bmatrix}
\underset{h \times p}{\mathbf{W}'}
\begin{bmatrix}
1 & 1 & 1 & 1 \\
1 & 0 & 0 & 0 \\
0 & 1 & 0 & 0 \\
0 & 0 & 1 & 0
\end{bmatrix}
\underset{p \times 1}{\mathbf{c}}
\begin{bmatrix}
1 \\
-1 \\
0 \\
0
\end{bmatrix}
\left. \right\}
$$

---

* Matrix $\mathbf{W}$ is obtained from (5.4-4), $\hat{\boldsymbol{\beta}}$ is obtained from Table 5.4-1, and the $\mathbf{c}_i'$'s are the coefficient vectors associated with the hypotheses $\mu_1 - \mu_2 = 0$, $\mu_3 - \mu_4 = 0$, and $(\mu_1 + \mu_2)/2 - (\mu_3 + \mu_4)/2 = 0$.

** Matrix $(\mathbf{X}'\mathbf{X})^{-1}$ and MSE are obtained from Table 5.4-1.

TABLE 5.6-1 (continued)

$$\hat{\sigma}^2_{\hat{\psi}_1} = 1.464 \left\{ \begin{bmatrix} 1 & -1 & 0 & 0 \end{bmatrix} \begin{bmatrix} 1/8 & 0 & 0 & 0 \\ 0 & 1/8 & 0 & 0 \\ 0 & 0 & 1/8 & 0 \\ 0 & 0 & 0 & 1/8 \end{bmatrix} \begin{bmatrix} 1 \\ -1 \\ 0 \\ 0 \end{bmatrix} \right\} = 1.464(2/8) = 0.366$$

$$\hat{\sigma}^2_{\hat{\psi}_2} = 1.464 \left\{ \begin{bmatrix} 0 & 0 & 1 & -1 \end{bmatrix} \begin{bmatrix} 1/8 & 0 & 0 & 0 \\ 0 & 1/8 & 0 & 0 \\ 0 & 0 & 1/8 & 0 \\ 0 & 0 & 0 & 1/8 \end{bmatrix} \begin{bmatrix} 0 \\ 0 \\ 1 \\ -1 \end{bmatrix} \right\} = 1.464(2/8) = 0.366$$

$$\hat{\sigma}^2_{\hat{\psi}_3} = 1.464 \left\{ \begin{bmatrix} 1/2 & 1/2 & -1/2 & -1/2 \end{bmatrix} \begin{bmatrix} 1/8 & 0 & 0 & 0 \\ 0 & 1/8 & 0 & 0 \\ 0 & 0 & 1/8 & 0 \\ 0 & 0 & 0 & 1/8 \end{bmatrix} \begin{bmatrix} 1/2 \\ 1/2 \\ -1/2 \\ -1/2 \end{bmatrix} \right\} = 1.464(1/8) = 0.183$$

(iii) Computation of test statistics:

t statistic	q statistic
$t = \dfrac{\hat{\psi}_1}{\hat{\sigma}_{\psi_1}} = \dfrac{-0.75}{\sqrt{0.366}} = \dfrac{-0.75}{0.605} = -1.24$	$q = \dfrac{\hat{\psi}_1}{\hat{\sigma}_{\overline{Y}}} = \dfrac{-0.75}{\sqrt{0.183}} = \dfrac{-0.75}{0.428} = -1.75$
$t = \dfrac{\hat{\psi}_2}{\hat{\sigma}_{\psi_2}} = \dfrac{-2.75}{\sqrt{0.366}} = \dfrac{-2.75}{0.605} = -4.55$	$q = \dfrac{\hat{\psi}_2}{\hat{\sigma}_{\overline{Y}}} = \dfrac{-2.75}{\sqrt{0.183}} = \dfrac{-2.75}{0.428} = -6.43$
$t = \dfrac{\hat{\psi}_3}{\hat{\sigma}_{\psi_3}} = \dfrac{-4.50}{\sqrt{0.183}} = \dfrac{-4.50}{0.428} = -10.51$	(not appropriate for a contrast involving more than two means)

5.7 ANOVA VIA THE REGRESSION MODEL APPROACH WITH EFFECT AND ORTHOGONAL CODING

Effect and orthogonal coding were described in Section 5.3. We will now illustrate the use of these coding schemes in testing hypotheses about the parameters of regression and experimental design models.

CORRESPONDENCE BETWEEN PARAMETERS OF THE REGRESSION MODEL WITH EFFECT CODING AND THE PARAMETERS OF THE EXPERIMENTAL DESIGN MODEL

The correspondence between the parameters of the regression model with effect coding and the parameters of the experimental design model for a type CR-4 design is as follows. Consider the ith observation in treatment level a_1.

$$\begin{aligned}
\text{Experimental} \quad & E(Y_{i1}) = \mu + \alpha_1 \\
\text{design model} \quad & \\
& = \mu_1 \\
\text{Regression} \quad & E(Y_i) = \beta_0 + \beta_1 1 + \beta_2 0 + \beta_3 0 \\
\text{model} \quad & \\
& = \beta_0 + \beta_1
\end{aligned}$$

Equating the two expectations for treatment level a_1, we find the following correspondence between the parameters:

(5.7-1a)
$$\begin{aligned}
E(Y_{i1}) = \beta_0 + \beta_1 &= \mu + \alpha_1 \\
&= \mu_1.
\end{aligned}$$

The correspondence for treatment levels a_2 through a_4 is

(5.7-1b)
$$\begin{aligned}
E(Y_{i2}) = \beta_0 + \beta_2 &= \mu + \alpha_2 \\
&= \mu_2
\end{aligned}$$

(5.7-1c)
$$\begin{aligned}
E(Y_{i3}) = \beta_0 + \beta_3 &= \mu + \alpha_3 \\
&= \mu_3
\end{aligned}$$

(5.7-1d)
$$\begin{aligned}
E(Y_{i4}) = \beta_0 - \beta_1 - \beta_2 - \beta_3 &= \mu + \alpha_4 \\
&= \mu_4.
\end{aligned}$$

The interpretation of β_0 is not obvious from (5.7-1a–d); by expressing β_0 as a function of the parameters $\mu_1, \ldots, \mu_4$ and $\beta_1, \ldots, \beta_3$ we can show that β_0 is equal to μ, the grand mean.

$$\begin{aligned}
\beta_0 &= \mu_1 - \beta_1 \\
\beta_0 &= \mu_2 - \beta_2 \\
\beta_0 &= \mu_3 - \beta_3 \\
\underline{\beta_0 = \mu_4 + \beta_1 + \beta_2 + \beta_3} \\
4\,\beta_0 &= \mu_1 + \mu_2 + \mu_3 + \mu_4
\end{aligned}$$

It follows that

(5.7-2)
$$\begin{aligned}
\beta_0 &= (\mu_1 + \mu_2 + \mu_3 + \mu_4)/4 \\
\beta_0 &= \mu.
\end{aligned}$$

Using (5.7-1) and (5.7-2) we see that

(5.7-3a)
$$\beta_1 = \mu_1 - \mu = \alpha_1$$

(5.7-3b)
$$\beta_2 = \mu_2 - \mu = \alpha_2$$

(5.7-3c)
$$\beta_3 = \mu_3 - \mu = \alpha_3$$

(5.7-3d)
$$-(\beta_1 + \beta_2 + \beta_3) = \mu_4 - \mu = \alpha_4.$$

From (5.7-3) we can see where effect coding got its name—the regression parameters are equal to the treatment effects in the experimental design model.

In Section 5.4 we constructed a $p \times h$ coefficient matrix $\mathbf{W}$ so that the product $\mathbf{W}\hat{\boldsymbol{\beta}}$ was equal to a vector of treatment level means for a type CR-p design. When effect coding is used, the $\mathbf{W}$ matrix has the following form.

(5.7-4)
$$\mathbf{W}_{p \times h} = \begin{bmatrix} 1 & 1 & 0 & 0 & 0 & \cdot & 0 \\ 1 & 0 & 1 & 0 & 0 & \cdot & 0 \\ 1 & 0 & 0 & 1 & 0 & \cdot & 0 \\ 1 & 0 & 0 & 0 & 1 & \cdot & 0 \\ \cdot & \cdot & \cdot & \cdot & \cdot & \cdot & \cdot \\ 1 & 0 & 0 & 0 & 0 & \cdot & 1 \\ 1 & -1 & -1 & -1 & -1 & \cdot & -1 \end{bmatrix} \begin{matrix} \text{1st row} \\ \text{2nd row} \\ \text{3rd row} \\ \text{4th row} \\ \\ p - \text{1th row} \\ p\text{th row} \end{matrix}$$

The pattern of ones and zeros in $\mathbf{W}$ follows that for the $\mathbf{X}$ matrix in Table 5.7-1.

COMPUTATIONAL PROCEDURES USING THE REGRESSION MODEL WITH EFFECT CODING

We will again use the hand-steadiness data in Table 4.3-1 to illustrate the computational procedures for this model. The statistical hypothesis that we want to test is

$$H_0: \quad \beta_1 = \beta_2 = \beta_3 = 0.$$

In matrix notation this hypothesis is

$$\underset{(h \times 1) \times h}{\mathbf{C}'} \quad \underset{h \times 1}{\boldsymbol{\beta}} \quad \underset{(h-1) \times 1}{\mathbf{0}}$$

$$\begin{bmatrix} 0 & 1 & 0 & 0 \\ 0 & 0 & 1 & 0 \\ 0 & 0 & 0 & 1 \end{bmatrix} \begin{bmatrix} \beta_0 \\ \beta_1 \\ \beta_2 \\ \beta_3 \end{bmatrix} = \begin{bmatrix} 0 \\ 0 \\ 0 \end{bmatrix}.$$

Computational procedures using the regression model with effect coding are illustrated in Table 5.7-1. These data were analyzed previously using the conventional ANOVA approach (Table 4.3-1) and dummy coding (Table 5.4-1). The values of the mean squares and F statistic in Table 5.7-1 are identical to those obtained earlier. Also, the null hypothesis can be rejected. From the relationships in (5.7-3) we also know that $\alpha_j \neq 0$ for some j in the experimental design model.

Procedures for computing the terms $\hat{\psi}_i$, $\hat{\sigma}_{\bar{Y}}$, and $\hat{\sigma}_{\psi_i}$ needed to evaluate multiple contrasts are illustrated in Table 5.7-2 on page 203. The null hypothesis

H_0: $\mu_1 - \mu_2 = 0$ is used for illustrative purposes. The results in Table 5.7-2 are identical to those presented previously in Section 4.4 and Table 5.6-1.

TABLE 5.7-1   Computational Procedures for a Type CR-4 Design Using a Regression Model with Effect Coding

(i)   Data and basic matrices ($N = 32$, $h = 4$):

$$
\begin{array}{c}
\underset{N\times 1}{\mathbf{y}} \qquad \underset{N\times h}{\mathbf{X}}
\end{array}
$$

		$\mathbf{x}_0$	$\mathbf{x}_1$	$\mathbf{x}_2$	$\mathbf{x}_3$
a_1	3	1	1	0	0
	6	1	1	0	0
	3	1	1	0	0
	3	1	1	0	0
	1	1	1	0	0
	2	1	1	0	0
	2	1	1	0	0
	2	1	1	0	0
a_2	4	1	0	1	0
	5	1	0	1	0
	4	1	0	1	0
	3	1	0	1	0
	2	1	0	1	0
	3	1	0	1	0
	4	1	0	1	0
	3	1	0	1	0
a_3	7	1	0	0	1
	8	1	0	0	1
	7	1	0	0	1
	6	1	0	0	1
	5	1	0	0	1
	6	1	0	0	1
	5	1	0	0	1
	6	1	0	0	1
a_4	7	1	−1	−1	−1
	8	1	−1	−1	−1
	9	1	−1	−1	−1
	8	1	−1	−1	−1
	10	1	−1	−1	−1
	10	1	−1	−1	−1
	9	1	−1	−1	−1
	11	1	−1	−1	−1

$$\sum_{i=1}^{N} Y_i = 172$$

$$
\underset{1\times N}{\mathbf{y}'} \qquad \underset{N\times 1}{\mathbf{y}} \qquad \underset{1\times 1}{(\mathbf{y}'\mathbf{y})}
$$

$$
\begin{bmatrix} 3 & 6 & 3 & \cdots & 11 \end{bmatrix}
\begin{bmatrix} 3 \\ 6 \\ 3 \\ \vdots \\ 11 \end{bmatrix} = 1160.0
$$

$$
\underset{h\times N}{\mathbf{X}'} \qquad \qquad \underset{N\times h}{\mathbf{X}}
$$

$$
\begin{bmatrix} 1 & 1 & 1 & \cdot & 1 \\ 1 & 1 & 1 & \cdot & -1 \\ 0 & 0 & 0 & \cdot & -1 \\ 0 & 0 & 0 & \cdot & -1 \end{bmatrix}
\begin{bmatrix} 1 & 1 & 0 & 0 \\ 1 & 1 & 0 & 0 \\ 1 & 1 & 0 & 0 \\ \cdot & \cdot & \cdot & \cdot \\ 1 & -1 & -1 & -1 \end{bmatrix} =
$$

$$
\underset{h\times h}{(\mathbf{X}'\mathbf{X})}
$$

$$
\begin{bmatrix} 32 & 0 & 0 & 0 \\ 0 & 16 & 8 & 8 \\ 0 & 8 & 16 & 8 \\ 0 & 8 & 8 & 16 \end{bmatrix}
$$

$$
\underset{h\times N}{\mathbf{X}'} \qquad \underset{N\times 1}{\mathbf{y}} \qquad \underset{h\times 1}{(\mathbf{X}'\mathbf{y})}
$$

$$
\begin{bmatrix} 1 & 1 & 1 & \cdot & 1 \\ 1 & 1 & 1 & \cdot & -1 \\ 0 & 0 & 0 & \cdot & -1 \\ 0 & 0 & 0 & \cdot & -1 \end{bmatrix}
\begin{bmatrix} 3 \\ 6 \\ 3 \\ \vdots \\ 11 \end{bmatrix} =
\begin{bmatrix} 172 \\ -50 \\ -44 \\ -22 \end{bmatrix}
$$

TABLE 5.7-1   (continued)

$$
\underset{h \times h}{(\mathbf{X'X})^{-1}} \qquad \underset{h \times 1}{(\mathbf{X'y})} \qquad \underset{h \times 1}{\hat{\boldsymbol{\beta}}}
$$

$$
\begin{bmatrix} 1/32 & 0 & 0 & 0 \\ 0 & 3/32 & -1/32 & -1/32 \\ 0 & -1/32 & 3/32 & -1/32 \\ 0 & -1/32 & -1/32 & 3/32 \end{bmatrix} \begin{bmatrix} 172 \\ -50 \\ -44 \\ -22 \end{bmatrix} = \begin{bmatrix} 5.375 \\ -2.625 \\ -1.875 \\ 0.875 \end{bmatrix}
$$

$$
\underset{1 \times h}{\hat{\boldsymbol{\beta}}'} \qquad \underset{h \times 1}{(\mathbf{X'y})} \qquad \underset{1 \times 1}{(\hat{\boldsymbol{\beta}}'\mathbf{X'y})}
$$

$$
[5.375 \quad \cdots \quad 0.875] \begin{bmatrix} 172 \\ -50 \\ -44 \\ -22 \end{bmatrix} = 1119.0
$$

(ii) Computation of sum of squares:

$$
\begin{aligned}
SSTO &= \mathbf{y'y} - (\Sigma Y_i)^2/N = 1160.0 - (172)^2/32 \\
&= 1160.0 - 924.5 = 235.5
\end{aligned}
$$

$$
\begin{aligned}
SSR &= \hat{\boldsymbol{\beta}}'\mathbf{X'y} - (\Sigma Y_i)^2/N = 1119.0 - (172)^2/32 \\
&= 1119.0 - 924.5 = 194.5
\end{aligned}
$$

$$
SSE = \mathbf{y'y} - \hat{\boldsymbol{\beta}}'\mathbf{X'y} = 1160.0 - 1119.0 = 41.0
$$

(iii) Computation of mean squares and F statistic:

$$
\begin{aligned}
MSR &= SSR/(h - 1) = 194.5/(4 - 1) = 64.833 \\
MSE &= SSE/(N - h) = 41.0/(32 - 4) = 1.464 \\
F &= MSR/MSE = 64.833/1.464 = 44.28
\end{aligned}
$$

$$
F_{.05;3,28} = 2.95
$$

TABLE 5.7-2   Computational Procedures for Testing Differences Among Means (Effect coding)

(i) Computation of means, $\bar{\mathbf{y}}$, and contrast, $\hat{\psi}_1$ ($h = 4$, $n = 8$, $p = 4$):*

$$
\underset{p \times h}{\mathbf{W}} \qquad \underset{h \times 1}{\hat{\boldsymbol{\beta}}} \qquad \underset{p \times 1}{\bar{\mathbf{y}}} \qquad \underset{1 \times p}{\mathbf{c}_1'} \qquad \underset{p \times 1}{\bar{\mathbf{y}}} \qquad \hat{\psi}_1
$$

$$
\begin{bmatrix} 1 & 1 & 0 & 0 \\ 1 & 0 & 1 & 0 \\ 1 & 0 & 0 & 1 \\ 1 & -1 & -1 & -1 \end{bmatrix} \begin{bmatrix} 5.375 \\ -2.625 \\ -1.875 \\ 0.875 \end{bmatrix} = \begin{bmatrix} 2.75 \\ 3.50 \\ 6.25 \\ 9.00 \end{bmatrix} \qquad [1 \quad -1 \quad 0 \quad 0] \begin{bmatrix} 2.75 \\ 3.50 \\ 6.25 \\ 9.00 \end{bmatrix} = -0.75
$$

---

*Matrix $\mathbf{W}$ is obtained from (5.7-4), $\hat{\boldsymbol{\beta}}$ is obtained from Table 5.7-1, and $\mathbf{c}'$ is the coefficient vector associated with the hypothesis $\mu_1 - \mu_2 = 0$.

TABLE 5.7-2   (continued)

(ii)   Computation of $\hat{\sigma}_{\bar{Y}}^2$ and $\hat{\sigma}_{\hat{\psi}_1}^2$:*

$$\hat{\sigma}_{\bar{Y}}^2 = \frac{MSE}{n} = \frac{1.464}{8} = 0.183$$

$$MSE \left\{ \underset{1 \times p}{\mathbf{c}'} \left[\underset{p \times h}{\mathbf{W}} \quad \underset{h \times h}{(\mathbf{X}'\mathbf{X})^{-1}} \quad \underset{h \times p}{\mathbf{W}'} \right] \underset{p \times 1}{\mathbf{c}} \right\}$$

$$\hat{\sigma}_{\hat{\psi}_1}^2 = 1.464 \left\{ [1 \quad -1 \quad 0 \quad 0] \begin{bmatrix} 1 & 1 & 0 & 0 \\ 1 & 0 & 1 & 0 \\ 1 & 0 & 0 & 1 \\ 1 & -1 & -1 & -1 \end{bmatrix} \begin{bmatrix} \frac{1}{32} & 0 & 0 & 0 \\ 0 & \frac{3}{32} & -\frac{1}{32} & -\frac{1}{32} \\ 0 & -\frac{1}{32} & \frac{3}{32} & -\frac{1}{32} \\ 0 & -\frac{1}{32} & -\frac{1}{32} & \frac{3}{32} \end{bmatrix} \begin{bmatrix} 1 & 1 & 1 & 1 \\ 1 & 0 & 0 & -1 \\ 0 & 1 & 0 & -1 \\ 0 & 0 & 1 & -1 \end{bmatrix} \begin{bmatrix} 1 \\ -1 \\ 0 \\ 0 \end{bmatrix} \right\}$$

$$= 1.464 \left\{ [1 \quad -1 \quad 0 \quad 0] \begin{bmatrix} \frac{1}{8} & 0 & 0 & 0 \\ 0 & \frac{1}{8} & 0 & 0 \\ 0 & 0 & \frac{1}{8} & 0 \\ 0 & 0 & 0 & \frac{1}{8} \end{bmatrix} \begin{bmatrix} 1 \\ -1 \\ 0 \\ 0 \end{bmatrix} \right\} = 1.464(2/8) = 0.366$$

(iii)   Computation of test statistics:

$$t = \frac{\hat{\psi}_1}{\hat{\sigma}_{\hat{\psi}_1}} = \frac{-0.75}{\sqrt{0.366}} = \frac{-0.75}{0.605} = -1.24 \qquad q = \frac{\hat{\psi}_1}{\hat{\sigma}_{\bar{Y}}} = \frac{-0.75}{\sqrt{0.183}} = \frac{-0.75}{0.428} = -1.75$$

*Matrix $(\mathbf{X}'\mathbf{X})^{-1}$ and MSE are obtained from Table 5.7-1.

It should be apparent from the foregoing that an analysis of variance can be performed in a variety of ways. As noted earlier, the solution to $\hat{\boldsymbol{\beta}} = (\mathbf{X}'\mathbf{X})^{-1}\mathbf{X}'\mathbf{y}$, where $\mathbf{X}$ is a matrix of indicator variable zeros and ones, depends on the particular coding scheme that is used. However, certain linear combinations of $\hat{\boldsymbol{\beta}}$ are unique and independent of the solution. For example, $\mathbf{W}\hat{\boldsymbol{\beta}}$ can be used to estimate population means, where $\mathbf{W}$ is based on the pattern of ones and zeros in the $\mathbf{X}$ matrix. A choice between dummy coding and effect coding is largely a matter of personal preference. For a type CR-p design, dummy coding is a simpler coding scheme than effect coding but the latter scheme has the advantage that

$$\beta_0 = \mu, \quad \beta_1 = \alpha_1, \quad \beta_2 = \alpha_2, \quad \ldots, \quad -(\beta_1 + \beta_2 + \cdots + \beta_{h-1}) = \alpha_p.$$

COMPUTATIONAL PROCEDURES USING THE REGRESSION MODEL WITH ORTHOGONAL CODING

We will now illustrate a third coding scheme for a structural matrix—*orthogonal coding*. As the name suggests, the independent variables of a regression model are coded so that the vectors corresponding to $X_0, X_{i1}, X_{i2}, \ldots, X_{i,h-1}$ are mutually orthogonal. The concept of orthogonality is not new; we have discussed it in connection with multiple comparisons among means (Sections 3.1 and 5.3) and trend analysis (Section 4.5).*

There is not just one orthogonal coding scheme but rather a variety of schemes. Each is subject only to the following restrictions on the columns of the **X** matrix: $\Sigma_{i=1}^{N} X_{ij} = 0$ for $j = 1, \ldots, h - 1$ and $\Sigma_{i=1}^{N} X_{ij}X_{ij'} = 0$ for all j and $j', j \neq j'$. If we want to test the overall null hypothesis H_0: $\alpha_1 = \alpha_2 = \alpha_3 = \alpha_4 = 0$ in a type CR-4 design, the first vector of the structural matrix can be assigned ones and the remaining $h - 1$ vectors coded as follows.

$$
\begin{array}{c}
 & \mathbf{X_0} \quad \mathbf{X_1} \quad \mathbf{X_2} \quad \mathbf{X_3} \\
a_1 \left\{ \begin{array}{l} \\ \\ \\ \end{array} \right. &
\left[\begin{array}{rrrr}
1 & 1 & 1 & 1 \\
1 & 1 & 1 & 1 \\
. & . & . & . \\
1 & -1 & 1 & 1 \\
1 & -1 & 1 & 1 \\
. & . & . & . \\
1 & 0 & -2 & 1 \\
1 & 0 & -2 & 1 \\
. & . & . & . \\
1 & 0 & 0 & -3 \\
1 & 0 & 0 & -3 \\
. & . & . & .
\end{array} \right]
\end{array}
$$

The correspondence between the parameters of the regression model and the parameters of the experimental design model for this coding scheme is

$$E(Y_{i1}) = \beta_0 + \beta_1 + \beta_2 + \beta_3 = \mu_1$$
$$E(Y_{i2}) = \beta_0 - \beta_1 + \beta_2 + \beta_3 = \mu_2$$
$$E(Y_{i3}) = \beta_0 - 2\beta_2 + \beta_3 = \mu_3$$
$$E(Y_{i4}) = \beta_0 - 3\beta_3 = \mu_4.$$

It can be shown that the β_0 parameter is equal to μ. The other β_j parameters do not have such a simple interpretation. However, the use of orthogonal coding does simplify the computation of sums of squares. Suppose, for example, that the last $h - 1$ columns of **X** contain orthogonal coefficients that define $p - 1$ trend components or $p - 1$ orthogonal contrasts among means. Then the sum of squares for the jth contrast, $SS \ \hat{\psi}_j$, is given by $SS \ \hat{\psi}_j = \hat{\beta}_j \Sigma_{i=1}^{N} Y_i X_{ij}$. We will illustrate the computation by performing a trend analysis on the hand-steadiness data in Table 4.3-1.

Recall from the discussion of trend analysis in Section 4.5 that the treatment sum of squares for a quantitative independent variable can be partitioned into $p - 1$ trend components—linear, quadratic, cubic, and so on. For the data in Table 4.3-1 we can code $\mathbf{x}_1$ in the structural matrix using the linear coefficients $(-3 \ -1 \ 1 \ 3)$ from Appendix E.12.** Similarly, $\mathbf{x}_2$ and $\mathbf{x}_3$ can be coded using the quadratic $(1 \ -1 \ -1 \ 1)$

---

* It is also discussed in Appendix D.1-8.

** The coefficients in Appendix E.12 are appropriate only for the equal n's case.

TABLE 5.7-3 Computational Procedures for a Type CR-4 Design Using a Regression Model with Orthogonal Coding

(i) Data and basic matrices ($N = 32$, $h = 4$):

$$\mathbf{y}_{N \times 1} \qquad \mathbf{X}_{N \times h}$$

$$
a_1 \left\{
\begin{array}{c}
3 \\ 6 \\ 3 \\ 3 \\ 1 \\ 2 \\ 2 \\ 2
\end{array}
\right.
\qquad
\begin{array}{cccc}
\mathbf{x}_0 & \mathbf{x}_1 & \mathbf{x}_2 & \mathbf{x}_3 \\
1 & -3 & 1 & -1 \\
1 & -3 & 1 & -1 \\
1 & -3 & 1 & -1 \\
1 & -3 & 1 & -1 \\
1 & -3 & 1 & -1 \\
1 & -3 & 1 & -1 \\
1 & -3 & 1 & -1 \\
1 & -3 & 1 & 1
\end{array}
$$

$$
a_2 \left\{
\begin{array}{c}
4 \\ 5 \\ 4 \\ 3 \\ 2 \\ 3 \\ 4 \\ 3
\end{array}
\right.
\qquad
\begin{array}{cccc}
1 & -1 & -1 & 3 \\
1 & -1 & -1 & 3 \\
1 & -1 & -1 & 3 \\
1 & -1 & -1 & 3 \\
1 & -1 & -1 & 3 \\
1 & -1 & -1 & 3 \\
1 & -1 & -1 & 3 \\
1 & -1 & -1 & 3
\end{array}
$$

$$
a_3 \left\{
\begin{array}{c}
7 \\ 8 \\ 7 \\ 6 \\ 5 \\ 6 \\ 5 \\ 6
\end{array}
\right.
\qquad
\begin{array}{cccc}
1 & 1 & -1 & -3 \\
1 & 1 & -1 & -3 \\
1 & 1 & -1 & -3 \\
1 & 1 & -1 & -3 \\
1 & 1 & -1 & -3 \\
1 & 1 & -1 & -3 \\
1 & 1 & -1 & -3 \\
1 & 1 & -1 & -3
\end{array}
$$

$$
a_4 \left\{
\begin{array}{c}
7 \\ 8 \\ 9 \\ 8 \\ 10 \\ 10 \\ 9 \\ 11
\end{array}
\right.
\qquad
\begin{array}{cccc}
1 & 3 & 1 & 1 \\
1 & 3 & 1 & 1 \\
1 & 3 & 1 & 1 \\
1 & 3 & 1 & 1 \\
1 & 3 & 1 & 1 \\
1 & 3 & 1 & 1 \\
1 & 3 & 1 & 1 \\
1 & 3 & 1 & 1
\end{array}
$$

$$\sum_{i=1}^{N} Y_i = 172$$

$$
\underset{1 \times N}{\mathbf{y}'} \quad \underset{N \times 1}{\mathbf{y}} \quad \underset{1 \times 1}{(\mathbf{y}'\mathbf{y})}
$$

$$
[3 \quad 6 \quad 3 \quad \cdots \quad 11]
\begin{bmatrix} 3 \\ 6 \\ 3 \\ \vdots \\ 11 \end{bmatrix} = 1160.0
$$

$$
\underset{h \times N}{\mathbf{X}'} \qquad \underset{N \times h}{\mathbf{X}} \qquad \underset{h \times h}{(\mathbf{X}'\mathbf{X})}
$$

$$
\begin{bmatrix}
1 & 1 & 1 & \cdot & 1 \\
-3 & -3 & -3 & \cdot & 3 \\
1 & 1 & 1 & \cdot & 1 \\
-1 & -1 & -1 & \cdot & 1
\end{bmatrix}
\begin{bmatrix}
1 & -3 & 1 & -1 \\
1 & -3 & 1 & -1 \\
1 & -3 & 1 & -1 \\
\cdot & \cdot & \cdot & \cdot \\
1 & 3 & 1 & 1
\end{bmatrix}
=
\begin{bmatrix}
32 & 0 & 0 & 0 \\
0 & 160 & 0 & 0 \\
0 & 0 & 32 & 0 \\
0 & 0 & 0 & 160
\end{bmatrix}
$$

$$
\underset{h \times N}{\mathbf{X}'} \qquad \underset{N \times 1}{\mathbf{y}} \qquad \underset{h \times 1}{(\mathbf{X}'\mathbf{y})}
$$

$$
\begin{bmatrix}
1 & 1 & 1 & \cdot & 1 \\
-3 & -3 & -3 & \cdot & 3 \\
1 & 1 & 1 & \cdot & 1 \\
-1 & -1 & -1 & \cdot & 1
\end{bmatrix}
\begin{bmatrix} 3 \\ 6 \\ 3 \\ \vdots \\ 11 \end{bmatrix}
=
\begin{bmatrix} 172 \\ 172 \\ 16 \\ -16 \end{bmatrix}
$$

$$
\underset{h \times h}{(\mathbf{X}'\mathbf{X})^{-1}} \quad \underset{h \times 1}{(\mathbf{X}'\mathbf{y})} \quad \underset{h \times 1}{\hat{\boldsymbol{\beta}}} \qquad \underset{1 \times h}{\hat{\boldsymbol{\beta}}'} \qquad \underset{h \times 1}{(\mathbf{X}'\mathbf{y})} \quad \underset{1 \times 1}{(\hat{\boldsymbol{\beta}}'\mathbf{X}'\mathbf{y})}
$$

$$
\begin{bmatrix}
1/32 & 0 & 0 & 0 \\
0 & 1/160 & 0 & 0 \\
0 & 0 & 1/32 & 0 \\
0 & 0 & 0 & 1/160
\end{bmatrix}
\begin{bmatrix} 172 \\ 172 \\ 16 \\ -16 \end{bmatrix}
=
\begin{bmatrix} 5.375 \\ 1.075 \\ .500 \\ -.100 \end{bmatrix}
\qquad
[5.375 \quad \cdots \quad -.100]
\begin{bmatrix} 172 \\ 172 \\ 16 \\ -16 \end{bmatrix}
= 1119.0
$$

TABLE 5.7-3 (continued)

(ii) Computation of sum of squares:

$$SSTO = \mathbf{y'y} - (\Sigma\, Y_i)^2/N = 1160.0 - (172)^2/32$$

$$= 1160.0 - 924.5 = 235.5$$

$$SSR = \hat{\boldsymbol{\beta}}'\mathbf{X'y} - (\Sigma\, Y_i)^2/N = 1119.0 - (172)^2/32$$

$$= 1119.0 - 924.5 = 194.5$$

$$SSE = \mathbf{y'y} - \hat{\boldsymbol{\beta}}'\, \mathbf{X'y} = 1160.0 - 1119.0 = 41.0$$

$$SS\hat{\psi}_{\text{lin}} = \hat{\beta}_1 \left(\sum_{i=1}^{N} Y_i X_{i1} \right)^* = 1.075(172) = 184.900$$

$$SS\hat{\psi}_{\text{quad}} = \hat{\beta}_2 \left(\sum_{i=1}^{N} Y_i X_{i2} \right)^{**} = 0.500(16) = 8.000$$

$$SS\hat{\psi}_{\text{cub}} = \hat{\beta}_3 \left(\sum_{i=1}^{N} Y_i X_{i3} \right)^\dagger = -0.100(-16) = 1.600$$

(iii) Computation of F test statistic:

$$F = \frac{SS\hat{\psi}_{\text{lin}}/1}{SSE/(N-h)} = \frac{184.900}{1.464} = 126.30$$

$$F = \frac{SS\hat{\psi}_{\text{quad}}/1}{SSE/(N-h)} = \frac{8.000}{1.464} = 5.46$$

$$F = \frac{SS\hat{\psi}_{\text{cub}}/1}{SSE/(N-h)} = \frac{1.600}{1.464} = 1.09$$

*This sum is the second element of $\mathbf{X'y}$.
**This sum is the third element of $\mathbf{X'y}$.
†This sum is the fourth element of $\mathbf{X'y}$.

and cubic $(-1\ 3\ -3\ 1)$ coefficients, respectively. This is shown in Table 5.7-3. The hypotheses to be tested at the .05 level of significance are

$$H_0: \quad \psi_{\text{lin}} = 0 \quad \text{versus} \quad H_1: \quad \psi_{\text{lin}} \neq 0$$
$$H_0: \quad \psi_{\text{quad}} = 0 \quad \text{versus} \quad H_1: \quad \psi_{\text{quad}} \neq 0$$
$$H_0: \quad \psi_{\text{cub}} = 0 \quad \text{versus} \quad H_1: \quad \psi_{\text{cub}} \neq 0.$$

These hypotheses are a priori and represent mutually orthogonal contrasts; accordingly each is tested at the .05 level.* The analysis of the data is presented in Table 5.7-3.

---

*   The rationale for adopting the individual hypothesis rather than the collection of hypotheses as the conceptual unit for making a type I error is discussed in Section 3.3.

The computations in Table 5.7-3 lead to the same values for $SS\hat{\psi}_{lin}$, $SS\hat{\psi}_{quad}$, and $SS\hat{\psi}_{cub}$ as those in Tables 4.5-2 and 4.5-3. The hypotheses $\psi_{lin} = 0$ and $\psi_{quad} = 0$ are rejected.

One feature of orthogonal coding deserves special mention. The $(X'X)$ matrix, unlike those for dummy and effect coding, is always a diagonal matrix. This simplifies determining its inverse $(X'X)^{-1}$, since the inverse of a diagonal matrix is obtained by replacing each diagonal element by its reciprocal.*

5.8   EXPERIMENTAL DESIGN MODEL APPROACH TO ANOVA

Experimental design models can be classified as either full rank linear models or less than full rank linear models. As we will see, the distinction is based on the rank of the structural matrix, X. The column (or row) rank of a matrix, denoted by r, is the number of linearly independent column (or row) vectors in the matrix. It makes no difference whether we consider column rank or row rank since the two are equal.

Let $x_0, x_1, \ldots, x_{h-1}$ denote column vectors of matrix X, and also let $a_0, a_1, \ldots, a_{h-1}$ denote scalars. Then $a_0 x_0 + a_1 x_1 + \cdots + a_{h-1} x_{h-1}$ is a linear combination of the x_j's, where $j = 0, \ldots, h - 1$. If the linear combination is equal to the null vector, 0, only when $a_0 = a_1 = \cdots = a_{h-1} = 0$, that is, $a_0 x_0 + a_1 x_1 + \cdots + a_{h-1} x_{h-1} = 0$, only if $a_j = 0$ for all j, then the x_j's are said to be *linearly independent*. If the linear combination is equal to the null vector for some $a_j \neq 0$, the vectors are *linearly dependent*. To state it another way, the vectors are linearly dependent if and only if one of them can be expressed as a linear combination of the others. For example, the column vectors of X_1

$$(5.8\text{-}1) \qquad X_1 = \begin{array}{c} \begin{array}{ccccc} x_0 & x_1 & x_2 & x_3 & x_4 \end{array} \\ \begin{bmatrix} 1 & 1 & 0 & 0 & 0 \\ 1 & 1 & 0 & 0 & 0 \\ 1 & 0 & 1 & 0 & 0 \\ 1 & 0 & 1 & 0 & 0 \\ 1 & 0 & 0 & 1 & 0 \\ 1 & 0 & 0 & 1 & 0 \\ 1 & 0 & 0 & 0 & 1 \\ 1 & 0 & 0 & 0 & 1 \end{bmatrix} \end{array}$$

are linearly dependent because

---

*   This is discussed in Appendix D.2-6.

$$
\underset{\mathbf{x}_0}{\begin{bmatrix} 1 \\ 1 \\ 1 \\ 1 \\ 1 \\ 1 \\ 1 \\ 1 \end{bmatrix}} = 1 \underset{\mathbf{x}_1}{\begin{bmatrix} 1 \\ 1 \\ 0 \\ 0 \\ 0 \\ 0 \\ 0 \\ 0 \end{bmatrix}} + 1 \underset{\mathbf{x}_2}{\begin{bmatrix} 0 \\ 0 \\ 1 \\ 1 \\ 0 \\ 0 \\ 0 \\ 0 \end{bmatrix}} + 1 \underset{\mathbf{x}_3}{\begin{bmatrix} 0 \\ 0 \\ 0 \\ 0 \\ 1 \\ 1 \\ 0 \\ 0 \end{bmatrix}} + 1 \underset{\mathbf{x}_4}{\begin{bmatrix} 0 \\ 0 \\ 0 \\ 0 \\ 0 \\ 0 \\ 1 \\ 1 \end{bmatrix}}.
$$

Furthermore, nonzero scalars can be found such that $a_0\mathbf{x}_0 + a_1\mathbf{x}_1 + a_2\mathbf{x}_2 + a_3\mathbf{x}_3 + a_4\mathbf{x}_4 = \mathbf{0}$.

$$
\underset{a_0\mathbf{x}_0}{1\begin{bmatrix} 1 \\ 1 \\ 1 \\ 1 \\ 1 \\ 1 \\ 1 \\ 1 \end{bmatrix}} - \underset{a_1\mathbf{x}_1}{1\begin{bmatrix} 1 \\ 1 \\ 0 \\ 0 \\ 0 \\ 0 \\ 0 \\ 0 \end{bmatrix}} - \underset{a_2\mathbf{x}_2}{1\begin{bmatrix} 0 \\ 0 \\ 1 \\ 1 \\ 0 \\ 0 \\ 0 \\ 0 \end{bmatrix}} - \underset{a_3\mathbf{x}_3}{1\begin{bmatrix} 0 \\ 0 \\ 0 \\ 0 \\ 1 \\ 1 \\ 0 \\ 0 \end{bmatrix}} - \underset{a_4\mathbf{x}_4}{1\begin{bmatrix} 0 \\ 0 \\ 0 \\ 0 \\ 0 \\ 0 \\ 1 \\ 1 \end{bmatrix}} = \underset{\mathbf{0}}{\begin{bmatrix} 0 \\ 0 \\ 0 \\ 0 \\ 0 \\ 0 \\ 0 \\ 0 \end{bmatrix}}
$$

Since the vectors are not linearly independent we know that the rank of $\mathbf{X}_1$ is less than five, the number of column vectors. Consider next the column vectors of $\mathbf{X}_2$.

(5.8-2)

$$
\mathbf{X}_2 = \begin{matrix} \mathbf{x}_1 \ \mathbf{x}_2 \ \mathbf{x}_3 \ \mathbf{x}_4 \\ \begin{bmatrix} 1 & 0 & 0 & 0 \\ 1 & 0 & 0 & 0 \\ 0 & 1 & 0 & 0 \\ 0 & 1 & 0 & 0 \\ 0 & 0 & 1 & 0 \\ 0 & 0 & 1 & 0 \\ 0 & 0 & 0 & 1 \\ 0 & 0 & 0 & 1 \end{bmatrix} \end{matrix}
$$

It can be shown that the linear combination $a_1\mathbf{x}_1 + a_2\mathbf{x}_2 + a_3\mathbf{x}_3 + a_4\mathbf{x}_4$ is equal to the null vector only if $a_1 = a_2 = a_3 = a_4 = 0$. Thus the column vectors of $\mathbf{X}_2$ are linearly independent and its rank is four. The rank r of an $N \times h$ matrix is always a unique number such that $0 < r \leq$ smaller of (N, h). For $h \leq N$, r cannot exceed h; if $r = h$, the matrix is of *full rank*. If $r < h$, the matrix is not of full rank. In this case we know that there are $h - r$ dependent column vectors in the matrix. Alternatively, if $N \leq h$ and the matrix is not of full rank, there are $N - r$ dependent row vectors in the matrix.

In Section 5.1 we saw that the system of equations represented by the experimental design model equation

(5.8-3) $$Y_{ij} = \mu + \alpha_j + \epsilon_{i(j)} \qquad (i = 1, \ldots, n; j = 1, \ldots, p)$$

can be written in matrix notation as

(5.8-4)
$$\underset{N \times 1}{\mathbf{y}} = \underset{N \times h}{\mathbf{X}} \; \underset{h \times 1}{\boldsymbol{\beta}} + \underset{N \times 1}{\boldsymbol{\epsilon}}.$$

The $N \times h$ structural matrix $\mathbf{X}$ has the following form.

$$\begin{bmatrix} 1 & 1 & 0 & \cdot & 0 \\ 1 & 1 & 0 & \cdot & 0 \\ 1 & 0 & 1 & \cdot & 0 \\ 1 & 0 & 1 & \cdot & 0 \\ \cdot & \cdot & \cdot & \cdot & \cdot \\ 1 & 0 & 0 & \cdot & 1 \\ 1 & 0 & 0 & \cdot & 1 \end{bmatrix}$$

The form of this matrix and that for $\mathbf{X}_1$ (5.8-1) are identical. Thus the structural matrix for model (5.8-3) is not of full rank and the model is classified as a less than full rank model. For such models the inverse $(\mathbf{X}'\mathbf{X})^{-1}$ does not exist and it is not possible to solve for $\hat{\boldsymbol{\beta}}$ by $\hat{\boldsymbol{\beta}} = (\mathbf{X}'\mathbf{X})^{-1}(\mathbf{X}'\mathbf{y})$ as we did for the regression model approach.* Three procedures are often used to solve for $\hat{\boldsymbol{\beta}}$ in less than full rank models:

1.   Solution by restricting the number of unknowns

2.   Solution by reparameterization

3.   Solution by use of a generalized inverse.

An introduction to the first and second procedures is given in Section 5.9. Detailed coverage of these methods can be found in Bock (1975), Finn (1974), Scheffé (1959), and Timm (1975). For an elementary introduction to the generalized inverse, see Searle (1966).

Consider now an alternative experimental design model for a completely randomized design:

(5.8-5) $$Y_{ij} = \mu_j + \epsilon_{i(j)} \qquad (i = 1, \ldots, n; j = 1, \ldots, p)$$

---

*   A consistent system of N equations in h unknowns for which $\mathbf{X}'\mathbf{X}\hat{\boldsymbol{\beta}} = \mathbf{X}'\mathbf{y}$, where $\mathbf{y}$ is a nonnull vector of scalars, has a unique solution if and only if the rank of $\mathbf{X}$ is equal to $h, (h < N)$. If the rank of $\mathbf{X}$ is less than h, the system has an infinite number of solutions. A system is consistent, has a solution, if and only if the rank of $\mathbf{X}$ is equal to the rank of the augmented matrix $[\mathbf{X} \quad \mathbf{y}]$ (this is a matrix formed by appending $\mathbf{y}$ to the $\mathbf{X}$ matrix). To state it another way, a system has a solution if and only if $\mathbf{y}$ is subject to the same linear dependencies as the rows of $\mathbf{X}$. This is discussed in Appendix D.2-7.

where Y_{ij} = the value of the dependent variable for subject i in treatment level j
  μ_j = a fixed unknown parameter representing the mean of population j
  $\epsilon_{i(j)}$ = a random error term that is independent of other $\epsilon_{i(j)}$'s and is normally
    distributed with mean equal to 0, $E(\epsilon_{i(j)}) = 0$, and variance equal to
    σ_ϵ^2, $V(\epsilon) = \sigma_\epsilon^2$.

Using the indicator variable 0,1 and letting $p = 4$ and $n = 2$, (5.8-5) can be written in matrix notation as

$$
\begin{bmatrix} Y_{11} \\ Y_{21} \\ Y_{12} \\ Y_{22} \\ Y_{13} \\ Y_{23} \\ Y_{14} \\ Y_{24} \end{bmatrix}
=
\begin{bmatrix} 1 & 0 & 0 & 0 \\ 1 & 0 & 0 & 0 \\ 0 & 1 & 0 & 0 \\ 0 & 1 & 0 & 0 \\ 0 & 0 & 1 & 0 \\ 0 & 0 & 1 & 0 \\ 0 & 0 & 0 & 1 \\ 0 & 0 & 0 & 1 \end{bmatrix}
\begin{bmatrix} \mu_1 \\ \mu_2 \\ \mu_3 \\ \mu_4 \end{bmatrix}
+
\begin{bmatrix} \epsilon_{1(1)} \\ \epsilon_{2(1)} \\ \epsilon_{1(2)} \\ \epsilon_{2(2)} \\ \epsilon_{1(3)} \\ \epsilon_{2(3)} \\ \epsilon_{1(4)} \\ \epsilon_{2(4)} \end{bmatrix}
$$

or

(5.8-6)
$$ \underset{N \times 1}{\mathbf{y}} = \underset{N \times p}{\mathbf{X}} \; \underset{p \times 1}{\boldsymbol{\mu}} + \underset{N \times 1}{\boldsymbol{\epsilon}} $$

where $\mathbf{y}$ = an $N \times 1$ vector of observations
  $\mathbf{X}$ = an $N \times p$ matrix of constants (0, 1)
  $\boldsymbol{\mu}$ = a $p \times 1$ vector of parameters $(\mu_1, \mu_2, \ldots, \mu_4)$
  $\boldsymbol{\epsilon}$ = an $N \times 1$ vector of *NID* errors with $E(\boldsymbol{\epsilon}) = \mathbf{0}$ and $V(\boldsymbol{\epsilon}) = \sigma_\epsilon^2 \mathbf{I}$.

The structural matrix for this model is identical to $\mathbf{X}_2$ (5.8-2) and is of full rank. Thus the inverse $(\mathbf{X}'\mathbf{X})^{-1}$ exists and it is possible to solve for $\hat{\boldsymbol{\mu}}$ by $\hat{\boldsymbol{\mu}} = (\mathbf{X}'\mathbf{X})^{-1}(\mathbf{X}'\mathbf{y})$.
    We have briefly introduced two experimental design models for a type CR-p design: $Y_{ij} = \mu + \alpha_j + \epsilon_{i(j)}$ and $Y_{ij} = \mu_j + \epsilon_{i(j)}$ $(i = 1, \ldots, n; j = 1, \ldots, p)$, where $\epsilon_{i(j)}$ is *NID*$(0, \sigma_\epsilon^2)$. The first model is classified as a less than full rank model; the second model is a full rank model. In the following sections we will describe procedures for computing sums of squares and testing hypotheses about parameters for these models.

5.9 ANOVA VIA THE LESS THAN FULL RANK EXPERIMENTAL DESIGN MODEL APPROACH

SOLUTION BY RESTRICTING THE NUMBER OF UNKNOWNS

We saw in the last section that a solution to the system of equations represented by

(5.9-1)
$$ Y_{ij} = \mu + \alpha_j + \epsilon_{i(j)} \qquad (i = 1, \ldots, n; j = 1, \ldots, p) $$

where $\epsilon_{i(j)}$ is $NID(0, \sigma_\epsilon^2)$, cannot be obtained from $\hat{\boldsymbol{\beta}} = (\mathbf{X}'\mathbf{X})^{-1}(\mathbf{X}'\mathbf{y})$ because the structural matrix is rank deficient $(r < h)$ and the inverse $(\mathbf{X}'\mathbf{X})^{-1}$ does not exist. Probably the most common solution to this problem is to bring $\mathbf{X}$ up to full rank by adding $h - r$ independent rows to the system in a manner that eliminates the linear dependencies among the columns of $\mathbf{X}$. The added rows represent independent equations that are usually interpreted as restrictions or side conditions on the $\alpha_1, \alpha_2, \ldots,$ α_p parameters. We will now show that the rank deficiency of $\mathbf{X}$ in (5.8-1) can be eliminated by adding one equation to the system representing the restriction $\alpha_1 + \alpha_2 + \cdots + \alpha_p = 0$.

The less than full rank model represented by (5.9-1) with p equal to 4 and n equal to 2 can be written as

(5.9-2)

$$
\underset{\underset{N\times 1}{}}{\mathbf{y}} \qquad \underset{\underset{N\times h}{}}{\mathbf{X}} \qquad \underset{\underset{h\times 1}{}}{\boldsymbol{\beta}} \qquad \underset{\underset{N\times 1}{}}{\boldsymbol{\epsilon}}
$$

$$
\begin{bmatrix} Y_{11} \\ Y_{21} \\ Y_{12} \\ Y_{22} \\ Y_{13} \\ Y_{23} \\ Y_{14} \\ Y_{24} \end{bmatrix} = \begin{bmatrix} \mathbf{x_0} & \mathbf{x_1} & \mathbf{x_2} & \mathbf{x_3} & \mathbf{x_4} \\ 1 & 1 & 0 & 0 & 0 \\ 1 & 1 & 0 & 0 & 0 \\ 1 & 0 & 1 & 0 & 0 \\ 1 & 0 & 1 & 0 & 0 \\ 1 & 0 & 0 & 1 & 0 \\ 1 & 0 & 0 & 1 & 0 \\ 1 & 0 & 0 & 0 & 1 \\ 1 & 0 & 0 & 0 & 1 \end{bmatrix} \begin{bmatrix} \mu \\ \alpha_1 \\ \alpha_2 \\ \alpha_3 \\ \alpha_4 \end{bmatrix} + \begin{bmatrix} \epsilon_{1(1)} \\ \epsilon_{2(1)} \\ \epsilon_{1(2)} \\ \epsilon_{2(2)} \\ \epsilon_{1(3)} \\ \epsilon_{2(3)} \\ \epsilon_{1(4)} \\ \epsilon_{2(4)} \end{bmatrix}.
$$

The restriction $\alpha_1 + \alpha_2 + \alpha_3 + \alpha_4 = 0$ to be added to (5.9-2) can be written in vector notation as

$$
\underset{\underset{(h-r)\times h}{}}{\mathbf{R}} \qquad \underset{\underset{h\times 1}{}}{\boldsymbol{\beta}} \qquad \underset{\underset{(h-r)\times 1}{}}{\boldsymbol{\theta}}
$$

$$
\begin{bmatrix} 0 & 1 & 1 & 1 & 1 \end{bmatrix} \begin{bmatrix} \mu \\ \alpha_1 \\ \alpha_2 \\ \alpha_3 \\ \alpha_4 \end{bmatrix} = [0].
$$

The choice of restrictions is arbitrary, provided that the resulting equations are linearly independent and independent of the equations already in the system. For our purposes the restriction $\alpha_1 + \alpha_2 + \cdots + \alpha_p = 0$ is a convenient choice since it is a characteristic of any fixed-effects experimental design model (see Section 4.1). Adding this restriction to (5.9-2) gives

$$
\begin{array}{cccc}
\underset{(N+1)\times 1}{\mathbf{y}} & \underset{(N+1)\times h}{\mathbf{X}} & \underset{h\times 1}{\boldsymbol{\beta}} & \underset{(N+1)\times 1}{\boldsymbol{\epsilon}}
\end{array}
$$

$$
\begin{bmatrix}
Y_{11} \\
Y_{21} \\
Y_{12} \\
Y_{22} \\
Y_{13} \\
Y_{23} \\
Y_{14} \\
Y_{24} \\
0
\end{bmatrix}
=
\begin{array}{ccccc}
\mathbf{x}_0 & \mathbf{x}_1 & \mathbf{x}_2 & \mathbf{x}_3 & \mathbf{x}_4 \\
\end{array}
\begin{bmatrix}
1 & 1 & 0 & 0 & 0 \\
1 & 1 & 0 & 0 & 0 \\
1 & 0 & 1 & 0 & 0 \\
1 & 0 & 1 & 0 & 0 \\
1 & 0 & 0 & 1 & 0 \\
1 & 0 & 0 & 1 & 0 \\
1 & 0 & 0 & 0 & 1 \\
1 & 0 & 0 & 0 & 1 \\
0 & 1 & 1 & 1 & 1
\end{bmatrix}
\begin{bmatrix}
\mu \\
\alpha_1 \\
\alpha_2 \\
\alpha_3 \\
\alpha_4
\end{bmatrix}
+
\begin{bmatrix}
\epsilon_{1(1)} \\
\epsilon_{2(1)} \\
\epsilon_{1(2)} \\
\epsilon_{2(2)} \\
\epsilon_{1(3)} \\
\epsilon_{2(3)} \\
\epsilon_{1(4)} \\
\epsilon_{2(4)} \\
0
\end{bmatrix}.
$$

(5.9-3)

In (5.9-3) we have added $[0]$ to $\mathbf{y}$, $[0 \quad 1 \quad 1 \quad 1 \quad 1]$ to $\mathbf{X}$, and $[0]$ to $\boldsymbol{\epsilon}$. By augmenting (5.9-2) with $h - r = 1$ equation, we have eliminated the rank deficiency of $\mathbf{X}$. The rank of $\mathbf{X}$ in (5.9-2) is four, the rank of $\mathbf{X}$ in (5.9-3) is five, which corresponds to the number of column vectors in $\mathbf{X}$. We can now obtain a solution to (5.9-3) in the usual way since the structural matrix is of full rank.

Let's review what we have done. The system of equations (5.9-2) can be written in matrix notation as

(5.9-4)
$$
\underset{N\times 1}{\mathbf{y}} = \underset{N\times h}{\mathbf{X}} \, \underset{h\times 1}{\boldsymbol{\beta}} + \underset{N\times 1}{\boldsymbol{\epsilon}}
$$

where $N = 8$, $h = 5$, and $r = 4$. Since $h - r = 1$, there is one dependent column vector in $\mathbf{X}$. To eliminate this deficiency we placed $h - r = 1$ restriction on the α_j's, namely, $\alpha_1 + \alpha_2 + \alpha_3 + \alpha_4 = 0$. The vectors and matrix in (5.9-4) are augmented to reflect this restriction and become

(5.9-5)
$$
\begin{bmatrix}
\underset{N\times 1}{\mathbf{y}} \\
\underset{(h-r)\times 1}{\boldsymbol{\theta}}
\end{bmatrix}
=
\begin{bmatrix}
\underset{N\times h}{\mathbf{X}} \\
\underset{(h-r)\times h}{\mathbf{R}}
\end{bmatrix}
\underset{h\times 1}{\boldsymbol{\beta}}
+
\begin{bmatrix}
\underset{N\times 1}{\boldsymbol{\epsilon}} \\
\underset{(h-r)\times 1}{\boldsymbol{\theta}}
\end{bmatrix}
$$

where $\boldsymbol{\theta} = [0]$ and $\mathbf{R} = [0 \quad 1 \quad 1 \quad 1 \quad 1]$. For convenience we will denote the augmented vector $\begin{bmatrix}\mathbf{y}\\\boldsymbol{\theta}\end{bmatrix}$ by $\mathbf{y}^*$, $\begin{bmatrix}\mathbf{X}\\\mathbf{R}\end{bmatrix}$ by $\mathbf{X}^*$, $\boldsymbol{\beta}$ by $\boldsymbol{\beta}^*$, since this term is subject to the restriction that the vector of unknown satisfies $\mathbf{R}\boldsymbol{\beta} = \boldsymbol{\theta}$, and $\begin{bmatrix}\boldsymbol{\epsilon}\\\boldsymbol{\theta}\end{bmatrix}$ by $\boldsymbol{\epsilon}^*$. Because $\mathbf{X}^*$ is of full rank, the least squares solution to

(5.9-6)
$$
\underset{(N+1)\times 1}{\mathbf{y}^*} = \underset{(N+1)\times h}{\mathbf{X}^*} \, \underset{h\times 1}{\boldsymbol{\beta}^*} + \underset{(N+1)\times 1}{\boldsymbol{\epsilon}^*}
$$

is obtained in the usual way, namely $\hat{\boldsymbol{\beta}}^* = (\mathbf{X}^{*\prime}\mathbf{X}^*)^{-1}(\mathbf{X}^{*\prime}\mathbf{y}^*)$.

TABLE 5.9-1 Computational Procedures for a Type CR-4 Design Using a Less than Full Rank Experimental Design Model (Solution by restricting the number of unknowns)

(i)   Data and basic matrices ($N = 32$, $h = 5$, $p = 4$):

$$
\mathbf{y}^*_{(N+1)\times 1} \qquad \mathbf{X}^*_{(N+1)\times h}
$$

$$
\mathbf{X_0}\ \mathbf{X_1}\ \mathbf{X_2}\ \mathbf{X_3}\ \mathbf{X_4}
$$

$$
a_1\left\{
\begin{bmatrix} 3 \\ 6 \\ 3 \\ 3 \\ 1 \\ 2 \\ 2 \\ 2 \end{bmatrix}
\begin{bmatrix} 1 & 1 & 0 & 0 & 0 \\ 1 & 1 & 0 & 0 & 0 \\ 1 & 1 & 0 & 0 & 0 \\ 1 & 1 & 0 & 0 & 0 \\ 1 & 1 & 0 & 0 & 0 \\ 1 & 1 & 0 & 0 & 0 \\ 1 & 1 & 0 & 0 & 0 \\ 1 & 1 & 0 & 0 & 0 \end{bmatrix}\right.
$$

$$
a_2\left\{
\begin{bmatrix} 4 \\ 5 \\ 4 \\ 3 \\ 2 \\ 3 \\ 4 \\ 3 \end{bmatrix}
\begin{bmatrix} 1 & 0 & 1 & 0 & 0 \\ 1 & 0 & 1 & 0 & 0 \\ 1 & 0 & 1 & 0 & 0 \\ 1 & 0 & 1 & 0 & 0 \\ 1 & 0 & 1 & 0 & 0 \\ 1 & 0 & 1 & 0 & 0 \\ 1 & 0 & 1 & 0 & 0 \\ 1 & 0 & 1 & 0 & 0 \end{bmatrix}\right.
$$

$$
a_3\left\{
\begin{bmatrix} 7 \\ 8 \\ 7 \\ 6 \\ 5 \\ 6 \\ 5 \\ 6 \end{bmatrix}
\begin{bmatrix} 1 & 0 & 0 & 1 & 0 \\ 1 & 0 & 0 & 1 & 0 \\ 1 & 0 & 0 & 1 & 0 \\ 1 & 0 & 0 & 1 & 0 \\ 1 & 0 & 0 & 1 & 0 \\ 1 & 0 & 0 & 1 & 0 \\ 1 & 0 & 0 & 1 & 0 \\ 1 & 0 & 0 & 1 & 0 \end{bmatrix}\right.
$$

$$
a_4\left\{
\begin{bmatrix} 7 \\ 8 \\ 9 \\ 8 \\ 10 \\ 10 \\ 9 \\ 11 \end{bmatrix}
\begin{bmatrix} 1 & 0 & 0 & 0 & 1 \\ 1 & 0 & 0 & 0 & 1 \\ 1 & 0 & 0 & 0 & 1 \\ 1 & 0 & 0 & 0 & 1 \\ 1 & 0 & 0 & 0 & 1 \\ 1 & 0 & 0 & 0 & 1 \\ 1 & 0 & 0 & 0 & 1 \\ 1 & 0 & 0 & 0 & 1 \end{bmatrix}\right.
$$

$$
\begin{bmatrix} 0 \end{bmatrix} \qquad \begin{bmatrix} 0 & 1 & 1 & 1 & 1 \end{bmatrix}
$$

$$
\sum_{i=1}^{N} Y_i = 172
$$

$$
\mathbf{y}^{*\prime}_{1\times(N+1)} \qquad \mathbf{y}^*_{(N+1)\times 1} \qquad (\mathbf{y}^{*\prime}\mathbf{y}^*)_{1\times 1}
$$

$$
\begin{bmatrix} 3 & 6 & 3 & \cdots & 0 \end{bmatrix}
\begin{bmatrix} 3 \\ 6 \\ 3 \\ \vdots \\ 0 \end{bmatrix} = 1160.0
$$

$$
\mathbf{X}^{*\prime}_{h\times(N+1)} \qquad \mathbf{X}^*_{(N+1)\times h} \qquad (\mathbf{X}^{*\prime}\mathbf{X}^*)_{h\times h}
$$

$$
\begin{bmatrix} 1 & 1 & 1 & \cdot & 0 \\ 1 & 1 & 1 & \cdot & 1 \\ 0 & 0 & 0 & \cdot & 1 \\ 0 & 0 & 0 & \cdot & 1 \\ 0 & 0 & 0 & \cdot & 1 \end{bmatrix}
\begin{bmatrix} 1 & 1 & 0 & 0 & 0 \\ 1 & 1 & 0 & 0 & 0 \\ 1 & 1 & 0 & 0 & 0 \\ \cdot & \cdot & \cdot & \cdot & \cdot \\ 0 & 1 & 1 & 1 & 1 \end{bmatrix} =
\begin{bmatrix} 32 & 8 & 8 & 8 & 8 \\ 8 & 9 & 1 & 1 & 1 \\ 8 & 1 & 9 & 1 & 1 \\ 8 & 1 & 1 & 9 & 1 \\ 8 & 1 & 1 & 1 & 9 \end{bmatrix}
$$

$$
\mathbf{X}^{*\prime}_{h\times(N+1)} \qquad \mathbf{y}^*_{(N+1)\times 1} \qquad (\mathbf{X}^{*\prime}\mathbf{y}^*)_{h\times 1}
$$

$$
\begin{bmatrix} 1 & 1 & 1 & \cdot & 0 \\ 1 & 1 & 1 & \cdot & 1 \\ 0 & 0 & 0 & \cdot & 1 \\ 0 & 0 & 0 & \cdot & 1 \\ 0 & 0 & 0 & \cdot & 1 \end{bmatrix}
\begin{bmatrix} 3 \\ 6 \\ 3 \\ \vdots \\ 0 \end{bmatrix} =
\begin{bmatrix} 172 \\ 22 \\ 28 \\ 50 \\ 72 \end{bmatrix}
$$

$$
(\mathbf{X}^{*\prime}\mathbf{X}^*)^{-1}_{h\times h} \qquad (\mathbf{X}^{*\prime}\mathbf{y}^*)_{h\times 1} \qquad \hat{\boldsymbol{\beta}}^*_{h\times 1}
$$

$$
\begin{bmatrix} 3/32 & -2/32 & -2/32 & -2/32 & -2/32 \\ -2/32 & 5/32 & 1/32 & 1/32 & 1/32 \\ -2/32 & 1/32 & 5/32 & 1/32 & 1/32 \\ -2/32 & 1/32 & 1/32 & 5/32 & 1/32 \\ -2/32 & 1/32 & 1/32 & 1/32 & 5/32 \end{bmatrix}
\begin{bmatrix} 172 \\ 22 \\ 28 \\ 50 \\ 72 \end{bmatrix} =
\begin{bmatrix} 5.375 \\ -2.625 \\ -1.875 \\ 0.875 \\ 3.625 \end{bmatrix}
$$

$$
\hat{\boldsymbol{\beta}}^{*\prime}_{1\times h} \qquad (\mathbf{X}^{*\prime}\mathbf{y}^*)_{h\times 1} \qquad (\hat{\boldsymbol{\beta}}^{*\prime}\mathbf{X}^{*\prime}\mathbf{y}^*)_{1\times 1}
$$

$$
\begin{bmatrix} 5.375 & \cdots & 3.625 \end{bmatrix}
\begin{bmatrix} 172 \\ 22 \\ 28 \\ 50 \\ 72 \end{bmatrix} = 1119.0
$$

TABLE 5.9-1 (continued)

(ii) Computation of sum of squares:

$$SSTO = \mathbf{y}^{\star\prime}\mathbf{y}^{\star} - \left(\sum_{i=1}^{N} Y_i\right)^2 \Big/ N = 1160.0 - (172)^2/32$$

$$= 1160.0 - 924.5 = 235.5$$

$$SSBG = \hat{\boldsymbol{\beta}}^{\star\prime}\mathbf{X}^{\star\prime}\mathbf{y}^{\star} - \left(\sum_{i=1}^{N} Y_i\right)^2 \Big/ N = 1119.0 - (172)^2/32$$

$$= 1119.0 - 924.5 = 194.5$$

$$SSWG = \mathbf{y}^{\star\prime}\mathbf{y}^{\star} - \hat{\boldsymbol{\beta}}^{\star\prime}\mathbf{X}^{\star\prime}\mathbf{y}^{\star} = 1160.0 - 1119.0 = 41.0$$

(iii) Computation of mean squares and F statistic:

$$MSBG = SSBG/(p - 1) = 194.5/(4 - 1) = 64.833$$

$$MSWG = SSWG/(N - p) = 41.0/(32 - 4) = 1.464$$

$$F = MSBG/MSWG = 64.833/1.464 = 44.28$$

$$F_{.05;3,28} = 2.95$$

The data for the sleep-deprivation experiment in Table 4.3-1 will be used to illustrate the computational procedures for obtaining a solution to (5.9-6). The statistical hypothesis that we will test at $\alpha = .05$ is

$$H_0: \quad \underset{\substack{\mathbf{C}' \\ p \times h}}{\begin{bmatrix} 0 & 1 & 0 & 0 & 0 \\ 0 & 0 & 1 & 0 & 0 \\ 0 & 0 & 0 & 1 & 0 \\ 0 & 0 & 0 & 0 & 1 \end{bmatrix}} \underset{\substack{\boldsymbol{\beta}^{\star} \\ h \times 1}}{\begin{bmatrix} \mu \\ \alpha_1 \\ \alpha_2 \\ \alpha_3 \\ \alpha_4 \end{bmatrix}} = \underset{\substack{\mathbf{0} \\ p \times 1}}{\begin{bmatrix} 0 \\ 0 \\ 0 \\ 0 \end{bmatrix}}.$$

The analysis is given in Table 5.9-1. The mean squares and F statistic are identical to those in Table 4.3-2, where the conventional approach to ANOVA was used. Since $F = MSBG/MSWG = 44.28$ exceeds $F_{.05;2,28} = 2.95$, the null hypothesis is rejected.

The treatment level means can be obtained by constructing a $p \times h$ coefficient matrix $\mathbf{W}$ from the structural matrix $\mathbf{X}$ so that $\mathbf{W}\boldsymbol{\beta}^{\star} = \bar{\mathbf{y}}$. For the data in Table 5.9-1, $\mathbf{W}$ is

(5.9-7)
$$\underset{p \times h}{\mathbf{W}} = \begin{bmatrix} 1 & 1 & 0 & 0 & 0 \\ 1 & 0 & 1 & 0 & 0 \\ 1 & 0 & 0 & 1 & 0 \\ 1 & 0 & 0 & 0 & 1 \end{bmatrix}.$$

Note that the ones and zeros in $\mathbf{W}$ correspond to the pattern in $\mathbf{X}$, the unaugmented structural matrix. Procedures for computing the terms $\hat{\psi}_i$, $\hat{\sigma}_{\bar{Y}}$, and $\hat{\sigma}_{\psi_i}$ needed to evaluate multiple contrasts are illustrated in Table 5.9-2. The null hypothesis H_0: $\mu_1 - \mu_2 = 0$ is used for illustrative purposes. The results in Table 5.9-2 are identical to those presented previously in Section 4.4 and Tables 5.6-1 and 5.7-2.

SOLUTION BY REPARAMETERIZATION

The method of reparameterization is another approach* to obtaining a solution to the system of equations represented by

(5.9-8)
$$Y_{ij} = \mu + \alpha_j + \epsilon_{i(j)} \qquad (i = 1, \ldots, n; j = 1, \ldots, p)$$

where $\epsilon_{i(j)}$ is $NID(0, \sigma_\epsilon^2)$. As we saw earlier the structural matrix for this system is rank deficient. The method of reparameterization, unlike the method of restricting the number of unknowns, does not let us solve for the original vector of parameters $\boldsymbol{\beta}' = [\mu \; \alpha_1 \; \alpha_2 \; \cdots \; \alpha_p]$. Instead we solve for only r linear combinations of the parameters, where r is the rank of the structural matrix. The r linear combinations represent a new set of unknowns, $\boldsymbol{\beta}^* = \mathbf{E}\boldsymbol{\beta}$, where $\boldsymbol{\beta}^*$ is the new $r \times 1$ vector of parameters, $\mathbf{E}$ is an $r \times h$ matrix of rank r whose rows are linearly dependent on the rows of the structural matrix $\mathbf{X}$, and $\boldsymbol{\beta}$ is the original $h \times 1$ vector of parameters. The choice of $\mathbf{E}$ is arbitrary, subject to the preceding restrictions. However, the rows of $\mathbf{E}$ should be chosen so that the new vector of parameters $\boldsymbol{\beta}^*$ has a useful interpretation.

For the hand-steadiness data in Table 4.3-1, consider the following two choices for $\mathbf{E}$. In these examples, $\boldsymbol{\beta}' = [\mu \; \alpha_1 \; \alpha_2 \; \alpha_3 \; \alpha_4]$.

(5.9-9)
$$\underset{r \times h}{\mathbf{E}_1} = \begin{bmatrix} 1 & \frac{1}{4} & \frac{1}{4} & \frac{1}{4} & \frac{1}{4} \\ 0 & 1 & 0 & 0 & -1 \\ 0 & 0 & 1 & 0 & -1 \\ 0 & 0 & 0 & 1 & -1 \end{bmatrix}$$

$$\underset{r \times h}{\mathbf{E}_1} \underset{h \times 1}{\boldsymbol{\beta}} = \begin{bmatrix} \mu + \dfrac{\alpha_1 + \alpha_2 + \alpha_3 + \alpha_4}{4} \\ \alpha_1 - \alpha_4 \\ \alpha_2 - \alpha_4 \\ \alpha_3 - \alpha_4 \end{bmatrix} \underset{r \times 1}{\boldsymbol{\beta}_1^*}$$

$$\underset{r \times h}{\mathbf{E}_2} = \begin{bmatrix} 1 & \frac{1}{4} & \frac{1}{4} & \frac{1}{4} & \frac{1}{4} \\ 0 & 1 & -1 & 0 & 0 \\ 0 & 0 & 1 & -1 & 0 \\ 0 & 0 & 0 & 1 & -1 \end{bmatrix}$$

$$\underset{r \times h}{\mathbf{E}_2} \underset{h \times 1}{\boldsymbol{\beta}} = \begin{bmatrix} \mu + \dfrac{\alpha_1 + \alpha_2 + \alpha_3 + \alpha_4}{4} \\ \alpha_1 - \alpha_2 \\ \alpha_2 - \alpha_3 \\ \alpha_3 - \alpha_4 \end{bmatrix} \underset{r \times 1}{\boldsymbol{\beta}_2^*}$$

It is easy to show that the rows of $\mathbf{E}_1$, for example, are linear combinations of the rows of $\mathbf{X}$ in Table 5.9-3.

---

* For an in-depth discussion of this approach see Bock (1975) and Graybill (1961); Tatsuoka (1975) provides an excellent shorter introduction.

TABLE 5.9-2 Computational Procedures for Testing Differences Among Means (Solution by restricting the number of unknowns)

(i) Computation of means, $\bar{y}$, and contrast, $\hat{\psi}_1$ ($h = 5$, $n = 8$, $p = 4$):*

$$
\underset{p \times h}{\mathbf{W}} \quad \underset{h \times 1}{\hat{\boldsymbol{\beta}}^*} \quad = \quad \underset{p \times 1}{\bar{\mathbf{y}}}
$$

$$
\begin{bmatrix} 1 & 1 & 0 & 0 & 0 \\ 1 & 0 & 1 & 0 & 0 \\ 1 & 0 & 0 & 1 & 0 \\ 1 & 0 & 0 & 0 & 1 \end{bmatrix} \begin{bmatrix} 5.375 \\ -2.625 \\ -1.875 \\ 0.875 \\ 3.625 \end{bmatrix} = \begin{bmatrix} 2.75 \\ 3.50 \\ 6.25 \\ 9.00 \end{bmatrix}
$$

$$
\underset{1 \times p}{\mathbf{c}'} \qquad \underset{p \times 1}{\bar{\mathbf{y}}} \qquad \underset{1 \times 1}{\hat{\psi}_1}
$$

$$
\begin{bmatrix} 1 & -1 & 0 & 0 \end{bmatrix} \begin{bmatrix} 2.75 \\ 3.50 \\ 6.25 \\ 9.00 \end{bmatrix} = -0.75
$$

(ii) Computation of $\hat{\sigma}_{\bar{Y}}^2$ and $\sigma_{\hat{\psi}_1}^2$:**

$\hat{\sigma}_{\bar{Y}}^2 = MSWG/n = 1.464/8 = 0.183$

$$
\underset{1 \times p}{\mathbf{c}'} \qquad\qquad \underset{p \times h}{\mathbf{W}} \qquad\qquad (\underset{h \times h}{\mathbf{X}^{*\prime}\mathbf{X}^*})^{-1} \qquad \underset{h \times p}{\mathbf{W}'} \qquad \underset{p \times 1}{\mathbf{c}}
$$

$$
\sigma_{\hat{\psi}_1}^2 = MSWG \left\{ \begin{bmatrix} 1 & -1 & 0 & 0 \end{bmatrix} \begin{bmatrix} 1 & 1 & 0 & 0 & 0 \\ 1 & 0 & 1 & 0 & 0 \\ 1 & 0 & 0 & 1 & 0 \\ 1 & 0 & 0 & 0 & 1 \end{bmatrix} \begin{bmatrix} 3/32 & -2/32 & -2/32 & -2/32 & -2/32 \\ -2/32 & 5/32 & 1/32 & 1/32 & 1/32 \\ -2/32 & 1/32 & 5/32 & 1/32 & 1/32 \\ -2/32 & 1/32 & 1/32 & 5/32 & 1/32 \\ -2/32 & 1/32 & 1/32 & 1/32 & 5/32 \end{bmatrix} \begin{bmatrix} 1 & 1 & 1 & 1 \\ 1 & 0 & 0 & 0 \\ 0 & 1 & 0 & 0 \\ 0 & 0 & 1 & 0 \\ 0 & 0 & 0 & 1 \end{bmatrix} \begin{bmatrix} 1 \\ -1 \\ 0 \\ 0 \end{bmatrix} \right\}
$$

$$
\sigma_{\hat{\psi}_1}^2 = 1.464 \left\{ \begin{bmatrix} 1 & -1 & 0 & 0 \end{bmatrix} \begin{bmatrix} 1/8 & 0 & 0 & 0 \\ 0 & 1/8 & 0 & 0 \\ 0 & 0 & 1/8 & 0 \\ 0 & 0 & 0 & 1/8 \end{bmatrix} \begin{bmatrix} 1 \\ -1 \\ 0 \\ 0 \end{bmatrix} \right\} = 1.464(2/8) = 0.366
$$

(iii) Computation of test statistics:

$$
t = \frac{\hat{\psi}_1}{\hat{\sigma}_{\hat{\psi}_1}} = \frac{-0.75}{\sqrt{0.366}} = \frac{-0.75}{0.605} = -1.24 \qquad q = \frac{\hat{\psi}_1}{\hat{\sigma}_{\bar{Y}}} = \frac{-0.75}{\sqrt{0.183}} = \frac{-0.75}{0.428} = -1.75
$$

* Matrix $\mathbf{W}$ is obtained from (5.9-7), $\hat{\boldsymbol{\beta}}^*$ is obtained from Table 5.9-1, and $\mathbf{c}'$ is the coefficient vector associated with the hypothesis $\mu_1 - \mu_2 = 0$.

** Matrix $(\mathbf{X}^{*\prime}\mathbf{X}^*)^{-1}$ and $MSWG$ are obtained from Table 5.9-1.

Rows of $\mathbf{X}$ *in Table* 5.9-3	*Rows of* $\mathbf{E}_1$

(1) (9) (17) (25)	(1)
$\tfrac{1}{4}[1\ 1\ 0\ 0\ 0] + \tfrac{1}{4}[1\ 0\ 1\ 0\ 0] + \tfrac{1}{4}[1\ 0\ 0\ 1\ 0] + \tfrac{1}{4}[1\ 0\ 0\ 0\ 1] = [1\ \tfrac{1}{4}\ \tfrac{1}{4}\ \tfrac{1}{4}\ \tfrac{1}{4}]$	$[1\ \tfrac{1}{4}\ \tfrac{1}{4}\ \tfrac{1}{4}\ \tfrac{1}{4}]$
(1) (25)	(2)
$1[1\ 1\ 0\ 0\ 0] - 1[1\ 0\ 0\ 0\ 1]$	$= [0\ 1\ 0\ 0\ -1]$
(9) (25)	(3)
$1[1\ 0\ 1\ 0\ 0] - 1[1\ 0\ 0\ 0\ 1]$	$= [0\ 0\ 1\ 0\ -1]$
(17) (25)	(4)
$1[1\ 0\ 0\ 1\ 0] - 1[1\ 0\ 0\ 0\ 1]$	$= [0\ 0\ 0\ 1\ -1]$

Both $\mathbf{E}_1$ and $\mathbf{E}_2$ would provide interpretable reparameterizations of $\boldsymbol{\beta}$; in the discussion that follows we will adopt $\mathbf{E}_1$ and henceforth refer to it as $\mathbf{E}$.

As we have seen, we cannot solve for $\hat{\boldsymbol{\beta}}$ in the normal equations

(5.9-10)
$$\underset{h\times N}{\mathbf{X}'}\ \underset{N\times h}{\mathbf{X}}\ \underset{h\times 1}{\hat{\boldsymbol{\beta}}} = \underset{h\times N}{\mathbf{X}'}\ \underset{N\times 1}{\mathbf{y}}$$

because the $\mathbf{X}$ matrix is rank deficient—$r < h$ and $(\mathbf{X}'\mathbf{X})^{-1}$ does not exist. However, the method of reparameterization enables us to solve for $\hat{\boldsymbol{\beta}}^\star$ in

(5.9-11)
$$\underset{r\times N}{\mathbf{Z}'}\ \underset{N\times r}{\mathbf{Z}}\ \underset{r\times 1}{\hat{\boldsymbol{\beta}}^\star} = \underset{r\times N}{\mathbf{Z}'}\ \underset{N\times 1}{\mathbf{y}}$$

where $\mathbf{X}$ has been replaced by $\mathbf{Z}$, an $N \times r$ matrix that is of full rank. The $\mathbf{Z}$ matrix is called a *basis matrix*. We will now see how $\mathbf{Z}$ is obtained.

If the rows of $\mathbf{E}$ are linearly dependent on the rows of $\mathbf{X}$, it can be shown (Timm, 1975, 42) that $\mathbf{X}$ is factorable into the product of $\mathbf{ZE}$, that is

(5.9-12)
$$\underset{N\times h}{\mathbf{X}} = \underset{N\times r}{\mathbf{Z}}\ \underset{r\times h}{\mathbf{E}}.$$

It is a simple matter to solve for $\mathbf{Z}$. We begin by postmultiplying both sides of (5.9-12) by $\mathbf{E}'$. Then postmultiplying by $(\mathbf{EE}')^{-1}$ gives the solution.

$$\underset{N\times h}{\mathbf{X}}\ \underset{h\times r}{\mathbf{E}'} = \underset{N\times r}{\mathbf{Z}}\ \underset{r\times h}{\mathbf{E}}\ \underset{h\times r}{\mathbf{E}'}$$

$$\underset{N\times r}{(\mathbf{XE}')}\underset{r\times r}{(\mathbf{EE}')^{-1}} = \underset{N\times r}{\mathbf{Z}}\ \underset{r\times r}{(\mathbf{EE}')}\underset{r\times r}{(\mathbf{EE}')^{-1}}$$

$$\underset{N\times r}{(\mathbf{XE}')}\underset{r\times r}{(\mathbf{EE}')^{-1}} = \underset{N\times r}{\mathbf{Z}}$$

Now that we know how to compute $\mathbf{Z}$ we can solve for $\hat{\boldsymbol{\beta}}^\star$ in (5.9-11) in the usual way, namely,

(5.9-13)
$$\underset{r\times 1}{\hat{\boldsymbol{\beta}}^\star} = \underset{r\times r}{(\mathbf{Z}'\mathbf{Z})^{-1}}\underset{r\times 1}{(\mathbf{Z}'\mathbf{y})}.$$

The data for the sleep-deprivation experiment in Table 4.3-1 will be used to illustrate the computational procedures for obtaining a solution to $\mathbf{y} = \mathbf{Z}\boldsymbol{\beta}^\star + \boldsymbol{\epsilon}$. A test of the null hypothesis $\alpha_1 = \alpha_2 = \alpha_3 = \alpha_4 = 0$ is equivalent to testing the null hypothesis

$$\alpha_1 - \alpha_4 = 0$$
$$\alpha_2 - \alpha_4 = 0$$
$$\alpha_3 - \alpha_4 = 0.$$

The latter hypothesis can be written in matrix notation as

(5.9-14)

$$H_0: \quad \underset{(p-1)\times r}{\mathbf{C'}} \quad \underset{r\times 1}{\boldsymbol{\beta^\star}} \quad \underset{(p-1)\times 1}{\mathbf{0}}$$

$$\begin{bmatrix} 0 & 1 & 0 & 0 \\ 0 & 0 & 1 & 0 \\ 0 & 0 & 0 & 1 \end{bmatrix} \begin{bmatrix} \mu + (\alpha_1 + \alpha_2 + \alpha_3 + \alpha_4)/4 \\ \alpha_1 - \alpha_4 \\ \alpha_2 - \alpha_4 \\ \alpha_3 - \alpha_4 \end{bmatrix} = \begin{bmatrix} 0 \\ 0 \\ 0 \end{bmatrix}.$$

The .05 level of significance is adopted. Table 5.9-3 presents the analysis. The sums of squares are identical to those in Table 4.3-2 where the traditional approach to ANOVA was used. Since $F = MSBG/MSWG = 44.28$ exceeds $F_{.05;3,28} = 2.95$, we can reject the null hypothesis.

The treatment level means can be obtained by constructing a $p \times r$ coefficient matrix $\mathbf{W}$ based on the pattern of ones and zeros in the basis matrix $\mathbf{Z}$. For the data in Table 5.9-3, $\mathbf{W}$ is

(5.9-15)

$$\underset{p\times r}{\mathbf{W}} = \begin{bmatrix} 1 & 3/4 & -1/4 & -1/4 \\ 1 & -1/4 & 3/4 & -1/4 \\ 1 & -1/4 & -1/4 & 3/4 \\ 1 & -1/4 & -1/4 & -1/4 \end{bmatrix}.$$

Procedures for computing the terms ψ_i, $\hat{\sigma}_{\bar{Y}}$, and σ_{ψ_i} needed to evaluate multiple contrasts are illustrated in Table 5.9-4. The null hypothesis H_0: $\mu_1 - \mu_2 = 0$ is used for illustrative purposes. The results in Table 5.9-4 are identical to those presented previously in Section 4.4 and Tables 5.6-1, 5.7-2, and 5.9-2.

The reader may feel that the procedure we have used to obtain a solution to $\hat{\boldsymbol{\beta}}^\star = (\mathbf{Z'Z})^{-1}(\mathbf{Z'y})$ is a bit arbitrary since $\mathbf{E}$, which was used to obtain $\mathbf{Z}$, was chosen subject only to the restrictions that it is an $r \times h$ matrix of rank r whose rows are linearly dependent on the rows of the structural matrix $\mathbf{X}$. This is correct and as a result the solution $\hat{\boldsymbol{\beta}}^\star$ is not unique. However, by choosing the rows of $\mathbf{W}$ as in (5.9-15), $\mathbf{W}\hat{\boldsymbol{\beta}}^\star$ has the same value regardless of the choice of $\mathbf{E}$.

It may be helpful to briefly review the problem addressed by the method of reparameterization and the way the problem was solved. We began with the system of equations (5.9-8) for a completely randomized design. The system can be written in matrix notation as

$$\underset{N\times 1}{\mathbf{y}} = \underset{N\times h}{\mathbf{X}} \underset{h\times 1}{\boldsymbol{\beta}} + \underset{N\times 1}{\boldsymbol{\epsilon}}.$$

TABLE 5.9-3 Computational Procedures for a Type CR-4 Design Using a Less than Full Rank Experimental Design Model (Solution by reparameterization)

(i) Data and basic matrices ($N = 32$, $h = 5$, $p = 4$, $r = 4$):[*]

$$
\begin{array}{c}
\underset{N \times 1}{\mathbf{y}} \qquad \underset{N \times h}{\mathbf{X}}
\end{array}
$$

$$
\begin{array}{cc}
 & \mathbf{x}_0\ \mathbf{x}_1\ \mathbf{x}_2\ \mathbf{x}_3\ \mathbf{x}_4 \\
a_1 \left\{ \begin{bmatrix} 3 \\ 6 \\ 3 \\ 3 \\ 1 \\ 2 \\ 2 \\ 2 \end{bmatrix} \right. &
\begin{bmatrix} 1 & 1 & 0 & 0 & 0 \\ 1 & 1 & 0 & 0 & 0 \\ 1 & 1 & 0 & 0 & 0 \\ 1 & 1 & 0 & 0 & 0 \\ 1 & 1 & 0 & 0 & 0 \\ 1 & 1 & 0 & 0 & 0 \\ 1 & 1 & 0 & 0 & 0 \\ 1 & 1 & 0 & 0 & 0 \end{bmatrix}
\end{array}
$$

$$
a_2 \left\{ \begin{bmatrix} 4 \\ 5 \\ 4 \\ 3 \\ 2 \\ 3 \\ 4 \\ 3 \end{bmatrix} \right.
\begin{bmatrix} 1 & 0 & 1 & 0 & 0 \\ 1 & 0 & 1 & 0 & 0 \\ 1 & 0 & 1 & 0 & 0 \\ 1 & 0 & 1 & 0 & 0 \\ 1 & 0 & 1 & 0 & 0 \\ 1 & 0 & 1 & 0 & 0 \\ 1 & 0 & 1 & 0 & 0 \\ 1 & 0 & 1 & 0 & 0 \end{bmatrix}
$$

$$
a_3 \left\{ \begin{bmatrix} 7 \\ 8 \\ 7 \\ 6 \\ 5 \\ 6 \\ 5 \\ 6 \end{bmatrix} \right.
\begin{bmatrix} 1 & 0 & 0 & 1 & 0 \\ 1 & 0 & 0 & 1 & 0 \\ 1 & 0 & 0 & 1 & 0 \\ 1 & 0 & 0 & 1 & 0 \\ 1 & 0 & 0 & 1 & 0 \\ 1 & 0 & 0 & 1 & 0 \\ 1 & 0 & 0 & 1 & 0 \\ 1 & 0 & 0 & 1 & 0 \end{bmatrix}
$$

$$
a_4 \left\{ \begin{bmatrix} 7 \\ 8 \\ 9 \\ 8 \\ 10 \\ 10 \\ 9 \\ 11 \end{bmatrix} \right.
\begin{bmatrix} 1 & 0 & 0 & 0 & 1 \\ 1 & 0 & 0 & 0 & 1 \\ 1 & 0 & 0 & 0 & 1 \\ 1 & 0 & 0 & 0 & 1 \\ 1 & 0 & 0 & 0 & 1 \\ 1 & 0 & 0 & 0 & 1 \\ 1 & 0 & 0 & 0 & 1 \\ 1 & 0 & 0 & 0 & 1 \end{bmatrix}
$$

$$
\sum_{i=1}^{N} Y_i = 172
$$

$$
\underset{1 \times N}{\mathbf{y}'} \qquad \underset{N \times 1}{\mathbf{y}} \qquad \underset{1 \times 1}{(\mathbf{y}'\mathbf{y})}
$$

$$
[3 \quad 6 \quad 3 \quad \cdots \quad 11] \begin{bmatrix} 3 \\ 6 \\ 3 \\ \vdots \\ 11 \end{bmatrix} = 1160.0
$$

$$
\underset{r \times h}{\mathbf{E}} \qquad\qquad \underset{h \times r}{\mathbf{E}'} \qquad\qquad \underset{r \times r}{(\mathbf{E}\mathbf{E}')}
$$

$$
\begin{bmatrix} 1 & \tfrac14 & \tfrac14 & \tfrac14 & \tfrac14 \\ 0 & 1 & 0 & 0 & -1 \\ 0 & 0 & 1 & 0 & -1 \\ 0 & 0 & 0 & 1 & -1 \end{bmatrix}
\begin{bmatrix} 1 & 0 & 0 & 0 \\ \tfrac14 & 1 & 0 & 0 \\ \tfrac14 & 0 & 1 & 0 \\ \tfrac14 & 0 & 0 & 1 \\ \tfrac14 & -1 & -1 & -1 \end{bmatrix}
=
\begin{bmatrix} 1\tfrac14 & 0 & 0 & 0 \\ 0 & 2 & 1 & 1 \\ 0 & 1 & 2 & 1 \\ 0 & 1 & 1 & 2 \end{bmatrix}
$$

[*] Matrix **E** is defined in (5.9-9).

TABLE 5.9-3 (continued)

$$
\underset{N \times h}{\mathbf{X}} \qquad \underset{h \times r}{\mathbf{E}'} \qquad \underset{r \times r}{(\mathbf{E}\mathbf{E}')^{-1}} \qquad \underset{N \times r}{\mathbf{Z}}
$$

$$
\begin{bmatrix}
1 & 1 & 0 & 0 & 0 \\
1 & 1 & 0 & 0 & 0 \\
1 & 1 & 0 & 0 & 0 \\
\cdot & \cdot & \cdot & \cdot & \cdot \\
1 & 0 & 0 & 0 & 1
\end{bmatrix}
\begin{bmatrix}
1 & 0 & 0 & 0 \\
1/4 & 1 & 0 & 0 \\
1/4 & 0 & 1 & 0 \\
1/4 & 0 & 0 & 1 \\
1/4 & -1 & -1 & -1
\end{bmatrix}
\begin{bmatrix}
4/5 & 0 & 0 & 0 \\
0 & 3/4 & -1/4 & -1/4 \\
0 & -1/4 & 3/4 & -1/4 \\
0 & -1/4 & -1/4 & 3/4
\end{bmatrix}
=
\begin{bmatrix}
1 & 3/4 & -1/4 & -1/4 \\
1 & 3/4 & -1/4 & -1/4 \\
1 & 3/4 & -1/4 & -1/4 \\
1 & 3/4 & -1/4 & -1/4 \\
1 & 3/4 & -1/4 & -1/4 \\
1 & 3/4 & -1/4 & -1/4 \\
1 & 3/4 & -1/4 & -1/4 \\
1 & 3/4 & -1/4 & -1/4 \\
1 & -1/4 & 3/4 & -1/4 \\
1 & -1/4 & 3/4 & -1/4 \\
1 & -1/4 & 3/4 & -1/4 \\
1 & -1/4 & 3/4 & -1/4 \\
1 & -1/4 & 3/4 & -1/4 \\
1 & -1/4 & 3/4 & -1/4 \\
1 & -1/4 & 3/4 & -1/4 \\
1 & -1/4 & 3/4 & -1/4 \\
1 & -1/4 & -1/4 & 3/4 \\
1 & -1/4 & -1/4 & 3/4 \\
1 & -1/4 & -1/4 & 3/4 \\
1 & -1/4 & -1/4 & 3/4 \\
1 & -1/4 & -1/4 & 3/4 \\
1 & -1/4 & -1/4 & 3/4 \\
1 & -1/4 & -1/4 & 3/4 \\
1 & -1/4 & -1/4 & 3/4 \\
1 & -1/4 & -1/4 & -1/4 \\
1 & -1/4 & -1/4 & -1/4 \\
1 & -1/4 & -1/4 & -1/4 \\
1 & -1/4 & -1/4 & -1/4 \\
1 & -1/4 & -1/4 & -1/4 \\
1 & -1/4 & -1/4 & -1/4 \\
1 & -1/4 & -1/4 & -1/4 \\
1 & -1/4 & -1/4 & -1/4
\end{bmatrix}
$$

$$
\underset{r \times N}{\mathbf{Z}'} \qquad \underset{N \times 1}{\mathbf{y}} \qquad \underset{r \times 1}{\mathbf{Z}'\mathbf{y}}
$$

$$
\begin{bmatrix}
1 & 1 & 1 & \cdot & 1 \\
3/4 & 3/4 & 3/4 & \cdot & -1/4 \\
-1/4 & -1/4 & -1/4 & \cdot & -1/4 \\
-1/4 & -1/4 & -1/4 & \cdot & -1/4
\end{bmatrix}
\begin{bmatrix}
3 \\ 6 \\ 3 \\ \vdots \\ \vdots \\ 11
\end{bmatrix}
=
\begin{bmatrix}
172 \\ -21 \\ -15 \\ 7
\end{bmatrix}
$$

$$
\underset{r \times r}{(\mathbf{Z}'\mathbf{Z})^{-1}} \qquad \underset{r \times 1}{(\mathbf{Z}'\mathbf{y})} \qquad \underset{r \times 1}{\hat{\boldsymbol{\beta}}^{\star}}
$$

$$
\begin{bmatrix}
1/32 & 0 & 0 & 0 \\
0 & 8/32 & 4/32 & 4/32 \\
0 & 4/32 & 8/32 & 4/32 \\
0 & 4/32 & 4/32 & 8/32
\end{bmatrix}
\begin{bmatrix}
172 \\ -21 \\ -15 \\ 7
\end{bmatrix}
=
\begin{bmatrix}
5.375 \\ -6.250 \\ -5.500 \\ -2.750
\end{bmatrix}
$$

$$
\underset{1 \times r}{\hat{\boldsymbol{\beta}}^{\star\prime}} \qquad \underset{r \times 1}{\mathbf{Z}'\mathbf{y}} \qquad \underset{1 \times 1}{(\hat{\boldsymbol{\beta}}^{\star\prime}\mathbf{Z}'\mathbf{y})}
$$

$$
[5.375 \quad \cdots \quad -2.750]
\begin{bmatrix}
172 \\ -21 \\ -15 \\ 7
\end{bmatrix}
= 1119.0
$$

TABLE 5.9-3   (continued)

(ii)   Computation of sum of squares:

$$SSTO = \mathbf{y'y} - \left(\sum_{i=1}^{N} Y_i\right)^2 \Big/ N = 1160.0 - (172)^2/32$$

$$= 1160.0 - 924.5 = 235.5$$

$$SSBG = \hat{\boldsymbol{\beta}}^{*\prime}\mathbf{Z'y} - \left(\sum_{i=1}^{N} Y_i\right)^2 \Big/ N = 1119.0 - (172)^2/32$$

$$= 1119.0 - 924.5 = 194.5$$

$$SSWG = \mathbf{y'y} - \hat{\boldsymbol{\beta}}^{*\prime}\mathbf{Z'y} = 1160.0 - 1119.0 = 41.0$$

(iii)   Computation of mean squares and F statistic:

$$MSBG = SSBG/(p - 1) = 194.5/(4 - 1) = 64.833$$

$$MSWG = SSWG/(N - p) = 41.0/(32 - 4) = 1.464$$

$$F = MSBG/MSWG = 64.833/1.464 = 44.28$$

$$F_{.05;3,28} = 2.95$$

Adopting the least squares approach, we sought a solution that minimized the sum of the squared errors

$$\hat{\boldsymbol{\epsilon}}'\hat{\boldsymbol{\epsilon}} = (\mathbf{y} - \mathbf{X}\hat{\boldsymbol{\beta}})'(\mathbf{y} - \mathbf{X}\hat{\boldsymbol{\beta}}).$$

However, the usual approach to solving for $\hat{\boldsymbol{\beta}}$ in the normal equations

$$\mathbf{X'X}\hat{\boldsymbol{\beta}} = \mathbf{X'y}$$

could not be used because $\mathbf{X}$ is not of full rank and hence the inverse $(\mathbf{X'X})^{-1}$ does not exist. Our solution to this problem was to reparameterize. That is, instead of trying to solve for the original parameters $\boldsymbol{\beta}$, we defined a new set $\boldsymbol{\beta}^*$ whose elements were linear combinations of the original parameters. The new parameters were given by

$$\boldsymbol{\beta}^* = \mathbf{E}\boldsymbol{\beta}$$

where $\mathbf{E}$ was of full row rank and its rows were linearly dependent on the rows of the structural matrix $\mathbf{X}$. Subject to these restrictions, we chose $\mathbf{E}$ so that the product $\mathbf{E}\boldsymbol{\beta} = \boldsymbol{\beta}^*$ enabled us to test the overall null hypothesis H_0: $\alpha_1 = \alpha_2 = \alpha_3 = \alpha_4 = 0$. The new vector of unknowns was

$$\boldsymbol{\beta}^* = \mathbf{E}\boldsymbol{\beta} = \begin{bmatrix} \mu + \dfrac{\alpha_1 + \alpha_2 + \alpha_3 + \alpha_4}{4} \\ \alpha_1 - \alpha_4 \\ \alpha_2 - \alpha_4 \\ \alpha_3 - \alpha_4 \end{bmatrix}.$$

TABLE 5.9-4 Computational Procedures for Testing Differences Among Means (Solution by reparameterization)

(i) Computation of means, $\bar{\mathbf{y}}$, and contrast, $\hat{\psi}_1$ ($n = 8$, $p = 4$, $r = 4$):*

$$
\underset{p\times r}{\mathbf{W}} \qquad \underset{r\times 1}{\hat{\boldsymbol{\beta}}^*} \qquad \underset{p\times 1}{\bar{\mathbf{y}}}
$$

$$
\begin{bmatrix} 1 & \tfrac{3}{4} & -\tfrac{1}{4} & -\tfrac{1}{4} \\ 1 & -\tfrac{1}{4} & \tfrac{3}{4} & -\tfrac{1}{4} \\ 1 & -\tfrac{1}{4} & -\tfrac{1}{4} & \tfrac{3}{4} \\ 1 & -\tfrac{1}{4} & -\tfrac{1}{4} & -\tfrac{1}{4} \end{bmatrix}
\begin{bmatrix} 5.375 \\ -6.250 \\ -5.500 \\ -2.750 \end{bmatrix}
=
\begin{bmatrix} 2.75 \\ 3.50 \\ 6.25 \\ 9.00 \end{bmatrix}
$$

$$
\underset{1\times p}{\mathbf{c}'} \qquad \underset{p\times 1}{\bar{\mathbf{y}}} \qquad \hat{\psi}_1
$$

$$
\begin{bmatrix} 1 & -1 & 0 & 0 \end{bmatrix}
\begin{bmatrix} 2.75 \\ 3.50 \\ 6.25 \\ 9.00 \end{bmatrix}
= -0.75
$$

(ii) Computation of $\hat{\sigma}_{\bar{y}}^2$ and $\sigma_{\psi_1}^2$:**

$$\hat{\sigma}_{\bar{y}}^2 = MSWG/n = 1.464/8 = 0.183$$

$$
MSWG \left\{ \underset{1\times p}{\mathbf{c}'} \qquad \underset{p\times r}{\mathbf{W}} \right.
$$

$$
\sigma_{\psi_1}^2 = 1.464 \begin{bmatrix} 1 & -1 & 0 & 0 \end{bmatrix}
\begin{bmatrix} \tfrac{1}{8} & 0 & 0 & 0 \\ 0 & \tfrac{1}{8} & 0 & 0 \\ 0 & 0 & \tfrac{1}{8} & 0 \\ 0 & 0 & 0 & \tfrac{1}{8} \end{bmatrix}
\begin{bmatrix} 1 & \tfrac{3}{4} & -\tfrac{1}{4} & -\tfrac{1}{4} \\ 1 & -\tfrac{1}{4} & \tfrac{3}{4} & -\tfrac{1}{4} \\ 1 & -\tfrac{1}{4} & -\tfrac{1}{4} & \tfrac{3}{4} \\ 1 & -\tfrac{1}{4} & -\tfrac{1}{4} & -\tfrac{1}{4} \end{bmatrix}
$$

$$
\underset{r\times r}{(\mathbf{Z}'\mathbf{Z})^{-1}} \qquad\qquad \underset{r\times p}{\mathbf{W}'}
$$

$$
\times \begin{bmatrix} \tfrac{1}{32} & 0 & 0 & 0 \\ 0 & \tfrac{8}{32} & \tfrac{4}{32} & \tfrac{4}{32} \\ 0 & \tfrac{4}{32} & \tfrac{8}{32} & \tfrac{4}{32} \\ 0 & \tfrac{4}{32} & \tfrac{4}{32} & \tfrac{8}{32} \end{bmatrix}
\begin{bmatrix} 1 & 1 & 1 & 1 \\ \tfrac{3}{4} & -\tfrac{1}{4} & -\tfrac{1}{4} & -\tfrac{1}{4} \\ -\tfrac{1}{4} & \tfrac{3}{4} & -\tfrac{1}{4} & -\tfrac{1}{4} \\ -\tfrac{1}{4} & -\tfrac{1}{4} & \tfrac{3}{4} & -\tfrac{1}{4} \end{bmatrix}
\begin{bmatrix} 1 \\ -1 \\ 0 \\ 0 \end{bmatrix}
$$

$$
\underset{p\times 1}{\mathbf{c}}
$$

$$
= 1.464 \begin{bmatrix} 1 & -1 & 0 & 0 \end{bmatrix}
\begin{bmatrix} -1 \\ -1 \\ 0 \\ 0 \end{bmatrix}
= 1.464(2/8) = 0.366
$$

(iii) Computation of test statistics:

$$t = \frac{\hat{\psi}_1}{\hat{\sigma}_{\psi_1}} = \frac{-0.75}{\sqrt{0.366}} = \frac{-0.75}{0.605} = -1.24 \qquad q = \frac{\hat{\psi}_1}{\hat{\sigma}_{\bar{y}}} = \frac{-0.75}{\sqrt{0.183}} = \frac{-0.75}{0.428} = -1.75$$

---

* Matrix $\mathbf{W}$ is obtained from (5.9-15), $\hat{\boldsymbol{\beta}}^*$ is obtained from Table 5.9-3, and $\mathbf{c}'$ is the coefficient vector associated with the hypothesis $\mu_1 - \mu_2 = 0$.

** Matrix $(\mathbf{Z}'\mathbf{Z})^{-1}$ and $MSWG$ are obtained from Table 5.9-3.

By selecting the rows of $\mathbf{E}$ so that they were linearly dependent on the rows of $\mathbf{X}$, the so-called *estimable condition,* we ensured that $\boldsymbol{\beta}^{\star}$ was estimable. Although we could not solve for $\hat{\boldsymbol{\beta}}' = [\hat{\mu} + \hat{\alpha}_1 + \hat{\alpha}_2 + \hat{\alpha}_3 + \hat{\alpha}_4]$ in the normal equations

$$\mathbf{X}'\mathbf{X}\hat{\boldsymbol{\beta}} = \mathbf{X}'\mathbf{y}$$

we were able to solve for $\hat{\boldsymbol{\beta}}^{\star}$ in

$$\mathbf{Z}'\mathbf{Z}\hat{\boldsymbol{\beta}}^{\star} = \mathbf{Z}'\mathbf{y}$$

because we had replaced the rank deficient structural matrix $\mathbf{X}$ by a basis matrix $\mathbf{Z}$ that is of full rank.

A test of the total null hypothesis $\mathbf{C}'\boldsymbol{\beta}^{\star} = \mathbf{0}$, where

$$\underset{p \times r}{\mathbf{C}'} = \begin{bmatrix} 1 & 0 & 0 & 0 \\ 0 & 1 & 0 & 0 \\ 0 & 0 & 1 & 0 \\ 0 & 0 & 0 & 1 \end{bmatrix}$$

is given by

$$F = \frac{\{(\mathbf{C}'\hat{\boldsymbol{\beta}}^{\star})'[\mathbf{C}'(\mathbf{Z}'\mathbf{Z})^{-1}\mathbf{C}]^{-1}(\mathbf{C}'\hat{\boldsymbol{\beta}}^{\star})\}/(p - 1)}{(\mathbf{y}'\mathbf{y} - \hat{\boldsymbol{\beta}}^{\star\prime}\mathbf{Z}'\mathbf{y})/(N - p)}.$$

For the null hypothesis specified by (5.9-14), where

$$\underset{(p-1) \times r}{\mathbf{C}'} = \begin{bmatrix} 0 & 1 & 0 & 0 \\ 0 & 0 & 1 & 0 \\ 0 & 0 & 0 & 1 \end{bmatrix}$$

this statistic simplifies to

$$F = \frac{\left[\hat{\boldsymbol{\beta}}^{\star\prime}\mathbf{Z}'\mathbf{y} - \left(\sum_{i=1}^{N} Y_i \right)^2 \Big/ N \right] \Big/ (p - 1)}{(\mathbf{y}'\mathbf{y} - \hat{\boldsymbol{\beta}}^{\star\prime}\mathbf{Z}'\mathbf{y})/(N - p)}$$

with $p - 1$ and $N - p$ degrees of freedom. In the next section we will describe a completely different approach to the analysis of variance—an approach that uses a full rank experimental design model.

5.10 ANOVA VIA THE FULL RANK EXPERIMENTAL DESIGN MODEL APPROACH

The experimental design model discussed in the last section,

(5.10-1) $$Y_{ij} = \mu + \alpha_j + \epsilon_{i(j)} \qquad (i = 1, \ldots, n; j = 1, \ldots, p)$$

where $\epsilon_{i(j)}$ is $NID(0, \sigma_\epsilon^2)$, contains the parameters μ, α_1, α_2, $\ldots$, α_p. An alternative model, called the full rank experimental design model,

$$Y_{ij} = \mu_j + \epsilon_{i(j)} \qquad (i = 1, \ldots, n; j = 1, \ldots, p)$$

where $\epsilon_{i(j)}$ is $NID(0, \sigma_\epsilon^2)$, contains the parameters, μ_1, μ_2, $\ldots$, μ_p. Although we described model (5.10-1) first, this is not the order in which the models evolved historically. According to Urquhart, Weeks, and Henderson (1973), Fisher's early development of ANOVA was conceptualized by his colleagues in terms of means. It was not until later that the means were given a kind of linear structure in terms of other parameters, for example,

$$\mu_j = \mu + \alpha_j.$$

Because of this structure, model (5.10-1) is referred to as an overparameterized model. As we have seen, the model is not of full rank, and this led to the development of three approaches to solving for $\hat{\boldsymbol{\beta}}$ in the normal equations $\mathbf{X'X\hat{\boldsymbol{\beta}}} = \mathbf{X'y}$: (1) solution by restricting the number of unknowns, (2) solution by reparameterization, and (3) solution by use of a generalized inverse.

The less than full rank model poses other problems for an experimenter, such as determining which parameters or parametric functions are estimable. These problems are avoided in the full rank experimental design model since the structural matrix is of full rank and all of the parameters are estimable.* An added advantage, as we will see in subsequent chapters, is that regardless of the complexity of the design an experimenter always knows exactly what hypothesis is being tested.**

We will use the data for the sleep-deprivation experiment in Table 4.3-1 to illustrate the computational procedures for the full rank experimental design model. The null hypothesis for these data is

(5.10-2) $$H_0: \quad \alpha_1 = \alpha_2 = \alpha_3 = \alpha_4 = 0.$$

An equivalent hypothesis in terms of population means for the full rank experimental design model is

---

*   For an in-depth discussion of the full rank experimental design model, the reader is referred to Carlson and Timm (1974), Speed (1969), Timm and Carlson (1975), and Urquhart, Weeks, and Henderson (1973).

**   This point is discussed by Hocking and Speed (1975) and by Timm and Carlson (1975).

(5.10-3) $$H_0: \quad \mu_1 - \mu_4 = 0, \quad \mu_2 - \mu_4 = 0, \quad \mu_3 - \mu_4 = 0.$$

In matrix notation this hypothesis is written as

$$H_0: \quad \underset{(p-1)\times h}{\mathbf{C}'} \quad \underset{h \times 1}{\boldsymbol{\mu}} \quad \underset{(p-1)\times 1}{\mathbf{0}}$$

$$H_0: \quad \begin{bmatrix} 1 & 0 & 0 & -1 \\ 0 & 1 & 0 & -1 \\ 0 & 0 & 1 & -1 \end{bmatrix} \begin{bmatrix} \mu_1 \\ \mu_2 \\ \mu_3 \\ \mu_4 \end{bmatrix} = \begin{bmatrix} 0 \\ 0 \\ 0 \end{bmatrix}.$$

Note that there are several ways of testing hypothesis (5.10-2). A second way is to test $H_0: \quad \mu_1 - \mu_2 = 0, \quad \mu_2 - \mu_3 = 0, \quad \mu_3 - \mu_4 = 0.$ This can be written as

$$H_0: \quad \underset{(p-1)\times h}{\mathbf{C}'} \quad \underset{h \times 1}{\boldsymbol{\mu}} \quad \underset{(p-1)\times 1}{\mathbf{0}}$$

$$H_0: \quad \begin{bmatrix} 1 & -1 & 0 & 0 \\ 0 & 1 & -1 & 0 \\ 0 & 0 & 1 & -1 \end{bmatrix} \begin{bmatrix} \mu_1 \\ \mu_2 \\ \mu_3 \\ \mu_4 \end{bmatrix} = \begin{bmatrix} 0 \\ 0 \\ 0 \end{bmatrix}.$$

In order for the hypothesis to be testable the $\mathbf{C}'$ matrix must be of full row rank. This means that each row of $\mathbf{C}'$ must be linearly independent of every other row. The maximum number of such rows is $p - 1$, which is why it is necessary to express the null hypothesis as (5.10-3), for example.*

Procedures for computing sums of squares and mean squares for the hand-steadiness data are illustrated in Table 5.10-1. The results in Table 5.10-1 are identical to those in Table 4.3-2 where the traditional approach to ANOVA was used. Since $F = 64.833/1.464 = 44.28$ exceeds $F_{.05;3,28} = 2.95$, the null hypothesis is rejected. Two formulas for computing $SSBG$ are illustrated in Table 5.10-1. When more complex designs are discussed, a modification of the formula $SSBG = (\mathbf{C}'\hat{\boldsymbol{\mu}})' [\mathbf{C}'(\mathbf{X}'\mathbf{X})^{-1}\mathbf{C}]^{-1} (\mathbf{C}'\hat{\boldsymbol{\mu}})$ will be used. Procedures for computing the terms $\hat{\psi}_i$, $\hat{\sigma}_{\bar{Y}}$, and $\hat{\sigma}_{\psi_i}$ needed to evaluate multiple contrasts are illustrated in Table 5.10-2. The null hypothesis $H_0: \quad \mu_1 - \mu_2 = 0$ is used for illustrative purposes. The results in Table 5.10-2 are identical to those presented previously in Section 4.4 and Tables 5.6-1, 5.7-2, 5.9-2, and 5.9-4.

---

*   The full rank experimental design model approach can be thought of as a reparameterization of (5.9-8). For the data in Table 5.9-3,

$$\mathbf{E} = \begin{bmatrix} 1 & 1 & 0 & 0 & 0 \\ 1 & 0 & 1 & 0 & 0 \\ 1 & 0 & 0 & 1 & 0 \\ 1 & 0 & 0 & 0 & 1 \end{bmatrix} \quad \text{and} \quad \mathbf{E}\boldsymbol{\beta} = \begin{bmatrix} \mu + \alpha_1 \\ \mu + \alpha_2 \\ \mu + \alpha_3 \\ \mu + \alpha_4 \end{bmatrix} = \begin{bmatrix} \mu_1 \\ \mu_2 \\ \mu_3 \\ \mu_4 \end{bmatrix}.$$

TABLE 5.10-1 Computational Procedures for a Type CR-4 Design Using a Full Rank Experimental Design Model

(i) Data and basic matrices ($N = 32$, $h = 4$, $p = 4$):

$$
\begin{array}{c}
\underset{N \times 1}{\mathbf{y}} \qquad\qquad \underset{N \times h}{\mathbf{X}}
\end{array}
$$

$$
\begin{array}{cc}
& \mathbf{x_1}\ \mathbf{x_2}\ \mathbf{x_3}\ \mathbf{x_4} \\
a_1 \left\{ \begin{array}{c} 3 \\ 6 \\ 3 \\ 3 \\ 1 \\ 2 \\ 2 \\ 2 \end{array} \right. &
\left[\begin{array}{cccc} 1 & 0 & 0 & 0 \\ 1 & 0 & 0 & 0 \\ 1 & 0 & 0 & 0 \\ 1 & 0 & 0 & 0 \\ 1 & 0 & 0 & 0 \\ 1 & 0 & 0 & 0 \\ 1 & 0 & 0 & 0 \\ 1 & 0 & 0 & 0 \end{array}\right. \\
a_2 \left\{ \begin{array}{c} 4 \\ 5 \\ 4 \\ 3 \\ 2 \\ 3 \\ 4 \\ 3 \end{array} \right. &
\begin{array}{cccc} 0 & 1 & 0 & 0 \\ 0 & 1 & 0 & 0 \\ 0 & 1 & 0 & 0 \\ 0 & 1 & 0 & 0 \\ 0 & 1 & 0 & 0 \\ 0 & 1 & 0 & 0 \\ 0 & 1 & 0 & 0 \\ 0 & 1 & 0 & 0 \end{array} \\
a_3 \left\{ \begin{array}{c} 7 \\ 8 \\ 7 \\ 6 \\ 5 \\ 6 \\ 5 \\ 6 \end{array} \right. &
\begin{array}{cccc} 0 & 0 & 1 & 0 \\ 0 & 0 & 1 & 0 \\ 0 & 0 & 1 & 0 \\ 0 & 0 & 1 & 0 \\ 0 & 0 & 1 & 0 \\ 0 & 0 & 1 & 0 \\ 0 & 0 & 1 & 0 \\ 0 & 0 & 1 & 0 \end{array} \\
a_4 \left\{ \begin{array}{c} 7 \\ 8 \\ 9 \\ 8 \\ 10 \\ 10 \\ 9 \\ 11 \end{array} \right. &
\left.\begin{array}{cccc} 0 & 0 & 0 & 1 \\ 0 & 0 & 0 & 1 \\ 0 & 0 & 0 & 1 \\ 0 & 0 & 0 & 1 \\ 0 & 0 & 0 & 1 \\ 0 & 0 & 0 & 1 \\ 0 & 0 & 0 & 1 \\ 0 & 0 & 0 & 1 \end{array}\right]
\end{array}
$$

$$\sum_{i=1}^{N} Y_i = 172$$

$$
\underset{1 \times N}{\mathbf{y'}} \qquad \underset{N \times 1}{\mathbf{y}} \qquad \underset{1 \times 1}{(\mathbf{y'y})}
$$

$$
[3 \ \ 6 \ \ 3 \ \cdots \ 11] \begin{bmatrix} 3 \\ 6 \\ 3 \\ \vdots \\ 11 \end{bmatrix} = 1160.0
$$

$$
\underset{h \times N}{\mathbf{X'}} \qquad\qquad \underset{N \times h}{\mathbf{X}} \qquad\qquad \underset{h \times h}{(\mathbf{X'X})}
$$

$$
\begin{bmatrix} 1 & 1 & 1 & \cdot & 0 \\ 0 & 0 & 0 & \cdot & 0 \\ 0 & 0 & 0 & \cdot & 0 \\ 0 & 0 & 0 & \cdot & 1 \end{bmatrix}
\begin{bmatrix} 1 & 0 & 0 & 0 \\ 1 & 0 & 0 & 0 \\ 1 & 0 & 0 & 0 \\ \cdot & \cdot & \cdot & \cdot \\ 0 & 0 & 0 & 1 \end{bmatrix}
=
\begin{bmatrix} 8 & 0 & 0 & 0 \\ 0 & 8 & 0 & 0 \\ 0 & 0 & 8 & 0 \\ 0 & 0 & 0 & 8 \end{bmatrix}
$$

$$
\underset{h \times N}{\mathbf{X'}} \qquad \underset{N \times 1}{\mathbf{y}} \qquad \underset{h \times 1}{(\mathbf{X'y})}
$$

$$
\begin{bmatrix} 1 & 1 & 1 & \cdot & 0 \\ 0 & 0 & 0 & \cdot & 0 \\ 0 & 0 & 0 & \cdot & 0 \\ 0 & 0 & 0 & \cdot & 1 \end{bmatrix}
\begin{bmatrix} 3 \\ 6 \\ 3 \\ \vdots \\ 11 \end{bmatrix}
=
\begin{bmatrix} 22 \\ 28 \\ 50 \\ 72 \end{bmatrix}
$$

$$
\underset{h \times h}{(\mathbf{X'X})^{-1}} \qquad \underset{h \times 1}{(\mathbf{X'y})} \qquad \underset{h \times 1}{\hat{\boldsymbol{\mu}}}
$$

$$
\begin{bmatrix} \tfrac{1}{8} & 0 & 0 & 0 \\ 0 & \tfrac{1}{8} & 0 & 0 \\ 0 & 0 & \tfrac{1}{8} & 0 \\ 0 & 0 & 0 & \tfrac{1}{8} \end{bmatrix}
\begin{bmatrix} 22 \\ 28 \\ 50 \\ 72 \end{bmatrix}
=
\begin{bmatrix} 2.75 \\ 3.50 \\ 6.25 \\ 9.00 \end{bmatrix}
$$

$$
\underset{1 \times h}{\hat{\boldsymbol{\mu}}'} \qquad \underset{h \times 1}{\mathbf{X'y}} \qquad \underset{1 \times 1}{(\hat{\boldsymbol{\mu}}'\mathbf{X'y})}
$$

$$
[2.75 \ \cdots \ 9.00] \begin{bmatrix} 22 \\ 28 \\ 50 \\ 72 \end{bmatrix} = 1119.0
$$

$$
\underset{(p-1) \times h}{\mathbf{C'}} \qquad \underset{h \times 1}{\hat{\boldsymbol{\mu}}} \qquad \underset{(p-1) \times 1}{(\mathbf{C'}\hat{\boldsymbol{\mu}})}
$$

$$
\begin{bmatrix} 1 & 0 & 0 & -1 \\ 0 & 1 & 0 & -1 \\ 0 & 0 & 1 & -1 \end{bmatrix}
\begin{bmatrix} 2.75 \\ 3.50 \\ 6.25 \\ 9.00 \end{bmatrix}
=
\begin{bmatrix} -6.25 \\ -5.50 \\ -2.75 \end{bmatrix}
$$

TABLE 5.10-1 (continued)

$$SSBG = \underset{1\times(p-1)}{[-6.25 \quad -5.50 \quad -2.75]} \left\{ \underset{(p-1)\times h}{\begin{bmatrix} 1 & 0 & 0 & -1 \\ 0 & 1 & 0 & -1 \\ 0 & 0 & 1 & -1 \end{bmatrix}} \underset{h\times h}{\begin{bmatrix} \tfrac{1}{8} & 0 & 0 & 0 \\ 0 & \tfrac{1}{8} & 0 & 0 \\ 0 & 0 & \tfrac{1}{8} & 0 \\ 0 & 0 & 0 & \tfrac{1}{8} \end{bmatrix}} \underset{h\times(p-1)}{\begin{bmatrix} 1 & 0 & 0 \\ 0 & 1 & 0 \\ 0 & 0 & 1 \\ -1 & -1 & -1 \end{bmatrix}} \right\}^{-1} \underset{(p-1)\times1}{\begin{bmatrix} -6.25 \\ -5.50 \\ -2.75 \end{bmatrix}}$$

$$\underset{1\times(p-1)}{(\mathbf{C'\hat{\mu}})'} \quad \underset{(p-1)\times(p-1)}{[\mathbf{C'(X'X)^{-1}C}]^{-1}} \quad \underset{(p-1)\times1}{(\mathbf{C'\hat{\mu}})}$$

$$= [-6.25 \quad -5.50 \quad -2.75]\begin{bmatrix} 6 & -2 & -2 \\ -2 & 6 & -2 \\ -2 & -2 & 6 \end{bmatrix}\begin{bmatrix} -6.25 \\ -5.50 \\ -2.75 \end{bmatrix} = 194.5$$

(ii) Computation of sum of squares:

$$SSTO = \mathbf{y'y} - \left(\sum_{i=1}^{N} Y_i\right)^2 \Big/ N = 1160.0 - (172)^2/32$$

$$= 1160.0 - 924.5 = 235.5$$

$$SSBG = \mathbf{\hat{\mu}'X'y} - \left(\sum_{i=1}^{N} Y_i\right)^2 \Big/ N = 1119.0 - (172)^2/32$$

$$= 1119.0 - 924.5 = 194.5$$

$$= (\mathbf{C'\hat{\mu}})'[\mathbf{C'(X'X)^{-1}C}]^{-1}(\mathbf{C'\hat{\mu}}) = 194.5$$

$$SSWG = \mathbf{y'y} - \mathbf{\hat{\mu}'X'y} = 1160.0 - 1119.0 = 41.0$$

(iii) Computation of mean squares and F statistic:

$$MSBG = SSBG/(p-1) = 194.5/(4-1) = 64.833$$

$$MSWG = SSWG/(N-p) = 41.0/(32-4) = 1.464$$

$$F = MSBG/MSWG = 64.833/1.464 = 44.28$$

$$F_{.05;3,28} = 2.95$$

TABLE 5.10-2 Computational Procedures for Testing Differences Among Means (Full rank experimental design model)

(i) Computation of contrast, $\hat{\psi}_1$ ($h = 4$, $n = 8$):*

$$\underset{1\times h}{\mathbf{c'}} \qquad \underset{h\times1}{\mathbf{\hat{\mu}}} \qquad \hat{\psi}_1$$

$$[1 \quad -1 \quad 0 \quad 0]\begin{bmatrix} 2.75 \\ 3.50 \\ 5.50 \\ 9.00 \end{bmatrix} = -0.75$$

TABLE 5.10-2   (continued)

(ii)   Computation of $\hat{\sigma}_{\bar{Y}}^2$ and $\hat{\sigma}_{\psi_1}^2$:**

$$\hat{\sigma}_{\bar{Y}}^2 = MSWG/n = 1.464/8 = 0.183$$

$$MSWG\begin{bmatrix} \underset{1 \times h}{\mathbf{c}'} & \underset{h \times h}{(\mathbf{X}'\mathbf{X})^{-1}} & \underset{h \times 1}{\mathbf{c}} \end{bmatrix}$$

$$\hat{\sigma}_{\psi_1}^2 = 1.464 \left\{ \begin{bmatrix} 1 & -1 & 0 & 0 \end{bmatrix} \begin{bmatrix} \frac{1}{8} & 0 & 0 & 0 \\ 0 & \frac{1}{8} & 0 & 0 \\ 0 & 0 & \frac{1}{8} & 0 \\ 0 & 0 & 0 & \frac{1}{8} \end{bmatrix} \begin{bmatrix} 1 \\ -1 \\ 0 \\ 0 \end{bmatrix} \right\}$$

$$= 1.464\,(2/8) = 0.366$$

(iii)   Computation of test statistics:

$$t = \frac{\hat{\psi}_1}{\hat{\sigma}_{\psi_1}} = \frac{-0.75}{\sqrt{0.366}} = \frac{-0.75}{0.605} = -1.24 \qquad q = \frac{\hat{\psi}_1}{\hat{\sigma}_{\bar{Y}}} = \frac{-0.75}{\sqrt{0.183}} = \frac{-0.75}{0.428} = -1.75$$

* $\hat{\boldsymbol{\mu}}$ is obtained from Table 5.10-1; $\mathbf{c}'$ is the coefficient vector associated with the hypothesis $\mu_1 - \mu_2 = 0$.

** Matrix $(\mathbf{X}'\mathbf{X})^{-1}$ and $MSWG$ are obtained from Table 5.10-1.

5.11   SUMMARY

This chapter has described six variations on the general linear model approach to ANOVA. All of them involve inverting a matrix, a task best left to a digital computer. Today this is no longer a deterent to the use of the approach because computers are widely available. Most computer software packages use the general linear model approach in performing an analysis of variance. Consequently a familiarity with the approach is helpful in interpreting ANOVA computer printouts.

A familiarity with the approach is also helpful in seeing similarities between analysis of variance and regression analysis, two procedures that are often viewed by experimenters as being quite distinct. In this regard, the general linear model approach is particularly attractive because it encompasses all ANOVA experimental design models and regression models. And the general linear model approach provides a much smoother transition to the study of multivariate methods than the traditional approach. Despite these advantages, it is also important to study the traditional approach because I believe this leads to a clearer understanding of basic concepts in the design of experiments—a point of view shared by others (Collier and Hummel, 1977, 101). Ideally one should understand both approaches.

As demonstrated in this chapter, a qualitative regression model can be used to estimate the parameters of the analysis of variance model and partition the total variation so as to test the ANOVA null hypothesis. This is accomplished by establishing a correspondence between the $p - 1$ treatment levels of a type CR-p design and the $X_{i1}, X_{i2}, \ldots, X_{i, h-1}$ independent variables of a regression model equation

$$Y_i = \beta_0 + \beta_1 X_{i1} + \beta_2 X_{i2} + \cdots + \beta_{h-1} X_{i,h-1} + \epsilon_i \quad (i = 1, \ldots, N).$$

Three schemes for coding the independent variable were discussed—dummy coding, effect coding, and orthogonal coding. Although the schemes provide different solutions to $\hat{\boldsymbol{\beta}}' = [\hat{\beta}_0 \, \hat{\beta}_1 \cdots \hat{\beta}_{h-1}]$ in the least squares normal equations, certain linear combinations of $\hat{\boldsymbol{\beta}}$ are unique and independent of the solution.

Experimental design models can be classified as either less than full rank models or full rank models. The distinction is based on whether the structural matrix $\mathbf{X}$ is of full rank. The solution $\hat{\boldsymbol{\beta}}' = [\hat{\mu} \, \hat{\alpha}_1 \cdots \hat{\alpha}_p]$ to the less than full rank model

$$Y_{ij} = \mu + \alpha_j + \epsilon_{i(j)} \quad (i = 1, \ldots, n; j = 1, \ldots, p)$$

where $\epsilon_{i(j)}$ is $NID(0, \sigma_\epsilon^2)$, cannot be obtained in the usual way since the inverse $(\mathbf{X}' \mathbf{X})^{-1}$ does not exist. Two ways around this problem were discussed: restricting the number of unknowns and reparameterizing. The first method adds a sufficient number of independent equations—side conditions—to the original system to remove its rank deficiency. The system is then of full rank and a unique solution is obtained with the imposed side conditions. In the second method, a new set of r unknowns, $\boldsymbol{\beta}^\star$, is defined whose elements are linear combinations of the original h parameters $\boldsymbol{\beta}$. The new parameters are related to the original parameters by $\boldsymbol{\beta}^\star = \mathbf{E}\boldsymbol{\beta}$, where $\mathbf{E}$ is of full row rank and its rows are linearly dependent on the rows of the structural matrix $\mathbf{X}$. Instead of solving for the original h parameters, one solves for r parameters in $\boldsymbol{\beta}^\star$. By solving for only r parameters where r is the rank of $\mathbf{X}$, the less than full rank system is reduced to full rank and a solution, $\boldsymbol{\beta}^\star$, is obtained in the usual way. The new vector $\boldsymbol{\beta}^\star$ is chosen so that it has a useful interpretation, subject, of course, to the given restrictions.

The full rank experimental design model

$$Y_{ij} = \mu_j + \epsilon_{i(j)} \quad (i = 1, \ldots, n; j = 1, \ldots, p)$$

where $\epsilon_{i(j)}$ is $NID(0, \sigma_\epsilon^2)$, circumvents the problem of rank deficiency. The solution $\hat{\boldsymbol{\beta}}' = \hat{\boldsymbol{\mu}}' = [\hat{\mu}_1 \, \hat{\mu}_2 \cdots \hat{\mu}_p]$ to the least squares normal equations can be obtained in the usual way since the structural matrix $\mathbf{X}$ is of full rank. Although this is helpful, a more important advantage of the full rank model is that an experimenter can always discern exactly what hypothesis is being tested. The importance of this will become evident when more complex designs are discussed.

5.12 REVIEW EXERCISES

1.   [5.1] What is the key difference between regression analysis and analysis of variance when the independent variable is (**a**) quantitative (**b**) qualitative?

†2. [5.1] Comment on the statement, "The linear model for a type CR-p design is $Y_{ij} = \mu + \alpha_j + \epsilon_{i(j)}$ $(i = 1, \ldots, n; j = 1, \ldots, p)$."

3. [5.1] Both of the model equations $Y_i = \beta_0 + \beta_1 X_i + \epsilon_i$ $(i = 1, \ldots, N)$ and $Y_i = \beta_0 + \beta_1 X_i + \beta_2 X_i^2 + \epsilon_i$ $(i = 1, \ldots, N)$ are described as linear; explain.

4. [5.1] Write the following systems of equations in terms of vectors and matrices.

†a) Regression model equations with two independent variables

$$Y_1 = \beta_0 + \beta_1 X_{11} + \beta_2 X_{12} + \epsilon_1$$
$$Y_2 = \beta_0 + \beta_1 X_{21} + \beta_2 X_{22} + \epsilon_2$$
$$Y_3 = \beta_0 + \beta_1 X_{31} + \beta_2 X_{32} + \epsilon_3$$

b) Regression model equations with three independent variables

$$Y_1 = \beta_0 + \beta_1 X_{11} + \beta_2 X_{12} + \beta_3 X_{13} + \epsilon_1$$
$$Y_2 = \beta_0 + \beta_1 X_{21} + \beta_2 X_{22} + \beta_3 X_{23} + \epsilon_2$$
$$Y_3 = \beta_0 + \beta_1 X_{31} + \beta_2 X_{32} + \beta_3 X_{33} + \epsilon_3$$
$$Y_4 = \beta_0 + \beta_1 X_{41} + \beta_2 X_{42} + \beta_3 X_{43} + \epsilon_4$$

5. [5.1] Write the following systems of equations in terms of vectors and matrices.

†a) Type CR-2 design

$$Y_{11} = \mu + \alpha_1 1 + \alpha_2 0 + \epsilon_{1(1)}$$
$$Y_{21} = \mu + \alpha_1 1 + \alpha_2 0 + \epsilon_{2(1)}$$
$$Y_{12} = \mu + \alpha_1 0 + \alpha_2 1 + \epsilon_{1(2)}$$
$$Y_{22} = \mu + \alpha_1 0 + \alpha_2 1 + \epsilon_{2(2)}$$

b) Type CR-3 design

$$Y_{11} = \mu + \alpha_1 1 + \alpha_2 0 + \alpha_3 0 + \epsilon_{1(1)}$$
$$Y_{21} = \mu + \alpha_1 1 + \alpha_2 0 + \alpha_3 0 + \epsilon_{2(1)}$$
$$Y_{12} = \mu + \alpha_1 0 + \alpha_2 1 + \alpha_3 0 + \epsilon_{1(2)}$$
$$Y_{22} = \mu + \alpha_1 0 + \alpha_2 1 + \alpha_3 0 + \epsilon_{2(2)}$$
$$Y_{13} = \mu + \alpha_1 0 + \alpha_2 0 + \alpha_3 1 + \epsilon_{1(3)}$$
$$Y_{23} = \mu + \alpha_1 0 + \alpha_2 0 + \alpha_3 1 + \epsilon_{2(3)}$$

†6. [5.2] What is special about least squares estimators?

7. [5.2] Write the partial joint null hypothesis using matrix notation for

†a) Exercise 4(a)   **b)** Exercise 4(b)   **†c)** Exercise 5(a)   **d)** Exercise 5(b).

8. [5.3] Write the **X** matrix for a qualitative regression model for the following ANOVA designs using (i) dummy coding and (ii) effect coding.

†a) Type CR-2 design with n equal to 2   **b)** Type CR-2 design with n equal to 3
†c) Type CR-3 design with n equal to 2   **d)** Type CR-3 design with n equal to 3

9. [5.4] Assume that the **X** matrix for a qualitative regression model has been coded for the following ANOVA designs using (i) dummy coding and (ii) effect coding. Indicate the correspondence between the parameters of the regression and experimental design models.

†a) Type CR-3 design   **†b)** Type CR-4 design   **c)** Type CR-5 design

d) Type CR-6 design

†10. The effects of an interpersonal skill training program for male psychiatric inpatients was investigated. The program was designed to increase performance competence in initiating conversations, dealing with rejection, and being more assertive and self-disclosing. Six patients were randomly assigned to one of three groups. Patients in group a_1, referred to as the interpersonal skill training group, received three one-hour training sessions covering 11 problem situations. Training techniques included behavior rehearsal, modeling, coaching, recorded response playback, and corrective feedback. Patients in group a_2, the pseudotherapy control group, covered the same 11 problem situations, but instead of receiving suggestions for specific behavioral changes, they were encouraged to explore their feelings about the problems and to seek insight into the psychological and historical reasons for these feelings. Patients in group a_3, referred to as the assessment control group, participated in the posttreatment assessment but received no training.

At the conclusion of the experiment, the patients were instructed to perform seven tasks including initiating a conversation with a male stranger (a confederate), asking the stranger to lunch, and terminating the conversation after ten minutes. The confederate confronted the patient with three "critical moments": not hearing the patient's name when introduced, responding to the lunch invitation with an excuse, and asking the patient, "Tell me about yourself." The dependent variable was the number of interaction tasks (out of seven) that the patient was able to complete. The following data were obtained. The number of subjects has been reduced so that the computations can be done without a computer. (Experiment suggested by Goldsmith, Jean B. and McFall, Richard M. Development and evaluation of an interpersonal skill-training program for psychiatric inpatients. *Journal of Abnormal Psychology*, 1975, *84*, 51–58.)

a_1	a_2	a_3
8	2	1
6	3	3

a) [5.3] Write a qualitative regression model for this type CR-3 design.

b) [5.3] Write the **X** matrix using (i) dummy coding, (ii) effect coding, and (iii) orthogonal coding (compare a_1 with a_2 and the average of a_1 and a_2 with a_3).

c) [5.4] Write the null hypothesis for β_1 and β_2 using matrix notation.

d) [5.4] Assume that dummy coding has been used; compute the following: $\mathbf{y'y}$, $\mathbf{X'X}$, $(\mathbf{X'X})^{-1}$, $\mathbf{X'y}$, $\hat{\boldsymbol{\beta}}$, and $\hat{\boldsymbol{\beta}}'\mathbf{X'y}$.

e) [5.4] Assume that dummy coding has been used; compute MSR, MSE, and $SSTO$. Construct an ANOVA table; let $\alpha = .05$.

f) [5.4] Determine the correspondence for this type CR-3 design between the parameters of the regression and experimental design models when dummy coding has been used.

g) [5.4] Compute the vector of treatment means using $\mathbf{W}\hat{\boldsymbol{\beta}} = \bar{\mathbf{y}}$; assume that dummy coding has been used.

h) [5.5] (i) Write a reduced model for computing $SSE(0)$, that is, a model containing no independent variables. (ii) Compute the following vectors and matrices for this reduced model (let the subscript 1 identify the reduced model): $\mathbf{X}_1$, $\mathbf{X}_1'\mathbf{X}_1$, $(\mathbf{X}_1'\mathbf{X}_1)^{-1}$, $\mathbf{X}_1'\mathbf{y}$, $\hat{\boldsymbol{\beta}}_1$, and

$\hat{\boldsymbol{\beta}}_1' \mathbf{X}_1' \mathbf{y}$. (iii) Compute $SSE(0)$ and $SSR(X_1\ X_2 \mid 0) = SSE(0) - SSE(X_1\ X_2)$, where $SSE(X_1\ X_2)$ is equal to SSE from Exercise 10(e). (iv) Compare $SSR(X_1\ X_2 \mid 0)$ with SSR from Exercise 10(e).

i) [5.6] Use Dunnett's statistic to test null hypotheses for contrasts involving the control group mean; compute $\sigma_{\hat{\psi}_i}^2$ using

$$MSE\{\mathbf{c}'[\mathbf{W}(\mathbf{X}'\mathbf{X})^{-1}\mathbf{W}']\mathbf{c}\}.$$

The critical value of Dunnett's statistic, $tD'_{.05/2;3,3}$, is approximately 4.

j) [5.7] Determine the correspondence for this type CR-3 design between the parameters of the regression and experimental design models when effect coding is used.

k) [5.7] Assume that effect coding has been used; compute the following: $\mathbf{X}'\mathbf{X}$, $(\mathbf{X}'\mathbf{X})^{-1}$, $\mathbf{X}'\mathbf{y}$, $\hat{\boldsymbol{\beta}}$, and $\hat{\boldsymbol{\beta}}'\mathbf{X}'\mathbf{y}$.

l) [5.7] (i) Assume that effect coding has been used; compute MSR, MSE, and $SSTO$. (ii) If Exercise 10(e) was done, compare the results of the two analyses.

m) [5.7] Compute the vector of treatment level means using $\mathbf{W}\hat{\boldsymbol{\beta}} = \bar{\mathbf{y}}$; assume that effect coding has been used.

11. The effects of externally imposed deadlines on an individual's task performance and subsequent interest in the task were investigated. The subjects, eight male undergraduate college students, were asked to play with five sets of enjoyable word games under one of four sets of instructions. In the no-deadline condition, treatment level a_1, subjects played with the game with no performance requirements or time constraints. In the two deadline conditions, the subjects were told (1) that they would be allowed to play with the game for 15 minutes, (2) to work as quickly as possible, and (3) that most students could finish the game within the time period. In the implied-deadline condition, treatment level a_2, no further instructions were given; in the explicit-deadline condition, treatment level a_3, the subjects were told that they were required to finish the game in the allotted time in order for their data to be useful. In the work-fast condition, treatment level a_4, subjects were asked to work as quickly as possible but were given no time limits or information about the performance of others.

Eight subjects were randomly assigned to four groups with two subjects in each group. The four conditions were randomly assigned to the groups. One of the dependent variables was the length of time required to complete the games. The following data were obtained. The number of subjects has been reduced so that the computations can be done without a computer. (Experiment suggested by Amabile, Teresa M., DeJong, William, and Lepper, Mark R. Effects of externally imposed deadlines on subsequent intrinsic motivation. *Journal of Personality and Social Psychology,* 1976, *34,* 92–98.)

a_1	a_2	a_3	a_4
13	11	9	10
14	9	10	11

a) [5.3] Write a qualitative regression model for this type CR-4 design.

b) [5.3] Write the **X** matrix using (i) dummy coding, (ii) effect coding, and (iii) orthogonal coding (compare a_1 with a_2, a_3 with a_4, and the average of a_1 and a_2 with the average of a_3 and a_4).

c) [5.4] Write the null hypothesis for β_1, β_2, and β_3 using matrix notation.

d) [5.4] Assume that dummy coding has been used; compute the following: $\mathbf{y'y}$, $\mathbf{X'X}$, $(\mathbf{X'X})^{-1}$, $\mathbf{X'y}$, $\hat{\boldsymbol{\beta}}$, and $\hat{\boldsymbol{\beta}}'\mathbf{X'y}$.

e) [5.4] Assume that dummy coding has been used; compute MSR, MSE, and $SSTO$. Construct an ANOVA table; let $\alpha = .05$.

f) [5.4] Determine the correspondence for this type CR-4 design between the parameters of the regression and experimental design models when dummy coding has been used.

g) [5.4] Compute the vector of treatment level means using $\mathbf{W}\hat{\boldsymbol{\beta}} = \bar{\mathbf{y}}$; assume that dummy coding has been used.

h) [5.5] (i) Write a reduced model for computing $SSE(0)$, that is, a model containing no independent variables. (ii) Compute the following vectors and matrices for this reduced model (let the subscript 1 identify the reduced model): $\mathbf{X}_1$, $\mathbf{X}_1'\mathbf{X}_1$, $(\mathbf{X}_1'\mathbf{X}_1)^{-1}$, $\mathbf{X}_1'\mathbf{y}$, $\hat{\boldsymbol{\beta}}_1$, and $\hat{\boldsymbol{\beta}}_1'\mathbf{X}_1'\mathbf{y}$. (iii) Compute $SSE(0)$ and $SSR(X_1\,X_2\,X_3 \mid 0) = SSE(0) - SSE(X_1\,X_2\,X_3)$, where $SSE(X_1\,X_2\,X_3)$ is equal to SSE from Exercise 11(e). (iv) Compare $SSR(X_1\,X_2\,X_3 \mid 0)$ with SSR from Exercise 11(e).

i) [5.6] Use Scheffé's statistic to test the following hypothesis:

$$H_0: \quad \mu_1 - (\mu_2 + \mu_3 + \mu_4)/3 = 0.$$

Compute $\sigma^2_{\hat{\psi}_i}$ using

$$MSE\{\mathbf{c}'[\mathbf{W}(\mathbf{X'X})^{-1}\mathbf{W}']\mathbf{c}\}.$$

j) [5.7] Determine the correspondence for this type CR-4 design between the parameters of the regression and experimental design models when effect coding is used.

k) [5.7] Assume that effect coding has been used; compute the following: $\mathbf{X'X}$, $(\mathbf{X'X})^{-1}$, $\mathbf{X'y}$, $\hat{\boldsymbol{\beta}}$, and $\hat{\boldsymbol{\beta}}'\mathbf{X'y}$.

l) [5.7] (i) Assume that effect coding has been used; compute MSR, MSE, and $SSTO$. (ii) If Exercise 11(e) was done, compare the results of the two analyses.

m) [5.7] Compute the vector of treatment level means using $\mathbf{W}\hat{\boldsymbol{\beta}} = \bar{\mathbf{y}}$; assume that effect coding has been used.

12.   [5.8] (i) Determine the rank of the following matrices. (ii) Which are of full rank?

†a)
$$\begin{bmatrix} 1 & 1 & 0 \\ 1 & 1 & 0 \\ 1 & 0 & 1 \\ 1 & 0 & 1 \end{bmatrix}$$
†b)
$$\begin{bmatrix} 1 & 1 \\ 1 & 1 \\ 1 & 0 \\ 1 & 0 \end{bmatrix}$$
c)
$$\begin{bmatrix} 1 & 1 \\ 1 & 1 \\ 1 & -1 \\ 1 & -1 \end{bmatrix}$$
d)
$$\begin{bmatrix} 1 & 0 \\ 1 & 0 \\ 0 & 1 \\ 0 & 1 \end{bmatrix}$$

e)
$$\begin{bmatrix} 1 & 1 & 0 & 0 \\ 1 & 1 & 0 & 0 \\ 1 & 0 & 1 & 0 \\ 1 & 0 & 1 & 0 \\ 1 & 0 & 0 & 1 \\ 1 & 0 & 0 & 1 \end{bmatrix}$$
†f)
$$\begin{bmatrix} 1 & 1 & 0 & 0 \\ 1 & 1 & 0 & 0 \\ 1 & 0 & 1 & 0 \\ 1 & 0 & 1 & 0 \\ 1 & 0 & 0 & 1 \\ 1 & 0 & 0 & 1 \\ 0 & 1 & 1 & 1 \end{bmatrix}$$
†g)
$$\begin{bmatrix} 1 & 0 & 0 \\ 1 & 0 & 0 \\ 0 & 1 & 0 \\ 0 & 1 & 0 \\ 0 & 0 & 1 \\ 0 & 0 & 1 \end{bmatrix}$$

†13.   [5.9] Exercise 10 described an experiment to determine the effects of an interpersonal skill training program for male psychiatric inpatients.

a) Write a less than full rank experimental design model for this type CR-3 design.

b) (i) Write the $\mathbf{X}$ matrix for the model in Exercise 13(a), subject to the restriction $\alpha_1 + \alpha_2 + \alpha_3 = 0$. (ii) Write the restriction using matrix notation.

c) Write the null hypothesis for the treatment using vectors and matrices.

d) Assume that the restriction in Exercise 13(b) has been used; compute the following: $\mathbf{y}^{*\prime}\mathbf{y}^*$, $\mathbf{X}^{*\prime}\mathbf{X}^*$, $(\mathbf{X}^{*\prime}\mathbf{X}^*)^{-1}$, $\mathbf{X}^{*\prime}\mathbf{y}^*$, $\hat{\boldsymbol{\beta}}^*$, and $\hat{\boldsymbol{\beta}}^{*\prime}\mathbf{X}^{*\prime}\mathbf{y}^*$. (This problem can be done without a computer; it involves inverting a 4×4 matrix.)

e) (i) Compute $MSBG$, $MSWG$, and $SSTO$, subject to the restriction that $\alpha_1 + \alpha_2 + \alpha_3 = 0$. (ii) If you did Exercise 10(e), compare the results with MSR, MSE, and $SSTO$ obtained there.

f) Compute the vector of treatment level means using $\mathbf{W}\hat{\boldsymbol{\beta}}^* = \bar{\mathbf{y}}$.

†14. [5.9] The method of reparameterization can be used to obtain a solution to a system of equations when the structural matrix is rank deficient. Suppose that we want to solve for

$$\hat{\boldsymbol{\beta}}^*_{3\times 1} = \begin{bmatrix} \mu + (\alpha_1 + \alpha_2 + \alpha_3)/3 \\ \alpha_1 - \alpha_3 \\ \alpha_2 - \alpha_3 \end{bmatrix}$$

in Exercise 10.

a) Compute the following matrices: $\mathbf{E}$, $(\mathbf{EE}')^{-1}$, $(\mathbf{Z}'\mathbf{Z})^{-1}$, $\mathbf{Z}'\mathbf{y}$, $\hat{\boldsymbol{\beta}}^*$. (This problem can be done without a computer, but the computations are very tedious.)

b) (i) Compute $MSBG$, $MSWG$, and $SSTO$ using the method of reparameterization. (ii) If you did Exercise 10(e) or 13(e), compare the results with $MSBG(MSR)$, $MSWG(MSE)$, and $SSTO$ obtained there.

c) Write the null hypothesis for the treatment using vectors and matrices.

d) Compute the vector of treatment level means using $\mathbf{W}\hat{\boldsymbol{\beta}}^* = \bar{\mathbf{y}}$.

15. [5.9] Exercise 11 described an experiment to determine the effects of externally imposed deadlines on an individual's task performance and subsequent interest in the task.

a) Write a less than full rank experimental design model for this type CR-4 design.

b) (i) Write the $\mathbf{X}$ matrix for the model in Exercise 15(a), subject to the restriction $\alpha_1 + \alpha_2 + \alpha_3 + \alpha_4 = 0$. (ii) Write the restriction using matrix notation.

c) Write the null hypothesis for the treatment using vectors and matrices.

d) Assume that the restriction in Exercise 15(b) has been used; compute the following: $\mathbf{y}^{*\prime}\mathbf{y}^*$, $\mathbf{X}^{*\prime}\mathbf{X}^*$, $(\mathbf{X}^{*\prime}\mathbf{X}^*)^{-1}$, $\mathbf{X}^{*\prime}\mathbf{y}^*$, $\hat{\boldsymbol{\beta}}^*$, and $\hat{\boldsymbol{\beta}}^{*\prime}\mathbf{X}^{*\prime}\mathbf{y}^*$. (This problem can be done without a computer, but the computations are fairly tedious. It involves inverting a 5×5 matrix.)

e) (i) Compute $MSBG$, $MSWG$, and $SSTO$, subject to the restriction that $\alpha_1 + \alpha_2 + \alpha_3 + \alpha_4 = 0$. (ii) If you did Exercise 11(e), compare the results with $MSBG(MSR)$, $MSWG(MSE)$, and $SSTO$ obtained there.

f) Compute the vector of treatment level means using $\mathbf{W}\hat{\boldsymbol{\beta}}^* = \bar{\mathbf{y}}$.

16. [5.9] For the data in Exercise 11, the method of reparameterization can be used to solve for

$$\hat{\boldsymbol{\beta}}^*_{4\times 1} = \begin{bmatrix} \mu + (\alpha_1 + \alpha_2 + \alpha_3 + \alpha_4)/4 \\ \alpha_1 - \alpha_4 \\ \alpha_2 - \alpha_4 \\ \alpha_3 - \alpha_4 \end{bmatrix}.$$

a) Compute the following matrices: $\mathbf{E}$, $(\mathbf{EE'})^{-1}$, $(\mathbf{Z'Z})^{-1}$, $\mathbf{Z'y}$, and $\hat{\boldsymbol{\beta}}^{*}$. (This problem can be done without a computer, but the computations are very tedious. It involves inverting a 5×5 matrix.)

b) (i) Compute $MSBG$, $MSWG$, and $SSTO$ using the method of reparameterization. (ii) If you did Exercise 11(e) or 15(e), compare the results with $MSBG(MSR)$, $MSWG(MSE)$, and $SSTO$ obtained there.

c) Write the null hypothesis for the treatment using vectors and matrices.

d) Compute the vector of treatment level means using $\mathbf{W}\hat{\boldsymbol{\beta}}^{*} = \overline{\mathbf{y}}$.

†**17.**   [5.10] A null hypothesis equivalent to H_0: $\alpha_1 = \alpha_2 = \alpha_3 = \alpha_4 = 0$ can be formulated for a full rank experimental design model in a variety of ways.

a) Write four alternative formulations of this null hypothesis using population means.

b) Write each null hypothesis using vectors and matrices.

18.   [5.10] **a)** Formulate three null hypotheses for a full rank experimental design model that are equivalent to H_0: $\alpha_1 = \alpha_2 = \alpha_3 = 0$.

b) Write each null hypothesis using vectors and matrices.

†**19.**   [5.10] **a)** For the data in Exercise 10, write a full rank experimental design model.

b) Using the data in Exercise 10, compute the following: $\mathbf{y'y}$, $\mathbf{X'X}$, $(\mathbf{X'X})^{-1}$, $\mathbf{X'y}$, $\hat{\boldsymbol{\mu}}$, and $\hat{\boldsymbol{\mu}}'\mathbf{X'y}$. (This problem can be done without a computer. It involves inverting a 3×3 diagonal matrix.)

c) (i) Compute $MSBG$, $MSWG$, and $SSTO$ using the full rank experimental design model. (ii) If you did Exercise 10(e), 13(e), or 14(b), compare the results with $MSBG(MSR)$, $MSWG(MSE)$, and $SSTO$ obtained there.

20.   [5.10] **a)** For the data in Exercise 11, write a full rank experimental design model.

b) Using the data in Exercise 11, compute the following: $\mathbf{y'y}$, $\mathbf{X'X}$, $(\mathbf{X'X})^{-1}$, $\mathbf{X'y}$, $\hat{\boldsymbol{\mu}}$, and $\hat{\boldsymbol{\mu}}'\mathbf{X'y}$. (This problem can be done without a computer. It involves inverting a 4×4 diagonal matrix.)

c) (i) Compute $MSBG$, $MSWG$, and $SSTO$ using the full rank experimental design model. (ii) If you did Exercise 11(e), 15(e), or 16(b), compare the results with $MSBG(MSR)$, $MSWG(MSE)$, and $SSTO$ obtained there.

6 RANDOMIZED BLOCK DESIGNS

6.1 DESCRIPTION OF THE RANDOMIZED BLOCK DESIGN

This chapter describes two designs—a randomized block design and a generalized randomized block design—that employ a blocking procedure to reduce error variance. In the behavioral sciences and education, differences among the experimental units may make a significant contribution to error variance and thereby mask or obscure the effects of a treatment. Similarly, administering the levels of a treatment under different environmental conditions—say at different times of the day, locations, and seasons of the year—may also mask treatment effects. Variation in the dependent variable attributable to such sources is called nuisance variation. We saw in Section 1.2 that there are three experimental approaches to controlling these undesired sources of variation:

1. Hold the nuisance variable constant, for example, use only 18-year-old females, and administer all treatment levels at the same time.

237

2.   Assign the treatment levels randomly to the experimental units so that known and unsuspected sources of variation among the units are distributed over the entire experiment and thus do not affect just one or a limited number of treatment levels.

3.   Include the nuisance variable as one of the factors in the experiment.

The latter approach uses a blocking procedure to isolate or partition out variation attributable to the nuisance variable so that it does not appear in estimates of treatment and error effects. The procedure involves forming n blocks of p homogeneous experimental units, where p is the number of levels of the treatment and the n blocks correspond to the levels of the nuisance variable. The blocks are formed so that the units in each block are more homogeneous with respect to the nuisance variable than those in different blocks.

To illustrate, suppose we plan to compare the effectiveness of three methods for memorizing German vocabulary, and we want to isolate the nuisance variable of IQ. We could assign the three students with the highest IQs to block one, the three with the next highest IQs to block two, and so on. The n blocks thus formed correspond to n arbitrarily determined levels of IQ. The three memorizing methods are then assigned to the experimental units randomly and independently for each block. The layout for this experiment is shown in Figure 6.1-1. Treatment A has three levels denoted by a_1, a_2, and a_3. This experiment illustrates an important feature of a randomized block design—the use of a blocking procedure to isolate a nuisance variable. The design is designated as a *type RB-p design,* where p stands for the number of treatment levels.

A randomized block design is appropriate for experiments that meet, in addition to the general assumptions of analysis of variance, the following three conditions.

1.   One treatment with $p \geq 2$ treatment levels and one nuisance variable with $n \geq 2$ levels.

2.   Formation of n blocks each containing p homogeneous experimental units. The variability among units within each block should be less than the variability among units in different blocks.

FIGURE 6.1-1   Layout for type RB-3 design; Y_{ij} denotes a score in one of the $i = 1, \ldots, n$ blocks and $j = 1, \ldots, p$ treatment levels. In this example, $p = 3$. The jth treatment level mean is computed from $\overline{Y}_{\cdot j} = \Sigma_{i=1}^{n} Y_{ij}/n$, the ith block mean is computed from $\overline{Y}_{i\cdot} = \Sigma_{j=1}^{p} Y_{ij}/p$, and the grand mean is computed from $\overline{Y}_{\cdot\cdot} = \Sigma_{i=1}^{n} \Sigma_{j=1}^{p} Y_{ij}/np$.

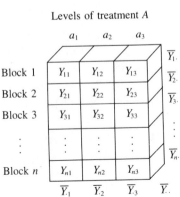

3.   Random assignment of treatment levels to experimental units within each block. The design requires n sets of p homogeneous experimental units, a total of $N = np$ units.* When a block consists of a single experimental unit who receives all p treatment levels, the order of presentation of the p levels is randomized independently for each experimental unit if the nature of the treatment permits. For this case the design requires a total of n experimental units who are each observed p times.

CRITERIA FOR BLOCKING

In forming blocks it is important to assign experimental units to blocks so that those within a block are as similar as possible with respect to the dependent variable while those in different blocks are less similar. Homogeneity of the experimental units can be achieved by using (1) litter mates or identical twins, (2) subjects matched by test scores or other relevant variables, (3) subjects matched by mutual selection, or (4) subjects as their own controls. If blocks are composed of litter mates or identical twins, homogeneity is achieved with respect to genetic characteristics. It is assumed that the behavior of subjects having identical or similar heredities will be more homogeneous than the behavior of subjects having dissimilar heredities. An alternative procedure is to match subjects on the basis of a variable that correlates with the dependent variable. This procedure was used in the experiment to compare the effectiveness of three methods for memorizing German vocabulary. It was assumed, with good reason, that IQ and memorizing ability are positively correlated. Another alternative is to use subjects who are matched by mutual selection, for example, husband and wife pairs, business partners, or club members. An experimenter must always ascertain that subjects matched in this way are in fact more similar with respect to the dependent variable than are unmatched subjects. Knowing a husband's attitudes on certain issues, for example, may provide considerable information about his wife's attitudes on the issues, and vice versa. However, knowing the husband's mechanical aptitude or his recreational interests may provide no information about his wife's aptitude or interests. The last alternative, use of subjects as their own controls, is appropriate for treatment levels that have relatively short-duration effects. The nature of the treatment should be such that the effects of one condition dissipate before the subject is observed under another condition. Otherwise the subsequent observations will reflect the cumulative effects of the preceding conditions. There is no such restriction, of course, when carryover effects, such as learning or fatigue, are the principal interest of the experimenter.

     The procedure used to form the blocks does not affect the computational analysis of the data. However, it does affect the interpretation of the results. The results of an experiment with repeated measurements generalize to a population of subjects who have been exposed to all treatment levels. The results of an experiment

---

*   A variation on this design, the generalized randomized block design described in Section 6.10, contains a total of $N = npw$ experimental units with np homogeneous units in each of w blocks (groups).

in which a block consists of p experimental units generalize to a population of subjects who have been exposed to only one treatment level. Some writers reserve the designation *randomized block design* for the latter case. They refer to an experiment with repeated measurements in which the order of administration of the treatment levels is randomized independently for each subject as a *subjects-by-treatments design*. A repeated measurements experiment in which the order of administration of the treatment levels is the same for all experimental units is referred to as a *subjects-by-trials design*.

When experimenters consider potential blocking variables they often overlook characteristics of the environmental setting. Blocking with respect to time of day, day of the week, season, location, and apparatus can significantly decrease error variance. The use of time is a particularly effective blocking variable since it often isolates a number of additional sources of variability: circadian body cycles, fatigue, changes in weather conditions, and drifts in calibration of equipment, to mention only a few. Ideas for appropriate blocking variables may come from one's experience in a research area, the literature, common sense, or intuition.

To reiterate, the purpose of blocking is to reduce the error variance and obtain a more precise estimate of treatment effects, thereby obtaining a more powerful test of a false null hypothesis. If the variation among experimental units within blocks is appreciably smaller than the variation among blocks, a randomized block design is more powerful than a completely randomized design. A measure of the relative efficiency of the two designs is described in Section 6.8.

EXPERIMENTAL DESIGN MODEL FOR A TYPE RB-p DESIGN

A score, Y_{ij}, in a randomized block design is a composite that reflects the effects of treatment j and block i plus all other sources of variation that affect Y_{ij}. These latter sources of variation are collectively referred to as *experimental error* or *error effects*. Our expectations about Y_{ij} can be expressed more formally by an experimental design model.

(6.1-1)
$$Y_{ij} = \mu + (\mu_{.j} - \mu) + (\mu_{i.} - \mu) + \epsilon_{ij}$$
$$= \mu + \alpha_j + \pi_i + \epsilon_{ij} \quad (i = 1, \ldots, n; j = 1, \ldots, p)$$

where Y_{ij} = a score for the experimental unit in block i and treatment level j

μ = the overall population mean

α_j = the effect of treatment level j and is subject to the restriction $\Sigma_{j=1}^{p} \alpha_j = 0$

π_j = the effect of block i that is $NID(0, \sigma_\pi^2)$

ϵ_{ij} = the experimental error that is $NID(0, \sigma_\epsilon^2)$; ϵ_{ij} is independent of π_i.

Model (6.1-1), which is called a mixed model or model III, is probably the most commonly used model for a type RB-p design. It is appropriate for the vocabulary-memorizing experiment described earlier. There we were only interested in three methods of memorizing German vocabulary, hence, the levels of treatment A were fixed. If we replicated the experiment, we would use the same three methods.

However, the n blocks corresponded to n arbitrarily determined levels of IQ. If we replicated the experiment with a new sample of subjects, the IQ levels would undoubtedly be different. Thus it seems appropriate to regard the blocks as a random variable.

The experimental error ϵ_{ij} in model (6.1-1) is often referred to as a *residual error*. This designation is an apt one since ϵ_{ij} is that portion of a score that remains after all the other terms in the model have been subtracted from it. For model (6.1-1), ϵ_{ij} is equal to

$$\epsilon_{ij} = Y_{ij} - [\mu + (\mu_{\cdot j} - \mu) + (\mu_{i \cdot} - \mu)]$$
$$= Y_{ij} - \mu - \mu_{\cdot j} + \mu - \mu_{i \cdot} + \mu$$
$$= Y_{ij} - \mu_{\cdot j} - \mu_{i \cdot} + \mu.$$

The values of the parameters μ, α_j, π_i, and ϵ_{ij} in model (6.1-1) are unknown, but they can be estimated from sample data as follows.

Statistic	Parameter Estimated
$\overline{Y}_{\cdot\cdot}$	μ
$\overline{Y}_{\cdot j} - \overline{Y}_{\cdot\cdot}$	α_j or $\mu_{\cdot j} - \mu$
$\overline{Y}_{i \cdot} - \overline{Y}_{\cdot\cdot}$	π_i or $\mu_{i \cdot} - \mu$
$Y_{ij} - \overline{Y}_{\cdot j} - \overline{Y}_{i \cdot} + \overline{Y}_{\cdot\cdot}$	ϵ_{ij}

PARTITION OF THE TOTAL SUM OF SQUARES

We will now show that a measure of the total variability among scores in an experiment, called the total sum of squares ($SSTO$), can be partitioned into three parts: sum of squares due to treatment A (SSA), sum of squares due to blocks ($SSBL$), and sum of squares due to experimental error or residual ($SSRES$). We begin by replacing the parameters in model (6.1-1) by statistics and rearranging terms as follows.

$$Y_{ij} - \mu = \quad \alpha_j \quad + \quad \pi_i \quad + \quad \epsilon_{ij}$$
$$Y_{ij} - \overline{Y}_{\cdot\cdot} = (\overline{Y}_{\cdot j} - \overline{Y}_{\cdot\cdot}) + (\overline{Y}_{i \cdot} - \overline{Y}_{\cdot\cdot}) + (Y_{ij} - \overline{Y}_{\cdot j} - \overline{Y}_{i \cdot} + \overline{Y}_{\cdot\cdot})$$
$$\begin{matrix} \text{(Total} & \text{(Treatment} & \text{(Block} & \text{(Residual)} \\ \text{deviation)} & \text{effect)} & \text{effect)} & \end{matrix}$$

Next we square both sides of the equation.

$$(Y_{ij} - \overline{Y}_{\cdot\cdot})^2 = [(\overline{Y}_{\cdot j} - \overline{Y}_{\cdot\cdot}) + (\overline{Y}_{i \cdot} - \overline{Y}_{\cdot\cdot}) + (Y_{ij} - \overline{Y}_{\cdot j} - \overline{Y}_{i \cdot} + \overline{Y}_{\cdot\cdot})]^2$$

This equation is for a single score, but we can treat each of the np scores in the same way and sum the resulting equations to obtain

$$\sum_{j=1}^{p} \sum_{i=1}^{n} (Y_{ij} - \overline{Y}_{\cdot\cdot})^2 = \sum_{j=1}^{p} \sum_{i=1}^{n} [(\overline{Y}_{\cdot j} - \overline{Y}_{\cdot\cdot}) + (\overline{Y}_{i \cdot} - \overline{Y}_{\cdot\cdot}) + (Y_{ij} - \overline{Y}_{\cdot j} - \overline{Y}_{i \cdot} + \overline{Y}_{\cdot\cdot})]^2.$$

The quantity on the left is the total sum of squares. It can be shown using elementary algebra and the summation rules in Appendix A that the term on the right is equal to $SSA + SSBL + SSRES$.*

$$(6.1\text{–}2)\quad \sum_{j=1}^{p}\sum_{i=1}^{n}(Y_{ij}-\overline{Y}_{..})^2 = n\sum_{j=1}^{p}(\overline{Y}_{.j}-\overline{Y}_{..})^2 + p\sum_{i=1}^{n}(\overline{Y}_{i.}-\overline{Y}_{..})^2 + \sum_{j=1}^{p}\sum_{i=1}^{n}(Y_{ij}-\overline{Y}_{.j}-\overline{Y}_{i.}+\overline{Y}_{..})^2$$

$$SSTO \quad = \quad SSA \quad + \quad SSBL \quad + \quad SSRES$$

The sum of squares formulas in (6.1-2) are not the most convenient for computational purposes. More convenient formulas are given in Table 6.2-1.** Mean squares are obtained by dividing each sum of squares by its degrees of freedom.

$$MSTO = SSTO/(np-1)$$
$$MSA = SSA/(p-1)$$
$$MSBL = SSBL/(n-1)$$
$$MSRES = SSRES/(n-1)(p-1)$$

The derivation of the expected values of the mean squares follows the procedures illustrated for the completely randomized design in Section 2.3. The derivation will not be given since no new principles are involved and the algebra is tedious. The results of the derivation are summarized in Table 6.2-2 and Section 6.3.

A test of the treatment A null hypothesis

$$H_0:\quad \alpha_j = 0 \quad\text{for all } j \qquad\text{or}\qquad \mu_{.1}=\mu_{.2}=\cdots=\mu_{.p}$$

versus

$$H_1:\quad \alpha_j \neq 0 \quad\text{for some } j \qquad\text{or}\qquad \mu_{.j}\neq\mu_{.j'} \quad\text{for some } j \text{ and } j' \ (j\neq j')$$

for model (6.1-1) is given by $F = MSA/MSRES$. If the null hypothesis is true, this ratio is distributed as F with $p-1$ and $(n-1)(p-1)$ degrees of freedom.[†] A test of the block null hypothesis

$$H_0:\quad \sigma_\pi^2 = 0$$

versus

$$H_1:\quad \sigma_\pi^2 > 0$$

is given by $F = MSBL/MSRES$. If the null hypothesis is true, this ratio is distributed as F with $n-1$ and $(n-1)(p-1)$ degrees of freedom. Recall that model (6.1-1) is a mixed model in which the blocks are regarded as random effects. Thus this null hypothesis concerns the population of blocks from which the n blocks in the experi-

---

*   See Exercise 4 for the derivation.

**   See Exercise 5 for the derivation of the computational formulas.

†   Assumptions underlying a randomized block design are discussed in detail in Section 6.4.

ment are a random sample. Ordinarily an experimenter is not particularly interested in testing this null hypothesis. One expects this null hypothesis to be false since the blocks in the experiment were formed so as to be dissimilar. Before examining the fixed and mixed models for a randomized block design, we will illustrate procedures for computing mean squares and testing the null hypotheses.

6.2 COMPUTATIONAL EXAMPLE FOR A TYPE RB-*p* DESIGN

Assume that an experimenter wants to evaluate the relative merits of four versions of an altimeter, an instrument used to display altitude. Eight helicopter pilots with from 500 to 3000 flight hours are available to serve as subjects. Accuracy of reading the altimeter at low altitude is of prime importance, so the dependent variable is amount of reading error. It is anticipated that the amount and type of previous flying experience of pilots may affect their performance with the experimental altimeters.

To isolate the nuisance variable of previous flying experience and other idiosyncratic characteristics of the pilots, a type RB-4 design with repeated measures is used. Each subject makes 100 readings under simulated flight conditions with each experimental altimeter. The mean of each pilot's reading errors constitutes the data subjected to statistical analysis. The sequence in which the four altimeters are presented in the experiment is randomized independently for each subject. The appropriate design model is a mixed model since the levels of treatment A (altimeters) are fixed while those for blocks (previous flying experience) are random. If we replicated the experiment, we would use the same four altimeters, but the pilots and the amount of their flying experience would be different.

The following statistical hypotheses are advanced:

$$H_0: \quad \alpha_j = 0 \quad \text{for all } j$$
$$H_1: \quad \alpha_j \neq 0 \quad \text{for some } j.$$

The level of significance adopted is .05. The data and computational procedures are given in Table 6.2-1. The results of the analysis are summarized in Table 6.2-2. Since $F = MSA/MSRES$ is significant, we conclude that the mean number of reading errors is not the same for all of the altimeters. Procedures for determining which population means differ are described in Section 6.5. One can also test the hypothesis $\sigma_\pi^2 = 0$ to determine if there are differences among the means of the hypothetical populations associated with previous flying experience. Contrary to expectations, $F = MSBL/MSRES$ is not significant. This suggests that previous flying experience was not an effective blocking variable. Thus we have used seven degrees of freedom in isolating that portion of the total variance due to block variation, but to little avail.

TABLE 6.2-1   Computational Procedures for a Type RB-4 Design

(i)   Data and notation [Y_{ij} denotes a score for the experimental unit in block i and treatment level j; $j = 1, \ldots, p$ treatment levels (a_j); $i = 1, \ldots, n$ blocks (s_i)]:

AS Summary Table

Entry is Y_{ij}

	a_1	a_2	a_3	a_4	$\sum_{j=1}^{p} Y_{ij}$
s_1	3	4	7	7	21
s_2	6	5	8	8	27
s_3	3	4	7	9	23
s_4	3	3	6	8	20
s_5	1	2	5	10	18
s_6	2	3	6	10	21
s_7	2	4	5	9	20
s_8	2	3	6	11	22
$\sum_{i=1}^{n} Y_{ij} =$	22	28	50	72	

(ii)   Computational symbols:

$$\sum_{j=1}^{p} \sum_{i=1}^{n} Y_{ij} = 3 + 6 + 3 + \cdots + 11 = 172.0$$

$$\left(\sum_{j=1}^{p} \sum_{i=1}^{n} Y_{ij} \right)^2 \bigg/ np = [Y] = (172.0)^2/(8)(4) = 924.5$$

$$\sum_{j=1}^{p} \sum_{i=1}^{n} Y_{ij}^2 = [AS] = (3)^2 + (6)^2 + (3)^2 + \cdots + (11)^2 = 1160.0$$

$$\sum_{j=1}^{p} \left(\sum_{i=1}^{n} Y_{ij} \right)^2 \bigg/ n = [A] = \frac{(22)^2}{8} + \frac{(28)^2}{8} + \cdots + \frac{(72)^2}{8} = 1119.0$$

$$\sum_{i=1}^{n} \left(\sum_{j=1}^{p} Y_{ij} \right)^2 \bigg/ p = [S] = \frac{(21)^2}{4} + \frac{(27)^2}{4} + \cdots + \frac{(22)^2}{4} = 937.0$$

(iii)   Computational formulas:

$SSTO = [AS] - [Y] = 1160.0 - 924.5 = 235.5$

$SSA = [A] - [Y] = 1119.0 - 924.5 = 194.5$

$SSBL = [S] - [Y] = 937.0 - 924.5 = 12.5$

$SSRES = [AS] - [A] - [S] + [Y] = 1160.0 - 1119.0 - 937.0 + 924.5 = 28.5$

TABLE 6.2-2   ANOVA Table for Type RB-4 Design

	Source	SS	df	MS	F	E(MS) Model III
1	Treatment levels (Altimeters)	194.500	$p - 1 = 3$	64.833	[1/3] 47.78*	$\sigma_\epsilon^2 + n \sum_{j=1}^{p} \alpha_j^2/(p-1)$
2	Blocks (Pilots and previous flying experience)	12.500	$n - 1 = 7$	1.786	[2/3] 1.32	$\sigma_\epsilon^2 + p\sigma_\pi^2$
3	Residual	28.500	$(n-1)(p-1) = 21$	1.357		σ_ϵ^2
4	Total	235.500	$np - 1 = 31$			

* $p < .01$

POWER OF THE ANALYSIS OF VARIANCE F TEST

Procedures for calculating power and determining the number of subjects necessary to achieve a specified power are described in Section 4.3. These procedures generalize directly to a randomized block design. However, instead of entering Table E.14 with $\nu_2 = p(n-1)$, the degrees of freedom are equal to $(n-1)(p-1)$.

ESTIMATES OF STRENGTH OF ASSOCIATION

We noted in Section 4.6 that a significant F for treatment effects indicates that there is an association between the independent and dependent variables. Indices of strength of association for a fixed-effects model are given by

$$\hat{\omega}_{Y|A}^2 = \frac{\sum_{j=1}^{p} \hat{\alpha}_j^2/p}{\hat{\sigma}_\epsilon^2 + \sum_{i=1}^{n} \hat{\pi}_i^2/n + \sum_{j=1}^{p} \hat{\alpha}_j^2/p} = \frac{SSA - (p-1)MSRES}{SSTO + MSRES}$$

$$\hat{\omega}_{Y|BL}^2 = \frac{\sum_{i=1}^{n} \hat{\pi}_i^2/n}{\hat{\sigma}_\epsilon^2 + \sum_{i=1}^{n} \hat{\pi}_i^2/n + \sum_{j=1}^{p} \hat{\alpha}_j^2/p} = \frac{SSBL - (n-1)MSRES}{SSTO + MSRES}.$$

Similar measures for a random-effects model are given by

$$\hat{\rho}_{IY|A} = \frac{\hat{\sigma}_{\alpha}^2}{\hat{\sigma}_{\epsilon}^2 + \hat{\sigma}_{\pi}^2 + \hat{\sigma}_{\alpha}^2} \qquad \text{and} \qquad \hat{\rho}_{IY|BL} = \frac{\hat{\sigma}_{\pi}^2}{\hat{\sigma}_{\epsilon}^2 + \hat{\sigma}_{\pi}^2 + \hat{\sigma}_{\alpha}^2}$$

where $\hat{\sigma}_{\epsilon}^2 = MSRES$
$\hat{\sigma}_{\pi}^2 = (MSBL - MSRES)/p$
$\hat{\sigma}_{\alpha}^2 = (MSA - MSRES)/n.$

Indices of strength of association for a mixed model in which the levels of treatment A are fixed and those for blocks are random are

$$\hat{\omega}_{Y|A}^2 = \frac{\sum_{j=1}^{p} \hat{\alpha}_j^2/p}{\hat{\sigma}_{\epsilon}^2 + \hat{\sigma}_{\pi}^2 + \sum_{j=1}^{p} \hat{\alpha}_j^2/p} \qquad \text{and} \qquad \hat{\rho}_{IY|BL} = \frac{\hat{\sigma}_{\pi}^2}{\hat{\sigma}_{\epsilon}^2 + \hat{\sigma}_{\pi}^2 + \sum_{j=1}^{p} \hat{\alpha}_j^2/p}$$

where $\hat{\sigma}_{\epsilon}^2 = MSRES$

$\hat{\sigma}_{\pi}^2 = (MSBL - MSRES)/p$

$$\sum_{j=1}^{p} \hat{\alpha}_j^2/p = \frac{p-1}{np} (MSA - MSRES).$$

The rationale underlying the formula for $\Sigma \hat{\alpha}_j^2/p$, for example, can be understood in terms of the $E(MS)$ for MSA and $MSBL$ shown in Table 6.2-2.

$$E\left[\frac{p-1}{np}(MSA - MSRES)\right] = \frac{p-1}{np}\left\{\left[\sigma_{\epsilon}^2 + n\sum_{j=1}^{p}\alpha_j^2/(p-1)\right] - \sigma_{\epsilon}^2\right\}$$

$$= \sum_{j=1}^{p}\alpha_j^2/p.$$

For the data in Table 6.2-2,

$$\hat{\omega}_{Y|A}^2 = \quad = \frac{5.951}{1.357 + .107 + 5.951} = .80$$

and

$$\hat{\rho}_{IY|BL} = \frac{.107}{1.357 + .107 + 5.951} = .01.$$

In words, the independent variable of type of altimeter accounts for 80% of the variance in the dependent variable (error in reading the altimeters), but the blocking variable accounts for only 1% of the variance. This is additional evidence for concluding that previous flying experience was not an effective blocking variable.

In Section 4.6 we noted that an F ratio provides no information concerning the size of treatment effects, only whether they are significant. Trivial effects can achieve statistical significance if the sample is sufficiently large. The interpretation of research results in terms of $\hat{\omega}^2$ or $\hat{\rho}_I$ is an important adjunct to significance tests.

6.3 ALTERNATIVE MODELS FOR A TYPE RB-*p* DESIGN

EXPECTED VALUES FOR OTHER MODELS

The expected values of the mean squares given in Table 6.2-2 are appropriate for the mixed model

(6.3-1)
$$Y_{ij} = \mu + \alpha_j + \pi_i + \epsilon_{ij} \qquad (i = 1, \ldots, n; j = 1, \ldots, p)$$

described in Section 6.1. There we assumed that the levels of treatment A were fixed but those for blocks were random. Suppose instead that the levels of treatment A and blocks are either both fixed or both random. The resulting models are referred to, respectively, as a fixed-effects model (model I) and a random-effects model (model II). For model I, $\Sigma_{j=1}^{p} \alpha_j = 0$ and $\Sigma_{i=1}^{n} \pi_i = 0$. For model II, we assume that α_j is $NID(0, \sigma_\alpha^2)$, π_i is $NID(0, \sigma_\pi^2)$, α_j is independent of π_i and ϵ_{ij}, and π_i is independent of ϵ_{ij}. The expected values of the mean squares for these two models are as follows.

Model I	*Model II*
$E(MSA) = \sigma_\epsilon^2 + n \sum_{j=1}^{p} \alpha_j^2/(p-1)$	$\sigma_\epsilon^2 + n\sigma_\alpha^2$
$E(MSBL) = \sigma_\epsilon^2 + p \sum_{i=1}^{n} \pi_i^2/(n-1)$	$\sigma_\epsilon^2 + p\sigma_\pi^2$
$E(MSRES) = \sigma_\epsilon^2$	σ_ϵ^2

The F statistics for testing the hypotheses

Model I H_0: $\alpha_j = 0$ for all j and H_0: $\pi_i = 0$ for all i

Model II H_0: $\sigma_\alpha^2 = 0$ and H_0: $\sigma_\pi^2 = 0$

are, respectively, $F = MSA/MSRES$ and $F = MSBL/MSRES$. Although the form of the F ratios is the same for model I and model II, the inferences are different—for model I, inferences apply only to the p treatment levels and n blocks in the experiment; for model II, inferences apply to the populations from which the p treatment levels and n blocks are random samples. The expected values for a mixed model in which the levels of treatment A are random but those for blocks are fixed are as follows.

Model III (A random, Blocks fixed)

$$E(MSA) = \sigma_\epsilon^2 + n\sigma_\alpha^2$$

$$E(MSBL) = \sigma_\epsilon^2 + p \sum_{i=1}^{n} \pi_i^2/(n-1)$$

$$E(MSRES) = \sigma_\epsilon^2$$

The appropriate null hypotheses for this model are

$$\textit{Model III} \qquad H_0: \; \sigma_\alpha^2 = 0 \qquad \text{and} \qquad H_0: \; \pi_i = 0 \quad \text{for all } i.$$

A NONADDITIVE MODEL

Model (6.3-1) contains no provision for nonadditivity of block and treatment effects, that is, we assume that in the population the scores for each block have the same trend over the p treatment levels. If this condition obtains, block and treatment effects are said to be *additive* (not to interact). If the assumption is not tenable, we need to include a term in model (6.3-1) that reflects the nonadditivity (interaction). A model should reflect all of the effects that are actually present in an experiment. Thus the appropriate model equation for a randomized block design if block and treatment effects are not additive is

(6.3-2) $\qquad Y_{ij} = \mu + \alpha_j + \pi_i + (\pi\alpha)_{ij} + \epsilon_{ij} \quad (i = 1, \ldots, n; j = 1, \ldots, p)$

where $(\pi\alpha)_{ij}$ is the interaction effect for block i and treatment level j.

The presence of interaction effects alters the expected values of the mean squares as shown in Table 6.3-1. An examination of this table reveals a problem in evaluating treatment A and blocks for model I. It is not possible to form an F ratio in which the expected value of the numerator contains one more term than that in the denominator. For example, the expected values of the numerator and denominator of the F statistic for testing the hypothesis $\alpha_j = 0$ for all j are

$$\frac{E(MSA)}{E(MSRES)} = \frac{\sigma_\epsilon^2 + n \sum_{j=1}^{p} \alpha_j^2/(p-1)}{\sigma_\epsilon^2 + \sum_{j=1}^{p} \sum_{i=1}^{n} (\pi\alpha)_{ij}^2/(n-1)(p-1)}.$$

The presence of the interaction term in the denominator negatively biases the test; consequently, an experimenter will reject too few false null hypotheses. The same problem occurs with model III when the levels of A are random and when blocks are random. For these cases, the expected values are, respectively,

$$\frac{E(MSA)}{E(MSRES)} = \frac{\sigma_\epsilon^2 + n\sigma_\alpha^2}{\sigma_\epsilon^2 + \sigma_{\pi\alpha}^2} \qquad \text{and} \qquad \frac{E(MSBL)}{E(MSRES)} = \frac{\sigma_\epsilon^2 + p\sigma_\pi^2}{\sigma_\epsilon^2 + \sigma_{\pi\alpha}^2}$$

If a negatively biased test is significant, an experimenter can feel confident that the null hypothesis is false. However, the interpretation of an insignificant ratio is ambiguous. An insignificant ratio may occur because the test is negatively biased, the test lacks adequate power, or the null hypothesis is in fact true. For model II and model III (A fixed), tests of treatment A are not negatively biased. However, the tests are less powerful if an interaction term appears in the model than if it does not. To see why this is so, consider a test of treatment A for model II. Expected values for the additive and nonadditive models are, respectively,

TABLE 6.3-1 Expected Values of Means Squares for Type RB-p Design When Treatment and Block Effects Are Not Additive

Mean Square	Model I A Fixed BL Fixed	Model II A Random BL Random	Model III A Fixed BL Random	Model III A Random BL Fixed
MSA	$\sigma_\epsilon^2 + n\sum_{j=1}^p \alpha_j^2/(p-1)$	$\sigma_\epsilon^2 + \sigma_{\pi\alpha}^2 + n\sigma_\alpha^2$	$\sigma_\epsilon^2 + \sigma_{\pi\alpha}^2 + n\sum_{j=1}^p \alpha_j^2/(p-1)$	$\sigma_\epsilon^2 + n\sigma_\alpha^2$
MSBL	$\alpha_\epsilon^2 + p\sum_{i=1}^n \pi_i^2/(n-1)$	$\sigma_\epsilon^2 + \sigma_{\pi\alpha}^2 + p\sigma_\pi^2$	$\sigma_\epsilon^2 + p\sigma_\pi^2$	$\sigma_\epsilon^2 + \sigma_{\pi\alpha}^2 + p\sum_{i=1}^n \pi_i^2/(n-1)$
MSRES	$\sigma_\epsilon^2 + \sum_{j=1}^p\sum_{i=1}^n (\pi\alpha)_{ij}^2/(n-1)(p-1)$	$\sigma_\epsilon^2 + \sigma_{\pi\alpha}^2$	$\sigma_\epsilon^2 + \sigma_{\pi\alpha}^2$	$\sigma_\epsilon^2 + \sigma_{\pi\alpha}^2$

$$\frac{E(MSA)}{E(MSRES)} = \frac{\sigma_\epsilon^2 + n\sigma_\alpha^2}{\sigma_\epsilon^2} \quad \text{and} \quad \frac{\sigma_\epsilon^2 + \sigma_{\pi\alpha}^2 + n\sigma_\alpha^2}{\sigma_\epsilon^2 + \sigma_{\pi\alpha}^2}.$$

Suppose that $\sigma_\epsilon^2 = 15$, $\sigma_{\pi\alpha}^2 = 10$, and $\sigma_\alpha^2 = 8$. The values of the F ratios for the additive and nonadditive models are

$$F = \frac{15 + 8}{15} = \frac{23}{15} = 1.53 \quad \text{and} \quad F = \frac{15 + 10 + 8}{15 + 10} = \frac{33}{25} = 1.32.$$

The addition of a constant, $\sigma_{\pi\alpha}^2 = 10$, to the numerator and denominator of the nonadditive model has reduced the value of the F ratio and resulted in a less powerful test of the false null hypothesis.

    We have just seen two reasons why an experimenter hopes an additive model rather than a nonadditive model describes the data—it avoids a negative bias for model I and model III (A random) and it results in a more powerful test of a false null hypothesis. Before turning to some underlying assumptions associated with a randomized block design, we will describe a test of the hypothesis that block and treatment effects are additive.

TEST FOR ADDITIVITY OF BLOCK AND TREATMENT EFFECTS

A preliminary test for choosing between the additive and nonadditive model equations, for example, between

$$Y_{ij} = \mu + \alpha_j + \pi_i + \epsilon_{ij} \quad (i = 1, \ldots, n; j = 1, \ldots, p)$$

and

$$Y_{ij} = \mu + \alpha_j + \pi_i + (\pi\alpha)_{ij} + \epsilon_{ij} \quad (i = 1, \ldots, n; j = 1, \ldots, p)$$

has been developed by Tukey (1949). The procedure involves partitioning the residual sum of squares into two components: a one-degree of freedom block × treatment-level interaction component that represents nonadditivity and a $(n - 1)(p - 1) - 1$ degree of freedom component that provides an error term for testing the significance of the nonadditivity component.

    To get a better feeling for the meaning of additivity and nonadditivity, consider the experiment represented in Table 6.2-1. Suppose that treatment level a_3 increased the population dependent measure by $\alpha_3 = .875$ and block s_2 increased it by $\pi_2 = 1.375$. The two effects would be additive if the increase in the dependent measure due to α_3 and π_2 together were $.875 + 1.375 = 2.250$. An additive model is one for which all α_j and π_i are additive. We can think of additivity in another way. If the population values for each of the eight blocks in Table 6.2-1 changed by X amount from a_1 to a_2, by X' amount from a_2 to a_3 and by X'' amount from a_3 to a_4, the block–treatment-level interaction would be zero and the additive model would apply. If we graphed the data, the scores for each block would be parallel. An interaction would appear as two or more nonparallel lines.

Tukey's test is sensitive to situations in which the scores for each block follow the same general trend but the amount of change from a_1 to a_2, or from a_2 to a_3, et cetera, is not the same for all blocks.* This is a limitation of the test since other forms of nonadditivity do occur. Nevertheless, it can be a useful preliminary test in designs having one score per cell. In such designs there is no within-cell error term, and a residual is used as an estimate of experimental error. If the test for nonadditivity is insignificant, it lends credence to the hypothesis that *MSRES* is an estimate of σ_ϵ^2 rather than $\sigma_\epsilon^2 + \sigma_{\pi\alpha}^2$ or $\sigma_\epsilon^2 + \sum_{j=1}^{p} \sum_{i=1}^{n} (\pi\alpha)_{ij}^2 / (n - 1)(p - 1)$.

Computational procedures for the test of nonadditivity are illustrated in Table 6.3-2. The level of significance adopted for the test, $\alpha = .10$, reflects a concern about committing a type II error. The use of a numerically large probability value (.05, .10, .25) will increase the power of the test if the null hypothesis is false. It is evident from the table that the test is significant. Thus the nonadditive model appears to underlie the data. This is confirmed by an inspection of Figure 6.3-1 where the residual for each observation, $\hat{\epsilon}_{ij} = Y_{ij} - \overline{Y}_{.j} - \overline{Y}_{i.} + \overline{Y}_{..}$, is plotted against the estimated value, $\hat{Y}_{ij} = \overline{Y}_{..} + (\overline{Y}_{.j} - \overline{Y}_{..}) + (\overline{Y}_{i.} - \overline{Y}_{..})$, which does not include the residual. If the additive model were appropriate, there would be no association between the size of the residuals and the estimated values. We see from Figure 6.3-1 that large positive and negative values of $\hat{\epsilon}_{ij}$ tend to be associated with large $\hat{Y}_{ij}$'s, producing a funnel shape.

A nonadditive model poses an additional interpretation problem for an experimenter. The significant test for nonadditivity is a signal to an experimenter that the effects of a treatment cannot be described for the population of blocks as a whole but instead must be described individually for each block. In other words, an experimenter cannot draw the same conclusions regarding treatment levels for all blocks.

Sometimes the use of a different criterion measure or one of the trans-

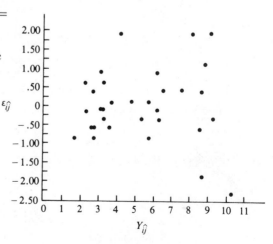

FIGURE 6.3-1 Plot of residual, $\hat{\epsilon}_{ij} = Y_{ij} - \overline{Y}_{.j} - \overline{Y}_{i.} + \overline{Y}_{..}$, versus estimated value, $\hat{Y}_{ij} = \overline{Y}_{..} + (\overline{Y}_{.j} - \overline{Y}_{..}) + (\overline{Y}_{i.} - \overline{Y}_{..})$. The presence of an association between the size of $\hat{\epsilon}_{ij}$ and $\hat{Y}_{ij}$, in this case a funnel shape, suggests that the additive model is not appropriate.

* More specifically, it tests the hypothesis that a special type of contrast-contrast interaction in which the coefficients of the contrasts are the $(\overline{Y}_{i.} - \overline{Y}_{..})$ and $(\overline{Y}_{.j} - \overline{Y}_{..})$ deviations is equal to zero. Contrast-contrast interactions are discussed in Section 8.6.

TABLE 6.3-2   Computational Procedures for Test for Nonadditivity

(i)   Data and notation [Y_{ij} denotes a score for the experimental unit in block i and treatment level j; $j = 1$, ..., p treatment levels (a_j); $i = 1$, ..., n blocks (s_i)]:

AS Summary Table
Entry is Y_{ij}

	a_1	a_2	a_3	a_4	$\overline{Y}_{i\cdot}$	$d_{i\cdot} = \overline{Y}_{i\cdot} - \overline{Y}_{\cdot\cdot}$
s_1	3	4	7	7	5.25	$-.125$
s_2	6	5	8	8	6.75	1.375
s_3	3	4	7	9	5.75	$.375$
s_4	3	3	6	8	5.00	$-.375$
s_5	1	2	5	10	4.50	$-.875$
s_6	2	3	6	10	5.25	$-.125$
s_7	2	4	5	9	5.00	$-.375$
s_8	2	3	6	11	5.50	$.125$

$$\overline{Y}_{\cdot j} = \quad 2.75 \quad 3.50 \quad 6.25 \quad 9.00 \qquad \overline{Y}_{\cdot\cdot} = \sum_{j=1}^{p}\sum_{i=1}^{n} Y_{ij}/np$$

$$d_{\cdot j} = \overline{Y}_{\cdot j} - \overline{Y}_{\cdot\cdot} = -2.625 \quad -1.875 \quad .875 \quad 3.625 \qquad = 172/(8)(4) = 5.375$$

(ii) d_{ij} Summary Table*

	a_1	a_2	a_3	a_4
s_1	.328125	.234375	$-.109375$	$-.453125$
s_2	-3.609375	-2.578125	1.203125	4.984375
s_3	$-.984375$	$-.703125$	.328125	1.359375
s_4	.984375	.703125	$-.328125$	-1.359375
s_5	2.296875	1.640625	$-.765625$	-3.171875
s_6	.328125	.234375	$-.109375$	$-.453125$
s_7	.984375	.703125	$-.328125$	-1.359375
s_8	$-.328125$	$-.234375$	.109375	.453125

*Entries in this table are product of $d_{i\cdot} \times d_{\cdot j}$. For example, $(d_{1\cdot})(d_{\cdot 1}) = (-.125)(-2.625) = .328125$, $(d_{2\cdot})(d_{\cdot 1}) = (1.375)(-2.625) = -3.609375$.

(iii)   Computational formulas:

$$SSNONADD = \frac{\left(\sum\limits_{j=1}^{p}\sum\limits_{i=1}^{n} d_{ij}Y_{ij}\right)^2}{\left(\sum\limits_{i=1}^{n} d_{i\cdot}^2\right)\left(\sum\limits_{j=1}^{p} d_{\cdot j}^2\right)} = \frac{[(.328125)(3) + (-3.609375)(6) + \cdots + (.453125)(11)]^2}{[(-.125)^2 + \cdots + (.125)^2][(-2.625)^2 + \cdots + (3.625)^2]}$$

$$= \frac{(-24.6875)^2}{(3.1250)(24.3125)} = \frac{609.4727}{75.9766} = 8.022$$

$$SSREM = SSRES - SSNONADD = 28.500 - 8.022 = 20.478$$

TABLE 6.3-2 (continued)

$dfNONADD = 1$

$dfREM = dfRES - 1 = 21 - 1 = 20$

$FNONADD = \dfrac{SSNONADD/dfNONADD}{SSREM/dfREM} = \dfrac{8.022/1}{20.478/20} = \dfrac{8.022}{1.024} = 7.83$

$F_{.10;1,20} = 2.97$

(iv) Computational checks:

$$\sum_{j=1}^{p} \sum_{i=1}^{n} d_{ij}^2 = \left(\sum_{i=1}^{n} d_{i\cdot}^2 \right) \left(\sum_{j=1}^{p} d_{\cdot j}^2 \right)$$

$$\sum_{j=1}^{p} \sum_{i=1}^{n} d_{ij}^2 = (.328125)^2 + \cdots + (.453125)^2 = 75.9766$$

$$\left(\sum_{i=1}^{n} d_{i\cdot}^2 \right) \left(\sum_{j=1}^{p} d_{\cdot j}^2 \right) = (3.1250)(24.3125) = 75.9766$$

$$np \left(\sum_{i=1}^{n} d_{i\cdot}^2 \right) \left(\sum_{j=1}^{p} d_{\cdot j}^2 \right) = (SSBL)(SSA)$$

$$np \left(\sum_{i=1}^{n} d_{i\cdot}^2 \right) \left(\sum_{j=1}^{p} d_{\cdot j}^2 \right) = (8)(4)(3.1250)(24.3125) = 2431.25$$

$$(SSBL)(SSA) = (12.5)(194.5) = 2431.25$$

---

formations described in Section 2.6 will eliminate or minimize the block–treatment-level interaction. Considering the advantages of the additive model, these procedures are worth considering.

*6.4 SOME ASSUMPTIONS UNDERLYING A TYPE RB-p DESIGN

COMPARISON OF ERROR TERMS FOR TYPE CR-p AND RB-p DESIGNS

In this section we will examine in more detail some of the assumptions underlying a type RB-p design. We begin by comparing the error terms for type CR-p and RB-p designs. Recall that the fixed-effects experimental design model for a type CR-p

---

*   Portions of this section assume a familiarity with the matrix operations in Appendix D. The essential ideas can be grasped without this background.

design is

(6.4-1)
$$Y_{ij} = \mu + \alpha_j + \epsilon_{i(j)} \qquad (i = 1, \ldots, n; j = 1, \ldots, p)$$

where $\Sigma_{j=1}^{p} \alpha_j = 0$ and $\epsilon_{i(j)}$ is $NID(0, \sigma_\epsilon^2)$. For this model the variance of the Y_{ij}'s for population a_j, denoted by $\sigma_{Y_j}^2$ or σ_j^2, is due only to variation among the $\epsilon_{i(j)}$'s since μ and α_j are both constants for the jth population. Earlier we referred to this variance as error variance and denoted it by $\sigma_{\epsilon_j}^2$. According to model (6.4-1), $\sigma_{\epsilon_1}^2 = \sigma_{\epsilon_2}^2 = \cdots = \sigma_{\epsilon_p}^2 = \sigma_\epsilon^2$, or equivalently, $\sigma_{Y_1}^2 = \sigma_{Y_2}^2 = \cdots = \sigma_{Y_p}^2 = \sigma_Y^2$. The best estimator of the common error variance, σ_ϵ^2, based on random samples from the p populations is the weighted mean

$$MSWG = \frac{(n_1 - 1)\hat{\sigma}_{\epsilon_1}^2 + \cdots + (n_p - 1)\hat{\sigma}_{\epsilon_p}^2}{(n_1 - 1) + \cdots + (n_p - 1)}.$$

Consider now the mixed model for a randomized block design described in Section 6.1. There we assumed that

$$Y_{ij} = \mu + \alpha_j + \pi_i + \epsilon_{ij} \qquad (i = 1, \ldots n; j = 1, \ldots, p)$$

where $\Sigma_{j=1}^{p} \alpha_j = 0$, π_i is $NID(0, \sigma_\pi^2)$, ϵ_{ij} is $NID(0, \sigma_\epsilon^2)$, and ϵ_{ij} is independent of π_i. For this model the variance of the Y_{ij}'s for population a_j is due to variation among the ϵ_{ij}'s and the π_i's. As in the type CR-p design, μ and α_j are constants for the jth population. Assuming that ϵ_{ij} and π_i are uncorrelated,

$$\sigma_{Y_j}^2 = \sigma_\epsilon^2 + \sigma_\pi^2 \qquad (j = 1, \ldots, p).$$

This follows from Rule B.2 in Appendix B concerning the variance of a random variable.

$$V(Y) = E\{[Y - E(Y)]^2\}$$
$$= E(Y^2) - [E(Y)]^2$$

Replacing Y on the right side of Rule B.2 by $\mu + \alpha_j + \pi_i + \epsilon_{ij}$ gives

$$\sigma_{Y_j}^2 = E(\mu + \alpha_j + \pi_i + \epsilon_{ij})^2 - [E(\mu + \alpha_j + \pi_i + \epsilon_{ij})]^2$$
$$= (\mu^2 + \alpha_j^2 + \sigma_\pi^2 + \sigma_\epsilon^2 + 2\mu\alpha_j) - (\mu^2 + \alpha_j^2 + 2\mu\alpha_j)$$

(6.4-2)
$$= \sigma_\epsilon^2 + \sigma_\pi^2.$$

Two scores Y_{ij} and $Y_{ij'}$, $j \neq j'$, are not independent since the term π_i is common to both of them. Thus, although we assume that the errors are independent, we do not assume that the scores within a block are independent. The dependence of the scores in population a_j and $a_{j'}$ is reflected in their covariance, denoted by $\sigma_{Y_jY_{j'}}$ or $\sigma_{jj'}$. We will now show that $\sigma_{Y_jY_{j'}} = \sigma_\pi^2$. According to rule B.15 in Appendix B,

$$COV(X, Y) = E\{[X - E(X)][Y - E(Y)]\}$$
$$= E(XY) - E(X)E(Y).$$

Replacing X on the right side of the rule by $\mu + \alpha_j + \pi_i + \epsilon_{ij}$ and replacing Y by $\mu + \alpha_{j'} + \pi_i + \epsilon_{ij'}$ gives

$$\sigma_{Y_j Y_{j'}} = E[(\mu + \alpha_j + \pi_i + \epsilon_{ij})(\mu + \alpha_{j'} + \pi_i + \epsilon_{ij'})]$$
$$- E(\mu + \alpha_j + \pi_i + \epsilon_{ij})E(\mu + \alpha_{j'} + \pi_i + \epsilon_{ij'})$$
$$= (\mu^2 + \alpha_j \alpha_{j'} + \sigma_\pi^2 + \mu\alpha_j + \mu\alpha_{j'}) - (\mu^2 + \alpha_j \alpha_{j'} + \mu\alpha_j + \mu\alpha_{j'})$$

(6.4-3)
$$= \sigma_\pi^2.$$

To summarize, for a randomized block design (1) $\sigma_{Y_j}^2 = \sigma_\epsilon^2 + \sigma_\pi^2$, (2) the dependence of Y_{ij} and $Y_{ij'}$ is due to the π_i term, and (3) $\sigma_{Y_j Y_{j'}} = \sigma_\pi^2$.

Earlier we saw for a completely randomized design that $\sigma_\epsilon^2 = \sigma_Y^2$. We will now show for a randomized block design that $\sigma_\epsilon^2 = \sigma_Y^2(1 - \rho)$, where ρ denotes the correlation between any two treatment levels. Assume that the potential observations under treatment level j have variance

$$\sigma_{Y_j}^2 = \sigma_Y^2 \qquad \text{for all } j$$

and that the covariances

$$\sigma_{Y_j Y_{j'}} = \text{a constant for all pairs of } j \text{ and } j' \; (j \neq j').$$

In other words, the variances are homogeneous and the covariances are homogeneous. Then the correlation between potential observations for any two treatment levels j and j' can be written as

$$\rho = \frac{\displaystyle\sum_{i=1}^{N} (Y_{ij} - \mu_{\cdot j})(Y_{ij'} - \mu_{\cdot j'})}{N} \Bigg/ \sqrt{\left[\frac{\displaystyle\sum_{i=1}^{N}(Y_{ij} - \mu_{\cdot j})^2}{N}\right]\left[\frac{\displaystyle\sum_{i=1}^{N}(Y_{ij'} - \mu_{\cdot j'})^2}{N}\right]} = \frac{\sigma_{Y_j Y_{j'}}}{\sigma_Y^2}$$

and

$$\sigma_{Y_j Y_{j'}} = \rho \sigma_Y^2.$$

Combining these results with (6.4-2) and (6.4-3) gives

$$\sigma_Y^2 = \sigma_\epsilon^2 + \sigma_\pi^2$$
$$= \sigma_\epsilon^2 + \rho\sigma_Y^2$$

and

$$\sigma_\epsilon^2 = \sigma_Y^2 - \rho\sigma_Y^2$$

(6.4-4)
$$= \sigma_Y^2(1 - \rho).$$

A comparison of this error term with that for a completely randomized design,

$$\text{CR-}p \qquad E(MSWG) = \sigma_\epsilon^2 = \sigma_Y^2$$
$$\text{RB-}p \qquad E(MSRES) = \sigma_\epsilon^2 = \sigma_Y^2(1 - \rho)$$

illustrates why *MSRES* is normally smaller than *MSWG*. The key factor is the correlation coefficient ρ. The more precise the matching of experimental units within blocks, the higher the correlation between observations in any two treatment levels and the smaller the error term σ_ϵ^2. The expected values of the treatment and block means squares for the mixed model described in Section 6.1 can also be expressed in terms of σ_Y^2 and ρ as follows.

$$E(MSA) = \sigma_\epsilon^2 + n \sum_{j=1}^{p} \alpha_j^2/(p-1) = \sigma_Y^2(1-\rho) + n \sum_{j=1}^{p} \alpha_j^2/(p-1)$$

$$E(MSBL) = \sigma_\epsilon^2 + p\sigma_\pi^2$$
$$= \sigma_Y^2 - \rho\sigma_Y^2 + p\rho\sigma_Y^2$$
$$= \sigma_Y^2[1 + (p-1)\rho]$$

THE COMPOUND SYMMETRY ASSUMPTION AND TYPE H MATRICES

We arrived at $\sigma_\epsilon^2 = \sigma_Y^2(1-\rho)$, equation (6.4-4), by assuming that all of the σ_j^2's are equal and that all of the $\sigma_{jj'}$'s are equal. These equality conditions define a special type of *variance-covariance matrix*. This matrix, denoted by Σ_S, is

$$\underset{p \times p}{\Sigma_S} = \begin{array}{c} \\ a_1 \\ a_2 \\ a_3 \\ \cdot \\ a_p \end{array} \begin{array}{cccccc} a_1 & a_2 & a_3 & \cdot & a_p \\ \begin{bmatrix} \sigma_1^2 & \sigma_{12} & \sigma_{13} & \cdot & \sigma_{1p} \\ \sigma_{21} & \sigma_2^2 & \sigma_{23} & \cdot & \sigma_{2p} \\ \sigma_{31} & \sigma_{32} & \sigma_3^2 & \cdot & \sigma_{3p} \\ \cdot & \cdot & \cdot & \cdot & \cdot \\ \sigma_{p1} & \sigma_{p2} & \sigma_{p3} & \cdot & \sigma_p^2 \end{bmatrix} \end{array}$$

where all of the variances on the main diagonal are equal and all of the covariances off the main diagonal are equal. Such a matrix is said to have *compound symmetry* and is called a *type S matrix*.

Compound symmetry is a sufficient condition for the ratio $F = MSA/MSRES$ to be distributed as F with $\nu_1 = p - 1$ and $\nu_2 = (n-1)(p-1)$ under the hypothesis of no treatment effects. It is not, however, a necessary condition. Huynh and Feldt (1970) and Rouanet and Lépine (1970) have shown that the F ratio will also be distributed as F if the variances of differences for all pairs of treatment levels are homogeneous, that is, if

$$\sigma_{Y_j - Y_{j'}}^2 = \sigma_j^2 + \sigma_{j'}^2 - 2\sigma_{jj'} = \text{a constant for all } j \text{ and } j' \ (j \neq j').$$

Matrices that satisfy this less restrictive condition are called *type H matrices*, Σ_H. All matrices having compound symmetry satisfy this condition and thus form a subset of the broader class of type H matrices. The following population matrix is an example of a type H matrix.

$$
\underset{4 \times 4}{\mathbf{\Sigma}_H} = \begin{array}{c} \\ a_1 \\ a_2 \\ a_3 \\ a_4 \end{array} \begin{array}{cccc} a_1 & a_2 & a_3 & a_4 \\ \begin{bmatrix} 5.0 & 2.5 & 5.0 & 7.5 \\ 2.5 & 10.0 & 7.5 & 10.0 \\ 5.0 & 7.5 & 15.0 & 12.5 \\ 7.5 & 10.0 & 12.5 & 20.0 \end{bmatrix} \end{array}
$$

We can verify this by computing the variances of differences for all $p(p - 1)/2 = 6$ pairs of treatment levels as follows.

$$
\begin{aligned}
\sigma^2_{Y_1 - Y_2} &= 5.0 + 10.0 - 2(2.5) = 10.0 \\
\sigma^2_{Y_1 - Y_3} &= 5.0 + 15.0 - 2(5.0) = 10.0 \\
\sigma^2_{Y_1 - Y_4} &= 5.0 + 20.0 - 2(7.5) = 10.0 \\
\sigma^2_{Y_2 - Y_3} &= 10.0 + 15.0 - 2(7.5) = 10.0 \\
\sigma^2_{Y_2 - Y_4} &= 10.0 + 20.0 - 2(10.0) = 10.0 \\
\sigma^2_{Y_3 - Y_4} &= 15.0 + 20.0 - 2(12.5) = 10.0
\end{aligned}
$$

Since $\sigma^2_{Y_j - Y_{j'}}$ equals a constant for all j and j' ($j \neq j'$), $\mathbf{\Sigma}_H$ is a type H matrix. The important point is that equal variances and equal covariances are not required in order for the F statistic to be distributed as F.

THE CIRCULARITY ASSUMPTION

The necessary and sufficient condition for $F = MSA/MSRES$ to be distributed as F with $\nu_1 = p - 1$ and $\nu_2 = (n - 1)(p - 1)$ under the hypothesis of no treatment effects is the *circularity assumption*

$$
\underset{(p-1) \times p}{\mathbf{C}^{\star\prime}} \quad \underset{p \times p}{\mathbf{\Sigma}} \quad \underset{p \times (p-1)}{\mathbf{C}^{\star}} = \lambda \underset{(p-1) \times (p-1)}{\mathbf{I}}
$$

where $\mathbf{C}^{\star\prime}$ = any orthonormal coefficient matrix that represents the overall null hypothesis

$\mathbf{\Sigma}$ = a population variance-covariance matrix

λ = a scalar number $\lambda > 0$

$\mathbf{I}$ = an identity matrix.*

The assumption is really quite simple although the matrix notation tends to obscure this. The preceding type H matrix will be used to illustrate the assumption.

To begin, we need to reformulate the overall null hypothesis for treatment A, which is

$$
\text{H}_0: \quad \mu_{.1} = \mu_{.2} = \mu_{.3} = \mu_{.4}.
$$

---

*   Matrix operations are discussed in Appendix D. Orthonormal vectors are defined in D.1-8; an identity matrix is defined in D.2-5. If $\mathbf{C}^{\star\prime} \mathbf{\Sigma} \mathbf{C}^{\star} = \lambda\mathbf{I}$, $\mathbf{C}^{\star\prime} \mathbf{\Sigma} \mathbf{C}^{\star}$ is said to be spherical. Hence the necessary and sufficient condition is also referred to as the sphericity assumption.

This can be done in a variety of ways as was shown in Section 5.10. For example, two alternative formulations are

$$H_0: \quad \mu_{.1} - \mu_{.2} = 0 \qquad\qquad H_0: \quad \mu_{.1} - \mu_{.2} = 0$$
$$\mu_{.3} - \mu_{.4} = 0 \qquad\qquad\qquad (\mu_{.1} + \mu_{.2})/2 - \mu_{.3} = 0$$
$$(\mu_{.1} + \mu_{.2})/2 - (\mu_{.3} + \mu_{.4})/2 = 0 \qquad (\mu_{.1} + \mu_{.2} + \mu_{.3})/3 - \mu_{.4} = 0.$$

Using matrix notation, these hypotheses are, respectively,

$$
\overset{\mathbf{C_1'}}{\begin{bmatrix} 1.0 & -1.0 & 0 & 0 \\ 0 & 0 & 1.0 & -1.0 \\ .5 & .5 & -.5 & -.5 \end{bmatrix}}
\overset{\boldsymbol{\mu}}{\begin{bmatrix} \mu_{.1} \\ \mu_{.2} \\ \mu_{.3} \\ \mu_{.4} \end{bmatrix}}
=
\overset{\mathbf{0}}{\begin{bmatrix} 0 \\ 0 \\ 0 \end{bmatrix}}
$$

and

$$
\overset{\mathbf{C_2'}}{\begin{bmatrix} 1.0 & -1.0 & 0 & 0 \\ .5 & .5 & -1.0 & 0 \\ .33 & .33 & .33 & -1 \end{bmatrix}}
\overset{\boldsymbol{\mu}}{\begin{bmatrix} \mu_{.1} \\ \mu_{.2} \\ \mu_{.3} \\ \mu_{.4} \end{bmatrix}}
=
\overset{\mathbf{0}}{\begin{bmatrix} 0 \\ 0 \\ 0 \end{bmatrix}}.
$$

The orthonormal form (see Appendix D.1-8) of the coefficient matrices $\mathbf{C_1'}$ and $\mathbf{C_2'}$ is

$$
\mathbf{C_1^{\star\prime}} = \begin{bmatrix} .7071 & -.7071 & 0 & 0 \\ 0 & 0 & .7071 & -.7071 \\ .5000 & .5000 & -.5000 & -.5000 \end{bmatrix}
$$

and

$$
\mathbf{C_2^{\star\prime}} = \begin{bmatrix} .7071 & -.7071 & 0 & 0 \\ .4082 & .4082 & -.8165 & 0 \\ .2887 & .2887 & .2887 & -.8660 \end{bmatrix}.
$$

To determine if $\boldsymbol{\Sigma}_H$ satisfies the circularity assumption, we compute

$$
\overset{\mathbf{C_1^{\star\prime}}}{\begin{bmatrix} .7071 & -.7071 & 0 & 0 \\ 0 & 0 & .7071 & -.7071 \\ .5000 & .5000 & -.5000 & -.5000 \end{bmatrix}}
\overset{\boldsymbol{\Sigma}_H}{\begin{bmatrix} 5.0 & 2.5 & 5.0 & 7.5 \\ 2.5 & 10.0 & 7.5 & 10.0 \\ 5.0 & 7.5 & 15.0 & 12.5 \\ 7.5 & 10.0 & 12.5 & 20.0 \end{bmatrix}}
$$

$$
\overset{\mathbf{C_1^{\star}}}{\begin{bmatrix} .7071 & 0 & .5000 \\ -.7071 & 0 & .5000 \\ 0 & .7071 & -.5000 \\ 0 & -.7071 & -.5000 \end{bmatrix}}
=
\overset{\lambda}{5}
\overset{\mathbf{I}}{\begin{bmatrix} 1 & 0 & 0 \\ 0 & 1 & 0 \\ 0 & 0 & 1 \end{bmatrix}}.
$$

Since $\mathbf{C}_1^{*\prime} \, \boldsymbol{\Sigma}_H \mathbf{C}_1^{*} = 5\mathbf{I}$, we conclude that the assumption is satisfied. The outcome of this procedure is not affected by the particular orthonormal matrix chosen to test the overall null hypothesis. We would have reached the same conclusion, for example, if we had used the $\mathbf{C}_2^{*\prime}$ coefficient matrix since $\mathbf{C}_2^{*\prime} \, \boldsymbol{\Sigma} \, \mathbf{C}_2^{*}$ is also equal to $5\mathbf{I}$. Rouanet and Lépine (1970) have shown that when $\mathbf{C}^{*\prime} \boldsymbol{\Sigma} \mathbf{C}^{*}$ satisfies the circularity assumption, the assumption is also satisfied for all subsets of contrasts. Furthermore, the circularity property is equivalent to the property that $\sigma^2_{Y_j - Y_{j'}}$ is equal to a constant for all j and j' ($j \neq j'$). However, the circularity assumption may hold for a subset of contrasts of interest (local circularity) although the variance-covariance matrix is not a type H matrix.*

In summary, the requirements for $F = MSA / MSRES$ to be distributed as F with $(p - 1)$ and $(n - 1)(p - 1)$ degrees of freedom given that the null hypothesis is true are somewhat less restrictive than was once believed. The necessary and sufficient condition is the circularity assumption $\mathbf{C}^{*\prime} \boldsymbol{\Sigma} \mathbf{C}^{*} = \lambda \mathbf{I}$.

A TEST FOR SPHERICITY

A statistic developed by Mauchley (1940) uses a sample estimate of the population variance-covariance matrix, $\boldsymbol{\Sigma}$, to test the hypothesis that $\mathbf{C}^{*\prime} \boldsymbol{\Sigma} \mathbf{C}^{*}$ satisfies the circularity assumption. If $\mathbf{C}^{*\prime} \boldsymbol{\Sigma} \mathbf{C}^{*} = \lambda \mathbf{I}$, then $\mathbf{C}^{*\prime} \hat{\boldsymbol{\Sigma}} \mathbf{C}^{*}$, where $\hat{\boldsymbol{\Sigma}}$ is the sample variance-covariance matrix, should estimate $\lambda \mathbf{I}$. A matrix whose diagonal elements are equal and off-diagonal elements are zero is said to be spherical. If $\mathbf{C}^{*\prime} \boldsymbol{\Sigma} \mathbf{C}^{*}$ is spherical, the χ^2 statistic defined in Table 6.4-1 is approximately distributed as chi square. Since $\chi^2 > \chi^2_{.05,5}$, the null hypothesis that $\mathbf{C}^{*\prime} \boldsymbol{\Sigma} \mathbf{C}^{*}$ is spherical is rejected. In other words, the circularity assumption for the altimeter data in Table 6.2-1 is not tenable. Under these conditions, the test of treatment effects, $F = MSA / MSRES$, is positively biased (Box, 1954b). Fortunately there is a correction for this bias, as we will see in the next section.

A THREE-STEP TESTING STRATEGY

As we have just seen, one can use a sample variance-covariance matrix to test the hypothesis that $\mathbf{C}^{*\prime} \boldsymbol{\Sigma} \mathbf{C}^{*}$ satisfies the circularity assumption. The routine use of this preliminary test, however, is not recommended. A more useful approach when the tenability of the circularity assumption is in doubt stems from the work of Box (1954b). He showed that the true distribution of the F statistic with $(p - 1)$ and $(n - 1)(p - 1)$ degrees of freedom for any arbitrary variance-covariance matrix can be approximated by an F statistic with reduced degrees of freedom. The modified degrees of freedom are $(p - 1)\theta$ and $(n - 1)(p - 1)\theta$, where θ is a number that depends on the degree of departure of the population variance-covariance matrix from the required form. When the matrix satisfies the circularity assumption, the value of θ is one; otherwise θ is less than one, with a minimum of $1/(p - 1)$.

In any practical situation, the value of θ is unknown. However, the work of

---

*   See Mendoza, Toothaker, and Crain (1976) for an example.

TABLE 6.4-1   Computation of Mauchley's Sphericity Test Statistic for Data in Table 6.2-1

(i)   Basic matrices [$\hat{\Sigma}$, the sample variance-covariance matrix, is obtained from Table 6.4-2 (i), $\mathbf{C}_1^{*\prime}$ is defined in Section 6.4]:

$$\underset{3 \times 4}{\mathbf{C}^{*\prime}}$$

$$\underset{4 \times 4}{\hat{\Sigma}}$$

$$\begin{bmatrix} .7071 & -.7071 & 0 & 0 \\ 0 & 0 & .7071 & -.7071 \\ .5000 & .5000 & -.5000 & -.5000 \end{bmatrix} \begin{bmatrix} 2.2143 & 1.1429 & 1.3571 & -1.1429 \\ 1.1429 & .8571 & .7143 & -.7143 \\ 1.3571 & .7143 & 1.0714 & -.7143 \\ -1.1429 & -.7143 & -.7143 & 1.7143 \end{bmatrix}$$

$$\underset{4 \times 3}{\mathbf{C}^{*}}$$

$$(\underset{3 \times 3}{\mathbf{C}^{*\prime}\, \hat{\Sigma}\, \mathbf{C}^{*}})$$

$$\begin{bmatrix} .7071 & 0 & .5000 \\ -.7071 & 0 & .5000 \\ 0 & .7071 & -.5000 \\ 0 & -.7071 & -.5000 \end{bmatrix} = \begin{bmatrix} .3928 & .5357 & .4041 \\ .5357 & 2.1071 & 1.6163 \\ .4041 & 1.6163 & 1.5715 \end{bmatrix}$$

$$|\mathbf{C}^{*\prime}\, \hat{\Sigma}\, \mathbf{C}^{*}| = .1793*$$
$$\text{Trace } (\mathbf{C}^{*\prime}\, \hat{\Sigma}\, \mathbf{C}^{*}) = 4.0714**$$

(ii)   Computation of χ^2 test statistic:

$$W = \frac{|\mathbf{C}^{*\prime}\hat{\Sigma}\mathbf{C}^{*}|}{\left[\dfrac{\text{Trace } (\mathbf{C}^{*\prime}\hat{\Sigma}\mathbf{C}^{*})}{(p-1)}\right]^{(p-1)}} = \frac{.1793}{\left[\dfrac{4.0663}{4-1}\right]^{(4-1)}} = .0717$$

$$d = 1 - [(2p^2 - 3p + 3)/6(n-1)(p-1)] = 1 - \{[2(4)^2 - 3(4) + 3]/6(8-1)(4-1)\}$$
$$= .8175$$

$$\chi^2 = -(n-1)\, d \ln(W) = -(8-1)\, .8715 \ln .0717$$
$$= 15.08$$

$$df = [p(p-1)/2] - 1 = [4(4-1)/2] - 1 = 5$$

$$\chi^2_{.05,5} = 11.07$$

*This number is the determinant of the $(\mathbf{C}^{*\prime}\hat{\Sigma}\mathbf{C}^{*})$ matrix. A determinant is a specified function of the elements of a square matrix. For a 3×3 matrix, the determinant is equal to $(e_{11}e_{22}e_{33} + e_{12}e_{23}e_{31} + e_{13}e_{21}e_{32}) - (e_{11}e_{23}e_{32} + e_{12}e_{21}e_{33} + e_{13}e_{22}e_{31})$, where e_{ij} is an element in row i and column j of the matrix.

**The trace of a matrix is the sum of the elements on the diagonal, for example, $e_{11} + e_{22} + e_{33}$, where the e's are diagonal elements.

Collier, Baker, Mandeville, and Hayes (1967), Huynh and Feldt (1976), and Stoloff (1970) indicates that a satisfactory estimate of θ can be obtained from the sample variance-covariance matrix. Collier et al. (1967) proposed a maximum likelihood estimator of θ. Their estimator $\hat{\theta}$, which is defined in Table 6.4-2, is biased, the extent of the bias increasing as θ approaches 1.0. In an effort to reduce this bias for large values of θ, Huynh and Feldt (1976) and Huynh (1978) have recommended using the estimator $\tilde{\theta}$ (see Table 6.4-2) when $\theta \geq .75$. Rogan, Keselman, and Mendoza (1979) have made a similar recommendation. The $\bar{\theta}$ estimator, which is computed from the ratio of two unbiased estimators, is also biased. Gary (1981) investigated the type I error for the two estimators in the context of a randomized block factorial design for the following values of θ: 1.0, .96, .69, and .52. He concluded that $\hat{\theta}$ provided better type I error protection than $\bar{\theta}$ for all values of θ investigated except 1.0. One reason for the liberalness of $\bar{\theta}$ is that it can assume values greater than one, in which case it is set equal to one. It can be shown that $\bar{\theta} \geq \hat{\theta}$; the two statistics are equal when $\hat{\theta} = 1/(p - 1)$. Pending further research, the use of $\hat{\theta}$ is recommended in the following three-step testing strategy for a randomized block design.

1. Compare the value of the *F* statistic with the critical value of *F* for $(p - 1)$ and $(n - 1)(p - 1)$ degrees of freedom. If *F* does not exceed this critical value, the analysis stops and *F* is declared not significant. If *F* exceeds the critical value, proceed to step 2.

2. Compare the value of the *F* statistic with the critical value of *F* for $(p - 1)$ $[1/(p - 1)] = 1$ and $[(n - 1)(p - 1)][1/(p - 1)] = (n - 1)$ degrees of freedom. If *F* exceeds this critical value, the test is declared significant. If *F* does not exceed this critical value, proceed to step 3. The use of 1 and $n - 1$ degrees of freedom for the test is referred to as the *Geisser-Greenhouse* (1958) *conservative F test* or simply as the *conservative F test*.

3. Compute the sample estimate $\hat{\theta}$ of θ (see Table 6.4-2 for the computational procedure) and the modified degrees of freedom $(p - 1)\hat{\theta}$ and $(n - 1)$ $(p - 1)\hat{\theta}$. Determine the critical value of *F* for these modified degrees of freedom. If *F* exceeds this critical value, the test is declared significant. Otherwise, the test is not significant. The use of a sample estimate of θ in determining the degrees of freedom for the *F* test is referred to as an *adjusted F test*.

The computation of $\hat{\theta}$ and $\bar{\theta}$ for the data in Table 6.2-1 is illustrated in Table 6.4-2. The value of the *F* statistic for treatment *A* from Table 6.2-1 is $F = 47.78$. This *F* exceeds the critical value for a conventional *F* test given in Table 6.4-2(iv). It also exceeds the value for the conservative *F* test. Hence, according to step 2 of the three-step testing procedure, *F* is declared significant. If the conservative *F* had not been significant, we would have proceeded to step 3 and computed $\hat{\theta}$. When $F \geq F(conservative)$, this is not necessary since it is always true that $F(conventional) \leq F(adjusted) \leq F(conservative)$. The value of $\hat{\theta}$, .419, is close to its lower bound, $1/(p - 1) = .333$, which is consistent with the earlier finding in Table 6.4-1 that the population variance-covariance matrix does not satisfy the circularity assumption.

In concluding this section, a comment about the use of Hotelling's (1931) exact T^2 test is in order. This procedure is sometimes recommended when the tenability of the circularity assumption is in doubt. Although it does provide better control of the type I error than an adjusted *F* test, its power is appreciably lower (Imhof, 1962). Thus its use is not recommended for experiments with fewer than 25 blocks.

TABLE 6.4-2   Computation of $\hat{\theta}$ and $\tilde{\theta}$ for Data in Table 6.2-1

(i)   Variance-covariance matrix [Elements on the main diagonal are $\hat{\sigma}_j^2$'s; elements off the main diagonal are $\hat{\sigma}_{jj'}$'s]:

$$\hat{\Sigma} = \begin{array}{c} \\ a_1 \\ a_2 \\ a_3 \\ a_4 \end{array} \begin{array}{cccc} a_1 & a_2 & a_3 & a_4 \\ \begin{bmatrix} 2.2143 & 1.1429 & 1.3571 & -1.1429 \\ 1.1429 & .8571 & .7143 & -.7143 \\ 1.3571 & .7143 & 1.0714 & -.7143 \\ -1.1429 & -.7143 & -.7143 & 1.7143 \end{bmatrix} \end{array}$$

$$\hat{\sigma}_j^2 = \frac{\sum\limits_{i=1}^{n} Y_{ij}^2 - \dfrac{\left(\sum\limits_{i=1}^{n} Y_{ij}\right)^2}{n}}{n-1}$$

$$\hat{\sigma}_{jj'} = \frac{\sum\limits_{i=1}^{n} Y_{ij}Y_{ij'} - \dfrac{\left(\sum\limits_{i=1}^{n} Y_{ij}\right)\left(\sum\limits_{i=1}^{n} Y_{ij'}\right)}{n}}{n-1}$$

(ii)   Computation of $\hat{\theta}$:

p = number of treatment levels = 4

$E_{jj'}$ = element in row j and column j' of $\hat{\Sigma}$ ($j = 1, \ldots, p; j' = 1, \ldots p$)

$\overline{E}_{..} = \Sigma\Sigma E_{jj'}/p^2$ = mean of all elements
   = $(2.2143 + 1.1429 + \cdots + 1.7143)/(4)^2 = 0.4464$

$\Sigma\Sigma E_{jj'}^2$ = sum of each element squared
   = $(2.2143)^2 + (1.1429)^2 + \cdots + (1.7143)^2 = 21.6941$

$\overline{E}_D$ = mean of elements on the main diagonal ($D = 11, 22, \ldots, pp$)
   = $(E_{11} + E_{22} + \cdots + E_{pp})/p = (2.2143 + .8571 + \cdots + 1.7143)/4 = 1.4643$

$\overline{E}_{j.}$ = mean of elements in row j
   $\overline{E}_{1.} = [2.2143 + 1.1429 + \cdots + (-1.1429)]/4 = 0.8929$
   $\overline{E}_{2.} = [1.1429 + .8571 + \cdots + (-.7143)]/4 = 0.5000$
   $\cdots\cdots\cdots\cdots\cdots\cdots\cdots\cdots\cdots\cdots\cdots$
   $\overline{E}_{4.} = [(-1.1429) + (-.7143) + \cdots + 1.7143]/4 = -0.2143$

$$\hat{\theta} = \frac{p^2(\overline{E}_D - \overline{E}_{..})^2}{\left(p-1\right)\left(\Sigma\Sigma E_{jj'}^2 - 2p\sum\limits_{j=1}^{p}\overline{E}_{j.}^2 + p^2\overline{E}_{..}^2\right)}$$

$$= \frac{(4)^2(1.4643 - .4464)^2}{(4-1)\{21.6941 - 2(4)[(.8929)^2 + (.5000)^2 + (.6071)^2 + (-.2143)^2] + (4)^2(.4464)^2\}}$$

$$= \frac{16.5779}{39.5650} = 0.4190$$

(iii)   Computation of $\tilde{\theta}$:

$$\tilde{\theta} = \frac{n(p-1)\hat{\theta} - 2}{(p-1)[n - 1 - (p-1)\hat{\theta}]} = \frac{8(4-1).4190 - 2}{(4-1)[8 - 1 - (4-1).4190]} = \frac{8.0560}{17.2290} = 0.468$$

(iv)   Critical values of F for treatment A:

$$dfA = (p-1)\hat{\theta} = (4-1).4190 = 1.257$$
$$(n-1)(p-1)\hat{\theta} = (8-1)(4-1).4190 = 8.799$$
$$F_{.05;3,21}(conventional) = 3.07$$
$$F_{.05;1,7}(conservative) = 5.59$$
$$F_{.05;1.257,8.799}(adjusted) \cong 4.94$$

6.5   PROCEDURES FOR TESTING DIFFERENCES AMONG MEANS

Tests of differences among means in a type RB-p design have the same general form as those illustrated in Section 4.4 for a completely randomized design. However, $MSWG$ in the formula should be replaced with $MSRES$ if the circularity assumption is satisfied or with $MSRES_i$ if it is not satisfied. Formulas using $MSRES$ are given first. An a priori t test for the ith contrast among means has the following form:

$$t = \frac{\hat{\psi}_i}{\hat{\sigma}_{\psi_i}} = \frac{c_1 \overline{Y}_{.1} + c_2 \overline{Y}_{.2} + \cdots + c_p \overline{Y}_{.p}}{\sqrt{MSRES\left(\dfrac{c_1^2}{n} + \dfrac{c_2^2}{n} + \cdots + \dfrac{c_p^2}{n}\right)}}$$

$$df = (n - 1)(p - 1).$$

To reject the null hypothesis that $c_1\mu_1 + c_2\mu_2 + \cdots + c_p\mu_p = 0$, the absolute value of t must equal or exceed $t_{\alpha/2,\nu}$ (or $t_{\alpha,\nu}$ for a one-tailed test in which case the value of t must be consistent with the alternative hypothesis).

A posteriori tests as described by Tukey (1953) and Scheffé (1959) are given by

$$q = \frac{\hat{\psi}_i}{\hat{\sigma}_{\overline{Y}}} = \frac{c_j \overline{Y}_{.j} + c_{j'} \overline{Y}_{.j'}}{\sqrt{\dfrac{MSRES}{n}}}$$

$$df = (n - 1)(p - 1)$$

and

$$F = \frac{\hat{\psi}_i^2}{\hat{\sigma}_{\psi_i}^2} = \frac{\left(c_1 \overline{Y}_{.1} + c_2 \overline{Y}_{.2} + \cdots + c_p \overline{Y}_{.p}\right)^2}{MSRES\left(\dfrac{c_1^2}{n} + \dfrac{c_2^2}{n} + \cdots + \dfrac{c_p^2}{n}\right)}$$

$$df = (p - 1), (n - 1)(p - 1)$$

respectively. The null hypothesis is rejected if $|q| \geq q_{\alpha;p,\nu}$ for Tukey's test. For Scheffé's test, the hypothesis is rejected if $F \geq F'_{\alpha;\nu_1,\nu_2}$, where

$$F'_{\alpha;\nu_1,\nu_2} = (p - 1)F_{\alpha;(p-1),(n-1)(p-1)}.$$

The use of the overall residual mean square, $MSRES$, in computing $\hat{\sigma}_{\psi_i}$ and $\hat{\sigma}_{\overline{Y}}$ is appropriate if $\mathbf{C^{*\prime}\Sigma C^*}$ satisfies the circularity assumption. If, as is the case for the data in Table 6.2-1, the assumption is not tenable, the overall mean square should be replaced by residual mean squares, $MSRES_i$, appropriate for the specific contrasts of interest. The error degrees of freedom for this case are equal to $n - 1$. There is reason to believe that tests of treatment contrasts are not robust with respect to violation of the circularity assumption (Boik, 1975, 1981). Maxwell (1980) found that Tukey's test using $MSRES_i$ with $n - 1$ degrees of freedom becomes somewhat liberal when p is large. In contrast, Dunn's test using $MSRES_i$ with $n - 1$ degrees of freedom controls the type I error rate at or less than α.

The computational procedures for the t (or Dunn) and Tukey test statistics are illustrated in Table 6.5-1. For purposes of illustration, assume that we are interested

TABLE 6.5-1   Computational Procedures for Testing Differences Among Means When the Circularity Assumption Is Not Tenable

(i) Computation of contrasts [Coefficients of a contrast are applied to data in Table 6.2-1(i)]:

	$\hat{\psi}_1$ at s_i	$\hat{\psi}_2$ at s_i	$\hat{\psi}_3$ at s_i
	$=(1)Y_{i1}+(-1)Y_{i2}+(0)Y_{i3}+(0)Y_{i4}$	$=(0)Y_{i1}+(0)Y_{i2}+(1)Y_{i3}+(-1)Y_{i4}$	$=(.5)Y_{i1}+(.5)Y_{i2}+(-.5)Y_{i3}+(-.5)Y_{i4}$
s_1	$(1)3+(-1)4+(0)7+(0)7=-1$	$(0)3+(0)4+(1)7+(-1)7=0$	$(.5)3+(.5)4+(-.5)7+(-.5)7=-3.5$
s_2	$(1)6+(-1)5+(0)8+(0)8=1$	$(0)6+(0)5+(1)8+(-1)8=0$	$(.5)6+(.5)5+(-.5)8+(-.5)8=-2.5$
s_3	$(1)3+(-1)4+(0)7+(0)9=-1$	$(0)3+(0)4+(1)7+(-1)9=-2$	$(.5)3+(.5)4+(-.5)7+(-.5)9=-4.5$
s_4	$(1)3+(-1)3+(0)6+(0)8=0$	$(0)3+(0)3+(1)6+(-1)8=-2$	$(.5)3+(.5)3+(-.5)6+(-.5)8=-4.0$
s_5	$(1)1+(-1)2+(0)5+(0)10=-1$	$(0)1+(0)2+(1)5+(-1)10=-5$	$(.5)1+(.5)2+(-.5)5+(-.5)10=-6.0$
s_6	$(1)2+(-1)3+(0)6+(0)10=-1$	$(0)2+(0)3+(1)6+(-1)10=-4$	$(.5)2+(.5)3+(-.5)6+(-.5)10=-5.5$
s_7	$(1)2+(-1)4+(0)5+(0)9=-2$	$(0)2+(0)4+(1)5+(-1)9=-4$	$(.5)2+(.5)4+(-.5)5+(-.5)9=-4.0$
s_8	$(1)2+(-1)3+(0)6+(0)11=-1$	$(0)2+(0)3+(1)6+(-1)11=-5$	$(.5)2+(.5)3+(-.5)6+(-.5)11=-6.0$

$$\sum_{i=1}^{n}\hat{\psi}_1 \text{ at } s_i = -6 \qquad \sum_{i=1}^{n}\hat{\psi}_2 \text{ at } s_i = -22 \qquad \sum_{i=1}^{n}\hat{\psi}_3 \text{ at } s_i = -36$$

$$\sum_{i=1}^{n}\hat{\psi}_1^2 \text{ at } s_i = 10 \qquad \sum_{i=1}^{n}\hat{\psi}_2^2 \text{ at } s_i = 90 \qquad \sum_{i=1}^{n}\hat{\psi}_3^2 \text{ at } s_i = 173$$

$$\hat{\psi}_1 = \left(\sum_{i=1}^{n}\hat{\psi}_1 \text{ at } s_i\right)/n = -6/8 \qquad \hat{\psi}_2 = \left(\sum_{i=1}^{n}\hat{\psi}_2 \text{ at } s_i\right)/n = -22/8 \qquad \hat{\psi}_3 = \left(\sum_{i=1}^{n}\hat{\psi}_3 \text{ at } s_i\right)/n = -36/8$$

$$= -.75 \qquad\qquad\qquad\qquad = -2.75 \qquad\qquad\qquad\qquad = -4.5$$

(ii) Computation of $\hat{\sigma}_{\psi_i}$ and $\hat{\sigma}_{\nabla}$:

$$SSRES_1 = \left[\sum_{i=1}^{n}\hat{\psi}_1^2 \text{ at } s_i - \frac{\left(\sum_{i=1}^{n}\hat{\psi}_1 \text{ at } s_i\right)^2}{n}\right] \Bigg/ \sum_{j=1}^{p}c_{1j}^2 = \left[10 - \frac{(-6)^2}{8}\right]\Bigg/[(1)^2+(-1)^2+(0)^2+(0)^2] = 2.75$$

$$SSRES_2 = \left[\sum_{i=1}^{n}\hat{\psi}_2^2 \text{ at } s_i - \frac{\left(\sum_{i=1}^{n}\hat{\psi}_2 \text{ at } s_i\right)^2}{n}\right] \Bigg/ \sum_{j=1}^{p}c_{2j}^2 = \left[90 - \frac{(-22)^2}{8}\right]\Bigg/[(0)^2+(0)^2+(1)^2+(-1)^2] = 14.75$$

$$SSRES_3 = \left[\sum_{i=1}^{n}\hat{\psi}_3^2 \text{ at } s_i - \frac{\left(\sum_{i=1}^{n}\hat{\psi}_3 \text{ at } s_i\right)^2}{n}\right] \Bigg/ \sum_{j=1}^{p}c_{3j}^2 = \left[173 - \frac{(-36)^2}{8}\right]\Bigg/[(.5)^2+(.5)^2+(-.5)^2+(-.5)^2] = 11.00$$

$MSRES_1 = SSRES_1/(n - 1) = 2.75/(8 - 1) = 0.393$

$MSRES_2 = SSRES_2/(n - 1) = 14.75/(8 - 1) = 2.107$

$MSRES_3 = SSRES_3/(n - 1) = 11.00/(8 - 1) = 1.571$

$$\hat{\sigma}_{\psi_1} = \sqrt{MSRES_1 \left(\frac{\sum_{j=1}^{p} c_{1j}^2}{n} \right)} = \sqrt{.393(2/8)} = 0.313$$

$$\hat{\sigma}_{\psi_2} = \sqrt{MSRES_2 \left(\frac{\sum_{j=1}^{p} c_{2j}^2}{n} \right)} = \sqrt{2.107(2/8)} = 0.726$$

$$\hat{\sigma}_{\psi_3} = \sqrt{MSRES_3 \left(\frac{\sum_{j=1}^{p} c_{3j}^2}{n} \right)} = \sqrt{1.571(1/8)} = 0.443$$

$$\hat{\sigma}_{\bar{y}_1} = \sqrt{MSRES_1/n} = \sqrt{.393/8} = 0.222$$

$$\hat{\sigma}_{\bar{y}_2} = \sqrt{MSRES_2/n} = \sqrt{2.107/8} = 0.513$$

(iii) Computation of test statistics:

t statistic	Tukey statistic
$t = \dfrac{\hat{\psi}_1}{\hat{\sigma}_{\psi_1}} = \dfrac{-.75}{0.313} = -2.40$	$q = \dfrac{\hat{\psi}_1}{\hat{\sigma}_{\bar{y}_1}} = \dfrac{-.75}{0.222} = -3.38$
$t = \dfrac{\hat{\psi}_2}{\hat{\sigma}_{\psi_2}} = \dfrac{-2.75}{0.726} = -3.79$	$q = \dfrac{\hat{\psi}_2}{\hat{\sigma}_{\bar{y}_2}} = \dfrac{-2.75}{0.513} = -5.36$
$t = \dfrac{\hat{\psi}_3}{\hat{\sigma}_{\psi_3}} = \dfrac{-4.5}{0.443} = -10.16$	$q_{.05;4,7} = 4.68$

$t_{.05/2,7} = 2.365$

in the following contrasts: $\psi_1 = \mu_{\cdot1} - \mu_{\cdot2}$, $\psi_2 = \mu_{\cdot3} - \mu_{\cdot4}$, and $\psi_3 = (\mu_{\cdot1} + \mu_{\cdot2})/2 - (\mu_{\cdot3} + \mu_{\cdot4})/2$. Since these contrasts are orthogonal, the t statistic can be used to test the respective null hypotheses. According to Table 6.5-1(iii), the absolute value of t exceeds the critical value for all three contrasts. Hence the null hypotheses, H_0: $\mu_{\cdot1} - \mu_{\cdot2} = 0$, H_0: $\mu_{\cdot3} - \mu_{\cdot4} = 0$, and H_0: $(\mu_{\cdot1} + \mu_{\cdot2})/2 - (\mu_{\cdot3} + \mu_{\cdot4})/2 = 0$, are rejected.

It is evident from Table 6.5-1 that when the circularity assumption is not tenable the procedures for testing differences among means are more complex. Less evident is the loss of power in testing a false null hypothesis that occurs when $MSRES_i$ is used in place of $MSRES$. In the present example $MSRES_i$ has 7 degrees of freedom, many fewer than the 21 associated with $MSRES$. The advantage of using $MSRES_i$ is that the resulting test is exact. The formula for $MSRES_i$ like that for $MSRES$ is an interaction formula. In the present example

$MSRES = 1.357$ (block $\times$ treatment A interaction with 21 degrees of freedom)

$MSRES_1 = 0.393$ (block $\times$ contrast 1 interaction with 7 degrees of freedom)

$MSRES_2 = 2.107$ (block $\times$ contrast 2 interaction with 7 degrees of freedom)

$MSRES_3 = 1.571$ (block $\times$ contrast 3 interaction with 7 degrees of freedom).

If the overall circularity assumption were tenable, these four residual mean squares would estimate the same population error variance. In this example, the use of $MSRES$ in place of $MSRES_1$ would have led to a negatively biased test and to not rejecting the hypothesis that $\mu_{\cdot1} - \mu_{\cdot2} = 0$.

6.6   TESTS FOR TRENDS IN THE DATA

If the treatment levels in an experiment differ quantitatively, additional insight concerning the experiment may be obtained by performing a trend analysis. Procedures for carrying out a trend analysis were described in Section 4.5 in connection with a completely randomized design. These procedures apply, with one modification, to a randomized block design. The modification has to do with the choice of an error term. If the circularity assumption is tenable, $MSRES$ can be used in the denominator of the F ratio for testing the significance of the linear contrast, quadratic contrast, and so on. For example,

$$F = \frac{MS\hat{\psi}_{\text{lin}}}{MSRES} \quad \text{and} \quad F = \frac{MS\hat{\psi}_{\text{quad}}}{MSRES}.$$

If the assumption is not tenable, a residual mean square appropriate for the specific trend contrast of interest should be used, for example, $MSRES_{\text{lin}}$, $MSRES_{\text{quad}}$, and so on. The procedures for computing $MS\hat{\psi}_{\text{lin}}$ and $MSRES_{\text{lin}}$ are shown in Table 6.6-1. These procedures generalize to $MS\hat{\psi}_{\text{quad}}$ and $MSRES_{\text{quad}}$, $MS\hat{\psi}_{\text{cub}}$ and $MSRES_{\text{cub}}$, and so on, by replacing the linear coefficients with the appropriate trend coefficients.

TABLE 6.6-1 Computational Procedures for Testing the Significance of the Linear Contrast When the Circularity Assumption Is Not Tenable

---

(i) Computation of linear contrasts [Coefficients of the linear contrast from Appendix Table E.12, $c_1 = -3$, $c_2 = -1$, $c_3 = 1$, $c_4 = 3$, are applied to the data in Table 6.2-1(i)]:

$\hat{\psi}_{\text{lin}}$ at $s_i = (-3)Y_{i1} + (-1)Y_{i2} + (1)Y_{i3} + (3)Y_{i4}$

$$
\begin{aligned}
s_1 &= (-3)3 + (-1)4 + (1)7 + (3)7 = 15 \\
s_2 &= (-3)6 + (-1)5 + (1)8 + (3)8 = 9 \\
s_3 &= (-3)3 + (-1)4 + (1)7 + (3)9 = 21 \\
s_4 &= (-3)3 + (-1)3 + (1)6 + (3)8 = 18 \\
s_5 &= (-3)1 + (-1)2 + (1)5 + (3)10 = 30 \\
s_6 &= (-3)2 + (-1)3 + (1)6 + (3)10 = 27 \\
s_7 &= (-3)2 + (-1)4 + (1)5 + (3)9 = 22 \\
s_8 &= (-3)2 + (-1)3 + (1)6 + (3)11 = 30
\end{aligned}
$$

$$\sum_{i=1}^{n} \hat{\psi}_{\text{lin}} \text{ at } s_i = 172$$

$$\sum_{i=1}^{n} \hat{\psi}_{\text{lin}}^2 \text{ at } s_i = 4084$$

(ii) Computation of sum of squares and F statistic:

$$SS\hat{\psi}_{\text{lin}} = \left(\sum_{i=1}^{n} \hat{\psi}_{\text{lin}} \text{ at } s_i\right)^2 / n \sum_{j=1}^{p} c_{\text{lin } j}^2$$

$$= (172)^2 / \{8[(-3)^2 + (-1)^2 + (1)^2 + (3)^2]\} = 184.900$$

$df = 1$

$$SSRES_{\text{lin}} = \left[\sum_{i=1}^{n} \hat{\psi}_{\text{lin}}^2 \text{ at } s_i - \frac{\left(\sum_{i=1}^{n} \hat{\psi}_{\text{lin}} \text{ at } s_i\right)^2}{n}\right] / \sum_{j=1}^{p} c_{\text{lin } j}^2$$

$$= [4084 - (172)^2 / 8] / [(-3)^2 + (-1)^2 + (1)^2 + (3)^2] = 19.300$$

$df = n - 1 = 7$

$MS\hat{\psi}_{\text{lin}} = SS\hat{\psi}_{\text{lin}} / 1 = 184.900$

$MSRES_{\text{lin}} = SSRES_{\text{lin}} / (n - 1) = 19.300 / 7 = 2.757$

$F = MS\hat{\psi}_{\text{lin}} / MSRES_{\text{lin}} = 184.900 / 2.757 = 67.07$

$F_{.05; 1, 7} = 5.59$

6.7 ESTIMATING MISSING OBSERVATIONS

In the conduct of experiments there are, unfortunately, occasions when one or more observations are missing—the equipment malfunctioned, a subject did not come for the last session, the wrong treatment condition was presented, and so on. The analysis procedures in Section 6.2 require that each treatment level appears once in each block. Thus, to carry out the statistical analysis the missing data must be replaced. Often it is not possible or feasible to go back and rerun the treatment condition. One alternative is to use an analysis procedure that permits missing data. Such procedures are described in Section 6.9. The alternative described here is to estimate the missing observation. A procedure due to Yates (1933) involves substituting for the missing observation one that minimizes the error sum of squares. We will use the following notation.

$$X_{ij} = \text{missing observation}$$

$$\sum_{j=1}^{p'} Y_{ij} = \text{sum of remaining observations in block containing } X_{ij}$$

$$\sum_{i=1}^{n'} Y_{ij} = \text{sum of remaining observations in column containing } X_{ij}$$

$$\sum_{j=1}^{p'} \sum_{i=1}^{n'} Y_{ij} = \text{sum of all remaining observations}$$

The value of X_{ij} that minimizes the sample error sum of squares,

$$\sum_{j=1}^{p} \sum_{i=1}^{n} \hat{\epsilon}_{ij}^2 = \sum_{j=1}^{p} \sum_{i=1}^{n} (Y_{ij} - \overline{Y}_{i\cdot} - \overline{Y}_{\cdot j} + \overline{Y}_{\cdot\cdot})^2$$

is the one for which

(6.7-1) $$\hat{\epsilon}_{ij} = X_{ij} - \overline{Y}_{i\cdot} - \overline{Y}_{\cdot j} + \overline{Y}_{\cdot\cdot} = 0.$$

The terms $\overline{Y}_{i\cdot}$, $\overline{Y}_{\cdot j}$, and $\overline{Y}_{\cdot\cdot}$ can be expressed as

$$\overline{Y}_{i\cdot} = \left(\sum_{j=1}^{p'} Y_{ij} + X_{ij} \right) / p$$

$$\overline{Y}_{\cdot j} = \left(\sum_{i=1}^{n'} Y_{ij} + X_{ij} \right) / n$$

$$\overline{Y}_{\cdot\cdot} = \left(\sum_{j=1}^{p'} \sum_{i=1}^{n'} Y_{ij} + X_{ij} \right) / np.$$

Substituting in (6.7-1) gives

$$X_{ij} - \frac{\sum\limits_{j=1}^{p'} Y_{ij} + X_{ij}}{p} - \frac{\sum\limits_{i=1}^{n'} Y_{ij} + X_{ij}}{n} + \frac{\sum\limits_{j=1}^{p'} \sum\limits_{i=1}^{n'} Y_{ij} + X_{ij}}{np} = 0.$$

Solving for X_{ij} we obtain Yates's formula for a missing observation:

$$X_{ij} - \frac{X_{ij}}{p} - \frac{X_{ij}}{n} + \frac{X_{ij}}{np} = \frac{\sum\limits_{j=1}^{p'} Y_{ij}}{p} + \frac{\sum\limits_{i=1}^{n'} Y_{ij}}{n} - \frac{\sum\limits_{j=1}^{p'} \sum\limits_{i=1}^{n'} Y_{ij}}{np}$$

$$X_{ij} = \frac{n \sum\limits_{j=1}^{p'} Y_{ij} + p \sum\limits_{i=1}^{n'} Y_{ij} - \sum\limits_{j=1}^{p'} \sum\limits_{i=1}^{n'} Y_{ij}}{(n-1)(p-1)}.$$

To illustrate the estimation procedure, suppose that observation Y_{73} in Table 6.2-1 is missing. An estimate of the observation is given by

$$X_{73} = \frac{8(15) + 4(45) - 167}{(8-1)(4-1)} = 6.3.$$

The estimate (6.3) is reasonably close to the original score which is 5. The value of the estimated observation should be inserted in the data matrix and the analysis of variance performed as shown in Table 6.2-1, but with one small modification. The degrees of freedom for *SSRES* and *SSTO* should be reduced by one.

If more than one observation is missing, an experimenter can *guesstimate* the values for all but one of the missing observations; this one is computed by the preceding formula. The computed value is inserted in the data matrix and another of the missing observations that was guesstimated is computed. This iterative process is repeated until the computed values for all missing observations have stabilized. Generally three cycles of the procedure are sufficient for this purpose. For each estimated missing observation, one degree of freedom should be subtracted from the *df* for *SSRES* and *SSTO*.

Yates's estimation procedure produces a slight positive bias in the *F* tests for treatment and block effects. An unbiased test for treatment effects can be obtained by computing a corrected treatment sum of squares as follows:

$$SSA\,(corrected) = SSTO' - SSRES - SSBL'$$

where *SSTO'* and *SSBL'* are computed from the data matrix prior to inserting the estimated observation(s); *SSRES* is computed from the data matrix after inserting the estimated observation(s). For these computations it is necessary to replace np by N and p by the number of observations in the respective blocks. For example, assume that Y_{73} in Table 6.2-1 is missing. As we saw earlier, $X_{73} = 6.3$. The corrected treatment sum of squares is *SSA(corrected)* = 235.355 − 27.334 − 12.355 = 195.666. This does not differ much from the uncorrected treatment sum of squares, *SSA(uncor-*

rected), which is equal to 196.933. If only one observation has been estimated, a simpler formula for *SSA(corrected)* is

$$SSA(corrected) = SSA(uncorrected) - \left[\sum_{j=1}^{p'} Y_{ij} - (p-1)X_{ij}\right]^2 / [p(p-1)].$$

A corrected block sum of squares is given by

$$SSBL(corrected) = SSTO' - SSRES - SSA'$$
$$= 235.355 - 27.334 - 196.141 = 11.880$$

where *SSTO'* and *SSRES* are computed as before and *SSA'* is computed from the data matrix prior to inserting the estimated observation(s). A simpler formula, if only one observation has been estimated, is

$$SSBL(corrected) = SSBL(uncorrected) - \left[\sum_{i=1}^{n'} Y_{ij} - (n-1)X_{ij}\right]^2 / [n(n-1)].$$

A *t* statistic for contrasts involving a treatment level with one missing observation is given by

$$t = \frac{c_j \overline{Y}_{.j} + c_{j'} \overline{Y}_{.j'}}{\sqrt{MSRES\left[\dfrac{2}{n} + \dfrac{p}{n(n-1)(p-1)}\right]}}.$$

If more than one observation has been estimated, the *t* ratio is given by

$$t = \frac{c_j \overline{Y}_{.j} + c_{j'} \overline{Y}_{.j'}}{\sqrt{MSRES\left(\dfrac{1}{e_j} + \dfrac{1}{e_{j'}}\right)}}$$

where e_j and $e_{j'}$ correspond to the number of *effective replications* in each treatment level. According to Taylor (1948), the value of e_j is computed for each block using the following rules.

1.   Assign 1 if block contains an observation for both e_j and $e_{j'}$.

2.   Assign $(p-2)/(p-1)$ if block contains an observation for e_j but not for $e_{j'}$.

3.   Assign 0 if the block does not contain an observation for e_j.

Assume, for example, that the values in Table 6.2-1 for Y_{31} and Y_{73} are missing and that estimates of these observations have been made. The values of e_1 and e_3 required for the comparison of $\overline{Y}_1$ with $\overline{Y}_3$ are, respectively,

$$e_1 = 1 + 1 + 0 + 1 + 1 + 1 + 2/3 + 1 = 6\tfrac{2}{3}$$
$$e_3 = 1 + 1 + 2/3 + 1 + 1 + 1 + 0 + 1 = 6\tfrac{2}{3}.$$

6.8 RELATIVE EFFICIENCY OF RANDOMIZED BLOCK DESIGN

A randomized block design enables an experimenter to isolate variation associated with a nuisance variable and thereby reduce the error term. If block effects representing the nuisance variable are appreciably greater than zero, a type RB-p design is more efficient than a type CR-p design. In the latter design, the error term $MSWG$ includes effects due to the nuisance variable. A measure of the relative efficiency of the two designs ignoring differences in degrees of freedom is given by

$$\text{Relative efficiency} = \frac{MSWG}{MSRES}$$

where $MSWG$ is obtained from a randomized block analysis by

$$MSWG = \frac{(n-1)MSBL + n(p-1)MSRES}{np-1}.$$

For the altimeter experiment in Section 6.2, an estimate of $MSWG$ is

$$MSWG = \frac{(8-1)1.786 + 8(4-1)1.357}{(8)(4)-1} = 1.454.$$

The efficiency of the randomized block design relative to that of a completely randomized design is

$$\text{Relative efficiency} = \frac{1.454}{1.357} = 1.07.$$

This indicates that if the blocking variable had not been used and treatment levels were assigned randomly to $np = 32$ experimental units in a completely randomized design, the resulting error variance would be 1.07 times as large as that for the randomized block design. A correction developed by Fisher (1935) for the smaller error degrees of freedom for the randomized block design can be incorporated in the efficiency index as follows.

$$\text{Relative efficiency} = \left(\frac{dfRES_{RB} + 1}{dfRES_{RB} + 3}\right)\left(\frac{dfWG_{CR} + 3}{dfWG_{CR} + 1}\right)\frac{MSWG}{MSRES}$$
$$= (0.98)(1.07) = 1.05$$

In this example, the efficiency of the randomized block design is not appreciably greater than that of a completely randomized design. This is another indication that the selection of previous flying experience as the nuisance variable was not a good choice.

The number of subjects n_j in each treatment level of a completely randomized design necessary to match the efficiency of a randomized block design is

$$n_j = \text{Relative efficiency} \times n = (1.05)(8) \cong 9$$

where n is the number of blocks in the randomized block design.

Partitioning of the total sum of squares and degrees of freedom for the two designs is shown in Figure 6.8-1. It is evident from this figure that $SSRES$ for a randomized block design will be smaller than $SSWG$ if the block effects are not equal to zero. If the block effects are equal to zero, $SSRES = SSWG$. In this case, the completely randomized design is more efficient than a randomized block design because the error term is based on more degrees of freedom, for example, $p(n - 1)$ as opposed to $(n - 1)(p - 1)$. It should be apparent that $SSBL$ in a type RB-p design must account for a sizable portion of $SSTO$ in order to compensate for the loss of $n - 1$ degrees of freedom in the $MSRES$ error term. If block effects are equal to zero, $MSBL$ and $MSRES$ can be pooled. The resulting error term has $(n - 1) + (n - 1)$ $(p - 1) = p(n - 1)$ degrees of freedom. The general question of whether or not to pool is discussed in Section 8.11.

An experimenter who uses a randomized block design hopes to be compensated for the additional effort required to form homogeneous blocks by obtaining a smaller error term and greater power to reject a false null hypothesis. The question of whether the total experimental effort involved in matching subjects or obtaining their repeated participation is less than the effort required to run more subjects in a completely randomized design should be raised for each proposed research project. Scarcity of subjects makes a randomized block design an attractive alternative to a completely randomized design when the nature of the treatment permits obtaining repeated measures on the same subjects.

FIGURE 6.8-1   Schematic partition of the total sum of squares and degrees of freedom for type CR-4 and RB-4 designs. A shaded square identifies the sum of squares used in computing the error term ($MSWG$ or $MSRES$) for the designs.

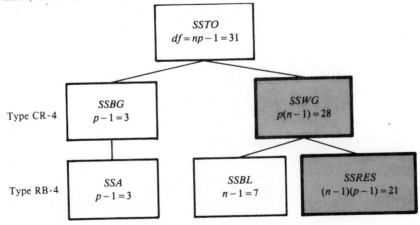

*6.9   GENERAL LINEAR MODEL APPROACH TO TYPE RB-*p* DESIGN

In Chapter 5 we described six approaches to analyzing data for a type CR-*p* design using the general linear model. Here two of the approaches—the regression model with dummy coding and the full rank experimental design model—will be applied to a randomized block design.

REGRESSION MODEL WITH DUMMY CODING

Consider the type RB-4 design for the altimeter experiment described in Section 6.2. A qualitative regression model with $h - 1 = (p - 1) + (n - 1) = 10$ independent variables $(X_{i1}\ X_{i2}, \ldots, X_{i10})$ can be formulated for this design as follows:

(6.9-1)
$$Y_i = \beta_0 X_0 + \overbrace{\beta_1 X_{i1} + \beta_2 X_{i2} + \beta_3 X_{i3}}^{p - 1 = 3 \text{ treatment levels}}$$

$$\overbrace{+ \beta_4 X_{i4} + \beta_5 X_{i5} + \beta_6 X_{i6} + \beta_7 X_{i7} + \beta_8 X_{i8} + \beta_9 X_{i9} + \beta_{10} X_{i10}}^{n - 1 = 7 \text{ blocks}} + \epsilon_i$$
$$(i = 1, \ldots, N)$$

where ϵ_i is $NID(0, \sigma_\epsilon^2)$. A test of H_0: $\beta_1 = \beta_2 = \beta_3 = 0$ provides an indirect test of the hypothesis H_0: $\alpha_1 = \alpha_2 = \alpha_3 = \alpha_4 = 0$, and a test of H_0: $\beta_4 = \cdots = \beta_{10} = 0$ provides an indirect test of H_0: $\pi_1 = \cdots = \pi_8 = 0$. This is accomplished by establishing a correspondence between the $h - 1 = 10$ qualitative independent variables of the regression model and the $n - 1 = 7$ blocks and $p - 1 = 3$ treatment levels of the type RB-4 design. The independent variables of the regression model are coded as follows.

$$X_{i1} = \begin{cases} 1 \text{ if an observation is in treatment level 1} \\ 0 \text{ otherwise} \end{cases}$$

$$X_{i2} = \begin{cases} 1 \text{ if an observation is in treatment level 2} \\ 0 \text{ otherwise} \end{cases}$$

$$X_{i3} = \begin{cases} 1 \text{ if an observation is in treatment level 3} \\ 0 \text{ otherwise} \end{cases}$$

$$X_{i4} = \begin{cases} 1 \text{ if an observation is in block 1} \\ 0 \text{ otherwise} \end{cases}$$

. . . .

---

*   This section assumes a familiarity with the matrix operations in Appendix D and Chapter 5. The reader who is interested only in the traditional approach to analysis of variance can, without loss of continuity, omit this section.

$$X_{i10} = \begin{cases} 1 \text{ if an observation is in block 7} \\ 0 \text{ otherwise} \end{cases}$$

For example, observation Y_{21} in block 2 and treatment level 1 is coded

$$Y_{21} = \beta_0 1 + \beta_1 1 + \beta_2 0 + \beta_3 0 + \beta_4 0 + \beta_5 1 + \beta_6 0$$
$$+ \beta_7 0 + \beta_8 0 + \beta_9 0 + \beta_{10} 0 + \epsilon_{21}$$
$$= \beta_0 + \beta_1 + \beta_5 + \epsilon_{21}.$$

This coding scheme results in the following correspondence between the parameters of the regression model and the parameters of the type RB-4 experimental design model.*

(6.9-2)

$$\beta_0 = \mu + \alpha_4 + \pi_8 \qquad \beta_6 = \pi_3 - \pi_8$$
$$\beta_1 = \alpha_1 - \alpha_4 \qquad \beta_7 = \pi_4 - \pi_8$$
$$\beta_2 = \alpha_2 - \alpha_4 \qquad \beta_8 = \pi_5 - \pi_8$$
$$\beta_3 = \alpha_3 - \alpha_4 \qquad \beta_9 = \pi_6 - \pi_8$$
$$\beta_4 = \pi_1 - \pi_8 \qquad \beta_{10} = \pi_7 - \pi_8$$
$$\beta_5 = \pi_2 - \pi_8$$

A test of the null hypothesis H_0: $\beta_1 = \beta_2 = \beta_3 = 0$,

$$H_0: \quad \underset{\substack{\mathbf{C}'_1 \\ (p-1)\times h}}{\begin{bmatrix} 0 & 1 & 0 & 0 & 0 & 0 & 0 & 0 & 0 & 0 & 0 \\ 0 & 0 & 1 & 0 & 0 & 0 & 0 & 0 & 0 & 0 & 0 \\ 0 & 0 & 0 & 1 & 0 & 0 & 0 & 0 & 0 & 0 & 0 \end{bmatrix}} \underset{\substack{\boldsymbol{\beta} \\ h\times 1}}{\begin{bmatrix} \beta_0 \\ \beta_1 \\ \beta_2 \\ \vdots \\ \beta_{10} \end{bmatrix}} = \underset{\substack{\mathbf{0} \\ (p-1)\times 1}}{\begin{bmatrix} 0 \\ 0 \\ 0 \end{bmatrix}}$$

involves a comparison of the full model (6.9-1) with the reduced model

$$Y_i = \beta_0 X_0 + \beta_4 X_{i4} + \beta_5 X_{i5} + \beta_6 X_{i6} + \beta_7 X_{i7} + \beta_8 X_{i8} + \beta_9 X_{i9} + \beta_{10} X_{i10} + \epsilon_i.$$

The regression sum of squares reflecting the contribution of X_1, X_2, and X_3 over and above that due to including $X_4, \ldots, X_{10}$ in the model is given by**

---

*   The correspondence can be shown by equating the regression and experimental design model expectations for selected values of Y_{ij}. For example, following the procedure in Section 5.4,

$$E(Y_{84}) = \beta_0 = \mu + \alpha_4 + \pi_8$$
$$E(Y_{81}) = \beta_0 + \beta_1 = \mu + \alpha_1 + \pi_8$$
$$\beta_1 = \mu + \alpha_1 + \pi_8 - \beta_0$$
$$= (\mu + \alpha_1 + \pi_8) - (\mu + \alpha_4 + \pi_8)$$
$$= \alpha_1 - \alpha_4.$$

**   The notation used here is introduced in Section 5.5.

$$SSR(X_1 \; X_2 \; X_3 | X_4 \cdots X_{10}) = \overbrace{SSE(X_4 \cdots X_{10})}^{SSE(R)} - \overbrace{SSE(X_1 \cdots X_{10})}^{SSE(F)}.$$

Similarly, a test of H_0: $\beta_4 = \beta_5 = \cdots = \beta_{10} = 0$,

$$H_0: \underset{(n-1)\times h}{\overset{\mathbf{C}_2'}{\begin{bmatrix} 0 & 0 & 0 & 0 & 1 & 0 & 0 & 0 & 0 & 0 & 0 \\ 0 & 0 & 0 & 0 & 0 & 1 & 0 & 0 & 0 & 0 & 0 \\ 0 & 0 & 0 & 0 & 0 & 0 & 1 & 0 & 0 & 0 & 0 \\ 0 & 0 & 0 & 0 & 0 & 0 & 0 & 1 & 0 & 0 & 0 \\ 0 & 0 & 0 & 0 & 0 & 0 & 0 & 0 & 1 & 0 & 0 \\ 0 & 0 & 0 & 0 & 0 & 0 & 0 & 0 & 0 & 1 & 0 \\ 0 & 0 & 0 & 0 & 0 & 0 & 0 & 0 & 0 & 0 & 1 \end{bmatrix}}} \; \underset{h\times 1}{\overset{\boldsymbol{\beta}}{\begin{bmatrix} \beta_0 \\ \beta_1 \\ \beta_2 \\ \cdot \\ \cdot \\ \beta_{10} \end{bmatrix}}} = \underset{(n-1)\times 1}{\overset{\mathbf{0}}{\begin{bmatrix} 0 \\ 0 \\ 0 \\ 0 \\ 0 \\ 0 \\ 0 \end{bmatrix}}}$$

involves a comparison of the full model (6.9-1) with the reduced model

$$Y_i = \beta_0 X_0 + \beta_1 X_{i1} + \beta_2 X_{i2} + \beta_3 X_{i3} + \epsilon_i.$$

The regression sum of squares reflecting the contribution of $X_4, \ldots, X_{10}$ is obtained from

$$SSR(X_4 \cdots X_{10} | X_1 \; X_2 \; X_3) = \overbrace{SSE(X_1 \; X_2 \; X_3)}^{SSE(R)} - \overbrace{SSE(X_1 \cdots X_{10})}^{SSE(F)}.$$

Because of the correspondence between the parameters of the regression model and the parameters of the type RB-4 experimental design model, a test of $\beta_1 = \beta_2 = \beta_3 = 0$ provides an indirect test of

$$H_0: \quad \alpha_j = 0 \quad \text{for all } j.$$

Similarly, a test of $\beta_4 = \cdots = \beta_{10} = 0$ provides an indirect test of

$$H_0: \quad \sigma_\pi^2 = 0$$

where blocks represent random effects.

The procedures for computing F statistics for testing the treatment and block null hypotheses are illustrated in Table 6.9-1; the results are summarized in Table 6.9-2. A comparison of Table 6.9-2 with Table 6.2-2, where the traditional computational procedures were used, reveals that both approaches lead to the same results.

REGRESSION MODEL WITH MISSING OBSERVATIONS

Experiments in which one or more scores are missing can be readily analyzed using the regression approach. One simply sets up the regression model for the observations that are available and then fits the model to these data. For purposes of illustration, the example in Section 6.7 involving one missing observation, Y_{73}, is reanalyzed using

TABLE 6.9-1 Computational Procedures for a Type RB-4 Design Using a Regression Model with Dummy Coding

(i) Data and basic matrices for full model ($N = 32$, $h = 11$, $n = 8$, $p = 4$):

$$\underset{1 \times N}{\mathbf{y}'} = [3 \quad 6 \quad 3 \quad \cdots \quad 11]$$

$$\underset{N \times 1}{\mathbf{y}} \quad \underset{1 \times 1}{(\mathbf{y}'\mathbf{y})}$$

$$[3 \quad 6 \quad 3 \quad \cdots \quad 11] \begin{bmatrix} 3 \\ 6 \\ 3 \\ \vdots \\ 11 \end{bmatrix} = 1160.0$$

$\mathbf{X}_0$ ($N \times h$): Treatment levels ($\mathbf{x}_1$ $\mathbf{x}_2$ $\mathbf{x}_3$), Blocks ($\mathbf{x}_4$ $\mathbf{x}_5$ $\mathbf{x}_6$ $\mathbf{x}_7$ $\mathbf{x}_8$ $\mathbf{x}_9$ $\mathbf{x}_{10}$), with $\mathbf{x}_0$ column and $\mathbf{y}$ ($N \times 1$):

$\mathbf{y}$		$\mathbf{x}_0$	$\mathbf{x}_1$	$\mathbf{x}_2$	$\mathbf{x}_3$	$\mathbf{x}_4$	$\mathbf{x}_5$	$\mathbf{x}_6$	$\mathbf{x}_7$	$\mathbf{x}_8$	$\mathbf{x}_9$	$\mathbf{x}_{10}$
3	a_1	1	1	0	0	1	0	0	0	0	0	0
6		1	1	0	0	0	1	0	0	0	0	0
3		1	1	0	0	0	0	1	0	0	0	0
3		1	1	0	0	0	0	0	1	0	0	0
1		1	1	0	0	0	0	0	0	1	0	0
2		1	1	0	0	0	0	0	0	0	1	0
2		1	1	0	0	0	0	0	0	0	0	1
2		1	1	0	0	0	0	0	0	0	0	0
4	a_2	1	0	1	0	1	0	0	0	0	0	0
5		1	0	1	0	0	1	0	0	0	0	0
4		1	0	1	0	0	0	1	0	0	0	0
3		1	0	1	0	0	0	0	1	0	0	0
2		1	0	1	0	0	0	0	0	1	0	0
3		1	0	1	0	0	0	0	0	0	1	0
4		1	0	1	0	0	0	0	0	0	0	1
3		1	0	1	0	0	0	0	0	0	0	0
7	a_3	1	0	0	1	1	0	0	0	0	0	0
8		1	0	0	1	0	1	0	0	0	0	0
7		1	0	0	1	0	0	1	0	0	0	0
6		1	0	0	1	0	0	0	1	0	0	0
5		1	0	0	1	0	0	0	0	1	0	0
6		1	0	0	1	0	0	0	0	0	1	0
5		1	0	0	1	0	0	0	0	0	0	1
6		1	0	0	1	0	0	0	0	0	0	0
7	a_4	1	0	0	0	1	0	0	0	0	0	0
8		1	0	0	0	0	1	0	0	0	0	0
9		1	0	0	0	0	0	1	0	0	0	0
8		1	0	0	0	0	0	0	1	0	0	0
10		1	0	0	0	0	0	0	0	1	0	0
10		1	0	0	0	0	0	0	0	0	1	0
9		1	0	0	0	0	0	0	0	0	0	1
11		1	0	0	0	0	0	0	0	0	0	0

$$\underset{h \times h}{(\mathbf{X}_0'\mathbf{X}_0)} = \begin{bmatrix} 32 & 8 & 8 & 8 & 4 & 4 & 4 & 4 & 4 & 4 & 4 \\ 8 & 8 & 0 & 0 & 1 & 1 & 1 & 1 & 1 & 1 & 1 \\ 8 & 0 & 8 & 0 & 1 & 1 & 1 & 1 & 1 & 1 & 1 \\ 8 & 0 & 0 & 8 & 1 & 1 & 1 & 1 & 1 & 1 & 1 \\ 4 & 1 & 1 & 1 & 4 & 0 & 0 & 0 & 0 & 0 & 0 \\ 4 & 1 & 1 & 1 & 0 & 4 & 0 & 0 & 0 & 0 & 0 \\ 4 & 1 & 1 & 1 & 0 & 0 & 4 & 0 & 0 & 0 & 0 \\ 4 & 1 & 1 & 1 & 0 & 0 & 0 & 4 & 0 & 0 & 0 \\ 4 & 1 & 1 & 1 & 0 & 0 & 0 & 0 & 4 & 0 & 0 \\ 4 & 1 & 1 & 1 & 0 & 0 & 0 & 0 & 0 & 4 & 0 \\ 4 & 1 & 1 & 1 & 0 & 0 & 0 & 0 & 0 & 0 & 4 \end{bmatrix}$$

$$\sum_{i=1}^{N} Y_i = 172$$

TABLE 6.9-1 (continued)

$$(\mathbf{X}_0'\mathbf{X}_0)^{-1}$$
$$h \times h$$

$$
\begin{bmatrix}
11/32 & -4/32 & -4/32 & -4/32 & -8/32 & -8/32 & -8/32 & -8/32 & -8/32 & -8/32 & -8/32 \\
-4/32 & 8/32 & 4/32 & 4/32 & 0 & 0 & 0 & 0 & 0 & 0 & 0 \\
-4/32 & 4/32 & 8/32 & 4/32 & 0 & 0 & 0 & 0 & 0 & 0 & 0 \\
-4/32 & 4/32 & 4/32 & 8/32 & 0 & 0 & 0 & 0 & 0 & 0 & 0 \\
-8/32 & 0 & 0 & 0 & 16/32 & 8/32 & 8/32 & 8/32 & 8/32 & 8/32 & 8/32 \\
-8/32 & 0 & 0 & 0 & 8/32 & 16/32 & 8/32 & 8/32 & 8/32 & 8/32 & 8/32 \\
-8/32 & 0 & 0 & 0 & 8/32 & 8/32 & 16/32 & 8/32 & 8/32 & 8/32 & 8/32 \\
-8/32 & 0 & 0 & 0 & 8/32 & 8/32 & 8/32 & 16/32 & 8/32 & 8/32 & 8/32 \\
-8/32 & 0 & 0 & 0 & 8/32 & 8/32 & 8/32 & 8/32 & 16/32 & 8/32 & 8/32 \\
-8/32 & 0 & 0 & 0 & 8/32 & 8/32 & 8/32 & 8/32 & 8/32 & 16/32 & 8/32 \\
-8/32 & 0 & 0 & 0 & 8/32 & 8/32 & 8/32 & 8/32 & 8/32 & 8/32 & 16/32
\end{bmatrix}
\begin{array}{c} (\mathbf{X}_0'\mathbf{y}) \\ h \times 1 \\ \begin{bmatrix} 172 \\ 22 \\ 28 \\ 50 \\ 21 \\ 27 \\ 23 \\ 20 \\ 18 \\ 21 \\ 20 \end{bmatrix} \end{array}
=
\begin{array}{c} \hat{\boldsymbol{\beta}}_0 \\ h \times 1 \\ \begin{bmatrix} 9.125 \\ -6.250 \\ -5.500 \\ -2.750 \\ -0.250 \\ 1.250 \\ 0.250 \\ -0.500 \\ -1.000 \\ -0.250 \\ -0.500 \end{bmatrix} \end{array}
$$

$$
\underset{1 \times h}{\hat{\boldsymbol{\beta}}_0'} \qquad \underset{h \times 1}{(\mathbf{X}_0'\mathbf{y})} \qquad \underset{1 \times 1}{(\hat{\boldsymbol{\beta}}_0'\mathbf{X}_0'\mathbf{y})}
$$

$$
\begin{bmatrix} 9.125 & \cdots & -0.500 \end{bmatrix}
\begin{bmatrix} 172 \\ 22 \\ \vdots \\ \vdots \\ 20 \end{bmatrix} = 1131.5
$$

Computation of sum of squares for full model:

$$
SSTO = \mathbf{y}'\mathbf{y} - \left(\sum_{i=1}^{N} Y_i \right)^2 \Big/ N = 1160.0 - (172)^2/32
$$

$$
= 1160.0 - 924.5 = 235.5
$$

$$
SSE = SSE(X_1 \cdots X_{10}) = \mathbf{y}'\mathbf{y} - \hat{\boldsymbol{\beta}}_0'\mathbf{X}_0'\mathbf{y}
$$

$$
= 1160.0 - 1131.5 = 28.5
$$

(ii)   Matrices for reduced model, $Y_i = \beta_0 X_0 + \beta_4 X_{i4} + \cdots + \beta_{10} X_{i10} + \epsilon_i$ ($h = 8$):

$$
\underset{h \times h}{(\mathbf{X}_1'\mathbf{X}_1)} =
\begin{bmatrix}
32 & 4 & 4 & 4 & 4 & 4 & 4 & 4 \\
4 & 4 & 0 & 0 & 0 & 0 & 0 & 0 \\
4 & 0 & 4 & 0 & 0 & 0 & 0 & 0 \\
4 & 0 & 0 & 4 & 0 & 0 & 0 & 0 \\
4 & 0 & 0 & 0 & 4 & 0 & 0 & 0 \\
4 & 0 & 0 & 0 & 0 & 4 & 0 & 0 \\
4 & 0 & 0 & 0 & 0 & 0 & 4 & 0 \\
4 & 0 & 0 & 0 & 0 & 0 & 0 & 4
\end{bmatrix}
$$

$$(\mathbf{X}_1'\mathbf{X}_1)^{-1}$$
$$h \times h$$

$$
\begin{bmatrix}
8/32 & -8/32 & -8/32 & -8/32 & -8/32 & -8/32 & -8/32 & -8/32 \\
-8/32 & 16/32 & 8/32 & 8/32 & 8/32 & 8/32 & 8/32 & 8/32 \\
-8/32 & 8/32 & 16/32 & 8/32 & 8/32 & 8/32 & 8/32 & 8/32 \\
-8/32 & 8/32 & 8/32 & 16/32 & 8/32 & 8/32 & 8/32 & 8/32 \\
-8/32 & 8/32 & 8/32 & 8/32 & 16/32 & 8/32 & 8/32 & 8/32 \\
-8/32 & 8/32 & 8/32 & 8/32 & 8/32 & 16/32 & 8/32 & 8/32 \\
-8/32 & 8/32 & 8/32 & 8/32 & 8/32 & 8/32 & 16/32 & 8/32 \\
-8/32 & 8/32 & 8/32 & 8/32 & 8/32 & 8/32 & 8/32 & 16/32
\end{bmatrix}
\begin{array}{c} (\mathbf{X}_1'\mathbf{y}) \\ h \times 1 \\ \begin{bmatrix} 172 \\ 21 \\ 27 \\ 23 \\ 20 \\ 18 \\ 21 \\ 20 \end{bmatrix} \end{array}
=
\begin{array}{c} \hat{\boldsymbol{\beta}}_1 \\ h \times 1 \\ \begin{bmatrix} 5.50 \\ -0.25 \\ 1.25 \\ 0.25 \\ -0.50 \\ -1.00 \\ -0.25 \\ -0.50 \end{bmatrix} \end{array}
$$

TABLE 6.9-1 (continued)

$$
\underset{1\times h}{\boldsymbol{\beta}'_1} \qquad \underset{h\times 1}{(\mathbf{X}'_1\mathbf{y})} \qquad \underset{1\times 1}{(\boldsymbol{\beta}'_1\mathbf{X}'_1\mathbf{y})}
$$

$$
[5.50 \ \cdots \ -0.50]\begin{bmatrix} 172 \\ 21 \\ \vdots \\ 20 \end{bmatrix} = 937.0
$$

Computation of sum of squares for reduced model:

$$SSE(X_4 \cdots X_{10}) = \mathbf{y}'\mathbf{y} - \hat{\boldsymbol{\beta}}'_1\mathbf{X}'_1\mathbf{y}$$

$$= 1160.0 - 937.0 = 223.0$$

$$SSR(X_1 X_2 X_3 | X_4 \cdots X_{10}) = SSE(X_4 \cdots X_{10}) - SSE(X_1 \cdots X_{10})$$

$$= 223.0 - 28.5 = 194.5$$

(iii)   Matrices for reduced model, $Y_i = \beta_0 X_0 + \beta_1 X_{i1} + \cdots + \beta_3 X_{i3} + \epsilon_i$ $(h = 4)$:

$$
\underset{h\times h}{(\mathbf{X}'_2\mathbf{X}_2)} = \begin{bmatrix} 32 & 8 & 8 & 8 \\ 8 & 8 & 0 & 0 \\ 8 & 0 & 8 & 0 \\ 8 & 0 & 0 & 8 \end{bmatrix}
\qquad
\underset{h\times h}{(\mathbf{X}'_2\mathbf{X}_2)^{-1}} \underset{h\times 1}{\mathbf{X}'_2\mathbf{y}}
\begin{bmatrix} \frac{1}{8} & -\frac{1}{8} & -\frac{1}{8} & -\frac{1}{8} \\ -\frac{1}{8} & \frac{2}{8} & \frac{1}{8} & \frac{1}{8} \\ -\frac{1}{8} & \frac{1}{8} & \frac{2}{8} & \frac{1}{8} \\ -\frac{1}{8} & \frac{1}{8} & \frac{1}{8} & \frac{2}{8} \end{bmatrix}
\begin{bmatrix} 172 \\ 22 \\ 28 \\ 50 \end{bmatrix} =
\underset{h\times 1}{\hat{\boldsymbol{\beta}}_2}
\begin{bmatrix} 9.00 \\ -6.25 \\ -5.50 \\ -2.75 \end{bmatrix}
$$

$$
\underset{1\times h}{\hat{\boldsymbol{\beta}}'_2} \qquad \underset{h\times 1}{(\mathbf{X}'_2\mathbf{y})} \qquad \underset{1\times 1}{(\hat{\boldsymbol{\beta}}'_2\mathbf{X}'_2\mathbf{y})}
$$

$$
[9.00 \ \cdots \ -2.75]\begin{bmatrix} 172 \\ 22 \\ 28 \\ 50 \end{bmatrix} = 1119.0
$$

Computation of sum of squares for reduced model:

$$SSE(X_1 X_2 X_3) = \mathbf{y}'\mathbf{y} - \hat{\boldsymbol{\beta}}'_2\mathbf{X}'_2\mathbf{y}$$

$$= 1160.0 - 1119.0 = 41.0$$

$$SSR(X_4 \cdots X_{10} | X_1 X_2 X_3) = SSE(X_1 X_2 X_3) - SSE(X_1 \cdots X_{10})$$

$$= 41.0 - 28.5 = 12.5$$

(iv)   Computation of mean squares:

$$MSR(X_1 X_2 X_3 | X_4 \cdots X_{10}) = SSR(X_1 X_2 X_3 | X_4 \cdots X_{10})/(p-1) = 194.5/(4-1) = 64.833$$

$$MSR(X_4 \cdots X_{10} | X_1 X_2 X_3) = SSR(X_4 \cdots X_{10} | X_1 X_2 X_3)/(n-1) = 12.5/(8-1) = 1.786$$

$$MSE = SSE/(N - n - p + 1) = 28.5/(32 - 8 - 4 + 1) = 1.357$$

TABLE 6.9-1 (continued)

(v) Computation of F ratios:

$F = MSR(X_1 X_2 X_3 | X_4 \cdots X_{10})/MSE = 64.833/1.357 = 47.78$

$F_{.05;3,21}(conventional) = 3.07$

$F_{.05;1,7}(conservative) = 5.59$

$F = MSR(X_4 \cdots X_{10} | X_1 X_2 X_3)/MSE = 1.786/1.357 = 1.32$

$F_{.05;7,21}(conventional) = 2.49$

TABLE 6.9-2 Analysis of Variance Table for Altimeter Data

Source	SS	df	MS	F
Regression for altimeters $(X_1 X_2 X_3 \mid X_4 \cdots X_{10})$	194.500	$p - 1 = 3$	64.833	47.78*
Regression for pilots and previous flying experience $(X_4 \cdots X_{10} \mid X_1 X_2 X_3)$	12.500	$n - 1 = 7$	1.786	1.32
Error	28.500	$N - n - p + 1 = 21$	1.357	
Total	235.500	$N - 1 = 31$		

*$p < .01$

the regression approach. The computational procedures are the same as those in Table 6.9-1; however, the number of rows in $\mathbf{y}$ and $\mathbf{X}_0$ is reduced from 32 to 31. The results of the analyses are given in Table 6.9-3. As this example illustrates, when one or more scores are missing

$$SSTO \neq SSR(X_1 X_2 X_3 \mid X_4 \cdots X_{10}) + SSR(X_4 \cdots X_{10} \mid X_1 X_2 X_3) + SSE.$$

A comparison of the values of the sums of squares in Table 6.9-3 with the corrected sums of squares in Section 7.6 where Yates's procedure was used reveals that, within rounding error, they are equal.

$$SSA(corrected) = 195.666 \qquad SSR(X_1 X_2 X_3 | X_4 \cdots X_{10}) = 195.667$$
$$SSBL(corrected) = 11.880 \qquad SSR(X_4 \cdots X_{10} | X_1 X_2 X_3) = 11.880$$
$$SSRES = 27.334 \qquad SSE(X_1 \cdots X_{10}) = 27.333$$

The nature of the null hypotheses that these sums of squares are used to test is discussed later in connection with the full rank experimental design model.

TABLE 6.9-3   Summary of Analysis When Observation Y_{73} Is Missing

$SSTO = \mathbf{y}'\mathbf{y} - (\Sigma Y_i)^2/N = 1135.000 - (167)^2/31 = 235.355$

$df = N - 1 = 30$

$SSE(X_1 \cdots X_{10}) = \mathbf{y}'\mathbf{y} - \hat{\boldsymbol{\beta}}_0'\mathbf{X}_0'\mathbf{y} = 1135.000 - 1107.667 = 27.333$

$df = N - n - p + 1 = 20$

$SSE(X_4 \cdots X_{10}) = \mathbf{y}'\mathbf{y} - \hat{\boldsymbol{\beta}}_1'\mathbf{X}_1'\mathbf{y} = 1135.000 - 912.000 = 223.000$

$SSR(X_1 X_2 X_3 \mid X_4 \cdots X_{10}) = 223.000 - 27.333 = 195.667$

$df = p - 1 = 3$

$SSE(X_1 X_2 X_3) = \mathbf{y}'\mathbf{y} - \hat{\boldsymbol{\beta}}_2'\mathbf{X}_2'\mathbf{y} = 1135.000 - 1095.787 = 39.213$

$SSR(X_4 \cdots X_{10} \mid X_1 X_2 X_3) = 39.213 - 27.333 = 11.880$

$df = n - 1 = 7$

$SSR(X_1 X_2 X_3 \mid X_4 \cdots X_{10}) + SSR(X_4 \cdots X_{10} \mid X_1 X_2 X_3) + SSE = 234.880$

TESTING DIFFERENCES AMONG MEANS

Procedures for estimating population means and computing $\hat{\sigma}_{\psi_i}$ and $\hat{\sigma}_{\bar{Y}}$ follow those described in Section 5.6. Unbiased estimators for $\boldsymbol{\mu}_{\cdot j}' = [\mu_{\cdot 1} \; \mu_{\cdot 2} \; \cdots \; \mu_{\cdot p}]$ and $\boldsymbol{\mu}_{i\cdot}' = [\mu_{1\cdot} \; \mu_{2\cdot} \; \cdots \; \mu_{n\cdot}]$ are, respectively, $\mathbf{W}_1\hat{\boldsymbol{\beta}}_0 = \bar{\mathbf{y}}_1$ and $\mathbf{W}_2\hat{\boldsymbol{\beta}}_0 = \bar{\mathbf{y}}_2$, where $\bar{\mathbf{y}}_1' = [\bar{Y}_{\cdot 1} \; \bar{Y}_{\cdot 2} \; \cdots \; \bar{Y}_{\cdot p}]$, $\bar{\mathbf{y}}_2' = [\bar{Y}_{1\cdot} \; \bar{Y}_{2\cdot} \; \cdots \; \bar{Y}_{n\cdot}]$,

$$
\mathbf{W}_1 \atop {p \times h} =
\begin{bmatrix}
1 & \overbrace{1 \;\; 0 \;\; 0 \;\; \cdot \;\; 0}^{p-1 \text{ terms}} & \overbrace{1/n \;\; 1/n \;\; \cdot \;\; 1/n}^{n-1 \text{ terms}} \\
1 & 0 \;\; 1 \;\; 0 \;\; \cdot \;\; 0 & 1/n \;\; 1/n \;\; \cdot \;\; 1/n \\
1 & 0 \;\; 0 \;\; 1 \;\; \cdot \;\; 0 & 1/n \;\; 1/n \;\; \cdot \;\; 1/n \\
\cdot & \cdot \;\; \cdot \;\; \cdot \;\; \cdot \;\; \cdot & \cdot \;\;\; \cdot \;\;\; \cdot \;\;\; \cdot \\
1 & 0 \;\; 0 \;\; 0 \;\; \cdot \;\; 0 & 1/n \;\; 1/n \;\; \cdot \;\; 1/n
\end{bmatrix}
$$

and

$$
\mathbf{W}_2 \atop {n \times h} =
\begin{bmatrix}
1 & \overbrace{1/p \;\; 1/p \;\; \cdot \;\; 1/p}^{p-1 \text{ terms}} & \overbrace{1 \;\; 0 \;\; 0 \;\; \cdot \;\; 0}^{n-1 \text{ terms}} \\
1 & 1/p \;\; 1/p \;\; \cdot \;\; 1/p & 0 \;\; 1 \;\; 0 \;\; \cdot \;\; 0 \\
1 & 1/p \;\; 1/p \;\; \cdot \;\; 1/p & 0 \;\; 0 \;\; 1 \;\; \cdot \;\; 0 \\
\cdot & \cdot \;\;\; \cdot \;\;\; \cdot \;\;\; \cdot & \cdot \;\; \cdot \;\; \cdot \;\; \cdot \;\; \cdot \\
1 & 1/p \;\; 1/p \;\; \cdot \;\; 1/p & 0 \;\; 0 \;\; 0 \;\; \cdot \;\; 0
\end{bmatrix}.
$$

The form of $\mathbf{W}$ is determined by the correspondence between the parameters of the regression model and the parameters of a type RB-p experimental design model. This correspondence for a type RB-4 design is given in (6.9-2). The $\mathbf{W}_1$ matrix for this design is

$$
\mathbf{W}_1 \atop 4\times 11
=
\begin{bmatrix}
1 & 1 & 0 & 0 & 1/8 & 1/8 & 1/8 & 1/8 & 1/8 & 1/8 & 1/8 \\
1 & 0 & 1 & 0 & 1/8 & 1/8 & 1/8 & 1/8 & 1/8 & 1/8 & 1/8 \\
1 & 0 & 0 & 1 & 1/8 & 1/8 & 1/8 & 1/8 & 1/8 & 1/8 & 1/8 \\
1 & 0 & 0 & 0 & 1/8 & 1/8 & 1/8 & 1/8 & 1/8 & 1/8 & 1/8
\end{bmatrix}.
$$

Unbiased estimators for $\sigma^2_{\psi_i}$ and $\sigma^2_{\bar{Y}}$ for treatment A are given by, respectively,

$$
\hat{\sigma}^2_{\psi_i} = MSE\{ \underset{1\times p}{\mathbf{c}'} \; [\underset{p\times h}{\mathbf{W}_1} (\underset{h\times h}{\mathbf{X}'_0\mathbf{X}_0})^{-1} \underset{h\times p}{\mathbf{W}'_1}] \; \underset{p\times 1}{\mathbf{c}} \}
$$

$$
= MSE\left[\frac{c_1^2}{n} + \frac{c_2^2}{n} + \cdots + \frac{c_p^2}{n} \right]
$$

$$
\hat{\sigma}^2_{\bar{Y}} = MSE/n
$$

where $\mathbf{c}'$ is a vector that defines the ith contrast. The corresponding formulas for blocks are

$$
\hat{\sigma}^2_{\psi_i} = MSE\{ \underset{1\times n}{\mathbf{c}'} \; [\underset{n\times h}{\mathbf{W}_2} (\underset{h\times h}{\mathbf{X}'_0\mathbf{X}_0})^{-1} \underset{h\times n}{\mathbf{W}'_2}] \; \underset{n\times 1}{\mathbf{c}} \}
$$

$$
= MSE\left[\frac{c_1^2}{p} + \frac{c_2^2}{p} + \cdots + \frac{c_n^2}{p} \right]
$$

$$
\hat{\sigma}^2_{\bar{Y}} = MSE/p.
$$

FULL RANK EXPERIMENTAL DESIGN MODEL

The less than full rank experimental design model for the randomized block design given earlier is

(6.9-3)
$$
Y_{ij} = \mu + \alpha_j + \pi_i + \epsilon_{ij} \qquad (i = 1, \ldots, n; j = 1, \ldots, p)
$$

where ϵ_{ij} is $NID(0, \sigma^2_\epsilon)$. This model does not contain a block $\times$ treatment A interaction term, $(\pi\alpha)_{ij}$, and if it did there is no way of obtaining separate estimates of $(\pi\alpha)_{ij}$ and ϵ_{ij}. Recall from Section 6.5 that the formula for $MSRES$, which is used to estimate σ^2_ϵ, is the formula for the block $\times$ treatment A interaction.

The full rank experimental design model corresponding to (6.9-3) is

(6.9-4)
$$
Y_{ij} = \mu_{ij} + \epsilon_{ij} \qquad (i = 1, \ldots, n; j = 1, \ldots, p)
$$

where ϵ_{ij} is $NID(0, \sigma^2_\epsilon)$ and μ_{ij} is subject to the restrictions that

$$
\mu_{ij} - \mu_{i'j} - \mu_{ij'} + \mu_{i'j'} = 0 \quad \text{for all } i, i', j, \text{ and } j'.
$$

These restrictions specify that all block $\times$ treatment level interaction effects equal zero. The restrictions, which are a part of the model, are necessary since it is not possible to estimate interaction effects separately from error effects. With the addition of these restrictions the model is called a *restricted full rank experimental design model*. As we will see, the analysis of models with restrictions is considerably more complex than that for unrestricted models.

The analysis procedures for a restricted full rank experimental design model will be illustrated using the altimeter example described in Section 6.2. We begin by describing in detail the restriction that all block $\times$ treatment A interaction effects equal zero. This restriction can be expressed in a number of ways.* One convenient way is to form interaction terms based on each set of crossed lines in Figure 6.9-1. The two μ_{ij}'s connected by a dashed line are subtracted from the μ_{ij}'s connected by a solid line. This produces $s = N - n - p + 1 = 21$ terms of the form $\mu_{ij} - \mu_{i'j} - \mu_{ij'} + \mu_{i'j'}$ that are then set equal to zero. For example,

(1) Rows 1 and 2, columns 1 and 2 $\mu_{11} - \mu_{21} - \mu_{12} + \mu_{22} = 0$

(2) Rows 2 and 3, columns 1 and 2 $\mu_{21} - \mu_{31} - \mu_{22} + \mu_{32} = 0$

(3) Rows 3 and 4, columns 1 and 2 $\mu_{31} - \mu_{41} - \mu_{32} + \mu_{42} = 0$

. . . .

(21) Rows 7 and 8, columns 3 and 4 $\mu_{73} - \mu_{83} - \mu_{74} + \mu_{84} = 0.$

FIGURE 6.9-1   An interaction term of the form $\mu_{ij} - \mu_{i'j} - \mu_{ij'} + \mu_{i'j'}$ is obtained from the crossed lines by subtracting the two μ_{ij}'s connected by a dashed line from the μ_{ij}'s connected by a solid line.

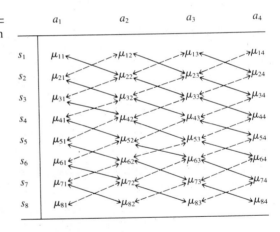

The complete set of interaction terms and restrictions can be written using matrix notation as $\mathbf{R'}\boldsymbol{\mu} = \boldsymbol{\theta}$; the matrix and vectors are specified in equation (6.9-5). This portentious equation simply states that all block $\times$ treatment A interaction effects equal zero—it is the restriction that is appended to model (6.9-4). Matrix $\mathbf{R'}$ in (6.9-5) is of full row rank s, $s \leq h$, where h is the number of μ_{ij}'s in the full rank experimental design model.

The null hypothesis for treatment A can be formulated in terms of treatment-level population means, H_0: $\mu_{.1} = \mu_{.2} = \mu_{.3} = \mu_{.4}$, or in terms of block–treatment-

---

*   See, for example, Timm and Carlson (1975, 18–19, 37).

$$\underset{s \times 1}{\boldsymbol{\theta}} = \underset{h \times 1}{\boldsymbol{\mu}} \quad [\mu_{11}\ \mu_{21}\ \mu_{31}\ \mu_{41}\ \mu_{51}\ \mu_{61}\ \mu_{71}\ \mu_{81}\ \mu_{12}\ \mu_{22}\ \cdots\ \mu_{74}\ \mu_{84}]$$

$$\underset{s \times h}{\mathbf{R}'}$$

Row: 1 2 3 4 5 6 7 8 9 10 ⋯ 20 21

(6.9-5)

level population means,

$$H_0: \quad \mu_{11} = \mu_{12} = \mu_{13} = \mu_{14}$$
$$\mu_{21} = \mu_{22} = \mu_{23} = \mu_{24}$$
$$\mu_{31} = \mu_{32} = \mu_{33} = \mu_{34}$$
$$\cdots\cdots\cdots\cdots\cdots$$
$$\mu_{81} = \mu_{82} = \mu_{83} = \mu_{84}.$$

The latter hypothesis specifies that, simultaneously, for each block, the treatment-level population means are equal. For the full rank experimental design model, the two hypotheses are expressed as

(6.9-6) $$H_0: \quad \mu_{\cdot 1} - \mu_{\cdot 4} = \mu_{\cdot 2} - \mu_{\cdot 4} = \mu_{\cdot 3} - \mu_{\cdot 4} = 0$$

and

(6.9-7) $$H_0: \quad \mu_{11} - \mu_{14} = \mu_{12} - \mu_{14} = \mu_{13} - \mu_{14} = 0$$
$$\mu_{21} - \mu_{24} = \mu_{22} - \mu_{24} = \mu_{23} - \mu_{24} = 0$$
$$\mu_{31} - \mu_{34} = \mu_{32} - \mu_{34} = \mu_{33} - \mu_{34} = 0$$
$$\cdots\cdots\cdots\cdots\cdots\cdots\cdots\cdots\cdots\cdots$$
$$\mu_{81} - \mu_{84} = \mu_{82} - \mu_{84} = \mu_{83} - \mu_{84} = 0.$$

In matrix notation these hypotheses can be written as $\mathbf{C}'_{i(A)}\boldsymbol{\mu} = \mathbf{0}$, where $\boldsymbol{\mu}$ is an $h \times 1$ vector of μ_{ij}'s and $\mathbf{0}$ is an $h \times 1$ vector of zeros.

The $\mathbf{C}'_{i(A)}$ matrices corresponding to (6.9-6) and (6.9-7) are, respectively,

and

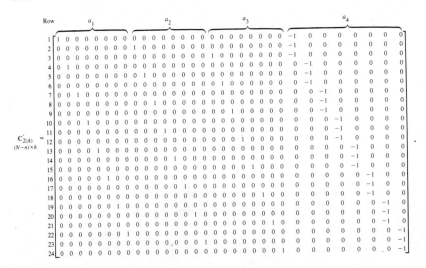

For convenience, matrix $\mathbf{C}'_{1(A)}$ can be multiplied by 8 since multiplying a coefficient matrix by a nonzero constant does not alter the nature of the hypothesis. When there are no missing observations, testing hypothesis (6.9-6) is equivalent to testing (6.9-7). However, as we will see, the hypotheses are not equivalent when one or more observations are missing.

We want to test $\mathbf{C}'\boldsymbol{\mu} = \mathbf{0}$ subject to the restrictions $\mathbf{R}'\boldsymbol{\mu} = \boldsymbol{\theta}$. To do this we form an augmented matrix $\mathbf{Q}'$ using the $\mathbf{R}'$ matrix and those rows of $\mathbf{C}'$ that are not identical to the rows of $\mathbf{R}'$, inconsistent with them, or linearly dependent on them. The resulting $\mathbf{Q}'$ matrix has $s + p - 1 > s$ linearly independent rows—the s rows of $\mathbf{R}'$ and $p - 1$ of the rows of $\mathbf{C}'$. That is,

$$\mathbf{Q}'_{(s+p-1)\times h} = \begin{bmatrix} \mathbf{R}' \\ {}_{s\times h} \\ \mathbf{C}' \\ {}_{(p-1)\times h} \end{bmatrix}.$$

Similarly, we can define the augmented matrix $\boldsymbol{\eta}$ as

$$\boldsymbol{\eta}_{(s+p-1)\times 1} = \begin{bmatrix} \boldsymbol{\theta} \\ {}_{s\times 1} \\ \mathbf{0} \\ {}_{(p-1)\times 1} \end{bmatrix}.$$

Ordinarily $\boldsymbol{\eta}$ is an $(s + p - 1) \times 1$ vector of zeros. The matrix equation $\mathbf{Q}'\boldsymbol{\mu} = \boldsymbol{\eta}$ combines the hypothesis that differences between means are equal to zero with the restriction that all interaction terms are equal to zero. It is a simple matter to combine the $s = 21$ rows of $\mathbf{R}'$ and the $p - 1 = 3$ rows of $\mathbf{C}'_{1(A)}$ to test hypothesis (6.9-6). However, combining the $s = 21$ rows of $\mathbf{R}'$ with the $N - n = 24$ rows of $\mathbf{C}'_{2(A)}$ to test (6.9-7) poses a problem since $N - n - p + 1 = 21$ rows of the resulting $\mathbf{Q}'$ matrix are linearly dependent. Identifying the 21 dependent rows in $\mathbf{Q}'$ could be tedious. Fortunately it turns out that, for matrices constructed like $\mathbf{R}'$ and $\mathbf{C}'_{2(A)}$, the last s rows of $\mathbf{Q}'$ are always dependent on the first $s + p - 1$ rows and should be deleted. To put it another way, $\mathbf{Q}'$ should consist of the s rows of $\mathbf{R}'$ and the first $p - 1$ rows of $\mathbf{C}'_{2(A)}$. It is a simple matter to show that the last 21 rows of $\mathbf{C}'_{2(A)}$ are linearly dependent on the s rows in $\mathbf{R}'$ and the first $p - 1$ rows of $\mathbf{C}'_{2(A)}$. We will let $1\mathbf{c}'$ denote the first row of $\mathbf{C}'_{2(A)}$, $1\mathbf{r}'$, the first row of $\mathbf{R}'$, and so on. Then

$$4\mathbf{c}' = 1\mathbf{c}' - 1\mathbf{r}' - 8\mathbf{r}' - 15\mathbf{r}'$$
$$5\mathbf{c}' = 2\mathbf{c}' - 8\mathbf{r}' - 15\mathbf{r}'$$
$$6\mathbf{c}' = 3\mathbf{c}' - 15\mathbf{r}'$$
$$7\mathbf{c}' = 1\mathbf{c}' - 1\mathbf{r}' - 2\mathbf{r}' - 8\mathbf{r}' - 9\mathbf{r}' - 15\mathbf{r}' - 16\mathbf{r}'$$
$$8\mathbf{c}' = 2\mathbf{c}' - 8\mathbf{r}' - 9\mathbf{r}' - 15\mathbf{r}' - 16\mathbf{r}'$$
$$9\mathbf{c}' = 3\mathbf{c}' - 15\mathbf{r}' - 16\mathbf{r}'$$
$$10\mathbf{c}' = 1\mathbf{c}' - 1\mathbf{r}' - 2\mathbf{r}' - 3\mathbf{r}' - 8\mathbf{r}' - 9\mathbf{r}' - 10\mathbf{r}' - 15\mathbf{r}' - 16\mathbf{r}' - 17\mathbf{r}'$$
$$\cdots\cdots\cdots\cdots\cdots\cdots\cdots\cdots\cdots\cdots\cdots\cdots\cdots\cdots\cdots\cdots$$
$$24\mathbf{c}' = 3\mathbf{c}' - 15\mathbf{r}' - 16\mathbf{r}' - 17\mathbf{r}' - 18\mathbf{r}' - 19\mathbf{r}' - 20\mathbf{r}' - 21\mathbf{r}'.$$

To summarize, the equation $\mathbf{Q}'\boldsymbol{\mu} = \boldsymbol{\eta}$ combines the null hypothesis, $\mathbf{C}'\boldsymbol{\mu} = \mathbf{0}$, and the restrictions, $\mathbf{R}'\boldsymbol{\mu} = \boldsymbol{\theta}$. If we want to test hypothesis (6.9-6), $\mathbf{Q}'_{1(A)}$ consists of $\mathbf{R}'$ and $\mathbf{C}'_{1(A)}$. If we want to test (6.9-7), $\mathbf{Q}'_{2(A)}$ consists of $\mathbf{R}'$ plus the first $p - 1 = 3$ rows of $\mathbf{C}'_{2(A)}$. The two hypotheses are equivalent when there are no missing observations.

The formulation of hypotheses for testing block effects follows that described for treatment A. Again, two null hypotheses can be tested:

(6.9-8) $$H_0: \quad \mu_{1.} - \mu_{8.} = \mu_{2.} - \mu_{8.} = \cdots = \mu_{7.} - \mu_{8.} = 0$$

and

(6.9-9)
$$H_0: \quad \mu_{11} - \mu_{81} = \mu_{21} - \mu_{81} = \cdots = \mu_{71} - \mu_{81} = 0$$
$$\mu_{12} - \mu_{82} = \mu_{22} - \mu_{82} = \cdots = \mu_{72} - \mu_{82} = 0$$
$$\mu_{13} - \mu_{83} = \mu_{23} - \mu_{83} = \cdots = \mu_{73} - \mu_{83} = 0$$
$$\mu_{14} - \mu_{84} = \mu_{24} - \mu_{84} = \cdots = \mu_{74} - \mu_{84} = 0.$$

These hypotheses can be written in matrix notation as $\mathbf{C}'\boldsymbol{\mu} = \mathbf{0}$, where the $\mathbf{C}'_{i(B)}$ matrices corresponding to (6.9-8) and (6.9-9) are, respectively,

```
                 a₁                        a₂                        a₃                        a₄
        ┌                                                                                                        ┐
        │ ¼  0  0  0  0  0  -¼  ¼  0  0  0  0  0  -¼  ¼  0  0  0  0  0  -¼  ¼  0  0  0  0  0  -¼ │
        │ 0  ¼  0  0  0  0  -¼  0  ¼  0  0  0  0  -¼  0  ¼  0  0  0  0  -¼  0  ¼  0  0  0  0  -¼ │
C'₁₍ᵦ₎= │ 0  0  ¼  0  0  0  -¼  0  0  ¼  0  0  0  -¼  0  0  ¼  0  0  0  -¼  0  0  ¼  0  0  0  -¼ │
(n-1)×h │ .....................................................................................................│
        │ 0  0  0  0  0  0  ¼  -¼  0  0  0  0  0  0  ¼  -¼  0  0  0  0  0  0  ¼  -¼  0  0  0  0  0  0  ¼  -¼ │
        └                                                                                                        ┘
```

and

```
                  a₁                       a₂                       a₃                       a₄
        ┌                                                                                                      ┐
        │ 1  0  0  0  0  0  0  -1  0  0  0  0  0  0  0  0  0  0  0  0  0  0  0  0  0  0  0  0  0  0  0  0 │
        │ 0  1  0  0  0  0  0  -1  0  0  0  0  0  0  0  0  0  0  0  0  0  0  0  0  0  0  0  0  0  0  0  0 │
C'₂₍ᵦ₎= │ 0  0  1  0  0  0  0  -1  0  0  0  0  0  0  0  0  0  0  0  0  0  0  0  0  0  0  0  0  0  0  0  0 │ .
(N-p)×h │ 0  0  0  1  0  0  0  -1  0  0  0  0  0  0  0  0  0  0  0  0  0  0  0  0  0  0  0  0  0  0  0  0 │
        │ ....................................................................................................│
        │ 0  0  0  0  0  0  0   0  0  0  0  0  0  0  0  0  0  0  0  0  0  0  0  0  0  0  0  0  0  0  1  -1 │
        └                                                                                                      ┘
```

To test $\mathbf{C}'_{1(B)}\boldsymbol{\mu} = \mathbf{0}$ or $\mathbf{C}'_{2(B)}\boldsymbol{\mu} = \mathbf{0}$ subject to the restrictions $\mathbf{R}'\boldsymbol{\mu} = \mathbf{0}$, we form augmented matrices $\mathbf{Q}'_{1(B)}$ or $\mathbf{Q}'_{2(B)}$ using the $\mathbf{R}'$ matrix and either $\mathbf{C}'_{1(B)}$ or the first $n - 1$ rows of $\mathbf{C}'_{2(B)}$. The resulting $\mathbf{Q}'_{i(B)}$ matrix is of order $(s + n - 1) \times h$.

For purposes of illustration, we will test hypotheses (6.9-7) and (6.9-9) for the altimeter experiment. The procedures for computing F statistics are illustrated in Table 6.9-4. The strategy for testing F statistics outlined in Section 6.4 is followed in Table 6.9-4. Table 6.9-4(i) contains two estimators for population block–treatment-level cell means: $\hat{\mu}$ and $\hat{\mu}_R$. The estimator $\hat{\mu}$, which is equal to $\mathbf{y}$ since each cell contains one observation, is not subject to the restriction that all block $\times$ treatment A interaction effects equal zero. The second estimator $\hat{\mu}_R$ yields estimates that are subject to the zero interaction restriction imposed on model (6.9-4).

TESTING DIFFERENCES AMONG MEANS

To test hypotheses about population means, we need to compute estimates of the contrasts of interest $\hat{\psi}_i$ and either the standard error of the contrast $\hat{\sigma}_{\psi_i}$ or the standard error of a mean $\hat{\sigma}_{\bar{Y}}$. The formulas for computing these estimates are as follows:

$$\hat{\psi}_i = \underset{1 \times h}{\mathbf{c}'} \, \underset{h \times 1}{\hat{\boldsymbol{\mu}}_R}, \text{ where } \mathbf{c}' \text{ is a vector that defines the } i\text{th contrast}$$

$$\hat{\sigma}_{\psi_i} = [MSRES(\mathbf{c}'\mathbf{A}\mathbf{c})]^{1/2}$$

$$\mathbf{A} = \mathbf{I} - \mathbf{R}(\mathbf{R}'\mathbf{R})^{-1}\mathbf{R}'$$

$$\hat{\sigma}_{\bar{Y}} = (MSRES/n)^{1/2} \text{ for treatment } A \text{ and } (MSRES/p)^{1/2} \text{ for blocks.}$$

FULL RANK EXPERIMENTAL DESIGN MODEL WITH MISSING OBSERVATIONS

The computational procedures in Table 6.9-4 can also be used when one or more observations are missing. Assume, for example, that Y_{73} is missing. The loss of this observation reduces N from 32 to 31 and the number of required restrictions in $\mathbf{R}'$ from $N - n - p + 1 = 21$ to 20. Since Y_{73} does not appear in the model, $\mathbf{R}'$ cannot contain any interaction term involving this observation. This eliminates the following interaction terms; the numbers in parentheses refer to the rows of the $\mathbf{R}'$ matrix in (6.9-5). The reader may find it helpful to refer to Figure 6.9-1.

$$(13) \quad \mu_{62} - \mu_{72} - \mu_{63} + \mu_{73}$$
$$(14) \quad \mu_{72} - \mu_{82} - \mu_{73} + \mu_{83}$$
$$(20) \quad \mu_{63} - \mu_{73} - \mu_{64} + \mu_{74}$$
$$(21) \quad \mu_{73} - \mu_{83} - \mu_{74} + \mu_{84}$$

After deleting these rows from $\mathbf{R}'$, 17 rows remain, three short of the required 20. The four deleted rows, (13), (14), (20), and (21), can be combined to obtain three new linearly independent restrictions that do not involve Y_{73}. These three restrictions, which are added to $\mathbf{R}'$ to bring it up to row rank $s = 20$, are

$$(13) + (14) \quad (\mu_{62} - \mu_{72} - \mu_{63} + \mu_{73}) + (\mu_{72} - \mu_{82} - \mu_{73} + \mu_{83})$$
$$= \mu_{62} - \mu_{63} - \mu_{82} + \mu_{83}$$
$$(13) + (20) \quad (\mu_{62} - \mu_{72} - \mu_{63} + \mu_{73}) + (\mu_{63} - \mu_{73} - \mu_{64} + \mu_{74})$$
$$= \mu_{62} - \mu_{72} - \mu_{64} + \mu_{74}$$
$$(14) + (21) \quad (\mu_{72} - \mu_{82} - \mu_{73} + \mu_{83}) + (\mu_{73} - \mu_{83} - \mu_{74} + \mu_{84})$$
$$= \mu_{72} - \mu_{82} - \mu_{74} + \mu_{84}.$$

The $\mathbf{Q}'$ matrix consists of the 20 rows of $\mathbf{R}'$ plus $\mathbf{C}'_{1(A)}$ or the first three rows of $\mathbf{C}'_{2(A)}$. The $\mathbf{C}'_{1(A)}$ and $\mathbf{C}'_{2(A)}$ matrices, like $\mathbf{R}'$, must be altered to reflect the missing observation, and this alters the hypotheses that are tested, as we will see next.

TABLE 6.9-4   Computational Procedures for a Type RB-4 Design Using a Restricted Full Rank Experimental Design Model

(i)   Data and matrix formulas for computing sums of squares ($\mathbf{R}'$, $\mathbf{Q}'_{2(A)}$ and $\mathbf{Q}'_{2(B)}$ are defined in the text, $N = 32$, $h = 32$, $n = 8$, $p = 4$, $s = 21$):

$$\underset{N\times 1}{\mathbf{y}} \qquad \underset{h\times 1}{\hat{\boldsymbol{\mu}}} \quad \underset{h\times 1}{[\mathbf{R}(\mathbf{R}'\mathbf{R})^{-1}\mathbf{R}'\hat{\boldsymbol{\mu}}]} \quad \underset{h\times 1}{\hat{\boldsymbol{\mu}}_R}$$

$$a_1 \begin{cases} \begin{bmatrix} 3 \\ 6 \\ 3 \\ 3 \\ 1 \\ 2 \\ 2 \\ 2 \end{bmatrix} & \begin{bmatrix} 0.375 \\ 1.875 \\ -0.125 \\ 0.625 \\ -0.875 \\ -0.625 \\ -0.375 \\ -0.875 \end{bmatrix} & \begin{bmatrix} 2.625 \\ 4.125 \\ 3.125 \\ 2.375 \\ 1.875 \\ 2.625 \\ 2.375 \\ 2.875 \end{bmatrix} \end{cases}$$

$$a_2 \begin{cases} \begin{bmatrix} 4 \\ 5 \\ 4 \\ 3 \\ 2 \\ 3 \\ 4 \\ 3 \end{bmatrix} & \begin{bmatrix} 0.625 \\ 0.125 \\ 0.125 \\ -0.125 \\ -0.625 \\ -0.375 \\ 0.875 \\ -0.625 \end{bmatrix} & \begin{bmatrix} 3.375 \\ 4.875 \\ 3.875 \\ 3.125 \\ 2.625 \\ 3.375 \\ 3.125 \\ 3.625 \end{bmatrix} \end{cases}$$

$$a_3 \begin{cases} \begin{bmatrix} 7 \\ 8 \\ 7 \\ 6 \\ 5 \\ 6 \\ 5 \\ 6 \end{bmatrix} & \begin{bmatrix} 0.875 \\ 0.375 \\ 0.375 \\ 0.125 \\ -0.375 \\ -0.125 \\ -0.875 \\ -0.375 \end{bmatrix} & \begin{bmatrix} 6.125 \\ 7.625 \\ 6.625 \\ 5.875 \\ 5.375 \\ 6.125 \\ 5.875 \\ 6.375 \end{bmatrix} \end{cases}$$

$$a_4 \begin{cases} \begin{bmatrix} 7 \\ 8 \\ 9 \\ 8 \\ 10 \\ 10 \\ 9 \\ 11 \end{bmatrix} & \begin{bmatrix} -1.875 \\ -2.375 \\ -0.375 \\ -0.625 \\ 1.875 \\ 1.125 \\ 0.375 \\ 1.875 \end{bmatrix} & \begin{bmatrix} 8.875 \\ 10.375 \\ 9.375 \\ 8.625 \\ 8.125 \\ 8.875 \\ 8.625 \\ 9.125 \end{bmatrix} \end{cases}$$

$$\sum_{i=1}^{N} Y_i = 172$$

$$\underset{1\times N}{\mathbf{y}'} \qquad \underset{N\times 1}{\mathbf{y}} \qquad \underset{1\times 1}{(\mathbf{y}'\mathbf{y})}$$

$$[3 \quad 6 \quad \cdots \quad 11] \begin{bmatrix} 3 \\ 6 \\ \vdots \\ 11 \end{bmatrix} = 1160.0$$

$$SSTO = \mathbf{y}'\mathbf{y} - (\Sigma\, Y_i)^2/N$$
$$= 1160.0 - (172)^2/32 = 235.5$$

$$SSRES = \underset{1\times s}{(\mathbf{R}'\hat{\boldsymbol{\mu}})'} \underset{s\times s}{(\mathbf{R}'\mathbf{R})^{-1}} \underset{s\times 1}{(\mathbf{R}'\hat{\boldsymbol{\mu}})} = 28.5$$

$$SSA = \underset{1\times(s+p-1)}{(\mathbf{Q}'_{2(A)}\hat{\boldsymbol{\mu}})'} \underset{(s+p-1)\times(s+p-1)}{(\mathbf{Q}'_{2(A)}\mathbf{Q}_{2(A)})^{-1}} \underset{(s+p-1)\times 1}{(\mathbf{Q}'_{2(A)}\hat{\boldsymbol{\mu}})} - SSRES$$
$$= 223.0 - 28.5 = 194.5$$

$$SSBL = \underset{1\times(s+n-1)}{(\mathbf{Q}'_{2(B)}\hat{\boldsymbol{\mu}})'} \underset{(s+n-1)\times(s+n-1)}{(\mathbf{Q}'_{2(B)}\mathbf{Q}_{2(B)})^{-1}} \underset{(s+n-1)}{(\mathbf{Q}'_{2(B)}\hat{\boldsymbol{\mu}})} - SSRES$$
$$= 41.0 - 28.5 = 12.5$$

Alternative formula for SSA:

$$SSA = \underset{1\times(N-n)}{(\mathbf{C}'_{2(A)}\hat{\boldsymbol{\mu}}_R)'} \underset{(N-n)\times(N-n)}{(\mathbf{C}'_{2(A)}\mathbf{A}\mathbf{C}_{2(A)})^{-1}} \underset{(N-n)\times 1}{(\mathbf{C}'_{2(A)}\hat{\boldsymbol{\mu}}_R)}$$

where $\mathbf{A} = \underset{h\times h}{\mathbf{I}} - \underset{h\times h}{[\mathbf{R}(\mathbf{R}'\mathbf{R})^{-1}\mathbf{R}']}$

Alternative formula for $SSBL$:

$$SSBL = \underset{1\times(N-p)}{(\mathbf{C}'_{2(B)}\hat{\boldsymbol{\mu}}_R)'} \underset{(N-p)\times(N-p)}{(\mathbf{C}'_{2(B)}\mathbf{A}\mathbf{C}_{2(B)})^{-1}} \underset{(N-p)\times 1}{(\mathbf{C}'_{2(B)}\hat{\boldsymbol{\mu}}_R)}$$

(ii)   Computation of mean squares:

$$MSA = SSA/(p-1) = 194.5/(4-1) = 64.833$$

$$MSBL = SSBL/(n-1) = 12.5/(8-1) = 1.786$$

$$MSRES = SSRES/(N-n-p+1) = 28.5/(32-8-4+1) = 1.357$$

TABLE 6.9-4 (continued)

(iii) Computation of F ratios:

$F = MSA/MSRES = 64.833/1.357 = 47.78$

$F_{.05;3,21}(conventional) = 3.07$

$F_{.05;1,7}(conservative) = 5.59$

$F = MSBL/MSRES = 1.786/1.357 = 1.32$

$F_{.05;7,21}(conventional) = 2.49$

We noted earlier that when there are no missing observations, $C'_{1(A)}$ and $C'_{2(A)}$ test equivalent hypotheses. This is not true when there are missing observations. Consider the type RB-3 design shown in Figure 6.9-2. Letting $n_{ij} = 0$ or 1 depending on whether an observation is or is not in the ijth cell, the full rank experimental design model is

$$n_{ij}Y_{ij} = n_{ij}\mu_{ij} + n_{ij}\epsilon_{ij} \qquad (i = 1, \ldots, n_{\cdot j}; j = 1, \ldots, p)$$

where ϵ_{ij} is $NID(0, \sigma_\epsilon^2)$ and $n_{ij}\mu_{ij}$ is subject to the restrictions that $\mu_{ij} - \mu_{i'j} - \mu_{ij'} + \mu_{i'j'}$ (or sums and differences of these functions to eliminate nonexistent μ_{ij}'s) $= 0$.

FIGURE 6.9-2 Layout for type RB-3 design showing correspondence between $\mu_{\cdot j}$ and $\mu + \alpha_j + \Sigma_{i=1}^{n_{\cdot j}} n_{ij}\pi_i/n_{\cdot j}$.

	a_1	a_2	a_3
s_1	$n_{11} = 1$ $\mu_{11} = \mu + \alpha_1 + \pi_1$	$n_{12} = 1$ $\mu_{12} = \mu + \alpha_2 + \pi_1$	$n_{13} = 1$ $\mu_{13} = \mu + \alpha_3 + \pi_1$
s_2	$n_{21} = 1$ $\mu_{21} = \mu + \alpha_1 + \pi_2$	$n_{22} = 1$ $\mu_{22} = \mu + \alpha_2 + \pi_2$	$n_{23} = 1$ $\mu_{23} = \mu + \alpha_3 + \pi_2$
s_3	$n_{31} = 1$ $\mu_{31} = \mu + \alpha_1 + \pi_3$	$n_{32} = 0$	$n_{33} = 1$ $\mu_{33} = \mu + \alpha_3 + \pi_3$
s_4	$n_{41} = 1$ $\mu_{41} = \mu + \alpha_1 + \pi_4$	$n_{42} = 1$ $\mu_{42} = \mu + \alpha_2 + \pi_4$	$n_{43} = 1$ $\mu_{43} = \mu + \alpha_3 + \pi_4$

Mean of μ_{ij}'s $\quad = \quad \displaystyle\sum_{i=1}^{n_{\cdot 1}} n_{i1}\mu_{i1}/n_{\cdot 1} = \mu_{\cdot 1} \qquad \displaystyle\sum_{i=1}^{n_{\cdot 2}} n_{i2}\mu_{i2}/n_{\cdot 2} = \mu_{\cdot 2} \qquad \displaystyle\sum_{i=1}^{n_{\cdot 3}} n_{i3}\mu_{i3}/n_{\cdot 3} = \mu_{\cdot 3}$

Mean of $(\mu + \alpha_j + \pi_i)$'s $= \mu + \alpha_1 + \dfrac{\pi_1 + \pi_2 + \pi_3 + \pi_4}{n_{\cdot 1}} \qquad \mu + \alpha_2 + \dfrac{\pi_1 + \pi_2 + \pi_4}{n_{\cdot 2}} \qquad \mu + \alpha_3 + \dfrac{\pi_1 + \pi_2 + \pi_3 + \pi_4}{n_{\cdot 3}}$

The corresponding classical fixed-effects less than full rank experimental design model is

$$n_{ij}Y_{ij} = n_{ij}(\mu + \alpha_j + \pi_i) + n_{ij}\epsilon_{ij} \qquad (i = 1, \ldots, n_{.j}; j = 1, \ldots, p)$$

where ϵ_{ij} is $NID(0, \sigma_\epsilon^2)$, $\sum_{j=1}^p \alpha_j = 0$, and $\sum_{i=1}^n \pi_i = 0$. For this type RB-3 design, we would probably be interested in testing the following null hypothesis concerning block–treatment-level population means.

(6.9-10a)
$$H_0: \quad \mu_{11} = \mu_{12} = \mu_{13}$$
$$\mu_{21} = \mu_{22} = \mu_{23}$$
$$\mu_{31} = \qquad \mu_{33}$$
$$\mu_{41} = \mu_{42} = \mu_{43}$$

An alternative way of expressing this hypothesis that is used with the full rank experimental design model is

(6.9-10b)
$$H_0: \quad \mu_{11} - \mu_{13} = \mu_{12} - \mu_{13} = 0$$
$$\mu_{21} - \mu_{23} = \mu_{22} - \mu_{23} = 0$$
$$\mu_{31} - \mu_{33} = 0$$
$$\mu_{41} - \mu_{43} = \mu_{42} - \mu_{43} = 0$$

or

(6.9-10c)
$$H_0: \quad n_{ij}\mu_{ij} - n_{ij'}\mu_{ij'} = 0 \quad \text{for all } i, j, \text{ and } j' \ (j \neq j').$$

In matrix notation, hypothesis (6.9-10b) is

$$
H_0: \;
\underset{\substack{\mathbf{C}'_{2(A)} \\ (N-n) \times h}}{
\begin{bmatrix}
1 & 0 & 0 & 0 & 0 & 0 & 0 & -1 & 0 & 0 & 0 \\
0 & 0 & 0 & 0 & 1 & 0 & 0 & -1 & 0 & 0 & 0 \\
0 & 1 & 0 & 0 & 0 & 0 & 0 & 0 & -1 & 0 & 0 \\
0 & 0 & 0 & 0 & 0 & 1 & 0 & 0 & -1 & 0 & 0 \\
0 & 0 & 1 & 0 & 0 & 0 & 0 & 0 & 0 & -1 & 0 \\
0 & 0 & 0 & 1 & 0 & 0 & 0 & 0 & 0 & 0 & -1 \\
0 & 0 & 0 & 0 & 0 & 0 & 1 & 0 & 0 & 0 & -1
\end{bmatrix}}
\underset{\substack{\boldsymbol{\mu} \\ h \times 1}}{
\begin{bmatrix}
\mu_{11} \\ \mu_{21} \\ \mu_{31} \\ \mu_{41} \\ \mu_{12} \\ \mu_{22} \\ \mu_{42} \\ \mu_{13} \\ \mu_{23} \\ \mu_{33} \\ \mu_{43}
\end{bmatrix}}
=
\underset{\substack{\mathbf{0} \\ (N-n) \times 1}}{
\begin{bmatrix}
0 \\ 0 \\ 0 \\ 0 \\ 0 \\ 0 \\ 0
\end{bmatrix}}.
$$

For the classical less than full rank experimental design model, it can be seen from Figure 6.9-2 that testing (6.9-10b) is equivalent to testing

(6.9-11a)

$$H_0: \quad \text{Block 1} \quad \alpha_1 - \alpha_3 = \alpha_2 - \alpha_3 = 0$$
$$\text{Block 2} \quad \alpha_1 - \alpha_3 = \alpha_2 - \alpha_3 = 0$$
$$\text{Block 3} \quad \alpha_1 - \alpha_3 = 0$$
$$\text{Block 4} \quad \alpha_1 - \alpha_3 = \alpha_2 - \alpha_3 = 0$$

or

(6.9-11b)

$$H_0: \quad \alpha_j = 0 \quad \text{for all } j.$$

A second hypothesis which is sometimes tested involves treatment-level population means.

(6.9-12a)

$$H_0: \quad \mu_{\cdot 1} = \mu_{\cdot 2} = \mu_{\cdot 3}$$

For the full rank experimental design model this hypothesis is expressed as

(6.9-12b)

$$H_0: \quad \mu_{\cdot 1} - \mu_{\cdot 3} = \mu_{\cdot 2} - \mu_{\cdot 3} = 0$$

or

(6.9-12c)

$$H_0: \quad \frac{1}{n_{\cdot j}} \sum_{i=1}^{n_{\cdot j}} n_{ij}\mu_{ij} - \frac{1}{n_{\cdot j'}} \sum_{i=1}^{n_{\cdot j'}} n_{ij'}\mu_{ij'} = 0 \quad \text{for all } j \text{ and } j'(j \neq j').$$

In matrix notation, hypothesis (6.9-12b) is

$$
H_0: \quad
\underset{\substack{\mathbf{C}'_{1(A)} \\ (p-1)\times 1}}{
\begin{bmatrix}
\tfrac{1}{4} & \tfrac{1}{4} & \tfrac{1}{4} & \tfrac{1}{4} & 0 & 0 & 0 & -\tfrac{1}{4} & -\tfrac{1}{4} & -\tfrac{1}{4} & -\tfrac{1}{4} \\
0 & 0 & 0 & 0 & \tfrac{1}{3} & \tfrac{1}{3} & \tfrac{1}{3} & -\tfrac{1}{4} & -\tfrac{1}{4} & -\tfrac{1}{4} & -\tfrac{1}{4}
\end{bmatrix}}
\underset{\substack{\boldsymbol{\mu} \\ h\times 1}}{
\begin{bmatrix}
\mu_{11} \\ \mu_{21} \\ \mu_{31} \\ \mu_{41} \\ \mu_{12} \\ \mu_{22} \\ \mu_{42} \\ \mu_{13} \\ \mu_{23} \\ \mu_{33} \\ \mu_{43}
\end{bmatrix}}
=
\underset{\substack{\mathbf{0} \\ (p-1)\times 1}}{
\begin{bmatrix} 0 \\ 0 \end{bmatrix}}.
$$

For the classical less than full rank experimental design model, it can be seen from Figure 6.9-2 that testing (6.9-12b) is equivalent to testing

(6.9-13a) $\quad H_0: \quad \alpha_1 - \alpha_3 = \alpha_2 + \dfrac{\pi_1 + \pi_2 + \pi_4}{n_{\cdot 2}} - \left(\alpha_3 + \dfrac{\pi_1 + \pi_2 + \pi_3 + \pi_4}{n_{\cdot 3}} \right) = 0$

or

(6.9-13b)
$$H_0: \quad \text{all } \alpha_j + \frac{1}{n_{\cdot j}} \sum_{i=1}^{n_{\cdot j}} n_{ij} \, \pi_i \text{ are equal.}$$

The point of all of this is that a test of (6.9-10c) corresponds to a test of the equality of the α_j's; however, a test of (6.9-12c) corresponds to a test of the equality of the $\alpha_j + (1/n_{\cdot j}) \sum_{i=1}^{n_{\cdot j}} n_{ij} \pi_i$'s, which is a different hypothesis. For the special case in which there are no missing observations,

$$\sum_{i=1}^{n_{\cdot j}} n_{ij} \pi_i = \sum_{i=1}^{n_{\cdot j'}} n_{ij'} \pi_i \quad \text{for all } j \text{ and } j'$$

and hypotheses (6.9-10c) and (6.9-12c) test equivalent hypotheses.

An important advantage of the full rank experimental model over the less than full rank experimental design model and regression model is that an experimenter always knows exactly what hypothesis is being tested. In fact, the null hypothesis and the restrictions on the parameters of the model are specified when the $\mathbf{Q'}$ matrix is formed. This advantage of the full rank model is worth pursuing a bit further.

In Section 6.7 we used Yates's procedure to estimate a missing observation, Y_{73}, and then computed a corrected treatment sum of squares, $SSA\,(corrected)$ = 195.666. We reanalyzed these data in Section 6.9 (Table 6.9-3) using a regression model and obtained the same value within rounding error for SSR $(X_1\,X_2\,X_3|X_4 \cdots X_{10})$. Earlier, we did not specify the null hypothesis that these sums of squares were used to test, although a test of treatment A in some sense was implied. By reanalyzing these data using the full rank experimental design model and comparing the results with those obtained earlier we can identify the null hypothesis that was tested. For hypothesis (6.9-10c), SSA = 195.667; for hypothesis (6.9-12c), SSA = 196.164. Thus, the hypothesis tested earlier was

$$H_0: \quad n_{ij}\mu_{ij} - n_{ij'}\mu_{ij'} = 0 \quad \text{for all } i, j, \text{ and } j'(j \neq j')$$

or, in terms of treatment effects,

$$H_0: \quad \alpha_j = 0 \quad \text{for all } j$$

and not

$$H_0: \quad \mu_{\cdot 1} = \mu_{\cdot 2} = \mu_{\cdot 3} = \mu_{\cdot 4}.^*$$

---

*   Some writers in discussing less than full rank experimental design models with missing observations refer to the sum of squares for testing hypothesis (6.9-10c) as SSA_{adj} or $R(\alpha|\mu, \pi)$ and that for testing (6.9-12c) as SSA_{unadj} or $R(\alpha|\mu)$. Unfortunately, these designations, like $SS(X_1 \cdots X_{p-1}|X_p \cdots X_{h-1})$, are of little help in identifying the actual hypothesis that the sum of squares is used to test. When such designations are used, the null hypothesis should be fully specified, preferably in terms of population means.

6.10 GENERALIZED RANDOMIZED BLOCK DESIGN

A generalized randomized block design, denoted by GRB-p, has the advantages of a randomized block design and, in addition, enables an experimenter to test a group $\times$ treatment A interaction. Wilk (1955) provided the first detailed description of the design. It has subsequently been discussed by Addelman (1969, 1970) and by Wilk and Kempthorne (1956). The design is appropriate for experiments that meet the following three conditions.

1. One treatment with $p \geq 2$ treatment levels and one nuisance variable with $w \geq 2$ levels.

2. Formation of w groups each containing np homogeneous experimental units, where $n \geq 2$.

3. Random assignment of treatment levels to the np units in each group so that each treatment level is assigned to exactly n units. The design requires w sets of np homogeneous experimental units and a total of $N = npw$ units.

EXPERIMENTAL DESIGN MODEL
FOR A TYPE GRB-p DESIGN

The classical experimental design model for a type GRB-p design is

$$Y_{ijz} = \mu + \alpha_j + \zeta_z + (\alpha\zeta)_{jz} + \epsilon_{i(jz)} \quad (i = 1, \ldots, n; j = 1, \ldots, p; z = 1, \ldots, w)$$

where
- Y_{ijz} = a score for experimental unit i in cell jz
- μ = the overall population mean
- α_j = the effect of treatment level j, subject to the restriction $\sum_{j=1}^{p} \alpha_j = 0$
- ζ_z = the effect of group z that is $NID(0, \sigma_\pi^2)$
- $(\alpha\zeta)_{jz}$ = the interaction of the jth treatment level and zth group and is $NID\{0, [(p-1)/p]\sigma_{\alpha\zeta}^2\}$, independent of ζ_z, and subject to the restriction $\sum_{j=1}^{p} (\alpha\zeta)_{jz} = 0$ for all z
- $\epsilon_{i(jz)}$ = the experimental error for unit i in cell jz and is $NID(0, \sigma_\epsilon^2)$, $\epsilon_{i(jz)}$ is independent of ζ_z and $(\alpha\zeta)_{jz}$.

This model, which is called a mixed model or model III, is indistinguishable in form from a mixed model for a completely randomized two-treatment factorial design (type CRF-pq design). However, the two designs differ in the number of treatments and in the way treatment levels (combinations) are assigned to experimental units. Many designs that are called completely randomized factorial designs are actually generalized randomized block designs. The type GRB-p design has only one treatment whose levels are randomly assigned to the experimental units within a group. The confusion between the designs is more likely to occur when one of the factors is an organismic variable. The use of an organismic variable always restricts the assignment of "treatment combinations" to the experimental units. Block diagrams of type GRB-4 and CRF-22 designs are shown in Figure 6.10-1.

FIGURE 6.10-1   Block diagrams of type GRB-4 and CRF-22 designs. The letters S_1, S_2, S_3, and S_4 denote groups (samples) of subjects. In the type GRB-4 design, the four treatment levels are assigned randomly to the np subjects within each group. In the type CRF-22 design, the four treatment combinations are assigned randomly to four random samples of n subjects.

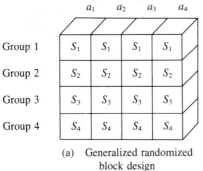

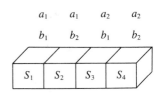

(a)   Generalized randomized   (b)   Completely randomized
         block design   factorial design

COMPUTATIONAL EXAMPLE FOR TYPE GRB-p DESIGN

For purposes of illustration, assume that 32 helicopter pilots are available to participate in the altimeter experiment described in Section 6.2. Eight of the pilots have less than 500 flight hours (group 1), eight have between 500 and 1499 hours (group 2), eight have between 1500 and 2499 hours (group 3), and eight, 2500 or more hours (group 4). Each level of treatment A is randomly assigned to two pilots in each group. Again we will assume a mixed model in which the levels of treatment A are fixed. The following statistical hypotheses are advanced:

$$H_0: \quad \alpha_j = 0 \quad \text{for all } j$$
$$H_0: \quad \sigma_\zeta^2 = 0$$
$$H_0: \quad \sigma_{\alpha\zeta}^2 = 0.$$

The .05 level of significance is adopted. The data and computational procedures are given in Table 6.10-1; the results of the analysis are summarized in Table 6.10-2.

According to Table 6.10-2, the hypothesis $\alpha_j = 0$ for all j is rejected. Since the group $\times$ treatment A interaction is insignificant, it is reasonable to believe that in the population the scores for each group have the same trend over the p treatment levels. The interpretation of the results is more complicated when the interaction is significant; for a discussion of the issues, see Sections 8.3 and 8.6. One feature in Table 6.10-2 deserves special comment. The proper error term for testing MSA is $MSG \times A$; $MSWCELL$ (mean square within cell) is used to test MSG and $MSG \times A$.

TABLE 6.10-1 Computational Procedures for a Type GRB-4 Design

Data and notation [Y_{ijz} denotes a score for the ith experimental unit in treatment level j and group z; $i = 1,$ ..., n experimental units (s_i); $j = 1, \ldots, p$ treatment levels (a_j); $z = 1, \ldots, w$ groups (g_z)]:

AGS Summary Table

Entry is Y_{ijz}

	a_1	a_2	a_3	a_4
g_1	3	4	7	7
	6	5	8	8
g_2	3	4	7	9
	1	2	5	10
g_3	2	3	6	10
	2	4	5	9
g_4	3	3	6	8
	2	3	6	11

AG Summary Table

Entry is $\sum_{i=1}^{n} Y_{ijz}$

	a_1	a_2	a_3	a_4	$\sum_{i=1}^{n}\sum_{j=1}^{p} Y_{ijz}$
$n = 2$					
g_1	9	9	15	15	48
g_2	4	6	12	19	41
g_3	4	7	11	19	41
g_4	5	6	12	19	42

$\sum_{i=1}^{n}\sum_{z=1}^{w} Y_{ijz} = 22 \qquad 28 \qquad 50 \qquad 72$

(ii) Computational symbols:

$$\sum_{i=1}^{n}\sum_{j=1}^{p}\sum_{z=1}^{w} Y_{ijz} = 3 + 6 + 3 + \cdots + 11 = 172.00$$

$$\sum_{i=1}^{n}\sum_{j=1}^{p}\sum_{z=1}^{w} Y_{ijz}^2 = [AGS] = (3)^2 + (6)^2 + (3)^2 + \cdots + (11)^2 = 1160.00$$

$$\left(\sum_{i=1}^{n}\sum_{j=1}^{p}\sum_{z=1}^{w} Y_{ijz}\right)^2/npw = [Y] = (172)^2/(2)(4)(4) = 924.50$$

$$\sum_{j=1}^{p}\left(\sum_{i=1}^{n}\sum_{z=1}^{w} Y_{ijz}\right)^2/nw = [A] = \frac{(22)^2}{(2)(4)} + \frac{(28)^2}{(2)(4)} + \cdots + \frac{(72)^2}{(2)(4)} = 1119.00$$

$$\sum_{z=1}^{w}\left(\sum_{i=1}^{n}\sum_{j=1}^{p} Y_{ijz}\right)^2/np = [G] = \frac{(48)^2}{(2)(4)} + \frac{(41)^2}{(2)(4)} + \cdots + \frac{(42)^2}{(2)(4)} = 928.75$$

$$\sum_{j=1}^{p}\sum_{z=1}^{w}\left(\sum_{i=1}^{n} Y_{ijz}\right)^2/n = [AG] = \frac{(9)^2}{2} + \frac{(4)^2}{2} + \frac{(4)^2}{2} + \cdots + \frac{(19)^2}{2} = 1141.00$$

(iii) Computational formulas:

$SSTO = [AGS] - [Y] = 1160.00 - 924.50 = 235.50$

$SSA = [A] - [Y] = 1119.00 - 924.50 = 194.50$

$SSG = [G] - [Y] = 928.75 - 924.50 = 4.25$

$SSG \times A = [AG] - [A] - [G] + [Y] = 1141.00 - 1119.00 - 928.75 + 924.50 = 17.75$

$SSWCELL = [AGS] - [AG] = 1160.00 - 1141.00 = 19.00$

The choice of F-ratio denominators is based on the principle that the expected value of the denominator should contain all of the terms in the numerator except the one being tested. The expected values for the fixed effects, random effects, and mixed (A random) models are given in Table 6.10-3. The circularity assumption (Section 6.4)

TABLE 6.10-2 ANOVA Table for Type GRB-4 Design

Source	SS	df	MS	F	E(MS) Model III (A Fixed)
1 Treatment (Altimeters)	194.500	$p - 1 = 3$	64.833	[⅓]**32.88*	$\sigma_\epsilon^2 + n\sigma_{\alpha\zeta}^2 + nw \sum_{j=1}^{p} \alpha_j^2/(p-1)$
2 Groups (Previous flying experience)	4.250	$w - 1 = 3$	1.417	[²⁄₄] 1.19	$\sigma_\epsilon^2 + np\sigma_\zeta^2$
3 Interaction of altimeters and previous flying experience	17.750	$(p-1)(w-1) = 9$	1.972	[³⁄₄] 1.66	$\sigma_\epsilon^2 + n\sigma_{\alpha\zeta}^2$
4 Within cell	19.000	$pw(n-1) = 16$	1.188		σ_ϵ^2
5 Total	235.500	$npw - 1 = 31$			

*$p < .01$

**The F ratio is obtained by dividing the MS is row 1 by the MS in row 3; this is indicated symbolically by [⅓].

TABLE 6.10-3 Expected Values of Mean Squares for Type GRB-p Design

Mean Square	Model I A Fixed Groups Fixed	Model II A Random Groups Random	Model III A Random Groups Fixed
MSA	$\sigma_\epsilon^2 + nw \sum_{j=1}^{p} \alpha_j^2/(p-1)$	$\sigma_\epsilon^2 + n\sigma_{\alpha\zeta}^2 + nw\sigma_\alpha^2$	$\sigma_\epsilon^2 + nw\sigma_\alpha^2$
MSG	$\sigma_\epsilon^2 + np \sum_{z=1}^{w} \zeta_z^2/(w-1)$	$\sigma_\epsilon^2 + n\sigma_{\alpha\zeta}^2 + np\sigma_\zeta^2$	$\sigma_\epsilon^2 + n\sigma_{\alpha\zeta}^2 + np \sum_{z=1}^{w} \zeta_z^2/(w-1)$
MSG × A	$\sigma_\epsilon^2 + n \sum_{j=1}^{p}\sum_{z=1}^{w} (\alpha\zeta)_{jz}^2/(p-1)(w-1)$	$\sigma_\epsilon^2 + n\sigma_{\alpha\zeta}^2$	$\sigma_\epsilon^2 + n\sigma_{\alpha\zeta}^2$
MSWCELL	σ_ϵ^2	σ_ϵ^2	σ_ϵ^2

is required for tests that use $MSG \times A$ as the error term. Homogeneity of population within-cell variances is required for those that use $MSWCELL$. Procedures for determining if the latter assumption is tenable are presented in Section 2.5. If the cell n's are unequal, the full rank experimental design model approach described in Section 8.13 should be used.

PROCEDURES FOR TESTING DIFFERENCES AMONG MEANS

Tests of differences among treatment A means have the same general form as those in Section 6.5 for a type RB-*p* design. The denominator of a contrast $\hat{\psi}_i$ for model I is

$$\hat{\sigma}_{\psi_i} = \sqrt{MSWCELL \left(\sum_{j=1}^{p} \frac{c_j^2}{nw} \right)} \quad \text{or} \quad \hat{\sigma}_{\bar{Y}} = \sqrt{\frac{MSWCELL}{nw}}.$$

For model III (A fixed), the denominator is

$$\hat{\sigma}_{\psi_i} = \sqrt{MSG \times A \left(\sum_{j=1}^{p} \frac{c_j^2}{nw} \right)} \quad \text{or} \quad \hat{\sigma}_{\bar{Y}} = \sqrt{\frac{MSG \times A}{nw}}.$$

The use of $MSWCELL$ or $MSG \times A$ in the preceding formulas is appropriate if the homogeneity of variance assumption or the circularity assumption, respectively, is tenable. If not, the procedures described in Chapter 2 or Section 6.5 should be used.

6.11 ADVANTAGES AND DISADVANTAGES OF TYPE RB-*p* AND GRB-*p* DESIGNS

The major advantages of the randomized block and generalized randomized block designs are as follows.

1. Designs provide greater power relative to a completely randomized design for many research applications. The designs permit an experimenter to isolate the effects of one nuisance variable and thereby reduce the experimental error and obtain a more precise estimate of treatment effects.

2. Flexibility; any number of treatment levels and blocks can be used in an experiment.

3. Simplicity in analysis of data.

4. The type GRB-*p* design also enables an experimenter to test the significance of the group $\times$ treatment A interaction. Furthermore, an appropriate error term is available for testing treatment A even though the interaction is significant.

The major disadvantages of the designs are as follows.

1.   If a large number of treatment levels are included in the experiment, it becomes difficult to form blocks or groups having minimum within-block (group) variability.

2.   In the fixed-effects and mixed (A random) models for a type RB-p design, a test of treatment effects is negatively biased if $(\pi\alpha)_{ij} > 0$.

3.   The designs involve somewhat more restrictive assumptions, for example, circularity, than the completely randomized design.

6.12   REVIEW EXERCISES

1.   [4.1, 6.1] Compare the randomization procedures for type CR-p and RB-p designs.

†2.   [6.1] What are the most common approaches to forming homogeneous blocks in a type RB-p design?

3.   [6.1] For each of the following experiments, suggest an appropriate blocking variable.

a) The effect of jogging on heart rate.

b) The effect of worktable height on assembly line productivity.

c) The effect of meaningfulness of nonsense syllables on the number of trials required to learn the syllables.

†4.   [6.1] Show that

$$\sum_{j=1}^{p} \sum_{i=1}^{n} [(\overline{Y}_{\cdot j} - \overline{Y}_{\cdot \cdot}) + (\overline{Y}_{i\cdot} - \overline{Y}_{\cdot \cdot}) + (Y_{ij} - \overline{Y}_{\cdot j} - \overline{Y}_{\cdot i} + \overline{Y}_{\cdot \cdot})]^2$$

$$= n \sum_{j=1}^{p} (\overline{Y}_{\cdot j} - \overline{Y}_{\cdot \cdot})^2 + p \sum_{i=1}^{n} (\overline{Y}_{i\cdot} - \overline{Y}_{\cdot \cdot})^2 + \sum_{j=1}^{p} \sum_{i=1}^{n} (Y_{ij} - \overline{Y}_{\cdot j} - \overline{Y}_{i\cdot} + \overline{Y}_{\cdot \cdot})^2.$$

†5.   [6.1] Derive the computational formulas for a type RB-p design from the deviation formulas in equation (6.1-2).

6.   It was hypothesized that brain-damaged patients would score lower on the Willner Unusual Meanings Vocabulary Test (*WUMV*), which measures knowledge of unusual meanings of familiar words, and the Willner-Sheerer Analogy Test (*WSA*) than on the vocabulary items of the Wechsler Adult Intelligence Scale (*WAIS*). A random sample of 12 brain-damaged patients took all three tests. The order of administration of the tests was randomized independently for each patient. The dependent variable was the subject's standard score on each test. According to the test norms, all three tests have a mean of 10 and a standard deviation of 3. The following data were obtained. (Experiment suggested by Willner, W. Impairment of knowledge of unusual meanings of familiar words in brain damage and schizophrenia. *Journal of Abnormal Psychology*, 1965, *70*, 405–411.)

	a_1 WAIS	a_2 WUMV	a_3 WSA
s_1	15	12	11
s_2	10	11	8
s_3	6	4	3
s_4	7	7	5
s_5	9	6	6
s_6	16	14	10
s_7	11	10	7
s_8	13	9	4
s_9	12	10	8
s_{10}	10	8	7
s_{11}	11	9	9
s_{12}	14	11	10

†a) [6.2] Test the hypothesis H_0: $\alpha_j = 0$ for all j; let $\alpha = .05$. Assume for this exercise and (b) through (d) that the additive model (Section 6.3) is appropriate and the circularity assumption (Section 6.4) is tenable.

†b) [6.2] Calculate the power of the test of H_0: $\alpha_j = 0$ for all j.

†c) [6.2] For the hypothesis H_0: $\alpha_j = 0$ for all j, determine the number of blocks required to achieve a power of approximately .88.

†d) [6.2] Calculate $\hat{\omega}^2_{Y|A}$ and $\hat{\rho}_{IY|BL}$.

†e) [6.3] Use Tukey's procedure to determine if the additive model is appropriate; let $\alpha = .10$.

†f) [6.3] Construct a figure like 6.3-1 for $\hat{\epsilon}_{ij}$ and $\hat{Y}_{ij}$. Is the figure consistent with Tukey's test for nonadditivity?

†g) [6.4] Compute the value of $\hat{\theta}$ and $F_{.05}(adjusted)$. The use of $F(adjusted)$ is not required for the test of H_0: $\alpha_j = 0$ for all j; explain why this is so.

†h) [6.5] The null hypothesis in Exercise 6(a) does not specifically address the experimenter's research interest. Formulate a more appropriate null hypothesis and test it using Dunn's statistic; use $MSRES$ in the formula and let $\alpha = .05$. The critical values in Dunn's table are for two-tailed tests. The one-tailed critical value can be obtained from the t table since $t'D_{.05;2,22} = t_{.05/2,22}$. Suggest at least one alternative explanation for the observed differences among the means of the three tests.

†i) [6.7] Assume that Y_{32} is missing. Use Yates's procedure to estimate the missing observation; compute $SSA(corrected)$.

†j) [6.8] (i) Determine the efficiency of the randomized block design relative to a completely randomized design; use Fisher's correction. (ii) How many subjects would be required in a completely randomized design to match the efficiency of the randomized block design?

k) Prepare a "results and discussion section" for the *Journal of Abnormal Psychology*.

7. [6.2] Six monkeys (*Macaca mulatta*) were trained to search a visual display for a target stimulus (3-inch circle). The effect of 0, 1, 2, 3, 4, and 5 irrelevant stimuli (geometric shapes) on probability of correct response was investigated. The subjects had free access to a manipulanda with which they could turn on a visual display for five seconds. If they pressed the panel

containing the target stimulus, a banana-flavored food pellet was delivered to a food cup. If the incorrect panel was pressed or if no response was made, there was a 5-second period of darkness and time-out during which pressing the manipulanda had no effect. It was hypothesized that irrelevant stimuli appearing with the target stimulus would decrease the monkey's response accuracy. The following data were obtained. (Experiment suggested by Latto, R. Visual search in monkeys. *Perceptual and Motor Skills*, 1971, *32*, 307–312.)

	a_1 (0)*	a_2 (1)	a_3 (2)	a_4 (3)	a_5 (4)	a_6 (5)
s_1	98	91	95	90	93	94
s_2	100	100	99	98	99	97
s_3	92	89	92	87	89	88
s_4	99	95	97	93	96	95
s_5	99	95	99	94	100	97
s_6	100	100	100	96	99	99

    *   Numbers in parentheses denote the number of irrelevant geometric shapes

†**a)** [6.2] Test the hypothesis H_0: $\alpha_j = 0$ for all j; let $\alpha = .01$. Assume for this exercise and (b) through (d) that the additive model (Section 6.3) is appropriate and the circularity assumption (Section 6.4) is tenable.

†**b)** [6.2] Calculate the power of the test of H_0: $\alpha_j = 0$ for all j.

†**c)** [6.2] For the hypothesis H_0: $\alpha_j = 0$ for all j, determine the number of blocks required to achieve a power of approximately .94.

†**d)** [6.2] Calculate $\hat{\omega}^2_{Y|A}$ and $\hat{\rho}_{IY|BL}$.

†**e)** [6.3] Use Tukey's procedure to determine if the additive model is appropriate; let $\alpha = .10$.

†**f)** [6.3] Construct a figure like 6.3-1 for $\hat{e}_{ij}$ and $\hat{Y}_{ij}$. Is the figure consistent with Tukey's test for nonadditivity?

†**g)** [6.4] Compute the value of $\hat{\theta}$ and $F_{.01}(adjusted)$. Test the hypothesis H_0: $\alpha_j = 0$ for all j using the adjusted F value.

†**h)** [6.5] Use Dunnett's statistic to determine which population means differ from the control population (treatment level a_1). (i) Compute the statistic using both *MSRES* and *MSRES$_i$*; use a one-tailed test, let $\alpha = .01$. (ii) Does the choice of error terms affect your decisions about differences among the population means? (iii) Based on Exercise 7(g), which error term is the most appropriate?

†**i)** [6.6] (i) Test the hypothesis that the relationship between the independent and dependent variables is nonlinear. Use *SSRES*$_{dep\ from\ lin}$ = *SSRES* − *SSRES*$_{lin}$ with $df_{dep\ from\ lin}$ = $dfRES$ − $dfRES_{lin}$. (ii) What percent of the total trend in the sample is accounted for by the nonlinear component?

†**j)** [6.7] Assume that Y_{52} is missing. Use Yates's procedure to estimate the missing observation; compute *SSA (corrected)*.

†**k)** [6.8] (i) Determine the efficiency of the randomized block design relative to a completely randomized design; use Fisher's correction. (ii) How many subjects would be required in a completely randomized design to match the efficiency of the randomized block design?

  l) Prepare a "results and discussion section" for *Perceptual and Motor Skills*.

8. Adaptation in the oral performance of stutterers was investigated. A random sample of ten stutterers each read a 300-word factual passage five successive times on each of three days. The days were separated by at least one, but not more than two weeks. The subjects were observed through a one-way window by four observers who timed their reading. It was hypothesized that the population mean overall reading rates would differ for the three days. The following data were obtained. (Experiment suggested by Cullman, W. L. Stability of adaptation in the oral performance of stutterers. *Journal of Speech and Hearing,* 1963, 6, 70–83.)

	a_1	a_2	a_3
	Day 1	Day 2	Day 3
s_1	200	225	235
s_2	185	200	217
s_3	173	197	210
s_4	196	233	228
s_5	198	212	239
s_6	212	236	251
s_7	191	229	248
s_8	178	235	240
s_9	180	199	220
s_{10}	189	215	233

a) [6.2] Test the hypothesis H_0: $\alpha_j = 0$ for all j; let $\alpha = .01$. Assume for this exercise and (b) through (c) that the additive model (Section 6.3) is appropriate and the circularity assumption (Section 6.4) is tenable.

b) [6.2] Determine the number of blocks required to achieve a power of approximately .97.

c) [6.2] Calculate $\hat{\omega}^2_{Y|A}$ and $\hat{\rho}_{IY|BL}$.

d) [6.3] Use Tukey's procedure to determine if the additive model is appropriate; let $\alpha = .10$.

e) [6.3] Construct a figure like 6.3-1 for $\hat{\epsilon}_{ij}$ and $\hat{Y}_{ij}$. Is the figure consistent with Tukey's test for nonadditivity?

f) [6.4] Compute the value of $\hat{\theta}$ and $F_{.01}(adjusted)$. The use of $F(adjusted)$ is not required for the test of H_0: $\alpha_j = 0$ for all j; explain why this is so.

g) [6.5] (i) Use Tukey's statistic to test all two-mean null hypotheses; let $\alpha = .01$. (ii) Justify the use of *MSRES* instead of *MSRES_i*.

h) [6.6] (i) Test the hypothesis that the relationship between the independent and dependent variables is nonlinear. Use both *SSRES* with $df = (n-1)(p-1)$ and $SSRES_{dep\ from\ lin} = SSRES - SSRES_{lin}$ with $df_{dep\ from\ lin} = df RES - df RES_{lin}$. (ii) Does the choice of error terms affect your decision about nonlinearity in the population? (iii) Based on Exercise 8(f), which error term is most appropriate? (iv) In this example, $MS\hat{\psi}_{dep\ from\ lin} = MS\hat{\psi}_{quad}$ and $MSRES_{dep\ from\ lin} = MSRES_{quad}$. Explain why this is true. (v) What percent of the total trend in the sample is accounted for by the nonlinear component?

i) [6.7] Assume that Y_{91} is missing. Use Yates's procedure to estimate the missing observation; compute *SSA*(*corrected*).

j) [6.8] (i) Determine the efficiency of the randomized block design relative to a completely randomized design; use Fisher's correction. (ii) How many subjects would be required in a completely randomized design to match the efficiency of the randomized block design?

k) Prepare a "results and discussion section" for the *Journal of Speech and Hearing.*

9. Pain thresholds to electrical stimulation were measured in ten male volunteers using the psychophysical method of limits. The independent variable was the rate of decrease in the noxious stimulus in the descending trials; five rates were used ranging from slow to fast. The subjects were instructed to say "Pain Gone" as soon as all of the pain had disappeared. Following a series of practice trials, the subjects were given four trials with each of the five rates. The order of presentation of the rates was randomized independently for each subject. The dependent variable was the mean voltage applied when the subject said "Pain Gone" and was measured in arbitrary units on a 0- to 100-point scale. (Experiment suggested by Horland, A. A. and Wolff, B. B. Changes in descending pain threshold related to rate of noxious stimulation. *Journal of Abnormal Psychology*, 1973, *81*, 39–45.)

	a_1 (slow)	a_2	a_3	a_4 (fast)
s_1	31	31	26	23
s_2	30	29	29	22
s_3	27	27	21	11
s_4	19	28	16	17
s_5	25	24	22	17
s_6	29	19	27	18
s_7	39	28	30	21
s_8	32	30	26	29
s_9	29	37	36	31
s_{10}	27	25	22	21

a) [6.2] Test the hypothesis H_0: $\alpha_j = 0$ for all j; let $\alpha = .05$. Assume for this exercise and (b) through (d) that the additive model (Section 6.3) is appropriate and the circularity assumption (Section 6.4) is tenable.

b) [6.2] Calculate the power of the test of H_0: $\alpha_j = 0$ for all j.

c) [6.2] For the hypothesis H_0: $\alpha_j = 0$ for all j, determine the number of blocks required to achieve a power of approximately .84.

d) [6.2] Calculate $\hat{\omega}_{Y|A}$ and $\hat{\rho}_{IY|BL}$.

e) [6.3] Use Tukey's procedure to determine if the additive model is appropriate; let $\alpha = .10$.

f) [6.3] Construct a figure like 6.3-1 for $\hat{\epsilon}_{ij}$ and $\hat{Y}_{ij}$. Is the figure consistent with Tukey's test for nonadditivity?

g) [6.4] Compute the value of $\hat{\theta}$ and $F(adjusted)$. The use of $F(adjusted)$ is not required for the test of H_0: $\alpha_j = 0$ for all j; explain why this is so.

h) [6.5] (i) Use Tukey's statistic to test all two-mean null hypotheses; let $\alpha = .05$. (ii) Justify the use of $MSRES$ instead of $MSRES_i$.

i) [6.6] (i) Assume that the levels of the independent variable differ by a constant amount. Test the hypothesis that the relationship between the independent and dependent variables is nonlinear. Use both $SSRES$ with $df = (n - 1)(p - 1)$ and $SSRES_{dep\ from\ lin} = SSRES - SSRES_{lin}$ with $df_{dep\ from\ lin} = dfRES - dfRES_{lin}$. (ii) Does the choice of error terms affect your decision about nonlinearity? (iii) Based on Exercise 9(g), which error term is most appropriate? (iv) What percent of the total trend in the sample is accounted for by the nonlinear component?

j) [6.7] Assume that Y_{64} is missing. Use Yates's procedure to estimate the missing observation; compute $SSA(corrected)$.

k) [6.8] (i) Determine the efficiency of the randomized block design relative to a completely randomized design; use Fisher's correction. (ii) How many subjects would be required in a completely randomized design to match the efficiency of the randomized block design?

l) Prepare a "results and discussion section" for the *Journal of Abnormal Psychology.*

10. The effect of strenuous to exhaustive physical exercise on performance of a discrimination task was investigated. Seven male subjects performed a line matching task while jogging at various speeds on a motor-driven treadmill. The task was self-paced—a new set of lines was presented immediately following the subject's response. The dependent variable was the number of responses made during successive 3-minute periods. The periods consisted of a pretest resting stage (level a_1), 2.5 mph exercise at 12% grade (level a_2), 3.4 mph exercise at 14% grade (level a_3), 4.2 mph exercise at 16% grade (level a_4), 5.0 mph exercise at 18% grade (level a_5), and posttest resting stage (level a_6). The following data were obtained. (Experiment suggested by McGlynn, G. H., Laughlin, N. T., and Bender, V. L. Effect of strenuous to exhaustive exercise on a discrimination task. *Perceptual and Motor Skills,* 1977, *44,* 1139–1147.)

	a_1 Pretest	a_2 2.5 mph	a_3 3.4 mph	a_4 4.2 mph	a_5 5.0 mph	a_6 Posttest
s_1	45	47	48	50	51	44
s_2	48	50	58	54	61	51
s_3	46	54	51	57	56	48
s_4	40	37	44	40	45	34
s_5	34	41	38	48	41	38
s_6	42	45	46	43	48	41
s_7	55	58	54	60	57	55

a) [6.2] Test the hypothesis H_0: $\alpha_j = 0$ for all j; let $\alpha = .01$. Assume for this exercise and (b) through (d) that the additive model (Section 6.3) is appropriate and the circularity assumption (Section 6.4) is tenable.

b) [6.2] Calculate the power of the test of H_0: $\alpha_j = 0$ for all j.

c) [6.2] For the hypothesis H_0: $\alpha_j = 0$ for all j, determine the number of blocks required to achieve a power of approximately .87.

d) [6.2] Calculate $\hat{\omega}_{Y|A}^2$ and $\hat{\rho}_{IY|BL}$.

e) [6.3] Use Tukey's procedure to determine if the additive model is appropriate; let $\alpha = .10$.

f) [6.3] Construct a figure like 6.3-1 for $\hat{\epsilon}_{ij}$ and $\hat{Y}_{ij}$. Is the figure consistent with Tukey's test for nonadditivity?

g) [6.4] Compute the value of $\hat{\theta}$ and $F_{.01}(adjusted)$; test the hypothesis H_0: $\alpha_j = 0$ for all j.

h) [6.5] Use Dunnett's statistic to determine which population means differ from the control population (treatment level a_1). (i) Compute the statistic using both $MSRES$ and $MSRES_i$; use a two-tailed test, let $\alpha = .01$. (ii) Does the choice of error terms affect your decision about differences among the population means? (iii) Based on Exercise 10(g), which error term is the most appropriate?

i) [6.7] Assume that Y_{71} is missing. Use Yates's procedure to estimate the missing observation; compute $SSA(corrected)$.

j) [6.8] (i) Determine the efficiency of the randomized block design relative to a completely randomized design; use Fisher's correction. (ii) How many subjects would be required in a completely randomized design to match the efficiency of the randomized block design?

k) Prepare a "results and discussion section" for *Perceptual and Motor Skills*.

11.   Following the procedures in Section 2.3 for a completely randomized design, derive the following $E(MS)$'s for a randomized block design. Assume an additive model.

†**a)** [6.3] $E(MSA)$, $E(MSBL)$, $E(MSRES)$; assume that treatment A and blocks are fixed.

b) [6.3] $E(MSA)$, $E(MSBL)$, $E(MSRES)$; assume that treatment A is fixed but blocks are random.

12.   [6.3] Explain why the interpretation of an insignificant F ratio for treatment A for the non-additive model may be ambiguous.

13.   [6.3] Why is the .10 level of significance rather than a higher level of significance often adopted for Tukey's test for nonadditivity of block and treatment effects?

†**14.**   [6.4] The error term for a randomized block design is normally smaller than that for a completely randomized design. **(a)** What determines the relative size of the two error terms? **(b)** Why might the error term for a randomized block design be larger than that for a completely randomized design?

15.   **a)** [6.4] Identify the type of matrix for the following.

†(i)

$$\Sigma = \begin{bmatrix} 5.0 & 2.5 & 5.0 \\ 2.5 & 10.0 & 7.5 \\ 5.0 & 7.5 & 15.0 \end{bmatrix}$$

†(ii)

$$\Sigma = \begin{bmatrix} 6.1 & 3.4 & 3.4 \\ 3.4 & 6.1 & 3.4 \\ 3.4 & 3.4 & 6.1 \end{bmatrix}$$

(iii)

$$\Sigma = \begin{bmatrix} 1.0 & .5 & 1.5 \\ .5 & 3.0 & 2.5 \\ 1.5 & 2.5 & 5.0 \end{bmatrix}$$

b) [6.4] Assume that we want to test the following overall null hypothesis

$$\begin{matrix} \mathbf{C'} & \boldsymbol{\mu} & \mathbf{0} \end{matrix}$$

$$\begin{bmatrix} 1 & -1 & 0 \\ \frac{1}{2} & \frac{1}{2} & -1 \end{bmatrix} \begin{bmatrix} \mu_1 \\ \mu_2 \\ \mu_3 \end{bmatrix} = \begin{bmatrix} 0 \\ 0 \\ 0 \end{bmatrix}.$$

Determine whether $\mathbf{C^{*'}} \Sigma \mathbf{C^*} = \lambda \mathbf{I}$ for the Σ's given in †(i), †(ii), and (iii).

16.   [6.5] In computing $\hat{\sigma}_{\psi_i}$ and $\hat{\sigma}_{\bar{Y}_i}$, either $MSRES$ or $MSRES_i$ may be used. What are the advantages and disadvantages of using $MSRES_i$ instead of $MSRES$?

†**17.**   [6.7] What condition do estimated observations obtained by Yates's procedure satisfy?

†**18.**   Exercise 6 described an experiment concerning the performance of brain-damaged patients on three psychological tests. Assume that instead of testing 12 patients, only two were tested. The following data were obtained. (This problem requires the inversion of small matrices. The computations are tedious, but they can be done without the aid of a computer.)

	a_1	a_2	a_3
s_1	15	12	11
s_2	10	11	8

a) [6.9] Write a regression model with $h - 1 = 3$ independent variables for this type RB-3 design.

b) [6.9] Specify the null hypothesis in terms of β's for (i) treatment levels and (ii) blocks.

c) [6.9] If dummy coding is used, how do the parameters of the regression model correspond to those of the type RB-3 less than full rank experimental design model?

d) [6.9] (i) Write the reduced regression model for testing the null hypothesis for treatment levels. (ii) Write the formula for the regression sum of squares for treatment levels in terms of the reduced and full models.

e) [6.9] Use the regression model to test the hypothesis that all treatment and block effects equal zero; let $\alpha = .05$.

f) [6.9] Write the (i) less than full rank experimental design model and the (ii) restricted full rank experimental design model for this type RB-3 design.

g) [6.9] Write the $\mathbf{Q}'$ matrix for the restricted full rank experimental design model for the type RB-3 design. Formulate the treatment null hypothesis for $\mathbf{Q}'$ in terms of (i) treatment-level population means, that is, $\mu_{.1} - \mu_{.3} = \mu_{.2} - \mu_{.3} = 0$ and (ii) block–treatment-level population means, that is,

$$\mu_{11} - \mu_{13} = \mu_{12} - \mu_{13} = 0$$
$$\mu_{21} - \mu_{23} = \mu_{22} - \mu_{23} = 0.$$

h) [6.9] (i) Use the restricted full rank experimental design model to test the hypothesis in Exercise 18(g) (ii); let $\alpha = .05$. (ii) Test the hypothesis $\mu_{11} - \mu_{21} = \mu_{12} - \mu_{22} = \mu_{13} - \mu_{23} = 0$; let $\alpha = .05$.

i) [6.9] Assume that Y_{21} is missing. Use the restricted full rank experimental design model to test

$$\mu_{11} - \mu_{13} = \mu_{12} - \mu_{13} = 0$$
$$\mu_{22} - \mu_{23} = 0$$

let $\alpha = .05$.

j) [6.7] (i) Assume that Y_{21} is missing. Use Yates's procedure to estimate the missing observation; compute $SSA\,(corrected)$. (ii) If you have done Exercise 18(i), compare $SSA\,(corrected)$ with SSA obtained in that exercise. Based on this comparison, what hypothesis is tested by $F = MSA\,(corrected)/MSRES$?

†19. a) [6.9] Exercise 7 described an experiment concerning the effect of irrelevant stimuli on probability of correct response. Write a regression model with $h - 1 = 10$ independent variables for this type RB-6 design.

b) [6.9] Specify the null hypothesis in terms of β's for (i) treatment levels and (ii) blocks.

c) [6.9] If dummy coding is used, how do the parameters of the regression model correspond to those of the type RB-6 less than full rank experimental design model.

d) [6.9] (i) Write the reduced regression model for testing the null hypothesis for treatment levels. (ii) Write the formula for the regression sum of squares for treatment levels in terms of the reduced and full models.

e) [6.9] Use the regression model approach to test the hypothesis that all treatment and block effects equal zero; let $\alpha = .01$. (This exercise requires the inversion of large matrices. A computer with a matrix inversion program should be used.)

f) [6.9] Compute $\bar{y}' = [\bar{Y}_{.1}\bar{Y}_{.2} \cdots \bar{Y}_{.6}]$ using $\bar{y} = W\hat{\beta}_0$.

20.   **a)** [6.9] Exercise 8 described an experiment concerning adaptation in the oral performance of stutterers. Write a regression model with $h - 1 = 11$ independent variables for this type RB-3 design.

b) [6.9] Specify the null hypothesis in terms of β's for (i) treatment levels and (ii) blocks.

c) [6.9] If dummy coding is used, how do the parameters of the regression model correspond to those of the type RB-3 less than full rank experimental design model.

d) [6.9] (i) Write the reduced regression model for testing the null hypothesis for treatment levels. (ii) Write the formula for the regression sum of squares for treatment levels in terms of the reduced and full models.

e) [6.9] Use the regression model to test the hypothesis that all treatment and block effects equal zero; let $\alpha = .01$. (This exercise requires the inversion of large matrices. A computer with a matrix inversion program should be used.)

f) [6.9] Write the (i) less than full rank experimental design model and the (ii) restricted full rank experimental design model for this type RB-3 design.

g) [6.9] Formulate the treatment null hypothesis in terms of (i) treatment-level population means, $\mu_{.j}$'s, and (ii) block-treatment-level population means, μ_{ij}'s.

h) [6.9] Use the restricted full rank experimental design model to test the hypothesis in Exercise 20(g) (ii); let $\alpha = .01$. (This exercise requires the inversion of large matrices. A computer with a matrix inversion program should be used.)

i) [6.9] Assume that Y_{91} is missing. (i) Use the restricted full rank experimental design model to test

$$\mu_{11} - \mu_{13} = \mu_{12} - \mu_{13} = 0$$
$$\mu_{21} - \mu_{23} = \mu_{22} - \mu_{23} = 0$$
$$\mu_{31} - \mu_{33} = \mu_{32} - \mu_{33} = 0$$
$$\cdots\cdots\cdots\cdots\cdots\cdots\cdots$$
$$\mu_{92} - \mu_{93} = 0$$
$$\mu_{10,1} - \mu_{10,3} = \mu_{10,2} - \mu_{10,3} = 0.$$

(ii) Use the regression model with dummy coding to test the null hypothesis for treatment A. Compare the results with those in Exercise 20(i) (i). (These exercises require the inversion of large matrices. A computer with a matrix inversion program should be used.) (iii) If you did Exercise 8(i), compare the value of SSA in Exercises 20(i) (i) and 20(i) (ii) with that for SSA (*corrected*).

21.   It was hypothesized that test anxiety and the resulting poorer performance among school children is due to previous failures and negative evaluations in school. A counterconditioning procedure was used to reduce test anxiety. Thirty children, 15 males and 15 females, were randomly assigned to one of three conditions. The children in the counterconditioning group were shown neutral words, followed by pictures of school-related scenes, followed by positive words. The children in a placebo condition saw the neutral words, followed by school scenes, but did not see the positive words. Those in the control group did not experience any of the treatment procedures. The dependent variable was performance on the Digit Span subtest of the Wechsler Intelligence Scale for Children. The following data were obtained. (Experiment suggested by Parish, T. S., Buntman, A. D., and Buntman, S. R. Effect of counterconditioning on test anxiety as indicated by digit span performance. *Journal of Educational Psychology,* 1976, *68,* 297–299.)

	a_1 Counter-conditioning	a_2 Placebo	a_3 Control
g_1 Males	13	8	6
	11	9	9
	8	12	11
	10	7	12
	12	11	8
g_2 Females	10	9	11
	8	8	7
	11	8	8
	14	10	4
	13	5	8

†a) [6.10] Test the hypothesis H_0: $\alpha_j = 0$ for all j, H_0: $\zeta_z = 0$ for all z and H_0: $(\alpha\zeta)_{jz} = 0$ for all jz; let $\alpha = .05$.

†b) [6.2] Calculate the power of the test of H_0: $\alpha_j = 0$ for all j.

†c) [6.2] For the hypothesis H_0: $\alpha_j = 0$ for all j, determine the number of subjects required to achieve a power of approximately .84.

†d) [6.2] Calculate $\hat{\omega}^2_{Y|A}$.

†e) [6.10] Use Tukey's statistic to test all two-mean null hypotheses; let $\alpha = .05$.

 f) Prepare a "results and discussion section" for the *Journal of Educational Psychology.*

7 LATIN SQUARE AND RELATED DESIGNS

7.1 DESCRIPTION OF LATIN SQUARE DESIGN

A Latin square design is an arrangement having p rows and p columns with p Latin letters assigned to the cells of the square so that each letter appears once in each row and once in each column. The design derives its name from an ancient puzzle that dealt with the number of different ways Latin letters could be arranged in a square matrix. If three treatment levels are designated by the letters A, B, and C, one of the 12 possible arrangements of a 3×3 Latin square is the following.

Each row and each column of the square represents a complete replication of the three treatment levels.

We saw in Chapter 6 that a randomized block design permits an experimenter to minimize the effects of one nuisance variable (variation among rows) in evaluating

308

treatment effects. A Latin square design extends this principle to two nuisance variables: variation associated with rows and variation associated with columns. As a result the design may be considerably more efficient than completely randomized and randomized block designs. For example, Cochran (1938, 1940) reported that during a seven-year period at Rothamsted Experimental Station and associated centers the efficiency of the Latin square design relative to type CR-p and RB-p designs was 222% and 137%, respectively.

The Latin square design has been used extensively in agricultural and industrial research. However, it has seen limited use in behavioral and educational research, probably because of the restrictive assumptions required. The design, which is designated as a type LS-p design, is appropriate for experiments that meet, in addition to the general assumptions of analysis of variance, the following conditions.

1. One treatment with $p \geq 2$ treatment levels and two nuisance variables with p levels each. The levels of one nuisance variable are assigned to the p rows of the square; the levels of the other nuisance variable are assigned to the p columns. A Latin square design with $p < 5$ is not practical because of the small number of degrees of freedom available for experimental error unless more than one experimental unit is assigned to each cell of the square or unless several squares are combined. This latter procedure is described in Chapters 12 and 13.

2. Formation of p rows and p columns each containing np homogeneous experimental units were $n \geq 1$. The variability among units within each row (column) should be less than the variability among units in different rows (columns).

3. It must be reasonable to assume that there are no interactions among the rows, columns, and treatment levels of the square. If this assumption is not fulfilled, a test of one or more of the corresponding effects is biased.

4. Random assignment of treatment levels to the p^2 cells of the square with the restriction that each treatment level must appear only once in a row and once in a column. The design requires a total of $N = np^2$ experimental units where $n \geq 1$.

7.2 CONSTRUCTION AND RANDOMIZATION OF LATIN SQUARES

A discussion of the construction of Latin square designs and extensive tables of designs are given by Fisher and Yates (1963). The purpose of this section is to acquaint the reader with the basic terminology and randomization procedures employed with type LS-p designs.

A Latin square is said to be a *standard square* if the first row and the first column are ordered alphabetically or numerically. The following are examples of standard squares.

```
A B      A B C      A B C D      A B C D      A B C D      A B C D
B A      B C A      B A D C      B C D A      B D A C      B A D C
         C A B      C D B A      C D A B      C A D B      C D A B
                    D C A B      D A B C      D C B A      D C B A
                      (1)          (2)          (3)          (4)
```

$\underbrace{}_{2 \times 2}$ $\underbrace{}_{3 \times 3}$ $\underbrace{\phantom{\text{4 × 4 squares}}}_{4 \times 4}$

```
A B C D E      A B C D E      A B C D E F
B C D E A      B A E C D      B C F A D E
C D E A B      C D A E B      C F B E A D
D E A B C      D E B A C      D E A B F C
E A B C D      E C D B A      E A D F C B
   (1)            (2)         F D E C B A
```

$\underbrace{\phantom{\text{5 × 5 squares}}}_{5 \times 5}$ $\underbrace{\phantom{\text{6 × 6 square}}}_{6 \times 6}$

Two squares are *conjugate* if the rows of one are identical to the columns of the other. The conjugate of square $5 \times 5(2)$ is

```
A B C D E
B A D E C
C E A B D.
D C E A B
E D B C A
```

A square is *self-conjugate* if the same square is obtained when the rows and columns are interchanged. For example, squares 2×2, 3×3, $4 \times 4(1, 2, 3,$ and $4)$, and $5 \times 5(1)$ are self-conjugate because the same arrangement of Latin letters results from the interchanging of rows and columns. Squares $5 \times 5(2)$ and 6×6 are not self-conjugate; this was just illustrated for square $5 \times 5(2)$.

Latin squares of any size can be constructed by a *one-step cyclic permutation* of a sequence of letters. This process involves successively moving the first letter in the sequence to the extreme right and simultaneously moving all other letters one position to the left. For example, from the sequence *ABCD* a one-step cyclic permutation gives *BCDA*, a second cyclic permutation gives *CDAB*, and a third gives *DABC*. The fourth cyclic permutation gives the original starting sequence *ABCD*. The Latin square constructed from these $p - 1$ one-step cyclic permutations of the letters *ABCD* is

```
A B C D
B C D A
C D A B
D A B C
```

ENUMERATION OF LATIN SQUARES

The letters in a given Latin square can be rearranged so as to produce a total of $p!(p-1)!$ Latin squares, including the original square. Enumeration of Latin squares of size 7×7 and smaller has been made by Fisher and Yates (1934), Norton (1939), and Sade (1951). The number of possible arrangements of Latin squares is as follows:

1. *2 × 2 Latin square.* One standard square (self-conjugate) and one non-standard square, which is obtained by interchanging either rows or columns of the standard square. Thus there is a total of two arrangements.

2. *3 × 3 Latin square.* One standard square (self-conjugate) and $(3!)(2!) = (3 \cdot 2 \cdot 1)(2 \cdot 1) = 12$ possible arrangements of this standard square.

3. *4 × 4 Latin square.* Four standard squares (all self-conjugate) with $(4!)(3!) = 144$ possible arrangements of each standard square. The total number of arrangements is equal to $4(144) = 576$.

4. *5 × 5 Latin square.* Twenty-five standard squares and their conjugates, plus 6 self-conjugate squares or 56 standard squares. The total number of possible arrangements is $56(5!)(4!) = 161{,}280$.

5. *6 × 6 Latin square.* 9408 standard squares with $9408(6!)(5!) = 812{,}851{,}200$ possible arrangements.

6. *7 × 7 Latin square.* 16,942,080 standard squares.

The number of possible standard squares increases sharply as the dimensions of the square increase.

RANDOMIZATION OF LATIN SQUARES

Rules for randomization of Latin squares have been described by Fisher and Yates (1963). Theoretically an experimenter who plans to use a Latin square design should randomly select a square from the population of all possible squares of the proper dimension. However, this is not practical for squares larger than 5×5. The following rules, which represent a compromise, are adequate for almost all research applications.

1. *2 × 2 Latin square.* Select one of two arrangements at random.

2. *3 × 3 Latin square.* Randomize the order of rows and columns of the standard square independently.

3. *4 × 4 Latin square.* Select one of the standard squares at random. Randomize the order of rows and columns of the standard square independently.

4. *5 × 5 and higher-order squares.* Arbitrarily select one of the standard squares. Randomize the order of rows and columns independently and assign the treatment levels randomly to the letters A, B, C, and so on.

For example, to select a 4×4 Latin square, draw three sets of four random digits (1,

2, 3, 4) from a table of random numbers. Assume that the digits are 2, 1, 3, 4; 3, 1, 2, 4; and 4, 3, 1, 2. Because the first digit is *two,* the second 4 × 4 square shown at the beginning of this section is selected. This square is

Columns

1 2 3 4

Rows		
1	A B C D
2	B C D A
3	C D A B
4	D A B C

The rows are ordered according to the second set of random digits (3, 1, 2, 4) to yield the following square.

Columns

1 2 3 4

Rows		
3	C D A B
1	A B C D
2	B C D A
4	D A B C

The final step in the randomization procedure is to order the columns according to the third set of random digits (4, 3, 1, 2). This gives the following square:

Columns

4 3 1 2

Rows		
3	B A C D
1	D C A B
2	A D B C
4	C B D A

The randomization procedures for Latin squares described here select a square at random from a set of all squares of the required size for 2 × 2, 3 × 3, and 4 × 4 squares. For 5 × 5 and higher-order squares, a square is selected at random from a set of squares. Although this procedure does not select higher-order squares from all possible squares, according to Cox (1958) it is suitable for both practical and theoretical purposes. Alternative randomization procedures are described by Federer (1955, 140–142) and Fisher and Yates (1963, 24).

7.3   COMPUTATIONAL EXAMPLE FOR TYPE LS-*p* DESIGN

Assume that we want to evaluate automobile tires by means of a road test. The independent variable is four kinds of rubber compounds, denoted by a_1, a_2, a_3, and

a_4, used in the construction of the tires. The dependent variable is thickness of tread remaining on each tire after 10,000 miles of driving. Random samples of tires made from each rubber compound are obtained from the production line. The tires are mounted on four cars according to the scheme shown in Table 7.3-1. The cars are designated by c_1, c_2, c_3, and c_4 and the wheel positions by b_1, right front, b_2, left front, b_3, right rear, and b_4, left rear. The rubber compounds and wheel positions represent fixed effects, the automobiles are regarded as random effects. The road test is repeated with a second random sample of drivers using the scheme in Table 7.3-1. Thus there are two observations in each cell. An alternative analysis described in Section 13.9 permits the isolation of variation attributable to the repetition of the road test.

An examination of Table 7.3-1 shows that each kind of tire is used equally often on each car and appears the same number of times at each wheel position. The type LS-4 design enables us to isolate the two nuisance variables of differences among automobiles and wheel positions while evaluating the effects of the rubber compounds. The Latin square in Table 7.3-1 is identical to the 4×4 square obtained by the randomization procedure illustrated in Section 7.2. To maintain a consistent notation, the treatment levels A, B, C, and D are denoted by a_1, a_2, a_3, and a_4. Treatment effects are denoted by $\alpha_j = \mu_{\cdot j \cdot \cdot} - \mu$, row effects by $\beta_k = \mu_{\cdot \cdot k \cdot} - \mu$, and column effects by $\gamma_l = \mu_{\cdot \cdot \cdot l} - \mu$, where μ is the grand mean.

The following statistical hypotheses are advanced:

$$H_0: \quad \alpha_j = 0 \quad \text{for all } j$$
$$H_1: \quad \alpha_j \neq 0 \quad \text{for some } j.$$

The level of significance adopted is .05. The data and computational procedures are given in Table 7.3-2. The results of the analysis are summarized in Table 7.3-3. The first test that should be performed is a test of the residual mean square, $F = MSRES/MSWCELL$. If this test is significant, it indicates that $MSRES$ and MSA or MSB or MSC or possibly all three of the latter mean squares estimate, in addition to main effects, one or more interaction components.* As a result, a test of A, B, or C, or a combination of these is positively biased. A test is said to be positively biased

---

TABLE 7.3-1   Randomization Plan for Tire Test

			Automobile			
			c_4	c_3	c_1	c_2
	R rear	b_3	a_2	a_1	a_3	a_4
Wheel	R front	b_1	a_4	a_3	a_1	a_2
Position	L front	b_2	a_1	a_4	a_2	a_3
	L rear	b_4	a_3	a_2	a_4	a_1

a_1, a_2, a_3, a_4 denote four kinds of rubber compounds used in the construction of the tires

---

*   Interaction components are discussed in Section 12.8.

if it yields too many significant results. If a test, say, of treatment A is significant, we have no way of knowing if this is due to treatment A or to a component of the BC or ABC interaction. Since the test of $MSRES$ in Table 7.3-3 is not significant, it is reasonable to assume that the rows, columns, and treatment levels do not interact and hence that tests of A, B, and C are not positively biased. We can conclude from the analysis in Table 7.3-3 that the treatment effects associated with the rubber compounds are not all equal to zero. Procedures for testing hypotheses concerning contrasts among the population means are described in Section 7.6.

When n is equal to or greater than 2 and the assumption of no interactions among treatment levels, rows, and columns is tenable, the analysis provides two

---

TABLE 7.3-2   Computational Procedures for a Type LS-4 Design

---

(i)   Data and notation [Y_{ijkl} denotes a score for the ith experimental unit in row k and column l; $i = 1, \ldots, n$ experimental units (s_i); $j = 1, \ldots, p$ treatment levels (a_j); $k = 1, \ldots, p$ rows (b_k); $l = 1, \ldots, p$ columns (c_l)]:

ABCS Summary Table

Entry is Y_{ijkl}

	c_1	c_2	c_3	c_4
b_1	a_1	a_2	a_3	a_4
	1	2	5	9
	2	3	6	8
b_2	a_2	a_3	a_4	a_1
	3	8	9	2
	4	6	8	3
b_3	a_3	a_4	a_1	a_2
	5	10	3	5
	7	11	2	4
b_4	a_4	a_1	a_2	a_3
	7	6	3	6
	10	3	4	7

ABC Summary Table

Entry is $\displaystyle\sum_{i=1}^{n} Y_{ijkl}$

	c_1	c_2	c_3	c_4	$\displaystyle\sum_{i=1}^{n}\sum_{l=1}^{p} Y_{ijkl}$
	$n = 2$				
b_1	a_1	a_2	a_3	a_4	
	3	5	11	17	36
b_2	a_2	a_3	a_4	a_1	
	7	14	17	5	43
b_3	a_3	a_4	a_1	a_2	
	12	21	5	9	47
b_4	a_4	a_1	a_2	a_3	
	17	9	7	13	46
$\displaystyle\sum_{i=1}^{n}\sum_{k=1}^{p} Y_{ijkl} =$	39	49	40	44	

A Summary Table

Entry is $\displaystyle\sum_{i=1}^{n}\sum_{k=1}^{p}\sum_{l=1}^{p} Y_{ijkl}$ for a_j

a_1	a_2	a_3	a_4
$np = 8$			
22	28	50	72

TABLE 7.3-2 (continued)

(ii) Computational symbols:

$$\sum_{i=1}^{n} \sum_{k=1}^{p} \sum_{l=1}^{p} Y_{ijkl} = 1 + 2 + \cdots + 7 = 172$$

$$\frac{\left(\sum_{i=1}^{n} \sum_{k=1}^{p} \sum_{l=1}^{p} Y_{ijkl}\right)^2}{np^2} = [Y] = \frac{(172)^2}{(2)(4)^2} = 924.500$$

$$\sum_{i=1}^{n} \sum_{k=1}^{p} \sum_{l=1}^{p} Y_{ijkl}^2 = [ABCS] = (1)^2 + (2)^2 + \cdots + (7)^2 = 1160.000$$

$$\sum_{k=1}^{p} \sum_{l=1}^{p} \frac{\left(\sum_{i=1}^{n} Y_{ijkl}\right)^2}{n} = [ABC] = \frac{(3)^2}{2} + \frac{(7)^2}{2} + \cdots + \frac{(13)^2}{2} = 1141.000$$

$$\sum_{j=1}^{p} \frac{\left(\sum_{i=1}^{n} \sum_{k=1}^{p} \sum_{l=1}^{p} Y_{ijkl}\right)^2}{np} = [A] = \frac{(22)^2}{(2)(4)} + \frac{(28)^2}{(2)(4)} + \cdots + \frac{(72)^2}{(2)(4)} = 1119.000$$

$$\sum_{k=1}^{p} \frac{\left(\sum_{i=1}^{n} \sum_{l=1}^{p} Y_{ijkl}\right)^2}{np} = [B] = \frac{(36)^2}{(2)(4)} + \frac{(43)^2}{(2)(4)} + \cdots + \frac{(46)^2}{(2)(4)} = 933.750$$

$$\sum_{l=1}^{p} \frac{\left(\sum_{i=1}^{n} \sum_{k=1}^{p} Y_{ijkl}\right)^2}{np} = [C] = \frac{(39)^2}{(2)(4)} + \frac{(49)^2}{(2)(4)} + \cdots + \frac{(44)^2}{(2)(4)} = 932.250$$

(iii) Computational formulas:

$$SSTO = [ABCS] - [Y] = 235.500$$
$$SSA = [A] - [Y] = 194.500$$
$$SSB = [B] - [Y] = 9.250$$
$$SSC = [C] - [Y] = 7.750$$
$$SSRES = [ABC] - [A] - [B] - [C] + 2[Y] = 5.000$$
$$SSWCELL = [ABCS] - [ABC] = 19.000$$

independent estimates of experimental error: *MSWCELL* and *MSRES*. Most writers recommend pooling these two mean squares to obtain a better estimate of the population error variance as follows:

$$MSRES_{\text{pooled}} = \frac{SSRES + SSWCELL}{(p - 1)(p - 2) + p^2(n - 1)} = \frac{5.000 + 19.000}{6 + 16} = 1.091.$$

This pooled mean square with 22 degrees of freedom replaces *MSWCELL* in computing the *F* statistics. The use of a pooled error estimate may be desirable when, as in our example, the within cell mean square is based on a small number of degrees

TABLE 7.3-3 ANOVA Table for Type LS-4 Design

Source	SS	df	MS	F	$E(MS)$ Model III
1 Rubber compounds (A)	194.500	$p - 1 = 3$	64.833	[1/5] 54.57*	$\sigma_\epsilon^2 + np \sum_{j=1}^{p} \alpha_j^2/(p - 1)$
2 Wheel positions (B)	9.250	$p - 1 = 3$	3.083	[2/5] 2.60	$\sigma_\epsilon^2 + np \sum_{k=1}^{p} \beta_k^2/(p - 1)$
3 Automobiles (C)	7.750	$p - 1 = 3$	2.583	[3/5] 2.17	$\sigma_\epsilon^2 + np\sigma_\gamma^2$
4 Residual	5.000	$(p - 1)(p - 2) = 6$	0.833	[4/5] 0.70	σ_ϵ^2
5 Within cell	19.000	$p^2(n - 1) = 16$	1.188		σ_ϵ^2
6 Total	235.500	$np^2 - 1 = 31$			

*$p < .01$

of freedom and the no interaction assumption is tenable. However, in our example the use of $MSRES_{pooled}$ would not have altered the outcome of the analysis. Issues surrounding the use of pooled error terms are discussed in Section 8.11.

POWER OF THE ANALYSIS OF VARIANCE *F* TEST

Procedures for calculating power and determining the number of subjects necessary to achieve a specified power are described in Section 4.3. These procedures generalize to a Latin square design. However, instead of entering Table E.14 with $\nu_2 = p(n - 1)$, the degrees of freedom are equal to those of the mean square error term.

ESTIMATES OF STRENGTH OF ASSOCIATION

Procedures for estimating strength of association, the proportion of the total variance accounted for by the treatment variable or by one of the nuisance variables, are described in Sections 4.6 and 6.2. These procedures will be illustrated for $\hat{\omega}^2_{Y|A}$. Recall that the rubber compounds and wheel positions represent fixed effects while the automobiles are regarded as random effects. Furthermore, we assume that treatment levels, rows, and columns do not interact. The formula for $\hat{\omega}^2_{Y|A}$ following the development in Sections 4.6 and 6.2 is

$$\hat{\omega}^2_{Y|A} = \frac{\sum_{j=1}^{p} \hat{\alpha}_j^2/p}{\hat{\sigma}_\epsilon^2 + \hat{\sigma}_\gamma^2 + \sum_{k=1}^{p} \hat{\beta}_k^2/p + \sum_{j=1}^{p} \hat{\alpha}_j^2/p}$$

where

$$\hat{\sigma}_\epsilon^2 = MSWCELL = 1.188$$
$$\hat{\sigma}_\gamma^2 = (MSC - MSWCELL)/np = (2.583 - 1.188)/(2)(4)$$
$$= 0.174$$
$$\sum_{k=1}^{p} \hat{\beta}_k^2/p = (MSB - MSWCELL)[(p - 1)/np^2]$$
$$= (3.083 - 1.188)[(4 - 1)/(2)(4)^2]$$
$$= 0.178$$
$$\sum_{j=1}^{p} \hat{\alpha}_j^2/p = (MSA - MSWCELL)[(p - 1)/np^2]$$
$$= (64.833 - 1.188)[(4 - 1)/(2)(4)^2]$$
$$= 5.967.$$

Substituting in the formula gives

$$\hat{\omega}^2_{Y|A} = \frac{5.967}{1.188 + 0.174 + 0.178 + 5.967} = 0.79.$$

This value of $\hat{\omega}^2$ is unusually high; it is rare to find treatments that account for more than 30% to 35% of the total variance.

7.4   COMPUTATIONAL PROCEDURES FOR $n = 1$

If the road test had not been repeated, there would be no within-cell error term. In this situation *MSRES* can be used as an error term for testing the treatment and nuisance effects if it can be assumed that *MSRES* estimates only experimental error. To state it another way, *MSRES* is an appropriate error term if and only if all of the interactions are equal to zero (i.e., $\sigma_{\alpha\beta}^2 = \sigma_{\alpha\gamma}^2 = \sigma_{\beta\gamma}^2 = \sigma_{\alpha\beta\gamma}^2 = 0$). Tukey's test for nonadditivity described in the next section provides a partial test of this assumption.

Computational procedures for the case where $n = 1$ are very similar to those shown in Table 7.3-2. However, there are several important differences: (1) a score is denoted by Y_{jkl}, (2) there is no *ABCS* Summary Table, [*ABCS*] term, and *MSWCELL* term, (3) $SSTO = [ABC] - [Y]$ and, (4) the summary tables and computational symbols should be modified by deleting n and $\sum_{i=1}^{n}$ wherever they appear.

In Section 7.1 it was observed that a Latin square design is not practical for $p < 5$ unless more than one observation per cell is obtained. This follows because, for 2×2, 3×3, and 4×4 Latin squares, the *MSRES* has only 1, 2, and 6 degrees of freedom, respectively. A 5×5 square has 12 degrees of freedom. Although marginal, this number of degrees of freedom is acceptable.

TUKEY'S TEST FOR NONADDITIVITY

As we have seen, if $n = 1$, there is no within-cell error term. In order to use *MSRES* as an error term, all interactions among rows, columns, and treatment levels must equal zero. A partial test of the additivity assumption when $n = 1$ can be made by means of Tukey's (1955) test for nonadditivity, which was described in Section 6.3 in connection with a type RB-*p* design. For purposes of illustration, assume that the data in the *ABC* Summary Table of Table 7.3-2 are based on n equal to one experimental unit in each cell. Computational procedures for Tukey's test are shown in Table 7.4-1.

It is apparent from the analysis that the assumption of addivity is tenable. A relatively low level of significance ($\alpha = .10$) was adopted for the *FNONADD* test. The selection of this numerically large probability value reflects a desire to increase the power of the test. A plot of the residual for each observation, $\hat{\epsilon}_{jkl} = Y_{jkl} - \overline{Y}_{j..} - \overline{Y}_{.k.} - \overline{Y}_{..l} + 2\overline{Y}_{...}$, against the estimated value, $\hat{Y}_{jkl} = \overline{Y}_{...} + (\overline{Y}_{j..} - \overline{Y}_{...}) + (\overline{Y}_{.k.} - \overline{Y}_{...}) + (\overline{Y}_{..l} - \overline{Y}_{...})$, is shown in Figure 7.4-1. The absence of an association between $\hat{\epsilon}_{jkl}$ and $\hat{Y}_{jkl}$ supports the tenability of the additive model.

If the test of nonadditivity is significant, it means that *MSRES* estimates components of the two and three factor interactions and that combinations of these components appear in *MSA*, *MSB*, and *MSC*. Under these conditions, tests of main and nuisance effects are biased in a complicated manner. For a discussion of this point see Scheffé (1959, 154–158), Wilk and Kempthorne (1957), and Section 13.7. If

$F = MSRES /MSWCELL$ or $FNONADD$ is significant, a transformation may be found that will produce additivity.

FIGURE 7.4-1   Plot of residual, $\hat{\epsilon}_{jkl} = Y_{jkl} - \overline{Y}_{\cdot\cdot\cdot} - d_j - d_k - d_l$, versus estimated value, $\hat{Y}_{jkl} = \overline{Y}_{\cdot\cdot\cdot} + d_j + d_k + d_l$, where $d_j = \overline{Y}_{j\cdot\cdot} - \overline{Y}_{\cdot\cdot\cdot}$, $d_k = \overline{Y}_{\cdot k\cdot} - \overline{Y}_{\cdot\cdot\cdot}$, and $d_l = \overline{Y}_{\cdot\cdot l} - \overline{Y}_{\cdot\cdot\cdot}$. The absence of an association between the size of $\hat{\epsilon}_{jkl}$ and Y_{jkl} suggests that the additive model is appropriate.

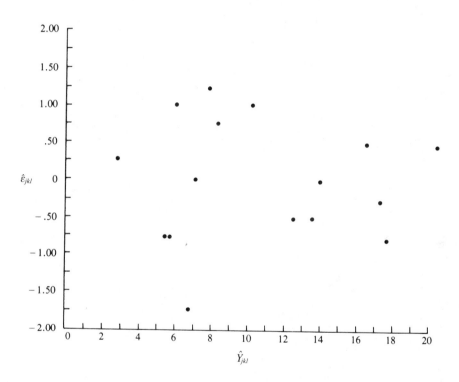

7.5   EXPERIMENTAL DESIGN MODEL FOR A TYPE LS-p DESIGN

The experimental design model for the experiment described in Section 7.3 is

(7.5-1) $Y_{ijkl} = \mu + \alpha_j + \beta_k + \gamma_l + \epsilon_{jkl} + \epsilon_{i(jkl)}$
$(i = 1, \ldots, n; j = 1, \ldots, p; k = 1, \ldots, p; l = 1, \ldots, p)$

TABLE 7.4-1 Computational Procedures for Test for Nonadditivity

(i) Data and notation [Y_{jkl} denotes a score for the experimental unit in row k and column l; $j = 1, \ldots,$
p treatment levels (a_j); $k = 1, \ldots, p$ rows (b_k), $l = 1, \ldots, p$ columns (c_l)]:

ABC Summary Table
Entry is Y_{jkl}

	c_1	c_2	c_3	c_4	$\overline{Y}_{.k.}$	$d_k = \overline{Y}_{.k.} - \overline{Y}_{...}$
b_1	a_1	a_2	a_3	a_4	9.00	-1.75
	3	5	11	17		
b_2	a_2	a_3	a_4	a_1	10.75	0
	7	14	17	5		
b_3	a_3	a_4	a_1	a_2	11.75	1.00
	12	21	5	9		
b_4	a_4	a_1	a_2	a_3	11.50	0.75
	17	9	7	13		

$$\overline{Y}_{..l} = \quad 9.75 \quad 12.25 \quad 10.00 \quad 11.00 \qquad \overline{Y}_{...} = \sum_{k=1}^{p} \sum_{l=1}^{p} Y_{jkl}/p^2$$

$$d_l = \overline{Y}_{..l} - \overline{Y}_{...} = -1.00 \quad 1.50 \quad -0.75 \quad 0.25 \qquad = 172/(4)^2 = 10.75$$

A Summary Table
Entry is $\sum_{k=1}^{p} \sum_{l=1}^{p} Y_{jkl}$ for a_j

	a_1	a_2	a_3	a_4
$p = 4$	22	28	50	72
$\overline{Y}_{j..} =$	5.50	7.00	12.50	18.00
$d_j = \overline{Y}_{j..} - \overline{Y}_{...} =$	-5.25	-3.75	1.75	7.25

$\hat{Y}_{jkl}$ *Table*
Entry is $\hat{Y}_{jkl} = \overline{Y}_{...} + d_j + d_k + d_l$
$e.g.$, $\hat{Y}_{111} = 10.75 + (-5.25) + (-1.75) + (-1.00)$
$\qquad = 2.75$

$\hat{e}_{jkl}$ *Table*
Entry is $\hat{e}_{jkl} = Y_{jkl} - \hat{Y}_{jkl}$
$e.g.$, $\hat{e}_{111} = 3 - 2.75$
$\qquad = 0.25$

	c_1	c_2	c_3	c_4			c_1	c_2	c_3	c_4
b_1	2.75	6.75	10.00	16.50		b_1	.25	-1.75	1.00	.50
b_2	6.00	14.00	17.25	5.75		b_2	1.00	0	$-.25$	$-.75$
b_3	12.50	20.50	5.75	8.25		b_3	$-.50$	.50	$-.75$	.75
b_4	17.75	7.75	7.00	13.50		b_4	$-.75$	1.25	0	$-.50$

Computational check: $\sum_{k=1}^{p} \sum_{l=1}^{p} \hat{Y}_{jkl}/p^2 = \overline{Y}_{...} = 10.75$

Computational check: The sum of entries in each row and column should equal zero.
Computational check:

$$\sum_{k=1}^{p} \sum_{l=1}^{p} \hat{e}_{jkl}^2 = SSRES = 10.00$$

TABLE 7.4-1 (continued)

R^2 Table

Entry is $R_{jkl}^2 = (\hat{Y}_{jkl} - \overline{Y}...)^2$

	c_1	c_2	c_3	c_4
b_1	64.0000	16.0000	.5625	33.0625
b_2	22.5625	10.5625	42.2500	25.0000
b_3	3.0625	95.0625	25.0000	6.2500
b_4	49.0000	9.0000	14.0625	7.5625

(ii) Computational formulas:

$$SSNONADD = \frac{\left(\sum_{k=1}^{p}\sum_{l=1}^{p}\hat{\epsilon}_{jkl}R_{jkl}^2\right)^2}{SSRES(R^2 \text{ Table})} = \frac{[(0.25)64.0000 + (1.00)(22.5625) + \cdots + (-0.50)\,7.5625]^2}{2937.8204}$$

$$= \frac{(1.0000)^2}{2937.8204} = .0003$$

[$SSRES(R^2$ Table) is computed from R^2 Table by the formula $[ABC] - [A] - [B] - [C] + 2[Y] = 20923.4063 - 16798.4297 - 12037.3281 - 11515.9531 + 2(11183.0625) = 2937.8204$.]

$dfNONADD = 1$

$SSREM = SSRES(ABC \text{ Table}) - SSNONADD = 10.000 - 0.0003 = 9.9997$

$dfREM = (p - 1)(p - 2) - 1 = (3)(2) - 1 = 5$

$$FNONADD = \frac{SSNONADD/dfNONADD}{SSREM/dfREM} = \frac{0.0003/1}{9.9997/5} = \frac{0.0003}{1.9999} = 0.0002$$

$F_{.10;1,5} = 4.06$

where Y_{ijkl} = a score for the ith experimental unit in row k and column l. The subscript j is redundant since the treatment level is known once k and l and the Latin square are specified

μ = the overall population mean

α_j = the effect of treatment level j and is subject to the restriction $\sum_{j=1}^{p}\alpha_j = 0$

β_k = the effect of row k and is subject to the restriction $\sum_{k=1}^{p}\beta_k = 0$

γ_l = the effect of column l that is $NID(0, \sigma_\gamma^2)$

ϵ_{jkl} = the residual error that represents the effect due to interactions among the treatment levels, rows, and columns. If all interactions are equal to zero, ϵ_{jkl} is $NID(0, \sigma_\epsilon^2)$ and independent of γ_l and $\epsilon_{i(jkl)}$

$\epsilon_{i(jkl)}$ = the within cell experimental error that is $NID(0, \sigma_\epsilon^2)$; $\epsilon_{i(jkl)}$ is independent of γ_l.

This is a mixed model since the automobiles (columns) are regarded as random effects. If we repeated the experiment, we undoubtedly would obtain a new sample of automobiles. The expected values of the mean squares for this model are given in Table 7.3-3. Expected values for other models are readily obtained by replacing a variance term by a sum of squared effects divided by its degrees of freedom when effects are fixed and vice versa when they are random. For example, if column effects in Table 7.3-3 were regarded as fixed, $np\sigma_\gamma^2$ would be replaced by $np\sum_{l=1}^{p} \gamma_l^2/(p-1)$ and subject to the restriction that $\sum_{l=1}^{p} \gamma_l = 0$.

The partition of the total sum of squares, which is the basis for the analysis in Table 7.3-3, is obtained by replacing the parameters in model (7.5-1) by statistics and rearranging terms as follows.

(7.5-2)

$$Y_{ijkl} - \mu = \alpha_j + \beta_k + \gamma_l$$

$$Y_{ijkl} - \overline{Y}_{\cdots} = (\overline{Y}_{\cdot j \cdot\cdot} - \overline{Y}_{\cdots}) + (\overline{Y}_{\cdot\cdot k\cdot} - \overline{Y}_{\cdots}) + (\overline{Y}_{\cdots l} - \overline{Y}_{\cdots})$$

(Total deviation)   (Treatment effect)   (Row effect)   (Column effect)

$$+ \qquad \epsilon_{jkl} \qquad + \qquad \epsilon_{i(jkl)}$$

$$+ (\overline{Y}_{\cdot jkl} - \overline{Y}_{\cdot j\cdot\cdot} - \overline{Y}_{\cdot\cdot k\cdot} - \overline{Y}_{\cdots l} + 2\overline{Y}_{\cdots}) + (Y_{ijkl} - \overline{Y}_{\cdot jkl})$$

(Residual effect)   (Within-cell effect)

Squaring both sides of (7.5-2) and summing over all of the observations following the procedures in Section 2.2 and 6.1 leads ultimately to the computational formulas in Table 7.3-2.* The derivation of the expected values of the mean squares follows the procedures illustrated in Section 2.3 for a completely randomized design.

7.6   PROCEDURES FOR TESTING DIFFERENCES AMONG MEANS

Tests of differences among means in a type LS-p design have the same general form as those illustrated in Chapter 3 and Sections 4.4 and 6.5. The denominator of the test statistics $\hat{\psi}_i/\hat{\sigma}_{\psi_i}$ and $\hat{\psi}_i/\hat{\sigma}_{\overline{Y}}$ is given by

---

*   See Exercise 10 for the derivation.

$$\hat{\sigma}_{\psi_i} = \sqrt{MSWCELL\left(\frac{c_1^2}{np} + \frac{c_2^2}{np} + \cdots + \frac{c_p^2}{np}\right)}$$

and

$$\hat{\sigma}_{\overline{Y}} = \sqrt{\frac{MSWCELL}{np}}.$$

A pooled mean square, $MSRES_{pooled}$, should be used in place of $MSWCELL$ if the pooled term was used in testing treatment effects and nuisance variables. If n is equal to one, the denominators of the test statistics become

$$\hat{\sigma}_{\psi_i} = \sqrt{MSRES\left(\frac{c_1^2}{p} + \frac{c_2^2}{p} + \cdots + \frac{c_p^2}{p}\right)} \quad \text{and} \quad \hat{\sigma}_{\overline{Y}} = \sqrt{\frac{MSRES}{p}}.$$

7.7   ESTIMATING MISSING OBSERVATIONS

Despite an experimenter's best efforts, occasionally one or more observations are missing. This presents a problem because the analysis procedures in Section 7.3 require that each treatment level appear n times in each row and column. To carry out the statistical analysis, it is necessary to replace the missing data or use an analysis procedure that is appropriate for unequal n's. If each cell of the data matrix is supposed to contain one observation, a formula described by Allan and Wishart (1930) and Yates (1933) can be used to estimate the missing observation(s). If n is supposed to be greater than one, analysis procedures described in Section 7.9 can be used.

The procedure for the case in which each cell is supposed to contain one observation is described next. It involves substituting for the missing observation one that minimizes the residual error sum of squares. The same procedure was used for a randomized block design (Section 6.7). The following notation is used.

X_{jkl} = missing observation

$\sum\limits_{j=1}^{p'} \sum\limits_{l=1}^{p'} Y_{jkl}$ for a_j = sum of remaining observations in treatment level containing X_{jkl}

$\sum\limits_{k=1}^{p'} Y_{jkl}$ = sum of remaining observations in column containing X_{jkl}

$\sum\limits_{l=1}^{p'} Y_{jkl}$ = sum of remaining observations in row containing X_{jkl}

$\sum\limits_{k=1}^{p'} \sum\limits_{l=1}^{p'} Y_{jkl}$ = sum of all remaining observations

The value of X_{jkl} that minimizes the sample residual error sum of squares $\sum_{k=1}^{P} \sum_{l=1}^{P} \hat{\epsilon}_{jkl}^2 = \sum_{k=1}^{P} \sum_{l=1}^{P} (Y_{jkl} - \overline{Y}_{j\cdot\cdot} - \overline{Y}_{\cdot k\cdot} - \overline{Y}_{\cdot\cdot l} + 2\overline{Y}_{\cdots})^2$ is the one for which

(7.7-1)
$$\hat{\epsilon}_{jkl} = X_{jkl} - \overline{Y}_{j\cdot\cdot} - \overline{Y}_{\cdot k\cdot} - \overline{Y}_{\cdot\cdot l} + 2\overline{Y}_{\cdots} = 0.$$

The terms $\overline{Y}_{j\cdot\cdot}$, $\overline{Y}_{\cdot k\cdot}$, $\overline{Y}_{\cdot\cdot l}$, and $\overline{Y}_{\cdots}$ can be expressed as

$$\overline{Y}_{j\cdot\cdot} = \left(\sum_{k=1}^{p'} \sum_{l=1}^{p'} Y_{jkl} \text{ for } a_j + X_{jkl} \right) \Big/ p$$

$$\overline{Y}_{\cdot k\cdot} = \left(\sum_{l=1}^{p'} Y_{jkl} + X_{jkl} \right) \Big/ p$$

$$\overline{Y}_{\cdot\cdot l} = \left(\sum_{k=1}^{p'} Y_{jkl} + X_{jkl} \right) \Big/ p$$

$$\overline{Y}_{\cdots} = \left(\sum_{k=1}^{p'} \sum_{l=1}^{p'} Y_{jkl} + X_{jkl} \right) \Big/ p^2.$$

Substituting in (7.7-1) gives

$$X_{jkl} - \frac{\sum_{k=1}^{p'} \sum_{l=1}^{p'} Y_{jkl} \text{ for } a_j + X_{jkl}}{p} - \frac{\sum_{l=1}^{p'} Y_{jkl} + X_{jkl}}{p} - \frac{\sum_{k=1}^{p'} Y_{jkl} + X_{jkl}}{p}$$

$$+ \frac{2\left(\sum_{k=1}^{p'} \sum_{l=1}^{p'} Y_{jkl} + X_{jkl} \right)}{p^2} = 0.$$

Solving for X_{jkl} we obtain the formula for a missing observation.

$$X_{jkl} - \frac{X_{jkl}}{\cdot\, p} - \frac{X_{jkl}}{p} - \frac{X_{jkl}}{p} + \frac{2X_{jkl}}{p^2}$$

$$= \frac{\sum_{k=1}^{p'} \sum_{l=1}^{p'} Y_{jkl} \text{ for } a_j}{p} + \frac{\sum_{l=1}^{p'} Y_{jkl}}{p} + \frac{\sum_{k=1}^{p'} Y_{jkl}}{p} - \frac{2\sum_{k=1}^{p'} \sum_{l=1}^{p'} Y_{jkl}}{p^2}$$

$$X_{jkl} = \frac{p\left(\sum_{k=1}^{p'} \sum_{l=1}^{p'} Y_{jkl} \text{ for } a_j + \sum_{k=1}^{p'} Y_{jkl} + \sum_{l=1}^{p'} Y_{jkl} \right) - 2\sum_{k=1}^{p'} \sum_{l=1}^{p'} Y_{jkl}}{(p-1)(p-2)}$$

If more than one observation is missing, an experimenter can *guesstimate* the values for all but one of the missing observations; this one is computed by the preceding formula. The computed value is then inserted in the data matrix and another of the missing observations that was guesstimated is computed. This iterative process is repeated until the computed values for all missing observations have stabilized. Generally three cycles of the procedure are sufficient for this purpose. For each estimated missing observation, one degree of freedom should be subtracted from the df for SSRES and SSTO.

The preceding procedure produces a slight positive bias in the F tests. An unbiased test for treatment effects has been described by Kempthorne (1952, 198) and is as follows. First, compute *SSRES* for the Latin square with the estimated missing values in the data matrix. Second, treat the data as a randomized block design, ignoring treatment A, and estimate missing values by the formula in Section 6.7 appropriate for this design. Third, compute *SSRES* for the randomized block design with the second set of estimated missing observations inserted in the data matrix. The corrected sum of squares for treatment A in the Latin square design is given by

$$SSA\,(corrected) = SSRES\,(RB\text{-}p) - SSRES\,(LS\text{-}p).$$

As noted previously, the degrees of freedom for *SSRES* must be reduced by the number of missing observations. When only one observation is missing, a simpler procedure for computing a corrected treatment sum of squares uses the following formula.

$SSA\,(corrected)$

$$= SSA - \frac{\left[\sum_{k=1}^{p'} \sum_{l=1}^{p'} Y_{jkl} - \sum_{k=1}^{p'} Y_{jkl} - \sum_{l=1}^{p'} Y_{jkl} - (p-1) \sum_{k=1}^{p'} \sum_{l=1}^{p'} Y_{jkl} \text{ for } a_j \right]^2}{[(p-1)(p-2)]^2}$$

A t statistic for contrasts involving a treatment level with one missing observation is given by

$$t = \frac{c_j \overline{Y}_{j..} + c_{j'} \overline{Y}_{j'..}}{\sqrt{MSRES \left[\dfrac{2}{p} + \dfrac{1}{(p-1)(p-2)} \right]}}.$$

If more than one observation has been estimated, the t statistic is given by

$$t = \frac{c_j \overline{Y}_{j..} + c_{j'} \overline{Y}_{j'..}}{\sqrt{MSRES \left[\dfrac{1}{e_j} + \dfrac{1}{e_{j'}} \right]}}$$

where e_j and $e_{j'}$ correspond to the number of *effective replications* in the jth and j'th treatment levels. An approximate rule proposed by Yates (1933) assigns a value to e_j and $e_{j'}$ for the jth and j'th treatment levels, respectively, according to the following scheme.

1. Assign 1 if the observation under consideration is present and the other observation is present in both the corresponding row and column.

2. Assign 2/3 if the observation under consideration is present and the other observation is missing in the corresponding row or column.

3. Assign 1/3 if the observation under consideration is present and the other observation is missing in both the corresponding row and column.

4. Assign 0 if the observation under consideration is missing.

The values of e_j and $e_{j'}$ are the sums of the values assigned for each of the j and j' cells, respectively. The application of similar rules is illustrated with a numerical example for a randomized block design in Section 6.7.

Estimation procedures when one or more rows, columns, or treatment levels are missing are discussed by DeLury (1946), Kempthorne (1952, 199–201), Yates (1936), and Yates and Hale (1939).

7.8 RELATIVE EFFICIENCY OF LATIN SQUARE DESIGN WITH $n = 1$

A Latin square design enables an experimenter to minimize the effects of two nuisance variables in evaluating treatment effects. If the row and column effects are appreciably greater than zero, a Latin square design is more powerful than a completely randomized or randomized block design. We will now see how to estimate the efficiency of a Latin square design relative to these designs.

RELATIVE EFFICIENCY OF LATIN SQUARE AND COMPLETELY RANDOMIZED DESIGNS

The relative efficiency of the type CR-p and LS-p designs, ignoring differences in degrees of freedom, is given by

(7.8-1)
$$\text{Relative efficiency} = \frac{MSWG}{MSRES}$$

where $MSWG$ for a completely randomized design is obtained from a Latin square analysis by

$$MSWG = \frac{MSB + MSC + (p - 1)MSRES}{(p + 1)}.$$

If the data in Table 7.3-2 were based on $n = 1$, the required mean squares for B, C, and residual would be, respectively, 6.167, 5.167, and 1.667. The value of $MSWG$ required in (7.8-1) is

$$MSWG = \frac{6.167 + 5.167 + (4 - 1)1.667}{4 + 1} = 3.267.$$

The efficiency of the Latin square design relative to that of a completely randomized design is

$$\text{Relative efficiency} = \frac{3.267}{1.667} = 1.96.$$

This indicates that if the blocking variables had not been used and treatment levels were assigned randomly to $np = 16$ experimental units in a completely randomized design, the resulting error variance would be 1.96 times as large as that of the Latin square design. A correction developed by Fisher (1935) for the smaller error degrees of freedom for the Latin square design can be incorporated in the efficiency index as follows.

Relative efficiency

$$= \left(\frac{df\,RES_{LS} + 1}{df\,RES_{LS} + 3}\right)\left(\frac{df\,WG_{CR} + 3}{df\,WG_{CR} + 1}\right)\frac{MSWG}{MSRES} = (0.90)(1.96) = 1.76.$$

RELATIVE EFFICIENCY OF LATIN SQUARE AND RANDOMIZED BLOCK DESIGNS

The efficiency of a Latin square design relative to a randomized block design may be estimated in two ways depending on whether the rows or columns of the squares are considered as replicates. In the present experiment, the column variable is automobiles and would be considered replications corresponding to blocks in a randomized block design. The relative efficiency of the two designs, considering columns as replications, is given by

$$\text{Relative efficiency} = \left(\frac{df\,RES_{LS} + 1}{df\,RES_{LS} + 3}\right)\left(\frac{(p - 1)^2 + 3}{(p - 1)^2 + 1}\right)\frac{MSRES_{RB}}{MSRES_{LS}}$$

where $MSRES_{RB}$ for a randomized block design is obtained from a Latin square analysis by

$$MSRES_{RB} = \frac{MSB + (p - 1)MSRES_{LS}}{p}$$

Again, considering the data in Table 7.3-2 to be based on $n = 1$, the relative efficiency is

$$\text{Relative efficiency} = \left(\frac{6 + 1}{6 + 3}\right)\left(\frac{(4 - 1)^2 + 3}{(4 - 1)^2 + 1}\right)\frac{2.792}{1.667} = (0.933)(1.675) = 1.56$$

where $MSRES_{RB} = [6.167 + (4 - 1)1.667]/4 = 2.792$. This estimate of the relative efficiency of the LS-p and RB-p designs includes a correction (0.933) for the smaller error degrees of freedom for the Latin square design.

If rows of the Latin square are considered as replications, $MSRES_{RB}$ is estimated by

$$MSRES_{RB} = \frac{MSC + (p - 1)MSRES_{LS}}{p} = \frac{5.167 + (4 - 1)1.667}{4} = 2.542.$$

The efficiency of a Latin square design relative to a randomized block design, considering rows as replications, is

$$\text{Relative efficiency} = \left(\frac{dfRES_{LS} + 1}{dfRES_{LS} + 3}\right)\left(\frac{(p-1)^2 + 3}{(p-1)^2 + 1}\right)\frac{MSRES_{RB}}{MSRES_{LS}}$$

$$= \left(\frac{6+1}{6+3}\right)\left(\frac{(4-1)^2 + 3}{(4-1)^2 + 1}\right)\frac{2.542}{1.667} = (0.933)(1.525) = 1.42.$$

SCHEMATIC COMPARISON OF TOTAL VARIATION FOR THREE DESIGNS

At the beginning of this section it was pointed out that a Latin square design is more powerful than either a completely randomized or randomized block design if the row and column effects in the Latin square design are appreciably greater than zero. This point can be illustrated by a schematic comparison of the total variation and degrees of freedom for the three designs.

Assume that an experiment has four treatment levels with four observations in each level. A partition of the total sum of squares and degrees of freedom for the three designs is presented in Figure 7.8-1. It is evident from this figure that $SSRES_{LS}$ will be smaller than $SSWG$ if either row or column effects are not equal to zero. Also, $SSRES_{LS}$ will be smaller than $SSRES_{RB}$ if column effects are not equal to zero. If column effects are equal to zero, $SSRES_{LS}$ is equal to $SSRES_{RB}$; and if row effects are also equal to zero, $SSRES_{LS}$ is equal to $SSWG$. Under these conditions, the completely randomized and randomized block designs are more efficient than a Latin square design because the error terms are based on more degrees of freedom.

FIGURE 7.8-1   Schematic comparison of total variation and degrees of freedom for three experimental designs. A shaded square identifies the variation used to estimate experimental error.

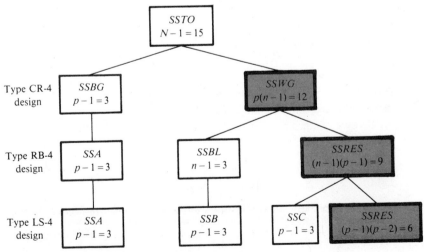

*7.9 GENERAL LINEAR MODEL APPROACH TO TYPE LS-*p* DESIGN

The general linear model approach introduced in Section 5.10 can be used to analyze data for a Latin square design. Two cases need to be distinguished: the restricted and unrestricted models. If each cell of a Latin square contains only one observation, a restricted full rank experimental design model should be used. If each cell contains two or more observations, either a restricted or an unrestricted model can be used. We will examine the unrestricted model first.

UNRESTRICTED FULL RANK MODEL

The unrestricted full rank experimental design model for a type LS-*p* design is

$$Y_{ijkl} = \mu_{jkl} + \epsilon_{i(jkl)} \qquad (i = 1, \ldots, n; j = 1, \ldots, p; k = 1, \ldots, p;$$
$$l = 1, \ldots, p)$$

where $\epsilon_{i(jkl)}$ is $NID(0, \sigma_\epsilon^2)$. The analysis procedures will be illustrated using the data in Table 7.3-2 for the tire experiment. In addition to computing $SSTO$ and $SSWCELL$, we need to formulate coefficient matrices that will let us compute SSA, SSB, SSC, and $SSRES$. As an aid to understanding the following null hypotheses and the coefficient matrices for these sources of variation, the reader may find it helpful to refer to Tables 7.9-1 and 7.9-2.

$$\text{Treatment } A \qquad H_0: \quad \mu_{1..} - \mu_{4..} = 0 \qquad \text{or} \qquad H_0: \quad \underset{(p-1)\times h}{\mathbf{C}'_A} \ \underset{h\times 1}{\boldsymbol{\mu}} = \underset{(p-1)\times 1}{\mathbf{0}_A}$$
$$\mu_{2..} - \mu_{4..} = 0$$
$$\mu_{3..} - \mu_{4..} = 0$$

where
$$\underset{3\times 16}{\mathbf{C}'_A} = \begin{bmatrix} 1 & 1 & 1 & 1 & 0 & 0 & 0 & 0 & 0 & 0 & 0 & 0 & -1 & -1 & -1 & -1 \\ 0 & 0 & 0 & 0 & 1 & 1 & 1 & 1 & 0 & 0 & 0 & 0 & -1 & -1 & -1 & -1 \\ 0 & 0 & 0 & 0 & 0 & 0 & 0 & 0 & 1 & 1 & 1 & 1 & -1 & -1 & -1 & -1 \end{bmatrix}$$

$$\underset{1\times 3}{\mathbf{0}'_A} = \begin{bmatrix} 0 & 0 & 0 \end{bmatrix}$$

and
$$\underset{1\times 16}{\boldsymbol{\mu}'} = \begin{bmatrix} \mu_{111} & \mu_{124} & \mu_{133} & \mu_{142} & \mu_{212} & \mu_{221} & \mu_{234} & \mu_{243} & \mu_{313} & \mu_{322} & \cdots & \mu_{441} \end{bmatrix}.$$

---

* This section assumes a familiarity with the matrix operations in Appendix D and the material on the general linear model in Chapter 5 and Section 6.9. The reader who is interested only in the traditional approach to the analysis of variance can, without loss of continuity, omit this section.

TABLE 7.9-1   Type LS-4 Design

ABC and A Summary Tables ($N = 32$ scores, $h = 16$ population means, $p = 4$ levels of treatment A):

ABC Summary Table

Entry is μ_{jkl}

A Summary Table

Entry is $\mu_{j..}$

	c_1	c_2	c_3	c_4	$\sum\limits_{l=1}^{p} \mu_{jkl}/p$		a_1	a_2	a_3	a_4
b_1	a_1 μ_{111}	a_2 μ_{212}	a_3 μ_{313}	a_4 μ_{414}	$\mu_{.1.}$		$\mu_{1..}$	$\mu_{2..}$	$\mu_{3..}$	$\mu_{4..}$
b_2	a_2 μ_{221}	a_3 μ_{322}	a_4 μ_{423}	a_1 μ_{124}	$\mu_{.2.}$					
b_3	a_3 μ_{331}	a_4 μ_{432}	a_1 μ_{133}	a_2 μ_{234}	$\mu_{.3.}$					
b_4	a_4 μ_{441}	a_1 μ_{142}	a_2 μ_{243}	a_3 μ_{344}	$\mu_{.4.}$					

$\sum\limits_{k=1}^{p} \mu_{jkl}/p = \quad \mu_{..1} \quad \mu_{..2} \quad \mu_{..3} \quad \mu_{..4}$

The coefficients, $c_{jkl} = \pm 1/p$ or 0, in $\mathbf{C}'_A$ have been multiplied by 4 to avoid fractions. This does not affect the nature of the hypothesis that is tested.

Rows (B) H_0: $\mu_{.1.} - \mu_{.4.} = 0$ or H_0: $\underset{(p-1)\times h}{\mathbf{C}'_B} \underset{h\times 1}{\boldsymbol{\mu}} = \underset{(p-1)\times 1}{\mathbf{0}_B}$

$\mu_{.2.} - \mu_{.4.} = 0$

$\mu_{.3.} - \mu_{.4.} = 0$

where $\underset{3\times16}{\mathbf{C}'_B} = \begin{bmatrix} 1 & 0 & 0 & -1 & 1 & 0 & 0 & -1 & 1 & 0 & 0 & -1 & 1 & 0 & 0 & -1 \\ 0 & 1 & 0 & -1 & 0 & 1 & 0 & -1 & 0 & 1 & 0 & -1 & 0 & 1 & 0 & -1 \\ 0 & 0 & 1 & -1 & 0 & 0 & 1 & -1 & 0 & 0 & 1 & -1 & 0 & 0 & 1 & -1 \end{bmatrix}$

and $\underset{1\times3}{\mathbf{0}'_B} = [0 \quad 0 \quad 0]$.

The coefficients, $c_{jkl} = \pm 1/p$ or 0, in $\mathbf{C}'_B$ have been multiplied by 4 to avoid fractions.

Columns (C) H_0: $\mu_{..1} - \mu_{..4} = 0$ or H_0: $\underset{(p-1)\times h}{\mathbf{C}'_C} \underset{h\times 1}{\boldsymbol{\mu}} = \underset{(p-1)\times 1}{\mathbf{0}_C}$

$\mu_{..2} - \mu_{..4} = 0$

$\mu_{..3} - \mu_{..4} = 0$

where $\underset{3\times16}{\mathbf{C}'_C} = \begin{bmatrix} 1 & -1 & 0 & 0 & 0 & 1 & -1 & 0 & 0 & 0 & 1 & -1 & -1 & 0 & 0 & 1 \\ 0 & -1 & 0 & 1 & 1 & 0 & -1 & 0 & 0 & 1 & 0 & -1 & -1 & 0 & 1 & 0 \\ 0 & -1 & 1 & 0 & 0 & 0 & -1 & 1 & 1 & 0 & 0 & -1 & -1 & 1 & 0 & 0 \end{bmatrix}$

and $\underset{1\times3}{\mathbf{0}'_C} = [0 \;\; 0 \;\; 0]$.

The coefficients, $c_{jkl} = \pm 1/p$ or 0, in $\mathbf{C}'_C$ have been multiplied by 4 to avoid fractions.

Residual $\quad H_0$: $\quad (\mu_{111} + \mu_{212} + \mu_{344} + \mu_{423}) - (\mu_{142} + \mu_{221} + \mu_{313} + \mu_{414}) = 0$

$\qquad\qquad\qquad (\mu_{124} + \mu_{221} + \mu_{313} + \mu_{432}) - (\mu_{111} + \mu_{234} + \mu_{322} + \mu_{423}) = 0$

$\qquad\qquad\qquad (\mu_{124} + \mu_{212} + \mu_{313} + \mu_{441}) - (\mu_{111} + \mu_{243} + \mu_{322} + \mu_{414}) = 0$

$\qquad\qquad\qquad (\mu_{133} + \mu_{221} + \mu_{322} + \mu_{414}) - (\mu_{124} + \mu_{212} + \mu_{331} + \mu_{423}) = 0$

$\qquad\qquad\qquad (\mu_{124} + \mu_{212} + \mu_{344} + \mu_{432}) - (\mu_{142} + \mu_{234} + \mu_{322} + \mu_{414}) = 0$

$\qquad\qquad\qquad (\mu_{133} + \mu_{221} + \mu_{313} + \mu_{441}) - (\mu_{111} + \mu_{243} + \mu_{331} + \mu_{423}) = 0$

or $\qquad\qquad H_0$: $\quad \underset{s\times h}{\mathbf{R}'} \; \underset{h\times1}{\boldsymbol{\mu}} = \underset{s\times1}{\boldsymbol{\theta}}$

where $\underset{6\times16}{\mathbf{R}'} = \begin{bmatrix} 1 & 0 & 0 & -1 & 1 & -1 & 0 & 0 & -1 & 0 & 0 & 1 & -1 & 1 & 0 & 0 \\ -1 & 1 & 0 & 0 & 0 & 1 & -1 & 0 & 1 & -1 & 0 & 0 & 0 & -1 & 1 & 0 \\ -1 & 1 & 0 & 0 & 1 & 0 & 0 & -1 & 1 & -1 & 0 & 0 & -1 & 0 & 0 & 1 \\ 0 & -1 & 1 & 0 & -1 & 1 & 0 & 0 & 0 & 1 & -1 & 0 & 1 & -1 & 0 & 0 \\ 1 & 0 & -1 & 0 & -1 & 0 & 1 & 0 & 1 & 0 & -1 & 0 & -1 & 0 & 1 & 0 \\ 0 & 1 & 0 & -1 & 0 & -1 & 0 & 1 & 0 & 1 & 0 & -1 & 0 & -1 & 0 & 1 \end{bmatrix}$

$\underset{1\times6}{\boldsymbol{\theta}'} = [0 \;\; 0 \;\; 0 \;\; 0 \;\; 0 \;\; 0]$

and $s = (p-1)(p-2)$. This hypothesis is the only one in the Latin square design that is unintelligible. In words, it states that the sources of variation corresponding to components of two- and three-treatment interactions are all equal to zero. An algorithm for obtaining the residual matrix coefficients for any Latin square design is described in the following section.

Formulas for computing sums of squares, F statistics, and means are given in Table 7.9-2. A comparison of these sums of squares with those in Table 7.3-2 where the traditional sum of squares approach is used reveals that they are identical.

Computational procedures for testing differences among means are similar to those shown in Table 5.10-2 for a type CR-*p* design. The required terms are $\hat{\psi}_i$ and $\hat{\sigma}_{\psi_i}$ or $\hat{\sigma}_{\bar{Y}}$. The computational formulas are

$$\hat{\psi}_i = \underset{1\times h}{\mathbf{c}'} \; \underset{h\times1}{\hat{\boldsymbol{\mu}}}$$

where $\mathbf{c}'$ is a coefficient vector that defines the contrast of interest

$$\hat{\sigma}_{\psi_i} = \left\{ MSWCELL \; [\underset{1\times h}{\mathbf{c}'} \; \underset{h\times h}{(\mathbf{X}'\mathbf{X})^{-1}} \; \underset{h\times1}{\mathbf{c}}] \right\}^{1/2}$$

or $\qquad\qquad \left[MSWCELL \left(\dfrac{c_1^2}{np} + \dfrac{c_2^2}{np} + \cdots + \dfrac{c_p^2}{np} \right) \right]^{1/2}$

$$\hat{\sigma}_{\bar{Y}} = (MSWCELL/np)^{1/2}.$$

TABLE 7.9-2 Computational Procedures for a Type LS-4 Design Using a Full Rank Experimental Design Model

(i) Data and basic matrices for computing sums of squares (C' and R' matrices are defined in the text; $N = 32$, $h = 16$, $p = 4$, $n = 2$, and $s = 6$):

	y ($N \times 1$)	X_1	X_2	X_3	X_4	X_5	X_6	X_7	X_8	X_9	X_{10}	X_{11}	X_{12}	X_{13}	X_{14}	X_{15}	X_{16}
$a_1b_1c_1$	1	1	0	0	0	0	0	0	0	0	0	0	0	0	0	0	0
	2	1	0	0	0	0	0	0	0	0	0	0	0	0	0	0	0
$a_1b_2c_4$	2	0	1	0	0	0	0	0	0	0	0	0	0	0	0	0	0
	3	0	1	0	0	0	0	0	0	0	0	0	0	0	0	0	0
$a_1b_3c_3$	3	0	0	1	0	0	0	0	0	0	0	0	0	0	0	0	0
	2	0	0	1	0	0	0	0	0	0	0	0	0	0	0	0	0
$a_1b_4c_2$	6	0	0	0	1	0	0	0	0	0	0	0	0	0	0	0	0
	3	0	0	0	1	0	0	0	0	0	0	0	0	0	0	0	0
$a_2b_1c_2$	2	0	0	0	0	1	0	0	0	0	0	0	0	0	0	0	0
	3	0	0	0	0	1	0	0	0	0	0	0	0	0	0	0	0
$a_2b_2c_1$	3	0	0	0	0	0	1	0	0	0	0	0	0	0	0	0	0
	4	0	0	0	0	0	1	0	0	0	0	0	0	0	0	0	0
$a_2b_3c_4$	5	0	0	0	0	0	0	1	0	0	0	0	0	0	0	0	0
	4	0	0	0	0	0	0	1	0	0	0	0	0	0	0	0	0
$a_2b_4c_3$	3	0	0	0	0	0	0	0	1	0	0	0	0	0	0	0	0
	4	0	0	0	0	0	0	0	1	0	0	0	0	0	0	0	0
$a_3b_1c_3$	5	0	0	0	0	0	0	0	0	1	0	0	0	0	0	0	0
	6	0	0	0	0	0	0	0	0	1	0	0	0	0	0	0	0
$a_3b_2c_2$	8	0	0	0	0	0	0	0	0	0	1	0	0	0	0	0	0
	6	0	0	0	0	0	0	0	0	0	1	0	0	0	0	0	0
$a_3b_3c_1$	5	0	0	0	0	0	0	0	0	0	0	1	0	0	0	0	0
	7	0	0	0	0	0	0	0	0	0	0	1	0	0	0	0	0
$a_3b_4c_4$	6	0	0	0	0	0	0	0	0	0	0	0	1	0	0	0	0
	7	0	0	0	0	0	0	0	0	0	0	0	1	0	0	0	0
$a_4b_1c_4$	9	0	0	0	0	0	0	0	0	0	0	0	0	1	0	0	0
	8	0	0	0	0	0	0	0	0	0	0	0	0	1	0	0	0
$a_4b_2c_3$	9	0	0	0	0	0	0	0	0	0	0	0	0	0	1	0	0
	8	0	0	0	0	0	0	0	0	0	0	0	0	0	1	0	0

TABLE 7.9-2 (continued)

$$
a_4b_3c_2 \left\{\begin{matrix} 10 \\ 11 \\ \text{—} \end{matrix}\right|\left\|\begin{matrix} 0 & 0 & 0 & 0 & 0 & 0 & 0 & 0 & 0 & 0 & 0 & 0 & 0 & 0 & 1 & 0 \\ 0 & 0 & 0 & 0 & 0 & 0 & 0 & 0 & 0 & 0 & 0 & 0 & 0 & 0 & 1 & 0 \\ - & - & - & - & - & - & - & - & - & - & - & - & - & - & - & - \end{matrix}\right|
$$
$$
a_4b_4c_1 \left\{\begin{matrix} 7 \\ 10 \end{matrix}\right|\left\|\begin{matrix} 0 & 0 & 0 & 0 & 0 & 0 & 0 & 0 & 0 & 0 & 0 & 0 & 0 & 0 & 0 & 1 \\ 0 & 0 & 0 & 0 & 0 & 0 & 0 & 0 & 0 & 0 & 0 & 0 & 0 & 0 & 0 & 1 \end{matrix}\right|
$$

$$
\sum_{i=1}^{N} Y_i = 172
$$

$$
\underset{1\times N}{\mathbf{y}'} \quad \underset{N\times 1}{\mathbf{y}'} \quad \underset{1\times 1}{(\mathbf{y}'\mathbf{y})} \qquad \underset{h\times N}{\mathbf{X}'} \qquad \underset{N\times h}{\mathbf{X}} \qquad \underset{h\times h}{\mathbf{X}'\mathbf{X}}
$$

$$
[1\ 2\ 2 \cdots 10]\begin{bmatrix} 1 \\ 2 \\ 2 \\ \vdots \\ 10 \end{bmatrix} = 1160.0 \quad \begin{bmatrix} 1 & 1 & 0 & 0 & \cdot & 0 \\ 0 & 0 & 1 & 1 & \cdot & 0 \\ 0 & 0 & 0 & 0 & \cdot & 0 \\ \cdot & \cdot & \cdot & \cdot & & \cdot \\ 0 & 0 & 0 & 0 & \cdot & 1 \end{bmatrix}\begin{bmatrix} 1 & 0 & 0 & \cdot & 0 \\ 1 & 0 & 0 & \cdot & 0 \\ 0 & 1 & 0 & \cdot & 0 \\ 0 & 1 & 0 & \cdot & 0 \\ \cdot & \cdot & \cdot & & \cdot \\ 0 & 0 & 0 & \cdot & 1 \end{bmatrix} = \begin{bmatrix} 2 & 0 & 0 & \cdot & 0 \\ 0 & 2 & 0 & \cdot & 0 \\ 0 & 0 & 2 & \cdot & 0 \\ 0 & 0 & 0 & \cdot & 0 \\ \cdot & \cdot & \cdot & & \cdot \\ 0 & 0 & 0 & \cdot & 2 \end{bmatrix}
$$

$$
\underset{h\times N}{\mathbf{X}'} \qquad \underset{N\times 1}{\mathbf{y}} \quad \underset{h\times 1}{(\mathbf{X}'\mathbf{y})} \qquad \underset{h\times h}{(\mathbf{X}'\mathbf{X})^{-1}} \qquad \underset{h\times 1}{(\mathbf{X}'\mathbf{y})} \quad \underset{h\times 1}{\hat{\boldsymbol{\mu}}}
$$

$$
\begin{bmatrix} 1 & 1 & 0 & 0 & \cdot & 0 \\ 0 & 0 & 1 & 1 & \cdot & 0 \\ 0 & 0 & 0 & 0 & \cdot & 0 \\ \cdot & \cdot & \cdot & \cdot & & \cdot \\ 0 & 0 & 0 & 0 & \cdot & 1 \end{bmatrix}\begin{bmatrix} 1 \\ 2 \\ 2 \\ \vdots \\ 10 \end{bmatrix} = \begin{bmatrix} 3 \\ 5 \\ 5 \\ \vdots \\ 17 \end{bmatrix} \quad \begin{bmatrix} \frac{1}{2} & 0 & 0 & \cdot & 0 \\ 0 & \frac{1}{2} & 0 & \cdot & 0 \\ 0 & 0 & \frac{1}{2} & \cdot & 0 \\ \cdot & \cdot & \cdot & & \cdot \\ 0 & 0 & 0 & \cdot & \frac{1}{2} \end{bmatrix}\begin{bmatrix} 3 \\ 5 \\ 5 \\ \vdots \\ 17 \end{bmatrix} = \begin{bmatrix} 1.5 \\ 2.5 \\ 2.5 \\ \vdots \\ 8.5 \end{bmatrix}
$$

$$
\underset{1\times h}{\hat{\boldsymbol{\mu}}'} \quad \underset{h\times 1}{(\mathbf{X}'\mathbf{y})} \quad \underset{1\times 1}{(\hat{\boldsymbol{\mu}}'\mathbf{X}'\mathbf{y})}
$$

$$
[1.5\ 2.5\ 2.5 \cdots 8.5]\begin{bmatrix} 3 \\ 5 \\ 5 \\ \vdots \\ 17 \end{bmatrix} = 1141.000
$$

(ii) Computation of sum of squares:

$$
SSTO = \mathbf{y}'\mathbf{y} - (\Sigma Y_i)^2/N = 1160.000 - (172)^2/32 = 235.500
$$

$$
SSA = \underset{1\times(p-1)}{(\mathbf{C}_A'\,\hat{\boldsymbol{\mu}})'}\ \underset{(p-1)\times(p-1)}{[\mathbf{C}_A'(\mathbf{X}'\mathbf{X})^{-1}\mathbf{C}_A]^{-1}}\ \underset{(p-1)\times 1}{(\mathbf{C}_A'\,\hat{\boldsymbol{\mu}})} = 194.500
$$

$$
SSB = \underset{1\times(p-1)}{(\mathbf{C}_B'\,\hat{\boldsymbol{\mu}})'}\ \underset{(p-1)\times(p-1)}{[\mathbf{C}_B'(\mathbf{X}'\mathbf{X})^{-1}\mathbf{C}_B]^{-1}}\ \underset{(p-1)\times 1}{(\mathbf{C}_B'\,\hat{\boldsymbol{\mu}})} = 9.250
$$

$$
SSC = \underset{1\times(p-1)}{(\mathbf{C}_C'\,\hat{\boldsymbol{\mu}})'}\ \underset{(p-1)\times(p-1)}{[\mathbf{C}_C'(\mathbf{X}'\mathbf{X})^{-1}\mathbf{C}_C]^{-1}}\ \underset{(p-1)\times 1}{(\mathbf{C}_C'\,\hat{\boldsymbol{\mu}})} = 7.750
$$

$$
SSRES = \underset{1\times s}{(\mathbf{R}'\,\hat{\boldsymbol{\mu}})'}\ \underset{s\times s}{[\mathbf{R}'(\mathbf{X}'\mathbf{X})^{-1}\mathbf{R}]^{-1}}\ \underset{s\times 1}{(\mathbf{R}'\,\hat{\boldsymbol{\mu}})} = 5.000
$$

$$
SSWCELL = \mathbf{y}'\mathbf{y} - \hat{\boldsymbol{\mu}}'\mathbf{X}'\mathbf{y} = 1160.000 - 1141.000 = 19.000
$$

TABLE 7.9-2   *(continued)*

(iii)   Computation of mean squares:

$$MSA = SSA/(p - 1) = 194.500/(4 - 1) = 64.833$$

$$MSB = SSB/(p - 1) = 9.250/(4 - 1) = 3.083$$

$$MSC = SSC/(p - 1) = 7.750/(4 - 1) = 2.583$$

$$MSRES = SSRES/(p - 1)(p - 2) = 5.000/(4 - 1)(4 - 2) = 0.833$$

$$MSWCELL = SSWCELL/p^2(n - 1) = 19.000/(4)^2(2 - 1) = 1.188$$

(iv)   Computation of F ratios:

$$F = MSA/MSWCELL = 54.57$$

$$F = MSB/MSWCELL = 2.60$$

$$F = MSC/MSWCELL = 2.17$$

$$F = MSRES/MSWCELL = 0.70$$

$$F_{.05;3,16} = 3.24$$

$$F_{.05;6,16} = 2.74$$

A coefficient vector for testing H_0: $\mu_{1..} - \mu_{2..} = 0$ for the data in Table 7.9-2 is

$$\mathbf{c}'_{1 \times 16} = [\tfrac{1}{4} \quad \tfrac{1}{4} \quad \tfrac{1}{4} \quad \tfrac{1}{4} \quad -\tfrac{1}{4} \quad -\tfrac{1}{4} \quad -\tfrac{1}{4} \quad -\tfrac{1}{4} \quad 0 \quad 0 \quad 0 \quad 0 \quad 0 \quad 0 \quad 0 \quad 0].$$

The unrestricted full rank model approach can be used to analyze data for a type LS-p design with missing observations and cells. The required modifications are similar to those presented in Section 8.13 for a completely randomized factorial design and will not be discussed here.

PROCEDURE FOR FORMING RESIDUAL MATRIX

The hypothesis that interactions involving A, B, and C are all equal to zero is expressed as

$$\mathbf{R}'_{s \times h} \underset{h \times 1}{\boldsymbol{\mu}} = \underset{s \times 1}{\boldsymbol{\theta}}$$

where $\mathbf{R}'$ is a coefficient matrix and $\boldsymbol{\theta}$ is a vector of zeros. We will now describe an algorithm for obtaining $\mathbf{R}'$. The 3×3 and 4×4 Latin squares in Figure 7.9-1 will be used for illustrative purposes.

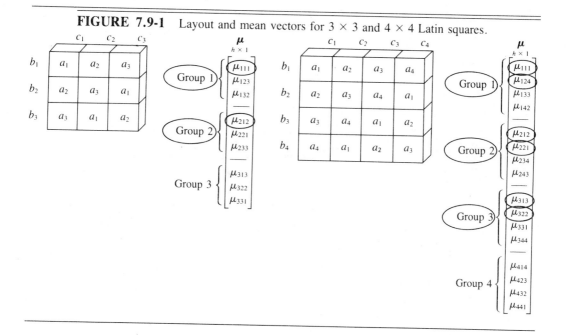

FIGURE 7.9-1 Layout and mean vectors for 3×3 and 4×4 Latin squares.

The algorithm will be illustrated for the 3×3 square first.*

Step 1   Partition $\boldsymbol{\mu}$ into p groups according to the levels of treatment A. (See Figure 7.9-1.)

Step 2   Form residual vectors, $\mathbf{r}_i'$, for $p - 2$ arbitrarily selected means in each of $p - 1$ arbitrarily selected groups. (In Figure 7.9-1 the $p - 1$ groups and $p - 2$ means within the groups that have been selected are circled.) Coefficients for a particular residual vector are determined as follows:

(a)   The circled mean is given the coefficient $(p^2 - 3p + 2)$.

(b)   Means having a subscript in common with the circled mean are given the coefficient $-(p - 2)$.

(c)   All other means are given the coefficient 2.

Example for μ_{111}: $\mathbf{r}_1' = \begin{bmatrix} 2 & -1 & -1 & -1 & -1 & 2 & -1 & 2 & -1 \end{bmatrix}$
Example for μ_{212}: $\mathbf{r}_2' = \begin{bmatrix} -1 & 2 & -1 & 2 & -1 & -1 & -1 & -1 & 2 \end{bmatrix}$

Step 3   If desired, linearly transform the residual vectors to reduce the coefficients to 1, 0, -1 form. For example,

---

*   I am indebted to H. Gary, Jr. for the algorithm.

$$\mathbf{r}_1^{*\prime} = [\mathbf{r}_1^\prime - \mathbf{r}_2^\prime]1/3 = [1 \quad -1 \quad 0 \quad -1 \quad 0 \quad 1 \quad 0 \quad 1 \quad -1]$$
$$\mathbf{r}_2^{*\prime} = [\mathbf{r}_1^\prime + 2\mathbf{r}_2^\prime]1/3 = [0 \quad 1 \quad -1 \quad 1 \quad -1 \quad 0 \quad -1 \quad 0 \quad 1].$$

The residual coefficient matrix for the 3×3 Latin square is

$$\mathbf{R}_{2\times9}^\prime = \begin{bmatrix} 1 & -1 & 0 & -1 & 0 & 1 & 0 & 1 & -1 \\ 0 & 1 & -1 & 1 & -1 & 0 & -1 & 0 & 1 \end{bmatrix}.$$

We can follow the same steps to obtain $\mathbf{R}^\prime$ for the 4×4 Latin square in Figure 7.9-1. Following step 2 we obtain $(p-1)(p-2) = 6$ residual vectors.

μ_{111}: $\mathbf{r}_1^\prime =$
$$[\ 6 \ -2 \ -2 \ -2 \ -2 \ -2 \ \ 2 \ \ 2 \ -2 \ \ 2 \ -2 \ \ 2 \ -2 \ \ 2 \ \ 2 \ -2]$$

μ_{124}: $\mathbf{r}_2^\prime =$
$$[-2 \ \ 6 \ -2 \ -2 \ \ 2 \ -2 \ -2 \ \ 2 \ \ 2 \ -2 \ \ 2 \ -2 \ -2 \ -2 \ \ 2 \ \ 2]$$

μ_{212}: $\mathbf{r}_3^\prime =$
$$[-2 \ \ 2 \ \ 2 \ -2 \ \ 6 \ -2 \ -2 \ -2 \ -2 \ -2 \ \ 2 \ \ 2 \ -2 \ \ 2 \ -2 \ \ 2]$$

μ_{221}: $\mathbf{r}_4^\prime =$
$$[-2 \ -2 \ \ 2 \ \ 2 \ -2 \ \ 6 \ -2 \ -2 \ \ 2 \ -2 \ -2 \ \ 2 \ \ 2 \ -2 \ \ 2 \ -2]$$

μ_{313}: $\mathbf{r}_5^\prime =$
$$[-2 \ \ 2 \ -2 \ \ 2 \ -2 \ \ 2 \ \ 2 \ -2 \ \ 6 \ -2 \ -2 \ -2 \ -2 \ -2 \ \ 2 \ \ 2]$$

μ_{322}: $\mathbf{r}_6^\prime =$
$$[\ 2 \ -2 \ \ 2 \ -2 \ -2 \ -2 \ \ 2 \ \ 2 \ -2 \ \ 6 \ -2 \ -2 \ \ 2 \ -2 \ -2 \ \ 2]$$

The vectors can be transformed to $1, 0, -1$ as follows. The transformation that is used must be one that will preserve the linear independence of the residual vectors.

$\mathbf{r}_1^{*\prime} = [\mathbf{r}_1^\prime + \mathbf{r}_3^\prime]1/4$
$$= [\ 1 \ \ 0 \ \ 0 \ -1 \ \ 1 \ -1 \ \ 0 \ \ 0 \ -1 \ \ 0 \ \ 0 \ \ 1 \ -1 \ \ 1 \ \ 0 \ \ 0]$$

$\mathbf{r}_2^{*\prime} = [\mathbf{r}_2^\prime + \mathbf{r}_4^\prime]1/4$
$$= [-1 \ \ 1 \ \ 0 \ \ 0 \ \ 0 \ \ 1 \ -1 \ \ 0 \ \ 1 \ -1 \ \ 0 \ \ 0 \ \ 0 \ -1 \ \ 1 \ \ 0]$$

$\mathbf{r}_3^{*\prime} = [\mathbf{r}_3^\prime + \mathbf{r}_5^\prime]1/4$
$$= [-1 \ \ 1 \ \ 0 \ \ 0 \ \ 1 \ \ 0 \ \ 0 \ -1 \ \ 1 \ -1 \ \ 0 \ \ 0 \ -1 \ \ 0 \ \ 0 \ \ 1]$$

$\mathbf{r}_4^{*\prime} = [\mathbf{r}_4^\prime + \mathbf{r}_6^\prime]1/4$
$$= [\ 0 \ -1 \ \ 1 \ \ 0 \ -1 \ \ 1 \ \ 0 \ \ 0 \ \ 0 \ \ 1 \ -1 \ \ 0 \ \ 1 \ -1 \ \ 0 \ \ 0]$$

$\mathbf{r}_5^{*\prime} = [\mathbf{r}_1^\prime + \mathbf{r}_5^\prime]1/4$
$$= [\ 1 \ \ 0 \ -1 \ \ 0 \ -1 \ \ 0 \ \ 1 \ \ 0 \ \ 1 \ \ 0 \ -1 \ \ 0 \ -1 \ \ 0 \ \ 1 \ \ 0]$$

$\mathbf{r}_6^{*\prime} = [\mathbf{r}_2^\prime + \mathbf{r}_6^\prime]1/4$
$$= [\ 0 \ \ 1 \ \ 0 \ -1 \ \ 0 \ -1 \ \ 0 \ \ 1 \ \ 0 \ \ 1 \ \ 0 \ -1 \ \ 0 \ -1 \ \ 0 \ \ 1]$$

The residual coefficient matrix for the 4×4 Latin square is

$$\underset{6 \times 16}{\mathbf{R}'} =$$

$$\begin{bmatrix} 1 & 0 & 0 & -1 & 1 & -1 & 0 & 0 & -1 & 0 & 0 & 1 & -1 & 1 & 0 & 0 \\ -1 & 1 & 0 & 0 & 0 & 1 & -1 & 0 & 1 & -1 & 0 & 0 & 0 & -1 & 1 & 0 \\ -1 & 1 & 0 & 0 & 1 & 0 & 0 & -1 & 1 & -1 & 0 & 0 & -1 & 0 & 0 & 1 \\ 0 & -1 & 1 & 0 & -1 & 1 & 0 & 0 & 0 & 1 & -1 & 0 & 1 & -1 & 0 & 0 \\ 1 & 0 & -1 & 0 & -1 & 0 & 1 & 0 & 1 & 0 & -1 & 0 & -1 & 0 & 1 & 0 \\ 0 & 1 & 0 & -1 & 0 & -1 & 0 & 1 & 0 & 1 & 0 & -1 & 0 & -1 & 0 & 1 \end{bmatrix}.$$

Graeco-Latin and hyper-Graeco-Latin square designs are described in Sections 7.10 and 7.11, respectively. Rules similar to those just described can be used to obtain $\mathbf{R}'$ for these designs. Let k denote the number of orthogonal Latin squares in a hyper-Graeco-Latin square design. The three-step algorithm is as follows.

Step 1   Partition $\boldsymbol{\mu}$ into p groups according to the levels of treatment A.

Step 2   Form residual vectors $\mathbf{r}_i'$ for $p - k - 1$ arbitrarily selected means in each of $p - 1$ arbitrarily selected groups. Circle the groups and means that are selected. Coefficients for a particular residual vector are determined as follows:

     (a)   The circled mean is given the coefficient $p^2 - (k + 2)p + (k + 1)$.

     (b)   Means having a subscript in common with the circled mean are given the coefficient $-(p - k - 1)$.

     (c)   Means not having a subscript in common with the circled mean are given the coefficient $(k + 1)$.

Step 3   If desired, linearly transform the residual vectors to reduce the coefficients to $1, 0, -1$ form.

RESTRICTED FULL RANK MODEL

A restricted full rank model should be used if the cells of a Latin square contain only one observation. For this case, a within-cell estimate of the population error variance cannot be obtained. Instead, the residual mean square is used to estimate the error variance under the assumption that interactions among A, B, and C are all equal to zero. A restricted model may also be used when $n > 1$ and an experimenter wants to pool *MSRES* and *MSWCELL* under the assumption that all interactions are equal to zero. The modifications of the computational formulas for this case follow those shown in Table 8.13-6.

The restricted full rank experimental design model when n is equal to 1 is

$$Y_{jkl} = \mu_{jkl} + \epsilon_{jkl} \qquad (j = 1, \ldots, p; k = 1, \ldots, p; l = 1, \ldots, p)$$

where ϵ_{jkl} is $NID(0, \sigma_\epsilon^2)$ and μ_{jkl} is subject to the restrictions that all interaction components involving A, B, and C are equal to zero. These restrictions can be expressed as

$$\underset{s \times h}{\mathbf{R}'} \underset{h \times 1}{\boldsymbol{\mu}} = \underset{s \times 1}{\boldsymbol{\theta}}$$

where $s = (p - 1)(p - 2)$. The coefficient matrices, $\mathbf{R}'$, $\mathbf{C}_A'$, and so on, for the restricted model are identical to those defined earlier for the unrestricted model.

We want to test the null hypotheses for treatment A and nuisance variables B and C, subject to the restrictions that $\mathbf{R}'\boldsymbol{\mu} = \boldsymbol{\theta}$. This is accomplished by forming augmented matrices of the form

$$\underset{(s+p-1) \times h}{\mathbf{Q}'} = \begin{bmatrix} \underset{s \times h}{\mathbf{R}'} \\ \underset{(p-1) \times h}{\mathbf{C}'} \end{bmatrix} \qquad \text{and} \qquad \underset{(s+p-1) \times 1}{\boldsymbol{\eta}} = \begin{bmatrix} \underset{s \times 1}{\boldsymbol{\theta}} \\ \underset{(p-1) \times 1}{\mathbf{0}} \end{bmatrix}$$

where $\mathbf{C}'$ is the coefficient matrix for one of the null hypotheses and $\mathbf{0}$ is the null vector associated with $\mathbf{C}'$.

The computational procedures for the restricted full rank experimental design model are illustrated in Table 7.9-3 using the data from the ABC Summary Table of Table 7.3-2. If, as in this example, the rows of $\mathbf{R}'$ are orthogonal to those of $\mathbf{C}_A'$, $\mathbf{C}_B'$, or $\mathbf{C}_C'$, the associated computational formula $(\mathbf{Q}'\hat{\boldsymbol{\mu}})' (\mathbf{Q}'\mathbf{Q})^{-1} (\mathbf{Q}'\hat{\boldsymbol{\mu}}) - SSRES$ simplifies to $(\mathbf{C}'\hat{\boldsymbol{\mu}}) (\mathbf{C}'\mathbf{C})^{-1} (\mathbf{C}'\hat{\boldsymbol{\mu}})$.*

Computational procedures for testing differences among means are similar to those shown in Section 6.9 for a randomized block design. The computation formulas are

$$\underset{1 \times h}{\hat{\psi}_1} = \underset{1 \times h}{\mathbf{c}'} \underset{h \times 1}{\hat{\boldsymbol{\mu}}_R} \qquad \text{where } \mathbf{c}' \text{ is a } 1 \times h \text{ coefficient vector that defines the}$$

contrast of interest

$$\underset{1 \times 1}{\hat{\sigma}_{\psi_i}} = [MSRES(\mathbf{c}'\mathbf{Ac})]^{1/2} \qquad \text{where } \underset{h \times h}{\mathbf{A}} = [\mathbf{I} - \underset{h \times h}{\mathbf{R}(\mathbf{R}'\mathbf{R})^{-1} \mathbf{R}'}]$$

$$= \left[MSRES\left(\frac{c_1^2}{p} + \frac{c_2^2}{p} + \cdots + \frac{c_p^2}{p}\right) \right]^{1/2}$$

$$\hat{\sigma}_{\bar{Y}} = (MSRES/p)^{1/2}.$$

The restricted full rank model approach can be used to analyze data for a type LS-p design with missing cells. The required modifications are similar to those presented in Section 6.9 for a randomized block design.

---

*   Hypotheses defined by the coefficient matrices $\mathbf{C}_1'$ and $\mathbf{C}_2'$ are said to be orthogonal if $\mathbf{C}_1' (\mathbf{X}'\mathbf{X})^{-1}$ $\cdot \mathbf{C}_2 = \mathbf{0}$. For designs with one observation per cell, this rule reduces to $\mathbf{C}_1'\mathbf{C}_2 = \mathbf{0}$ since $(\mathbf{X}'\mathbf{X})^{-1} = \mathbf{I}$, an identity matrix.

TABLE 7.9-3 Computational Procedures for a Type LS-4 Design Using a Restricted Full Rank Experimental Design Model

(i) Data and matrix formulas for computing sums of squares ($\mathbf{Q}'$ and $\mathbf{R}'$ matrices are defined in the text; $N = 16$, $h = 16$, $p = 4$, and $s = 6$):

$$
\begin{array}{c}
\mathbf{y} \\
{\scriptstyle N \times 1}
\end{array}
\qquad
\begin{array}{c}
\hat{\boldsymbol{\mu}} \\
{\scriptstyle h \times 1}
\end{array}
\quad
\begin{array}{c}
[\mathbf{R}(\mathbf{R}'\mathbf{R})^{-1}\mathbf{R}'\hat{\boldsymbol{\mu}}] \\
{\scriptstyle h \times 1}
\end{array}
\quad
\begin{array}{c}
\hat{\boldsymbol{\mu}}_R \\
{\scriptstyle h \times 1}
\end{array}
$$

$$
a_1 \left\{
\begin{bmatrix} 3 \\ 5 \\ 5 \\ 9 \end{bmatrix}
\right.
\quad
\begin{bmatrix} 0.25 \\ -0.75 \\ -0.75 \\ 1.25 \end{bmatrix}
\quad
\begin{bmatrix} 2.75 \\ 5.75 \\ 5.75 \\ 7.75 \end{bmatrix}
$$

$$
a_2 \left\{
\begin{bmatrix} 5 \\ 7 \\ 9 \\ 7 \end{bmatrix}
\right.
\quad
\begin{bmatrix} -1.75 \\ 1.00 \\ 0.75 \\ 0 \end{bmatrix}
\quad
\begin{bmatrix} 6.75 \\ 6.00 \\ 8.25 \\ 7.00 \end{bmatrix}
$$

$$
a_3 \left\{
\begin{bmatrix} 11 \\ 14 \\ 12 \\ 13 \end{bmatrix}
\right.
\quad
\begin{bmatrix} 1.00 \\ 0 \\ -0.50 \\ -0.50 \end{bmatrix}
\quad
\begin{bmatrix} 10.00 \\ 14.00 \\ 12.50 \\ 13.50 \end{bmatrix}
$$

$$
a_4 \left\{
\begin{bmatrix} 17 \\ 17 \\ 21 \\ 17 \end{bmatrix}
\right.
\quad
\begin{bmatrix} 0.50 \\ -0.25 \\ 0.50 \\ -0.75 \end{bmatrix}
\quad
\begin{bmatrix} 16.50 \\ 17.25 \\ 20.50 \\ 17.75 \end{bmatrix}
$$

$$
\sum_{i=1}^{N} Y_i = 172
$$

$$
\begin{array}{ccc}
\mathbf{y}' & \mathbf{y} & (\mathbf{y}'\mathbf{y}) \\
{\scriptstyle 1 \times N} & {\scriptstyle N \times 1} & {\scriptstyle 1 \times 1}
\end{array}
$$

$$
[3 \ \ 5 \ \cdots \ 17]
\begin{bmatrix} 3 \\ 5 \\ \vdots \\ 17 \end{bmatrix}
= 2282.00
$$

$$
SSTO = \mathbf{y}'\mathbf{y} - (\Sigma Y_i)^2/N = 2282.00 - (172)^2/16 = 433.00
$$

$$
SSRES = \underset{\scriptstyle 1 \times s}{(\mathbf{R}'\hat{\boldsymbol{\mu}})'} \ \underset{\scriptstyle s \times s}{(\mathbf{R}'\mathbf{R})^{-1}} \ \underset{\scriptstyle s \times 1}{(\mathbf{R}'\hat{\boldsymbol{\mu}})} = 10.00
$$

$$
SSA = \underset{\scriptstyle 1 \times (s+p-1)}{(\mathbf{Q}_A'\hat{\boldsymbol{\mu}})'} \ \underset{\scriptstyle (s+p-1) \times (s+p-1)}{(\mathbf{Q}_A'\mathbf{Q}_A)^{-1}} \ \underset{\scriptstyle (s+p-1) \times 1}{(\mathbf{Q}_A'\hat{\boldsymbol{\mu}})} - SSRES
$$
$$
= 399.00 - 10.00 = 389.00
$$

$$
SSB = \underset{\scriptstyle 1 \times (s+p-1)}{(\mathbf{Q}_B'\hat{\boldsymbol{\mu}})'} \ \underset{\scriptstyle (s+p-1) \times (s+p-1)}{(\mathbf{Q}_B'\mathbf{Q}_B)^{-1}} \ \underset{\scriptstyle (s+p-1) \times 1}{(\mathbf{Q}_B'\hat{\boldsymbol{\mu}})} - SSRES
$$
$$
= 28.50 - 10.00 = 18.50
$$

$$
SSC = \underset{\scriptstyle 1 \times (s+p-1)}{(\mathbf{Q}_C'\hat{\boldsymbol{\mu}})'} \ \underset{\scriptstyle (s+p-1) \times (s+p-1)}{(\mathbf{Q}_C'\mathbf{Q}_C)^{-1}} \ \underset{\scriptstyle (s+p-1) \times 1}{(\mathbf{Q}_C'\hat{\boldsymbol{\mu}})} - SSRES
$$
$$
= 25.50 - 10.00 = 15.50
$$

(ii) Computation of mean squares:

$$
MSA = SSA/(p - 1) = 389.00/3 = 129.67
$$

$$
MSB = SSB/(p - 1) = 18.50/3 = 6.17
$$

$$
MSC = SSC/(p - 1) = 15.50/3 = 5.17
$$

$$
MSRES = SSRES/(p - 1)(p - 2) = 10.00/(3)(2) = 1.67
$$

7.10 GRAECO-LATIN SQUARE DESIGN

We have seen that a Latin square design permits an experimenter to isolate variation due to two nuisance variables while evaluating treatment effects. A Graeco-Latin square design, denoted by the letters GLS-p, permits the isolation of three nuisance variables. A type GLS-p design consists of two superimposed orthogonal Latin squares. For example, if we denoted four levels of a treatment by the Latin letters A, B, C, and D, and four levels of a third nuisance variable by the Greek letters α, β, γ, and δ, the type GLS-4 design has the following form.

Columns
(Nuisance variable)

	1	2	3	4
1	$A\alpha$	$B\beta$	$C\gamma$	$D\delta$
2	$B\delta$	$A\gamma$	$D\beta$	$C\alpha$
3	$C\beta$	$D\alpha$	$A\delta$	$B\gamma$
4	$D\gamma$	$C\delta$	$B\alpha$	$A\beta$

Rows
(Nuisance
variable)

Two Latin squares are orthogonal if and only if when they are superimposed, each letter of one square occurs once and only once with each letter of the other square. An inspection of the preceding square reveals that it satisfies this requirement.

The computational procedures for a type GLS-p design are similar to those for a type LS-p design. The computational formulas, degrees of freedom, and expected values of mean squares for a fixed effects model are given in Table 7.10-1. The

TABLE 7.10-1 Computational Formulas for Type GLS-p Design

Source	df	Computational Formulas	$E(MS)$ Model I
A (Treatment)	$p - 1$	$[A] - [Y]$	$\sigma_\epsilon^2 + np \sum_{j=1}^{p} \alpha_j^2/(p - 1)$
B (Rows)	$p - 1$	$[B] - [Y]$	$\sigma_\epsilon^2 + np \sum_{k=1}^{p} \beta_k^2/(p - 1)$
C (Columns)	$p - 1$	$[C] - [Y]$	$\sigma_\epsilon^2 + np \sum_{l=1}^{p} \gamma_l^2/(p - 1)$
D (Third nuisance variable)	$p - 1$	$[D] - [Y]$	$\sigma_\epsilon^2 + np \sum_{m=1}^{p} \delta_m^2/(p - 1)$
Residual	$(p - 1)(p - 3)$	$[ABCD] - [A] - [B]$ $- [C] - [D] + 3[Y]$	σ_ϵ^2
Within cell	$p^2(n - 1)$	$[ABCDS] - [ABCD]$	σ_ϵ^2
Total	$np^2 - 1$	$[ABCDS] - [Y]$	

analysis follows that for the type LS-p design in Table 7.3-2; however, the *ABCS* and *ABC* Summary Tables are replaced by *ABCDS* and *ABCD* tables, respectively. Also, a *D* Summary Table for the third nuisance variable is required. The entry in this table is $\sum_{i=1}^{n} \sum_{k=1}^{p} \sum_{l=1}^{p} Y_{ijklm}$ for d_m, $m = 1, \ldots, p$.

Complete sets of orthogonal Latin squares of size 3, 4, 5, 7, 8, 9, and 10 are given by Fisher and Yates (1963, 25, 86–89). It has been shown (Fisher and Yates, 1934) that no orthogonal pair of 6×6 Latin squares exists. Orthogonal squares are known to exist for all odd prime numbers, powers of primes, and multiples of four. Bose, Shrikhande, and Parker (1960) proved that orthogonal squares also exist for squares of the size $4s + 2$, where $s > 1$.

Graeco-Latin square designs have not proven very useful in the behavioral sciences for several reasons. First, the design is not appropriate if any of the variables interact. In the planning stages of an experiment it is difficult to predict with any degree of confidence that all interactions among four variables equal zero. A second reason for the design's lack of popularity is the restriction that each variable must have the same number of levels. Achieving this balance can be difficult. Graeco-Latin squares are most useful as parts of more inclusive designs. These designs, which permit the isolation of certain interactions among variables, are described in Chapter 13.

7.11 HYPER-GRAECO-LATIN SQUARE DESIGNS

If one or more orthogonal Latin squares are superimposed on a Graeco-Latin square, the resulting square is called a hyper-Graeco-Latin square. The number of orthogonal Latin squares that can be combined in forming hyper-squares is limited. For example, no more than three orthogonal 4×4 squares can be combined. This is evident if the $p^2 - 1$ degrees of freedom for total sum of squares (disregarding within-cell degrees of freedom) is analyzed. For a 4×4 square, each mode of classification has $p - 1 = 3$ degrees of freedom. The five modes of classification—rows, columns, Latin letters, Greek letters, and Hebrew letters—account for the $(4)^2 - 1 = 15$ degrees of freedom. For a 5×5 square, the $p^2 - 1 = 24$ degrees of freedom are accounted for by six modes of classification. Hence, no more than four orthogonal 5×5 squares can be combined. The residual degrees of freedom and residual sum of squares formulas for hyper-Graeco-Latin squares with five and six modes of classification are, respectively,

$$df = (p - 1)(p - 4) \qquad SSRES = [ABCDE] - [A] - [B] - [C] - [D] \\ - [E] + 4[Y]$$

and

$$df = (p - 1)(p - 5) \qquad SSRES = [ABCDEF] - [A] - [B] - [C] - [D] \\ - [E] - [F] + 5[Y].$$

The sum of squares formulas for rows, columns, and so on have the same general pattern as the corresponding formulas in a Latin square design. For a discussion of hyper-Graeco-Latin squares, see Federer (1955, Ch. 15).

7.12 ADVANTAGES AND DISADVANTAGES OF TYPE LS-*p* DESIGN

The major advantages of the Latin square design are as follows.

1. Design provides greater power for many research applications than completely randomized and randomized block designs. The design permits an experimenter to isolate the effects of two nuisance variables and thereby reduce the experimental error and obtain a more precise estimate of treatment effects.

2. Simplicity in analysis of data.

The major disadvantages of the design are as follows.

1. The number of levels of each nuisance variable must equal the number of treatment levels. Because of this requirement, Latin squares larger than 8×8 are seldom used.

2. Squares smaller than 5×5 are not practical because of the small number of degrees of freedom unless each cell contains more than one experimental unit.

3. The assumption of the model that there are no interactions among any of the variables is quite restrictive. The complex designs described in Chapters 12 and 13 that use a Latin square as the building block do permit interactions among some variables.

4. Randomization is relatively complex.

7.13 REVIEW EXERCISES

1. [7.1] What is the principal advantage of a Latin square design over completely randomized and randomized block designs?

†2. [7.1] Why is a Latin square design rarely used in behavioral and educational research?

3. [7.2] **a)** Construct a 5×5 standard Latin square.

b) Is the square self-conjugate?

c) How many arrangements of this square are possible?

4.   [7.2] Randomize the Latin square obtained in Exercise 3(a); describe each of the steps in the randomization procedure.

5.   The effect of three levels of staff disciplinary intervention on patient morale in a large Veterans Administration hospital was investigated (a_1 = low intervention, a_2 = moderate intervention, a_3 = high intervention). The nuisance variables were length of the intervention (b_1 = 6 weeks, b_2 = 12 weeks, b_3 = 18 weeks) and degree of heterogeneity of social activities available (c_1 = low heterogeneity, c_2 = medium heterogeneity, c_3 = high heterogeneity). Twenty-seven chronic schizophrenic patients were randomly assigned to the cells of the following 3 × 3 Latin square. At the beginning and end of the study, each patient completed a 20-item morale inventory. The dependent variable was the difference between the pre- and posttest morale scores ($Y_{\text{posttest}} - Y_{\text{pretest}}$). The following data were obtained; the larger the number, the greater was the improvement in morale. (Experiment suggested by Jennings, R. D. Three levels of staff intervention and their effect on inpatient small group morale, leadership, and performance. *Journal of Abnormal Psychology*, 1968, *73*, 500–502.)

	c_1	c_2	c_3
	a_1	a_2	a_3
b_1	3	6	3
	6	3	5
	5	4	4
	a_2	a_3	a_1
b_2	4	6	5
	5	5	6
	2	6	5
	a_3	a_1	a_2
b_3	4	5	4
	3	4	3
	4	7	6

†a) [7.3] Test the hypotheses $\alpha_j = 0$ for all j, $\beta_k = 0$ for all k, and $\gamma_l = 0$ for all l; use *MSWCELL* as the error term and let $\alpha = .05$.

†b) [7.3] Calculate the power of the test of H_0: $\alpha_j = 0$ for all j.

†c) [7.3] For the hypothesis H_0: $\alpha_j = 0$ for all j, determine the number of subjects required to achieve a power of approximately .82.

†d) [7.3] Calculate $\hat{\omega}^2_{Y|A}$, $\hat{\omega}^2_{Y|B}$, and $\hat{\omega}^2_{Y|C}$.

e) Prepare a "results and discussion section" for the *Journal of Abnormal Psychology*.

†6.   The effect of five levels of associative strength of adjective-noun pairs on recognition threshold was investigated. Ten word pairs having associative strengths of .30–.39 (treatment level a_1), ten having associative strengths of .40–.49 (treatment level a_2), and so on were selected from Palermo and Jenkins's (1964) list. The adjective of a word pair was presented tachistoscopically for 100 milliseconds. Immediately afterwards, the noun was presented for 20 milliseconds. If the noun was not recognized, the exposure time for the noun on a subsequent trial was increased by 5 milliseconds. This was continued until the noun was recognized. The dependent variable was mean time in milliseconds required to recognize the ten nouns in a list. Twenty-five subjects were assigned to five rows of a Latin square so that those within a row were similar with respect to noun recognition thresholds. The columns of the Latin square corresponded to five graduate students who conducted the testing sessions. The following data were obtained.

	c_1	c_2	c_3	c_4	c_5
b_1	a_1	a_2	a_3	a_4	a_5
	72	62	66	51	40
b_2	a_2	a_3	a_4	a_5	a_1
	65	61	40	44	59
b_3	a_3	a_4	a_5	a_1	a_2
	55	46	35	63	54
b_4	a_4	a_5	a_1	a_2	a_3
	34	29	54	44	50
b_5	a_5	a_1	a_2	a_3	a_4
	51	49	43	30	25

a) [7.4] Test the hypotheses $\alpha_j = 0$ for all j, $\sigma_\beta^2 = 0$, and $\sigma_\gamma^2 = 0$; let $\alpha = .01$.

b) [7.3] Calculate the power of the test of H_0: $\alpha_j = 0$ for all j.

c) [7.3] Calculate $\hat{\omega}_{Y|A}^2$, $\hat{\rho}_{IY|B}$, and $\hat{\rho}_{IY|C}$.

d) [7.4] Use Tukey's procedure to determine if the additive model is appropriate; let $\alpha = .10$.

e) [7.6] Use Tukey's multiple comparison statistic to determine which treatment A means differ.

f) [4.5] (i) Test the hypothesis that the relationship between the independent and dependent variables is nonlinear; use

$$SS\hat{\psi}_{\text{lin}} = \left[\sum_{j=1}^{p} \left(c_{1j} \sum_{k=1}^{p} \sum_{l=1}^{p} Y_{jkl} \text{ for } a_j \right) \right]^2 \bigg/ p \sum_{j=1}^{p} c_{1j}^2$$

where c_{1j} denotes the linear trend coefficient for the jth treatment level. (ii) What percent of the treatment trend is accounted for by the nonlinear component?

g) [7.7] (i) Assume that Y_{322} is missing. Use Yates's procedure to estimate the missing observation; compute $SSA(\text{corrected})$ and the F statistic. (ii) Compute $\hat{\psi}_i/\hat{\sigma}_{\psi_i}$ for testing $\mu_1 - \mu_3 = 0$ as if this were an a priori hypothesis.

h) [7.8] Determine the efficiency of the randomized block design relative to a completely randomized design; use Fisher's correction.

7. The von Restorff effect, which is the facilitation in learning a particular item in a list due to making the item distinctive from other items, was investigated using different degrees of color saturation. Three degrees of color saturation were used; a_1 was high saturation, a_2 was medium saturation, a_3 was low saturation, and a_4 was a control condition in which no color was used. Sixteen first-grade students were assigned to one of four groups on the basis of their performance on the verbal subtests of the Wechsler Intelligence Scale for Children. The four children with the highest scores were assigned to row one of a 4 × 4 Latin square, the four with

the next highest scores to row two, and so on. The columns of the Latin square represented stimulus items selected from artificial alphabets that had low, medium, high, and very high resemblance to English letters. The children were required to associate nouns with the stimulus items using a paired associates learning paradigm. One stimulus item in each paired associate list was printed in color, except for the control condition, of course. A memory drum was used to present the lists. The dependent variable was the number of trials required to learn the noun associated with the colored stimulus or control stimulus. The following data were obtained.

	c_1 low	c_2 medium	c_3 high	c_4 very high
b_1	a_1 12	a_2 12	a_3 13	a_4 16
b_2	a_2 17	a_3 17	a_4 18	a_1 8
b_3	a_3 18	a_4 20	a_1 10	a_2 10
b_4	a_4 22	a_1 14	a_2 16	a_3 14

a) [7.4] Test the hypotheses $\alpha_j = 0$ for all j, $\sigma_\beta^2 = 0$, and $\gamma_l = 0$ for all l; let $\alpha = .01$.

b) [7.3] Calculate the power of the test of H_0: $\alpha_j = 0$ for all j.

c) [7.3] Calculate $\hat{\omega}_{Y|A}^2$, $\hat{\rho}_{IY|B}$, and $\hat{\omega}_{Y|C}^2$.

d) [7.4] Use Tukey's procedure to determine if the additive model is appropriate; let $\alpha = .05$.

e) [7.6] Use Dunnett's statistic to determine which treatment A population means differ from the mean of the control population; test nondirectional hypotheses.

f) [7.7] (i) Assume that Y_{221} is missing. Use Yates's procedure to estimate the missing observation; compute *SSA (corrected)* and the F statistic. (ii) Compute $\hat{\psi}_i / \hat{\sigma}_{\psi_i}$ for testing the hypothesis $\mu_2 - \mu_4 = 0$.

g) [7.8] Determine the efficiency of the randomized block design (assume no missing observations) relative to a completely randomized design; use Fisher's correction.

8. [7.4] If each cell of a Latin square design contains one experimental unit, it is necessary to use *MSRES* in testing hypotheses about treatment and nuisance effects. What assumption must be tenable in order to use the residual mean square as an error term? How can the tenability of this assumption be determined?

†9. [7.5] Write the expected values of the means squares for

a) A fixed-effects model, assume that n is greater than one.

b) A mixed model in which rows and columns are random, assume that n is equal to one.

†10. [7.5] It can be shown that

$$\sum_{i=1}^{n} \sum_{j=1}^{p} \sum_{k=1}^{p} \sum_{l=1}^{p} [(\overline{Y}_{\cdot j \cdot \cdot} - \overline{Y}_{\cdot \cdot \cdot}) + (\overline{Y}_{\cdot \cdot k \cdot} - \overline{Y}_{\cdot \cdot \cdot}) + (\overline{Y}_{\cdot \cdot \cdot l} - \overline{Y}_{\cdot \cdot \cdot})$$

$$+ (\overline{Y}_{\cdot jkl} - \overline{Y}_{\cdot j \cdot \cdot} - \overline{Y}_{\cdot \cdot k \cdot} - \overline{Y}_{\cdot \cdot \cdot l} + 2\overline{Y}_{\cdot \cdot \cdot}) + (Y_{ijkl} - \overline{Y}_{\cdot jkl})]^2$$

is equal to

$$np \sum_{j=1}^{p} (\overline{Y}_{\cdot j \cdot \cdot} - \overline{Y}_{\cdot \cdot \cdot \cdot})^2 + np \sum_{k=1}^{p} (\overline{Y}_{\cdot \cdot k \cdot} - \overline{Y}_{\cdot \cdot \cdot \cdot})^2 + np \sum_{l=1}^{p} (\overline{Y}_{\cdot \cdot \cdot l} - \overline{Y}_{\cdot \cdot \cdot \cdot})^2$$

$$+ n \sum_{j=1}^{p} \sum_{k=1}^{p} \sum_{l=1}^{p} (\overline{Y}_{\cdot jkl} - \overline{Y}_{\cdot j \cdot \cdot} - \overline{Y}_{\cdot \cdot k \cdot} - \overline{Y}_{\cdot \cdot \cdot l} + 2\overline{Y}_{\cdot \cdot \cdot \cdot})^2 + \sum_{i=1}^{n} \sum_{j=1}^{p} \sum_{k=1}^{p} \sum_{l=1}^{p} (Y_{ijkl} - \overline{Y}_{\cdot jkl})^2$$

plus ten cross-product terms that equal zero. In the derivation, remember that the subscripts j, k, and l are redundant; once the row and column, for example, are specified the treatment level is determined. Summation is performed over only np^2 scores rather than np^3 scores as $\sum_{i=1}^{n} \sum_{j=1}^{p} \sum_{k=1}^{p} \sum_{l=1}^{p}$ would suggest. Derive the computational formulas for *SSA* and *SSW-CELL* from the deviation formulas above.

11.   [7.6] Under what conditions is it appropriate to use a pooled error term in testing a hypothesis concerning a population contrast?

†12.   [7.7] What condition do estimated observations obtained by Yates's procedure satisfy?

13.   [7.7] Assume that data for the following Latin square where n is equal to one have been obtained and that scores Y_{221} and Y_{432} are missing.

	c_1	c_2	c_3	c_4
b_1	a_1	a_2	a_3	a_4
b_2	a_2	a_3	a_4	a_1
b_3	a_3	a_4	a_1	a_2
b_4	a_4	a_1	a_2	a_3

†a) Compute e_1 and e_2.

†b) Compute e_2 and e_3.

c) Compute e_2 and e_4.

†14.   [7.8] Suppose that the efficiency of a type LS-4 design relative to that for a type CR-4 design was found to be 1.35. Interpret this relative efficiency index.

†15.   Exercise 5 described an experiment concerning the effect of three levels of staff disciplinary intervention on patient morale.

a) Write a regression model with $h - 1 = 6$ independent variables for this type LS-3 design.

b) Specify the null hypothesis in terms of β_j's for treatment levels.

c) If dummy coding is used, how do the parameters of the regression model correspond to those of the type LS-4 experimental design model?

d) (i) Write the reduced regression model for testing the null hypothesis for treatment levels. (ii) Write the formula for the regression sum of squares for treatment levels in terms of the reduced and full models.

e) [7.9] Write an unrestricted full rank experimental design model for this type LS-3 design.

f) [7.9] Write the $\mathbf{C}'$ and $\mathbf{R}'$ matrices for testing the null hypotheses for A, B, C, and residual. Assume that $\boldsymbol{\mu}' = [\mu_{111} \ \mu_{123} \ \mu_{132} \ \mu_{212} \ \mu_{221} \ \cdots \ \mu_{331}]$.

g) [7.9] Use the full rank experimental design model approach to test the null hypotheses for A, B, C, and residual; use *MSWCELL* as the error term and let $\alpha = .05$. (This problem involves the inversion of 2 × 2 matrices and a 9 × 9 diagonal matrix. The computations can be done without the aid of a computer.)

†16. Exercise 6 described an experiment concerning the effect of five levels of associative strength of adjective-noun pairs on recognition threshold.

    a) Write a regression model with $h - 1 = 12$ independent variables for this type LS-5 design.

    b) Specify the null hypothesis in terms of β's for (i) treatment levels, (ii) rows, and (iii) columns.

    c) If dummy coding is used, how do the parameters of the regression model correspond to those of the type LS-5 experimental design model?

    d) (i) Write the reduced model for testing the null hypothesis for treatment levels. (ii) Write the formula for the regression sum of squares for treatment levels in terms of the reduced and full models.

    e) [7.9] Write a restricted full rank experimental design model for this type LS-5 design.

    f) [7.9] Write the $\mathbf{C}'$ and $\mathbf{R}'$ matrices for testing the null hypotheses for A, B, and C. Assume that $\boldsymbol{\mu}' = [\mu_{111} \ \mu_{125} \ \mu_{134} \ \mu_{143} \ \mu_{152} \ \mu_{212} \ \mu_{221} \ \cdots \ \mu_{551}]$. (*Note*: It is not necessary to reduce $\mathbf{R}'$ to the -1, 0, 1 form.)

    g) [7.9] Use the full rank experimental design model approach to test the overall null hypotheses for A, B, and C; let $\alpha = .01$. (This problem and 16(h) involve the inversion of large matrices. A computer with a matrix inversion program should be used.)

    h) [7.9] (i) Compute $\hat{\boldsymbol{\mu}}_R'$. (ii) Use Tukey's multiple comparison statistic to determine which treatment A population means differ.

17. Exercise 7 described an experiment that investigated the von Restorff effect.

    a) Write a regression model with $h - 1 = 9$ independent variables for this type LS-4 design.

    b) Specify the null hypotheses in terms of β's for (i) treatment levels, (ii) rows, and (iii) columns.

    c) If dummy coding is used, how do the parameters of the regression model correspond to those of the type LS-4 experimental design model?

    d) (i) Write the reduced model for testing the null hypothesis for treatment levels. (ii) Write the formula for the regression sum of squares for treatment levels in terms of the reduced and full models.

    e) [7.9] Write a restricted full rank experimental design model for this type LS-4 design.

    f) [7.9] Write the $\mathbf{C}'$ and $\mathbf{R}'$ matrices for testing the null hypotheses for A, B, and C. Assume that $\boldsymbol{\mu}' = [\mu_{111} \ \mu_{124} \ \mu_{133} \ \mu_{142} \ \mu_{212} \ \mu_{221} \ \cdots \ \mu_{441}]$.

    g) [7.9] Use the full rank experimental design model approach to test the overall null hypotheses for A, B, and C; let $\alpha = .01$. (This problem and 17(h) involve the inversion of large matrices. A computer with a matrix inversion program should be used.)

    h) [7.9] (i) Compute $\hat{\boldsymbol{\mu}}_R'$. (ii) Use Dunnett's multiple comparison statistic to determine which treatment A population means differ from the mean of the control population, $\mu_{4..}$; test nondirectional hypotheses.

18. [7.10] **†a)** Construct a 3 × 3 Graeco-Latin square.

b) Construct a 5×5 Graeco-Latin square.

†c) Construct a 4×4 hyper-Graeco-Latin square; use lowercase Latin letters for the third square.

†19. [7.10] A number of rules of thumb have been developed for determining the existence of various size orthogonal Latin squares. For squares of size 10×10 to 20×20, use the rules to support the existence of orthogonal squares; in each case cite the rule you used.

20. The effect of strength of association between two words on learning to read the second word in a two-word pair was investigated. A pretest was used to select 32 first-graders who were able to read the first word of each pair but not the second word. Following the pretest, word association training was given in which the response words were repeated 30 times after the stimulus words, treatment level a_1; 10 times, a_2; 5 times, a_3; and 0 times, a_4. The children were then shown the stimulus-response words printed on 5×8 index cards and given a series of reading training trials. The four two-word pairs, corresponding to the levels of variable B, were presented in four different sequences (variable C). Variable D represented the four assistants who conducted the reading training. The children were randomly assigned to the cells of the following 4×4 Graeco-Latin square. The dependent variable was the number of words recognized on a word-recognition test. (Experiment suggested by Samuels, S. J. and Wittrock, M. Word-association strength and learning to read. *Journal of Educational Psychology*, 1969, *60*, 248–252.)

	c_1	c_2	c_3	c_4
b_1	a_1d_1	a_2d_2	a_3d_3	a_4d_4
	4	3	3	0
	2	1	4	0
b_2	a_2d_4	a_1d_3	a_4d_2	a_3d_1
	4	3	1	2
	1	3	2	1
b_3	a_3d_2	a_4d_1	a_1d_4	a_2d_3
	2	1	2	3
	3	2	1	1
b_4	a_4d_3	a_3d_4	a_2d_1	a_1d_2
	1	1	2	2
	1	2	2	1

†a) [7.10] Test the hypotheses $\alpha_j = 0$ for all j, $\beta_k = 0$ for all k, $\sigma_\gamma^2 = 0$, and $\sigma_\delta^2 = 0$; use a pooled error term here and in (c) and (d). Let $\alpha = .05$.

†b) [7.3 and 7.10] Is it reasonable to assume that the treatment and nuisance variables do not interact? Explain.

†c) [7.3 and 7.10] Calculate the power of the test of H_0: $\alpha_j = 0$ for all j.

†d) [7.4 and 7.10] Use Dunnett's statistic to determine which population means differ from the mean of the control population; test nondirectional hypotheses.

e) Prepare a "results and discussion section" for the *Journal of Educational Psychology*.

†21. **a)** Write a regression model with $h - 1 = 12$ independent variables for the type GLS-4 design described in Exercise 20.

b) Specify the treatment null hypothesis in terms of β's.

c) If dummy coding is used, how do the parameters of the regression model correspond to those of the type GLS-4 experimental design model?

d) (i) Write the reduced model for testing the null hypothesis for treatment levels. (ii) Write the formula for the regression sum of squares for treatment levels in terms of the reduced and full models.

e) [7.9 and 7.10] Write a restricted full rank experimental design model for this type GLS-4 design.

f) [7.9 and 7.10] Write the $\mathbf{C}'$ and $\mathbf{R}'$ matrices for testing the null hypotheses for A, B, C, D, and residual. Assume that $\boldsymbol{\mu}' = [\mu_{1111} \ \mu_{1223} \ \mu_{1334} \ \mu_{1442} \ \mu_{2122} \ \mu_{2214} \cdot \cdot \cdot \mu_{4413}]$.

g) [7.9 and 7.10] Use the full rank experimental design model approach to test the hypotheses in 21(f). Use a pooled error term; let $\alpha = .05$.

h) [7.9 and 7.10] (i) Compute $\boldsymbol{\mu}'_R$. (ii) Use Dunnett's statistic to determine which population means differ from the mean of the control population; test nondirectional hypotheses.

22. Make a schematic comparison like Figure 7.8-1 for type CR-5, RB-5, LS-5, and GLS-5 designs.

8 COMPLETELY RANDOMIZED FACTORIAL DESIGN WITH TWO TREATMENTS

8.1 INTRODUCTION TO FACTORIAL EXPERIMENTS

A *factorial experiment* is one in which the effects of a particular combination of the levels of two or more treatments are investigated simultaneously. Factorial designs are constructed from the three basic *building block* designs described in Chapters 4, 6, and 7. They are the most widely used designs in the behavioral sciences, as an examination of recent volumes of behavioral science journals will verify. The design of factorial experiments using completely randomized designs is described here and in Chapter 9. Also described in Chapter 9 is a factorial experiment that uses randomized block designs. In Chapter 11, factorial experiments that combine both type CR-p and RB-p designs in the same experiment are described. These designs are referred to as split-plot factorial designs. Two additional variations of factorial experiments, a confounded factorial design and a fractional factorial design, are described in Chapters 12 and 13, respectively.

One assumption that has guided the writing of this and subsequent chapters is that complex designs appear less complex and are more easily understood when they are perceived as consisting of elementary building block designs. To help the reader identify the building block design, the letters CR, RB, or LS are used in the designation of factorial designs. One exception involves the split-plot factorial design whose designation is well established by contemporary usage.

8.2 DESCRIPTION OF COMPLETELY RANDOMIZED FACTORIAL DESIGN

The simplest factorial experiment, from the standpoint of data analysis and assignment of treatment levels to experimental units, is the *completely randomized factorial design*. A design with two treatments is designated as a type CRF-pq design, where p and q stand for the number of levels of treatments A and B, respectively, and F indicates that it is a factorial design. A completely randomized factorial design is appropriate for experiments that meet, in addition to the assumptions of a type CR-p design described in Chapter 4, the following conditions.

1. Two or more treatments with each treatment having two or more levels.

2. All levels of each treatment are investigated in combination with all levels of every other treatment. If there are p levels of one treatment and q levels of another treatment, the experiment contains pq treatment combinations.

3. Random assignment of treatment combinations to experimental units; each experimental unit must receive only one combination.

TREATMENT DESIGNATION SCHEME

This is a good place to describe in some detail the scheme that is used to designate treatments in factorial experiments. Treatments are designated by the capital letters A through G. A particular but unspecified level of a treatment is designated by lowercase letters a through g and a lowercase subscript, for example, a_j, b_k, and so on. If two unspecified levels of a treatment must be differentiated, a prime is used after one of the subscripts. For example, two levels of treatment A are a_j and $a_{j'}$. Specific levels of a treatment are designated by number subscripts, for example, a_1, a_2, and so on. Besides specifying a particular treatment level, it is also necessary to indicate the number of levels of a treatment. Seven lowercase letters, p, q, and so on, are used for this purpose. The complete designation scheme for treatments is as follows.

$$
\begin{array}{lll}
A & a_j & j = 1, \ldots, p \\
B & b_k & k = 1, \ldots, q \\
C & c_l & l = 1, \ldots, r \\
D & d_m & m = 1, \ldots, t
\end{array}
$$

$$E \quad e_o \quad o = 1, \ldots, u$$
$$F \quad f_h \quad h = 1, \ldots, v$$
$$G \quad g_z \quad z = 1, \ldots, w$$

The letter S is used in the notation scheme to designate blocks or samples of subjects (experimental units). The letters s_i refer to a particular but unspecified block or subject. There are n levels of s_i; that is, $i = 1, \ldots, n$. This designation scheme uses 24 letters of the alphabet. The remaining two letters, X and Y, are used to designate individual scores.

ASSIGNMENT OF TREATMENT COMBINATIONS TO EXPERIMENTAL UNITS

A block diagram of a type CRF-23 design is shown in Figure 8.2-1. All possible combinations of levels of the two treatments occur together in the experiment. The treatments are said to be *completely crossed*. This produces $pq = 6$ treatment combinations: $a_1b_1, a_1b_2, \ldots, a_2b_3$. It is assumed that a random sample of npq experimental units are available. The sample is randomly divided into pq subsamples of size n. The pq treatment combinations are then randomly assigned to the pq subsamples. If one or more of the treatments involve an organismic variable, for example sex, IQ, or year in school, it is necessary to deviate from this randomization procedure. Obviously one cannot assign the treatment combination "male and high reinforcement" to a subsample containing both males and females. In such cases, the randomization procedure for a generalized randomized block design in Section 6.10 is used. According to convention, the resulting design is called a completely randomized factorial design if at least two of the variables are regarded as treatments. In Figure 8.2-1 the letters $S_1, S_2, \ldots, S_6$ refer to six subsamples of n subjects. The minimum number of subjects required for a type CRF-23 design is equal to $n(2)(3) = 6$, where $n = 1$. However, it is desirable to have more than one experimental unit in each of the pq cells. If n is equal to 1, it is not possible to compute a within-cell estimate of experimental error. In this case, an interaction must be used to estimate experimental error under the often tenuous assumption that all interaction effects are equal to zero.

FIGURE 8.2-1   Block diagram of a type CRF-23 design and the two completely randomized designs from which it is constructed. The type CRF-23 design contains all possible combinations of the levels of treatments A and B from the type CR-2 and CR-3 designs. Furthermore, the six treatment combinations, $a_1b_1, a_1b_2, \ldots, a_2b_3$, are assigned randomly to the six random samples of subjects, $S_1, S_2, \ldots, S_6$.

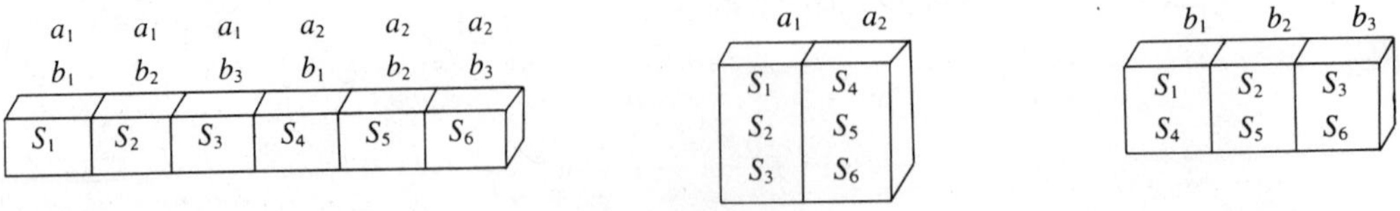

Type CRF-23 design   Type CR-2 design   Type CR-3 design

If $n = 5$, the preceding experiment would require $(5)(2)(3) = 30$ subjects. As the number of levels of a treatment increases, the number of subjects required increases rapidly. For example, if treatment A had three instead of two levels, the number of subjects that would be required is $(5)(3)(3) = 45$.

Recall that the number of subjects in each treatment level of a type CR-*p* design does not have to be equal. This is also true for type CRF-*pq* designs. However, the analysis and interpretation is greatly simplified if the n's in each cell are equal or if a particular proportionality, which is described in Section 8.13, exists among the cell n's. Accordingly, experimenters are strongly encouraged to have equal n's.

8.3 COMPUTATIONAL EXAMPLE FOR TYPE CRF-*pq* DESIGN

Assume that a statistical consultant has been called in to assist the police department of a large metropolitan city evaluate its human relations course for new officers. The independent variables are type of beat to which officers are assigned during the course, treatment A, and length of the course, treatment B. Treatment A has three levels: upper-class beat, a_1, middle-class beat, a_2, and inner-city beat, a_3. Treatment B also has three levels: five hours of human relations training, b_1, ten hours, b_2, and fifteen hours, b_3. The dependent variable is attitude toward minority groups following the course. A test developed and validated previously by the consultant is used to measure the dependent variable.

The research hypotheses leading to the experiment can be evaluated by testing the following null hypotheses.

$$H_0: \quad \alpha_j = 0 \quad \text{for all } j \qquad \text{or} \qquad \mu_{1.} = \mu_{2.} = \cdots = \mu_{p.}$$

$$H_1: \quad \alpha_j \neq 0 \quad \text{for some } j \qquad \text{or} \qquad \mu_{j.} \neq \mu_{j'.} \quad \text{for some } j \text{ and } j'(j \neq j')$$

$$H_0: \quad \beta_k = 0 \quad \text{for all } k \qquad \text{or} \qquad \mu_{.1} = \mu_{.2} = \cdots = \mu_{.q}$$

$$H_1: \quad \beta_k \neq 0 \quad \text{for some } k \qquad \text{or} \qquad \mu_{.k} \neq \mu_{.k'} \quad \text{for some } k \text{ and } k'(k \neq k')$$

$$H_0: \quad (\alpha\beta)_{jk} = 0 \quad \text{for all } j, k \qquad \text{or} \qquad \mu_{jk} - \mu_{j'k} - \mu_{jk'} + \mu_{j'k'} = 0$$
$$\text{for all } j, j', k, k'$$

$$H_1: \quad (\alpha\beta)_{jk} \neq 0 \quad \text{for some } j, k \qquad \text{or} \qquad \mu_{jk} - \mu_{j'k} - \mu_{jk'} + \mu_{j'k'} \neq 0$$
$$\text{for some } j, j', k, k'$$

The first two null hypotheses are the familiar hypotheses that all treatment effects are equal to zero. The third null hypothesis, that $(\alpha\beta)_{jk} = 0$ for all j and k, states that all population interaction effects for treatments A and B are equal to zero. Two treatments are said to interact if differences in performance under the levels of one treatment are different at two or more levels of the other treatment. An interaction can be thought of as the joint effects of treatments that are different from the sum of their individual effects. We will have more to say about interactions shortly. The level of significance adopted for all tests is .05.

To test the null hypotheses, a random sample of 45 police recruits was obtained. The sample was randomly divided into nine subsamples of five recruits each. The nine treatment combinations of the type CRF-33 design were then randomly assigned to the nine subsamples. All treatment levels of interest to the police chief are included in the experiment. Thus a fixed-effects model, model I, applies to the experiment. Before proceeding to test the statistical hypotheses, the raw data from the experiment were eyeballed and descriptive statistics—means and variances—were computed. This important step in the analysis of data is all too often omitted by novice experimenters. Such an examination may turn up suspected data recording errors and other anomalies requiring further investigation. Sometimes an examination of treatment means reveals only trivial effects that would be of no practical value even though they were statistically significant. An examination of means, variances, and a plot of residuals $\hat{\epsilon}$ versus estimated values $\hat{Y}$ may also point to assumptions of the model that require additional scrutiny.

The layout of the type CRF-33 design and computational procedures are shown in Table 8.3-1. The labels of the AB and ABS Summary Tables represent a convenient way of designating the sources of variation represented by scores in these tables—the more letters in the label, the more sources of variation. For example, both the ABS and AB Summary Tables contain information concerning variation among the levels of treatments A and B, but information concerning variation among subjects within cells, S, can only be obtained from the ABS Summary Table. To partition the total sum of squares in a factorial design, it is convenient to begin by constructing a summary table that contains all of the sources of variation that the design estimates. This summary table can be collapsed into smaller summary tables by summing the

---

TABLE 8.3-1   Computational Procedures for a Type CRF-33 Design

(i) Data and notation [Y_{ijk} denotes a score for experimental unit i in treatment combination jk; $i = 1, \ldots, n$ experimental units (s_i); $j = 1, \ldots, p$ levels of treatment $A(a_j)$; $k = 1, \ldots, q$ levels of treatment $B(b_k)$]:

ABS Summary Table
Entry is Y_{ijk}

a_1 b_1	a_1 b_2	a_1 b_3	a_2 b_1	a_2 b_2	a_2 b_3	a_3 b_1	a_3 b_2	a_3 b_3
24	44	38	30	35	26	21	41	42
33	36	29	21	40	27	18	39	52
37	25	28	39	27	36	10	50	53
29	27	47	26	31	46	31	36	49
42	43	48	34	22	45	20	34	64

TABLE 8.3-1 (continued)

AB Summary Table

Entry is $\sum\limits_{i=1}^{n} Y_{ijk}$

	b_1	b_2	b_3	$\sum\limits_{i=1}^{n}\sum\limits_{k=1}^{q} Y_{ijk}$
	$n = 5$			
a_1	165	175	190	530
a_2	150	155	180	485
a_3	100	200	260	560

$\sum\limits_{i=1}^{n}\sum\limits_{j=1}^{p} Y_{ijk} = 415 \qquad 530 \qquad 630$

(ii) Computational symbols:

$$\sum_{i=1}^{n}\sum_{j=1}^{p}\sum_{k=1}^{q} Y_{ijk} = 24 + 44 + \cdots + 64 = 1575.000$$

$$\frac{\left(\sum\limits_{i=1}^{n}\sum\limits_{j=1}^{p}\sum\limits_{k=1}^{q} Y_{ijk}\right)^2}{npq} = [Y] = \frac{(1575)^2}{(5)(3)(3)} = 55125.000$$

$$\sum_{i=1}^{n}\sum_{j=1}^{p}\sum_{k=1}^{q} Y_{ijk}^2 = [ABS] = (24)^2 + (44)^2 + \cdots + (64)^2 = 60345.000$$

$$\sum_{j=1}^{p}\frac{\left(\sum\limits_{i=1}^{n}\sum\limits_{k=1}^{q} Y_{ijk}\right)^2}{nq} = [A] = \frac{(530)^2}{(5)(3)} + \frac{(485)^2}{(5)(3)} + \frac{(560)^2}{(5)(3)} = 55315.000$$

$$\sum_{k=1}^{q}\frac{\left(\sum\limits_{i=1}^{n}\sum\limits_{j=1}^{p} Y_{ijk}\right)^2}{np} = [B] = \frac{(415)^2}{(5)(3)} + \frac{(530)^2}{(5)(3)} + \frac{(630)^2}{(5)(3)} = 56668.333$$

$$\sum_{j=1}^{p}\sum_{k=1}^{q}\frac{\left(\sum\limits_{i=1}^{n} Y_{ijk}\right)^2}{n} = [AB] = \frac{(165)^2}{5} + \frac{(175)^2}{5} + \cdots + \frac{(260)^2}{5} = 58095.000$$

(iii) Computational formulas:

$SSTO = [ABS] - [Y] = 5220.000$ $\qquad$ $SSAB = [AB] - [A] - [B] + [Y] = 1236.667$

$SSA = [A] - [Y] = 190.000$ $\qquad$ $SSWCELL = [ABS] - [AB] = 2250.000$

$SSB = [B] - [Y] = 1543.333$

entries over one or more sources of variation, a procedure that is continued until summary tables have been constructed from which all sums of squares can be computed. For a type CRF-pq design, only two summary tables are required: ABS and AB.

The analysis of variance is summarized in Table 8.3-2. The table contains three *a priori* families of contrasts, A, B, and AB; the overall null hypothesis for each family is evaluated at $\alpha = .05$ level of significance. The conceptual unit for error rate in this example is the individual family. The probability of falsely rejecting one or more overall null hypotheses (error rate experimentwise) is equal to $1 - (1 - .05)^3 = .14$. The relative merits of holding constant error rate per family or some larger conceptual unit, such as the experiment, have been widely debated. This issue is discussed in Section 3.3. It can be shown (see Box, Hunter, and Hunter, 1978, Ch. 7) that the three a priori families of contrasts are also mutually orthogonal. The practice adopted here is to make the family the conceptual unit for error rate if, as in Table 8.3-2, the families are a priori and orthogonal. Indeed, a type CRF-pq design can be regarded as a set of type CR-p designs with prespecified orthogonal families of comparisons, that is, A, B, and AB. According to the analysis in Table 8.3-2, the null hypotheses for treatment B and the AB interaction are rejected. Procedures for following up a significant interaction are discussed in the next section and in Section 8.6. If the AB interaction had not been significant, we would have performed multiple comparisons among the means. These procedures are discussed in Section 8.5.

GRAPHIC REPRESENTATION OF INTERACTION

A significant interaction is a signal that the interpretation of tests of treatments A and B must be qualified. This point can be clarified by examining graphs of the AB interaction in Figure 8.3-1(a and b). Differences among the sample treatment means

FIGURE 8.3-1   Graphs of the interaction between treatments A and B.

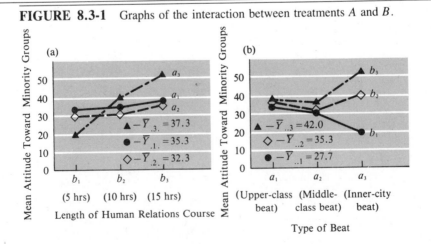

TABLE 8.3-2   ANOVA Table for Type CRF-33 Design

Source	SS	df		MS	F	$E(MS)$ Model I
1 A (Type of beat)	190.000	$p - 1 = 2$		95.000	[¹/₄] 1.52	$\sigma_\epsilon^2 + nq \sum_{j=1}^{p} \alpha_j^2/(p - 1)$
2 B (Training duration)	1543.333	$q - 1 = 2$		771.666	[²/₄] 12.35*	$\sigma_\epsilon^2 + np \sum_{k=1}^{q} \beta_k^2/(q - 1)$
3 AB	1236.667	$(p - 1)(q - 1) = 4$		309.167	[³/₄] 4.95*	$\sigma_\epsilon^2 + n \sum_{j=1}^{p} \sum_{k=1}^{q} (\alpha\beta)_{jk}^2/(p - 1)(q - 1)$
4 Within cell	2250.000	$pq(n - 1) = 36$		62.500		σ_ϵ^2
5 Total	5220.000	$npq - 1 = 44$				

*$p < .01$

shown in Figure 8.3-1(a) for a_1, a_2, and a_3 are quite small ($\overline{Y}_{.1.} - \overline{Y}_{.2.} = 3$, $\overline{Y}_{.1.} - \overline{Y}_{.3.} = -2$, and $\overline{Y}_{.2.} - \overline{Y}_{.3.} = -5$), which is consistent with the insignificant test for treatment A. But this tells only part of the story. If we examine differences among the a_1, a_2, and a_3 means at each level of treatment B, we discover some large differences. For example, the differences between a_1 and a_3 at b_1, b_2, and b_3 are $\overline{Y}_{.11} - \overline{Y}_{.31} = 13$; $\overline{Y}_{.12} - \overline{Y}_{.32} = -5$, and $\overline{Y}_{.13} - \overline{Y}_{.33} = -14$, respectively. The insignificant test of treatment A is misleading because differences among the means are not uniform across the levels of treatment B. The significant test for treatment B is also misleading. Differences among the treatment means for b_1, b_2, and b_3 in Figure 8.3-1(b) are fairly small at the first two levels of treatment A and are large only at a_3. As these examples illustrate, a significant interaction is a red flag signaling that tests of treatments provide only a partial and often misleading picture of what has happened in an experiment.

The first step in interpreting an interaction is to graph the means as in Figure 8.3-1. Additional examples of interactions are shown in Figure 8.3-2(a and b). As is customary, the means have been connected by straight lines. A significant interaction always appears as nonparallel lines. If treatments do not interact, the lines are parallel as in Figure 8.3-2(c). A test of H_0: $(\alpha\beta)_{jk} = 0$ for all j and k or $\mu_{jk} - \mu_{j'k} - \mu_{jk'} + \mu_{j'k'} = 0$ for all j, j', k, and k' helps an experimenter decide whether chance sampling error is a reasonable explanation for the presence of non-

FIGURE 8.3-2   Treatments A and B are said to interact if any portion of the lines connecting the means for a_j and $a_{j'}$ are not parallel. Examples (a) and (b) illustrate interaction between the two treatments. Example (c) illustrates treatments that do not interact. The necessary condition for the lines in (c) to be parallel is that $(\mu_{11} - \mu_{21}) - (\mu_{12} - \mu_{22}) = 0$ or $\mu_{11} - \mu_{21} - \mu_{12} + \mu_{22} = 0$.

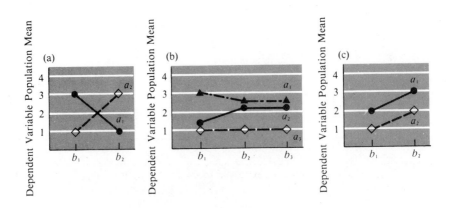

parallel lines connecting sample means. Additional steps in interpreting interactions are discussed in Section 8.6.*

POWER OF THE ANALYSIS OF VARIANCE *F* TEST

Procedures for estimating power and the number of subjects necessary to achieve a specified power are described in Section 4.3. These procedures generalize directly to a completely randomized factorial design. Formulas for computing $\hat{\phi}$ are as follows:

$$\hat{\phi} = \frac{\sqrt{\sum_{j=1}^{p} \hat{\alpha}_j^2 / p}}{\hat{\sigma}_\epsilon / \sqrt{nq}} \qquad \begin{array}{l} \nu_1 = p - 1 \\[2mm] \nu_2 = pq(n - 1) \end{array}$$

$$\hat{\phi} = \frac{\sqrt{\sum_{k=1}^{q} \hat{\beta}_k^2 / q}}{\hat{\sigma}_\epsilon / \sqrt{np}} \qquad \begin{array}{l} \nu_1 = q - 1 \\[2mm] \nu_2 = pq(n - 1) \end{array}$$

$$\hat{\phi} = \frac{\sqrt{\sum_{j=1}^{p} \sum_{k=1}^{q} (\hat{\alpha\beta})_{jk}^2 / (p - 1)(q - 1) + 1}}{\hat{\sigma}_\epsilon / \sqrt{n}} \qquad \begin{array}{l} \nu_1 = (p - 1)(q - 1) \\[2mm] \nu_2 = pq(n - 1). \end{array}$$

Note that the denominator of $\sum_{j=1}^{p} \hat{\alpha}_j^2$, $\sum_{k=1}^{q} \hat{\beta}_k^2$, and $\sum_{j=1}^{p} \sum_{k=1}^{q} (\hat{\alpha\beta})_{jk}^2$ is the degrees of freedom for the respective effects *plus one*.

8.4 EXPERIMENTAL DESIGN MODEL FOR A TYPE CRF-*pq* DESIGN

ASSUMPTIONS FOR FIXED-EFFECTS MODEL

A score, Y_{ijk}, in a completely randomized factorial design is a composite that reflects the effects of treatments A and B, the AB interaction, and all other sources of variation that affect Y_{ijk}. These latter sources of variation are collectively referred to as experimental error. Our conjecture about Y_{ijk} can be expressed by an experimental design model.

$$Y_{ijk} = \mu + (\mu_{j\cdot} - \mu) + (\mu_{\cdot k} - \mu) + (\mu_{jk} - \mu_{j\cdot} - \mu_{\cdot k} + \mu) + (Y_{ijk} - \mu_{jk})$$

(8.4-1) $$= \mu + \alpha_j + \beta_k + (\alpha\beta)_{jk} + \epsilon_{i(jk)}$$

$$(i = 1, \ldots, n; j = 1, \ldots, p; k = 1, \ldots, q)$$

---

*   Monlezun (1979) discusses extensions of graphical analysis for experiments with three treatments.

where Y_{ijk} = a score for the ith experimental unit in treatment combination $a_j b_k$

μ = the overall population mean

α_j = the effect of treatment level j and is subject to the restriction $\Sigma_{j=1}^p \alpha_j = 0$

β_k = the effect of treatment level k and is subject to the restriction $\Sigma_{k=1}^q \beta_k = 0$

$(\alpha\beta)_{jk}$ = the joint effect of treatment levels j and k (interaction of α_j and β_k) and is subject to the restrictions $\Sigma_{j=1}^p (\alpha\beta)_{jk} = 0$ and $\Sigma_{k=1}^q (\alpha\beta)_{jk} = 0$

$\epsilon_{i(jk)}$ = the experimental error that is $NID(0, \sigma_\epsilon^2)$.

Key features of the model can be summarized as follows. (1) The linear model is assumed to reflect all sources of variation that affect Y_{ijk}. (2) Treatments A and B and the AB interaction represent fixed effects; thus, (8.4-1) is a fixed-effects model (model I). (3) The error effect, $\epsilon_{i(jk)}$, is (a) independent of other ϵ's and is (b) normally distributed within each jk treatment population with (c) mean equal to zero and (d) variance equal to σ_ϵ^2. Assumption 3(d) can be restated as $\sigma_{11}^2 = \sigma_{21}^2 = \cdots = \sigma_{pq}^2$, the familiar homogeneity of variance assumption. Procedures for using sample data to determine the tenability of some of these assumptions are discussed in Section 2.5. Also discussed there are the consequences of violating the assumptions.

The values of the parameters μ, α_j, β_k, $(\alpha\beta)_{jk}$, and $\epsilon_{i(jk)}$ in model (8.4-1) are unknown, but they can be estimated from sample data as follows.

Statistic	*Parameter Estimated*
$\overline{Y}...$	μ
$\overline{Y}_{\cdot j \cdot} - \overline{Y}...$	α_j or $\mu_{j\cdot} - \mu$
$\overline{Y}_{\cdot\cdot k} - \overline{Y}...$	β_k or $\mu_{\cdot k} - \mu$
$\overline{Y}_{\cdot jk} - \overline{Y}_{\cdot j\cdot} - \overline{Y}_{\cdot\cdot k} + \overline{Y}...$	$(\alpha\beta)_{jk}$
$Y_{ijk} - \overline{Y}_{\cdot jk}$	$\epsilon_{i(jk)}$

PARTITION OF THE TOTAL SUM OF SQUARES

The partition of the total sum of squares, which is the basis for the analysis in Table 8.3-2, is obtained by replacing the parameters in model (8.4-1) by statistics and rearranging terms as follows.

(8.4-2)
$$Y_{ijk} - \mu = \alpha_j + \beta_k + (\alpha\beta)_{jk}$$
$$Y_{ijk} - \overline{Y}... = (\overline{Y}_{\cdot j\cdot} - \overline{Y}...) + (\overline{Y}_{\cdot\cdot k} - \overline{Y}...) + (\overline{Y}_{\cdot jk} - \overline{Y}_{\cdot j\cdot} - \overline{Y}_{\cdot\cdot k} + \overline{Y}...)$$
(Total   (*A* treatment   (*B* treatment   (Interaction effect)
deviation)   effect)   effect)

$$+ \; \epsilon_{i(jk)}$$
$$+ \; (Y_{ijk} - \overline{Y}_{\cdot jk})$$
(Within-cell
effect)

Squaring both sides of (8.4-2) and summing over all of the observations following the procedure in Sections 2.2 and 6.1 leads to the following partition of the total sum of squares.*

(8.4-3)
$$\sum_{i=1}^{n} \sum_{j=1}^{p} \sum_{k=1}^{q} (Y_{ijk} - \overline{Y}...)^2 = nq \sum_{j=1}^{p} (\overline{Y}_{\cdot j\cdot} - \overline{Y}...)^2 + np \sum_{k=1}^{q} (\overline{Y}_{\cdot\cdot k} - \overline{Y}...)^2$$
$$SSTO \qquad\qquad\quad = SSA \qquad\qquad\quad + SSB$$

$$+ \; n \sum_{j=1}^{p} \sum_{k=1}^{q} (\overline{Y}_{\cdot jk} - \overline{Y}_{\cdot j\cdot} - \overline{Y}_{\cdot\cdot k} + \overline{Y}...)^2 + \sum_{i=1}^{n} \sum_{j=1}^{p} \sum_{k=1}^{q} (Y_{ijk} - \overline{Y}_{\cdot jk})^2$$
$$+ \; SSAB \qquad\qquad\qquad\qquad\qquad + SSWCELL$$

The sum of squares formulas in (8.4-3) are not the most convenient for computational purposes. More convenient formulas are given in Table 8.3-1.** Mean squares are obtained by dividing each sum of squares by its degrees of freedom.

$$MSTO = SSTO/(npq - 1)$$
$$MSA = SSA/(p - 1)$$
$$MSB = SSB/(q - 1)$$
$$MSAB = SSAB/(p - 1)(q - 1)$$
$$MSWCELL = SSWCELL/pq(n - 1)$$

EXPECTED VALUES OF MEAN SQUARES FOR FIXED-EFFECTS, MIXED, AND RANDOM-EFFECTS MODELS

Random sampling was not used to determine the levels of treatments included in the police attitude experiment described in Section 8.3. Instead the particular levels were included because they were the only levels of interest to the experimenter. As discussed in Sections 2.3 and 4.7, the model for such experiments is called a fixed-effects model (model I) and the treatment and interaction effects are called fixed effects. The $E(MS)$'s given in Table 8.3-2 are appropriate for this model. A random-effects model (model II) results when the levels of all treatments in the experiment are selected randomly. The treatment and interaction effects are called random effects. If the levels for one or more, but not all, of the treatments are selected randomly, the model is a

---

*	See Exercise 8 for the derivation.
**	See Exercise 9 for the derivation.

mixed model (model III). The treatment effects are fixed or random depending on whether they were randomly sampled. Interaction effects are random effects if they involve one or more random effects, otherwise they are fixed effects.

A comparison of $E(MS)$'s for the fixed-effects, mixed, and random-effects models is shown in Table 8.4-1. Much confusion concerning the correct error term to use in testing main effects and interactions has arisen because of failure to distinguish among these three models. Recall from Section 4.7 that the numerator of an F ratio should contain, in addition to the expected value terms in the denominator, one additional term—the one being tested. Based on the $E(MS)$'s in Table 8.4-1, the F ratios for testing treatment A for models I and II are, respectively,

$$F = \frac{MSA}{MSWCELL} \quad \text{because} \quad \frac{E(MSA)}{E(MSWCELL)} = \frac{\sigma_\epsilon^2 + nq \sum_{j=1}^{p} \alpha_j^2 /(p-1)}{\sigma_\epsilon^2}$$

and

$$F = \frac{MSA}{MSAB} \quad \text{because} \quad \frac{E(MSA)}{E(MSAB)} = \frac{\sigma_\epsilon^2 + n\sigma_{\alpha\beta}^2 + nq\sigma_\alpha^2}{\sigma_\epsilon^2 + n\sigma_{\alpha\beta}^2}.$$

The proper error term for other tests can be determined from the table.

The type of effect, fixed versus random, also determines the nature of the null hypothesis that is tested. For example, the null hypotheses for treatment A where the α_j's are fixed effects or random effects are, respectively,

$$\text{H}_0\text{:} \quad \alpha_j = 0 \quad \text{for all } j \quad \text{and} \quad \text{H}_0\text{:} \quad \sigma_\alpha^2 = 0.$$

The null hypothesis for fixed effects concerns only the $j = 1, \ldots, p$ population treatment effects (α_j's) that are represented in the experiment. For random effects, the null hypothesis concerns all of the P population treatment effects, not just the p effects represented in the experiment ($p < P$). If the null hypothesis for a fixed-effects treatment is rejected, it can be concluded that at least two of the p population treatment effects are not equal. If the null hypothesis for a random-effects treatment is rejected, at least two of the treatment effects in the population of P effects are not equal. The computational procedures for models involving fixed and/or random effects are identical. The models differ in the selection of treatment levels, expected values of the MS's, and the nature of the null hypotheses that are tested.

ASSUMPTIONS FOR RANDOM-EFFECTS AND MIXED MODELS

The assumptions for a random-effects model are as follows.

$$Y_{ijk} = \mu + \alpha_j + \beta_k + (\alpha\beta)_{jk} + \epsilon_{i(jk)}$$
$$(i = 1, \ldots, n; j = 1, \ldots, p; k = 1, \ldots, q)$$

where μ is the overall population mean

TABLE 8.4-1   Expected Values of Mean Squares for a Type CRF-*pq* Design

Mean Square	Model I *A Fixed* *B Fixed*	Model II *A Random* *B Random*	Model III *A Fixed* *B Random*	Model III *A Random* *B Fixed*
1 MSA	$\sigma_\epsilon^2 + nq \sum_{j=1}^{p} \alpha_j^2/(p-1)$	$\sigma_\epsilon^2 + n\sigma_{\alpha\beta}^2 + nq\sigma_\alpha^2$	$\sigma_\epsilon^2 + n\sigma_{\alpha\beta}^2 + nq \sum_{j=1}^{p} \alpha_j^2/(p-1)$	$\sigma_\epsilon^2 + nq\sigma_\alpha^2$
2 MSB	$\sigma_\epsilon^2 + np \sum_{k=1}^{q} \beta_k^2/(q-1)$	$\sigma_\epsilon^2 + n\sigma_{\alpha\beta}^2 + np\sigma_\beta^2$	$\sigma_\epsilon^2 + np\sigma_\beta^2$	$\sigma_\epsilon^2 + n\sigma_{\alpha\beta}^2 + np \sum_{k=1}^{q} \beta_k^2/(q-1)$
3 MSAB	$\sigma_\epsilon^2 + n \sum_{j=1}^{p} \sum_{k=1}^{q} (\alpha\beta)_{jk}^2/(p-1)(q-1)$	$\sigma_\epsilon^2 + n\sigma_{\alpha\beta}^2$	$\sigma_\epsilon^2 + n\sigma_{\alpha\beta}^2$	$\sigma_\epsilon^2 + n\sigma_{\alpha\beta}^2$
4 MSWCELL	σ_ϵ^2	σ_ϵ^2	σ_ϵ^2	σ_ϵ^2

$$\alpha_j \text{ is } NID(0, \sigma_\alpha^2)$$

$$\beta_k \text{ is } NID(0, \sigma_\beta^2)$$

$$(\alpha\beta)_{jk} \text{ is } NID(0, \sigma_{\alpha\beta}^2)$$

$$\epsilon_{i(jk)} \text{ is } NID(0, \sigma_\epsilon^2) \text{ and independent of } \alpha_j, \beta_k, \text{ and } (\alpha\beta)_{jk}$$

As noted in Chapter 2, the assumption of independence is particularly important. The p values of the random variable α_j, for example, that occur in the experiment must be independent of each other and of β_k, $(\alpha\beta)_{jk}$, and $\epsilon_{i(jk)}$. The selection of the treatment levels by random sampling from a much larger population of levels is necessary for independence.

For a mixed model (A fixed), we assume that

1. α_j is the effect of treatment level j and is subject to the restriction $\sum_{j=1}^{p} \alpha_j = 0$.

2. β_k is $NID(0, \sigma_\beta^2)$.

3. $(\alpha\beta)_{jk}$ is normally distributed $(0, [(p-1)/p]\, \sigma_{\alpha\beta}^2)$ and is subject to the restriction that $\sum_{j=1}^{p} (\alpha\beta)_{jk} = 0$ for all k, β_k and $(\alpha\beta)_{jk}$ are independent as are $(\alpha\beta)_{jk}$ and $(\alpha\beta)_{j'k'}$ unless $k = k'$.

4. $\epsilon_{i(jk)}$ is $NID(0, \sigma_\epsilon^2)$, and is independent of β_k and $(\alpha\beta)_{jk}$.

In this model the variance of $(\alpha\beta)_{jk}$ is defined as $[(p-1)/p]\, \sigma_{\alpha\beta}^2$ rather than $\sigma_{\alpha\beta}^2$ in order to simplify the expected mean squares. For a discussion of alternative mixed models see Hocking (1973).

8.5   PROCEDURES FOR TESTING DIFFERENCES AMONG MEANS

Tests of differences among means in a type CRF-pq design have the same general form as those given in Chapter 3 and Section 4.4. The test statistics for making comparisons among treatment A means assuming a fixed-effects model have the form

$$\frac{\hat{\psi}_{i(A)}}{\hat{\sigma}_{\psi_{i(A)}}} = \frac{c_1 \overline{Y}_{.1.} + c_2 \overline{Y}_{.2.} + \cdots + c_p \overline{Y}_{.p.}}{\sqrt{MSWCELL\left(\dfrac{c_1^2}{nq} + \dfrac{c_2^2}{nq} + \cdots + \dfrac{c_p^2}{nq}\right)}}$$

$$\frac{\hat{\psi}_{i(A)}}{\hat{\sigma}_{\overline{Y}}} = \frac{c_j \overline{Y}_{.j.} + c_{j'} \overline{Y}_{.j'.}}{\sqrt{MSWCELL/nq}}$$

$$df = pq(n-1).$$

Test statistics for making comparisons among treatment B means have the form

$$\frac{\hat{\psi}_{i(B)}}{\hat{\sigma}_{\psi_{i(B)}}} = \frac{c_1\overline{Y}_{\cdot\cdot1} + c_2\overline{Y}_{\cdot\cdot2} + \cdots + c_q\overline{Y}_{\cdot\cdot q}}{\sqrt{MSWCELL\left(\dfrac{c_1^2}{np} + \dfrac{c_2^2}{np} + \cdots + \dfrac{c_q^2}{np}\right)}}$$

$$\frac{\hat{\psi}_{i(B)}}{\hat{\sigma}_{\overline{Y}}} = \frac{c_k\overline{Y}_{\cdot\cdot k} + c_{k'}\overline{Y}_{\cdot\cdot k'}}{\sqrt{MSWCELL/np}}$$

$$df = pq(n-1).$$

For random-effects and mixed models, the denominator of the F statistic used in testing the overall null hypotheses should be used in place of *MSWCELL* in $\hat{\psi}/\hat{\sigma}_{\psi_i}$ and $\hat{\psi}/\hat{\sigma}_{\overline{Y}}$.

Procedures for testing hypotheses about contrasts of the form

$$\hat{\psi}_{i(A)} \text{ at } b_k = c_1\mu_{1k} + c_2\mu_{2k} + \cdots + c_p\mu_{pk}$$

and

$$\hat{\psi}_{i(B)} \text{ at } a_j = c_1\mu_{j1} + c_2\mu_{j2} + \cdots + c_q\mu_{jq}$$

which are called *simple-effects contrasts* to distinguish them from main-effects contrasts, are discussed in Section 8.6.

8.6 MORE ON THE INTERPRETATION OF INTERACTIONS

We saw in Section 8.4 that if an interaction is significant, the interpretation of tests of main effects must be qualified. When this occurs an experimenter may, in an attempt to better understand the interaction, proceed to test hypotheses about simple main effects or hypotheses about treatment-contrast interactions. We will examine these two approaches next.

SIMPLE MAIN EFFECTS

We begin by distinguishing between main effects and simple main effects. For purposes of discussion we will assume a fixed-effects model. Main effects for treatments A and B are

$$\begin{aligned}
\alpha_1 &= \mu_{1\cdot} - \mu & \beta_1 &= \mu_{\cdot1} - \mu \\
\alpha_2 &= \mu_{2\cdot} - \mu & \beta_2 &= \mu_{\cdot2} - \mu \\
&\cdots\cdots\cdots & &\cdots\cdots\cdots \\
\alpha_j &= \mu_{j\cdot} - \mu & \beta_k &= \mu_{\cdot k} - \mu
\end{aligned}$$

$$\alpha_p = \mu_{p\cdot} - \mu \qquad \beta_q = \mu_{\cdot q} - \mu.$$

The null hypotheses can be stated as

$$H_0: \quad \mu_{1\cdot} = \mu_{2\cdot} = \cdots = \mu_{p\cdot} \qquad \text{and} \qquad H_0: \quad \mu_{\cdot 1} = \mu_{\cdot 2} = \cdots = \mu_{\cdot q}$$

or as

$$H_0: \quad \alpha_j = 0 \quad \text{for all } j \qquad \text{and} \qquad H_0: \quad \beta_k = 0 \quad \text{for all } k.$$

Simple main effects are defined as

$$\alpha_1 \text{ at } b_1 = \mu_{11} - \mu_{\cdot 1} \qquad \beta_1 \text{ at } a_1 = \mu_{11} - \mu_{1\cdot}$$
$$\alpha_2 \text{ at } b_1 = \mu_{21} - \mu_{\cdot 1} \qquad \beta_2 \text{ at } a_1 = \mu_{12} - \mu_{1\cdot}$$
$$\cdots\cdots\cdots\cdots\cdots\cdots$$
$$\alpha_j \text{ at } b_k = \mu_{jk} - \mu_{\cdot k} \qquad \beta_k \text{ at } a_j = \mu_{jk} - \mu_{j\cdot}$$
$$\cdots\cdots\cdots\cdots\cdots\cdots$$
$$\alpha_p \text{ at } b_q = \mu_{pq} - \mu_{\cdot q} \qquad \beta_q \text{ at } a_p = \mu_{pq} - \mu_{p\cdot}.$$

The null hypotheses of interest are often stated as

$$H_0: \quad \mu_{11} = \mu_{21} = \cdots = \mu_{p1} \qquad H_0: \quad \mu_{11} = \mu_{12} = \cdots = \mu_{1q}$$
$$H_0: \quad \mu_{12} = \mu_{22} = \cdots = \mu_{p2} \qquad H_0: \quad \mu_{21} = \mu_{22} = \cdots = \mu_{2q}$$
$$\cdots\cdots\cdots\cdots\cdots\cdots\cdots\cdots$$
$$H_0: \quad \mu_{1q} = \mu_{2q} = \cdots = \mu_{pq} \qquad H_0: \quad \mu_{p1} = \mu_{p2} = \cdots = \mu_{pq}$$

or as

$$H_0: \quad \alpha_j \text{ at } b_1 = 0 \quad \text{for all } j \qquad H_0: \quad \beta_k \text{ at } a_1 = 0 \quad \text{for all } k$$
$$H_0: \quad \alpha_j \text{ at } b_2 = 0 \quad \text{for all } j \qquad H_0: \quad \beta_k \text{ at } a_2 = 0 \quad \text{for all } k$$
$$\cdots\cdots\cdots\cdots\cdots\cdots\cdots\cdots$$
$$H_0: \quad \alpha_j \text{ at } b_q = 0 \quad \text{for all } j \qquad H_0: \quad \beta_k \text{ at } a_p = 0 \quad \text{for all } k.$$

When stated in the latter way the hypotheses are deceptively simple. The hypothesis H_0: α_j at $b_1 = 0$ for all j, for example, focuses on only the effects in treatment level b_1. However, as we will see, a test of this hypothesis is equivalent to testing $\alpha_j + (\alpha\beta)_{j1} = 0$ for all j.

Formulas for computing sums of squares and mean squares for simple main effects are similar to those for main effects.

Main-Effects MS's	*Simple Main-Effects MS's*

$$MSA = nq \sum_{j=1}^{p} (\overline{Y}_{\cdot j\cdot} - \overline{Y}_{\ldots})^2/(p-1) \qquad MSA \text{ at } b_k = n \sum_{j=1}^{p} (\overline{Y}_{\cdot jk} - \overline{Y}_{\cdot\cdot k})^2/(p-1)$$

$$MSB = np \sum_{k=1}^{q} (\overline{Y}_{\cdot\cdot k} - \overline{Y}_{\ldots})^2/(q-1) \qquad MSB \text{ at } a_j = n \sum_{k=1}^{q} (\overline{Y}_{\cdot jk} - \overline{Y}_{\cdot j\cdot})^2/(q-1)$$

More convenient computational formulas are given in Table 8.6-1. The police attitude data from Table 8.3-1 are used to illustrate the computations. The results for the computations are summarized in Table 8.6-2.

By now you are used to partitioning sums of squares and degrees of freedom. It can be shown using the data in Table 8.6-1 that

$$SSA + SSAB = SSA \text{ at } b_1 + SSA \text{ at } b_2 + \cdots + SSA \text{ at } b_q$$

$$= \sum_{k=1}^{q} SSA \text{ at } b_k$$

$$190.000 + 1236.667 = 463.333 + 203.333 + 760.000 = 1426.67$$

$$(p-1) + (p-1)(q-1) = (p-1) + (p-1) + \cdots + (p-1)$$

$$= q(p-1)$$

$$4 + 2 = 3(3-1) = 6$$

and

$$SSB + SSAB = SSB \text{ at } a_1 + SSB \text{ at } a_2 + \cdots + SSB \text{ at } a_p$$

$$= \sum_{j=1}^{p} SSB \text{ at } a_j$$

$$1543.333 + 1236.667 = 63.333 + 103.333 + 2613.333 = 2780.00$$

$$(q-1) + (p-1)(q-1) = (q-1) + (q-1) + \cdots + (q-1)$$

$$= p(q-1)$$

$$4 + 2 = 3(3-1) = 6.$$

The important point is that simple main-effects sums of squares represent a partition of a treatment sum of squares plus an interaction sum of squares. It can also be shown using the procedures illustrated in Section 2.3 that

$$E(MSA \text{ at } b_k) = \sigma_\epsilon^2 + n \sum_{j=1}^{p} (\alpha\beta)_{jk}^2/(p-1) + n \sum_{j=1}^{p} \alpha_j^2/(p-1)$$

$$E(MSB \text{ at } a_j) = \sigma_\epsilon^2 + n \sum_{k=1}^{q} (\alpha\beta)_{jk}^2/(q-1) + n \sum_{k=1}^{q} \beta_k^2/(q-1).$$

Hence a significant F ratio, say $F = (MSA \text{ at } b_1)/MSWCELL$, means that $(\alpha\beta)_{j1} \neq 0$ for some j, or $\alpha_j \neq 0$ for some j, or both.

We saw in Chapter 3 that there are a variety of approaches to evaluating multiple contrasts and differences of opinion among statisticians regarding the merits of the various approaches. One fundamental issue concerned the correct conceptual unit for error rate. We face the same issue in performing tests of simple main effects. In the absence of definitive research, the best we can do is to describe approaches that are consistent with the testing philosophy outlined in Chapter 3. A decision to perform simple main-effects tests is usually made following an examination and statistical analysis of data. Furthermore, SSA at b_1, SSA at b_2, . . . , SSB at a_p are not mutually

TABLE 8.6-1   Computational Procedures for Simple Main-Effects Sum of Squares

(i)   Data and notation [Y_{ijk} denotes a score for experimental unit i in treatment combination jk; $i = 1, \ldots, n$ experimental units (s_i); $j = 1, \ldots, p$ levels of treatment $A(a_j)$; $k = 1, \ldots, q$ levels of treatment $B(b_k)$]:

AB Summary Table

$$\text{Entry is } \sum_{i=1}^{n} Y_{ijk}$$

	b_1	b_2	b_3	$\sum_{i=1}^{n}\sum_{k=1}^{q} Y_{ijk}$
	$n = 5$			
a_1	165	175	190	530
a_2	150	155	180	485
a_3	100	200	260	560
$\sum_{i=1}^{n}\sum_{j=1}^{p} Y_{ijk} =$	415	530	630	

(ii)   Computation of SSA at b_k:

$$SSA \text{ at } b_1 = \sum_{j=1}^{p} \frac{\left(\sum_{i=1}^{n} Y_{ij1}\right)^2}{n} - \frac{\left(\sum_{i=1}^{n}\sum_{j=1}^{p} Y_{ij1}\right)^2}{np} = \frac{(165)^2}{5} + \frac{(150)^2}{5} + \frac{(100)^2}{5} - \frac{(415)^2}{(5)(3)} = 463.3333$$

$$SSA \text{ at } b_2 = \sum_{j=1}^{p} \frac{\left(\sum_{i=1}^{n} Y_{ij2}\right)^2}{n} - \frac{\left(\sum_{i=1}^{n}\sum_{j=1}^{p} Y_{ij2}\right)^2}{np} = \frac{(175)^2}{5} + \frac{(155)^2}{5} + \frac{(200)^2}{5} - \frac{(530)^2}{(5)(3)} = 203.3333$$

$$SSA \text{ at } b_3 = \sum_{j=1}^{p} \frac{\left(\sum_{i=1}^{n} Y_{ij3}\right)^2}{n} - \frac{\left(\sum_{i=1}^{n}\sum_{j=1}^{p} Y_{ij3}\right)^2}{np} = \frac{(190)^2}{5} + \frac{(180)^2}{5} + \frac{(260)^2}{5} - \frac{(630)^2}{(5)(3)} = 760.0000$$

Computational check: $\sum_{k=1}^{q} SSA \text{ at } b_k = SSA + SSAB = 1426.667$

(iii)   Computation of SSB at a_j:

$$SSB \text{ at } a_1 = \sum_{k=1}^{q} \frac{\left(\sum_{i=1}^{n} Y_{i1k}\right)^2}{n} - \frac{\left(\sum_{i=1}^{n}\sum_{k=1}^{q} Y_{i1k}\right)^2}{nq} = \frac{(165)^2}{5} + \frac{(175)^2}{5} + \frac{(190)^2}{5} - \frac{(530)^2}{(5)(3)} = 63.3333$$

$$SSB \text{ at } a_2 = \sum_{k=1}^{q} \frac{\left(\sum_{i=1}^{n} Y_{i2k}\right)^2}{n} - \frac{\left(\sum_{i=1}^{n}\sum_{k=1}^{q} Y_{i2k}\right)^2}{nq} = \frac{(150)^2}{5} + \frac{(155)^2}{5} + \frac{(180)^2}{5} - \frac{(485)^2}{(5)(3)} = 103.3333$$

TABLE 8.6-1 (continued)

$$SSB \text{ at } a_3 = \sum_{k=1}^{q} \frac{\left(\sum\limits_{i=1}^{n} Y_{i3k}\right)^2}{n} - \frac{\left(\sum\limits_{i=1}^{n}\sum\limits_{k=1}^{q} Y_{i3k}\right)^2}{nq} = \frac{(100)^2}{5} + \frac{(200)^2}{5} + \frac{(260)^2}{5} - \frac{(560)^2}{(5)(3)} = 2613.3333$$

Computational check: $\sum\limits_{j=1}^{p} SSB$ at $a_j = SSB + SSAB = 2780.000$

TABLE 8.6-2 ANOVA Table with Tests of Simple Main Effects

	Source	SS	df	MS		F
1	A (Type of beat)	190.000	$p - 1 = 2$	95.000	[1/10]	1.52
2	B (Training duration)	1543.333	$q - 1 = 2$	771.666	[2/10]	12.35*
3	AB	1236.667	$(p - 1)(q - 1) = 4$	309.167	[3/10]	4.95**
4	A at b_1	463.333	$p - 1 = 2$	231.666	[4/10]	3.71
5	A at b_2	203.333	$p - 1 = 2$	101.666	[5/10]	1.63
6	A at b_3	760.000	$p - 1 = 2$	380.000	[6/10]	6.08§
7	B at a_1	63.333	$q - 1 = 2$	31.666	[7/10]	0.51
8	B at a_2	103.333	$q - 1 = 2$	51.666	[8/10]	0.83
9	B at a_3	2613.333	$q - 1 = 2$	1306.666	[9/10]	20.91§
10	Within cell	2250.000	$pq(n - 1) = 36$	62.500		
11	Total	5220.000	$npq - 1 = 44$			

$*F_{.05;2,36} = 3.26$
$**F_{.05;4,36} = 2.63$
$§F_{.15/6;2,36} = 4.09$

orthogonal. One recommended procedure in such cases is to assign the same error rate to the collection of tests as that allotted to the family. In our example, the simple main-effects sum of squares represent a partition of three families, A, B, and AB, with an overall error rate of $.05 + .05 + .05 = .15$.* For a type CRF-pq design, the number of simple main-effects tests that can be performed is always equal to $p + q$. We can hold the error rate for the collection of the six simple main-effects tests in our example at or less than .15 by means of Dunn's procedure, which is based on the Bonferroni inequality. Alternatively, we can use a procedure discussed by Gabriel (1964, 1969) called the *simultaneous test procedure* to control the probability of making at least one type I error at or less than .15. The application of this procedure that we will describe is related to Scheffé's method and Roy's (1953) union inter-

---

* An alternative conception is presented by Marascuilo and Levin (1970) in which treatment B, say, is conceived of as being nested within A. Using this approach, the simple main-effects sums of squares represent a partition of two families: A and B within A. The overall error rate is $.05 + .05 = .10$.

section principle. We will illustrate the application of Dunn's procedure first.

Using Dunn's procedure, each simple main-effects F ratio is evaluated at the $\alpha_{PE}/(p + q) = .15/6 = .025$ level of significance. The critical value of the test statistic is $F_{.15/6;2,36} = 4.09$. On the basis of this criterion, two simple main-effects hypotheses are rejected—H_0: α_j at $b_3 = 0$ for all j and H_0: β_k at $a_3 = 0$ for all k (see Table 8.6-2). An experimenter would normally proceed next to evaluate contrasts among the means embedded in these two hypotheses. Scheffé's statistic can be used to test all possible simple-effects hypotheses of the form

$$H_0: \quad \psi_{i(A)} \text{ at } b_3 = c_j\mu_{j3} + c_{j'}\mu_{j'3} + c_{j''}\mu_{j''3} = 0$$
$$H_0: \quad \psi_{i(B)} \text{ at } a_3 = c_k\mu_{3k} + c_{k'}\mu_{3k'} + c_{k''}\mu_{3k''} = 0$$

where the c_j's and c_k's are coefficients that define the contrasts. The computational formulas for the F statistic are

$$F = \frac{\hat{\psi}_{i(A)}^2 \text{ at } b_3}{\hat{\sigma}_{\psi_{i(A)}}^2 \text{ at } b_3} = \frac{[c_j\overline{Y}_{.j3} + c_{j'}\overline{Y}_{.j'3} + c_{j''}\overline{Y}_{.j''3}]^2}{MSWCELL\left(\dfrac{c_j^2}{n} + \dfrac{c_{j'}^2}{n} + \dfrac{c_{j''}^2}{n}\right)}$$

$$F = \frac{\hat{\psi}_{i(B)}^2 \text{ at } a_3}{\hat{\sigma}_{\psi_{i(B)}}^2 \text{ at } a_3} = \frac{[c_k\overline{Y}_{.3k} + c_{k'}\overline{Y}_{.3k'} + c_{k''}\overline{Y}_{.3k''}]^2}{MSWCELL\left(\dfrac{c_k^2}{n} + \dfrac{c_{k'}^2}{n} + \dfrac{c_{k''}^2}{n}\right)}.$$

The critical value for Scheffé's test statistic is

$$\nu_1 F_{.025;\nu_1,\nu_2} = 2F_{.025;2,36} = 2(4.09) = 8.18.$$

The second approach to controlling error rate, the simultaneous test procedure (STP), is particularly well suited to data snooping. It shares several important characteristics of Scheffé's procedure. First, the probability of making at least one type I error among all of the subdecisions does not exceed the significance level of the omnibus test. Second, all hypotheses may be tested simultaneously and without reference to one another. Third, it leads to coherent tests in that if a contrast is declared significant, the omnibus set containing the contrast will also be significant, and if the omnibus set is not significant, no contrasts among subsets of the data will be significant. However, if the omnibus test is significant, the STP may not detect a significant one-degree of freedom contrast among subsets of the data.

According to the simultaneous test procedure, an F test statistic is significant if it exceeds $(\nu_1 F_{\alpha;\nu_1,\nu_2})/\nu_3$, where α is the significance level of the omnibus test, ν_1 is the degrees of freedom of the omnibus numerator mean square, ν_2 is the degrees of freedom of the error mean square, and ν_3 is the degrees of freedom of the mean square that is being tested.* For example, the critical value for $F = (MSA \text{ at } b_k)/MSWCELL$ is $(8F_{.15;8,36})/2 = 6.52$. The significance level of the omnibus test is .05 + .05 +

---

*   In Gabriel's (1964, 1969) descriptions of the procedure, a sum of squares for the hypothesis that is being tested is significant if it exceeds the critical value $MSWCELL$ $(\nu_1 F_{\alpha;\nu_1,\nu_2})$. The advantage of this formulation is that the critical value remains the same although the sums of squares being tested may have different degrees of freedom.

.05 = .15; the degrees of freedom for A, B, and AB are $2 + 2 + 4 = 8$; the degrees of freedom of $MSWCELL$ is 36; and the degrees of freedom of MSA at b_k is 2. Using this critical value, the null hypothesis H_0: β_k at $a_3 = 0$ for all k is rejected. As a follow up, all possible contrasts of the form

$$\hat{\psi}_{i(B)}^2 \text{ at } a_3 = [c_k \overline{Y}_{\cdot 3k} + c_{k'} \overline{Y}_{\cdot 3k'} + c_{k''} \overline{Y}_{\cdot 3k''}]^2$$

can be evaluated using Scheffé's test statistic with

$$\nu_1 F_{.15; \nu_1, \nu_2} = 8 F_{.15; 8, 36} = 8(1.63) = 13.04$$

as the critical value. The critical values for the STP and Scheffé's statistic are equivalent when ν_3 is equal to one. In that case

$$(\nu_1 F_{\alpha; \nu_1, \nu_2})/1 = \nu_1 F_{\alpha; \nu_1, \nu_2}.$$

    In concluding this discussion of simple main-effects tests, note that questions surrounding their use have been the subject of continuing debate among statisticians (Games, 1973; Levin and Marascuilo, 1972, 1973; Marascuilo and Levin, 1970, 1976). Boik (1975) has examined some apparent inconsistencies and interpretation problems that can occur when the tests are applied to designs with three or more treatments. Procedures for controlling error rate and achieving coherent tests have been discussed by Betz and Gabriel (1978). However, even with refinements, a basic limitation of the simple main-effects approach remains: it partitions a pooled term involving a treatment and an interaction. Tests of hypotheses about simple main-effects and contrasts involving cell means may be interesting, but they do not help us understand the interaction between two variables. The approach described next enables an experimenter to gain a better understanding of the nature and sources of nonadditivity in data.

TREATMENT-CONTRAST INTERACTIONS

Whenever two treatments interact, we know that some contrast for one treatment is different at two or more levels of the other treatment. In other words, there is at least one contrast that interacts with the other treatment. Such interactions are called *treatment-contrast interactions* to distinguish them from omnibus interactions.* We will now see how to follow up a significant omnibus interaction by partitioning it into specific meaningful treatment-contrast interactions.

    As usual we begin by graphing the interaction. A graph of the AB interaction for the police attitude experiment is given in Figure 8.3-1. The next step is to examine the data for interesting contrasts of the form

$$\hat{\psi}_{i(A)} = c_1 \overline{Y}_{\cdot 1} + c_2 \overline{Y}_{\cdot 2} + \cdots + c_p \overline{Y}_{\cdot p}$$

---

*   Boik (1975, 1979) and Marascuilo and Levin (1970) provide excellent surveys of the various types of interactions that can be tested in a factorial design. A more theoretical discussion is given by Gabriel, Putter, and Wax (1973). Harter's (1970) paper contains numerical examples but is narrower in scope.

and

$$\hat{\psi}_{i(B)} = c_1\overline{Y}_{\cdot\cdot1} + c_2\overline{Y}_{\cdot\cdot2} + \cdots + c_q\overline{Y}_{\cdot\cdot q}.$$

Suppose that the experimenter is interested in the following contrasts.

$$\hat{\psi}_{1(B)} = \overline{Y}_{\cdot\cdot1} - \overline{Y}_{\cdot\cdot2} \qquad \hat{\psi}_{1(A)} = \overline{Y}_{\cdot1\cdot} - \overline{Y}_{\cdot2\cdot}$$

$$\hat{\psi}_{2(B)} = \overline{Y}_{\cdot\cdot1} - \overline{Y}_{\cdot\cdot3} \qquad \hat{\psi}_{2(A)} = (1/2)\overline{Y}_{\cdot1\cdot} + (1/2)\overline{Y}_{\cdot2\cdot} - \overline{Y}_{\cdot3\cdot}$$

$$\hat{\psi}_{3(B)} = \overline{Y}_{\cdot\cdot2} - \overline{Y}_{\cdot\cdot3}$$

We know from the significant AB interaction that at least one contrast interacts with the other treatment. To determine whether any of these contrasts interact, we can test the following null hypotheses:

H_0: $\alpha\psi_{1(B)} = \delta$ for all j H_0: $\beta\psi_{1(A)} = \delta$ for all k

H_0: $\alpha\psi_{2(B)} = \delta$ for all j H_0: $\beta\psi_{2(A)} = \delta$ for all k

H_0: $\alpha\psi_{3(B)} = \delta$ for all j

where δ is a constant for a given hypothesis. Procedures for computing treatment-contrast mean squares and testing the associated null hypotheses are illustrated in Table 8.6-3. The simultaneous test procedure is used to control the probability of making at least one type I error at or less than .05 for the collection of all possible tests involving the omnibus interaction.

According to Table 8.6-3, we can reject H_0: $\alpha\psi_{2(B)} = \delta$ for all j and H_0: $\beta\psi_{2(A)} = \delta$ for all k. Contrast $\psi_{2(B)} = \mu_{j1} - \mu_{j3}$ is a comparison of 5 hours of human relations training versus 15 hours. We know from the F test that the value of this contrast is not the same at all levels of treatment A. Since this is so, we might want to determine if the contrast is different for the upper-class beat versus the middle-class beat, $\psi_{1(A)} = \mu_{1k} - \mu_{2k}$, and whether it is different for the average of the upper- and middle-class beats versus the inner-city beat, $\psi_{2(A)} = [(\mu_{1k} + \mu_{2k})/2 - \mu_{3k}]$. The appropriate null hypotheses are

$$H_0: \quad \psi_{1(A)}\psi_{2(B)} = 0 \qquad \text{and} \qquad H_0: \quad \psi_{2(A)}\psi_{2(B)} = 0.$$

Since these hypotheses involve two contrasts they are called *contrast-contrast interactions*. To follow up the other significant $B\hat{\psi}_{2(A)}$ interaction, we might want to test

$$H_0: \quad \psi_{2(A)}\psi_{1(B)} = 0, \qquad H_0: \quad \psi_{2(A)}\psi_{2(B)} = 0, \qquad \text{and} \qquad H_0: \quad \psi_{2(A)}\psi_{3(B)} = 0.$$

Note that the second hypothesis duplicates one of the preceding. Procedures for computing contrast-contrast mean squares are illustrated in Table 8.6-4. The results of all of the tests are summarized in Table 8.6-5.

Tests of treatment-contrast interactions and contrast-contrast interactions provide additional insight concerning the AB interaction. From row 5 of Table 8.6-5 we know that $\psi_{2(B)} = \mu_{j1} - \mu_{j3}$ is not the same at all levels of treatment A. We followed this up in rows 9 and 10. We found no evidence that $\psi_{2(B)}$ interacts with $\psi_{1(A)} = \mu_{1k} - \mu_{2k}$, upper-class beat versus middle-class beat. However, $\psi_{2(B)}$ does interact with $\psi_{2(A)} = [(\mu_{1k} + \mu_{2k})/2 - \mu_{3k}]$, average of upper- and middle-class beats versus the inner-city beat. These conclusions are consistent with the sample data in

Figure 8.3-1. According to row 8, $\psi_{2(A)} = [(\mu_{1k} + \mu_{2k})/2 - \mu_{3k}]$ is not the same at all levels of treatment B. In following this up, we found, as noted earlier, that the only interesting contrast that $\psi_{2(A)}$ interacts with is $\psi_{2(B)}$ (see row 10).

There is no reason for believing that two of the interesting contrasts interact with the other treatment. These are $\psi_{3(B)}$ (see row 6) and $\psi_{1(A)}$ (see row 7). If we are willing to act as if these null hypotheses are in fact true, and if the test of treatment A or B is significant, we can test the main-effects hypotheses H_0: $\psi_{3(B)} = \mu_{.2} - \mu_{.3} = 0$ and H_0: $\psi_{1(A)} = \mu_{1.} - \mu_{2.} = 0$.* Earlier we saw that the test for treatment B is significant; the test for treatment A is not significant. Scheffé's statistic for testing the null hypothesis for the treatment B contrast, $\hat{\psi}_{3(B)}$, is

$$F = \frac{[(1)35.333 + (-1)42.000]^2}{62.5\left[\dfrac{(1)^2}{(5)(3)} + \dfrac{(-1)^2}{(5)(3)}\right]} = \frac{(-6.667)^2}{8.333} = 5.33.$$

The critical value for Scheffé's test statistic is $2F_{.05;2,36} = 2(3.26) = 6.52$. This null hypothesis cannot be rejected. Many experimenters would be reluctant to test hypotheses such as H_0: $\mu_{.2} - \mu_{.3} = 0$ and H_0: $\mu_{1.} - \mu_{2.} = 0$, believing that the AB interaction renders such contrasts meaningless. As we have seen, this is not the case. The significant AB interaction indicates that there is at least one contrast that is different at two or more levels of the other treatment. It does not mean that all contrasts for one treatment interact with the other treatment.

Earlier we saw that the sum of simple main-effects sum of squares is not equal to the omnibus interaction sum of squares. It is a simple matter to show that for mutually orthogonal contrasts, the sum of treatment-contrast interaction sum of squares is equal to the omnibus interaction sum of squares. That is,

$$\sum_{i=1}^{p-1} SSB\hat{\psi}_{i(A)} = SSAB \qquad \text{and} \qquad \sum_{i=1}^{q-1} SSA\hat{\psi}_{i(B)} = SSAB.$$

To illustrate, we will use contrasts defined by the coefficient vectors

$$\mathbf{c}'_{1(A)} = [1 \quad -1 \quad 0] \qquad \text{and} \qquad \mathbf{c}'_{2(A)} = [{}^1/_2 \quad {}^1/_2 \quad -1]$$

in Table 8.6-3. These contrasts are orthogonal; hence,

$$SSB\hat{\psi}_{1(A)} + SSB\hat{\psi}_{2(A)} = SSAB$$
$$5.000 + 1231.667 = 1236.667.$$

The important point is that treatment-contrast interactions, unlike simple-main effects, represent a partition of the omnibus interaction.

---

*   The merits of behaving as if, for all practical purposes, the null hypothesis is true when the probability associated with the test statistic is large ($p > .25$) has been the subject of much debate. The practice is fairly common in some areas, such as performing preliminary tests on the appropriateness of a particular model (see Section 8.11).

TABLE 8.6-3   Computational Procedures for Treatment-Contrast Interactions (Data from Table 8.3-1)

(i)   Data and notation [$\overline{Y}_{\cdot jk}$ denotes a mean for the experimental units in treatment combination jk; $i = 1, \ldots, n$ experimental units (s_i); $j = 1, \ldots, p$ levels of treatment $A(a_j)$; $k = 1, \ldots, q$ levels of treatment $B(b_k)$]:

AB Summary Table
Entry is $\overline{Y}_{\cdot jk}$

	b_1 (5 hrs.)	b_2 (10 hrs.)	b_3 (15 hrs.)	$\overline{Y}_{\cdot j \cdot}$
	$n = 5$			
a_1 (Upper class)	$\overline{Y}_{\cdot 11} = 33$	$\overline{Y}_{\cdot 12} = 35$	$\overline{Y}_{\cdot 13} = 38$	35.333
a_2 (Middle class)	$\overline{Y}_{\cdot 21} = 30$	$\overline{Y}_{\cdot 22} = 31$	$\overline{Y}_{\cdot 23} = 36$	32.333
a_3 (Inner city)	$\overline{Y}_{\cdot 31} = 20$	$\overline{Y}_{\cdot 32} = 40$	$\overline{Y}_{\cdot 33} = 52$	37.333
$\overline{Y}_{\cdot \cdot k} =$	27.667	35.333	42.000	$\overline{Y}_{\cdots} = 35.000$

(ii)   Coefficients of interesting contrasts (Contrasts are defined in the text):

$$\mathbf{c}'_{1(B)} = \begin{bmatrix} 1 & -1 & 0 \end{bmatrix} \qquad \mathbf{c}'_{1(A)} = \begin{bmatrix} 1 & -1 & 0 \end{bmatrix}$$

$$\mathbf{c}'_{2(B)} = \begin{bmatrix} 1 & 0 & -1 \end{bmatrix} \qquad \mathbf{c}'_{2(A)} = \begin{bmatrix} 1/2 & 1/2 & -1 \end{bmatrix}$$

$$\mathbf{c}'_{3(B)} = \begin{bmatrix} 0 & 1 & -1 \end{bmatrix}$$

(iii)   Contrasts at each level of the other treatment:

$\hat{\psi}_{1(B)}$ at $a_1 = (1)33 + (-1)35 + (0)38 = -2$ $\qquad$ $\hat{\psi}_{1(A)}$ at $b_1 = (1)33 + (-1)30 + (0)20 = 3$

$\hat{\psi}_{1(B)}$ at $a_2 = (1)30 + (-1)31 + (0)36 = -1$ $\qquad$ $\hat{\psi}_{1(A)}$ at $b_2 = (1)35 + (-1)31 + (0)40 = 4$

$\hat{\psi}_{1(B)}$ at $a_3 = (1)20 + (-1)40 + (0)52 = -20$ $\qquad$ $\hat{\psi}_{1(A)}$ at $b_3 = (1)38 + (-1)36 + (0)52 = 2$

$\hat{\psi}_{2(B)}$ at $a_1 = (1)33 + (0)35 + (-1)38 = -5$ $\qquad$ $\hat{\psi}_{2(A)}$ at $b_1 = (1/2)33 + (1/2)30 + (-1)20 = 11.5$

$\hat{\psi}_{2(B)}$ at $a_2 = (1)30 + (0)31 + (-1)36 = -6$ $\qquad$ $\hat{\psi}_{2(A)}$ at $b_2 = (1/2)35 + (1/2)31 + (-1)40 = -7.0$

$\hat{\psi}_{2(B)}$ at $a_3 = (1)20 + (0)40 + (-1)52 = -32$ $\qquad$ $\hat{\psi}_{2(A)}$ at $b_3 = (1/2)38 + (1/2)36 + (-1)52 = -15.0$

$\hat{\psi}_{3(B)}$ at $a_1 = (0)33 + (1)35 + (-1)38 = -3$

$\hat{\psi}_{3(B)}$ at $a_2 = (0)30 + (1)31 + (-1)36 = -5$

$\hat{\psi}_{3(B)}$ at $a_3 = (0)20 + (1)40 + (-1)52 = -12$

(iv)   Computation of treatment-contrast interaction sum of squares:

$$SSA\hat{\psi}_{i(B)} = n \left[\sum_{j=1}^{p} \hat{\psi}^2_{i(B)} \text{ at } a_j - \frac{\left(\sum_{j=1}^{p} \hat{\psi}_{i(B)} \text{ at } a_j \right)^2}{p} \right] \Bigg/ \sum_{k=1}^{q} c^2_{i(B)}$$

$$df = (p - 1)1$$

TABLE 8.6-3 (continued)

$$SSA\hat{\psi}_{1(B)} = 5\left[(-2)^2 + (-1)^2 + (-20)^2 - \frac{(-23)^2}{3}\right] \Big/ [(1)^2 + (-1)^2 + (0)^2] = 571.667$$

$$df = 2$$

$$SSA\hat{\psi}_{2(B)} = 5\left[(-5)^2 + (-6)^2 + (-32)^2 - \frac{(-43)^2}{3}\right] \Big/ [(1)^2 + (0)^2 + (-1)^2] = 1171.667$$

$$df = 2$$

$$SSA\hat{\psi}_{3(B)} = 5\left[(-3)^2 + (-5)^2 + (-12)^2 - \frac{(-20)^2}{3}\right] \Big/ [(0)^2 + (1)^2 + (-1)^2] = 111.667$$

$$df = 2$$

$$SSB\hat{\psi}_{i(A)} = n\left[\sum_{k=1}^{q} \hat{\psi}_{i(A)}^2 \text{ at } b_k - \frac{\left(\sum_{k=1}^{q} \hat{\psi}_{i(A)} \text{ at } b_k\right)^2}{q}\right] \Big/ \sum_{j=1}^{p} c_{i(A)}^2$$

$$df = (q - 1)1$$

$$SSB\hat{\psi}_{1(A)} = 5\left[(3)^2 + (4)^2 + (2)^2 - \frac{(9)^2}{3}\right] \Big/ [(1)^2 + (-1)^2 + (0)^2] = 5.000$$

$$df = 2$$

$$SSB\hat{\psi}_{2(A)} = 5\left[(11.5)^2 + (-7.0)^2 + (-15.0)^2 - \frac{(-10.5)^2}{3}\right] \Big/ [(1/2)^2 + (1/2)^2 + (-1)^2] = 1231.667$$

$$df = 2$$

(v)   Computation of F ratios:

$$F = \frac{SSA\hat{\psi}_{1(B)}/(p-1)1}{MSWCELL} = \frac{571.667/2}{62.500} = 4.57$$

$$F = \frac{SSA\hat{\psi}_{2(B)}/(p-1)1}{MSWCELL} = \frac{1171.667/2}{62.500} = 9.37*$$

$$F = \frac{SSA\hat{\psi}_{3(B)}/(p-1)1}{MSWCELL} = \frac{111.667/2}{62.500} = 0.89$$

$$F = \frac{SSB\hat{\psi}_{1(A)}/(q-1)1}{MSWCELL} = \frac{5.000/2}{62.500} = 0.04$$

$$F = \frac{SSB\hat{\psi}_{2(A)}/(q-1)1}{MSWCELL} = \frac{1231.667/2}{62.500} = 9.85*$$

$$(\nu_1 F_{.05;\nu_1,\nu_2})/\nu_3 = (4F_{.05;4,36})/2 = 4(2.63)/2 = 5.26$$

---

*$p < .01$

TABLE 8.6-4   Computational Procedures for Contrast-Contrast Interactions (Data from Table 8.6-3)

(i)   Computation of contrast-contrast interactions:

$$SS\hat{\psi}_{i(A)}\hat{\psi}_{i(B)} = n\left(\sum_{j=1}^{p} c_{i(A)}\hat{\psi}_{i(B)} \text{ at } a_j\right)^2 \bigg/ \sum_{j=1}^{p} c_{i(A)}^2 \sum_{k=1}^{q} c_{i(B)}^2$$

$$\text{Alternative formula} = n\left(\sum_{k=1}^{q} c_{i(B)}\hat{\psi}_{i(A)} \text{ at } b_k\right)^2 \bigg/ \sum_{j=1}^{p} c_{i(A)}^2 \sum_{k=1}^{q} c_{i(B)}^2$$

$$df = 1$$

$$SS\hat{\psi}_{1(A)}\hat{\psi}_{2(B)} = 5[(1)(-5) + (-1)(-6) + (0)(-32)]^2/[(1)^2 + (-1)^2 + (0)^2][(1)^2 + (0)^2 + (-1)^2]$$

$$= 5(1)^2/(2)(2) = 1.250$$

$$SS\hat{\psi}_{2(A)}\hat{\psi}_{2(B)} = 5[(1/2)(-5) + (1/2)(-6) + (-1)(-32)]^2/[(1/2)^2 + (1/2)^2 + (-1)^2][(1)^2 + (0)^2 + (-1)^2]$$

$$= 5(26.5)^2/(1.5)(2) = 1170.417$$

$$SS\hat{\psi}_{2(A)}\hat{\psi}_{1(B)} = 5[(1/2)(-2) + (1/2)(-1) + (-1)(-20)]^2/[(1/2)^2 + (1/2)^2 + (-1)^2][(1)^2 + (-1)^2 + (0)^2]$$

$$= 5(18.5)^2/(1.5)(2) = 570.417$$

$$SS\hat{\psi}_{2(A)}\hat{\psi}_{3(B)} = 5[(1/2)(-3) + (1/2)(-5) + (-1)(-12)]^2/[(1/2)^2 + (1/2)^2 + (-1)^2][(0)^2 + (1)^2 + (-1)^2]$$

$$= 5(8)^2/(1.5)(2) = 106.667$$

(ii)   Alternative computational approach:

$$SS\hat{\psi}_{i(A)}\hat{\psi}_{i(B)} = n\left[\left(\sum_{j=1}^{p}\sum_{k=1}^{q} c_{jk}\overline{Y}_{\cdot jk}\right)^2 \bigg/ \sum_{j=1}^{p}\sum_{k=1}^{q} c_{jk}^2\right]$$ where the c_{jk}'s are elements of the matrix obtained by

taking the outer product of $\mathbf{c}_{i(A)}$ and $\mathbf{c}_{i(B)}$.

For example,

$$\mathbf{c}_{1(A)}\mathbf{c}_{2(B)}' = \begin{bmatrix} 1 \\ -1 \\ 0 \end{bmatrix} \begin{bmatrix} 1 & 0 & -1 \end{bmatrix} = \begin{bmatrix} 1 & 0 & -1 \\ -1 & 0 & 1 \\ 0 & 0 & 0 \end{bmatrix}.$$

$$SS\hat{\psi}_{1(A)}\hat{\psi}_{2(B)} = 5\{[(1)33 + (0)35 + \cdots + (0)52]^2/[(1)^2 + (0)^2 + \cdots + (0)^2]\}$$

$$= 5(1)^2/4 = 1.250$$

$$df = 1 \qquad \text{the rank of the coefficient matrix formed by } \mathbf{c}_{1(A)}\mathbf{c}_{2(B)}'$$

An experimenter may gain additional insight into the nature of the *AB* inter-action by identifying the coefficients of the one-degree of freedom contrast-contrast interaction sum of squares that accounts for the largest portion of the *AB* interaction. As an introduction to this procedure, we will begin by considering treatment sum of squares. It is easy to determine the coefficients of a contrast, say $\hat{\psi}_{i(A)}$, with one-degree of freedom such that $SS\hat{\psi}_{i(A)} = SSA$. The coefficients are given by $c_1 = \overline{Y}_{1\cdot} - \overline{Y}_{\cdots}$,

TABLE 8.6-5 ANOVA Table for Type CRF-33 Design

	Source	SS	df	MS		F
1	A (Type of beat)	190.000	$p - 1 = 2$	95.000	$[^1/_{13}]$	1.52
2	B (Training duration)	1543.333	$q - 1 = 2$	771.666	$[^2/_{13}]$	12.35*
3	AB	1236.667	$(p - 1)(q - 1) = 4$	309.167	$[^3/_{13}]$	4.95**
4	$A\hat{\psi}_{1(B)}$	571.667	$(p - 1)1 = 2$	285.834	$[^4/_{13}]$	4.57
5	$A\hat{\psi}_{2(B)}$	1171.667	$(p - 1)1 = 2$	585.834	$[^5/_{13}]$	9.37§
6	$A\hat{\psi}_{3(B)}$	111.667	$(p - 1)1 = 2$	55.834	$[^6/_{13}]$	0.89
7	$B\hat{\psi}_{1(A)}$	5.000	$(q - 1)1 = 2$	2.500	$[^7/_{13}]$	0.04
8	$B\hat{\psi}_{2(A)}$	1231.667	$(q - 1)1 = 2$	615.834	$[^8/_{13}]$	9.85§
9	$\hat{\psi}_{1(A)}\hat{\psi}_{2(B)}$	1.250	1	1.250	$[^9/_{13}]$	0.02
10	$\hat{\psi}_{2(A)}\hat{\psi}_{2(B)}$	1170.417	1	1170.417	$[^{10}/_{13}]$	18.73§§
11	$\hat{\psi}_{2(A)}\hat{\psi}_{1(B)}$	570.417	1	570.417	$[^{11}/_{13}]$	9.13
12	$\hat{\psi}_{2(A)}\hat{\psi}_{3(B)}$	106.667	1	160.667	$[^{12}/_{13}]$	1.71
13	Within cell	2250.000	$pq(n - 1) = 36$	62.500		
14	Total	5220.000	$npq - 1 = 44$			

$\mathbf{c}'_{1(B)} = [1 \quad -1 \quad 0]$, $\mathbf{c}'_{2(B)} = [1 \quad 0 \quad -1]$, $\mathbf{c}'_{3(B)} = [0 \quad 1 \quad -1]$
$\mathbf{c}'_{1(A)} = [1 \quad -1 \quad 0]$, $\mathbf{c}'_{2(A)} = [^1/_2 \quad ^1/_2 \quad -1]$

*$F_{.05;2,36} = 3.26$
**$F_{.05;4,36} = 2.63$
§$(4F_{.05;4,36})/2 = 4(2.63)/2 = 5.26$
§§$(4F_{.05;4,36})/1 = 4(2.63) = 10.52$

$c_2 = \overline{Y}_{.2.} - \overline{Y}_{...}, \ldots, c_p = \overline{Y}_{.p.} - \overline{Y}_{...}$. For the data in Table 8.6-3, the coefficients for $\hat{\psi}_{i(A)}$ are $\mathbf{c}'_{i(A)} = [.333333 \quad -2.666667 \quad 2.333333].$* The sum of squares for this contrast is given by

$$SS\hat{\psi}_{i(A)} = nq\left(\sum_{j=1}^{p} c_j\overline{Y}_{.j.}\right)^2 \Big/ \sum_{j=1}^{p} c_j^2$$
$$= (5)(3)(12.6667)^2/12.6667$$
$$= 190.000$$

which is equal to SSA.

Similarly, the coefficients of a contrast-contrast interaction can be found for which $SS\hat{\psi}_{i(A)}\hat{\psi}_{i(B)} = SSAB$. The coefficients are given by $c_{11} = \overline{Y}_{11} - \overline{Y}_{.1.} - $

---

* If each coefficient is divided by the largest coefficient, $c_i/\max c_i$, the sum of the absolute value of the coefficients is equal to two. For example, $|.333333/-2.666667| + |-2.666667/-2.666667| + |2.333333/-2.666667| = |-.125| + |1.000| + |-0.875| = 2.000$. This transformation has no affect on the value of the contrast sum of squares.

$\overline{Y}_{\cdot 1} + \overline{Y}_{\cdot\cdot\cdot}$, $c_{21} = \overline{Y}_{21} - \overline{Y}_{2\cdot} - \overline{Y}_{\cdot 1} + \overline{Y}_{\cdot\cdot\cdot}$, $\ldots$, $c_{pq} = \overline{Y}_{pq} - \overline{Y}_{p\cdot} - \overline{Y}_{\cdot q} + \overline{Y}_{\cdot\cdot\cdot}$. The matrix of coefficients for the data in Table 8.6-3 is

$$\mathbf{C} = \begin{bmatrix} 5.000000 & -.666666 & -4.333333 \\ 5.000000 & -1.666666 & -3.333333 \\ -10.000000 & 2.333334 & 7.666667 \end{bmatrix}.$$

For example, the coefficient in the $j = $ 1st row and $k = $ 1st column is

$$c_{11} = 33.000 - 35.333 - 27.667 + 35.000 = 5.000.$$

The sum of squares for the contrast-contrast interaction is given by

$$SS\hat{\psi}_{i(A)}\hat{\psi}_{i(B)} = n\left(\sum_{j=1}^{p} \sum_{k=1}^{q} c_{jk}\overline{Y}_{\cdot jk} \right)^2 \Bigg/ \sum_{j=1}^{p} \sum_{k=1}^{q} c_{jk}^2$$

$$= 5(247.3334)^2/247.3333$$

$$= 1236.668$$

which within rounding error is equal to $SSAB$. The degrees of freedom for this contrast-contrast interaction is equal to 2, the rank of the $\mathbf{C}$ matrix. The rank of $\mathbf{C}$ is generally equal to the minimum of $(p - 1)$ and $(q - 1)$.

With this introduction we are now ready to identify the coefficients of the *one-degree of freedom* contrast-contrast interaction that accounts for the maximum proportion of $SSAB$. Methods for determining these coefficients have been described by Gollob (1968a, 1968b), Mandel (1969, 1971) and Snee (1972). An approximate procedure which uses coefficients in the $\mathbf{C}$ matrix has been described by Bradu and Gabriel (1974). According to their procedure, one first determines the maximum $|c_{jk}|$ in the $\mathbf{C}$ matrix. For the preceding $\mathbf{C}$ matrix, max $|c_{jk}| = 10$. This coefficient occurs in column 1 and row 3. The coefficient vector in column 1 is defined as $\mathbf{c}_{i(A)}$; the vector in row 3 is defined as $\mathbf{c}_{i(B)}$. The outer product of these two vectors is

$$\mathbf{c}_{i(A)}\mathbf{c}'_{i(B)} = \begin{bmatrix} 5 \\ 5 \\ -10 \end{bmatrix} \begin{bmatrix} -10 & 2.333334 & 7.666667 \end{bmatrix}$$

$$= \begin{bmatrix} -50 & 11.666670 & 38.333335 \\ -50 & 11.666670 & 38.333335 \\ 100 & -23.333340 & -76.666670 \end{bmatrix}.$$

The sum of squares is

$$SS\hat{\psi}_{i(A)}\hat{\psi}_{i(B)} = n\left(\sum_{j=1}^{p} \sum_{k=1}^{q} c_{jk}\overline{Y}_{\cdot jk} \right)^2 \Bigg/ \sum_{j=1}^{p} \sum_{k=1}^{q} c_{jk}^2$$

$$= 5(-2463.333)^2/24633.335 = 1231.667$$

with $df = 1$, since the rank of the matrix formed by the outer product of $\mathbf{c}_{i(A)}\mathbf{c}'_{i(B)}$ is one. This one-degree of freedom contrast accounts for 99.6% of the AB interaction

$$\frac{SS\hat{\psi}_{i(A)}\hat{\psi}_{i(B)}}{SSAB} = \frac{1231.667}{1236.668} = .996.$$

As expected, the null hypothesis for this contrast-contrast interaction is rejected.

$$F = \frac{MS\hat{\psi}_{i(A)}\hat{\psi}_{i(B)}}{MSWCELL} = \frac{1231.667/1}{62.500} = 19.71$$

$(4F_{.05;4,36})/1 = 4(2.63) = 10.52$

To simplify the interpretation of this contrast-contrast interaction, the coefficients in the vectors can be divided by the largest coefficient, for example,

$\mathbf{c}'_{i(A)} = [5/-10 \quad 5/-10 \quad -10/-10] = [-1/2 \quad -1/2 \quad 1]$

$\mathbf{c}'_{i(B)} = [-10/-10 \quad 2.3333333/-10 \quad 7.6666667/-10] = [1.00 \quad -.23 \quad -.77]$.

In words, the most potent one-degree of freedom interaction between type of beat and hours of human relations training involves the average of the upper- and middle-class beats versus the inner-city beat, $\hat{\psi}_{i(A)} = \{[(-1/2)\overline{Y}_{1.} + (-1/2)\overline{Y}_{2.}]/2 + (1)\overline{Y}_{3.}\}$, and a weighted average of the 10- and 15-hour courses versus the 5-hour course, $\hat{\psi}_{i(B)} = \{(1)\overline{Y}_{.1} + [(-.23)\overline{Y}_{.2} + (-.77)\overline{Y}_{.3}]\}$. As we saw earlier, this interaction accounts for 99.6% of the AB interaction.

The presence of a significant interaction can certainly complicate an experimenter's life. As we have seen, the follow-up procedures are both complex and tedious. However, interactions among variables are common and experimenters must be prepared to deal with them. In this section we have described two follow-up approaches: tests of simple main effects and tests of treatment-contrast and contrast-contrast interactions. In the hands of a creative experimenter, the latter approach can provide useful insights into the nature and sources of nonadditivity in data.

8.7 TEST FOR TRENDS IN THE DATA

When one or more treatments in a factorial experiment are quantitative, additional insight concerning their effects can be obtained by partitioning the data into orthogonal trend contrasts. The general rationale for trend analysis is presented in Section 4.5.

TEST FOR PRESENCE OF TREND

A test of the hypothesis of no trend in the population means for treatment A is provided by

$$F = \frac{MSA}{MSWCELL}.$$

For the data in Table 8.3-1, treatment A is not a quantitative variable and therefore is not suitable for trend tests. A test of the hypothesis of no trend for treatment B is provided by

$$F = \frac{MSB}{MSWCELL}.$$

If the F ratio is significant, we conclude that the population dependent variable means are related in some manner to the independent variable. According to the analysis in Table 8.3-2, this F is significant. Thus there is a trend in the population. A graph of the trend for the sample data is shown in Figure 8.7-1. If the overall test for the

FIGURE 8.7-1   Overall trend of data for levels of treatment B.

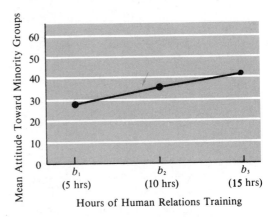

presence of a trend is not significant, further tests for trends with respect to the treatment should not be made unless an experimenter has advanced a priori hypotheses concerning specific trends. A posteriori procedures should be used for trend snooping, whereas a priori procedures can be used to carry out planned trend tests.

A factorial experiment enables an experimenter to make a trend test not previously described. This test provides an answer to the question: "Is the trend of population means for one treatment the same for all levels of a second treatment?" This test has the form

$$F = \frac{MSAB}{MSWCELL}.$$

According to the analysis of data presented in Table 8.3-2, the AB interaction is significant. A graph illustrating the differences in trends for treatment B at the three levels of treatment A appears in Figure 8.3-1. If, as in the present example, the AB interaction is significant, interest shifts from the overall treatment trend contrasts described in Section 4.5 to interaction trend contrasts.

PARTITION OF TREATMENTS AND INTERACTION INTO TREND CONTRASTS

We noted in Section 4.5 that $SSBG$ in a completely randomized design can be partitioned into $p - 1$ trend contrasts. Similarly, if both treatments of, say, a type CRF-33 design represent quantitative variables, the data can be partitioned into

$pq - 1 = 8$ trend contrasts as shown in Table 8.7-1. The reader may have difficulty visualizing the meaning of linear × linear, linear × quadratic, etc, trend contrasts. Suppose for illustrative purposes that treatment A in Table 8.3-1 is also a quantitative variable and that the levels are separated by equal intervals. A response surface for these data is shown in Figure 8.7-2. We know from the significant AB interaction that the profiles for treatment A at the q levels of B are not parallel and that the profiles for B at the p levels of A are not parallel. An examination of Figure 8.7-2 suggests that the profiles for treatment A are linear and quadratic in form, whereas those for treatment B are predominately linear. An experimenter might wonder if the linear and quadratic components of the treatment A trend interact with the linear component of treatment B; that is, are the linear × linear and quadratic × linear interactions significant? This type of analysis of the AB interaction provides the experimenter with an indication of the fit of differently shaped surfaces to the population means.

TABLE 8.7-1 Partition of Treatments and Interaction for a Type CRF-33 Design When Both Treatments Are Quantitative

Source	df	Source	df
A	2	AB	4
Linear	1	Linear × linear	1
Quadratic	1	Linear × quadratic	1
B	2	Quadratic × linear	1
Linear	1	Quadratic × quadratic	1
Quadratic	1		

FIGURE 8.7-2 Response surface for data in Table 8.3-1.

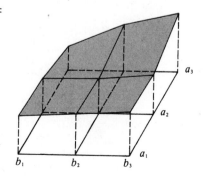

If only one of the treatments in a type CRF-33 design is quantitative, a different partition of the data is required. For the data in Table 8.3-1 where only treatment B is quantitative, the partition in Table 8.7-2 is appropriate. Profiles for treatment B at the three levels of A are shown in Figure 8.3-1. The question that the analysis in Table 8.7-2 is designed to answer is: "Which trend contrasts account for the differences in the profiles?" Procedures for carrying out these trend analyses are presented in the following sections.

TABLE 8.7-2   Partition of Treatments and Interaction for a Type CRF-33 Design When Only Treatment B Is Quantitative

Source	df	
A	2	
B	2	
Linear		1
Quadratic		1
AB	4	
Difference in linear trend		2
Difference in quadratic trend		2

TESTS OF TREND CONTRASTS FOR TREATMENTS

Although the AB interaction for the data in Table 8.3-1 is significant, an experimenter might want to know how much of the variation in the dependent variable is accounted for by the linear and quadratic trend components in treatment B. We will assume that an experimenter has advanced a priori hypotheses with respect to both trend components. Alternatively, an experimenter might simply want to test the hypothesis that the trend is nonlinear. Procedures for testing these hypotheses are identical to those in Section 4.5 and will not be repeated here. Results of the computations for treatment B are as follows.

$$SS\hat{\psi}_{\text{lin}(B)} = \frac{\left[\sum_{k=1}^{q}\left(c_{1k}\sum_{i=1}^{n}\sum_{j=1}^{p}Y_{ijk}\right)\right]^2}{np\sum_{k=1}^{q}c_{1k}^2} = \frac{(215)^2}{(5)(3)(2)} = 1540.833$$

$$df = 1$$

$$SS\hat{\psi}_{\text{quad}(B)} = \frac{\left[\sum_{k=1}^{q}\left(c_{2k}\sum_{i=1}^{n}\sum_{j=1}^{p}Y_{ijk}\right)\right]^2}{np\sum_{k=1}^{q}c_{2k}^2} = \frac{(-15)^2}{(5)(3)(6)} = 2.500$$

$$df = 1$$

$$F = \frac{MS\hat{\psi}_{\text{lin}(B)}}{MSWCELL} = \frac{1540.833}{62.500} = 24.65$$

$$F = \frac{MS\,\hat{\psi}_{\text{quad}(B)}}{MSWCELL} = \frac{2.500}{62.500} = 0.04$$

$$F_{.05;1,36} = 8.94$$

It is apparent from the foregoing that the linear contrast is significant but the quadratic is not. Therefore the overall trend for treatment B shown in Figure 8.7-1 can be adequately described by an equation of the form

$$\hat{\mu}_k = \beta_0 + \beta_1 X_k.$$

Procedures for fitting a linear equation to the trend in Figure 8.7-1 are described in Section 4.5. The linear component of the trend accounts for almost all of the variation in treatment B, as the following computations show.

$$\frac{SS\,\hat{\psi}_{\text{lin}(B)}}{SSB} = \frac{1540.833}{1543.333} \times 100 = 99.8\% \qquad \frac{SS\,\hat{\psi}_{\text{quad}(B)}}{SSB} = \frac{2.500}{1543.333} \times 100 = 0.2\%$$

PARTITION OF *AB* INTERACTION INTO TREND CONTRAST-CONTRAST INTERACTIONS

If, in a type CRF-33 design, the two treatments represent quantitative variables, the four degrees of freedom for the AB interaction can be partitioned as was shown in Table 8.7-1. Procedures for testing the hypotheses

$$H_0: \quad \psi_{\text{lin}(A)}\psi_{\text{lin}(B)} = 0 \qquad H_0: \quad \psi_{\text{lin}(A)}\psi_{\text{quad}(B)} = 0$$
$$H_0: \quad \psi_{\text{quad}(A)}\psi_{\text{lin}(B)} = 0 \qquad H_0: \quad \psi_{\text{quad}(A)}\psi_{\text{quad}(B)} = 0$$

are given in Table 8.7-3. The data are taken from Table 8.3-1; for purposes of illustration we assume that both variables are quantitative and that the levels of treatment A are also separated by equal intervals. The orthogonal coefficients in part (i) of Table 8.7-3 are obtained from Appendix E.12.

In evaluating the significance of the F ratios in Table 8.7-3 we again face the issue of the correct conceptual unit for error rate. This issue is examined in detail in Sections 3.3 and 8.6 and there is little that we can add here. Since the four contrast-contrast interactions in Table 8.7-3 are orthogonal, the question reduces to whether the tests are a priori or a posteriori. Critical values for both cases are given in the table.

PARTITION OF *AB* INTERACTION INTO TREATMENT×TREND-CONTRAST INTERACTIONS

If one but not both treatments in a type CRF-33 design represent a quantitative variable, the four degrees of freedom for the AB interaction can be partitioned as was shown in Table 8.7-2. A significant test of $F = MSAB/MSWCELL$ tells us that the trend for the quantitative variable is not the same for all levels of the qualitative variable.

TABLE 8.7-3   Computational Procedures for Testing Trend Constrast-Contrast Interactions (Data in part (iii) are from Table 8.3-1.)

(i) Coefficients for linear and quadratic contrasts:

	Treatment A			Treatment B		
Linear coefficients $(c_{1j}) =$	-1	0	1	$(c_{1k}) =$ -1	0	1
Quadratic coefficients $(c_{2j}) =$	1	-2	1	$(c_{2k}) =$ 1	-2	1

(ii) Products of coefficients for treatments A and B ($d_{jk(l \times l)} = c_{1j}c_{1k}$):

A lin $\times$ B lin

$$d_{11(l \times l)} = -1 \times -1 = 1$$
$$d_{12(l \times l)} = -1 \times 0 = 0$$
$$d_{13(l \times l)} = -1 \times 1 = -1$$
$$d_{21(l \times l)} = 0 \times -1 = 0$$
$$d_{22(l \times l)} = 0 \times 0 = 0$$
$$d_{23(l \times l)} = 0 \times 1 = 0$$
$$d_{31(l \times l)} = 1 \times -1 = -1$$
$$d_{32(l \times l)} = 1 \times 0 = 0$$
$$d_{33(l \times l)} = 1 \times 1 = 1$$

A lin $\times$ B quad

$$d_{11(l \times q)} = -1 \times 1 = -1$$
$$d_{12(l \times q)} = -1 \times -2 = 2$$
$$d_{13(l \times q)} = -1 \times 1 = -1$$
$$d_{21(l \times q)} = 0 \times 1 = 0$$
$$d_{22(l \times q)} = 0 \times -2 = 0$$
$$d_{23(l \times q)} = 0 \times 1 = 0$$
$$d_{31(l \times q)} = 1 \times 1 = 1$$
$$d_{32(l \times q)} = 1 \times -2 = -2$$
$$d_{33(l \times q)} = 1 \times 1 = 1$$

A quad $\times$ B lin

$$d_{11(q \times l)} = 1 \times -1 = -1$$
$$d_{12(q \times l)} = 1 \times 0 = 0$$
$$d_{13(q \times l)} = 1 \times 1 = 1$$
$$d_{21(q \times l)} = -2 \times -1 = 2$$
$$d_{22(q \times l)} = -2 \times 0 = 0$$
$$d_{23(q \times l)} = -2 \times 1 = -2$$
$$d_{31(q \times l)} = 1 \times -1 = -1$$
$$d_{32(q \times l)} = 1 \times 0 = 0$$
$$d_{33(q \times l)} = 1 \times 1 = 1$$

A quad $\times$ B quad

$$d_{11(q \times q)} = 1 \times 1 = 1$$
$$d_{12(q \times q)} = 1 \times -2 = -2$$
$$d_{13(q \times q)} = 1 \times 1 = 1$$
$$d_{21(q \times q)} = -2 \times 1 = -2$$
$$d_{22(q \times q)} = -2 \times -2 = 4$$
$$d_{23(q \times q)} = -2 \times 1 = -2$$
$$d_{31(q \times q)} = 1 \times 1 = 1$$
$$d_{32(q \times q)} = 1 \times -2 = -2$$
$$d_{33(q \times q)} = 1 \times 1 = 1$$

(iii) Treatment combination sums:

	a_jb_1	a_jb_2	a_jb_3
$\sum_{i=1}^{n} Y_{i1k} = $	165	175	190
$\sum_{i=1}^{n} Y_{i2k} = $	150	155	180
$\sum_{i=1}^{n} Y_{i3k} = $	100	200	260

(iv) Computation of sum of squares:

$$SS\hat{\psi}_{\text{lin}(A)}\hat{\psi}_{\text{lin}(B)} = \frac{\left(\sum_{j=1}^{p} \sum_{k=1}^{q} d_{jk(l \times l)} \sum_{i=1}^{n} Y_{ijk}\right)^2}{n \sum_{j=1}^{p} \sum_{k=1}^{q} d_{jk(l \times l)}^2} = \frac{[(1)165 + (0)175 + \cdots + (1)260]^2}{5[(1)^2 + (0)^2 + \cdots + (1)^2]}$$

$$= \frac{(135)^2}{5(4)} = 911.250$$

TABLE 8.7-3 (continued)

$$df = 1$$

$$SS\hat{\psi}_{\text{lin}(A)}\hat{\psi}_{\text{quad}(B)} = \frac{\left(\sum_{j=1}^{p}\sum_{k=1}^{q} d_{jk(l\times q)}\sum_{i=1}^{n} Y_{ijk}\right)^2}{n\sum_{j=1}^{p}\sum_{k=1}^{q} d_{jk(l\times q)}^2} = \frac{[(-1)165 + (2)175 + \cdots + (1)260]^2}{5[(-1)^2 + (2)^2 + \cdots + (1)^2]}$$

$$= \frac{(-45)^2}{5(12)} = 33.750$$

$$df = 1$$

$$SS\hat{\psi}_{\text{quad}(A)}\hat{\psi}_{\text{lin}(B)} = \frac{\left(\sum_{j=1}^{p}\sum_{k=1}^{q} d_{jk(q\times l)}\sum_{i=1}^{n} Y_{ijk}\right)^2}{n\sum_{j=1}^{p}\sum_{k=1}^{q} d_{jk(q\times l)}^2} = \frac{[(-1)165 + (0)175 + \cdots + (1)260]^2}{5[(-1)^2 + (0)^2 + \cdots + (1)^2]}$$

$$= \frac{(125)^2}{5(12)} = 260.417$$

$$df = 1$$

$$SS\hat{\psi}_{\text{quad}(A)}\hat{\psi}_{\text{quad}(B)} = \frac{\left(\sum_{j=1}^{p}\sum_{k=1}^{q} d_{jk(q\times q)}\sum_{i=1}^{n} Y_{ijk}\right)^2}{n\sum_{j=1}^{p}\sum_{k=1}^{q} d_{jk(q\times q)}^2} = \frac{[(1)165 + (-2)175 + \cdots + (1)260]^2}{5[(1)^2 + (-2)^2 + \cdots + (1)^2]}$$

$$= \frac{(-75)^2}{5(36)} = 31.250$$

$$df = 1$$

(v) Computation of F ratios:

$$F = \frac{(SS\hat{\psi}_{\text{lin}(A)}\hat{\psi}_{\text{lin}(B)})/1}{MSWCELL} = \frac{911.250}{62.500} = 14.58$$

$$F = \frac{(SS\hat{\psi}_{\text{lin}(A)}\hat{\psi}_{\text{quad}(B)})/1}{MSWCELL} = \frac{33.750}{62.500} = 0.54$$

$$F = \frac{(SS\hat{\psi}_{\text{quad}(A)}\hat{\psi}_{\text{lin}(B)})/1}{MSWCELL} = \frac{260.417}{62.500} = 4.17$$

$$F = \frac{(SS\hat{\psi}_{\text{quad}(A)}\hat{\psi}_{\text{quad}(B)})/1}{MSWCELL} = \frac{31.250}{62.500} = 0.50$$

Critical value for a priori orthogonal contrasts:

$$F_{.05;1,36} = 4.11$$

Critical value for Scheffé's procedure:

$$\nu_1 F_{.05;\nu_1,\nu_2} = 4F_{.05;4,36} = 4(2.63) = 10.52$$

The null hypotheses that we want to test are

$$H_0: \quad \alpha\psi_{\text{lin}(B)} = \delta \quad \text{for all } j \quad \text{and} \quad H_0: \quad \alpha\psi_{\text{quad}(B)} = \delta \quad \text{for all } j$$

where δ is a constant for a particular hypothesis. To determine whether the AB interaction is due to differences in the linear or quadratic trend components, the analysis illustrated in Table 8.7-4 can be performed. The data are taken from Table 8.3-1. The hypothesis $H_0: \quad \alpha\psi_{\text{lin}(B)} = \delta$ for all j is rejected, which means that treatment A interacts with the linear contrast. Differences in the linear contrast for the three levels of treatment A account for $SSA\hat{\psi}_{\text{lin}(B)}/SSAB = (1171.667/1236.667) \times 100 = 95\%$ of the AB interaction sum of squares.

TABLE 8.7-4   Computational Procedures for Testing Treatment $\times$ Trend-Contrast Interactions (Data in part (i) are from Table 8.3-1.)

(i)   Treatment combination sums and coefficients for linear and quadratic contrasts:

Treatment combination sums

	a_jb_1	a_jb_2	a_jb_3	Treatment B		
$\sum_{i=1}^{n} Y_{i1k} = 165$		175	190	Linear coefficients $(c_{1k}) = -1$	0	1
				Quadratic coefficients $(c_{2k}) = 1$	-2	1
$\sum_{i=1}^{n} Y_{i2k} = 150$		155	180			
$\sum_{i=1}^{n} Y_{i3k} = 100$		200	260			

(ii)   Computation of contrasts at each level of treatment A:

$$\hat{\psi}_{i(B)} \text{ at } a_j = c_{i1}\Sigma Y_{ij1} + c_{i2}\Sigma Y_{ij2} + c_{i3}\Sigma Y_{ij3}$$

$$\hat{\psi}_{1(B)} \text{ at } a_1 = (-1)165 + (0)175 + (1)190 = 25$$

$$\hat{\psi}_{1(B)} \text{ at } a_2 = (-1)150 + (0)155 + (1)180 = 30$$

$$\hat{\psi}_{1(B)} \text{ at } a_3 = (-1)100 + (0)200 + (1)260 = 160$$

$$\hat{\psi}_{2(B)} \text{ at } a_1 = (1)165 + (-2)175 + (1)190 = 5$$

$$\hat{\psi}_{2(B)} \text{ at } a_2 = (1)150 + (-2)155 + (1)180 = 20$$

$$\hat{\psi}_{2(B)} \text{ at } a_3 = (1)100 + (-2)200 + (1)260 = -40$$

(iii)   Computation of sum of squares:

$$SSA\hat{\psi}_{\text{lin}(B)} = \left[\sum_{j=1}^{p} \hat{\psi}_{1(B)}^2 \text{ at } a_j - \frac{\left(\sum_{j=1}^{p} \hat{\psi}_{1(B)} \text{ at } a_j \right)^2}{p} \right] \bigg/ n \sum_{k=1}^{q} c_{1k}^2$$

TABLE 8.7-4 (continued)

$$= \left[(25)^2 + (30)^2 + (160)^2 - \frac{(215)^2}{3} \right] / 5[(-1)^2 + (0)^2 + (1)^2] = 1171.667$$

$$df = (p - 1)1 = 2$$

$$SSA\hat{\psi}_{\text{quad}(B)} = \left[\sum_{j=1}^{p} \hat{\psi}_{2(B)}^2 \text{ at } a_j - \frac{\left(\sum_{j=1}^{p} \hat{\psi}_{2(B)} \text{ at } a_j \right)^2}{p} \right] / n \sum_{k=1}^{q} c_{2k}^2$$

$$= \left[(5)^2 + (20)^2 + (-40)^2 - \frac{(-15)^2}{3} \right] / 5[(1)^2 + (-2)^2 + (1)^2] = 65.000$$

$$df = (p - 1)1 = 2$$

(iv) Computation of F ratios:

$$F = \frac{SSA\hat{\psi}_{\text{lin}(B)}/(p - 1)1}{MSWCELL} = \frac{1171.667/2}{62.500} = 9.37$$

$$F = \frac{SSA\hat{\psi}_{\text{quad}(B)}/(p - 1)1}{MSWCELL} = \frac{65.000/2}{62.500} = 0.52$$

Critical value for a priori orthogonal contrasts:

$$F_{.05:2.36} = 3.26$$

Critical value for simultaneous test procedure:

$$(\nu_1 F_{.05:\nu_1,\nu_2})/\nu_3 = (4F_{.05:4.36})/2 = 4(2.63)/2 = 5.26$$

8.8 ESTIMATING STRENGTH OF ASSOCIATION IN A TYPE CRF-*pq* DESIGN

A significant F ratio indicates that some relationship exists between the dependent variable and independent variables. Although this information is useful, the point was made in Section 4.6 that an F ratio provides no information concerning the strength of the association. Indices of strength of association for type CRF-*pq* fixed-effects model are as follows (the data are from the police attitude experiment reported in Table 8.3-1).

$$\hat{\omega}_{Y|A}^2 = \frac{SSA - (p - 1)MSWCELL}{SSTO + MSWCELL} = \frac{190.000 - (3 - 1)62.500}{5220.000 + 62.500} = .012$$

$$\hat{\omega}_{Y|B}^2 = \frac{SSB - (q - 1)MSWCELL}{SSTO + MSWCELL} = \frac{1543.333 - (3 - 1)62.500}{5220.000 + 62.500} = .268$$

$$\hat{\omega}^2_{Y|AB} = \frac{SSAB - (p - 1)(q - 1)MSWCELL}{SSTO + MSWCELL}$$

$$= \frac{1236.667 - (3 - 1)(3 - 1)62.500}{5220.000 + 62.500} = .187$$

According to the preceding analysis, treatment B (number of hours of human relations training) accounts for the largest proportion of the variance in the dependent variable ($\hat{\omega}^2_{Y|B} = .27$), while treatment A (beat to which a trainee is assigned) accounts for a negligible portion ($\hat{\omega}^2_{Y|A} = .01$). Both treatments and the interaction combined account for $(.012 + .268 + .187) \times 100 = 47\%$ of the variance in change in attitudes toward minority groups, which is a sizeable proportion.

If the estimated value of ω^2 is negative, $\hat{\omega}^2$ is set equal to zero. The value of $\hat{\omega}^2$ will be positive if the F ratio for the associated mean square is significant. Information concerning the proportion of variance in the dependent variable accounted for by the treatments and interaction is most useful to an experimenter in interpreting the outcome of research. It cannot be emphasized enough that a trivial association may achieve statistical significance if the sample is made sufficiently large.

A comparable measure of association for a type CRF-pq design random-effects model is the intraclass correlation (ρ_I). Formulas for computing the intraclass correlation are as follows.

$$\hat{\rho}_{IY|A} = \frac{\hat{\sigma}^2_\alpha}{\hat{\sigma}^2_\epsilon + \hat{\sigma}^2_{\alpha\beta} + \hat{\sigma}^2_\beta + \hat{\sigma}^2_\alpha}$$

$$\hat{\rho}_{IY|B} = \frac{\hat{\sigma}^2_\beta}{\hat{\sigma}^2_\epsilon + \hat{\sigma}^2_{\alpha\beta} + \hat{\sigma}^2_\beta + \hat{\sigma}^2_\alpha}$$

$$\hat{\rho}_{IY|AB} = \frac{\hat{\sigma}^2_{\alpha\beta}}{\hat{\sigma}^2_\epsilon + \hat{\sigma}^2_{\alpha\beta} + \hat{\sigma}^2_\beta + \hat{\sigma}^2_\alpha}$$

Estimates of the different variance components can be obtained from the expected values of the mean squares given in Table 8.4-1. According to this table, the expected values of the mean squares are

$$E(MSA) = \sigma^2_\epsilon + n\sigma^2_{\alpha\beta} + nq\sigma^2_\alpha$$
$$E(MSB) = \sigma^2_\epsilon + n\sigma^2_{\alpha\beta} + np\sigma^2_\beta$$
$$E(MSAB) = \sigma^2_\epsilon + n\sigma^2_{\alpha\beta}$$
$$E(MSWCELL) = \sigma^2_\epsilon.$$

It follows that estimates of the variance components are as follows.

$$\hat{\sigma}^2_\alpha = \frac{MSA - MSAB}{nq}$$

$$\hat{\sigma}^2_\beta = \frac{MSB - MSAB}{np}$$

$$\hat{\sigma}_{\alpha\beta}^2 = \frac{MSAB - MSWCELL}{n}$$

$$\hat{\sigma}_{\epsilon}^2 = MSWCELL$$

If an estimated component is negative, that component is estimated to be zero.

Indices of strength of association for a mixed model in which the levels of treatment A are fixed effects but those for treatment B are random effects are as follows:

$$\hat{\omega}_{Y|A}^2 = \frac{\sum_{j=1}^{p} \hat{\alpha}_j^2/p}{\hat{\sigma}_{\epsilon}^2 + \hat{\sigma}_{\alpha\beta}^2 + \hat{\sigma}_{\beta}^2 + \sum_{j=1}^{p} \hat{\alpha}_j^2/p}$$

$$\hat{\rho}_{IY|B} = \frac{\hat{\sigma}_{\beta}^2}{\hat{\sigma}_{\epsilon}^2 + \hat{\sigma}_{\alpha\beta}^2 + \hat{\sigma}_{\beta}^2 + \sum_{j=1}^{p} \hat{\alpha}_j^2/p}$$

$$\hat{\rho}_{IY|AB} = \frac{\hat{\sigma}_{\alpha\beta}^2}{\hat{\sigma}_{\epsilon}^2 + \hat{\sigma}_{\alpha\beta}^2 + \hat{\sigma}_{\beta}^2 + \sum_{j=1}^{p} \hat{\alpha}_j^2/p}$$

where

$$\frac{\sum_{j=1}^{p} \hat{\alpha}_j^2}{p} = \frac{(p-1)(MSA - MSAB)}{npq}$$

$$\hat{\sigma}_{\beta}^2 = \frac{MSB - MSWCELL}{np}$$

$$\hat{\sigma}_{\alpha\beta}^2 = \frac{MSAB - MSWCELL}{n}$$

$$\hat{\sigma}_{\epsilon}^2 = MSWCELL.$$

The magnitude of treatment effects can be interpreted directly in terms of variance component estimates. This alternative approach to interpreting the outcome of research is discussed by Bross (1950), Scheffé (1959, 231–235), Searle (1971a, Ch. 11; 1971b, 1–76), and Tukey (1953).

8.9   RULES FOR DERIVING EXPECTED VALUES OF MEAN SQUARES

An experimenter may find it necessary to derive $E(MS)$'s for a complex design in order to determine the proper denominators for F ratios. A laborious procedure for deriving expected values was presented in Section 2.3. A simple set of rules is described in this section that leads to the same results. These rules apply to type CR-p and RB-p designs and all designs constructed from these building block designs. An equal number of experimental units in each treatment level or treatment combination is assumed. Rules similar to those presented here have been described by Bennett and Franklin (1954, 413) and Cornfield and Tukey (1956).

For purposes of illustration, expected values of mean squares will be derived for a type CRF-pq design. The procedure consists of six steps or rules.

Rule 1.   Write the experimental design model equation for the design.

$$Y_{ijk} = \mu + \alpha_j + \beta_k + (\alpha\beta)_{jk} + \epsilon_{i(jk)}$$
$$(i = 1, \ldots, n; j = 1, \ldots, p; k = 1, \ldots, q)$$

Rule 2.   Construct a two-way table. See Table 8.9-1 for details of the layout.

    a.   Row headings in part (i) of the table consist of the terms on the right side of the model equation with the exception of the grand mean μ.

    b.   Column headings in part (ii) consist of the subscripts used in the model, e.g., i, j, k. Below each subscript, indicate the number of levels over which summation takes place, e.g., n, p, q. The number of columns must equal the number of different subscripts in the model.

Rule 3.   Entries below each column heading in part (ii) are determined as follows.

    a.   If the column heading, i, j, k, and so on in part (ii) appears as a subscript of a row term in part (i), but not in parentheses, enter the sampling fraction—$1 - (n/N), 1 - (p/P), 1 - (q/Q)$, and so on—appropriate for that column in the row. The lowercase letter in the sampling fractions stands for the number of levels of the variable that are in the experiment; the uppercase letter, the number in the population.

    b.   If the column heading does not appear as a subscript of a row term, enter the number of levels, n, p, q, appropriate for that column in the row.

    c.   If the column heading appears in the subscript of a row term, but in parentheses, write the number 1 in that row.

Rule 4.   Entries in part (iii) for a given row consist of the variances for those terms in the model that contain *all* the subscripts of the row term. No distinction is made between subscripts in parentheses and those not in parentheses. For example, the subscript in row 1 is j. Variances for the terms in the model that contain the subscript j are σ_ϵ^2, $\sigma_{\alpha\beta}^2$, and σ_α^2.

Rule 5.   Coefficients of the variances in part (iii) for a given row are obtained by covering up the column(s) in part (ii) headed by subscript(s) that appear in the row (part i), but not including subscripts in parentheses, and multiplying the variance for an effect in part (iii) by the uncovered terms in part (ii) from the row for that effect. Consider row 1; its subscript is j so column j is covered leaving columns i and k exposed. The coefficients for σ_α^2 are n and q from row 1, the coefficients for $\sigma_{\alpha\beta}^2$ are n and $1 - (q/Q)$ from row 3, and the coefficients for σ_ϵ^2 are $1 - (n/N)$ and 1 from row 4. (According to Rule 6, $(1 - n/N)$ always equals 1.)

Rule 6. Sampling fractions in part (iii) equal 0 if the corresponding terms in the experimental design model represent fixed effects, and 1 if the terms are random effects. Values between 0 and 1 may be appropriate for the sampling fraction, but these two values are most often used in practice. Experimental error $\epsilon_{i(jk)}$ is considered to be a random variable; thus $(1 - n/N)$ is always equal to 1.

TABLE 8.9-1 Expected Values of Mean Squares for Type CRF-pq Design

(i) Effects	(ii)			(iii) $E(MS)$
	i n	j p	k q	
α_j	n	$1 - \dfrac{p}{P}$	q	$\sigma_\epsilon^2 + n\left(1 - \dfrac{q}{Q}\right)\sigma_{\alpha\beta}^2 + nq\sigma_\alpha^2$
β_k	n	p	$1 - \dfrac{q}{Q}$	$\sigma_\epsilon^2 + n\left(1 - \dfrac{p}{P}\right)\sigma_{\alpha\beta}^2 + np\sigma_\beta^2$
$(\alpha\beta)_{jk}$	n	$1 - \dfrac{p}{P}$	$1 - \dfrac{q}{Q}$	$\sigma_\epsilon^2 + \qquad n\sigma_{\alpha\beta}^2$
$\epsilon_{i(jk)}$	$1 - \dfrac{n}{N}$	1	1	σ_ϵ^2

Once the table is completed, expected values of mean squares for fixed-effects, random-effects, and mixed models can be readily determined. For example, if treatments A and B are fixed effects, their respective sampling fractions, $(1 - p/P)$ and $(1 - q/Q)$, are equal to zero and all products involving these sampling fractions equal zero. In writing out the $E(MS)$'s from part (iii), the variances for fixed effects should be replaced by sums of squared effects divided by their respective degrees of freedom. For example, since treatments A and B are fixed, σ_α^2 should be replaced by $\Sigma \alpha_j^2/(p - 1)$, and σ_β^2 by $\Sigma \beta_k^2/(q - 1)$. If, as in the example, the interaction involves all fixed effects, it too should be replaced by the sum of the squared interaction effects divided by degrees of freedom; that is, $\sigma_{\alpha\beta}^2$ is replaced by $\Sigma\Sigma(\alpha\beta)_{jk}^2/(p - 1)(q - 1)$. The resulting expected values of mean squares and the form of the F ratios for this fixed-effects model are as follows.

1 $E(MSA)$ $\quad = \sigma_\epsilon^2 + nq\Sigma\alpha_j^2/(p - 1)$ $\qquad F = [1/4]$
2 $E(MSB)$ $\quad = \sigma_\epsilon^2 + np\Sigma\beta_k^2/(q - 1)$ $\qquad F = [2/4]$
3 $E(MSAB)$ $\quad = \sigma_\epsilon^2 + n\Sigma\Sigma(\alpha\beta)_{jk}^2/(p - 1)(q - 1)$ $\qquad F = [3/4]$
4 $E(MSWCELL) = \sigma_\epsilon^2$

Let A be a fixed effect and B a random effect. Then the sampling fraction $(1 - q/Q)$ is equal to one; $(1 - p/P)$ is equal to zero as before. The expected values of mean squares and the form of the F ratio for this mixed model are as follows.

1 $E(MSA)$ $= \sigma_\epsilon^2 + n\sigma_{\alpha\beta}^2 + nq\Sigma\alpha_j^2/(p-1)$ $F = [1/3]$

2 $E(MSB)$ $= \sigma_\epsilon^2 + np\sigma_\beta^2$ $F = [2/4]$

3 $E(MSAB)$ $= \sigma_\epsilon^2 + n\sigma_{\alpha\beta}^2$ $F = [3/4]$

4 $E(MSWCELL) = \sigma_\epsilon^2$

A second example showing the application of the six rules appears in Table 8.9-2 for a type CRF-pqr design, which is described in Section 9.2. This example has three treatments. The experimental design model equation is

$$Y_{ijkl} = \mu + \alpha_j + \beta_k + \gamma_l + (\alpha\beta)_{jk} + (\alpha\gamma)_{jl} + (\beta\gamma)_{kl} + (\alpha\beta\gamma)_{jkl} + \epsilon_{i(jkl)}$$
$$(i = 1, \ldots, n; j = 1, \ldots, p; k = 1, \ldots, q; l = 1, \ldots, r).$$

If treatments A, B, and C are random effects, the expected values of mean squares and the form of the F ratios are as follows. For this random-effects model, all of the

1 $E(MSA)$ $= \sigma_\epsilon^2 + n\sigma_{\alpha\beta\gamma}^2 + nq\sigma_{\alpha\gamma}^2 + nr\sigma_{\alpha\beta}^2 + nqr\sigma_\alpha^2$ $F = [1/?]$

2 $E(MSB)$ $= \sigma_\epsilon^2 + n\sigma_{\alpha\beta\gamma}^2 + np\sigma_{\beta\gamma}^2 + nr\sigma_{\alpha\beta}^2 + npr\sigma_\beta^2$ $F = [2/?]$

3 $E(MSC)$ $= \sigma_\epsilon^2 + n\sigma_{\alpha\beta\gamma}^2 + np\sigma_{\beta\gamma}^2 + nq\sigma_{\alpha\gamma}^2 + npq\sigma_\gamma^2$ $F = [3/?]$

4 $E(MSAB)$ $= \sigma_\epsilon^2 + n\sigma_{\alpha\beta\gamma}^2 + nr\sigma_{\alpha\beta}^2$ $F = [4/7]$

5 $E(MSAC)$ $= \sigma_\epsilon^2 + n\sigma_{\alpha\beta\gamma}^2 + nq\sigma_{\alpha\gamma}^2$ $F = [5/7]$

6 $E(MSBC)$ $= \sigma_\epsilon^2 + n\sigma_{\alpha\beta\gamma}^2 + np\sigma_{\beta\gamma}^2$ $F = [6/7]$

7 $E(MSABC)$ $= \sigma_\epsilon^2 + n\sigma_{\alpha\beta\gamma}^2$ $F = [7/8]$

8 $E(MSWCELL) = \sigma_\epsilon^2$

sampling fractions are equal to one. An examination of the $E(MS)$'s for this design indicates that the model does not provide error terms for testing the three main effects. The error term for MSA, for example, should contain the following terms:

$$\sigma_\epsilon^2 + n\sigma_{\alpha\beta\gamma}^2 + nq\sigma_{\alpha\gamma}^2 + nr\sigma_{\alpha\beta}^2.$$

It is apparent that such an error term does not exist. One can approach this problem in several ways, depending on the outcome of tests of the interactions. If all of the interactions are significant, an experimenter would not ordinarily be interested in tests of main effects. Hence, in this case, the problem of testing main effects does not come up. A second approach involves performing preliminary tests on the model. Suppose that a test of, say, $F = MSAB/MSABC$ is not significant at the .25 level of significance. The inclusion of $(\alpha\beta)_{jk}$ in the experimental design model and, hence, $nr\sigma_{\alpha\beta}^2$ in the $E(MS)$ for treatment A are open to question. If it can be concluded that $nr\sigma_{\alpha\beta}^2 = 0$, the proper error term for testing treatment A is $MSAC$. A general discussion of preliminary tests on the model and pooling procedures appears in Section 8.11. A third approach involves piecing together an error term having the proper form for testing main effects. This procedure is described in Section 8.10.

TABLE 8.9-2 Expected Values of Mean Squares for Type CRF-pqr Design

(i)	(ii)				(iii)
Effects	i	j	k	l	$E(MS)$
	n	p	q	r	
α_j	n	$1 - \dfrac{p}{P}$	q	r	$\sigma^2_\epsilon + n\left(1 - \dfrac{q}{Q}\right)\left(1 - \dfrac{r}{R}\right)\sigma^2_{\alpha\beta\gamma} + nq\left(1 - \dfrac{r}{R}\right)\sigma^2_{\alpha\gamma} + n\left(1 - \dfrac{q}{Q}\right)r\sigma^2_{\alpha\beta} + nqr\sigma^2_\alpha$
β_k	n	p	$1 - \dfrac{q}{Q}$	r	$\sigma^2_\epsilon + n\left(1 - \dfrac{p}{P}\right)\left(1 - \dfrac{r}{R}\right)\sigma^2_{\alpha\beta\gamma} + np\left(1 - \dfrac{r}{R}\right)\sigma^2_{\beta\gamma} + n\left(1 - \dfrac{p}{P}\right)r\sigma^2_{\alpha\beta} + npr\sigma^2_\beta$
γ_l	n	p	q	$1 - \dfrac{r}{R}$	$\sigma^2_\epsilon + n\left(1 - \dfrac{p}{P}\right)\left(1 - \dfrac{q}{Q}\right)\sigma^2_{\alpha\beta\gamma} + np\left(1 - \dfrac{q}{Q}\right)\sigma^2_{\beta\gamma} + n\left(1 - \dfrac{p}{P}\right)q\sigma^2_{\alpha\gamma} + npq\sigma^2_\gamma$
$(\alpha\beta)_{jk}$	n	$1 - \dfrac{p}{P}$	$1 - \dfrac{q}{Q}$	r	$\sigma^2_\epsilon + n\left(1 - \dfrac{r}{R}\right)\sigma^2_{\alpha\beta\gamma} + nr\sigma^2_{\alpha\beta}$
$(\alpha\gamma)_{jl}$	n	$1 - \dfrac{p}{P}$	q	$1 - \dfrac{r}{R}$	$\sigma^2_\epsilon + n\left(1 - \dfrac{q}{Q}\right)\sigma^2_{\alpha\beta\gamma} + nq\sigma^2_{\alpha\gamma}$
$(\beta\gamma)_{kl}$	n	p	$1 - \dfrac{q}{Q}$	$1 - \dfrac{r}{R}$	$\sigma^2_\epsilon + n\left(1 - \dfrac{p}{P}\right)\sigma^2_{\alpha\beta\gamma} + np\sigma^2_{\beta\gamma}$
$(\alpha\beta\gamma)_{jkl}$	n	$1 - \dfrac{p}{P}$	$1 - \dfrac{q}{Q}$	$1 - \dfrac{r}{R}$	$\sigma^2_\epsilon + n\sigma^2_{\alpha\beta\gamma}$
$\epsilon_{i(jkl)}$	$1 - \dfrac{n}{N}$	1	1	1	σ^2_ϵ

USE OF PARENTHESES TO INDICATE NESTING

The experimental design models for some of the designs that have been presented included subscripts in parentheses. This information was used (Rule 3) in completing the two-way table for deriving the $E(MS)$'s. The use of parentheses around one or more subscripts indicates that the effect corresponding to the subscript not in parentheses is nested. Effects that are restricted to a single level of a treatment are said to be *nested* within that treatment. The symbol $\epsilon_{i(jk)}$ in the model for a type CRF-pq design indicates that the error effect for the ith experimental unit is nested within the jkth treatment combination. An examination of the *ABS* Summary Table in Table 8.3-1 shows that n different subjects appear in each cell of the table. Thus a particular sample of n experimental units is nested because it is restricted to a single treatment combination. Treatments A and B are *crossed* (not nested) in this design because every level of treatment A appears in combination with every level of treatment B.

8.10   QUASI F RATIOS

We saw in the previous section that mean squares having appropriate expected values for constructing F ratios are not always available. Under these conditions, an error term that does have the appropriate expected values can be pieced together. The resulting F ratios are called *quasi F ratios* and are designated by the symbols F' or F''.

    The error term for testing MSA in a type CRF-pqr design, assuming a random-effects model, should have the following expected value.

$$\sigma_\epsilon^2 + n\sigma_{\alpha\beta\gamma}^2 + nq\sigma_{\alpha\gamma}^2 + nr\sigma_{\alpha\beta}^2$$

Although a mean square with this expected value does not exist for this model, a composite mean square with the required expected value can be pieced together. The required error term, following the procedure suggested by Satterthwaite (1946), is given by a linear combination of the means squares $MSAB + MSAC - MSABC$. This can be shown as follows.

$$E(MSAB) = \sigma_\epsilon^2 + n\sigma_{\alpha\beta\gamma}^2 \qquad\qquad + nr\sigma_{\alpha\beta}^2$$
$$+\ E(MSAC) = \sigma_\epsilon^2 + n\sigma_{\alpha\beta\gamma}^2 + nq\sigma_{\alpha\gamma}^2$$
$$-\ E(MSABC) = -\sigma_\epsilon^2 - n\sigma_{\alpha\beta\gamma}^2$$

***

$$E(MSAB) + E(MSAC) - E(MSABC) = \sigma_\epsilon^2 + n\sigma_{\alpha\beta\gamma}^2 + nq\sigma_{\alpha\gamma}^2 + nr\sigma_{\alpha\beta}^2$$

A test of treatment A is given by the quasi F' ratio

$$F' = \frac{MSA}{MSAB + MSAC - MSABC}.$$

The F' ratio has the general form

$$F' = \frac{MS_1}{MS_2 + MS_3 - MS_4}.$$

The degrees of freedom for the denominator of this ratio is the nearest integral value of

$$df = \frac{(MS_2 + MS_3 - MS_4)^2}{MS_2^2/df_2 + MS_3^2/df_3 + MS_4^2/df_4}$$

where df_2, df_3, and df_4 are the degrees of freedom for the respective mean squares. The degrees of freedom for the numerator of the F' ratio is the regular df for MS_1.

One problem inherent in the F' ratio is the possibility of obtaining a negative denominator. Cochran (1951) proposed that Satterthwaite's method be modified by adding mean squares to the numerator rather than subtracting them from the denominator. This ratio, which is designated as F'', has the general form

$$F'' = \frac{MS_1 + MS_2}{MS_3 + MS_4}.$$

The degrees of freedom for numerator and denominator are equal to the nearest integral values, respectively, of

$$\frac{(MS_1 + MS_2)^2}{MS_1^2/df_1 + MS_2^2/df_2} \quad \text{and} \quad \frac{(MS_3 + MS_4)^2}{MS_3^2/df_3 + MS_4^2/df_4}.$$

A quasi F'' ratio for testing treatment A for the type CRF-*pqr* design just described is given by

$$F'' = \frac{MSA + MSABC}{MSAC + MSAB}.$$

It can be shown that this ratio has the required form in terms of expected values of mean squares.

$$\frac{E(MS_1) + E(MS_2)}{E(MS_3) + E(MS_4)} = \frac{2\sigma_\epsilon^2 + 2n\sigma_{\alpha\beta\gamma}^2 + nq\sigma_{\alpha\gamma}^2 + nr\sigma_{\alpha\beta}^2 + nqr\sigma_\alpha^2}{2\sigma_\epsilon^2 + 2n\sigma_{\alpha\beta\gamma}^2 + nq\sigma_{\alpha\gamma}^2 + nr\sigma_{\alpha\beta}^2}$$

Quasi F' and F'' ratios for testing treatments B and C in a type CRF-*pqr* design, assuming a random-effects model, have the following form.

$$F' = \frac{MSB}{MSAB + MSBC - MSABC} \qquad F'' = \frac{MSB + MSABC}{MSAB + MSBC}$$

$$F' = \frac{MSC}{MSAC + MSBC - MSABC} \qquad F'' = \frac{MSC + MSABC}{MSAC + MSBC}$$

The sampling distribution of a quasi F ratio when the null hypothesis is true is not central F, although the latter distribution may be used as an approximation. Conditions under which the approximation holds have been examined by Gaylor and Hopper (1969).

8.11 PRELIMINARY TESTS ON THE MODEL AND POOLING PROCEDURES

The selection of an experimental design and associated design model is largely based on an experimenter's subject-matter knowledge. Other factors influencing the selection of a design are discussed in Section 1.3. The model selected should include all sources of variation that an experimenter is interested in and that are expected to contribute to the total variation. All sources of variation not specifically included in the model as treatment or nuisance effects become a part of the experimental error.

A factorial design should be employed whenever an experimenter expects that interaction terms are important sources of variation. Unfortunately factors other than an interest in higher-order interactions may lead an experimenter to use a complex factorial design: availability of a particular computer software package or familiarity with the layout and analysis of the complex design. We have mentioned two situations. In the first, interactions are included in the model by choice. In the second, they are included for convenience. Experimenters in the second category typically have little commitment to the experimental design model that is adopted. Procedures discussed in this section are concerned with using data obtained in an experiment to make preliminary tests on the appropriateness of a particular model. Statisticians disagree on whether to adhere to the model specified at the beginning of an experiment, even though the data suggest that it is incorrect, or whether it is permissible or even desirable to modify the model along lines suggested by the data (Games, 1973; Levin and Marascuilo, 1972, 1973; Marascuilo and Levin, 1976; Scheffé, 1959, 126–127). The procedures recommended here represent a middle-of-the-road position with respect to the issue of preliminary tests and pooling. This position is similar to that adopted by Bozivich, Bancroft, and Hartley (1956) and by Green and Tukey (1960).

Assume that a type CRF-pqr design has been used in an experiment and that the treatment levels were selected randomly. The model equation for this design is

$$Y_{ijkl} = \mu + \alpha_j + \beta_k + \gamma_l + (\alpha\beta)_{jk} + (\alpha\gamma)_{jl} + (\beta\gamma)_{kl} + (\alpha\beta\gamma)_{jkl} + \epsilon_{i(jkl)}.$$

The experimental design model and associated expected values of the mean squares determine the form of the tests of the null hypotheses. If there is a question about whether interaction terms should appear in the model, preliminary tests can be performed. Such tests are designed to revise or confirm the specification of parameters included in the model.

The first component that is usually tested is the highest-order interaction; in this case the three-treatment interaction $\sigma^2_{\alpha\beta\gamma}$. According to the expected values of the mean squares for this design in Section 8.9, the F ratio for the test of $\sigma^2_{\alpha\beta\gamma} = 0$ has the form

$$F = \frac{MSABC}{MSWCELL}.$$

A type II error—failing to reject the hypothesis that $\sigma^2_{\alpha\beta\gamma} = 0$ when it is false—should be avoided in preliminary tests. The probability of a type II error can be made relatively small by adopting a numerically large value of α, say, $\alpha = .25$. If the test for $\sigma^2_{\alpha\beta\gamma} = 0$ is insignificant at the .25 level and an experimenter has no a priori reason for including this component in the model, it can be dropped. The revised model equation is

$$Y_{ijkl} = \mu + \alpha_j + \beta_k + \gamma_l + (\alpha\beta)_{jk} + (\alpha\gamma)_{jl} + (\beta\gamma)_{kl} + \epsilon_{ijkl}.$$

The term $\sigma^2_{\alpha\beta\gamma}$ does not appear in this model. Thus $E(MSABC) = \sigma^2_\epsilon$ instead of $\sigma^2_\epsilon + n\sigma^2_{\alpha\beta\gamma}$.

For this design we can compute two independent estimates of experimental error: MSWCELL and MSABC. When MSWCELL is based on a small number of degrees of freedom, it is often desirable to combine MSABC with MSWCELL to form a pooled error term that has more degrees of freedom. The resulting error term is called a *residual* error and is given by

$$MSRES = \frac{SSWCELL + SSABC}{pqr(n-1) + (p-1)(q-1)(r-1)}$$

with $df = pqr(n-1) + (p-1)(q-1)(r-1)$. The expected values of mean squares for the revised model are shown in Table 8.11-1.

The next step in the preliminary test of the model is to determine if the three two-treatment interactions should be retained. According to Table 8.11-1, the F ratios have the form

$$F = \frac{MSAB}{MSRES} \qquad F = \frac{MSAC}{MSRES} \qquad F = \frac{MSBC}{MSRES}.$$

TABLE 8.11-1   Expected Values of Mean Squares for Revised Model and F Ratios $(\sigma^2_{\alpha\beta\gamma} = 0)$

Source of Variation	$E(MS)$	F
1 MSA	$\sigma^2_\epsilon + nq\sigma^2_{\alpha\gamma} + nr\sigma^2_{\alpha\beta} + nqr\sigma^2_\alpha$	[1/?]
2 MSB	$\sigma^2_\epsilon + np\sigma^2_{\beta\gamma} + nr\sigma^2_{\alpha\beta} + npr\sigma^2_\beta$	[2/?]
3 MSC	$\sigma^2_\epsilon + np\sigma^2_{\beta\gamma} + nr\sigma^2_{\alpha\gamma} + npq\sigma^2_\gamma$	[3/?]
4 MSAB	$\sigma^2_\epsilon + nr\sigma^2_{\alpha\beta}$	[4/7]
5 MSAC	$\sigma^2_\epsilon + nq\sigma^2_{\alpha\gamma}$	[5/7]
6 MSBC	$\sigma^2_\epsilon + np\sigma^2_{\beta\gamma}$	[6/7]
7 $\left.\begin{array}{l} MSABC \\ MSWCELL \end{array}\right\} MSRES$	σ^2_ϵ	

Assume that tests of $\sigma_{\alpha\beta}^2 = 0$ and $\sigma_{\beta\gamma}^2 = 0$ are insignificant at the .25 level but $\sigma_{\alpha\gamma}^2 = 0$ is significant. The revised model equation under these conditions is

$$Y_{ijkl} = \mu + \alpha_j + \beta_k + \gamma_l + \alpha\gamma_{jl} + \epsilon_{ijkl}.$$

The expected values of mean squares for this revised model and F ratios are shown in Table 8.11-2. The pooled error term for testing $\sigma_{\beta}^2 = 0$ and $\sigma_{\alpha\gamma}^2 = 0$ is given by

$$MSRES = $$
$$\frac{SSAB \quad + \quad SSBC \quad + \quad SSABC \quad + \; SSWCELL}{(p-1)(q-1) + (q-1)(r-1) + (p-1)(q-1)(r-1) + pqr(n-1)}.$$

Alternative rules for carrying out preliminary tests and pooling have been described by Paull (1950) and by Bozivich, Bancroft, and Hartley (1956). Green and Tukey (1960) present a detailed example of the application of Paull's rule.

TABLE 8.11-2   Expected Values of Mean Squares for Revised Model and F Ratios ($\sigma_{\alpha\beta}^2 = 0$, $\sigma_{\beta\gamma}^2 = 0$, $\sigma_{\alpha\beta\gamma}^2 = 0$)

Source of Variation	$E(MS)$	F
1 MSA	$\sigma_\epsilon^2 + nq\sigma_{\alpha\gamma}^2 + nqr\sigma_\alpha^2$	[1/4]
2 MSB	$\sigma_\epsilon^2 + npr\sigma_\beta^2$	[2/5]
3 MSC	$\sigma_\epsilon^2 + nq\sigma_{\alpha\gamma}^2 + npq\sigma_\gamma^2$	[3/4]
4 $MSAC$	$\sigma_\epsilon^2 + nq\sigma_{\alpha\gamma}^2$	[4/5]
5 $\left.\begin{array}{l} MSAB \\ MSBC \\ MSABC \\ MSWCELL \end{array}\right\} MSRES$	σ_ϵ^2	

ADVANTAGES AND DISADVANTAGES OF PRELIMINARY TESTS AND POOLING

One advantage of carrying out preliminary tests and pooling is readily apparent. If interaction components can be deleted from the model and pooled, the resulting error term will have more degrees of freedom. However, the potential disadvantages of carrying out preliminary tests and pooling, such as pooling nonzero interaction components with experimental error, may outweigh the advantages if the original error term is based on an adequate number of degrees of freedom, say, 20–30. A problem with preliminary tests and pooling is that subsequent tests are carried out as if no preliminary tests had preceded them. The sampling distributions of statistics used in tests that are preceded by preliminary tests differ from those not preceded by preliminary tests. In general, the appropriate sampling distribution of the statistic at each stage of preliminary testing of the model is not available to the experimenter. The use of a sampling distribution that is not appropriate for sequential decisions probably introduces a slight positive bias, that is, too often rejecting a null hypothesis when it should not be rejected. Thus pooling introduces contingencies that are difficult to evaluate statistically. Adherence to the original model eliminates this problem, although tests associated with an incorrect model may be less powerful than tests based on a revised model.

An experimenter can take three positions with respect to conducting preliminary tests and pooling. These are (1) never pool, (2) always pool, and (3) pool only when evidence indicates that the initial model is incorrect. Both the first and third rules have much to recommend them. The author favors the third. For other discussions of this point, see Bennett and Franklin (1954, 392); Bozivich, Bancroft, and Hartley (1956); Mead, Bancroft, and Han (1975); Paull (1950); and Scheffé (1959, 126–127).

8.12 THE ANALYSIS OF COMPLETELY RANDOMIZED FACTORIAL DESIGNS WITH $n = 1$

Experimenters are often interested in conducting research that involves many treatments, each with many levels. A type CRF-354 design, which is not an unusually large experiment by current standards, has 60 treatment combinations. To compute a within-cell error term with two subjects per cell, a minimum of 120 subjects is required. However, a sample of this size may not be available or considerations, such as time or cost, may preclude the use of such a large sample. Under these conditions, several design alternatives are available to an experimenter. One alternative, which is described in this section, is to assign only one subject to each treatment combination. If this approach is followed, the highest-order interaction instead of the within-cell error term is used as an estimate of experimental error. Additional design alternatives are described in Chapters 11, 12, and 13; these designs permit an experimenter, by careful planning, to carry out multitreatment experiments with a minimum of subjects.

Consider the following model equations for a two-treatment factorial experiment.

$$Y_{ijk} = \mu + \alpha_j + \beta_k + (\alpha\beta)_{jk} + \epsilon_{i(jk)}$$

$$Y_{ijk} = \mu + \alpha_j + \beta_k + \epsilon_{ijk}$$

The first model includes a two-treatment interaction, the second does not. If the second model provides an adequate description of the sources of variation in the experiment, $MSAB$ and $MSWCELL$ both estimate experimental error. For experiments in which $n = 1$, a mean square within-cell error term cannot be computed, in which case $MSAB$ may be used in place of $MSWCELL$. The basic question to be answered is: "Which model is appropriate for the experiment?"

Tukey's test for nonadditivity described in Section 6.3 is helpful in deciding between the two models. The model in which

$$E(MSAB) = \sigma_\epsilon^2 + \sigma_{\alpha\beta}^2$$

is called a *nonadditive model*. The one in which

$$E(MSAB) = \sigma_\epsilon^2$$

is called an *additive model*. Computational procedures for Tukey's test for a randomized block design are shown in Table 6.3-1. These procedures generalize with slight modification to factorial experiments. A test of nonadditivity for a type CRF-pq design requires the substitution in Table 6.3-1 of an AB Summary Table in place of the AS Summary Table. The computational formulas are

$$SSNONADD = \frac{\left(\sum_{j=1}^{p} \sum_{k=1}^{q} d_{jk} \sum_{i=1}^{n} Y_{ijk} \right)^2}{\left(\sum_{j=1}^{p} d_{j\cdot}^2 \right)\left(\sum_{k=1}^{q} d_{\cdot k}^2 \right)}$$

$$SSREM = SSAB - SSNONADD$$
$$dfNONADD = 1$$
$$dfREM = dfAB - 1$$
$$FNONADD = \frac{MSNONADD}{MSREM}.$$

If *FNONADD* is not significant at a numerically high level of significance ($\alpha = .10-.25$), it lends credence to the assumption that $E(MSAB) = \sigma_\epsilon^2$. In this case with $n = 1$, *MSAB* is used as the error term for testing treatments A and B. If *FNONADD* is significant, a transformation that will eliminate the AB interaction may be found. For a discussion of Tukey's test and the effects of various transformations on nonadditivity, see Harter and Lum (1962).

Tukey's test for nonadditivity in Section 6.3 can be easily modified for factorial experiments with three or more treatments. For a type CRF-pqr design, the AS Summary Table in Table 6.3-1 is replaced by an ABC Summary Table. A d_{jkl} Summary Table must be substituted for the d_{ij} Summary Table. The required computational formulas are

$$SSNONADD = \frac{\left(\sum_{j=1}^{p} \sum_{k=1}^{q} \sum_{l=1}^{r} d_{jkl} \sum_{i=1}^{n} Y_{ijkl} \right)^2}{\left(\sum_{j=1}^{p} d_{j\cdot\cdot}^2 \right)\left(\sum_{k=1}^{q} d_{\cdot k\cdot}^2 \right)\left(\sum_{l=1}^{r} d_{\cdot\cdot l}^2 \right)}$$

$$SSREM = SSABC - SSNONADD$$
$$dfNONADD = 1$$

$$dfREM = dfABC - 1$$

$$FNONADD = \frac{MSNONADD}{MSREM}.$$

If *FNONADD* is insignificant at, say, $\alpha = .25$, the nonadditive model is rejected in favor of the additive model in which $E(MSABC) = \sigma_{\epsilon}^2$.

*8.13 ANALYSIS OF COMPLETELY RANDOMIZED FACTORIAL DESIGNS WITH UNEQUAL CELL *n*'s

It is usually desirable to have equal *n*'s in each cell of a completely randomized factorial design, although this is not always possible. Three cases will be examined: one in which the cell *n*'s are proportional, a second in which they are disproportional but every cell contains at least one observation, and a third in which they are disproportional with one or more empty cells. As we will see, the traditional computational procedures described in Section 8.3 can be used for the proportional case although the general linear model approach, preferably the full rank experimental design model introduced in Section 5.10, is recommended for all three cases.

GENERAL LINEAR MODEL APPROACHES TO THE ANALYSIS OF EXPERIMENTS WITH UNEQUAL CELL *n*'s AND AT LEAST ONE OBSERVATION PER CELL

The issue of the proper way to analyze factorial experiments with unequal cell *n*'s has been debated by applied statisticians for over a decade. The debate began in earnest in 1969 with the publication of a paper by Overall and Spiegel on three methods for using a regression model with effect coding to analyze data from a type CRF-*pq* design.* Before discussing these and other methods, we will review two schemes for coding the independent variables of a qualitative regression model.

For purposes of illustration, the police attitude data in Table 8.3-1 will be used. We will assume that observation $Y_{511} = 42$ is missing. A qualitative regression model with $h - 1 = (p - 1) + (q - 1) + (p - 1)(q - 1) = 8$ independent variables $(X_{i1}, X_{i2}, \ldots, X_{i,p-1}X_{i,p+q-2})$ can be formulated for this design as follows:

☆   This section assumes a familarity with the matrix operations in Appendix D, Chapter 5, and Section 6.9.

*   The key papers in the debate include Applebaum and Cramer (1974), Carlson and Timm (1974), Cramer and Applebaum (1980), Keren and Lewis (1977), Kutner (1974), Lewis and Keren (1977), O'Brien (1976), Overall and Spiegel (1969, 1973), Overall, Spiegel, and Cohen (1975), and Rawlings (1972, 1973).

$$\overbrace{\phantom{\beta_1 X_{i1} + \beta_2 X_{i2}}}^{A} \qquad \overbrace{\phantom{\beta_3 X_{i3} + \beta_4 X_{i4}}}^{B}$$

(8.13-1) $Y_i = \beta_0 X_0 + \beta_1 X_{i1} + \beta_2 X_{i2} + \beta_3 X_{i3} + \beta_4 X_{i4}$

$$\overbrace{\phantom{+ \beta_5 X_{i1}X_{i3} + \beta_6 X_{i1}X_{i4} + \beta_7 X_{i2}X_{i3} + \beta_8 X_{i2}X_{i4}}}^{AB}$$

$$+ \beta_5 X_{i1}X_{i3} + \beta_6 X_{i1}X_{i4} + \beta_7 X_{i2}X_{i3} + \beta_8 X_{i2}X_{i4} + \epsilon_i \qquad (i = 1, \ldots, N)$$

where ϵ_i is $NID(0, \sigma_\epsilon^2)$. In previous regression model examples (Sections 6.9 and 7.9) we used dummy coding. However, for experiments with interactions, we will find that effect coding leads to a simpler correspondence between the parameters of the regression model and those of the less than full rank experimental design model. The two coding schemes are as follows.

	Effect Coding	*Dummy Coding*
$X_{i1} = \begin{cases} \\ \\ \\ \end{cases}$	1 if observation is in a_1 -1 if observation is in a_p 0 otherwise	1 if observation is in a_1 0 otherwise

. . . .

| $X_{i,p-1} = \begin{cases} \\ \\ \\ \end{cases}$ | 1 if observation is in a_{p-1}
-1 if observation is in a_p
0 otherwise | 1 if observation is in a_{p-1}

0 otherwise |
| $X_{ip} = \begin{cases} \\ \\ \\ \end{cases}$ | 1 if observation is in b_1
-1 if observation is in b_{q-1}
0 otherwise | 1 if observation is in b_1

0 otherwise |

. . . .

| $X_{i,p+q-2} = \begin{cases} \\ \\ \\ \end{cases}$ | 1 if observation is in b_{q-1}
-1 if observation is in b_q
0 otherwise | 1 if observation is in b_{q-1}

0 otherwise |
| $X_{i1}X_{ip} = \begin{cases} \\ \\ \end{cases}$ | product of coded values
associated with a_1 and b_1 | product of coded values
associated with a_1 and b_1 |

. . . .

| $X_{i,p-1}X_{i,p+q-2} = \begin{cases} \\ \\ \\ \end{cases}$ | product of coded values
associated with a_{p-1} and
b_{q-1} | product of coded values
associated with a_{p-1} and
b_{q-1} |

The only new terms in the coding scheme are $X_{i1}X_{ip}, \ldots, X_{i,p-1}X_{i,p+q-2}$, which are used to code interactions. A type CRF-pq design requires $(p - 1)(q - 1)$ such terms, one for each of the AB interaction degrees of freedom. An example illustrating effect coding for the data in Table 8.3-1, assuming that Y_{511} is missing, is given in Table 8.13-1.

Procedures for determining the correspondence between h parameters of a regression model and the parameter of a less than full rank experimental design model are described in Sections 5.4 and 5.7. The correspondence for a type CRF-33 design with equal cell n's is as follows.

| | | Experimental Design Parameters | |
Regression Parameters		Effect Coding	Dummy Coding
β_0	$=$	μ	$\mu + \alpha_3 + \beta_3 + (\alpha\beta)_{33}$
β_1	$=$	α_1	$\alpha_1 - \alpha_3 + (\alpha\beta)_{13} - (\alpha\beta)_{33}$
β_2	$=$	α_2	$\alpha_2 - \alpha_3 + (\alpha\beta)_{23} - (\alpha\beta)_{33}$
β_3	$=$	β_1	$\beta_1 - \beta_3 + (\alpha\beta)_{31} - (\alpha\beta)_{33}$
β_4	$=$	β_2	$\beta_2 - \beta_3 + (\alpha\beta)_{32} - (\alpha\beta)_{33}$
β_5	$=$	$(\alpha\beta)_{11}$	$(\alpha\beta)_{11} - (\alpha\beta)_{31} - (\alpha\beta)_{13} + (\alpha\beta)_{33}$
β_6	$=$	$(\alpha\beta)_{12}$	$(\alpha\beta)_{12} - (\alpha\beta)_{32} - (\alpha\beta)_{13} + (\alpha\beta)_{33}$
β_7	$=$	$(\alpha\beta)_{21}$	$(\alpha\beta)_{21} - (\alpha\beta)_{31} - (\alpha\beta)_{23} + (\alpha\beta)_{33}$
β_8	$=$	$(\alpha\beta)_{22}$	$(\alpha\beta)_{22} - (\alpha\beta)_{32} - (\alpha\beta)_{23} + (\alpha\beta)_{33}$

(8.13-2)

It is apparent that effect coding leads to a simpler correspondence between the parameters of the two models. Hence effect coding is preferred when the experimental design model includes one or more interaction terms.

By now you are familiar with the general approach to testing hypotheses about the parameters of a regression model. Recall that a test of the hypothesis H_0: $\beta_1 = \beta_2 = 0$ for model (8.13-1) involves a comparison of the error sum of squares for the reduced model $SSE(X_3 \ X_4 \cdots X_2X_4)$ with the error sum of squares for the full model $SSE(X_1 \ X_2 \cdots X_2X_4)$. The test statistic is

$$F = \frac{[SSE(X_3 \ X_4 \cdots X_2X_4) - SSE(X_1 \ X_2 \cdots X_2X_4)]/(p - 1)}{SSE(X_1 \ X_2 \cdots X_2X_4)/(N - h)}$$
$$= \frac{MSR(X_1 \ X_2|X_3 \ X_4 \cdots X_2X_4)}{MSE}.$$

As we have seen, we test the null hypothesis for a particular set of parameters of the regression model because the test is relevant to some interesting null hypothesis

TABLE 8.13-1 Example Illustrating Effect Coding for the Data in Table 8.3-1 (Y_{511} Is Assumed to Be Missing.)

(i) Data and $\mathbf{X}_0$ matrix ($N = 44$, $h = 9$):

$$\mathbf{X}_0 \quad {}_{N \times h}$$

	y $_{N \times 1}$	$\mathbf{X}_0$	A $\mathbf{X}_1$	$\mathbf{X}_2$	B $\mathbf{X}_3$	$\mathbf{X}_4$	AB $\mathbf{X}_1\mathbf{X}_3$	$\mathbf{X}_1\mathbf{X}_4$	$\mathbf{X}_2\mathbf{X}_3$	$\mathbf{X}_2\mathbf{X}_4$
a_1b_1	24	1	1	0	1	0	1	0	0	0
	33	1	1	0	1	0	1	0	0	0
	37	1	1	0	1	0	1	0	0	0
	29	1	1	0	1	0	1	0	0	0
a_1b_2	44	1	1	0	0	1	0	1	0	0
	36	1	1	0	0	1	0	1	0	0
	25	1	1	0	0	1	0	1	0	0
	27	1	1	0	0	1	0	1	0	0
	43	1	1	0	0	1	0	1	0	0
a_1b_3	38	1	1	0	−1	−1	−1	−1	0	0
	29	1	1	0	−1	−1	−1	−1	0	0
	28	1	1	0	−1	−1	−1	−1	0	0
	47	1	1	0	−1	−1	−1	−1	0	0
	48	1	1	0	−1	−1	−1	−1	0	0
a_2b_1	30	1	0	1	1	0	0	0	1	0
	21	1	0	1	1	0	0	0	1	0
	39	1	0	1	1	0	0	0	1	0
	26	1	0	1	1	0	0	0	1	0
	34	1	0	1	1	0	0	0	1	0
a_2b_2	35	1	0	1	0	1	0	0	0	1
	40	1	0	1	0	1	0	0	0	1
	27	1	0	1	0	1	0	0	0	1
	31	1	0	1	0	1	0	0	0	1
	22	1	0	1	0	1	0	0	0	1
a_2b_3	26	1	0	1	−1	−1	0	0	−1	−1
	27	1	0	1	−1	−1	0	0	−1	−1
	36	1	0	1	−1	−1	0	0	−1	−1
	46	1	0	1	−1	−1	0	0	−1	−1
	45	1	0	1	−1	−1	0	0	−1	−1
a_3b_1	21	1	−1	−1	1	0	−1	0	−1	0
	18	1	−1	−1	1	0	−1	0	−1	0
	10	1	−1	−1	1	0	−1	0	−1	0
	31	1	−1	−1	1	0	−1	0	−1	0
	20	1	−1	−1	1	0	−1	0	−1	0
a_3b_2	41	1	−1	−1	0	1	0	−1	0	−1
	39	1	−1	−1	0	1	0	−1	0	−1
	50	1	−1	−1	0	1	0	−1	0	−1
	36	1	−1	−1	0	1	0	−1	0	−1
	34	1	−1	−1	0	1	0	−1	0	−1
a_3b_3	42	1	−1	−1	−1	−1	1	1	1	1
	52	1	−1	−1	−1	−1	1	1	1	1
	53	1	−1	−1	−1	−1	1	1	1	1
	49	1	−1	−1	−1	−1	1	1	1	1
	64	1	−1	−1	−1	−1	1	1	1	1

concerning the parameters of the experimental design model. For example, based on the correspondence in (8.13-2), we know that if $\beta_1 = \beta_2 = 0$, $\alpha_1 = \alpha_2 = \alpha_3 = 0$, since for a fixed-effects model $\Sigma_{j=1}^{p} \alpha_j = 0$. The choice of a reduced model for testing an interesting hypothesis for type CR-p, RB-p, and LS-p designs is straightforward. The situation is more complicated for a type CRF-33 design since a variety of reduced models can be defined that appear to be relevant to an interesting hypothesis. Some of these model equations and a simplified designation for the associated error sum of squares are as follows.

Model	Error Sum of Squares for Model
$\overbrace{\phantom{\beta_1 X_{i1} + \beta_2 X_{i2}}}^{A}\ \overbrace{\phantom{\beta_3 X_{i3} + \beta_4 X_{i4}}}^{B}$ $Y_i = \beta_0 X_0 + \beta_1 X_{i1} + \beta_2 X_{i2} + \beta_3 X_{i3} + \beta_4 X_{i4}$ $\overbrace{\phantom{\beta_5 X_{i1}X_{i3} + \beta_6 X_{i1}X_{i4} + \beta_7 X_{i2}X_{i3} + \beta_8 X_{i2}X_{i4}}}^{AB}$ $+\ \beta_5 X_{i1}X_{i3} + \beta_6 X_{i1}X_{i4} + \beta_7 X_{i2}X_{i3} + \beta_8 X_{i2}X_{i4} + \epsilon_i$	$SSE(A, B, AB)$
$\overbrace{\phantom{\beta_1 X_{i1} + \beta_2 X_{i2}}}^{A}\ \overbrace{\phantom{\beta_3 X_{i3} + \beta_4 X_{i4}}}^{B}$ $Y_i = \beta_0 X_0 + \beta_1 X_{i1} + \beta_2 X_{i2} + \beta_3 X_{i3} + \beta_4 X_{i4} + \epsilon_i$	$SSE(A, B)$
$\overbrace{\phantom{\beta_1 X_{i1} + \beta_2 X_{i2}}}^{A}$ $Y_i = \beta_0 X_0 + \beta_1 X_{i1} + \beta_2 X_{i2} + \epsilon_i$	$SSE(A)$
$\overbrace{\phantom{\beta_3 X_{i3} + \beta_4 X_{i4}}}^{B}$ $Y_i = \beta_0 X_0 + \beta_3 X_{i3} + \beta_4 X_{i4} + \epsilon_i$	$SSE(B)$
$\overbrace{\phantom{\beta_5 X_{i1}X_{i3} + \beta_6 X_{i1}X_{i4} + \beta_7 X_{i2}X_{i3} + \beta_8 X_{i2}X_{i4}}}^{AB}$ $Y_i = \beta_0 X_0 + \beta_5 X_{i1}X_{i3} + \beta_6 X_{i1}X_{i4} + \beta_7 X_{i2}X_{i3} + \beta_8 X_{i2}X_{i4}$ $+\ \epsilon_i$	$SSE(AB)$
$Y_i = \beta_0 X_0 + \epsilon_i$	$SSE(0)$
$\overbrace{\phantom{\beta_1 X_{i1} + \beta_2 X_{i2}}}^{A}$ $Y_i = \beta_0 X_0 + \beta_1 X_{i1} + \beta_2 X_{i2}$ $\overbrace{\phantom{\beta_5 X_{i1}X_{i3} + \beta_6 X_{i1}X_{i4} + \beta_7 X_{i2}X_{i3} + \beta_8 X_{i2}X_{i4}}}^{AB}$ $+\ \beta_5 X_{i1}X_{i3} + \beta_6 X_{i1}X_{i4} + \beta_7 X_{i2}X_{i3} + \beta_8 X_{i2}X_{i4} + \epsilon_i$	$SSE(A, AB)$
$\overbrace{\phantom{\beta_3 X_{i3} + \beta_4 X_{i4}}}^{B}$ $Y_i = \beta_0 X_0 + \beta_3 X_{i3} + \beta_4 X_{i4}$ $\overbrace{\phantom{\beta_5 X_{i1}X_{i3} + \beta_6 X_{i1}X_{i4} + \beta_7 X_{i2}X_{i3} + \beta_8 X_{i2}X_{i4}}}^{AB}$ $+\ \beta_5 X_{i1}X_{i3} + \beta_6 X_{i1}X_{i4} + \beta_7 X_{i2}X_{i3} + \beta_8 X_{i2}X_{i4} + \epsilon_i$	$SSE(B, AB)$

For the preceding set of models, there are a variety of ways of constructing tests of the regression parameters that are relevant to hypotheses about treatments A and B and the AB interaction. Consider the following model comparisons.

Treatment A

$$SSR(A \mid B, AB) = SSE(B, AB) - SSE(A, B, AB)$$
$$SSR(A \mid AB) = SSE(AB) - SSE(A, AB)$$
$$SSR(A \mid B) = SSE(B) - SSE(A, B)$$
$$SSR(A) = SSE(0) - SSE(A)$$

Treatment B

$$SSR(B \mid A, AB) = SSE(A, AB) - SSE(A, B, AB)$$
$$SSR(B \mid AB) = SSE(AB) - SSE(B, AB)$$
$$SSR(B \mid A) = SSE(A) - SSE(A, B)$$
$$SSR(B) = SSE(0) - SSE(B)$$

AB Interaction

$$SSR(AB \mid A, B) = SSE(A, B) - SSE(A, B, AB)$$
$$SSR(AB \mid A) = SSE(A) - SSE(A, AB)$$
$$SSR(AB \mid B) = SSE(B) - SSE(B, AB)$$
$$SSR(AB) = SSE(0) - SSE(AB)$$

If the cell n's are equal, the various ways of computing sums of squares lead to the same results. These results are also identical to those obtained using the traditional sums of squares formulas, SSA, SSB, and $SSAB$. That is,

$$SSA = SSR(A \mid B, AB) = SSR(A \mid AB) = SSR(A \mid B) = SSR(A)$$
$$SSB = SSR(B \mid A, AB) = SSR(B \mid AB) = SSR(B \mid A) = SSR(B)$$
$$SSAB = SSR(AB \mid A, B) = SSR(AB \mid A) = SSR(AB \mid B) = SSR(AB).$$

If the cell n's are unequal but the number in each of the jk cells satisfies

$$n_{jk} = \frac{n_{j\cdot} n_{\cdot k}}{N}$$

the cell n's are said to be *proportional,* otherwise they are *disproportional.* We will have more to say about this shortly. For the proportional case

$$SSA = SSR(A \mid B) = SSR(A) \neq SSR(A \mid B, AB) \neq SSR(A \mid AB)$$
$$SSB = SSR(B \mid A) = SSR(B) \neq SSR(B \mid A, AB) \neq SSR(B \mid AB)$$
$$SSAB = SSR(AB \mid A, B) \neq SSR(AB \mid A) \neq SSR(AB \mid B) \neq SSR(AB).$$

In the following section we will see that SSA and SSB test hypotheses concerning the equality of weighted means ($\bar{\mu}_{j\cdot} = \Sigma_{k=1}^{q} n_{jk}\mu_{jk}/n_{\cdot k}$ and $\bar{\mu}_{\cdot k} = \Sigma_{j=1}^{p} n_{jk}\mu_{jk}/n_{j\cdot}$, respectively); $SSR(A \mid B, AB)$ and $SSR(B \mid A, AB)$ test hypotheses concerning unweighted means ($\mu_{j\cdot} = \Sigma_{k=1}^{q} \mu_{jk}/q$ and $\mu_{\cdot k} = \Sigma_{j=1}^{p} \mu_{jk}/p$, respectively). For the disproportional case, the traditional sum of squares approach cannot be used. Furthermore, the various ways of constructing tests of the regression parameters all lead to different values for the sums of squares. The question naturally arises as to which values are correct. As we will see, this question does not have a simple answer.

There seems to be some consensus among applied statisticians that $SSR(AB \mid A, B)$ is a reasonable way to compute a sum of squares for the AB interaction. However, there is little agreement when it comes to computing sums of squares for treatments A and B. The alternative computational approaches give different results and obviously test different hypotheses. One way to choose among them is to determine which ones provide tests of interesting null hypotheses concerning population means. This is easier said than done. For the unequal n case, it is difficult to relate the hypotheses tested by the various regression model comparisons to hypotheses concerning population means. We will adopt an indirect approach to this problem. It consists of testing interesting null hypotheses using the full rank experimental design model and determining which regression model comparisons lead to identical F statistics and hence test equivalent hypotheses.* The proportional cell n case will be examined first.

ANALYSIS PROCEDURES FOR EXPERIMENTS WITH PROPORTIONAL CELL *n*'s

In some research situations, an experimenter may choose to have unequal cell n's. This might occur if the treatment levels require different kinds of equipment that are in limited supply and if the levels are to be administered to all subjects at the same time. Consider a type CRF-24 design. If 20 units of equipment associated with treatment level a_2 are available but only 12 units associated with a_1, an experimenter might decide to assign 32 subjects to the two levels of treatment A so as to use all available equipment. This means assigning 12 subjects to a_1, 20 to a_2, and 8 to each level of treatment B. This design is shown in Table 8.13-2(a).

We saw earlier that if the number of subjects in each of the jk treatment combinations satisfies

$$n_{jk} = \frac{n_j \cdot n_{\cdot k}}{N}$$

the cell n's are said to be *proportional*. The design in Table 8.13-2(a) satisfies the proportional requirement. The meaning of n_{jk} and so on should be clear from the example. Two other examples illustrating proportional cell n's are shown in Table 8.13-2. The advantage of having proportional cell n's is that the traditional sum of squares procedures described in Section 8.3 can, with slight modifications, be used to analyze the data. The modifications are shown in part (ii) of Table 8.13-2.

A set of data involving proportional cell n's is shown in Table 8.13-3. The traditional sum of squares approach has been used to compute the sums of squares in the table. The AB Summary Table contains two sets of estimates of population treatment-level means: $\hat{\mu}_{j\cdot}$ and $\hat{\mu}_{\cdot k}$ are simple averages of the respective cell means; $\hat{\bar{\mu}}_{j\cdot}$ and $\hat{\bar{\mu}}_{\cdot k}$ weight each cell mean proportional to the cell sample size. An experimenter

---

*   Searle (1971a, Ch. 7) provides a more elegant approach to this problem.

TABLE 8.13-2   Examples Illustrating Proportional Cell n's

(i)   Three examples:

(a) AB Summary Table

	b_1	b_2	b_3	b_4	$n_{j\cdot}$
a_1	$n_{11} = 3$	$n_{12} = 3$	$n_{13} = 3$	$n_{14} = 3$	12
a_2	$n_{21} = 5$	$n_{22} = 5$	$n_{23} = 5$	$n_{24} = 5$	20
$n_{\cdot k}$	$= 8$	8	8	8	$N = 32$

(b) AB Summary Table

	b_1	b_2	b_3	b_4	$n_{j\cdot}$
a_1	$n_{11} = 3$	$n_{12} = 4$	$n_{13} = 4$	$n_{14} = 5$	16
a_2	$n_{21} = 3$	$n_{22} = 4$	$n_{23} = 4$	$n_{24} = 5$	16
$n_{\cdot k}$	$= 6$	8	8	10	$N = 32$

(c) AB Summary Table

	b_1	b_2	b_3	b_4	$n_{j\cdot}$
a_1	$n_{11} = 1$	$n_{12} = 1$	$n_{13} = 2$	$n_{14} = 4$	8
a_2	$n_{21} = 3$	$n_{22} = 3$	$n_{23} = 6$	$n_{24} = 12$	24
$n_{\cdot k}$	$= 4$	4	8	16	$N = 32$

(ii)   Modified computational symbols when n's are proportional:

$$\frac{\left(\sum_{i=1}^{n_{jk}} \sum_{j=1}^{p} \sum_{k=1}^{q} Y_{ijk}\right)^2}{N} = [Y]$$

$$\sum_{i=1}^{n_{jk}} \sum_{j=1}^{p} \sum_{k=1}^{q} Y_{ijk}^2 = [ABS]$$

$$\sum_{j=1}^{p} \frac{\left(\sum_{i=1}^{n_{jk}} \sum_{k=1}^{q} Y_{ijk}\right)^2}{n_{j\cdot}} = [A]$$

$$\sum_{k=1}^{q} \frac{\left(\sum_{i=1}^{n_{jk}} \sum_{j=1}^{p} Y_{ijk}\right)^2}{n_{\cdot k}} = [B]$$

$$\sum_{j=1}^{p} \sum_{k=1}^{q} \frac{\left(\sum_{i=1}^{n_{jk}} Y_{ijk}\right)^2}{n_{jk}} = [AB]$$

$SSTO = [ABS] - [Y]$ $df = N - 1$

$SSA = [A] - [Y]$ $df = p - 1$

$SSB = [B] - [Y]$ $df = q - 1$

$SSWCELL = [ABS] - [AB]$ $df = N - pq$

$SSAB = [AB] - [A] - [B] + [Y]$ $df = (p - 1)(q - 1)$

$N = n_{11} + n_{12} + \cdots + n_{pq}$

TABLE 8.13-3 Data and Sums of Squares for Type CRF-23 Design (Cell n's proportional)

(i) Data:

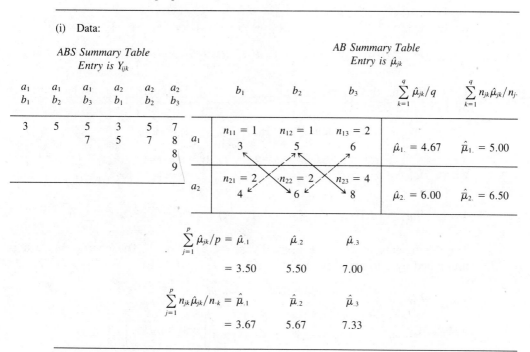

ABS Summary Table
Entry is Y_{ijk}

AB Summary Table
Entry is $\hat{\mu}_{jk}$

a_1 b_1	a_1 b_2	a_1 b_3	a_2 b_1	a_2 b_2	a_2 b_3
3	5	5	3	5	7
	7	5	7	8	8
				8	
				9	

AB Summary Table with columns b_1, b_2, b_3, $\sum_{k=1}^{q} \hat{\mu}_{jk}/q$, $\sum_{k=1}^{q} n_{jk}\hat{\mu}_{jk}/n_{j\cdot}$:

a_1: $n_{11} = 1$, 3; $n_{12} = 1$, 5; $n_{13} = 2$, 6; $\hat{\mu}_{1\cdot} = 4.67$; $\hat{\bar{\mu}}_{1\cdot} = 5.00$

a_2: $n_{21} = 2$, 4; $n_{22} = 2$, 6; $n_{23} = 4$, 8; $\hat{\mu}_{2\cdot} = 6.00$; $\hat{\bar{\mu}}_{2\cdot} = 6.50$

$$\sum_{j=1}^{p} \hat{\mu}_{jk}/p = \hat{\mu}_{\cdot 1} \quad \hat{\mu}_{\cdot 2} \quad \hat{\mu}_{\cdot 3}$$
$$= 3.50 \quad 5.50 \quad 7.00$$

$$\sum_{j=1}^{p} n_{jk}\hat{\mu}_{jk}/n_{\cdot k} = \hat{\bar{\mu}}_{\cdot 1} \quad \hat{\bar{\mu}}_{\cdot 2} \quad \hat{\bar{\mu}}_{\cdot 3}$$
$$= 3.67 \quad 5.67 \quad 7.33$$

(ii) Sum of squares using modified computational symbols from Table 8.13-2:

$$SSTO = 42.000 \qquad SSAB = 0.667$$
$$SSA = 6.000 \qquad SSWCELL = 8.000$$
$$SSB = 27.333$$

might be interested in testing one of the following sets of null hypotheses.

Unweighted Means Hypotheses	*Weighted Means Hypotheses for A and B*
H_0: $\mu_{1\cdot} - \mu_{2\cdot} = 0$	H_0: $\bar{\mu}_{1\cdot} - \bar{\mu}_{2\cdot} = 0$
H_0: $\mu_{\cdot 1} - \mu_{\cdot 3} = 0$ $\mu_{\cdot 2} - \mu_{\cdot 3} = 0$	H_0: $\bar{\mu}_{\cdot 1} - \bar{\mu}_{\cdot 3} = 0$ $\bar{\mu}_{\cdot 2} - \bar{\mu}_{\cdot 3} = 0$
H_0: $\mu_{jk} - \mu_{j'k} - \mu_{jk'} + \mu_{j'k'} = 0$ for all j, j', k, and k'	H_0: $\mu_{jk} - \mu_{j'k} - \mu_{jk'} + \mu_{j'k'} = 0$ for all j, j', k, and k'

It is a simple matter to test either set of null hypotheses using the unrestricted full rank model approach introduced in Section 5.10. The computational procedures will be presented later; for now we will focus on the null hypotheses. These hypotheses can be expressed in matrix notation as follows.

The unweighted means hypotheses are presented first.

$$\text{Treatment } A \qquad H_0: \quad \underset{(p-1)\times h}{\mathbf{C}'_{1(A)}} \quad \underset{h\times 1}{\boldsymbol{\mu}} = \underset{(p-1)\times 1}{\mathbf{0}_{1(A)}}$$

where $\mathbf{C}'_{1(A)} = [\frac{1}{3} \quad \frac{1}{3} \quad \frac{1}{3} \quad -\frac{1}{3} \quad -\frac{1}{3} \quad -\frac{1}{3}]$ or $[1 \quad 1 \quad 1 \quad -1 \quad -1 \quad -1]$,

$\boldsymbol{\mu}' = [\mu_{11} \quad \mu_{12} \quad \mu_{13} \quad \mu_{21} \quad \mu_{22} \quad \mu_{23}]$, and $\mathbf{0}_{1(A)} = [0]$.

The coefficients in $\mathbf{C}'_{1(A)}$ are given by $c_{jk} = \pm 1/q$. The coefficients can be multiplied by 3 to avoid fractions.

$$\text{Treatment } B \qquad H_0: \quad \underset{(q-1)\times h}{\mathbf{C}'_{1(B)}} \quad \underset{h\times 1}{\boldsymbol{\mu}} = \underset{(q-1)\times 1}{\mathbf{0}_{1(B)}}$$

where $\mathbf{C}'_{1(B)} = \begin{bmatrix} \frac{1}{2} & 0 & -\frac{1}{2} & \frac{1}{2} & 0 & -\frac{1}{2} \\ 0 & \frac{1}{2} & -\frac{1}{2} & 0 & \frac{1}{2} & -\frac{1}{2} \end{bmatrix}$ or $\begin{bmatrix} 1 & 0 & -1 & 1 & 0 & -1 \\ 0 & 1 & -1 & 0 & 1 & -1 \end{bmatrix}$,

and $\mathbf{0}_{1(B)} = \begin{bmatrix} 0 \\ 0 \end{bmatrix}$.

The coefficients in $\mathbf{C}'_{1(B)}$ are given by $c_{jk} = \pm 1/p$ or 0. The coefficients can be multiplied by 2 to avoid fractions.

$$AB \text{ Interaction} \qquad H_0: \quad \underset{(p-1)(q-1)}{\mathbf{C}'_{1(AB)}} \quad \underset{h\times 1}{\boldsymbol{\mu}} = \underset{(p-1)(q-1)\times h}{\mathbf{0}_{1(AB)}}$$

where $\mathbf{C}'_{1(AB)} = \begin{bmatrix} 1 & -1 & 0 & -1 & 1 & 0 \\ 0 & 1 & -1 & 0 & -1 & 1 \end{bmatrix}$ and $\mathbf{0}_{1(AB)} = \begin{bmatrix} 0 \\ 0 \end{bmatrix}$.

The coefficients, $c_{jk} = \pm 1$ or 0, in $\mathbf{C}'_{AB}$ are determined from the two sets of crossed lines in the AB Summary Table of Table 8.13-3 (See Section 6.9 for the construction of interaction coefficient matrices).

The weighted means hypotheses for treatments A and B are as follows.

$$\text{Treatment } A \qquad H_0: \quad \underset{(p-1)\times h}{\mathbf{C}'_{2(A)}} \quad \underset{h\times 1}{\boldsymbol{\mu}} = \underset{(p-1)\times 1}{\mathbf{0}_{2(A)}}$$

where $\mathbf{C}'_{2(A)} = [\frac{1}{4} \quad \frac{1}{4} \quad \frac{2}{4} \quad -\frac{1}{4} \quad -\frac{1}{4} \quad -\frac{2}{4}]$

$$\text{or} \qquad [1 \quad 1 \quad 2 \quad -1 \quad -1 \quad -2], \text{ and } \mathbf{0}_{2(A)} = [0].$$

The coefficients in $\mathbf{C}'_{2(A)}$ are given by $c_{jk} = \pm\, n_{jk}/n_{j.}$. Such coefficients are data dependent; that is, they depend on the number of observations, n_{jk}, that appear in each cell.

$$\text{Treatment } B \qquad H_0: \quad \underset{(q-1)\times h}{\mathbf{C}'_{2(B)}} \quad \underset{h\times 1}{\boldsymbol{\mu}} = \underset{(q-1)\times 1}{\mathbf{0}_{2(B)}}$$

where $\mathbf{C}'_{2(B)} = \begin{bmatrix} \frac{1}{3} & 0 & -\frac{1}{3} & \frac{2}{3} & 0 & -\frac{2}{3} \\ 0 & \frac{1}{3} & -\frac{1}{3} & 0 & \frac{2}{3} & -\frac{2}{3} \end{bmatrix}$ or $\begin{bmatrix} 1 & 0 & -1 & 2 & 0 & -2 \\ 0 & 1 & -1 & 0 & 2 & -2 \end{bmatrix}$, and $\quad \mathbf{0}_{2(B)} = \begin{bmatrix} 0 \\ 0 \end{bmatrix}.$

The coefficients in $\mathbf{C}'_{2(B)}$ are given by $c_{jk} = \pm n_{jk}/n_{\cdot k}$ or 0.

A comparison of the sums of squares for the data in Table 8.13-3 using the (1) unrestricted full rank model approach, (2) traditional approach, and (3) regression model approach is given in Table 8.13-4. From these comparisons one can discern the type of hypothesis—unweighted means or weighted means—tested by the traditional sum of squares approach and the various regression model approaches. An experimenter would most likely be interested in testing hypotheses involving unweighted means unless the sample n_{jk}'s were proportional to the population n_{jk}'s.

Experiments with proportional cell n's rarely happen by chance. Advanced planning is required if all n_{jk}'s are to satisfy $n_{jk} = (n_j \cdot n_{\cdot k})/N$. Experiments with disproportional cell n's, on the other hand, often begin as equal n experiments but, for one reason or another, data for some experimental units are not obtained. Common reasons for missing data are equipment failed, subject forgot to report for the experimental session or was ill, an animal subject died, or the wrong experimental condition was presented to a subject. These conditions inevitably lead to disproportional cell n's. Disproportional cell n's also occur in observational research and research with intact groups such as students in different classrooms. Also disproportional cell n's may occur when data are reanalyzed in terms of independent variables not originally considered in the experiment. As you may have surmised, factorial experiments with disproportional cell n's are quite common. The analysis of such experiments is the subject of the next two sections. We will consider first the case in which each cell contains at least one observation.

TABLE 8.13-4 Comparison of Sums of Squares for the Data in Table 8.13-3 Using Three Computational Approaches

Full rank model	$SS_{1(A)}$	$SS_{2(A)}$	$SS_{1(B)}$	$SS_{2(B)}$	$SS_{1(AB)}$
Traditional Approach		SSA		SSB	$SSAB$
Regression model	$SSR(A \mid B, AB)$	$SSR(A \mid B)$, $SSR(A)$	$SSR(B \mid A, B)$	$SSR(B \mid A)$, $SSR(B)$	$SSR(AB \mid A, B)$
Value of sum of squares	4.27	6.00	22.00	27.33	0.67
F statistic	3.20	4.50	8.25	10.25	0.50

GENERAL LINEAR MODEL APPROACHES TO THE ANALYSIS OF EXPERIMENTS WITH DISPROPORTIONAL CELL n's AND AT LEAST ONE OBSERVATION PER CELL

Overall and Spiegel (1969) have described three regression model approaches to analyzing data for a type CRF-pq design with one or more missing observations. Their approaches, denoted by I, II, and III, and a fourth approach, denoted by IV, are as follows.

Method I	*Method II*	*Method IIIa*	*Method IIIb*	*Method IV*
$SSR(A\,\vert\,B,\,AB)$	$SSR(A\,\vert\,B)$	$SSR(A)$	$SSR(A\,\vert\,B)$	$SSR(A)$
$SSR(B\,\vert\,A,\,AB)$	$SSR(B\,\vert\,A)$	$SSR(B\,\vert\,A)$	$SSR(B)$	$SSR(B)$
$SSR(AB\,\vert\,A,\,B)$	$SSR(AB\,\vert\,A,\,B)$	$SSR(AB\,\vert\,A,\,B)$	$SSR(AB\,\vert\,A,\,B)$	$SSR(AB\,\vert\,A,\,B)$
SSE	SSE	SSE	SSE	SSE

When cell n's are equal, the various hypotheses tested by these approaches are equivalent in terms of the parameters of the full rank experimental design model. Unfortunately, this is not the case for factorial experiments with unequal n's. To determine the experimental design model hypotheses that are tested when the cell n's are disproportional, we will adopt the indirect approach used earlier for the proportional n case. It consists of testing different null hypotheses using the full rank experimental design model and comparing the results with those for the regression model approaches. When this is done for a type CRF-pq design, the following results are obtained. The full rank experimental design model hypotheses are denoted by $1(A)$, $1(B)$, and so on.

Method I

$1(A)$ H_0: $\displaystyle\sum_{k=1}^{q} \mu_{jk}/q - \sum_{k=1}^{q} \mu_{j'k}/q = 0$ or $\mu_{j\cdot} - \mu_{j'\cdot} = 0$

$\hspace{10cm}$ for all j and j'

$1(B)$ H_0: $\displaystyle\sum_{j=1}^{p} \mu_{jk}/p - \sum_{j=1}^{p} \mu_{jk'}/p = 0$ or $\mu_{\cdot k} - \mu_{\cdot k'} = 0$

$\hspace{10cm}$ for all k and k'

$1(AB)$ H_0: $\mu_{jk} - \mu_{j'k} - \mu_{jk'} + \mu_{j'k'} = 0$ for all $j,\,j',\,k,\,k'$

Method II

$3(A)$ H_0: $\displaystyle\sum_{k=1}^{q} \left(n_{jk} - \frac{n_{jk}^2}{n_{\cdot k}} \right) \mu_{jk} - \sum_{j \neq j'} \sum_{k=1}^{q} \left(\frac{n_{jk} n_{j'k}}{n_{\cdot k}} \right) \mu_{j'k} = 0$

$\hspace{9cm}$ for $j = 1, \ldots, p - 1$

or   1(A) subject to restrictions that $\mu_{jk} - \mu_{j'k} - \mu_{jk'} + \mu_{j'k'} = 0$

for all j, j', k, k'

$3(B)$ H_0: $\displaystyle\sum_{j=1}^{p} \left(n_{jk} - \frac{n_{jk}^2}{n_{j\cdot}} \right) \mu_{jk} - \sum_{j=1}^{p} \sum_{k \neq k'} \left(\frac{n_{jk} n_{jk'}}{n_{j\cdot}} \right) \mu_{jk'} = 0$

for $k = 1, \ldots, q - 1$

or   1(B) subject to restrictions that $\mu_{jk} - \mu_{j'k} - \mu_{jk'} + \mu_{j'k'} = 0$

for all j, j', k, k'

$1(AB)$ H_0: $\mu_{jk} - \mu_{j'k} - \mu_{jk'} + \mu_{j'k'} = 0$ for all j, j', k, k'

Method IIIa

$2(A)$ H_0: $\displaystyle\sum_{k=1}^{q} n_{jk}\mu_{jk}/n_{j\cdot} - \sum_{k=1}^{q} n_{j'k}\mu_{j'k}/n_{j'} = 0$ or $\bar{\mu}_{j\cdot} - \bar{\mu}_{j'} = 0$

for all j and j'

$3(B)$ H_0: $\displaystyle\sum_{j=1}^{p} \left(n_{jk} - \frac{n_{jk}^2}{n_{j\cdot}} \right) \mu_{jk} - \sum_{j=1}^{p} \sum_{k \neq k'} \left(\frac{n_{jk} - n_{jk'}}{n_{j\cdot}} \right) \mu_{jk'} = 0$

for $k = 1, \ldots, q - 1$

or   1(B) subject to restrictions that $\mu_{jk} - \mu_{j'k} - \mu_{jk'} + \mu_{j'k'} = 0$

for all j, j', k, k'

$1(AB)$ H_0: $\mu_{jk} - \mu_{j'k} - \mu_{jk'} + \mu_{j'k'} = 0$ for all j, j', k, k'

Method IV

$2(A)$ H_0: $\displaystyle\sum_{k=1}^{q} n_{jk}\mu_{jk}/n_{j\cdot} - \sum_{k=1}^{q} n_{j'k}\mu_{j'k}/n_{j'} = 0$ or $\bar{\mu}_{j\cdot} - \bar{\mu}_{j'} = 0$

for all j and j'

$2(B)$ H_0: $\displaystyle\sum_{j=1}^{p} n_{jk}\mu_{jk}/n_{\cdot k} - \sum_{j=1}^{p} n_{jk'}\mu_{jk'}/n_{\cdot k'} = 0$ or $\bar{\mu}_{\cdot k} - \bar{\mu}_{\cdot k'} = 0$

for all k and k'

$1(AB)$ H_0: $\mu_{jk} - \mu_{j'k} - \mu_{jk'} + \mu_{j'k'} = 0$ for all j, j', k, k'

Hypotheses denoted by 1(A) and 1(B) involve unweighted cell means; those denoted by 2(A), 2(B), 3(A), and 3(B) involve weighted cell means, where the weights depend on the number of observations that appear in each cell of the experiment.*

Earlier we described a regression model with effect coding for a type CRF-pq design. We will now describe a full rank experimental design model for the same design. The full rank experimental design model corresponding to regression model (8.13-1) is

(8.13-3) $Y_{ijk} = \mu_{jk} + \epsilon_{i(jk)}$ $(i = 1, \ldots, n_{jk}; j = 1, \ldots, p; k = 1, \ldots, q)$

where $\epsilon_{i(jk)}$ is $NID(0, \sigma_\epsilon^2)$.** Model (8.13-3) is an unrestricted model. The computational procedures presented in Section 5.10 for a type CR-p design generalize directly to this CRF-pq design. We can require that the AB interaction effects equal zero. With the addition of the restrictions

$$\mu_{jk} - \mu_{j'k} - \mu_{jk'} + \mu_{j'k'} = 0 \quad \text{for all } j, j', k, k'$$

model (8.13-3) becomes a restricted full rank experimental design model. The computational procedures presented in Section 6.9 for a type RB-p design are used in analyzing this restricted full rank experimental design model.

For purposes of comparison, the police attitude data in Table 8.13-1 where Y_{511} is missing have been analyzed using the unrestricted and restricted full rank experimental design models and the four regression model approaches. The results are summarized in parts (i) and (ii) of Table 8.13-5. Part (iii) gives the coefficient matrices used in computing the sums of squares for the full rank models. An outline of the computational procedures for the full rank model is given in Table 8.13-6. Most likely an experimenter would be interested in testing hypotheses involving unweighted means, for example,

1(A) H_0: $\mu_{j\cdot} - \mu_{j'\cdot} = 0$ for all j and j'

1(B) H_0: $\mu_{\cdot k} - \mu_{\cdot k'} = 0$ for all k and k'

---

*   These hypotheses can also be expressed in terms of the classical less than full rank experimental design model equation $Y_{ijk} = \mu + \alpha_j + \beta_k + (\alpha\beta)_{jk} + \epsilon_{i(jk)}$ for a type CRF-pq design. It can be shown, for example, that 1(A), 2(A), and 3(A) correspond, respectively, to

H_0: $\alpha_j + \sum_{k=1}^{q} (\alpha\beta)_{jk}/q$ are equal for all j or H_0: $\alpha_j = 0$ for all j, with the addition of the usual

side conditions $\sum_{j=1}^{p} \alpha_j = \sum_{k=1}^{q} \beta_j = \sum_{j=1}^{p} (\alpha\beta)_{jk} = \sum_{k=1}^{q} (\alpha\beta)_{jk} = 0$

H_0: $\alpha_j + \sum_{k=1}^{q} n_{jk} [\beta_k + (\alpha\beta)_{jk}]/n_{j\cdot}$ are equal for all j or H_0: $\alpha_j = 0$ for all j, with the addition

of the usual side conditions $\bar{\alpha}_\cdot = \bar{\beta}_\cdot = \overline{(\alpha\beta)}_{j\cdot} = \overline{(\alpha\beta)}_{\cdot k} = 0$

H_0: $(n_{j\cdot} - \sum_{k=1}^{q} n_{jk}^2/n_{\cdot k})\alpha_j - \sum_{j \neq j'} \left(\sum_{k=1}^{q} n_{jk}n_{j'k}/n_{\cdot k} \right)\alpha_{j'} + \sum_{k=1}^{q} (n_{jk} - n_{jk}^2/n_{\cdot k})(\alpha\beta)_{jk}$

$- \sum_{j \neq j'} \sum_{k=1}^{q} (n_{jk}n_{j'k}/n_{\cdot k})(\alpha\beta)_{j'k} = 0$ for $j = 1, \ldots, p - 1$

(Searle 1971a, 307–310, 315).

**   For an in-depth discussion of the use of the full rank experimental design model with experiments having disproportional cell n's, see Timm and Carlson (1975).

1(AB) H_0: $\mu_{jk} - \mu_{j'k} - \mu_{jk'} + \mu_{j'k'} = 0$ for all j, j', k, k'.

These are the hypotheses tested by Method I.

Method II might be of some interest since it tests the same hypotheses after setting the AB interaction for the sample data in the experiment equal to zero. Note that this is a departure from current practice in the equal n case. Recall that for the equal n case the AB interaction is tested first; if it is not significant, treatments A and B are tested. Method II adds an additional step before testing treatments A and B—adjusting the sample data to make the AB interaction zero. A disadvantage of this procedure is that the adjustment is data dependent; that is, it depends on the sample n_{jk}'s.

The solution obtained by Method III is of less interest to researchers in the behavioral sciences and education. The outcome of Method III depends on the order in which the sums of squares for treatments A and B are computed. It requires an initial a priori ordering of the hypotheses and then estimating each treatment sum of squares adjusting for those preceding it in the ordering and ignoring those following it. For an experiment with two treatments there are two possible testing sequences: $SSR(A)$ followed by $SSR(B|A)$ as shown in Table 8.13-5 and $SSR(B)$ followed by $SSR(A|B)$. Method III is the only one of the regression approaches for which the sums of squares add up to the total sum of squares. However, considering the null hypotheses tested by the procedure, it is hard to imagine why an experimenter would choose to use it.

Method IV, like Methods II and III, tests hypotheses that are data dependent, which is usually a disadvantage. An exception occurs when the sample n_{jk}'s are proportional to the population N_{jk}'s in which case Method IV might be of interest. For a more detailed evaluation of the methods, the reader is referred to the excellent papers by Appelbaum and Cramer (1974), Carlson and Timm (1974), Cramer and Appelbaum (1980), and Timm and Carlson (1975). The simplest solution to the problem of unequal and disproportional cell n's is to use the full rank experimental design model approach. When it is used, one is never in doubt about the null hypotheses that are tested.

Before turning to the disproportional case with one or more empty cells, we will briefly mention three approximate procedures that are sometimes recommended. For a fuller discussion of these and other procedures see Gosslee and Lucas (1965); Searle (1971a, 316–369); and Speed, Hocking, and Hackney (1978). One procedure involves making the cell n's equal by estimating the missing observations. This is appropriate if most of the cell n's are equal. Recall that this procedure was recommended for randomized block and Latin square designs having missing observations. The procedure is satisfactory for experiments that do not involve interactions. However, if one or more interactions are substantial, the estimation formulas are not satisfactory.

Another procedure when cell n's are disproportional is to randomly set aside data in order to reduce all cell n's to the same size. A conventional sum of squares analysis can then be performed on the data. Questions concerning this method immediately come to mind. How similar should the cell n's be in order to use the procedure? How much data should one be willing to set aside? Unfortunately, definitive answers to these questions are not available.

TABLE 8.13-5   Comparison of Analyses Using Four Regression Model Approaches and the Full Rank Experimental Design Model Approach

(i) Regression model approaches:

Method I Source	df	SS	Method II Source	SS	Method IIIa Source	SS	Method IV Source	SS
$SSR(A\mid B, AB)$	2	188.09	$SSR(A\mid B)$	190.42	$SSR(A)$	187.51	$SSR(A)$	187.51
$SSR(B\mid A, AB)$	2	1641.29	$SSR(B\mid A)$	1716.21	$SSR(B\mid A)$	1716.21	$SSR(B)$	1713.30
$SSR(AB\mid A, B)$	4	1117.42	$SSR(AB\mid A, B)$	1117.42	$SSR(AB\mid A, B)$	1117.42	$SSR(AB\mid A, B)$	1117.42
SSE	35	2148.75	SSE	2148.75	SSE	2148.75	SSE	2148.75
$SSTO$	43	5169.89	$SSTO$	5169.89	$SSTO$	5169.89	$SSTO$	5169.89

(ii) Full rank model approach (The coefficient matrices, C', are defined in part iii):

Source	df	Hypothesis	SS	Method II Hypothesis	SS	Method IIIa Hypothesis	SS	Method IV Hypothesis	SS
SSA	2	$C'_{1(A)}\boldsymbol{\mu} = \mathbf{0}$	188.09	$C'_{1(A)}\boldsymbol{\mu} = \mathbf{0}^*$ or $C'_{3(A)}\boldsymbol{\mu} = \mathbf{0}$	190.42	$C'_{2(A)}\boldsymbol{\mu} = \mathbf{0}$	187.51	$C'_{2(A)}\boldsymbol{\mu} = \mathbf{0}$	187.51
SSB	2	$C'_{1(B)}\boldsymbol{\mu} = \mathbf{0}$	1641.29	$C'_{1(B)}\boldsymbol{\mu} = \mathbf{0}^*$ or $C'_{3(B)}\boldsymbol{\mu} = \mathbf{0}$	1716.21	$C'_{1(B)}\boldsymbol{\mu} = \mathbf{0}^*$ or $C'_{3(B)}\boldsymbol{\mu} = \mathbf{0}$	1716.21	$C'_{2(B)}\boldsymbol{\mu} = \mathbf{0}$	1713.30
$SSAB$	4	$C'_{1(AB)}\boldsymbol{\mu} = \mathbf{0}$	1117.42	$C'_{1(AB)}\boldsymbol{\mu} = \mathbf{0}$	1117.42	$C'_{1(AB)}\boldsymbol{\mu} = \mathbf{0}$	1117.42	$C'_{1(AB)}\boldsymbol{\mu} = \mathbf{0}$	1117.42
$SSWCELL$	35		2148.75		2148.75		2148.75		2148.75
$SSTO$	43		5169.89		5169.89		5169.89		5169.89

*Subject to the restriction that $\mu_{jk} - \mu_{j'k} - \mu_{jk'} + \mu_{j'k'} = 0$ for all j, j', k, and k'

TABLE 8.13-5 (continued)

(iii) Coefficient matrices for full rank model:

$$\mathbf{C}'_{1(A)} \quad (\text{Treatment } A)$$
$$\underset{(p-1)\times h}{}$$

$$\begin{bmatrix} 1 & 1 & 1 & 0 & 0 & 0 & -1 & -1 & -1 \\ 0 & 0 & 0 & 1 & 1 & 1 & -1 & -1 & -1 \end{bmatrix}$$

$$\mathbf{C}'_{1(B)} \quad (\text{Treatment } B)$$
$$\underset{(q-1)\times h}{}$$

$$\begin{bmatrix} 1 & 0 & -1 & 1 & 0 & -1 & 1 & 0 & -1 \\ 0 & 1 & -1 & 0 & 1 & -1 & 0 & 1 & -1 \end{bmatrix}$$

$$\mathbf{C}'_{1(AB)} \quad (AB \text{ Interaction})$$
$$\underset{(p-1)(q-1)\times h}{}$$

$$\begin{bmatrix} 1 & -1 & 0 & -1 & 1 & 0 & 0 & 0 & 0 \\ 1 & -1 & 0 & -1 & 0 & -1 & 0 & 0 & 0 \\ 0 & 0 & 0 & 1 & -1 & 0 & -1 & 1 & 0 \\ 0 & 0 & 0 & 1 & 0 & -1 & -1 & 0 & -1 \end{bmatrix}$$

$$\mathbf{C}'_{2(A)} \quad (\text{Treatment } A)$$
$$\underset{(p-1)\times h}{}$$

$$\begin{bmatrix} \dfrac{n_{11}}{n_{1\cdot}} & \dfrac{n_{12}}{n_{1\cdot}} & \dfrac{n_{13}}{n_{1\cdot}} & 0 & 0 & 0 & -\dfrac{n_{31}}{n_{3\cdot}} & -\dfrac{n_{32}}{n_{3\cdot}} & -\dfrac{n_{33}}{n_{3\cdot}} \\ 0 & 0 & 0 & \dfrac{n_{21}}{n_{2\cdot}} & \dfrac{n_{22}}{n_{2\cdot}} & \dfrac{n_{23}}{n_{2\cdot}} & -\dfrac{n_{31}}{n_{3\cdot}} & -\dfrac{n_{32}}{n_{3\cdot}} & -\dfrac{n_{33}}{n_{3\cdot}} \end{bmatrix}$$

$$\mathbf{C}'_{2(B)} \quad (\text{Treatment } B)$$
$$\underset{(q-1)\times h}{}$$

$$\begin{bmatrix} \dfrac{n_{11}}{n_{\cdot 1}} & 0 & -\dfrac{n_{13}}{n_{\cdot 3}} & \dfrac{n_{21}}{n_{\cdot 1}} & 0 & -\dfrac{n_{23}}{n_{\cdot 3}} & \dfrac{n_{31}}{n_{\cdot 1}} & 0 & -\dfrac{n_{33}}{n_{\cdot 3}} \\ 0 & \dfrac{n_{12}}{n_{\cdot 2}} & -\dfrac{n_{13}}{n_{\cdot 3}} & 0 & \dfrac{n_{22}}{n_{\cdot 2}} & -\dfrac{n_{23}}{n_{\cdot 3}} & 0 & \dfrac{n_{32}}{n_{\cdot 2}} & -\dfrac{n_{33}}{n_{\cdot 3}} \end{bmatrix}$$

$$\mathbf{C}'_{3(A)} \quad (\text{Treatment } A)$$
$$\underset{(p-1)\times h}{}$$

$$\begin{bmatrix} \left(n_{11} - \dfrac{n_{11}^2}{n_{\cdot 1}}\right) & \left(n_{12} - \dfrac{n_{12}^2}{n_{\cdot 2}}\right) & \left(n_{13} - \dfrac{n_{13}^2}{n_{\cdot 3}}\right) & -\dfrac{n_{11}n_{21}}{n_{\cdot 1}} & -\dfrac{n_{12}n_{22}}{n_{\cdot 2}} & -\dfrac{n_{13}n_{23}}{n_{\cdot 3}} & -\dfrac{n_{11}n_{31}}{n_{\cdot 1}} & -\dfrac{n_{12}n_{32}}{n_{\cdot 2}} & -\dfrac{n_{13}n_{33}}{n_{\cdot 3}} \\ -\dfrac{n_{21}n_{11}}{n_{\cdot 1}} & -\dfrac{n_{22}n_{12}}{n_{\cdot 2}} & -\dfrac{n_{23}n_{13}}{n_{\cdot 3}} & \left(n_{21} - \dfrac{n_{21}^2}{n_{\cdot 1}}\right) & \left(n_{22} - \dfrac{n_{22}^2}{n_{\cdot 2}}\right) & \left(n_{23} - \dfrac{n_{23}^2}{n_{\cdot 3}}\right) & -\dfrac{n_{21}n_{31}}{n_{\cdot 1}} & -\dfrac{n_{22}n_{32}}{n_{\cdot 2}} & -\dfrac{n_{21}n_{33}}{n_{\cdot 3}} \end{bmatrix}$$

$$\mathbf{C}'_{3(B)} \quad (\text{Treatment } B)$$
$$\underset{(q-1)\times h}{}$$

$$\begin{bmatrix} \left(n_{11} - \dfrac{n_{11}^2}{n_{1\cdot}}\right) & -\dfrac{n_{11}n_{12}}{n_{1\cdot}} & -\dfrac{n_{11}n_{13}}{n_{1\cdot}} & \left(n_{21} - \dfrac{n_{21}^2}{n_{2\cdot}}\right) & -\dfrac{n_{21}n_{22}}{n_{2\cdot}} & -\dfrac{n_{21}n_{23}}{n_{2\cdot}} & \left(n_{31} - \dfrac{n_{31}^2}{n_{3\cdot}}\right) & -\dfrac{n_{31}n_{32}}{n_{3\cdot}} & -\dfrac{n_{31}n_{33}}{n_{3\cdot}} \\ -\dfrac{n_{12}n_{11}}{n_{1\cdot}} & \left(n_{12} - \dfrac{n_{12}^2}{n_{1\cdot}}\right) & -\dfrac{n_{12}n_{13}}{n_{1\cdot}} & -\dfrac{n_{22}n_{21}}{n_{2\cdot}} & \left(n_{22} - \dfrac{n_{22}^2}{n_{2\cdot}}\right) & -\dfrac{n_{22}n_{23}}{n_{2\cdot}} & -\dfrac{n_{32}n_{31}}{n_{3\cdot}} & \left(n_{32} - \dfrac{n_{32}^2}{n_{3\cdot}}\right) & -\dfrac{n_{32}n_{33}}{n_{3\cdot}} \end{bmatrix}$$

TABLE 8.13-6   Outline of Computational Procedures for a Type CRF-33 Design Using a Full Rank Experimental Design Model Approach (Y_{511} Is Assumed to Be Missing.)

(i)   Data and basic matrix [$N = 44$, $h = 9$; the $\mathbf{C}'$'s are defined in Table 8.13-2(iii)]:

	$\mathbf{y}$ $N \times 1$	$\mathbf{X}$ $N \times h$								
		$\mathbf{x}_1$	$\mathbf{x}_2$	$\mathbf{x}_3$	$\mathbf{x}_4$	$\mathbf{x}_5$	$\mathbf{x}_6$	$\mathbf{x}_7$	$\mathbf{x}_8$	$\mathbf{x}_9$
a_1b_1	24	1	0	0	0	0	0	0	0	0
	33	1	0	0	0	0	0	0	0	0
	37	1	0	0	0	0	0	0	0	0
	29	1	0	0	0	0	0	0	0	0
a_1b_2	44	0	1	0	0	0	0	0	0	0
	36	0	1	0	0	0	0	0	0	0
	25	0	1	0	0	0	0	0	0	0
	27	0	1	0	0	0	0	0	0	0
	43	0	1	0	0	0	0	0	0	0
a_1b_3	38	0	0	1	0	0	0	0	0	0
	29	0	0	1	0	0	0	0	0	0
	28	0	0	1	0	0	0	0	0	0
	47	0	0	1	0	0	0	0	0	0
	48	0	0	1	0	0	0	0	0	0
a_2b_1	30	0	0	0	1	0	0	0	0	0
	21	0	0	0	1	0	0	0	0	0
	39	0	0	0	1	0	0	0	0	0
	26	0	0	0	1	0	0	0	0	0
	34	0	0	0	1	0	0	0	0	0
a_2b_2	35	0	0	0	0	1	0	0	0	0
	40	0	0	0	0	1	0	0	0	0
	27	0	0	0	0	1	0	0	0	0
	31	0	0	0	0	1	0	0	0	0
	22	0	0	0	0	1	0	0	0	0
a_2b_3	26	0	0	0	0	0	1	0	0	0
	27	0	0	0	0	0	1	0	0	0
	36	0	0	0	0	0	1	0	0	0
	46	0	0	0	0	0	1	0	0	0
	45	0	0	0	0	0	1	0	0	0
a_3b_1	21	0	0	0	0	0	0	1	0	0
	18	0	0	0	0	0	0	1	0	0
	10	0	0	0	0	0	0	1	0	0
	31	0	0	0	0	0	0	1	0	0
	20	0	0	0	0	0	0	1	0	0
a_3b_2	41	0	0	0	0	0	0	0	1	0
	39	0	0	0	0	0	0	0	1	0
	50	0	0	0	0	0	0	0	1	0
	36	0	0	0	0	0	0	0	1	0
	34	0	0	0	0	0	0	0	1	0

TABLE 8.13-6 (continued)

$$a_3b_3 \left\{ \begin{bmatrix} 42 \\ 52 \\ 53 \\ 49 \\ 64 \end{bmatrix} \quad \begin{bmatrix} 0 & 0 & 0 & 0 & 0 & 0 & 0 & 0 & 1 \\ 0 & 0 & 0 & 0 & 0 & 0 & 0 & 0 & 1 \\ 0 & 0 & 0 & 0 & 0 & 0 & 0 & 0 & 1 \\ 0 & 0 & 0 & 0 & 0 & 0 & 0 & 0 & 1 \\ 0 & 0 & 0 & 0 & 0 & 0 & 0 & 0 & 1 \end{bmatrix} \right. $$

$$\sum_{i=1}^{N} Y_i = 1575$$

Unrestricted Model

$$\hat{\boldsymbol{\mu}} = (\mathbf{X}'\mathbf{X})^{-1}(\mathbf{X}'\mathbf{y})$$

$$SSTO = \mathbf{y}'\mathbf{y} - (\Sigma Y_i)^2/N = 5169.8864$$

$$SSA = (\mathbf{C}'_{1(A)}\hat{\boldsymbol{\mu}})'[\mathbf{C}'_{1(A)}(\mathbf{X}'\mathbf{X})^{-1}\mathbf{C}_{1(A)}]^{-1}(\mathbf{C}'_{1(A)}\hat{\boldsymbol{\mu}})$$
$$= 188.09$$

$$SSB = (\mathbf{C}'_{1(B)}\hat{\boldsymbol{\mu}})'[\mathbf{C}'_{1(B)}(\mathbf{X}'\mathbf{X})^{-1}\mathbf{C}_{1(B)}]^{-1}(\mathbf{C}'_{1(B)}\hat{\boldsymbol{\mu}})$$
$$= 1641.29$$

$$SSAB = (\mathbf{C}'_{1(AB)}\hat{\boldsymbol{\mu}})'[\mathbf{C}'_{1(AB)}(\mathbf{X}'\mathbf{X})^{-1}\mathbf{C}_{1(AB)}]^{-1}(\mathbf{C}'_{1(AB)}\hat{\boldsymbol{\mu}})$$
$$= 1117.42$$

$$SSWCELL = \mathbf{y}'\mathbf{y} - \hat{\boldsymbol{\mu}}'(\mathbf{X}'\mathbf{y}) = 2148.75$$

Restricted Model
$$(\mu_{jk} - \mu_{j'k} - \mu_{jk'} + \mu_{j'k'} = 0 \quad \text{for all } j, j', k, k')$$

$$SSA = \{(\mathbf{Q}'_{1(A)}\hat{\boldsymbol{\mu}})'[\mathbf{Q}'_{1(A)}(\mathbf{X}'\mathbf{X})^{-1}\mathbf{Q}_{1(A)}]^{-1}(\mathbf{Q}'_{1(A)}\hat{\boldsymbol{\mu}}) - (\mathbf{R}'\hat{\boldsymbol{\mu}})'[\mathbf{R}'(\mathbf{X}'\mathbf{X})^{-1}\mathbf{R}]^{-1}(\mathbf{R}'\hat{\boldsymbol{\mu}})\}$$
$$= 1307.8368 - 1117.4167$$
$$= 190.42$$

$$SSB = \{(\mathbf{Q}'_{1(B)}\hat{\boldsymbol{\mu}})'[\mathbf{Q}'_{1(B)}(\mathbf{X}'\mathbf{X})^{-1}\mathbf{Q}_{1(B)}]^{-1}(\mathbf{Q}'_{1(B)}\hat{\boldsymbol{\mu}}) - (\mathbf{R}'\hat{\boldsymbol{\mu}})'[\mathbf{R}'(\mathbf{X}'\mathbf{X})^{-1}\mathbf{R}]^{-1}(\mathbf{R}'\hat{\boldsymbol{\mu}})\}$$
$$= 2833.6310 - 1117.4167$$
$$= 1716.21$$

$$\mathbf{R}' = \mathbf{C}'_{1(AB)} \quad \text{in Table 8.13-5}$$

$$\mathbf{Q}'_{1(A)} = \begin{bmatrix} \mathbf{R}' \\ \mathbf{C}'_{1(A)} \end{bmatrix} \quad \mathbf{Q}'_{1(B)} = \begin{bmatrix} \mathbf{R}' \\ \mathbf{C}'_{1(B)} \end{bmatrix}$$

$$\hat{\boldsymbol{\mu}}_R = \hat{\boldsymbol{\mu}} - (\mathbf{X}'\mathbf{X})^{-1}\mathbf{R}[\mathbf{R}'(\mathbf{X}'\mathbf{X})^{-1}\mathbf{R}]^{-1}(\mathbf{R}'\hat{\boldsymbol{\mu}})$$
$$\hat{\boldsymbol{\mu}}_R \text{ estimates population means subject to restrictions } \mathbf{R}'\boldsymbol{\mu} = \mathbf{0}$$

When each cell contains at least one observation, an *unweighted-means analysis* can be performed. The procedure consists of first computing cell means for each treatment combination. The cell means are then subjected to a conventional sum of squares analysis. The final step in the analysis is to multiply the treatment and interaction sums of squares by the harmonic mean of the cell n's; that is, by $\bar{n} = pq/(1/n_{11} + 1/n_{12} + \cdots + 1/n_{pq})$. Unfortunately the resulting mean squares do not have χ^2 distributions, and hence their ratios do not provide F statistics for testing null hypotheses. Reasonably satisfactory adjustments using amended degrees of freedom have been described by Gosslee and Lucas (1965). At a time when most researchers have access to computers and software packages for manipulating matrices, it is difficult to defend the use of these approximate procedures. The full rank experimental design model, in particular, is recommended because one always knows the null hypotheses that are tested and the restrictions, if any, that are imposed on the model.

ANALYSIS OF EXPERIMENTS WITH DISPROPORTIONAL CELL n's AND ONE OR MORE EMPTY CELLS

The full rank experimental design model approach can be used to analyze data for a type CRF-pq design with empty cells. However, this condition causes problems in interpreting the results. Consider the CRF-33 design in Table 8.13-7 where cell a_1b_3 contains no observations. An experimenter might be interested in one of the following

TABLE 8.13-7   Type CRF-33 Design with Disproportional Cell n's and an Empty Cell

ABS Summary Table
Entry is Y_{ijk}

a_1 b_1	a_1 b_2	a_1 b_3	a_2 b_1	a_2 b_2	a_2 b_3	a_3 b_1	a_3 b_2	a_3 b_3
24	44		30	35	26	21	41	42
33	36		21	40	27	18	39	52
37	25		39	27	36	10	50	53
29	27		26	31	46	31	36	49
	43		34	22	45	20	34	64

AB Summary Table
Entry is $\hat{\mu}_{jk}$

	b_1	b_2	b_3	$\sum_{k=1}^{q} \hat{\mu}_{jk}/q_{(j)}$	$\sum_{k=1}^{q} n_{jk}\hat{\mu}_{jk}/n_{j\cdot}$
a_1	$\hat{\mu}_{11}$	$\hat{\mu}_{12}$	Empty	$\hat{\mu}_{1\cdot} = 32.875$	$\hat{\bar{\mu}}_{1\cdot} = 33.111$
a_2	$\hat{\mu}_{21}$	$\hat{\mu}_{22}$	$\hat{\mu}_{23}$	$\hat{\mu}_{2\cdot} = 32.333$	$\hat{\bar{\mu}}_{2\cdot} = 32.333$
a_3	$\hat{\mu}_{31}$	$\hat{\mu}_{32}$	$\hat{\mu}_{33}$	$\hat{\mu}_{3\cdot} = 37.333$	$\hat{\bar{\mu}}_{3\cdot} = 37.333$

$$\sum_{j=1}^{p} \hat{\mu}_{jk}/p_{(k)} = \qquad \hat{\mu}_{\cdot 1} \qquad \hat{\mu}_{\cdot 2} \qquad \hat{\mu}_{\cdot 3}$$

$$= 26.917 \qquad 35.333 \qquad 44.000$$

$$\sum_{j=1}^{p} n_{jk}\hat{\mu}_{jk}/n_{\cdot k} = \qquad \hat{\bar{\mu}}_{\cdot 1} \qquad \hat{\bar{\mu}}_{\cdot 2} \qquad \hat{\bar{\mu}}_{\cdot 3}$$

$$= 26.643 \qquad 35.333 \qquad 44.000$$

sets of null hypotheses for treatments A and B.*

Average of Cell Means

$1(A)$ H_0: $\mu_{1.} - \mu_{3.} = 0$
$\mu_{2.} - \mu_{3.} = 0$

     or $(\mu_{11} + \mu_{12})/2 - (\mu_{31} + \mu_{32} + \mu_{33})/3 = 0$
$(\mu_{21} + \mu_{22} + \mu_{23})/3 - (\mu_{31} + \mu_{32} + \mu_{33})/3 = 0$

$1(B)$ H_0: $\mu_{.1} - \mu_{.2} = 0$
$\mu_{.2} - \mu_{.3} = 0$

     or $(\mu_{11} + \mu_{21} + \mu_{31})/3 - (\mu_{12} + \mu_{22} + \mu_{32})/3 = 0$
$(\mu_{12} + \mu_{22} + \mu_{32})/3 - (\mu_{23} + \mu_{33})/2 = 0$

Weighted Means

$2(A)$ H_0: $\bar{\mu}_{1.} - \bar{\mu}_{3.} = 0$
$\bar{\mu}_{2.} - \bar{\mu}_{3.} = 0$

     or $(4\mu_{11} + 5\mu_{12})/9 - (5\mu_{31} + 5\mu_{32} + 5\mu_{33})/15 = 0$
$(5\mu_{21} + 5\mu_{22} + 5\mu_{23})/15 - (5\mu_{31} + 5\mu_{32} + 5\mu_{33})/15 = 0$

$2(B)$ H_0: $\bar{\mu}_{.1} - \bar{\mu}_{.2} = 0$
$\bar{\mu}_{.2} - \bar{\mu}_{.3} = 0$

     or $(4\mu_{11} + 5\mu_{21} + 5\mu_{31})/14 - (5\mu_{12} + 5\mu_{22} + 5\mu_{32})/15 = 0$
$(5\mu_{12} + 5\mu_{22} + 5\mu_{32})/15 - (5\mu_{23} + 5\mu_{33})/10 = 0$

The coefficient matrices for the full rank model null hypotheses, $\mathbf{C'\mu} = \mathbf{0}$, are as follows.

$$\underset{(p-1)\times h}{\mathbf{C}'_{1(A)}} = \begin{bmatrix} 1/2 & 1/2 & 0 & 0 & 0 & -1/3 & -1/3 & -1/3 \\ 0 & 0 & 1/3 & 1/3 & 1/3 & -1/3 & -1/3 & -1/3 \end{bmatrix}$$

$$\underset{(p-1)\times h}{\mathbf{C}'_{2(A)}} = \begin{bmatrix} 4/9 & 5/9 & 0 & 0 & 0 & -5/15 & -5/15 & -5/15 \\ 0 & 0 & 5/15 & 5/15 & 5/15 & -5/15 & -5/15 & -5/15 \end{bmatrix}$$

The coefficients in $\mathbf{C}'_{1(A)}$ are given by $c_{jk} = \pm 1/q_{(j)}$ or 0; the coefficients in $\mathbf{C}'_{2(A)}$ are given by $c_{jk} = \pm n_{jk}/n_{j.}$ or 0.

$$\underset{(q-1)\times h}{\mathbf{C}'_{1(B)}} = \begin{bmatrix} 1/3 & -1/3 & 1/3 & -1/3 & 0 & 1/3 & -1/3 & 0 \\ 0 & 1/3 & 0 & 1/3 & -1/2 & 0 & 1/3 & -1/2 \end{bmatrix}$$

$$\underset{(q-1)\times h}{\mathbf{C}'_{2(B)}} = \begin{bmatrix} 4/14 & -5/15 & 5/14 & -5/15 & 0 & 5/14 & -5/15 & 0 \\ 0 & 5/15 & 0 & 5/15 & -5/10 & 0 & 5/15 & -5/10 \end{bmatrix}$$

The coefficients in $\mathbf{C}'_{1(B)}$ are given by $c_{jk} = \pm 1/p_{(k)}$ or 0. The coefficients in $\mathbf{C}'_{2(B)}$ are given by $c_{jk} = \pm n_{jk}/n_{.k}$ or 0. The vector of means is

$$\underset{1\times h}{\boldsymbol{\mu}'} = [\mu_{11}\ \mu_{12}\ \mu_{21}\ \mu_{22}\ \mu_{23}\ \mu_{31}\ \mu_{32}\ \mu_{33}].$$

---

*   Jennings and Ward (1982) provide an excellent discussion of the missing cell problem in the context of a type CRF-23 design.

The preceding hypotheses are not directly comparable to those for designs with no empty cells. The empty cell a_1b_3 affects selected contrasts. For example, contrasts $\mu_1. - \mu_3. = 0$ and $\bar{\mu}_1. - \bar{\mu}_3. = 0$ involve a comparison of the effects of treatment levels a_1 and a_3 as well as the effects of treatment b_3, since the latter effects do not appear in $\mu_1.$ and $\bar{\mu}_1.$ but do appear in $\mu_3.$ and $\bar{\mu}_3..$* Similarly, contrasts $\mu._2 - \mu._3 = 0$ and $\bar{\mu}._2 - \bar{\mu}._3 = 0$ involve the effects of treatment levels b_2 and b_3 as well as the effects of treatment level a_1, since the latter effects do not appear in $\mu._3$ and $\bar{\mu}._3$ but do appear in $\mu._2$ and $\bar{\mu}._2$.

The null hypothesis for the AB interaction must be modified since the interaction term involving μ_{13} is excluded from the model. This eliminates a test of $\mu_{12} - \mu_{13} - \mu_{22} + \mu_{23} = 0$. The AB interaction has $h - p - q + 1 = 8 - 3 - 3 + 1 = 3$ degrees of freedom. Three interaction functions of the form $\mu_{jk} - \mu_{j'k} - \mu_{jk'} + \mu_{j'k'} = 0$ can be constructed from the sets of crossed lines in the AB Summary Table of Table 8.13-7. This construction follows the procedure described in Section 6.9. The interaction null hypothesis is

$$H_0: \quad \mu_{11} - \mu_{12} - \mu_{21} + \mu_{22} = 0$$
$$\mu_{21} - \mu_{22} - \mu_{31} + \mu_{32} = 0$$
$$\mu_{22} - \mu_{23} - \mu_{32} + \mu_{33} = 0.$$

In matrix notation this hypothesis becomes $\mathbf{C}'_{1(AB)}\boldsymbol{\mu} = \mathbf{0}$, where

$$\underset{(h-p-q+1)\times h}{\mathbf{C}'_{1(AB)}} = \begin{bmatrix} 1 & -1 & -1 & 1 & 0 & 0 & 0 & 0 \\ 0 & 0 & 1 & -1 & 0 & -1 & 1 & 0 \\ 0 & 0 & 0 & 1 & -1 & 0 & -1 & 1 \end{bmatrix}.$$

Caution must be exercised in interpreting a test of this null hypothesis. If it is rejected, we conclude that at least one function of the form $\mu_{jk} - \mu_{j'k} - \mu_{jk'} + \mu_{j'k'} \neq 0$. However, failure to reject the hypothesis does not imply that all functions of the form $\mu_{jk} - \mu_{j'k} - \mu_{jk'} + \mu_{j'k'} = 0$ since we are unable to test $\mu_{12} - \mu_{13} - \mu_{22} + \mu_{23} = 0$. It should be apparent from the foregoing that an experimenter should make every effort to avoid empty cells.

8.14 ADVANTAGES AND DISADVANTAGES OF FACTORIAL DESIGNS

The major advantages of factorial designs are as follows.

1. All subjects are used in simultaneously evaluating the effects of two or more

---

*   In terms of the classical less than full rank model equation, $\mu_{jk} = \mu + \alpha_j + \beta_k + (\alpha\beta)_{jk}$, it can be shown that a test of $\mu_1. - \mu_3. = 0$ is equivalent to a test of $\{\alpha_1 + (\beta_1 + \beta_2)/2 + [(\alpha\beta)_{11} + (\alpha\beta)_{12}]/2\} - \{\alpha_2 + (\beta_1 + \beta_2 + \beta_3)/3 + [(\alpha\beta)_{31} + (\alpha\beta)_{32} + (\alpha\beta)_{33}]/3\} = 0$. Imposing the usual restrictions or side conditions $\sum_{k=1}^{q} \beta_k = 0$ and $\sum_{k=1}^{q} (\alpha\beta)_{jk} = 0$, this contrast becomes $\{\alpha_1 + (\beta_1 + \beta_2)/2 + [(\alpha\beta)_{11} + (\alpha\beta)_{12}]/2\} - \alpha_3 = 0$. Similarly, a test of $\bar{\mu}_1. - \bar{\mu}_3. = 0$ is equivalent to a test of $\{\alpha_1 + (4\beta_1 + 5\beta_2)/9 + [4(\alpha\beta)_{11} + 5(\alpha\beta)_{12}]/9\} - \alpha_3 = 0$.

treatments. The effects of each treatment are evaluated with the same precision as if the entire experiment had been devoted to that treatment alone. Factorial experiments thus permit efficient use of resources.

2. They enable an experimenter to evaluate interaction effects.

The major disadvantages of factorial designs are as follows.

1. If numerous treatments are included in the experiment, the number of subjects required may be prohibitive.

2. A factorial design lacks simplicity in interpretation of results if interaction effects are present. Unfortunately interactions among variables in the behavioral sciences and education are common.

3. The use of a factorial design commits an experimenter to a relatively large experiment. Small exploratory experiments may indicate much more promising lines of investigation than those originally envisioned. Relatively small experiments permit greater freedom in the pursuit of serendipity.

4. Factorial experiments are generally less efficient in determining optimum levels of treatments or treatment combinations than a series of smaller experiments, each based on the results of the preceding experiments.

8.15 REVIEW EXERCISES

1. [8.1] What is the major difference between factorial designs and type CR-p, RB-p, and LS-p designs?

2. [8.2] List the treatment combinations for the following factorial designs.
   †a) Type CRF-22   b) Type CRF-23   c) Type CRF-33

3. [8.2] How many subjects are required for the following designs? Assume that n is equal to two.
   †a) Type CRF-23   b) Type CRF-324   c) Type CRF-235

4. It was hypothesized that people who are required to evaluate someone they have just hurt tend to denigrate the victim as a means of justifying the harmful act. To investigate this hypothesis, white male college students were required to give a series of either painful or mild electric shocks (treatment A) as feedback for errors made by a white or black male confederate (treatment B) working at a learning task. Shocks were not actually delivered to the confederate, but he acted as if he were being shocked. Prior to and after administering the shocks, the subjects rated the confederates in terms of likability, intelligence, and personal adjustment. The dependent variable was the change in ratings from the pretest to the posttest. The four treatment combinations were randomly assigned to random subsamples of five subjects each. The following data were obtained. (Experiment suggested by Katz, I., Glass, D. C., and Cohn, S. Ambivalence, guilt, and the scapegoating of minority group victims. *Journal of Experimental Social Psychology*, 1973, *9*, 423–436.)

a_1	a_1	a_2	a_2
b_1	b_2	b_1	b_2
14	0	-6	9
10	-8	-2	-7
2	-4	-10	5
-2	8	-18	1
6	4	-14	-3

a_1 = mild shock
a_2 = strong shock
b_1 = black confederate
b_2 = white confederate

†a) [8.3] Test the hypotheses $\alpha_j = 0$ for all j, $\beta_k = 0$ for all k, and $(\alpha\beta)_{jk} = 0$ for all j and k; let $\alpha = .05$.

†b) [8.3] Graph the interaction between treatments A and B; interpret the graphs.

†c) [8.3] Calculate the power of the tests of H$_0$: $\alpha_j = 0$ for all j and H$_0$: $(\alpha\beta)_{jk} = 0$ for all j and k.

†d) [8.3] Determine the value of n necessary to achieve a power of approximately .85 for treatment A.

†e) [8.8] Calculate $\hat{\omega}_{Y|A}^2$ and $\hat{\omega}_{Y|AB}^2$.

f) Prepare a "results and discussion section" appropriate for the *Journal of Experimental Social Psychology*.

5.   It was hypothesized that persons who are less physically attractive believe that they have less control over reinforcements in their lives than those more attractive. To test this hypothesis, 36 male college students were shown one of six photographs and asked to fill out the Rotter I-E scale the way they thought the person in the photograph would. Individuals who score at the internal end of the I-E countinuum, low scores, perceive events as contingent on their behavior. Those at the external end, high scores, perceive events as the result of luck, chance, or powerful others. Thirty-six head and shoulder photographs were obtained from college yearbooks. Half of the photographs were of males and half of females. This variable is treatment A. Prior to the experiment, the photographs were assigned to one of three physical attractiveness categories: high, moderate, and low—treatment B. The 36 subjects were randomly assigned to six sub-samples containing six subjects each. The six sets of photographs were randomly assigned to the subsamples of subjects. The following data were obtained. (Experiment suggested by Miller, A. G. Social perception of internal-external control. *Perceptual and Motor Skills*, 1970, *30*, 103–109.)

a_1	a_1	a_1	a_2	a_2	a_2
b_1	b_2	b_3	b_1	b_2	b_3
9	8	14	10	6	14
10	8	9	9	12	13
9	12	11	6	9	10
13	7	6	15	9	13
5	8	10	10	10	12
8	4	10	11	8	16

a_1 = male
a_2 = female
b_1 = high attractiveness
b_2 = moderate attractiveness
b_3 = low attractiveness

a) [8.3] Test the hypotheses $\alpha_j = 0$ for all j, $\beta_k = 0$ for all k, and $(\alpha\beta)_{jk} = 0$ for all j and k; let $\alpha = .05$.

b) [8.3] (i) Graph the interaction between treatments A and B. (ii) Are the graphs consistent with the test of the AB interaction?

c) [8.3] Calculate the power of the tests of treatments A and B.

d) [8.3] Determine the value of n necessary to achieve a power of approximately .85 for treatment A.

e) [8.5] Use Tukey's statistic to determine which population means differ for treatment B.

f) [8.8] Calculate $\hat{\omega}_{Y|A}^2$ and $\hat{\omega}_{Y|B}^2$.

g) Prepare a "results and discussion section" appropriate for the journal of *Perceptual and Motor Skills*.

6. The experiment described in Exercise 5 was repeated using a random sample of 36 female college students. The following data were obtained. (Experiment suggested by Miller, A. G. Social perception of internal-external control. *Perceptual and Motor Skills*, 1970, *30*, 103–109.)

| a_1 | a_1 | a_1 | a_2 | a_2 | a_2 |
b_1	b_2	b_3	b_1	b_2	b_3
12	10	9	7	8	13
8	12	7	13	7	14
10	8	9	9	3	19
6	6	5	9	8	9
8	10	11	5	12	16
4	14	12	11	10	14

†**a)** [8.3] Test the hypotheses $\alpha_j = 0$ for all j, $\beta_k = 0$ for all k, and $(\alpha\beta)_{jk} = 0$ for all j and k; let $\alpha = .05$.

†**b)** [8.3 and 8.6] (i) Graph the AB interaction. (ii) Test the following hypotheses: H_0: $\alpha\psi_{1(B)} = \delta$ for all j, H_0: $\alpha\psi_{2(B)} = \delta$ for all j, and H_0: $\alpha\psi_{3(B)} = \delta$ for all j, where $\psi_{1(B)} = \mu_{\cdot 1} - \mu_{\cdot 2}$, $\psi_{2(B)} = \mu_{\cdot 1} - \mu_{\cdot 3}$, and $\psi_{3(B)} = \mu_{\cdot 2} - \mu_{\cdot 3}$. Use $(\nu_1 F_{.05:\nu_1,\nu_2})/\nu_3$ as the critical value for each test.

†**c)** [8.3] Calculate the power of the test of H_0: $\beta_k = 0$ for all k and $(\alpha\beta)_{jk} = 0$ for all j and k.

†**d)** [8.3] Determine the value of n necessary to achieve a power of approximately .85 for the AB interaction.

†**e)** [8.8] Calculate $\hat{\omega}_{Y|B}^2$ and $\hat{\omega}_{Y|AB}^2$.

f) Prepare a "results and discussion section" appropriate for the journal of *Perceptual and Motor Skills*.

7. The effectiveness of three counseling-analogue procedures and a control condition in helping undergraduate females to resolve personal anger conflicts was investigated. Subjects first took Gendlin's nine-item questionnaire that measures the degree to which people can use their emotional experience facilitatively. Those with scores above 2.5 were designated as high focusers, a_1; those with scores less than 2.5 were designated low focusers, a_2. The subjects were then asked to think of a recent unresolved, anger-arousing incident involving another person (the provocateur) and visualize it for two minutes. Following this, they filled out Thayer's Activation-Deactivation Adjective Check List, a self-report measure of anger, anxiety, and hostile or friendly attitudes toward the provocateur. Each of the 32 subjects was randomly assigned to one of four counseling-analogue conditions (treatment B). Subjects in the control condition b_1 answered a series of questions concerning the physical characteristics of the anger-arousing incident. Those in the intellectual-analysis condition b_2 were asked to

"coldly" and rationally answer questions that were designed to facilitate intellectual exploration of the incident. Those in the discharge condition b_3 pretended that the provocateur was sitting across from them and they expressed their anger toward him or her. Those in the role-play condition b_4 in addition switched chairs and role-played the part of the provocateur. Following the treatment condition, the subjects again filled out the adjective check list. The dependent variable was the amount of change in anger scores from the pretest to the posttest. The following data were obtained. (Experiment suggested by Bohart, A. C. Role playing and interpersonal-conflict reduction. *Journal of Counseling Psychology*, 1977, *24*, 15–24.)

| a_1 | a_1 | a_1 | a_1 | a_2 | a_2 | a_2 | a_2 |
b_1	b_2	b_3	b_4	b_1	b_2	b_3	b_4
5	−5	4	−1	−1	3	9	−1
7	−1	2	−5	3	−3	10	3
4	3	1	0	−1	2	7	−6
1	−2	−1	2	−5	−4	5	−3

a) [8.3] Test the hypotheses $\alpha_j = 0$ for all j, $\beta_k = 0$ for all k, and $(\alpha\beta)_{jk} = 0$ for all j and k; let $\alpha = .05$.

b) [8.3 and 8.6] (i) Graph the AB interaction. (ii) Test the following hypotheses:
H_0: $\alpha\psi_{1(B)} = \delta$ for all j, H_0: $\alpha\psi_{2(B)} = \delta$ for all j, and H_0: $\alpha\psi_{3(B)} = \delta$ for all j, where $\psi_{1(B)} = \mu_{\cdot 1} - \mu_{\cdot 2}$, $\psi_{2(B)} = \mu_{\cdot 1} - \mu_{\cdot 3}$, and $\psi_{3(B)} = \mu_{\cdot 1} - \mu_{\cdot 4}$. Use $(\nu_1 F_{.05;\nu_1,\nu_2})/\nu_3$ as the critical value for each test.

c) [8.3] Calculate the power of the tests of H_0: $\beta_k = 0$ for all k and H_0: $(\alpha\beta)_{jk} = 0$ for all j and k.

d) [8.8] Calculate $\hat{\omega}^2_{Y|B}$ and $\hat{\omega}^2_{Y|AB}$.

e) Prepare a "results and discussion section" appropriate for the *Journal of Counseling Psychology*.

†8.   [8.4] Show that $\sum_{i=1}^{n} \sum_{j=1}^{p} \sum_{k=1}^{q} [(\overline{Y}_{\cdot j\cdot} - \overline{Y}_{\cdots}) + (\overline{Y}_{\cdot\cdot k} - \overline{Y}_{\cdots}) + (\overline{Y}_{\cdot jk} - \overline{Y}_{\cdot j\cdot} - \overline{Y}_{\cdot\cdot k} + \overline{Y}_{\cdots}) + (Y_{ijk} - \overline{Y}_{\cdot jk})]^2 = nq \sum_{j=1}^{p} (\overline{Y}_{\cdot j\cdot} - \overline{Y}_{\cdots})^2 + np \sum_{k=1}^{q} (\overline{Y}_{\cdot\cdot k} - \overline{Y}_{\cdots})^2$
$+ n \sum_{j=1}^{p} \sum_{k=1}^{q} (\overline{Y}_{\cdot jk} - \overline{Y}_{\cdot j\cdot} - \overline{Y}_{\cdot\cdot k} + \overline{Y}_{\cdots})^2 + \sum_{i=1}^{n} \sum_{j=1}^{p} \sum_{k=1}^{q} (Y_{ijk} - \overline{Y}_{\cdot jk})^2$.

†9.   [8.4] Derive the computational formulas for *SSA* and *SSWCELL* from the deviation formulas in equation (8.4-3).

10.   [8.6] If the hypothesis α_j at $b_1 = 0$ for all j is rejected, the interpretation is ambiguous; explain.

11.   [8.6] List the similarities between Scheffé's procedure and the simultaneous test procedure.

†12.   [8.6] When is the critical value for Scheffé's procedure equal to that for the simultaneous test procedure?

13.   [8.6] Sometimes a meaningful interpretation can be made of a main effect, say $\mu_{j\cdot} - \mu_{j'\cdot}$, even though the associated treatment interacts with another treatment; explain.

†14.   [8.6] When does $\sum_{i=1}^{p-1} SSB\hat{\psi}_{i(A)} = \sum_{i=1}^{q-1} SSA\hat{\psi}_{i(B)} = SSAB$?

†15.   [8.9] Use the rules in Section 8.9 to derive the expected values of mean squares for the following model

$$Y_{ijk} = \mu + \alpha_j + \beta_{k(j)} + \epsilon_{i(jk)} \qquad (i = 1, \ldots, n; j = 1, \ldots, p; k = 1, \ldots, q)$$

where $\epsilon_{i(jk)}$ is $NID(0, \sigma_\epsilon^2)$, A is fixed, and B is random. Denote the means squares by MSA, MSB $w.A$, and $MSWCELL$.

†16. [8.10] Assume that the expected values of mean squares for a design are as follows.

$$E(MSA) = \sigma_\epsilon^2 + n\sigma_{\beta\gamma}^2 + nq\sigma_{\alpha\gamma}^2 + nr\sigma_\beta^2 + nqr\sigma_\alpha^2$$
$$E(MSB \ w.A) = \sigma_\epsilon^2 + n\sigma_{\beta\gamma}^2 + nr\sigma_\beta^2$$
$$E(MSC) = \sigma_\epsilon^2 + n\sigma_{\beta\gamma}^2 + npq\sigma_\gamma^2$$
$$E(MSAC) = \sigma_\epsilon^2 + n\sigma_{\beta\gamma}^2 + nq\sigma_{\alpha\gamma}^2$$
$$E(MSB \ w.A \times C) = \sigma_\epsilon^2 + n\sigma_{\beta\gamma}^2$$
$$E(MSWCELL) = \sigma_\epsilon^2$$

Construct F' and F'' ratios for testing treatment A.

†17. [8.10 and 8.11] Assume that the following data have been obtained for a type CRF-322 design (model II) with n equal to two.

$$
\begin{array}{ll}
MSA = 21.17 & MSAC = 4.21 \\
MSB = \ \ 8.04 & MSBC = 7.85 \\
MSC = 12.19 & MSABC = 1.61 \\
MSAB = \ \ 1.60 & MSWCELL = 1.56
\end{array}
$$

a) Perform preliminary tests on the model and pool where appropriate; write out the ANOVA table showing values of mean squares and degrees of freedom.

b) Indicate the expected values of the mean squares for the final model.

c) Construct an F' ratio for testing treatment C.

18. [8.13] Assume that for the data in Exercise 4, observations Y_{521} and Y_{522} are missing. Perform a conventional analysis for proportional cell n's.

†19. [8.13] Exercise 4 described an experiment in which subjects rated a confederate in terms of likeability, intelligence, and personal adjustment.

a) Write a regression model with $h - 1 = 3$ independent variables for this type CRF-22 design.

b) Specify the null hypotheses in terms of β's.

c) Indicate the correspondence between the $h = 4$ parameters of the regression model and the parameters of a type CRF-22 less than full rank experimental design model for (i) effect coding and (ii) dummy coding.

d) Use the regression model approach with effect coding to test the hypothesis that all treatment and interaction effects equal zero; let $\alpha = .05$. (This problem involves the inversion of small diagonal matrices. The computations can be done without the aid of a computer.)

e) Write a full rank experimental design model for this type CRF-22 design.

f) Write the $\mathbf{C}'$ matrices for testing the null hypotheses for treatments A and B and the AB interaction.

g) Use the full rank experimental design model approach to test the hypotheses in 19(f); let $\alpha = .05$. (This problem involves the inversion of small diagonal matrices. The computations can be done without the aid of a computer.)

20.   [8.13] Exercise 5 described an experiment in which male subjects were shown photographs and asked to fill out the Rotter I-E scale the way they thought the person in the photograph would.

   a) Assume that observation $Y_{211} = 10$ is missing. Use the full rank experimental design model approach to test the null hypotheses for this type CRF-23 design. (This problem involves the inversion of small matrices. The computations, while tedious, can be done without the aid of a computer.)

   b) Assume that observation $Y_{211} = 10$ is missing and that cell a_1b_2 is empty. Use the full rank experimental design model approach to test the null hypotheses. (This problem involves the inversion of small matrices. The computations, while tedious, can be done without the aid of a computer.)

   c) How does the empty cell a_1b_2 affect the interpretation of the AB interaction?

†21.   [8.13] Exercise 7 described an experiment involving four approaches to resolving personal anger conflicts.

   a) Assume that observation $Y_{113} = 4$ is missing. Use the full rank experimental design model approach to test the null hypotheses. (This problem involves the inversion of small matrices. The computations, while tedious, can be done without the aid of a computer.)

   b) Assume that observation $Y_{113} = 4$ is missing and that cell a_2b_2 is empty. Use the full rank experimental design model approach to test the hypotheses. (This problem involves the inversion of small matrices. The computations, while tedious, can be done without the aid of a computer.)

   c) How does the empty cell a_2b_2 affect the interpretation of the AB interaction?

9 COMPLETELY RANDOMIZED FACTORIAL DESIGN WITH THREE OR MORE TREATMENTS AND RANDOMIZED BLOCK FACTORIAL DESIGN

9.1 INTRODUCTION TO TYPE CRF-pqr DESIGN

This chapter describes the layout and analysis of a completely randomized factorial design with three or more treatments. Also discussed is a factorial design that is constructed from randomized block designs.

The traditional analysis procedures described in Section 8.3 as well as those for the full rank model in Section 8.13 can be easily extended to experiments having three or more treatments. We will begin with a three-treatment completely randomized factorial design. A block diagram of a type CRF-222 design appears in Figure 9.1-1.

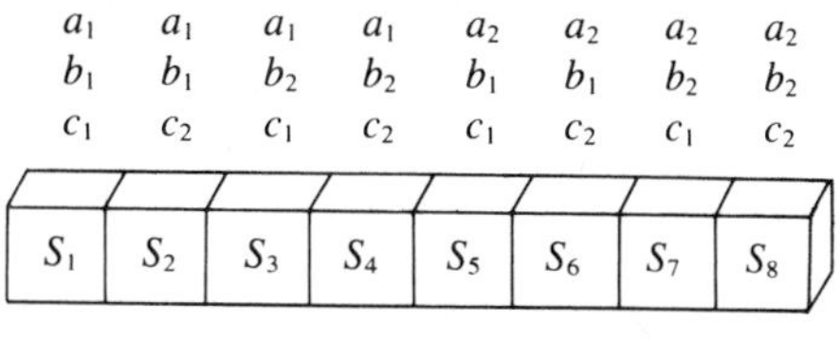

FIGURE 9.1-1 Block diagram of type CRF-222 design. The eight treatment combinations, $a_1b_1c_1$, $a_1b_1c_2$, . . . , $a_2b_2c_2$, are assigned randomly to eight random samples of subjects, $S_1, S_2, \ldots, S_8$.

a_1	a_1	a_1	a_1	a_2	a_2	a_2	a_2
b_1	b_1	b_2	b_2	b_1	b_1	b_2	b_2
c_1	c_2	c_1	c_2	c_1	c_2	c_1	c_2
S_1	S_2	S_3	S_4	S_5	S_6	S_7	S_8

The experimental design model is

$$Y_{ijkl} = \mu + \alpha_j + \beta_k + \gamma_l + (\alpha\beta)_{jk} + (\alpha\gamma)_{jl} + (\beta\gamma)_{kl} + (\alpha\beta\gamma)_{jkl} + \epsilon_{i(jkl)}$$
$$(i = 1, \ldots, n; j = 1, \ldots, p; k = 1, \ldots, q; l = 1, \ldots, r)$$

where $\epsilon_{i(jkl)}$ is $NID(0, \sigma_\epsilon^2)$. This design enables an experimenter to test hypotheses concerning treatments A, B, and C and four interactions AB, AC, BC, and ABC. General procedures for enumerating interactions in a factorial design are described in the following section.

ENUMERATION OF INTERACTIONS FOR ANY TYPE CRF DESIGN

The interaction terms associated with an experiment can be readily determined by writing down all combinations of treatment letters, ignoring the order of the letters. For example, treatments A, B, and C can be combined to form the following combinations:

$$AB$$
$$AC$$
$$BC$$
$$ABC.$$

Interactions involving two letters are called two-treatment interactions, first-order interactions, or double interactions. If three letters are involved, the interaction is a three-treatment interaction, second-order or triple interaction, and so on. Four treatment letters (A, B, C, D) can form the following combinations:

AB	ABC
AC	ABD
AD	ACD
BC	BCD
BD	$ABCD.$
CD	

A four-treatment experiment that is based on a completely randomized building block design is designated by the letters CRF-*pqrt*.

In general the number of two-, three-, and so on treatment interactions is given by

$$_kC_n = \frac{k!}{n!(k-n)!}$$

where k is the number of treatments and n is the number of letters in the interaction. For example, a four-treatment experiment has six two-treatment interactions,

$$_4C_2 = \frac{4!}{2!(4-2)!} = \frac{4 \cdot 3 \cdot 2 \cdot 1}{2 \cdot 1(2 \cdot 1)} = \frac{24}{4} = 6$$

four three-treatment interactions,

$$_4C_3 = \frac{4!}{3!(4-3)!} = \frac{4 \cdot 3 \cdot 2 \cdot 1}{3 \cdot 2 \cdot 1(1)} = \frac{24}{6} = 4$$

and one four-treatment interaction,

$$_4C_4 = \frac{4!}{4!(4-4)!} = \frac{4 \cdot 3 \cdot 2 \cdot 1}{4 \cdot 3 \cdot 2 \cdot 1} = \frac{24}{24} = 1.$$

9.2 COMPUTATIONAL EXAMPLE FOR TYPE CRF-*pqr* DESIGN

We will now illustrate the computational procedures for a type CRF-222 design. For purposes of illustration assume that an experimenter is interested in the effects of magnitude of reinforcement (treatment A), social deprivation (treatment B), and the sex of the adult who administers reinforcement (treatment C: c_1 denotes a male and c_2 denotes a female) on children's choice behavior. The task performed by the children consists of taking marbles from a bin, one at a time, and inserting them in one of two distinctively marked holes in the top of a box. The hole preferred by each child was determined during a preliminary familiarization session. The dependent variable is the number of marbles inserted in the nonpreferred hole during the last ten minutes of a 15-minute experimental session.

    Social reinforcement, treatment A, was administered by an adult during the experimental session: level a_1 consisted of smiles, nods, and praise statements by the experimenter when a child placed a marble in the nonpreferred hole; level a_2 consisted of smiles for using the nonpreferred hole but no nods or praise statements. Prior to the experimental session the children were subjected to two levels of social deprivation, treatment B. The deprivation consisted of being left alone in a waiting room for 40 minutes, b_1, or 80 minutes, b_2. The waiting room contained numerous toys with which the children could amuse themselves. However, during the deprivation period they were isolated from contact with adults and other children. All treatment levels of interest to the experimenter are included in the experiment. Thus a fixed-effects model (model I) applies to the experiment.

    Assume that a total of 32 children have been obtained by random sampling from a population. The children are randomly divided into eight subsamples of four each and the $2 \times 2 \times 2 = 8$ treatment combinations randomly assigned to the subsamples.

    The layout of this type of CRF-222 design and computational procedures are shown in Table 9.2-1. The analysis of variance is summarized in Table 9.2-2. As can

---

TABLE 9.2-1   Computational Procedures for a Type CRF-222 Design

---

(i)   Data and notation [Y_{ijkl} denotes a score for experimental unit i in treatment combination jkl; $i = 1, \ldots, n$ experimental units (s_i); $j = 1, \ldots, p$ levels of treatment A (a_j); $k = 1, \ldots, q$ levels of treatment B (b_k); $l = 1, \ldots, r$ levels of treatment C (c_l)]:

ABCS Summary Table
Entry is Y_{ijkl}

| a_1 | a_1 | a_1 | a_1 | a_2 | a_2 | a_2 | a_2 |
| b_1 | b_1 | b_2 | b_2 | b_1 | b_1 | b_2 | b_2 |
c_1	c_2	c_1	c_2	c_1	c_2	c_1	c_2
3	4	7	7	1	2	5	10
6	5	8	8	2	3	6	10
3	4	7	9	2	4	5	9
3	3	6	8	2	3	6	11

ABC Summary Table
Entry is $\sum\limits_{i=1}^{n} Y_{ijkl}$

| | b_1 | b_1 | b_2 | b_2 |
	c_1	c_2	c_1	c_2
a_1	$n = 4$ 15	16	28	32
a_2	7	12	22	40

AB Summary Table
Entry is $\sum\limits_{i=1}^{n}\sum\limits_{l=1}^{r} Y_{ijkl}$

	b_1	b_2	$\sum\limits_{i=1}^{n}\sum\limits_{k=1}^{q}\sum\limits_{l=1}^{r} Y_{ijkl}$
a_1	$nr = 8$ 31	60	91
a_2	19	62	81
$\sum\limits_{i=1}^{n}\sum\limits_{j=1}^{p}\sum\limits_{l=1}^{r} Y_{ijkl} = 50$	122		

AC Summary Table
Entry is $\sum\limits_{i=1}^{n}\sum\limits_{k=1}^{q} Y_{ijkl}$

	c_1	c_2
a_1	$nq = 8$ 43	48
a_2	29	52
$\sum\limits_{i=1}^{n}\sum\limits_{j=1}^{p}\sum\limits_{k=1}^{q} Y_{ijkl} = 72$	100	

BC Summary Table
Entry is $\sum\limits_{i=1}^{n}\sum\limits_{j=1}^{p} Y_{ijkl}$

	c_1	c_2
b_1	$np = 8$ 22	28
b_2	50	72

Table 9.2-1 (continued)

(ii)   Computational symbols:

$$\sum_{i=1}^{n}\sum_{j=1}^{p}\sum_{k=1}^{q}\sum_{l=1}^{r} Y_{ijkl} = 3 + 6 + \cdots + 11 = 172.000$$

$$\frac{\left(\sum_{i=1}^{n}\sum_{j=1}^{p}\sum_{k=1}^{q}\sum_{l=1}^{r} Y_{ijkl}\right)^2}{npqr} = [Y] = \frac{(172.000)^2}{(4)(2)(2)(2)} = 924.500$$

$$\sum_{i=1}^{n}\sum_{j=1}^{p}\sum_{k=1}^{q}\sum_{l=1}^{r} Y_{ijkl}^2 = [ABCS] = (3)^2 + (6)^2 + \cdots + (11)^2 = 1160.000$$

$$\sum_{j=1}^{p} \frac{\left(\sum_{i=1}^{n}\sum_{k=1}^{q}\sum_{l=1}^{r} Y_{ijkl}\right)^2}{nqr} = [A] = \frac{(91)^2}{(4)(2)(2)} + \frac{(81)^2}{(4)(2)(2)} = 927.625$$

$$\sum_{k=1}^{q} \frac{\left(\sum_{i=1}^{n}\sum_{j=1}^{p}\sum_{l=1}^{r} Y_{ijkl}\right)^2}{npr} = [B] = \frac{(50)^2}{(4)(2)(2)} + \frac{(122)^2}{(4)(2)(2)} = 1086.500$$

$$\sum_{l=1}^{r} \frac{\left(\sum_{i=1}^{n}\sum_{j=1}^{p}\sum_{k=1}^{q} Y_{ijkl}\right)^2}{npq} = [C] = \frac{(72)^2}{(4)(2)(2)} + \frac{(100)^2}{(4)(2)(2)} = 949.000$$

$$\sum_{j=1}^{p}\sum_{k=1}^{q} \frac{\left(\sum_{i=1}^{n}\sum_{l=1}^{r} Y_{ijkl}\right)^2}{nr} = [AB] = \frac{(31)^2}{(4)(2)} + \cdots + \frac{(62)^2}{(4)(2)} = 1095.750$$

$$\sum_{j=1}^{p}\sum_{l=1}^{r} \frac{\left(\sum_{i=1}^{n}\sum_{k=1}^{q} Y_{ijkl}\right)^2}{nq} = [AC] = \frac{(43)^2}{(4)(2)} + \cdots + \frac{(52)^2}{(4)(2)} = 962.250$$

$$\sum_{k=1}^{q}\sum_{l=1}^{r} \frac{\left(\sum_{i=1}^{n}\sum_{j=1}^{p} Y_{ijkl}\right)^2}{np} = [BC] = \frac{(22)^2}{(4)(2)} + \cdots + \frac{(72)^2}{(4)(2)} = 1119.000$$

$$\sum_{j=1}^{p}\sum_{k=1}^{q}\sum_{l=1}^{r} \frac{\left(\sum_{i=1}^{n} Y_{ijkl}\right)^2}{n} = [ABC] = \frac{(15)^2}{4} + \cdots + \frac{(40)^2}{4} = 1141.500$$

(iii)   Computational formulas:

$$SSTO = [ABCS] - [Y] = 235.500$$
$$SSA = [A] - [Y] = 3.125$$
$$SSB = [B] - [Y] = 162.000$$
$$SSC = [C] - [Y] = 24.500$$

Table 9.2-1 (continued)

$$SSAB = [AB] - [A] - [B] + [Y] = 6.125$$

$$SSAC = [AC] - [A] - [C] + [Y] = 10.125$$

$$SSBC = [BC] - [B] - [C] + [Y] = 8.000$$

$$SSABC = [ABC] - [AB] - [AC] - [BC] + [A] + [B] + [C] - [Y] = 3.125$$

$$SSWCELL = [ABCS] - [ABC] = 18.500$$

TABLE 9.2-2   ANOVA Table for Type CRF-222 Design

Source	df	MS	F	E(MS) Model I
1 A (Magnitude of reinforcement)	$p - 1 = 1$	3.125	[1/8] 4.05	$\sigma_\epsilon^2 + nqr \sum\limits_{j=1}^{p} \alpha_j^2/(p - 1)$
2 B (Social deprivation)	$q - 1 = 1$	162.000	[2/8] 210.12*	$\sigma_\epsilon^2 + npr \sum\limits_{k=1}^{q} \beta_k^2/(q - 1)$
3 C (Sex of reinforcer)	$r - 1 = 1$	24.500	[3/8] 31.78*	$\sigma_\epsilon^2 + npq \sum\limits_{l=1}^{r} \gamma_l^2/(r - 1)$
4 AB	$(p - 1)(q - 1) = 1$	6.125	[4/8] 7.94*	$\sigma_\epsilon^2 + nr \sum\limits_{j=1}^{p} \sum\limits_{k=1}^{q} (\alpha\beta)_{jk}^2/(p - 1)(q - 1)$
5 AC	$(p - 1)(r - 1) = 1$	10.125	[5/8] 13.13*	$\sigma_\epsilon^2 + nq \sum\limits_{j=1}^{p} \sum\limits_{l=1}^{r} (\alpha\gamma)_{jl}^2/(p - 1)(r - 1)$
6 BC	$(q - 1)(r - 1) = 1$	8.000	[6/8] 10.38*	$\sigma_\epsilon^2 + np \sum\limits_{k=1}^{q} \sum\limits_{l=1}^{r} (\beta\gamma)_{kl}^2/(q - 1)(r - 1)$
7 ABC	$(p - 1)(q - 1)(r - 1) = 1$	3.125	[7/8] 4.05	$\sigma_\epsilon^2 + n \sum\limits_{j=1}^{p} \sum\limits_{k=1}^{q} \sum\limits_{l=1}^{r} (\alpha\beta\gamma)_{jkl}^2/(p - 1)(q - 1)(r - 1)$
8 WCELL	$pqr(n - 1) = 24$	0.771		σ_ϵ^2
9 Total	$npqr - 1 = 31$			

*$p < .01$

be seen, all two-treatment interactions are significant. Under these conditions an experimenter would turn to the procedures in Section 8.6 in order to gain a better understanding of the way the variables interact.

SIMPLE-EFFECTS TESTS

If one chooses to test hypotheses about simple main effects, the following provide useful computational checks.

Simple Main-Effects Sum of Squares

$$\sum_{k=1}^{q} SSA \text{ at } b_k = SSA + SSAB \qquad \sum_{l=1}^{r} SSB \text{ at } c_l = SSB + SSBC$$

$$\sum_{l=1}^{r} SSA \text{ at } c_l = SSA + SSAC \qquad \sum_{j=1}^{p} SSC \text{ at } a_j = SSC + SSAC$$

$$\sum_{j=1}^{p} SSB \text{ at } a_j = SSB + SSAB \qquad \sum_{k=1}^{q} SSC \text{ at } b_k = SSC + SSBC$$

Simple Simple Main-Effects Sum of Squares

$$\sum_{k=1}^{q} \sum_{l=1}^{r} SSA \text{ at } b_k c_l = SSA + SSAB + SSAC + SSABC$$

$$\sum_{j=1}^{p} \sum_{l=1}^{r} SSB \text{ at } a_j c_l = SSB + SSAB + SSBC + SSABC$$

$$\sum_{j=1}^{p} \sum_{k=1}^{q} SSC \text{ at } a_j b_k = SSC + SSAC + SSBC + SSABC$$

Simple Interaction-Effects Sum of Squares

$$\sum_{l=1}^{r} SSAB \text{ at } c_l = SSAB + SSABC$$

$$\sum_{k=1}^{q} SSAC \text{ at } b_k = SSAC + SSABC$$

$$\sum_{j=1}^{p} SSBC \text{ at } a_j = SSBC + SSABC$$

9.3 PATTERNS UNDERLYING COMPUTATIONAL FORMULAS FOR TYPE CRF DESIGNS

KINDS OF COMPUTATIONAL FORMULAS USED IN ANOVA

The reader may feel that a bewildering array of computational formulas are used in analysis of variance. At first glance this appears to be true, but it can be shown that actually only four kinds of formulas are used. These can be classified as (1) between, (2) within, (3) interaction, and (4) total formulas. Consider the formulas for type CR-p, RB-p, LS-p, CRF-pq, and CRF-pqr designs. The between formulas have the form:

type CR-p design	$[A] - [Y]$
type RB-p design	$[A] - [Y], [S] - [Y]$
type LS-p design	$[A] - [Y], [B] - [Y], [C] - [Y]$
type CRF-pq design	$[A] - [Y], [B] - [Y]$
type CRF-pqr design	$[A] - [Y], [B] - [Y], [C] - [Y].$

The pattern that underlies these formulas is obvious. The within formulas have the form:

type CR-p design	$[AS] - [A]$
type RB-p design	no within sum of squares
type LS-p design	$[ABCS] - [ABC]$ for $n > 1$
type CRF-pq design	$[ABS] - [AB]$
type CRF-pqr design	$[ABCS] - [ABC].$

The third kind of formula, the one for computing an interaction sum of squares, has the following general form:

type CR-p design	no interaction sum of squares
type RB-p design	$[AS] - [A] - [S] + [Y]$
type LS-p design	$[ABC] - [A] - [B] - [C] + 2[Y]$
type CRF-pq design	$[AB] - [A] - [B] + [Y]$
type CRF-pqr design	$[AB] - [A] - [B] + [Y],$
	$[AC] - [A] - [C] + [Y],$
	$[BC] - [B] - [C] + [Y],$
	$[ABC] - [AB] - [AC] - [BC] + [A] + [B]$
	$\quad + [C] - [Y]$
type CRF-$pqrt$ design	$[ABCD] - [ABC] - [ABD] - [ACD]$
	$\quad - [BCD] + [AB] + [AC] + [AD]$
	$\quad + [BC] + [BD] + [CD] - [A] - [B]$
	$\quad - [C] - [D] + [Y].$

In general, an interaction involving k treatments is given by

$$k \text{ treatment interaction} = [k \text{ treatment letters}]$$
$$- \Sigma[\text{all combinations of } k - 1 \text{ treatment letters}]$$
$$+ \Sigma[\text{all combinations of } k - 2 \text{ treatment letters}]$$
$$- \cdots \pm \Sigma[\text{all } k - (k - 1) \text{ treatment letters}] \pm [Y].$$

If k is an even number, the sign of $[Y]$ is plus; if k is an odd number, the sign is negative. This is evident because the sign changes for each group of terms. In some designs, such as a randomized block design, the interaction formula is referred to as a *residual* formula. This designation is often used when an interaction mean square is used in lieu of a within-groups or a within-cell mean square to estimate experimental error.

The final kind of formula is a total sum of squares formula. This formula has the general form illustrated by the following examples:

type CR-p design	$[AS] - [Y]$
type RB-p design	$[AS] - [Y]$
type LS-p design	$[ABCS] - [Y]$
type CRF-pq design	$[ABS] - [Y]$
type CRF-pqr design	$[ABCS] - [Y]$.

Complex designs described in subsequent chapters should appear less complex when it is realized that all designs involve at most only four kinds of formulas and hence four kinds of variation.

UNDERSTANDING THE ABBREVIATED COMPUTATIONAL FORMULAS

The rationale underlying the computational formulas is relatively simple. Terms such as $[A]$, $[AB]$, and $[ABS]$ in a type CRF-pq design are given by

$$[A] = \sum_{j=1}^{p} \frac{\left(\sum_{i=1}^{n} \sum_{k=1}^{q} Y_{ijk} \right)^2}{nq}$$

$$[AB] = \sum_{j=1}^{p} \sum_{k=1}^{q} \frac{\left(\sum_{i=1}^{n} Y_{ijk} \right)^2}{n}$$

$$[ABS] = \sum_{i=1}^{n} \sum_{j=1}^{p} \sum_{k=1}^{q} Y_{ijk}^2 .$$

The *correction term* $[Y]$ is given by

$$[Y] = \frac{\left(\sum_{i=1}^{n} \sum_{j=1}^{p} \sum_{k=1}^{q} Y_{ijk} \right)^2}{npq} .$$

If the correction term $[Y]$ is subtracted from $[A]$, $[B]$, $[AB]$, or $[ABS]$, a sum of squares is obtained. Thus,

$$[A] - [Y] = SSA$$
$$[B] - [Y] = SSB$$
$$[AB] - [Y] = SSA + SSB + SSAB$$
$$[ABS] - [Y] = SSA + SSB + SSAB + SSWCELL(\text{subjects}) = SSTO.$$

The last two formulas require a word of explanation. Consider $[AB] - [Y]$ first. The term $[AB]$ may be thought of as including variation due to each letter, A and B, and the interaction of the two letters AB. It includes the AB interaction because treatments A and B are crossed in a type CRF-pq design. To obtain the AB interaction sum of squares it is necessary to subtract SSA and SSB from $([AB] - [Y])$.

$$SSAB = ([AB] - [Y]) - ([A] - [Y]) - ([B] - [Y])$$
$$= [AB] - [A] - [B] + [Y]$$

The term $[ABS]$ in the last formula includes variation due to A, B, S(subjects), and the AB interaction. The reader may wonder why $[ABS]$ does not also include variation due to the interactions AS, BS, and ABS. The answer is that only treatments A and B are crossed; subjects (S) are nested within the pq treatments combinations.

The sum of squares due to subjects, $SSWCELL$, is obtained by subtracting $[AB]$ from $[ABS]$, that is,

$$[ABS] - [AB] = SSWCELL.$$

*FORMULATING INTERACTION NULL HYPOTHESES FOR THE FULL RANK MODEL

We have seen how to formulate interaction null hypotheses for the full rank model when the experiment contains two treatments (see Sections 6.9 and 8.13). In this section we will see how to formulate interaction null hypotheses for experiments with three or more treatments. A type CRF-pqr design has $(p - 1)(q - 1)(r - 1)$ three-treatment interaction terms of the form

$$\mu_{jkl} - \mu_{j'kl} - \mu_{jk'l} + \mu_{j'k'l} - \mu_{jkl'} + \mu_{j'kl'} + \mu_{jk'l'} - \mu_{j'k'l'}.$$

This three-treatment interaction term looks a little forebidding, but we will see that the underlying pattern is quite simple. A three-treatment interaction is equal to zero if the interaction between any two treatments is the same at all levels of the third treatment. Consider the population means for a type CRF-233 design in Figure 9.3-1. The AB interaction at each level of treatment C has the form

---

_* This section assumes a familiarity with the matrix operations in Appendix D and Sections 6.9 and 8.13. The reader who is interested only in the traditional approach to the analysis of variance can, without loss of continuity, omit this section.

FIGURE 9.3-1   Two-treatment interaction terms of the form $\mu_{jkl} - \mu_{j'kl} - \mu_{jk'l} + \mu_{j'k'l}$ are obtained from the crossed lines by subtracting the two μ_{jkl}'s connected by a dashed line from the μ_{jkl}'s connected by a solid line. A three-treatment interaction term is obtained by subtracting a two-treatment interaction term at the third level of treatment C from the corresponding two-treatment interaction term at the first or second level of treatment C. For example, $(\mu_{111} - \mu_{211} - \mu_{121} + \mu_{221}) - (\mu_{113} - \mu_{213} - \mu_{123} + \mu_{223}) = \mu_{111} - \mu_{211} - \mu_{121} + \mu_{221} - \mu_{113} + \mu_{213} + \mu_{123} - \mu_{223}$.

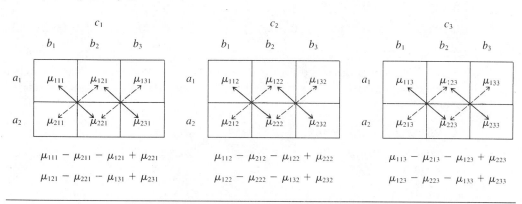

$$AB \text{ at } c_1 = \mu_{jk1} - \mu_{j'k1} - \mu_{jk'1} + \mu_{j'k'1} \qquad \text{for all } j, j', k, k'$$
$$AB \text{ at } c_2 = \mu_{jk2} - \mu_{j'k2} - \mu_{jk'2} + \mu_{j'k'2} \qquad \text{for all } j, j', k, k'$$
$$AB \text{ at } c_3 = \mu_{jk3} - \mu_{j'k3} - \mu_{jk'3} + \mu_{j'k'3} \qquad \text{for all } j, j', k, k'.$$

The ABC interaction is equal to zero if

$$(AB \text{ at } c_1) - (AB \text{ at } c_3) = 0$$

or

$$(\mu_{jk1} - \mu_{j'k1} - \mu_{jk'1} + \mu_{j'k'1}) - (\mu_{jk3} - \mu_{j'k3} - \mu_{jk'3} + \mu_{j'k'3}) = 0$$

and

$$(AB \text{ at } c_2) - (AB \text{ at } c_3) = 0$$

or

$$(\mu_{jk2} - \mu_{j'k2} - \mu_{jk'2} + \mu_{j'k'2}) - (\mu_{jk3} - \mu_{j'k3} - \mu_{jk'3} + \mu_{j'k'3}) = 0$$

for all j, j', k, k'. The $(p - 1)(q - 1)(r - 1) = 4$ interaction terms for the means in Figure 9.3-1 are as follows.

$$\mu_{111} - \mu_{211} - \mu_{121} + \mu_{221} - \mu_{113} + \mu_{213} + \mu_{123} - \mu_{223}$$
$$\mu_{112} - \mu_{212} - \mu_{122} + \mu_{222} - \mu_{113} + \mu_{213} + \mu_{123} - \mu_{223}$$
$$\mu_{121} - \mu_{221} - \mu_{131} + \mu_{231} - \mu_{123} + \mu_{223} + \mu_{133} - \mu_{233}$$
$$\mu_{122} - \mu_{222} - \mu_{132} + \mu_{232} - \mu_{123} + \mu_{223} + \mu_{133} - \mu_{233}$$

In matrix notation the $\mathbf{C}'$ matrix for testing the hypothesis that the ABC interaction is equal to zero, $\mathbf{C}'_{ABC}\,\boldsymbol{\mu} = \mathbf{0}$, is

$$\mathbf{C}'_{ABC} \atop \scriptstyle (p-1)(q-1)(r-1)\times h} =$$

$$
\begin{bmatrix}
1 & 0 & -1 & -1 & 0 & 1 & 0 & 0 & 0 & -1 & 0 & 1 & 1 & 0 & -1 & 0 & 0 & 0 \\
0 & 1 & -1 & 0 & -1 & 1 & 0 & 0 & 0 & 0 & -1 & 1 & 0 & 1 & -1 & 0 & 0 & 0 \\
0 & 0 & 0 & 1 & 0 & -1 & -1 & 0 & 1 & 0 & 0 & 0 & -1 & 0 & 1 & 1 & 0 & -1 \\
0 & 0 & 0 & 0 & 1 & -1 & 0 & -1 & 1 & 0 & 0 & 0 & 0 & -1 & 1 & 0 & 1 & -1
\end{bmatrix}
$$

where $\boldsymbol{\mu}' = \begin{bmatrix} \mu_{111} & \mu_{112} & \mu_{113} & \mu_{121} & \mu_{122} & \mu_{123} & \mu_{131} & \mu_{132} & \mu_{133} & \mu_{211} & \mu_{212} & \cdots & \mu_{233} \end{bmatrix}$.

The pattern described for three-treatment interactions generalizes to higher-order interactions. Consider a type CRF-2332 design. The ABC interaction at each level of treatment D has the form

$$ABC \text{ at } d_1 = \mu_{jkl1} - \mu_{j'kl1} - \mu_{jk'l1} + \mu_{j'k'l1} - \mu_{jkl'1} + \mu_{j'kl'1} + \mu_{jk'l'1} - \mu_{j'k'l'1}$$

$$ABC \text{ at } d_2 = \mu_{jkl2} - \mu_{j'kl2} - \mu_{jk'l2} + \mu_{j'k'l2} - \mu_{jkl'2} + \mu_{j'kl'2} + \mu_{jk'l'2} - \mu_{j'k'l'2}$$

for all j, j', k, k', l, l'. The $ABCD$ interaction is equal to zero if $(ABC \text{ at } d_1) - (ABC \text{ at } d_2) = 0$. There are $(p-1)(q-1)(r-1)(t-1) = 4$ such interaction terms.

Let us return to the type CRF-233 design in Figure 9.3-1. Suppose that observations Y_{i111}, Y_{i122}, and Y_{i222} are missing. The null hypothesis for the three-treatment interaction is unchanged. However, when there are missing observations, an experimenter has a choice among several null hypotheses that can be tested for each of the two-treatment interactions.* We will use the AB interaction to illustrate the two most interesting null hypotheses; they are

(9.3-1) H_0: $\mu_{jk\cdot} - \mu_{j'k\cdot} - \mu_{jk'\cdot} + \mu_{j'k'\cdot} = 0$ for all j, j', k, k'

where $\mu_{jk\cdot} = \Sigma_{l=1}^{r}\, \mu_{jkl}/r$, and

(9.3-2) H_0: $\bar{\mu}_{jk\cdot} - \bar{\mu}_{j'k\cdot} - \bar{\mu}_{jk'\cdot} + \bar{\mu}_{j'k'\cdot} = 0$ for all j, j', k, k'

where $\bar{\mu}_{jk\cdot} = \Sigma_{l=1}^{r}\, n_{jkl}\mu_{jkl}/n_{jk\cdot}$.

The $\mathbf{C}'$ matrices for testing these null hypotheses are, respectively,

$${\mathbf{C}'_{1(AB)} \atop \scriptstyle (p-1)(1-1)\times h} =$$

$$
\begin{bmatrix}
\frac{1}{r} & \frac{1}{r} & \frac{1}{r} & -\frac{1}{r} & -\frac{1}{r} & -\frac{1}{r} & 0 & 0 & 0 & -\frac{1}{r} & -\frac{1}{r} & -\frac{1}{r} & \frac{1}{r} & \frac{1}{r} & \frac{1}{r} & 0 & 0 & 0 \\
0 & 0 & 0 & \frac{1}{r} & \frac{1}{r} & \frac{1}{r} & -\frac{1}{r} & -\frac{1}{r} & -\frac{1}{r} & 0 & 0 & 0 & -\frac{1}{r} & -\frac{1}{r} & -\frac{1}{r} & \frac{1}{r} & \frac{1}{r} & \frac{1}{r}
\end{bmatrix}
$$

where $c_{jkl} = \pm 1/r$ or 0, and

---

*   I am indebted to James E. Carlson for providing unpublished material on this.

$$
\mathbf{C}'_{\substack{2(AB) \\ (p-1)(q-1)\times h}} = \begin{bmatrix} \dfrac{n_{111}}{n_{11\cdot}} & \dfrac{n_{112}}{n_{11\cdot}} & \dfrac{n_{113}}{n_{11\cdot}} & -\dfrac{n_{121}}{n_{12\cdot}} & -\dfrac{n_{122}}{n_{12\cdot}} & -\dfrac{n_{123}}{n_{12\cdot}} & 0 & 0 & 0 \\[3mm] 0 & 0 & 0 & \dfrac{n_{121}}{n_{12\cdot}} & \dfrac{n_{122}}{n_{12\cdot}} & \dfrac{n_{123}}{n_{12\cdot}} & -\dfrac{n_{131}}{n_{13\cdot}} & -\dfrac{n_{132}}{n_{13\cdot}} & -\dfrac{n_{133}}{n_{13\cdot}} \end{bmatrix}
$$

$$
\begin{bmatrix} -\dfrac{n_{211}}{n_{21\cdot}} & -\dfrac{n_{212}}{n_{21\cdot}} & -\dfrac{n_{213}}{n_{21\cdot}} & \dfrac{n_{221}}{n_{22\cdot}} & \dfrac{n_{222}}{n_{22\cdot}} & \dfrac{n_{223}}{n_{22\cdot}} & 0 & 0 & 0 \\[3mm] 0 & 0 & 0 & -\dfrac{n_{221}}{n_{22\cdot}} & -\dfrac{n_{222}}{n_{22\cdot}} & -\dfrac{n_{223}}{n_{22\cdot}} & \dfrac{n_{231}}{n_{23\cdot}} & \dfrac{n_{232}}{n_{23\cdot}} & \dfrac{n_{233}}{n_{23\cdot}} \end{bmatrix}
$$

where $c_{jkl} = \pm n_{jkl}/n_{jk\cdot}$ or 0. If, as is usually the case, the sample n_{jkl}'s are not proportional to the population N_{jkl}'s, an experimenter would ordinarily be interested in testing hypothesis (9.3-1). When there are no missing observations, hypotheses (9.3-1) and (9.3-2) are equivalent.

9.4 INTRODUCTION TO RANDOMIZED BLOCK FACTORIAL DESIGN

A randomized block factorial design uses a feature of a type RB-p design, namely the blocking technique, in isolating variation attributable to a nuisance variable while simultaneously evaluating two or more treatments and associated interactions. The building block for this factorial experiment is a randomized block design. An experiment with two treatments is designated as a type RBF-pq design. A comparison of this design with a type CRF-pq design is shown in Figure 9.4-1.

As indicated in the figure, the blocks can consist of n random samples of pq matched experimental units or a random sample of n experimental units who receive all pq treatment combinations. If matched subjects are used, the pq treatment combinations are randomly assigned to the units within a block with the restriction that each unit receives only one combination. If repeated measures are used, the order of presentation of the treatment combinations is randomized independently for each experimental unit. General considerations with respect to the use of matched subjects or repeated measures are discussed in Section 6.1 and apply to randomized block factorial designs. If the block size is quite large, other designs may be more appropriate. It is not always feasible to observe each subject under all pq combinations or to secure sets of pq matched subjects. Under these conditions a split-plot factorial design, described in Chapter 11, may be appropriate. It requires a subject to participate under only q of the pq treatment combinations. If sets of matched subjects are used, a set consists of q subjects instead of the pq subjects required in a type RBF-pq design.

FIGURE 9.4-1   Comparison of type RBF-23 and CRF-23 designs. In the type RBF-23 design, $s_1, s_2, \ldots, s_n$ denote n sets of $pq = 6$ matched subjects or if repeated measures are obtained n subjects who receive all pq treatment combinations. In the type CRF-23 design, $S_1, S_2, \ldots, S_n$ denote six random samples of n subjects, each.

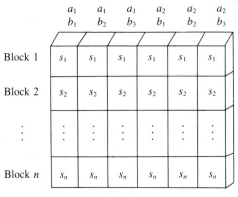

    (a) Type RBF-23 design   (b) Type CRF-23 design

Other designs that achieve a reduction in the number or treatment combinations to which subjects must be assigned are described in Chapters 12 and 13.

EXPERIMENTAL DESIGN MODEL FOR A TYPE RBF-pq DESIGN

A score Y_{ijk} in a two-treatment randomized block factorial design is a composite that is equal to the following terms in the experimental design model.

$$Y_{ijk} = \mu + (\mu_{\cdot j \cdot} - \mu) + (\mu_{\cdot \cdot k} - \mu) + (\mu_{\cdot jk} - \mu_{\cdot j \cdot} - \mu_{\cdot \cdot k} + \mu) + (\mu_{i \cdot \cdot} - \mu)$$
$$+ (Y_{ijk} - \mu_{i \cdot \cdot} - \mu_{\cdot jk} + \mu)$$

(9.4-1) $= \mu + \quad \alpha_j \quad + \quad \beta_k \quad + \quad (\alpha\beta)_{jk} \quad + \quad \pi_i$
$$+ \quad \epsilon_{ijk}$$

$$(i = 1, \ldots, n; j = 1, \ldots, p; k = 1, \ldots, q)$$

where μ = the overall mean

         α_j = the effect of treatment level j and is subject to the restriction $\sum_{j=1}^{p} \alpha_j = 0$

         β_k = the effect of treatment level k and is subject to the restriction $\sum_{k=1}^{q} \beta_k = 0$

         $(\alpha\beta)_{jk}$ = the joint effect of treatment levels j and k and is subject to the restrictions $\sum_{j=1}^{p} (\alpha\beta)_{jk} = 0$ and $\sum_{k=1}^{q} (\alpha\beta)_{jk} = 0$

$$\pi_i = \text{the effect of block } i \text{ that is } NID(0, \sigma_\pi^2)$$

$$\epsilon_{ijk} = \text{the experimental error that is } NID(0, \sigma_\epsilon^2); \epsilon_{ijk} \text{ is independent of } \pi_i.$$

Model (9.4-1) is called an additive model since all interactions involving blocks are assumed to equal zero. A nonadditive model will be examined later. Model (9.4-1) in which treatments are fixed but blocks are random (mixed model) is probably the most commonly used model for a type RBF-*pq* design. A residual mean square (*MSRES*), actually $MSAB \times BL$, is used to estimate experimental error since a within-cell mean square is not available. This involves somewhat restrictive assumptions. Before discussing these assumptions we will illustrate the layout and computational procedures for a type RBF-33 design.

9.5 COMPUTATIONAL EXAMPLE FOR A TYPE RBF-*pq* DESIGN

In Section 8.3 a completely randomized factorial design was used to evaluate the effects of (1) type of beat to which police officers are assigned during a human relations course and (2) length of the course on attitudes toward minority group members. For purposes of illustration we will modify the experiment slightly. Assume that the officers' pretest attitude scores were used to form five blocks such that the nine officers with the highest scores were assigned to block 1, the nine with the next highest scores, to block 2, and so on. The statistical hypotheses for this experiment are identical to those for the type CRF-33 design in Section 8.3. This randomized block factorial design also enables the experimenter to isolate the nuisance variable of differences in initial attitudes toward minority groups.

The layout of the type RBF-33 design and computational procedures are shown in Table 9.5-1. The analysis of variance is summarized in Table 9.5-2. According to this analysis, the *AB* interaction is significant. The procedures discussed in Section 8.6 can be used to gain a better understanding of the way the two variables interact.

A randomized block factorial design is more powerful than a completely randomized factorial design if the block effects in the former design are appreciably greater than zero. The reason for this is obvious if the two designs are compared with respect to the way the total variation is partitioned into the underlying sums of squares. This comparison is presented in Figure 9.5-1. In choosing between the two designs, one should consider not only their relative power but also the cost of forming blocks of matched subjects or obtaining repeated measures on the subjects versus the simpler procedure of randomly assigning treatment combinations to subjects.

TABLE 9.5-1 Computational Procedures for a Type RBF-33 Design

(i) Data and notation [Y_{ijk} denotes a score for an experimental unit in block i and treatment combination jk; $i = 1, \ldots, n$ blocks (s_i); $j = 1, \ldots, p$ levels of treatment A (a_j); $k = 1, \ldots, q$ levels of treatment B (b_k)]:

ABS Summary Table
Entry is Y_{ijk}

	a_1 b_1	a_1 b_2	a_1 b_3	a_2 b_1	a_2 b_2	a_2 b_3	a_3 b_1	a_3 b_2	a_3 b_3	$\sum\limits_{j=1}^{p}\sum\limits_{k=1}^{q} Y_{ijk}$
s_1	37	43	48	39	35	46	31	41	64	384
s_2	42	44	47	30	40	36	21	50	52	362
s_3	33	36	29	34	31	45	20	39	53	320
s_4	29	27	38	26	22	27	18	36	42	265
s_5	24	25	28	21	27	26	10	34	49	244

AB Summary Table

Entry is $\sum\limits_{i=1}^{n} Y_{ijk}$

	b_1	b_2	b_3	$\sum\limits_{i=1}^{n}\sum\limits_{k=1}^{q} Y_{ijk}$
	$n = 5$			
a_1	165	175	190	530
a_2	150	155	180	485
a_3	100	200	260	560
$\sum\limits_{i=1}^{n}\sum\limits_{j=1}^{p} Y_{ijk} =$	415	530	630	

(ii) Computational symbols:

$$\sum_{i=1}^{n}\sum_{j=1}^{p}\sum_{k=1}^{q} Y_{ijk} = 37 + 43 + \cdots + 49 = 1575.000$$

$$\frac{\left(\sum\limits_{i=1}^{n}\sum\limits_{j=1}^{p}\sum\limits_{k=1}^{q} Y_{ijk}\right)^2}{npq} = [Y] = \frac{(1575)^2}{(5)(3)(3)} = 55125.000$$

$$\sum_{i=1}^{n}\sum_{j=1}^{p}\sum_{k=1}^{q} Y_{ijk}^2 = [ABS] = (37)^2 + (43)^2 + \cdots + (49)^2 = 60345.000$$

$$\sum_{i=1}^{n} \frac{\left(\sum\limits_{j=1}^{p}\sum\limits_{k=1}^{q} Y_{ijk}\right)^2}{pq} = [S] = \frac{(384)^2}{(3)(3)} + \frac{(362)^2}{(3)(3)} + \cdots + \frac{(244)^2}{(3)(3)} = 56740.111$$

$$\sum_{j=1}^{p} \frac{\left(\sum\limits_{i=1}^{n}\sum\limits_{k=1}^{q} Y_{ijk}\right)^2}{nq} = [A] = \frac{(530)^2}{(5)(3)} + \frac{(485)^2}{(5)(3)} + \frac{(560)^2}{(5)(3)} = 55315.000$$

Table 9.5-1 (continued)

$$\sum_{k=1}^{q} \frac{\left(\sum_{i=1}^{n} \sum_{j=1}^{p} Y_{ijk}\right)^2}{np} = [B] = \frac{(415)^2}{(5)(3)} + \frac{(530)^2}{(5)(3)} + \frac{(630)^2}{(5)(3)} = 56668.333$$

$$\sum_{j=1}^{p} \sum_{k=1}^{q} \frac{\left(\sum_{i=1}^{n} Y_{ijk}\right)^2}{n} = [AB] = \frac{(165)^2}{5} + \frac{(175)^2}{5} + \cdots + \frac{(260)^2}{5} = 58095.000$$

(iii) Computational formulas:

$$SSTO = [ABS] - [Y] = 5220.000$$

$$SSBL = [S] - [Y] = 1615.111$$

$$SSTREAT = [AB] - [Y] = 2970.000$$

$$SSA = [A] - [Y] = 190.000$$

$$SSB = [B] - [Y] = 1543.333$$

$$SSAB = [AB] - [A] - [B] + [Y] = 1236.667$$

$$SSRES = [ABS] - [AB] - [S] + [Y] = 634.889$$

TABLE 9.5-2 ANOVA Table for Type RBF-33 Design

	Source	*SS*	*df*	*MS*	*F*	*E(MS)* *Model III*
1	Blocks	1615.111	$n - 1 =$ 4	403.778	[1/6] 20.35**	$\sigma_\epsilon^2 + pq\sigma_\pi^2$
2	Treatments	2970.000	$pq - 1 =$ 8	371.250		
3	A (Type of beat)	190.000	$p - 1 =$ 2	95.000	[3/6] 4.79*	$\sigma_\epsilon^2 + nq\sum_{j=1}^{p} \alpha_j^2/(p - 1)$
4	B (Training duration)	1543.333	$q - 1 =$ 2	771.666	[4/6] 38.89**	$\sigma_\epsilon^2 + np\sum_{k=1}^{q} \beta_k^2/(q - 1)$
5	AB	1236.667	$(p - 1)(q - 1) =$ 4	309.167	[5/6] 15.58**	$\sigma_\epsilon^2 + n\sum_{j=1}^{p} \sum_{k=1}^{q} (\alpha\beta)_{jk}^2/(p - 1)(q - 1)$
6	Residual	634.889	$(n - 1)(pq - 1) =$ 32	19.840		σ_ϵ^2
7	Total	5220.000	$npq - 1 =$ 44			

*$p < .05$

**$p < .01$

FIGURE 9.5-1   Schematic partition of the total sum of squares and degrees of freedom for type CRF-33 and RBF-33 designs. A shaded square identifies the sum of squares used in computing the error term (*MSWCELL* or *MSRES*) for the designs.

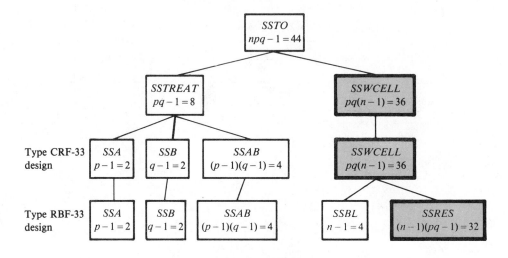

*9.6 EXPECTED VALUE OF MEAN SQUARES AND THE CIRCULARITY ASSUMPTION

The expected value of the mean squares for models I, II, and III for both the additive model and the nonadditive model, which is described later, can be obtained from Table 9.6-1. The terms $1 - (n/N)$, $1 - (p/P)$, and $1 - (q/Q)$ in the table become zero if the corresponding terms BL, A, and B represent fixed effects and one if they are random. If BL, A, and B represent fixed effects, the variance terms, σ_π^2, σ_α^2, and σ_β^2, should be replaced by $\Sigma \pi_i^2/(n - 1)$, $\Sigma \alpha_j^2/(p - 1)$, and $\Sigma \beta_k^2/(q - 1)$, respectively. If both A and B represent fixed effects, $\sigma_{\alpha\beta}^2$ should also be replaced by $\Sigma \Sigma (\alpha\beta)_{jk}^2/(p - 1)(q - 1)$.

The assumptions underlying F tests in a type RBF-pq design have been examined in detail by Rouanet and Lépine (1970). In summarizing their findings we will assume that the reader is familiar with the circularity assumption introduced in Section 6.4. As we have seen, a type RBF-pq design uses an interaction to estimate experimental error. Actually the residual mean square (*MSRES* or *MSAB* $\times$ *BL*) in Table 9.5-2 is a composite term that estimates σ_ϵ^2 plus $\sigma_{\alpha\pi}^2$, $\sigma_{\beta\pi}^2$, and $\sigma_{\alpha\beta\pi}^2$. If $\sigma_{\alpha\pi}^2$,

---

*   The latter portion of this section assumes a familiarity with the matrix operations in Appendix D and the material in Sections 6.4 and 6.9.

TABLE 9.6-1 Expected Value of Mean Squares for Type RBF-pq Design

Source	$E(MS)$ Additive Model
BL	$\sigma_\epsilon^2 + pq\sigma_\pi^2$
A	$\sigma_\epsilon^2 + n\left(1 - \dfrac{q}{Q}\right)\sigma_{\alpha\beta}^2 + nq\sigma_\alpha^2$
B	$\sigma_\epsilon^2 + n\left(1 - \dfrac{p}{P}\right)\sigma_{\alpha\beta}^2 + np\sigma_\beta^2$
AB	$\sigma_\epsilon^2 + n\sigma_{\alpha\beta}^2$
Res	σ_ϵ^2
	Nonadditive Model
BL	$\sigma_\epsilon^2 + \left(1 - \dfrac{p}{P}\right)\left(1 - \dfrac{q}{Q}\right)\sigma_{\alpha\beta\pi}^2 + p\left(1 - \dfrac{q}{Q}\right)\sigma_{\beta\pi}^2 + \left(1 - \dfrac{p}{P}\right)q\sigma_{\alpha\pi}^2 + pq\sigma_\pi^2$
A	$\sigma_\epsilon^2 + \left(1 - \dfrac{n}{N}\right)\left(1 - \dfrac{q}{Q}\right)\sigma_{\alpha\beta\pi}^2 + n\left(1 - \dfrac{q}{Q}\right)\sigma_{\alpha\beta}^2 + \left(1 - \dfrac{n}{N}\right)q\sigma_{\alpha\pi}^2 + nq\sigma_\alpha^2$
$A \times BL$	$\sigma_\epsilon^2 + \left(1 - \dfrac{q}{Q}\right)\sigma_{\alpha\beta\pi}^2 + q\sigma_{\alpha\pi}^2$
B	$\sigma_\epsilon^2 + \left(1 - \dfrac{n}{N}\right)\left(1 - \dfrac{p}{P}\right)\sigma_{\alpha\beta\pi}^2 + n\left(1 - \dfrac{p}{P}\right)\sigma_{\alpha\beta}^2 + \left(1 - \dfrac{n}{N}\right)p\sigma_{\beta\pi}^2 + np\sigma_\beta^2$
$B \times BL$	$\sigma_\epsilon^2 + \left(1 - \dfrac{p}{P}\right)\sigma_{\alpha\beta\pi}^2 + p\sigma_{\beta\pi}^2$
AB	$\sigma_\epsilon^2 + \left(1 - \dfrac{n}{N}\right)\sigma_{\alpha\beta\pi}^2 + n\sigma_{\alpha\beta}^2$
$A \times B \times BL$	$\sigma_\epsilon^2 + \sigma_{\alpha\beta\pi}^2$

$\sigma_{\beta\pi}^2$, and $\sigma_{\alpha\beta\pi}^2$ are not equal to zero, model equation (9.4-1) must be amended as follows.

(9.6-1) $\quad Y_{ijk} = \mu + \alpha_j + \beta_k + (\alpha\beta)_{jk} + \pi_i + (\alpha\pi)_{ji} + (\beta\pi)_{ki} + (\alpha\beta\pi)_{jki} + \epsilon_{ijk}$

If any of the interactions involving blocks is greater than zero, the model is referred to as a nonadditive model. Two preliminary tests may be helpful in choosing between the additive and nonadditive models. A nonsignificant test of H_0: $\sigma_{\alpha\pi}^2 = \sigma_{\beta\pi}^2 = \sigma_{\alpha\beta\pi}^2$, based on $MSA \times BL$, $MSB \times BL$, and $MSA \times B \times BL$, lends some credence to the hypothesis that the additive model applies. A nonsignificant test for nonadditivity also supports the hypothesis.

Two F statistics for testing treatment and interaction null hypotheses in a type RBF-pq design have been recommended in the literature. For convenience we will denote them by $F^\times$ and F^+. As we will see, a choice between $F^\times$ and F^+ should be based on the tenability of various circularity assumptions. The formulas for $F^\times$ and F^+ are as follows.

$$F^\times = \frac{MSA}{MSA \times BL} \qquad\qquad F^+ = \frac{MSA}{MSRES}$$

$$df = (p-1), (n-1)(p-1) \qquad df = (p-1), (n-1)(pq-1)$$

$$F^\times = \frac{MSB}{MSB \times BL} \qquad\qquad F^+ = \frac{MSB}{MSRES}$$

$$df = (q-1), (n-1)(q-1) \qquad df = (q-1), (n-1)(pq-1)$$

$$F^\times = \frac{MSAB}{MSA \times B \times BL} \qquad\qquad F^+ = \frac{MSAB}{MSRES}$$

$$df = (p-1)(q-1), \qquad\qquad df = (p-1)(q-1),$$
$$(n-1)(p-1)(q-1) \qquad\qquad (n-1)(pq-1)$$

Computational formulas for the denominators of the $F^\times$ statistics are as follows:

$$MSA \times BL = \frac{[AS] - [A] - [S] + [Y]}{(n-1)(p-1)} = 21.778$$

$$MSB \times BL = \frac{[BS] - [B] - [S] + [Y]}{(n-1)(q-1)} = 21.694$$

$$MSA \times B \times BL = \frac{[ABS] - [AB] - [AS] - [BS] + [A] + [B] + [S] - [Y]}{(n-1)(p-1)(q-1)}$$

$$= 17.944.$$

An F_{max} test for homogeneity of these three variances is not significant. This provides some support for using $MSRES$ to estimate σ_ϵ^2. From a casual inspection of the formula for $SSRES$, $[ABS] - [AB] - [S] + [Y]$, it is not evident that this formula is composed of three interaction formulas. This can be easily demonstrated by adding together the computational formulas for $SSA \times BL$, $SSB \times BL$, and $SSA \times B \times BL$.

$SSA \times BL$	$[AS] - [A] - [S] + [Y]$
$SSB \times BL$	$+ [BS] - [B] - [S] + [Y]$
$SSA \times B \times BL$	$+ [ABS] - [AB] - [AS] - [BS] + [A] + [B] + [S] + [Y]$
$SSAB \times BL$	$= [ABS] - [AB] - [S] + [Y]$

The reader who has followed the discussion concerning the general linear model in Chapter 5 and Sections 6.4, 6.9, and 8.13 is familiar with the matrix formulation of null hypotheses, for example,

$$\mathbf{C'\mu = 0}$$

where each row of $\mathbf{C'}$ sums to zero. If, in addition, the rows of $\mathbf{C'}$ are orthogonal and normalized, the coefficient matrix is said to be an orthonormal matrix and is denoted by $\mathbf{C}^{*\prime}$ (see Section 6.4). The hypothesis that all $\mu_{\cdot jk}$'s are equal is referred to as the *overall null hypothesis*. If it is true, all subhypotheses, $\mu_{\cdot 1\cdot} = \mu_{\cdot 2\cdot} = \cdots = \mu_{\cdot p\cdot}$, and so on, are also true. The statistic for testing the overall null hypothesis is $F^+ = MSTREAT/MSRES$. The necessary and sufficient condition for this F^+ statistic to be distributed as F with ν_1 and ν_2 degrees of freedom given that the overall null hypothesis is true is

$$\underset{(pq-1)\times pq}{\mathbf{C}^{*\prime}} \quad \underset{pq\times pq}{\boldsymbol{\Sigma}} \quad \underset{pq\times(pq-1)}{\mathbf{C}^{*}} = \lambda \quad \underset{(pq-1)\times(pq-1)}{\mathbf{I}}$$

where $\mathbf{C}^{*\prime}$ is an orthonormal matrix representing the overall null hypothesis that all $\mu_{\cdot jk}$'s are equal and $\boldsymbol{\Sigma}$ is the population variance-covariance matrix.* If $\boldsymbol{\Sigma}$ satisfies the circularity assumption for the overall null hypothesis, $\mathbf{C}^{*\prime}\boldsymbol{\mu} = \mathbf{0}$, it also satisfies the circularity assumption for all subhypotheses.** This means that $F^+ = MSA/MSRES$, $F^+ = MSB/MSRES$, . . . , and $F^\times = MSAB/MSA\times B\times BL$ are all distributed as F. The use of F^+ requires the strongest assumption, overall circularity, but it provides the most powerful tests of false null hypotheses. If overall circularity is not satisfied, it may still be true, for example, that

$$\underset{(p-1)\times pq}{\mathbf{C}^{*\prime}} \quad \underset{pq\times pq}{\boldsymbol{\Sigma}} \quad \underset{pq\times(p-1)}{\mathbf{C}^{*}} = \lambda \quad \underset{(p-1)\times(p-1)}{\mathbf{I}}$$

where $\mathbf{C}^{*\prime}$ is the coefficient matrix for testing the subhypothesis that all $\mu_{\cdot j\cdot}$'s are equal. When this condition, referred to as *local circularity,* is satisfied, $F^\times = MSA/MSA\times BL$ is distributed as F with ν_1 and ν_2 degrees of freedom. Tests using F^+ have more degrees of freedom than those using $F^\times$ and consequently are more powerful. Thus, F^+ is the preferred test statistic when $\boldsymbol{\Sigma}$ satisfies the circularity property for the overall null hypothesis. $F^\times$ can be used when $\boldsymbol{\Sigma}$ satisfies local circularity. If $\boldsymbol{\Sigma}$ does not satisfy local circularity for some subhypothesis, the three-step testing strategy in Section 6.4 can be used.

*9.7 ANALYSIS OF RANDOMIZED BLOCK DESIGN WITH MISSING OBSERVATIONS

A restricted full rank experimental design model can be used to compute sums of squares for a type RBF-pq design when one or more observations are missing. The

---

*   Mauchley's test for sphericity described in Section 6.4 uses the sample variance-covariance matrix $\hat{\boldsymbol{\Sigma}}$ to determine the tenability of the circularity assumption.

**   For a type RBF-pq design, this implies that $\sigma^2_{\alpha\pi} = \sigma^2_{\beta\pi} = \sigma^2_{\alpha\beta\pi}$.

☆   This section assumes a familiarity with the matrix operations in Appendix D and the full rank model that is discussed in Sections 6.9 and 8.13. The reader who is interested only in the traditional approach to analysis of variance can, without loss of continuity, omit this section.

restricted full rank model corresponding to model (9.4-1) is

(9.7-1) $Y_{ijk} = \mu_{ijk} + \epsilon_{ijk}$ $(i = 1, \ldots, n; j = 1, \ldots, p; k = 1, \ldots, q)$

where ϵ_{ijk} is $NID(0, \sigma_\epsilon^2)$ and μ_{ijk} is subject to the restrictions that $\mu_{i(jk)} - \mu_{i'(jk)} - \mu_{i(jk)'} + \mu_{i(jk)'} = 0$ for all i, i', (jk) and $(jk)'$. These restrictions specify that all block $\times$ ab treatment combination effects equal zero. We will assume that treatments A and B represent fixed effects and that blocks represent random effects.

The analysis procedures for this restricted full rank experimental design model will be illustrated for the data in Table 9.7-1 where Y_{221} is missing. We will assume that the loss of observation Y_{221} is not related to the nature of the a_2b_1 treatment combination. Using the procedure described in Section 6.9, the $s = N - n - pq + 1 = 2$ restrictions of the form $\mu_{i(jk)} - \mu_{i'(jk)} - \mu_{i(jk)'} + \mu_{i'(jk)'} = 0$ can be constructed from the crossed lines in the $\hat{\mu}_{ijk}$ Table of Table 9.7-1. The two restrictions are

$$\mu_{111} - \mu_{211} - \mu_{112} + \mu_{212} = 0$$
$$\mu_{112} - \mu_{212} - \mu_{122} + \mu_{222} = 0.$$

In matrix notation these restrictions are

TABLE 9.7-1   Type RBF-22 Design With Y_{221} Missing

ABS Summary Table
Entry is Y_{ijk}

	a_1 b_1	a_1 b_2	a_2 b_1	a_2 b_2
s_1	7	5	9	8
s_2	8	6		6

ABS Summary Table
Entry is $\hat{\mu}_{ijk}$

	a_1 b_1	a_1 b_2	a_2 b_1	a_2 b_2
s_1	$\hat{\mu}_{111}$	$\hat{\mu}_{112}$	$\hat{\mu}_{121}$	$\hat{\mu}_{122}$
s_2	$\hat{\mu}_{211}$	$\hat{\mu}_{212}$		$\hat{\mu}_{222}$

AB Summary Table
Entry is $\hat{\mu}_{\cdot jk}$

	b_1	b_2	$\sum_{k=1}^{q} \hat{\mu}_{\cdot jk}/q$
a_1	$n_{11} = 2$ 7.5	$n_{12} = 2$ 5.5	$\hat{\mu}_{\cdot 1 \cdot} = 6.5$
a_2	$n_{21} = 1$ 9	$n_{22} = 2$ 7	$\hat{\mu}_{\cdot 2 \cdot} = 8.0$

$\sum_{j=1}^{p} \hat{\mu}_{\cdot jk}/p = \hat{\mu}_{\cdot \cdot 1}$ $\hat{\mu}_{\cdot \cdot 2}$
$= 8.25$ 6.25

$$\underset{s\times h}{\mathbf{R}'} \ \underset{h\times 1}{\boldsymbol{\mu}} = \underset{s\times 1}{\boldsymbol{\theta}}$$

where $\quad \mathbf{R}' = \begin{bmatrix} 1 & -1 & -1 & 1 & 0 & 0 & 0 \\ 0 & 0 & 1 & -1 & 0 & -1 & 1 \end{bmatrix}$

$\boldsymbol{\mu}' = [\mu_{111} \quad \mu_{211} \quad \mu_{112} \quad \mu_{212} \quad \mu_{121} \quad \mu_{122} \quad \mu_{222}]$

and $\boldsymbol{\theta}$ is a vector of zeros.

An experimenter would most likely be interested in testing the following null hypotheses.

Treatment A $\quad$ H_0: $\quad \mu_{\cdot 1 \cdot} - \mu_{\cdot 2 \cdot} = 0 \quad$ or $\quad$ H_0: $\quad \underset{(p-1)\times h}{\mathbf{C}'_A} \ \underset{h\times 1}{\boldsymbol{\mu}} = \underset{(p-1)\times 1}{\mathbf{0}_A}$

where $\mathbf{C}'_A = [\frac{1}{4} \quad \frac{1}{4} \quad \frac{1}{4} \quad \frac{1}{4} \quad -\frac{1}{2} \quad -\frac{1}{4} \quad -\frac{1}{4}]$. The coefficients, c_{ijk}, are given by $\pm 1/n_{jk}q$.

Treatment B $\quad$ H_0: $\quad \mu_{\cdot\cdot 1} - \mu_{\cdot\cdot 2} = 0 \quad$ or $\quad$ H_0: $\quad \underset{(q-1)\times h}{\mathbf{C}'_B} \ \underset{h\times 1}{\boldsymbol{\mu}} = \underset{(q-1)\times 1}{\mathbf{0}_B}$

where $\mathbf{C}'_B = [\frac{1}{4} \quad \frac{1}{4} \quad -\frac{1}{4} \quad -\frac{1}{4} \quad \frac{1}{2} \quad -\frac{1}{4} \quad -\frac{1}{4}]$. The coefficients, c_{ijk}, are given by $\pm 1/n_{jk}p$.

AB Interaction $\quad$ H_0: $\quad \mu_{\cdot 11} - \mu_{\cdot 21} - \mu_{\cdot 12} + \mu_{\cdot 22} = 0$

or $\quad$ H_0: $\quad \underset{(p-1)(q-1)\times h}{\mathbf{C}'_{AB}} \ \underset{h\times 1}{\boldsymbol{\mu}} = \underset{(p-1)(q-1)\times 1}{\mathbf{0}_{AB}}$

where $\mathbf{C}'_{AB} = [\frac{1}{2} \quad \frac{1}{2} \quad -\frac{1}{2} \quad -\frac{1}{2} \quad -1 \quad \frac{1}{2} \quad \frac{1}{2}]$. The coefficients are given by $\pm 1/n_{jk}$.

We want to test $\mathbf{C}'_A\boldsymbol{\mu} = \mathbf{0}_A$, $\mathbf{C}'_B\boldsymbol{\mu} = \mathbf{0}_B$, and $\mathbf{C}'_{AB}\boldsymbol{\mu} = \mathbf{0}_{AB}$ subject to the restrictions $\mathbf{R}'\boldsymbol{\mu} = \boldsymbol{\theta}$. To do this we form the augmented matrices

$$\underset{(s+p-1)\times h}{\mathbf{Q}'_A} = \begin{bmatrix} \mathbf{R}' \\ \mathbf{C}'_A \end{bmatrix} \qquad \underset{(s+q-1)\times h}{\mathbf{Q}'_B} = \begin{bmatrix} \mathbf{R}' \\ \mathbf{C}'_B \end{bmatrix} \qquad \underset{[s+(p-1)(q-1)]\times h}{\mathbf{Q}'_{AB}} = \begin{bmatrix} \mathbf{R}' \\ \mathbf{C}'_{AB} \end{bmatrix}$$

$$\underset{(s+p-1)\times 1}{\boldsymbol{\eta}_A} = \begin{bmatrix} \boldsymbol{\theta} \\ \mathbf{0}_A \end{bmatrix} \qquad \underset{(s+q-1)\times 1}{\boldsymbol{\eta}_B} = \begin{bmatrix} \boldsymbol{\theta} \\ \mathbf{0}_B \end{bmatrix} \qquad \underset{[s+(p-1)(q-1)]\times 1}{\boldsymbol{\eta}_{AB}} = \begin{bmatrix} \boldsymbol{\theta} \\ \mathbf{0}_{AB} \end{bmatrix}.$$

The matrix equation $\mathbf{Q}'\boldsymbol{\mu} = \boldsymbol{\eta}$ combines the hypothesis that differences between treatment means are equal to zero with the restriction that all block $\times$ ab treatment combination effects equal zero. Procedures for computing sums of squares and F statistics are illustrated in Table 9.7-2. Also shown in Table 9.7-2 are procedures for computing sample treatment means, $\hat{\boldsymbol{\mu}}_R$, subject to the restrictions $\mathbf{R}'\boldsymbol{\mu} = \boldsymbol{\theta}$. One statistic for testing hypotheses about population means is $\hat{\psi}_i/\hat{\sigma}_{\psi_i}$, where

$$\hat{\psi}_i = \underset{1\times h}{\mathbf{c}'_i} \ \underset{h\times 1}{\hat{\boldsymbol{\mu}}_R}, \ \mathbf{c}'_i \text{ is a vector that defines the } i\text{th contrast}$$

$$\hat{\sigma}_{\psi_i} = \sqrt{MSRES(\mathbf{c}'_i \ \mathbf{A} \ \mathbf{c}_i)}$$

$$\underset{h\times h}{\mathbf{A}} = [\ \underset{h\times h}{\mathbf{I}} \ - \ \underset{h\times h}{\mathbf{R}(\mathbf{R}'\mathbf{R})^{-1} \ \mathbf{R}'}\ ].$$

TABLE 9.7-2 Computational Procedures for a Type RBF-22 Design Using a Restricted Full Rank Experimental Design Model (Y_{221} missing)

(i) Data and matrix formulas for computing sums of squares ($\mathbf{R}'$ and the $\mathbf{Q}''$s are defined in the text; $N = 7$, $h = 7$, $s = 2$):

$$
\begin{array}{c}
\underset{N\times1}{\mathbf{y}} \\
\underset{h\times1}{\hat{\boldsymbol{\mu}}} \quad \underset{h\times1}{[\mathbf{R}(\mathbf{R}'\mathbf{R})^{-1}\mathbf{R}'\hat{\boldsymbol{\mu}}]} \quad \underset{h\times1}{\hat{\boldsymbol{\mu}}_R}
\end{array}
$$

$$
\begin{array}{c}
a_1b_1 \\ \\ a_1b_2 \\ \\ a_2b_1 \\ \\ a_2b_2
\end{array}
\begin{bmatrix} 7 \\ 8 \\ \hline 5 \\ 6 \\ \hline 9 \\ \hline 8 \\ 6 \end{bmatrix}
-
\begin{bmatrix} -0.5 \\ 0.5 \\ \hline -0.5 \\ 0.5 \\ \hline 0 \\ \hline 1.0 \\ -1.0 \end{bmatrix}
=
\begin{bmatrix} 7.5 \\ 7.5 \\ \hline 5.5 \\ 5.5 \\ \hline 9.0 \\ \hline 7.0 \\ 7.0 \end{bmatrix}
$$

$$\sum_{i=1}^{N} Y_i = 49$$

$$
\underset{1\times N}{\mathbf{y}'} \quad \underset{N\times1}{\mathbf{y}} \quad \underset{1\times1}{(\mathbf{y}'\mathbf{y})}
$$

$$
\begin{bmatrix} 7 & 8 & \cdots & 6 \end{bmatrix}
\begin{bmatrix} 7 \\ 8 \\ \vdots \\ 6 \end{bmatrix} = 355
$$

$$
SSTO = \mathbf{y}'\mathbf{y} - (\Sigma Y_i)^2/N
$$

$$
= 355 - 343 = 12.00
$$

$$
SSRES = \underset{1\times s}{(\mathbf{R}'\hat{\boldsymbol{\mu}})'} \; \underset{s\times s}{(\mathbf{R}'\mathbf{R})^{-1}} \; \underset{s\times1}{(\mathbf{R}'\hat{\boldsymbol{\mu}})} = 3.0
$$

$$
SSA = \underset{1\times(s+p-1)}{(\mathbf{Q}_A'\hat{\boldsymbol{\mu}})'} \; \underset{(s+p-1)\times(s+p-1)}{(\mathbf{Q}_A'\mathbf{Q}_A)^{-1}} \; \underset{(s+p-1)\times1}{(\mathbf{Q}_A'\hat{\boldsymbol{\mu}})} - SSRES = 6.6 - 3.0 = 3.6
$$

$$
SSB = \underset{1\times(s+q-1)}{(\mathbf{Q}_B'\hat{\boldsymbol{\mu}})'} \; \underset{(s+q-1)\times(s+q-1)}{(\mathbf{Q}_B'\mathbf{Q}_B)^{-1}} \; \underset{(s+q-1)\times1}{(\mathbf{Q}_B'\hat{\boldsymbol{\mu}})} - SSRES = 9.4 - 3.0 = 6.4
$$

$$
SSAB = \underset{1\times[s+(p-1)(q-1)]}{(\mathbf{Q}_{AB}'\hat{\boldsymbol{\mu}})'} \; \underset{[s+(p-1)(q-1)]\times[s+(p-1)(q-1)]}{(\mathbf{Q}_{AB}'\mathbf{Q}_{AB})^{-1}} \; \underset{[s+(p-1)(q-1)]\times1}{(\mathbf{Q}_{AB}'\hat{\boldsymbol{\mu}})} - SSRES
$$

$$
= 3.0 - 3.0 = 0
$$

(ii) Computation of mean squares:

$$MSA = SSA/(p - 1) = 3.6/(2 - 1) = 3.6$$

$$MSB = SSB/(q - 1) = 6.4/(2 - 1) = 6.4$$

$$MSAB = SSAB/(p-1)(q-1) = 0/(2 - 1)(2 - 1) = 0$$

$$MSRES = SSRES/(N - n - pq + 1) = 3.0/[7 - 2 - (2)(2) + 1] = 1.5$$

(iii) Computation of F ratios:

$$F = MSA/MSRES = 3.6/1.5 = 2.40$$

$$F = MSB/MSRES = 6.4/1.5 = 4.27$$

$$F = MSAB/MSRES = 0/1.5 = 0$$

$$F_{.05;1,2} = 18.5$$

9.8 MINIMIZING TIME AND LOCATION EFFECTS BY THE USE OF A RANDOMIZED BLOCK FACTORIAL DESIGN

A useful application of a type RBF-pq design may not be apparent to the reader. Situations arise in behavioral and educational research in which all the pq treatment combinations can be administered to only one set of pq subjects at a time. At a later time or possibly at a different location, another set of subjects receives the same treatment combinations. In this type of experiment, the subjects in a set are matched in the sense that they receive the treatment combinations at the same time or at the same place. Differences among the blocks (sets) of subjects represent a time or place variable in addition to any other conditions that are not constant over the different treatment administrations. The use of a randomized block factorial design permits the isolation of these nuisance variables. An alternative analysis procedure, which is generally less desirable, is to pool subjects who receive the same treatment combinations and compute a within-cell error term. The data are analyzed as a completely randomized factorial design. The disadvantage of this procedure is that any variation due to the nuisance variables inflates the estimate of the experimental error. The two designs require the same number of subjects, but the randomized block factorial design permits the isolation of nuisance variables and is more congruent with the actual conduct of the experiment.

9.9 REVIEW EXERCISES

1. [9.1] Enumerate the treatments and interactions that can be tested in a

    †a) Type CRF-2332 design   b) Type CRF-34352 design.

2. [9.1] In a type CRF-354334 design, how many interactions involve

    †a) Two treatments?   †b) Three treatments?   c) Four treatments?

    d) Five treatments?   e) Six treatments?

3. [9.2] An experiment was performed to determine the relative effectiveness of two procedures for teaching an intermediate level biology laboratory: the conventional method in which students worked in pairs and performed each experiment by following a laboratory manual (treatment level a_1) and a second method in which the instructor performed the experiment while the students in pairs watched and recorded the results (treatment level a_2). Also investigated were (1) the effects of assigning members of the same sex or opposite sex (levels b_1 and b_2, respectively) to a lab pair and (2) achievement level in the introductory biology

laboratory (c_1 was medium achievement, c_2 was high achievement). Preenrollment data were used to select 40 male and 40 female students who scored between 91 and 100 on the introductory biology laboratory final and a like number who scored between 70 and 79. The students in each achievement category were assigned to a lab pair so that half the pairs contained members of the same sex and half, members of the opposite sex. The two teaching methods were randomly assigned to the lab pairs, subject to the restriction that each of the eight treatment combinations was assigned to ten lab pairs. The dependent variable was the mean performance of the members of a lab pair on the intermediate biology laboratory final examination. The following data were obtained.

a_1	a_1	a_1	a_1	a_2	a_2	a_2	a_2
b_1	b_1	b_2	b_2	b_1	b_1	b_2	b_2
c_1	c_2	c_1	c_2	c_1	c_2	c_1	c_2
93	99	53	81	96	78	47	83
91	98	79	99	82	92	49	59
84	96	77	98	84	90	56	67
79	89	68	97	77	83	56	60
76	86	66	90	80	69	69	69
79	89	64	100	74	66	58	71
74	84	66	90	72	73	60	71
72	97	55	87	77	70	58	73
60	75	62	80	65	78	65	75
82	77	70	78	63	81	62	82

†**a)** Test the null hypothesis for treatments and interactions; let $\alpha = .05$.

†**b)** Graph the interactions that are significant in part (a); interpret the interactions.

c) Prepare a "results and discussion section" for the *Journal of Educational Psychology*.

4.   [9.3] For a type CRF-343225 design, write the computational formulas using abbreviated notation for computing

†**a)** *SSABCD*   †**b)** *SSABCDE*

c) *SSCDE*   **d)** *SSBCDEF*.

†**5.**   [9.3] Write the **C′** matrices for a type CRF-232 design for testing the null hypotheses; let
$\mu' = [\mu_{111}\ \mu_{112}\ \mu_{121}\ \mu_{122}\ \mu_{131}\ \mu_{132}\ \mu_{211}\ \mu_{212}\ \mu_{221}\ \mu_{222}\ \mu_{231}\ \mu_{232}]$.

6.   [9.3] Write the **C′** matrices for a type CRF-233 design for testing the interaction null hypotheses; let $\mu' = [\mu_{111}\ \mu_{112}\ \mu_{113}\ \mu_{121}\ \mu_{122}\ \mu_{123}\ \mu_{131}\ \mu_{132}\ \mu_{133}\ \mu_{211}\ \mu_{212}\ \mu_{213} \cdots \mu_{233}]$.

†**7.**   Suppose that the following data have been obtained in a type RBF-23 design. Assume that the treatments represent fixed effects; blocks are random effects.

	a_1 b_1	a_1 b_2	a_1 b_3	a_2 b_1	a_2 b_2	a_2 b_3
s_1	4	3	8	3	3	5
s_2	2	4	6	2	1	2

a) [9.5] Use the traditional sum of squares approach to test the null hypotheses; let $\alpha = .05$.

b) Use Tukey's statistic to test all two-mean contrasts among the treatment B means. (*Note:* $\hat{\sigma}_{\bar{Y}_{.k}} = \sqrt{MSRES/np}$.)

c) [9.7] Use the full rank experimental design model approach to test the null hypotheses; let $\alpha = .05$.

8. The effect of vigorous physical exercise on performance of a line-matching task was investigated. Male college students were shown four vertical lines projected on a screen directly in front of them and at eye level while they exercised on a motor-driven treadmill. On the left of the screen there was a single line; on the right there were three lines numbered 1, 2, and 3. The subject was instructed to call out as quickly as possible the number of the line on the right that corresponded in length to the one on the left. Treatment A was the speed of the treadmill: $a_1 = 2.5$ mph, $a_2 = 3.4$ mph, and $a_3 = 4.3$ mph. Treatment B was the grade of the treadmill: $b_1 = 12\%$ grade and $b_2 = 18\%$ grade. The order in which the treatment combinations was administered to subjects was randomized independently for each subject. The dependent variable was the number of correct line matches during the last three minutes of each session. Assume that treatments represent fixed effects and blocks represent random effects. The following data were obtained.

	a_1 b_1	a_1 b_2	a_2 b_1	a_2 b_2	a_3 b_1	a_3 b_2
s_1	42	44	46	54	50	54
s_2	44	49	50	52	52	50
s_3	36	42	38	48	44	47
s_4	40	44	42	46	48	53
s_5	38	41	44	50	46	46

a) [9.5] Use the traditional sum of squares approach to test the null hypotheses for this type RBF-32 design; let $\alpha = .05$. Use *MSRES* as the error term.

b) [8.8] Compute $\hat{\rho}_I = \hat{\sigma}_\pi^2/(\hat{\sigma}_\epsilon^2 + \Sigma \hat{\alpha}_j^2/p + \Sigma \hat{\beta}_k^2/q + \Sigma\Sigma (\widehat{\alpha\beta})_{jk}^2/pq + \hat{\sigma}^2\pi)$. Was the use of a blocking variable effective? Explain.

c) [9.6] Would the use of $F^\times$ instead of F^+ ratios have led to different conclusions?

d) Use Tukey's statistic to determine which population means differ for treatment A. (*Note:* $\hat{\sigma}_{\bar{Y}_{j.}} = \sqrt{MSRES/nq}$.)

9. [9.6] In the context of a type RBF-33 design, use null hypotheses to distinguish overall circularity from local circularity.

†10. [9.7] Exercise 8 described an experiment in which subjects performed a line matching task while on a motor-driven treadmill. Assume that $Y_{211} = 44$ is missing.

a) Write a full rank experimental design model for this type RBF-32 design.

b) Write the $\mathbf{R}'$ matrix and the $\mathbf{C}'$ matrices for blocks, treatments A and B, and the AB interaction.

c) Use the full rank experimental design model approach to test the associated null hypotheses in part (b). (This problem involves the inversion of large matrices. A computer with a matrix inversion program should be used.)

10 HIERARCHICAL DESIGNS

10.1 INTRODUCTION TO HIERARCHICAL DESIGNS

A *hierarchical experiment* is one in which the levels of at least one treatment are nested within those of another treatment and all other treatments are completely crossed. If each level of treatment B appears with only *one* level of treatment A, B is said to be *nested* within A. This is denoted by $B(A)$. The distinction between nested and crossed treatments is illustrated in Figure 10.1-1.

Experimental designs with one or more nested treatments are particularly well suited to research in the behavioral sciences, education, and industry. To illustrate, consider an experiment to evaluate two types of programmed instruction materials (treatment levels a_1 and a_2) in four sixth-grade classes (treatment levels $b_1 \cdots b_4$). Each type of programmed material is randomly assigned to two classrooms. For administrative reasons, all children in a particular room must use the same type of programmed material. We will assume that each classroom contains the same number of children. This hierarchical experiment corresponds to the design illustrated in Figure 10.1-1(a). Each classroom b_k appears with only one level of programmed

FIGURE 10.1-1 Comparison of designs with nested and crossed treatments. The presence of S_i in a cell indicates that subjects are assigned to the associated treatment combination. In experiment (a), treatment B is nested within treatment A because b_1 and b_2 appear only with a_1 while b_3 and b_4 appear only with a_2.

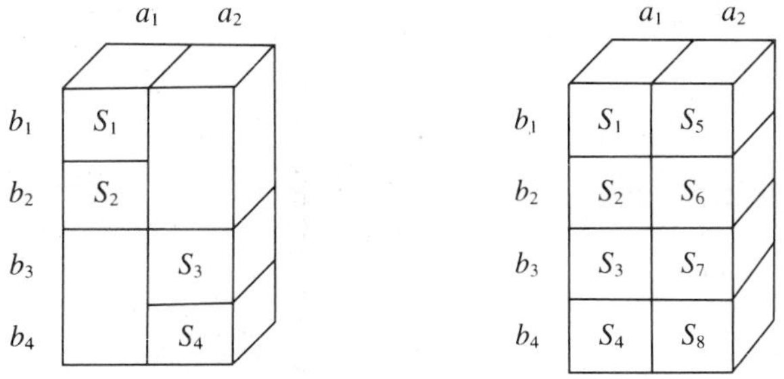

(a) Treatment B is nested within treatment A.

(b) Treatments A and B are crossed.

instruction. Thus treatment B is nested within A.

Nested treatments often occur in animal research where it is necessary to administer the same treatment level to all animals within the same cage or housing compound. Consider an experiment to investigate the effects of positive and negative ionization of air molecules (treatment levels a_1 and a_2, respectively) on the activity level of rats. The animals are radiated continuously for a three-week period, after which their activity level is measured in an open field situation. Sixteen rats are housed in four cages (treatment levels $b_1 \cdot \cdot \cdot b_4$) with four rats per cage. Each cage is equipped with ionizing equipment for producing the required condition within a cage. The cages are randomly assigned to the ionization conditions, two to each condition. In this example, cages are nested within the ionization conditions. This experiment also corresponds to the design shown in Figure 10.1-1(a). We assume that the activity level of a rat in the open field situation reflects the effects of (1) the ionization condition received, treatment A; (2) the cage in which the rat is housed, treatment $B(A)$; and (3) idiosyncratic characteristics of the rat, experimental error. This can be expressed more formally by means of the model equation

$$Y_{ijk} = \mu + \alpha_j + \beta_{k(j)} + \epsilon_{i(jk)} \quad (i = 1, \ldots, n; j = 1, \ldots, p; k = 1, \ldots, q)$$

where the symbol $\beta_{k(j)}$ indicates that the kth level of treatment B is nested within the jth level of A. Note that an $A \times B$ interaction term does not appear in the model.

In this and the programmed instruction experiment described earler, treatment $B(A)$ is really a nuisance variable. That is, the variable of classrooms or cages is included in the design and model equation because the experimenter suspects that it might affect the dependent variable and not because of an interest in the variable per se. One could argue that if the cages in the ionization experiment are identical this condition is constant for all subjects. This argument ignores such variables as differences in the location of the cages within a room and possible differences in dispersion of air ions within the cages as well as differences in ambient lighting, temperature, and humidity. Also ignored are differences in the social environment in the cages due to the presence of dominant or neurotic rats in certain cages, undetected infectious diseases, and so on. The use of a hierarchical design enables an experimenter to isolate the nuisance variable of cages, which might contribute significantly to the total variation in the rat's activity level.

A common error in the analysis of hierarchical experiments is to ignore the nuisance variable and treat the design as if it were a type CR-p design with $nq_{(j)}$ subjects in each level of treatment A. Here $q_{(j)}$ denotes the number of levels of treatment B that is nested within the jth level of treatment A. This may result in a biased test of treatment A. A correct analysis takes into account the effects of the nested treatment and makes the analysis congruent with the way the experiment was actually carried out.

Hierarchical designs constructed from completely randomized designs are appropriate for experiments that meet, in addition to the assumptions of a type CR-p design described in Chapter 4, the following conditions.

1.   Two or more treatments with each treatment having two or more levels.

2.   The levels of at least one treatment are nested within those of another treatment; all other treatments are completely crossed.

3.   Random assignment of the levels (combinations) of the nested treatment(s) to the unnested treatment (treatment combinations) and random assignment of the treatment combinations to the experimental units.

Hierarchical designs are often used in research situations where the third requirement cannot be met. For example, the programmed instruction experiment might be done in four schools using two sixth-grade classrooms in each school, a total of eight classrooms. In this design, the four schools can be randomly assigned to the two types of instruction materials but, obviously, classrooms in one school cannot be assigned to another school. Also, federal integration guidelines would probably preclude assigning the treatment combinations randomly to the students. Most observational studies by their very nature preclude the kind of randomization described in the third requirement. Campbell and Stanley (1966, 34) refer to designs in which the experimenter has limited control over randomization as *quasi-experimental designs*. The interpretation of the results of such experiments presents a real challenge. It seems that the difficulty of interpreting the outcome of research varies inversely with the degree of control an experimenter is able to exercise over randomization. Certainly observational studies are among the most difficult to interpret. Experimenters will find Campbell and Stanley's discussion and that in Cook and Campbell (1979) most helpful.

TYPES OF HIERARCHICAL DESIGNS

In this chapter we will describe hierarchical experiments based on completely randomized and randomized block designs. Depending on the building block design that is used, a two-treatment hierarchical design is denoted by CRH-$pq(A)$ or RBH-$pq(A)$, where $q(A)$ indicates that the q levels of treatment B are nested within treatment A. The design for the programmed instruction and ionization experiments described earlier is denoted by CRH-24(A). A completely randomized hierarchical design with three treatments can take any of the following forms: CRH-$pq(A)r(AB)$, CRPH-$pq(A)r$, CRPH-$pqr(A)$, CRPH-$pqr(B)$, CRPH-$pq(A)r(A)$, and CRPH-$pqr(AB)$. These designs are discussed in Section 10.6 so we will limit our comments here to the treatment designation scheme. Treatments $B(A)$ and $C(AB)$ in a type CRH-$pq(A)r(AB)$ design are completely nested: B is nested within A, and C is nested within A and B. The other designs involve partial nesting, which is indicated by the use of P in their designation, and are referred to as *completely randomized partial hierarchical designs*. In the type CRPH-$pq(A)r$ design, for example, treatment B is nested within A, but treatment C is crossed with A and $B(A)$. In the type CRPH-$pqr(AB)$ design, treatment C is nested within A and B, but treatments A and B are crossed.

Hierarchical designs are balanced if they have an equal number of experimental units within each treatment combination and an equal number of levels of the nested treatment within each level of the other treatment; otherwise they are unbalanced. The analysis and the interpretation of results are more complex for unbalanced designs than for balanced designs.* We will describe the analysis of balanced designs first using the traditional sum of squares approach. The analysis of unbalanced designs using the full rank experimental design model approach is discussed in Section 10.8.

10.2 COMPUTATIONAL EXAMPLE FOR TYPE CRH-$pq(A)$ DESIGN

Assume that the ionization experiment described earlier has been performed. However, instead of four cages, eight were used. The cages, treatment $B(A)$, were randomly divided into two subsamples of four each and the ionization conditions, positive (a_1) and negative (a_2), assigned randomly to the subsamples. Thirty-two rats were randomly divided into eight subsamples of four rats each and the $pq_{(j)} = 2(4) = 8$ treatment combinations assigned randomly to the subsamples. Here $q_{(j)}$ denotes the number of levels of treatment B that is nested within the jth level of treatment A. For a balanced type CRH-$pq(A)$ design, $q_{(j)}$ is constant for all j. The treatment effects for the ionization conditions are fixed, those for cages are assumed to be random. This is a common assumption for a nuisance variable. We are not interested in the eight cages used in the experiment, but rather in the population of cages from which the eight would have been a random sample if random sampling had been used.

---

*   For a discussion of the problems see Gill (1978, 185–199), Gower (1962), and Snedecor (1956, 271).

The research hypotheses leading to the experiment can be evaluated by testing the following null hypotheses.

$$H_0: \quad \alpha_j = 0 \quad \text{for all } j$$
$$H_1: \quad \alpha_j \neq 0 \quad \text{for some } j$$

$$H_0: \quad \sigma_\beta^2 = 0$$
$$H_1: \quad \sigma_\beta^2 \neq 0$$

If the latter null hypothesis is rejected, we conclude that $\sigma_\beta^2 \neq 0$ for a_1, or $\sigma_\beta^2 \neq 0$ for a_2, or both. The level of significance adopted is .05. The computational procedures are illustrated in Table 10.2-1. The results are summarized in Table 10.2-2. It is apparent that both null hypotheses can be rejected. From an examination of the data we know that rats exposed to negative air ions are more active than those exposed to positive ions. The finding that there are differences in activity level associated with the various cages is of little substantive interest since cages represent a nuisance variable. The significant F ratio for cages vindicates the decision to include this source of variation in the design.

TABLE 10.2-1 Computational Procedures for a Type CRH-28(A) Design

---

(i) Data and notation [Y_{ijk} denotes a score for experimental unit i in treatment combination jk; $i = 1$, . . . , n experimental units (s_i); $j = 1$, . . . , p levels of treatment A (a_j); $k = 1$, . . . , q levels of treatment B (b_k)]:

ABS Summary Table
Entry is Y_{ijk}

a_1	a_1	a_1	a_1	a_2	a_2	a_2	a_2
b_1	b_2	b_3	b_4	b_5	b_6	b_7	b_8
3	1	5	2	7	4	7	10
6	2	6	3	8	5	8	10
3	2	5	4	7	4	9	9
3	2	6	3	6	3	8	11

AB Summary Table

Entry is $\sum\limits_{i=1}^{n} Y_{ijk}$

	b_1	b_2	b_3	b_4	b_5	b_6	b_7	b_8	$\sum\limits_{i=1}^{n}\sum\limits_{k=1}^{q_{(j)}} Y_{ijk}$
a_1	$n = 4$ 15	7	22	12					56
a_2					28	16	32	40	116

Table 10.2-1 (continued)

(ii) Computational symbols:

$$\sum_{i=1}^{n}\sum_{j=1}^{p}\sum_{k=1}^{q_{(j)}} Y_{ijk} = 3 + 1 + \cdots + 11 = 172.000$$

$$\frac{\left(\sum_{i=1}^{n}\sum_{j=1}^{p}\sum_{k=1}^{q_{(j)}} Y_{ijk}\right)^2}{npq_{(j)}} = [Y] = \frac{(172)^2}{(4)(2)(4)} = 924.500$$

$$\sum_{i=1}^{n}\sum_{j=1}^{p}\sum_{k=1}^{q_{(j)}} Y_{ijk}^2 = [ABS] = (3)^2 + (1)^2 + \cdots + (11)^2 = 1160.00$$

$$\sum_{j=1}^{p}\frac{\left(\sum_{i=1}^{n}\sum_{k=1}^{q_{(j)}} Y_{ijk}\right)^2}{nq_{(j)}} = [A] = \frac{(56)^2}{(4)(4)} + \frac{(116)^2}{(4)(4)} = 1037.000$$

$$\sum_{j=1}^{p}\sum_{k=1}^{q_{(j)}}\frac{\left(\sum_{i=1}^{n} Y_{ijk}\right)^2}{n} = [AB] = \frac{(15)^2}{4} + \frac{(7)^2}{4} + \cdots + \frac{(40)^2}{4} = 1141.500$$

(iii) Computational formulas:

$$SSTO = [ABS] - [Y] = 235.500$$
$$SSA = [A] - [Y] = 112.500$$
$$SSB(A) = [AB] - [A] = 104.500$$
$$SSWCELL = [ABS] - [AB] = 18.500$$

TABLE 10.2-2   ANOVA Table for Type CRH-28(*A*) Design

	Source	SS	df	MS	F	E(MS) Model III (A Fixed)
1	A (Ionization)	112.500	$p - 1 = 1$	112.500	[½] 6.46*	$\sigma_\epsilon^2 + n\sigma_\beta^2 + nq_{(j)}\sum_{j=1}^{p}\alpha_j^2/(p-1)$
2	B(A) (Cages)	104.500	$p(q_{(j)} - 1) = 6$	17.417	[⅔] 22.59**	$\sigma_\epsilon^2 + n\sigma_\beta^2$
3	WCELL	18.500	$pq_{(j)}(n - 1) = 24$	0.771		σ_ϵ^2
4	Total	235.500	$npq_{(j)} - 1 = 31$			

*p < .05
**p < .01

Table 10.2-2 contains two error terms: one for testing MSA and another for testing $MSB(A)$. The test of MSA for model III (see Table 10.2-2) uses $MSB(A)$ in the denominator of the F ratio. In order for F to be distributed as central F when the null hypothesis is true, the population variance estimated by $MSB(A)$ should be composed of homogeneous sources of variation. It can be shown that $SSB(A)$ represents a pooled sum of squares.

Source	df
SSB at a_1	$q_{(1)} - 1$
SSB at a_2	$q_{(2)} - 1$
$\vdots$	$\vdots$
SSB at a_p	$q_{(p)} - 1$

$$SSB(A) = \sum_{j=1}^{p} SSB \text{ at } a_j \qquad p(q_{(j)} - 1)$$

Actually $SSB(A)$ is the pooled simple main effects of treatment B at each level of treatment A. The reader may recall from Section 8.6 that

$$\sum_{j=1}^{p} SSB \text{ at } a_j = SSB + SSAB.$$

Thus $SSB(A)$ in a hierarchical experiment is equivalent to $SSB + SSAB$ in the corresponding factorial experiment. Procedures for partitioning $SSB(A)$ into p sources of variation for testing the assumption of homogeneity are described in Section 8.6. The other error term estimator $MSWCELL$ should also be composed of homogeneous sources of variation. The procedures for testing this assumption are described in Section 2.5.

Typically the number of degrees of freedom associated with $MSB(A)$, the error term for testing MSA, is quite small. If there is reason for believing that $\sigma_\beta^2 = 0$, $MSB(A)$ and $MSWCELL$ are both estimators of the same population error variance and can be pooled. For a discussion of the issues involved in performing preliminary tests and pooling, see Section 8.11. Pooling is not appropriate for the ionization data in Table 10.2-1 because the hypothesis $\sigma_\beta^2 = 0$ is rejected.

AN INCORRECT ANALYSIS

A common error in analyzing experiments involving a nested nuisance variable is to ignore the variable and use an analysis appropriate for a type CR-p design. Instead of including the variable of, say, cages or classrooms in the design, the experimenter acts as if $nq_{(j)}$ experimental units were randomly assigned to each level of treatment A and the $nq_{(j)}$ units were all treated alike. If $\sigma_\beta^2 \neq 0$, as is so often the case, the test of treatment A using $MSWCELL$ as the error term is positively biased. An analysis of the data in Table 10.2-1 ignoring the variable of cages leads to the results shown in Table 10.2-3. The discrepancy between the incorrect analysis (CR-2) and the correct analysis (CRH-28(A)) is quite large. The two analyses lead to the same result for treatment A if $\sigma_\beta^2 = 0$ and the experimenter uses a pooled error term based on $MSB(A)$ and $MSWCELL$. Then the test of A is given by $F = MSA/MSRES = 112.500/4.100 = 27.44$, where

$$MSRES = (SSB(A) + SSWCELL)/[p(q_{(j)} - 1) + pq_{(j)}(n - 1)]$$
$$= (104.500 + 18.500)/(6 + 24)$$
$$= 4.100.$$

As noted earlier, the use of a pooled error term for these data is not appropriate.

TABLE 10.2-3 Comparison of Correct Analysis (CRH-28(A)) with an Incorrect Analysis (CR-2) for the Data in Table 10.2-1

Type CR-2 Design

	Source	Formula	SS	df		MS		F
1	Between Groups (Ionization)	$[A] - [Y]$	112.500	$p - 1 =$	1	112.500	$[\frac{1}{2}]$	27.44**
2	Within Groups	$[AS] - [A]$	123.000	$p(n - 1) =$	30	4.100		
3	Total	$[AS] - [Y]$	235.500	$np - 1 =$	31			

Type CRH-28(A) Design

	Source	Formula	SS	df		MS		F
1	A (Ionization)	$[A] - [Y]$	112.500	$p - 1 =$	1	112.500	$[\frac{1}{2}]$	6.46*
2	B (Cages)	$[AB] - [A]$	104.500	$p(q_{(j)} - 1) =$	6	17.417	$[\frac{2}{3}]$	22.59**
3	WCELL	$[ABS] - [AB]$	18.500	$pq_{(j)}(n - 1) =$	24	0.771		
4	Total	$[ABS] - [Y]$	235.500	$npq_{(j)} - 1 =$	31			

*$p < .05$

**$p < .01$

10.3 EXPERIMENTAL DESIGN MODEL FOR A TYPE CRH-$pq(A)$ DESIGN

ASSUMPTIONS FOR MIXED MODEL (A FIXED AND B RANDOM)

A score Y_{ijk} in a completely randomized hierarchical design is a composite that reflects the effects of treatments A and $B(A)$ plus all other sources of variation that affect Y_{ijk}. The experimental design model for a type CRH-$pq(A)$ design is as follows.

$$
\begin{aligned}
Y_{ijk} &= \mu + (\mu_{j\cdot} - \mu) + (\mu_{jk} - \mu_{j\cdot}) + (Y_{ijk} - \mu_{jk}) \\
&= \mu + \alpha_j \qquad + \beta_{k(j)} \qquad + \epsilon_{i(jk)} \\
& (i = 1, \ldots, n; j = 1, \ldots, p; k = 1, \ldots, q)
\end{aligned}
$$

(10.3-1)

where Y_{ijk} is a score for the ith experimental unit in treatment combination jk

μ is the overall mean

α_j is the effect of treatment level j and is subject to the restriction

$$
\sum_{j=1}^{p} \alpha_j = 0
$$

$\beta_{k(j)}$ is $NID(0, \sigma_\beta^2)$

$\epsilon_{i(jk)}$ is $NID(0, \sigma_\epsilon^2)$, and independent of $\beta_{k(j)}$.

The values of the parameters μ, α_j, $\beta_{k(j)}$, and $\epsilon_{i(jk)}$ in model (10.3-1) are unknown, but they can be estimated from sample data as follows.

Statistic	Parameter Estimated
$\overline{Y}_{\cdots}$	μ
$\overline{Y}_{\cdot j\cdot} - \overline{Y}_{\cdots}$	α_j or $\mu_{j\cdot} - \mu$
$\overline{Y}_{\cdot jk} - \overline{Y}_{\cdot j\cdot}$	$\beta_{k(j)}$ or $\mu_{jk} - \mu_{j\cdot}$
$Y_{ijk} - \overline{Y}_{\cdot jk}$	$\epsilon_{i(jk)}$

PARTITION OF THE TOTAL SUM OF SQUARES

The partition of the total sum of squares, which is the basis for the analysis in Table 10.2-2, is obtained by replacing the parameters in model (10.3-1) by statistics and rearranging terms as follows.

(10.3-2)

$$
\begin{aligned}
Y_{ijk} - \mu &= \alpha_j \qquad\qquad + \beta_{k(j)} \qquad\quad + \epsilon_{i(jk)} \\
Y_{ijk} - \overline{Y}_{\cdots} &= (\overline{Y}_{\cdot j\cdot} - \overline{Y}_{\cdots}) + (\overline{Y}_{\cdot jk} - \overline{Y}_{\cdot j\cdot}) + (Y_{ijk} - \overline{Y}_{\cdot jk})
\end{aligned}
$$

$\qquad\quad$ (Total $\qquad\quad$ (Treatment A $\quad$ (Treatment $B(A)$ $\quad$ (Within-cell
$\qquad\quad$ deviation) $\qquad\quad$ effect) $\qquad\qquad$ effect) $\qquad\qquad$ effect)

Squaring both sides of (10.3-2) and summing over all of the observations following the procedures in Sections 2.2 and 6.1 lead to the following partition of the total sum of squares.

(10.3-3)

$$\sum_{i=1}^{n}\sum_{j=1}^{p}\sum_{k=1}^{q(j)} (Y_{ijk} - \overline{Y}_{...})^2 = nq_{(j)}\sum_{j=1}^{p}(\overline{Y}_{.j.} - \overline{Y}_{...})^2$$

$$SSTO \qquad = \qquad SSA$$

$$+ n\sum_{j=1}^{p}\sum_{k=1}^{q(j)}(\overline{Y}_{.jk} - \overline{Y}_{.j.})^2 + \sum_{i=1}^{n}\sum_{j=1}^{p}\sum_{k=1}^{q(j)}(Y_{ijk} - \overline{Y}_{.jk})^2$$

$$+ \qquad SSB(A) \qquad + \qquad SSWCELL$$

More convenient formulas for computational purposes are given in Table 10.2-1. Mean squares are obtained by dividing each sum of squares by its degrees of freedom as shown in Table 10.2-2.

ASSUMPTIONS FOR OTHER MODELS AND EXPECTATIONS OF *MS*'S

The assumptions for a fixed-effects model are as follows.

$$Y_{ijk} = \mu + \alpha_j + \beta_{k(j)} + \epsilon_{i(jk)} \qquad (i = 1, \ldots, n; j = 1, \ldots, p; k = 1, \ldots, q)$$

where α_j is subject to the restriction $\displaystyle\sum_{j=1}^{p}\alpha_j = 0$

$\beta_{k(j)}$ is subject to the restriction $\displaystyle\sum_{k=1}^{q(j)}\beta_{k(j)} = 0$ for all j

$\epsilon_{i(jk)}$ is $NID(0, \sigma_\epsilon^2)$.

If treatments A and $B(A)$ represent random effects, the assumptions are as follows.

α_j is $NID(0, \sigma_\alpha^2)$

$\beta_{k(j)}$ is $NID(0, \sigma_\beta^2)$

$\epsilon_{i(jk)}$ is $NID(0, \sigma_\epsilon^2)$, and independent of α_j and $\beta_{k(j)}$.

Experiments in which treatment A is random but $B(A)$ is fixed are fairly rare. The assumptions for this mixed model are as follows.

α_j is $NID(0, \sigma_\alpha^2)$

$\beta_{k(j)}$ is subject to the restriction $\displaystyle\sum_{k=1}^{q(j)}\beta_{k(j)} = 0$ for all j

$\epsilon_{i(jk)}$ is $NID(0, \sigma_\epsilon^2)$, and independent of α_j.

The expected values for these models are given in Table 10.3-1. These expected values and those for other hierarchical designs are easily derived using the rules in Section 8.9.

TABLE 10.3-1   Expected Values of Mean Squares for Type CRH-$pq(A)$ Design

Mean Square	Model I A Fixed B(A) Fixed	Model II A Random B(A) Random	Model III* A Random B(A) Fixed
1 MSA	$\sigma_\epsilon^2 + nq_{(j)}\sum\limits_{j=1}^{p}\alpha_j^2/(p-1)$	$\sigma_\epsilon^2 + n\sigma_\beta^2 + nq_{(j)}\sigma_\alpha^2$	$\sigma_\epsilon^2 + nq_{(j)}\sigma_\alpha^2$
2 MSB(A)	$\sigma_\epsilon^2 + n\sum\limits_{j=1}^{p}\sum\limits_{k=1}^{q_{(j)}}\beta_{k(j)}^2/p(q_{(j)}-1)$	$\sigma_\epsilon^2 + n\sigma_\beta^2$	$\sigma_\epsilon^2 + n\sum\limits_{j=1}^{p}\sum\limits_{k=1}^{q_{(j)}}\beta_{k(j)}^2/p(q_{(j)}-1)$
3 MSWCELL	σ_ϵ^2	σ_ϵ^2	σ_ϵ^2

*Expected values for model III (A fixed) are given in Table 10.2-2.

10.4   PROCEDURES FOR TESTING DIFFERENCES AMONG MEANS

Tests of differences among means in a type CRH-$pq(A)$ design have the same general form as that given in Chapter 3. The test statistics for making comparisons among treatment A means for a mixed model (A fixed) have the form

$$\frac{\hat{\psi}_{i(A)}}{\hat{\sigma}_{\psi_{i(A)}}} = \frac{c_1\overline{Y}_{\cdot 1\cdot} + c_2\overline{Y}_{\cdot 2\cdot} + \cdots + c_p\overline{Y}_{\cdot p\cdot}}{\sqrt{MSB(A)\left(\dfrac{c_1^2}{nq_{(1)}} + \dfrac{c_2^2}{nq_{(2)}} + \cdots + \dfrac{c_p^2}{nq_{(p)}}\right)}}$$

$$\frac{\hat{\psi}_{i(A)}}{\hat{\sigma}_{\overline{Y}}} = \frac{c_j\overline{Y}_{\cdot j\cdot} + c_{j'}\overline{Y}_{\cdot j'\cdot}}{\sqrt{MSB(A)/nq_{(j)}}}.$$

For a fixed-effects model, the mean square in the denominator of the above ratios should be replaced by *MSWCELL*. The general rule is that the error term used in testing the overall null hypothesis should be used in testing specific hypotheses concerning the population means. The number of degrees of freedom for these ratios is equal to the degrees of freedom associated with the mean square used in the denominator: $p(q_{(j)}-1)$ or $pq_{(j)}(n-1)$.

An experimenter is ordinarily not interested in testing hypotheses concerning the nested treatment. However, if this is desired, the test statistics have the following form for both the fixed-effects model and the mixed model (B fixed).

$$\frac{\hat{\psi}_{i(B(A))}}{\hat{\sigma}_{\psi_{i(B(A))}}} = \frac{c_1\overline{Y}_{\cdot j1} + c_2\overline{Y}_{\cdot j2} + \cdots + c_q\overline{Y}_{\cdot jq}}{\sqrt{MSWCELL\left(\dfrac{c_1^2}{n} + \dfrac{c_2^2}{n} + \cdots + \dfrac{c_q^2}{n}\right)}}$$

$$\frac{\hat{\psi}_{i(B(A))}}{\hat{\sigma}_{\overline{Y}}} = \frac{c_k \overline{Y}_{\cdot jk} + c_{k'} \overline{Y}_{\cdot jk'}}{\sqrt{MSWCELL/n}}$$

10.5 ESTIMATING STRENGTH OF ASSOCIATION IN A TYPE CRH-$pq(A)$ DESIGN

Indices of strength of association for a fixed-effects model are given by

$$\hat{\omega}_{Y|A}^2 = \frac{SSA - (p-1)MSWCELL}{SSTO + MSWCELL}$$

$$\hat{\omega}_{Y|B(A)}^2 = \frac{SSB(A) - (q_{(j)} - 1)MSWCELL}{SSTO + MSWCELL}.$$

If an estimated value of ω^2 is negative, that estimate is set equal to zero. Comparable measures for a random-effects model are given by

$$\hat{\rho}_{IY|A} = \frac{\hat{\sigma}_{\alpha}^2}{\hat{\sigma}_{\epsilon}^2 + \hat{\sigma}_{\beta}^2 + \hat{\sigma}_{\alpha}^2}$$

$$\hat{\rho}_{IY|B(A)} = \frac{\hat{\sigma}_{\beta}^2}{\hat{\sigma}_{\epsilon}^2 + \hat{\sigma}_{\beta}^2 + \hat{\sigma}_{\alpha}^2}.$$

Estimates of the various variance components can be obtained from

$$\hat{\sigma}_{\alpha}^2 = \frac{MSA - MSB(A)}{nq_{(j)}}$$

$$\hat{\sigma}_{\beta}^2 = \frac{MSB(A) - MSWCELL}{n}$$

$$\hat{\sigma}_{\epsilon}^2 = MSWCELL.$$

If an estimated component is negative, that estimate is set equal to zero. The rationale underlying these formulas should be obvious from an examination of the $E(MS)$'s for the random-effects model in Table 10.3-1.

A mixed model (A fixed) was appropriate for the ionization experiment described in Sections 10.1 and 10.2. Measures of strength of association for this model are as follows:

$$\hat{\omega}_{Y|A}^2 = \frac{\sum_{j=1}^{p} \hat{\alpha}_j^2 / p}{\hat{\sigma}_{\epsilon}^2 + \hat{\sigma}_{\beta}^2 + \sum_{j=1}^{p} \hat{\alpha}_j^2 / p}$$

$$\hat{\rho}_{IY|B(A)} = \frac{\hat{\sigma}_{\beta}^2}{\hat{\sigma}_{\epsilon}^2 + \hat{\sigma}_{\beta}^2 + \sum\limits_{j=1}^{p} \hat{\alpha}_j^2 / p} \, .$$

Based on the $E(MS)$'s shown in Table 10.2-2, the terms in these formulas are given by

$$\frac{\sum\limits_{j=1}^{p} \hat{\alpha}_j^2}{p} = \frac{(p-1)[MSA - MSB(A)]}{npq_{(j)}} = \frac{(2-1)(112.500 - 17.417)}{(4)(2)(4)}$$
$$= 2.971$$

$$\hat{\sigma}_{\beta}^2 = \frac{MSB(A) - MSWCELL}{n} = \frac{17.417 - 0.771}{4}$$
$$= 4.162$$

$$\hat{\sigma}_{\epsilon}^2 = MSWCELL = 0.771.$$

The values of $\hat{\omega}_{Y|A}^2$ and $\hat{\rho}_{IY|B(A)}$ are

$$\hat{\omega}_{Y|A}^2 = 2.971/(0.771 + 4.162 + 2.971) = 0.38$$
$$\hat{\rho}_{IY|B(A)} = 4.162/(0.771 + 4.162 + 2.971) = 0.53.$$

Thus both the ionization treatment and the cages account for an appreciable portion of the variance in the dependent variable.

10.6 DESCRIPTION OF OTHER COMPLETELY RANDOMIZED HIERARCHICAL DESIGNS

This section gives a brief description of a variety of hierarchical designs that are constructed from completely randomized and randomized block designs. Suggestions for analyzing these designs using computer programs written for crossed treatments are given in the following section.

TYPE CRH-$pq(A)r(AB)$ DESIGN

Consider an experiment to evaluate the efficacy of a new drug. The new drug, a_1, and another drug used for the same purpose, a_2, are randomly assigned to n patients in eight wards, treatment C, in four hospitals, treatment B. The design is diagrammed in Figure 10.6-1. The experimental design model equation for this completely nested design in which hospitals are nested within the two drugs, and wards are nested within hospitals and drugs is as follows.

$$Y_{ijkl} = \mu + \alpha_j + \beta_{k(j)} + \gamma_{l(jk)} + \epsilon_{i(jkl)}$$
$$(i = 1, \ldots, n; j = 1, \ldots, p; k = 1, \ldots, q; l = 1, \ldots, r)$$

The computational formulas and so on for this design are given in Table 10.6-1. The sums of squares $SSB(A)$ and $SSC(AB)$ represent pooled simple main effects and pooled simple simple main effects, respectively. It can be shown that these sums of squares correspond to

$$SSB(A) = \sum_{j=1}^{p} SSB \text{ at } a_j \qquad = SSB + SSAB$$

$$SSC(AB) = \sum_{j=1}^{p} \sum_{k=1}^{q(j)} SSC \text{ at } a_j b_{k(j)} = SSC + SSAC + SSCB + SSABC$$

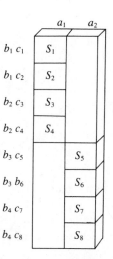

FIGURE 10.6-1 Block diagram of type CRH-24(A)8(AB) design

TABLE 10.6-1 Type CRH-$pq(A)r(AB)$ Design

Source	Formula	df	E(MS)*
1 A	$[A] - [Y]$	$p - 1$	$\sigma_\epsilon^2 + n\left(1 - \dfrac{r}{R}\right)\sigma_\gamma^2 + nr_{(jk)}\left(1 - \dfrac{q}{Q}\right)\sigma_\beta^2 + nq_{(j)}r_{(jk)}\sigma_\alpha^2$
2 $B(A)$	$[AB] - [A]$	$p(q_{(j)} - 1)$	$\sigma_\epsilon^2 + n\left(1 - \dfrac{r}{R}\right)\sigma_\gamma^2 + nr_{(jk)}\sigma_\beta^2$
3 $C(AB)$	$[ABC] - [AB]$	$pq_{(j)}(r_{(jk)} - 1)$	$\sigma_\epsilon^2 + n\sigma_\gamma^2$
4 $WCELL$	$[ABCS] - [ABC]$	$pq_{(j)}r_{(jk)}(n - 1)$	σ_ϵ^2
5 Total	$[ABCS] - [Y]$	$npq_{(j)}r_{(jk)} - 1$	

*The variances should be replaced by $\Sigma\alpha_j^2/(p - 1)$, $\Sigma\beta_k^2/(q - 1)$, and $\Sigma\gamma_l^2/(r - 1)$ if the corresponding treatments are fixed.

in a factorial design. In order for $MSB(A)$ and $MSC(AB)$ to serve as denominators of F ratios the mean squares should be composed of homogeneous sources of variation, for example,

$$MSC \text{ at } a_1 b_{1(1)} = MSC \text{ at } a_1 b_{2(1)} = \cdots = MSC \text{ at } a_p b_{q(p)}.$$

TYPE CRPH-$pq(A)r$, TYPE CRPH-$pqr(A)$, AND TYPE CRPH-$pqr(B)$ DESIGNS

The experimental design model equation for a type CRPH-$pq(A)r$ design in which treatment B is nested within A, but treatment C is crossed with A and $B(A)$ is

$$Y_{ijkl} = \mu + \alpha_j + \beta_{k(j)} + \gamma_l + (\alpha\gamma)_{jl} + (\beta\gamma)_{k(j)l} + \epsilon_{i(jkl)}$$
$$(i = 1, \ldots, n; j = 1, \ldots, p; k = 1, \ldots, q; l = 1, \ldots, r).$$

The design is diagrammed in Figure 10.6-2; computational formulas are given in Table 10.6-2.

It can be shown that $SSB(A)$ and $SSB(A) \times C$ correspond to

$$SSB(A) = \sum_{j=1}^{p} SSB \text{ at } a_j = SSB + SSAB$$

$$SSB(A) \times C = \sum_{j=1}^{p} SSBC \text{ at } a_j = SSBC + SSABC$$

in a factorial design.

One variation on this design is to nest treatment C within A and cross treatments A and B. The other variation is to nest treatment C within B and cross treatments A and B. The experimental design model equations are, respectively,

$$Y_{ijkl} = \mu + \alpha_j + \beta_k + \gamma_{l(j)} + (\alpha\beta)_{jk} + (\beta\gamma)_{kl(j)} + \epsilon_{i(jkl)}$$
$$Y_{ijkl} = \mu + \alpha_j + \beta_k + \gamma_{l(k)} + (\alpha\beta)_{jk} + (\alpha\gamma)_{jl(k)} + \epsilon_{i(jkl)}.$$

These designs are diagrammed in Figure 10.6-3; the computational formulas are given in Table 10-6.3.

FIGURE 10.6-2   Block diagram of type CRPH-24(A)2 design

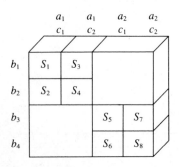

TABLE 10.6-2 Type CRPH-$pq(A)r$ Design

	Source	Formula	df	E(MS)*
1	A	$[A] - [Y]$	$p - 1$	$\sigma_\epsilon^2 + n\left(1 - \dfrac{q}{Q}\right)\left(1 - \dfrac{r}{R}\right)\sigma_{\beta\gamma}^2$ $+ nq_{(j)}\left(1 - \dfrac{r}{R}\right)\sigma_{\alpha\gamma}^2$ $+ nr\left(1 - \dfrac{q}{Q}\right)\sigma_\beta^2 + nq_{(j)}r\sigma_\alpha^2$
2	$B(A)$	$[AB] - [A]$	$p(q_{(j)} - 1)$	$\sigma_\epsilon^2 + n\left(1 - \dfrac{r}{R}\right)\sigma_{\beta\gamma}^2 + nr\sigma_\beta^2$
3	C	$[C] - [Y]$	$r - 1$	$\sigma_\epsilon^2 + n\left(1 - \dfrac{q}{Q}\right)\sigma_{\beta\gamma}^2$ $+ nq_{(j)}\left(1 - \dfrac{p}{P}\right)\sigma_{\alpha\gamma}^2 + npq_{(j)}\sigma_\gamma^2$
4	AC	$[AC] - [A] - [C] + [Y]$	$(p - 1)(r - 1)$	$\sigma_\epsilon^2 + n\left(1 - \dfrac{q}{Q}\right)\sigma_{\beta\gamma}^2 + nq_{(j)}\sigma_{\alpha\gamma}^2$
5	$B(A) \times C$	$[ABC] - [AB] - [AC] + [A]$	$p(q_{(j)} - 1)(r - 1)$	$\sigma_\epsilon^2 + n\sigma_{\beta\gamma}^2$
6	WCELL	$[ABCS] - [ABC]$	$pq_{(j)}r(n - 1)$	σ_ϵ^2
7	Total	$[ABCS] - [Y]$	$npq_{(j)}r - 1$	

* The variances should be replaced by $\Sigma\alpha_j^2/(p - 1)$, $\Sigma\beta_k^2/(q - 1)$, and $\Sigma\gamma_l^2/(r - 1)$ if the corresponding treatments are fixed.

FIGURE 10.6-3 Block diagrams of type CRPH-224(A) and type CRPH-224(B) designs

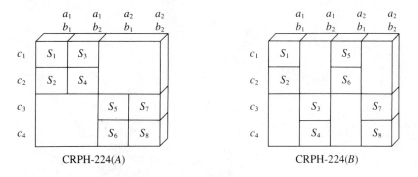

CRPH-224(A) CRPH-224(B)

TYPE CRPH-$pq(A)r(A)$ DESIGN

The experimental design model equation for a type CRPH-$pq(A)r(A)$ design in which both treatments B and C are nested within treatment A is

TABLE 10.6-3   Type CRPH-$pqr(A)$ and Type CRPH-$pqr(B)$ Designs

Type CRPH-pqr(A) Design

Source	Formula	df	E(MS)*
1 A	$[A] - [Y]$	$p - 1$	$\sigma_\epsilon^2 + n\left(1 - \dfrac{q}{Q}\right)\left(1 - \dfrac{r}{R}\right)\sigma_{\beta\gamma}^2$ $+ nr_{(j)}\left(1 - \dfrac{q}{Q}\right)\sigma_{\alpha\beta}^2$ $+ nq\left(1 - \dfrac{r}{R}\right)\sigma_\gamma^2 + nqr_{(j)}\sigma_\alpha^2$
2 B	$[B] - [Y]$	$q - 1$	$\sigma_\epsilon^2 + n\left(1 - \dfrac{r}{R}\right)\sigma_{\beta\gamma}^2$ $+ nr_{(j)}\left(1 - \dfrac{p}{P}\right)\sigma_{\alpha\beta}^2 + npr_{(j)}\sigma_\beta^2$
3 C(A)	$[AC] - [A]$	$p(r_{(j)} - 1)$	$\sigma_\epsilon^2 + n\left(1 - \dfrac{q}{Q}\right)\sigma_{\beta\gamma}^2 + nq\sigma_\gamma^2$
4 AB	$[AB] - [A] - [B] + [Y]$	$(p - 1)(q - 1)$	$\sigma_\epsilon^2 + n\left(1 - \dfrac{r}{R}\right)\sigma_{\beta\gamma}^2 + nr_{(j)}\sigma_{\alpha\beta}^2$
5 B × C(A)	$[ABC] - [AB] - [AC] + [A]$	$p(q - 1)(r_{(j)} - 1)$	$\sigma_\epsilon^2 + n\sigma_{\beta\gamma}^2$
6 WCELL	$[ABCS] - [ABC]$	$pqr_{(j)}(n - 1)$	σ_ϵ^2
7 Total	$[ABCS] - [Y]$	$npqr_{(j)} - 1$	

Type CRPH-pqr(B) Design

Source	Formula	df	E(MS)*
1 A	$[A] - [Y]$	$p - 1$	$\sigma_\epsilon^2 + n\left(1 - \dfrac{r}{R}\right)\sigma_{\alpha\gamma}^2$ $+ nr_{(k)}\left(1 - \dfrac{q}{Q}\right)\sigma_{\alpha\beta}^2 + nqr_{(k)}\sigma_\alpha^2$
2 B	$[B] - [Y]$	$q - 1$	$\sigma_\epsilon^2 + n\left(1 - \dfrac{p}{P}\right)\left(1 - \dfrac{r}{R}\right)\sigma_{\alpha\gamma}^2$ $+ nr_{(k)}\left(1 - \dfrac{p}{P}\right)\sigma_{\alpha\beta}^2$ $+ np\left(1 - \dfrac{r}{R}\right)\sigma_\gamma^2 + npr_{(k)}\sigma_\beta^2$
3 C(B)	$[BC] - [B]$	$q(r_{(k)} - 1)$	$\sigma_\epsilon^2 + n\left(1 - \dfrac{p}{P}\right)\sigma_{\alpha\gamma}^2 + np\sigma_\gamma^2$
4 AB	$[AB] - [A] - [B] + [Y]$	$(p - 1)(q - 1)$	$\sigma_\epsilon^2 + n\left(1 - \dfrac{r}{R}\right)\sigma_{\alpha\gamma}^2 + nr_{(k)}\sigma_{\alpha\beta}^2$
5 A × C(B)	$[ABC] - [AB] - [BC] + [B]$	$q(p - 1)(r_{(k)} - 1)$	$\sigma_\epsilon^2 + n\sigma_{\alpha\gamma}^2$
6 WCELL	$[ABCS] - [ABC]$	$pqr_{(k)}(n - 1)$	σ_ϵ^2
7 Total	$[ABCS] - [Y]$	$npqr_{(k)} - 1$	

* The variances should be replaced by $\Sigma\alpha_j^2/(p - 1)$, $\Sigma\beta_k^2/(q - 1)$, and $\Sigma\gamma_k^2/(r - 1)$ if the corresponding treatments are fixed.

$$Y_{ijkl} = \mu + \alpha_j + \beta_{k(j)} + \gamma_{l(j)} + (\beta\gamma)_{k(j)l(j)} + \epsilon_{i(jkl)}$$
$$(i = 1, \ldots, n; j = 1, \ldots, p; k = 1, \ldots, q; l = 1, \ldots, r).$$

This design is diagrammed in Figure 10.6-4; computational formulas are given in Table 10.6-4.

It can be shown that $SSB(A)$, $SSC(A)$, and $SSB(A) \times C(A)$ correspond to

FIGURE 10.6-4 Block diagram of type CRPH-24(A)4(A) design

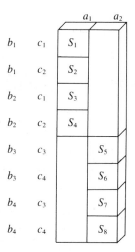

TABLE 10.6-4 Type CRPH-$pq(A)r(A)$ Design

Source	Formula	df	$E(MS)*$
1 A	$[A] - [Y]$	$p - 1$	$\sigma_\epsilon^2 + n\left(1 - \dfrac{q}{Q}\right)\left(1 - \dfrac{r}{R}\right)\sigma_{\beta\gamma}^2$ $+ nq_{(j)}\left(1 - \dfrac{r}{R}\right)\sigma_\gamma^2$ $+ nr_{(j)}\left(1 - \dfrac{q}{Q}\right)\sigma_\beta^2$ $+ nq_{(j)}r_{(j)}\sigma_\alpha^2$
2 B(A)	$[AB] - [A]$	$p(q_{(j)} - 1)$	$\sigma_\epsilon^2 + n\left(1 - \dfrac{r}{R}\right)\sigma_{\beta\gamma}^2 + nr_{(j)}\sigma_\beta^2$
3 C(A)	$[AC] - [A]$	$p(r_{(j)} - 1)$	$\sigma_\epsilon^2 + n\left(1 - \dfrac{q}{Q}\right)\sigma_{\beta\gamma}^2 + nq_{(j)}\sigma_\gamma^2$
4 B(A) × C(A)	$[ABC] - [AB] - [AC] + [A]$	$p(q_{(j)} - 1)(r_{(j)} - 1)$	$\sigma_\epsilon^2 + n\sigma_{\beta\gamma}^2$
5 WCELL	$[ABCS] - [ABC]$	$pq_{(j)}r_{(j)}(n - 1)$	σ_ϵ^2
6 Total	$[ABCS] - [Y]$	$npq_{(j)}r_{(j)} - 1$	

*The variances should be replaced by $\Sigma\alpha_j^2/(p - 1)$, $\Sigma\beta_k^2/(q - 1)$, and $\Sigma\gamma_l^2/(r - 1)$ if the corresponding treatments are fixed.

$$SSB(A) = \sum_{j=1}^{p} SSB \text{ at } a_j \quad = SSB + SSAB$$

$$SSC(A) = \sum_{j=1}^{p} SSC \text{ at } a_j \quad = SSC + SSAC$$

$$SSB(A) \times C(A) = \sum_{j=1}^{p} SSBC \text{ at } a_j = SSBC + SSABC$$

in a factorial design.

TYPE CRPH-*pqr*(*AB*) DESIGN

A type CRPH-*pqr*(*AB*) design is one in which treatment C is nested within both A and B, but treatments A and B are crossed. For example, suppose a new drug education program for junior high students is to be evaluated. We will denote the program by a_1 and the control condition by a_2. The levels of treament A are assigned to eight social studies classes, treatment C, in two schools, treatment B. This design is diagrammed in Figure 10.6-5.

The experimental design model equation is

$$Y_{ijkl} = \mu + \alpha_j + \beta_k + \gamma_{l(jk)} + (\alpha\beta)_{jk} + \epsilon_{i(jkl)}$$
$$(i = 1, \ldots, n; j = 1, \ldots, p; k = 1, \ldots, q; l = 1, \ldots, r).$$

The computational formulas are given in Table 10.6-5. The sum of squares $SSC(AB)$ corresponds to

$$SSC(AB) = \sum_{j=1}^{p} \sum_{k=1}^{q} SSC \text{ at } a_j b_k = SSC + SSAC + SSBC + SSABC$$

FIGURE 10.6-5   Block diagram of type CRPH-228(*AB*) design

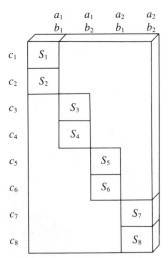

TABLE 10.6-5 Type CRPH-$pqr(AB)$ Design

Source	Formula	df	$E(MS)^*$
1 A	$[A] - [Y]$	$p - 1$	$\sigma_\epsilon^2 + nr_{(jk)}\left(1 - \dfrac{q}{Q}\right)\sigma_{\alpha\beta}^2 + n\left(1 - \dfrac{r}{R}\right)\sigma_\gamma^2$ $+ nqr_{(jk)}\sigma_\alpha^2$
2 B	$[B] - [Y]$	$q - 1$	$\sigma_\epsilon^2 + nr_{(jk)}\left(1 - \dfrac{p}{P}\right)\sigma_{\alpha\beta}^2 + n\left(1 - \dfrac{r}{R}\right)\sigma_\gamma^2$ $+ npr_{(jk)}\sigma_\beta^2$
3 $C(AB)$	$[ABC] - [AB]$	$pq(r_{(jk)} - 1)$	$\sigma_\epsilon^2 + n\sigma_\gamma^2$
4 AB	$[AB] - [A] - [B] + [Y]$	$(p - 1)(q - 1)$	$\sigma_\epsilon^2 + nr_{(jk)}\sigma_{\alpha\beta}^2$
5 $WCELL$	$[ABCS] - [ABC]$	$pqr_{(jk)}(n - 1)$	σ_ϵ^2
6 Total	$[ABCS] - [Y]$	$npqr_{(jk)} - 1$	

*The variances should be replaced by $\Sigma\alpha_j^2/(p - 1)$, $\Sigma\beta_k^2/(q - 1)$, and $\Sigma\gamma_l^2/(r - 1)$ if the corresponding treatments are fixed.

in a factorial design.

We have now described all of the three-treatment hierarchical designs that are based on a type CR-p design. The patterns that underlie hierarchical designs with four or more treatments are straightforward extensions of those just described. Accordingly, in the next section we will describe only one of the four-treatment completely randomized partial hierarchical designs.

TYPE CRPH-$pqrt(C)$ DESIGN

The experimental design model equation for a type CRPH-$pqrt(C)$ design in which treatments A, B, and C are crossed but D is nested within C is

$$Y_{ijklm} = \mu + \alpha_j + \beta_k + \gamma_l + \delta_{m(l)} + (\alpha\beta)_{jk} + (\alpha\gamma)_{jl} + (\alpha\delta)_{jm(l)}$$
$$+ (\beta\gamma)_{kl} + (\beta\delta)_{km(l)} + (\alpha\beta\gamma)_{jkl} + (\alpha\beta\delta)_{jkm(l)} + \epsilon_{i(jklm)}$$
$$(i = 1, \ldots, n; j = 1, \ldots, p; k = 1, \ldots, q; l = 1, \ldots, r;$$
$$m = 1, \ldots, t).$$

The design is diagrammed in Figure 10.6-6; computational formulas are given in Table 10.6-6. The sums of squares $SSD(C)$, $SSA \times D(C)$, $SSB \times D(C)$, and $SSAB \times D(C)$ correspond to

$$SSD(C) = \sum_{l=1}^{r} SSD \text{ at } c_l = SSD + SSCD$$

FIGURE 10.6-6 Block diagram of type CRPH-2224(C) design

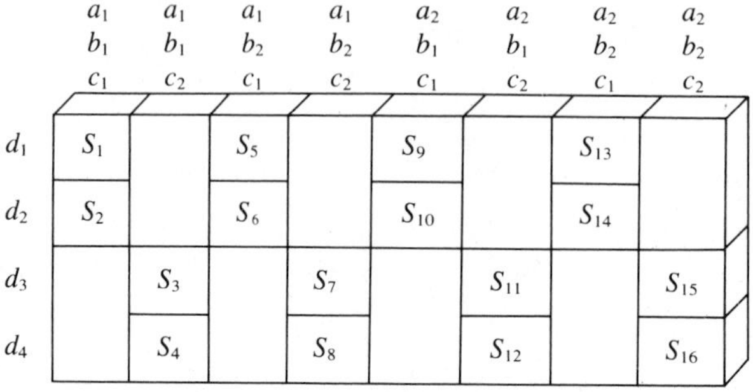

	a_1 b_1 c_1	a_1 b_1 c_2	a_1 b_2 c_1	a_1 b_2 c_2	a_2 b_1 c_1	a_2 b_1 c_2	a_2 b_2 c_1	a_2 b_2 c_2
d_1	S_1		S_5		S_9		S_{13}	
d_2	S_2		S_6		S_{10}		S_{14}	
d_3		S_3		S_7		S_{11}		S_{15}
d_4		S_4		S_8		S_{12}		S_{16}

TABLE 10.6-6 Type CRPH-$pqrl$(C) Design

	Source	Formula	df	$E(MS)$*
1	A	$[A] - [Y]$	$p - 1$	$\sigma_\epsilon^2 + n\left(1 - \dfrac{q}{Q}\right)\left(1 - \dfrac{t}{T}\right)\sigma_{\alpha\beta\delta}^2$ $+ nt_{(l)}\left(1 - \dfrac{q}{Q}\right)\left(1 - \dfrac{r}{R}\right)\sigma_{\alpha\beta\gamma}^2$ $+ nq\left(1 - \dfrac{t}{T}\right)\sigma_{\alpha\delta}^2$ $+ nqt_{(l)}\left(1 - \dfrac{r}{R}\right)\sigma_{\alpha\gamma}^2$ $+ nrt_{(l)}\left(1 - \dfrac{q}{Q}\right)\sigma_{\alpha\beta}^2 + nqrt_{(l)}\sigma_\alpha^2$
2	B	$[B] - [Y]$	$q - 1$	$\sigma_\epsilon^2 + n\left(1 - \dfrac{p}{P}\right)\left(1 - \dfrac{t}{T}\right)\sigma_{\alpha\beta\delta}^2$ $+ nt_{(l)}\left(1 - \dfrac{p}{P}\right)\left(1 - \dfrac{r}{R}\right)\sigma_{\alpha\beta\gamma}^2$ $+ np\left(1 - \dfrac{t}{T}\right)\sigma_{\beta\delta}^2$ $+ npt_{(l)}\left(1 - \dfrac{r}{R}\right)\sigma_{\beta\gamma}^2$ $+ nrt_{(l)}\left(1 - \dfrac{p}{P}\right)\sigma_{\alpha\beta}^2 + nprt_{(l)}\sigma_\beta^2$
3	C	$[C] - [Y]$	$r - 1$	$\sigma_\epsilon^2 + n\left(1 - \dfrac{p}{P}\right)\left(1 - \dfrac{q}{Q}\right)\left(1 - \dfrac{t}{T}\right)\sigma_{\alpha\beta\delta}^2$ $+ nt\left(1 - \dfrac{p}{P}\right)\left(1 - \dfrac{q}{Q}\right)\sigma_{\alpha\beta\gamma}^2$ $+ np\left(1 - \dfrac{q}{Q}\right)\left(1 - \dfrac{t}{T}\right)\sigma_{\beta\delta}^2$ $+ npt_{(l)}\left(1 - \dfrac{q}{Q}\right)\sigma_{\beta\gamma}^2$ $+ nq\left(1 - \dfrac{p}{P}\right)\left(1 - \dfrac{t}{T}\right)\sigma_{\alpha\delta}^2$

TABLE 10.6-6 (continued)

$$+ nqt_{(l)}\left(1 - \frac{p}{P}\right)\sigma^2_{\alpha\gamma}$$

$$+ npq\left(1 - \frac{t}{T}\right)\sigma^2_{\delta}$$

$$+ npqt_{(l)}\sigma^2_{\gamma}$$

4 $D(C)$ $[CD] - [C]$ $r(t_{(l)} - 1)$ $\sigma^2_{\epsilon} + n\left(1 - \frac{p}{P}\right)\left(1 - \frac{q}{Q}\right)\sigma^2_{\alpha\beta\delta}$

$$+ np\left(1 - \frac{q}{Q}\right)\sigma^2_{\beta\delta}$$

$$+ nq\left(1 - \frac{p}{P}\right)\sigma^2_{\alpha\delta} + npq\sigma^2_{\delta}$$

5 AB $[AB] - [A] - [B] + [Y]$ $(p - 1)(q - 1)$ $\sigma^2_{\epsilon} + n\left(1 - \frac{t}{T}\right)\sigma^2_{\alpha\beta\delta}$

$$+ nt_{(l)}\left(1 - \frac{r}{R}\right)\sigma^2_{\alpha\beta\gamma}$$

$$+ nrt_{(l)}\sigma^2_{\alpha\beta}$$

6 AC $[AC] - [A] - [C] + [Y]$ $(p - 1)(r - 1)$ $\sigma^2_{\epsilon} + n\left(1 - \frac{q}{Q}\right)\left(1 - \frac{t}{T}\right)\sigma^2_{\alpha\beta\delta}$

$$+ nt\left(1 - \frac{q}{Q}\right)\sigma^2_{\alpha\beta\gamma}$$

$$+ nq\left(1 - \frac{t}{T}\right)\sigma^2_{\alpha\delta}$$

$$+ nqt_{(l)}\sigma^2_{\alpha\gamma}$$

7 $A \times D(C)$ $[ACD] - [AC] - [CD] + [C]$ $(p - 1)r(t_{(l)} - 1)$ $\sigma^2_{\epsilon} + n\left(1 - \frac{q}{Q}\right)\sigma^2_{\alpha\beta\delta} + nq\sigma^2_{\alpha\delta}$

8 BC $[BC] - [B] - [C] + [Y]$ $(q - 1)(r - 1)$ $\sigma^2_{\epsilon} + n\left(1 - \frac{p}{P}\right)\left(1 - \frac{t}{T}\right)\sigma^2_{\alpha\beta\delta}$

$$+ nt_{(l)}\left(1 - \frac{p}{P}\right)\sigma^2_{\alpha\beta\gamma}$$

$$+ np\left(1 - \frac{t}{T}\right)\sigma^2_{\beta\delta}$$

$$+ npt_{(l)}\sigma^2_{\beta\gamma}$$

9 $B \times D(C)$ $[BCD] - [BC] - [CD] + [C]$ $(q - 1)r(t_{(l)} - 1)$ $\sigma^2_{\epsilon} + n\left(1 - \frac{p}{P}\right)\sigma^2_{\alpha\beta\delta} + np\sigma^2_{\beta\delta}$

10 ABC $[ABC] - [AB] - [AC] - [BC] + [A] + [B] + [C] - [Y]$ $(p - 1)(q - 1)(r - 1)$ $\sigma^2_{\epsilon} + n\left(1 - \frac{t}{T}\right)\sigma^2_{\alpha\beta\delta} + nt_{(l)}\sigma^2_{\alpha\beta\gamma}$

11 $AB \times D(C)$ $[ABCD] - [ABC] - [ACD] - [BCD] + [AC] + [BC] + [CD] - [C]$ $(p - 1)(q - 1)r(t_{(l)} - 1)$ $\sigma^2_{\epsilon} + n\sigma^2_{\alpha\beta\delta}$

12 $WCELL$ $[ABCDS] - [ABCD]$ $pqrt_{(l)}(n - 1)$ σ^2_{ϵ}

---

13 Total $[ABCDS] - [Y]$ $npqrt_{(l)} - 1$

---

* The variances should be replaced by $\Sigma\alpha^2_j/(p - 1)$, $\Sigma\beta^2_k/(q - 1)$, $\Sigma\gamma^2_l/(r - 1)$, and $\Sigma\delta^2_m/(t - 1)$ if the corresponding treatments are fixed.

$$SSA \times D(C) = \sum_{l=1}^{r} SSAD \text{ at } c_l = SSAD + SSACD$$

$$SSB \times D(C) = \sum_{l=1}^{r} SSBD \text{ at } c_l = SSBD + SSBCD$$

$$SSAB \times D(C) = \sum_{l=1}^{r} SSABD \text{ at } c_l = SSABD + SSABCD$$

in a factorial design.

TYPE RBH-$pq(A)$ DESIGN

Hierarchical designs can be constructed using a type RB-p design as the building block design. In this section we describe a type RBH-$pq(A)$ design in which the q levels of treatment B are nested within those of treatment A.

Suppose an educational researcher wants to compare the effectiveness of two ways of using reading pacers in increasing sixth-graders' reading speed. In condition a_1 the rate of presentation of reading material is increased by a small amount every five minutes. In condition a_2 the child controls the speed and is encouraged to increase it as much as possible. Four reading pacers are modified to provide the constant and self-paced conditions and the equipment installed in four study rooms. The rooms are not identical and there is no way to ensure that other extraneous conditions, such as ambient noise, room illumination, and so on, are constant. Accordingly the rooms are designated as levels of treatment $B(A)$. Twenty children are assigned to one of five blocks based on a pretest of their reading speed; subjects within a block have similar scores. The four combinations of treatments A and $B(A)$ are assigned randomly to the subjects within each block. A block diagram of this design is shown in Figure 10.6-7; computational formulas are given in Table 10.6-7.

If blocks do not interact with treatments A and $B(A)$, the experimental design model equation is

$$Y_{ijk} = \mu + \alpha_j + \beta_{k(j)} + \pi_i + \epsilon_{ijk}$$
$$(i = 1, \ldots, n; j = 1, \ldots, p; k = 1, \ldots, q).$$

If the block-treatment interactions, $\sigma_{\alpha\pi}^2$ and $\sigma_{\beta\pi}^2$, are not equal to zero, the model equation must be amended as follows.

$$Y_{ijk} = \mu + \alpha_j + \beta_{k(j)} + \pi_i + (\alpha\pi)_{jk} + (\beta\pi)_{k(j)i} + \epsilon_{ijk}$$

Computational procedures for the additive and nonadditive models are discussed in Section 9.6.

The construction of randomized block hierarchical and partial hierarchical designs with three or more treatments follows the pattern illustrated earlier for the type CR-p building block design.

FIGURE 10.6-7   Block diagram of type RBH-24(A) design

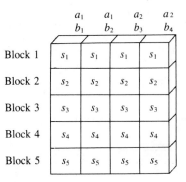

TABLE 10.6-7   Type RBH-$pq(A)$ Design

	Source	Formula	df	$E(MS)$* (additive model)
1	Blocks	$[S] - [Y]$	$n - 1$	$\sigma_\epsilon^2 + pq_{(j)}\sigma_\pi^2$
2	Treatments	$[AB] - [Y]$	$pq_{(j)} - 1$	
3	A	$[A] - [Y]$	$p - 1$	$\sigma_\epsilon^2 + n\left(1 - \dfrac{q}{Q}\right)\sigma_\beta^2 + nq_{(j)}\sigma_\alpha^2$
4	$B(A)$	$[AB] - [A]$	$p(q_{(j)} - 1)$	$\sigma_\epsilon^2 + n\sigma_\beta^2$
5	Residual	$[ABS] - [AB] - [S] + [Y]$	$(n - 1)(pq_{(j)} - 1)$	σ_ϵ^2
6	$A \times BL$	$[AS] - [A] - [S] + [Y]$	$(n - 1)(p - 1)$	
7	$B(A) \times BL$	$[ABS] - [AB] - [AS] + [A]$	$(n - 1)p(q_{(j)} - 1)$	
8	Total	$[ABS] - [Y]$	$npq_{(j)} - 1$	

* The variances should be replaced by $\Sigma\alpha_j^2/(p - 1)$ and $\Sigma\beta_k^2/(q - 1)$ if the corresponding treatments are fixed.

10.7 ANALYZING HIERARCHICAL DESIGNS USING COMPUTER PROGRAMS FOR CROSSED TREATMENTS

Computer programs for analyzing experiments with crossed treatments are widely available. Such is not the case for nested treatments. Fortunately, programs for crossed treatments can be used to analyze data for any balanced hierarchical design. We will describe the procedure using a type CRPH-$pq(A)r$ design. The first step is to renumber the levels of the nested treatment (B) as if they were crossed with A:

$b_{1(1)} = b_1$, $b_{2(1)} = b_2$, . . . , $b_{q(1)} = b_q$; $b_{1(2)} = b_1$, $b_{2(2)} = b_2$, and so on. The data can then be analyzed using a program appropriate for a three-treatment completely randomized factorial design. The computer printout will yield nine sums of squares as follows:

$$SSA \qquad SSAB \qquad SSABC$$
$$SSB \qquad SSAC \qquad SSWCELL$$
$$SSC \qquad SSBC \qquad SSTO.$$

These sums of squares can be combined as shown in Table 10-7.1 to obtain the sums of squares for the type CRPH-$pq(A)r$ design. Once these sums of squares have been computed, they are divided by the appropriate degrees of freedom to obtain mean squares. These steps and the subsequent tests of significance are performed by hand—a minor task compared to performing the entire analysis by hand or writing a new computer program.

TABLE 10.7-1   Correspondence Between the Sums of Squares for Type CRPH-$pq(A)r$ and Type CRF-pqr Designs

Sums of Squares		Degrees of Freedom	
CRPH-$pq(A)r$	CRF-pqr	CRPH-$pq(A)r$	CRF-pqr
SSA	$= SSA$	$p - 1$	$= p - 1$
$SSB(A)$	$= SSB + SSAB$	$p(q_{(j)} - 1)$	$= (q - 1) + (p - 1)(q - 1)$
SSC	$= SSC$	$r - 1$	$= r - 1$
$SSAC$	$= SSAC$	$(p - 1)(r - 1)$	$= (p - 1)(r - 1)$
$SSB(A) \times C$	$= SSBC + SSABC$	$p(q_{(j)} - 1)(r - 1)$	$= (p - 1)(r - 1) + (p - 1)(q - 1)(r - 1)$
$SSWCELL$	$= SSWCELL$	$pq_{(j)}r(n - 1)$	$= pqr(n - 1)$

*10.8 ANALYSIS OF UNBALANCED HIERARCHICAL DESIGNS

Data from unbalanced hierarchical designs (fixed-effects model) can be readily analyzed using the full rank experimental design model approach. Suppose that the ionization experiment described in Sections 10.1 and 10.2 is unbalanced because the equipment in cage 2 malfunctioned and two of the animals in other cages died. We will assume that the deaths were unrelated to the nature of the treatments. The data for this experiment are shown in Table 10.8-1. Note that all of the observations in cell a_1b_2 along with Y_{413} and Y_{427} are missing.

☆   This section assumes a familiarity with the matrix operations in Appendix D and Chapter 5. The reader who is interested only in the traditional approach to analysis of variance can, without loss of continuity, omit this section.

TABLE 10.8-1   Data for Ionization Experiment

			ABS Summary Table Entry is Y_{ijk}							

a_1	a_1	a_1	a_1	a_2	a_2	a_2	a_2
b_1	b_2	b_3	b_4	b_5	b_6	b_7	b_8
3	1	5	4	7	7	10	
6	2	6	5	8	8	10	
3	2	5	4	9	7	9	
3		6	3	8		11	

AB Summary Table
Entry is $\hat{\mu}_{jk}$

	a_1	a_2	$\hat{\mu}_{jk}$
$b_{1(1)}$	3.7500		3.7500
$b_{3(1)}$	1.6667		1.6667
$b_{4(1)}$	5.5000		5.5000
$b_{5(2)}$		4.0000	4.0000
$b_{6(2)}$		8.0000	8.0000
$b_{7(2)}$		7.3333	7.3333
$b_{8(2)}$		10.0000	10.0000

$$\sum_{k=1}^{q_{(j)}} \hat{\mu}_{jk}/q_{(j)} = \hat{\mu}_{1.} \qquad \hat{\mu}_{2.}$$

$$= 3.6389 \qquad 7.3333$$

The full rank experimental design model for this hierarchical design is

$$Y_{ijk} = \mu_{jk} + \epsilon_{i(jk)} \qquad (i = 1, \ldots, n_{jk}; j = 1, \ldots, p; k = 1, \ldots, q_{(j)})$$

where $\epsilon_{i(jk)}$ is $NID(0, \sigma_\epsilon^2)$. Assuming a fixed-effects model, we would be interested in testing the following null hypotheses for treatments A and $B(A)$, respectively,

$$H_0: \quad \frac{\mu_{11} + \mu_{13} + \mu_{14}}{3} - \frac{\mu_{25} + \mu_{26} + \mu_{27} + \mu_{28}}{4} = 0 \quad \text{or} \quad H_0: \quad \mu_{1.} - \mu_{2.} = 0$$

$$H_0: \quad \mu_{11} - \mu_{14} = \mu_{13} - \mu_{14} = 0$$
$$\mu_{25} - \mu_{28} = \mu_{26} - \mu_{28} = \mu_{27} - \mu_{28} = 0.$$

The null hypothesis for treatment A involves a simple average of the means at each level of treatment B. Alternatively, a hypothesis involving weighted means could be tested (see Section 8.13). In matrix notation the hypotheses for treatments A and $B(A)$ become

$$\mathbf{C}_A' \boldsymbol{\mu} = \mathbf{0}_A \qquad \text{and} \qquad \mathbf{C}_{B(A)}' \boldsymbol{\mu} = \mathbf{0}_{B(A)}$$

where

$$\underset{(p-1)\times h}{\mathbf{C}_A'} = [\tfrac{1}{3} \quad \tfrac{1}{3} \quad \tfrac{1}{3} \quad -\tfrac{1}{4} \quad -\tfrac{1}{4} \quad -\tfrac{1}{4} \quad -\tfrac{1}{4}]$$

$$\underset{\substack{\sum_{j=1}^{p}(q_{(j)}-1)\times h}}{\mathbf{C}_{B(A)}'} = \begin{bmatrix} 1 & 0 & -1 & 0 & 0 & 0 & 0 \\ 0 & 1 & -1 & 0 & 0 & 0 & 0 \\ 0 & 0 & 0 & 1 & 0 & 0 & -1 \\ 0 & 0 & 0 & 0 & 1 & 0 & -1 \\ 0 & 0 & 0 & 0 & 0 & 1 & -1 \end{bmatrix}$$

TABLE 10.8-2   Computational Procedures for Type CRH-$pq(A)$ Design Using a Full Rank Experimental Design Model

(i) Data and matrix formulas for computing sums of squares ($N = 26$, $h = 7$; the $\mathbf{C}''$s are defined in the text):

$$\underset{N \times 1}{\mathbf{y}} \qquad \underset{N \times h}{\mathbf{X}}$$

$$
\begin{array}{c}
\\
a_1\,b_1 \left\{\begin{array}{c} \\ \\ \\ \\ \end{array}\right.\\
\\
a_1\,b_3 \left\{\begin{array}{c} \\ \\ \\ \end{array}\right.\\
\\
a_1\,b_4 \left\{\begin{array}{c} \\ \\ \\ \\ \end{array}\right.\\
\\
a_2\,b_5 \left\{\begin{array}{c} \\ \\ \\ \\ \end{array}\right.\\
\\
a_2\,b_6 \left\{\begin{array}{c} \\ \\ \\ \\ \end{array}\right.\\
\\
a_2\,b_7 \left\{\begin{array}{c} \\ \\ \\ \end{array}\right.\\
\\
a_2\,b_8 \left\{\begin{array}{c} \\ \\ \\ \\ \end{array}\right.
\end{array}
$$

	$\mathbf{x}_1$	$\mathbf{x}_2$	$\mathbf{x}_3$	$\mathbf{x}_4$	$\mathbf{x}_5$	$\mathbf{x}_6$	$\mathbf{x}_7$
3	1	0	0	0	0	0	0
6	1	0	0	0	0	0	0
3	1	0	0	0	0	0	0
3	1	0	0	0	0	0	0
1	0	1	0	0	0	0	0
2	0	1	0	0	0	0	0
2	0	1	0	0	0	0	0
5	0	0	1	0	0	0	0
6	0	0	1	0	0	0	0
5	0	0	1	0	0	0	0
6	0	0	1	0	0	0	0
4	0	0	0	1	0	0	0
5	0	0	0	1	0	0	0
4	0	0	0	1	0	0	0
3	0	0	0	1	0	0	0
7	0	0	0	0	1	0	0
8	0	0	0	0	1	0	0
9	0	0	0	0	1	0	0
8	0	0	0	0	1	0	0
7	0	0	0	0	0	1	0
8	0	0	0	0	0	1	0
7	0	0	0	0	0	1	0
10	0	0	0	0	0	0	1
10	0	0	0	0	0	0	1
9	0	0	0	0	0	0	1
11	0	0	0	0	0	0	1

$$\sum_{i=1}^{n} Y_i = 152$$

$$
\underset{1 \times N}{\mathbf{y}'} \qquad \underset{N \times 1}{\mathbf{y}} \qquad \underset{1 \times 1}{(\mathbf{y}'\mathbf{y})}
$$

$$
\begin{bmatrix} 3 & 6 & \cdots & 11 \end{bmatrix}
\begin{bmatrix} 3 \\ 6 \\ \vdots \\ 11 \end{bmatrix} = 1082.00
$$

$$
\underset{h \times h}{(\mathbf{X}'\mathbf{X})^{-1}} \qquad \underset{h \times 1}{(\mathbf{X}'\mathbf{y})} \qquad \underset{h \times 1}{\hat{\boldsymbol{\mu}}}
$$

$$
\begin{bmatrix}
\tfrac{1}{4} & 0 & 0 & 0 & 0 & 0 & 0 \\
0 & \tfrac{1}{3} & 0 & 0 & 0 & 0 & 0 \\
0 & 0 & \tfrac{1}{4} & 0 & 0 & 0 & 0 \\
0 & 0 & 0 & \tfrac{1}{4} & 0 & 0 & 0 \\
0 & 0 & 0 & 0 & \tfrac{1}{4} & 0 & 0 \\
0 & 0 & 0 & 0 & 0 & \tfrac{1}{3} & 0 \\
0 & 0 & 0 & 0 & 0 & 0 & \tfrac{1}{4}
\end{bmatrix}
\begin{bmatrix} 15 \\ 5 \\ 22 \\ 16 \\ 32 \\ 22 \\ 40 \end{bmatrix}
=
\begin{bmatrix} 3.7500 \\ 1.6667 \\ 5.5000 \\ 4.0000 \\ 8.0000 \\ 7.3333 \\ 10.0000 \end{bmatrix}
$$

$$
\begin{aligned}
SSTO &= \mathbf{y}'\mathbf{y} - (\Sigma Y_i)^2/N = 1082.00 - 888.62 \\
&= 193.38
\end{aligned}
$$

$$
\begin{aligned}
SSA &= \underset{1\times(p-1)}{(\mathbf{C}_A'\hat{\boldsymbol{\mu}})'} \, \underset{(p-1)\times(p-1)}{[\mathbf{C}_A'(\mathbf{X}'\mathbf{X})^{-1}\mathbf{C}_A]^{-1}} \, \underset{(p-1)\times 1}{(\mathbf{C}_A'\hat{\boldsymbol{\mu}})} \\
&= 85.15
\end{aligned}
$$

$$
\begin{aligned}
SSB(A) &= \underset{1\times\Sigma(q_{(j)}-1)}{(\mathbf{C}_{B(A)}'\hat{\boldsymbol{\mu}})'} \, \underset{\Sigma(q_{(j)}-1)\times\Sigma(q_{(j)}-1)}{[\mathbf{C}_{B(A)}'(\mathbf{X}'\mathbf{X})^{-1}\mathbf{C}_{B(A)}]^{-1}} \, \underset{\Sigma(q_{(j)}-1)\times 1}{(\mathbf{C}_{B(A)}'\hat{\boldsymbol{\mu}})} \\
&= 99.89
\end{aligned}
$$

$$
\begin{aligned}
SSWCELL &= \mathbf{y}'\mathbf{y} - \hat{\boldsymbol{\mu}}'(\mathbf{X}'\mathbf{y}) = 1082.00 - 1066.92 \\
&= 15.08
\end{aligned}
$$

TABLE 10.8-3 ANOVA Table for Type CRH-27(A) Design

Source	SS	df	MS	F
1 A (Ionization)	85.15	$p - 1 = 1$	85.15*	$[\frac{1}{3}]$ 107.78
2 $B(A)$ (Cages)	99.89	$\sum\limits_{j=1}^{p} (q_{(j)} - 1) = 5$	19.98*	$[\frac{2}{3}]$ 25.29
3 $WCELL$	15.08	$\sum\limits_{j=1}^{p} \sum\limits_{k=1}^{q_{(j)}} (n_{jk} - 1) = 19$	0.79	
4 Total	193.38	$N - 1 = 25$		

* $p < .01$

$$\mathbf{0}'_A = [0] \qquad \mathbf{0}'_{B(A)} = [0 \quad 0 \quad 0 \quad 0 \quad 0]$$

and

$$\boldsymbol{\mu}' = [\mu_{11} \quad \mu_{13} \quad \mu_{14} \quad \mu_{25} \quad \mu_{26} \quad \mu_{27} \quad \mu_{28}].$$

The coefficients in $\mathbf{C}'_A$ are given by $c_{jk} = \pm 1/q_{(j)}$; the coefficients in $\mathbf{C}'_{B(A)}$ are given by $c_{jk} = \pm 1$ or 0. A test of $\mathbf{C}'_{B(A)}\boldsymbol{\mu} = \mathbf{0}_{B(A)}$ is a test that simultaneously the μ_{jk}'s within each level of treatment A are equal. Alternatively, one can test the hypothesis that the μ_{jk}'s within the jth level of treatment A are equal, that is,

$$H_0: \quad \mu_{jk} - \mu_{jk'} = 0 \quad \text{for all } k \text{ at a given } j.$$

Procedures for computing the sums of squares necessary to test $\mathbf{C}'_A\boldsymbol{\mu} = \mathbf{0}_{(A)}$ and $\mathbf{C}'_{B(A)}\boldsymbol{\mu} = \mathbf{0}_{B(A)}$ at $\alpha = .05$ are illustrated in Table 10.8-2. The results of the analysis are summarized in Table 10.8-3. It is apparent that both null hypotheses can be rejected.

The analysis of unbalanced hierarchical designs for mixed and random-effects models follows that for the fixed-effects model. For additional information, one should consult Gill (1978, 185–210), Henderson (1953, 226–252), or Searle (1971a, Chs. 10 and 11).

10.9 ADVANTAGES AND DISADVANTAGES OF HIERARCHICAL DESIGNS

The major advantages of hierarchical designs are as follows.

1.   All subjects are used in simultaneously evaluating the effects of two or more treatments.

2.   They enable an experimenter to isolate the effects of nuisance variables and

evaluate treatments that cannot be crossed with other treatments.

The major disadvantages of hierarchical designs are as follows.

1. If numerous treatments are included in the experiment, the number of subjects required may be prohibitive.

2. The power of certain tests for mixed and random-effects models tends to be low because of the small number of degrees of freedom associated with the error term.

3. If, as is often the case, the nesting of treatment levels and/or the assignment of treatment combinations to experimental units is not random, the interpretation of the results may be ambiguous.

10.10 REVIEW EXERCISES

1. [10.1] Distinguish between

   a) Hierarchical and partial hierarchical designs

   b) Balanced hierarchical and unbalanced hierarchical designs.

2. An experiment was designed to evaluate the effectiveness of a pattern-practice approach to modifying the nonstandard dialect of Chicano children living in San Antonio, Texas. The essential elements of the approach, denoted by a_1, involved pattern drill in imitating audio recorded speech models and immediate feedback by a teacher. The control condition, a_2, involved reading stories aloud to the teacher. The children were praised for a good performance but no corrections or guidance was given. Twenty-four 12-year-old Chicano children with similar scores on a standardized speech test were randomly assigned to six miniclasses with four children in each. The six miniclasses, treatment B, were randomly assigned to the levels of treatment A. The dependent variable was the difference between the children's pre- and posttest scores on the test. The following data were obtained.

a_1	a_1	a_1	a_2	a_2	a_2
b_1	b_2	b_3	b_4	b_5	b_6
5	10	7	1	8	6
2	15	12	1	4	2
10	11	14	0	1	6
11	8	15	6	3	10

†a) [10.2] Test the hypotheses H_0: $\alpha_j = 0$ for all j and H_0: $\sigma_\beta^2 = 0$; let $\alpha = .05$.

†b) [10.2] Is it reasonable to assume that the population variance estimated by $MSB(A)$ is composed of homogeneous sources of variation? Test the hypotheses H_0: σ_β^2 at $a_1 = \sigma_\beta^2$ at a_2; let $\alpha = .25$.

†c) [10.5] Calculate $\hat{\omega}^2_{Y|A}$ and $\hat{\rho}_{IY|B(A)}$.

d) Prepare a "results and discussion section" for the *Journal of Educational Psychology*.

†3.   An industrial psychologist was interested in decreasing the time required to assemble an electronic component. Three assembly fixtures, treatment A, including the one currently in use, a_1, were evaluated. Five operators at each of six workplaces within the plant, treatment $B(A)$, were selected randomly. The assembly fixtures were randomly assigned to the six workplaces and the operators at those workplaces. After a three-week familiarization period, the following data on the number of units assembled per hour were collected.

a_1 b_1	a_1 b_2	a_2 b_3	a_2 b_4	a_3 b_5	a_3 b_6
17	13	21	25	26	32
7	23	24	29	34	32
9	14	14	26	32	37
12	18	19	24	30	34
15	22	17	21	28	35

a) [10.2] Test the hypotheses H_0: $\alpha_j = 0$ for all j and H_0: $\sigma^2_\beta = 0$; let $\alpha = .05$.

b) [10.2] Is it reasonable to assume that the population variance estimated by $MSB(A)$ is composed of homogeneous sources of variation? Use Cochran's C statistic to test the hypothesis H_0: σ^2_β at $a_1 = \sigma^2_\beta$ at $a_2 = \sigma^2_\beta$ at a_3; let $\alpha = .05$.

c) [10.4] Use Tukey's statistic to determine which assembly fixture population means differ.

d) [10.5] Calculate $\hat{\omega}^2_{Y|A}$ and $\hat{\rho}_{IY|B(A)}$.

e) [10.1] On the basis of the analysis, assembly fixture a_3 is superior to a_1, the one currently in use. Is there any aspect of the design of the experiment that would make you question the conclusion? How could the design have been improved?

4.   The effects of early environment on the problem-solving ability of rats at maturity were investigated. Rats were raised in either a normal environment, a_1, enriched environment, a_2, or restricted environment, a_3. Thirty-six rats were randomly assigned to nine cages, four rats to a cage. The cages were randomly assigned to the levels of treatment A. The dependent variable was the number of trials required to learn a visual discrimination task. The following data were obtained.

a_1 b_1	a_1 b_2	a_1 b_3	a_2 b_4	a_2 b_5	a_2 b_6	a_3 b_7	a_3 b_8	a_3 b_9
14	13	11	11	11	5	15	11	12
12	11	11	10	9	7	12	13	9
11	9	10	11	7	8	15	15	12
11	11	8	8	9	8	14	13	11

a) [10.2] Test the hypotheses H_0: $\alpha_j = 0$ for all j and H_0: $\sigma^2_\beta = 0$; let $\alpha = .05$.

b) [10.2] Is it reasonable to assume that the population variance estimated by $MSB(A)$ is composed of homogeneous sources of variation? Use Cochran's C statistic to test the hypothesis H_0: σ^2_β at $a_1 = \sigma^2_\beta$ at $a_2 = \sigma^2_\beta$ at a_3; let $\alpha = .05$.

c) [10.4] Use Tukey's statistic to determine which treatment A population means differ.

d) [10.5] Calculate $\hat{\omega}^2_{Y|A}$ and $\hat{\rho}_{IY|B(A)}$.

5.   [10.6] Identify the following designs; assume in each case that the building block design is a type CR-p design.

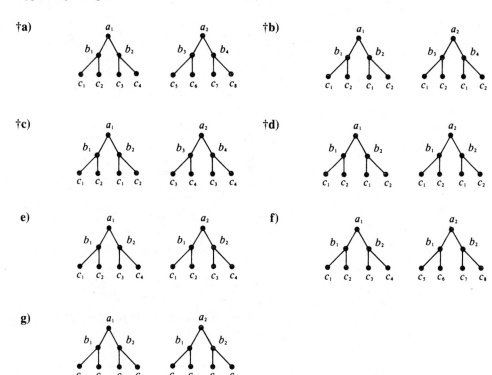

†a)   †b)

†c)   †d)

e)   f)

g)

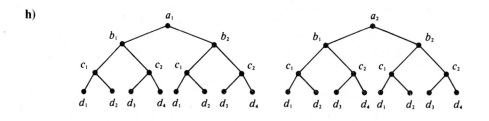

h)

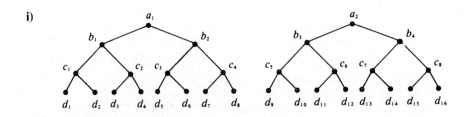

i)

†j)

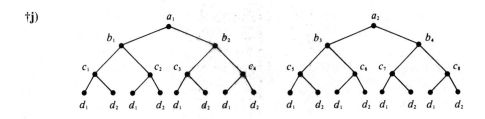

†k)

l)

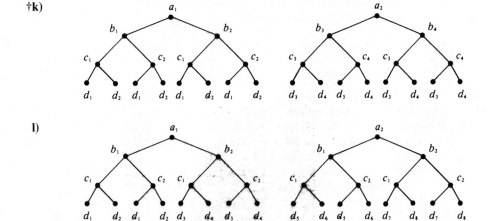

6.   Explain why the following is not a hierarchical design.

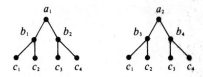

7.   [10.6] For each of the following designs (i) write the less than full rank experimental design model equation and (ii) construct a block diagram of the design. Use the minimum number of levels required for each treatment.

   †a) CRPH-*pqrt*(*B*)   †b) CRPH-*pqrt*(*AB*)   †c) CRH-*pq*(*A*)*r*(*AB*)*t*(*ABC*)

   d) CRPH-*pqrt*(*A*)   e) CRPH-*pq*(*A*)*r*(*A*)*t*(*A*)   f) CRPH-*pq*(*A*)*r*(*AB*)*t*

8.   [10.6] We have seen that there is a correspondence between the sum of squares for a nested treatment and the sums of squares in a factorial design. For example, $SSB(A) = SSB + SSAB$ and $SSC(AB) = SSC + SSAC + SSBC + SSABC$. (i) For each of the following, indicate the correspondence.

   †a) $SSC(B)$ †b) $SSB \times D(C)$ c) $SSD(BC)$

   †d) $SSA \times D(BC)$ e) $SSE(ABD)$ †f) $SSD(ABC)$

g) $SSA \times C(B)$ **h)** $SSD \times E(ABC)$ †**i)** $SSB(A) \times C(A)$

j) $SSC(AB) \times D$ **k)** $SSC(A) \times E(D)$ †**l)** $SSB(A) \times C(A) \times^* D(A)$

(ii) It is evident from part (i) that a simple rule governs the correspondence between the nested sum of squares and the sums of squares in the factorial design. State the rule.

9.   [10.7] Show by means of a table (see Table 10.7-1) how you would use a computer program for a type CRF-*pqrt* design to analyze data for a

     †**a)** Type CRPH-*pqrt*(*AB*) design   **b)** Type CRPH-*pq*(*A*)*r*(*A*)*t*(*A*) design.

†**10.**   [10.8] Exercise 2 described an experiment that was concerned with modifying the dialect of Chicano children. Suppose that observation $Y_{213} = 12$ is missing. Analyze these data using the full rank experimental design model approach; test hypotheses concerning simple averages of cell means. The mean squares can be obtained by analyzing the data as if a fixed-effects model were appropriate. The $E(MS)$'s for the mixed model are used to determine the form of the F ratio for testing null hypotheses. (This exercise requires the inversion of a 4 × 4 matrix. It can be done without a computer or calculator having a matrix inversion program but the computations are tedious.)

†**11.**   [10.8] Exercise 3 described an experiment that was concerned with decreasing the time required to assemble an electronic component. Suppose that observation $Y_{311} = 9$ and cell $a_1 b_2$ are missing. Analyze these data using the full rank experimental design model approach; test hypotheses concerning simple averages of cell means. The mean squares can be obtained by analyzing the data as if a fixed-effects model were appropriate. The $E(MS)$'s for the mixed model are used to determine the form of the F ratio for testing null hypotheses. (This exercise requires the inversion of 2 × 2 matrices. It can be done without a computer or calculator having a matrix inversion program.)

12.   [10.8] Exercise 4 described an experiment that investigated the effects of early environment on the problem-solving ability of rats. Analyze these data using the full rank experimental design model approach. For this balanced design, the mean squares for a mixed model can be obtained by analyzing the data as if a fixed-effects model were appropriate. The $E(MS)$'s for the mixed model are used to determine the form of the F ratio for testing H_0: $\alpha_j = 0$ for all j and H_0: $\sigma_\beta^2 = 0$. (This exercise requires the inversion of a 6 × 6 matrix. A computer with a matrix inversion program should be used.)

11 SPLIT-PLOT FACTORIAL DESIGN: DESIGN WITH GROUP-TREATMENT CONFOUNDING

11.1 INTRODUCTION TO SPLIT-PLOT FACTORIAL DESIGN

A split-plot factorial design with two or more completely crossed treatments is one of the most widely used designs in the behavioral sciences. As we will see, it contains features of two building block designs: a type CR-p design and a type RB-p design. A key feature of the randomized block design described in Chapter 6 is the blocking procedure that permits one to isolate the effects of a nuisance variable. This is accomplished by forming $i = 1, \ldots, n$ blocks of homogeneous experimental units, where the blocks represent the levels of the nuisance variable. A two-treatment split-plot factorial design extends this procedure to p samples or groups of such blocks. The levels of one treatment are randomly assigned to the p groups of blocks; the levels of a second treatment are randomly assigned to the experimental units within each block. An examination of Figure 11.1(a) should help to clarify the distinctive character of a split-plot factorial design. According to this figure, the subjects in sample S_1 of the split-plot factorial design receive treatment level a_1 in combination with all

489

FIGURE 11.1-1 Comparison of three factorial designs. The letters $S_1, S_2, \ldots, S_6$ denote groups (samples) of subjects.

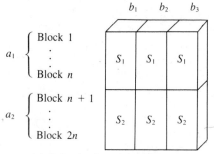

(a) Split-plot factorial design

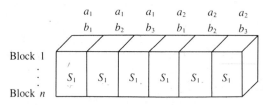

(b) Completely randomized factorial design

(c) Randomized block factorial design

levels of treatment B. Those in sample S_2 receive a_2 in combination with all levels of treatment B. By contrast, in the type CRF-23 design shown in Figure 11.1-1(b) each sample receives only one combination of the two treatments, whereas in the type RBF-23 design in Figure 11.1-1(c) S_1 receives all combinations of the two treatments. The block size for the randomized block factorial design is six, whereas that for the split-plot factorial design is only three. This difference can be important since the large block size required for many type RBF-pq designs precludes their use.

Split-plot factorial designs were originally used in agricultural research. The term *plot* refers to a plot of land that was subdivided or split into subplots. The use of these designs in educational research was popularized by E. F. Lindquist (1953) who referred to them as *mixed designs* because they contain a mixture of effects. Consider treatment A in Figure 11.1(a). The difference between levels a_1 and a_2 involves the difference between the blocks in S_1 and S_2 as well as the effects of treatment A. Such effects are called *between-blocks* effects. Consider now treatment B. Differences between any two levels of treatment B do not involve the difference between the blocks in S_1 and S_2 since the same blocks appear with all levels of the treatment. The effects of treatment B as well as those for the AB interaction are called *within-blocks* effects. The split-plot factorial design is the first one we have considered that contains a mixture of between- and within-blocks effects.

Looking at treatment A again it is apparent that the effects attributable to a_1 and a_2 are indistinguishable from those due to S_1 and S_2, the two groups of blocks. The groups and treatment A are said to be *completely confounded*. The effects of treatment B and the AB interaction are free of such confounding. Group-treatment confounding does not affect the interpretability of treatment effects, although, as we will see, it does affect precision and hence power. Group-treatment confounding occurs in several experimental designs including the completely randomized design described in Chapter 4. There, the groups also correspond to samples of n experimental units. Two more types of confounding—group-interaction confounding and treatment-interaction confounding—are described in Chapters 12 and 13, respectively. The type of confounding in a design is one of the factors used in classifying the design. Group-treatment confounding is characteristic of all split-plot factorial designs.

The advantage of a split-plot factorial design over a randomized block factorial design—smaller block size—involves a tradeoff that needs to be made explicit—less precision in estimating effects for the confounded treatment. The *precision* of an estimator is generally measured by the standard error of its sampling distribution: the smaller the standard error, the greater the precision. Type CRF-pq and type RBF-pq designs, assuming an additive model, use one error term in testing hypotheses involving A, B and AB. The two-treatment split-plot factorial design, however, uses two error terms: one for testing treatment A and a different and usually much smaller error term for treatment B and the AB interaction. As a result, the power of the tests for B and AB is greater than that for A.

The smaller block size of a split-plot factorial design is especially appealing. However, it is a poor design choice if an experimenter's primary interests involve treatments A and B. In this case, the experimenter should consider using a randomized block factorial design that does not involve confounding or one of the confounded factorial designs in Chapter 12 that involves group-interaction confounding. A split-plot factorial design is a good design choice if one's primary interests involve treatment B and the AB interaction.

The tests of treatments A and B in a split-plot factorial design, when viewed separately, resemble those in type CR-p and type RB-p designs, respectively. This analogy is discussed in Section 11.3.

We will denote a two-treatment split-plot factorial design by the letters SPF-$p \cdot q$. The lowercase letter preceding the dot stands for the number of levels of the between-blocks treatment (the confounded treatment); the letter after the dot stands for the number of levels of the within-blocks treatment (the unconfounded treatment). The design in Figure 11.1-1(a), for example, is a type SPF-2·3 design. The general designation for a design with one between-blocks treatment and two within-blocks treatments is SPF-$p \cdot qr$.

A split-plot factorial design is appropriate for experiments that meet, in addition to the assumptions of the experimental design model, the following conditions.

1. Two or more treatments with each treatment having two or more levels.

2. The number of combinations of treatment levels is greater than the desired size of each block.

3. If repeated measurements on the experimental units are obtained for treatment

B, each block contains one experimental unit, otherwise each block contains q experimental units.

4.   For the repeated measurements case, p samples of n experimental units from a population are randomly assigned to the levels of treatment A. The sequence of administration of the levels of treatment B to the experimental units in the np blocks is randomized independently for each block. An exception to this procedure is made when the nature of treatment B precludes randomization of the presentation order.

5.   For the two-treatment nonrepeated measurements case, p samples of n blocks each containing q matched experimental units from a population are randomly assigned to the levels of treatment A. Following this, the levels of treatment B are randomly assigned to the q experimental units within each block.

11.2 COMPUTATIONAL EXAMPLE FOR TYPE SPF-$p \cdot q$ DESIGN

Let us assume that an experimenter is interested in vigilance performance. An experiment has been designed to evaluate the relative effectiveness of two modes of signal presentation during a four-hour monitoring period. Treatment A is mode of signal presentation and has two levels: a_1 is an auditory signal (tone) and a_2 is a visual signal (light). Treatment B has four levels corresponding to successive hour monitoring periods: b_1 is the first hour, b_2 is the second hour, and so on. The research hypotheses leading to this experiment can be evaluated by means of statistical tests of the following null hypotheses:

$$H_0: \quad \alpha_j = 0 \quad \text{for all } j$$
$$H_1: \quad \alpha_j \neq 0 \quad \text{for some } j$$
$$H_0: \quad \beta_k = 0 \quad \text{for all } k$$
$$H_1: \quad \beta_k \neq 0 \quad \text{for some } k$$
$$H_0: \quad (\alpha\beta)_{jk} = 0 \quad \text{for all } j, k$$
$$H_1: \quad (\alpha\beta)_{jk} \neq 0 \quad \text{for some } j, k.$$

The experimenter's primary interest is in determining if the two treatments interact. The level of significance adopted for all tests is .05.

To test the preceding hypotheses, a random sample of eight subjects was obtained. The subjects were randomly divided into two subsamples (groups) of four each. The two levels of treatment A were then randomly assigned to the subsamples and observed under all four levels of treatment B. The dependent variable is response latency to the auditory and visual signals. Response latency scores were subjected to a logarithmic transformation as described in Section 2.5. The data that are analyzed are means of transformed latency scores for each successive hour during the four-hour monitoring session.

The layout of the type SPF-2·4 design and computational procedures are shown in Table 11.2-1. The analysis of variance is summarized in Table 11.2-2. According to this analysis, the null hypotheses for treatment B and the AB interaction are rejected. A graph of the AB interaction is given in Figure 11.2-1. An inspection of the graph indicates that response latency is shorter for visual signals than for auditory signals during the first three hours of monitoring but longer during the fourth hour. The best type of signal depends on the particular time period.

FIGURE 11.2-1   Graphs of the interaction between treatments A and B.

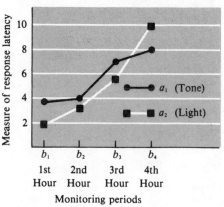
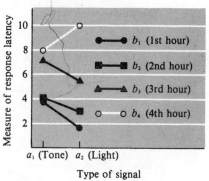

11.3   EXPERIMENTAL DESIGN MODEL FOR A TYPE SPF-$p \cdot q$ DESIGN

MIXED MODEL

A score Y_{ijk} in a split-plot factorial design is a composite as indicated in the following experimental design model.

$$
\begin{aligned}
Y_{ijk} &= \mu + (\mu_{\cdot j \cdot} - \mu) + (\mu_{ij \cdot} - \mu_{\cdot j \cdot}) + (\mu_{\cdot \cdot k} - \mu) + (\mu_{\cdot jk} - \mu_{\cdot j \cdot} - \mu_{\cdot \cdot k} + \mu) \\
&= \mu + \quad \alpha_j \quad + \quad \pi_{i(j)} \quad + \quad \beta_k \quad + \quad (\alpha\beta)_{jk} \\
&\qquad\quad + (Y_{ijk} - \mu_{\cdot jk} - \mu_{ij \cdot} + \mu_{\cdot j \cdot}) + \epsilon_{ijk} \\
&\qquad\quad + \qquad (\beta\pi)_{ki(j)} \qquad\qquad + \epsilon_{ijk} \\
&\qquad\qquad (i = 1, \ldots, n; j = 1, \ldots, p; k = 1, \ldots, q)
\end{aligned}
$$

(11.3-1)

TABLE 11.2-1   Computational Procedures for a Type SPF-2·4 Design

(i)   Data and notation [Y_{ijk} denotes a score for the experimental unit in block i and treatment combination jk; $i = 1, \ldots, n$ blocks (s_i) within each a_j; $j = 1, \ldots, p$ levels of treatment A (a_j); $k = 1, \ldots,$ q levels of treatment B (b_k)]:

ABS Summary Table
Entry is Y_{ijk}

		b_1	b_2	b_3	b_4	$\sum\limits_{k=1}^{q} Y_{ijk}$
a_1	s_1	3	4	7	7	21
	s_2	6	5	8	8	27
	s_3	3	4	7	9	23
	s_4	3	3	6	8	20
a_2	s_5	1	2	5	10	18
	s_6	2	3	6	10	21
	s_7	2	4	5	9	20
	s_8	2	3	6	11	22

AB Summary Table
Entry is $\sum\limits_{i=1}^{n} Y_{ijk}$

	b_1	b_2	b_3	b_4	$\sum\limits_{i=1}^{n}\sum\limits_{k=1}^{q} Y_{ijk}$
a_1	$n = 4$ 15	16	28	32	91
a_2	7	12	22	40	81
$\sum\limits_{i=1}^{n}\sum\limits_{j=1}^{p} Y_{ijk} =$	22	28	50	72	

(ii)   Computational symbols:

$$\sum_{i=1}^{n}\sum_{j=1}^{p}\sum_{k=1}^{q} Y_{ijk} = 3 + 4 + \cdots + 11 = 172.000$$

$$\frac{\left(\sum\limits_{i=1}^{n}\sum\limits_{j=1}^{p}\sum\limits_{k=1}^{q} Y_{ijk}\right)^2}{npq} = [Y] = \frac{(172.000)^2}{(4)(2)(4)} = 924.500$$

$$\sum_{i=1}^{n}\sum_{j=1}^{p}\sum_{k=1}^{q} Y_{ijk}^2 = [ABS] = (3)^2 + (4)^2 + \cdots + (11)^2 = 1160.000$$

$$\sum_{i=1}^{n}\sum_{j=1}^{p}\frac{\left(\sum\limits_{k=1}^{q} Y_{ijk}\right)^2}{q} = [AS] = \frac{(21)^2}{4} + \frac{(27)^2}{4} + \cdots + \frac{(22)^2}{4} = 937.000$$

TABLE 11.2-1 (continued)

$$\sum_{j=1}^{p} \frac{\left(\sum_{i=1}^{n}\sum_{k=1}^{q} Y_{ijk}\right)^2}{nq} = [A] = \frac{(91)^2}{(4)(4)} + \frac{(81)^2}{(4)(4)} = 927.625$$

$$\sum_{k=1}^{q} \frac{\left(\sum_{i=1}^{n}\sum_{j=1}^{p} Y_{ijk}\right)^2}{np} = [B] = \frac{(22)^2}{(4)(2)} + \frac{(28)^2}{(4)(2)} + \cdots + \frac{(72)^2}{(4)(2)} = 1119.000$$

$$\sum_{j=1}^{p}\sum_{k=1}^{q} \frac{\left(\sum_{i=1}^{n} Y_{ijk}\right)^2}{n} = [AB] = \frac{(15)^2}{4} + \frac{(16)^2}{4} + \cdots + \frac{(40)^2}{4} = 1141.500$$

(iii)   Computational formulas:

$$SSTO = [ABS] - [Y] = 235.500 \qquad SSWITHIN\ BL = [ABS] - [AS] = 223.000$$
$$SSBETWEEN\ BL = [AS] - [Y] = 12.500 \qquad\qquad SSB = [B] - [Y] = 194.500$$
$$SSA = [A] - [Y] = 3.125 \qquad\qquad SSAB = [AB] - [A] - [B] + [Y] = 19.375$$
$$SSBL(A) = [AS] - [A] = 9.375 \qquad SSB \times BL(A) = [ABS] - [AB] - [AS] + [A] = 9.125$$

---

where Y_{ijk} = a score for the ith experimental unit in treatment combination $a_j b_k$

μ = the overall population mean

α_j = the effect of treatment level j and is subject to the restriction $\sum_{j=1}^{p} \alpha_j = 0$

$\pi_{i(j)}$ = the effect of block i that is nested within a_j and is $NID(0, \sigma_\pi^2)$

$(\alpha\beta)_{jk}$ = the joint effect of treatment levels j and k and is subject to the restrictions $\sum_{j=1}^{p} (\alpha\beta)_{jk} = 0$ and $\sum_{k=1}^{q} (\alpha\beta)_{jk} = 0$

$(\beta\pi)_{ki(j)}$ = the joint effect of treatment level k and block i that is nested within a_j and is $NID(0, \sigma_{\beta\pi}^2)$

ϵ_{ijk} = the experimental error that is $NID(0, \sigma_\epsilon^2)$

The values of the parameters in model (11.3-1) are unknown, but they can be estimated from sample data as follows.

Statistic	Parameter Estimated
$\overline{Y}_{\cdot\cdot\cdot}$	μ
$\overline{Y}_{\cdot j\cdot} - \overline{Y}_{\cdot\cdot\cdot}$	α_j
$\overline{Y}_{ij\cdot} - \overline{Y}_{\cdot j\cdot}$	$\pi_{i(j)}$
$\overline{Y}_{\cdot\cdot k} - \overline{Y}_{\cdot\cdot\cdot}$	β_k
$\overline{Y}_{\cdot jk} - \overline{Y}_{\cdot j\cdot} - \overline{Y}_{\cdot\cdot k} + \overline{Y}_{\cdot\cdot\cdot}$	$(\alpha\beta)_{jk}$
$Y_{ijk} - \overline{Y}_{\cdot jk} - \overline{Y}_{ij\cdot} + \overline{Y}_{\cdot j\cdot}$	$(\beta\pi)_{ki(j)}$

The partition of the total sum of squares is obtained by the now familiar procedure of replacing the parameters in model (11.3-1) with statistics and rearranging terms (see Sections 2.2, 6.1, 7.5, and 8.4). Formulas for computing the various sums of squares are given in Table 11.2-1. The degrees of freedom for the sums of squares are given

TABLE 11.2-2 ANOVA Table for Type SPF-2·4 Design

Source	SS	df	MS	F	E(MS) Model III (A and B fixed; Blocks random)
1 Between blocks	12.500	$np - 1 = 7$			
2 A (Type of signal)	3.125	$p - 1 = 1$	3.125	[2/3] 2.00	$\sigma_\epsilon^2 + q\sigma_\pi^2 + nq\Sigma\alpha_j^2/(p - 1)$
3 Blocks w. A	9.375	$p(n - 1) = 6$	1.562		$\sigma_\epsilon^2 + q\sigma_\pi^2$
4 Within blocks	223.000	$np(q - 1) = 24$			
5 B (Monitoring periods)	194.500	$q - 1 = 3$	64.833	[5/7] 127.88*	$\sigma_\epsilon^2 + \sigma_{\beta\pi}^2 + np\Sigma\beta_k^2/(q - 1)$
6 AB	19.375	$(p - 1)(q - 1) = 3$	6.458	[6/7] 12.74*	$\sigma_\epsilon^2 + \sigma_{\beta\pi}^2 + n\Sigma\Sigma(\alpha\beta)_{jk}^2/(p - 1)(q - 1)$
7 B × blocks w. A	9.125	$p(n - 1)(q - 1) = 18$	0.507		$\sigma_\epsilon^2 + \sigma_{\beta\pi}^2$
8 Total	235.500	$npq - 1 = 31$			

*$p < .01$

in Table 11.2-2 along with the expected values of mean squares for the mixed model (A and B fixed, blocks random). Before presenting the $E(MS)$'s for other models, we will examine the nature of the two error terms in the type SPF-$p \cdot q$ design.

NATURE OF $\overset{M}{SSBL}(A)$ AND $\overset{M}{SSB} \times BL(A)$

Unlike the designs described previously, a type SPF-$p \cdot q$ design uses two error terms: $MSBL(A)$ for testing the between-blocks terms, and $MSB \times BL(A)$ for testing the within-blocks terms. We will show that $MSBL(A)$ corresponds to $MSWG$ in a completely randomized design and that $MSB \times BL(A)$ corresponds to $MSRES_{pooled}$, where the residual mean squares for p randomized block designs have been pooled. We will consider $MSBL(A)$ first.

The data in Table 11.2-1 can be laid out as a type CR-2 design if we ignore treatment B. This layout is shown in the following AS summary table and is appropriate since in the vigilance experiment the two levels of treatment A were randomly assigned to two random samples (groups) of four subjects each. The mean squares for the type CR-2 design are computed as follows.

AS Summary Table

a_1	a_2
$s_1 = 21$	$s_5 = 18$
$s_2 = 27$	$s_6 = 21$
$s_3 = 23$	$s_7 = 20$
$s_4 = 20$	$s_8 = 22$

$$[Y] = 3698.0$$
$$[AS] = 3748.0$$
$$[A] = 3710.5$$
$$MSBG = ([A] - [Y])/(p - 1) = 12.5/(2 - 1) = 12.50$$
$$MSWG = ([AS] - [A])/p(n - 1) = 37.5/[2(4 - 1)] = 6.25$$

The scores in the AS summary table are based on the sum of four observations. If we divide $MSBG$ and $MSWG$ by 4, we find that the resulting values are equal to MSA and $MSBL(A)$ in Table 11.2-2:

$$\frac{MSBG}{4} = MSA = 3.125$$

$$\frac{MSWG}{4} = MSBL(A) = 1.562.$$

Thus, $F = MSA/MSBL(A)$ in a split-plot factorial design is analogous to $F = MSBG/MSWG$ in a completely randomized design. In order for the F statistic to be distributed as the F distribution when the null hypothesis is true, $MSBL(a_1)$ and $MSBL(a_2)$ must estimate the same population variance. This is the familiar homogeneity of variance assumption that was introduced in Chapter 2 for a completely randomized design. Procedures for testing this assumption are described in Section 2.5.

Consider now the data at each level of a_j in Table 11.2-1. These data can be laid out as two type RB-4 designs as shown in the following *BS* Summary Tables. This is appropriate since in the vigilance experiment each set of blocks represents a random sample of n subjects for whom nq observations are obtained. The residual mean square for each of the type RB-4 designs can be computed as follows.

BS Summary Table Based on a_1

	b_1	b_2	b_3	b_4
s_1	3	4	7	7
s_2	6	5	8	8
s_3	3	4	7	9
s_4	3	3	6	8

$$[Y] = 517.562$$
$$[BS] = 585.000$$
$$[B] = 572.250$$
$$[S] = 524.750$$
$$MSRES(a_1) = \frac{[BS] - [B] - [S] + [Y]}{(n-1)(q-1)}$$
$$= 0.618$$

BS Summary Table Based on a_2

	b_1	b_2	b_3	b_4
s_5	1	2	5	10
s_6	2	3	6	10
s_7	2	4	5	9
s_8	2	3	6	11

$$[Y] = 410.062$$
$$[BS] = 575.000$$
$$[B] = 569.250$$
$$[S] = 412.250$$
$$MSRES(a_2) = \frac{[BS] - [B] - [S] + [Y]}{(n-1)(q-1)}$$
$$= 0.396$$

A pooled residual mean square is given by

$$MSRES_{pooled} = \frac{(n-1)(q-1)MSRES(a_1) + (n-1)(q-1)MSRES(a_2)}{(n-1)(q-1) + (n-1)(q-1)}$$
$$= \frac{(9)0.618 + (9)0.396}{9+9} = 0.507$$

which is equal to $MSB \times BL(A)$ in the split-plot factorial design. Thus $MSB \times BL(A)$ is simply the pooled interaction of treatment B and the blocks that are nested within each level of A.

We saw in Section 6.4 that when we are testing a within-blocks treatment the necessary and sufficient condition for the F statistic to be distributed as the F distribution given that the null hypothesis is true is the circularity condition. We will see in Section 11.4 how this applies to a split-plot factorial design.

EXPECTED VALUES OF MEAN SQUARES FOR A TYPE SPF-$p \cdot q$ DESIGN

The expected values of mean squares for models I, II, and III can be easily obtained from Table 11.3-1. The terms $1 - (n/N)$, $1 - (p/P)$, and $1 - (q/Q)$ in the table become zero if the corresponding terms BL, A, and B represent fixed effects and one if the corresponding terms are random. If, for example, the effects for blocks and treatment B are random but those for treatment A are fixed, the expected value of MSA is

$$E(MSA) = \sigma_\epsilon^2 + \sigma_{\beta\pi}^2 + n\sigma_{\alpha\beta}^2 + q\sigma_\pi^2 + nq \sum_{j=1}^{p} \alpha_j^2/(p-1).$$

The term σ_α^2 in the table has been replaced by $\Sigma_{j=1}^{p} \alpha_j^2/(p-1)$ since treatment A is fixed. It is apparent from Table 11.3-1 that the between-blocks error term $MSBL(A)$ does not include the required terms for testing MSA. An error term for testing MSA can be pieced together using the procedure described in Section 8.10. The quasi F' statistic for testing MSA is given by

$$F' = \frac{MSA}{MSBL(A) + MSAB - MSB \times BL(A)}.$$

The degrees of freedom for the numerator is given by $p - 1$; the degrees of freedom for the denominator is given by

$$\frac{[MSBL(A) + MSAB - MSB \times BL(A)]^2}{\dfrac{[MSBL(A)]^2}{df_{BL(A)}} + \dfrac{(MSAB)^2}{df_{AB}} + \dfrac{[MSB \times BL(A)]^2}{df_{B \times BL(A)}}}.$$

TABLE 11.3-1 Table for Determining $E(MS)$ for Type SPF-$p \cdot q$ Design

Source	$E(MS)$
A	$\sigma_\epsilon^2 + \left(1 - \dfrac{n}{N}\right)\left(1 - \dfrac{q}{Q}\right)\sigma_{\beta\pi}^2 + n\left(1 - \dfrac{q}{Q}\right)\sigma_{\alpha\beta}^2 + q\left(1 - \dfrac{n}{N}\right)\sigma_\pi^2 + nq\sigma_\alpha^2$
Blocks w. A	$\sigma_\epsilon^2 + \left(1 - \dfrac{q}{Q}\right)\sigma_{\beta\pi}^2 + q\sigma_\pi^2$
B	$\sigma_\epsilon^2 + \left(1 - \dfrac{n}{N}\right)\sigma_{\beta\pi}^2 + n\left(1 - \dfrac{p}{P}\right)\sigma_{\alpha\beta}^2 + np\sigma_\beta^2$
AB	$\sigma_\epsilon^2 + \left(1 - \dfrac{n}{N}\right)\sigma_{\beta\pi}^2 + n\sigma_{\alpha\beta}^2$
$B \times$ blocks w. A	$\sigma_\epsilon^2 + \sigma_{\beta\pi}^2$

11.4 SOME ASSUMPTIONS UNDERLYING A TYPE SPF-$p \cdot q$ DESIGN*

ASSUMPTIONS FOR BETWEEN-BLOCKS TESTS

A type SPF-$p \cdot q$ design involves two sets of assumptions: one set for the between-blocks test and another set for the within-blocks tests. The assumptions for the between-blocks test can be dispensed with in short order. As we saw in Section 11.3, the statistic $F = MSA/MSBL(A)$ in a split-plot factorial design is analogous to $F = MSBG/MSWG$ in a completely randomized design. It follows that the assumptions for the latter design discussed in Section 2.3 apply as well to the between-blocks test in the split-plot factorial design. These assumptions are discussed in detail in Section 2.5 and will not be repeated here.

ASSUMPTIONS FOR WITHIN-BLOCKS TESTS

The assumptions for the within-blocks tests include those described in Section 6.4 for a randomized block design and an assumption that we will refer to as the multisample circularity assumption. It states that

$$\mathbf{C}_B^{\star\prime} \mathbf{\Sigma}_{a_1} \mathbf{C}_B^{\star} = \mathbf{C}_B^{\star\prime} \mathbf{\Sigma}_{a_2} \mathbf{C}_B^{\star} = \cdots = \mathbf{C}_B^{\star\prime} \mathbf{\Sigma}_{a_p} \mathbf{C}_B^{\star} = \lambda \mathbf{I}$$

where $\mathbf{\Sigma}_{a_j}$ = a population variance-covariance matrix for each of the $j = 1, \ldots, p$ levels of treatment A

$\mathbf{C}_B^{\star\prime}$ = any orthonormal coefficient matrix that represents the null hypothesis for treatment B

λ = a scalar number ($\lambda > 0$)

$\mathbf{I}$ = an identity matrix.

For the vigilance experiment described in Section 11.2, the population variance-covariance matrices for a_1 and a_2 can be represented as follows.

$$
\mathbf{\Sigma}_{a_1} \; (q \times q) \qquad\qquad \mathbf{\Sigma}_{a_2} \; (q \times q)
$$

	b_1	b_2	b_3	b_4		b_1	b_2	b_3	b_4
b_1	σ_1^2	σ_{12}	σ_{13}	σ_{14}	b_1	σ_1^2	σ_{12}	σ_{13}	σ_{14}
b_2	σ_{21}	σ_2^2	σ_{23}	σ_{24}	b_2	σ_{21}	σ_2^2	σ_{23}	σ_{24}
b_3	σ_{31}	σ_{32}	σ_3^2	σ_{34}	b_3	σ_{31}	σ_{32}	σ_3^2	σ_{34}
b_4	σ_{41}	σ_{42}	σ_{43}	σ_4^2	b_4	σ_{41}	σ_{42}	σ_{43}	σ_4^2

---

* Portions of this section assume a familiarity with Section 6.4 and the matrix operations in Appendix D. The essential ideas can be grasped without this background.

Ordinarily these population matrices are unknown but they can be estimated from the data in Table 11.2-1. The estimators are denoted by $\hat{\boldsymbol{\Sigma}}_{a_1}$ and $\hat{\boldsymbol{\Sigma}}_{a_2}$. Formulas for computing the sample variances, $\hat{\sigma}_k^2$, and covariances, $\hat{\sigma}_{kk'}$, for the jth level of treatment A are, respectively,

$$\hat{\sigma}_k^2 = \left[\sum_{i=1}^n Y_{ijk}^2 - \left(\sum_{i=1}^n Y_{ijk} \right)^2 \Big/ n \right] \Big/ (n-1)$$

$$\hat{\sigma}_{kk'} = \left[\sum_{i=1}^n Y_{ijk} Y_{ijk'} - \left(\sum_{i=1}^n Y_{ijk} \right)\left(\sum_{i=1}^n Y_{ijk'} \right) \Big/ n \right] \Big/ (n-1).$$

The sample variance-covariance matrices for the data in Table 11.2-1 are

$$\hat{\boldsymbol{\Sigma}}_{a_1}$$

	b_1	b_2	b_3	b_4
b_1	2.2500	1.0000	1.0000	0
b_2	1.0000	.6667	.6667	0
b_3	1.0000	.6667	.6667	0
b_4	0	0	0	.6667

and

$$\hat{\boldsymbol{\Sigma}}_{a_2}$$

	b_1	b_2	b_3	b_4
b_1	.2500	.3333	.1667	0
b_2	.3333	.6667	0	-.3333
b_3	.1667	0	.3333	.3333
b_4	0	-.3333	.3333	.6667

We need to specify one more matrix in order to compute $\mathbf{C}_B^{*\prime} \, \boldsymbol{\Sigma}_{a_j} \mathbf{C}_B^*$. It is the ortho-normal coefficient matrix for testing the null hypothesis that all treatment B means are equal.[*] This matrix is

$$\mathbf{C}_B^{*\prime} = \begin{bmatrix} .7071 & -.7071 & 0 & 0 \\ 0 & 0 & .7071 & -.7071 \\ .5000 & .5000 & -.5000 & -.5000 \end{bmatrix}.$$

The null hypothesis for treatment B can be stated as

$$\mathbf{C}_B^{*\prime} \, \boldsymbol{\mu} = \mathbf{0}$$

where $\quad \boldsymbol{\mu}' = [\mu_{.1} \quad \mu_{.2} \quad \mu_{.3} \quad \mu_{.4}] \quad$ and $\quad \mathbf{0}' = [0 \quad 0 \quad 0].$

Sample estimates of $\mathbf{C}_B^{*\prime} \, \boldsymbol{\Sigma}_{a_j} \mathbf{C}_B^*$ for a_1 and a_2 are, respectively,

---

[*]   See Section 6.4.

$$\mathbf{C}_B^{\star\prime}\hat{\mathbf{\Sigma}}_{a_1}\mathbf{C}_B^{\star} \qquad\qquad \mathbf{C}_B^{\star\prime}\,\hat{\mathbf{\Sigma}}_{a_2}\mathbf{C}_B^{\star}$$

$$
\begin{bmatrix}
.4583 & .1666 & .4419 \\
.1666 & .6667 & .5893 \\
.4419 & .5893 & .7292
\end{bmatrix}
\quad\text{and}\quad
\begin{bmatrix}
.1250 & -.0833 & -.3241 \\
-.0833 & .1667 & .2946 \\
-.3241 & .2946 & .8958
\end{bmatrix}
$$

Given these sample matrices, the question arises as to the tenability of the multisample circularity assumption.

   If one wants to test the tenability of the multisample circularity assumption, Huynh and Feldt (1970) and Huynh and Mandeville (1979) have suggested that the assumption be tested in two steps. In the first step, one tests

(11.4-1) $$\mathrm{H}_0:\quad \mathbf{C}_B^{\star\prime}\,\mathbf{\Sigma}_{a_1}\mathbf{C}_B^{\star} = \mathbf{C}_B^{\star\prime}\,\mathbf{\Sigma}_{a_2}\mathbf{C}_B^{\star} = \cdots = \mathbf{C}_B^{\star\prime}\,\mathbf{\Sigma}_{a_p}\mathbf{C}_B^{\star}.$$

If this null hypothesis is not rejected, one tests

(11.4-2) $$\mathrm{H}_0:\quad \mathbf{C}_B^{\star\prime}\mathbf{\Sigma}_{\text{pooled}}\mathbf{C}_B^{\star} = \lambda\mathbf{I}$$

where $\mathbf{C}_B^{\star\prime}\,\mathbf{\Sigma}_{\text{pooled}}\,\mathbf{C}_B^{\star}$ is the pooled population matrix. The elements in $\mathbf{C}_B^{\star\prime}\,\mathbf{\Sigma}_{\text{pooled}}\,\mathbf{C}_B^{\star}$ are weighted averages of the corresponding elements in

$$\mathbf{C}_B^{\star\prime}\,\mathbf{\Sigma}_{a_1}\mathbf{C}_B^{\star},\;\ldots\;,\;\mathbf{C}_B^{\star\prime}\,\mathbf{\Sigma}_{a_p}\mathbf{C}_B^{\star}$$

where the weights are given by $(n_j - 1)/\Sigma_{j=1}^{p}\,(n_j - 1)$ $\quad(j = 1, \ldots, p)$.

   A statistic that can be used to test hypothesis (11.4-1) has been described by Box (1950). The test is a generalization of Bartlett's test for homogeneity of variances mentioned in Section 2.5, and, like Bartlett's test, is probably not robust with respect to nonnormality. The test statistic is

$$\chi^2 = (1 - E_1)M$$

where $$M = \left(\sum_{j=1}^{p} n_j - p\right) \ln\left|\mathbf{C}_B^{\star\prime}\,\hat{\mathbf{\Sigma}}_{\text{pooled}}\,\mathbf{C}_B^{\star}\right| - \sum_{j=1}^{p}\left[(n_j - 1)\ln\left|\mathbf{C}_B^{\star\prime}\,\hat{\mathbf{\Sigma}}_{a_j}\mathbf{C}_B^{\star}\right|\right]$$

$$E_1 = \frac{2q^2 + 3q - 1}{6(q + 1)(p - 1)}\left(\sum_{j=1}^{p}\frac{1}{n_j - 1} - \frac{1}{\displaystyle\sum_{j=1}^{p} n_j - p}\right)$$

q is the number of rows of $\mathbf{C}_B^{\star\prime}$ (q is 3 in our example), p is the number of levels of treatment A, and $\mathbf{C}_B^{\star\prime}\,\hat{\mathbf{\Sigma}}_{\text{pooled}}\,\mathbf{C}_B^{\star}$ is the pooled estimator of the common matrix. Also note that ln denotes the natural logarithm and $|\quad|$ denotes the determinant of a matrix.

   The χ^2 test statistic has a sampling distribution that is approximated by a chi-square distribution with

$$df_1 = \frac{q(q + 1)(p - 1)}{2}.$$

The approximation appears to be good if p and q do not exceed 4 or 5, and each n_j is perhaps 20 or more. When n_j is small or q and/or p are large, a more precise test is given by

$$F = \frac{M}{b}$$

with degrees of freedom equal to df_1 and

$$df_2 = (df_1 + 2)/(E_2 - E_1^2)$$

$$E_2 = \frac{(q - 1)(q + 2)}{6(p - 1)} \left[\sum_{j=1}^{p} \frac{1}{(n_j - 1)^2} - \frac{1}{\left(\sum_{j=1}^{p} n - p \right)^2} \right]$$

$$b = \frac{df_1}{1 - E_1 - (df_1/df_2)}.$$

The second hypothesis (11.4-2) can be tested following the procedures described in Table 6.4-1. Note that $\mathbf{C}^{*\prime} \hat{\mathbf{\Sigma}} \mathbf{C}^*$ should be replaced by $\mathbf{C}_B^{*\prime} \hat{\mathbf{\Sigma}}_{\text{pooled}} \mathbf{C}_B^*$. In Section 6.4 the routine testing of the circularity assumption was not recommended. This advice applies as well to the multisample circularity assumption. We saw in Section 6.4 that if the circularity assumption is not tenable the F test is positively biased. Fortunately the true distribution of the F statistic for any arbitrary variance-covariance matrix, assuming that

$$\mathbf{C}_B^{*\prime} \mathbf{\Sigma}_{a_1} \mathbf{C}_B^* = \mathbf{C}_B^{*\prime} \mathbf{\Sigma}_{a_2} \mathbf{C}_B^* = \cdots = \mathbf{C}_B^{*\prime} \mathbf{\Sigma}_{a_p} \mathbf{C}_B^*$$

can be approximated by an F statistic with reduced degrees of freedom. For a type SPF-$p \cdot q$ design, the degrees of freedom are

$$(q - 1)\theta \quad \text{and} \quad p(n - 1)(q - 1)\theta \qquad \text{for } F = MSB/MSB \times BL(A)$$
$$(p - 1)(q - 1)\theta \quad \text{and} \quad p(n - 1)(q - 1)\theta \qquad \text{for } F = MSAB/MSB \times BL(A)$$

where θ is a number that depends on the degree of departure of the population variance-covariance matrix from the required form. The computation of two estimates of θ is illustrated in Table 11.4-1.* The three-step testing strategy for the two within-blocks terms, treatment B and the AB interaction, is the same except for the degrees of freedom. The strategy is illustrated for treatment B; a fuller discussion is given in Section 6.4.

1. *Conventional F test.* First compare the value of the F statistic with $F_{\alpha; q-1, p(n-1)(q-1)}$. If F does not exceed this critical value, the analysis stops and F is declared not significant; if F exceeds the critical value, proceed to step 2.

2. *Conservative F test.* Compare the value of the F statistic with the critical

---

*   See Section 6.4 for comments about the estimators.

TABLE 11.4-1   Computation of $\hat{\theta}$ and $\tilde{\theta}$ for Data in Table 11.2-1.

(i)   Variance-covariance matrix [Elements in $\hat{\boldsymbol{\Sigma}}_{\text{pooled}}$ are weighted averages of the corresponding elements in $\hat{\boldsymbol{\Sigma}}_{a_1}$ and $\hat{\boldsymbol{\Sigma}}_{a_2}$, where the weights are given by $(n_j - 1)/\Sigma_{j=1}^{p}(n_j - 1)$ $(j = 1, \ldots, p)$]:

$$
\hat{\boldsymbol{\Sigma}}_{\substack{\text{pooled} \\ (q \times q)}} = \begin{array}{c} b_1 \\ b_2 \\ b_3 \\ b_4 \end{array}
\begin{array}{cccc}
\quad b_1 & \quad b_2 & \quad b_3 & \quad b_4 \\
\left[\begin{array}{cccc}
1.2500 & .6667 & .5834 & 0 \\
.6667 & .6667 & .3334 & -.1667 \\
.5834 & .3334 & .5000 & .1667 \\
0 & -.1667 & .1667 & .6667
\end{array}\right]
\end{array}
$$

(ii)   Computation of $\hat{\theta}$:

q = number of levels of treatment $B = 4$

$E_{kk'}$ = element in row k and column k' of $\hat{\boldsymbol{\Sigma}}_{\text{pooled}}$ $(k = 1, \ldots, q; k' = 1, \ldots, q)$

$\overline{E}_{..}$ = $\Sigma\Sigma E_{kk'}/q^2$ = mean of all elements

$= (1.2500 + .6667 + \cdots + .6667)/(4)^2 = .3906$

$\Sigma\Sigma E_{kk'}^2$ = sum of each element squared

$= (1.2500)^2 + (.6667)^2 + \cdots + (.6667)^2 = 4.6046$

$\overline{E}_D$ = mean of elements on the main diagonal $(D = 11, 22, \ldots, qq)$

$= (E_{11} + E_{22} + \cdots + E_{qq})/q = (1.2500 + .6667 + \cdots + .6667)/4 = .7708$

$\overline{E}_{k.}$ = mean of elements in row k

$\overline{E}_{1.} = [1.2500 + .6667 + \cdots + 0]/4 = .6250$

$\overline{E}_{2.} = [.6667 + .6667 + \cdots + (-.1667)]/4 = .3750$

$\overline{E}_{3.} = [.5834 + .3334 + \cdots + .1667]/4 = .3959$

$\overline{E}_{4.} = [0 + (-.1667) + \cdots + .6667]/4 = .1667$

$$
\hat{\theta} = \frac{q^2(\overline{E}_D - \overline{E}_{..})^2}{(q-1)\left(\Sigma\Sigma E_{kk'}^2 - 2q\sum_{k=1}^{q}\overline{E}_{k.}^2 + q^2\overline{E}_{..}^2\right)}
$$

$$
= \frac{(4)^2(.7708 - .3906)^2}{(4-1)\{4.6046 - 2(4)[(.6250)^2 + (.3750)^2 + \cdots + (.1667)^2] + (4)^2(.3906)^2\}}
$$

$$
= \frac{2.3128}{3.9585} = 0.5843
$$

(iii)   Computation of $\tilde{\theta}$:

$$
\tilde{\theta} = \frac{np(q-1)\hat{\theta} - 2}{(q-1)[np - p - (q-1)\hat{\theta}]}
$$

$$
= \frac{(4)(2)(4-1)0.5843 - 2}{(4-1)[(4)(2) - 2 - (4-1)0.5843]} = \frac{12.0232}{12.7413}
$$

$$
= 0.9436
$$

(iv)   Critical values of F for testing treatment B:

$$
F_{.05;3,18}(conventional) = 3.16
$$

TABLE 11.4-1 (continued)

$$F_{.05;1,6}(conservative) = 5.99$$
$$dfB = (q - 1)\hat{\theta} = (4 - 1)0.5843 = 1.753$$
$$dfB \times BL(A) = p(n - 1)(q - 1)\hat{\theta} = 2(4 - 1)(4 - 1)0.5843 = 10.517$$
$$F_{.05;1.753,10.517}(adjusted) \cong 4.16$$

(v)   Critical values of F for testing AB interaction:

$$F_{.05;3,18}(conventional) = 3.16$$
$$F_{.05;1,6}(conservative) = 5.99$$
$$dfAB = (p - 1)(q - 1)\hat{\theta} = (2 - 1)(4 - 1)0.5843 = 1.753$$
$$dfB \times BL(A) = p(n - 1)(q - 1)\hat{\theta} = 2(4 - 1)(4 - 1)0.5843 = 10.517$$
$$F_{.05;1.753,10.517}(adjusted) \cong 4.16$$

---

value for $(q - 1)[1/(q - 1)] = 1$ and $p(n - 1)(q - 1)[1/(q - 1)] = p(n - 1)$ degrees of freedom. If F exceeds $F_{\alpha;1,p(n-1)}$, the test is declared significant; if not, proceed to step 3.

3.   *Adjusted F test.* Compute the sample estimate, $\hat{\theta}$, of θ and the modified degrees of freedom $(q - 1)\hat{\theta}$ and $p(n - 1)(q - 1)\hat{\theta}$. Determine the critical value of F for these modified degrees of freedom. The test is declared significant if F exceeds $F_{\alpha;(q-1)\hat{\theta},p(n-1)(q-1)\hat{\theta}}$; if not, it is not significant.

The values of the F statistic for treatment B and the AB interaction in Table 11.2-2 are, respectively, 127.88 and 12.74. These F's exceed the critical value for the conventional F test in Table 11.4-1. They also exceed the value for the conservative F test. Hence, according to step 2 of the three-step testing strategy, both F's are declared significant. If one or both of the conservative F's had not been significant, we would have proceeded to the third step and performed an adjusted F test for the mean square(s) that was not declared significant at the second step. It is interesting to note also that the value of $\hat{\theta}$ is 0.5843, which is close to its lower bound, $1/(q - 1) = 0.3333$. This suggests that Σ departs appreciably from the required form.

11.5   PROCEDURES FOR TESTING HYPOTHESES ABOUT MEANS

CONTRASTS FOR MAIN-EFFECTS MEANS

In the following discussion, we assume that treatments A and B are fixed and blocks are random. Statistics for testing hypotheses about treatment A means,

$$H_0: \quad \psi_{i(A)} = c_1\mu_{.1.} + c_2\mu_{.2.} + \cdots + c_p\mu_{.p.} = 0$$

are as follows.

$$\frac{\hat{\psi}_{i(A)}}{\hat{\sigma}_{\psi_{i(A)}}} = \frac{c_1\overline{Y}_{1\cdot} + c_2\overline{Y}_{2\cdot} + \cdots + c_p\overline{Y}_{p\cdot}}{\sqrt{MSBL(A)\left(\dfrac{c_1^2}{nq} + \dfrac{c_2^2}{nq} + \cdots + \dfrac{c_p^2}{nq}\right)}}$$

$$\frac{\hat{\psi}_{i(A)}}{\hat{\sigma}_{\overline{Y}}} = \frac{c_j\overline{Y}_{j\cdot} + c_{j'}\overline{Y}_{j'\cdot}}{\sqrt{MSBL(A)/nq}}$$

$$df = p(n - 1)$$

Statistics for testing hypotheses about treatment B means,

$$\text{H}_0\text{:} \quad \hat{\psi}_{i(B)} = c_1\mu_{\cdot\cdot 1} + c_2\mu_{\cdot\cdot 2} + \cdots + c_q\mu_{\cdot\cdot q} = 0$$

are as follows.

$$\frac{\hat{\psi}_{i(B)}}{\hat{\sigma}_{\psi_{i(B)}}} = \frac{c_1\overline{Y}_{\cdot\cdot 1} + c_2\overline{Y}_{\cdot\cdot 2} + \cdots + c_q\overline{Y}_{\cdot\cdot q}}{\sqrt{MSB \times BL(A)\left(\dfrac{c_1^2}{np} + \dfrac{c_2^2}{np} + \cdots + \dfrac{c_q^2}{np}\right)}}$$

$$\frac{\hat{\psi}_{i(B)}}{\hat{\sigma}_{\overline{Y}}} = \frac{c_k\overline{Y}_{\cdot\cdot k} + c_{k'}\overline{Y}_{\cdot\cdot k'}}{\sqrt{MSB \times BL(A)/np}}$$

$$df = p(n - 1)(q - 1)$$

If $\mathbf{C}_B^{\star\prime} \mathbf{\Sigma}_{a_j} \mathbf{C}_B^{\star}$ ($j = 1, \ldots, p$) are equal for all j but $\mathbf{C}_B^{\star\prime} \mathbf{\Sigma}_{\text{pooled}} \mathbf{C}_B^{\star} \neq \lambda\mathbf{I}$, an exact test of

$$\text{H}_0\text{:} \quad \psi_{i(B)} = c_1\mu_{\cdot\cdot 1} + c_2\mu_{\cdot\cdot 2} + \cdots + c_q\mu_{\cdot\cdot q} = 0$$

can be obtained by replacing $MSB \times BL(A)$ in $\hat{\sigma}_{\psi_{i(B)}}$ or $\hat{\sigma}_{\overline{Y}}$ with a mean-square error term appropriate for the particular contrast. We will denote the error term for the ith contrast by $MS\hat{\psi}_{i(B)} \times BL(A)$; it has $p\,(n - 1)$ degrees of freedom. The computation of an error term for the contrast

$$\psi_{1(B)} = (1)\mu_{\cdot\cdot 1} + (-1)\mu_{\cdot\cdot 2} + (0)\mu_{\cdot\cdot 3} + (0)\mu_{\cdot\cdot 4}$$

is illustrated in Table 11.5-1. The data are from the vigilance experiment described in Section 11.2. According to Table 11.5-1, the value of the error term is 0.2917. If $\psi_{1(B)}$ represented an a posteriori contrast, Scheffé's test statistic could be used to test the null hypothesis $\psi_{1(B)} = 0$. The hypothesis cannot be rejected as the following computations show.

$$F = \frac{\hat{\psi}_{1(B)}^2}{\hat{\sigma}_{\psi_{1(B)}}^2} = \frac{[(1)2.75 + (-1)3.50 + (0)6.25 + (0)9.00]^2}{0.2917\left[\dfrac{(1)^2}{(4)(2)} + \dfrac{(-1)^2}{(4)(2)} + \dfrac{(0)^2}{(4)(2)} + \dfrac{(0)^2}{(4)(2)}\right]}$$

$$= \frac{(-.75)^2}{.073} = 7.71$$

$$F' = (4 - 1)F_{.05;4-1,2(4-1)} = 14.28$$

The use of mean square error terms appropriate for specific contrasts was introduced in Section 6.5. The advantage of using $MS\hat{\psi}_{i(B)} \times BL(A)$ in place of $MSB \times BL(A)$ is that the resulting test is exact. If the multisample circularity assumption is tenable,

---

TABLE 11.5-1 Procedures for Computing a Mean-Square Error Term Appropriate for a Specific Contrast Among Treatment B Means

---

(i) Computation of contrast for each block (Coefficients of the contrast, $c_1 = 1$, $c_2 = -1$, $c_3 = 0$, $c_4 = 0$, are applied to the data in Table 11.2-1):

$$(1)\ Y_{ij1} + (-1)Y_{ij2} + (0)Y_{ij3} + (0)Y_{ij4} = \hat{\psi}_{1(B)} \text{ at } a_j s_i$$

	s_1	(1) 3 + (−1)4 + (0)7 + (0) 7	−1
	s_2	(1) 6 + (−1)5 + (0)8 + (0) 8	1
a_1	s_3	(1) 3 + (−1)4 + (0)7 + (0) 9	−1
	s_4	(1) 3 + (−1)3 + (0)6 + (0) 8	0

$$\sum_{i=1}^{n} \hat{\psi}_{1(B)} \text{ at } a_1 s_i = -1$$

	s_5	(1) 1 + (−1)2 + (0)5 + (0)10	−1
	s_6	(1) 2 + (−1)3 + (0)6 + (0)10	−1
a_2	s_7	(1) 2 + (−1)4 + (0)5 + (0) 9	−2
	s_8	(1) 2 + (−1)3 + (0)6 + (0)11	−1

$$\sum_{i=1}^{n} \hat{\psi}_{1(B)} \text{ at } a_2 s_i = -5$$

$$\sum_{j=1}^{p} \left(\sum_{i=1}^{n} \hat{\psi}_{1(B)} \text{ at } a_j s_i \right)^2 = (-1)^2 + (-5)^2 = 26$$

$$\sum_{j=1}^{p} \sum_{i=1}^{n} \hat{\psi}_{1(B)}^2 \text{ at } a_j s_i = (-1)^2 + (1)^2 + \cdots + (-1)^2 = 10$$

(ii) Computation of $MS\hat{\psi}_{1(B)} \times BL(A)$:

$$SS\hat{\psi}_{1(B)} \times BL(A) = \left[\sum_{j=1}^{p} \sum_{i=1}^{n} \hat{\psi}_{1(B)}^2 \text{ at } a_j s_i - \frac{\sum_{j=1}^{p} \left(\sum_{i=1}^{n} \hat{\psi}_{1(B)} \text{ at } a_j s_i \right)^2}{n} \right] \Big/ \sum_{k=1}^{q} c_k^2$$

$$= \left[10 - \frac{(26)}{4} \right] \Big/ [(1)^2 + (-1)^2 + (0)^2 + (0)^2]$$

$$= 1.7500$$

$$MS\hat{\psi}_{1(B)} \times BL(A) = SS\hat{\psi}_{1(B)} \times BL(A)/p(n-1)$$

$$= 1.7500/[2(4-1)] = 0.2917$$

---

the use of $MSB \times BL(A)$ instead of $MS\hat{\psi}_{i(B)} \times BL(A)$ leads to a more powerful test. In the discussion that follows, the assumption

$$\mathbf{C}_B^{\star\prime} \, \Sigma_{a_j} \mathbf{C}_B^{\star} \qquad \text{are equal for all } j \quad (j = 1, \ldots, p)$$

is required whenever a within-blocks error term is pooled over the p levels of treatment A. The assumption

$$\mathbf{C}_B^{\star\prime} \, \Sigma_{\text{pooled}} \mathbf{C}_B^{\star} = \lambda \mathbf{I}$$

will be referred to as the circularity assumption.

CONTRASTS FOR SIMPLE-EFFECTS MEANS

Statistics for testing hypotheses about simple-effects means for treatment A at b_k,

$$H_0: \quad \psi_{i(A)} \text{ at } b_k = c_1\mu_{\cdot 1k} + c_2\mu_{\cdot 2k} + \cdots + c_p\mu_{\cdot pk} = 0$$

are as follows:

$$\frac{\hat{\psi}_{i(A)} \text{ at } b_k}{\hat{\sigma}_{\psi_{i(A)}} \text{ at } b_k} = \frac{c_1\overline{Y}_{\cdot 1k} + c_2\overline{Y}_{\cdot 2k} + \cdots + c_p\overline{Y}_{\cdot pk}}{\sqrt{MSWCELL\left[\dfrac{c_1^2}{n} + \dfrac{c_2^2}{n} + \cdots + \dfrac{c_p^2}{n}\right]}}$$

$$\frac{\hat{\psi}_{i(A)} \text{ at } b_k}{\hat{\sigma}_{\overline{Y}}} = \frac{c_j\overline{Y}_{\cdot jk} + c_{j'}\overline{Y}_{\cdot j'k}}{\sqrt{MSWCELL/n}}$$

where $\qquad MSWCELL$ is equal to $\dfrac{SSBL(A) + SSB \times BL(A)}{p(n-1) + p(n-1)(q-1)}$.

The use of $MSWCELL$ in the denominator of these ratios is discussed later. We know from the formula for $MSWCELL$ just given that it is composed of between-blocks and within-blocks variation. We also know that in a split-plot factorial design the corresponding population error variances are generally not homogeneous. Cochran and Cox (1957, 100, 298) point out that under these conditions test statistics, such as Student's t and Tukey's q, do not follow their distributions except in special cases. They propose a conservative test that can be used whenever error terms estimating different sources of variability are pooled. The critical value of the test statistic is equal to

$$h_\alpha' = \frac{h_1 MSB(A) + h_2 MSB \times BL(A)(q-1)}{MSB(A) + MSB \times BL(A)(q-1)}$$

where h_α' denotes the critical value of one of the test statistics, t, $t'D$, q, et cetera, and h_1 and h_2 are the tabled values of the test statistic at α level of significance for the df associated with $MSB(A)$ and $MSB \times BL(A)$, respectively. For example, if Tukey's statistic is used to test $\psi_{i(A)}$ at $b_k = 0$ for the data in Table 11.2-1, the critical value of $q = (\hat{\psi}_{i(A)} \text{ at } b_k)/\hat{\sigma}_{\overline{Y}}$ is

$$q'_{.05} = \frac{q_{.05,6}MSB(A) + q_{.05,18}MSB \times BL(A)(q - 1)}{MSB(A) + MSB \times BL(A)(q - 1)}$$

$$= \frac{(3.46)(1.562) + (2.97)(0.507)(4 - 1)}{1.562 + 0.507(4 - 1)}$$

$$= 3.22.$$

The value of h'_α will always be between those for h_1 and h_2 except when the degrees of freedom for the two error terms are equal; then $h'_\alpha = h_1 = h_2$. See Taylor (1950) for an extensive discussion of standard error formulas and approximate degrees of freedom.

Statistics for testing hypotheses about simple-effects means for treatment B at a_j,

$$H_0: \quad \psi_{i(B)} \text{ at } a_j = c_1\mu_{.j1} + c_2\mu_{.j2} + \cdots + c_q\mu_{.jq} = 0$$

are as follows.

$$\frac{\hat{\psi}_{i(B)} \text{ at } a_j}{\hat{\sigma}_{\psi_{i(B)}} \text{ at } a_j} = \frac{c_1\overline{Y}_{.j1} + c_2\overline{Y}_{.j2} + \cdots + c_q\overline{Y}_{.jq}}{\sqrt{MSB \times BL(A)\left(\dfrac{c_1^2}{n} + \dfrac{c_2^2}{n} + \cdots + \dfrac{c_q^2}{n}\right)}}$$

$$\frac{\hat{\psi}_{i(B)} \text{ at } a_j}{\hat{\sigma}_{\overline{Y}}} = \frac{c_k\overline{Y}_{.jk} + c_{k'}\overline{Y}_{.jk'}}{\sqrt{MSB \times BL(A)/n}}$$

$$df = p(n - 1)(q - 1)$$

If the circularity assumption is not tenable, an exact test of $\hat{\psi}_{i(B)}$ at $a_j = 0$ is obtained by replacing $MSB \times BL(A)$ in $\hat{\sigma}_{\psi_{i(B)}}$ or $\hat{\sigma}_{\overline{Y}}$ with $[SS\hat{\psi}_{i(B)} \times BL(a_j)]/(n - 1)$, where

$$SS\hat{\psi}_{i(B)} \times BL(a_j) = \left[\sum_{i=1}^{n} \hat{\psi}_{i(B)}^2 \text{ at } a_js_i - \frac{\left(\sum_{i=1}^{n} \hat{\psi}_{i(B)} \text{ at } a_js_i\right)^2}{n}\right] \Bigg/ \sum_{k=1}^{q} c_k^2.$$

The computation of this mean square error term is similar to that for $MS\hat{\psi}_{i(B)} \times BL(A)$ that is illustrated in Table 11.5-1.

In the following section we will see why $MSWCELL$ was used in testing $\psi_{i(A)}$ at $b_k = 0$ but $MSB \times BL(A)$ was used in testing $\psi_{i(B)}$ at $a_j = 0$.

RULE GOVERNING CHOICE OF ERROR TERM FOR SIMPLE-EFFECTS MEANS AND SIMPLE MAIN EFFECTS

An easy-to-follow rule governs the choice of error terms for testing hypotheses about simple-effects means and simple main effects (see Section 11.6) when the design contains between- and within-blocks effects. Recall from Section 8.6 that

$$\sum_{k=1}^{q} SSA \text{ at } b_k = SSA + SSAB$$

$$\sum_{j=1}^{p} SSB \text{ at } a_j = SSB + SSAB.$$

According to Table 11.2-2, when A and B represent fixed effects and blocks are random, the error terms for testing treatment and interaction null hypotheses are

$$MSBL(A) \qquad \text{for treatment } A$$
$$MSB \times BL(A) \qquad \text{for treatment } B$$
$$MSB \times BL(A) \qquad \text{for } AB \text{ interaction.}$$

The rule governing the choice of error terms states that if the treatment(s) and interaction(s) that equal the sum of simple main effects have different error terms, as in the case of ΣSSA at b_k, the error terms should be pooled in testing null hypotheses for the associated simple-effects means and simple main effects. If the treatment(s) and interaction(s) that equal the sum of simple main effects have the same error term, as in the case of ΣSSB at a_j, that error term should be used in testing null hypotheses for the associated simple-effects means and simple main effects. Thus, because A and AB have different error terms, the error term for testing, say, $\psi_{i(A)}$ at $b_k = 0$ is

$$MS_{\text{pooled}} = \frac{SSBL(A) + SSB \times BL(A)}{p(n-1) + p(n-1)(q-1)}.$$

It turns out that this pooled error term is actually $MSWCELL$, which, for purposes of testing treatments A and B, was partitioned into between- and within-blocks components.

Earlier we observed that $MSWCELL$ in a split-plot factorial design is likely to be composed of heterogeneous sources of variance. If the variances are heterogeneous, tests for simple-effects means and simple main effects for treatment A are biased. In general, the bias will be small if the degrees of freedom for $MSWCELL$ are greater than 30. An alternative to the conservative h' test procedure described earlier is to pool the within-cell variances from only those cells used to compute the simple-effects means or simple main effects. The degrees of freedom for this mean square error term are reduced to $v(n-1)$, where v is the number of variances that are pooled.

11.6 PROCEDURES FOR TESTING HYPOTHESES ABOUT SIMPLE MAIN EFFECTS AND TREATMENT-CONTRAST INTERACTIONS

If the AB interaction in a type SPF-$p \cdot q$ design is significant, an experimenter may be interested in testing hypotheses about simple main effects or treatment-contrast interactions. The relative merits of the two procedures are discussed in Section 8.6.

SIMPLE MAIN EFFECTS

Simple main effects procedures can be used to test the following kinds of hypotheses.

$$H_0: \quad \alpha_j \text{ at } b_k = 0 \quad \text{for all } j \qquad \text{or} \qquad \mu_{\cdot 1k} = \mu_{\cdot 2k} = \cdots = \mu_{\cdot pk}$$

$$H_0: \quad \beta_k \text{ at } a_j = 0 \quad \text{for all } k \qquad \text{or} \qquad \mu_{\cdot j1} = \mu_{\cdot j2} = \cdots = \mu_{\cdot jq}$$

Assuming a mixed model in which treatments A and B are fixed and blocks are random, the statistics for testing the preceding hypotheses are, respectively,

$$F = \frac{(SSA \text{ at } b_k)/(p - 1)}{MSWCELL}$$

with degrees of freedom equal to $p - 1$ and $[(p - 1) + p(n - 1)(q - 1)]$ and

$$F = \frac{(SSB \text{ at } a_j)/(q - 1)}{MSB \times BL(A)}$$

with degrees of freedom equal to $(q - 1)$ and $p(n - 1)(q - 1)$, where

$$SSA \text{ at } b_k = \sum_{j=1}^{p} \frac{\left(\sum_{i=1}^{n} Y_{ijk}\right)^2}{n} - \frac{\left(\sum_{i=1}^{n}\sum_{j=1}^{p} Y_{ijk}\right)^2}{np}$$

$$SSB \text{ at } a_j = \sum_{k=1}^{q} \frac{\left(\sum_{i=1}^{n} Y_{ijk}\right)^2}{n} - \frac{\left(\sum_{i=1}^{n}\sum_{k=1}^{q} Y_{ijk}\right)^2}{nq}.$$

The rationale for using $MSWCELL$ in testing H_0: α_j at $b_k = 0$ for all j is discussed in Section 11.5. As we have seen, this error term is usually composed of heterogeneous sources of variation. If the degrees of freedom for $MSWCELL$ are less than 30, it may be desirable to pool the within-cell variances from only those cells used in computing SSA at b_j. The resulting error term has $v(n - 1)$ degrees of freedom, where v is the number of within-cell variances that have been pooled.

The use of $MSB \times BL(A)$ in testing H_0: β_k at $a_j = 0$ is appropriate if the circularity assumption is tenable. If the assumption is not tenable, the three-step testing strategy described in Section 11.4 can be used.

TREATMENT-CONTRAST INTERACTIONS

If the AB interaction is significant, considerable insight into the sources of non-additivity in the data can be obtained by testing the following kinds of hypotheses:

$$H_0: \quad \alpha\psi_{i(B)} = \delta \text{ for all } j$$

$$H_0: \quad \beta\psi_{i(A)} = \delta \text{ for all } k$$

where

$$\psi_{i(B)} = c_1\mu_{\cdot j1} + c_2\mu_{\cdot j2} + \cdots + c_q\mu_{\cdot jq}$$

$$\psi_{i(A)} = c_1\mu_{\cdot 1k} + c_2\mu_{\cdot 2k} + \cdots + c_p\mu_{\cdot pk}$$

and δ is a constant for a given hypothesis. An examination of the sample data usually suggests a number of interesting $\psi_{i(B)}$'s and $\psi_{i(A)}$'s. Statistics for testing the preceding hypotheses are

$$F = \frac{SSA\hat{\psi}_{i(B)}/(p-1)}{MSB \times BL(A)}$$

with degrees of freedom equal to $(p-1)$ and $p(n-1)(q-1)$ and

$$F = \frac{SSB\hat{\psi}_{i(A)}/(q-1)}{MSB \times BL(A)}$$

with degrees of freedom equal to $(q-1)$ and $p(n-1)(q-1)$, where

$$SSA\hat{\psi}_{i(B)} = \left[\sum_{j=1}^{p} \left(\sum_{i=1}^{n} \hat{\psi}_{i(B)} \text{ at } a_j s_i \right)^2 - \frac{\left(\sum_{j=1}^{p} \sum_{i=1}^{n} \hat{\psi}_{i(B)} \text{ at } a_j s_i \right)^2}{p} \right] \Bigg/ n \sum_{k=1}^{q} c_k^2$$

$$SSB\hat{\psi}_{i(A)} = \left[\sum_{k=1}^{q} \hat{\psi}_{i(A)}^2 \text{ at } b_k - \frac{\left(\sum_{k=1}^{q} \hat{\psi}_{i(A)} \text{ at } b_k \right)^2}{q} \right] \Bigg/ n \sum_{j=1}^{p} c_j^2$$

$$\hat{\psi}_{i(B)} \text{ at } a_j s_i = c_1 Y_{ij1} + c_2 Y_{ij2} + \cdots + c_q Y_{ijq}$$

$$\hat{\psi}_{i(A)} \text{ at } b_k = c_1 Y_{11k} + c_1 Y_{21k} + \cdots + c_1 Y_{n1k} + \cdots + c_p Y_{1pk} + c_p Y_{2pk}$$
$$+ \cdots + c_p Y_{npk}.$$

The simultaneous test procedure or the Dunn-Šidák procedure, described in Section 3.4, should be used in conjunction with these tests since they are a posteriori.

If the circularity assumption is not tenable, $MSB \times BL(A)$ in the F statistics should be replaced by error terms appropriate for the specific treatment-contrast interactions, for example,

$$F = \frac{SSA\hat{\psi}_{i(B)}/(p-1)}{SS\hat{\psi}_{i(B)} \times BL(A)/p(n-1)}$$

$$F = \frac{SSB\hat{\psi}_{i(A)}/(q-1)}{SSB \times BL(\hat{\psi}_{i(A)})/p'(n-1)(q-1)}$$

where p' denotes the number of levels of treatment A involved in $\hat{\psi}_{i(A)}$. The computations will be illustrated for the vigilance data in Table 11.5-1. Suppose an experimenter is interested in the contrasts defined by

$$\mathbf{c}'_{1(B)} = [1 \quad -1 \quad 0 \quad 0] \quad \text{and} \quad \mathbf{c}'_{1(A)} = [1 \quad -1].*$$

We will compute $SSA\hat{\psi}_{1(B)}$ and $SS\hat{\psi}_{1(B)} \times BL(A)$ first; the required terms

---

*   In practice the contrast defined by $\mathbf{c}_{1(A)}$ would not be tested since the hypothesis $\beta\psi_{1(A)} = \delta$ for all k is equivalent to $\alpha\beta = 0$ for all j and k when, as in the vigilance experiment, p is equal to two.

$$\sum_{i=1}^{n} \hat{\psi}_{1(B)} \text{ at } a_1 s_i = -1 \quad \text{ and } \quad \sum_{i=1}^{n} \hat{\psi}_{1(B)} \text{ at } a_2 s_i = -5$$

are obtained from Table 11.5-1. The sum of squares for the interaction of treatment A and $\hat{\psi}_{1(B)}$ is given by

$$SSA\hat{\psi}_{1(B)} = \left\{ (-1)^2 + (-5)^2 - \frac{[(-1) + (-5)]^2}{2} \right\} \Big/ (4)[(1)^2 + (-1)^2 + (0)^2 + (0)^2]$$
$$= 1.0000.$$

Procedures for computing the error term $MS\hat{\psi}_{1(B)} \times BL(A)$ are illustrated in Table 11.5-1. The F statistic is

$$F = \frac{1.0000/1}{0.2917} = 3.43.$$

The critical value of F using the simultaneous test procedure is

$$[(p-1)(q-1)F_{\alpha;(p-1)(q-1),p(n-1)}]/(p-1) = (3F_{.05;3,6})/1 = 14.28.$$

The hypothesis $\alpha\psi_{1(B)} = \delta$ for all j cannot be rejected. Procedures for computing the terms needed to test the hypothesis $\beta\psi_{1(A)} = \delta$ for all k are illustrated in Table 11.6-1. According to the table, this hypothesis cannot be rejected.

   We conclude this section with a summary comment about the choice of error terms. In Section 8.6 we saw that treatment-contrast interactions, unlike simple main effects, represent a partition of an omnibus interaction. The error mean square that is used in testing the omnibus interaction should also be used in testing the various treatment-contrast interactions if the circularity assumption is tenable. If it is not tenable, the error mean square that is used to test the omnibus interaction should be partitioned into mean squares appropriate for the specific treatment-contrast interactions, that is into either $MS\hat{\psi}_{i(B)} \times BL(A)$ or $MSB \times BL(\psi_{i(A)})$.

11.7 TEST FOR TRENDS IN A TYPE SPF-$p \cdot q$ DESIGN

Procedures for performing trend analyses are described in Sections 4.5 and 8.7. These procedures generalize with slight modification to split-plot factorial designs.

TESTS OF TREND CONTRASTS AND TREATMENT × TREND-CONTRAST INTERACTIONS

The vigilance data in Table 11.2-1 are suitable for trend analysis because treatment B, successive hour monitoring periods, represents a quantitative variable. Furthermore, the levels of treatment B are separated by equal intervals. Thus the trend

TABLE 11.6-1   Procedures for Computing $MSB\hat{\psi}_{i(A)}$ and $MSB \times BL(\hat{\psi}_{i(A)})$

(i)   Computation of contrast (Coefficients of the contrast, $c_1 = 1$ and $c_2 = -1$, are applied to data in Table 11.2-1):

		$(1)Y_{i11}$	$(1)Y_{i12}$	$(1)Y_{i13}$	$(1)Y_{i14}$	$\sum\limits_{k=1}^{q} c_1 Y_{i1k}$
	s_1	(1) 3	(1) 4	(1) 7	(1) 7	21
	s_2	(1) 6	(1) 5	(1) 8	(1) 8	27
a_1	s_3	(1) 3	(1) 4	(1) 7	(1) 9	23
	s_4	(1) 3	(1) 3	(1) 6	(1) 8	20

$$\sum\limits_{i=1}^{n} c_1 Y_{i1k} = \qquad 15 \qquad\qquad 16 \qquad\qquad 28 \qquad\qquad 32 \qquad \sum\limits_{i=1}^{n}\sum\limits_{k=1}^{q} c_1 Y_{i1k} = \quad 91$$

		$(-1)Y_{i21}$	$(-1)Y_{i22}$	$(-1)Y_{i23}$	$(-1)Y_{i24}$	$\sum\limits_{k=1}^{q} c_2 Y_{i2k}$
	s_5	(−1) 1	(−1) 2	(−1) 5	(−1) 10	−18
	s_6	(−1) 2	(−1) 3	(−1) 6	(−1) 10	−21
a_2	s_7	(−1) 2	(−1) 4	(−1) 5	(−1) 9	−20
	s_8	(−1) 2	(−1) 3	(−1) 6	(−1) 11	−22

$$\sum\limits_{i=1}^{n} c_2 Y_{i2k} = \qquad -7 \qquad\qquad -12 \qquad\qquad -22 \qquad\qquad -40 \qquad \sum\limits_{i=1}^{n}\sum\limits_{k=1}^{q} c_2 Y_{i2k} = -81$$

$$\hat{\psi}_{1(A)} \text{ at } b_k = \sum\limits_{j=1}^{p}\sum\limits_{i=1}^{n} c_j Y_{ijk} = \qquad 8 \qquad\qquad 4 \qquad\qquad 6 \qquad\qquad -8 \qquad \sum\limits_{k=1}^{q} \hat{\psi}_{1(A)} \text{ at } b_k = \quad 10$$

(ii)   Computational symbols:

$$\sum\limits_{i=1}^{n}\sum\limits_{j=1}^{p}\sum\limits_{k=1}^{q} c_j Y_{ijk}^2 = [ABS] = (3)^2 + (4)^2 + \cdots + (-11)^2 = 1160.00$$

$$\sum\limits_{i=1}^{n}\sum\limits_{j=1}^{p} \frac{\left(\sum\limits_{k=1}^{q} c_j Y_{ijk}\right)^2}{q} = [AS] = \frac{(21)^2}{4} + \frac{(27)^2}{4} + \cdots + \frac{(-22)^2}{4} = 937.000$$

$$\sum\limits_{j=1}^{p} \frac{\left(\sum\limits_{i=1}^{n}\sum\limits_{k=1}^{q} c_j Y_{ijk}\right)^2}{nq} = [A] = \frac{(91)^2}{(4)(4)} + \frac{(-81)^2}{(4)(4)} = 927.625$$

$$\sum\limits_{j=1}^{p}\sum\limits_{k=1}^{q} \frac{\left(\sum\limits_{i=1}^{n} c_j Y_{ijk}\right)}{n} = [AB] = \frac{(15)^2}{4} + \frac{(16)^2}{4} + \cdots + \frac{(-40)^2}{4} = 1141.500$$

(iii)   Computational formulas:

$$SSB\hat{\psi}_{1(A)} = \left[\sum\limits_{k=1}^{q} \hat{\psi}_{1(A)}^2 \text{ at } b_k - \frac{\left(\sum\limits_{k=1}^{q} \hat{\psi}_{1(A)} \text{ at } b_k\right)^2}{q}\right] \Bigg/ n\sum\limits_{j=1}^{p} c_j^2$$

Table 11.6-1 (continued)

$$= \left[(8)^2 + (4)^2 + \cdots + (-8)^2 - \frac{(10)^2}{4} \right] \Big/ (4)[(1)^2 + (-1)^2]$$

$$= (180) - 25)/(4)(2) = 19.375$$

$$SSB \times BL(\hat{\psi}_{1(A)}) = [ABS] - [AB] - [AS] + [A] = 9.125$$

(iv) Computation of F ratio (p'denotes the number of levels of treatment A involved in $\hat{\psi}_{i(A)}$):

$$F = \frac{SSB\,\hat{\psi}_{1(A)}/(q-1)1}{SSB \times BL(\hat{\psi}_{1(A)})/p'(n-1)(q-1)} = \frac{.19.375/(4-1)1}{9.125/[2(4-1)(4-1)]} = 12.74$$

$$[(p-1)(q-1)F_{\alpha;(p-1)(q-1),p'(n-1)(q-1)}]/(q-1) = (3F_{.05;3,18})/3 = 3.16$$

---

coefficients in Appendix Table E.12 can be used. The significant test of $F = MSB/MSB \times BL(A)$ in Table 11.2-2 tells us that one or more trend contrasts for treatment B are greater than zero. If the test of $F = MSAB/MSB \times BL(A)$ were not also significant, an experimenter would be interested in testing the following trend-contrast null hypotheses:

$$H_0: \quad \psi_{\text{lin}(B)} = 0 \qquad H_0: \quad \psi_{\text{quad}(B)} = 0 \qquad H_0: \quad \psi_{\text{cub}(B)} = 0.$$

The test statistics are

$$F = \frac{MS\hat{\psi}_{\text{lin}(B)}}{MSB \times BL(A)} \qquad F = \frac{MS\hat{\psi}_{\text{quad}(B)}}{MSB \times BL(A)} \qquad F = \frac{MS\hat{\psi}_{\text{cub}(B)}}{MSB \times BL(A)}.$$

Because the test of $F = MSAB/MSB \times BL(A)$ in Table 11.2-2 is significant, interest shifts from tests of trend contrasts to tests of treatment $\times$ trend-contrast interactions. Rejection of the hypothesis $\alpha\beta = 0$ for all j and k tells us that the trend for treatment B is not the same for all levels of treatment A, type of signal. The following treatment $\times$ trend-contrast interaction null hypotheses are of interest:

$$H_0: \quad \alpha\psi_{\text{lin}(B)} = \delta \text{ for all } j \qquad H_0: \quad \alpha\psi_{\text{quad}(B)} = \delta \text{ for all } j$$
$$H_0: \quad \alpha\psi_{\text{cub}(B)} = \delta \text{ for all } j$$

where δ is a constant for a particular hypothesis. The test statistics are

$$F = \frac{MSA\hat{\psi}_{\text{lin}(B)}}{MSB \times BL(A)} \qquad F = \frac{MSA\hat{\psi}_{\text{quad}(B)}}{MSB \times BL(A)} \qquad F = \frac{MSA\hat{\psi}_{\text{cub}(B)}}{MSB \times BL(A)}.$$

The use of $MSB \times BL(A)$ in the denominator of the F statistics is appropriate if the circularity assumption is tenable. If it is not tenable, $MSB \times BL(A)$ should be replaced by error terms appropriate for the specific trend contrasts of interest, for example,

$$MS\hat{\psi}_{\text{lin}(B)} \times BL(A) \qquad MS\hat{\psi}_{\text{quad}(B)} \times BL(A) \qquad \text{and so on.}$$

The computational procedures for testing $\alpha\psi_{\text{lin}(B)} = \delta$ for all j are illustrated in Table 11.7-1. Procedures for testing $\psi_{\text{lin}(B)} = 0$ are also shown in the table. The mean squares required to test $\alpha\psi_{\text{quad}(B)} = \delta$ for all j and $\alpha\psi_{\text{cub}(B)} = \delta$ for all j are computed by replacing the linear trend coefficients in Table 11.7-1(i) by the quadratic and cubic coefficients, respectively. The results of the trend analyses are summarized in Table 11.7-2. The error rate for the collection of the three a posteriori treatment $\times$ trend-contrast interaction tests can be held at or less than .05 by means of Dunn's procedure or the simultaneous test procedure described in Section 8.6. If Dunn's procedure is

---

TABLE 11.7-1   Computational Procedures for Testing Trend Contrasts and Treatment $\times$ Trend-Contrast Interactions

(i)   Computation of linear contrasts (Coefficients of the linear contrast from Appendix E, $c_1 = -3$, $c_2 = -1$, $c_3 = 1$, $c_4 = 3$, are applied to the data in Table 11.2-1):

$$(-3)Y_{ij1} + (-1)Y_{ij2} + (1)Y_{ij3} + (3)Y_{ij4} = \hat{\psi}_{\text{lin}(B)} \text{ at } a_j s_i$$

	s_1	$(-3)\,3 + (-1)\,4 + (1)\,7 + (3)\quad 7$	15
	s_2	$(-3)\,6 + (-1)\,5 + (1)\,8 + (3)\quad 8$	9
a_1	s_3	$(-3)\,3 + (-1)\,4 + (1)\,7 + (3)\quad 9$	21
	s_4	$(-3)\,3 + (-1)\,3 + (1)\,6 + (3)\quad 8$	18

$$\sum_{i=1}^{n} \hat{\psi}_{\text{lin}(B)} \text{ at } a_1 s_i = \quad 63$$

	s_5	$(-3)\,1 + (-1)\,2 + (1)\,5 + (3)\quad 10$	30
	s_6	$(-3)\,2 + (-1)\,3 + (1)\,6 + (3)\quad 10$	27
a_2	s_7	$(-3)\,2 + (-1)\,4 + (1)\,5 + (3)\quad 9$	22
	s_8	$(-3)\,2 + (-1)\,3 + (1)\,6 + (3)\quad 11$	30

$$\sum_{i=1}^{n} \hat{\psi}_{\text{lin}(B)} \text{ at } a_2 s_i = 109$$

$$\sum_{j=1}^{p}\sum_{i=1}^{n} \hat{\psi}_{\text{lin}(B)} \text{ at } a_j s_i = 172$$

$$\sum_{j=1}^{p}\left(\sum_{i=1}^{n} \hat{\psi}_{\text{lin}(B)} \text{ at } a_j s_i\right)^2 = (63)^2 + (109)^2 = 15850$$

$$\sum_{j=1}^{p}\sum_{i=1}^{n} \hat{\psi}^2_{\text{lin}(B)} \text{ at } a_j s_i = (15)^2 + (9)^2 + \cdots + (30)^2 = 4084$$

(ii)   Computation of sum of squares:

$$SS\hat{\psi}_{\text{lin}(B)} = \left(\sum_{j=1}^{p}\sum_{i=1}^{n} \hat{\psi}_{\text{lin}(B)} \text{ at } a_j s_i\right)^2 \Big/ np \sum_{k=1}^{q} c_k^2$$

$$= (172)^2/(4)(2)[(-3)^2 + (-1)^2 + (1)^2 + (3)^2] = 184.900$$

$$df = 1$$

Table 11.7-1 (continued)

$$SSA\hat{\psi}_{\text{lin}(B)} = \left[\sum_{j=1}^{p} \left(\sum_{i=1}^{n} \hat{\psi}_{\text{lin}(B)} \text{ at } a_j s_i \right)^2 - \frac{\left(\sum_{j=1}^{p} \sum_{i=1}^{n} \hat{\psi}_{\text{lin}(B)} \text{ at } a_j s_i \right)^2}{p} \right] \Bigg/ n \sum_{k=1}^{p} c_k^2$$

$$= \left(15850 - \frac{(172)^2}{2} \right) \Bigg/ 4[(-3)^2 + (-1)^2 + (1)^2 + (3)^2] = 13.225$$

$$df = (p-1)1 = 1$$

$$SS\hat{\psi}_{\text{lin}(B)} \times BL(A) = \left[\sum_{j=1}^{p} \sum_{i=1}^{n} \hat{\psi}_{\text{lin}(B)}^2 \text{ at } a_j s_i - \frac{\sum_{j=1}^{p} \left(\sum_{i=1}^{n} \hat{\psi}_{\text{lin}(B)} \text{ at } a_j s_i \right)^2}{n} \right] \Bigg/ \sum_{k=1}^{q} c_k^2$$

$$= \left(4084 - \frac{15850}{4} \right) \Bigg/ [(-3)^2 + (-1)^2 + (1)^2 + (3)^2] = 6.075$$

$$df = p(n-1) = 2(4-1) = 6$$

(iii) Computation of F ratios:

$$F = \frac{SS\hat{\psi}_{\text{lin}(B)}/1}{SS\hat{\psi}_{\text{lin}(B)} \times BL(A)/p(n-1)} = \frac{184.900/1}{6.075/6} = 182.62$$

Critical value for a priori orthogonal contrasts:

$$F_{.05;1,p(n-1)} = 5.99$$

Critical value for three tests using Dunn's procedure:

$$F_{.05/3;1,p(n-1)} \cong 12.4$$

Critical value for simultaneous test procedure:

$$[(q-1)F_{.05;q-1,p(n-1)}]/1 = 14.28$$

$$F = \frac{SSA\hat{\psi}_{\text{lin}(B)}/(p-1)1}{SS\hat{\psi}_{\text{lin}(B)} \times BL(A)/p(n-1)} = \frac{13.225/1}{6.075/6} = 13.06$$

Critical value for a priori orthogonal contrasts:

$$F_{.05;(p-1)1,p(n-1)} = 5.99$$

Critical value for three tests using Dunn's procedure:

$$F_{.05/3;(p-1)1,p(n-1)} \cong 12.4$$

Critical value for simultaneous test procedure:

$$[(p-1)(q-1)F_{.05;(p-1)(q-1),p(n-1)}]/(p-1)1 = 14.28$$

TABLE 11.7-2   ANOVA Table for Trend Analysis

	Source	SS	df	MS		F
1	Between blocks	12.500	$np - 1 = 7$			
2	A (Type of signal)	3.125	$p - 1 = 1$	3.125	[2/3]	2.00
3	Blocks w. A	9.375	$p(n - 1) = 6$	1.562		
4	Within blocks	223.000	$np(q - 1) = 24$			
5	B (Monitoring periods)	194.500	$q - 1 = 3$	64.833	[5/10]	127.88*
6	AB	19.375	$(p - 1)(q - 1) = 3$	6.458	[6/10]	12.74*
7	$A \times \hat{\psi}_{\text{lin}(B)}$	13.225	$(p - 1)1 = 1$	13.225	[7/11]	13.06**
8	$A \times \hat{\psi}_{\text{quad}(B)}$	3.125	$(p - 1)1 = 1$	3.125	[8/12]	10.00
9	$A \times \hat{\psi}_{\text{cub}(B)}$	3.025	$(p - 1)1 = 1$	3.025	[9/13]	15.45**
10	$B \times$ blocks w. A	9.125	$p(n - 1)(q - 1) = 18$	0.507		
11	$\hat{\psi}_{\text{lin}(B)} \times$ blocks w. A	6.075	$p(n - 1) = 6$	1.0125		
12	$\hat{\psi}_{\text{quad}(B)} \times$ blocks w. A	1.875	$p(n - 1) = 6$	0.3125		
13	$\hat{\psi}_{\text{cub}(B)} \times$ blocks w. A	1.175	$p(n - 1) = 6$	0.1958		
14	Total	235.500	$npq - 1 = 31$			

*$p < .01$

**$p < .05$ based on Dunn's procedure

used, the hypotheses $\alpha\psi_{\text{lin}(B)} = \delta$ for all j and $\alpha\psi_{\text{cub}(B)} = \delta$ for all j can be rejected. The $A\hat{\psi}_{\text{lin}(B)}$ interaction accounts for $(13.225/19.375) \times 100 = 68.3\%$ of the variation due to the AB interaction while $A\hat{\psi}_{\text{cub}(B)}$ accounts for only $(3.025/19.375) \times 100 = 15.6\%$ of the variation.

The computational procedures are slightly different for testing hypotheses about trend contrasts and treatment × trend-contrast interactions if treatment A represents a quantitative variable but treatment B is a qualitative variable. Depending on whether treatment A or the AB interaction is significant, an experimenter might want to test one of the following sets of hypotheses:

$$H_0: \quad \psi_{\text{lin}(A)} = 0, \ H_0: \quad \psi_{\text{quad}(A)} = 0, \ \ldots, \ H_0: \quad \psi_{(p-1)\text{th trend}(A)} = 0$$

$$H_0: \quad \beta\psi_{\text{lin}(A)} = \delta \text{ for all } k, \ H_0: \quad \beta\psi_{\text{quad}(A)} = \delta \text{ for all } k, \ \ldots,$$
$$H_0: \quad \beta\psi_{(p-1)\text{th trend}(A)} = \delta \text{ for all } k$$

where δ is a constant for a particular hypothesis. The test statistics for the ith trend contrast have the form

$$F = \frac{SS\psi_{i(A)}/1}{SSBL(A)/p(n - 1)} \quad \text{and} \quad F = \frac{SSB\psi_{i(A)}/(q - 1)}{SSB \times BL(A)}$$

where

$$SS\hat{\psi}_{i(A)} = \left[\sum_{j=1}^{p}\left(\sum_{i=1}^{n}\sum_{k=1}^{q} c_j Y_{ijk}\right)\right]^2 \bigg/ nq\sum_{j=1}^{p} c_j^2$$

$$SSBL(A) = [AS] - [A] = \sum_{i=1}^{n} \sum_{j=1}^{p} \frac{\left(\sum_{k=1}^{q} c_j Y_{ijk} \right)^2}{q} - \sum_{j=1}^{p} \frac{\left(\sum_{i=1}^{n} \sum_{k=1}^{q} c_j Y_{ijk} \right)^2}{nq}$$

$$SSB\hat{\psi}_{i(A)} = \left[\sum_{k=1}^{q} \hat{\psi}_{i(A)}^2 \text{ at } b_k - \frac{\left(\sum_{k=1}^{q} \hat{\psi}_{i(A)} \text{ at } b_k \right)^2}{q} \right] \Bigg/ n \sum_{j=1}^{p} c_j^2$$

$$SSB \times BL(A) = [ABS] - [AB] - [AS] + [A]$$

$$= \sum_{i=1}^{n} \sum_{j=1}^{p} \sum_{k=1}^{q} c_j Y_{ijk}^2 - \sum_{j=1}^{p} \sum_{k=1}^{q} \frac{\left(\sum_{i=1}^{n} c_j Y_{ijk} \right)^2}{n} - \sum_{i=1}^{n} \sum_{j=1}^{p} \frac{\left(\sum_{k=1}^{q} c_j Y_{ijk} \right)^2}{q}$$

$$+ \sum_{j=1}^{p} \frac{\left(\sum_{i=1}^{n} \sum_{k=1}^{q} c_j Y_{ijk} \right)^2}{nq}$$

and c_j denotes the $j = 1, \ldots, p$ coefficients for the ith trend contrast. The layout that is used in computing these sums of squares is the same as that in Table 11.6-1. The reader may find it helpful to refer to this table.

TESTS OF TREND CONTRAST-CONTRAST INTERACTIONS

If both treatments in a type SPF-$p \cdot q$ design are quantitative variables, an experimenter can test the following kinds of hypotheses:

$$H_0: \quad \psi_{\text{lin}(A)} \psi_{\text{lin}(B)} = 0, \qquad H_0: \quad \psi_{\text{lin}(A)} \psi_{\text{quad}(B)} = 0,$$
$$H_0: \quad \psi_{\text{quad}(A)} \psi_{\text{lin}(B)} = 0, \qquad \text{and so on.}$$

Procedures for testing these hypotheses are given in Table 11.7-3. For purposes of illustration we assume that both variables in the table are quantitative and that the levels of each treatment are separated by equal intervals.

11.8 RELATIVE EFFICIENCY OF SPLIT-PLOT FACTORIAL DESIGN

We saw earlier that a split-plot factorial design is one alternative to a randomized block factorial design when the block size must be kept small. A numerical index of the relative efficiency of the two designs, disregarding differences in degrees of freedom,

TABLE 11.7-3 Computational Procedures for Testing Trend Contrast-Contrast Interactions

(i) Coefficients for linear and quadratic contrasts:

<table>
<tr><td align="center">Treatment A</td><td align="center">Treatment B</td></tr>
</table>

Treatment A	Treatment B
Linear coefficients $(c_{1j}) = -1 \quad 1$	Linear coefficients $(c_{1k}) = -3 \quad -1 \quad 1 \quad 3$
	Quadratic coefficients $(c_{2k}) = \quad 1 \quad -1 \quad -1 \quad 1$
	Cubic coefficients $(c_{3k}) = -1 \quad 3 \quad -3 \quad 1$

(ii) Products of coefficients for treatments A and B $(d_{jk(l \times l)} = c_{1j}c_{1k})$:

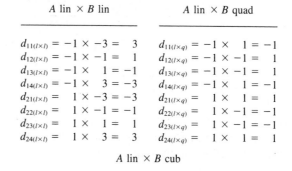

A lin $\times$ B lin

$$d_{11(l \times l)} = -1 \times -3 = 3$$
$$d_{12(l \times l)} = -1 \times -1 = 1$$
$$d_{13(l \times l)} = -1 \times 1 = -1$$
$$d_{14(l \times l)} = -1 \times 3 = -3$$
$$d_{21(l \times l)} = 1 \times -3 = -3$$
$$d_{22(l \times l)} = 1 \times -1 = -1$$
$$d_{23(l \times l)} = 1 \times 1 = 1$$
$$d_{24(l \times l)} = 1 \times 3 = 3$$

A lin $\times$ B quad

$$d_{11(l \times q)} = -1 \times 1 = -1$$
$$d_{12(l \times q)} = -1 \times -1 = 1$$
$$d_{13(l \times q)} = -1 \times -1 = 1$$
$$d_{14(l \times q)} = -1 \times 1 = -1$$
$$d_{21(l \times q)} = 1 \times 1 = 1$$
$$d_{22(l \times q)} = 1 \times -1 = -1$$
$$d_{23(l \times q)} = 1 \times -1 = -1$$
$$d_{24(l \times q)} = 1 \times 1 = 1$$

A lin $\times$ B cub

$$d_{11(l \times c)} = -1 \times -1 = 1$$
$$d_{12(l \times c)} = -1 \times 3 = -3$$
$$d_{13(l \times c)} = -1 \times -3 = 3$$
$$d_{14(l \times c)} = -1 \times 1 = -1$$
$$d_{21(l \times c)} = 1 \times -1 = -1$$
$$d_{22(l \times c)} = 1 \times 3 = 3$$
$$d_{23(l \times c)} = 1 \times -3 = -3$$
$$d_{24(l \times c)} = 1 \times 1 = 1$$

(iii) Computation of lin $\times$ lin contrast* (Coefficients of the contrast, $d_{11(l \times l)} = 3$, $d_{12(l \times l)} = 1$, ..., $d_{24(l \times l)} = 3$, are applied to data in Table 11.2-1):

		$(3)Y_{i11}$	$(1)Y_{i12}$	$(-1)Y_{i13}$	$(-3)Y_{i14}$	$= \hat{\psi}_{\text{lin}(A) \times \text{lin}(B)}$ at $a_1 s_i$
	s_1	(3) 3	(1) 4	(−1) 7	(−3) 7	−15
	s_2	(3) 6	(1) 5	(−1) 8	(−3) 8	− 9
a_1	s_3	(3) 3	(1) 4	(−1) 7	(−3) 9	−21
	s_4	(3) 3	(1) 3	(−1) 6	(−3) 8	−18

$$\sum_{i=1}^{n} \hat{\psi}_{\text{lin}(A) \times \text{lin}(B)} \text{ at } a_1 s_i = -63$$

Table 11.7-3 (continued)

		$(-3)Y_{i21}$	$(-1)Y_{i22}$	$(1)Y_{i23}$	$(3)Y_{i24}$	$= \hat{\psi}_{\text{lin}(A)\times\text{lin}(B)}$ at $a_2 s_i$
	s_5	$(-3)\,1$	$(-1)\,2$	$(1)\,5$	$(3)\,10$	30
	s_6	$(-3)\,2$	$(-1)\,3$	$(1)\,6$	$(3)\,10$	27
a_2	s_7	$(-3)\,2$	$(-1)\,4$	$(1)\,5$	$(3)\,\;9$	22
	s_8	$(-3)\,2$	$(-1)\,3$	$(1)\,6$	$(3)\,11$	30

$$\sum_{i=1}^{n} \hat{\psi}_{\text{lin}(A)\times\text{lin}(B)} \text{ at } a_2 s_i = \;\;109$$

$$\sum_{j=1}^{p}\sum_{i=1}^{n} \hat{\psi}_{\text{lin}(A)\times\text{lin}(B)} \text{ at } a_j s_i = \;-63 + 109 = 46$$

$$\sum_{j=1}^{p}\left(\sum_{i=1}^{n} \hat{\psi}_{\text{lin}(A)\times\text{lin}(B)} \text{ at } a_j s_i\right)^2 = (-63)^2 + (109)^2 = 15850$$

$$\sum_{j=1}^{p}\sum_{i=1}^{n} \hat{\psi}^2_{\text{lin}(A)\times\text{lin}(B)} \text{ at } a_j s_i = (-15)^2 + (-9)^2 + \cdots + (30)^2 = 4084$$

(iv) Computation of sum of squares:

$$SS\hat{\psi}_{\text{lin}(A)}\hat{\psi}_{\text{lin}(B)} = \left(\sum_{j=1}^{p}\sum_{i=1}^{n} \hat{\psi}_{\text{lin}(A)\times\text{lin}(B)} \text{ at } a_j s_i\right)^2 \Big/ n \sum_{j=1}^{p}\sum_{k=1}^{q} d^2_{jk(l\times l)}$$

$$= (46)^2/4[(3)^2 + (1)^2 + \cdots + (3)^2] = 13.225$$

$$SS\hat{\psi}_{\text{lin}(A)}\hat{\psi}_{\text{lin}(B)} \times BL(A) = \left[\sum_{j=1}^{p}\sum_{i=1}^{n} \hat{\psi}^2_{\text{lin}(A)\times\text{lin}(B)} \text{ at } a_j s_i - \frac{\sum_{j=1}^{p}\left(\sum_{i=1}^{n} \hat{\psi}_{\text{lin}(A)\times\text{lin}(B)} \text{ at } a_j s_i\right)^2}{n}\right] \Big/ \sum_{j=1}^{p}\sum_{k=1}^{q} d^2_{jk(l\times l)}$$

$$= \left(4084 - \frac{15850}{4}\right) \Big/ [(3)^2 + (1)^2 + \cdots + (3)^2] = 3.038$$

(v) Computation of F ratio:

$$F = \frac{SS\hat{\psi}_{\text{lin}(A)}\hat{\psi}_{\text{lin}(B)}/1}{SS\hat{\psi}_{\text{lin}(A)}\hat{\psi}_{\text{lin}(B)} \times BL(A)/p(n-1)} = \frac{13.225/1}{3.038/2(4-1)} = 26.12$$

Critical value for a priori orthogonal contrasts:

$$F_{.05;1,p(n-1)} = 5.99$$

Critical value for three tests using Dunn's procedure:

$$F_{.05/3;1,p(n-1)} \cong 12.4$$

Critical value for simultaneous test procedure:

$$[(p-1)(q-1)F_{.05;(p-1)(q-1),p(n-1)}]/1 = 14.28$$

*The computation of the lin × quad and lin × cub contrasts follows the pattern illustrated for the lin × lin contrast and will not be illustrated.

is given by the following formulas (Federer, 1955, 274). The data used in this example are from Table 11.2-2.

$$\begin{aligned}
\text{Relative efficiency} \atop \text{for } A &= \frac{[(p-1)MSBL(A) + p(q-1)MSB \times BL(A)]/(pq-1)}{MSBL(A)} \\[2mm]
&= \frac{[(2-1)1.562 + 2(4-1)0.507]/[(2)(4)-1]}{1.562} = \frac{0.658}{1.562} \\[2mm]
&= 0.42.
\end{aligned}$$

$$\begin{aligned}
\text{Relative efficiency} \atop \text{for } B \text{ and } AB &= \frac{[(p-1)MSBL(A) + p(q-1)MSB \times BL(A)]/(pq-1)}{MSB \times BL(A)} \\[2mm]
&= \frac{[(2-1)1.562 + 2(4-1)0.507]/[(2)(4)-1]}{0.507} = \frac{0.658}{0.507} \\[2mm]
&= 1.30.
\end{aligned}$$

In this example a test of treatment A is less than half as efficient in the split-plot factorial design as it is in the randomized block factorial design. On the other hand, the tests of B and AB are more efficient in the split-plot factorial design. The average standard error of a difference is the same for the two designs. The increased precision in testing the B and AB effects is obtained at the expense of the test of A effects.

To get a better feeling for the relative precision of the between- and within-blocks estimators in a type SPF-$p \cdot q$ design, it helps to examine the expected values of the two error terms. It can be shown for the mixed model where A and B are fixed and blocks are random that

$$E[MSBL(A)] = \sigma_\epsilon^2 + q\sigma_\pi^2 = \sigma_Y^2[1 + (q-1)\rho]$$
$$E[MSB \times BL(A)] = \sigma_\epsilon^2 + \sigma_{\beta\pi}^2 = \sigma_Y^2(1-\rho)$$

where σ_Y^2 is the population variance and ρ is the population correlation.* The larger ρ, the smaller the within-blocks error relative to the between-blocks error. Also, $MSB \times BL(A)$ is based on $p(n-1)(q-1)$ degrees of freedom while $MSBL(A)$ is based on only $p(n-1)$ degrees of freedom. For purposes of comparison, the expected value of $MSRES$ in a type RBF-pq design can be expressed as

$$E(MSRES) = \sigma_\epsilon^2 + \sigma_{(\alpha\beta)\times\pi}^2 = \sigma_Y^2\left(1 - \frac{q-1}{pq-1}\rho\right).$$

Hence, if ρ is greater than zero,

$$E[MSBL(A)] > E(MSRES) > E[MSB \times BL(A)].$$

From the foregoing it is apparent that a split-plot factorial design is an excellent design choice when an experimenter's primary interests involve treatment B

---

* See Section 6.4

and the *AB* interaction. However, a design that provides adequate power for testing *B* and *AB* may not provide adequate power for testing *A*. Of course the power of the *A* test can always be increased, if necessary, by securing additional blocks.

11.9 INTRODUCTION TO TYPE SPF-*pr·q* DESIGN

The general analysis procedures for a two-treatment split-plot factorial design can be extended to designs having three or more treatments. Numerical examples for the more frequently used split-plot factorial designs and rules for expanding the design to any combination of between- and within-blocks treatments are provided in subsequent sections.

DESCRIPTION OF TYPE SPF-*pr·q* DESIGN

The design described here and diagrammed in Figure 11.9-1 is a type SPF-22·4 design. It has two between-blocks treatments, *A* and *C*, and one within-blocks treatment, *B*. For the repeated measurements case, *pr* samples of *n* experimental units from a population are randomly assigned to the $a_j c_l$ treatment combinations. The sequence of administration of the levels of treatment *B* for the experimental units in the *npr* blocks is randomized independently for each block. For the nonrepeated measurements case, *pr* samples of *n* blocks each containing *q* matched experimental units

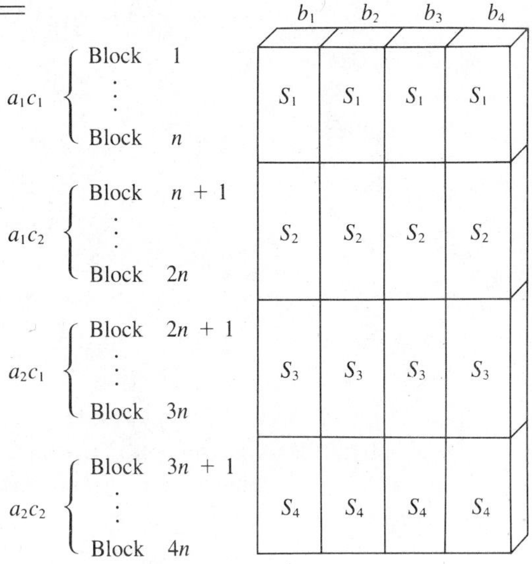

FIGURE 11.9-1 Type SPF-22·4 design; $S_1, S_2, \ldots, S_4$ denote samples of subjects who receive one combination of treatments *A* and *C* and all levels of treatment *B*.

from a population are randomly assigned to the $a_j c_l$ combinations. Following this, the levels of treatment B are randomly assigned to the q experimental units within each block. The experimental design model equation is

$$Y_{ijkl} = \mu + \alpha_j + \gamma_l + (\alpha\gamma)_{jl} + \pi_{i(jl)} + \beta_k + (\alpha\beta)_{jk} + (\beta\gamma)_{kl} + (\alpha\beta\gamma)_{jkl}$$
$$+ (\beta\pi)_{ki(jl)} + \epsilon_{ijkl}$$
$$(i = 1, \ldots, n; j = 1, \ldots, p; k = 1, \ldots, q; l = 1, \ldots, r).$$

COMPUTATIONAL EXAMPLE FOR TYPE SPF-$pr \cdot q$ DESIGN

The layout and computational procedures for a type SPF-22·4 design are illustrated in Table 11.9-1. The results of the analysis are summarized in Table 11.9-2.

SOME ASSUMPTIONS UNDERLYING A TYPE SPF-$pr \cdot q$ DESIGN

The between-blocks mean square error term $MSBL(AC)$ is composed of variation pooled over the pr combinations of treatments A and C. For the data in Table 11.9-1,

$$
\begin{array}{ll}
SSBL(AC) = 7.250 & pr(n-1) = 4 \\
SSBL(a_1 c_1) = 4.500 & n - 1 = 1 \\
SSBL(a_1 c_2) = 1.125 & n - 1 = 1 \\
SSBL(a_2 c_1) = 1.125 & n - 1 = 1 \\
SSBL(a_2 c_2) = 0.500 & n - 1 = 1
\end{array}
$$

where $SSBL(a_j c_l)$ is given by

$$SSBL(a_j c_l) = \sum_{i=1}^{n} \frac{\left(\sum_{k=1}^{q} Y_{ijkl} \right)^2}{q} - \frac{\left(\sum_{i=1}^{n} \sum_{k=1}^{q} Y_{ijkl} \right)^2}{nq}.$$

For example,

$$SSBL(a_1 c_1) = \sum_{i=1}^{n} \frac{\left(\sum_{k=1}^{q} Y_{i1k1} \right)^2}{q} - \frac{\left(\sum_{i=1}^{n} \sum_{k=1}^{q} Y_{i1k1} \right)^2}{nq}$$

$$= \frac{(21)^2}{4} + \frac{(27)^2}{4} - \frac{(48)^2}{(2)(4)} = 4.500.$$

In order for the between-blocks F statistics to be distributed as the F distribution when the null hypothesis is true, $MSBL(a_1 c_1)$, $MSBL(a_1 c_2)$, and so on must estimate the same population variance. This homogeneity of variance assumption can be tested by

$$F_{max} = \frac{\text{largest variance}}{\text{smallest variance}} = \frac{4.500/1}{0.500/1} = 9$$

$$F_{max,05;pr,n-1} > 142.$$

TABLE 11.9-1 Computational Procedures for a Type SPF-22·4 Design

(i) Data and notation [Y_{ijk} denotes a score for the experimental unit in block i and treatment combination jkl; $i = 1, \ldots, n$ blocks (s_i); $j = 1, \ldots, p$ levels of treatment A (a_j); $k = 1, \ldots, q$ levels of treatment B (b_k); $l = 1, \ldots, r$ levels of treatment C (c_l)]:

ABCS Summary Table
Entry is Y_{ijkl}

		b_1	b_2	b_3	b_4	$\sum_{k=1}^{q} Y_{ijkl}$
$a_1 c_1$	s_1	3	4	7	7	21
	s_2	6	5	8	8	27
$a_1 c_2$	s_3	3	4	7	9	23
	s_4	3	3	6	8	20
$a_2 c_1$	s_5	1	2	5	10	18
	s_6	2	3	6	10	21
$a_2 c_2$	s_7	2	4	5	9	20
	s_8	2	3	6	11	22

ABC Summary Table
Entry is $\sum_{i=1}^{n} Y_{ijkl}$

	b_1	b_2	b_3	b_4
$a_1 c_1$	$n = 2$			
	9	9	15	15
$a_1 c_2$	6	7	13	17
$a_2 c_1$	3	5	11	20
$a_2 c_2$	4	7	11	20

BC Summary Table
Entry is $\sum_{i=1}^{n} \sum_{j=1}^{p} Y_{ijkl}$

	b_1	b_2	b_3	b_4
c_1	$np = 4$			
	12	14	26	35
c_2	10	14	24	37

AB Summary Table
Entry is $\sum_{i=1}^{n} \sum_{l=1}^{r} Y_{ijkl}$

	b_1	b_2	b_3	b_4	$\sum_{i=1}^{n} \sum_{k=1}^{q} \sum_{l=1}^{r} Y_{ijkl}$
a_1	$nr = 4$				
	15	16	28	32	91
a_2	7	12	22	40	81
$\sum_{i=1}^{n} \sum_{j=1}^{p} \sum_{l=1}^{r} Y_{ijkl} =$	22	28	50	72	

AC Summary Table
Entry is $\sum_{i=1}^{n} \sum_{k=1}^{q} Y_{ijkl}$

	c_1	c_2
a_1	$nq = 8$	
	48	43
a_2	39	42
$\sum_{i=1}^{n} \sum_{j=1}^{p} \sum_{k=1}^{q} Y_{ijkl} =$	87	85

Table 11.9-1 (continued)

(ii)   Computational symbols:

$$\sum_{i=1}^{n} \sum_{j=1}^{p} \sum_{k=1}^{q} \sum_{l=1}^{r} Y_{ijkl} = 3 + 6 + \cdots + 11 = 172.000$$

$$\frac{\left(\sum_{i=1}^{n} \sum_{j=1}^{p} \sum_{k=1}^{q} \sum_{l=1}^{r} Y_{ijkl} \right)^2}{npqr} = [Y] = \frac{(172.000)^2}{(2)(2)(4)(2)} = 924.500$$

$$\sum_{i=1}^{n} \sum_{j=1}^{p} \sum_{k=1}^{q} \sum_{l=1}^{r} Y_{ijkl}^2 = [ABCS] = (3)^2 + (6)^2 + \cdots + (11)^2 = 1160.000$$

$$\sum_{i=1}^{n} \sum_{j=1}^{p} \sum_{l=1}^{r} \frac{\left(\sum_{k=1}^{q} Y_{ijkl} \right)^2}{q} = [ACS] = \frac{(21)^2}{4} + \frac{(27)^2}{4} + \cdots + \frac{(22)^2}{4} = 937.000$$

$$\sum_{j=1}^{p} \frac{\left(\sum_{i=1}^{n} \sum_{k=1}^{q} \sum_{l=1}^{r} Y_{ijkl} \right)^2}{nqr} = [A] = \frac{(91)^2}{(2)(4)(2)} + \frac{(81)^2}{(2)(4)(2)} = 927.625$$

$$\sum_{l=1}^{r} \frac{\left(\sum_{i=1}^{n} \sum_{j=1}^{p} \sum_{k=1}^{q} Y_{ijkl} \right)^2}{npq} = [C] = \frac{(87)^2}{(2)(2)(4)} + \frac{(85)^2}{(2)(2)(4)} = 924.625$$

$$\sum_{j=1}^{p} \sum_{l=1}^{r} \frac{\left(\sum_{i=1}^{n} \sum_{k=1}^{q} Y_{ijkl} \right)^2}{nq} = [AC] = \frac{(48)^2}{(2)(4)} + \frac{(43)^2}{(2)(4)} + \cdots + \frac{(42)^2}{(2)(4)} = 929.750$$

$$\sum_{k=1}^{q} \frac{\left(\sum_{i=1}^{n} \sum_{j=1}^{p} \sum_{l=1}^{r} Y_{ijkl} \right)^2}{npr} = [B] = \frac{(22)^2}{(2)(2)(2)} + \frac{(28)^2}{(2)(2)(2)} + \cdots + \frac{(72)^2}{(2)(2)(2)} = 1119.000$$

$$\sum_{j=1}^{p} \sum_{k=1}^{q} \frac{\left(\sum_{i=1}^{n} \sum_{l=1}^{r} Y_{ijkl} \right)^2}{nr} = [AB] = \frac{(15)^2}{(2)(2)} + \frac{(16)^2}{(2)(2)} + \cdots + \frac{(40)^2}{(2)(2)} = 1141.500$$

$$\sum_{k=1}^{q} \sum_{l=1}^{r} \frac{\left(\sum_{i=1}^{n} \sum_{j=1}^{p} Y_{ijkl} \right)^2}{np} = [BC] = \frac{(12)^2}{(2)(2)} + \frac{(10)^2}{(2)(2)} + \cdots + \frac{(37)^2}{(2)(2)} = 1120.500$$

$$\sum_{j=1}^{p} \sum_{k=1}^{q} \sum_{l=1}^{r} \frac{\left(\sum_{i=1}^{n} Y_{ijkl} \right)^2}{n} = [ABC] = \frac{(9)^2}{2} + \frac{(6)^2}{2} + \cdots + \frac{(20)^2}{2} = 1148.000$$

Table 11.9-1 (continued)

(iii) Computational formulas:

$$SSTO = [ABCS] - [Y] = 235.500$$

$$SSBETWEEN\ BL = [ACS] - [Y] = 12.500$$

$$SSA = [A] - [Y] = 3.125$$

$$SSC = [C] - [Y] = 0.125$$

$$SSAC = [AC] - [A] - [C] + [Y] = 2.000$$

$$SSBL(AC) = [ACS] - [AC] = 7.250$$

$$SSWITHIN\ BL = [ABCS] - [ACS] = 223.000$$

$$SSB = [B] - [Y] = 194.500$$

$$SSAB = [AB] - [A] - [B] + [Y] = 19.375$$

$$SSBC = [BC] - [B] - [C] + [Y] = 1.375$$

$$SSABC = [ABC] - [AB] - [AC] - [BC] + [A] + [B] + [C] - [Y] = 3.000$$

$$SSB \times BL(AC) = [ABCS] - [ABC] - [ACS] + [AC] = 4.750$$

TABLE 11.9-2   ANOVA Table for Type SPF-22·4 Design

	Source	SS	df	MS	F (A, B, and C fixed effects; Blocks random)	
1	Between blocks	12.500	$npr - 1 = 7$			
2	A	3.125	$p - 1 = 1$	3.125	[2/5]	1.72
3	C	0.125	$r - 1 = 1$	0.125	[3/5]	0.07
4	AC	2.000	$(p - 1)(r - 1) = 1$	2.000	[4/5]	1.10
5	Blocks w. AC	7.250	$pr(n - 1) = 4$	1.812		
6	Within blocks	223.000	$npr(q - 1) = 24$			
7	B	194.500	$q - 1 = 3$	64.833	[7/11]	163.72*
8	AB	19.375	$(p - 1)(q - 1) = 3$	6.458	[8/11]	16.31*
9	BC	1.375	$(q - 1)(r - 1) = 3$	0.458	[9/11]	1.16
10	ABC	3.000	$(p - 1)(q - 1)(r - 1) = 3$	1.000	[10/11]	2.53
11	$B \times$ blocks w. AC	4.750	$pr(n - 1)(q - 1) = 12$	0.396		
12	Total	235.500	$npqr - 1 = 31$			

*$p < .01$

The assumption of homogeneity of the between-blocks variances is tenable.

The within-blocks mean square error term $MSB \times BL(AC)$ is also a pooled term. For each of the pr combinations of treatments A and C, we can compute a $q \times q$ variance-covariance matrix $\hat{\Sigma}_{a_j c_l}$. The key assumption associated with testing the within-blocks null hypotheses is the multisample circularity assumption

$$\underset{[(q-1)\times q]}{\mathbf{C}_B^{*\prime}} \; \underset{(q\times q)}{\mathbf{\Sigma}_{a_j c_l}} \; \underset{[q\times(q-1)]}{\mathbf{C}_B^{*}} \; = \; \underset{[(q-1)\times(q-1)]}{\lambda \mathbf{I}} \qquad (j = 1, \ldots, p; l = 1, \ldots, r).$$

This assumption is discussed in Section 11.4.

If the tenability of the circularity assumption $\mathbf{C}_B^{*\prime} \, \mathbf{\Sigma}_{\text{pooled}} \, \mathbf{C}_B^{*} = \lambda \mathbf{I}$ is in doubt, the three-step testing strategy described in Section 11.4 can be used. The formula for computing $\hat{\theta}$ is the same as that in Table 11.4 for a type SPF-$p \cdot q$ design; the formula for $\tilde{\theta}$ is

$$\tilde{\theta} = \frac{npr(q-1)\hat{\theta} - 2}{(q-1)[npr - pr - (q-1)\hat{\theta}]}.$$

PROCEDURES FOR TESTING HYPOTHESES ABOUT MEANS

Procedures for testing hypotheses about means have been described in detail in Chapter 3 and Sections 8.5 and 11.5. These procedures generalize to a type SPF-$pr \cdot q$ design. The computational formulas are given in Table 11.9-3.

The use of $MSB \times BL(AC)$ as the error term in a test statistic is appropriate if the circularity assumption is tenable. If it is not tenable, an error term appropriate for the specific contrast should be used, that is, $MS\psi_{i(B)} \times BL(AC)$. The computation of such error terms is discussed in Section 11.5. When the error mean square in the denominator of a test statistic is composed of pooled mean squares, the critical value of the test statistic is approximated by h'_α as discussed in Section 11.5.

EXPECTED VALUES OF MEAN SQUARES
FOR A TYPE SPF-$pr \cdot q$ DESIGN

The expected values of mean squares for models I, II, and III can be determined from Table 11.9-4. The terms $1 - (n/N)$, $1 - (p/P)$, $1 - (q/Q)$, and $1 - (r/R)$ become zero if the corresponding terms BL, A, B, and C represent fixed effects and one if the corresponding terms are random. Also, the variances should be replaced by terms of the form $\Sigma\alpha_j^2/(p-1)$, $\Sigma\beta_k^2/(q-1)$, and $\Sigma\gamma_l^2/(r-1)$ if the corresponding treatments are fixed.

TABLE 11.9-3 Procedures for Testing Hypotheses About Means

(i) Main-effects means:

Treatment A

$$\frac{\hat{\psi}_{i(A)}}{\hat{\sigma}_{\psi_{i(A)}}} = \frac{c_1\overline{Y}_{.1..} + c_2\overline{Y}_{.2..} + \cdots + c_p\overline{Y}_{.p..}}{\sqrt{\dfrac{SSBL(AC)}{pr(n-1)}\left(\dfrac{c_1^2}{nqr} + \dfrac{c_2^2}{nqr} + \cdots + \dfrac{c_p^2}{nqr}\right)}}$$

$$df = pr(n-1)$$

Treatment B

$$\frac{\hat{\psi}_{i(B)}}{\hat{\sigma}_{\psi_{i(B)}}} = \frac{c_1\overline{Y}_{..1.} + c_2\overline{Y}_{..2.} + \cdots + c_q\overline{Y}_{..q.}}{\sqrt{\dfrac{SSB \times BL(AC)}{pr(n-1)(q-1)}\left(\dfrac{c_1^2}{npr} + \dfrac{c_2^2}{npr} + \cdots + \dfrac{c_q^2}{npr}\right)}}$$

$$df = pr(n-1)(q-1)$$

Treatment C

$$\frac{\hat{\psi}_{i(C)}}{\hat{\sigma}_{\psi_{i(C)}}} = \frac{c_1\overline{Y}_{...1} + c_2\overline{Y}_{...2} + \cdots + c_r\overline{Y}_{...r}}{\sqrt{\dfrac{SSBL(AC)}{pr(n-1)}\left(\dfrac{c_1^2}{npq} + \dfrac{c_2^2}{npq} + \cdots + \dfrac{c_r^2}{npq}\right)}}$$

$$df = pr(n-1)$$

(ii) Simple main-effects means:

Treatment A at b_k

$$\frac{\hat{\psi}_{i(A)} \text{ at } b_k}{\hat{\sigma}_{\psi_{i(A)}} \text{ at } b_k} = \frac{c_1\overline{Y}_{.1k.} + c_2\overline{Y}_{.2k.} + \cdots + c_p\overline{Y}_{.pk.}}{\sqrt{\dfrac{SSBL(AC) + SSB \times BL(AC)}{pr(n-1) + pr(n-1)(q-1)}\left(\dfrac{c_1^2}{nr} + \dfrac{c_2^2}{nr} + \cdots + \dfrac{c_p^2}{nr}\right)}}$$

$$df = pr(n-1) + pr(n-1)(q-1)$$

Treatment B at a_j

$$\frac{\hat{\psi}_{i(B)} \text{ at } a_j}{\hat{\sigma}_{\psi_{i(B)}} \text{ at } a_j} = \frac{c_1\overline{Y}_{.j1.} + c_2\overline{Y}_{.j2.} + \cdots + c_q\overline{Y}_{.jq.}}{\sqrt{\dfrac{SSB \times BL(AC)}{pr(n-1)(q-1)}\left(\dfrac{c_1^2}{nr} + \dfrac{c_2^2}{nr} + \cdots + \dfrac{c_q^2}{nr}\right)}}$$

$$df = pr(n-1)(q-1)$$

Treatment A at c_l

$$\frac{\hat{\psi}_{i(A)} \text{ at } c_l}{\hat{\sigma}_{\psi_{i(A)}} \text{ at } c_l} = \frac{c_1\overline{Y}_{.1\cdot l} + c_2\overline{Y}_{.2\cdot l} + \cdots + c_p\overline{Y}_{.p\cdot l}}{\sqrt{\dfrac{SSBL(AC)}{pr(n-1)}\left(\dfrac{c_1^2}{nq} + \dfrac{c_2^2}{nq} + \cdots + \dfrac{c_p^2}{nq}\right)}}$$

$$df = pr(n-1)$$

Table 11.9-3 (continued)

Treatment C at a_j

$$\frac{\hat{\psi}_{i(C)} \text{ at } a_j}{\hat{\sigma}_{\psi_{i(C)}} \text{ at } a_j} = \frac{c_1 \overline{Y}_{\cdot j \cdot 1} + c_2 \overline{Y}_{\cdot j \cdot 2} + \cdots + c_r \overline{Y}_{\cdot j \cdot r}}{\sqrt{\dfrac{SSBL(AC)}{pr(n-1)} \left(\dfrac{c_1^2}{nq} + \dfrac{c_2^2}{nq} + \cdots + \dfrac{c_r^2}{nq} \right)}}$$

$$df = pr(n-1)$$

Treatment B at c_l

$$\frac{\hat{\psi}_{i(B)} \text{ at } c_l}{\hat{\sigma}_{\psi_{i(B)}} \text{ at } c_l} = \frac{c_1 \overline{Y}_{\cdot \cdot 1l} + c_2 \overline{Y}_{\cdot \cdot 2l} + \cdots + c_q \overline{Y}_{\cdot \cdot ql}}{\sqrt{\dfrac{SSB \times BL(AC)}{pr(n-1)(q-1)} \left(\dfrac{c_1^2}{np} + \dfrac{c_2^2}{np} + \cdots + \dfrac{c_q^2}{np} \right)}}$$

$$df = pr(n-1)(q-1)$$

Treatment C at b_k

$$\frac{\hat{\psi}_{i(C)} \text{ at } b_k}{\hat{\sigma}_{\psi_{i(C)}} \text{ at } b_k} = \frac{c_1 \overline{Y}_{\cdot \cdot k1} + c_2 \overline{Y}_{\cdot \cdot k2} + \cdots + c_r \overline{Y}_{\cdot \cdot kr}}{\sqrt{\dfrac{SSBL(AC) + SSB \times BL(AC)}{pr(n-1) + pr(n-1)(q-1)} \left(\dfrac{c_1^2}{np} + \dfrac{c_2^2}{np} + \cdots + \dfrac{c_r^2}{np} \right)}}$$

$$df = pr(n-1) + pr(n-1)(q-1)$$

(iii) Simple-simple main-effects means:

Treatment A at $b_k c_l$

$$\frac{\hat{\psi}_{i(A)} \text{ at } b_k c_l}{\hat{\sigma}_{\psi_{i(A)}} \text{ at } b_k c_l} = \frac{c_1 \overline{Y}_{1kl} + c_2 \overline{Y}_{2kl} + \cdots + c_p \overline{Y}_{pkl}}{\sqrt{\dfrac{SSBL(AC) + SSB \times BL(AC)}{pr(n-1) + pr(n-1)(q-1)} \left(\dfrac{c_1^2}{n} + \dfrac{c_2^2}{n} + \cdots + \dfrac{c_p^2}{n} \right)}}$$

$$df = pr(n-1) + pr(n-1)(q-1)$$

Treatment B at $a_j c_l$

$$\frac{\hat{\psi}_{i(B)} \text{ at } a_j c_l}{\hat{\sigma}_{\psi_{i(B)}} \text{ at } a_j c_l} = \frac{c_1 \overline{Y}_{\cdot j 1l} + c_2 \overline{Y}_{\cdot j 2l} + \cdots + c_q \overline{Y}_{\cdot j ql}}{\sqrt{\dfrac{SSB \times BL(AC)}{pr(n-1)(q-1)} \left(\dfrac{c_1^2}{n} + \dfrac{c_2^2}{n} + \cdots + \dfrac{c_q^2}{n} \right)}}$$

$$df = pr(n-1)(q-1)$$

Treatment C at $a_j b_k$

$$\frac{\hat{\psi}_{i(C)} \text{ at } a_j b_k}{\hat{\sigma}_{\psi_{i(C)}} \text{ at } a_j b_k} = \frac{c_1 \overline{Y}_{\cdot jk1} + c_2 \overline{Y}_{\cdot jk2} + \cdots + c_r \overline{Y}_{\cdot jkr}}{\sqrt{\dfrac{SSBL(AC) + SSB \times BL(AC)}{pr(n-1) + pr(n-1)(q-1)} \left(\dfrac{c_1^2}{n} + \dfrac{c_2^2}{n} + \cdots + \dfrac{c_r^2}{n} \right)}}$$

$$df = pr(n-1) + pr(n-1)(q-1)$$

TABLE 11.9-4   Table for Determining $E(MS)$ for Type SPF-*pr·q* Design

Source	$E(MS)$
A	$\sigma_\epsilon^2 + \left(1 - \dfrac{n}{N}\right)\left(1 - \dfrac{q}{Q}\right)\sigma_{\beta\pi}^2 + n\left(1 - \dfrac{q}{Q}\right)\left(1 - \dfrac{r}{R}\right)\sigma_{\alpha\beta\gamma}^2 + nr\left(1 - \dfrac{q}{Q}\right)\sigma_{\alpha\beta}^2 + q\left(1 - \dfrac{n}{N}\right)\sigma_\pi^2$ $+ nq\left(1 - \dfrac{r}{R}\right)\sigma_{\alpha\gamma}^2 + nqr\sigma_\alpha^2$
C	$\sigma_\epsilon^2 + \left(1 - \dfrac{n}{N}\right)\left(1 - \dfrac{q}{Q}\right)\sigma_{\beta\pi}^2 + n\left(1 - \dfrac{p}{P}\right)\left(1 - \dfrac{q}{Q}\right)\sigma_{\alpha\beta\gamma}^2 + np\left(1 - \dfrac{q}{Q}\right)\sigma_{\beta\gamma}^2 + q\left(1 - \dfrac{n}{N}\right)\sigma_\pi^2$ $+ nq\left(1 - \dfrac{p}{P}\right)\sigma_{\alpha\gamma}^2 + npq\sigma_\gamma^2$
AC	$\sigma_\epsilon^2 + \left(1 - \dfrac{n}{N}\right)\left(1 - \dfrac{q}{Q}\right)\sigma_{\beta\pi}^2 + n\left(1 - \dfrac{q}{Q}\right)\sigma_{\alpha\beta\gamma}^2 + q\left(1 - \dfrac{n}{N}\right)\sigma_\pi^2 + nq\sigma_{\alpha\gamma}^2$
Blocks w. AC	$\sigma_\epsilon^2 + \left(1 - \dfrac{q}{Q}\right)\sigma_{\beta\pi}^2 + q\sigma_\pi^2$
B	$\sigma_\epsilon^2 + \left(1 - \dfrac{n}{N}\right)\sigma_{\beta\pi}^2 + n\left(1 - \dfrac{p}{P}\right)\left(1 - \dfrac{r}{R}\right)\sigma_{\alpha\beta\gamma}^2 + np\left(1 - \dfrac{r}{R}\right)\sigma_{\beta\gamma}^2 + nr\left(1 - \dfrac{p}{P}\right)\sigma_{\alpha\beta}^2$ $+ npr\sigma_\beta^2$
AB	$\sigma_\epsilon^2 + \left(1 - \dfrac{n}{N}\right)\sigma_{\beta\pi}^2 + n\left(1 - \dfrac{r}{R}\right)\sigma_{\alpha\beta\gamma}^2 + nr\sigma_{\alpha\beta}^2$
BC	$\sigma_\epsilon^2 + \left(1 - \dfrac{n}{N}\right)\sigma_{\beta\pi}^2 + n\left(1 - \dfrac{p}{P}\right)\sigma_{\alpha\beta\gamma}^2 + np\sigma_{\beta\gamma}^2$
ABC	$\sigma_\epsilon^2 + \left(1 - \dfrac{n}{N}\right)\sigma_{\beta\pi}^2 + n\sigma_{\alpha\beta\gamma}^2$
$B \times$blocks w. AC	$\sigma_\epsilon^2 + \sigma_{\beta\pi}^2$

11.10 COMPUTATIONAL PROCEDURES FOR TYPE SPF-*prt·q* DESIGN

The inclusion of a fourth treatment in a type SPF-*prt·q* design presents no new computational problems. The block diagram of a type SPF-222·4 design, shown in Figure 11.10-1, has three between-blocks treatments (A, C, and D) and one within-blocks treatment (B). The experimental design model equation is

FIGURE 11.10-1 Type SPF-222·4 design; $S_1, S_2, \ldots, S_8$ denote samples of subjects who receive one combination of treatments A, C, and D and all levels of treatment B.

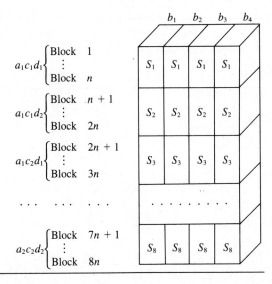

$$Y_{ijklm} = \mu + \alpha_j + \gamma_l + \delta_m + (\alpha\gamma)_{jl} + (\alpha\delta)_{jm} + (\gamma\delta)_{lm}$$
$$+ (\alpha\gamma\delta)_{jlm} + \pi_{i(jlm)} + \beta_k + (\alpha\beta)_{jk} + (\beta\gamma)_{kl} + (\beta\delta)_{km}$$
$$+ (\alpha\beta\gamma)_{jkl} + (\alpha\beta\delta)_{jkm} + (\beta\gamma\delta)_{klm} + (\alpha\beta\gamma\delta)_{jklm}$$
$$+ (\beta\pi)_{ki(jlm)} + \epsilon_{ijklm}$$

$(i = 1, \ldots, n; j = 1, \ldots, p; k = 1, \ldots, q; l = 1, \ldots, r; m = 1, \ldots, t).$

An outline of the computational procedures appears in Table 11.10-1. The F ratios are appropriate for a mixed model in which treatments A, B, C, and D are fixed and blocks are random.

PATTERN UNDERLYING COMPUTATIONAL FORMULAS FOR THE BETWEEN- AND WITHIN-BLOCKS ERROR MEAN SQUARES

In Section 9.3 we examined the pattern underlying the four kinds of formulas used in analysis of variance. Here we will illustrate the patterns for the between- and within-blocks error sums of squares and degrees of freedom for a split-plot factorial design having any number of between-blocks treatments. The patterns are as follows.

Design	Between-Blocks Error SS	df
SPF-$p \cdot q$	$SSBL(A) = [AS] - [A]$	$p(n-1)$
SPF-$pr \cdot q$	$SSBL(AC) = [ACS] - [AC]$	$pr(n-1)$
SPF-$prt \cdot q$	$SSBL(ACD) = [ACDS] - [ACD]$	$prt(n-1)$
SPF-$prtu \cdot q$	$SSBL(ACDE) = [ACDES] - [ACDE]$	$prtu(n-1)$

Design	Within-Blocks Error SS	df
SPF-$p \cdot q$	$SSB \times BL(A) = [ABS] - [AB] - [AS] + [A]$	$p(n-1)(q-1)$
SPF-$pr \cdot q$	$SSB \times BL(AC) = [ABCS] - [ABC] - [ACS] + [AC]$	$pr(n-1)(q-1)$
SPF-$prt \cdot q$	$SSB \times BL(ACD) = [ABCDS] - [ABCD] - [ACDS] + [ACD]$	$prt(n-1)(q-1)$
SPF-$prtu \cdot q$	$SSB \times BL(ACDE) = [ABCDES] - [ABCDE] - [ACDES] + [ACDE]$	$prtu(n-1)(q-1)$

TABLE 11.10-1   Outline of Computational Procedures for Type SPF-*prt·q* Design

(i)   Summary tables required:

AB Table	*BD* Table	*ACD* Table
AC Table	*CD* Table	*BCD* Table
AD Table	*ABC* Table	*ABCD* Table
BC Table	*ABD* Table	*ABCDS* Table

(ii)   Computational symbols:

AB Table $[A] = \sum_{j=1}^{p} \dfrac{\left(\sum_{i=1}^{n} \sum_{k=1}^{q} \sum_{l=1}^{r} \sum_{m=1}^{t} Y_{ijklm}\right)^2}{nqrt}$

CD Table $[CD] = \sum_{l=1}^{r} \sum_{m=1}^{t} \dfrac{\left(\sum_{i=1}^{n} \sum_{j=1}^{p} \sum_{k=1}^{q} Y_{ijklm}\right)^2}{npq}$

AB Table $[B] = \sum_{k=1}^{q} \dfrac{\left(\sum_{i=1}^{n} \sum_{j=1}^{p} \sum_{l=1}^{r} \sum_{m=1}^{t} Y_{ijklm}\right)^2}{nprt}$

ABC Table $[ABC] = \sum_{j=1}^{p} \sum_{k=1}^{q} \sum_{l=1}^{r} \dfrac{\left(\sum_{i=1}^{n} \sum_{m=1}^{t} Y_{ijklm}\right)^2}{nt}$

AC Table $[C] = \sum_{l=1}^{r} \dfrac{\left(\sum_{i=1}^{n} \sum_{j=1}^{p} \sum_{k=1}^{q} \sum_{m=1}^{t} Y_{ijklm}\right)^2}{npqt}$

ABD Table $[ABD] = \sum_{j=1}^{p} \sum_{k=1}^{q} \sum_{m=1}^{t} \dfrac{\left(\sum_{i=1}^{n} \sum_{l=1}^{r} Y_{ijklm}\right)^2}{nr}$

AD Table $[D] = \sum_{m=1}^{t} \dfrac{\left(\sum_{i=1}^{n} \sum_{j=1}^{p} \sum_{k=1}^{q} \sum_{l=1}^{r} Y_{ijklm}\right)^2}{npqr}$

ACD Table $[ACD] = \sum_{j=1}^{p} \sum_{l=1}^{r} \sum_{m=1}^{t} \dfrac{\left(\sum_{i=1}^{n} \sum_{k=1}^{q} Y_{ijklm}\right)^2}{nq}$

AB Table $[AB] = \sum_{j=1}^{p} \sum_{k=1}^{q} \dfrac{\left(\sum_{i=1}^{n} \sum_{l=1}^{r} \sum_{m=1}^{t} Y_{ijklm}\right)^2}{nrt}$

BCD Table $[BCD] = \sum_{k=1}^{q} \sum_{l=1}^{r} \sum_{m=1}^{t} \dfrac{\left(\sum_{i=1}^{n} \sum_{j=1}^{p} Y_{ijklm}\right)^2}{np}$

AC Table $[AC] = \sum_{j=1}^{p} \sum_{l=1}^{r} \dfrac{\left(\sum_{i=1}^{n} \sum_{k=1}^{q} \sum_{m=1}^{t} Y_{ijklm}\right)^2}{nqt}$

ABCD Table $[ABCD] = \sum_{j=1}^{p} \sum_{k=1}^{q} \sum_{l=1}^{r} \sum_{m=1}^{t} \dfrac{\left(\sum_{i=1}^{n} Y_{ijklm}\right)^2}{n}$

AD Table $[AD] = \sum_{j=1}^{p} \sum_{m=1}^{t} \dfrac{\left(\sum_{i=1}^{n} \sum_{k=1}^{q} \sum_{l=1}^{r} Y_{ijklm}\right)^2}{nqr}$

ACDS Table $[ACDS] = \sum_{i=1}^{n} \sum_{j=1}^{p} \sum_{l=1}^{r} \sum_{m=1}^{t} \dfrac{\left(\sum_{k=1}^{q} Y_{ijklm}\right)^2}{q}$

BC Table $[BC] = \sum_{k=1}^{q} \sum_{l=1}^{r} \dfrac{\left(\sum_{i=1}^{n} \sum_{j=1}^{p} \sum_{m=1}^{t} Y_{ijklm}\right)^2}{npt}$

ABCDS Table $[ABCDS] = \sum_{i=1}^{n} \sum_{j=1}^{p} \sum_{k=1}^{q} \sum_{l=1}^{r} \sum_{m=1}^{t} Y_{ijklm}^2$

BD Table $[BD] = \sum_{k=1}^{q} \sum_{m=1}^{t} \dfrac{\left(\sum_{i=1}^{n} \sum_{j=1}^{p} \sum_{l=1}^{r} Y_{ijklm}\right)^2}{npr}$

ABCDS Table $[Y] = \dfrac{\left(\sum_{i=1}^{n} \sum_{j=1}^{p} \sum_{k=1}^{q} \sum_{l=1}^{r} \sum_{m=1}^{t} Y_{ijklm}\right)^2}{npqrt}$

Table 11.10-1 (continued)

(iii) Computational formulas and F ratios:

	Sum of Squares	Degrees of Freedom	F ratio (A, B, C, and D fixed; Blocks random)
1	$SS\ b.\ BL = [ACDS] - [Y]$	$nprt - 1$	
2	$SSA = [A] - [Y]$	$p - 1$	$[2/9]$
3	$SSC = [C] - [Y]$	$r - 1$	$[3/9]$
4	$SSD = [D] - [Y]$	$t - 1$	$[4/9]$
5	$SSAC = [AC] - [A] - [C] + [Y]$	$(p - 1)(r - 1)$	$[5/9]$
6	$SSAD = [AD] - [A] - [D] + [Y]$	$(p - 1)(t - 1)$	$[6/9]$
7	$SSCD = [CD] - [C] - [D] + [Y]$	$(r - 1)(t - 1)$	$[7/9]$
8	$SSACD = [ACD] - [AC] - [AD] - [CD]$ $+ [A] + [C] + [D] - [Y]$	$(p - 1)(r - 1)(t - 1)$	$[8/9]$
9	$SSBL(ACD) = [ACDS] - [ACD]$	$prt(n - 1)$	
10	$SS\ w.\ BL = [ABCDS] - [ACDS]$	$nprt(q - 1)$	
11	$SSB = [B] - [Y]$	$q - 1$	$[11/19]$
12	$SSAB = [AB] - [A] - [B] + [Y]$	$(p - 1)(q - 1)$	$[12/19]$
13	$SSBC = [BC] - [B] - [C] + [Y]$	$(q - 1)(r - 1)$	$[13/19]$
14	$SSBD = [BD] - [B] - [D] + [Y]$	$(q - 1)(t - 1)$	$[14/19]$
15	$SSABC = [ABC] - [AB] - [AC] - [BC]$ $+ [A] + [B] + [C] - [Y]$	$(p - 1)(q - 1)(r - 1)$	$[15/19]$
16	$SSABD = [ABD] - [AB] - [AD] - [BD]$ $+ [A] + [B] + [D] - [Y]$	$(p - 1)(q - 1)(t - 1)$	$[16/19]$
17	$SSBCD = [BCD] - [BC] - [BD] - [CD]$ $+ [B] + [C] + [D] - [Y]$	$(q - 1)(r - 1)(t - 1)$	$[17/19]$
18	$SSABCD = [ABCD] - [ABC] - [ABD] - [ACD]$ $- [BCD] + [AB] + [AC] + [AD]$ $+ [BC] + [BD] + [CD] - [A]$ $- [B] - [C] - [D] + [Y]$	$(p - 1)(q - 1)(r - 1)(t - 1)$	$[18/19]$
19	$SSB \times BL(ACD) = [ABCDS] - [ABCD]$ $- [ACDS] + [ACD]$	$prt(n - 1)(q - 1)$	
20	$SSTO = [ABCDS] - [Y]$	$npqrt - 1$	

SOME ASSUMPTIONS UNDERLYING A TYPE SPF-*prt·q* DESIGN

The between-blocks mean square error term $MSBL(ACD)$ is composed of variation pooled over the *prt* combinations of treatments A, C, and D. We assume that $MSBL(a_1c_1d_1)$, $MSBL(a_1c_1d_2)$, . . . , $MSBL(a_pc_rd_t)$ all estimate the same population variance. This homogeneity of variance assumption can be tested by means of the F_{max} statistic or one of the other statistics described in Section 2.5.

The within-blocks mean square error term $MSB \times BL(ACD)$ is also a pooled term. For each of the *prt* combinations of treatments A, C, and D, we can construct a $q \times q$ variance-covariance matrix $\hat{\boldsymbol{\Sigma}}_{a_jc_ld_m}$. The key assumption associated with testing the within-blocks null hypotheses is the multisample circularity assumption

$$\underset{[(q-1)\times q]}{\mathbf{C}_B^{\star\prime}} \; \underset{(q\times q)}{\boldsymbol{\Sigma}_{a_jc_ld_m}} \; \underset{[q\times(q-1)]}{\mathbf{C}_B^{\star}} \; = \; \underset{[(q-1)\times(q-1)]}{\lambda\mathbf{I}} \qquad \begin{array}{l} (j = 1, \ldots, p; l = 1, \ldots, r; \\ m = 1, \ldots, t). \end{array}$$

If the tenability of the circularity assumption $\mathbf{C}_B^{\star\prime} \, \boldsymbol{\Sigma}_{\text{pooled}} \, \mathbf{C}_B^{\star} = \lambda\mathbf{I}$ is in doubt, the three-step testing strategy described in Section 11.4 can be used. The formula for computing $\hat{\theta}$ is the same as that in Table 11.4 for a type SPF-*p·q* design; the formula for $\tilde{\theta}$ is

$$\tilde{\theta} = \frac{nprt(q - 1)\hat{\theta} - 2}{(q - 1)[nprt - prt - (q - 1)\hat{\theta}]}.$$

11.11 COMPUTATIONAL PROCEDURES FOR TYPE SPF-*p·qr* DESIGN

A split-plot factorial design can be used in research situations involving two or more within-blocks treatments. One variation of this design, a type SPF-*p·qr* design, has one between-blocks treatment and two within-blocks treatments. An example of this design is diagrammed in Figure 11.1-1. For the repeated measurements case, p samples of n experimental units from a population are randomly assigned to the levels of treatment A. The sequence of administration of the b_kc_l treatment combinations to the experimental units in the np blocks is randomized independently for each block. For the nonrepeated measurements case, p samples of n blocks each containing qr matched experimental units from a population are randomly assigned to the levels of treatment A. Following this, the b_kc_l treatment combinations are randomly assigned to the qr experimental units within each block. The repeated measurements design requires np sets of qr matched subjects. The experimental design model equation is

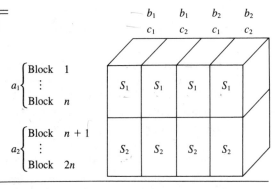

FIGURE 11.11-1   Type SPF-2·22 design; S_1 and S_2 denote samples of subjects who receive one level of treatment A and all combinations of treatments B and C.

$$Y_{ijkl} = \mu + \alpha_j + \pi_{i(j)} + \beta_k + (\alpha\beta)_{jk} + (\beta\pi)_{ki(j)} + \gamma_l + (\alpha\gamma)_{jl}$$
$$+ (\gamma\pi)_{li(j)} + (\beta\gamma)_{kl} + (\alpha\beta\gamma)_{jkl} + (\beta\gamma\pi)_{kli(j)} + \epsilon_{ijkl}$$
$$(i = 1, \ldots, n; j = 1, \ldots, p; k = 1, \ldots, q; l = 1, \ldots, r).$$

A type SPF-pr·q design is similar to the split-split-plot designs used in agricultural research. The essential difference is that in agricultural applications the levels of treatment A are assigned to plots (plots correspond to blocks of subjects in behavioral research). The plots are then subdivided for the levels of treatment B and subdivided again for the levels of treatment C. The levels of treatment B are randomly assigned to the split-plots and the levels of C are randomly assigned to the split-split-plots. This randomization procedure can be contrasted with that used in a type SPF-p·qr design, where the $b_k c_l$ treatment combinations are randomly assigned within each block. The two randomization procedures lead to different error terms for testing treatment C and all interactions involving C.

The computational procedures for a type SPF-p·qr design are illustrated in Table 11.11-1. The analysis is summarized in Table 11.11-2.

*SOME ASSUMPTIONS UNDERLYING A TYPE SPF-p·qr DESIGN

The between-blocks mean-square error term $MSBL(A)$ is composed of variation pooled over the p levels of treatment A. We assume that $MSBL(a_1), MSBL(a_2), \ldots, MSBL(a_p)$ all estimate the same population variance. This homogeneity of variance assumption can be tested by means of the F_{max} statistic or one of the other statistics described in Section 2.5.

Four within-blocks mean-square error terms can be computed for a type SPF-p·qr design: $MSB \times BL(A)$, $MSC \times BL(A)$, $MSB \times C \times BL(A)$,   and $MSBC \times BL(A)$, which will be described shortly. Mendoza, Toothaker, and Crain (1976) have examined the necessary and sufficient conditions for the within-blocks F ratios to be distributed as F, given that the associated null hypotheses are true. Their findings can be summarized as follows.

---

*   The latter portion of this section assumes a familiarity with Sections 6.4, 6.9, and 9.6.

TABLE 11.11-1 Computational Procedures for a Type SPF-2·22 Design

(i) Data and notation [Y_{ijkl} denotes a score for the experimental unit in block i and treatment combination jkl; $i = 1, \ldots, n$ blocks (s_i); $j = 1, \ldots, p$ levels of treatment A (a_j); $k = 1, \ldots, q$ levels of treatment B (b_k); $l = 1, \ldots, r$ levels of treatment C (c_l)]:

ABCS Summary Table

Entry is Y_{ijkl}

		b_1 c_1	b_1 c_2	b_2 c_1	b_2 c_2	$\sum\limits_{k=1}^{q}\sum\limits_{l=1}^{r} Y_{ijkl}$
	s_1	3	4	7	7	21
a_1	s_2	6	5	8	8	27
	s_3	3	4	7	9	23
	s_4	3	3	6	8	20
	s_5	1	2	5	10	18
a_2	s_6	2	3	6	10	21
	s_7	2	4	5	9	20
	s_8	2	3	6	11	22

ABC Summary Table

Entry is $\sum\limits_{i=1}^{n} Y_{ijkl}$

	b_1 c_1	b_1 c_2	b_2 c_1	b_2 c_2
a_1	$n = 4$ 15	16	28	32
a_2	7	12	22	40

AB Summary Table

Entry is $\sum\limits_{i=1}^{n}\sum\limits_{l=1}^{r} Y_{ijkl}$

	b_1	b_2	$\sum\limits_{i=1}^{n}\sum\limits_{k=1}^{q}\sum\limits_{l=1}^{r} Y_{ijkl}$
a_1	$nr = 8$ 31	60	91
a_2	19	62	81

$$\sum\limits_{i=1}^{n}\sum\limits_{j=1}^{p}\sum\limits_{l=1}^{r} Y_{ijkl} = 50 \quad 122$$

BC Summary Table

Entry is $\sum\limits_{i=1}^{n}\sum\limits_{j=1}^{p} Y_{ijkl}$

	c_1	c_2
b_1	$np = 8$ 22	28
b_2	50	72

AC Summary Table

Entry is $\sum\limits_{i=1}^{n}\sum\limits_{k=1}^{q} Y_{ijkl}$

	c_1	c_2
a_1	$nq = 8$ 43	48
a_2	29	52

$$\sum\limits_{i=1}^{n}\sum\limits_{j=1}^{p}\sum\limits_{k=1}^{q} Y_{ijkl} = 72 \quad 100$$

Table 11.11-1 (continued)

ABS Summary Table		

Entry is $\sum\limits_{l=1}^{r} Y_{ijkl}$

		b_1	b_2
		$r = 2$	
a_1	s_1	7	14
	s_2	11	16
	s_3	7	16
	s_4	6	14
a_2	s_5	3	15
	s_6	5	16
	s_7	6	14
	s_8	5	17

ACS Summary Table		

Entry is $\sum\limits_{k=1}^{q} Y_{ijkl}$

		c_1	c_2
		$q = 2$	
a_1	s_1	10	11
	s_2	14	13
	s_3	10	13
	s_4	9	11
a_2	s_5	6	12
	s_6	8	13
	s_7	7	13
	s_8	8	14

(ii) Computational symbols:

$$\sum_{i=1}^{n}\sum_{j=1}^{p}\sum_{k=1}^{q}\sum_{l=1}^{r} Y_{ijkl} = 3 + 4 + \cdots + 11 = 172.000$$

$$\frac{\left(\sum_{i=1}^{n}\sum_{j=1}^{p}\sum_{k=1}^{q}\sum_{l=1}^{r} Y_{ijkl}\right)^2}{npqr} = [Y] = \frac{(172.000)^2}{(4)(2)(2)(2)} = 924.500$$

$$\sum_{i=1}^{n}\sum_{j=1}^{p}\sum_{k=1}^{q}\sum_{l=1}^{r} Y_{ijkl}^2 = [ABCS] = (3)^2 + (4)^2 + \cdots + (11)^2 = 1160.000$$

$$\sum_{i=1}^{n}\sum_{j=1}^{p}\frac{\left(\sum_{k=1}^{q}\sum_{l=1}^{r} Y_{ijkl}\right)^2}{qr} = [AS] = \frac{(21)^2}{(2)(2)} + \frac{(27)^2}{(2)(2)} + \cdots + \frac{(22)^2}{(2)(2)} = 937.000$$

$$\sum_{j=1}^{p}\frac{\left(\sum_{i=1}^{n}\sum_{k=1}^{q}\sum_{l=1}^{r} Y_{ijkl}\right)^2}{nqr} = [A] = \frac{(91)^2}{(4)(2)(2)} + \frac{(81)^2}{(4)(2)(2)} = 927.625$$

$$\sum_{k=1}^{q}\frac{\left(\sum_{i=1}^{n}\sum_{j=1}^{p}\sum_{l=1}^{r} Y_{ijkl}\right)^2}{npr} = [B] = \frac{(50)^2}{(4)(2)(2)} + \frac{(122)^2}{(4)(2)(2)} = 1086.500$$

$$\sum_{j=1}^{p}\sum_{k=1}^{q}\frac{\left(\sum_{i=1}^{n}\sum_{l=1}^{r} Y_{ijkl}\right)^2}{nr} = [AB] = \frac{(31)^2}{(4)(2)} + \frac{(60)^2}{(4)(2)} + \cdots + \frac{(62)^2}{(4)(2)} = 1095.750$$

Table 11.11-1 (continued)

$$\sum_{i=1}^{n}\sum_{j=1}^{p}\sum_{k=1}^{q}\frac{\left(\sum_{l=1}^{r}Y_{ijkl}\right)^2}{r}=[ABS]=\frac{(7)^2}{2}+\frac{(14)^2}{2}+\cdots+\frac{(17)^2}{2}=1110.000$$

$$\sum_{l=1}^{r}\frac{\left(\sum_{i=1}^{n}\sum_{j=1}^{p}\sum_{k=1}^{q}Y_{ijkl}\right)^2}{npq}=[C]=\frac{(72)^2}{(4)(2)(2)}+\frac{(100)^2}{(4)(2)(2)}=949.000$$

$$\sum_{j=1}^{p}\sum_{l=1}^{r}\frac{\left(\sum_{i=1}^{n}\sum_{k=1}^{q}Y_{ijkl}\right)^2}{nq}=[AC]=\frac{(43)^2}{(4)(2)}+\frac{(48)^2}{(4)(2)}+\cdots+\frac{(52)^2}{(4)(2)}=962.250$$

$$\sum_{i=1}^{n}\sum_{j=1}^{p}\sum_{l=1}^{r}\frac{\left(\sum_{k=1}^{q}Y_{ijkl}\right)^2}{q}=[ACS]=\frac{(10)^2}{2}+\frac{(11)^2}{2}+\cdots+\frac{(14)^2}{2}=974.000$$

$$\sum_{k=1}^{q}\sum_{l=1}^{r}\frac{\left(\sum_{i=1}^{n}\sum_{j=1}^{p}Y_{ijkl}\right)^2}{np}=[BC]=\frac{(22)^2}{(4)(2)}+\frac{(28)^2}{(4)(2)}+\cdots+\frac{(72)^2}{(4)(2)}=1119.000$$

$$\sum_{j=1}^{p}\sum_{k=1}^{q}\sum_{l=1}^{r}\frac{\left(\sum_{i=1}^{n}Y_{ijkl}\right)^2}{n}=[ABC]=\frac{(15)^2}{4}+\frac{(16)^2}{4}+\cdots+\frac{(40)^2}{4}=1141.500$$

(iii) Computational formulas:

$$SSTO = [ABCS] - [Y] = 235.500$$
$$SSBETWEEN\ BL = [AS] - [Y] = 12.500$$
$$SSA = [A] - [Y] = 3.125$$
$$SSBL(A) = [AS] - [A] = 9.375$$
$$SSWITHIN\ BL = [ABCS] - [AS] = 223.000$$
$$SSB = [B] - [Y] = 162.000$$
$$SSAB = [AB] - [A] - [B] + [Y] = 6.125$$
$$SSB \times BL(A) = [ABS] - [AB] - [AS] + [A] = 4.875$$
$$SSC = [C] - [Y] = 24.500$$
$$SSAC = [AC] - [A] - [C] + [Y] = 10.125$$
$$SSC \times BL(A) = [ACS] - [AC] - [AS] + [A] = 2.375$$
$$SSBC = [BC] - [B] - [C] + [Y] = 8.000$$
$$SSABC = [ABC] - [AB] - [AC] - [BC] + [A] + [B] + [C] - [Y] = 3.125$$
$$SSB \times C \times BL(A) = [ABCS] - [ABC] - [ABS] - [ACS] + [AB] + [AC] + [AS] - [A] = 1.875$$

TABLE 11.11-2 ANOVA Table for Type SPF-2·22 Design

	Source	SS	df	MS	F (A, B, and C fixed effects; Blocks random)
1	Between blocks	12.500	$np - 1 = 7$		
2	A *Librarians*	3.125	$p - 1 = 1$	3.125	[2/3] 2.00
3	Blocks w. A	9.375	$p(n - 1) = 6$	1.562	
4	Within blocks	223.000	$np(qr - 1) = 24$		
5	B *Racial Comp*	162.000	$q - 1 = 1$	162.000	[5/7] 199.51*
6	AB	6.125	$(p - 1)(q - 1) = 1$	6.125	[6/7] 7.54*
7	B×blocks w. A	4.875	$p(n - 1)(q - 1) = 6$	0.812	
8	C *School Locat*	24.500	$r - 1 = 1$	24.500	[8/10] 61.87**
9	AC	10.125	$(p - 1)(r - 1) = 1$	10.125	[9/10] 25.57**
10	C×blocks w. A	2.375	$p(n - 1)(r - 1) = 6$	0.396	
11	BC	8.000	$(q - 1)(r - 1) = 1$	8.000	[11/13] 25.64**
12	ABC	3.125	$(p - 1)(q - 1)(r - 1) = 1$	3.125	[12/13] 10.02*
13	B×C×blocks w. A	1.875	$p(n - 1)(q - 1)(r - 1) = 6$	0.312	
14	Total	235.500	$npqr - 1 = 31$		

*p < .05
**p < .01

F Ratio	Necessary and Sufficient Conditions
$F = \dfrac{MSB}{MSB \times BL(A)}$ $F = \dfrac{MSAB}{MSB \times BL(A)}$	$\underset{(q-1)\times qr}{\mathbf{C}_B^{\star\prime}} \quad \underset{qr\times qr}{\mathbf{\Sigma}_{a_j}} \quad \underset{qr\times(q-1)}{\mathbf{C}_B^{\star}} = \underset{(q-1)\times(q-1)}{\lambda\mathbf{I}} \qquad (j = 1, \ldots, p)$
$F = \dfrac{MSC}{MSC \times BL(A)}$ $F = \dfrac{MSAC}{MSC \times BL(A)}$	$\underset{(r-1)\times qr}{\mathbf{C}_C^{\star\prime}} \quad \underset{qr\times qr}{\mathbf{\Sigma}_{a_j}} \quad \underset{qr\times(r-1)}{\mathbf{C}_C^{\star}} = \underset{(r-1)\times(r-1)}{\lambda\mathbf{I}} \qquad (j = 1, \ldots, p)$
$F = \dfrac{MSBC}{MSB \times C \times BL(A)}$ $F = \dfrac{MSABC}{MSB \times C \times BL(A)}$	$\underset{(q-1)(r-1)\times qr}{\mathbf{C}_{BC}^{\star\prime}} \quad \underset{qr\times qr}{\mathbf{\Sigma}_{a_j}} \quad \underset{qr\times(q-1)(r-1)}{\mathbf{C}_{BC}^{\star}} = \underset{(q-1)(r-1)\times(q-1)(r-1)}{\lambda\mathbf{I}} \qquad (j = 1, \ldots, p)$

The discussion that follows should help the reader get a better feeling for the matrices $\mathbf{C}_B^{*'}$, $\mathbf{\Sigma}_{a_j}$, and so on. For a type SPF-2·22 design, the population variance-covariance matrix, $\mathbf{\Sigma}_{a_j}$, has the form

$$
\underset{(2)(2)\times(2)(2)}{\mathbf{\Sigma}_{a_j}} =
\begin{matrix}
 & \begin{matrix} b_1 & b_1 & b_2 & b_2 \\ c_1 & c_2 & c_1 & c_2 \end{matrix} \\
\begin{matrix} b_1c_1 \\ b_1c_2 \\ b_2c_1 \\ b_2c_2 \end{matrix} &
\begin{bmatrix}
\sigma_1^2 & \sigma_{12} & \sigma_{13} & \sigma_{14} \\
\sigma_{21} & \sigma_2^2 & \sigma_{23} & \sigma_{24} \\
\sigma_{31} & \sigma_{32} & \sigma_3^2 & \sigma_{34} \\
\sigma_{41} & \sigma_{42} & \sigma_{43} & \sigma_4^2
\end{bmatrix}
\end{matrix}
$$

where, for example, σ_1^2 is the population variance for treatment combination b_1c_1 and σ_{12} is the population covariance for treatment combinations b_1c_1 and b_1c_2. The matrix $\mathbf{C}_B^{*'}$, which is an orthonormal coefficient matrix, is used to express the null hypothesis for treatment B. Recall from Section 6.9 that a null hypothesis can be expressed in matrix notation as

$$H_0: \quad \mathbf{C}^{*'}\boldsymbol{\mu} = \mathbf{0}.$$

The preceding multisample circularity assumptions are all examples of local circularity, which was introduced in Section 9.6. The overall multisample circularity assumption for a type SPF-2·22 design is

$$
\underset{(qr-1)\times qr}{\mathbf{C}^{*'}} \underset{qr \times qr}{\mathbf{\Sigma}_{a_j}} \underset{qr \times (qr-1)}{\mathbf{C}^{*}} = \underset{(qr-1)\times(qr-1)}{\lambda \mathbf{I}} \qquad (j = 1, \ldots, p)
$$

where $\mathbf{C}^{*'}$ is an orthonormal coefficient matrix for the within-blocks overall null hypothesis,

$$H_0: \quad \mu_{\cdot\cdot 11} = \mu_{\cdot\cdot 12} = \mu_{\cdot\cdot 21} = \mu_{\cdot\cdot 22}.^*$$

In matrix notation this hypothesis can be expressed as

$$
\underset{3\times 4}{\mathbf{C}^{*'}} \qquad \underset{4\times 1}{\boldsymbol{\mu}} \qquad \underset{3\times 1}{\mathbf{0}}
$$

$$
\begin{bmatrix}
.7071 & -.7071 & 0 & 0 \\
0 & 0 & .7071 & -.7071 \\
.5000 & .5000 & -.5000 & -.5000
\end{bmatrix}
\begin{bmatrix}
\mu_{\cdot\cdot 11} \\ \mu_{\cdot\cdot 12} \\ \mu_{\cdot\cdot 21} \\ \mu_{\cdot\cdot 22}
\end{bmatrix}
=
\begin{bmatrix}
0 \\ 0 \\ 0
\end{bmatrix}.
$$

If the overall multisample circularity assumption is tenable, a fourth mean-square error term,

$$MSBC \times BL(A) = \{[ABCS] - [ABC] - [AS] + [A]\}/p(n - 1)(qr - 1)$$

with $p(n - 1)(qr - 1)$ degrees of freedom can be used in testing all of the within-blocks null hypotheses. The resulting F tests are more powerful than tests that assume only local multisample circularity. However, overall multisample circularity is a much

---

*   This assumption implies that $\sigma_{\beta\pi}^2 = \sigma_{\gamma\pi}^2 = \sigma_{\beta\gamma\pi}^2 = \sigma_{[\beta\gamma]\pi}^2$, where $\sigma_{[\beta\gamma]\pi}^2$ is the interaction of the b_kc_l treatment combinations with blocks within each level of A.

more restrictive assumption. Consequently the usual practice is to partition $MSBC \times BL(A)$ into three within-blocks mean square error terms as follows; the data are from Table 11.11-2.

Sum of Squares		df	Mean Square
$SSBC \times BL(A)$	$= 9.125$	18	0.507
$SSB \times BL(A)$	$= 4.875$	6	0.812
$SSC \times BL(A)$	$= 2.375$	6	0.396
$SSB \times C \times BL(A)$	$= 1.875$	6	0.312

If one or more of the local multisample circularity assumptions are not tenable, the three-step testing strategy described in Section 11.6 can be used. The formulas for computing $\hat{\theta}$ and $\tilde{\theta}$ given in Table 11.4-1 can be used for a type SPF-$p \cdot qr$ design by replacing q with qr.

PROCEDURES FOR TESTING HYPOTHESES ABOUT MEANS

Procedures for testing hypotheses about means are similar to those described in Section 11.9. The formulas for testing hypotheses about main-effects means are as follows.

$$\frac{\hat{\psi}_{i(A)}}{\hat{\sigma}_{\psi_{i(A)}}} = \frac{c_1 \overline{Y}_{\cdot 1 \cdot \cdot} + c_2 \overline{Y}_{\cdot 2 \cdot \cdot} + \cdots + c_p \overline{Y}_{\cdot p \cdot \cdot}}{\sqrt{\dfrac{SSBL(A)}{p(n-1)} \left(\dfrac{c_1^2}{nqr} + \dfrac{c_2^2}{nqr} + \cdots + \dfrac{c_p^2}{nqr} \right)}}$$

$$df = p(n-1)$$

$$\frac{\hat{\psi}_{i(B)}}{\hat{\sigma}_{\psi_{i(B)}}} = \frac{c_1 \overline{Y}_{\cdot \cdot 1 \cdot} + c_2 \overline{Y}_{\cdot \cdot 2 \cdot} + \cdots + c_q \overline{Y}_{\cdot \cdot q \cdot}}{\sqrt{\dfrac{SSB \times BL(A)}{p(n-1)(q-1)} \left(\dfrac{c_1^2}{npr} + \dfrac{c_2^2}{npr} + \cdots + \dfrac{c_q^2}{npr} \right)}}$$

$$df = p(n-1)(q-1)$$

$$\frac{\hat{\psi}_{i(C)}}{\hat{\sigma}_{\psi_{i(C)}}} = \frac{c_1 \overline{Y}_{\cdot \cdot \cdot 1} + c_2 \overline{Y}_{\cdot \cdot \cdot 2} + \cdots + c_r \overline{Y}_{\cdot \cdot \cdot r}}{\sqrt{\dfrac{SSC \times BL(A)}{p(n-1)(r-1)} \left(\dfrac{c_1^2}{npq} + \dfrac{c_2^2}{npq} + \cdots + \dfrac{c_r^2}{npq} \right)}}$$

$$df = p(n-1)(r-1)$$

EXPECTED VALUES OF MEAN SQUARES FOR A TYPE SPF-$p \cdot qr$ DESIGN

The expected values of mean squares for models I, II, and III can be determined from Table 11.11-3. The terms $1 - (n/N)$, $1 - (p/P)$, $1 - (q/Q)$, and $1 - (r/R)$ become zero if the corresponding effects are fixed and one if they are random. Also the variances should be replaced by terms of the form $\Sigma\alpha_j^2/(p-1)$, $\Sigma\beta_k^2/(q-1)$, and $\Sigma\gamma_i^2/(r-1)$ if the corresponding treatments are fixed.

TABLE 11.11-3 Table for Determining $E(MS)$ for Type SPF-$p\cdot qr$ Design

Source	$E(MS)$
A	$\sigma_\epsilon^2 + \left(1-\dfrac{n}{N}\right)\left(1-\dfrac{q}{Q}\right)\left(1-\dfrac{r}{R}\right)\sigma_{\beta\gamma\pi}^2 + n\left(1-\dfrac{q}{Q}\right)\left(1-\dfrac{r}{R}\right)\sigma_{\alpha\beta\gamma}^2 + q\left(1-\dfrac{n}{N}\right)\left(1-\dfrac{r}{R}\right)\sigma_{\gamma\pi}^2$ $+ nq\left(1-\dfrac{r}{R}\right)\sigma_{\alpha\gamma}^2 + r\left(1-\dfrac{n}{N}\right)\left(1-\dfrac{q}{Q}\right)\sigma_{\beta\pi}^2 + nr\left(1-\dfrac{q}{Q}\right)\sigma_{\alpha\beta}^2 + qr\left(1-\dfrac{n}{N}\right)\sigma_\pi^2$ $+ nqr\sigma_\alpha^2$
$BL(A)$	$\sigma_\epsilon^2 + \left(1-\dfrac{q}{Q}\right)\left(1-\dfrac{r}{R}\right)\sigma_{\beta\gamma\pi}^2 + q\left(1-\dfrac{r}{R}\right)\sigma_{\gamma\pi}^2 + r\left(1-\dfrac{q}{Q}\right)\sigma_{\beta\pi}^2 + qr\sigma_\pi^2$
B	$\sigma_\epsilon^2 + \left(1-\dfrac{n}{N}\right)\left(1-\dfrac{r}{R}\right)\sigma_{\beta\gamma\pi}^2 + n\left(1-\dfrac{p}{P}\right)\left(1-\dfrac{r}{R}\right)\sigma_{\alpha\beta\gamma}^2 + np\left(1-\dfrac{r}{R}\right)\sigma_{\beta\gamma}^2 + r\left(1-\dfrac{n}{N}\right)\sigma_{\beta\pi}^2$ $+ nr\left(1-\dfrac{p}{P}\right)\sigma_{\alpha\beta}^2 + npr\sigma_\beta^2$
AB	$\sigma_\epsilon^2 + \left(1-\dfrac{n}{N}\right)\left(1-\dfrac{r}{R}\right)\sigma_{\beta\gamma\pi}^2 + n\left(1-\dfrac{r}{R}\right)\sigma_{\alpha\beta\gamma}^2 + r\left(1-\dfrac{n}{N}\right)\sigma_{\beta\pi}^2 + nr\sigma_{\alpha\beta}^2$
$B\times BL(A)$	$\sigma_\epsilon^2 + \left(1-\dfrac{r}{R}\right)\sigma_{\beta\gamma\pi}^2 + r\sigma_{\beta\pi}^2$
C	$\sigma_\epsilon^2 + \left(1-\dfrac{n}{N}\right)\left(1-\dfrac{q}{Q}\right)\sigma_{\beta\gamma\pi}^2 + n\left(1-\dfrac{p}{P}\right)\left(1-\dfrac{q}{Q}\right)\sigma_{\alpha\beta\gamma}^2 + np\left(1-\dfrac{q}{Q}\right)\sigma_{\beta\gamma}^2 + q\left(1-\dfrac{n}{N}\right)\sigma_{\gamma\pi}^2$ $+ nq\left(1-\dfrac{p}{P}\right)\sigma_{\alpha\gamma}^2 + npq\sigma_\gamma^2$
AC	$\sigma_\epsilon^2 + \left(1-\dfrac{n}{N}\right)\left(1-\dfrac{q}{Q}\right)\sigma_{\beta\gamma\pi}^2 + n\left(1-\dfrac{q}{Q}\right)\sigma_{\alpha\beta\gamma}^2 + q\left(1-\dfrac{n}{N}\right)\sigma_{\gamma\pi}^2 + nq\sigma_{\alpha\gamma}^2$
$C\times BL(A)$	$\sigma_\epsilon^2 + \left(1-\dfrac{q}{Q}\right)\sigma_{\beta\gamma\pi}^2 + q\sigma_{\gamma\pi}^2$
BC	$\sigma_\epsilon^2 + \left(1-\dfrac{n}{N}\right)\sigma_{\beta\gamma\pi}^2 + n\left(1-\dfrac{p}{P}\right)\sigma_{\alpha\beta\gamma}^2 + np\sigma_{\beta\gamma}^2$
ABC	$\sigma_\epsilon^2 + \left(1-\dfrac{n}{N}\right)\sigma_{\beta\gamma\pi}^2 + n\sigma_{\alpha\beta\gamma}^2$
$B\times C\times BL(A)$	$\sigma_\epsilon^2 + \sigma_{\beta\gamma\pi}^2$

11.12 COMPUTATIONAL PROCEDURES FOR TYPE SPF-*p·qrt* DESIGN

Figure 11.12-1 shows a block diagram for a type SPF-2·222 design. The experimental design model equation is

$$
\begin{aligned}
Y_{ijklm} = {} & \mu + \alpha_j + \pi_{i(j)} + \beta_k + (\alpha\beta)_{jk} + (\beta\pi)_{ki(j)} + \gamma_l + (\alpha\gamma)_{jl} \\
& + (\gamma\pi)_{li(j)} + \delta_m + (\alpha\delta)_{jm} + (\delta\pi)_{mi(j)} + (\beta\gamma)_{kl} + (\alpha\beta\gamma)_{jkl} \\
& + (\beta\gamma\pi)_{kli(j)} + (\beta\delta)_{km} + (\alpha\beta\delta)_{jkm} + (\beta\delta\pi)_{kmi(j)} + (\gamma\delta)_{lm} \\
& + (\alpha\gamma\delta)_{jlm} + (\gamma\delta\pi)_{lmi(j)} + (\beta\gamma\delta)_{klm} + (\alpha\beta\gamma\delta)_{jklm} \\
& + (\beta\gamma\delta\pi)_{klmi(j)} + \epsilon_{ijklm}
\end{aligned}
$$

$(i = 1, \ldots, n; j = 1, \ldots, p; k = 1, \ldots, q; l = 1, \ldots, r; m = 1, \ldots, t).$

An outline of the computational procedures appears in Table 11.12-1. The F ratios are appropriate for a mixed model in which treatments $A, B, C,$ and D are fixed and blocks are random.

FIGURE 11.12-1   Type SPF-2·222 design; S_1 and S_2 denote samples of subjects who receive one level of treatment A and all combinations of treatments $B, C,$ and D.

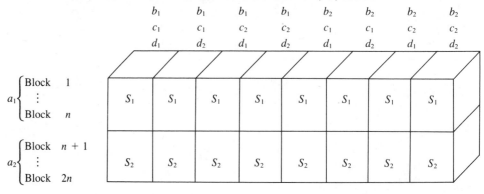

TABLE 11.12-1   Outline of Computational Procedures for Type SPF-*p·qrt* Design

(i)   Summary tables required:

AB Table	*ABC* Table	*ADS* Table
AC Table	*ABD* Table	*ABCD* Table
AD Table	*ACD* Table	*ABCS* Table
BC Table	*BCD* Table	*ABDS* Table
BD Table	*ABS* Table	*ACDS* Table
CD Table	*ACS* Table	*ABCDS* Table

TABLE 11.12-1 (continued)

(ii) Computational symbols:

AB Table $\quad [A] = \sum_{j=1}^{p} \dfrac{\left(\sum_{i=1}^{n} \sum_{k=1}^{q} \sum_{l=1}^{r} \sum_{m=1}^{t} Y_{ijklm} \right)^2}{nqrt}$

ABD Table $\quad [ABD] = \sum_{j=1}^{p} \sum_{k=1}^{q} \sum_{m=1}^{t} \dfrac{\left(\sum_{i=1}^{n} \sum_{l=1}^{r} Y_{ijklm} \right)^2}{nr}$

AB Table $\quad [B] = \sum_{k=1}^{q} \dfrac{\left(\sum_{i=1}^{n} \sum_{j=1}^{p} \sum_{l=1}^{r} \sum_{m=1}^{t} Y_{ijklm} \right)^2}{nprt}$

ACD Table $\quad [ACD] = \sum_{j=1}^{p} \sum_{l=1}^{r} \sum_{m=1}^{t} \dfrac{\left(\sum_{i=1}^{n} \sum_{k=1}^{q} Y_{ijklm} \right)^2}{nq}$

AC Table $\quad [C] = \sum_{l=1}^{r} \dfrac{\left(\sum_{i=1}^{n} \sum_{j=1}^{p} \sum_{k=1}^{q} \sum_{m=1}^{t} Y_{ijklm} \right)^2}{npqt}$

BCD Table $\quad [BCD] = \sum_{k=1}^{q} \sum_{l=1}^{r} \sum_{m=1}^{t} \dfrac{\left(\sum_{i=1}^{n} \sum_{j=1}^{p} Y_{ijklm} \right)^2}{np}$

AD Table $\quad [D] = \sum_{m=1}^{t} \dfrac{\left(\sum_{i=1}^{n} \sum_{j=1}^{p} \sum_{k=1}^{q} \sum_{l=1}^{r} Y_{ijklm} \right)^2}{npqr}$

ABS Table $\quad [ABS] = \sum_{i=1}^{n} \sum_{j=1}^{p} \sum_{k=1}^{q} \dfrac{\left(\sum_{l=1}^{r} \sum_{m=1}^{t} Y_{ijklm} \right)^2}{rt}$

AB Table $\quad [AB] = \sum_{j=1}^{p} \sum_{k=1}^{q} \dfrac{\left(\sum_{i=1}^{n} \sum_{l=1}^{r} \sum_{m=1}^{t} Y_{ijklm} \right)^2}{nrt}$

ACS Table $\quad [ACS] = \sum_{i=1}^{n} \sum_{j=1}^{p} \sum_{l=1}^{r} \dfrac{\left(\sum_{k=1}^{q} \sum_{m=1}^{t} Y_{ijklm} \right)^2}{qt}$

AC Table $\quad [AC] = \sum_{j=1}^{p} \sum_{l=1}^{r} \dfrac{\left(\sum_{i=1}^{n} \sum_{k=1}^{q} \sum_{m=1}^{t} Y_{ijklm} \right)^2}{nqt}$

ADS Table $\quad [ADS] = \sum_{i=1}^{n} \sum_{j=1}^{p} \sum_{m=1}^{t} \dfrac{\left(\sum_{k=1}^{q} \sum_{l=1}^{r} Y_{ijklm} \right)^2}{qr}$

AD Table $\quad [AD] = \sum_{j=1}^{p} \sum_{m=1}^{t} \dfrac{\left(\sum_{i=1}^{n} \sum_{k=1}^{q} \sum_{l=1}^{r} Y_{ijklm} \right)^2}{nqr}$

$ABCD$ Table $\quad [ABCD] = \sum_{j=1}^{p} \sum_{k=1}^{q} \sum_{l=1}^{r} \sum_{m=1}^{t} \dfrac{\left(\sum_{i=1}^{n} Y_{ijklm} \right)^2}{n}$

BC Table $\quad [BC] = \sum_{k=1}^{q} \sum_{l=1}^{r} \dfrac{\left(\sum_{i=1}^{n} \sum_{j=1}^{p} \sum_{m=1}^{t} Y_{ijklm} \right)^2}{npt}$

$ABCS$ Table $\quad [ABCS] = \sum_{i=1}^{n} \sum_{j=1}^{p} \sum_{k=1}^{q} \sum_{l=1}^{r} \dfrac{\left(\sum_{m=1}^{t} Y_{ijklm} \right)^2}{t}$

BD Table $\quad [BD] = \sum_{k=1}^{q} \sum_{m=1}^{t} \dfrac{\left(\sum_{i=1}^{n} \sum_{j=1}^{p} \sum_{l=1}^{r} Y_{ijklm} \right)^2}{npr}$

$ABDS$ Table $\quad [ABDS] = \sum_{i=1}^{n} \sum_{j=1}^{p} \sum_{k=1}^{q} \sum_{m=1}^{t} \dfrac{\left(\sum_{l=1}^{r} Y_{ijklm} \right)^2}{r}$

CD Table $\quad [CD] = \sum_{l=1}^{r} \sum_{m=1}^{t} \dfrac{\left(\sum_{i=1}^{n} \sum_{j=1}^{p} \sum_{k=1}^{q} Y_{ijklm} \right)^2}{npq}$

$ACDS$ Table $\quad [ACDS] = \sum_{i=1}^{n} \sum_{j=1}^{p} \sum_{l=1}^{r} \sum_{m=1}^{t} \dfrac{\left(\sum_{k=1}^{q} Y_{ijklm} \right)^2}{q}$

$ABCDS$ Table $\quad [AS] = \sum_{i=1}^{n} \sum_{j=1}^{p} \dfrac{\left(\sum_{k=1}^{q} \sum_{l=1}^{r} \sum_{m=1}^{t} Y_{ijklm} \right)^2}{qrt}$

$ABCDS$ Table $\quad [ABCDS] = \sum_{i=1}^{n} \sum_{j=1}^{p} \sum_{k=1}^{q} \sum_{l=1}^{r} \sum_{m=1}^{t} Y_{ijklm}^2$

TABLE 11.12-1 (continued)

$$ABC \text{ Table} \quad [ABC] = \sum_{j=1}^{p} \sum_{k=1}^{q} \sum_{l=1}^{r} \frac{\left(\sum_{i=1}^{n} \sum_{m=1}^{t} Y_{ijklm} \right)^2}{nt} \qquad ABCDS \text{ Table} \quad [Y] = \frac{\left(\sum_{i=1}^{n} \sum_{j=1}^{p} \sum_{k=1}^{q} \sum_{l=1}^{r} \sum_{m=1}^{t} Y_{ijklm} \right)^2}{npqrt}$$

(iii) Computational formulas and F ratios:

	Sum of Squares	Degrees of Freedom	F ratio (A, B, C, and D fixed; Blocks random)
1	$SS \text{ b. } BL = [AS] - [Y]$	$np - 1$	
2	$SSA = [A] - [Y]$	$p - 1$	$[2/3]$
3	$SSBL(A) = [AS] - [A]$	$p(n - 1)$	
4	$SS \text{ w. } BL = [ABCDS] - [AS]$	$np(qrt - 1)$	
5	$SSB = [B] - [Y]$	$q - 1$	$[5/7]$
6	$SSAB = [AB] - [A] - [B] + [Y]$	$(p - 1)(q - 1)$	$[6/7]$
7	$SSB \times BL(A) = [ABS] - [AB] - [AS] + [A]$	$p(n - 1)(q - 1)$	
8	$SSC = [C] - [Y]$	$r - 1$	$[8/10]$
9	$SSAC = [AC] - [A] - [C] + [Y]$	$(p - 1)(r - 1)$	$[9/10]$
10	$SSC \times BL(A) = [ACS] - [AC] - [AS] + [A]$	$p(n - 1)(r - 1)$	
11	$SSD = [D] - [Y]$	$t - 1$	$[11/13]$
12	$SSAD = [AD] - [A] - [D] + [Y]$	$(p - 1)(t - 1)$	$[12/13]$
13	$SSD \times BL(A) = [ADS] - [AD] - [AS] + [A]$	$p(n - 1)(t - 1)$	
14	$SSBC = [BC] - [B] - [C] + [Y]$	$(q - 1)(r - 1)$	$[14/16]$
15	$SSABC = [ABC] - [AB] - [AC] - [BC]$ $+ [A] + [B] + [C] - [Y]$	$(p - 1)(q - 1)(r - 1)$	$[15/16]$
16	$SSB \times C \times BL(A) = [ABCS] - [ABC] - [ABS]$ $- [ACS] + [AB] + [AC]$ $+ [AS] - [A]$	$p(n - 1)(q - 1)(r - 1)$	
17	$SSBD = [BD] - [B] - [D] + [Y]$	$(q - 1)(t - 1)$	$[17/19]$
18	$SSABD = [ABD] - [AB] - [AD] - [BD]$ $+ [A] + [B] + [D] - [Y]$	$(p - 1)(q - 1)(t - 1)$	$[18/19]$
19	$SSB \times D \times BL(A) = [ABDS] - [ABD] - [ABS]$ $- [ADS] + [AB] + [AD]$ $+ [AS] - [A]$	$p(n - 1)(q - 1)(t - 1)$	
20	$SSCD = [CD] - [C] - [D] + [Y]$	$(r - 1)(t - 1)$	$[20/22]$
21	$SSACD = [ACD] - [AC] - [AD] - [CD]$ $+ [A] + [C] + [D] - [Y]$	$(p - 1)(r - 1)(t - 1)$	$[21/22]$

Table 11.12-1 (continued)

	Sum of Squares	Degrees of Freedom	F ratio (A, B, C, and D fixed; Blocks random)
22	$SSC \times D \times BL(A) = [ACDS] - [ACD] - [ACS]$ $- [ADS] + [AC] + [AD]$ $+ [AS] - [A]$	$p(n - 1)(r - 1)(t - 1)$	
23	$SSBCD = [BCD] - [BC] - [BD] - [CD]$ $+ [B] + [C] + [D] - [Y]$	$(q - 1)(r - 1)(t - 1)$	$[23/25]$
24	$SSABCD = [ABCD] - [ABC] - [ABD]$ $- [ACD] - [BCD] + [AB]$ $+ [AC] + [AD] + [BC]$ $+ [BD] + [CD] - [A] - [B]$ $- [C] - [D] + [Y]$	$(p - 1)(q - 1)(r - 1)(t - 1)$	$[24/25]$
25	$SSB \times C \times D \times BL(A) = [ABCDS] - [ABCD] - [ABCS]$ $- [ABDS] - [ACDS] + [ABC]$ $+ [ABD] + [ACD] + [ABS]$ $+ [ACS] + [ADS] - [AB]$ $- [AC] - [AD] - [AS] + [A]$	$p(n - 1)(q - 1)(r - 1)(t - 1)$	
26	$SSTO = [ABCDS] - [Y]$	$npqrt - 1$	

11.13 COMPUTATIONAL PROCEDURES FOR TYPE SPF-*pr·qt* DESIGN

It is hoped that the reader has gained insight into the patterns underlying the analysis of all split-plot factorial designs. A block diagram for one more design, a type SPF-*pr·qt* design, is shown in Figure 11.13-1. The computational formulas for this design are given in Table 11.13-1. The experimental design model equation is

$$Y_{ijklm} = \mu + \alpha_j + \gamma_l + (\alpha\gamma)_{jl} + \pi_{i(jl)} + \beta_k + (\alpha\beta)_{jk} + (\beta\gamma)_{kl}$$
$$+ (\alpha\beta\gamma)_{jkl} + (\beta\pi)_{ki(jl)} + \delta_m + (\alpha\delta)_{jm} + (\gamma\delta)_{lm}$$
$$+ (\alpha\gamma\delta)_{jlm} + (\delta\pi)_{mi(jl)} + (\beta\delta)_{km} + (\alpha\beta\delta)_{jkm}$$
$$+ (\beta\gamma\delta)_{klm} + (\alpha\beta\gamma\delta)_{jklm} + (\beta\delta\pi)_{kmi(jk)} + \epsilon_{ijklm}$$

$(i = 1, \ldots, n; j = 1, \ldots, p; k = 1, \ldots, q; l = 1, \ldots, r; m = 1, \ldots, t).$

TABLE 11.13-1   Computational Formulas for Type SPF-$pr \cdot qt$ Design

(i)   Computational formulas:

	Sum of Squares	Degrees of Freedom	F ratio (A, B, C, and D fixed; Blocks random)
1	$SS \text{ b. } BL = [ACS] - [Y]$	$npr - 1$	
2	$SSA = [A] - [Y]$	$p - 1$	$[2/5]$
3	$SSC = [C] - [Y]$	$r - 1$	$[3/5]$
4	$SSAC = [AC] - [A] - [C] + [Y]$	$(p - 1)(r - 1)$	$[4/5]$
5	$SSBL(AC) = [ACS] - [AC]$	$pr(n - 1)$	
6	$SS \text{ w. } BL = [ABCDS] - [ACS]$	$npr(qt - 1)$	
7	$SSB = [B] - [Y]$	$q - 1$	$[7/11]$
8	$SSAB = [AB] - [A] - [B] + [Y]$	$(p - 1)(q - 1)$	$[8/11]$
9	$SSBC = [BC] - [B] - [C] + [Y]$	$(q - 1)(r - 1)$	$[9/11]$
10	$SSABC = [ABC] - [AB] - [AC] - [BC]$ $+ [A] + [B] + [C] - [Y]$	$(p - 1)(q - 1)(r - 1)$	$[10/11]$
11	$SSB \times BL(AC) = [ABCS] - [ABC] - [ACS] + [AC]$	$pr(n - 1)(q - 1)$	
12	$SSD = [D] - [Y]$	$t - 1$	$[12/16]$
13	$SSAD = [AD] - [A] - [D] + [Y]$	$(p - 1)(t - 1)$	$[13/16]$
14	$SSCD = [CD] - [C] - [D] + [Y]$	$(r - 1)(t - 1)$	$[14/16]$
15	$SSACD = [ACD] - [AC] - [AD] - [CD]$ $+ [A] + [C] + [D] - [Y]$	$(p - 1)(r - 1)(t - 1)$	$[15/16]$
16	$SSD \times BL(AC) = [ACDS] - [ACD] - [ACS] + [AC]$	$pr(n - 1)(t - 1)$	
17	$SSBD = [BD] - [B] - [D] + [Y]$	$(q - 1)(t - 1)$	$[17/21]$
18	$SSABD = [ABD] - [AB] - [AD] - [BD]$ $+ [A] + [B] + [D] - [Y]$	$(p - 1)(q - 1)(t - 1)$	$[18/21]$
19	$SSBCD = [BCD] - [BC] - [BD] - [CD]$ $+ [B] + [C] + [D] - [Y]$	$(q - 1)(r - 1)(t - 1)$	$[19/21]$
20	$SSABCD = [ABCD] - [ABC] - [ABD]$ $- [ACD] - [BCD] + [AB] + [AC]$ $+ [AD] + [BC] + [BD] + [CD]$ $- [A] - [B] - [C] - [D] + [Y]$	$(p - 1)(q - 1)(r - 1)(t - 1)$	$[20/21]$
21	$SSB \times D \times BL(AC) = [ABCDS] - [ABCD] - [ABCS]$ $- [ACDS] + [ABC] + [ACD]$ $+ [ACS] - [AC]$	$pr(n - 1)(q - 1)(t - 1)$	
22	$SSTO = [ABCDS] - [Y]$	$npqrt - 1$	

FIGURE 11.13-1 Type SPF-22·22 design; S_1, S_2, S_3, S_4 denote samples of subjects who receive one combination of treatments A and C and all combinations of treatments B and D.

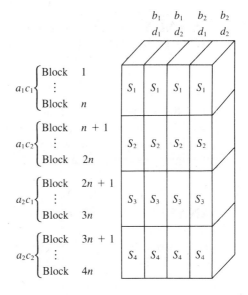

11.14 EVALUATION OF SEQUENCE EFFECTS

When repeated measurements are obtained, the procedure recommended in Section 11.1 is to randomize the order of administration of the levels independently for each experimental unit. If the administration of one treatment level affects a subject's performance on subsequent levels, the experiment is said to contain *carryover effects*. That portion of carryover effects attributable to the treatment levels being administered in a particular order is referred to as *sequence effects*. Randomization of the order of administration of the levels independently for each subject is an effective way to control sequence effects.

An alternative procedure is to include sequence effects as one of the treatments in the design. For example, a type SPF-2·3 design has 3! or 6 possible sequences in which treatment B can be administered. If we let the levels of treatment C correspond to the six sequences, the design can be analyzed as a type SPF-26·3 design. The experiment would require a minimum of 24 subjects, with two subjects assigned to each $a_j c_l$ treatment combination. When the number of levels of the repeated measurements treatment is greater than three, the number of subjects can be kept to a manageable size by using a random sample of the $q!$ possible sequences. The advantages of including sequence effects as one of the treatments is that (1) the effects are controlled and (2) an experimenter can determine if they affect the dependent variable.

*11.15 ANALYSIS OF A TYPE SPF-$p\cdot q$ DESIGN USING THE FULL RANK MODEL APPROACH

In this section, the full rank experimental design model approach is used to analyze data for a type SPF-$p\cdot q$ design. Three situations are examined: one in which the number of blocks in each level of treatment A is the same; a second in which the number of blocks is not the same; and a third in which there are missing observations within one or more blocks.

ANALYSIS WHEN $n_{1.} = n_{2.} = \cdots = n_{p.}$.

The restricted full rank experimental design model corresponding to the classical model (11.3-1) for a type SPF-$p\cdot q$ design is

(11.15-1) $Y_{ijk} = \mu_{ijk} + \epsilon_{ijk}$ $(i = 1, \ldots, n; j = 1, \ldots, p; k = 1, \ldots, q)$

where ϵ_{ijk} is $NID(0, \sigma_\epsilon^2)$ and μ_{ijk} is subject to the restrictions that for each j $\mu_{ijk} - \mu_{i'jk} - \mu_{ijk'} + \mu_{i'jk'} = 0$ for all i, i', k, and k'. These restrictions specify that for each level of treatment A the blocks $\times B$ treatment effects equal zero.

  The analysis procedures for this restricted full rank experimental design model will be illustrated for the data in Table 11.2-1. The following null hypotheses are of interest.

Treatment A

$$H_0:\ \mu_{\cdot 1 \cdot} - \mu_{\cdot 2 \cdot} = 0 \quad \text{or} \quad H_0:\ \underset{(p-1)\times h}{\mathbf{C}'_A}\ \underset{h\times 1}{\boldsymbol{\mu}} = \underset{(p-1)\times 1}{\mathbf{0}_A}$$

where $\underset{(1\times 32)}{\mathbf{C}'_A} =$

$$[1 \quad 1 \quad 1 \quad 1 \quad 1 \quad 1 \quad 1 \quad 1 \quad 1 \quad 1 \quad 1 \quad 1 \quad 1 \quad 1 \quad 1 \quad 1 \quad -1 \quad -1$$
$$-1 \quad -1 \quad -1 \quad -1 \quad -1 \quad -1 \quad -1 \quad -1 \quad -1 \quad -1 \quad -1 \quad -1 \quad -1 \quad -1]$$

   This section assumes a familiarity with Appendix D, Chapter 5, and Section 6.9. The reader who is interested only in the traditional approach to analysis of variance can, without loss of continuity, omit this section.

TABLE 11.15-1 Type SPF-2·4 Design

ABS Summary Table
Entry is μ_{ijk}

		b_1	b_2	b_3	b_4	$\sum\limits_{k=1}^{q} \mu_{ijk}/q$
a_1	s_1	μ_{111}	μ_{112}	μ_{113}	μ_{114}	$\mu_{11\cdot}$
	s_2	μ_{211}	μ_{212}	μ_{213}	μ_{214}	$\mu_{21\cdot}$
	s_3	μ_{311}	μ_{312}	μ_{313}	μ_{314}	$\mu_{31\cdot}$
	s_4	μ_{411}	μ_{412}	μ_{413}	μ_{414}	$\mu_{41\cdot}$
a_2	s_5	μ_{521}	μ_{522}	μ_{523}	μ_{524}	$\mu_{52\cdot}$
	s_6	μ_{621}	μ_{622}	μ_{623}	μ_{624}	$\mu_{62\cdot}$
	s_7	μ_{721}	μ_{722}	μ_{723}	μ_{724}	$\mu_{72\cdot}$
	s_8	μ_{821}	μ_{822}	μ_{823}	μ_{824}	$\mu_{82\cdot}$

AB Summary Table
Entry is $\mu_{\cdot jk}$

	b_1	b_2	b_3	b_4	$\sum\limits_{k=1}^{q} \mu_{\cdot jk}/q$
a_1	$\mu_{\cdot 11}$	$\mu_{\cdot 12}$	$\mu_{\cdot 13}$	$\mu_{\cdot 14}$	$\mu_{\cdot 1\cdot}$
a_2	$\mu_{\cdot 21}$	$\mu_{\cdot 22}$	$\mu_{\cdot 23}$	$\mu_{\cdot 24}$	$\mu_{\cdot 2\cdot}$

$$\sum_{j=1}^{p} \mu_{\cdot jk}/p = \mu_{\cdot\cdot 1} \qquad \mu_{\cdot\cdot 2} \qquad \mu_{\cdot\cdot 3} \qquad \mu_{\cdot\cdot 4}$$

$$\boldsymbol{\mu}' = [\mu_{111} \quad \mu_{211} \quad \mu_{311} \quad \mu_{411} \quad \mu_{112} \quad \mu_{212} \quad \mu_{312} \quad \mu_{412} \quad \mu_{113}$$
$$\mu_{213} \quad \mu_{313} \quad \mu_{413} \quad \cdot \quad \cdot \quad \cdot \quad \mu_{824}]$$

and $\mathbf{0}_A = [0]$. The coefficients, $c_{ijk} = \pm 1/nq$, in $\mathbf{C}'_A$ have been multiplied by 16 to avoid fractions. This does not affect the nature of the hypothesis that is tested. The reader may find it helpful to refer to Table 11.15-1 where the means for this hypothesis and those that follow are defined.

Treatment B

$$H_0: \quad \mu_{\cdot\cdot 1} - \mu_{\cdot\cdot 4} = 0 \qquad \text{or} \qquad H_0: \quad \underset{(q-1)\times h}{\mathbf{C}'_B} \quad \underset{h\times 1}{\boldsymbol{\mu}} = \underset{(q-1)\times 1}{\mathbf{0}_B}$$
$$\mu_{\cdot\cdot 2} - \mu_{\cdot\cdot 4} = 0$$
$$\mu_{\cdot\cdot 3} - \mu_{\cdot\cdot 4} = 0$$

where $C'_{(B)}$ =
(3×32)

$$
\begin{bmatrix}
1 & 1 & 1 & 1 & 0 & 0 & 0 & 0 & 0 & 0 & 0 & 0 & -1 & -1 & -1 & -1 \\
0 & 0 & 0 & 0 & 1 & 1 & 1 & 1 & 0 & 0 & 0 & 0 & -1 & -1 & -1 & -1 \\
0 & 0 & 0 & 0 & 0 & 0 & 0 & 0 & 1 & 1 & 1 & 1 & -1 & -1 & -1 & -1
\end{bmatrix}
$$

$$
\begin{matrix}
1 & 1 & 1 & 1 & 0 & 0 & 0 & 0 & 0 & 0 & 0 & 0 & -1 & -1 & -1 & -1 \\
0 & 0 & 0 & 0 & 1 & 1 & 1 & 1 & 0 & 0 & 0 & 0 & -1 & -1 & -1 & -1 \\
0 & 0 & 0 & 0 & 0 & 0 & 0 & 0 & 1 & 1 & 1 & 1 & -1 & -1 & -1 & -1
\end{matrix}
$$

and $0'_B = [0 \ \ 0 \ \ 0]$. The coefficients, $c_{ijk} = \pm 1/np$ or 0, in C'_B have been multiplied by 8 to avoid fractions.

AB Interaction

H_0: $\mu_{\cdot 11} - \mu_{\cdot 21} - \mu_{\cdot 12} + \mu_{\cdot 22} = 0$ or H_0: $\underset{(p-1)(q-1)\times h}{C'_{AB}}$ $\underset{h\times 1}{\mu}$ = $\underset{(p-1)(q-1)\times 1}{0_{AB}}$

$\mu_{\cdot 12} - \mu_{\cdot 22} - \mu_{\cdot 13} + \mu_{\cdot 23} = 0$

$\mu_{\cdot 13} - \mu_{\cdot 23} - \mu_{\cdot 14} + \mu_{\cdot 24} = 0$

where C'_{AB} =
(3×32)

$$
\begin{bmatrix}
1 & 1 & 1 & 1 & -1 & -1 & -1 & -1 & 0 & 0 & 0 & 0 & 0 & 0 & 0 & 0 \\
0 & 0 & 0 & 0 & 1 & 1 & 1 & 1 & -1 & -1 & -1 & -1 & 0 & 0 & 0 & 0 \\
0 & 0 & 0 & 0 & 0 & 0 & 0 & 0 & 1 & 1 & 1 & 1 & -1 & -1 & -1 & -1
\end{bmatrix}
$$

$$
\begin{matrix}
-1 & -1 & -1 & -1 & 1 & 1 & 1 & 1 & 0 & 0 & 0 & 0 & 0 & 0 & 0 & 0 \\
0 & 0 & 0 & 0 & -1 & -1 & -1 & -1 & 1 & 1 & 1 & 1 & 0 & 0 & 0 & 0 \\
0 & 0 & 0 & 0 & 0 & 0 & 0 & 0 & -1 & -1 & -1 & -1 & 1 & 1 & 1 & 1
\end{matrix}
$$

and $0'_{AB} = [0 \ \ 0 \ \ 0]$. The coefficients, $c_{ijk} = \pm 1/n$ or 0, in C'_{AB} have been multiplied by 4 to avoid fractions. In determining the coefficients in C'_{AB} the reader may find it helpful to refer to the *AB* Summary Table in Table 11.15-1. Each row of C'_{AB} represents one of the three sets of crossed lines in the *AB* Summary Table. The means connected by a dashed line are subtracted from those connected by a solid line.

We turn now to formulating coefficient matrices for error terms. The reader may recall from Section 11.2 that the error term used to test the null hypothesis for treatment *A* is the mean square blocks within levels of treatment *A*, *MSBL(A)*. This mean square is identical to that for testing the hypothesis that the block means within a given level of treatment *A* are equal, that is,

$$BL(A) \text{ Error}$$

$$\text{H}_0: \quad \mu_{11\cdot} - \mu_{41\cdot} = 0 \qquad \text{or} \qquad \text{H}_0: \quad \underset{p(n-1) \times h}{\mathbf{C}'_{BL(A)}} \quad \underset{h \times 1}{\boldsymbol{\mu}} \ = \ \underset{p(n-1) \times 1}{\mathbf{0}_{BL(A)}}$$

$$\mu_{21\cdot} - \mu_{41\cdot} = 0$$

$$\mu_{31\cdot} - \mu_{41\cdot} = 0$$

$$\mu_{52\cdot} - \mu_{82\cdot} = 0$$

$$\mu_{62\cdot} - \mu_{82\cdot} = 0$$

$$\mu_{72\cdot} - \mu_{82\cdot} = 0$$

where $\quad \underset{6 \times 32}{\mathbf{C}'_{BL(A)}} =$

$$
\begin{bmatrix}
1 & 0 & 0 & -1 & 1 & 0 & 0 & -1 & 1 & 0 & 0 & -1 & 1 & 0 & 0 & -1 \\
0 & 1 & 0 & -1 & 0 & 1 & 0 & -1 & 0 & 1 & 0 & -1 & 0 & 1 & 0 & -1 \\
0 & 0 & 1 & -1 & 0 & 0 & 1 & -1 & 0 & 0 & 1 & -1 & 0 & 0 & 1 & -1 \\
0 & 0 & 0 & 0 & 0 & 0 & 0 & 0 & 0 & 0 & 0 & 0 & 0 & 0 & 0 & 0 \\
0 & 0 & 0 & 0 & 0 & 0 & 0 & 0 & 0 & 0 & 0 & 0 & 0 & 0 & 0 & 0 \\
0 & 0 & 0 & 0 & 0 & 0 & 0 & 0 & 0 & 0 & 0 & 0 & 0 & 0 & 0 & 0 \\
\end{bmatrix}
$$

$$
\begin{bmatrix}
0 & 0 & 0 & 0 & 0 & 0 & 0 & 0 & 0 & 0 & 0 & 0 & 0 & 0 & 0 & 0 \\
0 & 0 & 0 & 0 & 0 & 0 & 0 & 0 & 0 & 0 & 0 & 0 & 0 & 0 & 0 & 0 \\
0 & 0 & 0 & 0 & 0 & 0 & 0 & 0 & 0 & 0 & 0 & 0 & 0 & 0 & 0 & 0 \\
1 & 0 & 0 & -1 & 1 & 0 & 0 & -1 & 1 & 0 & 0 & -1 & 1 & 0 & 0 & -1 \\
0 & 1 & 0 & -1 & 0 & 1 & 0 & -1 & 0 & 1 & 0 & -1 & 0 & 1 & 0 & -1 \\
0 & 0 & 1 & -1 & 0 & 0 & 1 & -1 & 0 & 0 & 1 & -1 & 0 & 0 & 1 & -1 \\
\end{bmatrix}
$$

and $\mathbf{0}'_{BL(A)} = [0 \quad 0 \quad 0 \quad 0 \quad 0 \quad 0]$. The coefficients, $c_{ijk} = \pm 1/q$ or 0, in $\mathbf{C}'_{BL(A)}$ have been multiplied by 4 to avoid fractions. The error term for testing the null hypotheses for treatment B and the AB interaction is the mean square interaction of treatment B and the blocks within levels of treatment A, $MSB \times BL(A)$. This mean square is identical to that for the $s = p(nq - n - q + 1) = 18$ restrictions on the μ_{ijk}'s. The restriction hypothesis is

$$B \times BL(A) \text{ Error}$$

$$\text{H}_0: \quad \mu_{111} - \mu_{211} - \mu_{112} + \mu_{212} = 0 \qquad \text{or} \qquad \text{H}_0: \quad \underset{s \times h}{\mathbf{R}'} \quad \underset{h \times 1}{\boldsymbol{\mu}} \ = \ \underset{s \times 1}{\boldsymbol{\theta}}$$

$$\mu_{211} - \mu_{311} - \mu_{212} + \mu_{312} = 0$$

$$\cdots\cdots\cdots\cdots\cdots\cdots\cdots\cdots$$

$$\mu_{313} - \mu_{413} - \mu_{314} + \mu_{414} = 0$$

$$\mu_{521} - \mu_{621} - \mu_{522} + \mu_{622} = 0$$

$$\mu_{621} - \mu_{721} - \mu_{622} + \mu_{722} = 0$$

$$\cdots\cdots\cdots\cdots\cdots\cdots\cdots\cdots$$

$$\mu_{723} - \mu_{823} - \mu_{724} + \mu_{824} = 0$$

where $\quad \mathbf{R}' = $
$$\underset{18\times32}{}$$

$$
\begin{bmatrix}
1 & -1 & 0 & 0 & -1 & 1 & 0 & 0 & 0 & 0 & 0 & 0 & 0 & 0 & 0 & 0 \\
0 & 1 & -1 & 0 & 0 & -1 & 1 & 0 & 0 & 0 & 0 & 0 & 0 & 0 & 0 & 0 \\
0 & 0 & 1 & -1 & 0 & 0 & -1 & 1 & 0 & 0 & 0 & 0 & 0 & 0 & 0 & 0 \\
0 & 0 & 0 & 0 & 1 & -1 & 0 & 0 & -1 & 1 & 0 & 0 & 0 & 0 & 0 & 0 \\
0 & 0 & 0 & 0 & 0 & 1 & -1 & 0 & 0 & -1 & 1 & 0 & 0 & 0 & 0 & 0 \\
0 & 0 & 0 & 0 & 0 & 0 & 1 & -1 & 0 & 0 & -1 & 1 & 0 & 0 & 0 & 0 \\
0 & 0 & 0 & 0 & 0 & 0 & 0 & 0 & 1 & -1 & 0 & 0 & -1 & 1 & 0 & 0 \\
0 & 0 & 0 & 0 & 0 & 0 & 0 & 0 & 0 & 1 & -1 & 0 & 0 & -1 & 1 & 0 \\
0 & 0 & 0 & 0 & 0 & 0 & 0 & 0 & 0 & 0 & 1 & -1 & 0 & 0 & -1 & 1 \\
0 & 0 & 0 & 0 & 0 & 0 & 0 & 0 & 0 & 0 & 0 & 0 & 0 & 0 & 0 & 0 \\
0 & 0 & 0 & 0 & 0 & 0 & 0 & 0 & 0 & 0 & 0 & 0 & 0 & 0 & 0 & 0 \\
\vdots & & & & & & & & & & & & & & & \\
0 & 0 & 0 & 0 & 0 & 0 & 0 & 0 & 0 & 0 & 0 & 0 & 0 & 0 & 0 & 0
\end{bmatrix}
$$

$$
\begin{bmatrix}
0 & 0 & 0 & 0 & 0 & 0 & 0 & 0 & 0 & 0 & 0 & 0 & 0 & 0 & 0 & 0 \\
0 & 0 & 0 & 0 & 0 & 0 & 0 & 0 & 0 & 0 & 0 & 0 & 0 & 0 & 0 & 0 \\
0 & 0 & 0 & 0 & 0 & 0 & 0 & 0 & 0 & 0 & 0 & 0 & 0 & 0 & 0 & 0 \\
0 & 0 & 0 & 0 & 0 & 0 & 0 & 0 & 0 & 0 & 0 & 0 & 0 & 0 & 0 & 0 \\
0 & 0 & 0 & 0 & 0 & 0 & 0 & 0 & 0 & 0 & 0 & 0 & 0 & 0 & 0 & 0 \\
0 & 0 & 0 & 0 & 0 & 0 & 0 & 0 & 0 & 0 & 0 & 0 & 0 & 0 & 0 & 0 \\
0 & 0 & 0 & 0 & 0 & 0 & 0 & 0 & 0 & 0 & 0 & 0 & 0 & 0 & 0 & 0 \\
0 & 0 & 0 & 0 & 0 & 0 & 0 & 0 & 0 & 0 & 0 & 0 & 0 & 0 & 0 & 0 \\
0 & 0 & 0 & 0 & 0 & 0 & 0 & 0 & 0 & 0 & 0 & 0 & 0 & 0 & 0 & 0 \\
1 & -1 & 0 & 0 & -1 & 1 & 0 & 0 & 0 & 0 & 0 & 0 & 0 & 0 & 0 & 0 \\
0 & 1 & -1 & 0 & 0 & -1 & 1 & 0 & 0 & 0 & 0 & 0 & 0 & 0 & 0 & 0 \\
\vdots & & & & & & & & & & & & & & & \\
0 & 0 & 0 & 0 & 0 & 0 & 0 & 0 & 0 & 0 & 1 & -1 & 0 & 0 & -1 & 1
\end{bmatrix}
$$

and $\boldsymbol{\theta}$ is an 18×1 vector of zeros. Each row of $\mathbf{R}'$ represents one of the 18 sets of crossed lines in the *ABS* Summary Table in Table 11.15-1.

We want to test the within-blocks null hypotheses, subject to the restrictions that $\mathbf{R}'\boldsymbol{\mu} = \boldsymbol{\theta}$. To do this we form augmented matrices of the form

$$
\mathbf{Q}'_w = \begin{bmatrix} \mathbf{R}' \\ \mathbf{C}'_w \end{bmatrix} \qquad \text{and} \qquad \boldsymbol{\eta}_w = \begin{bmatrix} \boldsymbol{\theta} \\ \mathbf{0}_w \end{bmatrix}
$$

where $\mathbf{C}'_w$ is the coefficient matrix for a within-blocks source of variation and $\mathbf{0}_w$ is the null vector associated with $\mathbf{C}'_w$. This same procedure was used with a randomized block design (see Section 6.9). Tests of the between-blocks null hypotheses resemble those for a completely randomized design and accordingly are not subject to the restrictions $\mathbf{R}'\boldsymbol{\mu} = \boldsymbol{\theta}$. If the rows of $\mathbf{R}'$ are orthogonal to the rows of a within-blocks coefficient matrix, the computational formula $(\mathbf{Q}'_w\hat{\boldsymbol{\mu}})'(\mathbf{Q}'_w\mathbf{Q}_w)^{-1}(\mathbf{Q}'_w\hat{\boldsymbol{\mu}}) - SSB \times BL(A)$ simplifies to $(\mathbf{C}'_w\hat{\boldsymbol{\mu}})'(\mathbf{C}'_w\mathbf{C}_w)^{-1}(\mathbf{C}'_w\hat{\boldsymbol{\mu}})$. Procedures for computing the sums of squares and F statistics are illustrated in Table 11.15-2. Also shown in Table 11.15-2 are procedures for computing within-blocks means $\hat{\boldsymbol{\mu}}_R$ subject to the restrictions $\mathbf{R}'\boldsymbol{\mu} = \boldsymbol{\theta}$. The results of the analysis in Table 11.15-2 are identical to those in Table 11.2-2 where the traditional sum of squares approach was used.

ANALYSIS WHEN $n_{j\cdot} \neq n_{j'\cdot}$ FOR SOME j AND j'

The restricted full rank experimental design model approach can be used to analyze data for a type SPF-$p\cdot q$ design having a different number of blocks within the levels of treatment A. Two cases need to be distinguished. In the first, the experimenter intended to have an equal number of blocks, but for one reason or another—equipment malfunctioned, appointment missed, wrong treatment condition administered—the number of blocks within the levels of A is not the same. In the second case, the experimenter deliberately selected an unequal number of blocks so that the number within the levels of A is proportional to the number in the corresponding populations. The two cases have implications for the kinds of hypotheses an experimenter would want to test since the means for treatment B are computed by "averaging" over the levels of treatment A. In the first case, the experimenter would want to weight the $\overline{Y}_{jk}$'s equally in computing the $\overline{Y}_{\cdot k}$'s. In the second case, the $\overline{Y}_{jk}$ should be weighted so that the $\overline{Y}_{\cdot k}$'s reflect the number of blocks and hence the size of the respective populations. The following example should help to clarify the distinction between the two cases.

Consider the data for a type SPF-2·3 design in Table 11.15-3. The following hypotheses are of interest.

Treatment A (treatment level means are simple average of cell means,
$$c_{ijk} = \pm 1/n_{jk}q)$$

$$\text{H}_0\text{:}\quad \mu_{\cdot 1\cdot} - \mu_{\cdot 2\cdot} = 0 \quad\text{or}\quad \text{H}_0\text{:}\quad \underset{(p-1)\times h}{\mathbf{C}'_A}\ \underset{h\times 1}{\boldsymbol{\mu}}\ =\ \underset{(p-1)\times 1}{\mathbf{0}_A}$$

where $\underset{1\times 15}{\mathbf{C}'_A} = [\tfrac{1}{6}\ \tfrac{1}{6}\ \tfrac{1}{6}\ \tfrac{1}{6}\ \tfrac{1}{6}\ \tfrac{1}{6}\ -\tfrac{1}{9}\ -\tfrac{1}{9}\ -\tfrac{1}{9}\ -\tfrac{1}{9}\ -\tfrac{1}{9}\ -\tfrac{1}{9}\ -\tfrac{1}{9}\ -\tfrac{1}{9}\ -\tfrac{1}{9}]$

$\underset{1\times 1}{\mathbf{0}_A} = [0]$

and $\underset{1\times 15}{\boldsymbol{\mu}'} = [\mu_{111}\ \mu_{211}\ \mu_{112}\ \mu_{212}\ \mu_{113}\ \mu_{213}\ \mu_{321}\ \mu_{421}\ \mu_{521}\ \mu_{322}\ \mu_{422}\ \cdots\ \mu_{523}]$.

Treatment B (means at a_1 and a_2 are weighted equally, $c_{ijk} = \pm 1/n_{jk}p$ or 0)

$$\text{H}_0\text{:}\quad \sum_{j=1}^{p} \mu_{\cdot j1}/p - \sum_{j=1}^{p} \mu_{\cdot j3}/p = 0$$

$$\sum_{j=1}^{p} \mu_{\cdot j2}/p - \sum_{j=1}^{p} \mu_{\cdot j3}/p = 0$$

TABLE 11.15-2 Computational Procedures for a Type SPF-$p \cdot q$ Design Using a Restricted Full Rank Experimental Design Model

(i) Data and matrix formulas for computing sums of squares ($\mathbf{C}'_A$, $\mathbf{C}'_{BL(A)}$, $\mathbf{R}'$, $\mathbf{Q}'_B$, and $\mathbf{Q}'_{AB}$ matrices are defined in the text; $N = 32$, $h = 32$, $p = 2$, $q = 4$, $n = 4$, $s = 18$):

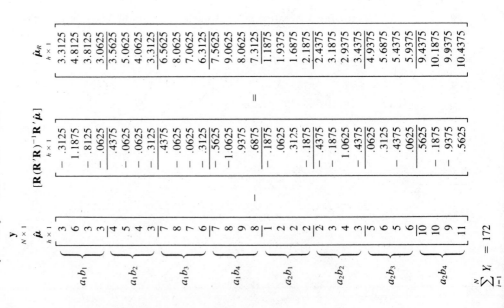

$$\underset{\substack{\mathbf{y} \\ N \times 1}}{\hat{\mu}} - [\mathbf{R}(\mathbf{R}'\mathbf{R})^{-1}\mathbf{R}'\hat{\mu}] = \hat{\mu}_R$$

	$\hat{\mu}$ ($h \times 1$)	$[\mathbf{R}(\mathbf{R}'\mathbf{R})^{-1}\mathbf{R}'\hat{\mu}]$ ($h \times 1$)	$\hat{\mu}_R$ ($h \times 1$)
a_1b_1	3	$-.3125$	3.3125
	6	1.1875	4.8125
	3	$-.8125$	3.8125
	4	$-.0625$	3.0625
a_1b_2	5	.4375	3.5625
	4	$-.0625$	5.0625
	3	$-.0625$	4.0625
		$-.3125$	3.3125
a_1b_3	7	.4375	6.5625
	8	$-.0625$	8.0625
	7	$-.0625$	7.0625
	6	$-.3125$	6.3125
a_1b_4	7	.5625	7.5625
	8	-1.0625	9.0625
	9	.9375	8.0625
	8	.6875	7.3125
a_2b_1	1	$-.1875$	1.1875
	2	.0625	1.9375
	2	.3125	1.6875
	2	$-.1875$	2.1875
a_2b_2	2	$-.4375$	2.4375
	3	$-.1875$	3.1875
	4	1.0625	2.9375
	3	$-.4375$	3.4375
a_2b_3	5	.0625	4.9375
	6	.3125	5.6875
	5	$-.4375$	5.4375
	6	$-.0625$	5.9375
a_2b_4	10	.5625	9.4375
	10	$-.1875$	10.1875
	9	$-.9375$	9.9375
	11	.5625	10.4375

$$\sum_{i=1}^{N} Y_i = 172$$

$$\underset{1 \times N}{\mathbf{y}'} \qquad \underset{N \times 1}{\mathbf{y}} \qquad \underset{1 \times 1}{(\mathbf{y}'\mathbf{y})}$$

$$[3 \quad 6 \quad \cdots \quad 11] \begin{bmatrix} 3 \\ 6 \\ \vdots \\ 11 \end{bmatrix} = 1160.000$$

$$SSTO = \mathbf{y}'\mathbf{y} - \left(\sum_{i=1}^{n} Y_i\right)^2 \Big/ N = 1160.000 - (172)^2/32 = 235.500$$

$$SSA = \underset{1 \times (p-1)}{(\mathbf{C}_A'\hat{\boldsymbol{\mu}})'} \; \underset{(p-1)\times(p-1)}{(\mathbf{C}_A'\mathbf{C}_A)^{-1}} \; \underset{(p-1)\times 1}{(\mathbf{C}_A'\hat{\boldsymbol{\mu}})} = 3.125$$

$$SSBL(A) = \underset{1 \times p(n-1)}{(\mathbf{C}_{BL(A)}'\hat{\boldsymbol{\mu}})'} \; \underset{p(n-1)\times p(n-1)}{(\mathbf{C}_{BL(A)}'\mathbf{C}_{BL(A)})^{-1}} \; \underset{p(n-1)\times 1}{(\mathbf{C}_{BL(A)}'\hat{\boldsymbol{\mu}})} = 9.375$$

$$SSB \times BL(A) = \underset{1 \times s}{(\mathbf{R}'\hat{\boldsymbol{\mu}})'} \; \underset{s \times s}{(\mathbf{R}'\mathbf{R})^{-1}} \; \underset{s \times 1}{(\mathbf{R}'\hat{\boldsymbol{\mu}})} = 9.125$$

$$SSB = \underset{1 \times (s+q-1)}{(\mathbf{Q}_B'\hat{\boldsymbol{\mu}})'} \; \underset{(s+q-1)\times(s+q-1)}{(\mathbf{Q}_B'\mathbf{Q}_B)^{-1}} \; \underset{(s+q-1)\times 1}{(\mathbf{Q}_B'\hat{\boldsymbol{\mu}})} - SSB \times BL(A)$$
$$= 203.625 - 9.125 = 194.500$$

$$SSAB = \underset{1 \times [s+(p-1)(q-1)]}{(\mathbf{Q}_{AB}'\hat{\boldsymbol{\mu}})'} \; \underset{[s+(p-1)(q-1)]\times[s+(p-1)(q-1)]}{(\mathbf{Q}_{AB}'\mathbf{Q}_{AB})^{-1}} \; \underset{[s+(p-1)(q-1)]\times 1}{(\mathbf{Q}_{AB}'\hat{\boldsymbol{\mu}})} - SSB \times BL(A)$$
$$= 28.500 - 9.125 = 19.375$$

(ii) Computation of mean squares:

$$MSA = SSA/(p-1) = 3.125$$

$$MSBL(A) = SSBL(A)/p(n-1) = 1.562$$

$$MSB = SSB/(q-1) = 64.833$$

$$MSAB = SSAB/(p-1)(q-1) = 6.458$$

$$MSB \times BL(A) = SSB \times BL(A)/p(n-1)(q-1) = 0.507$$

(iii) Computation of F ratios:

$$F = MSA/MSBL(A) = 2.00$$

$$F = MSB/MSB \times BL(A) = 127.88$$

$$F = MSAB/MSB \times BL(A) = 12.74$$

TABLE 11.15-3 Data for Type SPF-2·3 Design with $n_1 \neq n_2$.

ABS Summary Table
Entry is Y_{ijk}

		b_1	b_2	b_3
a_1	s_1	3	4	7
	s_2	6	5	8
a_2	s_3	3	4	7
	s_4	3	3	6
	s_5	1	2	5

ABS Summary Table
Entry is $\hat{\mu}_{ijk}$

		b_1	b_2	b_3	$\sum_{k=1}^{q} \hat{\mu}_{ijk}/q$
a_1	s_1	$\hat{\mu}_{111}$	$\hat{\mu}_{112}$	$\hat{\mu}_{113}$	$\hat{\mu}_{11\cdot}$
	s_2	$\hat{\mu}_{211}$	$\hat{\mu}_{212}$	$\hat{\mu}_{213}$	$\hat{\mu}_{21\cdot}$
a_2	s_3	$\hat{\mu}_{321}$	$\hat{\mu}_{322}$	$\hat{\mu}_{323}$	$\hat{\mu}_{32\cdot}$
	s_4	$\hat{\mu}_{421}$	$\hat{\mu}_{422}$	$\hat{\mu}_{423}$	$\hat{\mu}_{42\cdot}$
	s_5	$\hat{\mu}_{521}$	$\hat{\mu}_{522}$	$\hat{\mu}_{523}$	$\hat{\mu}_{52\cdot}$

$$\sum_{i=1}^{n} \sum_{j=1}^{p} \mu_{ijk}/n_{\cdot k} = \hat{\mu}_{\cdot\cdot 1} \quad \hat{\mu}_{\cdot\cdot 2} \quad \hat{\mu}_{\cdot\cdot 3}$$
$$= 3.2 \quad 3.6 \quad 6.6$$

AB Summary Table
Entry is $\hat{\mu}_{\cdot jk}$

	b_1	b_2	b_3	$\sum_{k=1}^{p} \hat{\mu}_{\cdot jk}/q$
a_1	$\hat{\mu}_{\cdot 11}$	$\hat{\mu}_{\cdot 12}$	$\hat{\mu}_{\cdot 13}$	$\hat{\mu}_{\cdot 1 \cdot} = 5.50$
a_2	$\hat{\mu}_{\cdot 21}$	$\hat{\mu}_{\cdot 22}$	$\hat{\mu}_{\cdot 23}$	$\hat{\mu}_{\cdot 2 \cdot} = 3.78$

$$\sum_{j=1}^{p} \hat{\mu}_{\cdot jk}/p = \hat{\mu}_{\cdot\cdot 1} \quad \hat{\mu}_{\cdot\cdot 2} \quad \hat{\mu}_{\cdot\cdot 3}$$
$$= 3.42 \quad 3.75 \quad 6.75$$

---

or

$$H_0: \quad \mu_{\cdot\cdot 1} - \mu_{\cdot\cdot 3} = 0 \qquad \text{or} \qquad H_0: \quad \underset{(q-1)\times h}{\mathbf{C}'_{1(B)}} \quad \underset{h\times 1}{\boldsymbol{\mu}} = \underset{(q-1)\times 1}{\mathbf{0}_{1(B)}}$$
$$\mu_{\cdot\cdot 2} - \mu_{\cdot\cdot 3} = 0$$

where $\underset{2\times 15}{\mathbf{C}'_{1(B)}} =$

$$\begin{bmatrix} 1/4 & 1/4 & 0 & 0 & -1/4 & -1/4 & 1/6 & 1/6 & 1/6 & 0 & 0 & 0 & -1/6 & -1/6 & -1/6 \\ 0 & 0 & 1/4 & 1/4 & -1/4 & -1/4 & 0 & 0 & 0 & 1/6 & 1/6 & 1/6 & -1/6 & -1/6 & -1/6 \end{bmatrix}$$

and $\quad \underset{1\times 2}{\mathbf{0}'_{1(B)}} = [0 \quad 0]$.

Treatment B (means at a_1 and a_2 are weighted proportional to the number of blocks, $c_{ijk} = \pm 1/n_{\cdot k}$ or 0)

$$H_0: \quad \sum_{j=1}^{p} n_{j1}\mu_{\cdot j1}/n_{\cdot 1} - \sum_{j=1}^{p} n_{j3}\mu_{\cdot j3}/n_{\cdot 3} = 0$$

$$\sum_{j=1}^{p} n_{j2}\mu_{\cdot j2}/n_{\cdot 2} - \sum_{j=1}^{p} n_{j3}\mu_{\cdot j3}/n_{\cdot 3} = 0$$

or

$$H_0: \quad \bar{\mu}_{\cdot\cdot 1} - \bar{\mu}_{\cdot\cdot 3} = 0 \qquad \text{or} \qquad H_0: \quad \underset{(q-1)\times h}{\mathbf{C}'_{2(B)}} \quad \underset{h\times 1}{\boldsymbol{\mu}} = \underset{(q-1)\times 1}{\mathbf{0}_{2(B)}}$$

$$\bar{\mu}_{\cdot\cdot 2} - \bar{\mu}_{\cdot\cdot 3} = 0$$

where $\underset{2\times 15}{\mathbf{C}'_{2(B)}} =$

$$\begin{bmatrix} 1/5 & 1/5 & 0 & 0 & -1/5 & -1/5 & 1/5 & 1/5 & 1/5 & 0 & 0 & 0 & -1/5 & -1/5 & -1/5 \\ 0 & 0 & 1/5 & 1/5 & -1/5 & -1/5 & 0 & 0 & 0 & 1/5 & 1/5 & 1/5 & -1/5 & -1/5 & -1/5 \end{bmatrix}$$

and $\underset{1\times 2}{\mathbf{0}'_{2(B)}} = [0 \quad 0]$.

The coefficients in $\mathbf{C}'_{2(B)}$ can be multiplied by 5 to eliminate fractions.

AB Interaction ($c_{ijk} = \pm 1/n_{jk}$ or 0)

$$H_0: \quad \mu_{\cdot 11} - \mu_{\cdot 21} - \mu_{\cdot 12} + \mu_{\cdot 22} = 0 \qquad \text{or} \qquad H_0: \quad \underset{(p-1)(q-1)\times h}{\mathbf{C}'_{AB}} \quad \underset{h\times 1}{\boldsymbol{\mu}} = \underset{(p-1)(q-1)\times 1}{\mathbf{0}_{AB}}$$

$$\mu_{\cdot 12} - \mu_{\cdot 22} - \mu_{\cdot 13} + \mu_{\cdot 23} = 0$$

where $\underset{2\times 15}{\mathbf{C}'_{AB}} =$

$$\begin{bmatrix} 1/2 & 1/2 & -1/2 & -1/2 & 0 & 0 & -1/3 & -1/3 & -1/3 & 1/3 & 1/3 & 1/3 & 0 & 0 & 0 \\ 0 & 0 & 1/2 & 1/2 & -1/2 & -1/2 & 0 & 0 & 0 & -1/3 & -1/3 & -1/3 & 1/3 & 1/3 & 1/3 \end{bmatrix}$$

and $\underset{1\times 2}{\mathbf{0}'_{AB}} = [0 \quad 0]$.

The coefficient matrices for computing the two error mean squares $MSBL(A)$ and $MSB \times BL(A)$ are based on the following hypotheses.

$BL(A)$ Error ($c_{ijk} = \pm 1/q$ or 0)

$$H_0: \quad \mu_{11\cdot} - \mu_{21\cdot} = 0 \qquad \text{or} \qquad H_0: \quad \underset{\substack{\sum_{j=1}^{p}(n_{j\cdot}-1)\times h}}{\mathbf{C}'_{BL(A)}} \quad \underset{h\times 1}{\boldsymbol{\mu}} = \underset{\substack{\sum_{j=1}^{p}(n_{j\cdot}-1)\times 1}}{\mathbf{0}_{BL(A)}}$$

$$\mu_{32\cdot} - \mu_{52\cdot} = 0$$

$$\mu_{42\cdot} - \mu_{52\cdot} = 0$$

where $\mathbf{C}'_{BL(A)} =$
$\underset{3\times 15}{}$

$$\begin{bmatrix} 1/3 & -1/3 & 1/3 & -1/3 & 1/3 & -1/3 & 0 & 0 & 0 & 0 & 0 & 0 & 0 & 0 & 0 \\ 0 & 0 & 0 & 0 & 0 & 0 & 1/3 & 0 & -1/3 & 1/3 & 0 & -1/3 & 1/3 & 0 & -1/3 \\ 0 & 0 & 0 & 0 & 0 & 0 & 0 & 1/3 & -1/3 & 0 & 1/3 & -1/3 & 0 & 1/3 & -1/3 \end{bmatrix}$$

and $\mathbf{0}'_{BL(A)} = [0 \ \ 0 \ \ 0]$.
$\underset{1\times 3}{}$

$$B \times BL(A) \text{ Error } (c_{ijk} = \pm 1 \text{ or } 0)$$

H_0: $\mu_{111} - \mu_{211} - \mu_{112} + \mu_{212} = 0$ or H_0: $\underset{s\times h}{\mathbf{R}'} \ \underset{h\times 1}{\boldsymbol{\mu}} = \underset{s\times 1}{\mathbf{0}}$

$\mu_{112} - \mu_{212} - \mu_{113} + \mu_{213} = 0$

$\mu_{321} - \mu_{421} - \mu_{322} + \mu_{422} = 0$

$\mu_{421} - \mu_{521} - \mu_{422} + \mu_{522} = 0$

$\cdot \ \cdot \ \cdot \ \cdot \ \cdot \ \cdot \ \cdot \ \cdot \ \cdot \ \cdot \ \cdot \ \cdot \ \cdot$

$\mu_{422} - \mu_{522} - \mu_{423} + \mu_{523} = 0$

where $\mathbf{R}' =$
$\underset{6\times 15}{}$

$$\begin{bmatrix} 1 & -1 & -1 & 1 & 0 & 0 & 0 & 0 & 0 & 0 & 0 & 0 & 0 & 0 & 0 \\ 0 & 0 & 1 & -1 & -1 & 1 & 0 & 0 & 0 & 0 & 0 & 0 & 0 & 0 & 0 \\ 0 & 0 & 0 & 0 & 0 & 0 & 1 & -1 & 0 & -1 & 1 & 0 & 0 & 0 & 0 \\ 0 & 0 & 0 & 0 & 0 & 0 & 0 & 1 & -1 & 0 & -1 & 1 & 0 & 0 & 0 \\ 0 & 0 & 0 & 0 & 0 & 0 & 0 & 0 & 1 & -1 & 0 & -1 & 1 & 0 \\ 0 & 0 & 0 & 0 & 0 & 0 & 0 & 0 & 0 & 1 & -1 & 0 & -1 & 1 \end{bmatrix}$$

$\boldsymbol{\theta}' = [0 \ \ 0 \ \ 0 \ \ 0 \ \ 0 \ \ 0]$
$\underset{1\times 6}{}$

and $s = \sum_{j=1}^{p} (n_{j\cdot}q - n_{j\cdot} - q + 1) = 2 + 4 = 6$. The sum of squares computational formulas are the same as those in Table 11.15-2. The results of the analyses are summarized in Table 11.15-4.

ANALYSIS WHEN ONE OR MORE SCORES ARE MISSING

The restricted full rank experimental design model can be used to test hypotheses when one or more scores are missing but all n_{jk}'s are greater than zero. To illustrate, consider the data in Table 11.15-5 where Y_{421} is missing. One would probably want to test the following null hypothesis for treatment A.

Treatment A H_0: $\mu_{\cdot 1\cdot} - \mu_{\cdot 2\cdot} = 0$ or H_0: $\underset{(p-1)\times h}{\mathbf{C}'_A} \ \underset{h\times 1}{\boldsymbol{\mu}} = \underset{(p-1)\times 1}{\mathbf{0}_A}$

where $\mathbf{C}'_A = [1/6 \ \ 1/6 \ \ 1/6 \ \ 1/6 \ \ 1/6 \ \ 1/6 \ \ -2/6 \ \ -1/6 \ \ -1/6 \ \ -1/6 \ \ -1/6]$
$\underset{1\times 11}{}$

$\boldsymbol{\mu}' = [\mu_{111} \ \ \mu_{211} \ \ \mu_{112} \ \ \mu_{212} \ \ \mu_{113} \ \ \mu_{213} \ \ \mu_{321} \ \ \mu_{322} \ \ \mu_{422} \ \ \mu_{323} \ \ \mu_{423}]$
$\underset{1\times 11}{}$

and $\mathbf{0}_A = [0]$. The coefficients in $\mathbf{C}'_A$ are given by $c_{ijk} = \pm 1/n_{jk}q$. Alternatively, one could test a hypothesis concerning weighted means, $\bar{\mu}_{\cdot j\cdot}$'s. The coefficient matrix for

TABLE 11.15-4 ANOVA Table for Type SPF-2·3 Design

Source	SS	df	MS	F (A and B fixed; Blocks random)
1 A	10.6779	$p - 1 = 1$	10.6779	3.08
2 Blocks w. A	10.3889	$\sum_{j=1}^{p} (n_{j \cdot} - 1) = 3$	3.4630	
3 B(1)	32.3556	$q - 1 = 2$	16.1778	54.60*
4 B(2)	34.5333	$q - 1 = 2$	17.2666	58.27*
5 AB	0.3555	$(p - 1)(q - 1) = 2$	0.1778	0.60
6 B × blocks w. A	1.7778	$N - \sum_{j=1}^{p} n_{j \cdot} - pq + p = 6$	0.2963	
7 Total	57.7333	$N - 1 = 14$		

*$p < .01$

computing the between-blocks error term $MSBL(A)$ is based on the following hypothesis

$BL(A)$ Error H_0: $\mu_{11 \cdot} - \mu_{21 \cdot} = 0$ or H_0: $\underset{\sum_{j=1}^{p} (n_{j \cdot}-1) \times h}{\mathbf{C}'_{BL(A)}}$ $\underset{h \times 1}{\boldsymbol{\mu}}$ $=$ $\underset{\sum_{j=1}^{p} (n_{j \cdot}-1) \times 1}{\mathbf{0}_{BL(A)}}$

$\mu_{32 \cdot} - \mu_{42 \cdot} = 0$

where $\underset{2 \times 11}{\mathbf{C}'_{BL(A)}} = \begin{bmatrix} \frac{1}{3} & -\frac{1}{3} & \frac{1}{3} & -\frac{1}{3} & \frac{1}{3} & -\frac{1}{3} & 0 & 0 & 0 & 0 & 0 \\ 0 & 0 & 0 & 0 & 0 & 0 & \frac{1}{3} & \frac{1}{3} & -\frac{1}{2} & \frac{1}{3} & -\frac{1}{2} \end{bmatrix}$

and $\underset{1 \times 2}{\mathbf{0}'_{BL(A)}} = [0 \quad 0]$.

The coefficients in $\mathbf{C}'_{BL(A)}$ are given by $c_{ijk} = \pm 1 / \sum_{k=1}^{q} n_{ijk}$ or 0, where n_{ijk} is 0 or 1. When scores are missing, an experimenter must decide whether a particular null hypothesis is a reasonable one to test. If A and B are fixed and blocks are random, it can be shown that testing $\mathbf{C}'_A \boldsymbol{\mu} = \mathbf{0}_A$ is equivalent to testing

$$H_0: \quad \alpha_1 - \alpha_2 = 0$$

for the classical less than full rank experimental design model.

For treatment B, one would probably want to test the following null hypothesis.

Treatment B H_0: $\mu_{\cdot \cdot 1} - \mu_{\cdot \cdot 3} = 0$ or H_0: $\underset{(q-1) \times h}{\mathbf{C}'_B}$ $\underset{h \times 1}{\boldsymbol{\mu}}$ $=$ $\underset{(q-1) \times 1}{\mathbf{0}_B}$

$\mu_{\cdot \cdot 2} - \mu_{\cdot \cdot 3} = 0$

where $\underset{2 \times 11}{\mathbf{C}'_B} = \begin{bmatrix} \frac{1}{4} & \frac{1}{4} & 0 & 0 & -\frac{1}{4} & -\frac{1}{4} & \frac{2}{4} & 0 & 0 & -\frac{1}{4} & -\frac{1}{4} \\ 0 & 0 & \frac{1}{4} & \frac{1}{4} & -\frac{1}{4} & -\frac{1}{4} & 0 & \frac{1}{4} & \frac{1}{4} & -\frac{1}{4} & -\frac{1}{4} \end{bmatrix}$

TABLE 11.15-5   Type SPF-2·3 Design with Y_{421} Missing

ABS Summary Table
Entry is Y_{ijk}

		b_1	b_2	b_3
a_1	s_1	3	4	7
	s_2	6	5	8
a_2	s_3	3	4	7
	s_4		3	6

ABS Summary Table
Entry is $\hat{\mu}_{ijk}$

		b_1	b_2	b_3	$\sum_{k=1}^{q} \hat{\mu}_{ijk} / \sum_{k=1}^{q} n_{ijk}$
a_1	s_1	$\hat{\mu}_{111}$	$\hat{\mu}_{112}$	$\hat{\mu}_{113}$	$\hat{\mu}_{11\cdot}$
	s_2	$\hat{\mu}_{211}$	$\hat{\mu}_{212}$	$\hat{\mu}_{213}$	$\hat{\mu}_{21\cdot}$
a_2	s_3	$\hat{\mu}_{321}$	$\hat{\mu}_{322}$	$\hat{\mu}_{323}$	$\hat{\mu}_{32\cdot}$
	s_4		$\hat{\mu}_{422}$	$\hat{\mu}_{423}$	$\hat{\mu}_{42\cdot}$

$$\sum_{i=1}^{n}\sum_{j=1}^{p} \hat{\mu}_{ijk}/n_{\cdot k} = \hat{\bar{\mu}}_{\cdot\cdot 1} \quad \hat{\bar{\mu}}_{\cdot\cdot 2} \quad \hat{\bar{\mu}}_{\cdot\cdot 3}$$

$$= 4.0 \quad 4.0 \quad 7.0$$

AB Summary Table
Entry is $\hat{\mu}_{\cdot jk}$

	b_1	b_2	b_3	$\sum_{k=1}^{q} \hat{\mu}_{\cdot jk}/q$	$\sum_{k=1}^{q} n_{jk}\hat{\mu}_{\cdot jk}/\sum_{k=1}^{q} n_{jk}$
a_1	$\hat{\mu}_{\cdot 11}$	$\hat{\mu}_{\cdot 12}$	$\hat{\mu}_{\cdot 13}$	$\hat{\mu}_{\cdot 1\cdot} = 5.50$	$\hat{\bar{\mu}}_{\cdot 1\cdot} = 5.50$
a_2	$\hat{\mu}_{\cdot 21}$	$\hat{\mu}_{\cdot 22}$	$\hat{\mu}_{\cdot 23}$	$\hat{\mu}_{\cdot 2\cdot} = 4.33$	$\hat{\bar{\mu}}_{\cdot 2\cdot} = 4.60$

$$\sum_{j=1}^{p} \hat{\mu}_{\cdot jk}/p = \hat{\mu}_{\cdot\cdot 1} \quad \hat{\mu}_{\cdot\cdot 2} \quad \hat{\mu}_{\cdot\cdot 3}$$

$$= 3.75 \quad 4.00 \quad 7.00$$

and $\mathbf{0}'_B = \begin{bmatrix} 0 & 0 \end{bmatrix}$.
$\scriptstyle 1\times 2$

The coefficients in $\mathbf{C}'_B$ are given by $c_{ijk} = \pm 1/n_{jk}p$ or 0. The restriction coefficient matrix for computing the error term $MSB \times BL(A)$ is based on the following null hypothesis.

$$B \times \text{ blocks w. } A \text{ Error}$$

$$\text{H}_0: \quad \mu_{111} - \mu_{211} - \mu_{112} + \mu_{212} = 0 \quad \text{or} \quad \text{H}_0: \quad \underset{s\times h}{\mathbf{R}'} \quad \underset{h\times 1}{\boldsymbol{\mu}} = \underset{s\times 1}{\boldsymbol{\theta}}$$

$$\mu_{112} - \mu_{212} - \mu_{113} + \mu_{213} = 0$$

$$\mu_{322} - \mu_{422} - \mu_{323} + \mu_{423} = 0$$

$$\text{where} \quad \underset{3\times 11}{\mathbf{R}'} = \begin{bmatrix} 1 & -1 & -1 & 1 & 0 & 0 & 0 & 0 & 0 & 0 & 0 \\ 0 & 0 & 1 & -1 & -1 & 1 & 0 & 0 & 0 & 0 & 0 \\ 0 & 0 & 0 & 0 & 0 & 0 & 0 & 1 & -1 & -1 & 1 \end{bmatrix} \quad \underset{3\times 1}{\boldsymbol{\theta}_R} = \begin{bmatrix} 0 \\ 0 \\ 0 \end{bmatrix}$$

and $s = N - \Sigma_{j=1}^{p} n_{j \cdot} - pq + p$. The coefficients in $\mathbf{R}'$ are given by $c_{ijk} = \pm 1$ or 0. The null hypothesis for the AB interaction is as follows.

$$AB \text{ Interaction}$$

$$\mathrm{H}_0: \quad \mu_{\cdot 11} - \mu_{\cdot 21} - \mu_{\cdot 12} + \mu_{\cdot 22} = 0 \qquad \text{or} \qquad \mathrm{H}_0: \quad \underset{(p-1)(q-1) \times h}{\mathbf{C}'_{AB}} \quad \underset{h \times 1}{\boldsymbol{\mu}} = \underset{(p-1)(q-1) \times 1}{\mathbf{0}_{AB}}$$

$$\mu_{\cdot 12} - \mu_{\cdot 22} - \mu_{\cdot 13} + \mu_{\cdot 23} = 0$$

where $\underset{2 \times 11}{\mathbf{C}'_{AB}} = \begin{bmatrix} \frac{1}{2} & \frac{1}{2} & -\frac{1}{2} & -\frac{1}{2} & 0 & 0 & -1 & \frac{1}{2} & \frac{1}{2} & 0 & 0 \\ 0 & 0 & \frac{1}{2} & \frac{1}{2} & -\frac{1}{2} & -\frac{1}{2} & 0 & -\frac{1}{2} & -\frac{1}{2} & \frac{1}{2} & \frac{1}{2} \end{bmatrix}$

and $\underset{1 \times 2}{\mathbf{0}'_{AB}} = \begin{bmatrix} 0 & 0 \end{bmatrix}$.

The coefficients in $\mathbf{C}'_{AB}$ are given by $c_{ijk} = \pm 1 / n_{jk}$ or 0.

We want to test the within-blocks null hypotheses, subject to the restrictions that $\mathbf{R}' \boldsymbol{\mu} = \boldsymbol{\theta}$. This is accomplished by forming augmented matrices of the form

$$\mathbf{Q}'_w = \begin{bmatrix} \mathbf{R}' \\ \mathbf{C}'_w \end{bmatrix} \qquad \text{and} \qquad \boldsymbol{\eta}_w = \begin{bmatrix} \boldsymbol{\theta} \\ \mathbf{0}_w \end{bmatrix}$$

where $\mathbf{C}'_w$ is the coefficient matrix for a within-blocks source of variation and $\mathbf{0}_w$ is the null vector associated with $\mathbf{C}'_w$. If the rows of $\mathbf{R}'$ are orthogonal to the rows of $\mathbf{C}'_w$, the within-blocks sum of squares formulas can be simplified as discussed previously. The formulas for computing the various sums of squares are the same as those in Table 11.15-2. The results of the analysis are summarized in Table 11.15-6.

TABLE 11.15-6 ANOVA Table for Type SPF-2·3 Design

	Source	SS	df		MS	F (A and B fixed; Blocks random)
1	A	3.500	$p - 1 =$	1	3.500	1.67
2	Blocks w. A	4.200	$\sum_{j=1}^{p} (n_{j \cdot} - 1) =$	2	2.100	
3	B	24.125	$q - 1 =$	2	12.062	27.17*
4	AB	0.125	$(p - 1)(q - 1) =$	2	0.062	0.14
5	B × blocks w. A	1.333	$N - \sum_{j=1}^{p} n_{j \cdot} - pq + p =$	3	0.444	
6	Total	32.909	$N - 1 =$	10		

*$p < .05$

11.16 ADVANTAGES AND DISADVANTAGES OF SPLIT-PLOT FACTORIAL DESIGNS

A split-plot factorial design is appropriate for many research problems. The most likely alternative design when repeated measures or matched subjects are employed is a randomized block factorial design. The advantages and disadvantages of a split-plot factorial design relative to a randomized block factorial design are as follows.

1. A split-plot factorial design requires a smaller block size than a randomized block factorial design.

2. A split-plot factorial design can be used in experiments where it is not possible to administer all treatment combinations within each block.

3. In a type SPF-$p \cdot q$ design, the power associated with tests of treatment B and the AB interaction is usually greater than that for treatment A. In a type RBF-pq design, the power of the three tests is the same. The power of within-blocks tests in a split-plot factorial design is greater than that for the corresponding tests in a randomized block factorial design. However, the power of between-blocks tests is usually less than that for the corresponding tests in a randomized block factorial design.

4. The analysis and randomization procedures for a split-plot factorial design are more complex than for a randomized block factorial design.

11.17 REVIEW EXERCISES

1. [11.1] Draw block diagrams (see Figure 11.1-1) for split-plot and randomized block factorial designs with the following levels of treatments A and B.

   †a) $p = 3$, $q = 2$ †b) $p = 3$, $q = 3$ c) $p = 4$, $q = 2$ d) $p = 4$, $q = 3$

2. [11.1] If repeated measurements are obtained, how many times must a subject be observed for the designs in Exercise 1?

†3. [11.1] How does the confounding of treatment A with groups of blocks affect the test of treatment A?

4. [11.1] Explain why a type RBF-23 design is a better choice than a type SPF-2 · 3 design if an experimenter is particularly interested in testing treatment A.

5. [11.1] Give a rule of thumb for determining whether a treatment is a between- or within-blocks treatment.

6. [11.2] Three procedures for inspecting printed circuit boards were investigated. The circuit boards were projected on a screen in front of an inspector. Condition a_1 was the standard condition in which the circuit board being inspected could be compared with the picture of a

perfect circuit board by pressing a button. Condition a_2 provided a perfect circuit board overlay in a contrasting color. Condition a_3 alternated temporally the circuit board being inspected with a picture of a perfect board every 200 ms. Three types of defects occurred in the circuit boards: condition b_1 was a filled hole in the board (fill defect), b_2 was a gap in one of the lines on the board (gap defect), and b_3 was a short between two lines of the circuit (short defect). Eighteen inspectors were randomly assigned to one of three groups with six in each group. The three levels of treatment A were randomly assigned to the groups. The subjects in each group received all three levels of treatment B. The order in which the levels of treatment B were presented was randomized independently for each subject. The dependent variable was the time in seconds required to detect a defect. The following data were obtained. (Experiment suggested by Liuzzo, J. G. and Drury, C. G. An evaluation of blink inspection. *Human Factors*, 1980, *22*, 201–210.)

		b_1	b_2	b_3
	s_1	10.6	5.9	8.6
	s_2	9.7	5.7	5.4
a_1	s_3	12.5	8.9	7.9
	s_4	9.6	6.8	3.1
	s_5	7.3	2.8	5.3
	s_6	6.7	4.2	2.3
	s_7	1.1	1.3	4.3
	s_8	3.2	6.3	5.4
a_2	s_9	5.4	4.2	7.3
	s_{10}	0.6	1.2	1.4
	s_{11}	2.4	3.1	2.1
	s_{12}	2.4	3.3	4.4
	s_{13}	7.8	6.9	4.7
	s_{14}	7.9	6.6	6.1
a_3	s_{15}	10.9	8.2	7.8
	s_{16}	4.8	5.1	1.8
	s_{17}	6.2	3.8	2.6
	s_{18}	8.4	9.9	4.9

†a) Test the null hypothesis for treatments A and B and the AB interaction; let $\alpha = .05$.

†b) Graph the AB interaction and interpret the graph.

c) Prepare a "results and discussion section" for the *Human Factors* journal.

7. [11.2] The performance of fifteen secretaries on three date-sorting tasks was compared at two times of day: 10 A.M., designated as treatment level b_1, and 4 P.M., designated as treatment level b_2. The task involved sorting a list of random dates written in British form (e.g., 30 1 74) into two, three, or four accounting periods—treatment levels a_1, a_2, and a_3, respectively. The secretaries were randomly assigned to one of three groups with five in each group. The levels of treatment A were randomly assigned to the groups. The tasks were performed for one hour on consecutive days—Thursday and Friday. The day on which a level of treatment B was presented was randomized independently for each secretary. The dependent variable Y was the

number of dates sorted minus 2500. The following data were obtained. (Experiment suggested by Monk, T. H. and Conrad, M. C. Time of day effects in a range of clerical tasks. *Human Factors*, 1979, *21*, 191–194.)

		b_1	b_2
a_1	s_1	171	189
	s_2	183	204
	s_3	145	154
	s_4	158	166
	s_5	196	179
a_2	s_6	213	249
	s_7	224	237
	s_8	198	224
	s_9	182	198
	s_{10}	172	214
a_3	s_{11}	200	212
	s_{12}	226	224
	s_{13}	213	196
	s_{14}	251	250
	s_{15}	238	239

a) Test the null hypothesis for treatments *A* and *B* and the *AB* interaction; let $\alpha = .05$.

b) Graph the *AB* interaction and interpret the graph.

c) Prepare a "results and discussion section" for the *Human Factors* journal.

8.   [11.3] Show that $\sum_{i=1}^{n} \sum_{j=1}^{p} \sum_{k=1}^{q} [(\overline{Y}_{.j.} - \overline{Y}_{...}) + (\overline{Y}_{ij.} - \overline{Y}_{.j.}) + (\overline{Y}_{..k} - \overline{Y}_{...}) +$
$(\overline{Y}_{.jk} - \overline{Y}_{.j.} - \overline{Y}_{..k} + \overline{Y}_{...}) + (Y_{ijk} - \overline{Y}_{.jk} - \overline{Y}_{ij.} + \overline{Y}_{.j.})]^2 = nq \sum_{j=1}^{p} (\overline{Y}_{.j.} - \overline{Y}_{...})^2 +$
$q \sum_{i=1}^{n} \sum_{j=1}^{p} (\overline{Y}_{ij.} - \overline{Y}_{.j.})^2 + np \sum_{k=1}^{q} (\overline{Y}_{..k} - \overline{Y}_{...})^2 + n \sum_{j=1}^{p} \sum_{k=1}^{q}$
$= 1 (\overline{Y}_{.jk} - \overline{Y}_{.j.} - \overline{Y}_{..k} + \overline{Y}_{...})^2 +$
$\sum_{i=1}^{n} \sum_{j=1}^{p} \sum_{k=1}^{q} (Y_{ijk} - \overline{Y}_{.jk} - \overline{Y}_{ij.} + \overline{Y}_{.j.})^2.$

†9.   [11.3] Derive the computational formula for $SSBL(A)$ from $q \sum_{i=1}^{n} \sum_{j=1}^{p} (\overline{Y}_{ij.} - \overline{Y}_{.j.})^2.$

10.   [11.4] †a) Use the F_{max} test to test the hypothesis H_0: $\sigma_1^2 = \sigma_2^2 = \sigma_3^2$. for the data in Exercise 6.

b) Test the multisample circularity assumption in two steps. The Mauchley test statistic in Table 6.4-1 should be modified as follows.

$$W = \frac{|\mathbf{C}_B^{\star\prime} \hat{\boldsymbol{\Sigma}}_{\text{pooled}} \mathbf{C}_B^{\star}|}{\left[\dfrac{\text{trace} (\mathbf{C}_B^{\star\prime} \hat{\boldsymbol{\Sigma}}_{\text{pooled}} \mathbf{C}_B^{\star})}{q-1}\right]^{q-1}}$$

$$d = 1 - [(2q^2 - 3q + 3)/6p(n-1)(q-1)]$$

$$\chi^2 = -p(n-1)d \ln(W)$$

$$df = [q/(q-1)/2] - 1$$

For this problem, let

$$\mathbf{C}_B^{\star'} = \begin{bmatrix} .7071 & -.7071 & 0 \\ .4082 & .4082 & -.8165 \end{bmatrix}.$$

The determinant of a 2×2 matrix is equal to $e_{11}e_{22} - e_{12}e_{21}$.

†**11.** [11.4] **a)** Compute $\hat{\theta}$ for the data in Exercise 6.

b) Perform an adjusted F test for treatment B and the AB interaction in Exercise 6.

c) How do the results of the adjusted test compare with the results of the conventional F test?

†**12.** [11.6] For the data in Exercise 6, test the following hypotheses:

a) H_0: $\alpha\psi_{1(B)} = \delta$ for all j **b)** H_0: $\alpha\psi_{2(B)} = \delta$ for all j

c) H_0: $\beta\psi_{1(A)} = \delta$ for all k **d)** H_0: $\beta\psi_{2(A)} = \delta$ for all k

where $\psi_{1(B)} = \mu_{\cdot j1} - \mu_{\cdot j2}$

$\psi_{2(B)} = (\mu_{\cdot j1} + \mu_{\cdot j2})/2 - \mu_{\cdot j3}$

$\psi_{1(A)} = \mu_{\cdot 1k} - \mu_{\cdot 2k}$

$\psi_{2(A)} = \mu_{\cdot 1k} - \mu_{\cdot 3k}.$

Compute error terms appropriate for the specific treatment-contrast interactions. Use the simultaneous test procedure and let $\alpha = .05$ for the collection of all possible tests.

13. [11.4] Use the $F_{\max}$ test to test the hypothesis H_0: $\sigma_{1\cdot}^2 = \sigma_{2\cdot}^2 = \sigma_{3\cdot}^2$ for the data in Exercise 7.

14. [11.4] **a)** Compute $\hat{\theta}$ for the data in Exercise 7.

b) What does this suggest regarding the hypothesis that $\mathbf{C}^{\star'}\mathbf{\Sigma}\mathbf{C}^{\star}$ is spherical?

15. [11.6] For the data in Exercise 7, test the following hypotheses:

a) H_0: $\beta\psi_{1(A)} = \delta$ for all k **b)** H_0: $\beta\psi_{2(A)} = \delta$ for all k

c) H_0: $\beta\psi_{3(A)} = \delta$ for all k

where $\psi_{1(A)} = \mu_{\cdot 1k} - \mu_{\cdot 2k}$

$\psi_{2(A)} = \mu_{\cdot 1k} - \mu_{\cdot 3k}$

$\psi_{3(A)} = \mu_{\cdot 2k} - \mu_{\cdot 3k}.$

Use the pooled error term $MSB \times BL(A)$ in testing these hypotheses. Also use the simultaneous test procedure and let $\alpha = .05$ for the collection of all possible tests.

†**16.** [11.8] For the data in Exercise 6, compute the relative efficiency of type SPF-3·3 and type RBF-33 designs for

a) Treatment A **b)** treatment B and AB interaction.

17. [11.8] For the data in Exercise 7, compute the relative efficiency of type SPF-3·2 and type RBF-32 designs for

a) Treatment A **b)** treatment B and AB interaction.

18. [11.9] and [11.10] Mauchley's test statistic for a type RB-p design is given in Table 6.4-1; the statistic for a type SPF-$p \cdot q$ design is given in Exercise 10(b). Extend the formulas for χ^2 and d for a

†**a)** Type SPF-$pr \cdot q$ design   **b)** type SPF-$prt \cdot q$ design.

†19. [11.11] List the key assumptions associated with the between- and within-blocks tests of a type SPF-$p \cdot qrt$ design.

20. [11.11] a) Specify the overall multisample circularity assumption for a type SPF-$p \cdot qr$ design.

b) If this assumption is tenable, what error term is used to test the within-blocks null hypotheses?

21. [11.14] Describe the modification necessary to evaluate sequence effects for the following designs:

a) Type SPF-$3 \cdot 2$ design   b) type SPF-$2 \cdot 3$ design   c) type SPF-$3 \cdot 5$ design.

†22. [11.15] Exercise 6 described an experiment to evaluate three procedures for inspecting printed circuit boards.

a) Write a full rank experimental design model for this type SPF-$3 \cdot 3$ design.

b) Specify the null hypotheses in terms of μ's for testing treatments A and B, the AB interaction, and the error terms $MSBL(A)$ and $MSB \times BL(A)$.

c) Indicate the dimensions of the following vectors and matrices:

$$\mathbf{y}, \; \boldsymbol{\mu}, \; \mathbf{C}'_A, \; \mathbf{C}'_B, \; \mathbf{Q}'_B, \; \mathbf{C}'_{AB}, \; \mathbf{Q}'_{AB}, \; \mathbf{C}'_{BL(A)}, \; \mathbf{R}'.$$

23. [11.15] Exercise 7 described an experiment to compare date-sorting performance at two times of day and three levels of task difficulty.

a) Write a full rank experimental design model for this type SPF-$3 \cdot 2$ design.

b) Specify the null hypotheses in terms of μ's for testing treatments A and B, the AB interaction, and the error terms $MSBL(A)$ and $MSB \times BL(A)$.

c) Indicate the dimensions of the following vectors and matrices:

$$\mathbf{y}, \; \boldsymbol{\mu}, \; \mathbf{C}'_A, \; \mathbf{C}'_B, \; \mathbf{Q}'_B, \; \mathbf{C}'_{AB}, \; \mathbf{Q}'_{AB}, \; \mathbf{C}'_{BL(A)}, \; \mathbf{R}'.$$

†24. [11.15] Female college students were exposed to either an aggressive or an erotic film, treatment levels a_1 and a_2, respectively, and then alternately provoked, treatment level b_1, and not provoked, treatment level b_2, in a series of games that provided opportunity for retaliation. Retaliation consisted in delivering a noxious noise to the opponent. The dependent variable was the duration of the noise presentation. Four students were randomly assigned to one of two groups. The levels of treatment A were randomly assigned to the groups. The order in which the levels of treatment B were presented was randomized independently for each subject. The following data were obtained.

		b_1	b_2
a_1	s_1	34	11
	s_2	28	7
a_2	s_3	16	10
	s_4	11	13

a) Write the $\mathbf{C}'$ and $\mathbf{R}'$ matrices for computing SSA, $SSBL(A)$, SSB, $SSAB$, and $SSB \times BL(A)$.

b) Use the full rank experimental design model approach to test the null hypotheses for A, B, and AB; let $\alpha = .05$. (This problem involves the inversion of 2×2 diagonal matrices. The computations are simple and can be done without the aid of a computer.)

25. [11.15] Assume that because of equipment failure data for subject 2 in Exercise 24 were lost.

a) Write the $\mathbf{C}'$ and $\mathbf{R}'$ matrices for computing SSA, $SSBL(A)$, SSB, $SSAB$, and $SSB \times BL(A)$. Assume that one is interested in unweighted means.

b) Use the full rank experimental design model approach to test the null hypotheses for A, B, and AB; let $\alpha = .05$. (This problem involves simple matrix multiplication and can be done without a computer.)

26. [11.15] Assume because of equipment failure that observation Y_{211} in Exercise 24 was lost.

†**a)** Write the $\mathbf{C}'$ and $\mathbf{R}'$ matrices for computing SSA, $SSBL(A)$, SSB, $SSAB$, and $SSB \times BL(A)$. Assume that one is interested in unweighted means for treatments A and B.

†**b)** Use the full rank experimental design model approach to test the null hypotheses for A, B, and AB; let $\alpha = .05$. (This problem involves simple matrix multiplication and can be done without a computer.)

c) Write the $\mathbf{C}'$ and $\mathbf{R}'$ matrices for computing SSA, $SSBL(A)$, SSB, $SSAB$, and $SSB \times BL(A)$. Assume that one is interested in weighted means for treatments A and B.

d) Use the full rank experimental design model approach to test the null hypotheses for A, B, and AB; let $\alpha = .05$. (This problem involves simple matrix multiplication and can be done without a computer.)

12 CONFOUNDED FACTORIAL DESIGNS: DESIGNS WITH GROUP-INTERACTION CONFOUNDING

12.1 INTRODUCTION TO CONFOUNDING IN FACTORIAL EXPERIMENTS

The designs in this chapter use the technique of group-interaction confounding to achieve a reduction in the number of treatment combinations that must be assigned to blocks. The technique was first described by Sir Ronald A. Fisher in 1926 and used as early as 1927 in agricultural research at the Rothamsted Experimental Station. The advantages of using a blocking procedure to isolate a nuisance variable have been discussed at length in Chapters 6, 7, 9, and 11. Unfortunately, when the number of treatment combinations is large, it may be difficult to secure enough homogeneous experimental units to form the blocks. Even a relatively small design, such as a type RBF-33 design, requires nine experimental units per block. Obtaining repeated observations on the same experimental units is no solution, for there is a practical limit to the number of times a subject can participate in an experiment. And the nature of the treatments often precludes obtaining more than one measurement per subject. The split-plot factorial design described in Chapter 11 provides one solution to the problem of unwieldy block size by assigning only a portion of the treatment combinations to each block. For example, a type SPF-3·4 design has 12 treatment combinations but only four are assigned to a block. This approach to reducing block size confounds the effects of treatment A with the effects of groups or samples of blocks. As a result, the test of treatment A is less powerful than those for treatment B and the AB interaction. This is satisfactory if an experimenter's primary interest is in testing B and AB. In most experiments, however, the treatments, not the interaction, are of primary interest. The designs described in this chapter, confounded factorial designs, are more appropriate for such experiments since they confound one or more interactions with groups. This reduces the number of treatment combinations within each block and, in addition, has an important advantage over the confounding scheme of a split-plot factorial design. If one of the interactions is expected to be negligible, it can be confounded with groups in order to estimate all treatment and remaining interaction effects with equal precision. Thus an experimenter can achieve a reduction in block size without sacrificing power in evaluating treatments and important interactions.

Confounded factorial designs can be constructed from either randomized block or Latin square building block designs. The letters RBCF and LSCF are used to designate a factorial design in which an interaction is *completely* confounded with groups or samples of blocks. If an interaction is *partially* confounded with groups, the design is designated by the letters RBPF. This type of design provides partial information with respect to the confounded interactions. The designation RBCF-p^k indicates that the design is restricted to the case in which k treatments each have p levels. For example, a type RBCF-3^2 design has two treatments, A and B, each having three levels.

A comparison among three factorial designs is shown in Figure 12.1-1. Suppose that repeated measurements are obtained on the subjects. In the type RBF-33 design, each subject in group S_1 receives all nine treatment combinations. In the type SPF-3·3 design, subjects in group S_1 receive only three treatment combinations— treatment level a_1 in combination with all levels of treatment B. In the type RBCF-3^2 design, subjects in S_1 also receive only three treatment combinations, but these include three levels of treatment A in combination with three levels of treatment B. As we will see, this scheme confounds the AB interaction or a component of the interaction with groups, but does not confound either treatment with groups. We will assume that all treatments in confounded factorial designs represent fixed effects and that blocks represent random effects. A block may correspond to a single experimental unit that is observed p times or to a set of p matched experimental units, each of which is observed once.

Type RBCF-p^k, RBPF-p^k, and LSCF-p^k designs are appropriate for experiments that meet, in addition to the assumptions of the experimental design model, the following conditions.

1.   Two or more treatments with each treatment having p levels, where $p \geq 2$. Exceptions to the general requirement that all treatments must have p levels are discussed in Sections 12.11 and 12.13.

2.   Treatment combinations are assigned to blocks so that only a fraction of them are represented within any one block. Variation between groups (samples of blocks) is confounded with one or more interactions. Thus an experimenter

FIGURE 12.1-1   Comparison of three factorial designs. The letters S_1, S_2, and S_3 denote groups (samples) of subjects.

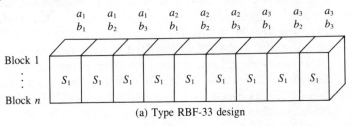

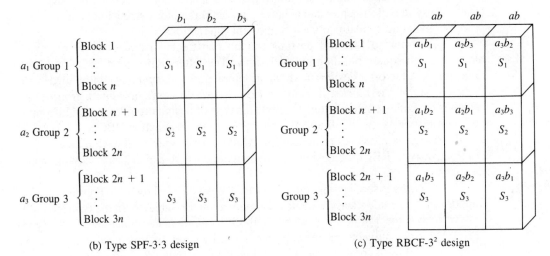

(a) Type RBF-33 design

(b) Type SPF-3·3 design

(c) Type RBCF-3^2 design

must be willing to sacrifice precision in estimating these interaction effects in order to achieve small block size. The interaction(s) that are confounded should be those that are thought to be negligible.

3. It must be possible to administer the levels of each treatment in every possible sequence. This requirement precludes the use of treatments whose levels consist of successive periods of time.

4. If repeated measurements are obtained, w samples of n experimental units from a population are randomly assigned to the w groups. The sequence of administration of the v treatment combinations within a block is randomized independently for each block.

5. For the nonrepeated measurements case, w samples of n blocks each containing v matched experimental units from a population are randomly assigned to the w groups. Following this, the v treatment combinations within a block are randomly assigned to the v matched experimental units within each block.

The construction of randomized block confounded factorial designs is described in the first part of this chapter. Confounding by means of a Latin square is covered in Section 12.13.

12.2 USE OF MODULAR ARITHMETIC IN CONSTRUCTING CONFOUNDED DESIGNS

The construction of type RBCF and RBPF designs requires a scheme for assigning treatment combinations to groups or samples of blocks so that variation between groups is confounded with one or more interactions or interaction components. Several schemes have been devised for this purpose (Yates, 1937; Kempthorne, 1952). One scheme, which is general in nature and relatively simple, uses modular arithmetic. We will now describe this scheme.

Let I and m be any integers, with $m > 0$. If I is divided by m, we obtain a quotient q and a remainder z, because

$$I = qm + z.$$

For example, let $I = 17$ and $m = 3$. Then $q = 5$ and $z = 2$, since

$$17 = 5(3) + 2.$$

In modular arithmetic the remainder is the term of interest. Consider now dividing $J = 5$ by $m = 3$. The remainder is equal to 2, since

$$5 = (1)(3) + 2.$$

Note that 17 and 5 leave the same remainder when divided by 3. Two integers I and J that leave the same remainder when divided by a positive integer m are said to be *congruent* with respect to the modulus m. This relation—congruence—can be written

$$I = J(\bmod m)$$

and read "I is congruent to J modulo m."

On reflection it should be apparent that any integer I is congruent to its remainder z; that is

$$I = z(\bmod m).$$

For example, if $I = 17$ and $m = 3$,

$$17 = 2(\bmod 3)$$

because when 17 and 2 are reduced modulo 3 (divided by the modulus 3) they leave the same remainder,

$$17 = 5(3) + 2 \qquad \text{and} \qquad 2 = 0(3) + 2.$$

The possible values of the remainder z are $0, 1, 2, \ldots, m - 1$. Thus an integer is always congruent to $0, 1, 2, \ldots,$ or $m - 1$, where m is the modulus. Consider the following examples.

$$0 = 0(\text{mod } 2) \qquad 0 = 0(\text{mod } 3)$$
$$1 = 1(\text{mod } 2) \qquad 1 = 1(\text{mod } 3)$$
$$2 = 0(\text{mod } 2) \qquad 2 = 2(\text{mod } 3)$$
$$3 = 1(\text{mod } 2) \qquad 3 = 0(\text{mod } 3)$$
$$4 = 0(\text{mod } 2) \qquad 4 = 1(\text{mod } 3)$$
$$5 = 1(\text{mod } 2) \qquad 5 = 2(\text{mod } 3)$$
$$6 = 0(\text{mod } 2) \qquad 6 = 0(\text{mod } 3)$$

MODULAR ADDITION AND MULTIPLICATION

Two operations of modular arithmetic are used in constructing confounded factorial designs: addition and multiplication. The operation of addition is illustrated by the following examples.

$$a_j + b_k = z(\text{mod } m)$$
$$0 + 0 = 0(\text{mod } 2)$$
$$1 + 0 = 1(\text{mod } 2)$$
$$0 + 1 = 1(\text{mod } 2)$$
$$1 + 1 = 0(\text{mod } 2)$$

$$0 + 0 = 0(\text{mod } 3)$$
$$0 + 1 = 1(\text{mod } 3)$$
$$0 + 2 = 2(\text{mod } 3)$$
$$1 + 1 = 2(\text{mod } 3)$$
$$1 + 2 = 0(\text{mod } 3)$$
$$2 + 2 = 1(\text{mod } 3)$$

To add two integers a_j and b_k, one obtains their sum and reduces it modulo m, that is, expresses it as a remainder with respect to the modulus m. This operation will be used later to confound an interaction with groups of blocks. We will let a_j, b_k, z, and m correspond to properties of an experimental design as follows:

> a_j and b_k denote levels of treatments A and B, respectively.
>
> z denotes a particular group of blocks.
>
> m denotes the number of levels of treatments A and B.

The second operation of modular arithmetic that is used in constructing factorial designs is multiplication. This operation is illustrated by the following examples.

$$(1)(1) = 1(\text{mod } 3)$$
$$(1)(2) = 2(\text{mod } 3)$$
$$(2)(2) = 1(\text{mod } 3)$$
$$(3)(2) = 0(\text{mod } 3)$$

To multiply two integers, one obtains their product and expresses it as a remainder with respect to the modulus m.

MODIFIED NOTATION SCHEME FOR TREATMENT LEVELS

Designs described in the first part of this chapter are limited to those of the form p^k, where p, the number of levels of each treatment, is a prime number. A prime number is any number divisible by no number smaller than itself other than one. Examples of prime numbers are 1, 2, 3, 5, 7, and 11. In order to use modular arithmetic in assigning treatment combinations to groups, we will have to use a new scheme for designating the levels of treatments, groups, and so on. According to this scheme, the first level of a treatment is designated by the subscript 0 instead of the usual 1. For example, the treatment levels of a type RBCF-3^2 design are designated by a_0, a_1, a_2, b_0, b_1, and b_2. The nine treatment combinations and their corresponding designations appear in Table 12.2-1. The digit in the first position indicates the level of treatment A; the digit in the second position indicates the level of treatment B. A design with three treatments requires three digits to designate a treatment combination. For example, if treatments A, B, and C are all at the first level, the designation is $a_0 b_0 c_0$ or 000.

This new notation scheme, carried to its logical conclusion, could lead to an odd looking notation for sums of observations. For example, the sum of $i = 0, \ldots, n - 1$ observations would be written

$$Y_0 + Y_1 + \cdots + Y_{n-1} = \sum_{i=0}^{n-1} Y_i.$$

In order to keep the notation for summation consistent with that used in Chapters 1–11, we will use the new notation only to identify the levels of treatments, blocks, and so on. We will revert back to the use of 1 for the first level of treatments and so on when summation is performed, that is, i will range over $1, \ldots, n$ and not $0, \ldots, n - 1$.* This dual notation scheme lets us write $\sum_{i=1}^{n} Y_i$ instead of $\sum_{i=0}^{n-1} Y_i$.

---

* In effect we are adding 1 to the initial and terminal values of the index of summation and 1 to the index of the summand, that is

$$\sum_{i=0}^{n-1} Y_i \qquad \text{becomes} \qquad \sum_{i=0+1}^{n-1+1} Y_{i+1}$$

where 0 and $n - 1$ are the initial and terminal values, respectively, of the index of summation and the subscript of Y is the index of the summand.

TABLE 12.2-1   Modified Notation Scheme for Factorial Designs

| $a_0\ a_0\ a_0$ | $a_1\ a_1\ a_1$ | $a_2\ a_2\ a_2$ |
$b_0\ b_1\ b_2$	$b_0\ b_1\ b_2$	$b_0\ b_1\ b_2$
00 01 02	10 11 12	20 21 22

ASSIGNMENT OF TREATMENT COMBINATIONS TO GROUPS

A factorial design with two levels of treatments A and B has four treatment combinations. Using the new notation for treatment levels, they are a_0b_0, a_0b_1, a_1b_0, and a_1b_1 or, more simply, 00, 01, 10, and 11. In order to use a randomized block factorial design, n blocks of size four would be required. Suppose it is only possible to observe a subject twice and that the experimenter's primary interest is in the two treatments. The block size can be reduced from four to two by confounding the AB interaction with groups of blocks. Modular arithmetic is used to determine which treatment combinations are assigned to each group of blocks. Let a_j denote the jth level of treatment A, and b_k, the kth level of treatment B. All treatment combinations satisfying the relation

$$a_j + b_k = z(\text{mod } 2)$$

where z is equal to 0, are assigned to group 0. Those satisfying the relation where z is equal to 1 are assigned to group 1. Modulus 2 is used because treatments A and B each have two levels. The range of z is 0 and 1, since all integers are congruent to 0, 1, . . . , or $m - 1$, and m is equal to 2 in this example.

Solving for a_j and b_k, we obtain

$$\left. \begin{array}{l} 0 + 0 = 0(\text{mod } 2) \\ 1 + 1 = 0(\text{mod } 2) \end{array} \right\} \quad \text{group 0 or } (AB)_0$$

$$\left. \begin{array}{l} 0 + 1 = 1(\text{mod } 2) \\ 1 + 0 = 1(\text{mod } 2) \end{array} \right\} \quad \text{group 1 or } (AB)_1.$$

Thus, treatment combinations 00 and 11 are assigned to group 0; combinations 01 and 10 are assigned to group 1. A block diagram for this type RBCF-2^2 design is shown in Figure 12.2-1. The notation $(AB)_z$ is an alternative way to denote the treatment combinations that are assigned to group z, $z = 0, 1$.

It can be shown that the arrangement in Figure 12.2-1 confounds the AB interaction with groups. Let μ_{ijkz} denote the population mean for block i, treatment combination a_jb_k, and group z. The AB interaction effect for the design in Figure 12.2-1 can be written as

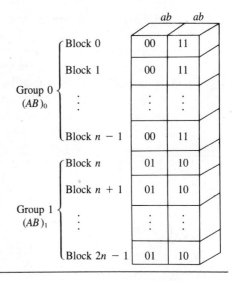

FIGURE 12.2-1 Block diagram of a type RBCF-2^2 design. The numbers in the cells denote the levels of treatments A and B. This design confounds the AB interaction with differences between groups.

$$\mu_{.000} - \mu_{.011} - \mu_{.101} + \mu_{.110} \quad \text{or} \quad (\mu_{.000} + \mu_{.110}) - (\mu_{.011} + \mu_{.101})$$

since by definition a two-treatment interaction effect has the form $\mu_{jk} - \mu_{jk'} - \mu_{j'k} + \mu_{j'k'}$. Consider now the contrast $\psi = \mu_{...0} - \mu_{...1}$ for groups 0 and 1,

$$\psi = (\mu_{.000} + \mu_{.110})/2 - (\mu_{.011} + \mu_{.101})/2.$$

This contrast involves the same means as the AB interaction effect. Since the effects of the AB interaction and groups are indistinguishable, they are said to be confounded.

We have now developed the basic concepts underlying the construction of randomized block completely confounded factorial designs where each treatment has two levels. Before describing partially confounded factorial designs and designs with three or five levels of each treatment, we will illustrate the computational procedures for type RBCF-2^2 and RBCF-2^3 designs.

12.3 COMPUTATIONAL PROCEDURES FOR TYPE RBCF-2^2 DESIGN

Assume that an experiment has been designed to evaluate the relative effectiveness of several procedures for using computer-aided instruction material. The material has been prepared to acquaint mechanics with servicing procedures for a new airplane engine. The criterion used to assess the effectiveness of the material was the number of simulated malfunctions in an engine that trainees were able to diagnose. The instruction material was presented to trainees by means of a CRT display. Treatment A consisted of two presentation rates for the material. Level a_0 was an unpaced rate in which trainees pressed the "return" key on the CRT unit when they were ready to view the next frame of information. Level a_1 was a paced presentation with 30 seconds between successive frames of information. The second variable investigated was the type of response that the trainees made to each frame of information. Two types of responses were investigated: one in which the trainee gave a verbal response to each frame of information, b_0, and a second in which the trainee typed a response using the keyboard of the CRT unit, b_1.

The research hypotheses leading to this experiment can be evaluated by means of statistical tests of the following null hypotheses.

$$H_0: \quad \alpha_j = 0 \quad \text{for all } j$$
$$H_1: \quad \alpha_j \neq 0 \quad \text{for some } j$$

$$H_0: \quad \beta_k = 0 \quad \text{for all } k$$
$$H_1: \quad \beta_k \neq 0 \quad \text{for some } k$$

$$H_0: \quad (\alpha\beta)_{jk} = 0 \quad \text{for all } j,k$$
$$H_1: \quad (\alpha\beta)_{jk} \neq 0 \quad \text{for some } j,k$$

The experimenter's primary interest is in evaluating hypotheses regarding the two treatments. The level of significance adopted for all tests is .05.

To evaluate these hypotheses, a random sample of 16 new trainees was obtained. Aptitude data were used to assign the trainees to eight blocks of size two so that those in a block had similar aptitude test scores. The eight blocks were randomly assigned to two groups, four blocks in each group. Following this, the a_jb_k treatment combinations appropriate for a group were randomly assigned to the matched trainees in each block. Ordinarily each group should contain at least 12 blocks in order to have an adequate number of degrees of freedom for experimental error.

The AB interaction was confounded with groups by using the relations

$$a_j + b_k = 0 (\text{mod } 2)$$
$$a_j + b_k = 1 (\text{mod } 2).$$

Treatment combinations 00 and 11 satisfy the first relation and were assigned to the blocks in group 0. Treatment combinations 01 and 10 satisfy the second relation and were assigned to the blocks in group 1. The layout of the type RBCF-2^2 design and computational procedures are shown in Table 12.3-1. The analysis of variance is summarized in Table 12.3-2. According to the analysis, the null hypotheses for both treatments are rejected. On the basis of the information in Tables 12.3-1 and 12.3-2, we can conclude that the paced presentation rate, a_0, is superior to the unpaced rate, a_1, and that typing a response at the CRT keyboard, b_1, is better than responding verbally, b_0.

The full rank experimental design model approach can also be used to analyze these data. This approach, which is useful if one or more scores is missing, is illustrated in Section 12.12.

12.4 EXPERIMENTAL DESIGN MODEL FOR A TYPE RBCF-2² DESIGN

A score Y_{ijkz} in a randomized block completely confounded factorial design is a composite that is equal to the following terms in the experimental design model.

$$Y_{ijkz} = \mu + \zeta_z + \pi_{i(z)} + \alpha_j + \beta_k + (\alpha\beta\pi)_{jki(z)} + \epsilon_{ijkz}$$
$$(i = 1, \ldots, n; j = 1, \ldots, p; k = 1, \ldots, q; z = 1, \ldots, w)$$

where

Y_{ijkz} = a score for the experimental unit in block i, treatment combination $a_j b_k$, and group z

ζ_z = the effect of group z and is subject to the restriction $\sum_{z=1}^{w} \zeta_z = 0$; the effects of groups and the AB interaction are completely confounded

$\pi_{i(z)}$ = the effect of block i that is $NID(0, \sigma_\pi^2)$

α_j = the effect of treatment level j and is subject to the restriction $\sum_{j=1}^{p} \alpha_j = 0$

β_k = the effect of treatment level k and is subject to the restriction $\sum_{k=1}^{q} \beta_k = 0$

$\alpha\beta\pi_{jki(z)}$ = the joint effect of $(\alpha\beta)_{jk}$ and $\pi_{i(z)}$

ϵ_{ijkz} = the experimental error that is $NID(0, \sigma_\epsilon^2)$; ϵ_{ijkz} is independent of $\pi_{i(z)}$. In this design, ϵ_{ijkz} cannot be estimated separately from $(\alpha\beta\pi)_{jki(z)}$.

Two sets of assumptions underlie the F tests for a type RBCF-2² design: one set for the between-blocks test and a second set for the within-blocks tests. The situation is similar to that described in Section 11.4 for a split-plot factorial design. The assumptions underlying the between-blocks test are the same as those for a completely randomized design (see Section 2.3). The key assumption is that the population variances for g_0 and g_1 are homogeneous. Sample estimators of the two variances are given by

$$\hat{\sigma}_{g_0}^2 = \frac{\sum_{i=1}^{n} \left(\sum_{jk=1}^{v} Y_{ijk0}\right)^2 - \frac{\left(\sum_{i=1}^{n}\sum_{jk=1}^{v} Y_{ijk0}\right)^2}{n}}{n-1}$$

and

$$\hat{\sigma}_{g_1}^2 = \frac{\sum_{i=1}^{n} \left(\sum_{jk=1}^{v} Y_{ijk1}\right)^2 - \frac{\left(\sum_{i=1}^{n}\sum_{jk=1}^{v} Y_{ijk1}\right)^2}{n}}{n-1}.$$

TABLE 12.3-1 Computational Procedures for a Type RBCF-2^2 Design

(i)   Data and notation [Y_{ijkz} denotes a score for the experimental unit in block i, treatment combination jk, and group z; $i = 1, \ldots, n$ blocks (s_i); $j = 1, \ldots, p$ levels of treatment A (a_j); $k = 1, \ldots, q$ levels of treatment B (b_k); $z = 1, \ldots, w$ groups (g_z); $jk = 1, \ldots, v$ combinations of $a_j b_k$ within a block]:

ABGS Summary Table
Entry is Y_{ijkz}

		ab	ab	$\sum\limits_{jk=1}^{v} Y_{ijkz}$	$\sum\limits_{i=1}^{n}\sum\limits_{jk=1}^{v} Y_{ijkz}$
		00	11		
	s_0	3	16	19	
g_0	s_1	5	14	19	
$(AB)_0$	s_2	6	17	23	81
	s_3	5	15	20	
		01	10		
	s_4	14	7	21	
g_1	s_5	14	6	20	
$(AB)_1$	s_6	16	7	23	91
	s_7	16	11	27	

AB Summary Table
Entry is $\sum\limits_{i=1}^{n} Y_{ijkz}$

	b_0	b_1	$\sum\limits_{i=1}^{n}\sum\limits_{k=1}^{q} Y_{ijkz}$
a_0	$n = 4$ 19	60	79
a_1	31	62	93
$\sum\limits_{i=1}^{n}\sum\limits_{j=1}^{p} Y_{ijkz} =$	50	122	

(ii)   Computational symbols:

$$\sum_{i=1}^{n}\sum_{jk=1}^{v}\sum_{z=1}^{w} Y_{ijkz} = 3 + 5 + \cdots + 11 = 172.000$$

$$\frac{\left(\sum\limits_{i=1}^{n}\sum\limits_{jk=1}^{v}\sum\limits_{z=1}^{w} Y_{ijkz}\right)^2}{nvw} = [Y] = \frac{(172.000)^2}{(4)(2)(2)} = 1849.000$$

Table 12.3-1 (continued)

$$\sum_{i=1}^{n} \sum_{jk=1}^{v} \sum_{z=1}^{w} Y_{ijkz}^2 = [ABGS] = (3)^2 + (5)^2 + \cdots + (11)^2 = 2220.000$$

$$\sum_{i=1}^{n} \sum_{z=1}^{w} \frac{\left(\sum_{jk=1}^{v} Y_{ijkz}\right)^2}{v} = [GS] = \frac{(19)^2}{2} + \frac{(19)^2}{2} + \cdots + \frac{(27)^2}{2} = 1875.000$$

$$\sum_{z=1}^{w} \frac{\left(\sum_{i=1}^{n} \sum_{jk=1}^{v} Y_{ijkz}\right)^2}{nv} = [G] = \frac{(81)^2}{(4)(2)} + \frac{(91)^2}{(4)(2)} = 1855.250$$

$$\sum_{j=1}^{p} \frac{\left(\sum_{i=1}^{n} \sum_{k=1}^{q} Y_{ijkz}\right)^2}{nq} = [A] = \frac{(79)^2}{(4)(2)} + \frac{(93)^2}{(4)(2)} = 1861.250$$

$$\sum_{k=1}^{q} \frac{\left(\sum_{i=1}^{n} \sum_{j=1}^{p} Y_{ijkz}\right)^2}{np} = [B] = \frac{(50)^2}{(4)(2)} + \frac{(122)^2}{(4)(2)} = 2173.000$$

$$\sum_{j=1}^{p} \sum_{k=1}^{q} \frac{\left(\sum_{i=1}^{n} Y_{ijkz}\right)^2}{n} = [AB] = \frac{(19)^2}{4} + \frac{(60)^2}{4} + \cdots + \frac{(62)^2}{4} = 2191.500$$

Computational formulas:

$$SSTO = [ABGS] - [Y] = 371.00 \qquad SSWITHIN\ BL = [ABGS] - [GS] = 345.00$$

$$SSBETWEEN\ BL = [GS] - [Y] = 26.00 \qquad SSA = [A] - [Y] = 12.25$$

$$SSG\ \text{or}\ SSAB = [G] - [Y] = 6.25 \qquad SSB = [B] - [Y] = 324.00$$

$$SSBL(G) = [GS] - [G] = 19.75 \qquad SSAB \times BL(G) = [ABGS] - [AB] - [GS] + [G]$$
$$= 8.75$$

For the data in Table 12.3-1, the values of $\hat{\sigma}_{g_0}^2$ and $\hat{\sigma}_{g_1}^2$ are

$$\hat{\sigma}_{g_0}^2 = \frac{1651 - \frac{(81)^2}{4}}{4 - 1} \qquad \hat{\sigma}_{g_1}^2 = \frac{2099 - \frac{(91)^2}{4}}{4 - 1}$$
$$= 3.583 \qquad\qquad = 9.583.$$

Procedures for testing the assumption of homogeneity of variance are described in Section 2.5.

    The assumptions for the within-blocks tests are somewhat less restrictive than those for a randomized block design (see Section 6.4) and the within-blocks tests for

TABLE 12.3-2 ANOVA Table for Type RBCF-2^2 Design

	Source	SS	df	MS	F	E(MS) Model III (A and B fixed, Blocks random)
1	Between blocks	26.000	$nw - 1 = 7$			
2	Groups or AB	6.250	$w - 1 = 1$	6.250	[2/3] 1.90	$\sigma_\epsilon^2 + v\sigma_\pi^2 + \dfrac{nv\sum\limits_{z=1}^{w}\zeta_z^2}{w-1}$
3	Blocks w. G	19.750	$w(n-1) = 6$	3.292		$\sigma_\epsilon^2 + v\sigma_\pi^2$
4	Within blocks	345.000	$nw(v-1) = 8$			
5	A (Presentation rate)	12.250	$p - 1 = 1$	12.250	[5/7] 8.40*	$\sigma_\epsilon^2 + \sigma_{\alpha\beta\pi}^2 + \dfrac{nw\sum\limits_{j=1}^{p}\alpha_j^2}{p-1}$
6	B (Response model)	324.000	$q - 1 = 1$	324.000	[6/7] 222.22**	$\sigma_\epsilon^2 + \sigma_{\alpha\beta\pi}^2 + \dfrac{nw\sum\limits_{k=1}^{q}\beta_k^2}{q-1}$
7	$AB \times BL(G)$	8.750	$w(n-1)(v-1) = 6$	1.458		$\sigma_\epsilon^2 + \sigma_{\alpha\beta\pi}^2$
8	$AB \times BL(g_0)$	4.375	$(n-1)(v-1) = 3$	1.458		
9	$AB \times BL(g_1)$	4.375	$(n-1)(v-1) = 3$	1.458		
10	Total	371.000	$nvw - 1 = 16$			

*$p < .05$
**$p < .01$

a split-plot factorial design (see Section 11.4). The mean square within-blocks error term $MSAB \times BL(G)$ is a pooled term that is equal to $[SSAB \times BL(g_0) + SSAB \times BL(g_1)]/w(n-1)(v-1)$. Since there are only two treatment combinations in each group of blocks, it can be shown that $MSAB \times BL(g_0)$ and $MSAB \times BL(g_1)$ are equal to the average of two variances minus a covariance. That is,

$$MSAB \times BL(g_0) = (\hat{\sigma}_{.000}^2 + \hat{\sigma}_{.110}^2)/2 - \hat{\sigma}_{(.000)(.110)}$$
$$= (1.5833 + 1.6667)/2 - .1667 = 1.458$$
$$MSAB \times BL(g_1) = (\hat{\sigma}_{.011}^2 + \hat{\sigma}_{.101}^2)/2 - \hat{\sigma}_{(.011)(.101)}$$
$$= (4.9167 + 1.3333)/2 - 1.6667 = 1.458.$$

The key assumption for the within-blocks tests is that the population variances estimated by $MSAB \times BL(g_0)$ and $MSAB \times BL(g_1)$ are equal.

The advantage of a type RBCF-2^2 design over a type RBF-22 design is that it enables an experimenter to reduce the block size from four to two. It accomplishes this without sacrificing power in testing treatments A and B since both are within-blocks treatments. A confounded factorial design is a better design choice than a split-plot factorial design if an experimenter is primarily interested in treatments A and B rather than the AB interaction.

12.5 LAYOUT AND ANALYSIS FOR A TYPE RBCF-2^3 DESIGN

The computational procedures for a randomized block completely confounded factorial design with two treatments, each having two levels, can be easily extended to three or more treatments. We will now describe the layout and analysis for a type RBCF-2^3 design. In this design there are four interactions: AB, AC, BC, and ABC. The size of a block can be reduced from eight to four by confounding one of these interactions with groups of blocks. The interaction that is chosen for this purpose should be one that is relatively unimportant or thought to be negligible. Usually this is the highest order interaction, in this example, the ABC interaction.

CONFOUNDING THE ABC INTERACTION WITH GROUPS

Modular arithmetic is used to determine which treatment combinations are assigned to each group of blocks. Let a_j denote the jth level of treatment A; b_k, the kth level of treatment B; and c_l, the lth level of treatment C. The ABC interaction can be confounded with groups by assigning the treatment combinations satisfying the relation

$$a_j + b_k + c_l = 0 \pmod 2$$

to group 0 and those satisfying

$$a_j + b_k + c_l = 1 \pmod 2$$

to group 1. Solving for a_j, b_k, and c_l, we obtain

$$
\left.
\begin{aligned}
000 &= 0 \pmod 2 \\
011 &= 0 \pmod 2 \\
101 &= 0 \pmod 2 \\
110 &= 0 \pmod 2
\end{aligned}
\right\} \text{ group 0 or } (ABC)_0
\qquad
\left.
\begin{aligned}
001 &= 1 \pmod 2 \\
010 &= 1 \pmod 2 \\
100 &= 1 \pmod 2 \\
111 &= 1 \pmod 2
\end{aligned}
\right\} \text{ group 1 or } (ABC)_1.
$$

A block diagram for this design is shown in Figure 12.5-1. To have an adequate number of degrees of freedom for experimental error, each group should contain at least five blocks.

The experimental design model equation for this design is

$$Y_{ijklz} = \mu + \zeta_z + \pi_{i(z)} + \alpha_j + \beta_k + \gamma_l + (\alpha\beta)_{jk} + (\alpha\gamma)_{jl} + (\beta\gamma)_{kl}$$
$$+ (\alpha\beta\gamma\pi)_{jkli(z)} + \epsilon_{ijklz} \quad (i = 1, \ldots, n; j = 1, \ldots, p;$$
$$k = 1, \ldots, q; l = 1, \ldots, r; z = 1, \ldots, w).$$

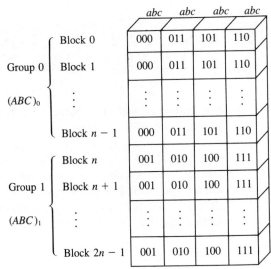

FIGURE 12.5-1   Block diagram of type RBCF-2^3 design. The numbers in the cells denote the levels of treatments A, B, and C, respectively. This design confounds the ABC interaction with differences between groups.

CONFOUNDING A TWO-TREATMENT INTERACTION WITH GROUPS

An experimenter could confound one of the two-treatment interactions, say AB, with groups by first determining the treatment combinations that satisfy the relation

$$a_j + b_k = z(\text{mod } 2) \quad z = 0, 1.$$

These combinations are

$$\begin{aligned}
00 &= 0\,(\text{mod } 2) \\
11 &= 0\,(\text{mod } 2)
\end{aligned}\Bigg\} \quad \text{group 0 } (AB)_0$$

$$\begin{aligned}
01 &= 1(\text{mod } 2) \\
10 &= 1(\text{mod } 2)
\end{aligned}\Bigg\} \quad \text{group 1 } (AB)_1.$$

Next, the levels 0 and 1 of treatment C are added to these combinations in a balanced manner.

add 0⟶   ⟶ add 1

$$\left.\begin{array}{l} 000 \\ 110 \end{array}\right. \quad \left.\begin{array}{l} 001 \\ 111 \end{array}\right\} \quad \text{group 0 } (AB)_0$$

$$\left.\begin{array}{l} 010 \\ 100 \end{array}\right. \quad \left.\begin{array}{l} 011 \\ 101 \end{array}\right\} \quad \text{group 1 } (AB)_1$$

The treatment combinations assigned to groups 0 and 1 are as follows.

	abc	abc	abc	abc
Group 0 $(AB)_0$	000	001	110	111
Group 1 $(AB)_1$	010	011	100	101

The experimental design model equation for this design is

$$Y_{ijklz} = \mu + \zeta_z + \pi_{i(z)} + \alpha_j + \beta_k + \gamma_l + (\alpha\gamma)_{jl} + (\beta\gamma)_{kl} + (\alpha\beta\gamma)_{jkl}$$
$$+ (\alpha\beta\gamma\pi)_{jkli(z)} + \epsilon_{ijklz} \quad (i = 1, \ldots, n; j = 1, \ldots, p;$$
$$k = 1, \ldots, q; l = 1, \ldots, r; z = 1, \ldots, w).$$

The procedures just described can be used to confound the AC interaction with groups. We begin with the treatment combinations that satisfy the relation

$$a_j + c_l = z(\text{mod } 2) \qquad z = 0, 1.$$

They are

$$\left.\begin{array}{l} 00 \\ 11 \end{array}\right\} \quad \text{group 0 } (AC)_0$$

$$\left.\begin{array}{l} 01 \\ 10 \end{array}\right\} \quad \text{group 1 } (AC)_1.$$

Next, the levels 0 and 1 of treatment B are added to these combinations in a balanced manner. For example,

add 0⟶   ⟶ add 1

$$\left.\begin{array}{l} 000 \\ 101 \end{array}\right. \quad \left.\begin{array}{l} 010 \\ 111 \end{array}\right\} \quad \text{group 0 } (AC)_0$$

$$\left.\begin{array}{l} 001 \\ 100 \end{array}\right. \quad \left.\begin{array}{l} 011 \\ 110 \end{array}\right\} \quad \text{group 1 } (AC)_1.$$

The treatment combinations assigned to groups 0 and 1 are as follows.

	abc	abc	abc	abc
Group 0 $(AC)_0$	000	010	101	111
Group 1 $(AC)_1$	001	011	100	110

The experimental design model equation for this design is

$$Y_{ijklz} = \mu + \zeta_z + \pi_{i(z)} + \alpha_j + \beta_k + \gamma_l + (\alpha\beta)_{jk} + (\beta\gamma)_{kl} + (\alpha\beta\gamma)_{jkl}$$
$$+ (\alpha\beta\gamma\pi)_{jkli(z)} + \epsilon_{ijklz} \qquad (i = 1, \ldots, n; j = 1, \ldots, p;$$
$$k = 1, \ldots, q; l = 1, \ldots, r; z = 1, \ldots, w).$$

COMPUTATIONAL EXAMPLE FOR TYPE RBCF-2³ DESIGN

The computational procedures for a type RBCF-2^3 design in which the ABC interaction is confounded with groups are shown in Table 12.5-1. The results of the analysis are summarized in Table 12.5-2. These computational procedures are easily extended to confounded factorial designs with four or more treatments.

TABLE 12.5-1 Computational Procedures for a Type RBCF-2^3 Design

---

(i) Data and notation [Y_{ijklz} denotes a score for the experimental unit in block i, treatment combination jkl, and group z; $i = 1, \ldots, n$ blocks (s_i); $j = 1, \ldots, p$ levels of treatment A (a_j); $k = 1, \ldots,$ q levels of treatment B (b_k); $l = 1, \ldots, r$ levels of treatment C (c_l); $z = 1, \ldots, w$ levels of groups (g_z); $jkl = 1, \ldots, v$ combinations of $a_j b_k c_l$ within a block]:

ABCGS Summary Table
Entry is Y_{ijklz}

		abc	abc	abc	abc	$\sum\limits_{jkl=1}^{v} Y_{ijklz}$	$\sum\limits_{i=1}^{n}\sum\limits_{jkl=1}^{v} Y_{ijklz}$
		000	011	101	110		
	s_0	3	7	2	5	17	
g_0	s_1	6	8	3	6	23	81
$(ABC)_0$	s_2	3	9	4	5	21	
	s_3	3	8	3	6	20	
		001	010	100	111		
	s_4	4	7	1	10	22	
g_1	s_5	5	8	2	10	25	91
$(ABC)_1$	s_6	4	7	2	9	22	
	s_7	3	6	2	11	22	

Table 12.5-1 (continued)

AB Summary Table

Entry is $\sum\limits_{i=1}^{n} \sum\limits_{z=1}^{w} Y_{ijklz}$

	b_0	b_1	$\sum\limits_{i=1}^{n}\sum\limits_{z=1}^{w}\sum\limits_{k=1}^{q} Y_{ijklz}$
a_0	$nw = 8$ 31	60	91
a_1	19	62	81
$\sum\limits_{i=1}^{n}\sum\limits_{z=1}^{w}\sum\limits_{j=1}^{p} Y_{ijklz} = 50$		122	

AC Summary Table

Entry is $\sum\limits_{i=1}^{n} \sum\limits_{z=1}^{w} Y_{ijklz}$

	c_0	c_1
a_0	$nw = 8$ 43	48
a_1	29	52
$\sum\limits_{i=1}^{n}\sum\limits_{z=1}^{w}\sum\limits_{j=1}^{p} Y_{ijklz} = 72$		100

BC Summary Table

Entry is $\sum\limits_{i=1}^{n} \sum\limits_{z=1}^{w} Y_{ijklz}$

	c_0	c_1
b_0	$nw = 8$ 22	28
b_1	50	72

ABC Summary Table

Entry is $\sum\limits_{i=1}^{n} Y_{ijklz}$

	b_0 c_0	b_0 c_1	b_1 c_0	b_1 c_1
a_0	$n = 4$ 15	16	28	32
a_1	7	12	22	40

(ii) Computational symbols:

$$\sum_{i=1}^{n} \sum_{jkl=1}^{v} \sum_{z=1}^{w} Y_{ijklz} = 3 + 6 + \cdots + 11 = 172.000$$

$$\frac{\left(\sum\limits_{i=1}^{n} \sum\limits_{jkl=1}^{v} \sum\limits_{z=1}^{w} Y_{ijklz}\right)^2}{nvw} = [Y] = \frac{(172.000)^2}{(4)(4)(2)} = 924.500$$

$$\sum_{i=1}^{n} \sum_{jkl=1}^{v} \sum_{z=1}^{w} Y_{ijklz}^2 = [ABCGS] = (3)^2 + (6)^2 + \cdots + (11)^2 = 1160.000$$

$$\sum_{i=1}^{n} \sum_{z=1}^{w} \frac{\left(\sum\limits_{jkl=1}^{v} Y_{ijklz}\right)^2}{v} = [GS] = \frac{(17)^2}{4} + \frac{(23)^2}{4} + \cdots + \frac{(22)^2}{4} = 934.000$$

Table 12.5-1 (continued)

$$\sum_{z=1}^{w} \frac{\left(\sum\limits_{i=1}^{n}\sum\limits_{jkl=1}^{v} Y_{ijklz}\right)^2}{nv} = [G] = \frac{(81)^2}{(4)(4)} + \frac{(91)^2}{(4)(4)} = 927.625$$

$$\sum_{j=1}^{p} \frac{\left(\sum\limits_{i=1}^{n}\sum\limits_{z=1}^{w}\sum\limits_{k=1}^{q} Y_{ijklz}\right)^2}{nwq} = [A] = \frac{(91)^2}{(4)(2)(2)} + \frac{(81)^2}{(4)(2)(2)} = 927.625$$

$$\sum_{k=1}^{q} \frac{\left(\sum\limits_{i=1}^{n}\sum\limits_{z=1}^{w}\sum\limits_{j=1}^{p} Y_{ijklz}\right)^2}{nwp} = [B] = \frac{(50)^2}{(4)(2)(2)} + \frac{(122)^2}{(4)(2)(2)} = 1086.500$$

$$\sum_{l=1}^{r} \frac{\left(\sum\limits_{i=1}^{n}\sum\limits_{z=1}^{w}\sum\limits_{j=1}^{p} Y_{ijklz}\right)^2}{nwp} = [C] = \frac{(72)^2}{(4)(2)(2)} + \frac{(100)^2}{(4)(2)(2)} = 949.000$$

$$\sum_{j=1}^{p}\sum_{k=1}^{q} \frac{\left(\sum\limits_{i=1}^{n}\sum\limits_{z=1}^{w} Y_{ijklz}\right)^2}{nw} = [AB] = \frac{(31)^2}{(4)(2)} + \frac{(60)^2}{(4)(2)} + \cdots + \frac{(62)^2}{(4)(2)} = 1095.750$$

$$\sum_{j=1}^{p}\sum_{l=1}^{r} \frac{\left(\sum\limits_{i=1}^{n}\sum\limits_{z=1}^{w} Y_{ijklz}\right)^2}{nw} = [AC] = \frac{(43)^2}{(4)(2)} + \frac{(48)^2}{(4)(2)} + \cdots + \frac{(52)^2}{(4)(2)} = 962.250$$

$$\sum_{k=1}^{q}\sum_{l=1}^{r} \frac{\left(\sum\limits_{i=1}^{n}\sum\limits_{z=1}^{w} Y_{ijklz}\right)^2}{nw} = [BC] = \frac{(22)^2}{(4)(2)} + \frac{(28)^2}{(4)(2)} + \cdots + \frac{(72)^2}{(4)(2)} = 1119.000$$

$$\sum_{j=1}^{p}\sum_{k=1}^{q}\sum_{l=1}^{r} \frac{\left(\sum\limits_{i=1}^{n} Y_{ijklz}\right)^2}{n} = [ABC] = \frac{(15)^2}{4} + \frac{(16)^2}{4} + \cdots + \frac{(40)^2}{4} = 1141.500$$

(iii)   Computational formulas:

$$SSTO = [ABCGS] - [Y] = 235.500 \qquad SSB = [B] - [Y] = 162.000$$

$$SSBETWEEN\ BL = [GS] - [Y] = 9.500 \qquad SSC = [C] - [Y] = 24.500$$

$$SSG = [G] - [Y] = 3.125 \qquad SSAB = [AB] - [A] - [B] + [Y] = 6.125$$

$$SSBL(G) = [GS] - [G] = 6.375 \qquad SSAC = [AC] - [A] - [C] + [Y] = 10.125$$

$$SSWITHIN\ BL = [ABCGS] - [GS] \qquad SSBC = [BC] - [B] - [C] + [Y] = 8.000$$

$$= 226.000 \qquad SSABC \times BL(G) = [ABCGS] - [ABC] - [GS] + [G]$$

$$SSA = [A] - [Y] = 3.125 \qquad = 12.125$$

TABLE 12.5-2 ANOVA Table for Type RBCF-2^3 Design

	Source	SS	df	MS	F	E(MS) Model III (A, B, and C fixed; Blocks random)
1	Between blocks	9.500	$nw - 1 = 7$			
2	Groups or ABC	3.125	$w - 1 = 1$	3.125	[2/3] 2.94	$\sigma_\epsilon^2 + v\sigma_\pi^2 + nv\Sigma\zeta_z^2/(w-1)$
3	Blocks w. G	6.375	$w(n-1) = 6$	1.063		$\sigma_\epsilon^2 + v\sigma_\pi^2$
4	Within blocks	226.000	$nw(v-1) = 24$			
5	A	3.125	$p - 1 = 1$	3.125	[5/11] 4.64*	$\sigma_\epsilon^2 + \sigma_{\alpha\beta\gamma\pi}^2 + 2nw\Sigma\alpha_j^2/(p-1)$
6	B	162.000	$q - 1 = 1$	162.000	[6/11] 240.36**	$\sigma_\epsilon^2 + \sigma_{\alpha\beta\gamma\pi}^2 + 2nw\Sigma\beta_k^2/(q-1)$
7	C	24.500	$r - 1 = 1$	24.500	[7/11] 36.35**	$\sigma_\epsilon^2 + \sigma_{\alpha\beta\gamma\pi}^2 + 2nw\Sigma\gamma_l^2/(r-1)$
8	AB	6.125	$(p-1)(q-1) = 1$	6.125	[8/11] 9.09**	$\sigma_\epsilon^2 + \sigma_{\alpha\beta\gamma\pi}^2 + nw\Sigma\Sigma(\alpha\beta)_{jk}^2/(p-1)(q-1)$
9	AC	10.125	$(p-1)(r-1) = 1$	10.125	[9/11] 15.02**	$\sigma_\epsilon^2 + \sigma_{\alpha\beta\gamma\pi}^2 + nw\Sigma\Sigma(\alpha\gamma)_{jl}^2/(p-1)(r-1)$
10	BC	8.000	$(q-1)(r-1) = 1$	8.000	[10/11] 11.87**	$\sigma_\epsilon^2 + \sigma_{\alpha\beta\gamma\pi}^2 + nw\Sigma\Sigma(\beta\gamma)_{kl}^2/(q-1)(r-1)$
11	$ABC \times BL(G)$	12.125	$w(n-1)(v-1) = 18$	0.674		$\sigma_\epsilon^2 + \sigma_{\alpha\beta\gamma\pi}^2$
12	$ABC \times BL(g_0)$	7.0625	$(n-1)(v-1) = 9$			
13	$ABC \times BL(g_1)$	5.0625	$(n-1)(v-1) = 9$			
14	Total	235.500	$nvz - 1 = 31$			

*$p < .05$
**$p < .01$

12.6   COMPLETE VERSUS PARTIAL CONFOUNDING

The two designs described thus far are examples of completely confounded factorial designs. In these designs the *AB* or *ABC* interaction was confounded with groups of blocks. As we have seen, confounded effects are usually evaluated with less precision than unconfounded, within-blocks, effects. In designs having more than two treatments, each with two levels, it is possible to confound different interactions in different groups of blocks and thereby obtain within-blocks information on these interactions. This procedure, which is called *partial confounding,* involves confounding one interaction in one group of blocks, a second interaction in a second group of blocks, and so on. For example, in a type RBPF-2^3 design the *AB* interaction can be confounded with the blocks in group 0, the *AC* interaction can be confounded with the blocks in group 1, the *BC* interaction can be confounded with the blocks in group 2, and the *ABC* interaction with the blocks in group 3. This design is diagrammed in Figure 12.6-1. Since the *AB* interaction is confounded only in group 0, within-blocks information on this interaction is available from groups 1, 2, and 3. It should be apparent from an examination of Figure 12.6-1 that within-blocks information from three of the four groups is also available for the *AC*, *BC*, and *ABC* interactions. The advantage of partial confounding is that the block size can be reduced and still obtain partial within-blocks information for each of the confounded interactions. If an interaction is known to be insignificant, complete confounding in which the interaction is confounded in all of the groups is preferable to partial confounding.

Federer (1955, 230) distinguishes between *balanced partial confounding* and *unbalanced partial confounding.* The former designation refers to designs in which all effects of a particular order, for example, all two-treatment interactions, are confounded with blocks an equal number of times. The type RBPF-2^3 design just de-

---

FIGURE 12.6-1 . Block diagram of a type RBPF-2^3 design. The numbers in the cells denote the levels of treatments *A*, *B*, and *C*. This design confounds a different interaction in each group of blocks.

		abc	abc	abc	abc
Group 0	Block 0 $(AB)_0$	000	001	110	111
	Block 1 $(AB)_1$	010	011	100	101
Group 1	Block 2 $(AC)_0$	000	010	101	111
	Block 3 $(AC)_1$	001	011	100	110
Group 2	Block 4 $(BC)_0$	000	011	100	111
	Block 5 $(BC)_1$	001	010	101	110
Group 3	Block 6 $(ABC)_0$	000	011	101	110
	Block 7 $(ABC)_1$	001	010	100	111

scribed, in which AB, AC, and BC were each confounded in one group of blocks, illustrates balanced partial confounding. If all effects of a particular order are confounded with blocks an unequal number of times, the arrangement is described as unbalanced partial confounding. For example, the AB and AC interactions could be confounded in groups 0 and 1, respectively. Since AB and AC are each confounded once but the BC interaction is not confounded, the design is said to involve unbalanced partial confounding.

12.7 COMPUTATIONAL PROCEDURES FOR TYPE RBPF-2³ DESIGN

The computational procedures for a randomized block partially confounded factorial design will be illustrated using the data in Table 12.5-1. We will assume that the AB, AC, BC, and ABC interactions have been confounded with the blocks in groups 0–3, respectively. The procedures described in Section 12.5 are used to determine which treatment combinations are assigned to blocks. The design is diagrammed in Figure 12.6-1.

If repeated measures are obtained on the experimental units, the design requires eight experimental units who are randomly assigned to four groups, each containing two units (blocks). The sequence of administration of the v treatment combinations within a block is randomized independently for each block. For the nonrepeated measurements case, 32 experimental units are required. Eight blocks each containing four matched experimental units are formed. The units are randomly assigned to four groups with two blocks in each group. The sets of treatment combinations are randomly assigned to the groups and then to the units within each block. More than eight blocks can be used in the design. However, the number of blocks in each group must always be a multiple of two.

The layout and computational procedures for this design are shown in Table 12.7-1; the analysis is summarized in Table 12.7-2. This analysis can be compared with that in Table 12.5-2 where the ABC interaction was completely confounded with groups. The advantage of partial confounding is that a within-blocks estimate of the ABC interaction can be computed. The reader has undoubtedly noted that this advantage is gained at the price of greater complexity in the analysis of the data. Furthermore the AB, AC, BC, and ABC interactions are computed from only three-fourths of the groups. The ratio 3/4 is called the *relative information* on the confounded effects (Yates, 1937). The choice between a completely confounded or a partially confounded design rests in part on the experimenter's expectations with respect to the interactions. If one is known to be insignificant, a completely confounded design is the better choice.

Numerous design possibilities are inherent in partial confounding for experi-

TABLE 12.7-1 Computational Procedures for a Type RBPF-2^3 Design

(i) Data and notation [Y_{ijklz} denotes a score for the experimental unit in block i, treatment combination jkl, and group z; $i = 1, \ldots, n$ blocks (s_i); $j = 1, \ldots, p$ levels of treatment A (a_j); $k = 1, \ldots,$ q levels of treatment B (b_k); $l = 1, \ldots, r$ levels of treatment C (c_l); $z = 1, \ldots, w$ levels of groups (g_z); $z = 1, \ldots w'$ levels of groups where an effect is not confounded; $jkl = 1, \ldots, v$ combinations of $a_j b_k c_l$ within a block]:

ABCGS Summary Table
Entry is Y_{ijklz}

			abc	abc	abc	abc	$\sum_{jkl=1}^{v} Y_{ijklz}$	$\sum_{i=1}^{n}\sum_{jkl=1}^{v} Y_{ijklz}$
			000	001	110	111		
	$(AB)_0$	s_0	3	4	5	10	22	
g_0								39
			010	011	100	101		
	$(AB)_1$	s_1	7	7	1	2	17	
			000	010	101	111		
	$(AC)_0$	s_2	6	8	3	10	27	
g_1								48
			001	011	100	110		
	$(AC)_1$	s_3	5	8	2	6	21	
			000	011	100	111		
	$(BC)_0$	s_4	3	9	2	9	23	
g_2								43
			001	010	101	110		
	$(BC)_1$	s_5	4	7	4	5	20	
			000	011	101	110		
	$(ABC)_0$	s_6	3	8	3	6	20	
g_3								42
			001	010	100	111		
	$(ABC)_1$	s_7	3	6	2	11	22	

Table 12.7-1 (continued)

AB Summary Table

Entry is $\sum_{i=1}^{n} \sum_{z=1}^{w} Y_{ijklz}$ from

$g_0,\ g_1,\ g_2,\ and\ g_3$

	b_0	b_1	$\sum_{i=1}^{n} \sum_{k=1}^{q} \sum_{z=1}^{w} Y_{ijklz}$
a_0	$nw = 8$ 31	60	91
a_1	19	62	81
$\sum_{i=1}^{n} \sum_{j=1}^{p} \sum_{z=1}^{w} Y_{ijklz} = 50$		122	

AC Summary Table

Entry is $\sum_{i=1}^{n} \sum_{z=1}^{w} Y_{ijklz}$ from

$g_0,\ g_1,\ g_2,\ and\ g_3$

	c_0	c_1
a_0	$nw = 8$ 43	48
a_1	29	52
$\sum_{i=1}^{n} \sum_{j=1}^{p} \sum_{z=1}^{w} Y_{ijklz} = 72$		100

AB′ Summary Table

Entry is $\sum_{i=1}^{n} \sum_{z=1}^{w'} Y_{ijklz}$ from

$g_1,\ g_2,\ and\ g_3$

	b_0	b_1	$\sum_{i=1}^{n} \sum_{k=1}^{q} \sum_{z=1}^{w'} Y_{ijklz}$
a_0	$nw' = 6$ 24	46	70
a_1	16	47	63
$\sum_{i=1}^{n} \sum_{j=1}^{p} \sum_{z=1}^{w'} Y_{ijklz} = 40$		93	

AC′ Summary Table

Entry is $\sum_{i=1}^{n} \sum_{z=1}^{w'} Y_{ijklz}$ from

$g_0,\ g_2,\ and\ g_3$

	c_0	c_1	$\sum_{i=1}^{n} \sum_{l=1}^{r} \sum_{z=1}^{w'} Y_{ijklz}$
a_0	$nw' = 6$ 29	35	64
a_1	21	39	60
$\sum_{i=1}^{n} \sum_{j=1}^{p} \sum_{z=1}^{w'} Y_{ijklz} = 50$		74	

AB″ Summary Table

Entry is $\sum_{i=1}^{n} \sum_{z=1}^{w'} Y_{ijklz}$ from

$g_0,\ g_1,\ and\ g_2$

	b_0	b_1	$\sum_{i=1}^{n} \sum_{k=1}^{q} \sum_{z=1}^{w'} Y_{ijklz}$
a_0	$nw' = 6$ 25	46	71
a_1	14	45	59
$\sum_{i=1}^{n} \sum_{j=1}^{p} \sum_{z=1}^{w'} Y_{ijklz} = 39$		91	

AC″ Summary Table

Entry is $\sum_{i=1}^{n} \sum_{z=1}^{w'} Y_{ijklz}$ from

$g_0,\ g_1,\ and\ g_2$

	c_0	c_1
a_0	$nw' = 6$ 34	37
a_1	21	38
$\sum_{i=1}^{n} \sum_{j=1}^{p} \sum_{z=1}^{w'} Y_{ijklz} = 55$		75

Table 12.7-1 (continued)

BC' Summary Table

Entry is $\sum\limits_{i=1}^{n} \sum\limits_{z=1}^{w'} Y_{ijklz}$ from

g_0, g_1, and g_3

ABC" Summary Table

Entry is $\sum\limits_{z=1}^{w'} Y_{ijklz}$ from

g_0, g_1, and g_2

	c_0	c_1	$\sum\limits_{i=1}^{n}\sum\limits_{l=1}^{r}\sum\limits_{z=1}^{w'} Y_{ijklz}$
b_0	$nw' = 6$ 17	20	37
b_1	38	54	92

$\sum\limits_{i=1}^{n}\sum\limits_{k=1}^{q}\sum\limits_{z=1}^{w'} Y_{ijklz} = 55 \qquad 74$

	b_0 c_0	b_0 c_1	b_1 c_0	b_1 c_1
a_0	$w' = 3$ 12	13	22	24
a_1	5	9	16	29

BC" Summary Table

Entry is $\sum\limits_{i=1}^{n}\sum\limits_{z=1}^{w'} Y_{ijklz}$ from

g_0, g_1, and g_2

	c_0	c_1
b_0	$nw' = 6$ 17	22
b_1	38	53

(ii)   Computational symbols:

$$\sum_{i=1}^{n}\sum_{jkl=1}^{v}\sum_{z=1}^{w} Y_{ijklz} = 3 + 4 + \cdots + 11 = 172.000$$

$$\frac{\left(\sum\limits_{i=1}^{n}\sum\limits_{jkl=1}^{v}\sum\limits_{z=1}^{w} Y_{ijklz}\right)^2}{nvw} = [Y] = \frac{(172.000)^2}{(2)(4)(4)} = 924.500$$

$$\sum_{i=1}^{n}\sum_{jkl=1}^{v}\sum_{z=1}^{w} Y_{ijklz}^2 = [ABCGS] = (3)^2 + (4)^2 + \cdots + (11)^2 = 1160.000$$

$$\sum_{i=1}^{n}\sum_{z=1}^{w} \frac{\left(\sum\limits_{jkl=1}^{v} Y_{ijklz}\right)^2}{v} = [GS] = \frac{(22)^2}{4} + \frac{(17)^2}{4} + \cdots + \frac{(22)^2}{4} = 939.000$$

$$\sum_{z=1}^{w} \frac{\left(\sum\limits_{i=1}^{n}\sum\limits_{jkl=1}^{v} Y_{ijklz}\right)^2}{nv} = [G] = \frac{(39)^2}{(2)(4)} + \frac{(48)^2}{(2)(4)} + \cdots + \frac{(42)^2}{(2)(4)} = 929.750$$

Table 12-7.1 (continued)

$$\sum_{j=1}^{p} \frac{\left(\sum_{i=1}^{n}\sum_{k=1}^{q}\sum_{z=1}^{w} Y_{ijklz}\right)^2}{nqw} = [A_{AB}]^* = \frac{(91)^2}{(2)(2)(4)} + \frac{(81)^2}{(2)(2)(4)} = 927.625$$

$$\sum_{j=1}^{p} \frac{\left(\sum_{i=1}^{n}\sum_{j=1}^{q}\sum_{z=1}^{w'} Y_{ijklz}\right)^2}{nqw'} = [A_{AB'}] = \frac{(70)^2}{(2)(2)(3)} + \frac{(63)^2}{(2)(2)(3)} = 739.083$$

$$\sum_{j=1}^{p} \frac{\left(\sum_{i=1}^{n}\sum_{l=1}^{r}\sum_{z=1}^{w'} Y_{ijklz}\right)^2}{nrw'} = [A_{AC'}] = \frac{(64)^2}{(2)(2)(3)} + \frac{(60)^2}{(2)(2)(3)} = 641.333$$

$$\sum_{j=1}^{p} \frac{\left(\sum_{i=1}^{n}\sum_{k=1}^{q}\sum_{z=1}^{w'} Y_{ijklz}\right)^2}{nqw'} = [A_{AB''}] = \frac{(71)^2}{(2)(2)(3)} + \frac{(59)^2}{(2)(2)(3)} = 710.167$$

$$\sum_{k=1}^{q} \frac{\left(\sum_{i=1}^{n}\sum_{j=1}^{p}\sum_{z=1}^{w} Y_{ijklz}\right)^2}{npw} = [B_{AB}] = \frac{(50)^2}{(2)(2)(4)} + \frac{(122)^2}{(2)(2)(4)} = 1086.500$$

$$\sum_{k=1}^{q} \frac{\left(\sum_{i=1}^{n}\sum_{j=1}^{p}\sum_{z=1}^{w'} Y_{ijklz}\right)^2}{npw'} = [B_{AB'}] = \frac{(40)^2}{(2)(2)(3)} + \frac{(93)^2}{(2)(2)(3)} = 854.083$$

$$\sum_{k=1}^{q} \frac{\left(\sum_{i=1}^{n}\sum_{j=1}^{p}\sum_{z=1}^{w'} Y_{ijklz}\right)^2}{npw'} = [B_{BC'}] = \frac{(37)^2}{(2)(2)(3)} + \frac{(92)^2}{(2)(2)(3)} = 819.417$$

$$\sum_{k=1}^{q} \frac{\left(\sum_{i=1}^{n}\sum_{j=1}^{p}\sum_{z=1}^{w'} Y_{ijklz}\right)^2}{npw'} = [B_{AB''}] = \frac{(39)^2}{(2)(2)(3)} + \frac{(91)^2}{(2)(2)(3)} = 816.833$$

$$\sum_{l=1}^{r} \frac{\left(\sum_{i=1}^{n}\sum_{j=1}^{p}\sum_{z=1}^{w} Y_{ijklz}\right)^2}{npw} = [C_{AC}] = \frac{(72)^2}{(2)(2)(4)} + \frac{(100)^2}{(2)(2)(4)} = 949.000$$

$$\sum_{l=1}^{r} \frac{\left(\sum_{i=1}^{n}\sum_{j=1}^{p}\sum_{z=1}^{w'} Y_{ijklz}\right)^2}{npw'} = [C_{AC'}] = \frac{(50)^2}{(2)(2)(3)} + \frac{(74)^2}{(2)(2)(3)} = 664.667$$

$$\sum_{l=1}^{r} \frac{\left(\sum_{i=1}^{n}\sum_{k=1}^{q}\sum_{z=1}^{w'} Y_{ijklz}\right)^2}{npw'} = [C_{BC'}] = \frac{(55)^2}{(2)(2)(3)} + \frac{(74)^2}{(2)(2)(3)} = 708.417$$

$$\sum_{l=1}^{r} \frac{\left(\sum_{i=1}^{n}\sum_{j=1}^{p}\sum_{z=1}^{w'} Y_{ijklz}\right)^2}{npw'} = [C_{AC''}] = \frac{(55)^2}{(2)(2)(3)} + \frac{(75)^2}{(2)(2)(3)} = 720.833$$

Table 12.7-1 (continued)

$$\sum_{j=1}^{p}\sum_{k=1}^{q}\frac{\left(\sum_{i=1}^{n}\sum_{z=1}^{w'}Y_{ijklz}\right)^2}{nw'} = [AB_{AB'}] = \frac{(24)^2}{(2)(3)} + \frac{(46)^2}{(2)(3)} + \cdots + \frac{(47)^2}{(2)(3)} = 859.500$$

$$\sum_{j=1}^{p}\sum_{k=1}^{q}\frac{\left(\sum_{i=1}^{n}\sum_{z=1}^{w'}Y_{ijklz}\right)^2}{nw'} = [AB_{AB''}] = \frac{(25)^2}{(2)(3)} + \frac{(46)^2}{(2)(3)} + \cdots + \frac{(45)^2}{(2)(3)} = 827.000$$

$$\sum_{j=1}^{p}\sum_{l=1}^{r}\frac{\left(\sum_{i=1}^{n}\sum_{z=1}^{w'}Y_{ijklz}\right)^2}{nw'} = [AC_{AC'}] = \frac{(29)^2}{(2)(3)} + \frac{(35)^2}{(2)(3)} + \cdots + \frac{(39)^2}{(2)(3)} = 671.333$$

$$\sum_{j=1}^{p}\sum_{l=1}^{r}\frac{\left(\sum_{i=1}^{n}\sum_{z=1}^{w'}Y_{ijklz}\right)^2}{nw'} = [AC_{AC''}] = \frac{(34)^2}{(2)(3)} + \frac{(37)^2}{(2)(3)} + \cdots + \frac{(38)^2}{(2)(3)} = 735.000$$

$$\sum_{k=1}^{q}\sum_{l=1}^{r}\frac{\left(\sum_{i=1}^{n}\sum_{z=1}^{w'}Y_{ijklz}\right)^2}{nw'} = [BC_{BC'}] = \frac{(17)^2}{(2)(3)} + \frac{(20)^2}{(2)(3)} + \cdots + \frac{(54)^2}{(2)(3)} = 841.500$$

$$\sum_{k=1}^{q}\sum_{l=1}^{r}\frac{\left(\sum_{i=1}^{n}\sum_{z=1}^{w'}Y_{ijklz}\right)^2}{nw'} = [BC_{BC''}] = \frac{(17)^2}{(2)(3)} + \frac{(22)^2}{(2)(3)} + \cdots + \frac{(53)^2}{(2)(3)} = 837.667$$

$$\sum_{j=1}^{p}\sum_{k=1}^{q}\sum_{l=1}^{r}\frac{\left(\sum_{z=1}^{w'}Y_{ijklz}\right)^2}{w'} = [ABC_{ABC''}] = \frac{(12)^2}{3} + \frac{(13)^2}{3} + \cdots + \frac{(29)^2}{3} = 858.667$$

$$\frac{\left(\sum_{i=1}^{n}\sum_{j=1}^{p}\sum_{k=1}^{q}\sum_{z=1}^{w'}Y_{ijklz}\right)^2}{npqw'} = [Y_{AB'}] = \frac{(133)^2}{(2)(2)(2)(3)} = 737.042$$

$$\frac{\left(\sum_{i=1}^{n}\sum_{j=1}^{p}\sum_{l=1}^{r}\sum_{z=1}^{w'}Y_{ijklz}\right)^2}{nprw'} = [Y_{AC'}] = \frac{(124)^2}{(2)(2)(2)(3)} = 640.667$$

$$\frac{\left(\sum_{i=1}^{n}\sum_{k=1}^{q}\sum_{l=1}^{r}\sum_{z=1}^{w'}Y_{ijklz}\right)^2}{nqrw'} = [Y_{BC'}] = \frac{(129)^2}{(2)(2)(2)(3)} = 693.375$$

$$\frac{\left(\sum_{j=1}^{p}\sum_{k=1}^{q}\sum_{l=1}^{r}\sum_{z=1}^{w'}Y_{ijklz}\right)^2}{pqrw'} = [Y_{ABC''}] = \frac{(130)^2}{(2)(2)(2)(3)} = 704.167$$

Table 12.7-1 (continued)

(iii)   Computational formulas:

$$SSTO = [ABCGS] - [Y] = 235.500$$

$$SSBETWEEN\ BL = [GS] - [Y] = 14.500$$

$$SSG = [G] - [Y] = 5.250$$

$$SSBL(G) = [GS] - [G] = 9.250$$

$$SSWITHIN\ BL = [ABCGS] - [GS] = 221.000$$

$$SSA = [A_{AB}] - [Y] = 3.125$$

$$SSB = [B_{AB}] - [Y] = 162.000$$

$$SSC = [C_{AC}] - [Y] = 24.500$$

$$SSAB' = [AB_{AB'}] - [A_{AB'}] - [B_{AB'}] + [Y_{AB'}] = 3.376$$

$$SSAC' = [AC_{AC'}] - [A_{AC'}] - [C_{AC'}] + [Y_{AC'}] = 6.000$$

$$SSBC' = [BC_{BC'}] - [B_{BC'}] - [C_{BC'}] + [Y_{BC'}] = 7.041$$

$$SSABC' = [ABC_{ABC''}] - [AB_{AB''}] - [AC_{AC''}] - [BC_{BC''}] + [A_{AB''}]$$

$$+ [B_{AB''}] + [C_{AC''}] - [Y_{ABC''}] = 2.666$$

$$SSRES = SSWITHIN\ BL - SSA - SSB - SSC - SSAB' - SSAC' - SSBC'$$

$$- SSABC' = 12.292$$

*The subscript identifies the summary table used in computing the source of variation.

menters who want to use their knowledge of a research area. For example, one could choose to confound the *AB* interaction in groups 0 and 1 and the *ABC* interaction in groups of 2 and 3. This is an example of unbalanced partial confounding, which was described in Section 12.6. If sufficient subjects are available for five groups, an experimenter could choose to confound *AB*, *AC*, and *BC* in groups 0 through 2, respectively, and *ABC* in groups 3 and 4. This design provides as much information on the *AB*, *AC*, and *BC* interactions as the completely confounded design summarized in Table 12.5-2. In addition, it provides three-fifths relative information on the *ABC* interaction.

COMPUTATION OF MEANS

Main-effects means for the data in Table 12.7-1 can be computed from summary tables based on all four groups since the treatments are not confounded with groups. For example,

TABLE 12.7-2   ANOVA Table for Type RBPF-2^3 Design

	Source	SS	df		MS	F		$E(MS)$ Model III (A, B, and C fixed; Blocks random)*
1	Between blocks	14.500	$nw - 1$	$= 7$				
2	Groups	5.250	$w - 1$	$= 3$				
3	Blocks w. groups	9.250	$w(n - 1)$	$= 4$				
4	AB' (between)	3.125	$(p - 1)(q - 1)$	$= 1$				
5	AC' (between)	4.500	$(p - 1)(r - 1)$	$= 1$				
6	BC' (between)	1.125	$(q - 1)(r - 1)$	$= 1$				
7	ABC' (between)	0.500	$(p - 1)(q - 1)(r - 1)$	$= 1$				
8	Within blocks	221.000	$nw(v - 1)$	$= 24$				
9	A	3.125	$p - 1$	$= 1$	3.125	$[^9\!/_{16}]$ 4.32		$\sigma_\epsilon^2 + nqw\,\Sigma\alpha_i^2/(p - 1)$
10	B	162.000	$q - 1$	$= 1$	162.000	$[^{10}\!/_{16}]$ 224.07§		$\sigma_\epsilon^2 + npw\,\Sigma\beta_k^2/(q - 1)$
11	C	24.500	$r - 1$	$= 1$	24.500	$[^{11}\!/_{16}]$ 33.89§		$\sigma_\epsilon^2 + npw\,\Sigma\gamma_l^2/(r - 1)$
12	AB' (within)	3.376	$(p - 1)(q - 1)$	$= 1$	3.376	$[^{12}\!/_{16}]$ 4.67**		$\sigma_\epsilon^2 + nw\,\Sigma\Sigma(\alpha\beta)_{jk}^2/(p - 1)(q - 1)$
13	AC' (within)	6.000	$(p - 1)(r - 1)$	$= 1$	6.000	$[^{13}\!/_{16}]$ 8.30**		$\sigma_\epsilon^2 + nw'\,\Sigma\Sigma(\alpha\gamma)_{jl}^2/(p - 1)(r - 1)$
14	BC' (within)	7.041	$(q - 1)(r - 1)$	$= 1$	7.041	$[^{14}\!/_{16}]$ 9.74§		$\sigma_\epsilon^2 + nw'\,\Sigma\Sigma(\beta\gamma)_{kl}^2/(q - 1)(r - 1)$
15	ABC' (within)	2.666	$(p - 1)(q - 1)(r - 1)$	$= 1$	2.666	$[^{15}\!/_{16}]$ 3.69		$\sigma_\epsilon^2 + w'\,\Sigma\Sigma\Sigma(\alpha\beta\gamma)_{jkl}^2/(p - 1)(q - 1)(r - 1)$
16	SSRES	12.292	$w(n - 1)(w' - 1)$ $+ w'(n - 1)(w' - 1)$	$= 17$	0.723			σ_ϵ^2
17	AG	4.125	$(p - 1)(w - 1)$	$= 3$				
18	BG	1.250	$(q - 1)(w - 1)$	$= 3$				
19	CG	1.750	$(r - 1)(w - 1)$	$= 3$				
20	$(AB) \times 3/4G$	3.250	$(n - 1)(w' - 1)$	$= 2$				
21	$(AC) \times 3/4G$	0.250	$(n - 1)(w' - 1)$	$= 2$				
22	$(BC) \times 3/4G$	0.583	$(n - 1)(w' - 1)$	$= 2$				
23	$(ABC) \times 3/4G$	1.084	$(n - 1)(w' - 1)$	$= 2$				
24	Total	235.500	$nvw - 1$	$= 31$				

*Interaction effects involving groups are assumed to equal zero.

**$p < .05$

§$p < .01$

$$\overline{Y}_{\cdot 1 \cdots} = \frac{\sum\limits_{i=1}^{n} \sum\limits_{k=1}^{q} \sum\limits_{z=1}^{w} Y_{i1klz}}{nqw} = \frac{91}{(2)(2)(4)} = 5.69$$

$$\overline{Y}_{\cdot 2 \cdots} = \frac{\sum\limits_{i=1}^{n} \sum\limits_{k=1}^{q} \sum\limits_{z=1}^{w} Y_{i2kz}}{nqw} = \frac{81}{(2)(2)(4)} = 5.06.$$

If an experimenter is interested in examining simple main-effects means or simple-simple main-effects means, it is necessary to compute them from adjusted cell totals based on the groups in which the interactions are not confounded. An *adjusted cell total* $\Sigma Y'_{\cdot jkl \cdot}$ is given by

$$\Sigma Y'_{\cdot jkl \cdot} = \frac{\Sigma(A)_j + \Sigma(B)_k + \Sigma(C)_l}{4}$$

$$+ \frac{\Sigma(AB)'_{j+k} + \Sigma(AC)'_{j+l} + \Sigma(BC)'_{k+l} + \Sigma(ABC)'_{j+k+l}}{3}$$

$$- \frac{3\left(\sum\limits_{i=1}^{n} \sum\limits_{jkl=1}^{v} \sum\limits_{z=1}^{w} Y_{ijklz} \right)}{4}$$

where

$\Sigma(A)_j$ = the sum of all observations at level a_j

$\Sigma(B)_k$ = the sum of all observations at level b_k

$\Sigma(C)_l$ = the sum of all observations at level c_l

$\Sigma(AB)'_{j+k}$ = the sum of all observations that satisfy the relation $a_j + b_k = j + k \pmod 2$ and are free of group effects (for example, are from groups g_1, g_2, and g_3)

$\Sigma(AC)'_{j+l}$ = the sum of all observations that satisfy the relation $a_j + c_l = j + l \pmod 2$ and are free of group effects

$\Sigma(BC)'_{k+l}$ = the sum of all observations that satisfy the relation $b_k + c_l = k + l \pmod 2$ and are free of group effects

$\Sigma(ABC)'_{j+k+l}$ = the sum of all observations that satisfy the relation $a_j + b_k + c_l = j + k + l \pmod 2$ and are free of group effects

$\sum\limits_{i=1}^{n} \sum\limits_{jkl=1}^{v} \sum\limits_{z=1}^{w} Y_{ijklz}$ = the sum of all observations.

The adjusted cell total for treatment combinations 000, for example, is given by

$$\Sigma Y'_{.000.} = \frac{\Sigma(A)_0 + \Sigma(B)_0 + \Sigma(C)_0}{4}$$

$$+ \frac{\Sigma(AB)'_{0+0} + \Sigma(AC)'_{0+0} + \Sigma(BC)'_{0+0} + \Sigma(ABC)'_{0+0+0}}{3}$$

$$- \frac{3\left(\sum_{i=1}^{n} \sum_{jkl=1}^{v} \sum_{z=1}^{w} Y_{ijklz}\right)}{4}$$

$$= \frac{91 + 50 + 72}{4}$$

$$+ \frac{[(24 + 47) + (29 + 39) + (17 + 54) + (12 + 24 + 9 + 16)]}{3}$$

$$- \frac{3(172)}{4}$$

$$= 14.5833.$$

The adjusted cell mean is given by

$$\overline{Y}'_{.000.} = \frac{\Sigma Y'_{.000.}}{w} = \frac{14.5833}{4} = 3.646.$$

The rest of the adjusted cell totals are given in Table 12.7-3. The unadjusted cell totals are included in the table for purposes of comparison.

Procedures for computing adjusted cell totals for other confounding schemes are described by Federer (1955; 173–176, 246).

TABLE 12.7-3   Adjusted and Unadjusted Cell Totals for the Data in Table 12.7-1

		b_0 c_0	b_0 c_1	b_1 c_0	b_1 c_1
a_0	Adjusted	$w = 4$ 14.5833	15.9167	27.9167	32.5833
	Unadjusted	15	16	28	32
a_1	Adjusted	7.7500	11.7500	21.7500	39.7500
	Unadjusted	7	12	22	40

12.8 COMPUTATIONAL PROCEDURES FOR TYPE RBCF-3^2 AND RBPF-3^2 DESIGNS

ASSIGNMENT OF TREATMENT COMBINATIONS TO GROUPS

In this section, we describe several confounding schemes for reducing the block size of a two-treatment factorial design where both treatments have three levels. The AB interaction can be partitioned into two orthogonal components as follows.

SS	df
AB	$(3 - 1)(3 - 1) = 4$
(AB)	2
(AB^2)	2

The two orthogonal components (AB) and (AB^2) have no special significance apart from their use in partitioning the interaction. Treatment combinations that satisfy the relations

$$a_j + b_k = 0 \pmod 3 \qquad (AB)_0$$
$$a_j + b_k = 1 \pmod 3 \qquad (AB)_1 \quad \text{group 0}$$
$$a_j + b_k = 2 \pmod 3 \qquad (AB)_2$$

constitute the (AB) component of the interaction. These treatment combinations are

00	12	21	$(AB)_0$
01	10	22	$(AB)_1$
02	11	20	$(AB)_2$.

The (AB^2) interaction component is composed of the treatment combinations that satisfy the relations

$$a_j + 2b_k = 0 \pmod 3 \qquad (AB^2)_0$$
$$a_j + 2b_k = 1 \pmod 3 \qquad (AB^2)_1 \quad \text{group 1}$$
$$a_j + 2b_k = 2 \pmod 3 \qquad (AB^2)_2$$

where the powers of A and B are used as the coefficients of a_j and b_k. These treatment combinations are

00	11	22	$(AB^2)_0$
02	10	21	$(AB^2)_1$
01	12	20	$(AB^2)_2$.

By convention, the power of A is always equal to one. This is necessary in order to uniquely define the interaction components since (A^2B) and (AB^2) define the same treatment combinations. That is,

$$2a_j + b_k = z \pmod 3 \qquad (z = 0, 1, 2)$$

defines the same nine combinations as

$$a_j + 2b_k = z \pmod 3 \qquad (z = 0, 1, 2).$$

A type RBCF-3^2 design is constructed by confounding either the (AB) or (AB^2) interaction component with groups. A block diagram in which the (AB) component is confounded with groups is shown in Figure 12.8-1(a). This design provides within-blocks information on the (AB^2) component of the AB interaction. An alternative design strategy is to confound (AB) in one group of blocks and (AB^2) in a second group of blocks as shown in Figure 12.8-1(b). This confounding scheme provides within-blocks information on both components of the AB interaction. The number of blocks in each group must equal three or a multiple of three.

FIGURE 12.8-1   Block diagrams of type RBCF-3^2 and RBPF-3^2 designs. The numbers in the cells denote the levels of treatments A and B. Design (a) confounds the (AB) component with groups; design (b) confounds the (AB) component with blocks in group 0 and the (AB^2) component with blocks in group 1.

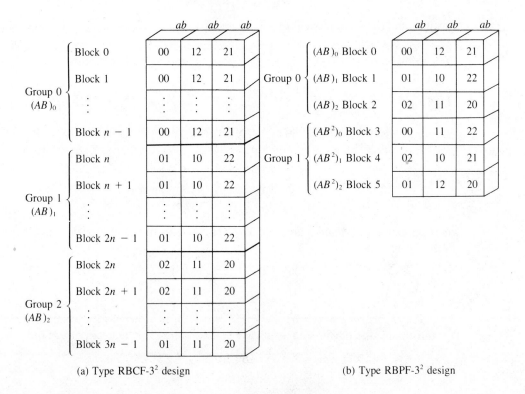

(a) Type RBCF-3^2 design   (b) Type RBPF-3^2 design

COMPUTATIONAL EXAMPLE FOR TYPE RBCF-3² DESIGN

The layout and computational procedures for a type RBCF-3² design are shown in Table 12.8-1; the analysis is summarized in Table 12.8-2. In this design the (AB) component is confounded with groups. Thus, within-blocks information is available for the $(AB)^2$ component. The experimental design model equation is

$$Y_{ijkz} = \mu + \zeta_{z(\text{between})} + \pi_{i(z)} + \alpha_j + \beta_k + (\alpha\beta)_{jk(\text{within})} + (\alpha\beta\pi)_{jki(z)} + \epsilon_{ijkz}$$
$$(i = 1, \ldots, n; j = 1, \ldots, p; k = 1, \ldots, q; z = 1, \ldots, w).$$

The assumptions underlying this design are similar to those for a type RBCF-2² design described in Section 12.4 and will not be repeated here.

COMPUTATION OF MEANS

In completely confounded designs, it is customary to compute main-effects and simple-effects contrasts from unadjusted cell totals. However, if the confounded effects, (AB) in our example, are equal to zero, an experimenter can adjust the sample cell totals to zero (AB) effects by the formula

$$\Sigma Y'_{\cdot jk\cdot} = \frac{\Sigma(A)_j + \Sigma(B)_k + \Sigma(AB^2)'_{j+2k}}{3} + \frac{\sum_{i=1}^{n}\sum_{jk=1}^{v}\sum_{z=1}^{w} Y_{ijkz}}{9} - \frac{\sum_{i=1}^{n}\sum_{jk=1}^{v}\sum_{z=1}^{w} Y_{ijkz}}{3}.$$

The meaning of the terms $\Sigma(A)_j$, $\Sigma(B)_k$, and so on should be apparent from the definitions given in Section 12.7. The adjustments to the cell totals affect the values of simple main-effects contrasts but not those for main-effects contrasts.

COMPUTATIONAL EXAMPLE FOR TYPE RBPF-3² DESIGN

The layout and computational procedures for a type RBPF-3² design where (AB) and (AB^2) are confounded with the blocks in group 0 and 1, respectively, are shown in Table 12.8-3. The analysis is summarized in Table 12.8-4. The summary in Table 12.8-4 should be compared with that in Table 12.8-2 where the (AB) component is completely confounded with groups. The advantage of the partially confounded design is that within-blocks information is available for both components of the AB interaction.

DIRECT COMPUTATION OF INTERACTION COMPONENTS

The computational procedures for $SSAB'$(between) and $SSAB'$(within) represent relatively simple analysis procedures, but they throw little light on the nature of these terms. Actually both sums of squares are pooled terms. For example,

$$SSAB'(\text{between}) = SS(AB) \text{ from } g_0 + SS(AB^2) \text{ from } g_1$$
$$SSAB'(\text{within}) = SS(AB^2) \text{ from } g_0 + SS(AB) \text{ from } g_1.$$

TABLE 12.8-1   Computational Procedures for a Type RBCF-3^2 Design

(i)   Data and notation [Y_{ijkz} denotes a score for the experimental unit in block i, treatment combination jk, and group z; $i = 1, \ldots, n$ blocks (s_i); $j = 1, \ldots, p$ levels of treatment A (a_j); $k = 1, \ldots,$ q levels of treatment B (b_k); $z = 1, \ldots, w$ levels of groups (g_z); $jk = 1, \ldots, v$ combinations of a_j and b_k within a block]:

ABGS Summary Table
Entry is Y_{ijkz}

		ab	ab	ab	$\sum\limits_{jk=1}^{v} Y_{ijkz}$	$\sum\limits_{i=1}^{n}\sum\limits_{jk=1}^{v} Y_{ijkz}$
		00	12	21		
g_0	s_0	3	7	4	14	25
$(AB)_0$	s_1	2	6	3	11	
		01	10	22		
g_1	s_2	4	3	6	13	21
$(AB)_1$	s_3	2	1	5	8	
		02	11	20		
g_2	s_4	7	3	3	13	24
$(AB)_2$	s_5	5	4	2	11	

AB Summary Table

Entry is $\sum\limits_{i=1}^{n} Y_{ijkz}$

	b_0	b_1	b_2	$\sum\limits_{i=1}^{n}\sum\limits_{k=1}^{q} Y_{ijkz}$
	$n = 2$			
a_0	5	6	12	23
a_1	4	7	13	24
a_2	5	7	11	23
$\sum\limits_{i=1}^{n}\sum\limits_{j=1}^{p} Y_{ijkz} = $ 14	20	36		

(ii)   Computational symbols:

$$\sum_{i=1}^{n}\sum_{jk=1}^{v}\sum_{z=1}^{w} Y_{ijkz} = 3 + 2 + \cdots + 2 = 70.000$$

$$\frac{\left(\sum\limits_{i=1}^{n}\sum\limits_{jk=1}^{v}\sum\limits_{z=1}^{w} Y_{ijkz}\right)^2}{nvw} = [Y] = \frac{(70.000)^2}{(2)(3)(3)} = 272.222$$

Table 12.8-1 (continued)

$$\sum_{i=1}^{n} \sum_{jk=1}^{v} \sum_{z=1}^{w} Y_{ijkz}^2 = [ABGS] = (3)^2 + (2)^2 + \cdots + (2)^2 = 326.000$$

$$\sum_{i=1}^{n} \sum_{z=1}^{w} \frac{\left(\sum_{jk=1}^{v} Y_{ijkz} \right)^2}{v} = [GS] = \frac{(14)^2}{3} + \frac{(11)^2}{3} + \cdots + \frac{(11)^2}{3} = 280.000$$

$$\sum_{z=1}^{w} \frac{\left(\sum_{i=1}^{n} \sum_{jk=1}^{v} Y_{ijkz} \right)^2}{nv} = [G] = \frac{(25)^2}{(2)(3)} + \frac{(21)^2}{(2)(3)} + \frac{(24)^2}{(2)(3)} = 273.667$$

$$\sum_{j=1}^{p} \frac{\left(\sum_{i=1}^{n} \sum_{k=1}^{q} Y_{ijkz} \right)^2}{nq} = [A] = \frac{(23)^2}{(2)(3)} + \frac{(24)^2}{(2)(3)} + \frac{(23)^2}{(2)(3)} = 272.333$$

$$\sum_{k=1}^{q} \frac{\left(\sum_{i=1}^{n} \sum_{j=1}^{p} Y_{ijkz} \right)^2}{np} = [B] = \frac{(14)^2}{(2)(3)} + \frac{(20)^2}{(2)(3)} + \frac{(36)^2}{(2)(3)} = 315.333$$

$$\sum_{j=1}^{p} \sum_{k=1}^{q} \frac{\left(\sum_{i=1}^{n} Y_{ijkz} \right)^2}{n} = [AB] = \frac{(5)^2}{2} + \frac{(6)^2}{2} + \cdots + \frac{(11)^2}{2} = 317.000$$

(iii) Computational formulas:

$$SSTO = [ABGS] - [Y] = 53.778$$
$$SSBETWEEN\ BL = [GS] - [Y] = 7.778$$
$$SSG = [G] - [Y] = 1.445$$
$$SSBL(G) = [GS] - [G] = 6.333$$
$$SSWITHIN\ BL = [ABGS] - [GS] = 46.000$$
$$SSA = [A] - [Y] = 0.111$$
$$SSB = [B] - [Y] = 43.111$$
$$SS(AB^2) = [AB] - [A] - [B] - [G] + 2[Y] = 0.111$$
$$SSAB \times BL(G) = [ABGS] - [AB] - [GS] + [G] = 2.667$$

The between-blocks interaction components are given by

$$SS(AB)\text{between} = \frac{\Sigma(AB)_0^2 + \Sigma(AB)_1^2 + \Sigma(AB)_2^2}{v} - \frac{\left(\sum_{i=1}^{n} \sum_{jk=1}^{v} Y_{ijk0} \right)^2}{nv}$$

$$SS(AB)^2\text{between} = \frac{\Sigma(AB^2)_0^2 + \Sigma(AB^2)_1^2 + \Sigma(AB^2)_2^2}{v} - \frac{\left(\sum_{i=1}^{n} \sum_{jk=1}^{v} Y_{ijk1} \right)^2}{nv}$$

TABLE 12.8-2 ANOVA Table for Type RBCF-3^2 Design

	Source	SS	df	MS	F	E(MS) (A and B fixed; Blocks random)
1	Between blocks	7.778	$nw - 1 = 5$			
2	Groups (AB)	1.445	$w - 1 = 2$	0.722	[⅔] 0.34	$\sigma_\epsilon^2 + v\sigma_\pi^2$ $+ nv\Sigma\zeta_z^2/(w - 1)$
3	Blocks w. groups	6.333	$w(n - 1) = 3$	2.111		$\sigma_\epsilon^2 + v\sigma_\pi^2$
4	Within groups	46.000	$nw(v - 1) = 12$			
5	A	0.111	$p - 1 = 2$	0.056	[⅝] 0.13	$\sigma_\epsilon^2 + \sigma_{\alpha\beta\pi}^2$ $+ nw\Sigma\alpha_j^2/(p - 1)$
6	B	43.111	$q - 1 = 2$	21.556	[⁹⁄₈] 48.55*	$\sigma_\epsilon^2 + \sigma_{\alpha\beta\pi}^2$ $+ nw\Sigma\beta_k^2/(q - 1)$
7	(AB^2) within	0.111	$(p - 1)(q - 1)$ $- (w - 1) = 2$	0.056	[⅞] 0.13	$\sigma_\epsilon^2 + \sigma_{\alpha\beta\pi}^2$ $+ n\Sigma\Sigma(\alpha\beta)_{jk}^2/$ $(p - 1)(q - 1)$
8	$AB \times BL(G)$	2.667	$w(n - 1)(v - 1) = 6$	0.444		$\sigma_\epsilon^2 + \sigma_{\alpha\beta\pi}^2$
9	$AB \times BL(g_0)$	0	$(n - 1)(v - 1) = 2$			
10	$AB \times BL(g_1)$	0.333	$(n - 1)(v - 1) = 2$			
11	$AB \times BL(g_2)$	2.333	$(n - 1)(v - 1) = 2$			
12	Total	53.778	$nvw - 1 = 17$			

$*p < .01$

where $\Sigma(AB)_z$ and $\Sigma(AB^2)_z$ are the sums of the observations in the treatment combinations that satisfy the relations

$$a_j + b_k = z(\mathrm{mod}\ 3) \quad \text{and} \quad a_j + 2b_k = z(\mathrm{mod}\ 3)$$

($z = 0, 1, 2$), respectively, and are confounded with blocks within a group. These treatment combinations and associated sums are as follows.

Group 0 Treatment Combinations			Sum	Group 1 Treatment Combinations			Sum
$(AB)_0 = 00$	12	21	$3 + 7 + 4 = 14$	$(AB^2)_0 = 00$	11	22	$2 + 4 + 5 = 11$
$(AB)_1 = 01$	10	22	$4 + 3 + 6 = 13$	$(AB^2)_1 = 02$	10	21	$5 + 1 + 3 = 9$
$(AB)_2 = 02$	11	20	$7 + 3 + 3 = 13$	$(AB^2)_2 = 01$	12	20	$2 + 6 + 2 = 10$

Computation of the between-blocks components of the AB interaction using the above sums gives

$$SS(AB)\text{between} = \frac{(14)^2 + (13)^2 + (13)^2}{3} - \frac{(40)^2}{(3)(3)} = 0.222$$

$$SS(AB^2)\text{between} = \frac{(11)^2 + (9)^2 + (10)^2}{3} - \frac{(30)^2}{(3)(3)} = 0.667.$$

The sum of these two components is equal to $SSAB'$(between). The two interaction components are completely confounded with blocks within groups. This can be easily verified by comparing the sources of variation that comprise $SS(AB)$between and $SS(AB^2)$between with those for $SSBL(G)$.

The within-blocks components of the AB interaction are given by

$$SS(AB^2)\text{within} = \frac{\Sigma(AB^2)_0^2 + \Sigma(AB^2)_1^2 + \Sigma(AB^2)_2^2}{v} - \frac{\left(\sum_{i=1}^{n}\sum_{jk=1}^{v} Y_{ijk0}\right)^2}{nv}$$

$$SS(AB)\text{within} = \frac{\Sigma(AB)_0^2 + \Sigma(AB)_1^2 + \Sigma(AB)_2^2}{v} - \frac{\left(\sum_{i=1}^{n}\sum_{jk=1}^{v} Y_{ijkl}\right)^2}{nv}$$

where $\Sigma(AB^2)_z$ and $\Sigma(AB)_z$ are not confounded with blocks within a group. The treatment combinations and associated sums are as follows.

| | Group 0 | | | | Group 1 | | |
Treatment Combination		Sum		Treatment Combination		Sum	
$(AB^2)_0 = 00$	11	22	$3 + 3 + 6 = 12$	$(AB)_0 = 00$	12	21	$2 + 6 + 3 = 11$
$(AB^2)_1 = 02$	10	21	$7 + 3 + 4 = 14$	$(AB)_1 = 01$	10	22	$2 + 1 + 5 = 8$
$(AB^2)_2 = 01$	12	20	$4 + 7 + 3 = 14$	$(AB)_2 = 02$	11	20	$5 + 4 + 2 = 11$

Computation of the two within-blocks components of the AB interaction gives

$$SS(AB^2)\text{within} = \frac{(12)^2 + (14)^2 + (14)^2}{3} - \frac{(40)^2}{(3)(3)} = 0.889$$

$$SS(AB)\text{within} = \frac{(11)^2 + (8)^2 + (11)^2}{3} - \frac{(30)^2}{(3)(3)} = 2.000.$$

It is customary to pool the two components as was done in Table 12.8-4 in carrying out a test of significance. If it is desirable to test either $SS(AB^2)$within or $SS(AB)$within separately, the procedure just described rather than the short-cut computational procedure in Table 12.8-3 must be used.

TABLE 12.8-3   Computational Procedures for a Type RBPF-3^2 Design

---

(i)   Data and notation [Y_{ijkz} denotes a score for the experimental unit in block i, treatment combination jk, and group z; $i = 1, \ldots, n$ blocks (s_i); $j = 1, \ldots, p$ levels of treatment A (a_j); $k = 1, \ldots, q$ levels of treatment B (b_k); $z = 1, \ldots, w$ levels of groups (g_z); $jk = 1, \ldots, v$ combinations of a_j and b_k within a block]:

ABGS Summary Table
Entry is Y_{ijkz}

			ab	ab	ab	$\sum\limits_{jk=1}^{v} Y_{ijkz}$	$\sum\limits_{i=1}^{n}\sum\limits_{jk=1}^{v} Y_{ijkz}$
	$(AB)_0$	s_0	00 3	12 7	21 4	14	
g_0	$(AB)_1$	s_1	01 4	10 3	22 6	13	40
	$(AB)_2$	s_2	02 7	11 3	20 3	13	
	$(AB^2)_0$	s_3	00 2	11 4	22 5	11	
g_1	$(AB^2)_1$	s_4	02 5	10 1	21 3	9	30
	$(AB^2)_2$	s_5	01 2	12 6	20 2	10	

AB Summary Table

Entry is $\sum\limits_{z=1}^{w} Y_{ijkz}$ from g_0 and g_1

	b_0	b_1	b_2	$\sum\limits_{k=1}^{q}\sum\limits_{z=1}^{w} Y_{ijkz}$
	$w = 2$			
a_0	5	6	12	23
a_1	4	7	13	24
a_2	5	7	11	23

$\sum\limits_{j=1}^{p}\sum\limits_{z=1}^{w} Y_{ijkz} =$ 14   20   36

AG Summary Table

Entry is $\sum\limits_{i=1}^{n} Y_{ijkz}$

	g_0	g_1
	$n = 3$	
a_0	14	9
a_1	13	11
a_2	13	10

BG Summary Table

Entry is $\sum\limits_{i=1}^{n} Y_{ijkz}$

	g_0	g_1
	$n = 3$	
b_0	9	5
b_1	11	9
b_2	20	16

Table 12.8-3 (continued)

(ii) Computational symbols:

$$\sum_{i=1}^{n} \sum_{jk=1}^{v} \sum_{z=1}^{w} Y_{ijkz} = 3 + 4 + \cdots + 2 = 70.000$$

$$\frac{\left(\sum_{i=1}^{n} \sum_{jk=1}^{v} \sum_{z=1}^{w} Y_{ijkz}\right)^2}{nvw} = [Y] = \frac{(70.000)^2}{(3)(3)(2)} = 272.222$$

$$\sum_{i=1}^{n} \sum_{jk=1}^{v} \sum_{z=1}^{w} Y_{ijkz}^2 = [ABGS] = (3)^2 + (4)^2 + \cdots + (2)^2 = 326.000$$

$$\sum_{i=1}^{n} \sum_{z=1}^{w} \frac{\left(\sum_{jk=1}^{v} Y_{ijkz}\right)^2}{v} = [GS] = \frac{(14)^2}{3} + \frac{(13)^2}{3} + \cdots + \frac{(10)^2}{3} = 278.667$$

$$\sum_{z=1}^{w} \frac{\left(\sum_{i=1}^{n} \sum_{jk=1}^{v} Y_{ijkz}\right)^2}{nv} = [G] = \frac{(40)^2}{(3)(3)} + \frac{(30)^2}{(3)(3)} = 277.778$$

$$\sum_{j=1}^{p} \frac{\left(\sum_{k=1}^{q} \sum_{z=1}^{w} Y_{ijkz}\right)^2}{qw} = [A] = \frac{(23)^2}{(3)(2)} + \frac{(24)^2}{(3)(2)} + \frac{(23)^2}{(3)(2)} = 272.333$$

$$\sum_{k=1}^{q} \frac{\left(\sum_{j=1}^{p} \sum_{z=1}^{w} Y_{ijkz}\right)^2}{qw} = [B] = \frac{(14)^2}{(3)(2)} + \frac{(20)^2}{(3)(2)} + \frac{(36)^2}{(3)(2)} = 315.333$$

$$\sum_{j=1}^{p} \sum_{z=1}^{w} \frac{\left(\sum_{i=1}^{n} Y_{ijkz}\right)^2}{n} = [AG] = \frac{(14)^2}{3} + \frac{(9)^2}{3} + \cdots + \frac{(10)^2}{3} = 278.667$$

$$\sum_{k=1}^{q} \sum_{z=1}^{w} \frac{\left(\sum_{i=1}^{n} Y_{ijkz}\right)^2}{n} = [BG] = \frac{(9)^2}{3} + \frac{(5)^2}{3} + \cdots + \frac{(16)^2}{3} = 321.333$$

(iii) Computational formulas:

$$SSTO = [ABGS] - [Y] = 53.778$$
$$SSBETWEEN\ BL = [GS] - [Y] = 6.445$$
$$SSG = [G] - [Y] = 5.556$$
$$SSBL(G) = [GS] - [G] = 0.889$$
$$SSWITHIN\ BL = [ABGS] - [GS] = 47.333$$

Table 12.8-3 (continued)

$$SSA = [A] - [Y] = 0.111$$

$$SSB = [B] - [Y] = 43.111$$

$$SSAB' = [ABGS] - [AG] - [BG] - [GS] + 2[G] = 2.889$$

$$SSAG = [AG] - [A] - [G] + [Y] = 0.778$$

$$SSBG = [BG] - [B] - [G] + [Y] = 0.444$$

$$SSRES = SSAG + SSBG = 1.222$$

TABLE 12.8-4   ANOVA Table for Type RBPF-3^2 Design

	Source	SS	df	MS	F	$E(MS)$ (A and B fixed; Blocks random)*
1	Between blocks	6.445	$nw - 1 = 5$			
2	Groups	5.556	$w - 1 = 1$			
3	Blocks w. groups [AB'(between)]	0.889	$w(n - 1) = 4$			
4	Within blocks	47.333	$nw(v - 1) = 12$			
5	A	0.111	$p - 1 = 2$	0.056	[⅝] 0.18	$\sigma_\epsilon^2 + qw\Sigma\alpha_j^2/(p - 1)$
6	B	43.111	$q - 1 = 2$	21.556	[⅞] 70.44**	$\sigma_\epsilon^2 + pw\Sigma\beta_k^2/(q - 1)$
7	AB'(within)	2.889	$(p - 1)(q - 1) = 4$	0.722	[⅞] 2.36	$\sigma_\epsilon^2 + (\frac{1}{2})w\Sigma\Sigma(\alpha\beta)_{jk}^2/(p - 1)(q - 1)$
8	Residual	1.222	4	0.306		σ_ϵ^2
9	AG	0.778	$(p - 1)(w - 1) = 2$			
10	BG	0.444	$(q - 1)(w - 1) = 2$			
11	Total	53.778	$nvw - 1 = 17$			

*Interaction effects involving blocks are assumed to equal zero.

**$p < .01$

COMPUTATION OF MEANS

Main-effects means for the data in Table 12.8-3 can be computed from the AB summary table. Simple main-effects means should be computed from adjusted cell totals based on the group in which the interaction component is not confounded. An adjusted cell total is given by

$$\Sigma Y'_{\cdot jk\cdot} = \frac{\Sigma(A)_j + \Sigma(B)_k}{3} + \frac{\Sigma(AB)'_{j+k} + \Sigma(AB^2)'_{j+2k}}{3/2} - \frac{\sum\limits_{i=1}^{n}\sum\limits_{jk=1}^{v}\sum\limits_{z=1}^{w} Y_{ijkz}}{3}$$

where
$\Sigma(A)_j$ = the sum of all observations at level a_j

$\Sigma(B)_k$ = the sum of all observations at level b_k

$\Sigma(AB)'_{j+k}$ = the sum of all observations that satisfy the relation $a_j + b_k = j + k \pmod 3$ and are free of block effects

$\Sigma(AB^2)'_{j+2k}$ = the sum of all observations that satisfy the relation $a_j + 2b_k = j + 2k \pmod 3$ and are free of block effects

$$\sum\limits_{i=1}^{n}\sum\limits_{jk=1}^{v}\sum\limits_{z=1}^{w} Y_{ijkz} = \text{the sum of all observations.}$$

The adjusted cell total for treatment combination 00, for example, is given by

$$\Sigma Y'_{\cdot 00\cdot} = \frac{\Sigma(A)_0 + \Sigma(B)_0}{3} + \frac{\Sigma(AB)'_{0+0} + \Sigma(AB^2)'_{0+(2)(0)}}{3/2} - \frac{\sum\limits_{i=1}^{n}\sum\limits_{jk=1}^{v}\sum\limits_{z=1}^{w} Y_{ijkz}}{3}$$

$$= \frac{23 + 14}{3} + \frac{[(2 + 6 + 3) + (3 + 3 + 6)]}{3/2} - \frac{70}{3}$$

$$= 4.333.$$

The adjusted cell mean is given by

$$\overline{Y}'_{\cdot 00\cdot} = \frac{\Sigma Y'_{\cdot 00\cdot}}{w} = \frac{4.333}{3} = 1.444.$$

The rest of the adjusted cell totals are given in Table 12.8-5. The unadjusted cell totals are included in the table for purposes of comparison.

TABLE 12.8-5   Adjusted and Unadjusted Cell Totals for the Data in Table 12.8-3

	a_0 b_0	a_0 b_1	a_0 b_2	a_1 b_0	a_1 b_1	a_1 b_2	a_2 b_0	a_2 b_1	a_2 b_2
Adjusted	4.333	5.667	13.000	4.000	6.667	13.333	5.667	7.667	9.667
Unadjusted	5	6	12	4	7	13	5	7	11

ALTERNATIVE LAYOUT FOR TYPE RBPF-3^2 DESIGN

The design in Table 12.8-3 provides only four degrees of freedom for testing the within-blocks null hypotheses. The degrees of freedom for experimental error can be increased by assigning $n > 1$ blocks to each $(AB)_0, (AB)_1, \ldots, (AB^2)_2$. This design is shown in Figure 12.8-2. The computational formulas and degrees of freedom for this design are given in Table 12.8-6. The analysis follows the pattern illustrated in Table 12.8-4 where one block is assigned to each $(AB)_0, (AB)_1, \ldots, (AB^2)_2$.

TABLE 12.8-6 Computational Formulas for Type RBPF-3^2 Design with $n > 1$

	Source	Formula	df	F	$E(MS)$ (A and B fixed; Blocks random)*
1	Between blocks	$[EGS]** - [Y]$	$nuw - 1$		
2	Groups	$[G] - [Y]$	$w - 1$		
3	AB'(between)	$[EG] - [G]$	$w(u - 1)$		
4	Blocks w. AB' (between)	$[EGS] - [EG]$	$uw(n - 1)$		
5	Within blocks	$[ABEGS] - [EGS]$	$nuw(v - 1)$		
6	A	$[A] - [Y]$	$p - 1$	[6/9]	$\sigma_\epsilon^2 + nqw\Sigma\alpha_j^2/(p - 1)$
7	B	$[B] - [Y]$	$q - 1$	[7/9]	$\sigma_\epsilon^2 + npw\Sigma\beta_k^2/(q - 1)$
8	AB'(within)	$[ABEG] - [AG] - [BG]$ $- [EG] + 2[G]$	$(p - 1)(q - 1)$	[8/9]	$\sigma_\epsilon^2 + (1/2)nw\Sigma\Sigma(\alpha\beta)_{jk}^2/$ $(p - 1)(q - 1)$
9	Residual	$SSWITHIN\ BL - SSA$ $- SSB - SSAB'$(within)			σ_ϵ^2
10	AG	$[AG] - [A] - [G] + [Y]$	$(p - 1)(w - 1)$		
11	BG	$[BG] - [B] - [G] + [Y]$	$(q - 1)(w - 1)$		
12	$AB \times BL(AB)_0$		$(n - 1)(v - 1)$		
13	$AB \times BL(AB)_1$		$(n - 1)(v - 1)$		
14	$AB \times BL(AB)_2$	$[ABEGS] - [ABEG]$ $- [EGS] + [EG]$	$(n - 1)(v - 1)$		
15	$AB \times BL(AB^2)_0$		$(n - 1)(v - 1)$		
16	$AB \times BL(AB^2)_1$		$(n - 1)(v - 1)$		
17	$AB \times BL(AB^2)_2$		$(n - 1)(v - 1)$		
18	Total	$[ABEGS] - [Y]$	$nuvw - 1$		

*Interaction effects involving blocks are assumed to equal zero.

**E denotes the levels of $(AB)_0, (AB)_1, \ldots, (AB^2)_2$; $o = 1, \ldots, u$ levels of E; the other terms are defined in Table 12.8-4.

FIGURE 12.8-2 Block diagram of type RBPF-3^2 design with $n > 1$ blocks assigned to each $(AB)_0$, $(AB)_1$, . . . , $(AB^2)_2$.

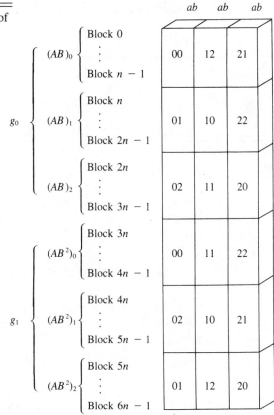

12.9 ANALYSIS PROCEDURES FOR HIGHER-ORDER CONFOUNDED DESIGNS

So far in this chapter we have described 2×2, $2 \times 2 \times 2$, and 3×3 experiments. The principles discussed in connection with these designs can be readily extended to any design of the form p^k, where p is a prime number. Experiments involving mixed primes, such as $3 \times 2 \times 2$ and $3 \times 3 \times 2$, are more difficult to lay out and analyze. A computational example for a $3 \times 2 \times 2$ design is given in Section 12.10. The purpose of this section is to extend the principles described earlier in connection with 2×2, $2 \times 2 \times 2$, and 3×3 designs to other unmixed designs. In the process we will see how to achieve further reductions in block size by confounding two or more interactions with groups of blocks.

TYPE RBCF-2^4 DESIGN WITH BLOCKS OF SIZE EIGHT

A type RBCF-2^4 design in which the $ABCD$ interaction is confounded with groups can be constructed using the relations

$$a_j + b_k + c_l + d_m = z(\text{mod } 2) \qquad (z = 0, 1).$$

Treatment combinations assigned to group 0 or $(ABCD)_0$ and group 1 or $(ABCD)_1$ are

group 0 $(ABCD)_0$ = 0000, 0011, 0101, 0110, 1001, 1010, 1100, 1111
group 1 $(ABCD)_1$ = 0001, 0010, 0100, 0111, 1000, 1011, 1101, 1110.

A complete block design, such as a type RBF-2222 design, requires blocks of size sixteen. By confounding the $ABCD$ interaction with groups of blocks, the block size can be reduced to eight. An experimenter can choose to confound an interaction other than $ABCD$ with groups if this seems desirable. Generally the highest-order interaction is confounded because it is of less interest and is less likely to be significant.

TYPE RBCF-2^4 DESIGN WITH BLOCKS OF SIZE FOUR

The block size of a type RBCF-2^4 design can be further reduced by using two defining relations instead of one to assign treatment combinations to blocks. For example, the two defining relations could be

$$a_j + b_k = z(\text{mod } 2) \qquad (z = 0, 1)$$
$$a_j + c_l + d_m = z(\text{mod } 2) \qquad (z = 0, 1).$$

Eight treatment combinations satisfy each of these sets of relations. For example,

$(ACD)_0$ = 0000, 0011, 1101, 1110, 0100, 0111, 1001, 1010
$(AB)_0$ = 0000, 0011, 1101, 1110, 0001, 0010, 1100, 1111.

If we examine these combinations closely we find that four of them satisfy both $(ACD)_0$ and $(AB)_0$. These combinations are

$$(ACD)_0 \, (AB)_0 = 0000, \quad 0011, \quad 1101, \quad 1110.$$

A different set of four treatment combinations satisfies the relations $(ACD)_0$ and $(AB)_1$, and so on. Thus we can assign four treatment combinations to blocks so that they satisfy, simultaneously, two defining relations. This procedure produces the type RBCF-2^4 design with blocks of size four shown in Table 12.9-1.

When two relations are simultaneously confounded with groups of blocks, their *generalized interaction(s)* or *treatment(s)* is also confounded. The generalized interaction(s) for any two defining relations symbolized by (X) and (Y) is given by the product $(X)(Y)^{m-z}$, reduced modulo m, where m is the modulus and z assumes values $m - 1$, $m - 2$, $\cdots$, $m - (m - 1)$. For treatments with two levels, $m - z = 2 - 1 = 1$ and $(X)(Y)^{2-1} = (X)(Y)$. If (ACD) is substituted for (X) and (AB) for (Y), the generalized interaction is given by

$$(ACD)(AB) = (A^2BCD) \text{ reduced modulo } 2 = (BCD).$$

TABLE 12.9-1   Layout of Type RBCF-2^4 Design in Blocks of Size Four

		abcd	*abcd*	*abcd*	*abcd*
		0000	0011	1101	1110
	s_0				
g_0	s_1				
$(ACD)_0 (AB)_0$	$\vdots$				
	s_{n-1}				
		0100	0111	1001	1010
	s_n				
g_1	s_{n+1}				
$(ACD)_0 (AB)_1$	$\vdots$				
	s_{2n-1}				
		0001	0010	1100	1111
	s_{2n}				
g_2	s_{2n+1}				
$(ACD)_1 (AB)_0$	$\vdots$				
	s_{3n-1}				
		0101	0110	1000	1011
	s_{3n}				
g_3	s_{3n+1}				
$(ACD)_1 (AB)_1$	$\vdots$				
	s_{4n-1}				

Hence, if (ACD) and (AB) are simultaneously confounded with groups of blocks, another interaction, (BCD), is also confounded with groups of blocks. Care must be used in choosing the defining contrasts so that no main effects are confounded with groups of blocks. Such confounding would have occurred if we had chosen $(ABCD)$ and (ACD) as the defining contrasts since then treatment B would have been the generalized treatment,

$$(ABCD)(ACD) = (A^2BC^2D^2) \text{ reduced modulo } 2 = B.$$

A summary of the computational formulas for a type RBCF-2^4 design with (ACD) and (AB) as the defining contrasts is given in Table 12.9-2.

TABLE 12.9-2 Computational Formulas for Type RBCF-2^4 Design in Blocks of Size Four

	Source	Formula	df	F	$E(MS)$ (A, B, C, and D fixed; Blocks random)
1	Between blocks	$[GS] - [Y]$	$nw - 1$		
2	AB, ACD, BCD (Groups)	$[G] - [Y]$	$w - 1$	$[2/3]$	
3	Blocks w. G	$[GS] - [G]$	$w(n - 1)$		
4	Within blocks	$[ABCDGS] - [GS]$	$nw(v - 1)$		
5	A	$[A] - [Y]$	$p - 1$	$[5/17]$	$\sigma_\epsilon^2 + \sigma_{\alpha\beta\gamma\delta\pi}^2 + 2nw\Sigma\alpha_j^2 / (p - 1)$
6	B	$[B] - [Y]$	$q - 1$	$[6/17]$	$\sigma_\epsilon^2 + \sigma_{\alpha\beta\gamma\delta\pi}^2 + 2nw\Sigma\beta_k^2 / (q - 1)$
7	C	$[C] - [Y]$	$r - 1$	$[7/17]$	$\sigma_\epsilon^2 + \sigma_{\alpha\beta\gamma\delta\pi}^2 + 2nw\Sigma\gamma_l^2 / (r - 1)$
8	D	$[D] - [Y]$	$l - 1$	$[8/17]$	$\sigma_\epsilon^2 + \sigma_{\alpha\beta\gamma\delta\pi}^2 + 2nw\Sigma\delta_m^2 / (t - 1)$
9	AC	$[AC] - [A] - [C] + [Y]$	$(p - 1)(r - 1)$	$[9/17]$	$\sigma_\epsilon^2 + \sigma_{\alpha\beta\gamma\delta\pi}^2 + nw\Sigma\Sigma(\alpha\gamma)_{jl}^2 / (p - 1)(r - 1)$
10	AD	$[AD] - [A] - [D] + [Y]$	$(p - 1)(t - 1)$	$[10/17]$	$\sigma_\epsilon^2 + \sigma_{\alpha\beta\gamma\delta\pi}^2 + nw\Sigma\Sigma(\alpha\delta)_{jm}^2 / (p - 1)(t - 1)$
11	BC	$[BC] - [B] - [C] + [Y]$	$(q - 1)(r - 1)$	$[11/17]$	$\sigma_\epsilon^2 + \sigma_{\alpha\beta\gamma\delta\pi}^2 + nw\Sigma\Sigma(\beta\gamma)_{kl}^2 / (q - 1)(r - 1)$
12	BD	$[BD] - [B] - [D] + [Y]$	$(q - 1)(t - 1)$	$[12/17]$	$\sigma_\epsilon^2 + \sigma_{\alpha\beta\gamma\delta\pi}^2 + nw\Sigma\Sigma(\beta\delta)_{km}^2 / (q - 1)(t - 1)$
13	CD	$[CD] - [C] - [D] + [Y]$	$(r - 1)(t - 1)$	$[13/17]$	$\sigma_\epsilon^2 + \sigma_{\alpha\beta\gamma\delta\pi}^2 + nw\Sigma\Sigma(\gamma\delta)_{lm}^2 / (r - 1)(t - 1)$
14	ABC	$[ABC] - [AB] - [AC]$ $- [BC] + [A] + [B]$ $+ [C] - [Y]$	$(p - 1)(q - 1)$ $(r - 1)$	$[14/17]$	$\sigma_\epsilon^2 + \sigma_{\alpha\beta\gamma\delta\pi}^2 + 2n\Sigma\Sigma\Sigma(\alpha\beta\gamma)_{jkl}^2 / (p - 1)(q - 1)(r - 1)$
15	ABD	$[ABD] - [AB] - [AD]$ $- [BD] + [A] + [B]$ $+ [D] - [Y]$	$(p - 1)(q - 1)$ $(t - 1)$	$[15/17]$	$\sigma_\epsilon^2 + \sigma_{\alpha\beta\gamma\delta\pi}^2 + 2n\Sigma\Sigma\Sigma(\alpha\beta\delta)_{jkm}^2 / (p - 1)(q - 1)(t - 1)$
16	$ABCD$	$[ABCD] - [ABC]$ $- [ABD] - [ACD]$ $- [BCD] + [AB]$ $+ [AC] + [AD]$ $+ [BC] + [BD] + [CD]$ $- [A] - [B] - [C]$ $- [D] + [Y]$	$(p - 1)(q - 1)$ $(r - 1)(t - 1)$	$[16/17]$	$\sigma_\epsilon^2 + \sigma_{\alpha\beta\gamma\delta\pi}^2 + n\Sigma\Sigma\Sigma\Sigma(\alpha\beta\gamma\delta)_{jklm}^2 / (p - 1)(q - 1)(r - 1)(t - 1)$
17	$ABCD \times BL(G)$	$[ABCDGS] - [ABCD]$ $- [GS] + [G]$	$w(n - 1)(v - 1)$		$\sigma_\epsilon^2 + \sigma_{\alpha\beta\gamma\delta\pi}^2$
18	Total	$[ABCDGS] - [Y]$	$nvw - 1$		

TYPE RBPF-3^3 DESIGN WITH BLOCKS OF SIZE NINE OR THREE

A type RBPF-3^3 design can be laid out in blocks of size nine by confounding one interaction component with each group of blocks. The design can also be laid out in blocks of size three by confounding two interaction components with each group of blocks. We will describe the layout using blocks of size nine first.

We saw in Section 12.8 that the AB interaction in a type RBPF-3^2 design can be partitioned into two orthogonal components as follows.

SS	df
AB	$(3-1)(3-1) = 4$
(AB)	2
(AB^2)	2

Similarly, the ABC interaction in a type RBPF-3^3 design can be partitioned into four orthogonal components as follows.

SS	df
ABC	$(3-1)(3-1)(3-1) = 8$
(ABC)	2
(ABC^2)	2
(AB^2C)	2
(AB^2C^2)	2

The ABC interaction can be partially confounded by assigning treatment combinations to blocks according to the following defining relations.

$$\text{Group 0 } (ABC)_z = a_j + b_k + c_l = z(\text{mod } 3) \qquad (z = 0, 1, 2)$$
$$\text{Group 1 } (ABC^2)_z = a_j + b_k + 2c_l = z(\text{mod } 3) \qquad (z = 0, 1, 2)$$
$$\text{Group 2 } (AB^2C)_z = a_j + 2b_k + c_l = z(\text{mod } 3) \qquad (z = 0, 1, 2)$$
$$\text{Group 3 } (AB^2C^2)_z = a_j + 2b_k + 2c_l = z(\text{mod } 3) \qquad (z = 0, 1, 2)$$

The design layout is shown in Table 12.9-3. This design provides within-blocks information on treatments A, B, C, and the AB, AC, and BC interactions from all four groups. It provides $\frac{3}{4}$ relative information on each component of the ABC interaction.

The design shown in Table 12.9-3 requires blocks of size nine. If this block size is considered too large, further confounding can be used to reduce the block size to three. This is accomplished by assigning treatment combinations to groups so that they simultaneously satisfy two defining relations instead of one. For example, the two defining relations might be

$$(ABC)_z = a_j + b_k + c_l = z(\text{mod } 3) \qquad (z = 0, 1, 2)$$
$$(AB^2)_z = a_j + 2b_k = z(\text{mod } 3) \qquad (z = 0, 1, 2).$$

TABLE 12.9-3   Layout of Type RBPF-3^3 Design in Blocks of Size Nine

			abc	abc	abc	abc	abc	abc	abc	abc	abc
						Treatment Combinations					
g_0	$(ABC)_0$	s_0	000	012	021	102	111	120	201	210	222
	$(ABC)_1$	s_1	001	010	022	100	112	121	202	211	220
	$(ABC)_2$	s_2	002	011	020	101	110	122	200	212	221
g_1	$(ABC^2)_0$	s_3	000	011	022	101	112	120	202	210	221
	$(ABC^2)_1$	s_4	002	010	021	100	111	122	201	212	220
	$(ABC^2)_2$	s_5	001	012	020	102	110	121	200	211	222
g_2	$(AB^2C)_0$	s_6	000	011	022	102	110	121	201	212	220
	$(AB^2C)_1$	s_7	001	012	020	100	111	122	202	210	221
	$(AB^2C)_2$	s_8	002	010	021	101	112	120	200	211	222
g_3	$(AB^2C^2)_0$	s_9	000	012	021	101	110	122	202	211	220
	$(AB^2C^2)_1$	s_{10}	002	011	020	100	112	121	201	210	222
	$(AB^2C^2)_2$	s_{11}	001	010	022	102	111	120	200	212	221

If these defining relations are confounded in groups of blocks, two generalized interactions are also confounded. These are

$$(ABC)(AB^2)^{3-1} = (ABC)(AB^2)^2 = (A^3B^5C) = (BC^2) \text{ reduced modulo 3}$$
$$(ABC)(AB^2)^{3-2} = (ABC)(AB^2) = (A^2B^3C) = (AC^2) \text{ reduced modulo 3.}$$

Actually, $(A^3B^5C) = (B^2C)$ and $(A^2B^3C) = (A^2C)$ when reduced modulo 3. As discussed earlier, the power of the first term should always equal 1 in order to uniquely define the interaction components. This is achieved by simply squaring the terms (B^2C) and (A^2C) and reducing them modulo 3. For example,

$$(B^2C)^2 = (B^4C^2) = (BC^2) \text{ reduced modulo 3}$$
$$(A^2C)^2 = (A^4C^2) = (AC^2) \text{ reduced modulo 3.}$$

It can be shown that (B^2C) and (BC^2) define the same treatment combinations; the same can be said for (A^2C) and (AC^2).

As we have seen, if (ABC) and (AB^2) are simultaneously confounded in groups of blocks, two other interaction components are also confounded: (BC^2) and (AC^2). Care must be exercised in choosing defining relations so that main effects are not confounded in groups of blocks. Kempthorne (1952, 299) lists 13 systems for confounding a $3 \times 3 \times 3$ design in blocks of size three; only four of these do not confound a treatment in groups of blocks. One layout for a type RBPF-3^3 design in blocks of size three that provides within-blocks information on all main effects and interaction components is shown in Table 12.9-4. A summary of the computational formulas for this design is given in Table 12.9-5.

Computation of $SSAB'$(within), $SSAC'$(within), $SSBC'$(within), and $SSABC'$ (within) follows the procedure outlined in Section 12.8. For example, the (AB) component is computed from the two groups in which it is not confounded. These groups are g_0 and g_1.

$$SS(AB)\text{within} = \frac{\Sigma(AB)_0^2 + \Sigma(AB)_1^2 + \Sigma(AB)_2^2}{18} - \frac{\left(\sum_{i=1}^{n}\sum_{jk=1}^{v} Y_{ijkl0} + \sum_{i=1}^{n}\sum_{jk=1}^{v} Y_{ijkl1}\right)^2}{54}$$

The (ABC^2) component is computed from groups g_0, g_2, and g_3 according to the formula

$$SS(ABC^2)\text{within} = \frac{\Sigma(ABC^2)_0 + \Sigma(ABC^2)_1 + \Sigma(ABC^2)_2}{27}$$

$$- \frac{\left(\sum_{i=1}^{n}\sum_{jkl=1}^{v} Y_{ijk0} + \sum_{i=1}^{n}\sum_{jkl=1}^{v} Y_{ijkl1} + \sum_{i=1}^{n}\sum_{jkl=1}^{v} Y_{ijkl2}\right)^2}{81}.$$

Adjusted treatment totals for the ABC Summary Table are given by

$$\Sigma Y'_{jkl\cdot} =$$

$$\frac{\Sigma(A)_j + \Sigma(B)_k + \Sigma(C)_l}{9}$$

$$+ \frac{2[\Sigma(AB)'_{j+k} + \Sigma(AB^2)'_{j+2k} + \Sigma(AC)'_{j+l} + \Sigma(AC^2)'_{j+2l} + \Sigma(BC)'_{k+l} + \Sigma(BC^2)'_{k+2l}]}{9}$$

$$+ \frac{4[\Sigma(ABC)'_{j+k+l} + \Sigma(ABC^2)'_{j+k+2l} + \Sigma(AB^2C)'_{j+2k+l} + \Sigma(AB^2C^2)'_{j+2k+2l}]}{27}$$

$$- \frac{4\left(\sum_{i=1}^{n}\sum_{jkl=1}^{v}\sum_{z=1}^{w} Y_{ijklz}\right)}{9}.$$

Adjusted treatment totals for a two-treatment summary table, for example AB, have the form

$$\Sigma Y'_{jk\cdot\cdot} = \frac{\Sigma(A)_j + \Sigma(B)_k}{3} + \frac{2[\Sigma(AB)'_{j+k} + \Sigma(AB^2)'_{j+2k}]}{3} - \frac{\sum_{i=1}^{n}\sum_{jkl=1}^{v}\sum_{z=1}^{w} Y_{ijklz}}{3}.$$

The symbols in the preceding formulas are defined in Section 12.7.

TABLE 12.9-4   Layout of Type RBPF-3^3 Design in Blocks of Size Three

				Treatment Combinations		
				abc	abc	abc
		$(ABC)_0$	$(AB^2)_0$	s_0	000	111	222
		$(ABC)_0$	$(AB^2)_1$	s_1	021	102	210
		$(ABC)_0$	$(AB^2)_2$	s_2	012	120	201
	g_0	$(ABC)_1$	$(AB^2)_0$	s_3	001	112	220
		$(ABC)_1$	$(AB^2)_1$	s_4	022	100	211
		$(ABC)_1$	$(AB^2)_2$	s_5	010	121	202
		$(ABC)_2$	$(AB^2)_0$	s_6	002	110	221
		$(ABC)_2$	$(AB^2)_1$	s_7	020	101	212
		$(ABC)_2$	$(AB^2)_2$	s_8	011	122	200
		$(ABC^2)_0$	$(AB^2)_0$	s_9	000	112	221
		$(ABC^2)_0$	$(AB^2)_1$	s_{10}	022	101	210
		$(ABC^2)_0$	$(AB^2)_2$	s_{11}	011	120	202
	g_1	$(ABC^2)_1$	$(AB^2)_0$	s_{12}	002	111	220
		$(ABC^2)_1$	$(AB^2)_1$	s_{13}	021	100	212
		$(ABC^2)_1$	$(AB^2)_2$	s_{14}	010	122	201
		$(ABC^2)_2$	$(AB^2)_0$	s_{15}	001	110	222
		$(ABC^2)_2$	$(AB^2)_1$	s_{16}	020	102	211
		$(ABC^2)_2$	$(AB^2)_2$	s_{17}	012	121	200
		$(AB^2C)_0$	$(AB)_0$	s_{18}	000	121	212
		$(AB^2C)_0$	$(AB)_1$	s_{19}	011	102	220
		$(AB^2C)_0$	$(AB)_2$	s_{20}	022	110	201
	g_2	$(AB^2C)_1$	$(AB)_0$	s_{21}	001	122	210
		$(AB^2C)_1$	$(AB)_1$	s_{22}	012	100	221
		$(AB^2C)_1$	$(AB)_2$	s_{23}	020	111	202
		$(AB^2C)_2$	$(AB)_0$	s_{24}	002	120	211
		$(AB^2C)_2$	$(AB)_1$	s_{25}	010	101	222
		$(AB^2C)_2$	$(AB)_2$	s_{26}	021	112	200
		$(AB^2C^2)_0$	$(AB)_0$	s_{27}	000	122	211
		$(AB^2C^2)_0$	$(AB)_1$	s_{28}	012	101	220
		$(AB^2C^2)_0$	$(AB)_2$	s_{29}	021	110	202
	g_3	$(AB^2C^2)_1$	$(AB)_0$	s_{30}	002	121	210
		$(AB^2C^2)_1$	$(AB)_1$	s_{31}	011	100	222
		$(AB^2C^2)_1$	$(AB)_2$	s_{32}	020	112	201

Table 12.9-4 (continued)

| | | | Treatment Combinations | | |
			abc	*abc*	*abc*
$(AB^2C^2)_2$	$(AB)_0$	s_{33}	001	120	212
$(AB^2C^2)_2$	$(AB)_1$	s_{34}	010	102	221
$(AB^2C^2)_2$	$(AB)_2$	s_{35}	022	111	200

TYPE RBPF-5^2 AND RBCF-5^2 DESIGNS WITH BLOCKS OF SIZE FIVE

The procedures that have been described for type RBCF-p^k and RBPF-p^k designs where p is equal to two or three are applicable to designs in which p is any prime number. If p is equal to five, the AB interaction degrees of freedom for type RBCF-5^2 and RBPF-5^2 designs can be partitioned as follows.

SS	df
AB	$(5 - 1)(5 - 1) = 16$
(AB)	4
(AB^2)	4
(AB^3)	4
(AB^4)	4

A balanced partially confounded design can be laid out in four groups or a multiple of four groups. Combinations of treatments A and B are assigned to blocks within each group by means of the following relations.

$$g_0 \ (AB)_z = a_j + b_k = z(\text{mod } 5) \qquad (z = 0, \ldots, 4)$$
$$g_1 \ (AB^2)_z = a_j + 2b_k = z(\text{mod } 5) \qquad (z = 0, \ldots, 4)$$
$$g_2 \ (AB^3)_z = a_j + 3b_k = z(\text{mod } 5) \qquad (z = 0, \ldots, 4)$$
$$g_3 \ (AB^4)_z = a_j + 4b_k = z(\text{mod } 5) \qquad (z = 0, \ldots, 4)$$

Each group contains five blocks corresponding to $z = 0, 1, 2, 3,$ and 4. A completely confounded design can be constructed by using any one of the four sets of relations just defined.

It is not possible to describe in this chapter all or even a majority of the designs for which the technique of confounding is useful. The approach that has been adopted is to present selected examples illustrating basic principles and procedures applicable to a broad range of designs. A listing of confounded designs, together with pertinent information, is given in Table 12.9-6. The table includes references that

TABLE 12.9-5   Computational Formulas for Type RBPF-3^3 Design in Blocks of Size Three

	Source	Formula	df	F	E(MS) (A, B, and C fixed; Blocks random)*
1	Between blocks	$[GS] - [Y]$	$nw - 1 = 35$		
2	Groups	$[G] - [Y]$	$w - 1 = 3$		
3	Blocks w. groups	$[GS] - [G]$	$w(n - 1) = 32$		
4	Within blocks	$[ABCG] - [GS]$	$nw(v - 1) = 72$		
5	A	$[A] - [Y]$	$p - 1 = 2$	[5/12]	$\sigma_\epsilon^2 + qrw \sum \alpha_j^2/(p - 1)$
6	B	$[B] - [Y]$	$q - 1 = 2$	[6/12]	$\sigma_\epsilon^2 + prw \sum \beta_k^2/(q - 1)$
7	C	$[C] - [Y]$	$r - 1 = 2$	[7/12]	$\sigma_\epsilon^2 + pqw \sum \gamma_i^2/(r - 1)$
8	AB'(within) $= (AB) + (AB^2)$	**	$(p - 1)(q - 1) = 4$	[8/12]	$\sigma_\epsilon^2 + \left(\frac{1}{2}\right)rw \sum\sum(\alpha\beta)_{jk}^2/(p - 1)(q - 1)$
9	AC'(within) $= (AC) + (AC^2)$	**	$(p - 1)(r - 1) = 4$	[9/12]	$\sigma_\epsilon^2 + \left(\frac{1}{2}\right)qw \sum\sum(\alpha\gamma)_{jk}^2/(p - 1)(r - 1)$
10	BC'(within) $= (BC) + (BC^2)$	**	$(q - 1)(r - 1) = 4$	[10/12]	$\sigma_\epsilon^2 + \left(\frac{1}{2}\right)pw \sum\sum(\beta\gamma)_{ki}^2/(q - 1)(r - 1)$
11	ABC'(within) $= (ABC) + (ABC^2)$ $+ (AB^2C)$ $+ (AB^2C^2)$	**	$(p - 1)(q - 1)(r - 1) = 8$	[11/12]	$\sigma_\epsilon^2 + \left(\frac{3}{4}\right)w \sum\sum(\alpha\beta\gamma)_{jkl}^2/(p - 1)(q - 1)(r - 1)$

12	Residual	$SSWITHIN\ BL - SSA - SSB - SSC$ $- SSAB' - SSAC' - SSBC'$ $- SSABC'$	$= 46$	σ_ϵ^2
13	AG		$(p-1)(w-1) = 6$	
14	BG		$(q-1)(w-1) = 6$	
15	CG		$(r-1)(w-1) = 6$	
16	$(AB) \times \frac{1}{2}G$		$(3-1)(w'-1) = 2$	
17	$(AB^2) \times \frac{1}{2}G$		$(3-1)(w'-1) = 2$	
18	$(AC) \times \frac{1}{2}G$		$(3-1)(w'-1) = 2$	
19	$(AC^2) \times \frac{1}{2}G$		$(3-1)(w'-1) = 2$	
20	$(BC) \times \frac{1}{2}G$		$(3-1)(w'-1) = 2$	
21	$(BC^2) \times \frac{1}{2}G$		$(3-1)(w'-1) = 2$	
22	$(ABC) \times \frac{3}{4}G$		$(3-1)(w''-1) = 4$	
23	$(ABC^2) \times \frac{3}{4}G$		$(3-1)(w''-1) = 4$	
24	$(AB^2C) \times \frac{3}{4}G$		$(3-1)(w''-1) = 4$	
25	$(AB^2C^2) \times \frac{3}{4}G$		$(3-1)(w''-1) = 4$	
26	Total	$[ABCGS] - [Y]$	$nvw - 1 = 107$	

*Interaction effects involving blocks are assumed to equal zero.

**See text

TABLE 12.9-6 Reference Information for Confounded Factorial Designs

Type Design	Number of Treatment Combinations	Number of Observations per Block	Number of Groups (Replications) for Balanced Design*	Interaction(s) Confounded and Relative Information**	Reference[§]
‡RBCF-2^2	4	2	x	AB	(4) p. 61, (7) p. 216
‡RBCF-2^3	8	4	x	any interaction	(1) p. 220, (2) p. 427, (3) p. 233, (7) p. 217, (10) p. 619, p. 635
‡RBPF-2^3	8	4	4	$AB(\frac{3}{4})$, $AC(\frac{3}{4})$, $BC(\frac{3}{4})$, $ABC(\frac{3}{4})$	(1) p. 220, (3) p. 244, (4) p. 62, (5) p. 201, (6) p. 275, (7) p. 231, (10) p. 620, p. 622
‡RBCF-2^4	16	8	x	any interaction	(1) p. 220, (2) p. 429, (5) p. 193
‡RBCF-2^4	16	4	x	AB, ACD, BCD; or AB, CD, $ABCD$	(1) p. 220, (2) p. 428, (5) p. 194
RBPF-2^4	16	4	6	all two-factor interactions $(\frac{2}{3})$, all three-factor interactions $(\frac{1}{2})$	(1) p. 220
RBCF-2^4	16	2	x	AB, AC, BC, AD, BD, CD, $ABCD$	(6) p. 261
RBPF-2^4	16	2	4	main effects $(\frac{3}{4})$, two-factor interactions $(\frac{1}{2})$, three-factor interactions $(\frac{1}{4})$	(6) p. 278
RBCF-2^5	32	16	x	any interaction	(2) p. 430
RBCF-2^5	32	8	x	ABC, ADE, $BCDE$; or AB, CDE, $ABCDE$	(1) p. 220, (2) p. 429, (4) p. 72, (7) p. 222

Design				Interactions confounded	References
RBPF-2^5	32	8	5	all three-factor interactions ($\frac{4}{5}$), all four-factor interactions ($\frac{4}{5}$)	(1) p. 220
RBCF-2^5	32	4	x	AB, AC, BC, DE, $ABDE$, $ACDE$, $BCDE$; or AB, CD, ACE, BDE, ADE, BCE, $ABCD$	(2) p. 428
‡RBCF-3^2	9	3	x	AB or AB^2	(2) p. 435, (3) p. 239, (7) p. 226
‡RBPF-3^2	9	3	2	AB components ($\frac{1}{2}$)	(3) p. 251, (4) p. 99, (6) p. 300, (10) p. 639, p. 640
RBCF-3^3	27	9	x	any interaction component	(4) p. 102, (7) p. 227
‡RBPF-3^3	27	9	4	ABC components ($\frac{3}{4}$)	(1) p. 222, (2) p. 438, (3) p. 251
‡RBPF-3^3	27	3	4	all two-factor interactions ($\frac{1}{2}$), ABC ($\frac{3}{4}$)	(6) p. 302, (10) p. 650
RBCF-3^4	81	9	x	ABC, AB^2D^2, AC^2D, BC^2D^2	(2) p. 436, (3) p. 252, (10) p. 655
RBPF-3^4	81	9	4	all three-factor interactions ($\frac{3}{4}$)	(7) p. 230
RBPF-4^2	16	4	3	AB components ($\frac{2}{3}$)	(1) p. 223, (6) p. 306
RBCF-4^3	64	16	x	ABC interaction	(1) p. 225
‡RBCF-5^2	25	5	x	AB, AB^2, AB^3, or AB^4	(8) p. 121
‡RBPF-5^2	25	5	4	AB components ($\frac{3}{4}$)	(9) p. 57

*The symbol x indicates that any number of groups (replicates) can be used; for a balanced design, groups (replicates) should be a multiple of the number that appears in the column.

**If a fraction does not follow the interaction, it is completely confounded.

§Information about a design can be found in the following references: (1) Cochran and Cox (1957), (2) Davies (1956), (3) Federer (1955), (4) Gill (1978, Vol. 2), (5) Johnson and Leone (1964), (6) Kempthorne (1952), (7) Montgomery (1976), (8) Nair (1938), (9) Nair (1940), and (10) Winer (1971).

‡Design described in this chapter.

describe the layout and main features of each design. Many of the designs listed in Table 12.9-6 are also described by Yates (1937).

12.10 ALTERNATIVE NOTATION AND COMPUTATIONAL SYSTEMS

There are almost as many notation systems in experimental design as there are books on the subject. Fortunately a common thread runs through most systems. A reader can, with a little effort, learn to be comfortable with each system. One specialized notation system, however, would be difficult to follow without a brief note of explanation. This system is widely used in designating treatment combinations when each treatment has only two levels. In this system the symbol for a treatment combination contains only those letters for which the treatment is at the higher level. The absence of a letter indicates that the treatment is at the lower level. If all treatments are at the lower level, the symbol (1) is used. The terms *lower* and *higher* may refer to positions on a scale or may be an arbitrary distinction between levels. A comparison of this specialized notation with the notation used in this chapter is as follows.

Treatment = 000   100   010   001   110   101   011   111
combination

Specialized = (1)   *a*   *b*   *c*   *ab*   *ac*   *bc*   *abc*
notation

Although this notation system has certain advantages over the notation in this chapter, it has not been used here because it is appropriate only for treatments having two levels. The notation in this chapter is a more general system in that it is appropriate for treatments with any number of levels. Furthermore, the notation in this chapter lends itself to the application of modular arithmetic.

Section 12.2 described a notation system for designating components of an interaction. An alternative scheme is used by Yates (1937) and many other writers. A comparison of this scheme and the one used in this book is

$$(AB^2) = AB(I), \qquad (AB^2C^2) = ABC(W), \qquad (ABC^2) = ABC(Y)$$
$$(AB) = AB(J), \qquad (AB^2C) = ABC(X), \qquad (ABC) = ABC(Z).$$

This alternative system is widely used but is less descriptive than that used in this chapter.

The computational procedures introduced in Chapters 4 and 5 have been adhered to throughout the book. These procedures were chosen not because they are the most efficient, but because they are relatively simple, illustrate fundamental principles clearly, and are easily adapted to the designs described in this book. Other computational schemes may be more efficient for the reader's purpose. Yates (1937) has devised a simple technique for obtaining treatment and interaction sums of squares when all treatments have the same number of levels. This technique is illustrated by Box, Hunter, and Hunter (1978, 322–324). Yates has also devised a simplified procedure for computing I and J interaction components. This procedure is described by Kempthorne (1952, 307) and Winer (1971, 612–618).

12.11 COMPUTATIONAL PROCEDURES FOR TYPE RBPF-32² DESIGN

Confounded designs in which the number of levels of each treatment are not equal are called *mixed designs*. Examples of mixed designs are $3 \times 2 \times 2$, $3 \times 3 \times 2$, and $4 \times 3 \times 2$ designs. These designs generally entail a more complex analysis than unmixed designs. And the choice of block size is much more restricted for the mixed designs. In order not to confound main effects, the block size must be an integral multiple of the number of levels of each treatment. Thus, for a $3 \times 2 \times 2$ design, the block size must equal six. This permits a_0, a_1, and a_2 to each occur twice in a block and b_0, b_1, and c_0, c_1 to occur three times. All 12 treatment combinations of a $3 \times 2 \times 2$ design can be assigned to two blocks as shown in the following illustration.

	abc	abc	abc	abc	abc	abc
Block 0	000	011	101	110	201	210
Block 1	001	010	100	111	200	211

A tabulation of the occurrence of a_j, b_k, c_l, a_jb_k, a_jc_l, and b_kc_l in the two blocks is as follows.

	Blocks 0	1		Blocks 0	1		Blocks 0	1		Blocks 0	1
a_0	2	2	a_0b_0	1	1	a_0c_0	1	1	b_0c_0	1	2
a_1	2	2	a_0b_1	1	1	a_0c_1	1	1	b_0c_1	2	1
a_2	2	2	a_1b_0	1	1	a_1c_0	1	1	b_1c_0	2	1
b_0	3	3	a_1b_1	1	1	a_1c_1	1	1	b_1c_1	1	2
b_1	3	3	a_2b_0	1	1	a_2c_0	1	1			
c_0	3	3	a_2b_1	1	1	a_2c_1	1	1			
c_1	3	3									

Each level of a_j, b_k, and c_l as well as each combination of a_jb_k and a_jc_l occurs equally often in blocks 0 and 1. This is not true for b_kc_l. As a result, the BC interaction is partially confounded with blocks. Furthermore the ABC interaction is also confounded with blocks because different combinations of $a_j b_k c_l$ occur in the blocks. On reflection it should be evident that the BC interaction, which involves four treatment combinations, must be confounded in a $3 \times 2 \times 2$ design because the block size is not a multiple of four. Another mixed design, a type RBPF-$3^2 2$ design, is also laid out in blocks of size six. In this design it is the AB and ABC interactions that are confounded. The block size of six allows all $a_j c_l$ and $b_k c_l$ treatment combinations to occur within a block but only six of the nine $a_j b_k$ combinations.

A balanced type RBPF-32^2 design can be laid out in three groups of two blocks each. The layout is shown in Table 12.11-1. The symbol $a_0(BC)_0$ stands for treatment combinations 000 and 011, where a_0 is equal to zero and $(BC)_0$ satisfies the relation

$$(BC)_0 = b_k + c_l = 0(\text{mod } 2) = 00 \text{ and } 11.$$

Treatment combinations denoted by $a_1(BC)_1$ are 101 and 110, where a_1 is equal to one and $(BC)_1$ satisfies the relation

$$(BC)_1 = b_k + c_l = 1(\text{mod } 2) = 01 \text{ and } 10.$$

Analysis procedures for a type RBPF-32^2 design are illustrated in Table 12.11-2. The procedures for adjusting $SSBC$ and $SSABC$ require special comment. We

---

TABLE 12.11-1   General Layout of Type RBPF-32^2 Design

		abc	*abc*	*abc*	*abc*	*abc*	*abc*
		$a_0(BC)_0$		$a_1(BC)_1$		$a_2(BC)_1$	
	Block 0	000	011	101	110	201	210
g_0							
		$a_0(BC)_1$		$a_1(BC)_0$		$a_2(BC)_0$	
	Block 1	001	010	100	111	200	211
		$a_0(BC)_1$		$a_1(BC)_0$		$a_2(BC)_1$	
	Block 2	001	010	100	111	201	210
g_1							
		$a_0(BC)_0$		$a_1(BC)_1$		$a_2(BC)_0$	
	Block 3	000	011	101	110	200	211
		$a_0(BC)_1$		$a_1(BC)_1$		$a_2(BC)_0$	
	Block 4	001	010	101	110	200	211
g_2							
		$a_0(BC)_0$		$a_1(BC)_0$		$a_2(BC)_1$	
	Block 5	000	011	100	111	201	210

saw earlier (Table 12.11-1) that the effects of blocks 0 and 1, 2 and 3, and so on, are confounded with $(BC)_0$ and $(BC)_1$ as well as the ABC interaction. Consequently the BC and ABC interactions must be adjusted for block effects. The rationale underlying these adjustments has been discussed by Federer (1955, 255), Kempthorne (1952, 348), Li (1944), Nair (1938), and Yates (1937). We will simply illustrate the procedure here. If the BC effects were not confounded with blocks, they could be estimated by taking the difference between two components, $\Sigma(BC)_0$ and $\Sigma(BC)_1$:

$$36BC\text{unadj} = \Sigma(BC)_0 - \Sigma(BC)_1$$

where $\Sigma(BC)_0$ is equal to the sum of the treatment combinations that satisfy $b_k + c_l = 0(\text{mod } 2)$ and $\Sigma(BC)_1$ is equal to the sum of the treatment combinations that satisfy $b_k + c_l = 1(\text{mod } 2)$. These sums can be obtained from the BC Summary

---

TABLE 12.11-2 Computational Procedures for a Type RBPF-32² Design

---

(i) Data and notation [Y_{ijklz} denotes a score for the experimental unit in block i, treatment combination jkl, and group z; $i = 1, \ldots, n$ blocks (s_i); $j = 1, \ldots, p$ levels of treatment A (a_j); $k = 1, \ldots, q$ levels of treatment B (b_k); $l = 1, \ldots, r$ levels of treatment C (c_l); $z = 1, \ldots, w$ levels of groups (g_z); $jkl = 1, \ldots, v$ combinations of a_j, b_k, and c_l within a block]:

ABCGS Summary Table
Entry is Y_{ijklz}

		abc	abc	abc	abc	abc	abc	$\sum\limits_{jkl=1}^{v} Y_{ijklz}$	$\sum\limits_{i=1}^{n}\sum\limits_{jkl=1}^{v} Y_{ijklz}$
g_0	s_0	000 3	011 7	101 4	110 7	201 3	210 6	30	
	s_1	001 4	010 7	100 3	111 9	200 3	211 8	34	64
g_1	s_2	001 2	010 5	100 2	111 10	201 3	210 6	28	
	s_3	000 1	011 10	101 3	110 6	200 2	211 11	33	61
g_2	s_4	001 3	010 6	101 4	110 7	200 2	211 9	31	
	s_5	000 2	011 8	100 3	111 8	201 3	210 7	31	62

Table 12.11-2 (continued)

AB Summary Table

Entry is $\sum\limits_{i=1}^{n} \sum\limits_{z=1}^{w} Y_{ijklz}$

	b_0	b_1	$\sum\limits_{i=1}^{n} \sum\limits_{k=1}^{q} \sum\limits_{z=1}^{w} Y_{ijklz}$
a_0	$nw = 6$ 15	43	58
a_1	19	47	66
a_2	16	47	63

$\sum\limits_{i=1}^{n} \sum\limits_{j=1}^{p} \sum\limits_{z=1}^{w} Y_{ijklz} = 50 \quad 137$

AC Summary Table

Entry is $\sum\limits_{i=1}^{n} \sum\limits_{z=1}^{w} Y_{ijklz}$

	c_0	c_1
a_0	$nw = 6$ 24	34
a_1	28	38
a_2	26	37

$\sum\limits_{i=1}^{n} \sum\limits_{j=1}^{p} \sum\limits_{z=1}^{w} Y_{ijklz} = 78 \quad 109$

ABC Summary Table

Entry is $\sum\limits_{z=1}^{w} Y_{ijklz}$

	b_0 c_0	b_0 c_1	b_1 c_0	b_1 c_1
a_0	$w = 3$ 6	9	18	25
a_1	8	11	20	27
a_2	7	9	19	28

BC Summary Table

Entry is $\sum\limits_{i=1}^{n} \sum\limits_{j=1}^{p} \sum\limits_{z=1}^{w} Y_{ijklz}$

	c_0	c_1
b_0	$pw = 9$ 21	29
b_1	57	80

(ii)   Computational symbols:

$$\sum\limits_{i=1}^{n} \sum\limits_{jkl=1}^{v} \sum\limits_{z=1}^{w} Y_{ijklz} = 3 + 4 + \cdots + 7 = 187.000$$

$$\frac{\left(\sum\limits_{i=1}^{n} \sum\limits_{jkl=1}^{v} \sum\limits_{z=1}^{w} Y_{ijklz} \right)^2}{nvw} = [Y] = \frac{(187.000)^2}{(2)(6)(3)} = 971.361$$

$$\sum\limits_{i=1}^{n} \sum\limits_{jkl=1}^{v} \sum\limits_{z=1}^{w} Y_{ijklz}^2 = [ABCGS] = (3)^2 + (4)^2 + \cdots + (7)^2 = 1239.000$$

Table 12.11-2 (continued)

$$\sum_{i=1}^{n}\sum_{z=1}^{w}\frac{\left(\sum_{jkl=1}^{v}Y_{ijklz}\right)^2}{v}=[GS]=\frac{(30)^2}{6}+\frac{(34)^2}{6}+\cdots+\frac{(31)^2}{6}=975.167$$

$$\sum_{z=1}^{w}\frac{\left(\sum_{i=1}^{n}\sum_{jkl=1}^{v}Y_{ijklz}\right)^2}{nv}=[G]=\frac{(64)^2}{(2)(6)}+\frac{(61)^2}{(2)(6)}+\frac{(62)^2}{(2)(6)}=971.750$$

$$\sum_{j=1}^{p}\frac{\left(\sum_{i=1}^{n}\sum_{k=1}^{q}\sum_{z=1}^{w}Y_{ijklz}\right)^2}{nqw}=[A]=\frac{(58)^2}{(2)(2)(3)}+\frac{(66)^2}{(2)(2)(3)}+\frac{(63)^2}{(2)(2)(3)}=974.083$$

$$\sum_{k=1}^{q}\frac{\left(\sum_{i=1}^{n}\sum_{j=1}^{p}\sum_{z=1}^{w}Y_{ijklz}\right)^2}{npw}=[B]=\frac{(50)^2}{(2)(3)(3)}+\frac{(137)^2}{(2)(3)(3)}=1181.611$$

$$\sum_{l=1}^{r}\frac{\left(\sum_{i=1}^{n}\sum_{j=1}^{p}\sum_{z=1}^{w}Y_{ijklz}\right)^2}{npw}=[C]=\frac{(78)^2}{(2)(3)(3)}+\frac{(109)^2}{(2)(3)(3)}=998.056$$

$$\sum_{j=1}^{p}\sum_{k=1}^{q}\frac{\left(\sum_{i=1}^{n}\sum_{z=1}^{w}Y_{ijklz}\right)^2}{nw}=[AB]=\frac{(15)^2}{(2)(3)}+\frac{(43)^2}{(2)(3)}+\cdots+\frac{(47)^2}{(2)(3)}=1184.833$$

$$\sum_{j=1}^{p}\sum_{l=1}^{r}\frac{\left(\sum_{i=1}^{n}\sum_{z=1}^{w}Y_{ijklz}\right)^2}{nw}=[AC]=\frac{(24)^2}{(2)(3)}+\frac{(34)^2}{(2)(3)}+\cdots+\frac{(37)^2}{(2)(3)}=1000.833$$

$$\sum_{k=1}^{q}\sum_{l=1}^{r}\frac{\left(\sum_{i=1}^{n}\sum_{j=1}^{p}\sum_{z=1}^{w}Y_{ijklz}\right)^2}{pw}=[BC]=\frac{(21)^2}{(3)(3)}+\frac{(29)^2}{(3)(3)}+\cdots+\frac{(80)^2}{(3)(3)}=1214.556$$

$$\sum_{j=1}^{p}\sum_{k=1}^{q}\sum_{l=1}^{r}\frac{\left(\sum_{z=1}^{w}Y_{ijklz}\right)^2}{w}=[ABC]=\frac{(6)^2}{3}+\frac{(9)^2}{3}+\cdots+\frac{(28)^2}{3}=1218.333$$

(iii) Computational formulas:

$$SSTO = [ABCGS] - [Y] = 267.639$$
$$SSBETWEEN\ BL = [GS] - [Y] = 3.806$$
$$SSG = [G] - [Y] = 0.389$$
$$SSBL(G) = [GS] - [G] = 3.417$$
$$SSWITHIN\ BL = [ABCGS] - [GS] = 263.833$$

Table 12.11-2 (continued)

$$SSA = [A] - [Y] = 2.722$$
$$SSB = [B] - [Y] = 210.250$$
$$SSC = [C] - [Y] = 26.695$$
$$SSAB = [AB] - [A] - [B] + [Y] = 0.500$$
$$SSAC = [AC] - [A] - [C] + [Y] = 0.055$$
$$SSBC = [BC] - [B] - [C] + [Y] = 6.250$$
$$SSB(adj) = (\text{see text}) = 4.500$$
$$SSABC = [ABC] - [AB] - [AC] - [BC] + [A] + [B] + [C] - [Y] = 0.500$$
$$SSABC(adj) = (\text{see text}) = 0.033$$
$$SSRES = [ABCGS] - [ABC] - [GS] + [Y] = 16.861$$
$$SSRES(adj) = SSRES + SSBC + SSABC - SSBC(adj) - SSABC(adj) = 19.078$$

Table. They are

$$00 + 11 = 21 + 80 = 101$$
$$01 + 10 = 29 + 57 = \quad 86.$$

The unadjusted sum of squares for the BC interaction is given by

$$SSBC = \frac{(BC\text{unadj})^2}{36} - \frac{[\Sigma(BC)_0 - \Sigma(BC)_1]^2}{36} = \frac{(101 - 86)^2}{36} = \frac{(15)^2}{36} = 6.25.$$

The reader can verify that $SSBC = 6.25$ is the same value that is obtained using the more familiar formula

$$SSBC = [BC] - [B] - [C] + [Y] = 6.25.$$

The BC effects can be adjusted for block effects by

$$BC\text{adj} = [\Sigma(BC)_0 - \Sigma(BC)_1] + \frac{(\Sigma S_0 - \Sigma S_1 + \Sigma S_2 - \Sigma S_3 + \Sigma S_4 - \Sigma S_5)}{w}$$

$$= (101 - 86) + \frac{(30 - 34 + 28 - 33 + 31 - 31)}{3}$$

$$= 15 - \frac{9}{3} = 12$$

where ΣS_i denotes the sum of all observations in block i. The adjusted BC effects are based on 32 effective replications as opposed to 36 for the unadjusted BC effects. Federer (1955, 260) illustrates the calculation of the number of effective replications. The adjusted sum of squares for the BC interaction is given by

$$SSBC\text{adj} = \frac{(BC\text{adj})^2}{32} = \frac{(12)^2}{32} = 4.50.$$

The same general procedure is used to adjust the *ABC* interaction for block effects. If the *ABC* interaction were not confounded, it could be computed as follows:

$$SSABC = \sum_{j=1}^{p} \frac{[\Sigma a_j(BC)_0 - \Sigma a_j(BC)_1]^2}{12} - SSBC.$$

Computation of the *ABC* interaction by this formula is analogous to summing the *BC* interaction over the *p* levels of treatment *A*. The numbers required for this computation are obtained from the *ABC* Summary Table.

$$\Sigma a_0(BC) - \Sigma a_0(BC)_1 = (000 + 011) - (001 + 010)$$
$$= (6 + 25) - (9 + 18) = 4$$
$$\Sigma a_1(BC)_0 - \Sigma a_1(BC)_1 = (100 + 111) - (101 + 110)$$
$$= (8 + 27) - (11 + 20) = 4$$
$$\Sigma a_2(BC)_0 - \Sigma a_2(BC)_1 = (200 + 211) - (201 + 210)$$
$$= (7 + 28) - (9 + 19) = 7$$
$$\sum_{j=1}^{p} \frac{[\Sigma a_j(BC)_0 - \Sigma a_j(BC)_1]^2}{12} = \frac{(4)^2}{12} + \frac{(4)^2}{12} + \frac{(7)^2}{12} = 6.75$$

If *SSBC* is subtracted from 6.75, the remainder is *SSABC*,

$$SSABC = 6.75 - 6.25 = 0.50.$$

The *SSABC* adj can be obtained by adjusting each $[\Sigma a_j(BC)_0 - \Sigma a_j(BC)_1]$ for block effects and subtracting *BC* adj.

$$3a_0BC\text{adj} = 3[\Sigma a_0(BC)_0 - \Sigma a_0(BC)_1] + (\Sigma S_1 - \Sigma S_0 + \Sigma S_2 - \Sigma S_3 + \Sigma S_4 - \Sigma S_5)$$
$$= 3(4) + (34 - 30 + 28 - 33 + 31 - 31) = 11$$
$$3a_1BC\text{adj} = 3[\Sigma a_1(BC)_0 - \Sigma a_1(BC)_1] + (\Sigma S_0 - \Sigma S_1 + \Sigma S_3 - \Sigma S_2 + \Sigma S_4 - \Sigma S_5)$$
$$= 3(4) + (30 - 34 + 33 - 28 + 31 - 31) = 13$$
$$3a_2BC\text{adj} = 3[\Sigma a_2(BC)_0 - \Sigma a_2(BC)_1] + (\Sigma S_0 - \Sigma S_1 + \Sigma S_2 - \Sigma S_3 + \Sigma S_5 - \Sigma S_4)$$
$$= 3(7) + (30 - 34 + 28 - 33 + 31 - 31) = 12$$

A check on the adjusted a_jBC components is given by

$$\sum_{j=1}^{p} (3a_jBC\text{adj}) = 3(BC\text{adj})$$
$$11 + 13 + 12 = (3)(12) = 36.$$

The adjusted sum of squares for the *ABC* interaction is

$SSABC\text{adj} =$

$$\frac{(3a_0BC\text{adj} - BC\text{adj})^2 + (3a_1BC\text{adj} - BC\text{adj})^2 + (3a_2BC\text{adj} - BC\text{adj})^2}{60}$$

$$= \frac{(11 - 12)^2 + (13 - 12)^2 + (12 - 12)^2}{60} = .033.$$

The results of the analysis in Table 12.11-2 and the adjustment just described are summarized in Table 12.11-3. It is apparent that the null hypotheses for treatments B and C and the BC interaction can be rejected. In view of the significant BC interaction, an experimenter might want to test hypotheses regarding simple effects contrasts. These tests should be carried out with adjusted $b_k c_l$ cell means in order to remove the confounding effects of the blocks. Procedures for adjusting the $b_k c_l$ cell means are described next.

TABLE 12.11-3 ANOVA Table for Type RBPF-32² Design

	Source	SS	df	MS	F	E(MS) (A, B, and C fixed; Blocks random)*
1	Between blocks	3.806	$nw - 1 = 5$			
2	Groups	0.389	$w - 1 = 2$			
3	Blocks w. groups	3.417	$w(n - 1) = 3$			
4	Within blocks	263.833	$nw(v - 1) = 30$			
5	A	2.722	$p - 1 = 2$	1.361	[5/12] 1.36	$\sigma_\epsilon^2 + qrw\Sigma\alpha_j^2/(p - 1)$
6	B	210.250	$q - 1 = 1$	210.250	[6/12] 209.41§	$\sigma_\epsilon^2 + prw\Sigma\beta_k^2/(q - 1)$
7	C	26.695	$r - 1 = 1$	26.695	[7/12] 26.59§	$\sigma_\epsilon^2 + pqw\Sigma\gamma_l^2/(r - 1)$
8	AB	0.500	$(p - 1)(q - 1) = 2$	0.250	[8/12] 0.25	$\sigma_\epsilon^2 + rw\Sigma\Sigma(\alpha\beta)_{jk}^2/ (p - 1)(q - 1)$
9	AC	0.055	$(p - 1)(r - 1) = 2$	0.028	[9/12] 0.03	$\sigma_\epsilon^2 + qw\Sigma\Sigma(\alpha\gamma)_{jl}^2/ (p - 1)(r - 1)$
10	BC(adj)	4.500	$(q - 1)(r - 1) = 1$	4.500	[10/12] 4.48**	$\sigma_\epsilon^2 + \left(\frac{8}{9}\right)pw\Sigma\Sigma(\beta\gamma)_{kl}^2/ (q - 1)(r - 1)$
11	ABC(adj)	0.033	$(p - 1)(q - 1)$ $(r - 1) = 2$	0.016	[11/12] 0.02	$\sigma_\epsilon^2 + \left(\frac{5}{9}\right)w\Sigma\Sigma\Sigma(\alpha\beta\gamma)_{jkl}^2/ (p - 1)(q - 1)$ $(r - 1)$
12	Residual(adj)	19.078	$= 19$	1.004		σ_ϵ^2
13	Total	267.639	$nvw - 1 = 35$			

*Interaction effects involving blocks are assumed to equal zero

**$p < .05$

§$p < .01$

PROCEDURES FOR TESTING DIFFERENCES AMONG MEANS

In a type RBPF-32^2 design, treatments A, B, and C, as well as the AB and AC interactions are not confounded with blocks. Comparisons among means for main effects and simple main effects involving AB and AC follow procedures described previously for factorial designs. However, because the BC interaction is partially confounded with blocks, cell means must be adjusted before testing simple-effects contrasts. The required adjustment for ΣY_{ij00z} and ΣY_{ij11z} is given by

$$\Sigma Y_{ij00z}(\text{adj}) \quad \text{or} \quad \Sigma Y_{ij11z}(\text{adj}) = \left[\frac{1}{6}(\Sigma S_0 + \Sigma S_2 + \Sigma S_4) + \frac{1}{3}(\Sigma S_1 + \Sigma S_3 + \Sigma S_5) \right]$$

$$- \frac{B(\text{adj})}{32} - \frac{\displaystyle\sum_{i=1}^{n}\sum_{jkl=1}^{v}\sum_{z=1}^{w} Y_{ijklz}}{4}$$

$$= \left[\frac{89}{6} + \frac{98}{3} \right] - \frac{12}{32} - \frac{187}{4} = 0.375.$$

The adjusted means are given by

$$\overline{Y}_{\cdot\cdot00\cdot}(\text{adj}) = \frac{\Sigma Y_{ij00z}}{pw} - \frac{\Sigma Y_{ij00z}(\text{adj})}{pw} = \frac{21}{(3)(3)} - \frac{0.375}{(3)(3)} = 2.29$$

$$\overline{Y}_{\cdot\cdot11\cdot}(\text{adj}) = \frac{\Sigma Y_{ij11z}}{pw} - \frac{\Sigma Y_{ij11z}(\text{adj})}{pw} = \frac{80}{(3)(3)} - \frac{0.375}{(3)(3)} = 8.85.$$

The adjustment for $\overline{Y}_{\cdot\cdot01\cdot}$ and $\overline{Y}_{\cdot\cdot10\cdot}$ is numerically equal to the adjustment for $\overline{Y}_{\cdot\cdot00\cdot}$ and $\overline{Y}_{\cdot\cdot11\cdot}$ but opposite in sign.

A t statistic for adjusted $b_k c_l$ cell means that takes into account the effective number of observations is

$$t = \frac{\hat{\psi}_{i(B)} \text{ at } c_l}{\hat{\sigma}_{\psi_{i(B)}} \text{ at } c_l} = \frac{c_k \overline{Y}_{\cdot\cdot kl\cdot} + c_{k'} \overline{Y}_{\cdot\cdot k'l\cdot}}{\sqrt{\dfrac{2MSRES}{pw} - \left(\dfrac{9}{8}\right)}}.$$

For a description of the adjustment procedure for simple-simple-effects contrasts, see Cochran and Cox (1957, 210).

OTHER MIXED DESIGNS

Mixed confounded factorial designs present many special computational problems. A complete presentation of these designs is beyond the scope of this book. General references include the classic monograph by Yates (1937) and the work of Li (1944). Additional references are cited in Table 12.11-4. This table presents relevant information for a variety of mixed confounded factorial designs.

TABLE 12.11-4 Reference Information for Mixed Confounded Factorial Designs

Type Design	Number of Treatment Combinations	Number of Observations per Block	Number of Replications for Balanced Design	Interaction(s) Confounded and Relative Information*	Reference**
§RBPF-32²	12	6	3	$BC(\frac{8}{9})$, $ABC(\frac{5}{9})$	(1) p. 224, (2) p. 253, (3) p. 348, (5) p. 661
RBPF-32³	24	6	3	$BC(\frac{8}{9})$, $BD(\frac{8}{9})$, $CD(\frac{8}{9})$, $ABC(\frac{5}{9})$, $ABD(\frac{5}{9})$, $ACD(\frac{5}{9})$	(1) p. 225
RBPF-3²2	18	6	4	$AB(\frac{7}{9})$, $ABC(\frac{5}{18})$	(1) p. 224, (2) p. 254, (3) p. 355, (5) p. 671 (6) p. 63
RBPF-3³2	54	6	4	$AB(\frac{7}{9})$, $AC(\frac{7}{9})$, $AD(\frac{7}{9})$, $ABD(\frac{5}{18})$, $ACD(\frac{5}{18})$, $BCD(\frac{5}{18})$, $ABC(\frac{1}{4})$	(1) p. 226, (4) p. 460
RBPF-42²	16	8	3	$ABC(\frac{2}{3})$	(1) p. 227
RBPF-432	24	12	9	$AC(\frac{26}{27})$, $AB(\frac{23}{27})$	(4) p. 475
RBPF-43²	36	12	2	$BC(\frac{7}{8})$, $ABC(\frac{7}{18})$	(4) p. 465
RBCF-4²2	32	16	x	ABC	(4) p. 486
RBPF-4²3	48	12	3	$AB(\frac{26}{27})$, $ABC(\frac{23}{27})$	(4) p. 467
RBPF-42³	32	8	3	$ABC(\frac{1}{3})$, $ABD(\frac{1}{3})$, $ACD(\frac{1}{3})$	(4) p. 469
RBPF-52²	20	10	5	$BC(\frac{24}{25})$, $ABC(\frac{19}{25})$	

*If a fraction does not follow the interaction, it is completely confounded with between-blocks variation.

**Information about a design can be found in the following references: (1) Cochran and Cox (1957), (2) Federer (1955), (3) Kempthorne (1952), (4) Li (1944), (5) Winer (1971), and (6) Yates (1937).

§Design described in this chapter.

*12.12 ANALYSIS OF CONFOUNDED FACTORIAL DESIGNS USING THE FULL RANK MODEL APPROACH

The full rank model approach introduced in Chapter 5 can be used to analyze data from confounded factorial experiments. The basic procedures have been discussed in Sections 8.13 and 11.15 and will not be repeated here. The reader who has followed the development of the full rank model knows that the crucial step in using the model is the formulation of coefficient matrices for testing hypotheses of interest. We will discuss the formulation of coefficient matrices for two of the designs described earlier, a type RBCF-2^2 design and a type RBPF-3^2 design.

COMPUTATIONAL PROCEDURES FOR A TYPE RBCF-2^2 DESIGN

The less than full rank experimental design model equation for a type RBCF-2^2 design is

$$Y_{ijkz} = \mu + \zeta_z + \pi_{i(z)} + \alpha_j + \beta_k + (\alpha\beta\pi)_{jki(z)} + \epsilon_{ijkz}$$
$$(i = 1, \ldots, n; j = 1, \ldots, p; k = 1, \ldots, q; z = 1, \ldots, w).$$

The restricted full rank experimental design model for this design is

$$Y_{ijkz} = \mu_{ijkz} + \epsilon_{ijkz}$$
$$(i = 1, \ldots, n; j = 1, \ldots, p; k = 1, \ldots, q; z = 1, \ldots, w)$$

where ϵ_{ijkz} is $NID(0, \sigma_\epsilon^2)$ and μ_{ijkz} is subject to the restrictions that for each z

$$\mu_{i(jk)z} - \mu_{i'(jk)z} - \mu_{i(jk)'z} + \mu_{i'(jk)'z} = 0 \qquad \text{for all } i, i', (jk), \text{ and } (jk)'.$$

These restrictions specify that for each group, the block × ab treatment combination effects equal zero. The number of treatment combinations per block is denoted by v.

The analysis procedures for this restricted full rank experimental design model will be illustrated using the data in Table 12.3-1. We need to formulate coefficient matrices that will let us compute sums of squares for

Between blocks	Within blocks
Groups (AB)	A
Blocks w. G	B
	$AB \times BL(G)$.

As in previous examples of the full rank model approach, we begin by specifying null hypotheses for each of these sources of variation. These hypotheses serve as guides

---

* This section assumes a familiarity with the matrix operations in Appendix D, Chapter 5, and Section 6.9. The reader who is interested only in the traditional approach to analysis of variance can, without loss of continuity, omit this section.

for formulating the coefficient matrices. The reader should refer to Table 12.12-1 as an aid in understanding the various hypotheses.

Between blocks H_0: $\mu_{0\cdot\cdot0} - \mu_{7\cdot\cdot1} = 0$ or H_0: $\underset{(nw-1)\times h}{C'_{BL}}\ \underset{h\times1}{\mu} = \underset{(nw-1)\times1}{0_{BL}}$

$\mu_{1\cdot\cdot0} - \mu_{7\cdot\cdot1} = 0$

$\mu_{2\cdot\cdot0} - \mu_{7\cdot\cdot1} = 0$

$\cdots\cdots\cdots$

$\mu_{6\cdot\cdot1} - \mu_{7\cdot\cdot1} = 0$

where $\underset{7\times16}{C'_{BL}} =$

$$\begin{bmatrix} 1 & 0 & 0 & 0 & 1 & 0 & 0 & 0 & 0 & 0 & 0 & -1 & 0 & 0 & 0 & -1 \\ 0 & 1 & 0 & 0 & 0 & 1 & 0 & 0 & 0 & 0 & 0 & -1 & 0 & 0 & 0 & -1 \\ 0 & 0 & 1 & 0 & 0 & 0 & 1 & 0 & 0 & 0 & 0 & -1 & 0 & 0 & 0 & -1 \\ 0 & 0 & 0 & 1 & 0 & 0 & 0 & 1 & 0 & 0 & 0 & -1 & 0 & 0 & 0 & -1 \\ 0 & 0 & 0 & 0 & 0 & 0 & 0 & 0 & 1 & 0 & 0 & -1 & 1 & 0 & 0 & -1 \\ 0 & 0 & 0 & 0 & 0 & 0 & 0 & 0 & 0 & 1 & 0 & -1 & 0 & 1 & 0 & -1 \\ 0 & 0 & 0 & 0 & 0 & 0 & 0 & 0 & 0 & 0 & 1 & -1 & 0 & 0 & 1 & -1 \end{bmatrix}$$

$\underset{1\times16}{\mu'} = [\mu_{0000}\quad \mu_{1000}\quad \mu_{2000}\quad \mu_{3000}\quad \mu_{0110}\quad \mu_{1110}\quad \mu_{2110}\quad \mu_{3110}\quad \mu_{4011}$

$\mu_{5011}\quad \mu_{6011}\quad \mu_{7011}\quad \mu_{4101}\quad \mu_{5101}\quad \mu_{6101}\quad \mu_{7101}]$

and $\underset{1\times7}{0'_{BL}} = [0\ \ 0\ \ 0\ \ 0\ \ 0\ \ 0\ \ 0]$. The coefficients, $c_{ijkz} = \pm1/v$ or 0, in C'_{BL} have been multiplied by 2 to avoid fractions. This does not affect the nature of the hypothesis that is tested.

Groups (AB) H_0: $\mu_{\cdot\cdot\cdot0} - \mu_{\cdot\cdot\cdot1}$ or H_0: $\underset{(z-1)\times h}{C'_G}\ \underset{h\times1}{\mu}\ \underset{(z-1)\times1}{0_G}$

where $\underset{1\times16}{C'_G} =$

$[1\quad 1\quad 1\quad 1\quad 1\quad 1\quad 1\quad 1\quad -1\quad -1\quad -1\quad -1\quad -1\quad -1\quad -1\quad -1]$

and $\underset{1\times1}{0_G} = \lfloor 0\rfloor$.

The coefficients, $c_{ijkz} = \pm1/nv$, in C'_G have been multiplied by 8 to avoid fractions.

Blocks w. G H_0: $\mu_{0\cdot\cdot0} - \mu_{3\cdot\cdot0} = 0$ or H_0: $\underset{w(n-1)\times h}{C'_{BL(G)}}\ \underset{h\times1}{\mu} = \underset{w(n-1)\times1}{0_{BL(G)}}$

$\mu_{1\cdot\cdot0} - \mu_{3\cdot\cdot0} = 0$

$\mu_{2\cdot\cdot0} - \mu_{3\cdot\cdot0} = 0$

$\mu_{4\cdot\cdot1} - \mu_{7\cdot\cdot1} = 0$

$\mu_{5\cdot\cdot1} - \mu_{7\cdot\cdot1} = 0$

$\mu_{6\cdot\cdot1} - \mu_{7\cdot\cdot1} = 0$

where $\underset{6\times16}{C'_{BL(G)}} =$

$$\begin{bmatrix} 1 & 0 & 0 & -1 & 1 & 0 & 0 & -1 & 0 & 0 & 0 & 0 & 0 & 0 & 0 & 0 \\ 0 & 1 & 0 & -1 & 0 & 1 & 0 & -1 & 0 & 0 & 0 & 0 & 0 & 0 & 0 & 0 \\ 0 & 0 & 1 & -1 & 0 & 0 & 1 & -1 & 0 & 0 & 0 & 0 & 0 & 0 & 0 & 0 \\ 0 & 0 & 0 & 0 & 0 & 0 & 0 & 0 & 1 & 0 & 0 & -1 & 1 & 0 & 0 & -1 \\ 0 & 0 & 0 & 0 & 0 & 0 & 0 & 0 & 0 & 1 & 0 & -1 & 0 & 1 & 0 & -1 \\ 0 & 0 & 0 & 0 & 0 & 0 & 0 & 0 & 0 & 0 & 1 & -1 & 0 & 0 & 1 & -1 \end{bmatrix}$$

TABLE 12.12-1 Type RBCF-2^2 Design

ABGS Summary Table
Entry is μ_{ijkz}

		ab	ab	$\displaystyle\sum_{jk=1}^{v} \mu_{ijkz}/v$	$\displaystyle\sum_{i=1}^{n}\sum_{jk=1}^{v} \mu_{ijkz}/nv$
	s_0	μ_{0000}	μ_{0110}	$\mu_{0\cdot\cdot0}$	
g_0	s_1	μ_{1000}	μ_{1110}	$\mu_{1\cdot\cdot0}$	$\mu_{\cdot\cdot\cdot0}$
	s_2	μ_{2000}	μ_{2110}	$\mu_{2\cdot\cdot0}$	
	s_3	μ_{3000}	μ_{3110}	$\mu_{3\cdot\cdot0}$	
	s_4	μ_{4011}	μ_{4101}	$\mu_{4\cdot\cdot1}$	
g_1	s_5	μ_{5011}	μ_{5101}	$\mu_{5\cdot\cdot1}$	$\mu_{\cdot\cdot\cdot1}$
	s_6	μ_{6011}	μ_{6101}	$\mu_{6\cdot\cdot1}$	
	s_7	μ_{7011}	μ_{7101}	$\mu_{7\cdot\cdot1}$	

AB Summary Table
Entry is μ_{ijkz}

	b_0	b_1	$\displaystyle\sum_{i=1}^{n}\sum_{k=1}^{q} \mu_{ijkz}/nq$
a_0	μ_{0000} μ_{1000} μ_{2000} μ_{3000}	μ_{4011} μ_{5011} μ_{6011} μ_{7011}	$\mu_{\cdot0\cdot\cdot}$
a_1	μ_{4101} μ_{5101} μ_{6101} μ_{7101}	μ_{0110} μ_{1110} μ_{2110} μ_{3110}	$\mu_{\cdot1\cdot\cdot}$

$$\sum_{i=1}^{n}\sum_{j=1}^{p} \mu_{ijkz}/np = \mu_{\cdot\cdot0\cdot} \qquad \mu_{\cdot\cdot1\cdot}$$

and $\quad \mathbf{0}'_{BL(G)} = \begin{bmatrix} 0 & 0 & 0 & 0 & 0 & 0 \end{bmatrix}.$
$\underset{1\times6}{}$

The coefficients, $c_{ijkz} = \pm 1/v$ or 0, in $\mathbf{C}'_{BL(G)}$ have been multiplied by 2 to avoid fractions.

Within blocks H_0:

$$\mu_{0000} - \mu_{0110} = 0$$
$$\mu_{1000} - \mu_{1110} = 0$$
$$\mu_{2000} - \mu_{2110} = 0$$
$$\mu_{3000} - \mu_{3110} = 0$$
$$\mu_{4011} - \mu_{4101} = 0$$
$$\vdots \qquad \vdots$$
$$\mu_{7011} - \mu_{7101} = 0$$

or H_0: $\underset{nw(v-1)\times h}{\mathbf{C}'_{\text{w}.BL}} \underset{h\times 1}{\boldsymbol{\mu}} = \underset{nw(v-1)\times 1}{\mathbf{O}_{\text{w}.BL}}$

where $\underset{8\times 16}{\mathbf{C}'_{\text{w}.BL}} =$

$$
\begin{bmatrix}
1 & 0 & 0 & 0 & -1 & 0 & 0 & 0 & 0 & 0 & 0 & 0 & 0 & 0 & 0 & 0 \\
0 & 1 & 0 & 0 & 0 & -1 & 0 & 0 & 0 & 0 & 0 & 0 & 0 & 0 & 0 & 0 \\
0 & 0 & 1 & 0 & 0 & 0 & -1 & 0 & 0 & 0 & 0 & 0 & 0 & 0 & 0 & 0 \\
0 & 0 & 0 & 1 & 0 & 0 & 0 & -1 & 0 & 0 & 0 & 0 & 0 & 0 & 0 & 0 \\
0 & 0 & 0 & 0 & 0 & 0 & 0 & 0 & 1 & 0 & 0 & 0 & -1 & 0 & 0 & 0 \\
0 & 0 & 0 & 0 & 0 & 0 & 0 & 0 & 0 & 1 & 0 & 0 & 0 & -1 & 0 & 0 \\
0 & 0 & 0 & 0 & 0 & 0 & 0 & 0 & 0 & 0 & 1 & 0 & 0 & 0 & -1 & 0 \\
0 & 0 & 0 & 0 & 0 & 0 & 0 & 0 & 0 & 0 & 0 & 1 & 0 & 0 & 0 & -1 \\
\end{bmatrix}
$$

and $\underset{1\times 8}{\mathbf{0}'_{\text{w}.BL}} = [0 \ 0 \ 0 \ 0 \ 0 \ 0 \ 0 \ 0]$.

The coefficients in $\mathbf{C}'_{\text{w}.BL}$ are given by $c_{ijkz} = \pm 1$ or 0.

Treatment A H_0: $\mu_{\cdot 0 \cdots} - \mu_{\cdot 1 \cdots} = 0$ or H_0: $\underset{(p-1)\times h}{\mathbf{C}'_A} \underset{h\times 1}{\boldsymbol{\mu}} = \underset{(p-1)\times 1}{\mathbf{0}_A}$

where $\underset{1\times 16}{\mathbf{C}'_A} =$
$$[1 \ \ 1 \ \ 1 \ \ 1 \ \ -1 \ \ -1 \ \ -1 \ \ -1 \ \ 1 \ \ 1 \ \ 1 \ \ 1 \ \ -1 \ \ -1 \ \ -1 \ \ -1]$$

and $\underset{1\times 1}{\mathbf{0}_A} = [0]$.

The coefficients, $c_{ijkz} = \pm 1/nq$, in $\mathbf{C}'_A$ have been multiplied by 8 to avoid fractions.

Treatment B H_0: $\mu_{\cdots 0 \cdot} - \mu_{\cdots 1 \cdot} = 0$ or H_0: $\underset{(q-1)\times h}{\mathbf{C}'_B} \underset{h\times 1}{\boldsymbol{\mu}} = \underset{(q-1)\times 1}{\mathbf{0}_B}$

where $\mathbf{C}_B' =$
$$\underset{1 \times 16}{}$$
$$[1 \quad 1 \quad 1 \quad 1 \quad -1 \quad -1 \quad -1 \quad -1 \quad -1 \quad -1 \quad -1 \quad -1 \quad 1 \quad 1 \quad 1 \quad 1]$$

and $\quad \underset{1 \times 1}{\mathbf{0}_B} = [0].$

The coefficients, $c_{ijkz} = \pm 1/np$, in $\mathbf{C}_B'$ have been multiplied by 8 to avoid fractions.

The $AB \times BL(G)$ null hypothesis corresponds to the $s = w(n-1)$ $(v-1) = 6$ restrictions on the μ_{ijkz}'s. We will denote the coefficient matrix by $\mathbf{R}'$. Construction of $\mathbf{R}'$ is facilitated by using the sets of crossed lines in the $ABGS$ Summary Table of Table 12.12-1. Each row of the $\mathbf{R}'$ matrix represents one of the six sets of crossed lines. Recall from Sections 6.9 and 9.1 that interaction terms of the form

$$\mu_{i(jk)z} - \mu_{i'(jk)z} - \mu_{i(jk)'z} + \mu_{i'(jk)'z}$$

can be obtained from the crossed lines by subtracting the two $\mu_{i(jk)z}$'s connected by a dashed line from the $\mu_{i(jk)z}$'s connected by a solid line.

$AB \times BL(G)$ Residual

$$\begin{array}{llll}
\mathrm{H}_0: & \mu_{0000} - \mu_{1000} - \mu_{0110} + \mu_{1110} = 0 & \text{or} & \mathrm{H}_0: \quad \underset{s \times h}{\mathbf{R}'} \underset{h \times 1}{\boldsymbol{\mu}} = \underset{s \times 1}{\boldsymbol{\theta}} \\
& \mu_{1000} - \mu_{2000} - \mu_{1110} + \mu_{2110} = 0 \\
& \mu_{2000} - \mu_{3000} - \mu_{2110} + \mu_{3110} = 0 \\
& \mu_{4011} - \mu_{5011} - \mu_{4101} + \mu_{5101} = 0 \\
& \mu_{5011} - \mu_{6011} - \mu_{5101} + \mu_{6101} = 0 \\
& \mu_{6011} - \mu_{7011} - \mu_{6101} + \mu_{7101} = 0
\end{array}$$

where $\quad \underset{6 \times 16}{\mathbf{R}'} =$

$$\begin{bmatrix}
1 & -1 & 0 & 0 & -1 & 1 & 0 & 0 & 0 & 0 & 0 & 0 & 0 & 0 & 0 & 0 \\
0 & 1 & -1 & 0 & 0 & -1 & 1 & 0 & 0 & 0 & 0 & 0 & 0 & 0 & 0 & 0 \\
0 & 0 & 1 & -1 & 0 & 0 & -1 & 1 & 0 & 0 & 0 & 0 & 0 & 0 & 0 & 0 \\
0 & 0 & 0 & 0 & 0 & 0 & 0 & 0 & 1 & -1 & 0 & 0 & -1 & 1 & 0 & 0 \\
0 & 0 & 0 & 0 & 0 & 0 & 0 & 0 & 0 & 1 & -1 & 0 & 0 & -1 & 1 & 0 \\
0 & 0 & 0 & 0 & 0 & 0 & 0 & 0 & 0 & 0 & 1 & -1 & 0 & 0 & -1 & 1
\end{bmatrix}$$

and $\quad \underset{1 \times 6}{\boldsymbol{\theta}'} = [0 \quad 0 \quad 0 \quad 0 \quad 0 \quad 0].$

The coefficients in $\mathbf{R}'$ are given by $c_{ijkz} = \pm 1$ or 0.

We want to test the within-blocks null hypotheses, subject to the restrictions that $\mathbf{R}'\boldsymbol{\mu} = \boldsymbol{\theta}$. To do this we form augmented matrices of the form

$$\mathbf{Q}_w' = \begin{bmatrix} \mathbf{R}' \\ \mathbf{C}_w' \end{bmatrix} \quad \text{and} \quad \boldsymbol{\eta}_w = \begin{bmatrix} \boldsymbol{\theta} \\ \mathbf{0}_w \end{bmatrix}$$

where $\mathbf{C}_w'$ is the coefficient matrix for a within-blocks source of variation and $\mathbf{0}_w$ is the null vector associated with $\mathbf{C}_w'$. If the rows of $\mathbf{R}'$ are orthogonal to the rows of a within-blocks coefficient matrix, that is, if

$$\mathbf{R}' \, \mathbf{C}_w = \mathbf{0}$$

then the sum of squares computational formula $(\mathbf{Q}_w' \hat{\boldsymbol{\mu}})' \, (\mathbf{Q}_w' \mathbf{Q}_w)^{-1} \, (\mathbf{Q}_w' \hat{\boldsymbol{\mu}}) - SSAB$ $\times BL(G)$ in Table 12.12-2 simplifies to $(\mathbf{C}_w' \hat{\boldsymbol{\mu}})' \, (\mathbf{C}_w' \mathbf{C}_w)^{-1} \, (\mathbf{C}_w' \hat{\boldsymbol{\mu}})$. Tests of the between-blocks null hypotheses are not subject to the restrictions $\mathbf{R}' \boldsymbol{\mu} = \boldsymbol{\theta}$. Procedures for computing the sums of squares and restricted means, $\hat{\boldsymbol{\mu}}_R$, are shown in Table 12.12-2. A comparison of these sums of squares with those in Table 12.3-1, where the traditional sum of squares approach is used, reveals that they are identical.

TABLE 12.12-2 Computational Procedures for a Type RBCF-2^2 Design Using a Restricted Full Rank Experimental Design Model (Data are from Table 12.3-1.)

(i) Data and matrix formulas for computing sums of squares: ($\mathbf{C}_{BL}'$, $\mathbf{C}_G'$, $\mathbf{C}_{BL(G)}'$, $\mathbf{R}'$, $\mathbf{Q}_{w.BL}'$, $\mathbf{Q}_A'$, and $\mathbf{Q}_B'$ are defined in the text; $N = 16$, $h = 16$, $p = 2$, $q = 2$, $n = 4$, $s = 6$):

$$
\begin{array}{c}
\underset{N\times 1}{\mathbf{y}}
\end{array}
$$

$$
\underset{h\times 1}{\hat{\boldsymbol{\mu}}}
\qquad
\underset{h\times 1}{\mathbf{R}(\mathbf{R}'\mathbf{R})^{-1}\mathbf{R}'\hat{\boldsymbol{\mu}}}
\qquad
\underset{h\times 1}{\hat{\boldsymbol{\mu}}_R}
$$

$$
a_0 b_0 \left\{ \begin{bmatrix} 3 \\ 5 \\ 6 \\ 5 \\ - \end{bmatrix} \right.
\begin{bmatrix} -1.125 \\ 0.875 \\ -0.125 \\ 0.375 \\ \hline \end{bmatrix}
\begin{bmatrix} 4.125 \\ 4.125 \\ 6.125 \\ 4.625 \\ \hline \end{bmatrix}
$$

$$
a_1 b_1 \left\{ \begin{bmatrix} 16 \\ 14 \\ 17 \\ 15 \\ - \end{bmatrix} \right.
\;-\;
\begin{bmatrix} 1.125 \\ -0.875 \\ 0.125 \\ -0.375 \\ \hline \end{bmatrix}
\;=\;
\begin{bmatrix} 14.875 \\ 14.875 \\ 16.875 \\ 15.375 \\ \hline \end{bmatrix}
$$

$$
a_0 b_1 \left\{ \begin{bmatrix} 14 \\ 14 \\ 16 \\ 16 \\ - \end{bmatrix} \right.
\begin{bmatrix} -0.125 \\ 0.375 \\ 0.875 \\ -1.125 \\ \hline \end{bmatrix}
\begin{bmatrix} 14.125 \\ 13.625 \\ 15.125 \\ 17.125 \\ \hline \end{bmatrix}
$$

$$
a_1 b_0 \left\{ \begin{bmatrix} 7 \\ 6 \\ 7 \\ 11 \end{bmatrix} \right.
\begin{bmatrix} 0.125 \\ -0.375 \\ -0.875 \\ 1.125 \end{bmatrix}
\begin{bmatrix} 6.875 \\ 6.375 \\ 7.875 \\ 9.875 \end{bmatrix}
$$

$$\sum_{i=1}^{N} Y_i = 172$$

$$
\underset{1\times N}{\mathbf{y}'} \qquad \underset{N\times 1}{\mathbf{y}} \qquad \underset{1\times 1}{(\mathbf{y}'\mathbf{y})}
$$

$$
[3 \; 5 \cdots 16]
\begin{bmatrix} 3 \\ 5 \\ \vdots \\ 16 \end{bmatrix}
= 2220.00
$$

Table 12.12-2 (continued)

$$SSTO = \mathbf{y'y} - (\Sigma Y_i)^2/N = 2220.00 - \frac{(172)^2}{16} = 371.00$$

$$SSBL = (\mathbf{C'_{BL}\hat{\mu}})' \ (\mathbf{C'_{BL}C_{BL}})^{-1} \ (\mathbf{C'_{BL}\hat{\mu}}) = 26.00$$
$$\scriptstyle 1\times(nw-1) \quad (nw-1)\times(nw-1) \quad (nw-1)\times 1$$

$$SSG = (\mathbf{C'_G\hat{\mu}})' \ (\mathbf{C'_G C_G})^{-1} \ (\mathbf{C'_G\hat{\mu}}) = 6.25$$
$$\scriptstyle 1\times(z-1) \quad (z-1)\times(z-1) \quad (z-1)\times 1$$

$$SSBL(G) = (\mathbf{C'_{BL(G)}\hat{\mu}})' \ (\mathbf{C'_{BL(G)}C_{BL(G)}})^{-1} \ (\mathbf{C'_{BL(G)}\hat{\mu}}) = 19.75$$
$$\scriptstyle 1\times w(n-1) \quad w(n-1)\times w(n-1) \quad w(n-1)\times 1$$

$$SSAB \times BL(G) = (\mathbf{R'\hat{\mu}})' \ (\mathbf{R'R'})^{-1} \ (\mathbf{R'\hat{\mu}}) = 8.75$$
$$\scriptstyle 1\times s \quad s\times s \quad s\times 1$$

$$SS \text{ w. } BL = (\mathbf{Q'_{w.BL}\hat{\mu}})' \ (\mathbf{Q'_{w.BL}Q_{w.BL}})^{-1} \ (\mathbf{Q'_{w.BL}\hat{\mu}}) - SSAB \times BL(G)$$
$$\scriptstyle 1\times(s+nw-1) \quad (s+nw-1)\times(s+nw-1) \quad (s+nw-1)\times 1$$
$$= 353.75 - 8.75 = 345.00$$

$$SSA = (\mathbf{Q'_A\hat{\mu}})' \ (\mathbf{Q'_A Q_A})^{-1} \ (\mathbf{Q'_A\hat{\mu}}) - SSAB \times BL(G)$$
$$\scriptstyle 1\times(s+p-1) \quad (s+p-1)\times(s+p-1) \quad (s+p-1)\times 1$$
$$= 21.00 - 8.75 = 12.25$$

$$SSB = (\mathbf{Q'_B\hat{\mu}})' \ (\mathbf{Q'_B Q_B})^{-1} \ (\mathbf{Q'_B\hat{\mu}}) - SSAB \times BL(G)$$
$$\scriptstyle 1\times(s+q-1) \quad (s+q-1)\times(s+q-1) \quad (s+q-1)\times 1$$
$$= 332.75 - 8.75 = 324.00$$

An important advantage of the full rank model approach is that it can be used to specify the null hypotheses tested when the number of blocks within each group is not equal and when there are missing observations. The use of the full rank model approach in these cases is described in detail for a type SPF-$p \cdot q$ design in Section 11.15. The procedures generalize to a type RBCF-p^k design.

COMPUTATIONAL PROCEDURES FOR A TYPE RBPF-3^2 DESIGN

The restricted full rank experimental design model for a type RBPF-3^2 design is

$$Y_{ijkz} = \mu_{ijkz} + \epsilon_{ijkz}$$
$$(i = 1, \ldots, n; j = 1, \ldots, p; k = 1, \ldots, q; z = 1, \ldots, w)$$

where ϵ_{ijkz} is $NID(0, \sigma_\epsilon^2)$ and μ_{ijkz} is subject to the restrictions that

$$\mu_{ijkz} - \mu_{i'jkz} - \mu_{ij'kz} + \mu_{i'j'kz} = 0 \quad \text{for all } i, i', j, \text{ and } j'$$
$$\mu_{ijkz} - \mu_{i'jkz} - \mu_{ijk'z} + \mu_{i'jk'z} = 0 \quad \text{for all } i, i', k, \text{ and } k'.$$

These restrictions specify that the block $\times$ treatment A interaction effects and the block $\times$ treatment B interaction effects equal zero.

The analysis procedures for this restricted full rank experimental design model will be illustrated using the data in Table 12.8-3. To perform the analysis, we need to formulate coefficient matrices for computing the following sums of squares.

Between blocks	A
Groups	B
Blocks w. G	AB' (within)
Within blocks	Residual

The null hypothesis and associated coefficient matrix for each of these sources of variation are as follows. It may be helpful to refer to Table 12.12-3 where the various means are defined.

Between blocks H_0: $\mu_{0 \cdot \cdot 0} - \mu_{5 \cdot \cdot 1} = 0$ or H_0: $\underset{(nw-1) \times h}{\mathbf{C}'_{BL}} \ \underset{h \times 1}{\boldsymbol{\mu}} = \underset{(nw-1) \times 1}{\mathbf{0}_{BL}}$

$\mu_{1 \cdot \cdot 0} - \mu_{5 \cdot \cdot 1} = 0$

$\mu_{2 \cdot \cdot 0} - \mu_{5 \cdot \cdot 1} = 0$

$\mu_{3 \cdot \cdot 0} - \mu_{5 \cdot \cdot 1} = 0$

$\mu_{4 \cdot \cdot 0} - \mu_{5 \cdot \cdot 1} = 0$

TABLE 12.12-3   Type RBPF-3^2 Design

ABGS Summary Table
Entry is μ_{ijkz}

			ab	ab	ab	$\sum_{jk=1}^{v} \mu_{ijkz}/v$	$\sum_{i=1}^{n}\sum_{jk=1}^{v} \mu_{ijkz}/nv$
	$(AB)_0$	s_0	μ_{0000}	μ_{0120}	μ_{0210}	$\mu_{0 \cdot \cdot 0}$	
g_0	$(AB)_1$	s_1	μ_{1010}	μ_{1100}	μ_{1220}	$\mu_{1 \cdot \cdot 0}$	$\mu_{\cdot \cdot \cdot 0}$
	$(AB)_2$	s_2	μ_{2020}	μ_{2110}	μ_{2200}	$\mu_{2 \cdot \cdot 0}$	
	$(AB^2)_0$	s_3	μ_{3001}	μ_{3111}	μ_{3221}	$\mu_{3 \cdot \cdot 1}$	
g_1	$(AB^2)_1$	s_4	μ_{4021}	μ_{4101}	μ_{4211}	$\mu_{4 \cdot \cdot 1}$	$\mu_{\cdot \cdot \cdot 1}$
	$(AB^2)_2$	s_5	μ_{5011}	μ_{5121}	μ_{5201}	$\mu_{5 \cdot \cdot 1}$	

AG Summary Table
Entry is $\mu_{\cdot j \cdot z}$

	a_0	a_1	a_2
g_0	$\mu_{\cdot 0 \cdot 0}$	$\mu_{\cdot 1 \cdot 0}$	$\mu_{\cdot 2 \cdot 0}$
g_1	$\mu_{\cdot 0 \cdot 1}$	$\mu_{\cdot 1 \cdot 1}$	$\mu_{\cdot 2 \cdot 1}$

$\sum_{z=1}^{w} \mu_{\cdot j \cdot z}/w = \mu_{\cdot 0 \cdot \cdot}$ $\mu_{\cdot 1 \cdot \cdot}$ $\mu_{\cdot 2 \cdot \cdot}$

BG Summary Table
Entry is $\mu_{\cdot \cdot kz}$

	b_0	b_1	b_2
g_0	$\mu_{\cdot \cdot 00}$	$\mu_{\cdot \cdot 10}$	$\mu_{\cdot \cdot 20}$
g_1	$\mu_{\cdot \cdot 01}$	$\mu_{\cdot \cdot 11}$	$\mu_{\cdot \cdot 21}$

$\sum_{z=1}^{w} \mu_{\cdot \cdot kz}/w = \mu_{\cdot \cdot 0 \cdot}$ $\mu_{\cdot \cdot 1 \cdot}$ $\mu_{\cdot \cdot 2 \cdot}$

where $\underset{5\times 18}{\mathbf{C}'_{BL}} =$

$$\begin{bmatrix} 1 & 0 & 0 & 1 & 0 & 0 & 1 & 0 & 0 & 0 & 0 & -1 & 0 & 0 & -1 & 0 & 0 & -1 \\ 0 & 1 & 0 & 0 & 1 & 0 & 0 & 1 & 0 & 0 & 0 & -1 & 0 & 0 & -1 & 0 & 0 & -1 \\ 0 & 0 & 1 & 0 & 0 & 1 & 0 & 0 & 1 & 0 & 0 & -1 & 0 & 0 & -1 & 0 & 0 & -1 \\ 0 & 0 & 0 & 0 & 0 & 0 & 0 & 0 & 0 & 1 & 0 & -1 & 1 & 0 & -1 & 1 & 0 & -1 \\ 0 & 0 & 0 & 0 & 0 & 0 & 0 & 0 & 0 & 0 & 1 & -1 & 0 & 1 & -1 & 0 & 1 & -1 \end{bmatrix}$$

$\underset{1\times 5}{\mathbf{0}'_{BL}} = [0 \quad 0 \quad 0 \quad 0 \quad 0]$

and $\underset{1\times 18}{\boldsymbol{\mu}'} = [\mu_{0000} \quad \mu_{1010} \quad \mu_{2020} \quad \mu_{0120} \quad \mu_{1100} \quad \mu_{2110} \quad \mu_{0210} \quad \mu_{1220}$
$\qquad\qquad \mu_{2200} \quad \mu_{3001} \quad \mu_{4021} \quad \mu_{5011} \quad \mu_{3111} \quad \mu_{4101} \quad \mu_{5121} \quad \mu_{3221} \quad \mu_{4211} \quad \mu_{5201}]$.

The coefficients, $c_{ijkz} = \pm 1/v$ or 0, in $\mathbf{C}'_{BL}$ have been multiplied by 3 to avoid fractions.

$\qquad$ Groups $\qquad$ H_0: $\mu_{\cdots 0} - \mu_{\cdots 1} = 0$ $\qquad$ or $\qquad$ H_0: $\underset{(z-1)\times h}{\mathbf{C}'_G} \; \underset{h\times 1}{\boldsymbol{\mu}} = \underset{(z-1)\times 1}{\mathbf{0}_G}$

where $\underset{1\times 18}{\mathbf{C}'_G} =$
$[1 \quad 1 \quad 1 \quad 1 \quad 1 \quad 1 \quad 1 \quad 1 \quad 1 \quad -1 \quad -1 \quad -1 \quad -1 \quad -1 \quad -1 \quad -1 \quad -1 \quad -1]$

and $\qquad \mathbf{0}_G = [0]$.

The coefficients, $c_{ijkz} = \pm 1/nv$, in $\mathbf{C}'_G$ have been multiplied by 9 to avoid fractions.

$\qquad$ Blocks w. G $\qquad$ H_0: $\mu_{0\cdot\cdot 0} - \mu_{2\cdot\cdot 0} = 0$ $\qquad$ or $\qquad$ H_0: $\underset{w(n-1)\times h}{\mathbf{C}'_{BL}} \; \underset{h\times 1}{\boldsymbol{\mu}} = \underset{w(n-1)\times 1}{\mathbf{0}_{BL(G)}}$
$\qquad\qquad\qquad\qquad\qquad \mu_{1\cdot\cdot 0} - \mu_{2\cdot\cdot 0} = 0$
$\qquad\qquad\qquad\qquad\qquad \mu_{3\cdot\cdot 1} - \mu_{5\cdot\cdot 1} = 0$
$\qquad\qquad\qquad\qquad\qquad \mu_{4\cdot\cdot 1} - \mu_{5\cdot\cdot 1} = 0$

where $\underset{4\times 18}{\mathbf{C}'_{BL(G)}} =$

$$\begin{bmatrix} 1 & 0 & -1 & 1 & 0 & -1 & 1 & 0 & -1 & 0 & 0 & 0 & 0 & 0 & 0 & 0 & 0 & 0 \\ 0 & 1 & -1 & 0 & 1 & -1 & 0 & 1 & -1 & 0 & 0 & 0 & 0 & 0 & 0 & 0 & 0 & 0 \\ 0 & 0 & 0 & 0 & 0 & 0 & 0 & 0 & 0 & 1 & 0 & -1 & 1 & 0 & -1 & 1 & 0 & -1 \\ 0 & 0 & 0 & 0 & 0 & 0 & 0 & 0 & 0 & 0 & 1 & -1 & 0 & 1 & -1 & 0 & 1 & -1 \end{bmatrix}$$

and $\qquad \underset{1\times 4}{\mathbf{0}'_{BL(G)}} = [0 \quad 0 \quad 0 \quad 0]$.

The coefficients, $c_{ijkz} = \pm 1/v$ or 0, in $\mathbf{C}'_{BL(G)}$ have been multiplied by 3 to avoid fractions.

$\qquad$ Within blocks $\quad H_0$: $\mu_{0000} - \mu_{0210} = 0$ $\qquad$ or $\qquad$ H_0: $\underset{nw(v-1)\times h}{\mathbf{C}'_{w.BL}} \; \underset{h\times 1}{\boldsymbol{\mu}} = \underset{nw(v-1)\times 1}{\mathbf{0}_{w.BL}}$
$\qquad\qquad\qquad\qquad\qquad \mu_{0120} - \mu_{0210} = 0$
$\qquad\qquad\qquad\qquad\qquad \mu_{1010} - \mu_{1220} = 0$
$\qquad\qquad\qquad\qquad\qquad \mu_{1100} - \mu_{1220} = 0$
$\qquad\qquad\qquad\qquad\qquad \mu_{2020} - \mu_{2200} = 0$

$$\mu_{2110} - \mu_{2200} = 0$$

$$\vdots$$

$$\mu_{5121} - \mu_{5201} = 0$$

where $\underset{12 \times 18}{\mathbf{C}'_{\text{w}.BL}} =$

$$
\begin{bmatrix}
1 & 0 & 0 & 0 & 0 & 0 & -1 & 0 & 0 & 0 & 0 & 0 & 0 & 0 & 0 & 0 & 0 & 0 \\
0 & 0 & 0 & 1 & 0 & 0 & -1 & 0 & 0 & 0 & 0 & 0 & 0 & 0 & 0 & 0 & 0 & 0 \\
0 & 1 & 0 & 0 & 0 & 0 & 0 & -1 & 0 & 0 & 0 & 0 & 0 & 0 & 0 & 0 & 0 & 0 \\
0 & 0 & 0 & 0 & 1 & 0 & 0 & -1 & 0 & 0 & 0 & 0 & 0 & 0 & 0 & 0 & 0 & 0 \\
0 & 0 & 1 & 0 & 0 & 0 & 0 & 0 & -1 & 0 & 0 & 0 & 0 & 0 & 0 & 0 & 0 & 0 \\
0 & 0 & 0 & 0 & 0 & 1 & 0 & 0 & -1 & 0 & 0 & 0 & 0 & 0 & 0 & 0 & 0 & 0 \\
0 & 0 & 0 & 0 & 0 & 0 & 0 & 0 & 0 & 1 & 0 & 0 & 0 & 0 & 0 & -1 & 0 & 0 \\
0 & 0 & 0 & 0 & 0 & 0 & 0 & 0 & 0 & 0 & 0 & 0 & 1 & 0 & 0 & -1 & 0 & 0 \\
0 & 0 & 0 & 0 & 0 & 0 & 0 & 0 & 0 & 0 & 1 & 0 & 0 & 0 & 0 & 0 & -1 & 0 \\
0 & 0 & 0 & 0 & 0 & 0 & 0 & 0 & 0 & 0 & 0 & 0 & 0 & 1 & 0 & 0 & -1 & 0 \\
0 & 0 & 0 & 0 & 0 & 0 & 0 & 0 & 0 & 0 & 0 & 1 & 0 & 0 & 0 & 0 & 0 & -1 \\
0 & 0 & 0 & 0 & 0 & 0 & 0 & 0 & 0 & 0 & 0 & 0 & 0 & 0 & 1 & 0 & 0 & -1
\end{bmatrix}
$$

and $\underset{1 \times 12}{\mathbf{0}'_{\text{w}.BL}} = [0 \quad 0 \quad 0 \quad 0 \quad 0 \quad 0 \quad 0 \quad 0 \quad 0 \quad 0 \quad 0 \quad 0].$

The coefficients in $\mathbf{C}'_{\text{w}.BL}$ are given by $c_{ijkz} = \pm 1$ or 0.

Treatment A H_0: $\mu_{.0..} - \mu_{.2..} = 0$ or H_0: $\underset{(p-1) \times h}{\mathbf{C}'_A} \underset{h \times 1}{\boldsymbol{\mu}} = \underset{(p-1) \times 1}{\mathbf{0}_A}$

$$\mu_{.1..} - \mu_{.2..} = 0$$

where $\underset{2 \times 18}{\mathbf{C}'_A} =$

$$
\begin{bmatrix}
1 & 1 & 1 & 0 & 0 & 0 & -1 & -1 & -1 & 1 & 1 & 1 & 0 & 0 & 0 & -1 & -1 & -1 \\
0 & 0 & 0 & 1 & 1 & 1 & -1 & -1 & -1 & 0 & 0 & 0 & 1 & 1 & 1 & -1 & -1 & -1
\end{bmatrix}
$$

and $\underset{1 \times 2}{\mathbf{0}'_A} = [0 \quad 0].$

The coefficients, $c_{ijkz} = \pm 1/nq$ or 0, in $\mathbf{C}'_A$ have been multiplied by 9 to avoid fractions.

Treatment B H_0: $\mu_{..0.} - \mu_{..2.} = 0$ or H_0: $\underset{(q-1) \times h}{\mathbf{C}'_B} \underset{h \times 1}{\boldsymbol{\mu}} = \underset{(q-1) \times 1}{\mathbf{0}_B}$

$$\mu_{..1.} - \mu_{..2.} = 0$$

where $\underset{2 \times 18}{\mathbf{C}'_B} =$

$$
\begin{bmatrix}
1 & 0 & -1 & -1 & 1 & 0 & 0 & -1 & 1 & 1 & -1 & 0 & 0 & 1 & -1 & -1 & 0 & 1 \\
0 & 1 & -1 & -1 & 0 & 1 & 1 & -1 & 0 & 0 & -1 & 1 & 1 & 0 & -1 & -1 & 1 & 0
\end{bmatrix}
$$

and $\underset{1 \times 2}{\mathbf{0}'_B} = [0 \quad 0].$

The coefficients, $c_{ijkz} = \pm 1/np$ or 0, in $\mathbf{C}'_B$ have been multiplied by 9 to avoid fractions.

     The AB interaction components, (AB) and (AB^2), are computed from the groups in which they are not confounded with blocks. We saw in Section 12.8 that

(AB) is not confounded in group 1 and (AB^2) is not confounded in group 0. The treatment combinations that are used to compute (AB^2) from group 0 and (AB) from group 1 are

$$
\begin{array}{llll}
(AB^2)_0 = 00 & 11 & 22 & \qquad (AB)_0 = 00 \quad 12 \quad 21 \\
(AB^2)_1 = 02 & 10 & 21 & \qquad (AB)_1 = 01 \quad 10 \quad 22 \\
(AB)^2_2 = 01 & 12 & 20 & \qquad (AB)_2 = 02 \quad 11 \quad 20.
\end{array}
$$

The null hypothesis for AB'(within) can be expressed as

$$
\text{H}_0: \quad (\mu_{0000} + \mu_{2110} + \mu_{1220}) - (\mu_{1010} + \mu_{0120} + \mu_{2200}) = 0
$$
$$
(\mu_{2020} + \mu_{1100} + \mu_{0210}) - (\mu_{1010} + \mu_{0120} + \mu_{2200}) = 0
$$
$$
(\mu_{3001} + \mu_{5121} + \mu_{4211}) - (\mu_{4021} + \mu_{3111} + \mu_{5201}) = 0
$$
$$
(\mu_{5011} + \mu_{4101} + \mu_{3221}) - (\mu_{4021} + \mu_{3111} + \mu_{5201}) = 0
$$

$$
\text{or} \quad \text{H}_0: \quad \underset{(p-1)(q-1)\times h}{\mathbf{C}'_{AB}} \underset{h\times 1}{\boldsymbol{\mu}} = \underset{(p-1)(q-1)\times 1}{\mathbf{0}_{AB}}
$$

where $\underset{4\times 18}{\mathbf{C}'_{AB}} =$

$$
\begin{bmatrix}
1 & -1 & 0 & -1 & 0 & 1 & 0 & 1 & -1 & 0 & 0 & 0 & 0 & 0 & 0 & 0 & 0 & 0 \\
0 & -1 & 1 & -1 & 1 & 0 & 1 & 0 & -1 & 0 & 0 & 0 & 0 & 0 & 0 & 0 & 0 & 0 \\
0 & 0 & 0 & 0 & 0 & 0 & 0 & 0 & 0 & 1 & -1 & 0 & -1 & 0 & 1 & 0 & 1 & -1 \\
0 & 0 & 0 & 0 & 0 & 0 & 0 & 0 & 0 & -1 & 1 & -1 & 1 & 0 & 1 & 0 & -1 \\
\end{bmatrix}
$$

and $\underset{1\times 4}{\mathbf{0}'_{AB}} = [0 \quad 0 \quad 0 \quad 0].$

The coefficients in $\mathbf{C}'_{AB}$ are given by $c_{ijkz} = \pm 1$ or 0.

The residual sum of squares is the pooled interactions of $A \times G$ and $B \times G$. The crossed lines in the AG and BG Summary Tables in Table 12.12-2 can be used to construct the null hypothesis for these sources of variation. We will denote this coefficient matrix by $\mathbf{R}'$ since it represents the $s = (p - 1)(w - 1) + (q - 1)$ $(w - 1) = 4$ restrictions that were placed on the μ_{ijkz}'s.

$$
\text{Residual} \quad \text{H}_0: \quad \mu_{\cdot 0 \cdot 0} - \mu_{\cdot 0 \cdot 1} - \mu_{\cdot 1 \cdot 0} + \mu_{\cdot 1 \cdot 1} = 0 \quad \text{or} \quad \text{H}_0: \underset{s\times h}{\mathbf{R}'} \underset{h\times 1}{\boldsymbol{\mu}} = \underset{s\times 1}{\boldsymbol{\theta}}
$$
$$
\mu_{\cdot 1 \cdot 0} - \mu_{\cdot 1 \cdot 1} - \mu_{\cdot 2 \cdot 0} + \mu_{\cdot 2 \cdot 1} = 0
$$
$$
\mu_{\cdot \cdot 00} - \mu_{\cdot \cdot 01} - \mu_{\cdot \cdot 10} + \mu_{\cdot \cdot 11} = 0
$$
$$
\mu_{\cdot \cdot 10} - \mu_{\cdot \cdot 11} - \mu_{\cdot \cdot 20} + \mu_{\cdot \cdot 21} = 0
$$

where $\underset{4\times 18}{\mathbf{R}'} =$

$$
\begin{bmatrix}
1 & 1 & 1 & -1 & -1 & -1 & 0 & 0 & 0 & -1 & -1 & -1 & 1 & 1 & 1 & 0 & 0 & 0 \\
0 & 0 & 0 & 1 & 1 & 1 & -1 & -1 & -1 & 0 & 0 & 0 & -1 & -1 & -1 & 1 & 1 & 1 \\
1 & -1 & 0 & 0 & 1 & -1 & -1 & 0 & 1 & -1 & 0 & 1 & 1 & -1 & 0 & 0 & 1 & -1 \\
0 & 1 & -1 & -1 & 0 & 1 & 1 & -1 & 0 & 0 & 1 & -1 & -1 & 0 & 1 & 1 & -1 & 0 \\
\end{bmatrix}
$$

and $\underset{1\times 4}{\boldsymbol{\theta}'} = [0 \quad 0 \quad 0 \quad 0].$

The coefficients, $c_{ijkz} = \pm 1/n$ or 0, in $\mathbf{R}'$ have been multiplied by 3 to avoid fractions.

We want to test the within-blocks null hypotheses, subject to the restrictions that $\mathbf{R}'\boldsymbol{\mu} = \boldsymbol{\theta}$. This is accomplished by forming augmented matrices of the form

$$\mathbf{Q}'_w = \begin{bmatrix} \mathbf{R}' \\ \mathbf{C}'_w \end{bmatrix} \quad \text{and} \quad \boldsymbol{\eta}_w = \begin{bmatrix} \boldsymbol{\theta} \\ \mathbf{0}_w \end{bmatrix}$$

where $\mathbf{C}'_w$ is the coefficient matrix for a within-blocks source of variation and $\mathbf{0}_w$ is the null vector associated with $\mathbf{C}'_w$. The null hypothesis that is tested has the form

$$\mathbf{Q}'_w \boldsymbol{\mu} = \boldsymbol{\eta}_w.$$

If the rows of $\mathbf{R}'$ are orthogonal to the rows of $\mathbf{C}'_w$, the sum of squares computational formula $(\mathbf{Q}'_w \hat{\boldsymbol{\mu}})' (\mathbf{Q}'_w \mathbf{Q}_w)^{-1} (\mathbf{Q}'_w \hat{\boldsymbol{\mu}}) - SSRES$ in Table 12.12-4 simplifies to $(\mathbf{C}'_w \hat{\boldsymbol{\mu}})'$ $\cdot (\mathbf{C}'_w \mathbf{C}_w)^{-1} (\mathbf{C}'_w \hat{\boldsymbol{\mu}})$. Procedures for computing the sums of squares and restricted means, $\hat{\boldsymbol{\mu}}_R$, are shown in Table 12.12-4. A comparison of these sums of squares with those in Table 12.8-3, where the traditional sums of squares approach is used, reveals that they are identical. We have now illustrated all of the basic procedures involved in using the full rank model to test hypotheses concerning fixed effects for randomized block confounded factorial designs.

12.13 GROUP-INTERACTION CONFOUNDING BY MEANS OF A LATIN SQUARE

An alternative confounding scheme that is much simpler than that already described uses a Latin square building block. Confounded factorial designs based on a Latin square are not limited to experiments in which the number of treatment levels is a prime number or a power of a prime number. This confounding scheme does have other restrictions that are described in subsequent paragraphs.

Consider the Latin Square in Figure 12.13-1. A comparison of the treatment combinations in this square with those in group 1 of Figure 12.8-2, which would form a type RBCF-3^2 design, reveals that they are identical. These treatment combinations are based on the relation $a_j + 2b_k = z \pmod 3$ or (AB^2). From this it is apparent that a confounded factorial design based on a Latin square also confounds a component of the AB interaction with groups of blocks. Although both designs use the same treatment combinations and reduce the block size from nine to three, they involve different construction procedures and assumptions.

A confounded factorial design based on a Latin square is denoted by the letters $LSCF$. The design in Figure 12.13-1 has two treatments and one nuisance variable. A factorial design based on a Latin square is classified as a confounded factorial design if the levels of either rows or columns correspond to an additional treatment instead of a nuisance variable. This point is developed further in Section 13.7.

TABLE 12.12-4 Computational Procedures for a Type RBPF-3^2 Design Using a Full Rank Experimental Design Model (Data are from Table 12.8-3.)

(i) Data and matrix formulas for computing sums of squares (C'_{BL}, C'_G, $C'_{BL(G)}$, R', $Q'_{w.BL}$, Q'_A, Q'_B, and Q'_{AB} are defined in the text; $N = 18$, $h = 18$, $p = 3$, $q = 3$, $n = 3$, $s = 4$):

$\underset{N\times 1}{\mathbf{y}}$	$\underset{h\times 1}{\hat{\boldsymbol{\mu}}}$	$\underset{h\times 1}{[\mathbf{R(R'R)^{-1}R'\hat{\boldsymbol\mu}}]}$	$\underset{h\times 1}{\hat{\boldsymbol\mu}_R}$
3	3	0.3889	2.6111
4	4	0.0556	3.9444
7	7	0.3889	6.6111
7	7	−0.1111	7.1111
3	3	−0.1111	3.1111
3	3	−0.4444	3.4444
4	4	−0.2778	4.2778
6	6	0.0556	5.9444
3	3	0.0556	2.9444
2	2	−0.3889	2.3889
5	5	−0.3889	5.3889
2	2	−0.0556	2.0556
4	4	0.4444	3.5556
1	1	0.1111	0.8889
6	6	0.1111	5.8889
5	5	−0.0556	5.0556
3	3	0.2778	2.7222
2	2	−0.0556	2.0556

(g_0 spans the first nine rows; g_1 spans the last nine rows.)

$$\sum_{i=1}^{N} Y_{ijkz} = 70$$

$$\underset{1\times N}{\mathbf{y}'} = [\,3 \quad 4 \quad \cdots \quad 2\,]$$

$$\underset{1\times 1}{(\mathbf{y'y})} = 326.000$$

$$SSTO = \mathbf{y'y} - (\Sigma Y_i)^2/N = 326.000 - \frac{(70)^2}{8} = 53.778$$

$$SSBL = \underset{1\times(nw-1)}{(\mathbf{C}'_{BL}\hat{\boldsymbol\mu})'} \; \underset{(nw-1)\times(nw-1)}{(\mathbf{C}'_{BL}\mathbf{C}_{BL})^{-1}} \; \underset{(nw-1)\times 1}{(\mathbf{C}'_{BL}\hat{\boldsymbol\mu})} = 6.445$$

$$SSG = \underset{1\times(z-1)}{(\mathbf{C}'_G\hat{\boldsymbol\mu})'} \; \underset{(z-1)\times(z-1)}{(\mathbf{C}'_G\mathbf{C}_G)^{-1}} \; \underset{(z-1)\times 1}{(\mathbf{C}'_G\hat{\boldsymbol\mu})} = 5.556$$

$$SSBL(G) = \underset{1\times w(n-1)}{(\mathbf{C}'_{BL(G)}\hat{\boldsymbol\mu})'} \; \underset{w(n-1)\times w(n-1)}{(\mathbf{C}'_{BL(G)}\mathbf{C}_{BL(G)})^{-1}} \; \underset{w(n-1)\times 1}{(\mathbf{C}'_{BL(G)}\hat{\boldsymbol\mu})} = 0.889$$

$$SSRES = \underset{1\times s}{(\mathbf{R}'\hat{\boldsymbol\mu})'} \; \underset{s\times s}{(\mathbf{R}'\mathbf{R})^{-1}} \; \underset{s\times 1}{(\mathbf{R}'\hat{\boldsymbol\mu})} = 1.222$$

$$SSw.BL = \underset{1\times nw(c-1)}{(\mathbf{Q}'_{w.BL}\hat{\boldsymbol\mu})'} \; \underset{nw(c-1)\times nw(c-1)}{(\mathbf{Q}'_{w.BL}\mathbf{Q}_{w.BL})^{-1}} \; \underset{nw(c-1)\times 1}{(\mathbf{Q}'_{w.BL}\hat{\boldsymbol\mu})} - SSRES$$
$$= 48.555 - 1.222 = 47.333$$

$$SSA = \underset{1\times(p-1)}{(\mathbf{Q}'_A\hat{\boldsymbol\mu})'} \; \underset{(p-1)\times(p-1)}{(\mathbf{Q}'_A\mathbf{Q}_A)^{-1}} \; \underset{(p-1)\times 1}{(\mathbf{Q}'_A\hat{\boldsymbol\mu})} - SSRES$$
$$= 1.333 - 1.222 = 0.111$$

$$SSB = \underset{1\times(q-1)}{(\mathbf{Q}'_B\hat{\boldsymbol\mu})'} \; \underset{(q-1)\times(q-1)}{(\mathbf{Q}'_B\mathbf{Q}_B)^{-1}} \; \underset{(q-1)\times 1}{(\mathbf{Q}'_B\hat{\boldsymbol\mu})} - SSRES$$
$$= 44.333 - 1.222 = 43.111$$

$$SSAB' = \underset{1\times(p-1)(q-1)}{(\mathbf{Q}'_{AB}\hat{\boldsymbol\mu})'} \; \underset{(p-1)(q-1)\times(p-1)(q-1)}{(\mathbf{Q}'_{AB}\mathbf{Q}_{AB})^{-1}} \; \underset{(p-1)(q-1)\times 1}{(\mathbf{Q}'_{AB}\hat{\boldsymbol\mu})} - SSRES$$
$$= 4.111 - 1.222 = 2.889$$

FIGURE 12.13-1   Block diagram of type LSCF-3^2 design. The numbers in the cells denote the level of treatment A.

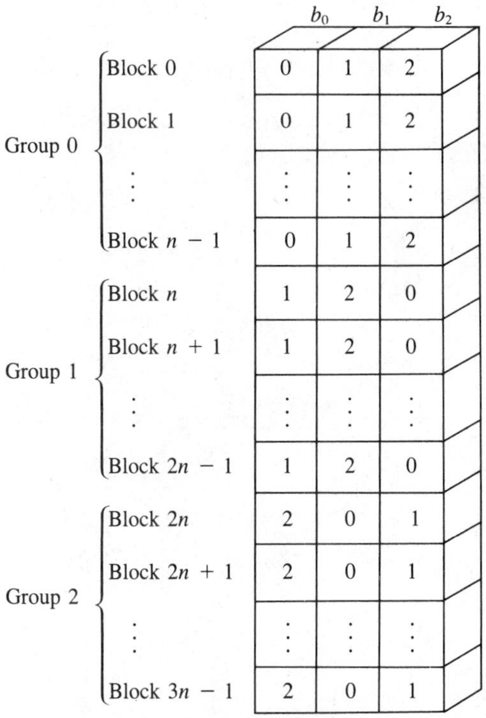

The experimental design model for the design in Figure 12.13-1 is

$$Y_{ijkz} = \mu + \zeta_z + \pi_{i(z)} + \alpha_j + \beta_k + (\alpha\beta)_{jk(\text{within})} + (\alpha\beta\pi)_{jki(z)} + \epsilon_{ijkz}$$
$$(i = 1, \ldots, n; j = 1, \ldots, p; k = 1, \ldots, p; z = 1, \ldots p)$$

where

Y_{ijkz} = a score for the experimental unit in block i, treatment combination $a_j b_k$, and group z

ζ_z = the effect of group z

$\pi_{i(z)}$ = the effect of block i that is $NID(0, \sigma_\pi^2)$

α_j = the effect of treatment level j and is subject to the restriction $\Sigma_{j=1}^{p} \alpha_j = 0$

β_k = the effect of treatment level k and is subject to the restriction $\Sigma_{k=1}^{p} \beta_k = 0$

$(\alpha\beta)_{jk}$ = the joint effect of treatment levels j and k that is not confounded with groups and is subject to the restrictions $\Sigma_{j=1}^{p} (\alpha\beta)_{jk} = 0$ and $\Sigma_{k=1}^{q} (\alpha\beta)_{jk} = 0$

$(\alpha\beta\pi)_{jki(z)}$ = the joint effect of $(\alpha\beta)_{jk}$ and $\pi_{i(z)}$

ϵ_{ijkz} = the experimental error that is $NID(0, \sigma_\epsilon^2)$; ϵ_{ijkz} is independent of $\pi_{i(z)}$. (In this design, ϵ_{ijkz} cannot be estimated separately from $(\alpha\beta\pi)_{jki(z)}$.)

The reader has probably noted an unusual feature of this model—the presence of an AB' interaction component in a design developed from a Latin square building block. This interaction component corresponds to $SS(AB)$ within described in Section 12.8. The (AB^2) component of the AB interaction is confounded with groups. It is assumed that the remaining interactions are equal to zero; that is, $\alpha\zeta$, $\beta\zeta$, and $\alpha\beta\zeta = 0$. Benjamin (1965) has explored in detail the assumptions underlying this design.

A type LSCF-p^2 design is restricted to experiments with two treatments and one nuisance variable. Any number of levels of the two treatments and the nuisance variable can be used in the design as long as they all have the same number of levels. The designation LSCF-3^2 instead of LSCF-33, for example, is used to convey this.

If repeated measurements are obtained on the experimental units, w samples of n units from a population are randomly assigned to the w groups (nuisance variable). The sequence of administration of the v treatment combinations within a block is randomized independently for each block. For the nonrepeated measurements case, w samples of n block, each containing v matched experimental units from a population, are randomly assigned to the w groups. Following this, the v treatment combinations within a block are randomly assigned to the v matched experimental units within each block.

For purposes of comparison, the data in Table 12.8-1 will be analyzed by means of a type LSCF-3^2 design. This example provides only six degrees of freedom for experimental error, which is inadequate. The design should have at least five blocks per group. This would provide 24 degrees of freedom for experimental error. It is assumed in Latin square confounded factorial designs that treatments represent fixed effects and blocks are random. The layout and computational procedures are presented in Table 12.13-1. The results are summarized in Table 12.13-2. According to this summary, the null hypothesis can be rejected for treatment B. The same decision was reached earlier using type RBCF-3^2 and type RBPF-3^2 designs; see the ANOVA summaries in Tables 12.8-2 and 12.8-4, respectively. All three designs require the same number of observations. Is one design a better choice than the others? The choice is really between a (1) type RBPF-3^2 design and (2) type RBCF-3^2 and type LSCF-3^2 designs because the latter designs are essentially equivalent. Consider the interaction and error degrees of freedom for the three designs summarized as follows.

Source	df for Type RBPF-3^2 Design	df for Type RBCF-3^2 Design	df for Type LSCF-3^2 Design
$SSAB$(within)	$(p-1)(q-1) = 4$	$(p-1)(q-1) - (w-1) = 2$	$(p-1)(p-2) = 2$
$SSRES$(within)	$2(3-1)(w-1) = 4$	$w(n-1)(v-1) = 6$	$p(n-1)(p-1) = 6$

The type RBCF-3^2 and type LSCF-3^2 designs provide more degrees of freedom for estimating experimental error, but do not provide within-blocks information on both components of the AB interaction. An experimenter who is more interested in the AB interaction than one of the treatments should consider a fourth design. A type SPF-$3 \cdot 3$ design, which is diagrammed in Figure 12.13-2, provides full information on the AB interaction, but confounds treatment A with groups of blocks. It should be apparent

---

TABLE 12.13-1 Computational Procedures for a Type LSCF-3^2 Design

---

(i) Data and notation [Y_{ijkz} denotes a score for the experimental unit in block i, treatment combination jk, and group z; $i = 1, \ldots, n$ blocks (s_i); $j = 1, \ldots, p$ levels of treatment A (a_j); $k = 1, \ldots,$ p levels of treatment B (b_k); $z = 1, \ldots, p$ levels of groups (g_z); $jk = 1, \ldots, p$ combinations of a_j and b_k within a block]:

ABGS Summary Table
Entry is Y_{ijkz}

		b_0	b_1	b_2	$\sum_{jk=1}^{p} Y_{ijkz}$	$\sum_{i=1}^{n}\sum_{jk=1}^{p} Y_{ijkz}$
		a_0	a_1	a_2		
g_0	s_0	3	3	6	12	23
	s_1	2	4	5	11	
		a_1	a_2	a_0		
g_1	s_2	3	4	7	14	23
	s_3	1	3	5	9	
		a_2	a_0	a_1		
g_2	s_4	3	4	7	14	24
	s_5	2	2	6	10	

AB Summary Table
Entry is $\sum_{i=1}^{n} Y_{ijkz}$

	b_0	b_1	b_2	$\sum_{i=1}^{n}\sum_{k=1}^{p} Y_{ijkz}$
a_0	$n = 2$ 5	6	12	23
a_1	4	7	13	24
a_2	5	7	11	23
$\sum_{i=1}^{n}\sum_{j=1}^{p} Y_{ijkz} = $	14	20	36	

Table 12.13-1 (continued)

(ii)   Computational symbols:

$$\sum_{i=1}^{n} \sum_{jk=1}^{p} \sum_{z=1}^{p} Y_{ijkz} = 3 + 2 + \cdots + 6 = 70.000$$

$$\frac{\left(\sum_{i=1}^{n} \sum_{jk=1}^{p} \sum_{z=1}^{p} Y_{ijkz}\right)^2}{np^2} = [Y] = \frac{(70.000)^2}{(2)(3)^2} = 272.222$$

$$\sum_{i=1}^{n} \sum_{jk=1}^{p} \sum_{z=1}^{p} Y_{ijkz}^2 = [ABGS] = (3)^2 + (2)^2 + \cdots + (6)^2 = 326.000$$

$$\sum_{i=1}^{n} \sum_{z=1}^{p} \frac{\left(\sum_{jk=1}^{p} Y_{ijkz}\right)^2}{p} = [GS] = \frac{(12)^2}{3} + \frac{(11)^2}{3} + \cdots + \frac{(10)^2}{3} = 279.333$$

$$\sum_{z=1}^{p} \frac{\left(\sum_{i=1}^{n} \sum_{jk=1}^{p} Y_{ijkz}\right)^2}{np} = [G] = \frac{(23)^2}{(2)(3)} + \frac{(23)^2}{(2)(3)} + \frac{(24)^2}{(2)(3)} = 272.333$$

$$\sum_{j=1}^{p} \frac{\left(\sum_{i=1}^{n} \sum_{k=1}^{p} Y_{ijkz}\right)^2}{np} = [A] = \frac{(23)^2}{(2)(3)} + \frac{(24)^2}{(2)(3)} + \frac{(23)^2}{(2)(3)} = 272.333$$

$$\sum_{k=1}^{p} \frac{\left(\sum_{i=1}^{n} \sum_{j=1}^{p} Y_{ijkz}\right)^2}{np} = [B] = \frac{(14)^2}{(2)(3)} + \frac{(20)^2}{(2)(3)} + \frac{(36)^2}{(2)(3)} = 315.333$$

$$\sum_{j=1}^{p} \sum_{k=1}^{p} \frac{\left(\sum_{i=1}^{n} Y_{ijkz}\right)^2}{n} = [AB] = \frac{(5)^2}{2} + \frac{(6)^2}{2} + \cdots + \frac{(11)^2}{2} = 317.000$$

(iii)   Computational formulas:

$$SSTO = [ABGS] - [Y] = 53.778$$
$$SSBETWEEN\ BL = [GS] - [Y] = 7.111$$
$$SSG = [G] - [Y] = 0.111$$
$$SSBL(G) = [GS] - [G] = 7.000$$
$$SSWITHIN\ BL = [ABGS] - [GS] = 46.667$$
$$SSA = [A] - [Y] = 0.111$$
$$SSB = [B] - [Y] = 43.111$$
$$SSAB'(\text{within}) = [AB] - [A] - [B] - [G] + 2[Y] = 1.445$$
$$SSAB \times BL(G) = [ABGS] - [AB] - [GS] + [G] = 2.000$$

TABLE 12.13-2 ANOVA Table for Type LSCF-3^2 Design

	Source	SS	df	MS	F	$E(MS)$ (A and B fixed; Blocks random)
1	Between blocks	7.111	$np - 1 = 5$			
2	Groups	0.111	$p - 1 = 2$	0.056	[2/3] 0.02	$\sigma_\epsilon^2 + p\sigma_\pi^2 + np\Sigma\zeta_z^2/(p-1)$
3	Blocks w. groups	7.000	$p(n-1) = 3$	2.333		$\sigma_\epsilon^2 + p\sigma_\pi^2$
4	Within blocks	46.667	$np(p-1) = 12$			
5	A	0.111	$p - 1 = 2$	0.056	[5/8] 0.17	$\sigma_\epsilon^2 + np\Sigma\alpha_j^2/(p-1)$
6	B	43.111	$p - 1 = 2$	21.556	[6/8] 64.73*	$\sigma_\epsilon^2 + np\Sigma\beta_k^2/(p-1)$
7	AB'(within)	1.445	$(p-1)(p-2) = 2$	0.722	[7/8] 2.17	$\sigma_\epsilon^2 + n\Sigma\Sigma(\alpha\beta)_{jk}^2/(p-1)(p-2)$
8	AB ×blocks w. groups	2.000	$p(n-1)(p-1) = 6$	0.333		σ_ϵ^2
9	Total	53.778	$np^2 - 1 = 17$			

*$p < .01$

that subtle variations in the assignment of treatment combinations to experimental units can markedly affect the nature and power of tests of significance.

COMPUTATIONAL PROCEDURES FOR TYPE LSCF-$r \cdot p^2$ DESIGN

The versatility of a Latin square in constructing confounded factorial designs is illustrated in Figure 12.13-3. This design is composed of r LSCF-p^2 designs. The levels of treatment C are assigned to r different $p \times p$ Latin squares. Assumptions of the type LSCF-3^2 design also apply to this design. There are no restrictions on the number of levels of treatment C, and interactions of C with A and B are permissible. Computational formulas for this design are shown in Table 12.13-3. The $E(MS)$ in Table 12.13-3 are appropriate for an experiment in which treatments A, B, and C are fixed effects. Blocks are assumed to be random. The AB components of the AB interaction can be tested separately for the various squares or pooled prior to the test. If the components are pooled, the resulting sum of squares represents a component of the ABC interaction.

FIGURE 12.13-2   Block diagram of type SPF-3 · 3 design. This design involves the same number of observations as the type RBCF-3^2, RBPF-3^2, and LSCF-3^2 designs discussed in the text.

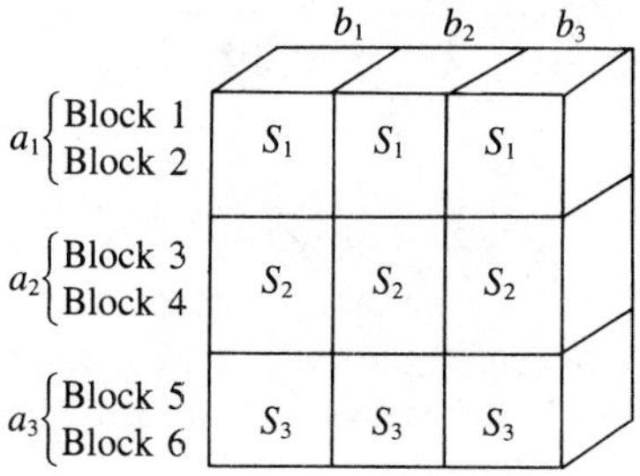

FIGURE 12.13-3   Block diagram of type LSCF-2 · 3^2 design. The r levels of treatment C are assigned to r different $p \times p$ Latin squares. The numbers in the cells denote the level of treatment A.

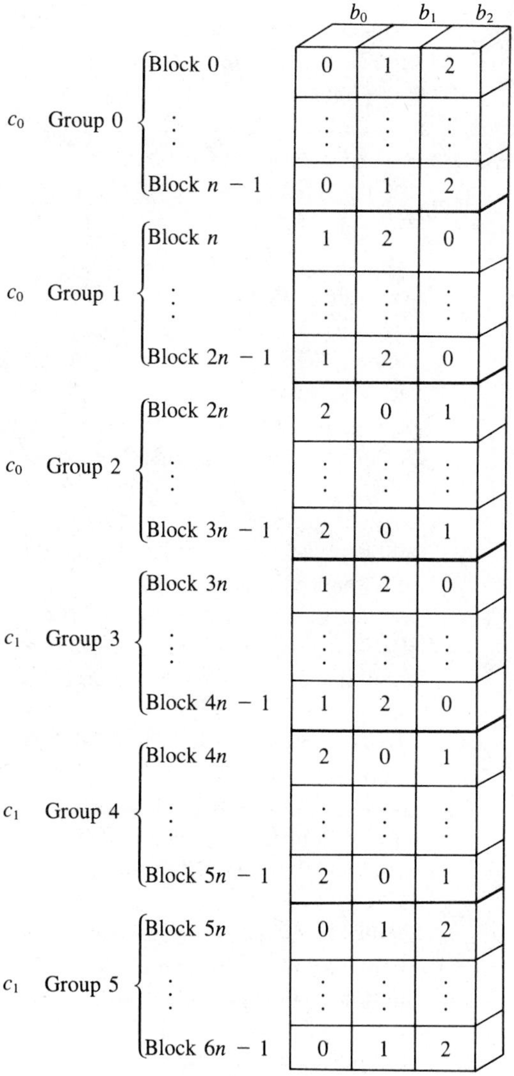

12.14   DOUBLE CONFOUNDING

It is possible to arrange treatment combinations in a Latin square in such a way that unimportant interactions are confounded with between-row variation and between-column variation. This procedure is called *double confounding*. Yates (1937) proposed the name *quasi-Latin squares* for these designs. They are useful for situations in which it is desirable to eliminate row and column variation from the experimental error and at the same time use relatively small rows and columns. They differ from a regular Latin square design in that each treatment combination does not appear once in each row and column. For a discussion of these designs, see Cochran and Cox (1957, 317), Federer (1955, 472), and Yates (1937).

12.15   ADVANTAGES AND DISADVANTAGES OF CONFOUNDING IN FACTORIAL DESIGNS

An examination of current literature in the behavioral sciences reveals an almost total absence of experiments using the confounded factorial designs described in this chapter. This observation is in sharp contrast to the observation that a split-plot factorial design, which involves a different form of confounding, is one of the most popular designs. How can one account for the relative popularity of these two designs, both of which accomplish the same objective of reducing block size? The answer to this question is probably to be found in an examination of the relative merits of the two designs and in the gradual evolution that is taking place in design sophistication among experimenters. Experimenters today are well aware of the potential advantage of a randomized block factorial design in reducing residual error variation by minimizing the effects of subject heterogeneity. As pointed out previously, however, type RBF-pq designs of only modest size may require a prohibitively large block size. For example, a type RBF-33 design requires blocks of size nine. A split-plot factorial design provides a solution to this problem by confounding treatment A with between-blocks variation. Thus a type SPF-3·3 design requires blocks of size three. An alternative solution is provided by type RBCF-3^2, RBPF-3^2, and LSCF-3^2 designs. These designs confound the AB interaction with between-blocks variation in order to achieve a block size of three. In all four designs, the advantage of reduced block size is achieved at a price. In the case of a type SPF-3·3 design, the price an experimenter pays is loss of power in evaluating a treatment. In the cases of type RBCF-3^2, RBPF-3^2, and LSCF-3^2 designs, power is lost in evaluating an interaction. The particular research application will determine which kind of information can be sacrificed and, indeed, if either form of confounding is appropriate. This discussion serves to emphasize a theme that runs throughout this book. The selection of the *best* design requires an intimate knowledge of a research area as well as a knowledge of the advantages and disadvantages of alternative designs.

TABLE 12.13-3 Computational Formulas for Type LSCF-$r \cdot p^2$ Design

	Source	Formula	df	F	$E(MS)$ (A, B, and C fixed; Blocks random)
1	Between blocks	$[CGS] - [Y]$	$npr - 1$		
2	C	$[C] - [Y]$	$r - 1$	$[2/3]$	$\sigma_\epsilon^2 + p\sigma_\pi^2 + np\Sigma\zeta_\zeta^2/(p-1) + np^2\Sigma\gamma_l^2/(p-1)$
3	Groups w. C	$[CG] - [C]$	$r(p-1)$	$[3/4]$	$\sigma_\epsilon^2 + p\sigma_\pi^2 + np\Sigma\zeta_\zeta^2/(p-1)$
4	Blocks w. groups	$[CGS] - [CG]$	$pr(n-1)$		$\sigma_\epsilon^2 + p\sigma_\pi^2$
5	Within blocks	$[ABCGS] - [CGS]$	$npr(p-1)$		
6	A	$[A] - [Y]$	$p-1$	$[6/12]$	$\sigma_\epsilon^2 + npr\Sigma\alpha_j^2/(p-1)$
7	B	$[B] - [Y]$	$p-1$	$[7/12]$	$\sigma_\epsilon^2 + npr\Sigma\beta_k^2/(p-1)$
8	AC	$[AC] - [A] - [C] + [Y]$	$(p-1)(r-1)$	$[8/12]$	$\sigma_\epsilon^2 + np\Sigma\Sigma(\alpha\gamma)_{jl}^2/(p-1)(r-1)$
9	BC	$[BC] - [B] - [C] + [Y]$	$(p-1)(r-1)$	$[9/12]$	$\sigma_\epsilon^2 + np\Sigma\Sigma(\beta\gamma)_{kl}^2/(p-1)(r-1)$
10	AB'(square c_0)	$[AB] - [A] - [B] - [G] + 2[Y]$ terms computed from c_0	$(p-1)(p-2)$	$[10/12]$	$\sigma_\epsilon^2 + n\Sigma\Sigma(\alpha\beta)_{jk}^2/(p-1)(p-2)$
11	AB'(square c_1)	$[AB] - [A] - [B] - [G] + 2[Y]$ terms computed from c_1	$(p-1)(p-2)$	$[11/12]$	$\sigma_\epsilon^2 + n\Sigma\Sigma(\alpha\beta)_{jk}^2/(p-1)(p-2)$
12	$AB \times BL(G)$	$[ABCGS] - [ABC] - [CGS] + [CG]$	$pr(n-1)(p-1)$		σ_ϵ^2
13	Total	$[ABCGS] - [Y]$	$np^2r - 1$		

With the foregoing discussion as a general frame of reference, the advantages and disadvantages of confounding interactions with groups of blocks can be stated as follows.

The advantages of type RBCF, RBPF, and LSCF designs are as follows.

1.   These designs can be used in experiments in which it is not possible to administer all treatment combinations within each block. This condition may exist because of lack of sufficient homogeneous subjects to form complete blocks or because only a portion of the treatment combinations can be administered in a relatively homogeneous time interval. If repeated observations are obtained on subjects, the sheer number of treatment combinations may preclude administration of all combinations to each subject.

2.   All main effects can be evaluated with equal precision in these designs.

3.   An experimenter is able to tailor the experimental design so that only interactions believed to be insignificant are partially or completely sacrificed.

The disadvantages of type RBCF, RBPF, and LSCF designs are as follows.

1.   Information is lost with respect to one or more interactions.

2.   The general layout and analysis of type RBCF and RBPF designs are more complex than the layout and analysis of type SPF and RBF designs. This is also true in making comparisons among means when partially confounded interactions are significant.

3.   Type RBCF, RBPF, and LSCF designs, with their greater complexity of layout and analysis, may actually involve more work for the experimenter than designs using more subjects, such as a completely randomized factorial design. The availability of subjects, cost of assigning homogeneous subjects to blocks, and cost of running subjects must be taken into account in the choice of a design.

12.16   REVIEW EXERCISES

1.   [12.2] Which of the following pairs of numbers are congruent?

   †a) 12 and 5 reduced modulo 2   †b) 6 and 10 reduced modulo 2

   c) 7 and 9 reduced modulo 2   †d) 6 and 12 reduced modulo 3

   e) 4 and 10 reduced modulo 3   †f) 9 and 24 reduced modulo 5

   g) 6 and 12 reduced modulo 5   h) 11 and 26 reduced modulo 5

2.   [12.2] Fill in the blank.

   †a) $2 + 4 =$ ____ (mod 5)   b) $6 + 1 =$ ____ (mod 2)   †c) $2 + 3 =$ ____ (mod 3)

   d) $1 + 1 =$ ____ (mod 2)   †e) $(2)(3) =$ ____ (mod 3)   f) $(3)(1) =$ ____ (mod 5)

†g) $1 + 3 =$ _____ (mod 2)   **h)** $3 + 3 =$ _____ (mod 3)   **†i)** $(1)(4) =$ _____ (mod 5)
j) $(3)(3) =$ _____ (mod 2)

3.   [12.2] Determine the treatment combinations assigned to groups $z = 0, \ldots, m - 1$ for the following relations.

    †a) $a_j + b_k = z \pmod 2$ **b)** $a_j + b_k = z \pmod 3$
    †c) $a_j + b_k = z \pmod 5$ **†d)** $a_j + b_k + c_l = z \pmod 2$
    e) $a_j + b_k + c_l = z \pmod 3$

4.   The effects of input sequence on processing temporal relationships in simple stories was investigated. After hearing stories presented in either chronological order (treatment level a_0) or flashback sequence (treatment level a_1), subjects made a decision about the underlying order of occurrence of two events. Two delay periods were used: b_0 was a 10-second delay after the hearing and b_1 was a 45-second delay. The dependent variable was time to make the decision. Ten pairs of monozygotic male twins were randomly assigned to two groups, five pairs in each group. The $a_j b_k$ treatment combinations appropriate for a group were randomly assigned to the twins in each block. The following data for this type RBCF-2^2 design were obtained.

		ab	ab
		00	11
	s_0	17	23
	s_1	15	14
g_0	s_2	13	18
	s_3	19	21
	s_4	16	19
		01	10
	s_5	15	18
	s_6	17	15
g_1	s_7	18	21
	s_8	12	15
	s_9	20	19

    †a) [12.3] Test the hypotheses $\alpha_j = 0$ for all j and $\beta_k = 0$ for all k; let $\alpha = .05$.

    †b) [12.3] Does this confounded factorial design appear to be a good choice for this experiment? Why?

    †c) [12.4] Test the homogeneity of variance hypothesis that $\sigma_{1.}^2 = \alpha_{2.}^2$. Use the F_{max} statistic; let $\alpha = .05$.

    †d) [12.4] Test the hypothesis that the population variances estimated by $MSAB \times BL(g_0)$ and $MSAB \times BL(g_1)$ are homogeneous. Use the F_{max} statistic; let $\alpha = .05$.

    e) Prepare a "results and discussion section" appropriate for the *Journal of Verbal Learning and Verbal Behavior*.

5.   [12.5] **a)** Construct a block diagram for a type RBCF-2^3 design in which the BC interaction is confounded with groups of blocks.

    b) Write the experimental design model equation for this design.

6.   [12.6] †a) Construct a block diagram for a type RBPF-2^3 design in which the AB and AC interactions are confounded in groups 0 and 1, respectively, and the ABC interaction is confounded in group 2.

   b) Write the computational formulas for this design using the abbreviated notation.

   †c) What is the relative information on the AB, AC, and ABC interactions?

7.   Observers made magnitude estimations of the annoyance value of aircraft noise while viewing TV. The basic noise spectrum was that of a commercial jet aircraft (DC-10) during takeoff, designated as treatment level a_0. This spectrum was modified by decreasing it by 6dB in the octave band centered at 1600 Hz, treatment level a_1, or in the octave band centered at 800 Hz, treatment level a_2. The three noise spectra were presented at three overall intensities: 88dBA, treatment level b_0; 85dBA, treatment level b_1; and 82dBA, treatment level b_2. Six male observers with normal audiograms were randomly assigned to two groups. The $a_j b_k$ treatment combinations of a type RBPF-3^2 design were randomly assigned to the subjects in each group. The order in which the combinations were presented was randomized independently for each subject. The following data were obtained.

			ab	*ab*	*ab*
			00	12	21
	s_0		18	6	9
			01	10	22
g_0	s_1		11	10	6
			02	11	20
	s_2		8	7	15
			00	11	22
	s_3		21	8	8
			02	10	21
g_1	s_4		13	13	9
			01	12	20
	s_5		13	7	13

   a) [12.8] Test the hypotheses $\alpha_j = 0$ for all j, $\beta_k = 0$ for all k, and $(\alpha\beta)'_{jk} = 0$ for all j and k; let $\alpha = .05$.

   b) [12.8] Does a partially confounded factorial design appear to be a good choice for this experiment? Why?

   c) [12.8] Why is a type RBPF-3^2 design a better choice than a type RBCF-3^2 design for this experiment?

   d) [12.8] Compute adjusted and unadjusted cell totals.

   e) Use Tukey's test statistic to determine which population means differ for treatments. A and B.

   f) Prepare a "results and discussion section" appropriate for the *Journal of Auditory Research*.

8.   [12.9] a) Determine the treatment combinations assigned to groups for a type RBCF-2^5 design with $a_j + b_k + c_l = z \pmod 2$ and $c_l + d_m + e_o = z \pmod 2$ as the defining relations.

b) What is the generalized interaction for this design?

†9. [12.9] Determine the treatment combinations assigned to groups for a type RBCF-5^2 design with $a_j + b_k = z \pmod 5$ as the defining relation.

10. [12.9] Determine the treatment combinations assigned to groups for a type RBPF-5^2 design with balanced partial confounding.

11. [12.9] What is the generalized interaction(s) when the following defining relations are used to assign treatment combinations to groups in a confounded factorial design?

  †a) Type RBCF-2^4 design based on (AC) and (BCD)

  b) Type RBCF-2^4 design based on (AD) and (BC)

  †c) Type RBCF-3^3 design based on (ABC) and (BC^2)

  d) Type RBCF-3^3 design based on (ABC) and (AC)

  †e) Type RBCF-3^3 design based on (ABC^2) and (AB^2)

  †f) Type RBCF-2^5 design based on (ABC), (ACD), and (BE)

  †g) Type RBCF-5^3 design based on (ABC) and (AB^2)

  h) Type RBCF-5^3 design based on (ABC) and (AB)

  i) Type RBCF-5^3 design based on (ABC) and (AB^3)

  j) Type RBCF-2^5 design based on (ABD), (ABE), and (CD)

†12. [12.12] Exercise 4 described an experiment for investigating the effects of input sequence on processing temporal relationships in simple stories.

  a) Write the (i) less than full rank experimental design model and (ii) restricted full rank experimental design model for this type RBCF-2^2 design.

  b) Specify the null hypotheses in terms of μ's for the full rank model for treatments A and B, groups, and the error terms $MSBL(G)$ and $MSAB \times BL(G)$.

  c) Indicate the dimensions of the following vectors and matrices: $\mathbf{y}$, $\boldsymbol{\mu}$, $\mathbf{C}'_{BL}$, $\mathbf{C}'_G$, $\mathbf{C}'_{BL(G)}$, $\mathbf{R}'$, $\mathbf{C}'_{w.BL}$, $\mathbf{C}'_A$, $\mathbf{C}'_B$, $\mathbf{Q}'_{w.BL}$, $\mathbf{Q}'_A$, and $\mathbf{Q}'_B$.

  d) Use the full rank experimental design model approach to test the null hypotheses for groups, treatment A, and treatment B; let $\alpha = .05$. (This problem involves inverting 8×8 matrices. A computer or calculator with a matrix inversion program should be used. Note that $\mathbf{R}'\mathbf{C}_w = \mathbf{0}$ for both within-blocks coefficient matrices.)

†13. [12.12] Assume that the following data have been obtained in Exercise 4.

		ab	ab
		00	11
g_0	s_0	17	23
	s_1	15	14
		01	10
g_1	s_2	15	18
	s_3	17	15

a) Write coefficient matrices for the following (let $\mathbf{y}' = [17\ 15\ 23\ 14\ 15\ 17\ 18\ 15]$): $\mathbf{C}'_{BL}$, $\mathbf{C}'_G$, $\mathbf{C}'_{BL(G)}$, $\mathbf{R}'$, $\mathbf{C}'_{w.BL}$, $\mathbf{C}'_A$, $\mathbf{C}'_B$.

b) Use the full rank experimental design model approach to compute SSG, $SSBL(G)$, $SSAB \times BL(G)$, SSA, and SSB. (This problem involves inverting 2×2 diagonal matrices and can be done without a computer. Note that $\mathbf{R}'\mathbf{C}'_w = \mathbf{0}$ for the within-blocks coefficient matrices.)

c) Test the null hypotheses for groups, treatment A, and treatment B; let $\alpha = .05$.

14. [12.12] Exercise 7 described an experiment to determine the annoyance value of three aircraft noise spectra.

a) Specify the null hypotheses in terms of μ's for the full rank model for groups, treatments A and B, AB'(within) interaction, and the error terms $MSBL(G)$ and $MSRES$.

b) Use the full rank experimental design model approach to test the null hypotheses for groups, treatments A and B, and AB'(within); let $\alpha = .05$. (This problem involves inverting 4×4 matrices. The computations can be done without a computer but they are tedious. Note that $\mathbf{R}'\mathbf{C}_w = \mathbf{0}$ for the three within-blocks coefficient matrices.)

c) Compute $\hat{\boldsymbol{\mu}}_R$ and use Tukey's statistic to determine which population means differ. (This problem involves manipulating an 18×18 matrix and should be done with a computer.)

†15. [12.13] The data of Exercise 7 can be analyzed as a type LSCF-3^2 design. The layout is as follows.

		b_0	b_1	b_2
		a_0	a_1	a_2
g_0	s_0	18	7	6
	s_1	21	8	8
		a_1	a_2	a_0
g_1	s_2	10	9	8
	s_3	13	9	13
		a_2	a_0	a_1
g_2	s_4	15	11	6
	s_5	13	13	7

a) Compare the randomization procedure for this design assuming the use of repeated measurements with that for a type RBPF-3^2 design.

b) Test the null hypotheses for groups, treatments A and B, and the AB interaction; let $\alpha = .05$.

c) Use Tukey's statistic to evaluate all pairwise comparisons among means.

d) Why might an experimenter prefer to analyze this experiment using a type LSCF-3^2 design rather than a type RBPF-3^2 design?

16. [12.13] Twelve elementary school teachers viewed a videotape of three fourth-grade boys after which they completed a referral form containing items from the Children's Personality Questionnaire, Form A, and the California Test of Personality, Elementary Series. The teachers were randomly assigned to one of six groups. Prior to viewing the videotape, teachers in three of the groups were told that the three boys were learning disabled (treatment level c_1). Teachers in the other three groups were told that the boys were normal (treatment level c_0). Additional fictitious information concerning the boys—occupation of father and classroom behavior—was also provided. The categories professional (level a_0), blue collar (level a_1), and unemployed (level a_2) were combined in two 3×3 Latin squares with the categories "no disruptive behavior" (b_0), "disruptive behavior" (b_1), and "disruptive, acting out behavior" (b_2). The treatment combinations in the rows of the two Latin squares were assigned randomly to the six groups. The dependent variable in this type LSCF-2.3^2 design was the number of referral items checked by the teachers. The following data were obtained.

		b_0	b_1	b_2
		a_0	a_1	a_2
c_0	s_0	13	8	5
	s_1	10	5	3
		a_1	a_2	a_0
c_0	s_2	4	10	9
	s_3	8	7	12
		a_2	a_0	a_1
c_0	s_4	5	8	15
	s_5	7	10	12
		a_1	a_2	a_0
c_1	s_6	7	11	19
	s_7	4	10	16
		a_2	a_0	a_1
c_1	s_8	8	12	20
	s_9	3	14	16
		a_0	a_1	a_2
c_1	s_{10}	4	11	13
	s_{11}	8	9	18

a) Test the null hypotheses for this experiment.

b) Graph the relevant interactions.

c) Prepare a "results and discussion section" appropriate for the *Learning Disability Quarterly*.

13 FRACTIONAL FACTORIAL DESIGNS: DESIGNS WITH TREATMENT-INTERACTION CONFOUNDING

13.1 INTRODUCTION TO FRACTIONAL FACTORIAL DESIGNS

The main properties of factorial designs have been considered in Chapters 8, 9, 11, and 12. One of the characteristics of these designs is that *all* treatment combinations appear in the experiment. Fractional factorial designs described in this chapter represent a departure from this characteristic. These designs include only a *fraction;* for example, ½, ⅓, ¼, and so on, of the treatment combinations of a complete factorial design. A $3 \times 3 \times 3 \times 3$ completely randomized factorial design has 81 treatment combinations. By using a one-third fractional replication, only 27 of these combinations need be included in the experiment.

Fractional factorial designs represent a relatively recent development in the evolution of experimental designs. The theory of fractional factorial designs was developed for 2^k and 3^k designs by Finney (1945, 1946a) and extended by Kempthorne (1947) to designs of the type p^k, where p is any prime number and k designates the number of treatments. Fractional factorial designs have found their greatest use in industrial research. Only limited application of these designs has been made in agricultural research, which historically has provided the impetus for the development of most designs in use today.

Fractional factorial designs have much in common with confounded factorial designs. These latter designs achieve a reduction in the number of treatment combinations that must be included within blocks. A fractional factorial design extends this principle to the entire experiment. The reduction in size of an experiment is not obtained without paying a price, however. Considerable ambiguity may exist in interpreting the results of fractionally replicated experiments. This arises because every sum of squares can be given two or more designations. For example, a sum of squares might be attributed to the effects of treatment A and the $BCDE$ interaction. The two or more designations given to the same sum of squares are called *aliases*. In a one-half replication of a 2^k factorial design, all sums of squares have two aliases; a one-fourth fractional replication results in all sums of squares having four aliases. This means, in the case of a one-fourth replication, that four names can be given to the same sum of squares. It is imperative that careful attention be given to the alias pattern of a proposed design in order to minimize confusion in interpreting tests of significance. Treatments are customarily aliased with higher-order interactions that are assumed to equal zero. This helps to minimize but does not eliminate ambiguity in interpreting the outcome of an experiment.

A fractional factorial design is appropriate for experiments that meet, in addition to the general assumptions of analysis of variance, the following conditions:

1. The experiment contains many treatments that result in a prohibitively large number of treatment combinations. Fractional replication is rarely used for experiments with less than four or five treatments.

2. The number of treatment levels should, if possible, be equal for all treatments. With the exception of fractional factorial experiments using a Latin square building block design, analysis procedures for experiments involving mixed numbers of treatment levels are relatively complex.

3. An experimenter should have some a priori reason for believing that a number of higher-order interactions are zero or small relative to main effects. In practice, fractional factorial designs, with the exception of those based on a Latin square, are most often used with treatments having either two or three levels. The use of a restricted number of levels increases the likelihood that interactions will be insignificant.

4. Fractional factorial designs are most useful for exploratory research and for situations that permit *follow-up* experiments to be performed. Thus a larger number of treatments can be investigated efficiently in an initial experiment, with subsequent experiments designed to focus on the most promising lines of investigation or to clarify the interpretation of the original analysis.

Among the disadvantages of a complete factorial design listed in Section 8.14 was that the design committed an experimenter to a relatively large experiment. In many research situations, a sequence of relatively small experiments, each based on the results of the preceding experiment, is more efficient than a single experiment in determining optimum levels of treatments. A fractional factorial design lends itself to this kind of sequential experimentation. Fractional factorial experiments using three *building block designs* are described in this chapter. The building block designs are the completely randomized, randomized block, and Latin square designs. These fractional factorial designs are designated by the letters CRFF, RBFF, and LSFF, respectively.

The use of a fractional factorial design can lead to a sizable reduction in the number of treatment combinations that must be included in an experiment. This is accomplished by *confounding* main effects with higher-order interactions. This form of confounding completes a cycle that began in Chapter 11. That chapter described split-plot factorial designs that involve *group-treatment* confounding. Chapter 12 described confounded factorial designs involving *group-interaction* confounding. Here we will discuss *treatment-interaction* confounding. Note that whenever confounding is used, some information is lost. However, if certain information concerning the outcome of the experiment is of negligible interest, an experimenter can employ confounding so as to sacrifice only this information. The advantage of confounding in terms of a reduction in experimental effort may more than compensate for the lost information.

13.2 GENERAL PROCEDURES FOR CONSTRUCTING TYPE CRFF-2^k DESIGNS

ONE-HALF REPLICATION DESIGNS

Procedures for constructing fractional factorial designs are closely related to those for confounded factorial designs. Recall that the treatment combinations in a 2^4 completely confounded factorial design can be assigned to two groups by means of the family of relations $(ABCD)_z$ modulo 2, where $z = 0, 1$. A 2^4 one-half fractional factorial design consists of the treatment combinations contained in either group $(ABCD)_0$ or $(ABCD)_1$. The interaction used to divide the treatment combinations into the two groups is called the *defining contrast*.

ALIAS PATTERN FOR FRACTIONAL FACTORIAL DESIGNS

All effects in a fractional factorial design have two or more aliases. The alias pattern for a one-half fractional factorial design can be determined by multiplying each of the effects in the experiment by the defining contrast and expressing the product modulo *m*. Computation of the alias pattern for a type CRFF-2^4 design is shown in Table 13.2-1. It is apparent from this table that all main effects have three-treatment interactions as aliases and that two-treatment interactions have other two-treatment interactions as aliases. This particular design is not satisfactory because of the ambiguity

TABLE 13.2-1 Alias Pattern for Type CRFF-2^4 Design with $ABCD$ as the Defining Contrast

(1) Source	(2) Defining Contrast	(3) Product (1) × (2)	(4) Alias (mod 2)
A	$ABCD$	A^2BCD	BCD
B	$ABCD$	AB^2CD	ACD
C	$ABCD$	ABC^2D	ABD
D	$ABCD$	$ABCD^2$	ABC
AB	$ABCD$	A^2B^2CD	CD
AC	$ABCD$	A^2BC^2D	BD
AD	$ABCD$	A^2BCD^2	BC
BC	$ABCD$	AB^2C^2D	AD
BD	$ABCD$	AB^2CD^2	AC
CD	$ABCD$	ABC^2D^2	AB
ABC	$ABCD$	$A^2B^2C^2D$	D
ABD	$ABCD$	$A^2B^2CD^2$	C
ACD	$ABCD$	$A^2BC^2D^2$	B
BCD	$ABCD$	$AB^2C^2D^2$	A

associated with interpreting significant two-treatment interactions. It is not possible to determine from this design whether a significant two-treatment interaction is associated with, for example, the AB or CD interaction. This question can be answered by conducting a second experiment that includes the treatment combinations not included in the first experiment. If the two experiments are combined, a regular type CRF-2222 design is obtained.

ONE-FOURTH REPLICATION AND ASSOCIATED ALIAS PATTERN

A one-fourth fractional replication can be obtained by using two defining contrasts to select treatment combinations that are included in an experiment. In this case, effects have four aliases. Two of the aliases are associated with the product of a treatment and the two defining contrasts. The third alias is associated with the product of a treatment and the *generalized interaction* of the two defining contrasts. The fourth alias is the treatment itself. Assume that the two defining contrasts are ABC and ACD. The generalized interaction of ABC and ACD, following procedures described in Section 12.9, is given by

$$(ABC)(ACD) = A^2BC^2D \text{ reduced modulo } 2 = BD.$$

Aliases of effects for a type CRFF-2^4 design are shown in Table 13.2-2.

A one-fourth replication of a type CRFF-2^4 design is an unsatisfactory design, for some main effects are aliased with other main effects. In this example, the alias of treatment B is treatment D. If an experiment contains a large number of treatments,

TABLE 13.2-2   Alias Pattern for One-Fourth Type CRFF-2^4 Design with ABC and ACD as the Defining Contrasts

Source	Defining Contrasts and Generalized Interaction	Products	Aliases (mod 2)
A	ABC, ACD, BD	A^2BC, A^2CD, ABD	BC, CD, ABD
B	ABC, ACD, BD	$AB^2C, ABCD, B^2D$	$AC, ABCD, D$
C	ABC, ACD, BD	ABC^2, AC^2D, BCD	AB, AD, BCD
D	ABC, ACD, BD	$ABCD, ACD^2, BD^2$	$ABCD, AC, B$
$\vdots$	$\vdots$ $\vdots$ $\vdots$	$\vdots$ $\vdots$ $\vdots$	$\vdots$ $\vdots$ $\vdots$
BCD	ABC, ACD, BD	$AB^2C^2D, ABC^2D^2, B^2CD^2$	AD, AB, C

a one-fourth replication is a practical design. For example, an experiment with eight treatments at two levels each can be designed for a one-fourth replication so that no treatments or two-treatment interactions are aliased with lower-order interactions. The number of treatment combinations can be reduced from 256 to 64 by this procedure. Care must be used in the selection of defining contrasts so as to avoid aliasing main effects with other main effects.

RELATION BETWEEN ONE-HALF REPLICATION OF 2^4 DESIGN AND COMPLETE 2^3 DESIGN

A careful examination of the alias pattern in Table 13.2-1 reveals an interesting feature of this one-half fractional factorial design. The incomplete four-treatment design contains all of the treatment combinations of a complete three-treatment design. That is, if any one of the four treatments is ignored, the experiment becomes a complete three-treatment factorial design. This point can be made clearer by a rearrangement of Table 13.2-1. For purposes of illustration, the effects associated with treatment D are ignored in the following source column.

Source	Alias
A	BCD
B	ACD
C	ABD
AB	CD
AC	BD
BC	AD
ABC	D

An examination of the source column reveals that all treatments and interactions of a type CRF-222 design are present. The significance of this arrangement will become apparent in the following section, which describes general analysis procedures for fractional factorial designs.

13.3 COMPUTATIONAL PROCEDURES FOR TYPE CRFF-2^4 DESIGN

Computational procedures for a one-half replication of a 2^4 design are identical to those for a complete replication of a 2^3 design. This means that, by ignoring one of the treatments, the analysis of an incomplete design can be carried out as if all the treatment combinations were included in the experiment. This procedure can be extended to one-fourth and one-eighth replications. Thus a 2^k incomplete design is analogous to a 2^{k-1} complete design, where k refers to the number of treatments. Similarly, one-fourth and one-eighth incomplete designs are analogous to 2^{k-2} and 2^{k-3} complete designs, respectively.

 The first step in laying out a type CRFF-2^4 design is to determine the treatment combinations to be included in the experiment. This is facilitated by the use of modular arithmetic. The convention adopted in Chapter 12 of using the subscript zero to designate the first level of a treatment will be followed in this chapter. It is customary in laying out a fractional factorial design to use the highest-order interaction as the defining contrast because it is less likely to be significant than lower-order interactions. Treatment combinations satisfying either $(ABCD)_0$ or $(ABCD)_1$ can be used as the defining contrast. Assume that $(ABCD)_0$ is selected by a toss of a coin. Treatment combinations that satisfy this relation are as follows:

$$a_j + b_k + c_l + d_m = 0 \pmod 2$$
$$= 0000,\ 1100,\ 1010,\ 1001,\ 0110,\ 0101,\ 0011,\ 1111.$$

The successive numbers refer to the level of treatments A, B, C, and D, respectively. Thus, 1010 stands for $a_1 b_0 c_1 d_0$.

 General procedures for analyzing fractional factorial designs are illustrated in Table 13.3-1. This example has the virtue of simplicity, but, as noted previously, a type CRFF-2^4 design is unsatisfactory because of ambiguity in interpreting two-treatment interactions. It is assumed that the following experimental design model equation is appropriate for this design:

$$Y_{ijklm} = \mu + \alpha_j + \beta_k + \gamma_l + \delta_m + (\alpha\beta)_{jk}(\gamma\delta)_{lm} + (\alpha\gamma)_{jl}(\beta\delta)_{km}$$
$$+ (\beta\gamma)_{kl}(\alpha\delta)_{jm} + \epsilon_{i(jklm)} \qquad (i = 1, \ldots, n; j = 1, \ldots, p;$$
$$k = 1, \ldots, q; l = 1, \ldots, r; m = 1, \ldots, t).$$

The notation $(\alpha\beta)_{jk}(\gamma\delta)_{lm}$ indicates that these two interactions are aliases which cannot be differentiated in this design. A fixed-effects model is assumed to be appropriate for the CRFF designs described in this chapter. For computational purposes, treatment D is ignored. The choice of which treatment to ignore is arbitrary. The analysis is summarized in Table 13.3-2.

 According to the analysis of variance table, the null hypothesis for treatments B and C can be rejected. The summary in Table 13.3-2 illustrates a major problem inherent in fractional factorial designs—how to interpret significant MS's that are aliased with other MS's of the same order. There is no way to determine from the

TABLE 13.3-1 Computational Procedures for a Type CRFF-2^4 Design

(i) Data and notation [Y_{ijklm} denotes a score for experimental unit i in treatment combination $jklm$; $i = 1,$ $\ldots, n$ experimental units (s_i); $j = 1, \ldots, p$ levels of treatment A (a_j); $k = 1, \ldots, q$ levels of treatment B (b_k); $l = 1, \ldots, r$ levels of treatment C (c_l); $m = 1, \ldots, t$ levels of treatment D (d_m); for purposes of analysis, treatment D is ignored]:

$ABCS = (\frac{1}{2}ABCDS)$ *Summary Table*

Entry is Y_{ijkl}

a_0	a_0	a_0	a_0	a_1	a_1	a_1	a_1
b_0	b_0	b_1	b_1	b_0	b_0	b_1	b_1
c_0	c_1	c_0	c_1	c_0	c_1	c_0	c_1
d_0	d_1	d_1	d_0	d_1	d_0	d_0	d_1
3	4	7	7	2	2	5	9
6	3	6	8	2	3	6	11

ABC Summary Table

Entry is $\sum_{i=1}^{n} Y_{ijkl}$

	b_0	b_0	b_1	b_1
	c_0	c_1	c_0	c_1
a_0	$n = 2$ 9	7	13	15
a_1	4	5	11	20

AB Summary Table

Entry is $\sum_{i=1}^{n} \sum_{l=1}^{r} Y_{ijkl}$

	b_0	b_1	$\sum_{i=1}^{n} \sum_{k=1}^{q} \sum_{l=1}^{r} Y_{ijkl}$
a_0	$nr = 4$ 16	28.	44
a_1	9	31	40
$\sum_{i=1}^{n} \sum_{j=1}^{p} \sum_{l=1}^{r} Y_{ijkl} = 25$		59	

AC Summary Table

Entry is $\sum_{i=1}^{n} \sum_{k=1}^{q} Y_{ijkl}$

	c_0	c_1
a_0	$nq = 4$ 22	22
a_1	15	25
$\sum_{i=1}^{n} \sum_{j=1}^{p} \sum_{k=1}^{q} Y_{ijkl} = 37$		47

BC Summary Table

Entry is $\sum_{i=1}^{n} \sum_{j=1}^{p} Y_{ijkl}$

	c_0	c_1
b_0	$np = 4$ 13	12
b_1	24	35

Table 13.13-1 (continued)

(ii)   Computational symbols:

$$\sum_{i=1}^{n}\sum_{j=1}^{p}\sum_{k=1}^{q}\sum_{l=1}^{r} Y_{ijkl} = 3 + 6 + \cdots + 11 = 84.00$$

$$\frac{\left(\sum_{i=1}^{n}\sum_{j=1}^{p}\sum_{k=1}^{q}\sum_{l=1}^{r} Y_{ijkl}\right)^2}{npqr} = [Y] = \frac{(84.00)^2}{(2)(2)(2)(2)} = 441.00$$

$$\sum_{i=1}^{n}\sum_{j=1}^{p}\sum_{k=1}^{q}\sum_{l=1}^{r} Y_{ijkl}^2 = [ABCS] = (3)^2 + (6)^2 + \cdots + (11)^2 = 552.00$$

$$\sum_{j=1}^{p}\frac{\left(\sum_{i=1}^{n}\sum_{k=1}^{q}\sum_{l=1}^{r} Y_{ijkl}\right)^2}{nqr} = [A] = \frac{(44)^2}{(2)(2)(2)} + \frac{(40)^2}{(2)(2)(2)} = 442.000$$

$$\sum_{k=1}^{q}\frac{\left(\sum_{i=1}^{n}\sum_{j=1}^{p}\sum_{l=1}^{r} Y_{ijkl}\right)^2}{npr} = [B] = \frac{(25)^2}{(2)(2)(2)} + \frac{(59)^2}{(2)(2)(2)} = 513.25$$

$$\sum_{l=1}^{r}\frac{\left(\sum_{i=1}^{n}\sum_{j=1}^{p}\sum_{k=1}^{q} Y_{ijkl}\right)^2}{npq} = [C] = \frac{(37)^2}{(2)(2)(2)} + \frac{(47)^2}{(2)(2)(2)} = 447.25$$

$$\sum_{j=1}^{p}\sum_{k=1}^{q}\frac{\left(\sum_{i=1}^{n}\sum_{l=1}^{r} Y_{ijkl}\right)^2}{nr} = [AB] = \frac{(16)^2}{(2)(2)} + \frac{(28)^2}{(2)(2)} + \cdots + \frac{(31)^2}{(2)(2)} = 520.50$$

$$\sum_{j=1}^{p}\sum_{l=1}^{r}\frac{\left(\sum_{i=1}^{n}\sum_{k=1}^{q} Y_{ijkl}\right)^2}{nq} = [AC] = \frac{(22)^2}{(2)(2)} + \frac{(22)^2}{(2)(2)} + \cdots + \frac{(25)^2}{(2)(2)} = 454.50$$

$$\sum_{k=1}^{q}\sum_{l=1}^{r}\frac{\left(\sum_{i=1}^{n}\sum_{j=1}^{p} Y_{ijkl}\right)^2}{np} = [BC] = \frac{(13)^2}{(2)(2)} + \frac{(12)^2}{(2)(2)} + \cdots + \frac{(35)^2}{(2)(2)} = 528.50$$

$$\sum_{j=1}^{p}\sum_{k=1}^{q}\sum_{l=1}^{r}\frac{\left(\sum_{i=1}^{n} Y_{ijkl}\right)^2}{n} = [ABC] = \frac{(9)^2}{2} + \frac{(7)^2}{2} + \cdots + \frac{(20)^2}{2} = 543.00$$

(iii)   Computational formulas:

$$SSTO = [ABCS] - [Y] = 111.00$$

$$SSA = [A] - [Y] = 1.00$$

$$SSB = [B] - [Y] = 72.25$$

$$SSC = [C] - [Y] = 6.25$$

Table 13.3-1 (continued)

$$SSAB = [AB] - [A] - [B] + [Y] = 6.25$$

$$SSAC = [AC] - [A] - [C] + [Y] = 6.25$$

$$SSBC = [BC] - [B] - [C] + [Y] = 9.00$$

$$SSABC = [ABC] - [AB] - [AC] - [BC] + [A] + [B] + [C] - [Y] = 1.00$$

$$SSWCELL = [ABCS] - [ABC] = 9.00$$

TABLE 13.3-2 ANOVA Table for Type CRFF-2^4 Design

	Source and Alias	SS	df	MS	F (Model I)
1	$A(BCD)$	1.000	$p - 1 = 1$	1.000	[$1/8$] N.S.
2	$B(ACD)$	72.250	$q - 1 = 1$	72.250	[$2/8$] 64.22**
3	$C(ABD)$	6.250	$r - 1 = 1$	6.250	[$3/8$] 5.56*
4	$D(ABC)$	1.000	$t - 1 = 1$	1.000	[$4/8$] N.S.
5	$AB(CD)$	6.250	$(p - 1)(q - 1) = 1$	6.250	[$5/8$] 5.56*
6	$AC(BD)$	6.250	$(p - 1)(r - 1) = 1$	6.250	[$6/8$] 5.56*
7	$BC(AD)$	9.000	$(q - 1)(r - 1) = 1$	9.000	[$7/8$] 8.00*
8	WCELL	9.000	$pqr(n - 1) = 8$	1.125	
9	Total	111.000	$npqr - 1 = 15$		

*$p < .05$
**$p < .01$

analysis just performed whether, for example, the *AB* or *CD* interaction is significant or whether both are significant. The experiment provides only one mean square that is a linear function of *AB* and *CD*. Confusion with respect to interpreting *AB* and *CD* can be resolved by carrying out the other half of the experiment and combining it with the first half. Davies (1956, 471) and Daniel (1956; 1976, Ch. 14) discuss the general problem of carrying out follow-up experiments to clarify the interpretation of fractional factorial experiments. Bennett and Franklin (1954, 597) interject a word of caution concerning *combined* experiments. They point out that some bias is introduced in the significance test for combined experiments if a decision to carry out the second half of an experiment is based on results obtained in the first half. The discussion in Section 8.11 concerning preliminary tests on the model and pooling procedures is relevant to this issue. In actual practice, it is customary to combine fractional factorial experiments as if no test of significance had preceded the joint analysis. Finney (1960, 139) states, "Undoubtedly the danger of bias arises, although the nature of this cannot be made clear without much fuller consideration of sequential experimentation, but many experimenters will, probably rightly, regard the risk as a reasonable price to pay for the advantages and economies gained."

It can be shown that the sums of squares computed for the AB and ABC interactions, for example, are indistinguishable from the sums of squares computed for the CD interaction and treatment D. The information contained in the $ABCS$ Summary Table can be used to construct the following CD Summary Table where treatment B has been ignored.

CD Summary Table

	d_0	d_1	$\sum\limits_{i=1}^{n}\sum\limits_{j=1}^{p}\sum\limits_{m=1}^{t} Y_{ijlm}$
c_0	$np = 4$ 20	17	37
c_1	20	27	47
$\sum\limits_{i=1}^{n}\sum\limits_{j=1}^{p}\sum\limits_{l=1}^{r} Y_{ijlm} = $	40	44	

$$[D] = \frac{(40)^2}{(2)(2)(2)} + \frac{(44)^2}{(2)(2)(2)} = 442.00$$

$$[CD] = \frac{(20)^2}{(2)(2)} + \cdots + \frac{(27)^2}{(2)(2)} = 454.50$$

$$SSD = [D] - [Y] = 1.00 = SSABC$$

$$SSCD = [CD] - [C] - [D] + [Y] = 6.25 = SSAB$$

A comparison of SSD with $SSABC$ and of $SSCD$ with $SSAB$ shows that the alias pairs are equal. This example could be extended to all alias pairs in Table 13.3-2.

Each sum of squares in a fractional factorial design can be given two or more names. In practice, an experimenter attempts to arrange the alias pattern so that one name is a more reasonable designation for effects than the other alias name. This is accomplished by aliasing main effects and lower-order interactions with higher-order interactions.

The example in Table 13.3-1 contains a total of 16 subjects with two observations in each cell. This was necessary in order to provide a within-cell estimate of experimental error. In larger fractional factorial designs, it is customary to obtain only one observation under the treatment combinations. The higher-order interactions are pooled in order to obtain an estimate of experimental error. A complete factorial design could be carried out with the same number of subjects used in this fractional factorial design. If, for example, 16 subjects were used in a type CRF-2222 design, it would not be possible to compute a within-cell error term. Under these conditions, the two- and three-treatment interactions could be pooled to form a residual error term. If these pooled interactions are insignificant, a complete factorial design would have been a better design choice for the data in Table 13.3-1 than the fractional factorial design. On the other hand, if some of the interactions are significant, the present analysis offers the advantage of a larger number of degrees of freedom for experimental error and a within-cell error term.

Extensive tables of plans (Cochran and Cox, 1957, 276 and Davies, 1956, 484) are available for 1/2, 1/4, 1/8, and so on fractional factorial designs that minimize undesirable alias patterns. Of particular interest are the extensive tables published by the National Bureau of Standards (1957). This latter publication provides plans for 1/2 through 1/256 replication of experiments with up to 16 treatments.

13.4 COMPUTATIONAL PROCEDURES FOR TYPE CRFF-3^4 DESIGN

If each treatment has three instead of two levels, the experiment can be designed so that only 1/3 or 1/9 or 1/27, etc. of the treatment combinations must be run. Computational procedures described in previous sections of this chapter are applicable to these experiments. However, the analysis and interpretation of fractional factorial designs are more complicated for the 3^k series than for the 2^k series. Effects have three aliases in a one-third replication, instead of two as in a one-half replication of the 2^k series. If a one-ninth replication is used, effects have nine aliases; whereas in a one-fourth replication of the 2^k series, effects have only four aliases. In general, fractional replication is less satisfactory for 3^k experiments than for 2^k experiments. One reason, in addition to greater complexity, is that a 3^k experiment must have a relatively large number of treatments ($k > 5$) in order to provide useful estimates of two-treatment interactions. Another problem with 3^k experiments that is not present in 2^k experiments concerns the interpretation of interactions. In a 3^k experiment, the AB interaction, for example, is partitioned into two components (AB) and (AB^2). It is customary in 3^k confounded factorial designs to pool these two estimates. This is rarely possible in a fractional factorial design because the components have different aliases. Thus, although the (AB) and (AB^2) components are orthogonal in the sense of representing nonoverlapping portions of the AB interaction, their physical interpretation is unclear. Extensive tables of fractional factorial designs for the 3^k series have been prepared by Connor and Zelen (1959).

Assume that an experimenter wants to carry out a one-third replication of a type CRFF-3^4 design. One of the components of the $ABCD$ interaction can be selected as the defining contrast. This interaction, which has 16 degrees of freedom, can be partitioned according to procedures described in Section 12.9 as follows.

Interaction	df
$ABCD$	16
$(ABCD)$	2
$(ABCD^2)$	2
(ABC^2D)	2
(AB^2CD)	2
(ABC^2D^2)	2
(AB^2CD^2)	2
(AB^2C^2D)	2
$(AB^2C^2D^2)$	2

Any one of the eight components of the interaction can be used to partition the treatment combinations into three sets. If the $(ABCD^2)$ component is selected as the defining contrast, three sets of treatment combinations are given by

$$a_j + b_k + c_l + 2d_m = z(\text{mod } 3) \qquad (z = 0, 1, 2).$$

Only one of the sets of relations, $(ABCD^2)_0$, $(ABCD^2)_1$, or $(ABCD^2)_2$, is used in the experiment. A table of random numbers can be employed to determine which of these three sets is adopted. Let us assume that the relation $(ABCD^2)_0$ has been selected. A complete experiment has 81 treatment combinations, but this number can be reduced to 27 by the use of the relation $(ABCD^2)_0$.

The pattern of aliases for treatments having three levels can be determined by procedures similar to those discussed in Section 12.9. If X and Y symbolize the defining contrast and treatment (or interaction), respectively, the alias pattern is given by $(X)(Y)$ and $(X)(Y)^2$, reduced modulo 3. For example, the aliases of treatment A are

$$(ABCD^2) \times A = A^2BCD^2 = (A^2BCD^2)^2 \text{ reduced modulo } 3 = (AB^2C^2D)$$
$$(ABCD^2) \times A^2 = A^3BCD^2 \text{ reduced modulo } 3 = (BCD^2).$$

The aliases associated with effects are given in Table 13.4-1 for a type CRFF-3^4 design in which $(ABCD^2)$ is the defining contrast. Treatments and two-treatment interaction components that are not aliased with other treatments or two-treatment interaction components are considered *measurable*. An asterisk designates the measurable effects in the type CRFF-3^4 design of Table 13.4-1.

The layout and computational procedures for the design are simplified by ignoring one of the treatments, say treatment D. It should be apparent from Table 13.4-1 that all main effects and interaction components can be obtained from summary tables that involve only treatments A, B, and C. The procedures described in Section 12.8 must be used to compute the *interaction component* sums of squares. The example used to illustrate the computation in Table 13.4-2 has one observation in each of the 27 treatment combinations. The levels of treatment D occurring in each cell of the ABC Summary Table are included for illustrative purposes only; this information is not used in the analysis. The reader can verify that the treatment combinations in the 27 cells satisfy the relation

$$a_j + b_k + c_l + 2d_m = 0(\text{mod } 3).$$

TABLE 13.4-1 Alias Pattern for Type CRFF-3^4 Design with $(ABCD^2)$ as the Defining Contrast

Source		Aliases			df
$A*$	$=$	(AB^2C^2D)	$=$	(BCD^2)	$p - 1 = 2$
$B*$	$=$	(AB^2CD^2)	$=$	(ACD^2)	$q - 1 = 2$
$C*$	$=$	(ABC^2D^2)	$=$	(ABD^2)	$r - 1 = 2$
(AB)	$=$	(ABC^2D)	$=$	(CD^2)	$(p - 1)(q - 1)/2 = 2$
$(AB^2)*$	$=$	(AC^2D)	$=$	(BC^2D)	$(p - 1)(q - 1)/2 = 2$
(AC)	$=$	(AB^2CD)	$=$	(BD^2)	$(p - 1)(r - 1)/2 = 2$
$(AC^2)*$	$=$	(AB^2D)	$=$	(BC^2D^2)	$(p - 1)(r - 1)/2 = 2$
(BC)	$=$	$(AB^2C^2D^2)$	$=$	(AD^2)	$(q - 1)(r - 1)/2 = 2$
$(BC^2)*$	$=$	(AB^2D^2)	$=$	(AC^2D^2)	$(q - 1)(r - 1)/2 = 2$
(ABC)	$=$	$(ABCD)$	$=$	$D*$	$(p - 1)(q - 1)(r - 1)/4 = 2$
(ABC^2)	$=$	(ABD)	$=$	$(CD)*$	$(p - 1)(q - 1)(r - 1)/4 = 2$
(AB^2C)	$=$	(ACD)	$=$	$(BD)*$	$(p - 1)(q - 1)(r - 1)/4 = 2$
(AB^2C^2)	$=$	$(AD)*$	$=$	(BCD)	$(p - 1)(q - 1)(r - 1)/4 = 2$

*Effects that are considered measurable.

TABLE 13.4-2 Computational Procedures for a Type CRFF-3^4 Design

(i) Data and notation [Y_{ijklm} denotes a score for experimental unit i in treatment combination $jklm$; $i = 1, \ldots, n$ experimental units (s_i); $j = 1, \ldots, p$ levels of treatment A (a_j); $k = 1, \ldots, q$ levels of treatment B (b_k); $l = 1, \ldots, r$ levels of treatment C (c_l); $m = 1, \ldots, t$ levels of treatment D (d_m); for purposes of analysis, treatment D is ignored]:

$ABC = (1/3\ ABCD)$ Summary Table
Entry is Y_{jkl}

		b_0 c_0	b_0 c_1	b_0 c_2	b_1 c_0	b_1 c_1	b_1 c_2	b_2 c_0	b_2 c_1	b_2 c_2
a_0	$n = 1$	d_0 3	d_1 4	d_2 3	d_1 3	d_2 5	d_0 10	d_2 5	d_0 9	d_1 11
a_1		d_1 4	d_2 3	d_0 5	d_2 4	d_0 6	d_1 10	d_0 5	d_1 8	d_2 12
a_2		d_2 3	d_0 3	d_1 7	d_0 6	d_1 7	d_2 8	d_1 6	d_2 9	d_0 13

Table 13.4-2 (continued)

AB Summary Table

Entry is $\sum_{l=1}^{r} Y_{jkl}$

	b_0	b_1	b_2	$\sum_{k=1}^{q}\sum_{l=1}^{r} Y_{jkl}$
a_0	$r = 3$ 10	18	25	53
a_1	12	20	25	57
a_2	13	21	28	62
$\sum_{j=1}^{p}\sum_{l=1}^{r} Y_{jkl} = $	35	59	78	

AC Summary Table

Entry is $\sum_{k=1}^{q} Y_{jkl}$

	c_0	c_1	c_2
a_0	$q = 3$ 11	18	24
a_1	13	17	27
a_2	15	19	28
$\sum_{j=1}^{p}\sum_{k=1}^{q} Y_{jkl} = $ 39	54	79	

BC Summary Table

Entry is $\sum_{j=1}^{p} Y_{jkl}$

	c_0	c_1	c_2
b_0	$p = 3$ 10	10	15
b_1	13	18	28
b_2	16	26	36

(ii) Computational symbols:

$$\sum_{j=1}^{p}\sum_{k=1}^{q}\sum_{l=1}^{r} Y_{jkl} = 3 + 4 + \cdots + 13 = 172.00$$

$$\frac{\left(\sum_{j=1}^{p}\sum_{k=1}^{q}\sum_{l=1}^{r} Y_{jkl}\right)^2}{pqr} = [Y] = \frac{(172.00)^2}{(3)(3)(3)} = 1095.70$$

$$\sum_{j=1}^{p}\sum_{k=1}^{q}\sum_{l=1}^{r} Y_{jkl}^2 = [ABC] = (3)^2 + (4)^2 + \cdots + (13)^2 = 1332.00$$

Table 13.4-2 (continued)

$$\sum_{j=1}^{p} \frac{\left(\sum_{k=1}^{q} \sum_{l=1}^{r} Y_{jkl}\right)^2}{qr} = [A] = \frac{(53)^2}{(3)(3)} + \frac{(57)^2}{(3)(3)} + \frac{(62)^2}{(3)(3)} = 1100.22$$

$$\sum_{k=1}^{q} \frac{\left(\sum_{j=1}^{p} \sum_{l=1}^{r} Y_{jkl}\right)^2}{pr} = [B] = \frac{(35)^2}{(3)(3)} + \frac{(59)^2}{(3)(3)} + \frac{(78)^2}{(3)(3)} = 1198.89$$

$$\sum_{l=1}^{r} \frac{\left(\sum_{j=1}^{p} \sum_{k=1}^{q} Y_{jkl}\right)^2}{pq} = [C] = \frac{(39)^2}{(3)(3)} + \frac{(54)^2}{(3)(3)} + \frac{(79)^2}{(3)(3)} = 1186.44$$

$$[D] = \frac{[\Sigma(ABC)_0]^2}{(3)(3)} + \frac{[\Sigma(ABC)_1]^2}{(3)(3)} + \frac{[\Sigma(ABC)_2]^2}{(3)(3)} = \frac{(60)^2}{(3)(3)} + \frac{(60)^2}{(3)(3)} + \frac{(52)^2}{(3)(3)}$$
$$= 1100.44, \quad \text{where } (ABC)_z = [a_j + b_k + c_l = z(\text{mod } 3)]$$

$$[(AB^2)] = \frac{[\Sigma(AB^2)_0]^2}{(3)(3)} + \frac{[\Sigma(AB^2)_1]^2}{(3)(3)} + \frac{[\Sigma(AB^2)_2]^2}{(3)(3)} = \frac{(58)^2}{(3)(3)} + \frac{(58)^2}{(3)(3)} + \frac{(56)^2}{(3)(3)}$$
$$= 1096.00, \quad \text{where } (AB^2)_z = [a_j + 2b_k = z(\text{mod } 3)]$$

$$[(AC^2)] = \frac{[\Sigma(AC^2)_0]^2}{(3)(3)} + \frac{[\Sigma(AC^2)_1]^2}{(3)(3)} + \frac{[\Sigma(AC^2)_2]^2}{(3)(3)} = \frac{(56)^2}{(3)(3)} + \frac{(60)^2}{(3)(3)} + \frac{(56)^2}{(3)(3)}$$
$$= 1096.89, \quad \text{where } (AC^2)_z = [a_j + 2c_l = z(\text{mod } 3)]$$

$$[(AD)] = \frac{[\Sigma(AB^2C^2)_0]^2}{(3)(3)} + \frac{[\Sigma(AB^2C^2)_1]^2}{(3)(3)} + \frac{[\Sigma(AB^2C^2)_2]^2}{(3)(3)} = \frac{(61)^2}{(3)(3)} + \frac{(57)^2}{(3)(3)} + \frac{(54)^2}{(3)(3)}$$
$$= 1098.44, \quad \text{where } (AB^2C^2)_z = [a_j + 2b_k + 2c_l = z(\text{mod } 3)]$$

$$[(BC^2)] = \frac{[\Sigma(BC^2)_0]^2}{(3)(3)} + \frac{[\Sigma(BC^2)_1]^2}{(3)(3)} + \frac{[\Sigma(BC^2)_2]^2}{(3)(3)} = \frac{(64)^2}{(3)(3)} + \frac{(54)^2}{(3)(3)} + \frac{(54)^2}{(3)(3)}$$
$$= 1103.11, \quad \text{where } (BC^2)_z = [b_k + 2c_l = z(\text{mod } 3)]$$

$$[(BD)] = \frac{[\Sigma(AB^2C)_0]^2}{(3)(3)} + \frac{[\Sigma(AB^2C)_1]^2}{(3)(3)} + \frac{[\Sigma(AB^2C)_2]^2}{(3)(3)} = \frac{(53)^2}{(3)(3)} + \frac{(63)^2}{(3)(3)} + \frac{(56)^2}{(3)(3)}$$
$$= 1101.56, \quad \text{where } (AB^2C)_z = [a_j + 2b_k + c_l = z(\text{mod } 3)]$$

$$[(CD)] = \frac{[\Sigma(ABC^2)_0]^2}{(3)(3)} + \frac{[\Sigma(ABC^2)_1]^2}{(3)(3)} + \frac{[\Sigma(ABC^2)_2]^2}{(3)(3)} = \frac{(59)^2}{(3)(3)} + \frac{(54)^2}{(3)(3)} + \frac{(59)^2}{(3)(3)}$$
$$= 1097.56, \quad \text{where } (ABC^2)_z = [a_j + b_k + 2c_l = z(\text{mod } 3)]$$

(iii) Computational formulas:

$$SSTO = [ABC] - [\dot{Y}] = 236.30 \qquad SS(AC^2) = [(AC^2)] - [Y] = 1.19$$
$$SSA = [A] - [Y] = 4.52 \qquad SS(AD) = [(AD)] - [Y] = 2.74$$
$$SSB = [B] - [Y] = 103.19 \qquad SS(BC^2) = [(BC^2)] - [Y] = 7.41$$

Table 13.4-2 (continued)

$$SSC = [C] - [Y] = 90.74 \qquad SS(BD) = [(BD)] - [Y] = 5.86$$

$$SSD = [D] - [Y] = 4.74 \qquad SS(CD) = [(CD)] - [Y] = 1.86$$

$$SS(AB^2) = [(AB^2)] - [Y] = 0.30 \qquad SSRES = SSTO - SSA - SSB - SSC - \cdots - SS(CD)$$

$$= 13.75$$

This example does not provide for a within-cell error term. If higher-order interaction components can be assumed insignificant, they can be pooled to form a residual error term.

    The results of the analysis are summarized in Table 13.4-3. The residual error term is based on only six degrees of freedom. If a larger number of degrees of freedom is desired, all of the remaining two- and three-treatment interaction components, with the exception of those aliased with treatments, namely, (BCD^2), (ACD^2), (ABD^2), and (ABC), can be pooled with the residual error term under the assumption that they are insignificant. If 54 subjects are available for the experiment, a within-cell error term can be computed. This requires two observations under each of the 27 treatment combinations. None of the interaction components in Table 13.4-3 is significant. This lends support to the assumption that the residual error term is an estimate only of random error.

    If the (CD) component, for example, had been significant, an experimenter would be faced with the problem of how to interpret this result. Several courses of action can be pursued in carrying out follow-up experiments designed to clarify the interpretation of aliased effects. The experimenter might choose to *complete* the

TABLE 13.4-3 ANOVA Table for Type CRFF-3⁴ Design

	Source	SS	df	MS	F (Model I)
1	A	4.52	2	2.26	$[1/11]$
2	B	103.19	2	51.60	$[2/11]$ 7.50*
3	C	90.74	2	45.37	$[3/11]$ 6.59*
4	D	4.74	2	2.37	$[4/11]$
5	(AB^2)	0.30	2	0.15	$[5/11]$
6	(AC^2)	1.19	2	0.60	$[6/11]$
7	(AD)	2.74	2	1.37	$[7/11]$
8	(BC^2)	7.41	2	3.70	$[8/11]$
9	(BD)	5.86	2	2.93	$[9/11]$
10	(CD)	1.86	2	0.93	$[10/11]$
11	Residual	13.75	6	6.88	
12	Total	236.30	26		

*$p < .05$.

experiment by running the remaining two-thirds of the treatment combinations. Another alternative would be to carry out a complete factorial experiment involving only treatments B, C, and D under the assumption, supported by this analysis, that treatment A is of no consequence.

Comparisons among means in a fractional factorial design can be carried out following procedures described in Section 8.5 for a complete factorial design. An alternative design for this 3^4 experiment uses a Latin square as the building block design. This design, which is described by Winer (1971, 704), uses three balanced Latin squares, with n subjects assigned to each cell. Fractional factorial designs based on a Latin square are described in Sections 13.7 through 13.12.

13.5 GENERAL PROCEDURES FOR CONSTRUCTING TYPE RBFF-2^k DESIGNS

The *building block* for fractional factorial designs described in the previous sections is a completely randomized design. Designs described here use a randomized block design as the building block and are designated by the letters RBFF. A one-half fractional factorial design, for example, can be laid out in two groups, four groups, and so on, by means of confounding procedures described in Chapter 12. No new principles are involved in these designs. The assumptions underlying a type RBF-pq design (Section 9.6) are also required for a type RBFF-2^k design.

General procedures for laying out fractional factorial designs in groups will be illustrated by means of a one-half replication of a type RBFF-2^5 design. The first step is to choose a defining contrast. Let the $ABCDE$ interaction be the defining contrast. The 32 treatment combinations of a complete factorial design can be reduced to 16 by the use of either relation $(ABCDE)_0$ or $(ABCDE)_1$. To assign the 16 treatment combinations to two groups of eight combinations each, it is necessary to confound an interaction other than the defining contrast with between-groups variation. The interaction selected as the confounding interaction should be one considered to be insignificant. Procedures for confounding an interaction with groups are described in Section 12.2. If the AB interaction is confounded with between-groups variation and $(ABCDE)_0$ is used as the defining contrast, one obtains the design shown in Table 13.5-1. In this design, treatment combinations in group 0 satisfy the two relations

$$(ABCDE)_0 = a_j + b_k + c_l + d_m + e_o = 0(\text{mod } 2)$$
$$(AB)_0 = a_j + b_k = 0(\text{mod } 2).$$

Treatment combinations in group 1 satisfy the relations

$$(ABCDE)_0 = a_j + b_k + c_l + d_m + e_o = 0(\text{mod } 2)$$
$$(AB)_1 = a_j + b_k = 1(\text{mod } 2).$$

All treatments and interactions except AB, its alias CDE, and the defining contrast $ABCDE$ are within-group effects. The analysis of the experiment is carried out as if the experiment were a complete four-treatment experiment. The alias pattern and computational formulas for this design appear in Table 13.5-2.

For purposes of computation, treatment E can be ignored. Only summary tables involving treatments A, B, C, and D are required for the analysis. All main effects are aliased with four-treatment interactions. If it can be assumed that the two- and three-treatment interactions are insignificant, they can be pooled to form a residual error term with nine degrees of freedom. This design is not satisfactory if an experimenter is interested in evaluating two-treatment interactions.

If an experiment contains six treatments, a one-half replication can be laid out in blocks of size eight. In this design, only 32 of the 64 treatment combinations are included in the experiment. This design permits an experimenter to evaluate all two-treatment interactions except one. Higher-order interactions are pooled to form a residual error term. To lay out the 32 treatment combinations in blocks of size eight, it is necessary to confound two interactions as described in Section 12.9 with between-groups variation.

A procedure for estimating missing observations in fractional factorial designs was described by Draper and Stoneman (1964), who illustrate the estimating procedure for one, two, and three missing observations. Alternatively the full rank experimental design model approach in Section 13.12 can be used.

TABLE 13.5-1   Layout of Type RBFF-2^5 Design in Two Groups

				Treatment Combinations				
	abcde	*abcde*	*abcde*	*abcde*	*abcde*	*abcde*	*abcde*	*abcde*
Group 0	00000	00110	00101	00011	11000	11110	11101	11011
Group 1	01111	01100	01010	01001	10100	10010	10001	10111

13.6 OTHER TYPES OF CRFF AND RBFF DESIGNS

Fractional factorial designs in which all treatments are at four levels can be easily constructed from 2^k plans. Cochran and Cox (1957) describe procedures for laying out these designs. Johnson and Leone (1964, 216) describe a 4×2^k fractional factorial design.

Designs with mixed treatments at two and three levels present special problems with respect to layout and analysis. Kempthorne (1952, 419) discusses some of the problems inherent in these designs. Connor and Young (1961) present plans for $2^k \times 3^l$ experiments, with k and l equal to one through nine treatments. Addelman (1963) has described general procedures for constructing complex fractional factorial

TABLE 13.5-2　Alias Pattern and Computational Formulas for Type RBFF-2^5 Design

Source	Alias	df	Computational Formulas	F (Model I)
1 Groups (AB)	CDE	1	$[AB] - [A] - [B] + [Y]$	$[2/16]$
2 A	BCDE	1	$[A] - [Y]$	$[3/16]$
3 B	ACDE	1	$[B] - [Y]$	$[4/16]$
4 C	ABDE	1	$[C] - [Y]$	$[5/16]$
5 D	ABCE	1	$[D] - [Y]$	$[6/16]$
6 E	ABCD	1	$[ABCD] - [ABC] - [ABD] - [ACD] - [BCD] + [AB] + [AC]$ $+ [AD] + [BC] + [BD] + [CD] - [A] - [B] - [C] - [D] + [Y]$	
7 AC	BDE	1	$[AC] - [A] - [C] + [Y]$	
8 AD	BCE	1	$[AD] - [A] - [D] + [Y]$	
9 BC	ADE	1	$[BC] - [B] - [C] + [Y]$	
10 BD	ACE	1	$[BD] - [B] - [D] + [Y]$	
11 CD	ABE	1	$[CD] - [C] - [D] + [Y]$	
12 ABC	DE	1	$[ABC] - [AB] - [AC] - [BC] + [A] + [B] + [C] - [Y]$	
13 ABD	CE	1	$[ABD] - [AB] - [AD] - [BD] + [A] + [B] + [D] - [Y]$	
14 ACD	BE	1	$[ACD] - [AC] - [AD] - [CD] + [A] + [C] + [D] - [Y]$	
15 BCD	AE	1	$[BCD] - [BC] - [BD] - [CD] + [B] + [C] + [D] - [Y]$	
16 Residual = pooled two- and three-treatment interactions				
17 Total		15	$[ABCD] - [Y]$	

designs. Other useful sources for fractional factorial designs are Addelman and Kempthorne (1961), Hahn and Shapiro (1966), and Margolin (1968).

13.7 INTRODUCTION TO LATIN SQUARE FRACTIONAL FACTORIAL DESIGN

LATIN SQUARE DESIGN AS A $1/p$ FRACTIONAL FACTORIAL DESIGN

The classic application of a Latin square design in agricultural research involves one treatment with *nuisance* variables assigned to rows and columns of the square. If the Latin square contains two treatments and one nuisance variable, it is called a Latin square confounded factorial design. If both variables assigned to rows and columns represent treatments instead of nuisance variables, the design is called a fractional factorial design (type LSFF design). Recall that the term *factorial experiment* refers to the simultaneous evaluation of two or more *crossed* treatments. Thus the Latin square in Figure 13.7-1 may be designated as a Latin square design or as a Latin square confounded factorial design or as a Latin square fractional factorial design, depending on the nature of the variables assigned to the rows and columns.

FIGURE 13.7-1 3×3 standard square

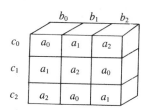

If a_j, b_k, and c_l represent three treatments, it can be shown that this Latin square is equivalent to a one-third replication of a 3^3 factorial experiment. Assume that the $(AB^2C^2)_0$ component has been selected as the defining contrast for a 3^3 factorial experiment. The treatment combinations are as follows:

$$(AB^2C^2)_0 = a_j + 2b_k + 2c_l = 0 (\text{mod } 3)$$
$$= 000, 021, 012, 101, 110, 122, 202, 211, 220$$

where the numbers refer to the level of treatments A, B, and C, respectively. A comparison of these treatment combinations with those in Figure 13.7-1 reveals that they are identical. Thus a standard 3×3 Latin square corresponds to the treatment combinations in a one-third replication of a 3^3 factorial experiment, with $(AB^2C^2)_0$ as the defining contrast. A complete 3^3 factorial experiment contains 27 treatment combinations. The Latin square, or one-third replication, contains only nine of these combinations.

In Section 7.2 it was noted that a 3×3 Latin square has 12 arrangements. These 12 arrangements can be generated by the family of relations

$$(ABC)_0 \qquad (AB^2C)_0$$
$$(ABC)_1 \qquad (AB^2C)_1$$
$$(ABC)_2 \qquad (AB^2C)_2$$
$$(ABC^2)_0 \qquad (AB^2C^2)_0$$
$$(ABC^2)_1 \qquad (AB^2C^2)_1$$
$$(ABC^2)_2 \qquad (AB^2C^2)_2.$$

A 2×2 Latin square has two arrangements. These correspond to the relations $(ABC)_0$ and $(ABC)_1$. A 2×2 Latin square is analogous to a one-half replication of a 2^3 factorial experiment.

If a one-third replication of a 3^3 factorial experiment corresponds to a 3×3 Latin square, the reader may wonder where $1/9$, $1/27$, and so on replications fit into the scheme. A $1/9$ replication of a 3^4 factorial experiment is a 3×3 Graeco-Latin square, and a $1/27$ replication of a 3^5 factorial experiment is a 3×3 hyper-Graeco-Latin square. The use of hyper-squares, where the maximum number of orthogonal squares has been superimposed on a Latin square, is an extreme form of fractionation. It can be shown that these designs represent the smallest fractional replication in which main effects are not aliased with one another.

THE IMPORTANCE OF ADDITIVITY OF ROW, COLUMN, AND SQUARE EFFECTS

To draw valid inferences from experimental designs constructed from Latin squares, interactions among row, column, and square variables must be zero. This requirement is much more likely to be met if the row and column variables represent nuisance or classification variables than if they represent additional treatments, as in the case of a fractional factorial design. In Chapter 7, which deals with Latin square designs, this *additivity requirement* was stated without explanation. A clearer presentation of the problem associated with nonadditivity of effects can be made by using the alias concept developed for fractional factorial designs.

Consider the 3×3 Latin square in Figure 13.7-1. This square can be constructed by using the $(AB^2C^2)_0$ component of the three-treatment interaction as the defining contrast. The aliases associated with treatment A are given by

$$(AB^2C^2)A = A^2B^2C^2 = (A^2B^2C^2)^2 \text{ reduced modulo } 3 = ABC$$
$$(AB^2C^2)A^2 = A^3B^2C^2 = (A^3B^2C^2)^2 \text{ reduced modulo } 3 = BC.$$

Thus treatment A is aliased with the (BC) and (ABC) components of the two- and three-treatment interactions, respectively. Similarly, it can be shown that

$$B = (AC^2) = (ABC^2)$$
$$C = (AB^2) = (AB^2C).$$

Hence in a 3×3 Latin square design or a one-third replication of a 3^3 factorial design,

A, B, and C are indistinguishable from components of the two- and three-treatment interactions. It can also be shown that the components of the two-treatment interactions not aliased with main effects are aliased with each other. That is,

$$(AB) = (AC) = (BC^2).$$

A general principle can be stated with respect to the use of Latin square designs or Latin squares as building blocks for more complex designs. Main effects will always be aliased with interaction effects. Thus, to interpret F ratios for main effects that are significant, it is necessary to assume that the aliased interactions (or components) are zero.

In large Latin squares, a relatively small portion of all two-treatment interaction components are aliased with treatments. Therefore, significant two-treatment interaction effects bias main effects tests less in large Latin squares than in small Latin squares. The situation is more complex if a within-cell error term is not available. Under these conditions, a residual error term is sometimes obtained by pooling all interaction MS's not aliased with main effects. If the pooled two- and three-treatment interaction components are not zero, they will negatively bias tests of main effects.

GENERAL INTRODUCTION TO TYPE LSFF-p^k DESIGNS

Subsequent sections of this chapter describe a number of complex designs that use a Latin square as the building block design. For convenience, these designs are designated as type LSFF designs. Additional examples of Latin square fractional factorial designs can be found in the thorough coverage by Winer (1971, Ch. 9).

The layout of type LSFF designs is much simpler than the layout of fractional factorial designs based on completely randomized and randomized block designs. The latter designs are generally restricted to experiments having two or three levels of each treatment; type LSFF designs do not have this restriction. In addition, type LSFF designs are not limited to experiments in which the number of levels of each treatment is a prime number. The use of mixed levels for treatments poses no computational problem if, of course, at least three of the treatments or classification variables have the same number of levels. The number of levels of rows, columns, and Latin letters constituting the square must be equal. It should be emphasized again that the use of a Latin square as a building block design requires a highly restrictive set of assumptions with respect to interactions. If interactions among variables that constitute the Latin square are not zero, main effects will be aliased with interaction effects.

13.8 COMPUTATIONAL PROCEDURES FOR TYPE LSFF-$p \cdot p^2$ DESIGN

If the design illustrated in Figure 13.7-1 of the previous section contains three treatments, it can be classified as a fractional factorial design. It is assumed that the $p^2 = 9$

cells of the square in Figure 13.7-1 contain nine random samples of n subjects ($n > 1$) from a common population. Computational procedures for this type LSFF-p^3 are identical to those for a regular Latin square design and have been described in Section 7.3.

A type LSFF-p^3 design can be easily modified for research situations in which it is possible to use matched subjects or repeated measures on the same subjects. The modified design is diagrammed in Figure 13.8-1 and is designated by the letters LSFF-$p \cdot p^2$. In this design, treatment A is a between-blocks treatment, whereas B and C are within-blocks treatments. This is indicated in the designation scheme by placing a dot after the between-blocks treatment. This same procedure was followed in the designation scheme for split-plot factorial designs. The use of the letter p in the designation for this design indicates that treatments A, B, and C must have the same number of levels. If repeated measurements are obtained, the design requires p random samples of n subjects from a common population. The p samples are randomly assigned to the levels of treatment A. The sequence of administration of the $b_k c_l$ treatment combinations is randomized independently for each subject. If matched subjects are used, p random samples of n sets of p matched subjects are required. The $b_k c_l$ treatment combinations are randomly assigned to the p subjects within each matched set.

FIGURE 13.8-1   Block diagram of type LSFF-3·3^2 design

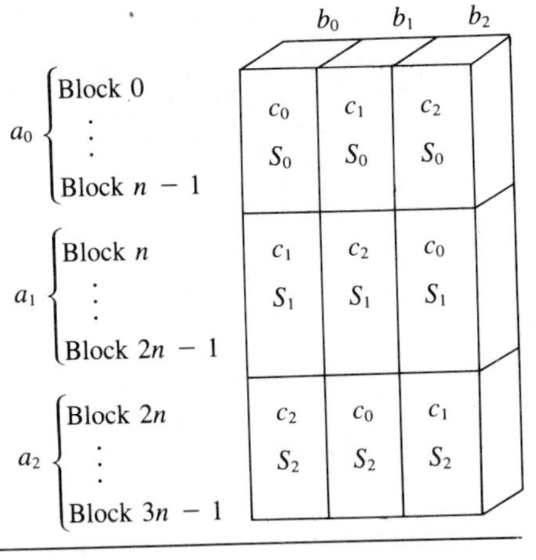

The experimental design model for this design is

$$Y_{ijkl} = \mu + \alpha_j + \pi_{i(j)} + \beta_k + \gamma_l + \text{res} + \beta\gamma\pi_{kli(j)} + \epsilon_{ijkl} \qquad (i = 1, \ldots, n;$$
$$j = 1, \ldots, p; k = 1, \ldots, p; l = 1, \ldots, p)$$

where Y_{ijkl} = a score for the experimental unit in block i and treatment combination $a_j b_k c_l$

μ = the overall mean

α_j = the effect of treatment level j and is subject to the restriction $\sum_{j=1}^{p} \alpha_j = 0$

$\pi_{i(j)}$ = the effect of block i that is $NID(0, \sigma_\pi^2)$

β_k = the effect of treatment level k and is subject to the restriction $\sum_{k=1}^{p} \beta_k = 0$

γ_l = the effect of treatment level l and is subject to the restriction $\sum_{l=1}^{p} \gamma_l = 0$

res = the effect that represents nonadditivity of effects α_j, β_k, and γ_l

$\beta\gamma\pi_{kli(j)}$ = the effect that represents nonadditivity of effects $(\beta\gamma)_{kl}$ and $\pi_{i(j)}$

ϵ_{ijkl} = the experimental error that is $NID(0, \sigma_\epsilon^2)$; ϵ_{ijkl} is independent of $\pi_{i(j)}$.

It is assumed that treatments A, B, and C represent fixed effects.

The layout of the type LSFF-3·3^2 design and computational formulas are shown in Table 13.8-1. The analysis of the data is summarized in Table 13.8-2. According to this summary, the null hypothesis can be rejected for treatment C but not for treatments A and B. A partial check on the assumption that all interactions among treatments are insignificant is given by the ratio

$$F = \frac{MSRES}{MSBC \times BL(A)}.$$

In this example, it seems safe to conclude that interactions among treatments are zero. Under this condition, the experimenter may wish to pool $MSRES$ with $MSBC \times BL(A)$ to obtain a better estimate of experimental error. Note that the test of treatment A is less powerful than the tests of treatments B and C. This same point has been made repeatedly with respect to the between-blocks treatments of a split-plot factorial design.

13.9 COMPUTATIONAL PROCEDURES FOR TYPE LSFF-p^3t DESIGN

In Section 7.3, an example involving a road test of four automobile tires was used to illustrate a possible application of a Latin square design. The levels of C, B, and A corresponded, respectively, to automobiles, wheel positions, and rubber compounds used in the tire construction. Two road tests using the same four automobiles were run, one after the other. In the analysis described in Section 7.3, the data were treated as if the two replications of the experiment were carried out under identical conditions. This is unrealistic because the two tests were separated by an interval of time and the cars were older during the second test. The within-cell error term included not only random error but also variation attributable to temperature and other climatic changes as well as variation associated with mechanical wear of the automobiles. An alternative design described in this section treats the two replications of the road test as two levels of a nuisance variable (D).

TABLE 13.8-1   Computational Procedures for a Type LSFF-$3\cdot3^2$ Design

(i)   Data and notation [Y_{ijkl} denotes a score for experimental unit i in treatment combination jkl; $i = 1,$ $\ldots, n$ experimental units (s_i); $j = 1, \ldots, p$ levels of treatment A (a_j); $k = 1, \ldots, p$ levels of treatment B_k; $l = 1, \ldots, p$ levels of treatment C (c_l)]:

ABCS Summary Table
Entry is Y_{ijkl}

		b_0	b_1	b_2	$\sum\limits_{k=1}^{p} Y_{ijkl}$
		c_0	c_1	c_2	
a_0	s_0	3	4	6	13
	s_1	1	4	6	11
		c_2	c_0	c_1	
a_1	s_2	7	3	3	13
	s_3	5	2	3	10
		c_1	c_2	c_0	
a_2	s_4	4	7	3	14
	s_5	2	5	2	9

ABC Summary Table
Entry is $\sum\limits_{i=1}^{n} Y_{ijkl}$

	b_0	b_1	b_2	$\sum\limits_{i=1}^{n}\sum\limits_{k=1}^{p} Y_{ijkl}$
	$n = 2$			
	c_0	c_1	c_2	
a_0	4	8	12	24
	c_2	c_0	c_1	
a_1	12	5	6	23
	c_1	c_2	c_0	
a_2	6	12	5	23

BC Summary Table
Entry is $\sum\limits_{i=1}^{n} Y_{ijkl}$

	b_0	b_1	b_2	$\sum\limits_{i=1}^{n}\sum\limits_{k=1}^{p} Y_{ijkl}$
c_0	$n = 2$ 4	5	5	14
c_1	6	8	6	20
c_2	12	12	12	36
$\sum\limits_{i=1}^{n}\sum\limits_{l=1}^{p} Y_{ijkl} =$	22	25	23	

Table 13.8-1 (continued)

(ii)   Computational symbols:

$$\sum_{i=1}^{n}\sum_{j=1}^{p}\sum_{k=1}^{p} Y_{ijkl} = 3 + 4 + \cdots + 2 = 70.00$$

$$\frac{\left(\sum\limits_{i=1}^{n}\sum\limits_{j=1}^{p}\sum\limits_{k=1}^{p} Y_{ijkl}\right)^2}{np^2} = [Y] = \frac{(70.00)^2}{(2)(3)^2} = 272.22$$

$$\sum_{i=1}^{n}\sum_{j=1}^{p}\sum_{k=1}^{p} Y_{ijkl}^2 = [ABCS] = (3)^2 + (4)^2 + \cdots + (2)^2 = 326.00$$

$$\sum_{i=1}^{n}\sum_{j=1}^{p}\frac{\left(\sum\limits_{k=1}^{p} Y_{ijkl}\right)^2}{p} = [AS] = \frac{(13)^2}{3} + \frac{(11)^2}{3} + \cdots + \frac{(9)^2}{3} = 278.67$$

$$\sum_{j=1}^{p}\frac{\left(\sum\limits_{i=1}^{n}\sum\limits_{k=1}^{p} Y_{ijkl}\right)^2}{np} = [A] = \frac{(24)^2}{(2)(3)} + \frac{(23)^2}{(2)(3)} + \frac{(23)^2}{(2)(3)} = 272.33$$

$$\sum_{k=1}^{p}\frac{\left(\sum\limits_{i=1}^{n}\sum\limits_{l=1}^{p} Y_{ijkl}\right)^2}{np} = [B] = \frac{(22)^2}{(2)(3)} + \frac{(25)^2}{(2)(3)} + \frac{(23)^2}{(2)(3)} = 273.00$$

$$\sum_{l=1}^{p}\frac{\left(\sum\limits_{i=1}^{n}\sum\limits_{k=1}^{p} Y_{ijkl}\right)^2}{np} = [C] = \frac{(14)^2}{(2)(3)} + \frac{(20)^2}{(2)(3)} + \frac{(36)^2}{(2)(3)} = 315.33$$

$$\sum_{j=1}^{p}\sum_{k=1}^{p}\frac{\left(\sum\limits_{i=1}^{n} Y_{ijkl}\right)^2}{p} = [ABC] = \frac{(4)^2}{2} + \frac{(8)^2}{2} + \cdots + \frac{(5)^2}{2} = 317.00$$

(iii)   Computational formulas:

$$SSTO = [ABCS] - [Y] = 53.78$$

$$SSBETWEEN\ BL = [AS] - [Y] = 6.45$$

$$SSA = [A] - [Y] = 0.11$$

$$SSBL(A) = [AS] - [A] = 6.34$$

$$SSWITHIN\ BL = [ABCS] - [AS] = 47.33$$

$$SSB = [B] - [Y] = 0.78$$

$$SSC = [C] - [Y] = 43.11$$

$$SSRES = [ABC] - [A] - [B] - [C] + 2[Y] = 0.78$$

$$SSBC \times BL(A) = [ABCS] - [ABC] - [AS] + [A] = 2.66$$

TABLE 13.8-2 ANOVA Table for Type LSFF-3·3² Design

	Source	SS	df		MS	F	E(MS) Model III
1	Between blocks	6.45	$np - 1$	$= 5$			
2	A	0.11	$p - 1$	$= 2$	0.06	[2/3] N.S.	$\sigma_\epsilon^2 + p\sigma_\pi^2 + np\Sigma\alpha_j^2/(p-1)$
3	BL(A)	6.34	$p(n - 1)$	$= 3$	2.11		$\sigma_\epsilon^2 + p\sigma_\pi^2$
4	Within blocks	47.33	$np(p - 1)$	$= 12$			
5	B	0.78	$p - 1$	$= 2$	0.39	[5/8] N.S.	$\sigma_\epsilon^2 + np\Sigma\beta_k^2/(p-1)$
6	C	43.11	$p - 1$	$= 2$	21.56	[6/8] 49.00*	$\sigma_\epsilon^2 + np\Sigma\gamma_i^2/(p-1)$
7	Residual	0.78	$(p - 1)(p - 2)$	$= 2$	0.39	[7/8] N.S.	$\sigma_\epsilon^2 + n\Sigma\text{res}^2/(p-1)(p-2)$
8	BC×BL(A)	2.66	$p(n - 1)(p - 1)$	$= 6$	0.44		σ_ϵ^2
9	Total	53.78	$np^2 - 1$	$= 17$			

*$p < .01$.

FIGURE 13.9-1 Block diagram of type LSFF-$4^3 2$ design

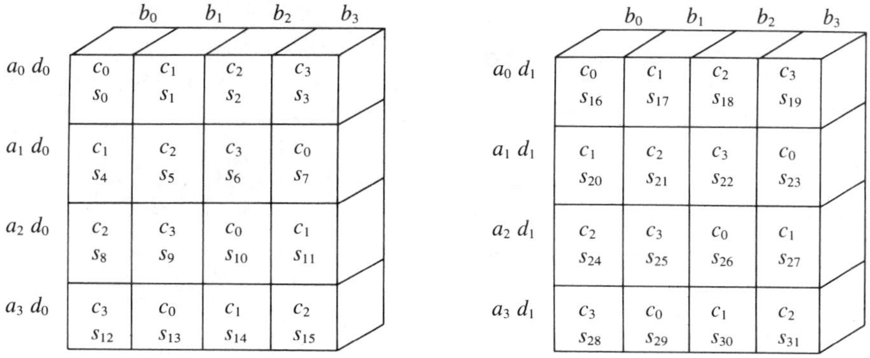

A type LSFF-$4^3 2$ design, in which treatment D corresponds to replications, is diagrammed in Figure 13.9-1. The experimental design model equation for this design is

$$Y_{ijklm} = \mu + \alpha_j + \beta_k + \gamma_l + \delta_m + (\alpha\delta)_{jm} + (\beta\delta)_{km} + (\gamma\delta)_{lm} + \text{res} + \epsilon_{jklm}$$
$$(j = 1, \ldots, p;\ k = 1, \ldots, p;\ l = 1, \ldots, p;\ m = 1, \ldots, t).$$

The terms in the model, with the exception of δ_m, $(\alpha\delta)_{jm}$, $(\beta\delta)_{km}$, and $(\gamma\delta)_{lm}$, are defined in Section 13.8. The term δ_m refers to the effect of treatment level m, which is a constant for all subjects within population m. This design permits an experimenter to evaluate the AD, BD, and CD interactions. Each cell of the design diagrammed in Figure 13.9-1 contains one observation. This design can be easily modified for the case in which each cell contains n observations. The required modifications are described in a subsequent paragraph.

The layout of the type LSFF-$4^3 2$ design, computational tables, and formulas are shown in Table 13.9-1. The analysis is summarized in Table 13.9-2. According to the analysis in Table 13.9-2, the effects of rubber compounds, wheel positions, and replications are significant. The superiority of the present analysis, in which the two road tests were treated as a fourth variable, compared to the analysis summarized in Table 7.3-3 is readily apparent. In the latter analysis, the within-cell variation included, in addition to random error, the effects of replications (D).

A type LSFF-$4^3 2$ design corresponds to a one-fourth replication of a type CRF-4442 design. If only one subject is assigned to each treatment combination, the latter design requires 128 subjects compared with 32 for the fractional factorial design. For research situations in which an experimenter believes that only one treatment is likely to interact with the other treatments, a type LSFF-p^3t design is a better choice than a complete factorial design. The example that was presented employed the same Latin square for both levels of treatment D. If it suits the experimenter's purpose, balanced sets of squares or independently randomized squares can be used. The number of levels of treatments A, B, and C must be equal. The only restriction on the number of levels of treatment D is that there must be at least two levels.

Comparisons among means follow the general procedures described in Section 8.5.

TABLE 13.9-1 Computational Procedures for a Type LSFF-$4^3 2$ Design

(i) Data and notation [Y_{jklm} denotes a score for the experimental unit in treatment combination $jklm$; $j = 1, \ldots, p$ levels of treatment A (a_j); $k = 1, \ldots, p$ levels of treatment B (b_k); $l = 1, \ldots,$ p levels of treatment C (c_l); $m = 1, \ldots, t$ levels of treatment D (d_m)]:

ABCD Summary Table
Entry is Y_{jklm}

		b_0	b_1	b_2	b_3
d_0	a_0	c_0 2	c_3 3	c_2 3	c_1 6
	a_1	c_1 2	c_0 4	c_3 5	c_2 4
	a_2	c_2 5	c_1 8	c_0 7	c_3 6
	a_3	c_3 9	c_2 9	c_1 11	c_0 10
d_1	a_0	c_0 1	c_3 2	c_2 2	c_1 3
	a_1	c_1 3	c_0 3	c_3 4	c_2 3
	a_2	c_2 6	c_1 6	c_0 5	c_3 7
	a_3	c_3 8	c_2 8	c_1 10	c_0 7

AD Summary Table

Entry is $\sum\limits_{k=1}^{p} Y_{jklm}$

	d_0	d_1	$\sum\limits_{k=1}^{p}\sum\limits_{m=1}^{t} Y_{jklm}$
	$p = 4$		
a_0	14	8	22
a_1	15	13	28
a_2	26	24	50
a_3	39	33	72

BD Summary Table

Entry is $\sum\limits_{j=1}^{p} Y_{jklm}$

	d_0	d_1	$\sum\limits_{j=1}^{p}\sum\limits_{m=1}^{t} Y_{jklm}$
	$p = 4$		
b_0	18	18	36
b_1	24	19	43
b_2	26	21	47
b_3	26	20	46

CD Summary Table

Entry is $\sum\limits_{j=1}^{p} Y_{jklm}$

	d_0	d_1	$\sum\limits_{j=1}^{p}\sum\limits_{m=1}^{t} Y_{jklm}$
	$p = 4$		
c_0	23	16	39
c_1	27	22	49
c_2	21	19	40
c_3	23	21	44
$\sum\limits_{j=1}^{p}\sum\limits_{l=1}^{p} Y_{jklm} = 94$	78		

Table 13.9-1 (continued)

(ii)   Computational symbols:

$$\sum_{j=1}^{p} \sum_{k=1}^{p} \sum_{m=1}^{t} Y_{jklm} = 2 + 3 + \cdots + 7 = 172.00$$

$$\frac{\left(\sum\limits_{j=1}^{p} \sum\limits_{k=1}^{p} \sum\limits_{m=1}^{t} Y_{jklm}\right)^2}{p^2 t} = [Y] = \frac{(172.00)^2}{(4)^2 2} = 924.50$$

$$\sum_{j=1}^{p} \sum_{k=1}^{p} \sum_{m=1}^{t} Y_{jklm}^2 = [ABCD] = (2)^2 + (3)^2 + \cdots + (7)^2 = 1160.00$$

$$\sum_{j=1}^{p} \frac{\left(\sum\limits_{k=1}^{p} \sum\limits_{m=1}^{t} Y_{jklm}\right)^2}{pt} = [A] = \frac{(22)^2}{(4)(2)} + \frac{(28)^2}{(4)(2)} + \cdots + \frac{(72)^2}{(4)(2)} = 1119.00$$

$$\sum_{k=1}^{p} \frac{\left(\sum\limits_{j=1}^{p} \sum\limits_{m=1}^{t} Y_{jklm}\right)^2}{pt} = [B] = \frac{(36)^2}{(4)(2)} + \frac{(43)^2}{(4)(2)} + \cdots + \frac{(46)^2}{(4)(2)} = 933.75$$

$$\sum_{l=1}^{p} \frac{\left(\sum\limits_{j=1}^{p} \sum\limits_{m=1}^{t} Y_{jklm}\right)^2}{pt} = [C] = \frac{(39)^2}{(4)(2)} + \frac{(49)^2}{(4)(2)} + \cdots + \frac{(44)^2}{(4)(2)} = 932.25$$

$$\sum_{m=1}^{t} \frac{\left(\sum\limits_{j=1}^{p} \sum\limits_{k=1}^{p} Y_{jklm}\right)^2}{p^2} = [D] = \frac{(94)^2}{(4)^2} + \frac{(78)^2}{(4)^2} = 932.50$$

$$\sum_{j=1}^{p} \sum_{m=1}^{t} \frac{\left(\sum\limits_{k=1}^{p} Y_{jklm}\right)^2}{p} = [AD] = \frac{(14)^2}{4} + \frac{(8)^2}{4} + \cdots + \frac{(33)^2}{4} = 1129.00$$

$$\sum_{k=1}^{p} \sum_{m=1}^{t} \frac{\left(\sum\limits_{j=1}^{p} Y_{jklm}\right)^2}{p} = [BD] = \frac{(18)^2}{4} + \frac{(18)^2}{4} + \cdots + \frac{(20)^2}{4} = 944.50$$

$$\sum_{l=1}^{p} \sum_{m=1}^{t} \frac{\left(\sum\limits_{j=1}^{p} Y_{jklm}\right)^2}{p} = [CD] = \frac{(23)^2}{4} + \frac{(16)^2}{4} + \cdots + \frac{(21)^2}{4} = 942.50$$

Table 13.9-1 (continued)

(iii) Computational formulas:

$$SSTO = [ABCD] - [Y] = 235.50$$
$$SSA = [A] - [Y] = 194.50$$
$$SSB = [B] - [Y] = 9.25$$
$$SSC = [C] - [Y] = 7.75$$
$$SSD = [D] - [Y] = 8.00$$
$$SSAD = [AD] - [A] - [D] + [Y] = 2.00$$
$$SSBD = [BD] - [B] - [D] + [Y] = 2.75$$
$$SSCD = [CD] - [C] - [D] + [Y] = 2.25$$
$$SSRES = [ABCD] - [AD] - [BD] - [CD] + 2[D] = 9.00$$

TABLE 13.9-2 ANOVA Table for Type LSFF-$4^3 2$ Design

	Source	SS	df	MS	F	E(MS) Model I
1	Rubber compounds (A)	194.50	$p - 1 = 3$	64.83	[$\frac{1}{8}$] 86.44**	$\sigma_\epsilon^2 + pt\Sigma\alpha_j^2/(p - 1)$
2	Wheel positions (B)	9.25	$p - 1 = 3$	3.08	[$\frac{2}{8}$] 4.11*	$\sigma_\epsilon^2 + pt\Sigma\beta_k^2/(p - 1)$
3	Automobiles (C)	7.75	$p - 1 = 3$	2.58	[$\frac{3}{8}$] 3.44	$\sigma_\epsilon^2 + pt\Sigma\gamma_l^2/(p - 1)$
4	Replications (D)	8.00	$t - 1 = 1$	8.00	[$\frac{4}{8}$] 10.67**	$\sigma_\epsilon^2 + p^2\Sigma\delta_m^2/(t - 1)$
5	AD	2.00	$(p - 1)(t - 1) = 3$	0.67	[$\frac{5}{8}$] 0.89	$\sigma_\epsilon^2 + p\Sigma\Sigma(\alpha\delta)_{jm}^2/(p - 1)(t - 1)$
6	BD	2.75	$(p - 1)(t - 1) = 3$	0.92	[$\frac{6}{8}$] 1.23	$\sigma_\epsilon^2 + p\Sigma\Sigma(\beta\delta)_{km}^2/(p - 1)(t - 1)$
7	CD	2.25	$(p - 1)(t - 1) = 3$	0.75	[$\frac{7}{8}$] 1.00	$\sigma_\epsilon^2 + p\Sigma\Sigma(\gamma\delta)_{lm}^2/(p - 1)(t - 1)$
8	Residual	9.00	$t(p - 1)(p - 2) = 12$	0.75		σ_ϵ^2
9	Total	235.50	$p^2 t - 1 = 31$			

*$p < .05$
**$p < .01$

TYPE LSFF-p^3t DESIGN WITH MORE THAN ONE SCORE PER CELL

The analysis shown in Table 13.9-1 can be modified for the case in which more than one observation is obtained within each cell. An *ABCDS* Summary Table must be constructed. The following modifications of the computational symbols as illustrated for $[A]$, $[CD]$, and $[ABCD]$ are required:

$$[A] = \sum_{j=1}^{p} \frac{\left(\sum_{i=1}^{n}\sum_{k=1}^{p}\sum_{m=1}^{t} Y_{ijklm}\right)^2}{npt}, \qquad [CD] = \sum_{l=1}^{p}\sum_{m=1}^{t} \frac{\left(\sum_{i=1}^{n}\sum_{j=1}^{p} Y_{ijklm}\right)^2}{np},$$

$$[ABCD] = \sum_{j=1}^{p}\sum_{k=1}^{p}\sum_{m=1}^{t} \frac{\left(\sum_{i=1}^{n} Y_{ijklm}\right)^2}{n}.$$

A within-cell error term is used as the denominator of F ratios. The formula is

$$SSWCELL = [ABCDS] - [ABCD]$$

with degrees of freedom equal to $p^2t(n-1)$.

COMPUTATIONAL PROCEDURES FOR TYPE LSFF-$pt \cdot p^2$ DESIGN

The design that has been described can be modified for the case in which matched subjects or repeated measures on the same subjects are obtained. A diagram of this design is shown in Figure 13.9-2. The letters S_i refer to pt random samples of n blocks

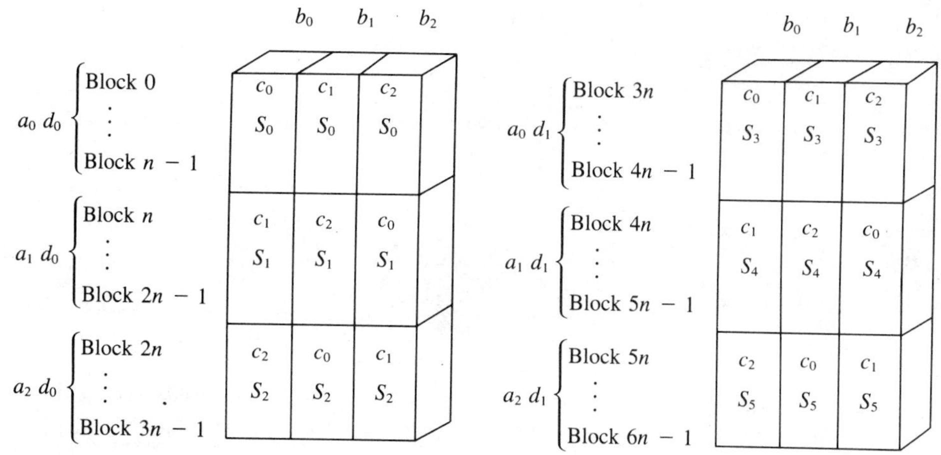

FIGURE 13.9-2   Block diagram of type LSFF-$32 \cdot 3^2$ design

of subjects if repeated measures are obtained or to pt random samples of n blocks of p homogeneous subjects if matched subjects are used. In the former case, administration of the $b_k c_l$ combinations is randomized independently for each subject.

The experimental design model equation for this design is

$$Y_{ijklm} = \mu + \alpha_j + \delta_m + (\alpha\delta)_{jm} + \pi_{i(jm)} + \beta_k + \gamma_l + (\beta\delta)_{km} + (\gamma\delta)_{lm} + \text{res}$$
$$+ (\beta\gamma\pi)_{kli(jm)} + \epsilon_{ijklm} \quad (i = 1, \ldots, n; j = 1, \ldots, p;$$
$$k = 1, \ldots, p; l = 1, \ldots, p; m = 1, \ldots, t).$$

Computational formulas and degrees of freedom for this design appear in Table 13.9-3.

TABLE 13.9-3   Computational Formulas for Type LSFF-$pt \cdot p^2$ Design

	Source	df	Computational Formulas	F	E(MS) Model III (A, B, C, and D fixed effects; Subjects random)
1	Between blocks	$npt - 1$	$[ADS] - [Y]$		
2	A	$p - 1$	$[A] - [Y]$	$[2/5]$	$\sigma_\epsilon^2 + p\sigma_\pi^2 + npt\Sigma\alpha_j^2/(p-1)$
3	D	$t - 1$	$[D] - [Y]$	$[3/5]$	$\sigma_\epsilon^2 + p\sigma_\pi^2 + np^2\Sigma\delta_m^2/(t-1)$
4	AD	$(p-1)(t-1)$	$[AD] - [A] - [D] + [Y]$	$[4/5]$	$\sigma_\epsilon^2 + p\sigma_\pi^2 + np\Sigma\Sigma(\alpha\delta)_{jm}^2/(p-1)(t-1)$
5	BL(AD)	$pt(n-1)$	$[ADS] - [AD]$		$\sigma_\epsilon^2 + p\sigma_\pi^2$
6	Within blocks	$npt(p-1)$	$[ABCDS] - [ADS]$		
7	B	$p - 1$	$[B] - [Y]$	$[7/12]$	$\sigma_\epsilon^2 + npt\Sigma\beta_k^2/(p-1)$
8	C	$p - 1$	$[C] - [Y]$	$[8/12]$	$\sigma_\epsilon^2 + npt\Sigma\gamma_l^2/(p-1)$
9	BD	$(p-1)(t-1)$	$[BD] - [B] - [D] + [Y]$	$[9/12]$	$\sigma_\epsilon^2 + np\Sigma\Sigma(\beta\delta)_{km}^2/(p-1)(t-1)$
10	CD	$(p-1)(t-1)$	$[CD] - [C] - [D] + [Y]$	$[10/12]$	$\sigma_\epsilon^2 + np\Sigma\Sigma(\gamma\delta)_{lm}^2/(p-1)(t-1)$
11	Residual	$t(p-1)(p-2)$	$[ABCD] - [AD] - [BD]$ $- [CD] + 2[D]$	$[11/12]$	$\sigma_\epsilon^2 + n\Sigma\Sigma \text{ res}^2/(p-)(p-2)$
12	BC × BL(AD)	$pt(n-1)(p-1)$	$[ABCDS] - [ABCD]$ $- [ADS] + [AD]$		σ_ϵ^2
13	Total	$np^2t - 1$	$[ABCDS] - [Y]$		

13.10   COMPUTATIONAL PROCEDURES FOR TYPE LSFF-p^4u DESIGN

The design described in this section is appropriate for experiments having five treatments. If four of the treatments have p levels each, the fifth treatment must have p^2

levels. For example, if the four treatments have two levels each, the fifth treatment must have $2^2 = 4$ levels. A block diagram for a type LSFF-$2^4 4$ design is shown in Figure 13.10-1. The design corresponds to a one-fourth replication of a type CRF-22224 design.

The experimental design model equation for this design is

$$Y_{ijklmo} = \mu + \alpha_j + \gamma_l + (\alpha\gamma)_{jl} + \beta_k + \delta_m + (\beta\delta)_{km} + \eta_o + \text{res} + \epsilon_{i(jklmo)}$$

$$(i = 1, \ldots, n; j = 1, \ldots, p; k = 1, \ldots, p; l = 1, \ldots, p;$$

$$m = 1, \ldots, p; o = 1, \ldots, u).$$

The Greek letter η_o designates one of the u levels of treatment E. It is assumed that all interactions not appearing in the model are zero. The design requires np^4 subjects who are randomly assigned to the p^4 cells. Computational formulas and degrees of freedom for this design appear in Table 13.10-1.

FIGURE 13.10-1 Block diagram
of type LSFF-$2^4 4$ design

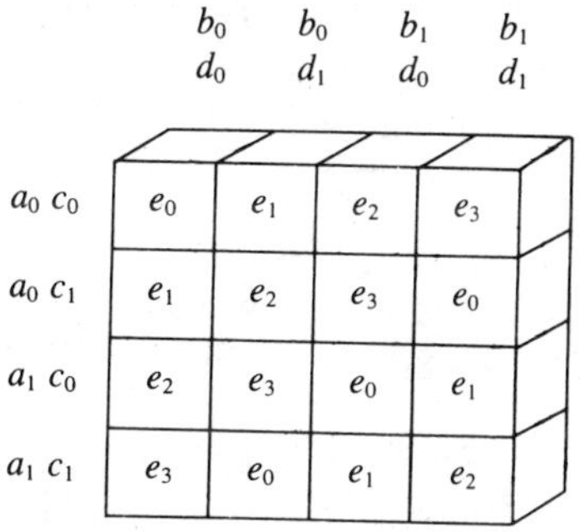

TABLE 13.10-1 Computational Formulas for Type LSFF-p^4u Design

Source	df	Computational Formulas	F	E(MS) Model I
1 A	$p - 1$	$[A] - [Y]$	[1/9]	$\sigma_\epsilon^2 + np^2 \sum \alpha_j^2/(p-1)$
2 C	$p - 1$	$[C] - [Y]$	[2/9]	$\sigma_\epsilon^2 + np^2 \sum \gamma_l^2/(p-1)$
3 AC	$(p-1)(p-1)$	$[AC] - [A] - [C] + [Y]$	[3/9]	$\sigma_\epsilon^2 + np\sum\sum(\alpha\gamma)_{jl}^2/(p-1)(p-1)$
4 B	$p - 1$	$[B] - [Y]$	[4/9]	$\sigma_\epsilon^2 + np^2 \sum \beta_k^2/(p-1)$
5 D	$p - 1$	$[D] - [Y]$	[5/9]	$\sigma_\epsilon^2 + np^2 \sum \delta_m^2/(p-1)$
6 BD	$(p-1)(p-1)$	$[BD] - [B] - [D] + [Y]$	[6/9]	$\sigma_\epsilon^2 + np\sum\sum(\beta\delta)_{km}^2/(p-1)(p-1)$
7 E	$u - 1$	$[E] - [Y]$	[7/9]	$\sigma_\epsilon^2 + np^2 \sum \eta_o^2/(u-1)$
8 Residual	$(p^2-1)(p^2-2)$	$[ABCDE] - [AC]$ $- [BD] - [E] + [2Y]$	[8/9]	$\sigma_\epsilon^2 + n\sum\sum \text{res}^2/(p^2-1)(p^2-2)$
9 WCELL	$p^4(n-1)$	$[ABCDES] - [ABCDE]$		σ_ϵ^2
10 Total	$np^4 - 1$	$[ABCDES] - [Y]$		

13.11 COMPUTATIONAL PROCEDURES FOR TYPE GLSFF-p^3 DESIGN

A Graeco-Latin square can be used as a building block for fractional factorial designs. The type GLSFF-3^3 design diagrammed in Figure 13.11-1 is appropriate for experiments that use repeated measures or matched subjects. The letters g_z designate groups of n blocks of subjects to whom the rows of the square are randomly assigned. There are w levels of g_z. This design requires w random samples of n subjects if repeated measures are obtained or nw sets of p matched subjects if matching is used. In the latter case, the $a_j c_l$ treatment combinations are randomly assigned to the matched subjects. If repeated measures are obtained, the sequence of administration of the $a_j c_l$ combinations is randomized independently for each subject.

The experimental design model equation for this design is

$$Y_{ijklz} = \mu + \zeta_z + \pi_{i(z)} + \alpha_j + \beta_k + \gamma_l + \text{res} + \alpha\beta\gamma\pi_{jkli(z)} + \epsilon_{ijklz}$$
$$(i = 1, \ldots, n; j = 1, \ldots, p;$$
$$k = 1, \ldots, p; l = 1, \ldots, p; z = 1, \ldots, w)$$

where ζ_z and $\pi_{i(z)}$ designate the effect of group z and the effect of person i who is nested within group z, respectively. This design requires the same number of observations as the type LSFF-$p \cdot p^2$ design summarized in Table 13.8-2. The advantage of a type GLSFF-p^3 design relative to a type LSFF-$p \cdot p^2$ design is that all treatment effects in the former design are within-subjects effects. Computational formulas for a type GLSFF-p^3 design are presented in Table 13.11-1. It is assumed that the treatments represent fixed effects and that the blocks are random effects.

FIGURE 13.11-1   Block diagram of type GLSFF-3^3 design

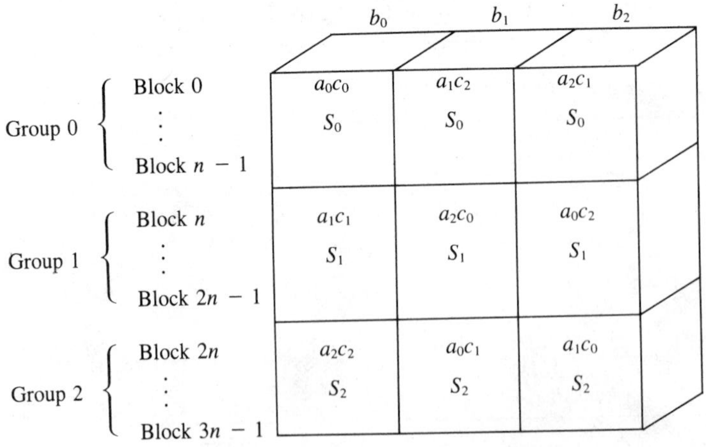

TABLE 13.11-1 Computational Formulas for Type GLSFF-p^3 Design

	Source	df	Computational Formulas	F	$E(MS)$ Model III (A, B, and C fixed effects; Subjects random)
1	Between blocks	$np - 1$	$[GS] - [Y]$		
2	Groups	$p - 1$	$[G] - [Y]$	[2/3]	$\sigma_\epsilon^2 + p\sigma_\pi^2 + np\Sigma\zeta_z^2/(w - 1)$
3	Blocks w. groups	$p(n - 1)$	$[GS] - [G]$		$\sigma_\epsilon^2 + p\sigma_\pi^2$
4	Within blocks	$np(p - 1)$	$[ABCGS] - [GS]$		
5	A	$p - 1$	$[A] - [Y]$	[5/9]	$\sigma_\epsilon^2 + np\Sigma\alpha_j^2/(p - 1)$
6	B	$p - 1$	$[B] - [Y]$	[6/9]	$\sigma_\epsilon^2 + np\Sigma\beta_k^2/(p - 1)$
7	C	$p - 1$	$[C] - [Y]$	[7/9]	$\sigma_\epsilon^2 + np\Sigma\gamma_l^2/(p - 1)$
8	Residual	$(p - 1)(p - 3)$	$[ABCG] - [A] - [B]$ $- [C] - [G] + 3[Y]$	[8/9]	$\sigma_\epsilon^2 + n\Sigma\Sigma\text{res}^2/(p - 1)$ $(p - 3)$
9	$ABC \times$ blocks w. groups	$p(n - 1)(p - 1)$	$[ABCGS] - [ABCG]$ $- [GS] + [G]$		σ_ϵ^2
10	Total	$np^2 - 1$	$[ABCGS] - [Y]$		

☆13.12 ANALYSIS OF FRACTIONAL FACTORIAL DESIGNS USING THE FULL RANK MODEL APPROACH

The full rank model approach can be used to analyze data from fractional factorial experiments. We will illustrate the procedure for two of the designs described earlier—a type CRFF-2^4 design and a type LSFF-$3 \cdot 3^2$ design.

COMPUTATIONAL PROCEDURES FOR A TYPE CRFF-2^4 DESIGN

The full rank experimental design model for a type CRFF-2^4 design is

$$Y_{ijkl} = \mu_{jkl} + \epsilon_{i(jkl)}$$
$$(i = 1, \ldots, n; j = 1, \ldots, p; k = 1, \ldots, q; l = 1, \ldots, r)$$

☆ This section assumes a familiarity with the matrix operations in Appendix D, Chapter 5, and Sections 6.9 and 7.9. The reader who is interested only in the traditional approach to analysis of variance can, without loss of continuity, omit this section.

where $\epsilon_{i(jkl)}$ is $NID(0, \sigma_\epsilon^2)$. The analysis procedures for this full rank experimental design model will be illustrated using the data in Table 13.3-1. In addition to computing $SSTO$ and $SSWCELL$, we need to formulate coefficient matrices that will let us compute sums of squares for

$$
\begin{array}{ll}
A(BCD) & AC(BD) \\
B(ACD) & BC(AD) \\
C(ABD) & ABC(D). \\
AB(CD) &
\end{array}
$$

As in previous examples using the full rank model approach, we begin by specifying null hypotheses for each of these sources of variation. These hypotheses serve as guides for formulating the coefficient matrices. The reader should refer to Table 13.12-1 as an aid in understanding the various hypotheses.

Treatment A H_0: $\mu_{0..} - \mu_{1..} = 0$ or H_0: $\underset{(p-1)\times h}{\mathbf{C}'_A} \ \underset{h\times 1}{\boldsymbol{\mu}} = \underset{(p-1)\times 1}{\mathbf{0}_A}$

where $\underset{1\times 8}{\mathbf{C}'_A} = \begin{bmatrix} 1 & 1 & 1 & 1 & -1 & -1 & -1 & -1 \end{bmatrix}$

$\underset{1\times 8}{\boldsymbol{\mu}'} = \begin{bmatrix} \mu_{000} & \mu_{001} & \mu_{010} & \mu_{011} & \mu_{100} & \mu_{101} & \mu_{110} & \mu_{111} \end{bmatrix}$

and $\underset{1\times 1}{\mathbf{0}_A} = \begin{bmatrix} 0 \end{bmatrix}$.

TABLE 13.12-1   Type CRFF-2^4 Design

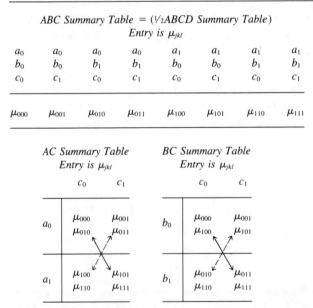

ABC Summary Table $= (\frac{1}{2}ABCD$ *Summary Table*)
Entry is μ_{jkl}

a_0	a_0	a_0	a_0	a_1	a_1	a_1	a_1
b_0	b_0	b_1	b_1	b_0	b_0	b_1	b_1
c_0	c_1	c_0	c_1	c_0	c_1	c_0	c_1
μ_{000}	μ_{001}	μ_{010}	μ_{011}	μ_{100}	μ_{101}	μ_{110}	μ_{111}

AC Summary Table
Entry is μ_{jkl}

	c_0	c_1
a_0	μ_{000} μ_{010}	μ_{001} μ_{011}
a_1	μ_{100} μ_{110}	μ_{101} μ_{111}

BC Summary Table
Entry is μ_{jkl}

	c_0	c_1
b_0	μ_{000} μ_{100}	μ_{001} μ_{101}
b_1	μ_{010} μ_{110}	μ_{011} μ_{111}

$\displaystyle\sum_{j=1}^{p}\sum_{k=1}^{q}\mu_{jkl}/pq = \mu_{..0}$ $\mu_{..1}$

Table 13.12-1 (*continued*)

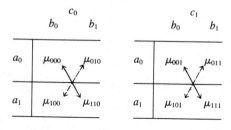

AB Summary Table
Entry is μ_{jkl}

	b_0	b_1	$\sum_{k=1}^{q} \sum_{l=1}^{r} \mu_{jkl}/qr$
a_0	μ_{000} μ_{010} μ_{001} μ_{011}		$\mu_{0\cdot\cdot}$
a_1	μ_{100} μ_{110} μ_{101} μ_{111}		$\mu_{1\cdot\cdot}$

$\sum_{j=1}^{p} \sum_{l=1}^{r} \mu_{jkl}/pr = \mu_{\cdot 0 \cdot} \qquad \mu_{\cdot 1 \cdot}$

AB at c_l Summary Table
Entry is μ_{jkl}

c_0

	b_0	b_1
a_0	μ_{000}	μ_{010}
a_1	μ_{100}	μ_{110}

c_1

	b_0	b_1
a_0	μ_{001}	μ_{011}
a_1	μ_{101}	μ_{111}

The coefficients, $c_{jkl} = \pm 1/qr$, in $\mathbf{C}_A'$ have been multiplied by 4 to avoid fractions. This does not affect the nature of the hypothesis that is tested.

$\qquad$ Treatment B $\qquad$ H$_0$: $\quad \mu_{\cdot 0 \cdot} - \mu_{\cdot 1 \cdot} = 0 \qquad$ or $\qquad$ H$_0$: $\quad \underset{(q-1)\times h}{\mathbf{C}_B'} \; \underset{h \times 1}{\boldsymbol{\mu}} \; = \; \underset{(q-1)\times 1}{\mathbf{0}_B}$

where $\underset{1 \times 8}{\mathbf{C}_B'} = \begin{bmatrix} 1 & 1 & -1 & -1 & 1 & 1 & -1 & -1 \end{bmatrix}$

and $\quad \underset{1 \times 1}{\mathbf{0}_B} = [0]$.

The coefficients, $c_{jkl} = \pm 1/pr$, in $\mathbf{C}_B'$ have been multiplied by 4 to avoid fractions.

$\qquad$ Treatment C $\qquad$ H$_0$: $\quad \mu_{\cdot\cdot 0} - \mu_{\cdot\cdot 1} = 0 \qquad$ or $\qquad$ H$_0$: $\quad \underset{(r-1)\times h}{\mathbf{C}_C'} \; \underset{h \times 1}{\boldsymbol{\mu}} \; = \; \underset{(r-1)\times 1}{\mathbf{0}_C}$

where $\underset{1 \times 8}{\mathbf{C}_C'} = \begin{bmatrix} 1 & -1 & 1 & -1 & 1 & -1 & 1 & -1 \end{bmatrix}$

and $\quad \underset{1 \times 1}{\mathbf{0}_C} = [0]$.

The coefficients, $c_{jkl} = \pm 1/pq$, in $\mathbf{C}_C'$ have been multiplied by 4 to avoid fractions.

$\qquad$ AB Interaction $\qquad$ H$_0$: $\quad \mu_{00\cdot} - \mu_{10\cdot} - \mu_{01\cdot} + \mu_{11\cdot} = 0$

or

$$\text{H}_0: \quad \underset{(p-1)(q-1)\times h}{\mathbf{C}_{AB}'} \; \underset{h \times 1}{\boldsymbol{\mu}} \; = \; \underset{(p-1)(q-1)\times 1}{\mathbf{0}_{AB}}$$

where $\underset{1 \times 8}{\mathbf{C}_{AB}'} = \begin{bmatrix} 1 & 1 & -1 & -1 & -1 & -1 & 1 & 1 \end{bmatrix}$

and $\quad \underset{1 \times 1}{\mathbf{0}_{AB}} = [0]$.

The coefficients, $c_{jkl} = \pm 1/r$, in $\mathbf{C}_{AB}'$ have been multiplied by 2 to avoid fractions.

$\qquad$ AC Interaction $\qquad$ H$_0$: $\quad \mu_{0\cdot 0} - \mu_{1\cdot 0} - \mu_{0\cdot 1} + \mu_{1\cdot 1} = 0$

TABLE 13.12-2 Outline of Computational Procedures for a Type CRFF-2^4 Design Using a Full Rank Experimental Design Model

(i) Data and basic matrix (the C''s are defined in the text; $N = 16$, $h = 8$, $n = 2$, $p = 2$, $q = 2$, $r = 2$):

$$\begin{array}{c} \mathbf{y} \\ N \times 1 \end{array} \qquad \begin{array}{c} \mathbf{X} \\ N \times h \end{array}$$

	y	X_1	X_2	X_3	X_4	X_5	X_6	X_7	X_8
$a_0b_0c_0$	3	1	0	0	0	0	0	0	0
	6	1	0	0	0	0	0	0	0
$a_0b_0c_1$	4	0	1	0	0	0	0	0	0
	3	0	1	0	0	0	0	0	0
$a_0b_1c_0$	7	0	0	1	0	0	0	0	0
	6	0	0	1	0	0	0	0	0
$a_0b_1c_1$	7	0	0	0	1	0	0	0	0
	8	0	0	0	1	0	0	0	0
$a_1b_0c_0$	2	0	0	0	0	1	0	0	0
	2	0	0	0	0	1	0	0	0
$a_1b_0c_1$	2	0	0	0	0	0	1	0	0
	3	0	0	0	0	0	1	0	0
$a_1b_1c_0$	5	0	0	0	0	0	0	1	0
	6	0	0	0	0	0	0	1	0
$a_1b_1c_1$	9	0	0	0	0	0	0	0	1
	11	0	0	0	0	0	0	0	1

$$\sum_{i=1}^{N} Y_i = 84$$

$$\hat{\boldsymbol{\mu}} = (\mathbf{X}'\mathbf{X})^{-1}(\mathbf{X}'\mathbf{y})$$

$$\begin{aligned} SSTO &= \mathbf{y}'\mathbf{y} - (\Sigma Y_i)^2/N \\ &= 552 - (84)^2/16 = 111.00 \end{aligned}$$

$$\begin{aligned} SSWCELL &= \mathbf{y}'\mathbf{y} - \hat{\boldsymbol{\mu}}'(\mathbf{X}'\mathbf{y}) \\ &= 552 - 543 = 9.00 \end{aligned}$$

$$\begin{aligned} SSA &= (\mathbf{C}_A'\hat{\boldsymbol{\mu}})' \, [\mathbf{C}_A'(\mathbf{X}'\mathbf{X})^{-1} \, \mathbf{C}_A]^{-1} \, (\mathbf{C}_A'\hat{\boldsymbol{\mu}}) \\ &= 1.00 \end{aligned}$$

$$\begin{aligned} SSB &= (\mathbf{C}_B'\hat{\boldsymbol{\mu}})' \, [\mathbf{C}_B'(\mathbf{X}'\mathbf{X})^{-1} \, \mathbf{C}_B]^{-1} \, (\mathbf{C}_B'\hat{\boldsymbol{\mu}}) \\ &= 72.25 \end{aligned}$$

$$\begin{aligned} SSC &= (\mathbf{C}_C'\hat{\boldsymbol{\mu}})' \, [\mathbf{C}_C'(\mathbf{X}'\mathbf{X})^{-1} \, \mathbf{C}_C]^{-1} \, (\mathbf{C}_C'\hat{\boldsymbol{\mu}}) \\ &= 6.25 \end{aligned}$$

$$\begin{aligned} SSAB &= (\mathbf{C}_{AB}'\hat{\boldsymbol{\mu}})' \, [\mathbf{C}_{AB}'(\mathbf{X}'\mathbf{X})^{-1} \, \mathbf{C}_{AB}]^{-1} \, (\mathbf{C}_{AB}'\hat{\boldsymbol{\mu}}) \\ &= 6.25 \end{aligned}$$

Table 13.12-2 (continued)

$$SSAC = (\mathbf{C}'_{AC}\hat{\boldsymbol{\mu}})' \, [\mathbf{C}'_{AC}(\mathbf{X}'\mathbf{X})^{-1} \, \mathbf{C}_{AC}]^{-1} \, (\mathbf{C}'_{AC}\hat{\boldsymbol{\mu}})$$
$$= 6.25$$

$$SSBC = (\mathbf{C}'_{BC}\hat{\boldsymbol{\mu}})' \, [\mathbf{C}'_{BC}(\mathbf{X}'\mathbf{X})^{-1} \, \mathbf{C}_{BC}]^{-1} \, (\mathbf{C}'_{BC}\hat{\boldsymbol{\mu}})$$
$$= 9.00$$

$$SSABC = (\mathbf{C}'_{ABC}\hat{\boldsymbol{\mu}})' \, [\mathbf{C}'_{ABC}(\mathbf{X}'\mathbf{X})^{-1} \, \mathbf{C}_{ABC}]^{-1} \, (\mathbf{C}'_{ABC}\hat{\boldsymbol{\mu}})$$
$$= 1.00$$

or

$$H_0: \quad \underset{(p-1)(r-1)\times h}{\mathbf{C}'_{AC}} \quad \underset{h\times 1}{\boldsymbol{\mu}} = \underset{(p-1)(r-1)\times 1}{\mathbf{0}_{AC}}$$

where $\underset{1\times 8}{\mathbf{C}'_{AC}} = [1 \quad -1 \quad 1 \quad -1 \quad -1 \quad 1 \quad -1 \quad 1]$

and $\quad \underset{1\times 1}{\mathbf{0}_{AC}} = [0].$

The coefficients, $c_{jkl} = \pm 1/q$, in $\mathbf{C}'_{AC}$ have been multiplied by 2 to avoid fractions.

    BC Interaction $\qquad H_0: \quad \mu_{\cdot 00} - \mu_{\cdot 10} - \mu_{\cdot 01} + \mu_{\cdot 11} = 0$

or

$$H_0: \quad \underset{(q-1)(r-1)\times h}{\mathbf{C}'_{BC}} \quad \underset{h\times 1}{\boldsymbol{\mu}} = \underset{(q-1)(r-1)\times 1}{\mathbf{0}_{BC}}$$

where $\underset{1\times 8}{\mathbf{C}'_{BC}} = [1 \quad -1 \quad -1 \quad 1 \quad 1 \quad -1 \quad -1 \quad 1]$

and $\quad \underset{1\times 1}{\mathbf{0}_{BC}} = [0].$

The coefficients, $c_{jkl} = \pm 1/p$, in $\mathbf{C}'_{BC}$ have been multiplied by 2 to avoid fractions.

    ABC Interaction $\qquad H_0: \quad (\mu_{000} - \mu_{100} - \mu_{010} - \mu_{110})$
$$-(\mu_{001} - \mu_{101} - \mu_{011} + \mu_{111}) = 0$$

or

$$H_0: \quad \underset{(p-1)(q-1)(r-1)\times h}{\mathbf{C}'_{ABC}} \quad \underset{h\times 1}{\boldsymbol{\mu}} = \underset{(p-1)(q-1)(r-1)\times 1}{\mathbf{0}_{ABC}}$$

where $\underset{1\times 8}{\mathbf{C}'_{ABC}} = [1 \quad -1 \quad -1 \quad 1 \quad -1 \quad 1 \quad 1 \quad -1]$

and $\quad \underset{1\times 1}{\mathbf{0}_{ABC}} = [0].$

The coefficients in $\mathbf{C}'_{ABC}$ are given by $c_{jkl} = \pm 1$. Procedures for constructing coefficient matrices for interactions involving three or more treatments are discussed in detail in Section 9.3.

    The formulas for computing sums of squares are given in Table 13.12-2. A comparison of these sums of squares with those in Table 13.3-1, where the traditional sum of squares approach is used, reveals that they are identical. The full rank model approach can also be used when there are missing scores and missing cells. The procedures are discussed in detail for a type CRF-*pq* design in Section 8.13.

COMPUTATIONAL PROCEDURES
FOR A TYPE LSFF-3·3² DESIGN

The restricted full rank experimental design model for a type LSFF-3·3² design is

$$Y_{ijkl} = \mu_{ijkl} + \epsilon_{ijkl}$$
$$(i = 1, \ldots, n; j = 1, \ldots, p; k = 1, \ldots, p; l = 1, \ldots, p)$$

where ϵ_{ijkl} is $NID(0, \sigma_\epsilon^2)$ and μ_{ijkl} is subject to the restrictions that for each j

$$\mu_{ij(kl)} - \mu_{i'j(kl)} - \mu_{ij(kl)'} + \mu_{i'j(kl)'} = 0 \quad \text{for all } i, i', (jk) \text{ and } (jk)'$$

and interaction components involving A, B, and C are equal to zero. The analysis procedures for this restricted full rank experimental design model will be illustrated using the data in Table 13.8-1. In addition to computing $SSTO$, we need to formulate coefficient matrices that will let us compute

$SSBL$	SSB
SSA	SSC
$SSBL(A)$	$SSRES$
$SSw.BL$	$SSBC \times BL(A)$.

As an aid to understanding the following null hypotheses and coefficient matrices for these sources of variation, the reader may find it helpful to refer to Table 13.12-3.

Between Blocks $\quad$ H$_0$: $\quad \mu_{00\cdot\cdot} - \mu_{52\cdot\cdot} = 0 \quad$ or $\quad$ H$_0$: $\quad \underset{(np-1)\times h}{\mathbf{C}'_{BL}} \; \underset{h\times 1}{\boldsymbol{\mu}} = \underset{(np-1)\times 1}{\mathbf{0}_{BL}}$

$$\mu_{10\cdot\cdot} - \mu_{52\cdot\cdot} = 0$$
$$\mu_{21\cdot\cdot} - \mu_{52\cdot\cdot} = 0$$
$$\mu_{31\cdot\cdot} - \mu_{52\cdot\cdot} = 0$$
$$\mu_{42\cdot\cdot} - \mu_{52\cdot\cdot} = 0$$

where $\underset{5\times 18}{\mathbf{C}'_{BL}} =$

$$\begin{bmatrix} 1 & 0 & 1 & 0 & 1 & 0 & 0 & 0 & 0 & 0 & 0 & 0 & 0 & -1 & 0 & -1 & 0 & -1 \\ 0 & 1 & 0 & 1 & 0 & 1 & 0 & 0 & 0 & 0 & 0 & 0 & 0 & -1 & 0 & -1 & 0 & -1 \\ 0 & 0 & 0 & 0 & 0 & 0 & 1 & 0 & 1 & 0 & 1 & 0 & 0 & -1 & 0 & -1 & 0 & -1 \\ 0 & 0 & 0 & 0 & 0 & 0 & 0 & 1 & 0 & 1 & 0 & 1 & 0 & -1 & 0 & -1 & 0 & -1 \\ 0 & 0 & 0 & 0 & 0 & 0 & 0 & 0 & 0 & 0 & 0 & 0 & 1 & -1 & 1 & -1 & 1 & -1 \end{bmatrix}$$

$$\underset{1\times 5}{\mathbf{0}'_{BL}} = \begin{bmatrix} 0 & 0 & 0 & 0 & 0 \end{bmatrix}$$

and $\underset{1\times 18}{\boldsymbol{\mu}'} = \begin{bmatrix} \mu_{0000} & \mu_{1000} & \mu_{0011} & \mu_{1011} & \mu_{0022} & \mu_{1022} & \mu_{2102} & \mu_{3102} & \mu_{2110} & \cdots & \mu_{5200} \end{bmatrix}$.

The coefficients, $c_{ijkl} = \pm 1/p$ or 0, in $\mathbf{C}'_{BL}$ have been multiplied by 3 to avoid fractions. This does not affect the nature of the hypothesis that is tested.

Treatment $A \quad$ H$_0$: $\quad \mu_{\cdot 0\cdot\cdot} - \mu_{\cdot 2\cdot\cdot} = 0 \quad$ or $\quad$ H$_0$: $\quad \underset{(p-1)\times h}{\mathbf{C}'_A} \; \underset{h\times 1}{\boldsymbol{\mu}} = \underset{(p-1)\times 1}{\mathbf{0}_A}$

$$\mu_{\cdot 1\cdot\cdot} - \mu_{\cdot 2\cdot\cdot} = 0$$

TABLE 13.12-3 Type LSFF-$3 \cdot 3^2$ Design

ABCS Summary Table
Entry is μ_{ijkl}

		b_0	b_1	b_2	$\sum\limits_{k=1}^{p} \mu_{ijkl}/p$
		c_0	c_1	c_2	
a_0	s_0	μ_{0000}	μ_{0011}	μ_{0022}	$\mu_{00\cdot\cdot}$
	s_1	μ_{1000}	μ_{1011}	μ_{1022}	$\mu_{10\cdot\cdot}$
		c_2	c_0	c_1	
a_1	s_2	μ_{2102}	μ_{2110}	μ_{2121}	$\mu_{21\cdot\cdot}$
	s_3	μ_{3102}	μ_{3110}	μ_{3121}	$\mu_{31\cdot\cdot}$
		c_1	c_2	c_0	
a_2	s_4	μ_{4201}	μ_{4212}	μ_{4220}	$\mu_{42\cdot\cdot}$
	s_5	μ_{5201}	μ_{5212}	μ_{5220}	$\mu_{52\cdot\cdot}$

AC Summary Table
Entry is $\mu_{\cdot j\cdot l}$

	c_0	c_1	c_2
a_0	$\mu_{\cdot 0\cdot 0}$	$\mu_{\cdot 0\cdot 1}$	$\mu_{\cdot 0\cdot 2}$
a_1	$\mu_{\cdot 1\cdot 0}$	$\mu_{\cdot 1\cdot 1}$	$\mu_{\cdot 1\cdot 2}$
a_2	$\mu_{\cdot 2\cdot 0}$	$\mu_{\cdot 2\cdot 1}$	$\mu_{\cdot 2\cdot 2}$
$\sum\limits_{j=1}^{p} \mu_{\cdot j\cdot l}/p =$	$\mu_{\cdot\cdot\cdot 0}$	$\mu_{\cdot\cdot\cdot 1}$	$\mu_{\cdot\cdot\cdot 2}$

ABC Summary Table
Entry is $\mu_{\cdot jkl}$

	b_0	b_1	b_2	$\sum\limits_{k=1}^{p} \mu_{\cdot jkl}/p$
	c_0	c_1	c_2	
a_0	$\mu_{\cdot 000}$	$\mu_{\cdot 011}$	$\mu_{\cdot 022}$	$\mu_{\cdot 0\cdot\cdot}$
	c_2	c_0	c_1	
a_1	$\mu_{\cdot 102}$	$\mu_{\cdot 110}$	$\mu_{\cdot 121}$	$\mu_{\cdot 1\cdot\cdot}$
	c_1	c_2	c_0	
a_2	$\mu_{\cdot 201}$	$\mu_{\cdot 212}$	$\mu_{\cdot 220}$	$\mu_{\cdot 2\cdot\cdot}$

$$\sum_{j=1}^{p} \mu_{\cdot jkl}/p = \mu_{\cdot\cdot 0\cdot} \qquad \mu_{\cdot\cdot 1\cdot} \qquad \mu_{\cdot\cdot 2\cdot}$$

where $\underset{2\times18}{\mathbf{C}'_A} =$

$$\begin{bmatrix} 1 & 1 & 1 & 1 & 1 & 1 & 0 & 0 & 0 & 0 & 0 & 0 & -1 & -1 & -1 & -1 & -1 & -1 \\ 0 & 0 & 0 & 0 & 0 & 0 & 1 & 1 & 1 & 1 & 1 & 1 & -1 & -1 & -1 & -1 & -1 & -1 \end{bmatrix}$$

and $\underset{1\times2}{\mathbf{0}'_A} = [0 \quad 0]$.

The coefficients, $c_{ijkl} = \pm 1/np$ or 0, in $\mathbf{C}'_A$ have been multiplied by 6 to avoid fractions.

Blocks Within A H_0: $\mu_{00\cdot\cdot} - \mu_{10\cdot\cdot} = 0$ or H_0: $\underset{p(n-1)\times h}{\mathbf{C}'_{BL(A)}} \underset{h\times1}{\boldsymbol{\mu}} = \underset{p(n-1)\times1}{\mathbf{0}_{BL(A)}}$

$\mu_{21\cdot\cdot} - \mu_{31\cdot\cdot} = 0$

$\mu_{42\cdot\cdot} - \mu_{52\cdot\cdot} = 0$

where $\underset{3\times18}{\mathbf{C}'_{BL(A)}} = \begin{bmatrix} 1 & -1 & 1 & -1 & 1 & -1 & 0 & 0 & 0 & 0 & 0 & 0 & 0 & 0 & 0 & 0 & 0 & 0 \\ 0 & 0 & 0 & 0 & 0 & 0 & 1 & -1 & 1 & -1 & 1 & -1 & 0 & 0 & 0 & 0 & 0 & 0 \\ 0 & 0 & 0 & 0 & 0 & 0 & 0 & 0 & 0 & 0 & 0 & 0 & 1 & -1 & 1 & -1 & 1 & -1 \end{bmatrix}$

and $\underset{1\times3}{\mathbf{0}'_{BL(A)}} = [0 \quad 0 \quad 0]$.

The coefficients, $c_{ijkl} = \pm 1/p$ or 0, in $\mathbf{C}'_{BL(A)}$ have been multiplied by 3 to avoid fractions.

Within Blocks H_0: $\mu_{0000} - \mu_{0022} = 0$ or H_0: $\underset{np(p-1)\times h}{\mathbf{C}'_{\mathrm{w}.BL}} \underset{h\times1}{\boldsymbol{\mu}} = \underset{np(p-1)\times1}{\mathbf{0}_{\mathrm{w}.BL}}$

$\mu_{0011} - \mu_{0022} = 0$

$\mu_{1000} - \mu_{1022} = 0$

$\mu_{1011} - \mu_{1022} = 0$

$\mu_{2102} - \mu_{2121} = 0$

$\cdots\cdots\cdots\cdots\cdots$

$\mu_{5212} - \mu_{5220} = 0$

where $\underset{12\times18}{\mathbf{C}'_{\mathrm{w}.BL}} =$

$$\begin{bmatrix} 1 & 0 & 0 & 0 & -1 & 0 & 0 & 0 & 0 & 0 & 0 & 0 & 0 & 0 & 0 & 0 & 0 & 0 \\ 0 & 0 & 1 & 0 & -1 & 0 & 0 & 0 & 0 & 0 & 0 & 0 & 0 & 0 & 0 & 0 & 0 & 0 \\ 0 & 1 & 0 & 0 & 0 & -1 & 0 & 0 & 0 & 0 & 0 & 0 & 0 & 0 & 0 & 0 & 0 & 0 \\ 0 & 0 & 0 & 1 & 0 & -1 & 0 & 0 & 0 & 0 & 0 & 0 & 0 & 0 & 0 & 0 & 0 & 0 \\ 0 & 0 & 0 & 0 & 0 & 0 & 1 & 0 & 0 & 0 & -1 & 0 & 0 & 0 & 0 & 0 & 0 & 0 \\ 0 & 0 & 0 & 0 & 0 & 0 & 0 & 0 & 1 & 0 & -1 & 0 & 0 & 0 & 0 & 0 & 0 & 0 \\ 0 & 0 & 0 & 0 & 0 & 0 & 0 & 0 & 1 & 0 & 0 & -1 & 0 & 0 & 0 & 0 & 0 & 0 \\ 0 & 0 & 0 & 0 & 0 & 0 & 0 & 0 & 0 & 1 & 0 & -1 & 0 & 0 & 0 & 0 & 0 & 0 \\ \cdot & \cdot & \cdot & \cdot & \cdot & \cdot & \cdot & \cdot & \cdot & \cdot & \cdot & \cdot & \cdot & \cdot & \cdot & \cdot & \cdot & \cdot \\ 0 & 0 & 0 & 0 & 0 & 0 & 0 & 0 & 0 & 0 & 0 & 0 & 0 & 0 & 0 & 1 & 0 & -1 \end{bmatrix}$$

and $\underset{1\times12}{\mathbf{0}'_{\mathrm{w}.BL}} = [0 \quad 0 \quad 0 \quad 0 \quad 0 \quad 0 \quad 0 \quad 0 \quad 0 \quad 0 \quad 0 \quad 0]$.

The coefficients in $\mathbf{C}'_{\mathrm{w}.BL}$ are given by $c_{ijkl} = \pm 1$ or 0.

Treatment B H_0: $\mu_{\cdot\cdot0\cdot} - \mu_{\cdot\cdot2\cdot} = 0$ or H_0: $\underset{(p-1)\times h}{\mathbf{C}'_B} \underset{h\times1}{\boldsymbol{\mu}} = \underset{(p-1)\times1}{\mathbf{0}_B}$

$\mu_{\cdot\cdot1\cdot} - \mu_{\cdot\cdot2\cdot} = 0$

where $\mathbf{C}'_B = \begin{bmatrix} 1 & 1 & 0 & 0 & -1 & -1 & 1 & 1 & 0 & 0 & -1 & -1 & 1 & 1 & 0 & 0 & -1 & -1 \\ 0 & 0 & 1 & 1 & -1 & -1 & 0 & 0 & 1 & 1 & -1 & -1 & 0 & 0 & 1 & 1 & -1 & -1 \end{bmatrix}$
$\underset{2\times 18}{}$

and $\underset{1\times 2}{\mathbf{0}'_B} = [0 \quad 0]$.

The coefficients, $c_{ijkl} = \pm 1/np$ or 0, in $\mathbf{C}'_B$ have been multiplied by 6 to avoid fractions.

Treatment C H_0: $\mu_{\cdot\cdot\cdot 0} - \mu_{\cdot\cdot\cdot 2} = 0$ or H_0: $\underset{(p-1)\times h}{\mathbf{C}'_C} \underset{h\times 1}{\boldsymbol{\mu}} = \underset{(p-1)\times 1}{\mathbf{0}_C}$

$\mu_{\cdot\cdot\cdot 1} - \mu_{\cdot\cdot\cdot 2} = 0$

where $\mathbf{C}'_C = \begin{bmatrix} 1 & 1 & 0 & 0 & -1 & -1 & -1 & -1 & 1 & 1 & 0 & 0 & 0 & 0 & -1 & -1 & 1 & 1 \\ 0 & 0 & 1 & 1 & -1 & -1 & -1 & -1 & 0 & 0 & 1 & 1 & 1 & 1 & -1 & -1 & 0 & 0 \end{bmatrix}$
$\underset{2\times 18}{}$

and $\underset{1\times 2}{\mathbf{0}'_C} = [0 \quad 0]$.

The coefficients, $c_{ijkl} = \pm 1/np$ or 0, in $\mathbf{C}'_C$ have been multiplied by 6 to avoid fractions.

Residual H_0: $(\mu_{\cdot 000} + \mu_{\cdot 121} + \mu_{\cdot 212}) - (\mu_{\cdot 022} + \mu_{\cdot 110} + \mu_{\cdot 201}) = 0$

$(\mu_{\cdot 022} + \mu_{\cdot 110} + \mu_{\cdot 201}) - (\mu_{\cdot 011} + \mu_{\cdot 102} + \mu_{\cdot 220}) = 0$

or

H_0: $\underset{(p-1)(p-2)\times h}{\mathbf{R}'} \underset{h\times 1}{\boldsymbol{\mu}} = \underset{(p-1)(p-2)\times 1}{\boldsymbol{\theta}}$

where $\underset{2\times 18}{\mathbf{R}'} =$

$\begin{bmatrix} 1 & 1 & 0 & 0 & -1 & -1 & 0 & 0 & -1 & -1 & 1 & 1 & -1 & -1 & 1 & 1 & 0 & 0 \\ 0 & 0 & -1 & -1 & 1 & 1 & -1 & -1 & 1 & 1 & 0 & 0 & 1 & 1 & 0 & 0 & -1 & -1 \end{bmatrix}$

and $\underset{1\times 2}{\boldsymbol{\theta}'} = [0 \quad 0]$.

The coefficients, $c_{ijkl} = \pm 1/n$ or 0, in $\mathbf{R}'$ have been multiplied by 2 to avoid fractions. This hypothesis is the only one in the Latin square design that is unintelligible. In words, it states that the source of variation corresponding to the (AB^2), (AC), and (BC) interaction components is equal to zero. These three components are aliases for this source of variation. An algorithm for obtaining the residual coefficient matrix for any Latin square is described in Section 7.9.

$BC \times BL(A)$ Residual H_0: $\mu_{0000} - \mu_{1000} - \mu_{0011} + \mu_{1011} = 0$

$\mu_{0011} - \mu_{1011} - \mu_{0022} + \mu_{1022} = 0$

$\mu_{2102} - \mu_{3102} - \mu_{2110} + \mu_{3110} = 0$

$\cdots\cdots\cdots\cdots\cdots\cdots\cdots$

$\mu_{4212} - \mu_{5212} - \mu_{4220} + \mu_{5220} = 0$

or

H_0: $\underset{p(n-1)(p-1)\times h}{\mathbf{R}'_{BC\times BL(A)}} \underset{h-1}{\boldsymbol{\mu}} = \underset{p(n-1)(p-1)\times 1}{\mathbf{0}_{BC\times BL(A)}}$

TABLE 13.12-4   Computational Procedures for a Type LSFF-$3 \cdot 3^2$ Design Using a Restricted Full Rank Experimental Design Model

(i) Data and matrix formulas for computing sums of squares (the $\mathbf{Q}'$, $\mathbf{R}'$, and $\mathbf{S}'$ matrices are defined in the text; $N = 18$, $h = 18$, $n = 2$, $p = 3$, $s = (p-1)(p-2) = 2$, $v = (p-1)(p-2) + p(n-1)(p-1) = 8$):

$$\begin{array}{c} \mathbf{y} \\ {\scriptstyle N \times 1} \\ \hat{\boldsymbol{\mu}} \\ {\scriptstyle h \times 1} \end{array} \qquad [\mathbf{S}(\mathbf{S}'\mathbf{S})^{-1}\mathbf{S}'\hat{\boldsymbol{\mu}}] \qquad \hat{\boldsymbol{\mu}}_S \qquad \hat{\boldsymbol{\mu}} - [\mathbf{R}(\mathbf{R}'\mathbf{R})^{-1}\mathbf{R}'\hat{\boldsymbol{\mu}}] = \hat{\boldsymbol{\mu}}_R$$

	$\mathbf{y}$, $\hat{\boldsymbol{\mu}}$	$[\mathbf{S}(\mathbf{S}'\mathbf{S})^{-1}\mathbf{S}'\hat{\boldsymbol{\mu}}]$	$\hat{\boldsymbol{\mu}}_S$	$\hat{\boldsymbol{\mu}}_R$
$a_0 b_0 c_0$	3	0.4444	2.5556	3.2222
	1	−0.8889	1.8889	1.2222
$a_0 b_1 c_1$	4	−0.0556	4.0556	3.7222
	4	0.6111	3.3889	3.7222
$a_0 b_2 c_2$	6	−0.3889	6.3889	6.0556
	6	0.2779	5.7222	6.0556
$a_1 b_0 c_2$	7	0.7778	6.2222	6.7222
	5	−0.2222	5.2222	4.7222
$a_1 b_1 c_0$	3	−0.0556	3.0556	3.0556
	2	−0.0556	2.0556	2.0556
$a_1 b_2 c_1$	3	−0.7222	3.7222	3.2222
	3	0.2778	2.7222	3.2222
$a_2 b_0 c_1$	4	0.1111	3.8889	4.0556
	2	−0.2222	2.2222	2.0556
$a_2 b_1 c_2$	7	−0.0556	7.0556	7.2222
	5	−0.3889	5.3889	5.2222
$a_2 b_2 c_0$	3	−0.0556	3.0556	2.7222
	2	0.6111	1.3889	1.7222

with the operator $-$ before the $[\mathbf{S}(\mathbf{S}'\mathbf{S})^{-1}\mathbf{S}'\hat{\boldsymbol{\mu}}]$ column and $=$ before the $\hat{\boldsymbol{\mu}}_S$ column.

$$\sum_{i=1}^{N} Y_i = 70$$

$$\begin{array}{ccc} \mathbf{y}' & \mathbf{y} & (\mathbf{y}'\mathbf{y}) \end{array}$$

$$\begin{bmatrix} 3 & 1 & \cdots & 2 \end{bmatrix} \begin{bmatrix} 3 \\ 1 \\ \vdots \\ 2 \end{bmatrix} = 326$$

$$SSTO = \mathbf{y}'\mathbf{y} - (\Sigma Y_i)^2/N$$
$$= 326 - (70)^2/18$$
$$= 53.78$$

$$SSb.RES = (\mathbf{R}'\hat{\boldsymbol{\mu}})'\,(\mathbf{R}'\mathbf{R})^{-1}\,(\mathbf{R}'\hat{\boldsymbol{\mu}}) = 0.78$$

Table 13.12-4 (continued)

$$SSBL = (\mathbf{Q}'_{BL}\hat{\boldsymbol{\mu}})' (\mathbf{Q}'_{BL}\mathbf{Q}_{BL})^{-1} (\mathbf{Q}'_{BL}\hat{\boldsymbol{\mu}}) - SSb.RES = 6.45$$

$$SSA = (\mathbf{Q}'_A\hat{\boldsymbol{\mu}})' (\mathbf{Q}'_A\mathbf{Q}_A)^{-1} (\mathbf{Q}'_A\hat{\boldsymbol{\mu}}) - SSb.RES = 0.11$$

$$SSBL(A) = (\mathbf{Q}'_{BL(A)}\hat{\boldsymbol{\mu}})' (\mathbf{Q}'_{BL(A)}\mathbf{Q}_{BL(A)})^{-1} (\mathbf{Q}'_{BL(A)}\hat{\boldsymbol{\mu}}) - SSb.RES = 6.34$$

$$SSw.RES = (\mathbf{S}'\hat{\boldsymbol{\mu}})' (\mathbf{S}'\mathbf{S})^{-1} (\mathbf{S}'\hat{\boldsymbol{\mu}}) = 3.44$$

$$SSw.BL = (\mathbf{Q}'_{w.BL}\hat{\boldsymbol{\mu}})' (\mathbf{Q}'_{w.BL}\mathbf{Q}_{w.BL})^{-1} (\mathbf{Q}'_{w.BL}\hat{\boldsymbol{\mu}}) - SSw.RES = 47.33$$

$$SSB = (\mathbf{Q}'_B\hat{\boldsymbol{\mu}})' (\mathbf{Q}'_B\mathbf{Q}_B)^{-1} (\mathbf{Q}'_B\hat{\boldsymbol{\mu}}) - SSw.RES = 0.78$$

$$SSC = (\mathbf{Q}'_C\hat{\boldsymbol{\mu}})' (\mathbf{Q}'_C\mathbf{Q}_C)^{-1} (\mathbf{Q}'_C\hat{\boldsymbol{\mu}}) - SSw.RES = 43.11$$

$$SSRES = SSb.RES = 0.78$$

$$SSBC \times BL(A) = SSw.RES - SSb.RES = 2.66$$

where $\mathbf{R}'_{BC \times BL(A)} =$
$\underset{6 \times 18}{}$

$$\begin{bmatrix}
1 & -1 & -1 & 1 & 0 & 0 & 0 & 0 & 0 & 0 & 0 & 0 & 0 & 0 & 0 & 0 & 0 & 0 \\
0 & 0 & 1 & -1 & -1 & 1 & 0 & 0 & 0 & 0 & 0 & 0 & 0 & 0 & 0 & 0 & 0 & 0 \\
0 & 0 & 0 & 0 & 0 & 0 & 1 & -1 & -1 & 1 & 0 & 0 & 0 & 0 & 0 & 0 & 0 & 0 \\
0 & 0 & 0 & 0 & 0 & 0 & 0 & 1 & -1 & -1 & 1 & 0 & 0 & 0 & 0 & 0 & 0 & 0 \\
0 & 0 & 0 & 0 & 0 & 0 & 0 & 0 & 0 & 0 & 0 & 1 & -1 & -1 & 1 & 0 & 0 \\
0 & 0 & 0 & 0 & 0 & 0 & 0 & 0 & 0 & 0 & 0 & 0 & 1 & -1 & -1 & 1
\end{bmatrix}$$

and $\underset{1 \times 6}{\mathbf{0}'_{BC \times BL(A)}} = [0 \quad 0 \quad 0 \quad 0 \quad 0 \quad 0]$

The coefficients in $\mathbf{R}'_{BC \times BL(A)}$ are given by $c_{ijkl} = \pm 1$ or 0.

Construction of $\mathbf{R}'_{BC \times BL(A)}$ is facilitated by using the sets of crossed lines in the *ABCS* Summary Table of Table 13.12-3. The $BC \times BL(A)$ null hypothesis corresponds to the first set of restrictions on the μ_{ijkl}'s. The residual null hypothesis described earlier corresponds to the second set of restrictions on the μ_{ijkl}'s.

We want to test the between-blocks null hypotheses, subject to the restrictions that $\mathbf{R}'\boldsymbol{\mu} = \boldsymbol{\theta}$. This is accomplished by forming augmented matrices of the form

$$\mathbf{Q}'_b = \begin{bmatrix} \mathbf{R}' \\ \mathbf{C}'_b \end{bmatrix} \quad \text{and} \quad \boldsymbol{\eta}_b = \begin{bmatrix} \boldsymbol{\theta} \\ \mathbf{0}_b \end{bmatrix}$$

where $\mathbf{C}'_b$ is the coefficient matrix for a between-blocks source of variation and $\mathbf{0}_b$ is the null vector associated with $\mathbf{C}'_b$. The null hypotheses for the within-blocks null hypotheses are tested subject to two sets of restrictions, $\mathbf{R}'\boldsymbol{\mu} = \boldsymbol{\theta}$ and $\mathbf{R}'_{BC \times BL(A)}\boldsymbol{\mu} = \mathbf{0}_{BC \times BL(A)}$. This is accomplished by forming augmented matrices of the form

$$\mathbf{Q}'_w = \begin{bmatrix} \mathbf{R}' \\ \mathbf{R}'_{BC \times BL(A)} \\ \mathbf{C}'_w \end{bmatrix} \quad \text{and} \quad \boldsymbol{\eta}_w = \begin{bmatrix} \boldsymbol{\theta} \\ \mathbf{0}_{BC \times BL(A)} \\ \mathbf{0}_w \end{bmatrix}$$

where $\mathbf{C}'_w$ is the coefficient matrix for a within-blocks source of variation and $\mathbf{0}_w$ is the null vector associated with $\mathbf{C}'_w$. If the rows of $\mathbf{R}'$ are mutually orthogonal to the rows of a between-blocks coefficient matrix, the computational formula $(\mathbf{Q}'_b\hat{\boldsymbol{\mu}})'(\mathbf{Q}'_b\mathbf{Q}_b)^{-1}(\mathbf{Q}'_b\hat{\boldsymbol{\mu}}) - SSb.RES$ simplifies to $(\mathbf{C}'_b\hat{\boldsymbol{\mu}})'(\mathbf{C}'_b\mathbf{C}_b)^{-1}(\mathbf{C}'_b\hat{\boldsymbol{\mu}})$. Similarly, if the rows of $\mathbf{R}'$ and $\mathbf{R}'_{BC\times BL(A)}$ are mutually orthogonal to the rows of a within-blocks coefficient matrix, $(\mathbf{Q}'_w\hat{\boldsymbol{\mu}})'(\mathbf{Q}'_w\mathbf{Q}_w)^{-1}(\mathbf{Q}'_w\hat{\boldsymbol{\mu}}) - SSw.RES$ simplifies to $(\mathbf{C}'_w\hat{\boldsymbol{\mu}})'(\mathbf{C}'_w\mathbf{C}_w)^{-1}(\mathbf{C}'_w\hat{\boldsymbol{\mu}})$. For convenience, we will use S to denote the augmented matrix of restrictions,

$$\underset{v\times h}{\mathbf{S}'} = \begin{bmatrix} \mathbf{R}' \\ \mathbf{R}'_{BC\times BL(A)} \end{bmatrix}$$

where $v = (p-1)(p-2) + p(n-1)(p-1)$.

Formulas for computing sums of squares and means subject to the restriction conditions are given in Table 13.12-4. A comparison of these sums of squares with those in Table 13.8-1, where the traditional sum of squares approach is used, reveals that they are identical.

13.13   ADVANTAGES AND DISADVANTAGES OF FRACTIONAL FACTORIAL DESIGNS

This chapter has described fractional factorial designs based on three building block designs: completely randomized, randomized block, and Latin square designs. Their chief advantages are as follows.

1.   They permit the evaluation of a large number of treatments but employ only a fraction of the total possible number of treatment combinations.

2.   Fractional factorial designs lend themselves to sequential research programs in which flexibility in the pursuit of promising lines of investigation is essential.

The chief disadvantages of these designs are as follows.

1.   Interpretation of the statistical analysis is complicated because treatments are aliased with interactions. These designs require the assumption of zero interactions among some or all treatments. This assumption is often unrealistic in the behavioral sciences.

2.   The requirement of these designs that all or most treatments have the same number of levels restricts their usefulness.

3.   The layout and computational procedures are more complex for fractional factorial designs than for conventional factorial designs.

13.14 REVIEW EXERCISES

1. [13.1] What is the distinguishing characteristic of fractional factorial designs?

2. [13.1] Compare the confounding in split-plot factorial, confounded factorial, and fractional factorial designs.

†3. [13.2] Determine the alias pattern for a type CRFF-2^5 design with $(ABCDE)$ as the defining contrast.

†4. [13.2] Determine the alias pattern for a type CRFF-2^5 design with (ABC) and (CDE) as the defining contrasts.

5. [13.2] Determine the alias pattern for a type CRFF-2^5 design with (ABC) and (BDE) as the defining contrasts.

†6. [13.3] Determine the treatment combinations in a type CRFF-2^5 design with $(ABCDE)_0$ as the defining contrast.

7. [13.3] Determine the treatment combinations in a type CRFF-2^5 design with $(ABC)_0$ and $(CDE)_0$ as the defining contrasts.

†8. [13.3] The effectiveness of a new medication for motion sickness was compared with that for dimenhydrinate, which is widely used to prevent nausea and vomiting. The medications represented the two levels of treatment A and were denoted, respectively, by a_0 and a_1. Thirty-two male volunteers who reported that they were susceptible to motion sickness participated in a three-hour bus ride over hilly terrain. The subjects in treatment condition b_0 had an unrestricted view of the road, those in treatment condition b_1 were able to see the road directly ahead, but their view to either side was restricted. The subjects in treatment condition c_0 listened to the sound of a waterfall on headphones, those in c_1 listened to a reading of *Soldiers of the Night* by David Schoenbrum. The subjects in treatment condition d_0 used a seat and shoulder belt; those in d_1 did not. The subjects sat in either the front or back of the bus, treatment conditions e_0 and e_1, respectively. The subjects were randomly assigned to 16 groups of two each. The treatment combinations of a type CRFF-2^5 design were randomly assigned to the groups. The dependent variable was the subject's rating of how they felt after the bus trip; the higher the rating, the more comfortable they felt. The following data were obtained.

a_0	a_0	a_0	a_0	a_0	a_0	a_0	a_0	a_1	a_1	a_1	a_1	a_1	a_1	a_1	a_1
b_0	b_0	b_0	b_0	b_1	b_1	b_1	b_1	b_0	b_0	b_0	b_0	b_1	b_1	b_1	b_1
c_0	c_0	c_1	c_1	c_0	c_0	c_1	c_1	c_0	c_0	c_1	c_1	c_0	c_0	c_1	c_1
d_0	d_1	d_0	d_1	d_0	d_1	d_0	d_1	d_0	d_1	d_0	d_1	d_0	d_1	d_0	d_1
e_1	e_0	e_0	e_1	e_0	e_1	e_1	e_0	e_0	e_1	e_1	e_0	e_1	e_0	e_0	e_1
10	12	18	13	15	9	9	11	13	11	14	14	10	10	13	10
15	18	13	19	11	13	16	16	18	17	20	18	14	17	17	16

a) Test the null hypotheses for this experiment; let $\alpha = .05$.

b) Is there any ambiguity in interpreting the outcome of this experiment? Explain.

9.   [13.4] For the following type CRFF-3^5 designs, indicate the aliases for treatment A.

   †**a)** Defining contrast is $(ABCDE)$.

   †**b)** Defining contrast is $(ABCDE^2)$.

   c) Defining contrast is (AB^2CDE).

   d) Defining contrast is (ABC^2D^2E).

   †**e)** Defining contrasts are (ABC) and (BCD). (Hint: don't forget the aliases for the two generalized interactions.)

   f) Defining contrasts are (ABC) and (CDE).

10.   [13.2 and 13.4] Indicate the number of aliases of a treatment for the following designs.

   †**a)** CRFF-2^4 with (ABC) as the defining contrast.

   b) CRFF-2^4 with (BCD) as the defining contrast.

   †**c)** CRFF-2^5 with (ABC) and (BCD) as the defining contrasts.

   d) CRFF-2^5 with (AB) and (CDE) as the defining contrasts.

   †**e)** CRFF-2^5 with (ABC), (BCD), and (DE) as the defining contrasts.

   †**f)** CRFF-3^4 with $(ABCD)$ as the defining contrast.

   g) CRFF-3^5 with (ABC) and (CDE^2) as the defining contrasts.

   †**h)** CRFF-3^5 with (AB^2C), (BC^2DE), and (ADE^2) as the defining contrasts.

   †**i)** CRFF-5^2 with (AB^3) as the defining contrast.

   j) CRFF-5^3 with (AB^2) and (BC^3) as the defining contrasts. (Note: this design has four generalized interactions: XY, XY^2, XY^3, and XY^4.)

11.   [13.2 and 13.4] Indicate the number of treatment combinations in each of the designs in Exercise 10.

†**12.**   [13.5] **a)** Determine the treatment combinations assigned to groups for a type RBFF-2^4 design with $(ABCD)_0$ as the defining contrast and the ABC interaction confounded with between-groups variation.

   b) Determine the alias pattern for this design.

   c) Is the ABC interaction a good choice to confound with groups? Why?

13.   [13.5] **a)** Determine the treatment combinations assigned to groups for a type RBFF-2^4 design with $(ABCD)_1$ as the defining contrast and the AB interaction confounded with between-groups variation.

   b) Determine the alias pattern for this design.

   c) Is the AB interaction a good choice to confound with groups? Why?

14.   [13.7] (i) Determine by trial and error the defining contrast for the following Latin squares. (ii) For each square determine the alias pattern.

†**a)**

	b_0	b_1
c_0	a_0	a_1
c_1	a_1	a_0

b)

	b_0	b_1
c_0	a_1	a_0
c_1	a_0	a_1

†c)

	b_0	b_1	b_2
c_0	a_0	a_2	a_1
c_1	a_2	a_1	a_0
c_2	a_1	a_0	a_2

d)

	b_0	b_1	b_2
c_0	a_0	a_2	a_1
c_1	a_1	a_0	a_2
c_2	a_2	a_1	a_0

†e)

	b_0	b_1	b_2
c_0	a_1	a_2	a_0
c_1	a_0	a_1	a_2
c_2	a_2	a_0	a_1

f)

	b_0	b_1	b_2
c_0	a_2	a_1	a_0
c_1	a_0	a_2	a_1
c_2	a_1	a_0	a_2

15. [13.8] Sixteen ten-year-olds read two stories over a period of two weeks. The main character in one of the stories was a boy (treatment condition b_0); the main character in the other story was a girl (treatment condition b_1). The age of the main character was given as either 10 years (treatment condition c_0) or 14 years (treatment condition c_1). Treatment A was the sex of the participant in the experiment; boys were designated by a_0 and girls by a_1. The treatment combinations of a 2×2 Latin square based on $(ABC)_0$ as the defining contrast were used in the experiment. The sequence of administration of the $b_k c_l$ combinations was randomized independently for each child. After reading a story, a child responded to 20 questions that were designed to assess their identification with the main character in the story. Two examples are, "Would you like to be (character's name)?" and "Would you like to do the things (character's name) did?" The following data for this type LSFF-$2\cdot2^2$ design were obtained.

		b_0	b_1			b_0	b_1
		c_0	c_1			c_1	c_0
	s_0	15	10		s_8	16	19
	s_1	14	10		s_9	14	14
	s_2	12	8		s_{10}	14	12
a_0	s_3	10	9	a_1	s_{11}	9	11
	s_4	12	6		s_{12}	14	14
	s_5	14	11		s_{13}	15	13
	s_6	17	15		s_{14}	16	15
	s_7	13	14		s_{15}	15	14

a) Test the null hypotheses for treatments A, B, and C; let $\alpha = .05$.

b) Is a type LSFF-$2\cdot2^2$ design a good choice for this experiment? Explain.

16. [13.9] Thirty-six pilots made simulated approaches and landings using the standard red/white 2-bar Visual Approach Slope Indicator, VASI, (treatment level a_0), 3-bar VASI (treatment level a_1), and the Austrian T-VASIS (treatment level a_2). The pilots flew Convair 580 aircraft simulators with a computer-generated visual system simulating a slightly reduced night-time airport scene (treatment level c_0), moderately reduced scene (treatment level c_1), and very reduced scene (treatment level c_2). Treatment B was the glidepath: $2°$ (treatment level b_0), $3°$ (treatment level b_1), and $4°$ (treatment level b_2). Two 580 simulators were used; they are denoted by d_0 and d_1. The dependent variable was a weighted measure of accuracy—the higher the score the more accurate an approach and landing. The 36 pilots were randomly assigned to two groups of 18 each. The groups were further randomly subdivided into nine subsamples containing two pilots each. The treatment combinations of a 3×3 Latin square based on $(ABC)_0$ as the defining contrast were randomly assigned to the nine subsamples of pilots in each group. The following data for this type LSFF-$3^3 2$ design were obtained.

		b_0	b_1	b_2				b_0	b_1	b_2
		c_0	c_2	c_1				c_0	c_2	c_1
	a_0	7	4	8			a_0	12	3	5
		8	9	5				11	7	3
		c_2	c_1	c_0				c_2	c_1	c_0
d_0	a_1	9	9	8	d_1	a_1		4	4	7
		6	8	11				8	7	10
		c_1	c_0	c_2				c_1	c_0	c_2
	a_2	11	10	7			a_2	8	14	5
		9	14	10				8	8	9

†a) Test the null hypotheses for treatments and interactions; let $\alpha = .05$.

†b) Does the assumption of no interactions among A, B, and C appear to be tenable? Explain.

†c) Use Tukey's statistic to test all two-mean null hypotheses for treatments A and C.

d) Prepare a "results and discussion section" appropriate for *Human Factors*.

17.   [13.12] Exercise 8 described an experiment to evaluate a new medication for preventing motion sickness.

†a) Specify the null hypotheses in terms of μ's for the full rank experimental design model; ignore treatment E.

b) Specify the coefficient matrices, $\mathbf{C}'$, for the null hypotheses in part (a).

†c) Use the full rank experimental design model approach to test the null hypotheses; let $\alpha = .05$. (This problem involves inverting a diagonal matrix and obtaining the inner product of 16×16 vectors. The computations can be done without a computer but they are tedious.)

18.   [13.12] Exercise 15 described an experiment to evaluate children's responses to stories with male and female characters.

a) Specify the null hypotheses for between-blocks variation, treatments A, B, and C, and within-blocks variation in terms of μ's for the full rank experimental design model.

b) Specify the coefficient matrices for computing $SSBL$, SSA, $SSBL(A)$, $SS\text{w.}BL$, SSB, SSC, and $SSBC \times BL(A)$. (Note: for a 2×2 Latin square, it is not possible to compute $SSRES$ and there is no $\mathbf{R}'$ matrix. The within-blocks null hypotheses are tested subject to the restrictions that $\mathbf{R}'_{BC \times BL(A)}\boldsymbol{\mu} = \mathbf{0}_{BC \times BL(A)}.$)

c) Use the full rank experimental design model approach to test the null hypothesis for treatments A, B, and C; let $\alpha = .05$. (This problem involves inverting large matrices and should be done with a computer.)

14 ANALYSIS OF COVARIANCE

14.1 INTRODUCTION TO ANALYSIS OF COVARIANCE

The emphasis in previous chapters has been on the use of *experimental control* to reduce variability due to experimental error and obtain unbiased estimates of treatment effects. Experimental control can take various forms, such as random assignment of subjects to treatment levels, stratification of subjects into homogeneous blocks, and refinement of techniques for measuring a dependent variable. An alternative approach to reducing experimental error and obtaining unbiased estimates of treatment effects involves the use of *statistical control*. This latter approach also enables an experimenter to remove potential sources of bias from an experiment, biases that are difficult or impossible to eliminate by experimental control.

The statistical control described in this chapter is analysis of covariance; it combines regression analysis with analysis of variance. The procedure involves measuring one or more concomitant variables (also called covariates) in addition to the dependent variable. The concomitant variable represents a source of variation that has not been controlled in the experiment and one that is believed to affect the dependent variable. Through analysis of covariance, the dependent variable can be adjusted so as to remove the effects of the uncontrolled source of variation represented by the concomitant variable. The potential advantages are (1) a reduction in experimental error and, hence, increased power and (2) a reduction in bias caused by differences among experimental units where those differences are not attributable to the manipulation of the independent variable.

APPLICATIONS OF ANALYSIS OF COVARIANCE

The goal of analysis of covariance, ANCOVA, is to reduce experimental error or obtain adjusted estimates of population means in addition to reducing experimental error. Experimental error can be reduced if a portion of the error variance $\hat{\sigma}_\epsilon^2$ associated with the dependent variable is predictable from a knowledge of the concomitant variable. Removing this predictable portion from $\hat{\sigma}_\epsilon^2$ results in a smaller error variance and, hence, a more powerful test of a false null hypothesis. The rationale for this is developed in considerable detail in Section 14.2. Here we will simply introduce some basic notions and describe several applications of ANCOVA. Consider an analysis of covariance experiment with two treatment levels a_1 and a_2. The dependent variable is denoted by Y, the concomitant variable by X. The relationship between X and Y for a_1 and a_2 might look like that shown in Figure 14.1-1. Each individual in the experiment contributes one point, determined by his or her X and Y scores, to the graph. The points form two scatterplots—one for each treatment level. These are represented in the figure by ellipses. Through each ellipsis a line is drawn representing the regression of Y on X. In the typical ANCOVA model it is assumed that the regression lines are

FIGURE 14.1-1   Scatterplots for the two treatment levels. The size of the error variance in ANOVA is determined by the dispersion of the marginal distributions. The size of the error variance in ANCOVA is determined by the dispersion of the conditional distributions. The higher the value of the correlation between X and Y, the greater is the reduction in the error variance due to using analysis of covariance.

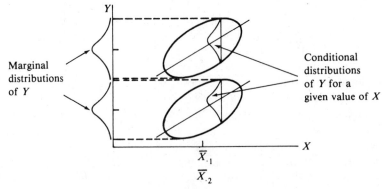

linear and parallel. The size of $\hat{\sigma}_{\epsilon}^{2}$, the error variance, in ANOVA is determined by the dispersion of the marginal distributions (see Figure 14.1-1). The size of $\hat{\sigma}_{\epsilon}^{2}$ in ANCOVA is determined by the dispersion of the conditional distributions. The higher the value of the correlation between X and Y, the narrower are the ellipses and the greater is the reduction in $\hat{\sigma}_{\epsilon}^{2}$ due to using analysis of covariance.

Figure 14.1-1 depicts an example in which the concomitant-variable means are equal. If experimental units are randomly assigned to treatment levels, one can expect the concomitant-variable means on the average to be equal. If random assignment is not used, differences among the means can be sizable as in Figure 14.1-2. This figure illustrates some of the effects of unequal concomitant-variable means. In parts (a) and (b), the absolute difference between adjusted dependent-variable means $\left|\overline{Y}_{\text{adj}\cdot1} - \overline{Y}_{\text{adj}\cdot2}\right|$ is smaller than that between unadjusted means $\left|\overline{Y}_{\cdot1} - \overline{Y}_{\cdot2}\right|$. In part (c), the absolute difference between adjusted means is larger than that between unadjusted means. Whenever the regression lines have the same slope and intercept, as in part (b), the difference between adjusted means will equal zero.

Analysis of covariance is often used in three kinds of research situations

---

FIGURE 14.1-2 When the concomitant-variable means, $\overline{X}_{\cdot1}$ and $\overline{X}_{\cdot2}$, differ, the absolute difference between adjusted means for the dependent variable can be less than that between unadjusted means, as in (a) and (b), or larger, as in (c).

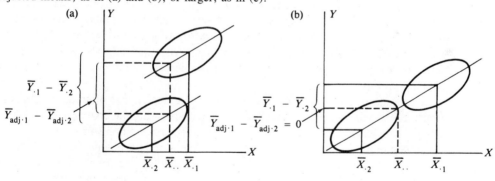

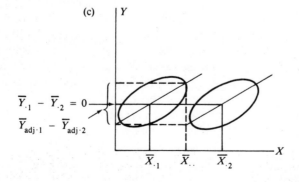

involving unequal concomitant-variable means. One of these involves the use of *intact groups,* a situation that is common in educational and industrial research. This situation can be illustrated by an experiment designed to evaluate four methods of teaching arithmetic. It is impractical for administrative reasons to assign four different teaching methods to the same classroom. An alternative is to assign the four teaching methods randomly to four different classrooms, a design with a serious defect. If differences in learning ability or similar characteristics exist among the classes prior to the introduction of the four teaching methods, these extraneous variables will bias the evaluation. In this example, it is possible to administer a test of intelligence prior to the beginning of the experiment in order to obtain an estimate of learning ability. If the classes differ in intelligence, one can, if certain assumptions are tenable, adjust the dependent variable, achievement in arithmetic, for difference in the concomitant variable, learning ability.

A note of caution concerning the use of *intact groups* is needed here. Experiments of this type are always subject to interpretation difficulties that are not present when random assignment is used in forming the experimental groups. Even when analysis of covariance is skillfully used, we can never be certain that some variable that has been overlooked will not bias the evaluation of an experiment.* This problem is absent in properly randomized experiments because the effects of all uncontrolled variables are distributed among the groups in such a way that they can be taken into account in the test of significance. The use of intact groups removes this safeguard.

Analysis of covariance is not limited to the use of only one concomitant variable. In the teaching example, differences in arithmetic achievement may also exist among the children prior to the beginning of the experiment. If concomitant measures of arithmetic achievement and intelligence are obtained prior to the introduction of the treatment, the dependent variable can be adjusted for both potential sources of bias.

A second situation in which analysis of covariance is often used is illustrated in the following example. It may become apparent during the course of an experiment that the subjects in the p treatment groups were not equated on some relevant variable at the beginning of the experiment although random assignment was employed. For example, an experiment might be designed to evaluate the effects of different drugs on stimulus generalization in rats. At the beginning of the experiment the rats are randomly assigned to p experimental groups and a bar-pressing response is shaped by operant conditioning procedures. If the p groups require different amounts of training to establish a stable bar-pressing response, this suggests that differences in learning ability exist among the groups. An experimenter may find, at the conclusion of the experiment, that amount of stimulus generalization is related to the amount of training necessary to establish the stable bar-pressing response. If certain assumptions are tenable, the generalization scores of the groups can be adjusted for differences in learning ability. Thus unsuspected differences present at the beginning of the experiment can be statistically controlled by analysis of covariance.

---

*   For discussions of this problem see Reichardt in Cook and Campbell (1979, Ch. 4) and Weisberg (1979). Huitema (1980, Ch. 7) and Smith (1957) also provide excellent treatments of the interpretation problems in ANCOVA.

Analysis of covariance may be useful in yet another research situation. In the example concerning the evaluation of four methods of teaching arithmetic, a third variable may bias the evaluation. This variable is the number of hours spent in study by students in the four classrooms. Variations in the daily schedules of the classrooms may provide more study periods for students in one class than for students in other classes. It would be difficult to experimentally control the amount of time available for arithmetic study. A practical alternative is to record each day the amount of study time available to the students. This information can then be used to make appropriate adjustments in the dependent variable. In this example, variation in the concomitant variable of study time did not occur until after the beginning of the experiment. It is assumed that amount of study time available during school is not influenced by the treatment. A teacher might allocate additional study periods for arithmetic because students were unable to master the material using the assigned teaching method. Under this condition, it would be incorrect to adjust the dependent variable for this concomitant variable.

Statistical control and experimental control are not mutually exclusive approaches for reducing bias and increasing precision. It may be convenient to control some variables by experimental control and others by statistical control. In general, an experimenter should attempt to use experimental control whenever possible. Statistical control is based on a series of assumptions, discussed in Section 14.4, that may prove untenable in a particular experiment.

SELECTION OF CONCOMITANT VARIABLE

Concomitant variables in analysis of covariance should be selected with care. Effects eliminated by a covariance adjustment must be irrelevant to the objectives of the experiment. Analysis of covariance can be used in conjunction with each of the experimental designs described in this book. A design based on a completely randomized building block design is denoted by the letters CRAC-p; one based on a Latin square building block design is denoted by LSAC-p, and so on. A covariance adjustment is appropriate for experiments that meet, in addition to the assumptions discussed in Section 14.4, the following conditions.

1. The experiment contains one or more extraneous sources of variation believed to affect the dependent variable and considered irrelevant to the objectives of the experiment.

2. Experimental control of the extraneous sources of variations is either not possible or not feasible.

3. It is possible to obtain a measure of the extraneous variation that does not include effects attributable to the treatment. Any one of the following situations will generally meet this third condition:
   a. The concomitant observations are obtained prior to presentation of the treatment levels, or
   b. The concomitant observations are obtained after the presentation of the treatment levels but before the treatment levels have had an opportunity to affect the concomitant variable, or

c. It can be assumed that the concomitant variable is unaffected by the treatment.

If the third condition is not satisfied and the concomitant variable is influenced by the treatment, the adjustment made on the dependent variable is biased. Consider the drug example cited previously. If administering the drug affected both the learning of the bar-pressing response (covariate) and the amount of generalization (dependent variable), it would not be possible to adjust the generalization scores correctly. An adjustment of the generalization scores for differences in learning of the bar-pressing response would also remove the effects of drugs (treatment) from the dependent variable. This follows because both the covariate and the dependent variable reflect the effects of the treatment.

Analysis of covariance can also be used as a general procedure for estimating missing observations. The rationale underlying this application of analysis of covariance is discussed by Coons (1957).

14.2 RATIONALE UNDERLYING COVARIATE ADJUSTMENT

If dependent and concomitant variables are designated by Y and X, respectively, how can Y be adjusted so as to be free of variation due to X? The adjustment used in analysis of covariance is based on regression analysis. Although the regression of Y on X need not be linear, only the linear case is considered. The discussion is limited to one covariate; multiple covariates are discussed in Section 14.6.

CALCULATION OF ADJUSTED TOTAL SUM OF SQUARES

The regression equation for predicting Y from X is

(14.2-1) $$\hat{Y}_{ij} = \hat{\beta}_T(X_{ij} - \overline{X}_{..}) + \overline{Y}_{..}$$

where $\hat{Y}_{ij}$ = the predicted score

$\hat{\beta}_T$ = the linear regression coefficient computed for the $i = 1, \ldots, n$ and $j = 1, \ldots, p$ pairs of observations

X_{ij} = the covariate for subject i in treatment level j

$\overline{X}_{..}$ = the mean of the covariate

$\overline{Y}_{..}$ = the mean of the dependent variable.

The sum of squares of residuals about this regression line is equal to

(14.2-2)
$$\sum_{i=1}^{n} \sum_{j=1}^{p} (Y_{ij} - \hat{Y}_{ij})^2$$

where $(Y_{ij} - \hat{Y}_{ij})$ is the deviation of the ijth score from the predicted score. This sum of squares represents the variation among the dependent scores which is *not* associated with the linear regression of Y on X. This is the sum of squares that is of interest in analysis of covariance. If $[\hat{\beta}_T(X_{ij} - \overline{X}_{..}) + \overline{Y}_{..}]$ is substituted for $\hat{Y}_{ij}$ in equation (14.2-2), the residual sum of squares can be shown to equal

$$\sum_{i=1}^{n} \sum_{j=1}^{p} (Y_{ij} - \hat{Y}_{ij})^2 = \sum_{i=1}^{n} \sum_{j=1}^{p} [(Y_{ij} - \overline{Y}_{..}) - \hat{\beta}_T(X_{ij} - \overline{X}_{..})]^2$$

$$= \sum_{i=1}^{n} \sum_{j=1}^{p} (Y_{ij} - \overline{Y}_{..})^2 - 2\hat{\beta}_T \sum_{i=1}^{n} \sum_{j=1}^{p} (X_{ij} - \overline{X}_{..})(Y_{ij} - \overline{Y}_{..})$$

$$+ \hat{\beta}_T^2 \sum_{i=1}^{n} \sum_{j=1}^{p} (X_{ij} - \overline{X}_{..})^2$$

but

$$\hat{\beta}_T = \frac{\displaystyle\sum_{i=1}^{n} \sum_{j=1}^{p} (X_{ij} - \overline{X}_{..})(Y_{ij} - \overline{Y}_{..})}{\displaystyle\sum_{i=1}^{n} \sum_{j=1}^{p} (X_{ij} - \overline{X}_{..})^2}$$

and

$$\hat{\beta}_T \sum_{i=1}^{n} \sum_{j=1}^{p} (X_{ij} - \overline{X}_{..})^2 = \sum_{i=1}^{n} \sum_{j=1}^{p} (X_{ij} - \overline{X}_{..})(Y_{ij} - \overline{Y}_{..}).$$

Thus,

$$\sum_{i=1}^{n} \sum_{j=1}^{p} (Y_{ij} - \hat{Y}_{ij})^2 = \sum_{i=1}^{n} \sum_{j=1}^{p} (Y_{ij} - \overline{Y}_{..})^2 - 2\hat{\beta}_T^2 \sum_{i=1}^{n} \sum_{j=1}^{p} (X_{ij} - \overline{X}_{..})^2$$

$$+ \hat{\beta}_T^2 \sum_{i=1}^{n} \sum_{j=1}^{p} (X_{ij} - \overline{X}_{..})^2$$

(14.2-3)
$$= \sum_{i=1}^{n} \sum_{j=1}^{p} (Y_{ij} - \overline{Y}_{..})^2 - \hat{\beta}_T^2 \sum_{i=1}^{n} \sum_{j=1}^{p} (X_{ij} - \overline{X}_{..})^2.$$

The term $[\hat{\beta}_T^2 \sum_{i=1}^{n} \sum_{j=1}^{p} (X_{ij} - \overline{X}_{..})^2]$ represents an adjustment that is made to the total sum of squares for Y, $[\sum_{i=1}^{n} \sum_{j=1}^{p} (Y_{ij} - \overline{Y}_{..})^2]$, which removes the linear effect of the covariate. This adjustment will always reduce the sum of the squares for Y if $\hat{\beta}_T \neq 0$. Consequently the sum of squares on the left side of equation (14.2-3) is sometimes called the *reduced* sum of squares. This sum of squares will be referred to in this chapter as an *adjusted* total sum of squares and designated by the symbol $T_{yy(adj)}$ or T_{adj} when it is clear from the context that it refers to the dependent variable. Unadjusted sums of squares for Y and X are designated by the symbols T_{yy} and T_{xx}, respectively. The use of T_{yy}, T_{xx}, and so on instead of SS to stand for a sum of squares is common practice in discussions of analysis of covariance.

As stated earlier the slope, $\hat{\beta}_T$, of the regression line used in predicting Y from X is given by

$$\hat{\beta}_T = \frac{\sum_{i=1}^{n} \sum_{j=1}^{p} (X_{ij} - \overline{X}_{..})(Y_{ij} - \overline{Y}_{..})}{\sum_{i=1}^{n} \sum_{j=1}^{p} (X_{ij} - \overline{X}_{..})^2} = \frac{T_{xy}}{T_{xx}}.$$

The term T_{xy} is called the sum of squares for the cross product of X and Y. It can be shown that the coefficient $\hat{\beta}_T$ provides the best fitting line according to the least squares criterion for the np pairs of observations. A brief discussion of the method of least squares is given in Section 5.2. An adjusted *total sum of squares*, using the abbreviated notation just described, can be written

(14.2-4)
$$T_{yy(adj)} = T_{yy} - \hat{\beta}_T^2 T_{xx} = T_{yy} - \frac{T_{xy}^2}{T_{xx}}.$$

The degrees of freedom for $T_{yy(adj)}$ are $np - 2$. One additional degree of freedom has been lost because of the linear restriction imposed on the sum of squares whereby the deviations are computed from the regression line.

CALCULATION OF ADJUSTED WITHIN-GROUPS SUM OF SQUARES

The total sum of squares $\sum_{i=1}^{n} \sum_{j=1}^{p} (Y_{ij} - \overline{Y}_{..})^2$ for a completely randomized design can be partitioned into between- and within-groups sums of squares as shown in Section 2.2.

$$\sum_{i=1}^{n} \sum_{j=1}^{p} (Y_{ij} - \overline{Y}_{..})^2 = \sum_{i=1}^{n} \sum_{j=1}^{p} (Y_{ij} - \overline{Y}_{.j})^2 + n \sum_{j=1}^{p} (\overline{Y}_{.j} - \overline{Y}_{..})^2$$

$$T_{yy} \qquad = \qquad S_{yy} \qquad + \qquad A_{yy}$$

The abbreviated designation for each sum of squares appears below the formulas.

Similarly, the total sum of squares for X and the cross product of X and Y can be partitioned into

$$\sum_{i=1}^{n} \sum_{j=1}^{p} (X_{ij} - \overline{X}_{..})^2 = \sum_{i=1}^{n} \sum_{j=1}^{p} (X_{ij} - \overline{X}_{.j})^2 + n \sum_{j=1}^{p} (\overline{X}_{.j} - \overline{X}_{..})^2$$

$$T_{xx} \qquad = \qquad S_{xx} \qquad + \qquad A_{xx}$$

and

$$\sum_{i=1}^{n} \sum_{j=1}^{p} (X_{ij} - \overline{X}_{..})(Y_{ij} - \overline{Y}_{..}) =$$

$$T_{xy} \qquad =$$

$$\sum_{i=1}^{n} \sum_{j=1}^{p} (X_{ij} - \overline{X}_{.j})(Y_{ij} - \overline{Y}_{.j}) + n \sum_{j=1}^{p} (\overline{X}_{.j} - \overline{X}_{..})(\overline{Y}_{.j} - \overline{Y}_{..})$$

$$S_{xy} \qquad + \qquad A_{xy}$$

An adjusted within-groups sum of squares, following the procedures for computing $T_{yy(\text{adj})}$, is given by

$$S_{yy(\text{adj})} = S_{yy} - \hat{\beta}_W^2 S_{xx} = S_{yy} - \frac{S_{xy}^2}{S_{xx}}$$

where $\hat{\beta}_W$ is the *within-groups* regression coefficient. The within-groups regression coefficient can be computed by

$$\hat{\beta}_W = \frac{\displaystyle\sum_{i=1}^{n} \sum_{j=1}^{p} (X_{ij} - \overline{X}_{.j})(Y_{ij} - \overline{Y}_{.j})}{\displaystyle\sum_{i=1}^{n} \sum_{j=1}^{p} (X_{ij} - \overline{X}_{.j})^2} = \frac{S_{xy}}{S_{xx}}.$$

The degrees of freedom for $S_{yy(\text{adj})}$ are $np - p - 1$.

INTERPRETATION OF $\hat{\beta}_T$, $\hat{\beta}_W$, AND $\hat{\beta}_B$

If data consist of np paired observations for p treatment levels, a number of different regression lines can be identified. Two regression lines, with slopes $\hat{\beta}_T$ and $\hat{\beta}_W$, have been mentioned in this section. A third regression line can also be identified—a *between-groups* regression line with slope $\hat{\beta}_B$. The interpretation of these three regression lines can be clarified by a simple numerical example and diagrams. The data in Table 14.2-1 will be used for purposes of illustration. Data for the three treatment levels have been plotted in Figure 14.2-1. If all 12 points in Figure 14.2-1 are plotted as if they represent one treatment level instead of three levels, a single regression line with slope equal to $\hat{\beta}_T$ can be drawn. The slope $\hat{\beta}_T$ of this single line is given by

$$\hat{\beta}_T = \frac{\displaystyle\sum_{i=1}^{n} \sum_{j=1}^{p} (X_{ij} - \overline{X}_{..})(Y_{ij} - \overline{Y}_{..})}{\displaystyle\sum_{i=1}^{n} \sum_{j=1}^{p} (X_{ij} - \overline{X}_{..})^2} = \frac{T_{xy}}{T_{xx}} = \frac{11.72}{16.06} = .73.$$

This regression line is shown in Figure 14.2-2. For convenience the X and Y scores

TABLE 14.2-1   Data from Experiment

a_{y1}	a_{x1}	a_{y2}	a_{x2}	a_{y3}	a_{x3}
1.0	1.0	2.6	4.5	4.8	3.0
1.5	2.0	2.0	2.0	4.0	2.0
2.0	4.0	2.3	3.0	5.3	4.0
1.8	3.0	2.5	4.0	6.0	5.0
Mean = 1.6	2.5	2.4	3.4	5.0	3.5

Grand mean $\overline{Y}.. = 3.0, \overline{X}.. = 3.1$

FIGURE 14.2-1   Regression lines for three treatment levels.

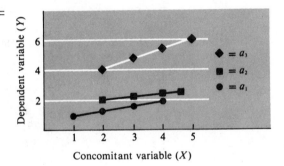

FIGURE 14.2-2   Total overall regression line with $\hat{\beta}_T = 0.73$.

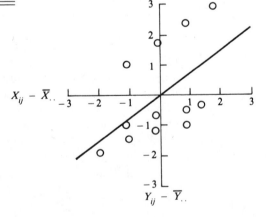

are expressed as deviations from their respective grand means; that is, $X_{ij} - \overline{X}..$ and $Y_{ij} - \overline{Y}..$.

The within-groups regression coefficient for each of the treatment levels shown in Figure 14.2-1 is given by

$$\hat{\beta}_{Wj} = \frac{\sum_{i=1}^{n} (X_{ij} - \overline{X}_{\cdot j})(Y_{ij} - \overline{Y}_{\cdot j})}{\sum_{i=1}^{n} (X_{ij} - \overline{X}_{\cdot j})^2} = \frac{S_{xyj}}{S_{xxj}}.$$

These three regression coefficients are, respectively,

$$\hat{\beta}_{W1} = \frac{S_{xy1}}{S_{xx1}} = \frac{1.65}{5.00} = .33$$

$$\hat{\beta}_{W2} = \frac{S_{xy2}}{S_{xx2}} = \frac{.88}{3.69} = .24$$

$$\hat{\beta}_{W3} = \frac{S_{xy3}}{S_{xx3}} = \frac{3.25}{5.00} = .65.$$

By computing a weighted mean of these three coefficients, one obtains the within-groups regression coefficient $\hat{\beta}_W$.

$$\hat{\beta}_W = \frac{(S_{xx1})(\hat{\beta}_{W1}) + (S_{xx2})(\hat{\beta}_{W2}) + (S_{xx3})(\hat{\beta}_{W3})}{S_{xx1} + S_{xx2} + S_{xx3}} = \frac{5.78}{13.69} = .42$$

The weights in this formula are the corresponding values of S_{xxj}. The formula is less convenient for computational purposes than the formula $\hat{\beta}_W = S_{xy}/S_{xx}$ given previously, but it helps to clarify the nature of $\hat{\beta}_W$. The regression lines corresponding to $\hat{\beta}_{W1}$, $\hat{\beta}_{W2}$, $\hat{\beta}_{W3}$, and $\hat{\beta}_W$ are plotted in Figure 14.2-3.

One of the assumptions underlying the adjustment of the within-groups sum of squares is that the within-groups regression coefficients are all estimates of the same

FIGURE 14.2-3 Regression lines corresponding to $\hat{\beta}_{W1}$, $\hat{\beta}_{W2}$, $\hat{\beta}_{W3}$, and $\hat{\beta}_W$.

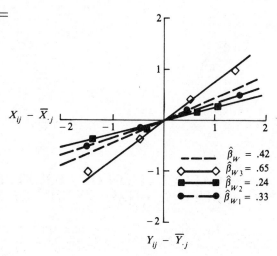

common population regression coefficient. That is,

$$\beta_{W1} = \beta_{W2} = \beta_{W3} = \beta_W.$$

Section 14.4 describes procedures for testing this assumption.

The final regression coefficient that needs to be defined is $\hat{\beta}_B$. This coefficient is given by

$$\hat{\beta}_B = \frac{n \sum_{j=1}^{p} (\overline{X}_{.j} - \overline{X}_{..})(\overline{Y}_{.j} - \overline{Y}_{..})}{n \sum_{j=1}^{p} (\overline{X}_{.j} - \overline{X}_{..})^2} = \frac{A_{xy}}{A_{xx}} = \frac{5.95}{2.37} = 2.51$$

and refers to the line that fits the means of the three treatment levels. This regression line is shown in Figure 14.2-4, where the mean of each treatment level is expressed as a deviation from the grand mean; that is, $\overline{X}_{.j} - \overline{X}_{..}$ and $\overline{Y}_{.j} - \overline{Y}_{..}$.

We noted that the total sum of squares for Y, X, and XY can be partitioned into between- and within-groups sums of squares. The total regression coefficient $\hat{\beta}_T$ is equal to the weighted mean of $\hat{\beta}_W$ and $\hat{\beta}_B$, where the last two terms are weighted by the corresponding sum of squares for X. That is

$$\hat{\beta}_T = \frac{S_{xx}\hat{\beta}_W + A_{xx}\hat{\beta}_B}{S_{xx} + A_{xx}} = \frac{(13.69)(.42) + (2.37)(2.51)}{13.69 + 2.37} = .73.$$

FIGURE 14.2-4   Regression line corresponding to $\hat{\beta}_B = 2.51$.

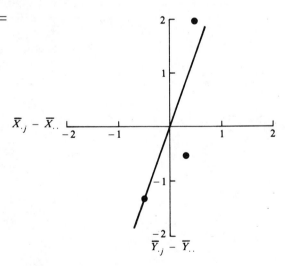

CALCULATION OF ADJUSTED BETWEEN-GROUPS SUM OF SQUARES

The adjusted between-groups sum of squares is given by

$$A_{yy(adj)} = T_{yy(adj)} - S_{yy(adj)}.$$

The reader may wonder why $A_{yy(adj)}$ is obtained by subtraction and not by

$$A_{yy(adj)} = A_{yy} - \hat{\beta}_B^2 A_{xx} = A_{yy} - \frac{A_{xy}^2}{A_{xx}}.$$

This latter formula is analogous to the formulas used to compute $T_{yy(adj)}$ and $S_{yy(adj)}$. Upon reflection, it is apparent that $\hat{\beta}_B$ is affected by differences among the means for the dependent variable. The adjustment that is made to the dependent variable, however, must be independent of the differences to be tested. Consequently $\hat{\beta}_B$ is unsatisfactory for making this adjustment. The coefficient $\hat{\beta}_W$ could be considered for this adjustment, for it is independent of the differences to be tested. However, $\hat{\beta}_W$ should not be used for adjusting both numerator and denominator of an F ratio. This is a consequence of the requirement that an F ratio must be the ratio of two *independent* chi-square variables, each divided by its respective degrees of freedom. The computation of $A_{yy(adj)}$ by subtraction circumvents the preceding problems. The degrees of freedom for $A_{yy(adj)}$ are $p - 1$, and not $p - 2$, because the between-groups regression line did not enter into the calculation of the adjusted sum of squares.

LINEAR MODEL FOR ANALYSIS OF COVARIANCE

The experimental design model equation for a completely randomized analysis of covariance design is

(14.2-5) $Y_{ij} = \mu + \alpha_j + \beta_W(X_{ij} - \overline{X}_{..}) + \epsilon_{i(j)}$ $(i = 1, \ldots, n; j = 1, \ldots, p)$

where μ, α_j, and $\epsilon_{i(j)}$ are the familiar terms for a completely randomized design. The terms β_W, X_{ij}, and $\overline{X}_{..}$ refer, respectively, to the linear regression coefficient, the value of the covariate for subject i in treatment level j, and the grand mean of the covariate scores. The difference between two observations Y_{ij} and $Y_{ij'}$ is an estimate of

(14.2-6)
$$Y_{ij} - Y_{ij'} = [\mu + \alpha_j + \beta_W(X_{ij} - \overline{X}_{..}) + \epsilon_{i(j)}] - [\mu + \alpha_{j'} + \beta_W(X_{ij'} - \overline{X}_{..}) + \epsilon_{i(j')}]$$
$$= \alpha_j - \alpha_{j'} + \beta_W(X_{ij} - X_{ij'}) + \epsilon_{i(j)} - \epsilon_{i(j')}.$$

It is apparent that unless the two covariates X_{ij} and $X_{ij'}$ are equal, the difference $(X_{ij} - X_{ij'})$ will affect the observed difference between Y_{ij} and $Y_{ij'}$. The terms in the linear model equation for analysis of covariance can be rearranged by subtracting $[\beta_W(X_{ij} - \overline{X}_{..})]$ from $Y_{ij} = \mu + \alpha_j + \epsilon_{i(j)}$. The difference is an *adjusted score*. Thus,

(14.2-7) $$Y_{adjij} = Y_{ij} - \beta_W(X_{ij} - \overline{X}_{..}) = \mu + \alpha_j + \epsilon_{i(j)}.$$

The adjusted score is free of the effects of the covariate. Furthermore, the adjusted score provides an estimate of the familiar terms of the linear model equation for a completely randomized design. However, $\epsilon_{i(j)}$ in ANCOVA will ordinarily be smaller than $\epsilon_{i(j)}$ in ANOVA. The magnitude of the reduction is discussed in Section 14.3.

14.3 LAYOUT AND COMPUTATIONAL PROCEDURES FOR TYPE CRAC-p DESIGN

Assume that the experiment described in Section 14.1 for evaluating four methods of teaching arithmetic has been carried out. Thirty-two students were randomly assigned to four classrooms with eight students in each room. The teaching methods were randomly assigned to the four classrooms. An intelligence test was administered to each student at the beginning of the experiment. These data are used to adjust arithmetic achievement scores obtained at the conclusion of the experiment for differences in intelligence among the students. If much larger samples had been used, differences in intelligence could be expected to be negligible due to random assignment and an ANOVA would have been appropriate.

The research hypothesis can be evaluated by means of a statistical test of the following null hypothesis:

$$H_0: \quad \alpha_j = 0 \quad \text{for all } j$$
$$H_1: \quad \alpha_j \neq 0 \quad \text{for some } j.$$

The level of significance adopted for the test is .05.

The layout of the design and computational formulas are shown in Table 14.3-1. The analysis is summarized in Table 14.3-2. The sums of squares in Table 14.3-2 have been adjusted for intellectual differences among the students. It is obvious

---

TABLE 14.3-1   Computational Procedures for a Type CRAC-4 Design

(i)   Data and notation [Y_{ij} and X_{ij} denote, respectively, a dependent score and a covariate score for experimental unit i in treatment level j; $i = 1, \ldots, n$ experimental units (s_i) and $j = 1, \ldots, p$ levels of treatment A (a_j)]:

AS Summary Table
Entries are Y_{ij} and X_{ij}

a_{y1}	a_{x1}	a_{y2}	a_{x2}	a_{y3}	a_{x3}	a_{y4}	a_{x4}
3	42	4	47	7	61	7	65
6	57	5	49	8	65	8	74
3	33	4	42	7	64	9	80
3	47	3	41	6	56	8	73
1	32	2	38	5	52	10	85
2	35	3	43	6	58	10	82
2	33	4	48	5	53	9	78
2	39	3	45	6	54	11	89

$\sum\limits_{i=1}^{n} Y_{ij}, \sum\limits_{i=1}^{n} X_{ij} = 22 \quad 318 \qquad 28 \quad 353 \qquad 50 \quad 463 \qquad 72 \quad 626$

Table 14.3-1 (continued)

(ii)   Computational symbols:

$$\sum_{i=1}^{n}\sum_{j=1}^{p} Y_{ij} = 3 + 6 + 3 + \cdots + 11 = 172.00$$

$$\frac{\left(\sum_{i=1}^{n}\sum_{j=1}^{p} Y_{ij}\right)^2}{np} = [Y] = \frac{(172)^2}{32} = 924.50$$

$$\sum_{i=1}^{n}\sum_{j=1}^{p} Y_{ij}^2 = [AS_y] = (3)^2 + (6)^2 + (3)^2 + \cdots + (11)^2 = 1160.00$$

$$\sum_{j=1}^{p}\frac{\left(\sum_{i=1}^{n} Y_{ij}\right)^2}{n} = [A_y] = \frac{(22)^2}{8} + \frac{(28)^2}{8} + \cdots + \frac{(72)^2}{8} = 1119.00$$

$$\sum_{i=1}^{n}\sum_{j=1}^{p} X_{ij} = 42 + 57 + 33 + \cdots + 89 = 1760.00$$

$$\frac{\left(\sum_{i=1}^{n}\sum_{j=1}^{p} X_{ij}\right)^2}{np} = [X] = \frac{(1760)^2}{32} = 96800.00$$

$$\sum_{i=1}^{n}\sum_{j=1}^{p} X_{ij}^2 = [AS_x] = (42)^2 + (57)^2 + (33)^2 + \cdots + (89)^2 = 105202.00$$

$$\sum_{j=1}^{p}\frac{\left(\sum_{i=1}^{n} X_{ij}\right)^2}{n} = [A_x] = \frac{(318)^2}{8} + \frac{(353)^2}{8} + \cdots + \frac{(626)^2}{8} = 103997.25$$

$$\frac{\left(\sum_{i=1}^{n}\sum_{j=1}^{p} X_{ij}\right)\left(\sum_{i=1}^{n}\sum_{j=1}^{p} Y_{ij}\right)}{np} = [XY] = \frac{(172)(1760)}{32} = 9460.00$$

$$\sum_{i=1}^{n}\sum_{j=1}^{p} X_{ij}Y_{ij} = [AS_{xy}] = (3)(42) + (6)(57) + (3)(33) + \cdots + (11)(89) = 10840.00$$

$$\sum_{j=1}^{p}\frac{\left(\sum_{i=1}^{n} X_{ij}\right)\left(\sum_{i=1}^{n} Y_{ij}\right)}{n} = [A_{xy}] = \frac{(22)(318)}{8} + \cdots + \frac{(72)(626)}{8} = 10637.75$$

(iii)   Computational formulas:

$$T_{yy} = [AS_y] - [Y] = 235.500$$
$$A_{yy} = [A_y] - [Y] = 194.500$$
$$S_{yy} = [AS_y] - [A_y] = 41.000$$
$$T_{xx} = [AS_x] - [X] = 8402.000$$
$$A_{xx} = [A_x] - [X] = 7197.250$$
$$S_{xx} = [AS_x] - [A_x] = 1204.750$$
$$T_{xy} = [AS_{xy}] - [XY] = 1380.000$$

$$A_{xy} = [A_{xy}] - [XY] = 1177.750$$
$$S_{xy} = [AS_{xy}] - [A_{xy}] = 202.250$$
$$T_{adj} = T_{yy} - \frac{(T_{xy})^2}{T_{xx}} = 8.840$$
$$S_{adj} = S_{yy} - \frac{(S_{xy})^2}{S_{xx}} = 7.047$$
$$A_{adj} = T_{adj} - S_{adj} = 1.793$$

TABLE 14.3-2 ANCOVA Table for Type CRAC-4 Design

	Source	SS	df	MS	F	$E(MS)$ Model I
1	Between groups, A_{adj} (Teaching methods)	1.793	$p - 1 = 3$	0.598	[½] 2.29 N.S.	$\sigma_\epsilon^2 + n\Sigma\alpha_j^2/(p - 1)$
2	Within groups, S_{adj}	7.047	$np - p - 1 = 27$	0.261		σ_ϵ^2
3	Total$_{adj}$	8.840	$np - 2 = 30$			

that the null hypothesis cannot be rejected. The reader may wonder what conclusion would have been drawn if the arithmetic achievement scores had not been adjusted for the nuisance variable of intelligence. The required sums of squares for this test also appear in Table 14.3-1. The F ratio is

$$F = \frac{A_{yy}/(p - 1)}{S_{yy}/(np - p)} = \frac{194.5/3}{41.0/28} = 44.28$$

$$F_{.05;3,28} = 2.95.$$

This F ratio is significant beyond the .05 level. Thus, if the covariate adjustment had not been used, the experimenter would have drawn an erroneous conclusion with respect to the four teaching methods. We infer from the F ratio for adjusted mean squares that there are no differences in arithmetic achievement among the four population means apart from differences in intelligence. A test of the hypothesis that the intelligence test scores of students in the four populations are equal is given by

$$F = \frac{A_{xx}/(p - 1)}{S_{xx}/(np - p)} = \frac{7197.250/3}{1204.750/28} = 55.76$$

$$F_{.05;3,28} = 2.95.$$

The null hypothesis with respect to intelligence scores can be rejected beyond the .05 level.

CORRELATION COEFFICIENTS FOR BETWEEN, WITHIN, AND TOTAL SUM OF SQUARES

Three correlation coefficients can be computed for the paired observations in Table 14.3-1:

$$r_T = \frac{T_{xy}}{\sqrt{T_{xx}T_{yy}}} = \frac{1380.00}{\sqrt{(8402.00)(235.50)}} = .98$$

$$r_B = \frac{A_{xy}}{\sqrt{A_{xx}A_{yy}}} = \frac{1177.75}{\sqrt{(7197.25)(194.50)}} = .99$$

$$r_W = \frac{S_{xy}}{\sqrt{S_{xx}S_{yy}}} = \frac{202.25}{\sqrt{(1204.75)(41.00)}} = .91$$

where r_T refers to the overall correlation between X and Y, r_B refers to the correlation between the treatment level means for X and Y, and r_W is the weighted average correlation between X and Y for the four treatment levels.

If r_B is larger than r_W, the reduction in the variation attributable to the treatment can be large relative to the reduction in the error variation. Under this condition, the F ratio in analysis of covariance will be smaller than the corresponding F ratio in analysis of variance. The data summarized in Table 14.3-2 illustrate this situation. In general, if r_B is negative and r_W is positive, the F ratio in analysis of covariance will be larger than the corresponding F ratio in analysis of variance.

The reduction in the error term MSS_{adj} that occurs from the use of analysis of covariance is determined by the size of r_W. The larger this correlation, the greater the reduction in the error term. An alternative method of computing an adjusted mean square for experimental error illustrates this fact. If σ_ϵ^2 is the experimental error when no covariance adjustment is used, this term reduces approximately to

$$\sigma_\epsilon^2(1 - \rho_W^2)\left(1 + \frac{1}{f_e - 2}\right)$$

by the use of analysis of covariance. Here ρ_W^2 is the population within-groups correlation coefficient, and f_e is the degrees of freedom associated with σ_ϵ^2 (Cochran, 1957). For the data in Table 14.3-1, an estimate based on the preceding formula yields

$$1.464[1 - (.91)^2]\left(1 + \frac{1}{28 - 2}\right) = 1.464(.1719)(1.0385) = 0.261$$

which is equal to MSS_{adj}. It is apparent from the preceding formula that the reduction in the error term due to the covariance adjustment is primarily a function of the size of r_W, which is used to estimate ρ_W.

14.4 ASSUMPTIONS OF THE MODEL FOR TYPE CRAC-*p* DESIGN

Let Y_{ij} be a measure for a randomly selected subject in treatment population j. For the fixed-effects experimental design model, it is assumed that

$$Y_{adjij} = Y_{ij} - \beta_W(X_{ij} - \overline{X}_{..}) = \mu + \alpha_j + \epsilon_{i(j)} \qquad (i = 1, \ldots, n; j = 1, \ldots, p)$$

where Y_{adjij} = the adjusted criterion measure
Y_{ij} = the unadjusted criterion measure
β_W = the common population linear regression coefficient for the treatment levels

X_{ij} = the covariate measure for subject i in treatment population j
$\overline{X}_{..}$ = the covariate sample mean
μ = the overall population mean
α_j = the effect of treatment j and is subject to the restriction
   $\sum_{j=1}^{p} \alpha_j = 0$
$\epsilon_{i(j)}$ = the experimental error that is $NID(0, \sigma_\epsilon^2)$.

In order for the F ratio in Table 14.3-2 to be distributed as the F distribution, the assumptions of a completely randomized design described in Section 2.5 must be met. In addition to these assumptions, the following assumptions must also be tenable:

1.   The population within-groups regression coefficients are homogeneous; that is, $\beta_1 = \beta_2 = \cdots = \beta_p = \beta_w$ for the p treatment levels.

2. X is measured without error.

3.   The residuals (deviations from regression) are NID with mean equal to zero and common variance.

This implies that the proper form of regression equation has been used. In the present example, it is assumed that the regression is linear. Of course, analysis of covariance is not appropriate unless the effects eliminated by covariate adjustment are irrelevant to the objectives of the experiment.

In general, tests of significance in the analysis of covariance are robust with respect to violation of the assumptions of normality and homogeneity of variance (Atiqullah, 1964). Less is known concerning the effects of violation of the added regression assumptions.* Departure from linearity results in biased estimates of treatment effects and reduced efficiency of the covariance analysis. Lord (1960) has shown that error of measurement in the covariate may either obscure true differences among the dependent variable means or create the illusion of differences where none exists. According to Snedecor and Cochran (1967), such errors are minor if the variance of the measurement error is small relative to the variance of X. Atiqullah (1964) has reported that heterogeneity of the β_j's results in loss of power and makes the interpretation of the test of treatment A difficult.

A test of the hypothesis that

$$\beta_1 = \beta_2 = \cdots = \beta_p$$

is given by

$$F = \frac{S_2/(p-1)}{S_1/p(n-2)}$$

which is distributed as the F distribution, with $p-1$ and $p(n-2)$ degrees of freedom. A numerically large level of significance ($\alpha = .10$ or $.25$) should be used

---

*   Excellent summaries of the ANCOVA assumptions are provided by Elashoff (1969), Glass, Peckham, and Sanders (1972), and Huitema (1980, Ch. 6). The special problems posed by quasi-experimental designs are examined in detail by Reichardt in Cook and Campbell (1979, Ch. 4), Cronbach et al. (1977), and Weisberg (1979).

for this test in order to avoid a type II error; that is, accepting the hypothesis of homogeneity of regression coefficients when, in fact, the hypothesis is false.

The statistics S_1 and S_2 are computed as follows:

$$S_1 = S_{yy} - \sum_{j=1}^{p} \frac{(S_{xyj})^2}{S_{xxj}}$$

where

$$S_{xyj} = [AS_{xyj}] - [A_{xyj}] \quad \text{and} \quad S_{xxj} = [AS_{xj}] - [A_{xj}]$$

and

$$S_2 = \sum_{j=1}^{p} \frac{(S_{xyj})^2}{S_{xxj}} - \frac{(S_{xy})^2}{S_{xx}}.$$

Table 14.3-1 provides the data for computing S_1 and S_2.

$$S_1 = 41 - \left[\frac{(79.50)^2}{529.50} + \frac{(20.50)^2}{100.88} + \frac{(34.25)^2}{174.88} + \frac{(68.00)^2}{399.50} \right]$$

$$= 41 - 34.38 = 6.62$$

$$S_2 = 34.38 - 33.95 = 0.43.$$

The F ratio is equal to

$$F = \frac{0.43/3}{6.62/24} = 0.52$$

which is less than the tabled value, $F_{.10;3,24} = 2.33$. Thus the assumption of homogeneity of regression coefficients is tenable. If this key assumption had not been tenable, an alternative procedure, such as the Johnson and Neyman (1936) technique which does not assume homogeneity of regression coefficients, could be used. Huitema (1980, Ch. 13) provides an excellent description of the procedure. Other procedures are discussed by Robson and Atkinson (1960), Searle (1971a, Section 8.2b), and Steel and Federer (1955).

The rationale underlying the test of homogeneity of regression coefficients has been discussed by Kendall (1948, 237–242). It can be shown that S_1 plus S_2 constitute the adjusted sum of squares for experimental error,

$$S_{adj} = 6.62 + .43 = 7.05.$$

S_1 corresponds to the variation of the individual observations around the unpooled within-groups regression lines. S_2 is the variation of the p within-groups regression coefficients around the pooled within-groups regression coefficient; that is, $\Sigma(\hat{\beta}_{Wj} - \hat{\beta}_W)^2$. The larger this latter source of variation relative to S_1, the less tenable the assumption of homogeneity of the within-groups regression coefficients.

The adjusted treatment sum of squares, like the adjusted error sum of squares, can be partitioned into two components. Tests associated with these components are described by Kendall (1948, 237–242). A test of the hypothesis that the regression for the p treatment means is linear, assuming that the slopes within the groups are the same, is given by

$$F = \frac{S_3/(p-2)}{(S_1 + S_2)/(np - p - 1)}$$

with degrees of freedom equal to $p - 2$ and $np - p - 1$. S_3 is computed from

$$S_3 = A_{yy} - \frac{A_{xy}^2}{A_{xx}}.$$

An approximate test of the hypothesis that the overall regression line is linear is given by

$$F = \frac{(S_2 + S_3 + S_4)/2(p-1)}{S_1/p(n-2)}$$

with degrees of freedom equal to $2(p - 1)$ and $p(n - 2)$. The terms S_2 and S_3 have already been defined; S_4 is given by

$$S_4 = \frac{A_{xy}^2}{A_{xx}} + \frac{S_{xy}^2}{S_{xx}} - \frac{T_{xy}^2}{T_{xx}}.$$

If the relationship between X and Y is not linear, it may be possible to transform the variables so that the resulting relationship is linear. Transformations are discussed in Section 2.6. Sometimes an appropriate transformation cannot be found, in which case a polynomial ANCOVA model can be used. This approach is described by Huitema (1980, Ch. 9).

14.5 PROCEDURES FOR TESTING DIFFERENCES AMONG MEANS IN TYPE CRAC-p DESIGN

Before comparisons among means can be made, the means must be adjusted for the concomitant variable. An adjusted mean for treatment level j is given by

$$\overline{Y}_{\text{adj}\cdot j} = \overline{Y}_{\cdot j} - \hat{\beta}_W(\overline{X}_{\cdot j} - \overline{X}_{\cdot\cdot})$$

where $\hat{\beta}_W = S_{xy}/S_{xx}$ and $\overline{X}_{\cdot\cdot}$ is the mean of all X observations. Although the treatment mean square in Table 14.3-2 is not significant, these data will be used to illustrate the computation of adjusted means. Adjusted means for levels a_1 and a_2 are given by

$$\overline{Y}_{\text{adj}\cdot 1} = \frac{22}{8} - \frac{202.25}{1204.75}\left(\frac{318}{8} - \frac{1760}{32}\right) = 5.31$$

$$\overline{Y}_{\mathrm{adj}\cdot 2} = \frac{28}{8} - \frac{202.25}{1204.75}\left(\frac{353}{8} - \frac{1760}{32}\right) = 5.33.$$

The values required for the computation are contained in Table 14.3-1. The difference $\overline{Y}_{\mathrm{adj}\cdot 1} - \overline{Y}_{\mathrm{adj}\cdot 2}$ can be interpreted as the value of the contrast that would be expected if both treatment levels had the same concomitant-variable mean.

A PRIORI CONTRASTS

Hypotheses concerning a priori orthogonal contrasts can be tested using

$$t = \frac{c_j \overline{Y}_{\mathrm{adj}\cdot j} + c_{j'} \overline{Y}_{\mathrm{adj}\cdot j'}}{\sqrt{MSS_{\mathrm{adj}}\left[\dfrac{1}{n_j} + \dfrac{1}{n_{j'}} + (\overline{X}_{\cdot j} - \overline{X}_{\cdot j'})^2/S_{xx}\right]}}$$

with $np - p - 1$ degrees of freedom. The computation will be illustrated for the hypothesis $\mu_{\mathrm{adj}1} - \mu_{\mathrm{adj}2} = 0$.

$$t = \frac{(1)5.31 + (-1)5.33}{\sqrt{0.261\,[2/8 + (39.750 - 44.125)^2/1204.75]}} = \frac{-0.02}{\sqrt{0.069}} = -0.08$$

The null hypothesis for this contrast cannot be rejected.

Hypotheses for a priori nonorthogonal contrasts can be tested using either the Dunn or the Dunn-Šidák procedures. The test statistic for both procedures is

$$tD = tDS = \frac{c_1 \overline{Y}_{\mathrm{adj}\cdot 1} + c_2 \overline{Y}_{\mathrm{adj}\cdot 2} + \cdots + c_p \overline{Y}_{\mathrm{adj}\cdot p}}{\sqrt{MS_{\mathrm{error}}\left[\dfrac{c_1^2}{n_1} + \dfrac{c_2^2}{n_2} + \cdots + \dfrac{c_p^2}{n_p}\right]}}$$

with $np - p - 1$ degrees of freedom, where

$$MS_{\mathrm{error}} = MSS_{\mathrm{adj}}\left[1 + \frac{A_{xx}/(p-1)}{S_{xx}}\right].$$

MS_{error} represents the average effective error per unit (Finney, 1946b).

The preceding formula is appropriate if the p treatment levels have been randomly assigned to the experimental units. If, instead, the treatment levels have been assigned to intact groups, it is likely that the group covariate means differ. For this case, the following formula should be used with the Dunn and Dunn-Šidák procedures:

$$tD = tDS =$$

$$\frac{c_1 \overline{Y}_{\mathrm{adj}\cdot 1} + c_2 \overline{Y}_{\mathrm{adj}\cdot 2} + \cdots + c_p \overline{Y}_{\mathrm{adj}\cdot p}}{\sqrt{MSS_{\mathrm{adj}}\left[\dfrac{c_1^2}{n_1} + \dfrac{c_2^2}{n_2} + \cdots + \dfrac{c_p^2}{n_p} + \dfrac{(c_1\overline{X}_{\cdot 1} + c_2\overline{X}_{\cdot 2} + \cdots + c_p\overline{X}_{\cdot p})^2}{S_{xx}}\right]}}$$

with $np - p - 1$ degrees of freedom. This formula is a generalization of the t formula given earlier.

A POSTERIORI CONTRASTS

Tests of a posteriori contrasts using Tukey's statistic

$$q = \frac{c_j \overline{Y}_{\text{adj} \cdot j} + c_{j'} \overline{Y}_{\text{adj} \cdot j'}}{\sqrt{MS_{\text{error}}/n}}$$

where MS_{error} is as just defined, use one of two sampling distributions. This statistic is referred to the studentized range distribution if the covariate is a fixed effect and the experimental units have been randomly assigned to the p treatment levels. If as is usually the case the covariate is a random effect, the statistic is denoted by qBP and is referred to the *generalized studentized range distribution*. The resulting test is called the Bryant-Paulson procedure (Bryant and Paulson, 1976). In order to reject the null hypothesis $\mu_{\text{adj}j} - \mu_{\text{adj}j'} = 0$, qBP must exceed $q^{\circ}_{\alpha;C,p,\nu}$, where C denotes the number of covariates, p is the number of treatment levels, and ν is the degrees of freedom associated with the MS error term. Values of $q^{\circ}_{\alpha;C,p,\nu}$ are given in Appendix Table E.19.

If the covariate is a random effect but the experimental units have not been randomly assigned to the p treatment levels, the following formula should be used for each contrast.

$$qBP = \frac{c_j \overline{Y}_{\text{adj} \cdot j} + c_{j'} \overline{Y}_{\text{adj} \cdot j'}}{\sqrt{MSS_{\text{adj}} \left[\frac{2}{n} + (\overline{X}_{\cdot j} - \overline{X}_{\cdot j'})^2 / S_{xx} \right] \Big/ 2}}$$

This statistic with $np - p - 1$ degrees of freedom is also referred to the generalized studentized range distribution.

The test statistic for Scheffé's test is

$$F = \frac{[c_1 \overline{Y}_{\text{adj} \cdot 1} + c_2 \overline{Y}_{\text{adj} \cdot 2} + \cdots + c_p \overline{Y}_{\text{adj} \cdot p}]^2}{MS_{\text{error}} \left[\dfrac{c_1^2}{n_1} + \dfrac{c_2^2}{n_2} + \cdots + \dfrac{c_p^2}{n_p} \right]}.$$

The value of F that is significant is given by $F' = (p - 1)F_{\alpha;p-1,np-p-1}$. This statistic is appropriate if the p treatment levels have been randomly assigned to the experimental units. If, say, the treatment levels have been randomly assigned to intact groups and this results in unequal covariate means, the following statistic should be used.

$$F = \frac{[c_1 \overline{Y}_{\text{adj} \cdot 1} + c_2 \overline{Y}_{\text{adj} \cdot 2} + \cdots + c_p \overline{Y}_{\text{adj} \cdot p}]^2}{MSS_{\text{adj}} \left[\dfrac{c_1^2}{n_1} + \dfrac{c_2^2}{n_2} + \cdots + \dfrac{c_p^2}{n_p} + \dfrac{(c_1 \overline{X}_{\cdot 1} + c_2 \overline{X}_{\cdot 2} + \cdots + c_p \overline{X}_{\cdot p})^2}{S_{xx}} \right]}$$

The critical value for this F statistic is $F' = (p - 1)F_{\alpha;p-1,np-p-1}$.

14.6 ANALYSIS WITH TWO COVARIATES

The procedures described for one covariate can be extended to experiments containing two or more covariates. That is, if two covariates designated as X and Z are available, it is possible to adjust the dependent variable Y so as to remove the extraneous variation associated with X and Z. An example illustrating the use of two covariates is presented in this section. The computation for the two-covariate case is relatively simple. If more than two covariates are used, the computations should be performed using the regression model approach in Section 14.11.

The model equation presented in Section 14.4 for a type CRAC-p design is

$$Y_{\text{adj}ij} = Y_{ij} - \beta_W(X_{ij} - \overline{X}_{..}) = \mu + \alpha_j + \epsilon_{i(j)}.$$

This equation can be generalized to two covariates as follows:

$$Y_{\text{adj}ij} = Y_{ij} - \beta_{Wyx}(X_{ij} - \overline{X}_{..}) - \beta_{Wyz}(Z_{ij} - \overline{Z}_{..}) = \mu + \alpha_j + \epsilon_{i(j)}.$$

If the two covariates selected for measurement in the experiment represent good choices, the inclusion of a third covariate generally will not add appreciably to the precision of the experiment. The analogous situation occurs in multiple correlation in which the inclusion of additional predictors reaches a point of diminishing returns; that is, there is some point beyond which the addition of predictors to, say, a test battery does not appreciably improve prediction.

Assume that in the experiment analyzed in Section 14.3, two covariates are measured prior to the beginning of the experiment—intelligence and arithmetic achievement. These covariates are designated by the letters X and Z, respectively. The layout of the design, computational tables, and formulas are shown in Table 14.6-1. The data in Table 14.6-1 for Y and X are identical to the data in Table 14.3-1. All the computational symbols in this latter table are required in the analysis. To simplify the presentation, only the new symbols required for the analysis appear in Table 14.6-1. The analysis of the experiment is summarized in Table 14.6-2. The inclusion of a second covariate in the experiment has reduced the experimental error and provided a better estimate of treatment effects relative to the use of only one covariate. As a result, the null hypothesis can be rejected. When both intelligence and initial arithmetic achievement are taken into account, the four teaching methods produce different results with respect to the dependent variable. The assumptions for this analysis are similar to those described in Section 14.4 for the case of one covariate.

CONTRASTS AMONG MEANS

Tests of differences among means are carried out with adjusted means. An adjusted mean for treatment level j is given by

$$\overline{Y}_{\text{adj} \cdot j} = \overline{Y}_{\cdot j} - \hat{\beta}_{Wyx}(\overline{X}_{\cdot j} - \overline{X}_{..}) - \hat{\beta}_{Wyz}(\overline{Z}_{\cdot j} - \overline{Z}_{..}).$$

TABLE 14.6-1 Computational Procedures for a Type CRAC-4 Design with Two Covariates

(i) Data and notation [Y_{ij} denotes a dependent score; X_{ij} and Z_{ij} denote covariate scores for experimental unit i in treatment level j; $i = 1, \ldots, n$ experimental units (S_i); $j = 1, \ldots, p$ levels of treatment A (a_j)]:

AS Summary Table

	a_{y1}	a_{x1}	a_{z1}		a_{y2}	a_{x2}	a_{z2}		a_{y3}	a_{x3}	a_{z3}		a_{y4}	a_{x4}	a_{z4}
	3	42	3		4	47	4		7	61	5		7	65	2
	6	57	5		5	49	6		8	65	7		8	74	4
	3	33	4		4	42	5		7	64	5		9	80	5
	3	47	4		3	41	2		6	56	4		8	73	5
	1	32	0		2	38	1		5	52	2		10	85	6
	2	35	1		3	43	2		6	58	3		10	82	6
	2	33	0		4	48	5		5	53	3		9	78	5
$\sum_{i=1}^{n} Y_{ij}, \sum_{i=1}^{n} X_{ij},$	2	39	2		3	45	3		6	54	4		11	89	7
or $\sum_{i=1}^{n} Z_{ij} =$	22	318	19		28	353	28		50	463	33		72	626	40

(ii) Computational symbols:*

$$\sum_{i=1}^{n} \sum_{j=1}^{p} Z_{ij} = 3 + 5 + 4 + \cdots + 7 = 120.000$$

$$\frac{\left(\sum_{i=1}^{n} \sum_{j=1}^{p} Z_{ij}\right)^2}{np} = [Z] = \frac{(120)^2}{32} = 450.000$$

$$\sum_{i=1}^{n} \sum_{j=1}^{p} Z_{ij}^2 = [AS_z] = (3)^2 + (5)^2 + (4)^2 + \cdots + (7)^2 = 560.000$$

$$\sum_{j=1}^{p} \frac{\left(\sum_{i=1}^{n} Z_{ij}\right)^2}{n} = [A_z] = \frac{(19)^2}{8} + \frac{(28)^2}{8} + \cdots + \frac{(40)^2}{8} = 479.250$$

$$\frac{\left(\sum_{i=1}^{n} \sum_{j=1}^{p} Y_{ij}\right)\left(\sum_{i=1}^{n} \sum_{j=1}^{p} Z_{ij}\right)}{np} = [YZ] = \frac{(172)(120)}{32} = 645.000$$

$$\sum_{i=1}^{n} \sum_{j=1}^{p} Y_{ij} Z_{ij} = [AS_{yz}] = (3)(3) + (6)(5) + \cdots + (11)(7) = 768.000$$

*Only symbols associated with Z are defined in this table. See Table 14.3-1 for other required symbols.

Table 14.6-1 (continued)

$$\sum_{j=1}^{p} \frac{\left(\sum_{i=1}^{n} Y_{ij}\right)\left(\sum_{i=1}^{n} Z_{ij}\right)}{n} = [A_{yz}] = \frac{(22)(19)}{8} + \cdots + \frac{(72)(40)}{8} = 716.500$$

$$\frac{\left(\sum_{i=1}^{n}\sum_{j=1}^{p} X_{ij}\right)\left(\sum_{i=1}^{n}\sum_{j=1}^{p} Z_{ij}\right)}{np} = [XZ] = \frac{(1760)(120)}{32} = 6600.000$$

$$\sum_{i=1}^{n}\sum_{j=1}^{p} X_{ij}Z_{ij} = [AS_{xz}] = (42)(3) + (57)(5) + \cdots + (89)(7) = 7280.000$$

$$\sum_{j=1}^{p} \frac{\left(\sum_{i=1}^{n} X_{ij}\right)\left(\sum_{i=1}^{n} Z_{ij}\right)}{n} = [A_{xz}] = \frac{(318)(19)}{8} + \cdots + \frac{(626)(40)}{8} = 7030.625$$

(iii) Computational formulas:

$$T_{yy} = [AS_y] - [Y] = 235.500 \qquad S_{yy} = [AS_y] - [A_y] = 41.000$$

$$T_{xx} = [AS_x] - [X] = 8402.000 \qquad S_{xx} = [AS_x] - [A_x] = 1204.750$$

$$T_{zz} = [AS_z] - [Z] = 110.000 \qquad S_{zz} = [AS_z] - [A_z] = 80.750$$

$$T_{xy} = [AS_{xy}] - [XY] = 1380.000 \qquad S_{xy} = [AS_{xy}] - [A_{xy}] = 202.250$$

$$T_{yz} = [AS_{yz}] - [YZ] = 123.000 \qquad S_{yz} = [AS_{yz}] - [A_{yz}] = 51.500$$

$$T_{xz} = [AS_{xz}] - [XZ] = 680.000 \qquad S_{xz} = [AS_{xz}] - [A_{xz}] = 249.375$$

$$\hat{\beta}_{Tyx} = \frac{T_{zz}T_{xy} - T_{xz}T_{yz}}{T_{xx}T_{zz} - T_{xz}^2} = 0.148 \qquad \hat{\beta}_{Wyx} = \frac{S_{zz}S_{xy} - S_{xz}S_{yz}}{S_{xx}S_{zz} - S_{xz}^2} = 0.099$$

$$\hat{\beta}_{Tyz} = \frac{T_{xx}T_{yz} - T_{xz}T_{xy}}{T_{xx}T_{zz} - T_{xz}^2} = 0.206 \qquad \hat{\beta}_{Wyz} = \frac{S_{xx}S_{yz} - S_{xz}S_{xy}}{S_{xx}S_{zz} - S_{xz}^2} = 0.331$$

$$T_{adj} = T_{yy} - \hat{\beta}_{Tyx}T_{xy} - \hat{\beta}_{Tyz}T_{yz} = 5.922 \qquad S_{adj} = S_{yy} - \hat{\beta}_{Wyx}S_{xy} - \hat{\beta}_{Wyz}S_{yz} = 3.931$$

$$A_{adj} = T_{adj} - S_{adj} = 1.991$$

TABLE 14.6-2   ANCOVA Table for Type CRAC-4 Design with Two Covariates

Source	SS	df	MS	F
1 Between groups, A_{adj} (Teaching methods)	1.991	$p - 1 = 3$	0.664	$[\frac{1}{2}]$ 4.40*
2 Within groups, S_{adj}	3.931	$np - p - 2 = 26$	0.151	
3 Total$_{adj}$	5.922	$np - 3 = 29$		

*$p < .05$.

Hypotheses concerning a priori orthogonal contrasts can be tested using

$$t = \frac{c_j \overline{Y}_{\text{adj} \cdot j} + c_{j'} \overline{Y}_{\text{adj} \cdot j'}}{\sqrt{MSS_{\text{adj}} \left[\dfrac{2}{n} + \dfrac{S_{zz}(\overline{X}_{\cdot j} - \overline{X}_{\cdot j'})^2 - 2S_{xz}(\overline{X}_{\cdot j} - \overline{X}_{\cdot j'})(\overline{Z}_{\cdot j} - \overline{Z}_{\cdot j'}) + S_{xx}(\overline{Z}_{\cdot j} - \overline{Z}_{\cdot j'})^2}{S_{xx}S_{zz} - S_{xz}^2} \right]}}$$

with $np - p - 2$ degrees of freedom. A priori nonorthogonal contrasts can be tested using either the Dunn or the Dunn-Šidák procedure. The test statistic for both procedures is

$$tD = tDS = \frac{c_1 \overline{Y}_{\text{adj} \cdot 1} + c_2 \overline{Y}_{\text{adj} \cdot 2} + \cdots + c_p \overline{Y}_{\text{adj} \cdot p}}{\sqrt{MS_{\text{error}} \left[\dfrac{c_1^2}{n_1} + \dfrac{c_2^2}{n_2} + \cdots + \dfrac{c_p^2}{n_p} \right]}}$$

with $np - p - 2$ degrees of freedom, where

$$MS_{\text{error}} = MSS_{\text{adj}} \left[1 + \frac{A_{xx}S_{zz} - 2A_{xz}S_{xz} + A_{zz}S_{xx}}{(p - 1)(S_{xx}S_{zz}) - S_{xz}^2} \right].$$

A posteriori contrasts can be tested using the statistic for the Bryant-Paulson and Tukey procedures or the Scheffé statistic. The formulas are, respectively,

$$qBP = q = \frac{c_j \overline{Y}_{\text{adj} \cdot j} + c_{j'} \overline{Y}_{\text{adj} \cdot j'}}{\sqrt{MS_{\text{error}}/n}}$$

$$F = \frac{[c_1 \overline{Y}_{\text{adj} \cdot 1} + c_2 \overline{Y}_{\text{adj} \cdot 2} + \cdots + c_p \overline{Y}_{\text{adj} \cdot p}]^2}{MS_{\text{error}} \left[\dfrac{c_1^2}{n_1} + \dfrac{c_2^2}{n_2} + \cdots + \dfrac{c_p^2}{n_p} \right]}$$

with $np - p - 2$ degrees of freedom. The term MS_{error} was just defined. The value of F that is significant is equal to $F' = (p - 1)F_{\alpha; p-1, np-p-2}$. Recall from Section 14.5 that the Bryant-Paulson procedure should be used when the covariate is a random effect; Tukey's procedure is appropriate when the covariate is a fixed effect.

14.7 ANALYSIS OF COVARIANCE FOR RANDOMIZED BLOCK DESIGN

The computational procedures described for a type CRAC-p design can easily be extended to other experimental designs, such as the type RBAC-p design described here. The experimental design model equation for this design is

$$Y_{\text{adj}ij} = Y_{ij} - \beta_W(X_{ij} - \overline{X}_{..}) = \mu + \alpha_j + \pi_i + \epsilon_{ij} \quad (i = 1, \ldots, n; j = 1, \ldots, p).$$

The main features of the analysis for a type RBAC-p design are presented in Table 14.7-1. The formulas for the unadjusted sums of squares are those for a randomized block design but the adjusted formulas follow the pattern shown in Table 14.3-1 for a completely randomized analysis of covariance design. For example,

$$\hat{\beta}_W = \frac{E_{xy}}{E_{xx}}, \quad A_{yy} = [A_y] - [Y], \quad S_{yy} = [S_y] - [Y],$$

$$E_{yy} = [AS_y] - [A_y] - [S_y] + [Y]$$

where
$$[A_y] = \sum_{j=1}^{p} \frac{\left(\sum_{i=1}^{n} Y_{ij}\right)^2}{n}, \quad [S_y] = \sum_{i=1}^{n} \frac{\left(\sum_{j=1}^{p} Y_{ij}\right)^2}{p},$$

$$[Y] = \frac{\left(\sum_{i=1}^{n} \sum_{j=1}^{p} Y_{ij}\right)^2}{np}, \quad [AS_y] = \sum_{i=1}^{n} \sum_{j=1}^{p} Y_{ij}^2.$$

The procedures shown in Table 14.7-1 are analogous to the adjustment procedures for a completely randomized analysis of covariance design. This is more apparent if the adjustment procedure for a type CRAC-p design is presented in the following way:

$$A_{\text{adj}} = T_{\text{adj}} - S_{\text{adj}} = \left[T_{yy} - \frac{(T_{xy})^2}{T_{xx}}\right] - S_{\text{adj}}$$

$$= \left[(A_{yy} + S_{yy}) - \frac{(A_{xy} + S_{xy})^2}{A_{xx} + S_{xx}}\right] - S_{\text{adj}}.$$

This follows because $T_{yy} = A_{yy} + S_{yy}$, $T_{xy} = A_{xy} + S_{xy}$, and $T_{xx} = A_{xx} + S_{xx}$, where A and S designate treatment and error sums of squares, respectively. The adjustment formula must be modified for a randomized block analysis of covariance, for in this design T_{yy} includes a block sum of squares in addition to treatment and error sums of squares. We will see in subsequent sections that whenever the total sum of squares contains variation other than treatment and error sums of squares, a subtotal consisting of only these latter sources of variation must be used in the analysis of covariance.

The assumptions associated with tests of significance are those for a randomized block design (Section 6.4) and the analysis of covariance (Section 14.4). Procedures for making comparisons among adjusted treatment means are, with one exception, identical to those described in Section 14.5 for a type CRAC-p design. The proper error term MSE_{adj} must be substituted for MSS_{adj}.

Computational procedures for the case in which an observation for Y or X is missing are described by Bartlett (1936) and Federer (1955, 513–517).

TABLE 14.7-1 Computational Procedures for Type RBAC-p Design

| Source | Sum of Squares | | | df | Adjusted Sum of Squares | F |
	yy	xy	xx			
1 Treatment (A)	A_{yy}	A_{xy}	A_{xx}	$p - 1$	$A_{\text{adj}} = (A_{yy} + E_{yy}) - \dfrac{(A_{xy} + E_{xy})^2}{A_{xx} + E_{xx}} - E_{\text{adj}}$	[1/3]
2 Blocks (S)	S_{yy}	S_{xy}	S_{xx}	$n - 1$		
3 Residual (E)	E_{yy}	E_{xy}	E_{xx}	$(p - 1)(n - 1) - 1$	$E_{\text{adj}} = E_{yy} - \dfrac{(E_{xy})^2}{E_{xx}}$	
4 Total	tot_{yy}	tot_{xy}	tot_{xx}	$np - 2$		

14.8 ANALYSIS OF COVARIANCE FOR LATIN SQUARE DESIGN

Computational procedures for a type LSAC-p design are shown in Table 14.8-1. The experimental design model equation for this design is

$$Y_{\text{adj}\,jkl} = Y_{jkl} - \beta_W(X_{jkl} - \overline{X}...) = \mu + \alpha_j + \beta_k + \gamma_l + \epsilon_{jkl}$$
$$(j = 1, \ldots, p; k = 1, \ldots, p; l = 1, \ldots, p).$$

The assumptions associated with this analysis are those for a Latin square design described in Sections 7.1 and 7.5 and the analysis of covariance described in Section 14.4. Comparisons among adjusted means can be made following the procedures presented in Section 14.5 for a type CRAC-p design.

Chapter 7 describes a Latin square design that contains more than one observation per cell. The proper error term for testing treatment effects in this design is *MSWCELL* rather than *MSRES*. If more than one observation per cell is obtained in a type LSAC-p design, three terms are required in addition to those shown in Table 14.8-1. These are the within-cell sums of squares associated with Y, XY, and X. These sums of squares are used in computing an adjusted within-cell error term and in adjusting the treatment sum of squares.

Numerical examples illustrating the computation for a type LSAC-p design are given by Federer (1955, 489–495) and Snedecor and Cochran (1967).

14.9 ANALYSIS OF COVARIANCE FOR FACTORIAL EXPERIMENTS

The analysis of covariance for a factorial experiment is a straightforward generalization of the procedures discussed in connection with a completely randomized analysis of covariance design. General analysis procedures for type CRFAC-pq and SPFAC-$p \cdot q$ designs are presented in this section.

ANALYSIS OF COVARIANCE FOR TYPE CRFAC-pq DESIGN

Computational procedures for a type CRFAC-pq design appear in Table 14.9-1. The model equation for this design is

$$Y_{\text{adj}\,ijk} = Y_{ijk} - \beta_W(X_{ijk} - \overline{X}...) = \mu + \alpha_j + \beta_k + (\alpha\beta)_{jk} + \epsilon_{i(jk)}$$
$$(i = 1, \ldots, n; j = 1, \ldots, p; k = 1, \ldots, q).$$

It is apparent from this model equation that an adjusted observation estimates the

TABLE 14.8-1 Computational Procedures for Type LSAC-p Design

Source	Sum of Squares yy	xy	xx	df	Adjusted Sum of Squares	F
1 Treatment (A)	A_{yy}	A_{xy}	A_{xx}	$p - 1$	$A_{adj} = (A_{yy} + E_{yy}) - \dfrac{(A_{xy} + E_{xy})^2}{A_{xx} + E_{xx}} - E_{adj}$	[1/4]
2 Rows (B)	B_{yy}	B_{xy}	B_{xx}	$p - 1$		
3 Columns (C)	C_{yy}	C_{xy}	C_{xx}	$p - 1$		
4 Residual (E)	E_{yy}	E_{xy}	E_{xx}	$(p - 1)(p - 2) - 1$	$E_{adj} = E_{yy} - \dfrac{(E_{xy})^2}{E_{xx}}$	
5 Total	tot_{yy}	tot_{xy}	tot_{xx}	$p^2 - 2$		

TABLE 14.9-1 Computational Procedures for Type CRFAC-pq Design

Source	Sum of Squares yy	xy	xx	df	Adjusted Sum of Squares	F (Model 1)
1 A	A_{yy}	A_{xy}	A_{xx}	$p - 1$	$A_{adj} = (A_{yy} + E_{yy}) - \dfrac{(A_{xy} + E_{xy})^2}{A_{xx} + E_{xx}} - E_{adj}$	[1/4]
2 B	B_{yy}	B_{xy}	B_{xx}	$q - 1$	$B_{adj} = (B_{yy} + E_{yy}) - \dfrac{(B_{xy} + E_{xy})^2}{B_{xx} + E_{xx}} - E_{adj}$	[2/4]
3 AB	AB_{yy}	AB_{xy}	AB_{xx}	$(p - 1)(q - 1)$	$AB_{adj} = (AB_{yy} + E_{yy}) - \dfrac{(AB_{xy} + E_{xy})^2}{AB_{xx} + E_{xx}} - E_{adj}$	[3/4]
4 W. cell (E)	E_{yy}	E_{xy}	E_{xx}	$pq(n - 1) - 1$	$E_{adj} = E_{yy} - \dfrac{(E_{xy})^2}{E_{xx}}$	
5 Total	tot_{yy}	tot_{xy}	tot_{xx}	$npq - 2$		

familiar terms of the linear model equation for a type CRF-pq design. In order for the F ratios in Table 14.9-1 to be distributed as the F distribution, the assumptions of a type CRF-pq design described in Section 8.4 must be met. In addition, the following assumptions must also be tenable: (1) the jk population within-cell regression coefficients are homogeneous, (2) X is measured without error, and (3) the residuals (deviations from regression) are *NID* with mean equal to zero and common variance. A test of the hypothesis that

$$H_0: \quad \beta_{Wjk} = \beta_W \quad \text{for all } j, k$$

is given by

$$F = \frac{S_2/(pq - 1)}{S_1/pq(n - 2)}$$

where $S_1 = E_{yy} - \sum_{j=1}^{p} \sum_{k=1}^{q} \frac{(E_{xyjk})^2}{E_{xxjk}}$

$$S_2 = \sum_{j=1}^{p} \sum_{k=1}^{q} \frac{(E_{xyjk})^2}{E_{xxjk}} - \frac{(E_{xy})^2}{E_{xx}}.$$

This F ratio is distributed as the F distribution with $pq - 1$ and $pq(n - 2)$ degrees of freedom. A numerically large level of significance should be used ($\alpha = .10$ or $.25$) in order to avoid a type II error.

Hypotheses concerning a priori orthogonal contrasts can be tested using

$$t = \frac{c_j \overline{Y}_{\text{adj} \cdot j \cdot} + c_{j'} \overline{Y}_{\text{adj} \cdot j' \cdot}}{\sqrt{MSWCELL_{\text{adj}} \left[\frac{2}{nq} + \frac{(\overline{X}_{\cdot j \cdot} - \overline{X}_{\cdot j' \cdot})^2}{E_{xx}} \right]}}$$

and

$$t = \frac{c_k \overline{Y}_{\text{adj} \cdot \cdot k} + c_{k'} \overline{Y}_{\text{adj} \cdot \cdot k'}}{\sqrt{MSWCELL_{\text{adj}} \left[\frac{2}{np} + \frac{(\overline{X}_{\cdot \cdot k} - \overline{X}_{\cdot \cdot k'})^2}{E_{xx}} \right]}}.$$

The degrees of freedom for these statistics and those that follow is $pq(n - 1) - 1$.

The test statistic for simple main-effects contrasts has the form

$$t = \frac{c_{jk} \overline{Y}_{\text{adj} \cdot jk} + c_{(jk)'} \overline{Y}_{\text{adj} \cdot (jk)'}}{\sqrt{MSWCELL_{\text{adj}} \left[\frac{2}{n} + \frac{(\overline{X}_{\cdot jk} - \overline{X}_{\cdot (jk)'})^2}{E_{xx}} \right]}}.$$

Adjusted means for A_y, B_y, and AB_y are given by

$$\overline{Y}_{\text{adj} \cdot j \cdot} = \overline{Y}_{\cdot j \cdot} - \hat{\beta}_W(\overline{X}_{\cdot j \cdot} - \overline{X}_{\cdots})$$
$$\overline{Y}_{\text{adj} \cdot \cdot k} = \overline{Y}_{\cdot \cdot k} - \hat{\beta}_W(\overline{X}_{\cdot \cdot k} - \overline{X}_{\cdots})$$
$$\overline{Y}_{\text{adj} \cdot jk} = \overline{Y}_{\cdot jk} - \hat{\beta}_W(\overline{X}_{\cdot jk} - \overline{X}_{\cdots})$$

where $\hat{\beta}_W = E_{xy}/E_{xx}$. A priori nonorthogonal contrasts can be tested using either the

Dunn or Dunn-Šidák procedures. The test statistics for both procedures are as follows:

$$tD = tDS = \frac{c_1\overline{Y}_{adj\cdot1\cdot} + c_2\overline{Y}_{adj\cdot2\cdot} + \cdots + c_p\overline{Y}_{adj\cdot p\cdot}}{\sqrt{MS_{error1}\left[\dfrac{c_1^2}{nq} + \dfrac{c_2^2}{nq} + \cdots + \dfrac{c_p^2}{nq}\right]}}$$

$$tD = tDS = \frac{c_1\overline{Y}_{adj\cdot\cdot1} + c_2\overline{Y}_{adj\cdot\cdot2} + \cdots + c_q\overline{Y}_{adj\cdot\cdot q}}{\sqrt{MS_{error2}\left[\dfrac{c_1^2}{np} + \dfrac{c_2^2}{np} + \cdots + \dfrac{c_q^2}{np}\right]}}$$

$$tD = tDS = \frac{c_{11}\overline{Y}_{adj\cdot11} + c_{12}\overline{Y}_{adj\cdot12} + \cdots + c_{pq}\overline{Y}_{adj\cdot pq}}{\sqrt{MS_{error3}\left[\dfrac{c_{11}^2}{n} + \dfrac{c_{12}^2}{n} + \cdots + \dfrac{c_{pq}^2}{n}\right]}}$$

where $MS_{error1} = MSWCELL_{adj}\left[1 + \dfrac{A_{xx}/(p-1)}{E_{xx}}\right]$

$MS_{error2} = MSWCELL_{adj}\left[1 + \dfrac{B_{xx}/(q-1)}{E_{xx}}\right]$

$MS_{error3} = MSWCELL_{adj}\left[1 + \dfrac{AB_{xx}/(p-1)(q-1)}{E_{xx}}\right].$

If the treatment combinations have not been randomly assigned to the experimental units, the denominators of the Dunn and Dunn-Šidák statistics should be replaced with

$$\sqrt{MSWCELL_{adj}\left[\frac{c_1^2}{nq} + \frac{c_2^2}{nq} + \cdots + \frac{c_p^2}{nq} + \frac{(c_1\overline{X}_{\cdot1\cdot} + c_2\overline{X}_{\cdot2\cdot} + \cdots + c_p\overline{X}_{\cdot p\cdot})^2}{E_{xx}}\right]}$$

$$\sqrt{MSWCELL_{adj}\left[\frac{c_1^2}{np} + \frac{c_2^2}{np} + \cdots + \frac{c_q^2}{np} + \frac{(c_1\overline{X}_{\cdot\cdot1} + c_2\overline{X}_{\cdot\cdot2} + \cdots + c_q\overline{X}_{\cdot\cdot q})^2}{E_{xx}}\right]}$$

$$\sqrt{MSWCELL_{adj}\left[\frac{c_{11}^2}{n} + \frac{c_{12}^2}{n} + \cdots + \frac{c_{pq}^2}{n} + \frac{(c_{11}\overline{X}_{\cdot11} + c_{12}\overline{X}_{\cdot12} + \cdots + c_{pq}\overline{X}_{\cdot pq})^2}{E_{xx}}\right]}.$$

A posteriori contrasts can be tested using statistics for the Bryant-Paulson and Tukey procedures or the Scheffé statistic. If treatment combinations have been randomly assigned to the experimental units, the following formulas can be used for the Bryant-Paulson and Tukey procedures.

$$qBP = q = \frac{c_j\overline{Y}_{adj\cdot j\cdot} + c_{j'}\overline{Y}_{adj\cdot j'\cdot}}{\sqrt{MS_{error1}/nq}}$$

$$qBP = q = \frac{c_k\overline{Y}_{adj\cdot\cdot k} + c_{k'}\overline{Y}_{adj\cdot\cdot k'}}{\sqrt{MS_{error2}/np}}$$

$$qBP = q = \frac{c_{jk}\overline{Y}_{adj\cdot jk} + c_{(jk)'}\overline{Y}_{adj\cdot(jk)'}}{\sqrt{MS_{error3}/n}}$$

If the treatment combinations have not been randomly assigned to the experimental units, the denominators of these statistics should be replaced with

$$\sqrt{MSWCELL_{\text{adj}}\left[\frac{2}{nq} + (\overline{X}_{\cdot j\cdot} - \overline{X}_{\cdot j'\cdot})^2/E_{xx}\right]}\Big/2$$

$$\sqrt{MSWCELL_{\text{adj}}\left[\frac{2}{np} + (\overline{X}_{\cdot\cdot k} - \overline{X}_{\cdot\cdot k'})^2/E_{xx}\right]}\Big/2$$

$$\sqrt{MSWCELL_{\text{adj}}\left[\frac{2}{n} + (\overline{X}_{\cdot jk} - \overline{X}_{\cdot(jk)'})^2/E_{xx}\right]}\Big/2.$$

Formulas for Scheffé's statistic are equal to the square of the formulas for the Dunn and Dunn-Šidák procedures. The critical values for Scheffé's test are as follows.

$$F' = (p - 1)F_{\alpha;p-1;pq(n-1)-1}$$
$$F' = (q - 1)F_{\alpha;q-1;pq(n-1)-1}$$
$$F' = (p - 1)(q - 1)F_{\alpha;(p-1)(q-1),pq(n-1)-1}$$

The analysis of factorial experiments with unequal numbers of observations per cell has been discussed by Das (1953), Federer (1955, 515–517), and Hazel (1946). The regression model approach described in Sections 8.13 and 14.11 can be used to analyze analysis of covariance factorial experiments with unequal n's.

ANALYSIS OF COVARIANCE FOR TYPE SPFAC-$p\cdot q$ DESIGN

The collection of concomitant observations in a split-plot factorial design can take one of two forms, as shown in Figure 14.9-1. In Figure 14.9-1(a), a single covariate

FIGURE 14.9-1 Type SPFAC-2·3 designs illustrating two forms for the concomitant variable.

		b_{x1}	b_{y1}	b_{y2}	b_{y3}
a_1	s_1	X_1	Y_{11}	Y_{12}	Y_{13}
	s_2	X_2	Y_{21}	Y_{22}	Y_{23}
a_2	s_3	X_3	Y_{31}	Y_{32}	Y_{33}
	s_4	X_4	Y_{41}	Y_{42}	Y_{43}

(a)

		b_{x1}	b_{y1}	b_{x2}	b_{y2}	b_{x3}	b_{y3}
a_1	s_1	X_{11}	Y_{11}	X_{12}	Y_{12}	X_{13}	Y_{13}
	s_2	X_{21}	Y_{21}	X_{22}	Y_{22}	X_{23}	Y_{23}
a_2	s_3	X_{31}	Y_{31}	X_{32}	Y_{32}	X_{33}	Y_{33}
	s_4	X_{41}	Y_{41}	X_{42}	Y_{42}	X_{43}	Y_{43}

(b)

measure is associated with all the criterion measures for a subject. In this case it is assumed that the covariate is obtained prior to administration of any of the treatment combinations. In Figure 14.9-1(b), each criterion measure for a subject is paired with a unique covariate measure. The design in part (a) can be considered a special case of part (b), where the covariate is identical for each criterion measure. For this reason, only computational procedures for the design in part (b) are given.

If the regression for between-blocks variation, β_B, is different from the regression for within-blocks variation, β_W, the following model equation for the design in part (b) is appropriate:

(14.9-1)
$$Y_{\text{adj}ijk} = Y_{ijk} - \beta_B(\overline{X}_{\cdot jk} - \overline{X}_{\dots}) - \beta_W(X_{ijk} - \overline{X}_{\cdot jk})$$
$$= \mu + \alpha_j + \pi_{i(j)} + \beta_k + (\alpha\beta)_{jk} + (\beta\pi)_{ki(j)} + \epsilon_{ijk}.$$

If $\beta_B = \beta_W = \beta$, the model equation can be simplified as follows:

(14.9-2)
$$Y_{\text{adj}ijk} = Y_{ijk} - \beta(X_{ijk} - \overline{X}_{\dots})$$
$$= \mu + \alpha_j + \pi_{i(j)} + \beta_k + (\alpha\beta)_{jk} + (\beta\pi)_{ki(j)} + \epsilon_{ijk}.$$

Computational procedures for a type SPFAC-$p \cdot q$ design, in which experimental design model equation (14.9-1) is assumed to be appropriate, are shown in Table 14.9-2.

Adjusted means for A, B, and AB, respectively, are given by

$$\overline{Y}_{\text{adj}\cdot j\cdot} = \overline{Y}_{\cdot j\cdot} - \hat{\beta}_B(\overline{X}_{\cdot j\cdot} - \overline{X}_{\dots})$$
$$\overline{Y}_{\text{adj}\cdot\cdot k} = \overline{Y}_{\cdot\cdot k} - \hat{\beta}_W(\overline{X}_{\cdot\cdot k} - \overline{X}_{\dots})$$
$$\overline{Y}_{\text{adj}\cdot jk} = \overline{Y}_{\cdot jk} - \hat{\beta}_B(\overline{X}_{\cdot j\cdot} - \overline{X}_{\dots}) - \hat{\beta}_W(\overline{X}_{\cdot jk} - \overline{X}_{\cdot j\cdot})$$

where $\hat{\beta}_B = \dfrac{S_{xy}}{S_{xx}}$ and $\hat{\beta}_W = \dfrac{E_{xy}}{E_{xx}}$.

Formulas for carrying out comparisons among adjusted means are similar to those for the type CRFAC-pq design described previously. The proper error term, $MSBL(A)_{\text{adj}}$ or $MSB \times BL(A)_{\text{adj}}$, must be inserted in the formulas. Comparisons among simple-effects means that are at different levels of treatment A require a modified adjustment formula. The formula is described by Winer (1971, 800–801).

If it can be assumed that $\beta_B = \beta_W$, the latter regression coefficient can be used to adjust the between-blocks and within-blocks variation. The simplified model (14.9-2) described earlier is appropriate for this case. A test of the hypothesis that $\beta_B = \beta_W$ is given by

$$t' = \frac{\hat{\beta}_B - \hat{\beta}_W}{\sqrt{\dfrac{MSS_{\text{adj}}}{S_{xx}} + \dfrac{MSE_{\text{adj}}}{E_{xx}}}}.$$

Because the variances in the denominator of this t' ratio are likely to be heterogeneous, t' is not distributed as the t distribution. The sampling distribution of t' is approximately distributed as the t distribution with

TABLE 14.9-2 Computational Procedures for Type SPFAC-$p \cdot q$ Design

Source	Sum of Squares			df	Adjusted Sum of Squares	F(Model III) (A and B fixed effects; Blocks random)
	yy	xy	xx			
1 A	A_{yy}	A_{xy}	A_{xx}	$p - 1$	$A_{\text{adj}} = (A_{yy} + S_{yy}) - \dfrac{(A_{xy} + S_{xy})^2}{A_{xx} + S_{xx}} - S_{\text{adj}}$	[1/2]
2 Blocks w. A	S_{yy}	S_{xy}	S_{xx}	$p(n - 1) - 1$	$S_{\text{adj}} = S_{yy} - \dfrac{(S_{xy})^2}{S_{xx}}$	
3 B	B_{yy}	B_{xy}	B_{xx}	$q - 1$	$B_{\text{adj}} = (B_{yy} + E_{yy}) - \dfrac{(B_{xy} + E_{xy})^2}{B_{xx} + E_{xx}} - E_{\text{adj}}$	[3/5]
4 AB	AB_{yy}	AB_{xy}	AB_{xx}	$(p - 1)(q - 1)$	$AB_{\text{adj}} = (AB_{yy} + E_{yy}) - \dfrac{(AB_{xy} + E_{xy})^2}{AB_{xx} + E_{xx}} - E_{\text{adj}}$	[4/5]
5 $B \times$ blocks w. A	E_{yy}	E_{xy}	E_{xx}	$p(n - 1)(q - 1) - 1$	$E_{\text{adj}} = E_{yy} - \dfrac{(E_{xy})^2}{E_{xx}}$	
6 Total	tot_{yy}	tot_{xy}	tot_{xx}	$npq - 3$		

$$df = \frac{\left(\dfrac{MSS_{adj}}{S_{xx}} + \dfrac{MSE_{adj}}{E_{xx}}\right)^2}{\dfrac{\left(\dfrac{MSS_{adj}}{S_{xx}}\right)^2}{df_S} + \dfrac{\left(\dfrac{MSE_{adj}}{E_{xx}}\right)^2}{df_E}}$$

where df_S and df_E refer to the degrees of freedom for MSS_{adj} and MSE_{adj}, respectively. A numerically large level of significance ($\alpha = .10$ or $.25$) should be used for this test in order to avoid a type II error. If the hypothesis that $\beta_B = \beta_W$ is tenable, the adjustment procedures shown in Table 14.9-2 for between-blocks sums of squares can be modified to reflect this. The adjustment procedures for within-blocks sums of squares are unchanged. Formulas for computing adjusted sums of squares for treatment A and blocks within groups are

$$A_{adj} = A_{yy} - 2\hat{\beta}_W A_{xy} + \hat{\beta}_W^2 A_{xx}$$
$$SSBL(A)_{adj} \text{ or } (S_{adj}) = S_{yy} - 2\hat{\beta}_W S_{xy} + \hat{\beta}_W^2 S_{xx}.$$

Adjusted means for A and AB, respectively, are given by

$$\overline{Y}_{adj \cdot j \cdot} = \overline{Y}_{\cdot j \cdot} - \hat{\beta}_W (\overline{X}_{\cdot j \cdot} - \overline{X}_{\dots})$$
$$\overline{Y}_{adj \cdot jk} = \overline{Y}_{\cdot jk} - \hat{\beta}_W (\overline{X}_{\cdot jk} - \overline{X}_{\dots}).$$

The formula for computing adjusted means for treatment B is unchanged.

We noted earlier that the design in Figure 14.9-1(a), which involves a single covariate measure for all criterion measures, is a special case of the design in Figure 14.9-1(b). Only the between-blocks terms are adjusted for the covariate. This follows because the within-blocks adjustments are equal to zero. That is, B_{xx}, AB_{xx}, E_{xx}, B_{xy}, AB_{xy}, and E_{xy} equal zero.

ANALYSIS OF COVARIANCE
FOR OTHER TYPES OF DESIGNS

The general procedures involved in applying analysis of covariance to different experimental designs should be apparent to the reader. Limitations due to space prevent a discussion of all analysis of covariance designs. Federer (1955, Ch. 16) has described the application of analysis of covariance to a variety of designs.

14.10   COVARIANCE VERSUS STRATIFICATION

If concomitant variables are obtained prior to presentation of a treatment, several alternative research strategies should be considered by an experimenter. The alternative described in this chapter is to statistically remove the variation in the dependent variable that is associated with the variation in the covariate. Another alternative is to use the covariate to form homogeneous blocks by assigning subjects with the highest covariate to one block, subjects with a lower covariate to a different block, and so on. If this type of stratification is employed, the data are analyzed by means of a randomized block design in which variation associated with blocks is isolated from error variation. In general, this design strategy is preferable to analysis of covariance if an experimenter's principal interest is in reducing the experimental error rather than in removing bias from estimates of treatment effects. Note that a randomized block analysis requires somewhat less restrictive assumptions than a covariance analysis.

The analysis of covariance design assumes that the correct form of regression equation has been fitted and that the within-treatment regression coefficients are homogeneous. The analysis is greatly simplified if the relationship between Y and X is linear. A randomized block design is essentially a function-free regression scheme and is appropriate even though the relationship between Y and X is nonlinear. Of course, it is assumed that the interaction of blocks and treatment levels is zero in a randomized block design. This assumption is similar to the assumption of homogeneity of regression coefficients in a covariance analysis.

According to Cochran (1957), covariance and stratification are almost equally effective in removing the effects of an extraneous source of variation from the experimental error if the relationship between X and Y is linear. If it is possible to assign subjects to blocks so that the X values are equal within a block, the use of stratification in a randomized block design reduces the error term to approximately

$$\sigma_\epsilon^2(1 - \rho_W^2)$$

where σ_ϵ^2 is the experimental error variance when no stratification is used and ρ_W is the within-groups correlation coefficient. The corresponding reduction due to analysis of covariance was given in Section 14.3 as

$$\sigma_\epsilon^2(1 - \rho_W^2)\left(1 + \frac{1}{f_e - 2}\right).$$

Feldt (1958) found that an analysis of covariance design is more precise than a randomized block design when the true correlation between X and Y is greater than 0.6. The randomized block design is superior when the correlation is less than 0.4.

A disadvantage of a randomized block design is that stratification of subjects into homogeneous blocks may not be feasible for administrative reasons or stratification may be impossible because the covariate is not available at the beginning of the experiment.

Another alternative research strategy is to use stratification of subjects with respect to X in addition to using analysis of covariance to adjust for any residual variation associated with X. It is doubtful if the increased precision of this procedure justifies the additional computational labor. The greatest gain from the simultaneous use of blocking and covariance occurs when the two techniques are used to control two different sources of variation, X and Z, which are believed to affect the dependent variable. The randomized block and Latin square analysis of covariance designs are normally used in this manner. Thus, in a type RBAC-p design, the variation associated with X can be removed by covariance adjustment while Z is removed experimentally by forming homogeneous blocks based on this source of variation. A general discussion of the merits of stratification versus analysis of covariance appears in Cox (1957), Feldt (1958), and Finney (1957).

ANALYSIS OF DIFFERENCE SCORES

In analysis of covariance, the relationship between Y and X is estimated from the data. As we noted, the computations required for analysis of covariance can be involved and laborious. Under certain conditions, the use of difference scores in a conventional analysis of variance achieves some of the same advantages as analysis of covariance without the laborious computations. This procedure is applicable to situations in which the concomitant variable is of the same nature as the dependent variable. For example, the covariate might be an initial score X_{ij} on a test and the dependent variable might be a score Y_{ij} obtained on the test following the introduction of a treatment. The difference score is defined as

$$D_{ij} = Y_{ij} - X_{ij}$$

and reflects the change in a subject's test performance that is attributable to treatment j. The analysis of variance is performed with the difference score. The ANOVA model equation for difference scores, assuming that subjects have been randomly assigned to p treatment levels, is

$$Y_{ij} - X_{ij} = \mu + \alpha_j + \epsilon_{i(j)}.$$

If the regression of Y on X is linear and $\beta_W = 1.0$, the analysis of difference scores gives the same estimate of treatment effects as analysis of covariance. In this case, the ANOVA model for difference scores and the ANCOVA model are essentially equivalent. To illustrate, the model equation for a type CRAC-p design was given earlier as

$$Y_{ij} = \mu + \alpha_j + \beta_W(X_{ij} - \overline{X}_{..}) + \epsilon_{i(j)}.$$

If β_W is equal to one, the model can be written as

$$Y_{ij} - X_{ij} + \overline{X}_{..} = \mu + \alpha_j + \epsilon_{i(j)}$$

which except for the constant $\overline{X}_{..}$ is the same as the ANOVA model for difference scores. The use of a difference score assumes a particular form for the residual relation between Y and X. This assumption may or may not be tenable for a particular experiment. The analysis of covariance does not require an a priori assumption concerning the value of β_W but instead determines the most suitable value of β_W from the data. According to Cox (1957) and Gourlay (1953), a moderate departure of β_W from the required value of one does not result in a serious loss of precision when difference scores are used. Reichardt in Cook and Campbell (1979) provides an excellent summary of the relative merits of analysis of covariance, stratification, and difference scores.

*14.11 GENERAL LINEAR MODEL APPROACH TO TYPE CRAC-*p* DESIGN

If a computer and appropriate software packages are available, the simplest way to estimate the parameters of the covariance model equation for a type CRAC-*p* design,

$$Y_{ij} = \mu + \alpha_j + \beta_W(X_{ij} - \overline{X}_{..}) + \epsilon_{i(j)} \qquad (i = 1, \ldots, n; j = 1, \ldots, p)$$

and test interesting hypotheses about these parameters is through the regression model approach introduced in Chapter 5. Consider the type CRAC-4 design for evaluating four methods of teaching arithmetic described in Section 14.3. A qualitative regression model with $h = (p - 1) + 1 = 4$ independent variables (X_{i1}, X_{i2}, X_{i3}, and X_{i4}) can be formulated for this design as follows:

(14.11-1) $$Y_i = \beta_0 X_0 + \beta_1 X_{i1} + \beta_2 X_{i2} + \beta_3 X_{i3} + \beta_4 X_{i4} + \epsilon_i \qquad (i = 1, \ldots, N)$$

where ϵ_i is $NID(0, \sigma_\epsilon^2)$. The independent variables of the regression model are coded as follows.

$$X_{i1} = \begin{cases} 1 \text{ if an observation is in treatment level } a_1 \\ 0 \text{ otherwise} \end{cases}$$

$$X_{i2} = \begin{cases} 1 \text{ if an observation is in treatment level } a_2 \\ 0 \text{ otherwise} \end{cases}$$

$$X_{i3} = \begin{cases} 1 \text{ if an observation is in treatment level } a_3 \\ 0 \text{ otherwise} \end{cases}$$

$$X_{i4} = X_{ij} - \overline{X}_{..}$$

It can be shown, following the procedure in Section 5.4, that this coding scheme results in the following correspondence between the parameters of the regression model and those of the type CRAC-4 experimental design model.

$$\begin{aligned} \beta_0 &= \mu + \alpha_4 & \beta_3 &= \alpha_3 - \alpha_4 \\ \beta_1 &= \alpha_1 - \alpha_4 & \beta_4 &= \beta_W \\ \beta_2 &= \alpha_2 - \alpha_4 \end{aligned}$$

It follows that a test of the null hypothesis H_0: $\beta_1 = \beta_2 = \beta_3 = 0$ provides an indirect test of H_0: $\alpha_1 = \alpha_2 = \alpha_3 = \alpha_4 = 0$ in the type CRAC-4 design. The test, following the procedures in Section 5.5, involves a comparison of the full model equation (14.11-1) with the reduced model equation

---

*   This section assumes a familiarity with the matrix operations in Appendix D and Chapter 5. The reader who is interested only in the traditional approach to analysis of covariance can, without loss of continuity, omit this section.

$$Y_i = \beta_0 X_0 + \beta_4 X_{i4} + \epsilon_i \qquad (i = 1, \ldots, N).$$

The regression sum of squares reflecting the contribution of X_1, X_2, and X_3 over and above that due to including X_4 in the model is given by

$$SSR(X_1\ X_2\ X_3|X_4) = \overbrace{SSE(X_4)}^{SSE(R)} - \overbrace{SSE(X_1\ X_2\ X_3\ X_4)}^{SSE(F)}.$$

Procedures for computing $SSR(X_1\ X_2\ X_3|X_4)$ and other terms required in testing the hypothesis H_0: $\beta_1 = \beta_2 = \beta_3 = 0$ are illustrated in Table 14.11-1. The data are from Table 14.3-1. A comparison of the mean squares in part (iii) of Table 14.11-1 with those in Table 14.3-2, where the traditional computational procedures were used, reveals that both approaches lead to the same results: $MSR(X_1\ X_2\ X_3|X_4) = A_{adj}$, $MSE = S_{adj}$, and $F = 2.29$.

TEST FOR HOMOGENEITY OF WITHIN-GROUP REGRESSION COEFFICIENTS

We saw in Section 14.2 that an important assumption of ANCOVA is that

$$\beta_{W1} = \beta_{W2} = \cdots = \beta_{Wp} = \beta_W.$$

We can formulate a regression model for the data in Table 14.11-1 that allows for $p - 1 = 3$ different within-group regression coefficients as follows

(14.11-2) $$\begin{aligned} Y_i = {} & \beta_0 X_0 + \beta_1 X_{i1} + \beta_2 X_{i2} + \beta_3 X_{i3} + \beta_4 X_{i4} \\ & + \beta_5 X_{i1} X_{i4} + \beta_6 X_{i2} X_{i4} + \beta_7 X_{i3} X_{i4} + \epsilon_i \qquad (i = 1, \ldots, N) \end{aligned}$$

where ϵ_i is $NID(0, \sigma_\epsilon^2)$. This model includes three interaction-like terms, $\beta_5 X_{i1} X_{i4}$, $\beta_6 X_{i2} X_{i4}$, and $\beta_7 X_{i3} X_{i4}$, that allow for unequal within-group regression coefficients. The structural matrix for this model contains three more columns than the structural matrix in Table 14.11-1. These columns are obtained by multiplying the coefficients for X_4 by those for X_1, X_2, and X_3. This procedure was used in Table 8.13 to compute interaction terms for a type CRF-pq design. A test of the hypothesis of no interaction

$$H_0: \quad \beta_5 = \beta_6 = \beta_7 = 0$$

is equivalent to testing the hypothesis that the within-group regression coefficients are equal. To test $\beta_5 = \beta_6 = \beta_7 = 0$, we compare the fit of model (14.11-2), the full model, with model (14.11-1), now called the reduced model. The test statistic is

$$F = \frac{MSR(X_1 X_4\ X_2 X_4\ X_3 X_4 | X_1\ X_2\ X_3\ X_4)}{MSE}$$

where $MSR(X_1 X_4\ X_2 X_4\ X_3 X_4 | X_1\ X_2\ X_3\ X_4)$
$= SSE(X_1 X_4\ X_2 X_4\ X_3 X_4) - SSE(X_1\ X_2 \cdots X_3 X_4)$

with $p - 1 = 3$ and $N - 2p = 24$ degrees of freedom.

ANALYSIS WITH SEVERAL COVARIATES

The regression model can be easily modified for the case in which there are several covariates. The model for the data in Table 14.6-1, where two covariates were used, is

$$Y_i = \beta_0 X_0 + \beta_1 X_{i1} + \beta_2 X_{i2} + \beta_3 X_{i3} + \beta_4 X_{i4} + \beta_5 X_{i5} + \epsilon_i \qquad (i = 1, \ldots, N)$$

where ϵ_i is $NID(0, \sigma_\epsilon^2)$. The first four independent variables of the regression model are coded as before; the fifth independent variable is coded $X_{i5} = Z_{ij} - \overline{Z}_{..}$. The regression sum of squares reflecting the contribution of X_1, X_2, and X_3 over and above that due to including X_4 and X_5 in the model is given by

$$SSR(X_1\, X_2\, X_3 | X_4\, X_5) = \overbrace{SSE(X_4\, X_5)}^{SSE(R)} - \overbrace{SSE(X_1\, X_2 \cdots X_5)}^{SSE(F)}.$$

The computation of these terms follows the procedures shown in Table 14.11-1.

14.12 ADVANTAGES AND DISADVANTAGES OF ANALYSIS OF COVARIANCE

The major advantages of analysis of covariance are as follows.

1.   Enables an experimenter to reduce the experimental error and remove one or more potential sources of bias that are difficult or impossible to eliminate by experimental control from estimates of experimental effects.

2.   Analysis of covariance provides approximately the same reduction in experimental error as the use of stratification of subjects. Covariance analysis can be used after the data have been collected. Stratification of subjects into homogeneous blocks must be carried out at the beginning of the experiment.

The major disadvantages of analysis of covariance are as follows.

1.   Computations required are more laborious than those for a corresponding analysis of variance.

2.   Analysis of covariance requires a somewhat restrictive set of assumptions that may prove untenable in a particular research application.

3.   Computational formulas for estimating missing observations and carrying out comparisons among means for some analysis of covariance designs are relatively complex.

TABLE 14.11-1 Computational Procedures for Type CRAC-4 Design Using a Regression Model with Dummy Coding

(i) Data and basic matrices for full model ($N = 32$, $h = 5$, $n = 8$, $p = 4$):

$$\mathbf{X}_0 \atop N \times h$$

	$\mathbf{y}$ $N \times 1$	$\mathbf{x}_0$	$\mathbf{x}_1$	$\mathbf{x}_2$	$\mathbf{x}_3$	$\mathbf{x}_4$		$\mathbf{y}'$ $1 \times N$		$\mathbf{y}$ $N \times 1$	$(\mathbf{y}'\mathbf{y})$ 1×1
	3	1	1	0	0	−13				3	
	6	1	1	0	0	2		[3 6 3 $\cdots$ 11]		6	
	3	1	1	0	0	−22				3	= 1160
	3	1	1	0	0	−8				:	
a_1	1	1	1	0	0	−23				:	
	2	1	1	0	0	−20				11	
	2	1	1	0	0	−22					
	2	1	1	0	0	−16					
	4	1	0	1	0	− 8					
	5	1	0	1	0	− 6					
	4	1	0	1	0	−13					
	3	1	0	1	0	−14					
a_2	2	1	0	1	0	−17					
	3	1	0	1	0	−12					
	4	1	0	1	0	− 7					
	3	1	0	1	0	−10					
	7	1	0	0	1	6					
	8	1	0	0	1	10					
	7	1	0	0	1	9					
	6	1	0	0	1	1					
a_3	5	1	0	0	1	− 3					
	6	1	0	0	1	3					
	5	1	0	0	1	− 2					
	6	1	0	0	1	− 1					
	7	1	0	0	0	10					
	8	1	0	0	0	19					
	9	1	0	0	0	25					
	8	1	0	0	0	18					
a_4	10	1	0	0	0	30					
	10	1	0	0	0	27					
	9	1	0	0	0	23					
	11	1	0	0	0	34					

$\Sigma Y_i = 172$

$$(\mathbf{X}_0' \mathbf{X}_0) \atop h \times h = \begin{bmatrix} 32 & 8 & 8 & 8 & 0 \\ 8 & 8 & 0 & 0 & -122 \\ 8 & 0 & 8 & 0 & -87 \\ 8 & 0 & 0 & 8 & 23 \\ 0 & -122 & -87 & 23 & 8402 \end{bmatrix} \qquad (\mathbf{X}_0' \mathbf{y}) \atop h \times 1 = \begin{bmatrix} 172 \\ 22 \\ 28 \\ 50 \\ 1380 \end{bmatrix}$$

Table 14.11-1 (continued)

$$
\underset{h \times h}{(\mathbf{X}_0'\mathbf{X}_0)^{-1}} \qquad\qquad\qquad \underset{h \times 1}{(\mathbf{X}_0'\mathbf{y})} \qquad \underset{h \times 1}{\hat{\boldsymbol{\beta}}_0}
$$

$$
\begin{bmatrix}
0.5737 & -0.8680 & -0.7836 & -0.5182 & -0.0193 \\
-0.8680 & 1.4803 & 1.2155 & 0.7761 & 0.0320 \\
-0.7836 & 1.2155 & 1.2166 & 0.7021 & 0.0283 \\
-0.5182 & 0.7761 & 0.7021 & 0.5946 & 0.0169 \\
-0.0193 & 0.0320 & 0.0283 & 0.0169 & 0.0008
\end{bmatrix}
\begin{bmatrix}
172 \\ 22 \\ 28 \\ 50 \\ 1380
\end{bmatrix}
=
\begin{bmatrix}
5.0969 \\ 0.2133 \\ 0.2288 \\ 0.6705 \\ 0.1679
\end{bmatrix}
$$

$$
\underset{1 \times h}{\hat{\boldsymbol{\beta}}_0'} \qquad\qquad \underset{h \times 1}{(\mathbf{X}_0'\mathbf{y})} \qquad \underset{1 \times 1}{(\hat{\boldsymbol{\beta}}_0'\mathbf{X}_0'\mathbf{y})}
$$

$$
[5.0969 \quad \cdots \quad 0.1679]
\begin{bmatrix}
172 \\ 22 \\ 28 \\ 50 \\ 1380
\end{bmatrix}
= 1152.9533
$$

Computation of sum of squares for full model:

$$
SSTO = \mathbf{y}'\mathbf{y} - (\Sigma Y_i)^2/N = 1160 - (172)^2/32 = 235.5000
$$

$$
\begin{aligned}
SSE = SSE(X_1\,X_2\,X_3\,X_4) = \mathbf{y}'\mathbf{y} - \hat{\boldsymbol{\beta}}_0'\mathbf{X}_0'\mathbf{y} \\
= 1160.0000 - 1152.9533 = 7.0467
\end{aligned}
$$

(ii)   Matrices for reduced model; $Y_i = \beta_0 X_0 + \beta_4 X_{i4} + \epsilon_i$ ($h = 2$):

$$
\underset{h \times h}{(\mathbf{X}_1'\mathbf{X}_1)^{-1}} \qquad \underset{h \times 1}{(\mathbf{X}_1'\mathbf{y})} \qquad \underset{h \times 1}{\hat{\boldsymbol{\beta}}_1}
$$

$$
\underset{h \times h}{(\mathbf{X}_1'\mathbf{X}_1)} = \begin{bmatrix} 32 & 0 \\ 0 & 8402 \end{bmatrix}
\qquad
\begin{bmatrix} 1/32 & 0 \\ 0 & 1/8402 \end{bmatrix}
\begin{bmatrix} 172 \\ 1380 \end{bmatrix}
=
\begin{bmatrix} 5.3750 \\ 0.1642 \end{bmatrix}
$$

$$
\underset{1 \times h}{\hat{\boldsymbol{\beta}}_1'} \qquad \underset{h \times 1}{\mathbf{X}_1'\mathbf{y}} \qquad \underset{1 \times 1}{(\hat{\boldsymbol{\beta}}_1'\mathbf{X}_1'\mathbf{y})}
$$

$$
[5.3750 \quad 0.1642]
\begin{bmatrix} 172 \\ 1380 \end{bmatrix}
= 1151.1603
$$

Computation of sum of squares for reduced model:

$$
\begin{aligned}
SSE(X_4) &= \mathbf{y}'\mathbf{y} - \hat{\boldsymbol{\beta}}_1'\mathbf{X}_1'\mathbf{y} \\
&= 1160 - 1151.1603 = 8.8397 \\
SSR(X_1\,X_2\,X_3|X_4) &= SSE(X_4) - SSE(X_1\,X_2\,X_3\,X_4) \\
&= 8.8397 - 7.0467 = 1.7930
\end{aligned}
$$

(iii)   Computation of mean squares and F ratio:

$$
\begin{aligned}
MSR(X_1\,X_2\,X_3|X_4) &= SSR(X_1\,X_2\,X_3|X_4)/(p - 1) = 1.7930/(4 - 1) = 0.5977 \\
MSE &= SSE(X_1\,X_2\,X_3\,X_4)/(N - p - 1) = 7.0467/(32 - 4 - 1) = 0.2610 \\
F &= MSR(X_1\,X_2\,X_3|X_4)/MSE = 2.29
\end{aligned}
$$

14.13 REVIEW EXERCISES

1.   [14.1] For each of the following experiments, identify the (i) independent variable, (ii) dependent variable, and (iii) concomitant variable.

     †a) The effect of four diets on the weight gain of rats was investigated. Data were recorded for the amount of each diet that was consumed.

     †b) The effect of three reading programs on the reading performance of fourth-graders was investigated. A reading achievement test was administered prior to beginning the programs.

     c) The effect of four workspace arrangements on employee productivity was investigated. Production records of the employees prior to the experiment were obtained.

2.   [14.1] Analysis of covariance is sometimes used to statistically equate intact groups that are known to differ on one or more relevant variables. What interpretation problem does this pose?

3.   [14.1] In designing an ANCOVA, how can one be certain that the measurement of the concomitant variable does not include effects attributable to the treatment?

4.   [14.2] Why is $\sum_{i=1}^{n} \sum_{j=1}^{p} (Y_{ij} - \hat{Y}_{ij})^2$ referred to as a reduced sum of squares?

†5.   [14.2] Show that $T_{xy} = S_{xy} + A_{xy}$ by replacing $(X_{ij} - \bar{X}..)$ in the formula for T_{xy} with $(X_{ij} - \bar{X}._j) + (\bar{X}._j - \bar{X}..)$ and $(Y_{ij} - \bar{Y}..)$ with $(Y_{ij} - \bar{Y}._j) + (\bar{Y}._j - \bar{Y}..)$.

†6.   [14.2] Compute the following coefficients and construct figures like Figure 14.2-2, 3, and 4.

a_{y1}	a_{x1}	a_{y2}	a_{x2}	a_{y3}	a_{x3}
2	3	5	5	0	1
4	6	3	3	1	3
5	9	6	5	4	5
1	2	2	2	3	3

     a) $\hat{\beta}_{w1}$, $\hat{\beta}_{w2}$, and $\hat{\beta}_{w3}$ b) $\hat{\beta}_B$ c) $\hat{\beta}_T$

7.   [14.2] For the data in Exercise 6, compute

     a) $\hat{\beta}_w = \dfrac{S_{xx1}\hat{\beta}_{w1} + S_{xx2}\hat{\beta}_{w2} + S_{xx3}\hat{\beta}_{w3}}{S_{xx1} + S_{xx2} + S_{xx3}}$ b) $\hat{\beta}_T = \dfrac{S_{xx}\hat{\beta}_w + A_{xx}\hat{\beta}_B}{S_{xx} + A_{xx}}$.

8.   Exercise 1 in Section 4.9 described an experiment to evaluate approaches to learning problem-solving strategies. In the experiment, 30 sixth-graders were randomly assigned to one of the two approaches or a control condition. Assume that one of Wallach and Kogan's tests of associative thinking was administered to the children at the beginning of the experiment. The following data were obtained; Y denotes the dependent variable and X denotes the concomitant variable.

a_{y1}	a_{x1}	a_{y2}	a_{x2}	a_{y3}	a_{x3}
11	21	11	21	7	21
12	23	14	24	18	26
19	25	10	21	16	25

13	23	9	20	11	21
17	23	12	23	9	22
15	24	13	24	10	23
17	24	10	23	13	25
14	20	8	21	14	24
13	22	14	25	12	24
16	24	11	24	12	23

†a) [14.3] Use ANCOVA to test H_0: $\alpha_j = 0$ for all j; let $\alpha = .05$.

†b) [14.3] (i) Compare the results for ANCOVA with those for ANOVA. (ii) Does the test of associative thinking appear to be an effective concomitant variable?

†c) [14.3] (i) Compute r_B and r_W. (ii) What does the relative size of r_B and r_W tell you about the effectiveness of the concomitant variable?

†d) [14.4] Test the hypothesis that $\beta_{w1} = \beta_{w2} = \beta_{w3}$, that is, that the within-group regression coefficients are homogeneous.

†e) [14.5] Use the Bryant-Paulson statistic to determine which population means differ.

f) Prepare a "results and discussion section" appropriate for the *Journal of Counseling Psychology*.

9. Exercise 3 in Section 4.9 described an experiment to evaluate the effect of written instructions designed to maximize a subject's attention to hypnotic facilitative information. The subjects, 36 hypnotically naive male and female college students, were randomly assigned to one of four groups with nine subjects in each group. Following the assignment, the subjects took the Harvard Group Scale of Hypnotic Susceptibility. The following data were obtained; Y denotes a subject's score on the Stanford Hypnotic Susceptibility Scale, Form C, and X denotes a subject's score on the Harvard Scale.

a_{y1}	a_{x1}	a_{y2}	a_{x2}	a_{y3}	a_{x3}	a_{y4}	a_{x4}
4	6	10	11	4	4	4	5
7	8	6	8	6	7	2	4
5	8	3	5	5	7	5	6
6	7	4	6	2	4	7	8
10	9	7	9	10	8	5	6
11	10	8	10	9	9	1	4
9	10	5	8	7	7	3	3
7	7	9	10	6	5	6	6
8	8	7	9	7	5	4	5

a) [14.3] Use ANCOVA to test H_0: $\alpha_j = 0$ for all j; let $\alpha = .05$.

b) [14.3] (i) Compare the results for ANCOVA with those for ANOVA. (ii) Does the Harvard Group Scale of Hypnotic Susceptibility appear to be an effective concomitant variable?

c) [14.3] (i) Compute r_B and r_W. (ii) What does the size of these coefficients tell you about the effectiveness of the concomitant variable?

d) [14.4] Test the hypothesis of homogeneity of the within-group regression coefficients.

e) [14.5] Use the Bryant-Paulson statistic to determine which population means differ.

f) Prepare a "results and discussion section" appropriate for the *Journal of Abnormal Psychology*.

10. Exercise 7 in Section 6.12 described an experiment to determine the effect of 0, 1, 2, 3, 4, and 5 irrelevant stimuli on probability of correct response in a visual search task. During the five-second viewing time, the length of time each animal actually scanned the display was recorded. The following data were obtained; Y denotes the probability of correct response and X denotes scanning time in seconds.

	a_{y1}	a_{x1}	a_{y2}	a_{x2}	a_{y3}	a_{x3}	a_{y4}	a_{x4}	a_{y5}	a_{x5}	a_{y6}	a_{x6}
s_1	98	1.5	91	2.5	95	2.6	90	2.3	93	2.7	94	2.9
s_2	100	2.8	100	4.1	99	4.0	98	4.6	99	4.6	97	3.6
s_3	92	1.2	89	1.8	92	2.7	87	2.7	89	3.1	88	3.2
s_4	99	3.0	95	2.6	97	3.3	93	3.8	96	3.0	95	4.7
s_5	99	2.6	95	3.2	99	3.6	94	3.3	100	4.5	97	4.8
s_6	100	3.1	100	3.6	100	4.1	96	4.4	99	5.0	99	5.0

†**a)** [14.7] Use ANCOVA to test H_0: $\alpha_j = 0$ for all j; let $\alpha = .01$.

†**b)** [14.3] (i) Compute r_B for treatment A. (ii) What does this tell you about the relationship between probability of detection and scanning time? (iii) Compute r_B for blocks. (iv) Interpret this relationship.

†**c)** [14.7] (i) Compare the results for ANCOVA for treatment A with those for ANOVA. (ii) Does scanning time appear to be an effective concomitant variable?

†**d)** [14.7] (i) Compute adjusted treatment level means. (ii) Use Dunnett's statistic to determine which population means differ from the control population mean (treatment level a_1). Use a one-tailed test and let $\alpha = .01$. Compute the statistic using MSE_{adj} and remember to replace S_{xx} with E_{xx}.

e) Prepare a "results and discussion section" appropriate for *Perceptual and Motor Skills*.

11. Exercise 4 in Section 8.15 described an experiment in which college students administered a series of either painful or mild electric shocks as feedback for errors made by a white or black male confederate. Assume that at the beginning of the experiment the subjects took the Test Anxiety Scale for Adults (TASA). The following data were obtained; Y denotes the change in ratings from the pretest to the posttest on likability, intelligence, and personal adjustment of the confederate, X denotes the TASA score.

a_{y1}	a_{x1}	a_{y1}	a_{x1}	a_{y2}	a_{x2}	a_{y2}	a_{x2}
b_{y1}	b_{x1}	b_{y2}	b_{x2}	b_{y1}	b_{x1}	b_{y2}	b_{x2}
14	21	0	16	-6	15	9	19
10	18	-8	16	-2	17	-7	14
2	17	-4	15	-10	14	5	18
-2	18	8	19	-18	13	1	15
6	17	4	17	-14	15	-3	15

a_1 = mild shock b_1 = black confederate
a_2 = strong shock b_2 = white confederate

†**a)** [14.9] Use ANCOVA to test H_0: $\alpha_j = 0$ for all j, H_0: $\beta_k = 0$ for all k, and

H_0: $(\alpha\beta)_{jk} = 0$ for all j, k; let $\alpha = .05$.

†b) [14.9] Compare the results for ANCOVA with those for ANOVA.

†c) [14.3] (i) Compute r_B for treatments A and B and the AB interaction; also compute r_W. (ii) Are these correlations consistent with the results in Exercise 11(b)?

†d) [14.9] Test the hypothesis of homogeneity of within-cell regression coefficients; let $\alpha = .25$.

†e) [14.9] Compute the four adjusted and four unadjusted treatment combination means.

f) (i) Graph the AB interaction in terms of adjusted and unadjusted treatment combination means. (ii) How does the use of adjusted instead of unadjusted treatment combination means alter your interpretation of the AB interaction?

g) Prepare a "results and discussion section" appropriate for the *Journal of Experimental Social Psychology*.

12. Exercise 5 in Section 8.15 described an experiment to test the hypothesis that persons who are less physically attractive believe that they have less control over reinforcements in their lives than those more attractive. Assume that ratings by three judges of the physical attractiveness of the participants in the experiment were also obtained during the experiment. The following data were obtained; Y denotes a score on the Rotter I-E scale, X denotes the mean physical attractiveness rating of a participant.

a_{y1} a_{x1} b_{y1} b_{x1}	a_{y1} a_{x1} b_{y2} b_{x2}	a_{y1} a_{x1} b_{y3} b_{x3}	a_{y2} a_{x2} b_{y1} b_{x1}	a_{y2} a_{x2} b_{y2} b_{x2}	a_{y2} a_{x2} b_{y3} b_{x3}
9 7	8 6	14 4	10 7	6 8	14 5
10 6	8 7	9 4	9 7	12 5	13 7
9 6	12 5	11 5	6 9	9 6	10 6
13 5	7 6	6 6	15 5	9 6	13 6
5 7	8 5	10 6	10 6	10 7	12 5
8 8	4 6	10 5	11 7	8 6	16 4

a) [14.9] Use ANCOVA to test H_0: $\alpha_j = 0$ for all j, H_0: $\beta_k = 0$ for all k, and H_0: $(\alpha\beta)_{jk} = 0$ for all j, k; let $\alpha = .05$.

b) [14.9] Compare the results for ANCOVA with those for ANOVA.

c) [14.3] (i) Compute r_B for treatments A and B and the AB interaction; also compute r_W. (ii) Are these correlations consistent with the results in Exercise 12(b)?

d) [14.9] Test the hypothesis of homogeneity of within-cell regression coefficients; let $\alpha = .25$.

e) [14.9] Compute adjusted means for treatments A and B.

f) [14.9] (i) Use the Bryant-Paulson statistic to determine which population means differ for treatment B. (ii) If you did Exercise 5(e) in Section 8.15, compare the two sets of tests for treatment B. (iii) Do the results of the ANCOVA support the original research hypothesis?

g) Prepare a "results and discussion section" appropriate for *Perceptual and Motor Skills*.

†13. a) [14.11] For the data in Exercise 8 of this section, write a regression model equation.

b) [14.11] Assume that dummy coding is used and $X_{i3} = X_{ij} - \bar{X}...$ Give the correspondence between the parameters of the regression model and those of the type CRAC-3 experimental design model.

c) [14.11] Write the reduced model equation for testing H_0: $\beta_1 = \beta_2 = 0$.

d) [14.11] Use the general linear model approach to test the null hypothesis in Exercise 13(c). (This exercise requires the inversion of a 4×4 matrix. A computer with a matrix inversion program should be used.)

e) [14.11] Write full and reduced regression model equations for testing the hypothesis of homogeneity of within-group regression coefficients.

f) [14.11] Use the regression model approach to test the homogeneity of within-group regression coefficients; let $\alpha = .25$. (This exercise requires the inversion of a 6×6 matrix. A computer with a matrix inversion program should be used.)

14.   **a)** [14.11] For the data in Exercise 9 of this section, write a regression model equation.

b) [14.11] Assume that dummy coding is used and $X_{i4} = X_{ij} - \overline{X}_{...}$. Give the correspondence between the parameters of the regression model and those of the type CRAC-4 design.

c) [14.11] Write the reduced model equation for testing H_0: $\beta_1 = \beta_2 = \beta_3 = 0$.

d) [14.11] Use the general linear model approach to test the null hypothesis in Exercise 14(c). (This exercise requires the inversion of a 5×5 matrix. A computer with a matrix inversion program should be used.)

e) [14.11] Write full and reduced regression model equations for testing the hypothesis of homogeneity of within-group regression coefficients.

f) [14.11] Use the regression model approach to test the homogeneity of within-group regression coefficients; let $\alpha = .25$. (This exercise requires the inversion of a 8×8 matrix. A computer with a matrix inversion program should be used.)

A RULES OF SUMMATION

This appendix describes the summation operator Σ and some of the more important rules governing the operator. Let Y denote a numerical variable. Specific values of this variable can be identified by means of a number subscript, for example,

$$Y_1, \ Y_2, \ Y_3, \ Y_4.$$

Thus, Y_1 (read as "Y sub one") denotes the value of Y for element 1, Y_2 denotes the value of Y for element 2, and so on. A specific but unspecified value of Y is denoted by a letter subscript, for example Y_i.

Suppose that i ranges over the positive integers 1 through n

$$Y_1, \ Y_2, \ Y_3, \ \ldots, \ Y_n.$$

The sum of Y for $i = 1, \ldots, n$ can be written as

$$Y_1 + Y_2 + Y_3 + \cdots + Y_n = \sum_{i=1}^{n} Y_i.$$

The symbol Σ (capital Greek sigma) like $+$ is an operator symbol that says to perform the operation of addition. However, $+$ indicates the addition of only two values while

$\Sigma_{i=1}^{n}$ says to add n values of the variable beginning with Y_1 and ending with Y_n. In the notation $\Sigma_{i=1}^{n}$, i is called the index of summation; other terms can be defined as follows.

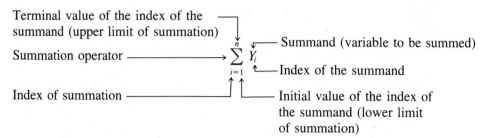

Terminal value of the index of the summand (upper limit of summation)

Summation operator

Index of summation

Summand (variable to be summed)

Index of the summand

Initial value of the index of the summand (lower limit of summation)

When the initial and terminal values of the index of the summand are obvious, the notation is often abbreviated as ΣY_i.

Consider the following matrix of scores that has $i = 1, \ldots, n$ rows, where $n = 4$, and $j = 1, \ldots, p$ columns, where $p = 3$.

	Col 1	Col 2	Col 3	$\sum\limits_{j=1}^{p} Y_{ij}$
Row 1	$Y_{11} = 4$	$Y_{12} = 5$	$Y_{13} = 1$	$\sum\limits_{j=1}^{p} Y_{1j} = 10$
Row 2	$Y_{21} = 3$	$Y_{22} = 2$	$Y_{23} = 6$	$\sum\limits_{j=1}^{p} Y_{2j} = 11$
Row 3	$Y_{31} = 1$	$Y_{32} = 8$	$Y_{33} = 2$	$\sum\limits_{j=1}^{p} Y_{3j} = 11$
Row 4	$Y_{41} = 5$	$Y_{42} = 3$	$Y_{43} = 1$	$\sum\limits_{j=1}^{p} Y_{4j} = 9$
	$\sum\limits_{i=1}^{n} Y_{i1} = 13$	$\sum\limits_{i=1}^{n} Y_{i2} = 18$	$\sum\limits_{i=1}^{n} Y_{i3} = 10$	

Two subscripts are required to identify a specific Y. It is customary to denote an unspecified row by the subscript i and an unspecified column by the subscript j. Thus, Y_{ij} denotes the value of Y in the ith row and the jth column. It is important to note the order of the subscripts because $Y_{21} = 3$ identifies a different score than $Y_{12} = 5$. The term $\Sigma_{i=1}^{n} Y_{i2}$ denotes the sum of $i = 1, \ldots, n$ scores in the second column.

$$\sum_{i=1}^{n} Y_{i2} = 5 + 2 + 8 + 3 = 18$$

The term $\Sigma_{j=1}^{p} Y_{3j}$ denotes the sum of $j = 1, \ldots, p$ scores in the third row.

$$\sum_{j=1}^{p} Y_{3j} = 1 + 8 + 2 = 11$$

The sum of all $np = (4)(3) = 12$ scores is denoted by double summation. For example, $\sum_{j=1}^{p} \sum_{i=1}^{n} Y_{ij}$ indicates that the $i = 1, \ldots, n$ scores are summed in each column and then summed over the $j = 1, \ldots, p$ columns. Thus,

$$\sum_{j=1}^{p} \sum_{i=1}^{n} Y_{ij} = (4 + 3 + 1 + 5) + (5 + 2 + 8 + 3) + (1 + 6 + 2 + 1)$$
$$= 13 + 18 + 10 = 41.$$

The innermost summation is ordinarily performed first, followed by the next summation on the left and so on. However, the order is of no great consequence since the same end result is obtained regardless of the order in which the summations are performed.

The mean of the four scores in the second column of the matrix of scores can be written as

$$\overline{Y}_{2} = \frac{1}{n} \sum_{i=1}^{n} Y_{i2} = \frac{1}{4} (5 + 2 + 8 + 3) = \frac{18}{4} = 4.5.$$

The dot in the symbol for the mean indicates that summation and averaging over the subscript replaced by the dot have occurred.

The most frequently used summation rules are described here. An understanding of these rules is necessary in order to follow the elementary derivations in this book.

RULE A.1   The sum of a constant. Let c be a constant; the sum over $i = 1, \ldots, n$ of the constant can be written as the product of the upper limit of summation, n, and c. That is

$$\sum_{i=1}^{n} c = \overbrace{c + c + \cdots + c}^{n \text{ terms}} = nc.$$

For example, let $c = 2$ and $i = 1, \ldots, 3$; then

$$\sum_{i=1}^{3} 2 = 2 + 2 + 2 = 3(2) = 6.$$

RULE A.2   The sum of a variable. Let Y be a variable with values $Y_1, Y_2, \ldots, Y_n$; the sum over $i = 1, \ldots, n$ of the variable is

$$\sum_{i=1}^{n} Y_i = Y_1 + Y_2 + \cdots + Y_n.$$

For example, let $Y_1 = 2$, $Y_2 = 3$, and $Y_3 = 4$; then

$$\sum_{i=1}^{3} Y_i = 2 + 3 + 4 = 9.$$

RULE A.3   The sum of the product of a constant and a variable. The sum $\sum_{i=1}^{n} cY_i$ can be written as the product of the constant and the sum of the variable; that is,

$$\sum_{i=1}^{n} cY_i = c \sum_{i=1}^{n} Y_i.$$

For example, let $c = 2$, and $Y_1 = 2$, $Y_2 = 3$, and $Y_3 = 4$; then

$$\sum_{i=1}^{n} cY_i = (2)(2) + (2)(3) + (2)(4)$$

$$= c \sum_{i=1}^{n} Y_i$$

$$= 2(2 + 3 + 4) = 18.$$

RULE A.4   If several operations, including summation, are to be performed on a numerical variable, the sequence of operations unless modified by the mathematical punctuation is (1) exponentiation, (2) multiplication and division, and (3) addition (summation) and subtraction. For example,

$$\sum_{i=1}^{n} Y_i^2 = Y_1^2 + Y_2^2 + \cdots + Y_n^2$$

$$\neq (Y_1 + Y_2 + \cdots + Y_n)^2$$

$$\left(\sum_{i=1}^{n} Y_i \right)^2 = (Y_1 + Y_2 + \cdots + Y_n)^2$$

$$\sum_{i=1}^{n} X_i Y_i = X_1 Y_1 + X_2 Y_2 + \cdots + X_n Y_n$$

$$\neq \sum_{i=1}^{n} X_i \sum_{i=1}^{n} Y_i.$$

RULE A.5   Distribution of summation. If the only operation to be performed before summation is addition or subtraction, the summation can be distributed among the separate terms. Let X and Y be two variables; then

$$\sum_{i=1}^{n} (X_i + Y_i) = \sum_{i=1}^{n} X_i + \sum_{i=1}^{n} Y_i.$$

For example, let $X_1 = 2$, $X_2 = 3$, $X_3 = 4$, $Y_1 = 5$, $Y_2 = 6$, and $Y_3 = 7$; then

$$\sum_{i=1}^{n} (X_i + Y_i) = (2 + 5) + (3 + 6) + (4 + 7)$$

$$= \sum_{i=1}^{n} X_i + \sum_{i=1}^{n} Y_i$$

$$= (2 + 3 + 4) + (5 + 6 + 7) = 27.$$

This rule applies to any number of terms. For example, let X, Y, and Z be variables and a, b, and c be constants; then according to Rules A.1, A.2, A.3, and A.5,

$$\sum_{i=1}^{n} (X_i + aY_i - bZ_i + c) = \sum_{i=1}^{n} X_i + a \sum_{i=1}^{n} Y_i - b \sum_{i=1}^{n} Z_i + nc.$$

RULE A.6 If one variable (say X_j) of a product of two variables ($X_j Y_{ij}$) being summed involves only the outside index of summation, this variable can be factored out of the inside summation sign. For example,

$$\sum_{j=1}^{p} \sum_{i=1}^{n} X_j Y_{ij} = \sum_{j=1}^{p} X_j \sum_{i=1}^{n} Y_{ij}.$$

This can be shown as follows:

$$\sum_{j=1}^{p} \sum_{i=1}^{n} X_j Y_{ij} = \sum_{j=1}^{p} (X_j Y_{1j} + X_j Y_{2j} + \cdots + X_j Y_{nj})$$

$$= \sum_{j=1}^{p} X_j (Y_{1j} + Y_{2j} + \cdots + Y_{nj})$$

$$= \sum_{j=1}^{p} X_j \sum_{i=1}^{n} Y_{ij}.$$

Note that

$$\sum_{j=1}^{p} X_j \sum_{i=1}^{n} Y_{ij} \neq \left(\sum_{j=1}^{p} X_j \right) \left(\sum_{i=1}^{n} Y_{ij} \right).$$

B RULES OF EXPECTATION, VARIANCE, AND COVARIANCE

Appendix A described the summation operator and rules of summation. Here we will describe the expected value operator and rules of expectation, variance, and covariance. The expected value of a random variable can be thought of as the long-run average value of the variable. More specifically, the expected value of a discrete random variable is obtained by multiplying each possible value of the variable by its associated probability and summing the product. That is,

$$E(Y) = \sum_{i=1}^{N} Y_i p(Y_i)$$

where E denotes the expected value operator and $\sum_{i=1}^{N} p(Y_i) = 1$. For a continuous random variable,

$$E(Y) = \int_{-\infty}^{\infty} Y f(Y) d(Y)$$

where $\int_{-\infty}^{\infty} f(Y) d(Y) = 1$.

The symbol E is similar to Σ in that both are instruction or operator symbols. We saw that Σ instructs us to obtain the sum of the term that follows it; E instructs us to obtain the expected value of the term that follows it. A number of rules of expectation, variance, and covariance will now be described.

Rule B.1 *Expectation of random variable.* Consider a population with mean equal to μ_Y. The expected value of the random variable Y is the mean of the population, that is,

$$E(Y) = \mu_Y.$$

Rule B.2 *Variance of random variable.* The variance of the random variable Y, $V(Y)$, can be expressed in terms of expectation:

$$\begin{aligned} V(Y) &= E[Y - E(Y)]^2 \\ &= E\{Y^2 - 2E(Y)Y + [E(Y)]^2\} \\ &= E(Y^2) - 2E(Y)E(Y) + [E(Y)]^2 \\ &= E(Y^2) - [E(Y)]^2 = \sigma_Y^2 \end{aligned}$$

where $E(Y)$ is the mean of population Y, a constant, and σ_Y^2 denotes the variance of population Y. This rule uses Rules B.3, B.4, and B.6 that follow.

Rule B.3 *Expectation of a constant.* The expected value of a constant, c, is

$$E(c) = c.$$

Rule B.4 *Expectation of the product of a constant and random variable.* If c is a constant and Y is a random variable, the expected value of cY is

$$E(cY) = cE(Y).$$

According to Rule B.1, this can be written as

$$E(cY) = c\mu_Y.$$

RULE B.5 *Expectation of a constant plus a random variable.* If c is a constant and Y is a random variable, the expected value of $(c + Y)$ is

$$E(c + Y) = c + E(Y).$$

According to Rule B.1, this can be written as

$$E(c + Y) = c + \mu_Y.$$

RULE B.6 *Expectation of sum of random variables.* If X and Y are random variables, the expected values of $(X + Y)$ and $(X - Y)$ are, respectively,

$$E(X + Y) = E(X) + E(Y)$$
$$E(X - Y) = E(X) - E(Y).$$

In other words, expectation can be distributed over an expression that has the form of a sum. This rule holds for any finite number of random variables.

RULE B.7 *Expectation of product of random variables.* If X and Y are random variables that are statistically independent, the expected value of (XY) is

$$E(XY) = E(X)E(Y).$$

If $E(XY) \neq E(X)E(Y)$, the variables are not independent.

RULE B.8   Expectation of sum of random variables from a common population.   If Y_1, $Y_2, \ldots, Y_n$ are random variables from a common population or from a population with the same mean, μ_Y, the expected value of the sum of the random variables is

$$E\left(\sum_{i=1}^{n} Y_i\right) = E(Y_1) + E(Y_2) + \cdots + E(Y_n) = n\mu_Y.$$

This rule follows from Rules B.1 and B.6.

RULE B.9   *Expectation of square of random variable.*   The expected value of the square of a random variable Y is

$$E(Y^2) = \mu_Y^2 + \sigma_Y^2.$$

This rule follows from Rule B.2. According to Rule B.2

$$E(Y^2) - [E(Y)]^2 = \sigma_Y^2$$
$$E(Y^2) = [E(Y)]^2 + \sigma_Y^2$$

but

$$E(Y) = \mu_Y. \qquad \text{(Rule B.1)}$$

Thus,

$$E(Y^2) = \mu_Y^2 + \sigma_Y^2.$$

RULE B.10.   *Expectation of sum of squared random variables from a common population.*   If $Y_1^2, Y_2^2, \ldots, Y_n^2$ are random variables from a population with mean equal to μ_Y and variance equal to σ_Y^2, the expected value of the sum of the squared random variables is

$$E\left(\sum_{i=1}^{n} Y_i^2\right) = E(Y_1^2 + Y_2^2 + \cdots + Y_n^2) = n(\mu_Y^2 + \sigma_Y^2).$$

This follows from Rules B.6 and B.9. We saw in Rule B.9 that $E(Y^2) = \mu_Y^2 + \sigma_Y^2$. Since there are n such terms, the expected value of these terms is $n(\mu_Y^2 + \sigma_Y^2)$.

RULE B.11.   *Expectation of the square of a sum of random variables from a common population.*   If $Y_1, Y_2, \ldots, Y_n$ are independent random variables from the same population with mean equal to μ_Y and variance equal to σ_Y^2, the expected value of the square of the sum of these random variables is

$$E\left(\sum_{i=1}^{n} Y_i\right)^2 = E(Y_1 + Y_2 + \cdots + Y_n)^2 = n(n\mu_Y^2 + \sigma_Y^2).$$

This can be shown as follows:

$$E(Y_1 + Y_2 + \cdots + Y_n)^2 = E[Y_1(Y_1 + Y_2 + \cdots + Y_n)$$
$$+ \, Y_2(Y_1 + Y_2 + \cdots + Y_n) + \cdots + Y_n(Y_1 + Y_2 + \cdots + Y_n)].$$

It is evident from the above expansion that there are n terms of the form Y_i^2 and $n(n-1)$ terms of the form $Y_iY_{i'}$, where $i \neq i'$. Consider first the n terms of the form Y_i^2; according to Rule B.10,

$$E(Y_1^2 + Y_2^2 + \cdots + Y_n^2) = n(\mu_Y^2 + \sigma_Y^2).$$

If the random variables Y_i and $Y_{i'}$ are independent for all i and i',

$$E(Y_iY_{i'}) = E(Y_i) \, E(Y_{i'}) \qquad \text{(Rule B.7)}$$

and

$$E(Y_i) = \mu_Y \qquad \text{and} \qquad E(Y_{i'}) = \mu_Y.$$

Thus,

$$E(Y_i) \, E(Y_{i'}) = (\mu_Y)(\mu_Y) = \mu_Y^2.$$

Because there are $n(n-1)$ terms of the form $Y_iY_{i'}$, their expected value is equal to

$$n(n-1)\mu_Y^2.$$

Combining these results we obtain

$$E\left(\sum_{i=1}^{n} Y_i\right)^2 = n(\mu_Y^2 + \sigma_Y^2) + n(n-1)\mu_Y^2$$

$$= n\mu_Y^2 + n\sigma_Y^2 + n^2\mu_Y^2 - n\mu_Y^2$$

$$= n^2 \, \mu_Y^2 + n\sigma_Y^2$$

$$= n(n \, \mu_Y^2 + \sigma_Y^2).$$

RULE B.12. *Variance of a constant.* The variance of a constant, c, is

$$V(c) = 0.$$

RULE B.13. *Variance of the product of a constant and random variable.* If c is a constant and Y is a random variable with variance equal to σ_Y^2, the variance of cY, $V(cY)$, is

$$V(cY) = c^2\sigma_Y^2.$$

This can be shown as follows. We begin by replacing Y in Rule B.2 by cY.

$$V(cY) = E(cY)^2 - [E(cY)]^2$$
$$= E(c^2Y^2) - [E(cY)]^2$$
$$= c^2E(Y^2) - [cE(Y)]^2 \qquad \text{(Rule B.4)}$$
$$= c^2\{E(Y^2) - [E(Y)]^2\}$$
$$= c^2\sigma_Y^2 \qquad \text{(Rule B.2)}$$

RULE B.14.   *Variance of a constant plus a random variable.*   If c is a constant and Y is a random variable, the variance of $(c + Y)$, $V(c + Y)$, is

$$V(c + Y) = \sigma_Y^2.$$

This can be shown as follows. We begin by replacing Y in Rule B.2 by $(c + Y)$.

$$
\begin{aligned}
V(c + Y) &= E(c + Y)^2 - [E(c + Y)]^2 \\
&= E(c^2 + 2cY + Y^2) - [c + E(Y)]^2 &\text{(Rule B.5)} \\
&= [c^2 + 2cE(Y) + E(Y^2)] \\
&\quad - \{c^2 + 2cE(Y) + [E(Y)]^2\} &\text{(Rules B.3, B.4, B.6)} \\
&= E(Y)^2 - [E(Y)]^2 = \sigma_Y^2 &\text{(Rule B.2)}
\end{aligned}
$$

RULE B.15.   *Covariance of random variables.*   If X and Y are random variables, the covariance of X and Y, $COV(X, Y)$, is

$$
\begin{aligned}
COV(X, Y) &= E\{[X - E(X)][Y - E(Y)]\} \\
&= E[XY - E(X)Y - E(Y)X + E(X)E(Y)] \\
&= E(XY) - E(X)E(Y) - E(Y)E(X) + E(X)E(Y) &\text{(Rules B.3, B.4, and B.6)} \\
&= E(XY) - E(X)E(Y).
\end{aligned}
$$

If X and Y are statistically independent, their covariance is zero. This can be shown as follows. According to Rule B.15

$$
\begin{aligned}
COV(X, Y) &= E(XY) - E(X)E(Y) \\
&= E(X)E(Y) - E(X)E(Y) &\text{(Rule B.7)} \\
&= 0.
\end{aligned}
$$

ORTHOGONAL COEFFICIENTS FOR UNEQUAL INTERVALS AND UNEQUAL n'S

C

Orthogonal coefficients for $p = 3$ to 10 treatment levels are given in Table E.12 of Appendix E. These coefficients can be used only if the treatment levels are separated by equal intervals and the sample n's are equal. This appendix describes procedures for deriving coefficients for any set of ordered treatment levels. Let us denote the numerical value of the jth treatment level by X_j. Orthogonal coefficients c_{ij} for the ith trend component are functions of X_j and have the form

$$c_{1j} = a_1 + X_j \qquad \text{(linear)}$$
$$c_{2j} = a_2 + b_2 X_j + X_j^2 \qquad \text{(quadratic)}$$
$$c_{3j} = a_3 + b_3 X_j + c_3 X_j^2 + X_j^3 \qquad \text{(cubic)}$$
$$c_{4j} = a_4 + b_4 X_j + c_4 X_j^2 + d_4 X_j^3 + X_j^4 \qquad \text{(quartic)}$$

$$\cdots\cdots\cdots\cdots\cdots\cdots\cdots\cdots\cdots\cdots\cdots\cdots\cdots\cdots$$

$$c_{p-1,j} = a_{p-1} + b_{p-1} X_j + c_{p-1} X_j^2 + d_{p-1} X_j^3 + \cdots + X_j^{p-1} \qquad (p - \text{1th power}).$$

Derivation of the coefficients c_{ij} will be illustrated for an experiment in which the numerical values of the p equal to four treatment levels are as follows: $X_1 = 0$, $X_2 = 5$, $X_3 = 10$, and $X_4 = 20$. The respective n's for the treatment levels are 5, 5, 4, and 4. For convenience, the values of X_j can be transformed by dividing each one by 5. The transformed values of X_j' are shown in column 3 of Table C.1-1. This transformation does not affect the ratios among the numerical values.

COEFFICIENTS FOR LINEAR COMPONENT

The procedure that is shown is similar to that described by Gaito (1965). Another version of the same procedure for the case of equal n's has been described by Grandage (1958). Both procedures involve solving a series of simultaneous equations. This becomes increasingly laborious for higher-order components. However, it is seldom necessary to go beyond the quartic component because the higher-order components are generally pooled in testing their significance and can be obtained in a type CR-p design from $SSBG - (SSLOWER\text{-}ORDER\ TRENDS)$. The details of the computational procedure for the linear coefficients are shown in Table C.1-1. Four equations corresponding to the four treatment levels are given in column 5 of Table C.1-1. The product of n_j times each equation is shown in column 6. According to the orthogonality condition, $\Sigma n_j c_{ij} = 0$. Thus

$$0 = 18a_1 + 29.$$

Solving for a_1 gives

$$18a_1 = -29$$
$$a_1 = -29/18.$$

The value of $a_1 = -29/18$ is substituted in the equations in column 5 to obtain column 7. The resulting coefficients for the linear component are fractions. For convenience, the fractions can be multiplied by a constant chosen so as to provide integers reduced to lowest terms. This has been done in column 8. The coefficients for the linear trend are -29, -11, 7, and 43.

COEFFICIENTS FOR QUADRATIC COMPONENT

The quadratic coefficients are derived from $c_{2j} = a_2 + b_2 X_j' + X_j'^2$. The product of n_j times each of the four equations that are based on $c_{2j} = a_2 + b_2 X_j' + X_j'^2$ are

TABLE C.1-1 Computation of Coefficients for Linear Component

(1) Treatment Level	(2) Numerical Value of Treatment Level X_j	(3) Transformed Value X_j'	(4) n_j	(5) $c_{1j} = a_1 + X_j'$	(6) $n_j c_{1j} = n_j a_1 + n_j X_j'$	(7) Linear Coefficients $c_{1j} = -29/18 + X_j'$	(8) Col(7) × 18
1	0	0	5	$c_{11} = a_1 + 0$	$5c_{11} = 5a_1 + 5(0)$	$c_{11} = -29/18 + 0 = -29/18$	−29
2	5	1	5	$c_{12} = a_1 + 1$	$5c_{12} = 5a_1 + 5(1)$	$c_{21} = -29/18 + 1 = -11/18$	−11
3	10	2	4	$c_{13} = a_1 + 2$	$4c_{13} = 4a_1 + 4(2)$	$c_{31} = -29/18 + 2 = 7/18$	7
4	20	4	4	$c_{14} = a_1 + 4$	$4c_{14} = 4a_1 + 4(4)$	$c_{41} = -29/18 + 4 = 43/18$	43

$$0 = 18a_1 + 29$$
$$a_1 = -29/18$$

$$5c_{21} = 5a_2 + 5b_2 0 + 5(0)^2$$
$$5c_{22} = 5a_2 + 5b_2 1 + 5(1)^2$$
$$4c_{23} = 4a_2 + 4b_2 2 + 4(2)^2$$
$$4c_{24} = 4a_2 + 4b_2 4 + 4(4)^2$$
$$\overline{\phantom{4c_{24} = }}$$
(C.1-1) $$0 = 18a_2 + 29b_2 + 85.$$

According to the orthogonality condition, $\Sigma n_j c_{2j} = 0$, and is so designated in equation (C.1-1). This equation contains two unknowns. The orthogonality condition also specifies that $\Sigma n_j c_{1j} c_{2j} = 0$. The value of b_2 can be determined from the four equations that result from the product of n_j, c_{1j}, and c_{2j}. These equations are

$$n_j c_{1j} c_{2j} = \quad n_j c_{1j} a_2 + \quad n_j c_{1j} b_2 X'_j + \quad n_j c_{1j} X'^2_j$$
$$5(-29)c_{21} = 5(-29)a_2 + 5(-29)b_2 0 + 5(-29)(0)^2$$
$$5(-11)c_{22} = 5(-11)a_2 + 5(-11)b_2 1 + 5(-11)(1)^2$$
$$4(7)c_{23} = 4(7)a_2 + 4(7)b_2 2 + 4(7)(2)^2$$
$$4(43)c_{24} = 4(43)a_2 + 4(43)b_2 4 + 4(43)(4)^2$$
$$\overline{\phantom{4(43)c_{24} = }}$$
$$0 = \quad 0a_2 + \quad 689b_2 + \quad 2809.$$

Solving for b_2 gives

$$689b_2 = -2809$$
$$b_2 = -2809/689 = -4.0769.$$

The value of $b_2 = -4.0769$ can be substituted in equation (C.1-1). Solving for a_2 gives

$$18a_2 = -29(-4.0769) - 85$$
$$18a_2 = 33.2301$$
$$a_2 = 1.8461.$$

The values of a_2 and b_2 can be substituted in the original equation $c_{2j} = a_2 + b_2 X'_j + X'^2_j$ to determine the quadratic coefficients. These coefficients are

$$c_{21} = 1.8461 + (-4.0769)(0) + (0)^2 = \quad 1.8461$$
$$c_{22} = 1.8461 + (-4.0769)(1) + (1)^2 = -1.2308$$
$$c_{23} = 1.8461 + (-4.0769)(2) + (2)^2 = -2.3077$$
$$c_{24} = 1.8461 + (-4.0769)(4) + (4)^2 = \quad 1.5385.$$

The calculations can be checked by determining if, in fact,

$$\Sigma n_j c_{2j} = 0 \quad \text{and} \quad \Sigma n_j c_{1j} c_{2j} = 0.$$

COEFFICIENTS FOR CUBIC AND HIGHER-ORDER COMPONENTS

Cubic coefficients are derived from

(C.1-2)
$$c_{3j} = a_3 + b_3 X_j' + c_3 X_j'^2 + X_j'^3.$$

This equation involves three unknowns a_3, b_3, and c_3. We solved for the two unknowns a_2 and b_2 in computing the quadratic coefficients by making use of the relations $\Sigma n_j c_{1j} = 0$ and $\Sigma n_j c_{1j} c_{2j} = 0$. The three unknowns in equation (C.1-2) can be calculated from equations based on the orthogonal conditions $\Sigma n_j c_{3j} = 0$, $\Sigma n_j c_{1j} c_{3j} = 0$, and $\Sigma n_j c_{2j} c_{3j} = 0$.

Quartic coefficients are derived from

$$c_{4j} = a_4 + b_4 X_j' + c_4 X_j'^2 + d_4 X_j'^3 + X_j'^4.$$

In solving for the four unknowns in this equation, the orthogonal conditions $\Sigma n_j c_{4j} = 0$, $\Sigma n_j c_{1j} c_{4j} = 0$, $\Sigma n_j c_{2j} c_{4j} = 0$, and $\Sigma n_j c_{3j} c_{4j}$ are used. This procedure can be extended to the p-1th power component.

Computation of orthogonal coefficients is simplified somewhat if the n's are equal. In this case,

$$\Sigma c_{1j} = 0, \qquad \Sigma c_{1j} c_{2j} = 0, \qquad \Sigma c_{1j} c_{3j} = 0, \qquad \text{and so on.}$$

Thus the n's can be dispensed with in all of the computations.

D MATRIX ALGEBRA

Matrix algebra is widely used in intermediate and advanced statistical methods. A familiarity with the subject is essential for understanding the general linear model approach to ANOVA, which is introduced in Chapter 5. Here we discuss only those aspects of elementary matrix algebra that are necessary for following the discussion in this book. More extensive presentations are given by Graybill (1969), Green and Carroll (1976), Hohn (1964), Searle (1966), and Timm (1975, Ch. 1).

D.1 VECTOR ALGEBRA

A *vector* is an ordered set of real numbers. The entries in a vector are called its *elements*; the number of elements is the *order* or *dimension* of the vector. We will denote vectors by lowercase English and Greek boldface letters; for example,

$$\underset{5\times 1}{\mathbf{y}} = \begin{bmatrix} 3 \\ 6 \\ 3 \\ 3 \\ 1 \end{bmatrix} \quad \underset{1\times 5}{\mathbf{y}'} = [3 \quad 6 \quad 3 \quad 3 \quad 1] \quad \underset{h\times 1}{\boldsymbol{\beta}} = \begin{bmatrix} \beta_0 \\ \beta_1 \\ \cdot \\ \cdot \\ \cdot \\ \beta_{h-1} \end{bmatrix} \quad \underset{1\times 4}{\mathbf{x}'} = [1 \quad -1 \quad 0 \quad 0].$$

A letter without a prime denotes a *column vector*, with a prime, a *row vector*. The numbers below a letter indicate the order or dimension of the vector. The first number indicates the number of rows and the second number, the number of columns. Thus, $\underset{5\times 1}{\mathbf{y}}$ is a five-element column vector; $\underset{1\times 5}{\mathbf{y}'}$ is a five-element row vector. The vector $\mathbf{y}'$ is also called the *transpose* of $\mathbf{y}$. The transpose of a vector is obtained by writing the ordered elements of a column vector as a row vector or vice versa. For example, the transpose of $\boldsymbol{\beta}$ is

$$\underset{1\times h}{\boldsymbol{\beta}'} = [\beta_0 \quad \beta_1 \quad \cdots \quad \beta_{h-1}] \quad \text{and} \quad \underset{1\times h}{(\boldsymbol{\beta}')'} = \underset{h\times 1}{\boldsymbol{\beta}} = \begin{bmatrix} \beta_0 \\ \beta_1 \\ \cdot \\ \cdot \\ \cdot \\ \beta_{h-1} \end{bmatrix}.$$

The transpose of a transposed vector is the original vector: $(\boldsymbol{\beta}')' = \boldsymbol{\beta}$.

We can think of vectors in a number of ways. So far we have thought of them as ordered *n*-tuples of numbers. We can also think of vectors as the coordinate representations of points in an *n*-dimensional geometric space. A two-element vector is a point in two-dimensional space, a three-element vector is a point in three-dimensional space, and so on. For example, let $\mathbf{v}' = [2 \quad 3]$ and $\mathbf{w}' = [2 \quad 3 \quad 3]$; these are represented as points in Figure D.1-1. We can also think of vectors as

FIGURE D.1-1 Vectors as the coordinate representations of points in a geometric space.

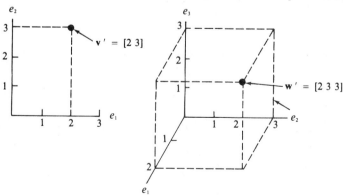

directed line segments in an *n*-dimensional geometric space. This conception of $\mathbf{v}'$ and $\mathbf{w}'$ is shown in Figure D.1-2. If, as in Figure D.1, the point of origin of a directed line segment is the origin of the coordinate system, it is called a *position vector*.

Vectors, like real numbers, obey a set of rules. We will describe some of these rules in the following sections.

FIGURE D.1-2 Vectors as directed line segments in a geometric space. Vector $\mathbf{y}'$, for example, is the line segment from the origin $[0 \quad 0]$ to the point $[2 \quad 3]$.

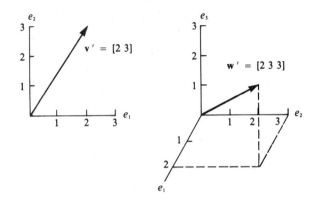

D.1-1 VECTOR EQUALITY

Two vectors of the same order are equal if and only if they have equal corresponding elements. Let

$$\mathbf{x}_{n\times 1} = \begin{bmatrix} X_1 \\ X_2 \\ \vdots \\ X_n \end{bmatrix} \quad \text{and} \quad \mathbf{y}_{n\times 1} = \begin{bmatrix} Y_1 \\ Y_2 \\ \vdots \\ Y_n \end{bmatrix}$$

then $\mathbf{x} = \mathbf{y}$ if and only if $X_i = Y_i$ ($i = 1, 2, \ldots, n$). For example,

$$\mathbf{x}_{3\times 1} = \begin{bmatrix} 3 \\ 6 \\ 1 \end{bmatrix} = \mathbf{y}_{3\times 1} = \begin{bmatrix} 3 \\ 6 \\ 1 \end{bmatrix}$$

but

$$\mathbf{x}_{3\times 1} = \begin{bmatrix} 3 \\ 6 \\ 1 \end{bmatrix} \neq \mathbf{z}_{3\times 1} = \begin{bmatrix} 3 \\ 1 \\ 6 \end{bmatrix}.$$

For two vectors to be equal, they must be of the same order, either both $n \times 1$ vectors or both $1 \times n$ vectors.

$$\mathbf{x}_{3\times 1} = \begin{bmatrix} 3 \\ 6 \\ 1 \end{bmatrix} \neq \mathbf{x}'_{1\times 3} = \begin{bmatrix} 3 & 6 & 1 \end{bmatrix}$$

D.1-2 VECTOR ADDITION AND SUBTRACTION

The addition of two or more vectors of the same order consists of adding their corresponding elements, that is, adding the first element of one vector to the first element of the other vector, and so on. Let $\mathbf{x}$ and $\mathbf{y}$ denote two $n \times 1$ vectors. Then

$$\mathbf{x}_{n\times 1} + \mathbf{y}_{n\times 1} = \begin{bmatrix} X_1 \\ X_2 \\ \cdot \\ \cdot \\ X_n \end{bmatrix} + \begin{bmatrix} Y_1 \\ Y_2 \\ \cdot \\ \cdot \\ Y_n \end{bmatrix} = \begin{bmatrix} X_1 + Y_1 \\ X_2 + Y_2 \\ \cdot \\ \cdot \\ X_n + Y_n \end{bmatrix}.$$

EXAMPLE 1 Let $\mathbf{x}' = \begin{bmatrix} 2 & 4 \end{bmatrix}$ and $\mathbf{y}' = \begin{bmatrix} 3 & 2 \end{bmatrix}$. Then

$$\mathbf{x}' + \mathbf{y}' = \begin{bmatrix} 2 & 4 \end{bmatrix} + \begin{bmatrix} 3 & 2 \end{bmatrix} = \begin{bmatrix} 2 + 3 & 4 + 2 \end{bmatrix} = \begin{bmatrix} 5 & 6 \end{bmatrix}.$$

This is shown geometrically in Figure D.1-3.

FIGURE D.1-3 Addition of $\mathbf{x}' = \begin{bmatrix} 2 & 4 \end{bmatrix}$ and $\mathbf{y}' = \begin{bmatrix} 3 & 2 \end{bmatrix}$. The sum $(\mathbf{x}' + \mathbf{y}') = \begin{bmatrix} 5 & 6 \end{bmatrix}$ is represented by the diagonal of a parallelogram determined by $\mathbf{x}'$ and $\mathbf{y}'$.

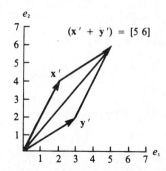

FIGURE D.1-4   Subtraction of $\mathbf{y}' = [3 \quad 2]$ from $\mathbf{x}' = [2 \quad 4]$ and vice versa. The difference vector $(\mathbf{x}' - \mathbf{y}') = [-1 \quad 2]$ is shown as a vector emanating from the origin with the same
length as that of the dashed line connecting $\mathbf{x}'$ and $\mathbf{y}'$. The direction of $(\mathbf{x}' - \mathbf{y}')$ is the same as
from $\mathbf{y}'$ to $\mathbf{x}'$. The difference vector $(\mathbf{y}' - \mathbf{x}') = [1 \quad -2]$ also begins at the origin, but its direction
is from $\mathbf{x}'$ to $\mathbf{y}'$.

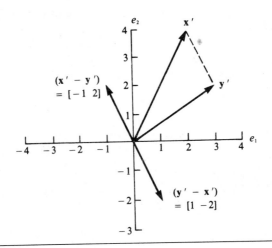

The subtraction of one vector from another of the same order consists of
changing the sign of the elements of the vector to be subtracted and proceeding as in
vector addition. Let $\mathbf{x}$ and $\mathbf{y}$ denote two $n \times 1$ vectors. Then

$$\underset{n \times 1}{\mathbf{x}} - \underset{n \times 1}{\mathbf{y}} = \begin{bmatrix} X_1 \\ X_2 \\ \vdots \\ X_n \end{bmatrix} - \begin{bmatrix} Y_1 \\ Y_2 \\ \vdots \\ Y_n \end{bmatrix} = \begin{bmatrix} X_1 + (-Y_1) \\ X_2 + (-Y_2) \\ \vdots \\ X_n + (-Y_n) \end{bmatrix}.$$

EXAMPLE 2   Let $\mathbf{x}' = [2 \quad 4]$ and $\mathbf{y}' = [3 \quad 2]$. Then

$$\mathbf{x}' - \mathbf{y}' = [2 \quad 4] - [3 \quad 2] = [2 - 3 \quad 4 - 2] = [-1 \quad 2].$$

This is shown geometrically in Figure D.1-4.

Vector addition exhibits the following properties. Let $\mathbf{x}$, $\mathbf{y}$, and $\mathbf{z}$ denote
vectors of the same order.

1. $\mathbf{x} + \mathbf{y}$ is a uniquely defined vector.

2. $(\mathbf{x} + \mathbf{y}) + \mathbf{z} = \mathbf{x} + (\mathbf{y} + \mathbf{z})$ (Associative law)

3. $\mathbf{x} + \mathbf{0} = \mathbf{x}$, where $\mathbf{0}$ is an $n \times 1$ vector of zeros called a *null* or *zero vector*.   (Identity law)

4. $\mathbf{x} + (-\mathbf{x}) = \mathbf{0}$, where $\mathbf{0}$ is an $n \times 1$ vector of zeros. (Inverse law)

5. $\mathbf{x} + \mathbf{y} = \mathbf{y} + \mathbf{x}$ (Commutative law)

D.1-3 SCALAR MULTIPLICATION OF A VECTOR

Scalar multiplication of a vector consists of multiplying each element of the vector by a scalar (real number). Let s denote a scalar and $\mathbf{x}$, an $n \times 1$ vector. Then

$$
s \underset{n \times 1}{\mathbf{x}} = s \begin{bmatrix} X_1 \\ X_2 \\ \cdot \\ \cdot \\ \cdot \\ X_n \end{bmatrix} = \begin{bmatrix} sX_1 \\ sX_2 \\ \cdot \\ \cdot \\ \cdot \\ sX_n \end{bmatrix}.
$$

EXAMPLE 1 Let $s = 2$ and $\mathbf{y} = \begin{bmatrix} 3 \\ 6 \\ 3 \\ 3 \\ 1 \end{bmatrix}$. Then

$$
s\mathbf{y} = 2\begin{bmatrix} 3 \\ 6 \\ 3 \\ 3 \\ 1 \end{bmatrix} = \begin{bmatrix} 2 \times 3 \\ 2 \times 6 \\ 2 \times 3 \\ 2 \times 3 \\ 2 \times 1 \end{bmatrix} = \begin{bmatrix} 6 \\ 12 \\ 6 \\ 6 \\ 2 \end{bmatrix}.
$$

EXAMPLE 2 Let $s = 2$ and $\mathbf{z}' = [2 \quad 3]$. Then

$$
s\mathbf{z}' = 2[2 \quad 3] = [2 \times 2 \quad 2 \times 3] = [4 \quad 6].
$$

This is shown geometrically in Figure D.1-5.

FIGURE D.1-5 Multiplication of a vector $\mathbf{z}' = [2 \quad 3]$ by a scalar $s = 2$.

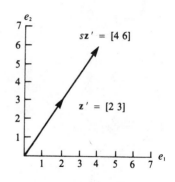

Scalar multiplication of a vector exhibits the following properties. Let $\mathbf{x}$ and $\mathbf{y}$ denote vectors of the same order and s and t, scalars.

1. $s\mathbf{x}$ is a uniquely defined vector.

2. $s(t\mathbf{x}) = (st)\mathbf{x}$ (Associative law)

3. $s\mathbf{x} = \mathbf{x}s$ (Commutative law)

4. $s(\mathbf{x} + \mathbf{y}) = s\mathbf{x} + s\mathbf{y}$ (Distributive law for vectors)

5. $(s + t)\mathbf{x} = s\mathbf{x} + t\mathbf{x}$ (Distributive law for scalars)

D.1-4 LINEAR COMBINATIONS OF VECTORS

A vector $\mathbf{x}_0$ of the form

$$\mathbf{x}_0 = s_1\mathbf{x}_1 + s_2\mathbf{x}_2 + \cdots + s_p\mathbf{x}_p$$

is called a *linear combination* of the vectors $\mathbf{x}_1, \mathbf{x}_2, \ldots, \mathbf{x}_p$, where $s_1, s_2, \ldots, s_p$ are arbitrary scalars. The subscripts $0, 1, 2, \ldots$ are used to distinguish the different vectors and different scalars.

EXAMPLE 1 $\mathbf{z} = \begin{bmatrix} 13 \\ 14 \end{bmatrix}$ is a linear combination of $\mathbf{x} = \begin{bmatrix} 2 \\ 4 \end{bmatrix}$ and $\mathbf{y} = \begin{bmatrix} 3 \\ 2 \end{bmatrix}$ because

$$\begin{bmatrix} 13 \\ 14 \end{bmatrix} = 2\begin{bmatrix} 2 \\ 4 \end{bmatrix} + 3\begin{bmatrix} 3 \\ 2 \end{bmatrix}.$$

EXAMPLE 2 $\mathbf{x}_0$, as defined here, is a linear combination of $\mathbf{x}_1, \mathbf{x}_2, \mathbf{x}_3$, and $\mathbf{x}_4$ because

$$
\begin{matrix} \mathbf{x}_0 \\ \begin{bmatrix} 1 \\ 1 \\ 1 \\ 1 \\ 1 \\ 1 \\ 1 \\ 1 \end{bmatrix} \end{matrix}
= 1
\begin{matrix} \mathbf{x}_1 \\ \begin{bmatrix} 1 \\ 1 \\ 0 \\ 0 \\ 0 \\ 0 \\ 0 \\ 0 \end{bmatrix} \end{matrix}
+ 1
\begin{matrix} \mathbf{x}_2 \\ \begin{bmatrix} 0 \\ 0 \\ 1 \\ 1 \\ 0 \\ 0 \\ 0 \\ 0 \end{bmatrix} \end{matrix}
+ 1
\begin{matrix} \mathbf{x}_3 \\ \begin{bmatrix} 0 \\ 0 \\ 0 \\ 0 \\ 1 \\ 1 \\ 1 \\ 0 \end{bmatrix} \end{matrix}
+ 1
\begin{matrix} \mathbf{x}_4 \\ \begin{bmatrix} 0 \\ 0 \\ 0 \\ 0 \\ 0 \\ 0 \\ 1 \\ 1 \end{bmatrix} \end{matrix}.
$$

D.1-5 LINEAR INDEPENDENCE

The vectors $\mathbf{x}_1, \mathbf{x}_2, \ldots, \mathbf{x}_n$ of the same order are *linearly dependent* if there exists scalars $s_1, s_2, \ldots, s_n$ not all zero such that

$$s_1\mathbf{x}_1 + s_2\mathbf{x}_2 + \cdots + s_n\mathbf{x}_n = \mathbf{0}.$$

If the preceding equation is satisfied only when all the scalars are zero, the vectors are said to be *linearly independent*.

EXAMPLE 1 Vectors x_0, x_1, x_2, and x_3, as defined here, are linearly dependent since nonzero scalars can be found such that the linear combination of vectors is equal to a null vector.

$$
1\begin{bmatrix} 1 \\ 1 \\ 1 \\ 1 \\ 1 \\ 1 \end{bmatrix}
- 1\begin{bmatrix} 1 \\ 1 \\ 0 \\ 0 \\ 0 \\ 0 \end{bmatrix}
- 1\begin{bmatrix} 0 \\ 0 \\ 1 \\ 1 \\ 0 \\ 0 \end{bmatrix}
- 1\begin{bmatrix} 0 \\ 0 \\ 0 \\ 0 \\ 1 \\ 1 \end{bmatrix}
= \begin{bmatrix} 0 \\ 0 \\ 0 \\ 0 \\ 0 \\ 0 \end{bmatrix}
$$

$$
\begin{matrix} x_0 & x_1 & x_2 & x_3 & 0 \end{matrix}
$$

EXAMPLE 2 Vectors $z' = [13 \quad 14]$, $x' = [2 \quad 4]$, and $y' = [3 \quad 2]$ are also linearly dependent.

$$
\begin{matrix} z' & x' & y' & 0' \end{matrix}
$$
$$
1[13 \quad 14] - 2[2 \quad 4] - 3[3 \quad 2] = [0 \quad 0]
$$

EXAMPLE 3 Vectors z_1, z_2, and z_3, as defined here, are linearly independent

$$
s_1\begin{bmatrix} 1 \\ 0 \\ 0 \end{bmatrix}
+ s_2\begin{bmatrix} 0 \\ 1 \\ 0 \end{bmatrix}
+ s_3\begin{bmatrix} 0 \\ 0 \\ 1 \end{bmatrix}
= \begin{bmatrix} s_1 \\ s_2 \\ s_3 \end{bmatrix}
= \begin{bmatrix} 0 \\ 0 \\ 0 \end{bmatrix}
$$

$$
\begin{matrix} z_1 & z_2 & z_3 & s & 0 \end{matrix}
$$

if and only if $s_1 = s_2 = s_3 = 0$. The vectors are independent because the only value of s_1, s_2, and s_3 for which the linear combination is equal to 0 is $s_1 = s_2 = s_3 = 0$.

EXAMPLE 4 Vectors x_0, x_1, and x_2, as defined here, are also linearly independent since a linear combination of the vectors equals zero only when all the scalars are zero.

$$
0\begin{bmatrix} 1 \\ 1 \\ 1 \\ 1 \\ 1 \\ 1 \end{bmatrix}
+ 0\begin{bmatrix} 1 \\ 1 \\ 0 \\ 0 \\ 0 \\ 0 \end{bmatrix}
+ 0\begin{bmatrix} 0 \\ 0 \\ 1 \\ 1 \\ 0 \\ 0 \end{bmatrix}
= \begin{bmatrix} 0 \\ 0 \\ 0 \\ 0 \\ 0 \\ 0 \end{bmatrix}
$$

$$
\begin{matrix} x_0 & x_1 & x_2 & 0 \end{matrix}
$$

D.1-6 INNER PRODUCT OF TWO VECTORS

The *inner product* (*scalar, minor,* or *dot product*) of two vectors **x** and **y** of the same dimension is the scalar quantity

$$\underset{1 \times n}{\mathbf{x'}} \underset{n \times 1}{\mathbf{y}} = [X_1 \quad X_2 \quad \cdots \quad X_n] \begin{bmatrix} Y_1 \\ Y_2 \\ \cdot \\ \cdot \\ \cdot \\ Y_n \end{bmatrix} = X_1 Y_1 + X_2 Y_2 + \cdots + X_n Y_n$$

$$= \sum_{i=1}^{n} X_i Y_i.$$

The inner product is also written **xy**, where it is understood that the first vector (premultiplying vector) is treated as a row vector and the second vector (post-multiplying vector), as a column vector.

EXAMPLE 1   Let $\mathbf{x'} = [2 \quad 4]$ and $\mathbf{y'} = [3 \quad 2]$. Then

$$\underset{1 \times 2}{\mathbf{x'}} \underset{2 \times 1}{\mathbf{y}} = [2 \quad 4] \begin{bmatrix} 3 \\ 2 \end{bmatrix} = (2 \times 3) + (4 \times 2) = 6 + 8 = 14.$$

Note: As will become evident in Section D.2-4,

$$\mathbf{xy'} = \begin{bmatrix} 2 \\ 4 \end{bmatrix} [3 \quad 2] = \begin{bmatrix} 6 & 4 \\ 12 & 8 \end{bmatrix}.$$

This product is called the *outer product* (*matrix* or *major product*). Unless indicated otherwise, **xy** refers to the inner product, **x'y**, rather than to **xy'**.

EXAMPLE 2   Let $\mathbf{y'} = [3 \quad 6 \quad 3 \quad 3 \quad 1]$. Then

$$\mathbf{y'y} = [3 \quad 6 \quad 3 \quad 3 \quad 1] \begin{bmatrix} 3 \\ 6 \\ 3 \\ 3 \\ 1 \end{bmatrix}$$

$$= (3 \times 3) + (6 \times 6) + (3 \times 3) + (3 \times 3) + (1 \times 1)$$

$$= 9 + 36 + 9 + 9 + 1 = 64.$$

Note that $\underset{1 \times n}{\mathbf{y'}} \underset{n \times 1}{\mathbf{y}} = \sum_{i=1}^{n} Y_i Y_i = \sum_{i=1}^{n} Y_i^2$. Premultiplying a vector by its transpose gives the sum of the squared elements (sum of squares).

EXAMPLE 3 Let $\mathbf{1}' = \begin{bmatrix} 1 & 1 & 1 & 1 & 1 \end{bmatrix}$ and $\mathbf{y}' = \begin{bmatrix} 3 & 6 & 3 & 3 & 1 \end{bmatrix}$. Then

$$\underset{1\times 5}{\mathbf{1}'} \ \underset{5\times 1}{\mathbf{y}} = \begin{bmatrix} 1 & 1 & 1 & 1 & 1 \end{bmatrix} \begin{bmatrix} 3 \\ 6 \\ 3 \\ 3 \\ 1 \end{bmatrix} = 3 + 6 + 3 + 3 + 1 = 16.$$

The vector $\mathbf{1}$ is called a *sum vector*. The inner product of $\mathbf{1}'$ and $\mathbf{y}$, where $\mathbf{1}'$ is a $1 \times n$ vector and $\mathbf{y}$ is an $n \times 1$ vector, is equal to $\sum_{i=1}^{n} Y_i$, the sum of the elements of $\mathbf{y}$.

Inner products exhibit the following properties. Let $\mathbf{x}$, $\mathbf{y}$, and $\mathbf{z}$ denote vectors of the same dimension and s and t, scalars.

1. $(s\mathbf{x}')(t\mathbf{y}) = st(\mathbf{x}'\mathbf{y})$ (Associative law)
2. $\mathbf{x}'\mathbf{y} = \mathbf{y}'\mathbf{x}$ (Commutative law)
3. $\mathbf{x}'(\mathbf{y} + \mathbf{z}) = \mathbf{x}'\mathbf{y} + \mathbf{x}'\mathbf{z}$ (Distributive law)
4. $\mathbf{x}'\mathbf{x} \geq 0$; $\mathbf{x}'\mathbf{x} = 0$, implies that $\mathbf{x} = \mathbf{0}$

D.1-7 LENGTH OF A VECTOR AND DISTANCE BETWEEN VECTORS

The length of an $n \times 1$ vector $\mathbf{x}$, denoted by $\|\mathbf{x}\|$, is defined as

$$\|\mathbf{x}\| = (\mathbf{x}'\mathbf{x})^{1/2} = \left(\sum_{i=1}^{n} X_i^2 \right)^{1/2}.$$

EXAMPLE 1 Let $\mathbf{v}' = \begin{bmatrix} 3 & 2 \end{bmatrix}$. Then

$$\|\mathbf{v}\| = \left\{ \begin{bmatrix} 3 & 2 \end{bmatrix} \begin{bmatrix} 3 \\ 2 \end{bmatrix} \right\}^{1/2} = [(3)^2 + (2)^2]^{1/2} = (13)^{1/2} = 3.6.$$

This is shown geometrically in Figure D.1-6.

FIGURE D.1-6 The length of $\mathbf{v}' = \begin{bmatrix} 3 & 2 \end{bmatrix}$, denoted by $\|\mathbf{v}\|$, is the distance from the origin $\mathbf{0}'$ to $\mathbf{v}'$.

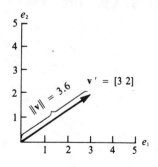

FIGURE D.1-7   The distance between vectors $\mathbf{v}' = \begin{bmatrix} 5 & 2 \end{bmatrix}$ and $\mathbf{w}' = \begin{bmatrix} 3 & 3 \end{bmatrix}$, denoted by $\|\mathbf{v} - \mathbf{w}\|$, is given by $[(2)^2 + (-1)^2]^{1/2}$. This follows from the Pythagorean theorem, which states that in a right triangle the square of the hypotenuse ($\|\mathbf{v} - \mathbf{w}\|^2$ in our example) is equal to the sum of the squares of the other two sides.

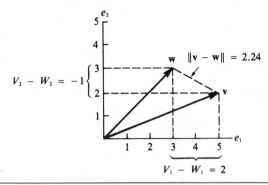

The distance between two $n \times 1$ vectors $\mathbf{x}$ and $\mathbf{y}$, denoted by $\|\mathbf{x} - \mathbf{y}\|$, is given by

$$\|\mathbf{x} - \mathbf{y}\| = \{[\mathbf{x} - \mathbf{y}]'[\mathbf{x} - \mathbf{y}]\}^{1/2} = \left[\sum_{i=1}^{n} (X_i - Y_i)^2 \right]^{1/2}.$$

EXAMPLE 2   Let $\mathbf{v}' = \begin{bmatrix} 5 & 2 \end{bmatrix}$ and $\mathbf{w}' = \begin{bmatrix} 3 & 3 \end{bmatrix}$. Then

$$\|\mathbf{v} - \mathbf{w}\| = \left\{ \begin{bmatrix} 5 - 3 & 2 - 3 \end{bmatrix} \begin{bmatrix} 5 - 3 \\ 2 - 3 \end{bmatrix} \right\}^{1/2} = [(2)^2 + (-1)^2]^{1/2} = (5)^{1/2} = 2.24.$$

This is shown geometrically in Figure D.1-7.

D.1-8 ORTHOGONAL AND ORTHONORMAL VECTORS

Two nonnull vectors $\mathbf{x}$ and $\mathbf{y}$ of the same order are *orthogonal*, denoted by $\mathbf{x} \perp \mathbf{y}$, if and only if their inner product is zero; that is,

$$\mathbf{x} \perp \mathbf{y} \quad \text{if and only if} \quad \mathbf{x}'\mathbf{y} = \sum_{i=1}^{n} X_i Y_i = 0.$$

EXAMPLE 1   Let $\mathbf{v}' = \begin{bmatrix} 1 & 2 \end{bmatrix}$ and $\mathbf{w}' = \begin{bmatrix} -2 & 1 \end{bmatrix}$. Then

$$\mathbf{v}'\mathbf{w} = \begin{bmatrix} 1 & 2 \end{bmatrix} \begin{bmatrix} -2 \\ 1 \end{bmatrix} = (1 \times -2) + (2 \times 1) = -2 + 2 = 0$$

which means that the vectors are orthogonal. This is shown geometrically in Figure D.1-8.

FIGURE D.1-8 The angle θ between two orthogonal vectors is always equal to 90°. In other words, the interior angle is always a right angle.

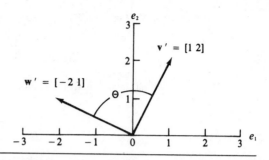

EXAMPLE 2 Let $\mathbf{v}' = [1 \quad -1 \quad 0 \quad 0]$ and $\mathbf{w}' = [0 \quad 0 \quad 1 \quad -1]$. Then

$$\mathbf{v}'\mathbf{w} = [1 \quad -1 \quad 0 \quad 0]\begin{bmatrix} 0 \\ 0 \\ 1 \\ -1 \end{bmatrix} = 0 + 0 + 0 + 0 = 0$$

which means that the vectors are orthogonal.

EXAMPLE 3 Let $\mathbf{v}' = [1 \quad -1 \quad 0 \quad 0]$ and $\mathbf{w}' = [0 \quad 1 \quad -1 \quad 0]$. Then

$$\mathbf{v}'\mathbf{w} = [1 \quad -1 \quad 0 \quad 0]\begin{bmatrix} 0 \\ 1 \\ -1 \\ 0 \end{bmatrix} = 0 - 1 + 0 + 0 = -1$$

which means that the vectors are not orthogonal.

Two nonnull vectors $\mathbf{x}$ and $\mathbf{y}$ of the same order are *orthonormal* if and only if their inner product is zero and each vector has unit length; that is

$$\mathbf{x}^\star \perp \mathbf{y}^\star \qquad \text{if and only if} \qquad \mathbf{x}'\mathbf{y} = 0 \qquad \text{and} \qquad \|\mathbf{x}\| = \|\mathbf{y}\| = 1$$

where * indicates that the vector has unit length. A vector for which $\|\mathbf{x}\| \neq 1$ can be made to have unit length, *normalized,* by multiplying each of its elements by $1/\|\mathbf{x}\|$.

EXAMPLE 4 Let $\mathbf{v}' = [4 \quad 3]$. Then

$$\mathbf{v}^\star = \frac{1}{\|\mathbf{v}\|}\mathbf{v} = \frac{1}{5}\begin{bmatrix} 4 \\ 3 \end{bmatrix} = \begin{bmatrix} 4/5 \\ 3/5 \end{bmatrix}.$$

We can verify that $\mathbf{v}^\star$ is a normalized vector as follows.

$$\|\mathbf{v}^\star\| = \left\{[4/5 \quad 3/5]\begin{bmatrix} 4/5 \\ 3/5 \end{bmatrix}\right\}^{1/2} = [(4/5)^2 + (3/5)^2]^{1/2}$$
$$= (25/25)^{1/2} = 1$$

EXAMPLE 5   Let $\mathbf{v}' = [0 \quad 1]$ and $\mathbf{w}' = [-1 \quad 0]$. Then

$$\mathbf{v}'\mathbf{w} = [0 \quad 1]\begin{bmatrix} -1 \\ 0 \end{bmatrix} = 0, \quad \|\mathbf{v}\| = 1, \quad \text{and } \|\mathbf{w}\| = 1$$

which means that the vectors are orthonormal.

D.2   MATRIX ALGEBRA

A *matrix* of order $n \times m$ is a rectangular array of real numbers arranged in n rows and m columns. We will denote matrices by uppercase English and Greek boldface letters; for example,

$$\mathbf{X}_{6\times 3} = \begin{bmatrix} 1 & 1 & 0 \\ 1 & 1 & 0 \\ 1 & 0 & 1 \\ 1 & 0 & 1 \\ 1 & 0 & 0 \\ 1 & 0 & 0 \end{bmatrix} \quad \mathbf{P}_{4\times 4} = \begin{bmatrix} 32 & 8 & 8 & 8 \\ 8 & 8 & 0 & 0 \\ 8 & 0 & 8 & 0 \\ 8 & 0 & 0 & 8 \end{bmatrix} \quad \mathbf{C}'_{3\times 4} = \begin{bmatrix} 0 & 1 & 0 & 0 \\ 0 & 0 & 1 & 0 \\ 0 & 0 & 0 & 1 \end{bmatrix}.$$

A prime after a letter denotes the *transpose* of a matrix. A transpose is obtained by interchanging rows and columns, so that the ith row of the original matrix becomes the ith column of the transposed matrix. For example, the transpose of $\mathbf{X}$ in the preceding examples is

$$\mathbf{X}'_{3\times 6} = \begin{bmatrix} 1 & 1 & 1 & 1 & 1 & 1 \\ 1 & 1 & 0 & 0 & 0 & 0 \\ 0 & 0 & 1 & 1 & 0 & 0 \end{bmatrix}$$

and the transpose of $\mathbf{C}'$ is

$$(\mathbf{C}')'_{3\times 4} = \mathbf{C}_{4\times 3} = \begin{bmatrix} 0 & 0 & 0 \\ 1 & 0 & 0 \\ 0 & 1 & 0 \\ 0 & 0 & 1 \end{bmatrix}.$$

An element in the ith row and jth column of matrix $\mathbf{C}$ is denoted by C_{ij}; for example, $C_{21} = 1$.

D.2-1 MATRIX EQUALITY

Two matrices of the same order are equal if and only if each element of one matrix is equal to the corresponding element of the other matrix. That is,

$$\underset{n \times m}{\mathbf{X}} = \underset{n \times m}{\mathbf{Y}} \quad \text{if and only if} \quad X_{ij} = Y_{ij} \quad (i = 1, \ldots, n; j = 1, \ldots, m).$$

EXAMPLE 1   Matrices **V** and **W**, as defined here, are equal

$$\underset{2 \times 3}{\mathbf{V}} = \begin{bmatrix} 1 & 3 & 5 \\ 1 & 3 & -2 \end{bmatrix} = \underset{2 \times 3}{\mathbf{W}} = \begin{bmatrix} 1 & 3 & 5 \\ 1 & 3 & -2 \end{bmatrix}$$

since $V_{ij} = W_{ij}$ for all i, j.

D.2-2 SCALAR MULTIPLICATION OF A MATRIX

Scalar multiplication of a matrix consists of multiplying each element of the matrix by a scalar. Let s denote a scalar and **X**, an $n \times m$ matrix, then

$$s \underset{n \times m}{\mathbf{X}} = s \begin{bmatrix} X_{11} & X_{12} & \cdot & X_{1m} \\ X_{21} & X_{22} & \cdot & X_{2m} \\ \cdot & \cdot & \cdot & \cdot \\ X_{n1} & X_{n2} & \cdot & X_{nm} \end{bmatrix} = \begin{bmatrix} sX_{11} & sX_{12} & \cdot & sX_{1m} \\ sX_{21} & sX_{22} & \cdot & sX_{2m} \\ \cdot & \cdot & \cdot & \cdot \\ sX_{n1} & sX_{n2} & \cdot & sX_{nm} \end{bmatrix}.$$

EXAMPLE 1   Let $s = 2$ and $\mathbf{V} = \begin{bmatrix} 1 & 3 \\ 4 & 2 \end{bmatrix}$. Then

$$\underset{2 \times 2}{s \mathbf{V}} = \begin{bmatrix} 2 \times 1 & 2 \times 3 \\ 2 \times 4 & 2 \times 2 \end{bmatrix} = \begin{bmatrix} 2 & 6 \\ 8 & 4 \end{bmatrix}.$$

Scalar multiplication of a matrix exhibits the following properties. Let s and t denote scalars and **X**, a matrix.

1. $s(t\mathbf{X}) = (st)\mathbf{X}$ (Associative law)
2. $s\mathbf{X} = \mathbf{X}s$ (Commutative law)
3. $(s + t)\mathbf{X} = s\mathbf{X} + t\mathbf{X}$ (Distributive law for scalars)

D.2-3 MATRIX ADDITION AND SUBTRACTION

The addition of two or more matrices of the same order consists of adding their corresponding elements; that is, adding the ijth element of one matrix to the ijth element of the other matrix for all ij. For example,

$$\underset{n \times m}{\mathbf{X}} + \underset{n \times m}{\mathbf{Y}} = \begin{bmatrix} X_{11} + Y_{11} & X_{12} + Y_{12} & \cdot & X_{1m} + Y_{1m} \\ X_{21} + Y_{21} & X_{22} + Y_{22} & \cdot & X_{2m} + Y_{2m} \\ \cdots\cdots\cdots\cdots\cdots\cdots\cdots\cdots\cdots\cdots\cdots \\ X_{n1} + Y_{n1} & X_{n2} + Y_{n2} & \cdot & X_{nm} + Y_{nm} \end{bmatrix}.$$

Two matrices of the same order are said to be *conformable for addition;* that is, they can be added.

EXAMPLE 1   The sum of matrices **V** and **W**, defined as follows, is

$$\underset{2\times3}{\mathbf{V}} + \underset{2\times3}{\mathbf{W}} = \begin{bmatrix} 1 & 3 & 5 \\ 1 & 3 & -2 \end{bmatrix} + \begin{bmatrix} 2 & 1 & 3 \\ 2 & 3 & 3 \end{bmatrix}$$

$$= \begin{bmatrix} 1+2 & 3+1 & 5+3 \\ 1+2 & 3+3 & -2+3 \end{bmatrix} = \begin{bmatrix} 3 & 4 & 8 \\ 3 & 6 & 1 \end{bmatrix}.$$

The subtraction of one matrix from another of the same order consists of changing the sign of the elements of the matrix to be subtracted and proceeding as in matrix addition.

EXAMPLE 2   For **V** and **W** defined in Example 1,

$$\mathbf{V} - \mathbf{W} = \mathbf{V} + (-1)\mathbf{W} = \begin{bmatrix} 1 & 3 & 5 \\ 1 & 3 & -2 \end{bmatrix} + \begin{bmatrix} -2 & -1 & -3 \\ -2 & -3 & -3 \end{bmatrix}$$

$$= \begin{bmatrix} 1+(-2) & 3+(-1) & 5+(-3) \\ 1+(-2) & 3+(-3) & -2+(-3) \end{bmatrix}$$

$$= \begin{bmatrix} -1 & 2 & 2 \\ -1 & 0 & -5 \end{bmatrix}.$$

Matrix addition exhibits the following properties. Let s denote a scalar and **X**, **Y**, **Z**, and **0**, matrices of the same order.

1. $(\mathbf{X} + \mathbf{Y}) + \mathbf{Z} = \mathbf{X} + (\mathbf{Y} + \mathbf{Z})$ (Associative law)
2. $\mathbf{X} + \mathbf{Y} = \mathbf{Y} + \mathbf{X}$ (Commutative law)
3. $s(\mathbf{X} + \mathbf{Y}) = s\mathbf{X} + s\mathbf{Y}$ (Distributive law for matrices)
4. $\mathbf{X} + \mathbf{0} = \mathbf{X}$, where **0** is a matrix whose elements are zero.   (Identity law)
5. $\mathbf{X} + (-\mathbf{X}) = \mathbf{0}$, where **0** is a matrix whose elements are zero.   (Inverse law)

D.2-4 MATRIX MULTIPLICATION

Matrix multiplication is defined for matrices in which the column order of the left-hand matrix equals the row order of the right-hand matrix. That is, if

$$\underset{n\times m}{\mathbf{X}} \quad \underset{m\times p}{\mathbf{Y}}$$

the matrices are said to be *conformable for multiplication.* The product of an $n \times m$

matrix and an $m \times p$ matrix is a matrix of order $n \times p$; that is,

$$\underset{n \times m}{\mathbf{X}} \ \underset{m \times p}{\mathbf{Y}} = \underset{n \times p}{\mathbf{Z}} \ .$$

The elements of $\mathbf{Z}$, Z_{ij}, are the sum of the products of the ith row of $\mathbf{X}$ with the jth column of $\mathbf{Y}$; in other words,

$$Z_{ij} = \sum_{k=1}^{m} X_{ik}Y_{kj} \qquad (i = 1, \ldots, n; j = 1, \ldots, p).$$

To illustrate, let $\mathbf{X}$ and $\mathbf{Y}$ denote matrices as follows:

$$\underset{2 \times 3}{\mathbf{X}} = \begin{bmatrix} X_{11} & X_{12} & X_{13} \\ X_{21} & X_{22} & X_{23} \end{bmatrix} \quad \text{and} \quad \underset{3 \times 2}{\mathbf{Y}} = \begin{bmatrix} Y_{11} & Y_{12} \\ Y_{21} & Y_{22} \\ Y_{31} & Y_{32} \end{bmatrix}.$$

Then

$$\underset{2 \times 3 \ 3 \times 2}{\mathbf{X} \ \mathbf{Y}} = \begin{bmatrix} \begin{array}{|ccc|} \hline X_{11} & X_{12} & X_{13} \\ \hline X_{21} & X_{22} & X_{23} \\ \hline \end{array} \end{bmatrix} \begin{bmatrix} \begin{array}{|c|c|} \hline Y_{11} \\ Y_{21} \\ Y_{31} \\ \hline \end{array} & \begin{array}{|c|} Y_{12} \\ Y_{22} \\ Y_{32} \\ \hline \end{array} \end{bmatrix} = \begin{bmatrix} Z_{11} & Z_{12} \\ Z_{21} & Z_{22} \end{bmatrix} = \underset{2 \times 2}{\mathbf{Z}}$$

where

$$\begin{bmatrix} \overbrace{X_{11}Y_{11} + X_{12}Y_{21} + X_{13}Y_{31}}^{Z_{11}} & \overbrace{X_{11}Y_{12} + X_{12}Y_{22} + X_{13}Y_{32}}^{Z_{12}} \\ \underbrace{X_{21}Y_{11} + X_{22}Y_{21} + X_{23}Y_{31}}_{Z_{21}} & \underbrace{X_{21}Y_{12} + X_{22}Y_{22} + X_{23}Y_{32}}_{Z_{22}} \end{bmatrix}.$$

Note that element Z_{11} is the sum of the products of the elements of row 1 of $\mathbf{X}$ and column 1 of $\mathbf{Y}$. Similarly, Z_{12} is the sum of the products of the elements of row 1 of $\mathbf{X}$ and column 2 of $\mathbf{Y}$. In vector terminology, element Z_{ij} is the inner product of the ith row vector of $\mathbf{X}$ and the jth column vector of $\mathbf{Y}$.

EXAMPLE 1 Let $\mathbf{X}$ and $\mathbf{Y}$ be two matrices as follows.

$$\begin{aligned} \underset{2 \times 3 \ 3 \times 2}{\mathbf{X} \ \mathbf{Y}} &= \begin{bmatrix} 2 & 1 & 3 \\ 0 & 2 & 2 \end{bmatrix} \begin{bmatrix} 3 & 1 \\ 1 & 2 \\ 4 & 1 \end{bmatrix} \\ &= \begin{bmatrix} (2 \times 3) + (1 \times 1) + (3 \times 4) & (2 \times 1) + (1 \times 2) + (3 \times 1) \\ (0 \times 3) + (2 \times 1) + (2 \times 4) & (0 \times 1) + (2 \times 2) + (2 \times 1) \end{bmatrix} \\ &= \begin{bmatrix} 19 & 7 \\ 10 & 6 \end{bmatrix} \end{aligned}$$

$$\underset{2\times3}{\mathbf{Y'}} \underset{3\times2}{\mathbf{X'}} = \begin{bmatrix} 3 & 1 & 4 \\ 1 & 2 & 1 \end{bmatrix} \begin{bmatrix} 2 & 0 \\ 1 & 2 \\ 3 & 2 \end{bmatrix}$$

$$= \begin{bmatrix} (3 \times 2) + (1 \times 1) + (4 \times 3) & (3 \times 0) + (1 \times 2) + (4 \times 2) \\ (1 \times 2) + (2 \times 1) + (1 \times 3) & (1 \times 0) + (2 \times 2) + (1 \times 2) \end{bmatrix}$$

$$= \begin{bmatrix} 19 & 10 \\ 7 & 6 \end{bmatrix}$$

EXAMPLE 2   Let **W** be a 4×4 matrix and $\boldsymbol{\beta}$ a 4×1 vector as follows.

$$\underset{4\times4}{\mathbf{W}} \underset{4\times1}{\boldsymbol{\beta}} = \begin{bmatrix} 1 & 1 & 0 & 0 \\ 1 & 0 & 1 & 0 \\ 1 & 0 & 0 & 1 \\ 1 & 0 & 0 & 0 \end{bmatrix} \begin{bmatrix} 2 \\ 3 \\ 1 \\ 4 \end{bmatrix} = \begin{bmatrix} (1 \times 2) + (1 \times 3) + (0 \times 1) + (0 \times 4) \\ (1 \times 2) + (0 \times 3) + (1 \times 1) + (0 \times 4) \\ (1 \times 2) + (0 \times 3) + (0 \times 1) + (1 \times 4) \\ (1 \times 2) + (0 \times 3) + (0 \times 1) + (0 \times 4) \end{bmatrix}$$

$$= \begin{bmatrix} 5 \\ 3 \\ 6 \\ 2 \end{bmatrix}$$

EXAMPLE 3   Let **X** be a 3×3 matrix and **c**, a 3×1 vector as follows.

$$\underset{1\times3}{\mathbf{c'}} \underset{3\times3}{\mathbf{X}} \underset{3\times1}{\mathbf{c}} = \begin{bmatrix} 2 & -1 & -1 \end{bmatrix} \begin{bmatrix} 5 & 3 & 1 \\ 3 & 2 & 0 \\ 1 & 0 & 2 \end{bmatrix} \begin{bmatrix} 2 \\ -1 \\ -1 \end{bmatrix}$$

$$= \begin{bmatrix} 6 & 4 & 0 \end{bmatrix} \begin{bmatrix} 2 \\ -1 \\ -1 \end{bmatrix}$$

$$= 8$$

Matrix multiplication exhibits the following properties. Let **X**, **Y**, and **Z** denote matrices that are conformable for multiplication and s, a scalar.

1. $(\mathbf{XY})\mathbf{Z} = \mathbf{X}(\mathbf{YZ})$ (Associative law)
2. $s(\mathbf{XY}) = (s\mathbf{X})\mathbf{Y}$ (Associative law)
3. $\mathbf{X}(\mathbf{Y} + \mathbf{Z}) = \mathbf{XY} + \mathbf{XZ}$ (Left distributive law)
4. $(\mathbf{X} + \mathbf{Y})\mathbf{Z} = \mathbf{XZ} + \mathbf{YZ}$ (Right distributive law)

As these properties indicate, there are similarities between the multiplication of scalars and the multiplication of matrices. However, there are also important differences:

Scalars	*Matrices*
1. $xy = yx$	**1.** $\mathbf{XY} \neq \mathbf{YX}$, in general.
2. If $xy = xz$ and $x \neq 0$, then $y = z$.	**2.** If $\mathbf{XY} = \mathbf{XZ}$ and $\mathbf{X} \neq \mathbf{0}$, it is not necessarily true that $\mathbf{Y} = \mathbf{Z}$.
3. If $xy = 0$, either $x = 0$ or $y = 0$, or both $x, y = 0$.	**3.** If $\mathbf{XY} = \mathbf{0}$, it is not necessarily true that $\mathbf{X} = \mathbf{0}$, $\mathbf{Y} = \mathbf{0}$ or both $\mathbf{X}$, $\mathbf{Y} = \mathbf{0}$
4. If $xy = 0$, then $yx = 0$.	**4.** If $\mathbf{XY} = \mathbf{0}$, it is not necessarily true that $\mathbf{YX} = \mathbf{0}$.

D.2-5 SOME SPECIAL TYPES OF MATRICES

1. *Augmented matrix*—a matrix that results from adjoining (uniting) two matrices.

EXAMPLE 1   Let $\mathop{\mathbf{X}}\limits_{3 \times 3} = \begin{bmatrix} 2 & 1 & 3 \\ 3 & 2 & 2 \\ 1 & 0 & 3 \end{bmatrix}$ and $\mathop{\mathbf{I}}\limits_{3 \times 3} = \begin{bmatrix} 1 & 0 & 0 \\ 0 & 1 & 0 \\ 0 & 0 & 1 \end{bmatrix}$. Then the augmented matrix,

denoted by $\mathop{[\mathbf{X} \quad \mathbf{I}]}\limits_{3 \times 6}$, is $\left[\begin{array}{ccc|ccc} 2 & 1 & 3 & 1 & 0 & 0 \\ 3 & 2 & 2 & 0 & 1 & 0 \\ 1 & 0 & 3 & 0 & 0 & 1 \end{array} \right]$.

EXAMPLE 2   Let $\mathbf{X}$ and $\mathbf{I}$ be defined as in Example 1. Then

$$\mathop{\begin{bmatrix} \mathbf{X} \\ \hline \mathbf{I} \end{bmatrix}}\limits_{6 \times 3} = \left[\begin{array}{ccc} 2 & 1 & 3 \\ 3 & 2 & 2 \\ 1 & 0 & 3 \\ \hline 1 & 0 & 0 \\ 0 & 1 & 0 \\ 0 & 0 & 1 \end{array} \right].$$

2. *Square matrix*—a matrix having the same number of rows and columns (a matrix of order $n \times n$).

EXAMPLE 3 $\mathop{\mathbf{X}}\limits_{3 \times 3} = \begin{bmatrix} 4 & 3 & 2 \\ 2 & 2 & 5 \\ 1 & 6 & 0 \end{bmatrix}$

Elements for which $i = j$ are called *diagonal elements* ($X_{11} = 4$, $X_{22} = 2$, $X_{33} = 0$); elements for which $i \neq j$ are called *off-diagonal elements* ($X_{12} = 3$, $X_{21} = 2$, and so on).

3.   *Symmetric matrix*—a square matrix that is equal to its transpose ($\mathbf{X} = \mathbf{X}'$).

$$\underset{4\times4}{\mathbf{X}} = \begin{bmatrix} 32 & 8 & 8 & 8 \\ 8 & 8 & 0 & 0 \\ 8 & 0 & 8 & 0 \\ 8 & 0 & 0 & 8 \end{bmatrix} = \underset{4\times4}{\mathbf{X}'} = \begin{bmatrix} 32 & 8 & 8 & 8 \\ 8 & 8 & 0 & 0 \\ 8 & 0 & 8 & 0 \\ 8 & 0 & 0 & 8 \end{bmatrix}$$

EXAMPLE 4

4.   *Diagonal matrix*—a square matrix in which all off-diagonal elements are zero. Elements on the diagonal may or may not be zero.

EXAMPLE 5 $\underset{3\times3}{\mathbf{D}} = \begin{bmatrix} 2 & 0 & 0 \\ 0 & 3 & 0 \\ 0 & 0 & 1 \end{bmatrix}$

5.   *Scalar matrix*—a diagonal matrix in which all diagonal elements are equal.

EXAMPLE 6 $\underset{3\times3}{\mathbf{X}} = \begin{bmatrix} 2 & 0 & 0 \\ 0 & 2 & 0 \\ 0 & 0 & 2 \end{bmatrix}$

6.   *Identity matrix* (denoted by $\mathbf{I}$)—a scalar matrix in which the diagonal elements are equal to unity.

EXAMPLE 7 $\underset{3\times3}{\mathbf{I}} = \begin{bmatrix} 1 & 0 & 0 \\ 0 & 1 & 0 \\ 0 & 0 & 1 \end{bmatrix}$

Multiplying a matrix by $\mathbf{I}$ leaves the matrix unchanged, that is, $\mathbf{XI} = \mathbf{IX} = \mathbf{X}$.

7.   *Null matrix* (denoted by $\mathbf{0}$)—a matrix in which all elements are zero.

EXAMPLE 8 $\underset{2\times3}{\mathbf{0}} = \begin{bmatrix} 0 & 0 & 0 \\ 0 & 0 & 0 \end{bmatrix}$

D.2-6 MATRIX INVERSION

Thus far we have discussed matrix addition, subtraction, and multiplication, but not matrix division. Division in the usual sense does not exist in matrix algebra. Instead of dividing matrix $\mathbf{W}$ by $\mathbf{X}$, we multiply $\mathbf{W}$ by a matrix called the inverse of $\mathbf{X}$, denoted by $\mathbf{X}^{-1}$. The inverse described here is only defined for square matrices.*

Recall from scalar algebra that we can divide a by b by multiplying a by the multiplicative inverse of b, that is,

$$\frac{a}{b} = a \times b^{-1} = a \times \frac{1}{b}.$$

---

*   Several special types of inverses exist for square and rectangular matrices; for a discussion of these, see Green and Carroll (1976) and Searle (1971a).

This inverse b^{-1} has the property that

$$bb^{-1} = b^{-1}b = 1.$$

In matrix algebra, a square matrix $\mathbf{X}^{-1}$ such that

$$\mathbf{X}\mathbf{X}^{-1} = \mathbf{X}^{-1}\mathbf{X} = \mathbf{I}$$

is said to be the inverse of $\mathbf{X}$. As we will see, it is not always possible to find a matrix $\mathbf{X}^{-1}$ such that $\mathbf{X}\mathbf{X}^{-1} = \mathbf{X}^{-1}\mathbf{X} = \mathbf{I}$. When an inverse does exist, however, it is unique. A square matrix for which an inverse exists is said to be *nonsingular*; one for which an inverse does not exist is called *singular*. Consider the following nonsingular matrix

$$\underset{4\times 4}{\mathbf{X}} = \begin{bmatrix} 32 & 8 & 8 & 8 \\ 8 & 8 & 0 & 0 \\ 8 & 0 & 8 & 0 \\ 8 & 0 & 0 & 8 \end{bmatrix}.$$

The inverse of this matrix is

$$\underset{4\times 4}{\mathbf{X}^{-1}} = \begin{bmatrix} \frac{1}{8} & -\frac{1}{8} & -\frac{1}{8} & -\frac{1}{8} \\ -\frac{1}{8} & \frac{2}{8} & \frac{1}{8} & \frac{1}{8} \\ -\frac{1}{8} & \frac{1}{8} & \frac{2}{8} & \frac{1}{8} \\ -\frac{1}{8} & \frac{1}{8} & \frac{1}{8} & \frac{2}{8} \end{bmatrix}$$

since

$$\overset{\mathbf{X}}{\begin{bmatrix} 32 & 8 & 8 & 8 \\ 8 & 8 & 0 & 0 \\ 8 & 0 & 8 & 0 \\ 8 & 0 & 0 & 8 \end{bmatrix}} \overset{\mathbf{X}^{-1}}{\begin{bmatrix} \frac{1}{8} & -\frac{1}{8} & -\frac{1}{8} & -\frac{1}{8} \\ -\frac{1}{8} & \frac{2}{8} & \frac{1}{8} & \frac{1}{8} \\ -\frac{1}{8} & \frac{1}{8} & \frac{2}{8} & \frac{1}{8} \\ -\frac{1}{8} & \frac{1}{8} & \frac{1}{8} & \frac{2}{8} \end{bmatrix}}$$

$$= \overset{\mathbf{X}^{-1}}{\begin{bmatrix} \frac{1}{8} & -\frac{1}{8} & -\frac{1}{8} & -\frac{1}{8} \\ -\frac{1}{8} & \frac{2}{8} & \frac{1}{8} & \frac{1}{8} \\ -\frac{1}{8} & \frac{1}{8} & \frac{2}{8} & \frac{1}{8} \\ -\frac{1}{8} & \frac{1}{8} & \frac{1}{8} & \frac{2}{8} \end{bmatrix}} \overset{\mathbf{X}}{\begin{bmatrix} 32 & 8 & 8 & 8 \\ 8 & 8 & 0 & 0 \\ 8 & 0 & 8 & 0 \\ 8 & 0 & 0 & 8 \end{bmatrix}} = \overset{\mathbf{I}}{\begin{bmatrix} 1 & 0 & 0 & 0 \\ 0 & 1 & 0 & 0 \\ 0 & 0 & 1 & 0 \\ 0 & 0 & 0 & 1 \end{bmatrix}}.$$

We turn now to procedures for obtaining an inverse. We will consider diagonal matrices first since they can be disposed of very quickly. The inverse of a diagonal matrix is obtained by replacing each element on the diagonal by its reciprocal.

EXAMPLE 1 Let $\underset{2\times 2}{\mathbf{D}} = \begin{bmatrix} 2 & 0 \\ 0 & 3 \end{bmatrix}$. Then

$$\underset{2\times 2}{\mathbf{D}^{-1}} = \begin{bmatrix} \frac{1}{2} & 0 \\ 0 & \frac{1}{3} \end{bmatrix}.$$

We can verify that $\mathbf{D}^{-1}$ is the inverse of $\mathbf{D}$ by computing $\mathbf{D}\mathbf{D}^{-1}$ or $\mathbf{D}^{-1}\mathbf{D}$. The product must equal $\mathbf{I}$.

$$\underset{2\times2}{\mathbf{D}}\ \underset{2\times2}{\mathbf{D}^{-1}} = \begin{bmatrix} 2 & 0 \\ 0 & 3 \end{bmatrix}\begin{bmatrix} \frac{1}{2} & 0 \\ 0 & \frac{1}{3} \end{bmatrix} = \begin{bmatrix} 1 & 0 \\ 0 & 1 \end{bmatrix} = \mathbf{I}$$

Obtaining the inverse of a nondiagonal matrix is more complicated and is best left to a computer. If this option is not available, the following rather tedious procedure, Gauss' matrix inversion technique, can be used.* Let $\mathbf{X}$ and $\mathbf{I}$ be $n \times n$ matrices and $[\mathbf{X}\mathbf{I}]$, the $n \times 2n$ augmented matrix that results form adjoining $\mathbf{I}$ to $\mathbf{X}$.

$$\underset{n\times2n}{[\mathbf{X}\mathbf{I}]} = \begin{bmatrix} X_{11}X_{12} & \cdot & X_{1n} & \vdots & 1 & 0 & \cdot & 0 \\ X_{21}X_{22} & \cdot & X_{2n} & \vdots & 0 & 1 & \cdot & 0 \\ \cdot \quad \cdot & \cdot & \cdot & \vdots & \cdot & \cdot & \cdot & \cdot \\ X_{n1}X_{n2} & \cdot & X_{nn} & \vdots & 0 & 0 & \cdot & 1 \end{bmatrix}$$

By means of a series of elementary row operations that will be described shortly, we can transform the augmented matrix $[\mathbf{X}\mathbf{I}]$ into $[\mathbf{I}\mathbf{X}^{-1}]$. As we will see, the row operations that change $\mathbf{X}$ into $\mathbf{I}$ also change $\mathbf{I}$ into $\mathbf{X}^{-1}$. The strategy, then, is to choose row operations that will successively transform $\mathbf{X}$ into an identity matrix. These same operations performed on $\mathbf{I}$ will transform it into the inverse of $\mathbf{X}$. The following three *row operations* can be performed on the augmented matrix.

1.   Multiplication of a row by a nonzero real number.
2.   Replacement of one row by the sum of that row and a constant times another row.
3.   Interchange of two rows.

These operations are illustrated in Table D.2-1.

The reader can verify that $\mathbf{X}^{-1}$ is indeed the inverse of $\mathbf{X}$ by showing that $\mathbf{X}\mathbf{X}^{-1}$ (or $\mathbf{X}^{-1}\mathbf{X}$) $= \mathbf{I}$. The number of row operations required to obtain an inverse if one exists depends on the peculiarities of the matrix and one's skill in using row operations. We could have shortened the procedure in Table D.2-1 by combining the steps in rows $i1$ and $i2$ (stages one and two), for example, letting $R_{11} = [R_{10} - (R_{20} + R_{30} + R_{40})]/8$, $R_{21} = (R_{20} - R_{11})/8$, and so on. The computation of an inverse for another matrix, denoted by $\mathbf{Y}$, is illustrated in Table D.2-2.

As mentioned earlier, inverses do not exist for all square matrices. If row operations cannot be found that transform a matrix into an identity matrix, the matrix does not have an inverse. In this case, at some stage of the procedure a row of zeros will appear on the left side of the augmented matrix. This is illustrated in Table D.2-3 for the singular matrix $\mathbf{Z}$. Try as one will, one can not find a series of row operations that will transform $\mathbf{Z}$ into an identity matrix. Apart from not having an inverse, how

---

*   Other procedures for obtaining an inverse are described by Green and Carroll (1976, Ch. 4), Hohn (1964, Ch. 3), and Searle (1966, Ch. 4).

TABLE D.2-1 Steps in Obtaining the Inverse of a Matrix

Row (R_{ij})*	X				I			
R_{10}	32	8	8	8	1	0	0	0
R_{20}	8	8	0	0	0	1	0	0
R_{30}	8	0	8	0	0	0	1	0
R_{40}	8	0	0	8	0	0	0	1
$R_{11} = R_{10} - (R_{20} + R_{30} + R_{40})$	8	0	0	0	1	-1	-1	-1
$R_{21} = R_{20} - R_{11}$	0	8	0	0	-1	2	1	1
$R_{31} = R_{30} - R_{11}$	0	0	8	0	-1	1	2	1
$R_{41} = R_{40} - R_{11}$	0	0	0	8	-1	1	1	2
$R_{12} = R_{11}/8$	1	0	0	0	$1/8$	$-1/8$	$-1/8$	$-1/8$
$R_{22} = R_{21}/8$	0	1	0	0	$-1/8$	$2/8$	$1/8$	$1/8$
$R_{32} = R_{31}/8$	0	0	1	0	$-1/8$	$1/8$	$2/8$	$1/8$
$R_{42} = R_{41}/8$	0	0	0	1	$-1/8$	$1/8$	$1/8$	$2/8$
	I				$\mathbf{X}^{-1}$			

*R_{ij} denotes the ith row at the jth stage; R_{i0} is the original matrix, R_{i1} is the transformed matrix at stage 1, and so on.

does matrix **Z** differ from **X** and **Y**? On close examination we see that column 1 of **Z** is equal to the sum of columns 2, 3, 4, and 5; that is, column 1 is a linear combination of 2, 3, 4, and 5. Also, row 1 is equal to the sum of rows 2, 3, 4, and 5. Accordingly, the column (row) vectors of **Z** are said to be linearly dependent.* This is not true for matrices **X** and **Y**, since none of their column (row) vectors can be expressed as linear combinations of the other column (row) vectors. These vectors are linearly independent. The number of linearly independent column (or row) vectors in a matrix is referred to as the *rank* of the matrix and denoted by r.** If r is equal to n in the case of square matrices or to the smaller of n and m in the case of rectangular matrices, the matrix is said to be of *full rank;* otherwise it is not of full rank. Matrices **X** and **Y** are of full rank since r, the number of linearly independent columns (rows), is equal to n, the number of columns (rows). Matrix **Z** is not of full rank since $r = 4$ is less than $n = 5$.

---

*	The concept of linear dependence is discussed in more detail in Appendix D.1-5 and Section 5.8.
**	The rank of a matrix can also be thought of as the order of the largest square submatrix for which an inverse exists. The concept of rank is discussed in more detail in Section 5.8.

TABLE D.2-2 Steps in Obtaining the Inverse of a Matrix

Row $(R_{ij})^*$	Y				I			
R_{10}	32	0	0	0	1	0	0	0
R_{20}	0	16	8	8	0	1	0	0
R_{30}	0	8	16	8	0	0	1	0
R_{40}	0	8	8	16	0	0	0	1
$R_{11} = R_{10}/32$	1	0	0	0	$1/32$	0	0	0
$R_{21} = (R_{20} - R_{30})/8$	0	1	-1	0	0	$1/8$	$-1/8$	0
$R_{31} = (R_{30} - R_{40})/8$	0	0	1	-1	0	0	$1/8$	$-1/8$
$R_{41} = (R_{40} - 8R_{21})/16$	0	0	1	1	0	$-1/16$	$1/16$	$1/16$
$R_{12} = R_{11}$	1	0	0	0	$1/32$	0	0	0
$R_{22} = R_{21} + R_{31}$	0	1	0	-1	0	$1/8$	0	$-1/8$
$R_{32} = (R_{31} + R_{41})/2$	0	0	1	0	0	$-1/32$	$3/32$	$-1/32$
$R_{42} = R_{41} - R_{32}$	0	0	0	1	0	$-1/32$	$-1/32$	$3/32$
$R_{13} = R_{12}$	1	0	0	0	$1/32$	0	0	0
$R_{23} = R_{22} + R_{42}$	0	1	0	0	0	$3/32$	$-1/32$	$-1/32$
$R_{33} = R_{32}$	0	0	1	0	0	$-1/32$	$3/32$	$-1/32$
$R_{43} = R_{42}$	0	0	0	1	0	$-1/32$	$-1/32$	$3/32$
	I				Y^{-1}			

$^*R_{ij}$ denotes the ith row at the jth stage.

The availability of an inverse is important since it is used in solving for unknowns, $\boldsymbol{\beta}$, in matrix equations such as

$$X'X\boldsymbol{\beta} = X'y.$$

Before seeing how this is done, let us consider some properties of inverses, transposes, and matrix rank. Let U, V, and W denote matrices that are conformable for addition or multiplication; X, Y, and Z denote nonsingular matrices of the same order; and s, a scalar.

1. The inverse of X is unique.
2. The inverse matrix is nonsingular.
3. $XX^{-1} = X^{-1}X = I$ (The inverse commutes with X.)
4. $(X^{-1})^{-1} = X$
5. $(XYZ)^{-1} = Z^{-1}Y^{-1}X^{-1}$

TABLE D.2-3 Steps Showing That an Inverse Does Not Exist

Row (R_{ij})*	Z					I				
R_{10}	8	2	2	2	2	1	0	0	0	0
R_{20}	2	2	0	0	0	0	1	0	0	0
R_{30}	2	0	2	0	0	0	0	1	0	0
R_{40}	2	0	0	2	0	0	0	0	1	0
R_{50}	2	0	0	0	2	0	0	0	0	1
$R_{11} = [R_{10} - (R_{20} + R_{30} + R_{40})]/2$	1	0	0	0	1	½	-½	-½	-½	0
$R_{21} = (R_{20} - R_{30})/2$	0	1	-1	0	0	0	½	-½	0	0
$R_{31} = (R_{30} - R_{40})/2$	0	0	1	-1	0	0	0	½	-½	0
$R_{41} = (R_{40} - R_{50})/2$	0	0	0	1	-1	0	0	0	½	-½
$R_{51} = (R_{50} - 2R_{11})$	0	0	0	0	0	-1	1	1	1	1

*R_{ij} denotes the ith row at the jth stage.

6. $(s\mathbf{X})^{-1} = (1/s)\mathbf{X}^{-1}$

7. $(\mathbf{U}')' = \mathbf{U}$

8. $\mathbf{I}' = \mathbf{I}$, where $\mathbf{I}$ is an identity matrix.

9. $(\mathbf{U} + \mathbf{V} + \mathbf{W})' = \mathbf{U}' + \mathbf{V}' + \mathbf{W}'$ and $(\mathbf{U} - \mathbf{V} - \mathbf{W})' = \mathbf{U}' - \mathbf{V}' - \mathbf{W}'$

10. $(\mathbf{UVW})' = \mathbf{W}'\mathbf{V}'\mathbf{U}'$

11. If $\mathbf{X}$ is symmetric, that is, $\mathbf{X}' = \mathbf{X}$, so is its inverse $(\mathbf{X}^{-1})' = \mathbf{X}^{-1}$.

12. The product of a matrix with its transpose is symmetric. Thus $(\mathbf{UU}')' = (\mathbf{U}')'\mathbf{U}' = \mathbf{UU}'$ and $(\mathbf{U}'\mathbf{U})' = \mathbf{U}'(\mathbf{U}')' = \mathbf{U}'\mathbf{U}$.

13. $(\mathbf{X}')^{-1} = (\mathbf{X}^{-1})'$

14. If $\mathbf{U} = \mathbf{0}$, the rank of $\mathbf{U}$, denoted by $r(\mathbf{U})$, is zero $(r = 0)$.

15. If $\underset{n \times m}{\mathbf{U}} \neq \mathbf{0}$, then $0 < r(\mathbf{U}) \le$ minimum of n and m.

16. If $\mathbf{I}$ is an identity matrix, $r(\underset{n \times n}{\mathbf{I}}) = n$.

17. If the rank of an $n \times m$ matrix is equal to the smaller of n and m, the matrix is of full rank, otherwise it is less than full rank.

18. The $r(\mathbf{U}) = r(\mathbf{U}')$.

19. The $r(\underset{n \times m}{\mathbf{U}} \ \underset{m \times p}{\mathbf{V}}) \le$ minimum of $r(\mathbf{U})$ and $r(\mathbf{V})$.

20. The rank of a matrix is not changed by premultiplying or postmultiplying it by a nonsingular matrix, that is, $r(\mathbf{U}) = r(\mathbf{XU}) = r(\mathbf{UX})$.

D.2-7 SOLVING SIMULTANEOUS LINEAR EQUATIONS

The inverse of a matrix is used to solve a system of N simultaneous equations in h unknowns $(\beta_0, \beta_1, \ldots, \beta_{h-1})$.

$$\beta_0 + \beta_1 X_{11} + \beta_2 X_{12} + \cdots + \beta_{h-1} X_{1,h-1} = Y_1$$
$$\beta_0 + \beta_1 X_{21} + \beta_2 X_{22} + \cdots + \beta_{h-1} X_{2,h-1} = Y_2$$
$$\cdots\cdots\cdots\cdots\cdots\cdots\cdots\cdots\cdots\cdots\cdots\cdots$$
$$\beta_0 + \beta_1 X_{N1} + \beta_2 X_{N2} + \cdots + \beta_{h-1} X_{N,h-1} = Y_N$$

Using matrix notation, this can be written as

(D.2-1)

$$\underset{N \times h}{\mathbf{X}} \quad \underset{h \times 1}{\boldsymbol{\beta}} \quad \underset{N \times 1}{\mathbf{y}}$$

$$\begin{bmatrix} 1 & X_{11} & X_{12} & \cdot & X_{1,h-1} \\ 1 & X_{21} & X_{22} & \cdot & X_{2,h-1} \\ \cdots & \cdots & \cdots & \cdots & \cdots \\ 1 & X_{N1} & X_{N2} & \cdot & X_{N,h-1} \end{bmatrix} \begin{bmatrix} \beta_0 \\ \beta_1 \\ \beta_2 \\ \cdot \\ \cdot \\ \cdot \\ \beta_{h-1} \end{bmatrix} = \begin{bmatrix} Y_1 \\ Y_2 \\ \cdot \\ \cdot \\ \cdot \\ Y_N \end{bmatrix}.$$

The system of equations represented by D.2-1 is *nonhomogeneous* if $Y_i \neq 0$ for some i and *homogeneous* if $Y_i = 0$ for all i. The systems in this book will always be nonhomogeneous. Systems of equations are also classified as consistent or inconsistent. A system is *consistent* if the rank of the augmented matrix $[\mathbf{Xy}]$ is equal to the rank of $\mathbf{X}$, that is if $r[\mathbf{Xy}] = r(\mathbf{X})$; otherwise the system is *inconsistent*. If the system is inconsistent, it does not have a solution. We will now see how the inverse is used in obtaining a solution to a nonhomogeneous system of equations. We begin by examining a parallel between scalar algebra and matrix algebra.

In scalar algebra we can solve for b in the equation

$$ab = c$$

if $a \neq 0$ by multiplying both sides of the equation by the inverse of a, for example,

$$\left(\frac{1}{a}\right)ab = \left(\frac{1}{a}\right)c$$

$$b = \frac{c}{a}.$$

In matrix algebra we solve for $\mathbf{B}$ in the equation

$$\underset{n \times n}{\mathbf{A}} \ \underset{n \times n}{\mathbf{B}} = \underset{n \times n}{\mathbf{C}}$$

in the same way. Assuming that $\mathbf{A}^{-1}$ exists, we premultiply both sides of the equation by $\mathbf{A}^{-1}$:

$$\mathbf{A}^{-1}\mathbf{AB} = \mathbf{A}^{-1}\mathbf{C}$$
$$\mathbf{IB} = \mathbf{A}^{-1}\mathbf{C} \quad \text{(since } \mathbf{A}^{-1}\mathbf{A} = \mathbf{I}\text{)}$$
$$\mathbf{B} = \mathbf{A}^{-1}\mathbf{C} \quad \text{(since } \mathbf{IB} = \mathbf{B}\text{)}.$$

To solve for $\mathbf{B}$ in

$$\underset{n \times n}{\mathbf{A}} \ \underset{n \times n}{\mathbf{B}} \ \underset{n \times n}{\mathbf{C}} = \underset{n \times n}{\mathbf{D}}$$

we premultiply both sides by $\mathbf{A}^{-1}$ and postmultiply both sides by $\mathbf{C}^{-1}$, assuming that $\mathbf{A}^{-1}$ and $\mathbf{C}^{-1}$ exist:

$$\mathbf{A}^{-1}\mathbf{ABCC}^{-1} = \mathbf{A}^{-1}\mathbf{DC}^{-1}$$
$$\mathbf{IBI} = \mathbf{A}^{-1}\mathbf{DC}^{-1}$$
$$\mathbf{B} = \mathbf{A}^{-1}\mathbf{DC}^{-1}.$$

The procedure for solving systems of nonhomogeneous equations such as

(D.2-2)
$$\underset{N \times h}{\mathbf{X}} \ \underset{h \times 1}{\boldsymbol{\beta}} = \underset{N \times 1}{\mathbf{y}} \quad (Y_i \neq 0 \text{ for some } i)$$

depends on whether $\mathbf{X}$ is square or rectangular and whether $\mathbf{X}$ is of full rank or less than full rank. If the number of equations equals the number of unknowns ($N = h$), $\mathbf{X}$ is square. If in addition, $\mathbf{X}$ is of full rank ($r(\mathbf{X}) = N = h$), a unique solution to

(D.2-2) is obtained by premultiplying both sides of the equation by $\mathbf{X}^{-1}$:

$$\underset{(h \times N)}{\mathbf{X}^{-1}} \; \underset{N \times h}{\mathbf{X}} \; \underset{(h \times 1)}{\boldsymbol{\beta}} = \underset{(h \times N)}{\mathbf{X}^{-1}} \; \underset{(N \times 1)}{\mathbf{y}}$$

$$\underset{(h \times 1)}{\boldsymbol{\beta}} = \underset{(h \times N)}{\mathbf{X}^{-1}} \; \underset{(N \times 1)}{\mathbf{y}} \; .$$

EXAMPLE 1   The system of two simultaneous equations in two unknowns (β_0, β_1)

$$\beta_0 1 + \beta_1 1 = 5$$
$$\beta_0 1 + \beta_1 0 = 2$$

can be written in matrix notation as

$$\underset{2 \times 2}{\mathbf{X}} \quad \underset{2 \times 1}{\boldsymbol{\beta}} \quad \underset{2 \times 1}{\mathbf{y}}$$

$$\begin{bmatrix} 1 & 1 \\ 1 & 0 \end{bmatrix} \begin{bmatrix} \beta_0 \\ \beta_1 \end{bmatrix} = \begin{bmatrix} 5 \\ 2 \end{bmatrix}.$$

The inverse of $\mathbf{X}$ is

$$\mathbf{X}^{-1} = \begin{bmatrix} 0 & 1 \\ 1 & -1 \end{bmatrix}.$$

The solution is given by

$$\underset{2 \times 1}{\boldsymbol{\beta}} \qquad \underset{2 \times 2}{\mathbf{X}^{-1}} \quad \underset{2 \times 1}{\mathbf{y}}$$

$$\begin{bmatrix} \beta_0 \\ \beta_1 \end{bmatrix} = \begin{bmatrix} 0 & 1 \\ 1 & -1 \end{bmatrix} \begin{bmatrix} 5 \\ 2 \end{bmatrix} = \begin{bmatrix} 2 \\ 3 \end{bmatrix}.$$

Thus, $\beta_0 = 2$ and $\beta_1 = 3$.

If $N > h$ in (D.2-2), $\mathbf{X}$ is rectangular and an inverse does not exist. A unique solution can still be obtained if $\mathbf{X}$ is of full rank. The procedure is as follows:

$$\underset{N \times h}{\mathbf{X}} \underset{h \times 1}{\boldsymbol{\beta}} = \underset{N \times 1}{\mathbf{y}}$$

$$\underset{h \times h}{(\mathbf{X}'\mathbf{X})} \underset{h \times 1}{\boldsymbol{\beta}} = \underset{h \times N}{\mathbf{X}'} \underset{N \times 1}{\mathbf{y}} \qquad \text{(Premultiply both sides by } \mathbf{X}'\text{; the matrix } (\mathbf{X}'\mathbf{X}) \text{ is square and has an inverse.)}$$

$$\underset{h \times h}{(\mathbf{X}'\mathbf{X})^{-1}} \underset{h \times h}{(\mathbf{X}'\mathbf{X})} \underset{h \times 1}{\boldsymbol{\beta}} = \underset{h \times h}{(\mathbf{X}'\mathbf{X})^{-1}} \underset{h \times N}{\mathbf{X}'} \underset{N \times 1}{\mathbf{y}} \qquad \text{(Premultiply both sides by } (\mathbf{X}'\mathbf{X})^{-1}.)$$

$$\underset{h \times h}{\mathbf{I}} \underset{h \times 1}{\boldsymbol{\beta}} = \underset{h \times h}{(\mathbf{X}'\mathbf{X})^{-1}} \underset{h \times N}{\mathbf{X}'} \underset{N \times 1}{\mathbf{y}}$$

$$\underset{h \times 1}{\boldsymbol{\beta}} = \underset{h \times h}{(\mathbf{X}'\mathbf{X})^{-1}} \underset{h \times N}{\mathbf{X}'} \underset{N \times 1}{\mathbf{y}} \; .$$

EXAMPLE 2   The system of three simultaneous equations in two unknowns (β_0, β_1)

$$\beta_0 1 + \beta_1 1 = 5$$
$$\beta_0 1 + \beta_1 1 = 5$$
$$\beta_0 1 + \beta_1 0 = 2$$

can be written in matrix notation as

$$
\underset{\substack{\mathbf{X} \\ 3\times 2}}{\begin{bmatrix} 1 & 1 \\ 1 & 1 \\ 1 & 0 \end{bmatrix}} \underset{\substack{\boldsymbol{\beta} \\ 2\times 1}}{\begin{bmatrix} \beta_0 \\ \beta_1 \end{bmatrix}} = \underset{\substack{\mathbf{y} \\ 3\times 1}}{\begin{bmatrix} 5 \\ 5 \\ 2 \end{bmatrix}}.
$$

The following are required in the solution.

$$
\underset{2\times 3}{\mathbf{X}'} \ \underset{3\times 2}{\mathbf{X}} = \begin{bmatrix} 1 & 1 & 1 \\ 1 & 1 & 0 \end{bmatrix} \begin{bmatrix} 1 & 1 \\ 1 & 1 \\ 1 & 0 \end{bmatrix} = \underset{\substack{(\mathbf{X}'\mathbf{X}) \\ 2\times 2}}{\begin{bmatrix} 3 & 2 \\ 2 & 2 \end{bmatrix}}
$$

$$
\underset{2\times 2}{(\mathbf{X}'\mathbf{X})^{-1}} = \begin{bmatrix} 1 & -1 \\ -1 & 1.5 \end{bmatrix}
$$

$$
\underset{2\times 3}{\mathbf{X}'} \ \underset{3\times 1}{\mathbf{y}} = \begin{bmatrix} 1 & 1 & 1 \\ 1 & 1 & 0 \end{bmatrix} \begin{bmatrix} 5 \\ 5 \\ 2 \end{bmatrix} = \underset{\substack{(\mathbf{X}'\mathbf{y}) \\ 2\times 1}}{\begin{bmatrix} 12 \\ 10 \end{bmatrix}}
$$

The solution is given by

$$
\underset{\substack{\boldsymbol{\beta} \\ 2\times 1}}{\begin{bmatrix} \beta_0 \\ \beta_1 \end{bmatrix}} = \underset{\substack{(\mathbf{X}'\mathbf{X})^{-1} \\ 2\times 2}}{\begin{bmatrix} 1 & -1 \\ -1 & 1.5 \end{bmatrix}} \underset{\substack{(\mathbf{X}'\mathbf{y}) \\ 2\times 1}}{\begin{bmatrix} 12 \\ 10 \end{bmatrix}} = \begin{bmatrix} 2 \\ 3 \end{bmatrix}.
$$

If **X** in (D.2-2) is of less than full rank, a unique solution does not exist. Procedures for obtaining solutions for this case are described in Section 5.9.

E TABLES

TABLE E.1 Squares, Square Roots, and Reciprocals

N	N^2	$\sqrt{N}$	$\sqrt{10N}$	$1/N$	N	N^2	$\sqrt{N}$	$\sqrt{10N}$	$1/N$ .0
					50	2 500	7.071 068	22.36068	2000000
1	1	1.000 000	3.162 278	1.0000000	51	2 601	7.141 428	22.58318	1960784
2	4	1.414 214	4.472 136	.5000000	52	2 704	7.211 103	22.80351	1923077
3	9	1.732 051	5.477 226	.3333333	53	2 809	7.280 110	23.02173	1886792
4	16	2.000 000	6.324 555	.2500000	54	2 916	7.348 469	23.23790	1851852
5	25	2.236 068	7.071 068	.2000000	55	3 025	7.416 198	23.45208	1818182
6	36	2.449 490	7.745 967	.1666667	56	3 136	7.483 315	23.66432	1785714
7	49	2.645 751	8.366 600	.1428571	57	3 249	7.549 834	23.87467	1754386
8	64	2.828 427	8.944 272	.1250000	58	3 364	7.615 773	24.08319	1724138
9	81	3.000 000	9.486 833	.1111111	59	3 481	7.681 146	24.28992	1694915
10	100	3.162 278	10.00000	.1000000	60	3 600	7.745 967	24.49490	1666667
11	121	3.316 625	10.48809	.09090909	61	3 721	7.810 250	24.69818	1639344
12	144	3.464 102	10.95445	.08333333	62	3 844	7.874 008	24.89980	1612903
13	169	3.605 551	11.40175	.07692308	63	3 969	7.937 254	25.09980	1587302
14	196	3.741 657	11.83216	.07142857	64	4 096	8.000 000	25.29822	1562500
15	225	3.872 983	12.24745	.06666667	65	4 225	8.062 258	25.49510	1538462
16	256	4.000 000	12.64911	.06250000	66	4 356	8.124 038	25.69047	1515152
17	289	4.123 106	13.03840	.05882353	67	4 489	8.185 353	25.88436	1492537
18	324	4.242 641	13.41641	.05555556	68	4 624	8.246 211	26.07681	1470588
19	361	4.358 899	13.78405	.05263158	69	4 761	8.306 624	26.26785	1449275
20	400	4.472 136	14.14214	.05000000	70	4 900	8.366 600	26.45751	1428571
21	441	4.582 576	14.49138	.04761905	71	5 041	8.426 150	26.64583	1408451
22	484	4.690 416	14.83240	.04545455	72	5 184	8.485 281	26.83282	1388889
23	529	4.795 832	15.16575	.04347826	73	5 329	8.544 004	27.01851	1369863
24	576	4.898 979	15.49193	.04166667	74	5 476	8.602 325	27.20294	1351351
25	625	5.000 000	15.81139	.04000000	75	5 625	8.660 254	27.38613	1333333
26	676	5.099 020	16.12452	.03846154	76	5 776	8.717 798	27.56810	1315789
27	729	5.196 152	16.43168	.03703704	77	5 929	8.774 964	27.74887	1298701
28	784	5.291 503	16.73320	.03571429	78	6 084	8.831 761	27.92848	1282051
29	841	5.385 165	17.02939	.03448276	79	6 241	8.888 194	28.10694	1265823
30	900	5.477 226	17.32051	.03333333	80	6 400	8.944 272	28.28427	1250000
31	961	5.567 764	17.60682	.03225806	81	6 561	9.000 000	28.46050	1234568
32	1 024	5.656 854	17.88854	.03125000	82	6 724	9.055 385	28.63564	1219512
33	1 089	5.744 563	18.16590	.03030303	83	6 889	9.110 434	28.80972	1204819
34	1 156	5.830 952	18.43909	.02941176	84	7 056	9.165 151	28.98275	1190476
35	1 225	5.916 080	18.70829	.02857143	85	7 225	9.219 544	29.15476	1176471
36	1 296	6.000 000	18.97367	.02777778	86	7 396	9.273 618	29.32576	1162791
37	1 369	6.082 763	19.23538	.02702703	87	7 569	9.327 379	29.49576	1149425
38	1 444	6.164 414	19.49359	.02631579	88	7 744	9.380 832	29.66479	1136364
39	1 521	6.244 998	19.74842	.02564103	89	7 921	9.433 981	29.83287	1123596
40	1 600	6.324 555	20.00000	.02500000	90	8 100	9.486 833	30.00000	1111111
41	1 681	6.403 124	20.24846	.02439024	91	8 281	9.539 392	30.16621	1098901
42	1 764	6.480 741	20.49390	.02380952	92	8 464	9.591 663	30.33150	1086957
43	1 849	6.557 439	20.73644	.02325581	93	8 649	9.643 651	30.49590	1075269
44	1 936	6.633 250	20.97618	.02272727	94	8 836	9.695 360	30.65942	1063830
45	2 025	6.708 204	21.21320	.02222222	95	9 025	9.746 794	30.82207	1052632
46	2 116	6.782 330	21.44761	.02173913	96	9 216	9.797 959	30.98387	1041667
47	2 209	6.855 655	21.67948	.02127660	97	9 409	9.848 858	31.14482	1030928
48	2 304	6.928 203	21.90890	.02083333	98	9 604	9.899 495	31.30495	1020408
49	2 401	7.000 000	22.13594	.02040816	99	9 801	9.949 874	31.46427	1010101
50	2 500	7.071 068	22.36068	.02000000	100	10 000	10.00000	31.62278	1000000

TABLE E.2 Random Numbers

	1	2	3	4	5	6	7	8	9	10	11	12	13	14	15	16	17	18	19	20	21	22	23	24	25
1	10	27	53	96	23	71	50	54	36	23	54	31	04	82	98	04	14	12	15	09	26	78	25	47	47
2	28	41	50	61	88	64	85	27	20	18	83	36	36	05	56	39	71	65	09	62	94	76	62	11	89
3	34	21	42	57	02	59	19	18	97	48	80	30	03	30	98	05	24	67	70	07	84	97	50	87	46
4	61	81	77	23	23	82	82	11	54	08	53	28	70	58	96	44	07	39	55	43	42	34	43	39	28
5	61	15	18	13	54	16	86	20	26	88	90	74	80	55	09	14	53	90	51	17	52	01	63	01	59
6	91	76	21	64	64	44	91	13	32	97	75	31	62	66	54	84	80	32	75	77	56	08	25	70	29
7	00	97	79	08	06	37	30	28	59	85	53	56	68	53	40	01	74	39	59	73	30	19	99	85	48
8	36	46	18	34	94	75	20	80	27	77	78	91	69	16	00	08	43	18	73	68	67	69	61	34	25
9	88	98	99	60	50	65	95	79	42	94	93	62	40	89	96	43	56	47	71	66	46	76	29	67	02
10	04	37	59	87	21	05	02	03	24	17	47	97	81	56	51	92	34	86	01	82	55	51	33	12	91
11	63	62	06	34	41	94	21	78	55	09	72	76	45	16	94	29	95	81	83	83	79	88	01	97	30
12	78	47	23	53	90	34	41	92	45	71	09	23	70	70	07	12	38	92	79	43	14	85	11	47	23
13	87	68	62	15	43	53	14	36	59	25	54	47	33	70	15	59	24	48	40	35	50	03	42	99	36
14	47	60	92	10	77	88	59	53	11	52	66	25	69	07	04	48	68	64	71	06	61	65	70	22	12
15	56	88	87	59	41	65	28	04	67	53	95	79	88	37	31	50	41	06	94	76	81	83	17	16	33
16	02	57	45	86	67	73	43	07	34	48	44	26	87	93	29	77	09	61	67	84	06	69	44	77	75
17	31	54	14	13	17	48	62	11	90	60	68	12	93	64	28	46	24	79	16	76	14	60	25	51	01
18	28	50	16	43	36	28	97	85	58	99	67	22	52	76	23	24	70	36	54	54	59	28	61	71	96
19	63	29	62	66	50	02	63	45	52	38	67	63	47	54	75	83	24	78	43	20	92	63	13	47	48
20	45	65	58	26	51	76	96	59	38	72	86	57	45	71	46	44	67	76	14	55	44	88	01	62	12
21	39	65	36	63	70	77	45	85	50	51	74	13	39	35	22	30	53	36	02	95	49	34	88	73	61
22	73	71	98	16	04	29	18	94	51	23	76	51	94	84	86	79	93	96	38	63	08	58	25	58	94
23	72	20	56	20	11	72	65	71	08	86	79	57	95	13	91	97	48	72	66	48	09	71	17	24	89
24	75	17	26	99	76	89	37	20	70	01	77	31	61	95	46	26	97	05	73	51	53	33	18	72	87
25	37	48	60	82	29	81	30	15	39	14	48	38	75	93	29	06	87	37	78	48	45	56	00	84	47
26	68	08	02	80	72	83	71	46	30	49	89	17	95	88	29	02	39	56	03	46	97	74	06	56	17
27	14	23	98	61	67	70	52	85	01	50	01	84	02	78	43	10	62	98	19	41	18	83	99	47	99
28	49	08	96	21	44	25	27	99	41	28	07	41	08	34	66	19	42	74	39	91	41	96	53	78	72
29	78	37	06	08	43	63	61	62	42	29	39	68	95	10	96	09	24	23	00	62	56	12	80	73	16
30	37	21	34	17	68	68	96	83	23	56	32	84	60	15	31	44	73	67	34	77	91	15	79	74	58
31	14	29	09	34	04	87	83	07	55	07	76	58	30	83	64	87	29	25	58	84	86	50	60	00	25
32	58	43	28	06	36	49	52	83	51	14	47	56	91	29	34	05	87	31	06	95	12	45	57	09	09
33	10	43	67	29	70	80	62	80	03	42	10	80	21	38	84	90	56	35	03	09	43	12	74	49	14
34	44	38	88	39	54	86	97	37	44	22	00	95	01	31	76	17	16	29	56	63	38	78	94	49	81
35	90	69	59	19	51	85	39	52	85	13	07	28	37	07	61	11	16	36	27	03	78	86	72	04	95
36	41	47	10	25	62	97	05	31	03	61	20	26	36	31	62	68	69	86	95	44	84	95	48	46	45
37	91	94	14	63	19	75	89	11	47	11	31	56	34	19	09	79	57	92	36	59	14	93	87	81	40
38	80	06	54	18	66	09	18	94	06	19	98	40	07	17	81	22	45	44	84	11	24	62	20	42	31
39	67	72	77	63	48	84	08	31	55	58	24	33	45	77	58	80	45	67	93	82	75	70	16	08	24
40	59	40	24	13	27	79	26	88	86	30	01	31	60	10	39	53	58	47	70	93	85	81	56	39	38
41	05	90	35	89	95	01	61	16	96	94	50	78	13	69	36	37	68	53	37	31	71	26	35	03	71
42	44	43	80	69	98	46	68	05	14	82	90	78	50	05	62	77	79	13	57	44	59	60	10	39	66
43	61	81	31	96	82	00	57	25	60	59	46	72	60	18	77	55	66	12	62	11	08	99	55	64	57
44	42	88	07	10	05	24	98	65	63	21	47	21	61	88	32	27	80	30	21	60	10	92	35	36	12
45	77	94	30	05	39	28	10	99	00	27	12	73	73	99	12	49	99	57	94	82	96	88	57	17	91
46	78	83	19	76	16	94	11	68	84	26	23	54	20	86	85	23	86	66	99	07	36	37	34	92	09
47	87	76	59	61	81	43	63	64	61	61	65	76	36	95	90	18	48	27	45	68	27	23	65	30	72
48	91	43	05	96	47	55	78	99	95	24	37	55	85	78	78	01	48	41	19	10	35	19	54	07	73
49	84	97	77	72	73	09	62	06	65	72	87	12	49	03	60	41	15	20	76	27	50	47	02	29	16
50	87	41	60	76	83	44	88	96	07	80	83	05	83	38	96	73	70	66	81	90	30	56	10	48	59

Table E.2 is taken from Table 33 of Fisher and Yates, *Statistical Tables for Biological, Agricultural and Medical Research*, published by Oliver and Boyd Ltd., Edinburgh, by permission of the authors and publishers.

TABLE E.2 (continued)

	1	2	3	4	5	6	7	8	9	10	11	12	13	14	15	16	17	18	19	20	21	22	23	24	25
1	22	17	68	65	84	68	95	23	92	35	87	02	22	57	51	61	09	43	95	06	58	24	82	03	47
2	19	36	27	59	46	13	79	93	37	55	39	77	32	77	09	85	52	05	30	62	47	83	51	62	74
3	16	77	23	02	77	09	61	87	25	21	28	06	24	25	93	16	71	13	59	78	23	05	47	47	25
4	78	43	76	71	61	20	44	90	32	64	97	67	63	99	61	46	38	03	93	22	69	81	21	99	21
5	03	28	28	26	08	73	37	32	04	05	69	30	16	09	05	88	69	58	28	99	35	07	44	75	47
6	93	22	53	64	39	07	10	63	76	35	87	03	04	79	88	08	13	13	85	51	55	34	57	72	69
7	78	76	58	54	74	92	38	70	96	92	52	06	79	79	45	82	63	18	27	44	69	66	92	19	09
8	23	68	35	26	00	99	53	93	61	28	52	70	05	48	34	56	65	05	61	86	90	92	10	70	80
9	15	39	25	70	99	93	86	52	77	65	15	33	59	05	28	22	87	26	07	47	86	96	98	29	06
10	58	71	96	30	24	18	46	23	34	27	85	13	99	24	44	49	18	09	79	49	74	16	32	23	02
11	57	35	27	33	72	24	53	63	94	09	41	10	76	47	91	44	04	95	49	66	39	60	04	59	81
12	48	50	86	54	48	22	06	34	72	52	82	21	15	65	20	33	29	94	71	11	15	91	29	12	03
13	61	96	48	95	03	07	16	39	33	66	98	56	10	56	79	77	21	30	27	12	90	49	22	23	62
14	36	93	89	41	26	29	70	83	63	51	99	74	20	52	36	87	09	41	15	09	98	60	16	03	03
15	18	87	00	42	31	57	90	12	02	07	23	47	37	17	31	54	08	01	88	63	39	41	88	92	10
16	88	56	53	27	59	33	35	72	67	47	77	34	55	45	70	08	18	27	38	90	16	95	86	70	75
17	09	72	95	84	29	49	41	31	06	70	42	38	06	45	18	64	84	73	31	65	52	53	37	97	15
18	12	96	88	17	31	65	19	69	02	83	60	75	86	90	68	24	64	19	35	51	56	61	87	39	12
19	85	94	57	24	16	92	09	84	38	76	22	00	27	69	85	29	81	94	78	70	21	94	47	90	12
20	38	64	43	59	98	98	77	87	68	07	91	51	67	62	44	40	98	05	93	78	23	32	65	41	18
21	53	44	09	42	72	00	41	86	79	79	68	47	22	00	20	35	55	31	51	51	00	83	63	22	55
22	40	76	66	26	84	57	99	99	90	37	36	63	32	08	58	37	40	13	68	97	87	64	81	07	83
23	02	17	79	18	05	12	59	52	57	02	22	07	90	47	03	28	14	11	30	79	20	69	22	40	98
24	95	17	82	06	53	31	51	10	96	46	92	06	88	07	77	56	11	50	81	69	40	23	72	51	39
25	35	76	22	42	92	96	11	83	44	80	34	68	35	48	77	33	42	40	90	60	73	96	53	97	86
26	26	29	13	56	41	85	47	04	66	08	34	72	57	59	13	82	43	80	46	15	38	26	61	70	04
27	77	80	20	75	82	72	82	32	99	90	63	95	73	76	63	89	73	44	99	05	48	67	26	43	18
28	46	40	66	44	52	91	36	74	43	53	30	82	13	54	00	78	45	63	98	35	55	03	36	67	68
29	37	56	08	18	09	77	53	84	46	47	31	91	18	95	58	24	16	74	11	53	44	10	13	85	57
30	61	65	61	68	66	37	27	47	39	19	84	83	70	07	48	53	21	40	06	71	95	06	79	88	54
31	93	43	69	64	07	34	18	04	52	35	56	27	09	24	86	61	85	53	83	45	19	90	70	99	00
32	21	96	60	12	99	11	20	99	45	18	48	13	93	55	34	18	37	79	49	90	65	97	38	20	46
33	95	20	47	97	97	27	37	83	28	71	00	06	41	41	74	45	89	09	39	84	51	67	11	52	49
34	97	86	21	78	73	10	65	81	92	59	58	76	17	14	97	04	76	62	16	17	17	95	70	45	80
35	69	92	06	34	13	59	71	74	17	32	27	55	10	24	19	23	71	82	13	74	63	52	52	01	41
36	04	31	17	21	56	33	73	99	19	87	26	72	39	27	67	53	77	57	68	93	60	61	97	22	61
37	61	06	98	03	91	87	14	77	43	96	43	00	65	98	50	45	60	33	01	07	98	99	46	50	47
38	85	93	85	86	88	72	87	08	62	40	16	06	10	89	20	23	21	34	74	97	76	38	03	29	63
39	21	74	32	47	45	73	96	07	94	52	09	65	90	77	47	25	76	16	19	33	53	05	70	53	30
40	15	69	53	82	80	79	96	23	53	10	65	39	07	16	29	45	33	02	43	70	02	87	40	41	45
41	02	89	08	04	49	20	21	14	68	86	87	63	93	95	17	11	29	01	95	80	35	14	97	35	33
42	87	18	15	89	79	85	43	01	72	73	08	61	74	51	69	89	74	39	82	15	94	51	33	41	67
43	98	83	71	94	22	59	97	50	99	52	08	52	85	08	40	87	80	61	65	31	91	51	80	32	44
44	10	08	58	21	66	72	68	49	29	31	89	85	84	46	06	59	73	19	85	23	65	09	29	75	63
45	47	90	56	10	08	88	02	84	27	83	42	29	72	23	19	66	56	45	65	79	20	71	53	20	25
46	22	85	61	68	90	49	64	92	85	44	16	40	12	89	88	50	14	49	81	06	01	82	77	45	12
47	67	80	43	79	33	12	83	11	41	16	25	58	19	68	70	77	02	54	00	52	53	43	37	15	26
48	27	62	50	96	72	79	44	61	40	15	14	53	40	65	39	27	31	58	50	28	11	39	03	34	25
49	33	78	80	87	15	38	30	06	38	21	14	47	47	07	26	54	96	87	53	32	40	36	40	96	76
50	13	13	92	66	99	47	24	49	57	74	32	25	43	62	17	10	97	11	69	84	99	63	22	32	98

TABLE E.3   Areas under the Standard Normal Distribution

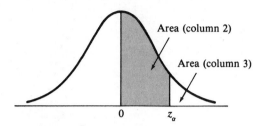

(1)	(2) Area between Mean and	(3) Area beyond	(1)	(2) Area between Mean and	(3) Area beyond	(1)	(2) Area between Mean and	(3) Area beyond
z_α	z_α	z_α	z_α	z_α	z_α	z_α	z_α	z_α
0.00	0.0000	0.5000	0.30	0.1179	0.3821	0.60	0.2257	0.2743
0.01	0.0040	0.4960	0.31	0.1217	0.3783	0.61	0.2291	0.2709
0.02	0.0080	0.4920	0.32	0.1255	0.3745	0.62	0.2324	0.2676
0.03	0.0120	0.4880	0.33	0.1293	0.3707	0.63	0.2357	0.2643
0.04	0.0160	0.4840	0.34	0.1331	0.3669	0.64	0.2389	0.2611
0.05	0.0199	0.4801	0.35	0.1368	0.3632	0.65	0.2422	0.2578
0.06	0.0239	0.4761	0.36	0.1406	0.3594	0.66	0.2454	0.2546
0.07	0.0279	0.4721	0.37	0.1443	0.3557	0.67	0.2486	0.2514
0.08	0.0319	0.4681	0.38	0.1480	0.3520	0.68	0.2517	0.2483
0.09	0.0359	0.4641	0.39	0.1517	0.3483	0.69	0.2549	0.2451
0.10	0.0398	0.4602	0.40	0.1554	0.3446	0.70	0.2580	0.2420
0.11	0.0438	0.4562	0.41	0.1591	0.3409	0.71	0.2611	0.2389
0.12	0.0478	0.4522	0.42	0.1628	0.3372	0.72	0.2642	0.2358
0.13	0.0517	0.4483	0.43	0.1664	0.3336	0.73	0.2673	0.2327
0.14	0.0557	0.4443	0.44	0.1700	0.3300	0.74	0.2704	0.2296
0.15	0.0596	0.4404	0.45	0.1736	0.3264	0.75	0.2734	0.2266
0.16	0.0636	0.4364	0.46	0.1772	0.3228	0.76	0.2764	0.2236
0.17	0.0675	0.4325	0.47	0.1808	0.3192	0.77	0.2794	0.2206
0.18	0.0714	0.4286	0.48	0.1844	0.3156	0.78	0.2823	0.2177
0.19	0.0753	0.4247	0.49	0.1879	0.3121	0.79	0.2852	0.2148
0.20	0.0793	0.4207	0.50	0.1915	0.3085	0.80	0.2881	0.2119
0.21	0.0832	0.4168	0.51	0.1950	0.3050	0.81	0.2910	0.2090
0.22	0.0871	0.4129	0.52	0.1985	0.3015	0.82	0.2939	0.2061
0.23	0.0910	0.4090	0.53	0.2019	0.2981	0.83	0.2967	0.2033
0.24	0.0948	0.4052	0.54	0.2054	0.2946	0.84	0.2995	0.2005
0.25	0.0987	0.4013	0.55	0.2088	0.2912	0.85	0.3023	0.1977
0.26	0.1026	0.3974	0.56	0.2123	0.2877	0.86	0.3051	0.1949
0.27	0.1064	0.3936	0.57	0.2157	0.2843	0.87	0.3078	0.1922
0.28	0.1103	0.3897	0.58	0.2190	0.2810	0.88	0.3106	0.1894
0.29	0.1141	0.3859	0.59	0.2224	0.2776	0.89	0.3133	0.1867

TABLE E.3 (continued)

(1) z_α	(2) Area between Mean and z_α	(3) Area beyond z_α	(1) z_α	(2) Area between Mean and z_α	(3) Area beyond z_α	(1) z_α	(2) Area between Mean and z_α	(3) Area beyond z_α
0.90	0.3159	0.1841	1.35	0.4115	0.0885	1.80	0.4641	0.0359
0.91	0.3186	0.1814	1.36	0.4131	0.0869	1.81	0.4649	0.0351
0.92	0.3212	0.1788	1.37	0.4147	0.0853	1.82	0.4656	0.0344
0.93	0.3238	0.1762	1.38	0.4162	0.0838	1.83	0.4664	0.0336
0.94	0.3264	0.1736	1.39	0.4177	0.0823	1.84	0.4671	0.0329
0.95	0.3289	0.1711	1.40	0.4192	0.0808	1.85	0.4678	0.0322
0.96	0.3315	0.1685	1.41	0.4207	0.0793	1.86	0.4686	0.0314
0.97	0.3340	0.1660	1.42	0.4222	0.0778	1.87	0.4693	0.0307
0.98	0.3365	0.1635	1.43	0.4236	0.0764	1.88	0.4699	0.0301
0.99	0.3389	0.1611	1.44	0.4251	0.0749	1.89	0.4706	0.0294
1.00	0.3413	0.1587	1.45	0.4265	0.0735	1.90	0.4713	0.0287
1.01	0.3438	0.1562	1.46	0.4279	0.0721	1.91	0.4719	0.0281
1.02	0.3461	0.1539	1.47	0.4292	0.0708	1.92	0.4726	0.0274
1.03	0.3485	0.1515	1.48	0.4306	0.0694	1.93	0.4732	0.0268
1.04	0.3508	0.1492	1.49	0.4319	0.0681	1.94	0.4738	0.0262
1.05	0.3531	0.1469	1.50	0.4332	0.0668	1.95	0.4744	0.0256
1.06	0.3554	0.1446	1.51	0.4345	0.0655	1.96	0.4750	0.0250
1.07	0.3577	0.1423	1.52	0.4357	0.0643	1.97	0.4756	0.0244
1.08	0.3599	0.1401	1.53	0.4370	0.0630	1.98	0.4761	0.0239
1.09	0.3621	0.1379	1.54	0.4382	0.0618	1.99	0.4767	0.0233
1.10	0.3643	0.1357	1.55	0.4394	0.0606	2.00	0.4772	0.0228
1.11	0.3665	0.1335	1.56	0.4406	0.0594	2.01	0.4778	0.0222
1.12	0.3686	0.1314	1.57	0.4418	0.0582	2.02	0.4783	0.0217
1.13	0.3708	0.1292	1.58	0.4429	0.0571	2.03	0.4788	0.0212
1.14	0.3729	0.1271	1.59	0.4441	0.0559	2.04	0.4793	0.0207
1.15	0.3749	0.1251	1.60	0.4452	0.0548	2.05	0.4798	0.0202
1.16	0.3770	0.1230	1.61	0.4463	0.0537	2.06	0.4803	0.0197
1.17	0.3790	0.1210	1.62	0.4474	0.0526	2.07	0.4808	0.0192
1.18	0.3810	0.1190	1.63	0.4484	0.0516	2.08	0.4812	0.0188
1.19	0.3830	0.1170	1.64	0.4495	0.0505	2.09	0.4817	0.0183
			1.645	0.4500	0.0500			
1.20	0.3849	0.1151	1.65	0.4505	0.0495	2.10	0.4821	0.0179
1.21	0.3869	0.1131	1.66	0.4515	0.0485	2.11	0.4826	0.0174
1.22	0.3888	0.1112	1.67	0.4525	0.0475	2.12	0.4830	0.0170
1.23	0.3907	0.1093	1.68	0.4535	0.0465	2.13	0.4834	0.0166
1.24	0.3925	0.1075	1.69	0.4545	0.0455	2.14	0.4838	0.0162
1.25	0.3944	0.1056	1.70	0.4554	0.0446	2.15	0.4842	0.0158
1.26	0.3962	0.1038	1.71	0.4564	0.0436	2.16	0.4846	0.0154
1.27	0.3980	0.1020	1.72	0.4573	0.0427	2.17	0.4850	0.0150
1.28	0.3997	0.1003	1.73	0.4582	0.0418	2.18	0.4854	0.0146
1.29	0.4015	0.0985	1.74	0.4591	0.0409	2.19	0.4857	0.0143
1.30	0.4032	0.0968	1.75	0.4599	0.0401	2.20	0.4861	0.0139
1.31	0.4049	0.0951	1.76	0.4608	0.0392	2.21	0.4864	0.0136
1.32	0.4066	0.0934	1.77	0.4616	0.0384	2.22	0.4868	0.0132
1.33	0.4082	0.0918	1.78	0.4625	0.0375	2.23	0.4871	0.0129
1.34	0.4099	0.0901	1.79	0.4633	0.0367	2.24	0.4875	0.0125

TABLE E.3 (continued)

(1) z_α	(2) Area between Mean and z_α	(3) Area beyond z_α	(1) z_α	(2) Area between Mean and z_α	(3) Area beyond z_α	(1) z_α	(2) Area between Mean and z_α	(3) Area beyond z_α
2.25	0.4878	0.0122	2.64	0.4959	0.0041	3.00	0.4987	0.0013
2.26	0.4881	0.0119	2.65	0.4960	0.0040	3.01	0.4987	0.0013
2.27	0.4884	0.0116	2.66	0.4961	0.0039	3.02	0.4987	0.0013
2.28	0.4887	0.0113	2.67	0.4962	0.0038	3.03	0.4988	0.0012
2.29	0.4890	0.0100	2.68	0.4963	0.0037	3.04	0.4988	0.0012
2.30	0.4893	0.0107	2.69	0.4964	0.0036	3.05	0.4989	0.0011
2.31	0.4896	0.0104	2.70	0.4965	0.0035	3.06	0.4989	0.0011
2.32	0.4898	0.0102	2.71	0.4966	0.0034	3.07	0.4989	0.0011
2.33	0.4901	0.0099	2.72	0.4967	0.0033	3.08	0.4990	0.0010
2.34	0.4904	0.0096	2.73	0.4968	0.0032	3.09	0.4990	0.0010
2.35	0.4906	0.0094	2.74	0.4969	0.0031	3.10	0.4990	0.0010
2.36	0.4909	0.0091	2.75	0.4970	0.0030	3.11	0.4991	0.0009
2.37	0.4911	0.0089	2.76	0.4971	0.0029	3.12	0.4991	0.0009
2.38	0.4913	0.0087	2.77	0.4972	0.0028	3.13	0.4991	0.0009
2.39	0.4916	0.0084	2.78	0.4973	0.0027	3.14	0.4992	0.0008
2.40	0.4918	0.0082	2.79	0.4974	0.0026	3.15	0.4992	0.0008
2.41	0.4920	0.0080	2.80	0.4974	0.0026	3.16	0.4992	0.0008
2.42	0.4922	0.0078	2.81	0.4975	0.0025	3.17	0.4992	0.0008
2.43	0.4925	0.0075	2.82	0.4976	0.0024	3.18	0.4993	0.0007
2.44	0.4927	0.0073	2.83	0.4977	0.0023	3.19	0.4993	0.0007
2.45	0.4929	0.0071	2.84	0.4977	0.0023	3.20	0.4993	0.0007
2.46	0.4931	0.0069	2.85	0.4978	0.0022	3.21	0.4993	0.0007
2.47	0.4932	0.0068	2.86	0.4979	0.0021	3.22	0.4994	0.0006
2.48	0.4934	0.0066	2.87	0.4979	0.0021	3.23	0.4994	0.0006
2.49	0.4936	0.0064	2.88	0.4980	0.0020	3.24	0.4994	0.0006
2.50	0.4938	0.0062	2.89	0.4981	0.0019	3.25	0.4994	0.0006
2.51	0.4940	0.0060	2.90	0.4981	0.0019	3.30	0.4995	0.0005
2.52	0.4941	0.0059	2.91	0.4982	0.0018	3.35	0.4996	0.0004
2.53	0.4943	0.0057	2.92	0.4982	0.0018	3.40	0.4997	0.0003
2.54	0.4945	0.0055	2.93	0.4983	0.0017	3.45	0.4997	0.0003
2.55	0.4946	0.0054	2.94	0.4984	0.0016	3.50	0.4998	0.0002
2.56	0.4948	0.0052	2.95	0.4984	0.0016	3.60	0.4998	0.0002
2.57	0.4949	0.0051	2.96	0.4985	0.0015	3.70	0.4999	0.0001
2.576	0.4950	0.0050	2.97	0.4985	0.0015	3.80	0.4999	0.0001
2.58	0.4951	0.0049	2.98	0.4986	0.0014	3.90	0.49995	0.00005
2.59	0.4952	0.0048	2.99	0.4986	0.0014	4.00	0.49997	0.00003
2.60	0.4953	0.0047						
2.61	0.4955	0.0045						
2.62	0.4956	0.0044						
2.63	0.4957	0.0043						

TABLE E.4 Percentage Points of Student's t Distribution

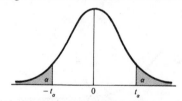

df	α .25 2α .50	.20 .40	.15 .30	.10 .20	.05 .10	.025 .05	.01 .02	.005 .01	.0005 .001
1	1.000	1.376	1.963	3.078	6.314	12.706	31.821	63.657	636.619
2	.816	1.061	1.386	1.886	2.920	4.303	6.965	9.925	31.598
3	.765	.978	1.250	1.638	2.353	3.182	4.541	5.841	12.924
4	.741	.941	1.190	1.533	2.132	2.776	3.747	4.604	8.610
5	.727	.920	1.156	1.476	2.015	2.571	3.365	4.032	6.869
6	.718	.906	1.134	1.440	1.943	2.447	3.143	3.707	5.959
7	.711	.896	1.119	1.415	1.895	2.365	2.998	3.499	5.408
8	.706	.889	1.108	1.397	1.860	2.306	2.896	3.355	5.041
9	.703	.883	1.100	1.383	1.833	2.262	2.821	3.250	4.781
10	.700	.879	1.093	1.372	1.812	2.228	2.764	3.169	4.587
11	.697	.876	1.088	1.363	1.796	2.201	2.718	3.106	4.437
12	.695	.873	1.083	1.356	1.782	2.179	2.681	3.055	4.318
13	.694	.870	1.079	1.350	1.771	2.160	2.650	3.012	4.221
14	.692	.868	1.076	1.345	1.761	2.145	2.624	2.977	4.140
15	.691	.866	1.074	1.341	1.753	2.131	2.602	2.947	4.073
16	.690	.865	1.071	1.337	1.746	2.120	2.583	2.921	4.015
17	.689	.863	1.069	1.333	1.740	2.110	2.567	2.898	3.965
18	.688	.862	1.067	1.330	1.734	2.101	2.552	2.878	3.922
19	.688	.861	1.066	1.328	1.729	2.093	2.539	2.861	3.883
20	.687	.860	1.064	1.325	1.725	2.086	2.528	2.845	3.850
21	.686	.859	1.063	1.323	1.721	2.080	2.518	2.831	3.819
22	.686	.858	1.061	1.321	1.717	2.074	2.508	2.819	3.792
23	.685	.858	1.060	1.319	1.714	2.069	2.500	2.807	3.767
24	.685	.857	1.059	1.318	1.711	2.064	2.492	2.797	3.745
25	.684	.856	1.058	1.316	1.708	2.060	2.485	2.787	3.725
26	.684	.856	1.058	1.315	1.706	2.056	2.479	2.779	3.707
27	.684	.855	1.057	1.314	1.703	2.052	2.473	2.771	3.690
28	.683	.855	1.056	1.313	1.701	2.048	2.467	2.763	3.674
29	.683	.854	1.055	1.311	1.699	2.045	2.462	2.756	3.659
30	.683	.854	1.055	1.310	1.697	2.042	2.457	2.750	3.646
40	.681	.851	1.050	1.303	1.684	2.021	2.423	2.704	3.551
60	.679	.848	1.046	1.296	1.671	2.000	2.390	2.660	3.460
120	.677	.845	1.041	1.289	1.658	1.980	2.358	2.617	3.373
∞	.674	.842	1.036	1.282	1.645	1.960	2.326	2.576	3.291

Table E.4 is taken from Table 3 of Fisher and Yates, *Statistical Tables for Biological, Agricultural and Medical Research*, published by Oliver and Boyd Ltd., Edinburgh, by permission of the authors and publishers.

TABLE E.5   Upper Percentage Points of the F Distribution

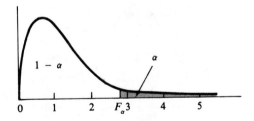

df for Denominator	α	1	2	3	4	5	6	7	8	9	10	11	12
						df for Numerator							
1	.25	5.83	7.50	8.20	8.58	8.82	8.98	9.10	9.19	9.26	9.32	9.36	9.41
	.10	39.9	49.5	53.6	55.8	57.2	58.2	58.9	59.4	59.9	60.2	60.5	60.7
	.05	161	200	216	225	230	234	237	239	241	242	243	244
2	.25	2.57	3.00	3.15	3.23	3.28	3.31	3.34	3.35	3.37	3.38	3.39	3.39
	.10	8.53	9.00	9.16	9.24	9.29	9.33	9.35	9.37	9.38	9.39	9.40	9.41
	.05	18.5	19.0	19.2	19.2	19.3	19.3	19.4	19.4	19.4	19.4	19.4	19.4
	.01	98.5	99.0	99.2	99.2	99.3	99.3	99.4	99.4	99.4	99.4	99.4	99.4
3	.25	2.02	2.28	2.36	2.39	2.41	2.42	2.43	2.44	2.44	2.44	2.45	2.45
	.10	5.54	5.46	5.39	5.34	5.31	5.28	5.27	5.25	5.24	5.23	5.22	5.22
	.05	10.1	9.55	9.28	9.12	9.01	8.94	8.89	8.85	8.81	8.79	8.76	8.74
	.01	34.1	30.8	29.5	28.7	28.2	27.9	27.7	27.5	27.3	27.2	27.1	27.1
4	.25	1.81	2.00	2.05	2.06	2.07	2.08	2.08	2.08	2.08	2.08	2.08	2.08
	.10	4.54	4.32	4.19	4.11	4.05	4.01	3.98	3.95	3.94	3.92	3.91	3.90
	.05	7.71	6.94	6.59	6.39	6.26	6.16	6.09	6.04	6.00	5.96	5.94	5.91
	.01	21.2	18.0	16.7	16.0	15.5	15.2	15.0	14.8	14.7	14.5	14.4	14.4
5	.25	1.69	1.85	1.88	1.89	1.89	1.89	1.89	1.89	1.89	1.89	1.89	1.89
	.10	4.06	3.78	3.62	3.52	3.45	3.40	3.37	3.34	3.32	3.30	3.28	3.27
	.05	6.61	5.79	5.41	5.19	5.05	4.95	4.88	4.82	4.77	4.74	4.71	4.68
	.01	16.3	13.3	12.1	11.4	11.0	10.7	10.5	10.3	10.2	10.1	9.96	9.89
6	.25	1.62	1.76	1.78	1.79	1.79	1.78	1.78	1.78	1.77	1.77	1.77	1.77
	.10	3.78	3.46	3.29	3.18	3.11	3.05	3.01	2.98	2.96	2.94	2.92	2.90
	.05	5.99	5.14	4.76	4.53	4.39	4.28	4.21	4.15	4.10	4.06	4.03	4.00
	.01	13.7	10.9	9.78	9.15	8.75	8.47	8.26	8.10	7.98	7.87	7.79	7.72
7	.25	1.57	1.70	1.72	1.72	1.71	1.71	1.70	1.70	1.69	1.69	1.69	1.68
	.10	3.59	3.26	3.07	2.96	2.88	2.83	2.78	2.75	2.72	2.70	2.68	2.67
	.05	5.59	4.74	4.35	4.12	3.97	3.87	2.79	3.73	3.68	3.64	3.60	3.57
	.01	12.2	9.55	8.45	7.85	7.46	7.19	6.99	6.84	6.72	6.62	6.54	6.47
8	.25	1.54	1.66	1.67	1.66	1.66	1.65	1.64	1.64	1.63	1.63	1.63	1.62
	.10	3.46	3.11	2.92	2.81	2.73	2.67	2.62	2.59	2.56	2.54	2.52	2.50
	.05	5.32	4.46	4.07	3.84	3.69	3.58	3.50	3.44	3.39	3.35	3.31	3.28
	.01	11.3	8.65	7.59	7.01	6.63	6.37	6.18	6.03	5.91	5.81	5.73	5.67
9	.25	1.51	1.62	1.63	1.63	1.62	1.61	1.60	1.60	1.59	1.59	1.58	1.58
	.10	3.36	3.01	2.81	2.69	2.61	2.55	2.51	2.47	2.44	2.42	2.40	2.38
	.05	5.12	4.26	3.86	3.63	3.48	3.37	3.29	3.23	3.18	3.14	3.10	3.07
	.01	10.6	8.02	6.99	6.42	6.06	5.80	5.61	5.47	5.35	5.26	5.18	5.11

TABLE E.5 (continued)

15	20	24	30	40	50	60	100	120	200	500	∞	α	df for Denominator
					df for Numerator								
9.49	9.58	9.63	9.67	9.71	9.74	9.76	9.78	9.80	9.82	9.84	9.85	.25	
61.2	61.7	62.0	62.3	62.5	62.7	62.8	63.0	63.1	63.2	63.3	63.3	.10	1
246	248	249	250	251	252	252	253	253	254	254	254	.05	
3.41	3.43	3.43	3.44	3.45	3.45	3.46	3.47	3.47	3.48	3.48	3.48	.25	
9.42	9.44	9.45	9.46	9.47	9.47	9.47	9.48	9.48	9.49	9.49	9.49	.10	2
19.4	19.4	19.5	19.5	19.5	19.5	19.5	19.5	19.5	19.5	19.5	19.5	.05	
99.4	99.4	99.5	99.5	99.5	99.5	99.5	99.5	99.5	99.5	99.5	99.5	.01	
2.46	2.46	2.46	2.47	2.47	2.47	2.47	2.47	2.47	2.47	2.47	2.47	.25	
5.20	5.18	5.18	5.17	5.16	5.15	5.15	5.14	5.14	5.14	5.14	5.13	.10	3
8.70	8.66	8.64	8.62	8.59	8.58	8.57	8.55	8.55	8.54	8.53	8.53	.05	
26.9	26.7	26.6	26.5	26.4	26.4	26.3	26.2	26.2	26.2	26.1	26.1	.01	
2.08	2.08	2.08	2.08	2.08	2.08	2.08	2.08	2.08	2.08	2.08	2.08	.25	
3.87	3.84	3.83	3.82	3.80	3.80	3.79	3.78	3.78	3.77	3.76	3.76	.10	4
5.86	5.80	5.77	5.75	5.72	5.70	5.69	5.66	5.66	5.65	5.64	5.63	.05	
14.2	14.0	13.9	13.8	13.7	13.7	13.7	13.6	13.6	13.5	13.5	13.5	.01	
1.89	1.88	1.88	1.88	1.88	1.88	1.87	1.87	1.87	1.87	1.87	1.87	.25	
3.24	3.21	3.19	3.17	3.16	3.15	3.14	3.13	3.12	3.12	3.11	3.10	.10	5
4.62	4.56	4.53	4.50	4.46	4.44	4.43	4.41	4.40	4.39	4.37	4.36	.05	
9.72	9.55	9.47	9.38	9.29	9.24	9.20	9.13	9.11	9.08	9.04	9.02	.01	
1.76	1.76	1.75	1.75	1.75	1.75	1.74	1.74	1.74	1.74	1.74	1.74	.25	
2.87	2.84	2.82	2.80	2.78	2.77	2.76	2.75	2.74	2.73	2.73	2.72	.10	6
3.94	3.87	3.84	3.81	3.77	3.75	3.74	3.71	3.70	3.69	3.68	3.67	.05	
7.56	7.40	7.31	7.23	7.14	7.09	7.06	6.99	6.97	6.93	6.90	6.88	.01	
1.68	1.67	1.67	1.66	1.66	1.66	1.65	1.65	1.65	1.65	1.65	1.65	.25	
2.63	2.59	2.58	2.56	2.54	2.52	2.51	2.50	2.49	2.48	2.48	2.47	.10	7
3.51	3.44	3.41	3.38	3.34	3.32	3.30	3.27	3.27	3.25	3.24	3.23	.05	
6.31	6.16	6.07	5.99	5.91	5.86	5.82	5.75	5.74	5.70	5.67	5.65	.01	
1.62	1.61	1.60	1.60	1.59	1.59	1.59	1.58	1.58	1.58	1.58	1.58	.25	
2.46	2.42	2.40	2.38	2.36	2.35	2.34	2.32	2.32	2.31	2.30	2.29	.10	8
3.22	3.15	3.12	3.08	3.04	3.02	3.01	2.97	2.97	2.95	2.94	2.93	.05	
5.52	5.36	5.28	5.20	5.12	5.07	5.03	4.96	4.95	4.91	4.88	4.86	.01	
1.57	1.56	1.56	1.55	1.55	1.54	1.54	1.53	1.53	1.53	1.53	1.53	.25	
2.34	2.30	2.28	2.25	2.23	2.22	2.21	2.19	2.18	2.17	2.17	2.16	.10	9
3.01	2.94	2.90	2.86	2.83	2.80	2.79	2.76	2.75	2.73	2.72	2.71	.05	
4.96	4.81	4.73	4.65	4.57	4.52	4.48	4.42	4.40	4.36	4.33	4.31	.01	

TABLE E.5　(continued)

| df for Denominator | α | \multicolumn{12}{c}{df for Numerator} |
		1	2	3	4	5	6	7	8	9	10	11	12
10	.25	1.49	1.60	1.60	1.59	1.59	1.58	1.57	1.56	1.56	1.55	1.55	1.54
	.10	3.29	2.92	2.73	2.61	2.52	2.46	2.41	2.38	2.35	2.32	2.30	2.28
	.05	4.96	4.10	3.71	3.48	3.33	3.22	3.14	3.07	3.02	2.98	2.94	2.91
	.01	10.0	7.56	6.55	5.99	5.64	5.39	5.20	5.06	4.94	4.85	4.77	4.71
11	.25	1.47	1.58	1.58	1.57	1.56	1.55	1.54	1.53	1.53	1.52	1.52	1.51
	.10	3.23	2.86	2.66	2.54	2.45	2.39	2.34	2.30	2.27	2.25	2.23	2.21
	.05	4.84	3.98	3.59	3.36	3.20	3.09	3.01	2.95	2.90	2.85	2.82	2.79
	.01	9.65	7.21	6.22	5.67	5.32	5.07	4.89	4.74	4.63	4.54	4.46	4.40
12	.25	1.46	1.56	1.56	1.55	1.54	1.53	1.52	1.51	1.51	1.50	1.50	1.49
	.10	3.18	2.81	2.61	2.48	2.39	2.33	2.28	2.24	2.21	2.19	2.17	2.15
	.05	4.75	3.89	3.49	3.26	3.11	3.00	2.91	2.85	2.80	2.75	2.72	2.69
	.01	9.33	6.93	5.95	5.41	5.06	4.82	4.64	4.50	4.39	4.30	4.22	4.16
13	.25	1.45	1.55	1.55	1.53	1.52	1.51	1.50	1.49	1.49	1.48	1.47	1.47
	.10	3.14	2.76	2.56	2.43	2.35	2.28	2.23	2.20	2.16	2.14	2.12	2.10
	.05	4.67	3.81	3.41	3.18	3.03	2.92	2.83	2.77	2.71	2.67	2.63	2.60
	.01	9.07	6.70	5.74	5.21	4.86	4.62	4.44	4.30	4.19	4.10	4.02	3.96
14	.25	1.44	1.53	1.53	1.52	1.51	1.50	1.49	1.48	1.47	1.46	1.46	1.45
	.10	3.10	2.73	2.52	2.39	2.31	2.24	2.19	2.15	2.12	2.10	2.08	2.05
	.05	4.60	3.74	3.34	3.11	2.96	2.85	2.76	2.70	2.65	2.60	2.57	2.53
	.01	8.86	6.51	5.56	5.04	4.69	4.46	4.28	4.14	4.03	3.94	3.86	3.80
15	.25	1.43	1.52	1.52	1.51	1.49	1.48	1.47	1.46	1.46	1.45	1.44	1.44
	.10	3.07	2.70	2.49	2.36	2.27	2.21	2.16	2.12	2.09	2.06	2.04	2.02
	.05	4.54	3.68	3.29	3.06	2.90	2.79	2.71	2.64	2.59	2.54	2.51	2.48
	.01	8.68	6.36	5.42	4.89	4.56	4.32	4.14	4.00	3.89	3.80	3.73	3.67
16	.25	1.42	1.51	1.51	1.50	1.48	1.47	1.46	1.45	1.44	1.44	1.44	1.43
	.10	3.05	2.67	2.46	2.33	2.24	2.18	2.13	2.09	2.06	2.03	2.01	1.99
	.05	4.49	3.63	3.24	3.01	2.85	2.74	2.66	2.59	2.54	2.49	2.46	2.42
	.01	8.53	6.23	5.29	4.77	4.44	4.20	4.03	3.89	3.78	3.69	3.62	3.55
17	.25	1.42	1.51	1.50	1.49	1.47	1.46	1.45	1.44	1.43	1.43	1.42	1.41
	.10	3.03	2.64	2.44	2.31	2.22	2.15	2.10	2.06	2.03	2.00	1.98	1.96
	.05	4.45	3.59	3.20	2.96	2.81	2.70	2.61	2.55	2.49	2.45	2.41	2.38
	.01	8.40	6.11	5.18	4.67	4.34	4.10	3.93	3.79	3.68	3.59	3.52	3.46
18	.25	1.41	1.50	1.49	1.48	1.46	1.45	1.44	1.43	1.42	1.42	1.41	1.40
	.10	3.01	2.62	2.42	2.29	2.20	2.13	2.08	2.04	2.00	1.98	1.96	1.93
	.05	4.41	3.55	3.16	2.93	2.77	2.66	2.58	2.51	2.46	2.41	2.37	2.34
	.01	8.29	6.01	5.09	4.58	4.25	4.01	3.84	3.71	3.60	3.51	3.43	3.37
19	.25	1.41	1.49	1.49	1.47	1.46	1.44	1.43	1.42	1.41	1.41	1.40	1.40
	.10	2.99	2.61	2.40	2.27	2.18	2.11	2.06	2.02	1.98	1.96	1.94	1.91
	.05	4.38	3.52	3.13	2.90	2.74	2.63	2.54	2.48	2.42	2.38	2.34	2.31
	.01	8.18	5.93	5.01	4.50	4.17	3.94	3.77	3.63	3.52	3.43	3.36	3.30
20	.25	1.40	1.49	1.48	1.46	1.45	1.44	1.43	1.42	1.41	1.40	1.39	1.39
	.10	2.97	2.59	2.38	2.25	2.16	2.09	2.04	2.00	1.96	1.94	1.92	1.89
	.05	4.35	3.49	3.10	2.87	2.71	2.60	2.51	2.45	2.39	2.35	2.31	2.28
	.01	8.10	5.85	4.94	4.43	4.10	3.87	3.70	3.56	3.46	3.37	3.29	3.23

TABLE E.5 (continued)

15	20	24	30	40	50	60	100	120	200	500	∞	α	df for Denominator
1.53	1.52	1.52	1.51	1.51	1.50	1.50	1.49	1.49	1.49	1.48	1.48	.25	
2.24	2.20	2.18	2.16	2.13	2.12	2.11	2.09	2.08	2.07	2.06	2.06	.10	10
2.85	2.77	2.74	2.70	2.66	2.64	2.62	2.59	2.58	2.56	2.55	2.54	.05	
4.56	4.41	4.33	4.25	4.17	4.12	4.08	4.01	4.00	3.96	3.93	3.91	.01	
1.50	1.49	1.49	1.48	1.47	1.47	1.47	1.46	1.46	1.46	1.45	1.45	.25	
2.17	2.12	2.10	2.08	2.05	2.04	2.03	2.00	2.00	1.99	1.98	1.97	.10	11
2.72	2.65	2.61	2.57	2.53	2.51	2.49	2.46	2.45	2.43	2.42	2.40	.05	
4.25	4.10	4.02	3.94	3.86	3.81	3.78	3.71	3.69	3.66	3.62	3.60	.01	
1.48	1.47	1.46	1.45	1.45	1.44	1.44	1.43	1.43	1.43	1.42	1.42	.25	
2.10	2.06	2.04	2.01	1.99	1.97	1.96	1.94	1.93	1.92	1.91	1.90	.10	12
2.62	2.54	2.51	2.47	2.43	2.40	2.38	2.35	2.34	2.32	2.31	2.30	.05	
4.01	3.86	3.78	3.70	3.62	3.57	3.54	3.47	3.45	3.41	3.38	3.36	.01	
1.46	1.45	1.44	1.43	1.42	1.42	1.42	1.41	1.41	1.40	1.40	1.40	.25	
2.05	2.01	1.98	1.96	1.93	1.92	1.90	1.88	1.88	1.86	1.85	1.85	.10	13
2.53	2.46	2.42	2.38	2.34	2.31	2.30	2.26	2.25	2.23	2.22	2.21	.05	
3.82	3.66	3.59	3.51	3.43	3.38	3.34	3.27	3.25	3.22	3.19	3.17	.01	
1.44	1.43	1.42	1.41	1.41	1.40	1.40	1.39	1.39	1.39	1.38	1.38	.25	
2.01	1.96	1.94	1.91	1.89	1.87	1.86	1.83	1.83	1.82	1.80	1.80	.10	14
2.46	2.39	2.35	2.31	2.27	2.24	2.22	2.19	2.18	2.16	2.14	2.13	.05	
3.66	3.51	3.43	3.35	3.27	3.22	3.18	3.11	3.09	3.06	3.03	3.00	.01	
1.43	1.41	1.41	1.40	1.39	1.39	1.38	1.38	1.37	1.37	1.36	1.36	.25	
1.97	1.92	1.90	1.87	1.85	1.83	1.82	1.79	1.79	1.77	1.76	1.76	.10	15
2.40	2.33	2.29	2.25	2.20	2.18	2.16	2.12	2.11	2.10	2.08	2.07	.05	
3.52	3.37	3.29	3.21	3.13	3.08	3.05	2.98	2.96	2.92	2.89	2.87	.01	
1.41	1.40	1.39	1.38	1.37	1.37	1.36	1.36	1.35	1.35	1.34	1.34	.25	
1.94	1.89	1.87	1.84	1.81	1.79	1.78	1.76	1.75	1.74	1.73	1.72	.10	16
2.35	2.28	2.24	2.19	2.15	2.12	2.11	2.07	2.06	2.04	2.02	2.01	.05	
3.41	3.26	3.18	3.10	3.02	2.97	2.93	2.86	2.84	2.81	2.78	2.75	.01	
1.40	1.39	1.38	1.37	1.36	1.35	1.35	1.34	1.34	1.34	1.33	1.33	.25	
1.91	1.86	1.84	1.81	1.78	1.76	1.75	1.73	1.72	1.71	1.69	1.69	.10	17
2.31	2.23	2.19	2.15	2.10	2.08	2.06	2.02	2.01	1.99	1.97	1.96	.05	
3.31	3.16	3.08	3.00	2.92	2.87	2.83	2.76	2.75	2.71	2.68	2.65	.01	
1.39	1.38	1.37	1.36	1.35	1.34	1.34	1.33	1.33	1.32	1.32	1.32	.25	
1.89	1.84	1.81	1.78	1.75	1.74	1.72	1.70	1.69	1.68	1.67	1.66	.10	18
2.27	2.19	2.15	2.11	2.06	2.04	2.02	1.98	1.97	1.95	1.93	1.92	.05	
3.23	3.08	3.00	2.92	2.84	2.78	2.75	2.68	2.66	2.62	2.59	2.57	.01	
1.38	1.37	1.36	1.35	1.34	1.33	1.33	1.32	1.32	1.31	1.31	1.30	.25	
1.86	1.81	1.79	1.76	1.73	1.71	1.70	1.67	1.67	1.65	1.64	1.63	.10	19
2.23	2.16	2.11	2.07	2.03	2.00	1.98	1.94	1.93	1.91	1.89	1.88	.05	
3.15	3.00	2.92	2.84	2.76	2.71	2.67	2.60	2.58	2.55	2.51	2.49	.01	
1.37	1.36	1.35	1.34	1.33	1.33	1.32	1.31	1.31	1.30	1.30	1.29	.25	
1.84	1.79	1.77	1.74	1.71	1.69	1.68	1.65	1.64	1.63	1.62	1.61	.10	20
2.20	2.12	2.08	2.04	1.99	1.97	1.95	1.91	1.90	1.88	1.86	1.84	.05	
3.09	2.94	2.86	2.78	2.69	2.64	2.61	2.54	2.52	2.48	2.44	2.42	.01	

TABLE E.5   (continued)

df for Denominator	α	1	2	3	4	5	6	7	8	9	10	11	12
		\multicolumn{12}{df for Numerator}											
22	.25	1.40	1.48	1.47	1.45	1.44	1.42	1.41	1.40	1.39	1.39	1.38	1.37
	.10	2.95	2.56	2.35	2.22	2.13	2.06	2.01	1.97	1.93	1.90	1.88	1.86
	.05	4.30	3.44	3.05	2.82	2.66	2.55	2.46	2.40	2.34	2.30	2.26	2.23
	.01	7.95	5.72	4.82	4.31	3.99	3.76	3.59	3.45	3.35	3.26	3.18	3.12
24	.25	1.39	1.47	1.46	1.44	1.43	1.41	1.40	1.39	1.38	1.38	1.37	1.36
	.10	2.93	2.54	2.33	2.19	2.10	2.04	1.98	1.94	1.91	1.88	1.85	1.83
	.05	4.26	3.40	3.01	2.78	2.62	2.51	2.42	2.36	2.30	2.25	2.21	2.18
	.01	7.82	5.61	4.72	4.22	3.90	3.67	3.50	3.36	3.26	3.17	3.09	3.03
26	.25	1.38	1.46	1.45	1.44	1.42	1.41	1.39	1.38	1.37	1.37	1.36	1.35
	.10	2.91	2.52	2.31	2.17	2.08	2.01	1.96	1.92	1.88	1.86	1.84	1.81
	.05	4.23	3.37	2.98	2.74	2.59	2.47	2.39	2.32	2.27	2.22	2.18	2.15
	.01	7.72	5.53	4.64	4.14	3.82	3.59	3.42	3.29	3.18	3.09	3.02	2.96
28	.25	1.38	1.46	1.45	1.43	1.41	1.40	1.39	1.38	1.37	1.36	1.35	1.34
	.10	2.89	2.50	2.29	2.16	2.06	2.00	1.94	1.90	1.87	1.84	1.81	1.79
	.05	4.20	3.34	2.95	2.71	2.56	2.45	2.36	2.29	2.24	2.19	2.15	2.12
	.01	7.64	5.45	4.57	4.07	3.75	3.53	3.36	3.23	3.12	3.03	2.96	2.90
30	.25	1.38	1.45	1.44	1.42	1.41	1.39	1.38	1.37	1.36	1.35	1.35	1.34
	.10	2.88	2.49	2.28	2.14	2.05	1.98	1.93	1.88	1.85	1.82	1.79	1.77
	.05	4.17	3.32	2.92	2.69	2.53	2.42	2.33	2.27	2.21	2.16	2.13	2.09
	.01	7.56	5.39	4.51	4.02	3.70	3.47	3.30	3.17	3.07	2.98	2.91	2.84
40	.25	1.36	1.44	1.42	1.40	1.39	1.37	1.36	1.35	1.34	1.33	1.32	1.31
	.10	2.84	2.44	2.23	2.09	2.00	1.93	1.87	1.83	1.79	1.76	1.73	1.71
	.05	4.08	3.23	2.84	2.61	2.45	2.34	2.25	2.18	2.12	2.08	2.04	2.00
	.01	7.31	5.18	4.31	3.83	3.51	3.29	3.12	2.99	2.89	2.80	2.73	2.66
60	.25	1.35	1.42	1.41	1.38	1.37	1.35	1.33	1.32	1.31	1.30	1.29	1.29
	.10	2.79	2.39	2.18	2.04	1.95	1.87	1.82	1.77	1.74	1.71	1.68	1.66
	.05	4.00	3.15	2.76	2.53	2.37	2.25	2.17	2.10	2.04	1.99	1.95	1.92
	.01	7.08	4.98	4.13	3.65	3.34	3.12	2.95	2.82	2.72	2.63	2.56	2.50
120	.25	1.34	1.40	1.39	1.37	1.35	1.33	1.31	1.30	1.29	1.28	1.27	1.26
	.10	2.75	2.35	2.13	1.99	1.90	1.82	1.77	1.72	1.68	1.65	1.62	1.60
	.05	3.92	3.07	2.68	2.45	2.29	2.17	2.09	2.02	1.96	1.91	1.87	1.83
	.01	6.85	4.79	3.95	3.48	3.17	2.96	2.79	2.66	2.56	2.47	2.40	2.34
200	.25	1.33	1.39	1.38	1.36	1.34	1.32	1.31	1.29	1.28	1.27	1.26	1.25
	.10	2.73	2.33	2.11	1.97	1.88	1.80	1.75	1.70	1.66	1.63	1.60	1.57
	.05	3.89	3.04	2.65	2.42	2.26	2.14	2.06	1.98	1.93	1.88	1.84	1.80
	.01	6.76	4.71	3.88	3.41	3.11	2.89	2.73	2.60	2.50	2.41	2.34	2.27
∞	.25	1.32	1.39	1.37	1.35	1.33	1.31	1.29	1.28	1.27	1.25	1.24	1.24
	.10	2.71	2.30	2.08	1.94	1.85	1.77	1.72	1.67	1.63	1.60	1.57	1.55
	.05	3.84	3.00	2.60	2.37	2.21	2.10	2.01	1.94	1.88	1.83	1.79	1.75
	.01	6.63	4.61	3.78	3.32	3.02	2.80	2.64	2.51	2.41	2.32	2.25	2.18

TABLE E.5 (continued)

15	20	24	30	40	50	60	100	120	200	500	∞	α	df for Denominator
					df for Numerator								
1.36	1.34	1.33	1.32	1.31	1.31	1.30	1.30	1.30	1.29	1.29	1.28	.25	
1.81	1.76	1.73	1.70	1.67	1.65	1.64	1.61	1.60	1.59	1.58	1.57	.10	22
2.15	2.07	2.03	1.98	1.94	1.91	1.89	1.85	1.84	1.82	1.80	1.78	.05	
2.98	2.83	2.75	2.67	2.58	2.53	2.50	2.42	2.40	2.36	2.33	2.31	.01	
1.35	1.33	1.32	1.31	1.30	1.29	1.29	1.28	1.28	1.27	1.27	1.26	.25	
1.78	1.73	1.70	1.67	1.64	1.62	1.61	1.58	1.57	1.56	1.54	1.53	.10	24
2.11	2.03	1.98	1.94	1.89	1.86	1.84	1.80	1.79	1.77	1.75	1.73	.05	
2.89	2.74	2.66	2.58	2.49	2.44	2.40	2.33	2.31	2.27	2.24	2.21	.01	
1.34	1.32	1.31	1.30	1.29	1.28	1.28	1.26	1.26	1.26	1.25	1.25	.25	
1.76	1.71	1.68	1.65	1.61	1.59	1.58	1.55	1.54	1.53	1.51	1.50	.10	26
2.07	1.99	1.95	1.90	1.85	1.82	1.80	1.76	1.75	1.73	1.71	1.69	.05	
2.81	2.66	2.58	2.50	2.42	2.36	2.33	2.25	2.23	2.19	2.16	2.13	.01	
1.33	1.31	1.30	1.29	1.28	1.27	1.27	1.26	1.25	1.25	1.24	1.24	.25	
1.74	1.69	1.66	1.63	1.59	1.57	1.56	1.53	1.52	1.50	1.49	1.48	.10	28
2.04	1.96	1.91	1.87	1.82	1.79	1.77	1.73	1.71	1.69	1.67	1.65	.05	
2.75	2.60	2.52	2.44	2.35	2.30	2.26	2.19	2.17	2.13	2.09	2.06	.01	
1.32	1.30	1.29	1.28	1.27	1.26	1.26	1.25	1.24	1.24	1.23	1.23	.25	
1.72	1.67	1.64	1.61	1.57	1.55	1.54	1.51	1.50	1.48	1.47	1.46	.10	30
2.01	1.93	1.89	1.84	1.79	1.76	1.74	1.70	1.68	1.66	1.64	1.62	.05	
2.70	2.55	2.47	2.39	2.30	2.25	2.21	2.13	2.11	2.07	2.03	2.01	.01	
1.30	1.28	1.26	1.25	1.24	1.23	1.22	1.21	1.21	1.20	1.19	1.19	.25	
1.66	1.61	1.57	1.54	1.51	1.48	1.47	1.43	1.42	1.41	1.39	1.38	.10	40
1.92	1.84	1.79	1.74	1.69	1.66	1.64	1.59	1.58	1.55	1.53	1.51	.05	
2.52	2.37	2.29	2.20	2.11	2.06	2.02	1.94	1.92	1.87	1.83	1.80	.01	
1.27	1.25	1.24	1.22	1.21	1.20	1.19	1.17	1.17	1.16	1.15	1.15	.25	
1.60	1.54	1.51	1.48	1.44	1.41	1.40	1.36	1.35	1.33	1.31	1.29	.10	60
1.84	1.75	1.70	1.65	1.59	1.56	1.53	1.48	1.47	1.44	1.41	1.39	.05	
2.35	2.20	2.12	2.03	1.94	1.88	1.84	1.75	1.73	1.68	1.63	1.60	.01	
1.24	1.22	1.21	1.19	1.18	1.17	1.16	1.14	1.13	1.12	1.11	1.10	.25	
1.55	1.48	1.45	1.41	1.37	1.34	1.32	1.27	1.26	1.24	1.21	1.19	.10	120
1.75	1.66	1.61	1.55	1.50	1.46	1.43	1.37	1.35	1.32	1.28	1.25	.05	
2.19	2.03	1.95	1.86	1.76	1.70	1.66	1.56	1.53	1.48	1.42	1.38	.01	
1.23	1.21	1.20	1.18	1.16	1.14	1.12	1.11	1.10	1.09	1.08	1.06	.25	
1.52	1.46	1.42	1.38	1.34	1.31	1.28	1.24	1.22	1.20	1.17	1.14	.10	200
1.72	1.62	1.57	1.52	1.46	1.41	1.39	1.32	1.29	1.26	1.22	1.19	.05	
2.13	1.97	1.89	1.79	1.69	1.63	1.58	1.48	1.44	1.39	1.33	1.28	.01	
1.22	1.19	1.18	1.16	1.14	1.13	1.12	1.09	1.08	1.07	1.04	1.00	.25	
1.49	1.42	1.38	1.34	1.30	1.26	1.24	1.18	1.17	1.13	1.08	1.00	.10	∞
1.67	1.57	1.52	1.46	1.39	1.35	1.32	1.24	1.22	1.17	1.11	1.00	.05	
2.04	1.88	1.79	1.70	1.59	1.52	1.47	1.36	1.32	1.25	1.15	1.00	.01	

TABLE E.6 Upper Percentage Points of the χ^2 Distribution

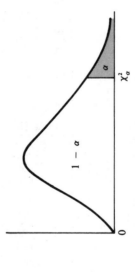

df	.99	.98	.95	.90	.80	.70	.50	.30	.20	.10	.05	.02	.01	.001
1	$.0^3157$	$.0^3628$	.00393	.0158	.0642	.148	.455	1.074	1.642	2.706	3.841	5.412	6.635	10.827
2	.0201	.0404	.103	.211	.446	.713	1.386	2.408	3.219	4.605	5.991	7.824	9.210	13.815
3	.115	.185	.352	.584	1.005	1.424	2.366	3.665	4.642	6.251	7.815	9.837	11.345	16.266
4	.297	.429	.711	1.064	1.649	2.195	3.357	4.878	5.989	7.779	9.488	11.668	13.277	18.467
5	.554	.752	1.145	1.610	2.343	3.000	4.351	6.064	7.289	9.236	11.070	13.388	15.086	20.515
6	.872	1.134	1.635	2.204	3.070	3.828	5.348	7.231	8.558	10.645	12.592	15.033	16.812	22.457
7	1.239	1.564	2.167	2.833	3.822	4.671	6.346	8.383	9.803	12.017	14.067	16.622	18.475	24.322
8	1.646	2.032	2.733	3.490	4.594	5.527	7.344	9.524	11.030	13.362	15.507	18.168	20.090	26.125
9	2.088	2.532	3.325	4.168	5.380	6.393	8.343	10.656	12.242	14.684	16.919	19.679	21.666	27.877
10	2.558	3.059	3.940	4.865	6.179	7.267	9.342	11.781	13.442	15.987	18.307	21.161	23.209	29.588

11	3.053	3.609	4.575	5.578	6.989	8.148	10.341	12.899	14.631	17.275	19.675	22.618	24.725	31.264
12	3.571	4.178	5.226	6.304	7.807	9.034	11.340	14.011	15.812	18.549	21.026	24.054	26.217	32.909
13	4.107	4.765	5.892	7.042	8.634	9.926	12.340	15.119	16.985	19.812	22.362	25.472	27.688	34.528
14	4.660	5.368	6.571	7.790	9.467	10.821	13.339	16.222	18.151	21.064	23.685	26.873	29.141	36.123
15	5.229	5.985	7.261	8.547	10.307	11.721	14.339	17.322	19.311	22.307	24.996	28.259	30.578	37.697
16	5.812	6.614	7.962	9.312	11.152	12.624	15.338	18.418	20.465	23.542	26.296	29.633	32.000	39.252
17	6.408	7.255	8.672	10.085	12.002	13.531	16.338	19.511	21.615	24.769	27.587	30.995	33.409	40.790
18	7.015	7.906	9.390	10.865	12.857	14.440	17.338	20.601	22.760	25.989	28.869	32.346	34.805	42.312
19	7.633	8.567	10.117	11.651	13.716	15.352	18.338	21.689	23.900	27.204	30.144	33.687	36.191	43.820
20	8.260	9.237	10.851	12.443	14.578	16.266	19.337	22.775	25.038	28.412	31.410	35.020	37.566	45.315
21	8.897	9.915	11.591	13.240	15.445	17.182	20.337	23.858	26.171	29.615	32.671	36.343	38.932	46.797
22	9.542	10.600	12.338	14.041	16.314	18.101	21.337	24.939	27.301	30.813	33.924	37.659	40.289	48.268
23	10.196	11.293	13.091	14.848	17.187	19.021	22.337	26.018	28.429	32.007	35.172	38.968	41.638	49.728
24	10.856	11.992	13.848	15.659	18.062	19.943	23.337	27.096	29.553	33.196	36.415	40.270	42.980	51.179
25	11.524	12.697	14.611	16.473	18.940	20.867	24.337	28.172	30.675	34.382	37.652	41.566	44.314	52.620
26	12.198	13.409	15.379	17.292	19.820	21.792	25.336	29.246	31.795	35.563	38.885	42.856	45.642	54.052
27	12.879	14.125	16.151	18.114	20.703	22.719	26.336	30.319	32.912	36.741	40.113	44.140	46.963	55.476
28	13.565	14.847	16.928	18.939	21.588	23.647	27.336	31.391	34.027	37.916	41.337	45.419	48.278	56.893
29	14.256	15.574	17.708	19.768	22.475	24.577	28.336	32.461	35.139	39.087	42.557	46.693	49.588	58.302
30	14.953	16.306	18.493	20.599	23.364	25.508	29.336	33.530	36.250	40.256	43.773	47.962	50.892	59.703

For $v > 30$, the expression $\sqrt{2\chi^2} - \sqrt{2v - 1}$ may be used as a normal deviate with unit variance.

Table E.6 is taken from Table 4 of Fisher and Yates, *Statistical Tables for Biological, Agricultural and Medical Research*, published by Oliver and Boyd Ltd., Edinburgh, by permission of the authors and publishers.

TABLE E.7 Percentage Points of the Studentized Range

Error df	α	Number of Means (p) or Number of Steps Between Ordered Means (r)									
		2	3	4	5	6	7	8	9	10	11
2	.05	6.08	8.33	9.80	10.9	11.7	12.4	13.0	13.5	14.0	14.4
	.01	14.0	19.0	22.3	24.7	26.6	28.2	29.5	30.7	31.7	32.6
3	.05	4.50	5.91	6.82	7.50	8.04	8.48	8.85	9.18	9.46	9.72
	.01	8.26	10.6	12.2	13.3	14.2	15.0	15.6	16.2	16.7	17.8
4	.05	3.93	5.04	5.76	6.29	6.71	7.05	7.35	7.60	7.83	8.03
	.01	6.51	8.12	9.17	9.96	10.6	11.1	11.5	11.9	12.3	12.6
5	.05	3.64	4.60	5.22	5.67	6.03	6.33	6.58	6.80	6.99	7.17
	.01	5.70	6.98	7.80	8.42	8.91	9.32	9.67	9.97	10.24	10.48
6	.05	3.46	4.34	4.90	5.30	5.63	5.90	6.12	6.32	6.49	6.65
	.01	5.24	6.33	7.03	7.56	7.97	8.32	8.61	8.87	9.10	9.30
7	.05	3.34	4.16	4.68	5.06	5.36	5.61	5.82	6.00	6.16	6.30
	.01	4.95	5.92	6.54	7.01	7.37	7.68	7.94	8.17	8.37	8.55
8	.05	3.26	4.04	4.53	4.89	5.17	5.40	5.60	5.77	5.92	6.05
	.01	4.75	5.64	6.20	6.62	6.96	7.24	7.47	7.68	7.86	8.03
9	.05	3.20	3.95	4.41	4.76	5.02	5.24	5.43	5.59	5.74	5.87
	.01	4.60	5.43	5.96	6.35	6.66	6.91	7.13	7.33	7.49	7.65
10	.05	3.15	3.88	4.33	4.65	4.91	5.12	5.30	5.46	5.60	5.72
	.01	4.48	5.27	5.77	6.14	6.43	6.67	6.87	7.05	7.21	7.36
11	.05	3.11	3.82	4.26	4.57	4.82	5.03	5.20	5.35	5.49	5.61
	.01	4.39	5.15	5.62	5.97	6.25	6.48	6.67	6.84	6.99	7.13
12	.05	3.08	3.77	4.20	4.51	4.75	4.95	5.12	5.27	5.39	5.51
	.01	4.32	5.05	5.50	5.84	6.10	6.32	6.51	6.67	6.81	6.94
13	.05	3.06	3.73	4.15	4.45	4.69	4.88	5.05	5.19	5.32	5.43
	.01	4.26	4.96	5.40	5.73	5.98	6.19	6.37	6.53	6.67	6.79
14	.05	3.03	3.70	4.11	4.41	4.64	4.83	4.99	5.13	5.25	5.36
	.01	4.21	4.89	5.32	5.63	5.88	6.08	6.26	6.41	6.54	6.66
15	.05	3.01	3.67	4.08	4.37	4.59	4.78	4.94	5.08	5.20	5.31
	.01	4.17	4.84	5.25	5.56	5.80	5.99	6.16	6.31	6.44	6.55
16	.05	3.00	3.65	4.05	4.33	4.56	4.74	4.90	5.03	5.15	5.26
	.01	4.13	4.79	5.19	5.49	5.72	5.92	6.08	6.22	6.35	6.46
17	.05	2.98	3.63	4.02	4.30	4.52	4.70	4.86	4.99	5.11	5.21
	.01	4.10	4.74	5.14	5.43	5.66	5.85	6.01	6.15	6.27	6.38
18	.05	2.97	3.61	4.00	4.28	4.49	4.67	4.82	4.96	5.07	5.17
	.01	4.07	4.70	5.09	5.38	5.60	5.79	5.94	6.08	6.20	6.31
19	.05	2.96	3.59	3.98	4.25	4.47	4.65	4.79	4.92	5.04	5.14
	.01	4.05	4.67	5.05	5.33	5.55	5.73	5.89	6.02	6.14	6.25
20	.05	2.95	3.58	3.96	4.23	4.45	4.62	4.77	4.90	5.01	5.11
	.01	4.02	4.64	5.02	5.29	5.51	5.69	5.84	5.97	6.09	6.19
24	.05	2.92	3.53	3.90	4.17	4.37	4.54	4.68	4.81	4.92	5.01
	.01	3.96	4.55	4.91	5.17	5.37	5.54	5.69	5.81	5.92	6.02
30	.05	2.89	3.49	3.85	4.10	4.30	4.46	4.60	4.72	4.82	4.92
	.01	3.89	4.45	4.80	5.05	5.24	5.40	5.54	5.65	5.76	5.85
40	.05	2.86	3.44	3.79	4.04	4.23	4.39	4.52	4.63	4.73	4.82
	.01	3.82	4.37	4.70	4.93	5.11	5.26	5.39	5.50	5.60	5.69
60	.05	2.83	3.40	3.74	3.98	4.16	4.31	4.44	4.55	4.65	4.73
	.01	3.76	4.28	4.59	4.82	4.99	5.13	5.25	5.36	5.45	5.53
120	.05	2.80	3.36	3.68	3.92	4.10	4.24	4.36	4.47	4.56	4.64
	.01	3.70	4.20	4.50	4.71	4.87	5.01	5.12	5.21	5.30	5.37
∞	.05	2.77	3.31	3.63	3.86	4.03	4.17	4.29	4.39	4.47	4.55
	.01	3.64	4.12	4.40	4.60	4.76	4.88	4.99	5.08	5.16	5.23

TABLE E.7 (continued)

\ Number of Means (p) or Number of Steps Between Ordered Means (r)										
12	13	14	15	16	17	18	19	20	α	Error df
14.7	15.1	15.4	15.7	15.9	16.1	16.4	16.6	16.8	.05	2
33.4	34.1	34.8	35.4	36.0	36.5	37.0	37.5	37.9	.01	
9.72	10.2	10.3	10.5	10.7	10.8	11.0	11.1	11.2	.05	3
17.5	17.9	18.2	18.5	18.8	19.1	19.3	19.5	19.8	.01	
8.21	8.37	8.52	8.66	8.79	8.91	9.03	9.13	9.23	.05	4
12.8	13.1	13.3	13.5	13.7	13.9	14.1	14.2	14.4	.01	
7.32	7.47	7.60	7.72	7.83	7.93	8.03	8.12	8.21	.05	5
10.70	10.89	11.08	11.24	11.40	11.55	11.68	11.81	11.93	.01	
6.79	6.92	7.03	7.14	7.24	7.34	7.43	7.51	7.59	.05	6
9.48	9.65	9.81	9.95	10.08	10.21	10.32	10.43	10.54	.01	
6.43	6.55	6.66	6.76	6.85	6.94	7.02	7.10	7.17	.05	7
8.71	8.86	9.00	9.12	9.24	9.35	9.46	9.55	9.65	.01	
6.18	6.29	6.39	6.48	6.57	6.65	6.73	6.80	6.87	.05	8
8.18	8.31	8.44	8.55	8.66	8.76	8.85	8.94	9.03	.01	
5.98	6.09	6.19	6.28	6.36	6.44	6.51	6.58	6.64	.05	9
7.78	7.91	8.03	8.13	8.23	8.33	8.41	8.49	8.57	.01	
5.83	5.93	6.03	6.11	6.19	6.27	6.34	6.40	6.47	.05	10
7.49	7.60	7.71	7.81	7.91	7.99	8.08	8.15	8.23	.01	
5.71	5.81	5.90	5.98	6.06	6.13	6.20	6.27	6.33	.05	11
7.25	7.36	7.46	7.56	7.65	7.73	7.81	7.88	7.95	.01	
5.61	5.71	5.80	5.88	5.95	6.02	6.09	6.15	6.21	.05	12
7.06	7.17	7.26	7.36	7.44	7.52	7.59	7.66	7.73	.01	
5.53	5.63	5.71	5.79	5.86	5.93	5.99	6.05	6.11	.05	13
6.90	7.01	7.10	7.19	7.27	7.35	7.42	7.48	7.55	.01	
5.46	5.55	5.64	5.71	5.79	5.85	5.91	5.97	6.03	.05	14
6.77	6.87	6.96	7.05	7.13	7.20	7.27	7.33	7.39	.01	
5.40	5.49	5.57	5.65	5.72	5.78	5.85	5.90	5.96	.05	15
6.66	6.76	6.84	6.93	7.00	7.07	7.14	7.20	7.26	.01	
5.35	5.44	5.52	5.59	5.66	5.73	5.79	5.84	5.90	.05	16
6.56	6.66	6.74	6.82	6.90	6.97	7.03	7.09	7.15	.01	
5.31	5.39	5.47	5.54	5.61	5.67	5.73	5.79	5.84	.05	17
6.48	6.57	6.66	6.73	6.81	6.87	6.94	7.00	7.05	.01	
5.27	5.35	5.43	5.50	5.57	5.63	5.69	5.74	5.79	.05	18
6.41	6.50	6.58	6.65	6.73	6.79	6.85	6.91	6.97	.01	
5.23	5.31	5.39	5.46	5.53	5.59	5.65	5.70	5.75	.05	19
6.34	6.43	6.51	6.58	6.65	6.72	6.78	6.84	6.89	.01	
5.20	5.28	5.36	5.43	5.49	5.55	5.61	5.66	5.71	.05	20
6.28	6.37	6.45	6.52	6.59	6.65	6.71	6.77	6.82	.01	
5.10	5.18	5.25	5.32	5.38	5.44	5.49	5.55	5.59	.05	24
6.11	6.19	6.26	6.33	6.39	6.45	6.51	6.56	6.61	.01	
5.00	5.08	5.15	5.21	5.27	5.33	5.38	5.43	5.47	.05	30
5.93	6.01	6.08	6.14	6.20	6.26	6.31	6.36	6.41	.01	
4.90	4.98	5.04	5.11	5.16	5.22	5.27	5.31	5.36	.05	40
5.76	5.83	5.90	5.96	6.02	6.07	6.12	6.16	6.21	.01	
4.81	4.88	4.94	5.00	5.06	5.11	5.15	5.20	5.24	.05	60
5.60	5.67	5.73	5.78	5.84	5.89	5.93	5.97	6.01	.01	
4.71	4.78	4.84	4.90	4.95	5.00	5.04	5.09	5.13	.05	120
5.44	5.50	5.56	5.61	5.66	5.71	5.75	5.79	5.83	.01	
4.62	4.68	4.74	4.80	4.85	4.89	4.93	4.97	5.01	.05	∞
5.29	5.35	5.40	5.45	5.49	5.54	5.57	5.61	5.65	.01	

TABLE E.8 Percentage Points of the Duncan New Multiple Range Test

Error df	Protection Level	Number of Means for Range Being Tested (r)													
		2	3	4	5	6	7	8	9	10	12	14	16	18	20
1	.05	18.0	18.0	18.0	18.0	18.0	18.0	18.0	18.0	18.0	18.0	18.0	18.0	18.0	18.0
	.01	90.0	90.0	90.0	90.0	90.0	90.0	90.0	90.0	90.0	90.0	90.0	90.0	90.0	90.0
2	.05	6.09	6.09	6.09	6.09	6.09	6.09	6.09	6.09	6.09	6.09	6.09	6.09	6.09	6.09
	.01	14.0	14.0	14.0	14.0	14.0	14.0	14.0	14.0	14.0	14.0	14.0	14.0	14.0	14.0
3	.05	4.50	4.50	4.50	4.50	4.50	4.50	4.50	4.50	4.50	4.50	4.50	4.50	4.50	4.50
	.01	8.26	8.5	8.6	8.7	8.8	8.9	8.9	9.0	9.0	9.0	9.1	9.2	9.3	9.3
4	.05	3.93	4.01	4.02	4.02	4.02	4.02	4.02	4.02	4.02	4.02	4.02	4.02	4.02	4.02
	.01	6.51	6.8	6.9	7.0	7.1	7.1	7.2	7.2	7.3	7.3	7.4	7.4	7.5	7.5
5	.05	3.64	3.74	3.79	3.83	3.83	3.83	3.83	3.83	3.83	3.83	3.83	3.83	3.83	3.83
	.01	5.70	5.96	6.11	6.18	6.26	6.33	6.40	6.44	6.5	6.6	6.6	6.7	6.7	6.8
6	.05	3.46	3.58	3.64	3.68	3.68	3.68	3.68	3.68	3.68	3.68	3.68	3.68	3.68	3.68
	.01	5.24	5.51	5.65	5.73	5.81	5.88	5.95	6.00	6.0	6.1	6.2	6.2	6.3	6.3
7	.05	3.35	3.47	3.54	3.58	3.60	3.61	3.61	3.61	3.61	3.61	3.61	3.61	3.61	3.61
	.01	4.95	5.22	5.37	5.45	5.53	5.61	5.69	5.73	5.8	5.8	5.9	5.9	6.0	6.0
8	.05	3.26	3.39	3.47	3.52	3.55	3.56	3.56	3.56	3.56	3.56	3.56	3.56	3.56	3.56
	.01	4.74	5.00	5.14	5.23	5.32	5.40	5.47	5.51	5.5	5.6	5.7	5.7	5.8	5.8
9	.05	3.20	3.34	3.41	3.47	3.50	3.52	3.52	3.52	3.52	3.52	3.52	3.52	3.52	3.52
	.01	4.60	4.86	4.99	5.08	5.17	5.25	5.32	5.36	5.4	5.5	5.5	5.6	5.7	5.7
10	.05	3.15	3.30	3.37	3.43	3.46	3.47	3.47	3.47	3.47	3.47	3.47	3.47	3.47	3.48
	.01	4.48	4.73	4.88	4.96	5.06	5.13	5.20	5.24	5.28	5.36	5.42	5.48	5.54	5.55
11	.05	3.11	3.27	3.35	3.39	3.43	3.44	3.45	3.46	3.46	3.46	3.46	3.46	3.47	3.48
	.01	4.39	4.63	4.77	4.86	4.94	5.01	5.06	5.12	5.15	5.24	5.28	5.34	5.38	5.39
12	.05	3.08	3.23	3.33	3.36	3.40	3.42	3.44	3.44	3.46	3.46	3.46	3.46	3.47	3.48
	.01	4.32	4.55	4.68	4.76	4.84	4.92	4.96	5.02	5.07	5.13	5.17	5.22	5.24	5.26
13	.05	3.06	3.21	3.30	3.35	3.38	3.41	3.42	3.44	3.45	3.45	3.46	3.46	3.47	3.47
	.01	4.26	4.48	4.62	4.69	4.74	4.84	4.88	4.94	4.98	5.04	5.08	5.13	5.14	5.15

df	α														
14	.05	3.03	3.18	3.27	3.33	3.37	3.39	3.41	3.42	3.44	3.45	3.46	3.46	3.47	3.47
	.01	4.21	4.42	4.55	4.63	4.70	4.78	4.83	4.87	4.91	4.96	5.00	5.04	5.06	5.07
15	.05	3.01	3.16	3.25	3.31	3.36	3.38	3.40	3.42	3.43	3.44	3.45	3.46	3.47	3.47
	.01	4.17	4.37	4.50	4.58	4.64	4.72	4.77	4.81	4.84	4.90	4.94	4.97	4.99	5.00
16	.05	3.00	3.15	3.23	3.30	3.34	3.37	3.39	3.41	3.43	3.44	3.45	3.46	3.47	3.47
	.01	4.13	4.34	4.45	4.54	4.60	4.67	4.72	4.76	4.79	4.84	4.88	4.91	4.93	4.94
17	.05	2.98	3.13	3.22	3.28	3.33	3.36	3.38	3.40	3.42	3.44	3.45	3.46	3.47	3.47
	.01	4.10	4.30	4.41	4.50	4.56	4.63	4.68	4.72	4.75	4.80	4.83	4.86	4.88	4.89
18	.05	2.97	3.12	3.21	3.27	3.32	3.35	3.37	3.39	3.41	3.43	3.45	3.46	3.47	3.47
	.01	4.07	4.27	4.38	4.46	4.53	4.59	4.64	4.68	4.71	4.76	4.79	4.82	4.84	4.85
19	.05	2.96	3.11	3.19	3.26	3.31	3.35	3.37	3.39	3.41	3.43	3.44	3.46	3.47	3.47
	.01	4.05	4.24	4.35	4.43	4.50	4.56	4.61	4.64	4.67	4.72	4.76	4.79	4.81	4.82
20	.05	2.95	3.10	3.18	3.25	3.30	3.34	3.36	3.38	3.40	3.43	3.44	3.46	3.46	3.47
	.01	4.02	4.22	4.33	4.40	4.47	4.53	4.58	4.61	4.65	4.69	4.73	4.76	4.78	4.79
22	.05	2.93	3.08	3.17	3.24	3.29	3.32	3.35	3.37	3.39	3.42	3.44	3.45	3.46	3.47
	.01	3.99	4.17	4.28	4.36	4.42	4.48	4.53	4.57	4.60	4.65	4.68	4.71	4.74	4.75
24	.05	2.92	3.07	3.15	3.22	3.28	3.31	3.34	3.37	3.38	3.41	3.44	3.45	3.46	3.47
	.01	3.96	4.14	4.24	4.33	4.39	4.44	4.49	4.53	4.57	4.62	4.64	4.67	4.70	4.72
26	.05	2.91	3.06	3.14	3.21	3.27	3.30	3.34	3.36	3.38	3.41	3.43	3.45	3.46	3.47
	.01	3.93	4.11	4.21	4.30	4.36	4.41	4.46	4.50	4.53	4.58	4.62	4.65	4.67	4.69
28	.05	2.90	3.04	3.13	3.20	3.26	3.30	3.33	3.35	3.37	3.40	3.43	3.45	3.46	3.47
	.01	3.91	4.08	4.18	4.28	4.34	4.39	4.43	4.47	4.51	4.56	4.60	4.62	4.65	4.67
30	.05	2.89	3.04	3.12	3.20	3.25	3.29	3.32	3.35	3.37	3.40	3.43	3.44	3.46	3.47
	.01	3.89	4.06	4.16	4.22	4.32	4.36	4.41	4.45	4.48	4.54	4.58	4.61	4.63	4.65
40	.05	2.86	3.01	3.10	3.17	3.22	3.27	3.30	3.33	3.35	3.39	3.42	3.44	3.46	3.47
	.01	3.82	3.99	4.10	4.17	4.24	4.30	4.34	4.37	4.41	4.46	4.51	4.54	4.57	4.59
60	.05	2.83	2.98	3.08	3.14	3.20	3.24	3.28	3.31	3.33	3.37	3.40	3.43	3.45	3.47
	.01	3.76	3.92	4.03	4.12	4.17	4.23	4.27	4.31	4.34	4.39	4.44	4.47	4.50	4.53
100	.05	2.80	2.95	3.05	3.12	3.18	3.22	3.26	3.29	3.32	3.36	3.40	3.42	3.45	3.47
	.01	3.71	3.86	3.93	4.06	4.11	4.17	4.21	4.25	4.29	4.35	4.38	4.42	4.45	4.48
∞	.05	2.77	2.92	3.02	3.09	3.15	3.19	3.23	3.26	3.29	3.34	3.38	3.41	3.44	3.47
	.01	3.64	3.80	3.90	3.98	4.04	4.09	4.14	4.17	4.20	4.26	4.31	4.34	4.38	4.41

Abridged from D. B. Duncan, Multiple range and multiple F tests, *Biometrics*, 1955, *11*, 1–42, with permission of the editor and the author.

TABLE E.9 Percentage Points for the Comparison of $p - 1$ Treatment Means with a Control

One-Tailed Comparisons

Error df	α	Number of Treatment Means, Including Control (p)								
		2	3	4	5	6	7	8	9	10
5	.05	2.02	2.44	2.68	2.85	2.98	3.08	3.16	3.24	3.30
	.01	3.37	3.90	4.21	4.43	4.60	4.73	4.85	4.94	5.03
6	.05	1.94	2.34	2.56	2.71	2.83	2.92	3.00	3.07	3.12
	.01	3.14	3.61	3.88	4.07	4.21	4.33	4.43	4.51	4.59
7	.05	1.89	2.27	2.48	2.62	2.73	2.82	2.89	2.95	3.01
	.01	3.00	3.42	3.66	3.83	3.96	4.07	4.15	4.23	4.30
8	.05	1.86	2.22	2.42	2.55	2.66	2.74	2.81	2.87	2.92
	.01	2.90	3.29	3.51	3.67	3.79	3.88	3.96	4.03	4.09
9	.05	1.83	2.18	2.37	2.50	2.60	2.68	2.75	2.81	2.86
	.01	2.82	3.19	3.40	3.55	3.66	3.75	3.82	3.89	3.94
10	.05	1.81	2.15	2.34	2.47	2.56	2.64	2.70	2.76	2.81
	.01	2.76	3.11	3.31	3.45	3.56	3.64	3.71	3.78	3.83
11	.05	1.80	2.13	2.31	2.44	2.53	2.60	2.67	2.72	2.77
	.01	2.72	3.06	3.25	3.38	3.48	3.56	3.63	3.69	3.74
12	.05	1.78	2.11	2.29	2.41	2.50	2.58	2.64	2.69	2.74
	.01	2.68	3.01	3.19	3.32	3.42	3.50	3.56	3.62	3.67
13	.05	1.77	2.09	2.27	2.39	2.48	2.55	2.61	2.66	2.71
	.01	2.65	2.97	3.15	3.27	3.37	3.44	3.51	3.56	3.61
14	.05	1.76	2.08	2.25	2.37	2.46	2.53	2.59	2.64	2.69
	.01	2.62	2.94	3.11	3.23	3.32	3.40	3.46	3.51	3.56
15	.05	1.75	2.07	2.24	2.36	2.44	2.51	2.57	2.62	2.67
	.01	2.60	2.91	3.08	3.20	3.29	3.36	3.42	3.47	3.52
16	.05	1.75	2.06	2.23	2.34	2.43	2.50	2.56	2.61	2.65
	.01	2.58	2.88	3.05	3.17	3.26	3.33	3.39	3.44	3.48
17	.05	1.74	2.05	2.22	2.33	2.42	2.49	2.54	2.59	2.64
	.01	2.57	2.86	3.03	3.14	3.23	3.30	3.36	3.41	3.45
18	.05	1.73	2.05	2.21	2.32	2.41	2.48	2.53	2.58	2.62
	.01	2.55	2.84	3.01	3.12	3.21	3.27	3.33	3.38	3.42
19	.05	1.73	2.03	2.20	2.31	2.40	2.47	2.52	2.57	2.61
	.01	2.54	2.83	2.99	3.10	3.18	3.25	3.31	3.36	3.40
20	.05	1.72	2.03	2.19	2.30	2.39	2.46	2.51	2.56	2.60
	.01	2.53	2.81	2.97	3.08	3.17	3.23	3.29	3.34	3.38
24	.05	1.71	2.01	2.17	2.28	2.36	2.43	2.48	2.53	2.57
	.01	2.49	2.77	2.92	3.03	3.11	3.17	3.22	3.27	3.31
30	.05	1.70	1.99	2.15	2.25	2.33	2.40	2.45	2.50	2.54
	.01	2.46	2.72	2.87	2.97	3.05	3.11	3.16	3.21	3.24
40	.05	1.68	1.97	2.13	2.23	2.31	2.37	2.42	2.47	2.51
	.01	2.42	2.68	2.82	2.92	2.99	3.05	3.10	3.14	3.18
60	.05	1.67	1.95	2.10	2.21	2.28	2.35	2.39	2.44	2.48
	.01	2.39	2.64	2.78	2.87	2.94	3.00	3.04	3.08	3.12
120	.05	1.66	1.93	2.08	2.18	2.26	2.32	2.37	2.41	2.45
	.01	2.36	2.60	2.73	2.82	2.89	2.94	2.99	3.03	3.06
∞	.05	1.64	1.92	2.06	2.16	2.23	2.29	2.34	2.38	2.42
	.01	2.33	2.56	2.68	2.77	2.84	2.89	2.93	2.97	3.00

Table reproduced from A multiple comparison procedure for comparing several treatments with a control. *Journal of the American Statistical Association*, 1955, *50*, 1096–1121, with permission of the author, C. W. Dunnett, and the editor.

TABLE E.9 (continued)

Two-Tailed Comparisons

| Error df | α | \multicolumn{9}{c}{Number of Treatment Means, Including Control (p)} |
		2	3	4	5	6	7	8	9	10
5	.05	2.57	3.03	3.29	3.48	3.62	3.73	3.82	3.90	3.97
	.01	4.03	4.63	4.98	5.22	5.41	5.56	5.69	5.80	5.89
6	.05	2.45	2.86	3.10	3.26	3.39	3.49	3.57	3.64	3.71
	.01	3.71	4.21	4.51	4.71	4.87	5.00	5.10	5.20	5.28
7	.05	2.36	2.75	2.97	3.12	3.24	3.33	3.41	3.47	3.53
	.01	3.50	3.95	4.21	4.39	4.53	4.64	4.74	4.82	4.89
8	.05	2.31	2.67	2.88	3.02	3.13	3.22	3.29	3.35	3.41
	.01	3.36	3.77	4.00	4.17	4.29	4.40	4.48	4.56	4.62
9	.05	2.26	2.61	2.81	2.95	3.05	3.14	3.20	3.26	3.32
	.01	3.25	3.63	3.85	4.01	4.12	4.22	4.30	4.37	4.43
10	.05	2.23	2.57	2.76	2.89	2.99	3.07	3.14	3.19	3.24
	.01	3.17	3.53	3.74	3.88	3.99	4.08	4.16	4.22	4.28
11	.05	2.20	2.53	2.72	2.84	2.94	3.02	3.08	3.14	3.19
	.01	3.11	3.45	3.65	3.79	3.89	3.98	4.05	4.11	4.16
12	.05	2.18	2.50	2.68	2.81	2.90	2.98	3.04	3.09	3.14
	.01	3.05	3.39	3.58	3.71	3.81	3.89	3.96	4.02	4.07
13	.05	2.16	2.48	2.65	2.78	2.87	2.94	3.00	3.06	3.10
	.01	3.01	3.33	3.52	3.65	3.74	3.82	3.89	3.94	3.99
14	.05	2.14	2.46	2.63	2.75	2.84	2.91	2.97	3.02	3.07
	.01	2.98	3.29	3.47	3.59	3.69	3.76	3.83	3.88	3.93
15	.05	2.13	2.44	2.61	2.73	2.82	2.89	2.95	3.00	3.04
	.01	2.95	3.25	3.43	3.55	3.64	3.71	3.78	3.83	3.88
16	.05	2.12	2.42	2.59	2.71	2.80	2.87	2.92	2.97	3.02
	.01	2.92	3.22	3.39	3.51	3.60	3.67	3.73	3.78	3.83
17	.05	2.11	2.41	2.58	2.69	2.78	2.85	2.90	2.95	3.00
	.01	2.90	3.19	3.36	3.47	3.56	3.63	3.69	3.74	3.79
18	.05	2.10	2.40	2.56	2.68	2.76	2.83	2.89	2.94	2.98
	.01	2.88	3.17	3.33	3.44	3.53	3.60	3.66	3.71	3.75
19	.05	2.09	2.39	2.55	2.66	2.75	2.81	2.87	2.92	2.96
	.01	2.86	3.15	3.31	3.42	3.50	3.57	3.63	3.68	3.72
20	.05	2.09	2.38	2.54	2.65	2.73	2.80	2.86	2.90	2.95
	.01	2.85	3.13	3.29	3.40	3.48	3.55	3.60	3.65	3.69
24	.05	2.06	2.35	2.51	2.61	2.70	2.76	2.81	2.86	2.90
	.01	2.80	3.07	3.22	3.32	3.40	3.47	3.52	3.57	3.61
30	.05	2.04	2.32	2.47	2.58	2.66	2.72	2.77	2.82	2.86
	.01	2.75	3.01	3.15	3.25	3.33	3.39	3.44	3.49	3.52
40	.05	2.02	2.29	2.44	2.54	2.62	2.68	2.73	2.77	2.81
	.01	2.70	2.95	3.09	3.19	3.26	3.32	3.37	3.41	3.44
60	.05	2.00	2.27	2.41	2.51	2.58	2.64	2.69	2.73	2.77
	.01	2.66	2.90	3.03	3.12	3.19	3.25	3.29	3.33	3.37
120	.05	1.98	2.24	2.38	2.47	2.55	2.60	2.65	2.69	2.73
	.01	2.62	2.85	2.97	3.06	3.12	3.18	3.22	3.26	3.29
∞	.05	1.96	2.21	2.35	2.44	2.51	2.57	2.61	2.65	2.69
	.01	2.58	2.79	2.92	3.00	3.06	3.11	3.15	3.19	3.22

Table reproduced from New tables for multiple comparisons with a control. *Biometrics*, 1964, *20*, 482–491, with permission of the author, C. W. Dunnett, and the editor.

TABLE E.10 Upper Percentage Points of the F_{max} Statistic

$$F_{max} = (\hat{\sigma}^2_{largest})/(\hat{\sigma}^2_{smallest})$$

df for $\hat{\sigma}^2_j$	α	\multicolumn{11}{c}{p = Number of Variances}
		2	3	4	5	6	7	8	9	10	11	12
2	.05	39.0	87.5	142	202	266	333	403	475	550	626	704
	.01	199	448	729	1036	1362	1705	2063	2432	2813	3204	3605
3	.05	15.4	27.8	39.2	50.7	62.0	72.9	83.5	93.9	104	114	124
	.01	47.5	85	120	151	184	216	249	281	310	337	361
4	.05	9.60	15.5	20.6	25.2	29.5	33.6	37.5	41.4	44.6	48.0	51.4
	.01	23.2	37.	49.	59.	69.	79.	89.	97.	106.	113.	120.
5	.05	7.15	10.8	13.7	16.3	18.7	20.8	22.9	24.7	26.5	28.2	29.9
	.01	14.9	22.	28.	33.	38.	42.	46.	50.	54.	57.	60.
6	.05	5.82	8.38	10.4	12.1	13.7	15.0	16.3	17.5	18.6	19.7	20.7
	.01	11.1	15.5	19.1	22.	25.	27.	30.	32.	34.	36.	37.
7	.05	4.99	6.94	8.44	9.70	10.8	11.8	12.7	13.5	14.3	15.1	15.8
	.01	8.89	12.1	14.5	16.5	18.4	20.	22.	23.	24.	26.	27.
8	.05	4.43	6.00	7.18	8.12	9.03	9.78	10.5	11.1	11.7	12.2	12.7
	.01	7.50	9.9	11.7	13.2	14.5	15.8	16.9	17.9	18.9	19.8	21.
9	.05	4.03	5.34	6.31	7.11	7.80	8.41	8.95	9.45	9.91	10.3	10.7
	.01	6.54	8.5	9.9	11.1	12.1	13.1	13.9	14.7	15.3	16.0	16.6
10	.05	3.72	4.85	5.67	6.34	6.92	7.42	7.87	8.28	8.66	9.01	9.34
	.01	5.85	7.4	8.6	9.6	10.4	11.1	11.8	12.4	12.9	13.4	13.9
12	.05	3.28	4.16	4.79	5.30	5.72	6.09	6.42	6.72	7.00	7.25	7.48
	.01	4.91	6.1	6.9	7.6	8.2	8.7	9.1	9.5	9.9	10.2	10.6
15	.05	2.86	3.54	4.01	4.37	4.68	4.95	5.19	5.40	5.59	5.77	5.93
	.01	4.07	4.9	5.5	6.0	6.4	6.7	7.1	7.3	7.5	7.8	8.0
20	.05	2.46	2.95	3.29	3.54	3.76	3.94	4.10	4.24	4.37	4.49	4.59
	.01	3.32	3.8	4.3	4.6	4.9	5.1	5.3	5.5	5.6	5.8	5.9
30	.05	2.07	2.40	2.61	2.78	2.91	3.02	3.12	3.21	3.29	3.36	3.39
	.01	2.63	3.0	3.3	3.4	3.6	3.7	3.8	3.9	4.0	4.1	4.2
60	.05	1.67	1.85	1.96	2.04	2.11	2.17	2.22	2.26	2.30	2.33	2.36
	.01	1.96	2.2	2.3	2.4	2.4	2.5	2.5	2.6	2.6	2.7	2.7
∞	.05	1.00	1.00	1.00	1.00	1.00	1.00	1.00	1.00	1.00	1.00	1.00
	.01	1.00	1.00	1.00	1.00	1.00	1.00	1.00	1.00	1.00	1.00	1.00

This table is abridged from Table 31 in *Biometrika Tables for Statisticians*, vol. 1, 2nd ed. New York: Cambridge, 1958. Edited by E. S. Pearson and H. O. Hartley. Reproduced with the kind permission of the editors and the trustees of *Biometrika*.

TABLE E.11 Upper Percentage Points of Cochran's Test for Homogeneity of Variance

$$C = \frac{\text{largest } \hat{\sigma}_j^2}{\Sigma \, \hat{\sigma}_j^2}$$

df for $\hat{\sigma}_j^2$	α	\multicolumn{11}{c}{Number of Variances (p)}
		2	3	4	5	6	7	8	9	10	15	20
1	.05	.9985	.9669	.9065	.8412	.7808	.7271	.6798	.6385	.6020	.4709	.3894
	.01	.9999	.9933	.9676	.9279	.8828	.8376	.7945	.7544	.7175	.5747	.4799
2	.05	.9750	.8709	.7679	.6838	.6161	.5612	.5157	.4775	.4450	.3346	.2705
	.01	.9950	.9423	.8643	.7885	.7218	.6644	.6152	.5727	.5358	.4069	.3297
3	.05	.9392	.7977	.6841	.5981	.5321	.4800	.4377	.4027	.3733	.2758	.2205
	.01	.9794	.8831	.7814	.6957	.6258	.5685	.5209	.4810	.4469	.3317	.2654
4	.05	.9057	.7457	.6287	.5441	.4803	.4307	.3910	.3584	.3311	.2419	.1921
	.01	.9586	.8335	.7212	.6329	.5635	.5080	.4627	.4251	.3934	.2882	.2288
5	.05	.8772	.7071	.5895	.5065	.4447	.3974	.3595	.3286	.3029	.2195	.1735
	.01	.9373	.7933	.6761	.5875	.5195	.4659	.4226	.3870	.3572	.2593	.2048
6	.05	.8534	.6771	.5598	.4783	.4184	.3726	.3362	.3067	.2823	.2034	.1602
	.01	:9172	.7606	.6410	.5531	.4866	.4347	.3932	.3592	.3308	.2386	.1877
7	.05	.8332	.6530	.5365	.4564	.3980	.3535	.3185	.2901	.2666	.1911	.1501
	.01	.8988	.7335	.6129	.5259	.4608	.4105	.3704	.3378	.3106	.2228	.1748
8	.05	.8159	.6333	.5175	.4387	.3817	.3384	.3043	.2768	.2541	.1815	.1422
	.01	.8823	.7107	.5897	.5037	.4401	.3911	.3522	.3207	.2945	.2104	.1646
9	.05	.8010	.6167	.5017	.4241	.3682	.3259	.2926	.2659	.2439	.1736	.1357
	.01	.8674	.6912	.5702	.4854	.4229	.3751	.3373	.3067	.2813	.2002	.1567
16	.05	.7341	.5466	.4366	.3645	.3135	.2756	.2462	.2226	.2032	.1429	.1108
	.01	.7949	.6059	.4884	.4094	.3529	.3105	.2779	.2514	.2297	.1612	.1248
36	.05	.6602	.4748	.3720	.3066	.2612	.2278	.2022	.1820	.1655	.1144	.0879
	.01	.7067	.5153	.4057	.3351	.2858	.2494	.2214	.1992	.1811	.1251	.0960
144	.05	.5813	.4031	.3093	.2513	.2119	.1833	.1616	.1446	.1308	.0889	.0675
	.01	.6062	.4230	.3251	.2644	.2229	.1929	.1700	.1521	.1376	.0934	.0709

Reprinted from Chapter 15 of *Techniques of Statistical Analysis,* edited by C. Eisenhart, M. W. Hastay, and W. A. Wallis, McGraw-Hill Book Company, 1947.

TABLE E.12   Coefficients of Orthogonal Polynomials

p	Polynomial	Coefficients										Σc_{ij}^2
3	Linear	−1	0	1								2
	Quadratic	1	−2	1								6
	Linear	−3	−1	1	3							20
4	Quadratic	1	−1	−1	1							4
	Cubic	−1	3	−3	1							20
	Linear	−2	−1	0	1	2						10
5	Quadratic	2	−1	−2	−1	2						14
	Cubic	−1	2	0	−2	1						10
	Quartic	1	−4	6	−4	1						70
	Linear	−5	−3	−1	1	3	5					70
6	Quadratic	5	−1	−4	−4	−1	5					84
	Cubic	−5	7	4	−4	−7	5					180
	Quartic	1	−3	2	2	−3	1					28
	Linear	−3	−2	−1	0	1	2	3				28
7	Quadratic	5	0	−3	−4	−3	0	5				84
	Cubic	−1	1	1	0	−1	−1	1				6
	Quartic	3	−7	1	6	1	−7	3				154
	Linear	−7	−5	−3	−1	1	3	5	7			168
	Quadratic	7	1	−3	−5	−5	−3	1	7			168
8	Cubic	−7	5	7	3	−3	−7	−5	7			264
	Quartic	7	−13	−3	9	9	−3	−13	7			616
	Quintic	−7	23	−17	−15	15	17	−23	7			2184
	Linear	−4	−3	−2	−1	0	1	2	3	4		60
	Quadratic	28	7	−8	−17	−20	−17	−8	7	28		2772
9	Cubic	−14	7	13	9	0	−9	−13	−7	14		990
	Quartic	14	−21	−11	9	18	9	−11	−21	14		2002
	Quintic	−4	11	−4	−9	0	9	4	−11	4		468
	Linear	−9	−7	−5	−3	−1	1	3	5	7	9	330
	Quadratic	6	2	−1	−3	−4	−4	−3	−1	2	6	132
10	Cubic	−42	14	35	31	12	−12	−31	−35	−14	42	8580
	Quartic	18	−22	−17	3	18	18	3	−17	−22	18	2860
	Quintic	−6	14	−1	−11	−6	6	11	1	−14	6	780

Table E.12 is taken from Table 23 of Fisher and Yates, *Statistical Tables for Biological, Agricultural and Medical Research,* published by Oliver and Boyd Ltd., Edinburgh, by permission of the authors and publishers.

TABLE E.13   Arcsin Transformation

$$\phi = 2 \arcsin \sqrt{Y}$$

Y	ϕ	Y	ϕ	Y	ϕ	Y	ϕ	Y	ϕ
.001	.0633	.041	.4078	.36	1.2870	.76	2.1177	.971	2.7993
.002	.0895	.042	.4128	.37	1.3078	.77	2.1412	.972	2.8053
.003	.1096	.043	.4178	.38	1.3284	.78	2.1652	.973	2.8115
.004	.1266	.044	.4227	.39	1.3490	.79	2.1895	.974	2.8177
.005	.1415	.045	.4275	.40	1.3694	.80	2.2143	.975	2.8240
.006	.1551	.046	.4323	.41	1.3898	.81	2.2395	.976	2.8305
.007	.1675	.047	.4371	.42	1.4101	.82	2.2653	.977	2.8371
.008	.1791	.048	.4418	.43	1.4303	.83	2.2916	.978	2.8438
.009	.1900	.049	.4464	.44	1.4505	.84	2.3186	.979	2.8507
.010	.2003	.050	.4510	.45	1.4706	.85	2.3462	.980	2.8578
.011	.2101	.06	.4949	.46	1.4907	.86	2.3746	.981	2.8650
.012	.2195	.07	.5355	.47	1.5108	.87	2.4039	.982	2.8725
.013	.2285	.08	.5735	.48	1.5308	.88	2.4341	.983	2.8801
.014	.2372	.09	.6094	.49	1.5508	.89	2.4655	.984	2.8879
.015	.2456	.10	.6435	.50	1.5708	.90	2.4981	.985	2.8960
.016	.2537	.11	.6761	.51	1.5908	.91	2.5322	.986	2.9044
.017	.2615	.12	.7075	.52	1.6108	.92	2.5681	.987	2.9131
.018	.2691	.13	.7377	.53	1.6308	.93	2.6062	.988	2.9221
.019	.2766	.14	.7670	.54	1.6509	.94	2.6467	.989	2.9315
.020	.2838	.15	.7954	.55	1.6710	.95	2.6906	.990	2.9413
.021	.2909	.16	.8230	.56	1.6911	.951	2.6952	.991	2.9516
.022	.2978	.17	.8500	.57	1.7113	.952	2.6998	.992	2.9625
.023	.3045	.18	.8763	.58	1.7315	.953	2.7045	.993	2.9741
.024	.3111	.19	.9021	.59	1.7518	.954	2.7093	.994	2.9865
.025	.3176	.20	.9273	.60	1.7722	.955	2.7141	.995	3.0001
.026	.3239	.21	.9521	.61	1.7926	.956	2.7189	.996	3.0150
.027	.3301	.22	.9764	.62	1.8132	.957	2.7238	.997	3.0320
.028	.3363	.23	1.0004	.63	1.8338	.958	2.7288	.998	3.0521
.029	.3423	.24	1.0239	.64	1.8546	.959	2.7338	.999	3.0783
.030	.3482	.25	1.0472	.65	1.8755	.960	2.7389		
.031	.3540	.26	1.0701	.66	1.8965	.961	2.7440		
.032	.3597	.27	1.0928	.67	1.9177	.962	2.7492		
.033	.3654	.28	1.1152	.68	1.9391	.963	2.7545		
.034	.3709	.29	1.1374	.69	1.9606	.964	2.7598		
.035	.3764	.30	1.1593	.70	1.9823	.965	2.7652		
.036	.3818	.31	1.1810	.71	2.0042	.966	2.7707		
.037	.3871	.32	1.2025	.72	2.0264	.967	2.7762		
.038	.3924	.33	1.2239	.73	2.0488	.968	2.7819		
.039	.3976	.34	1.2451	.74	2.0715	.969	2.7876		
.040	.4027	.35	1.2661	.75	2.0944	.970	2.7934		

TABLE E.14 Power Function for Analysis of Variance

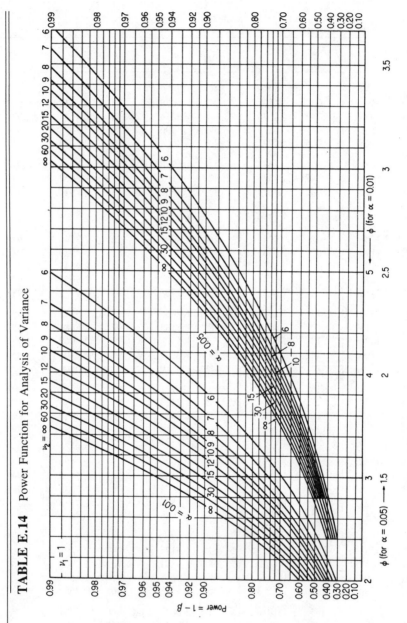

Reproduced with permission from E. S. Pearson and H. O. Hartley, Charts of the power function for analysis of variance tests, derived from the non-central *F*-distribution, *Biometrika*, 1951, *38*, 112–130.

TABLE E.14 (continued)

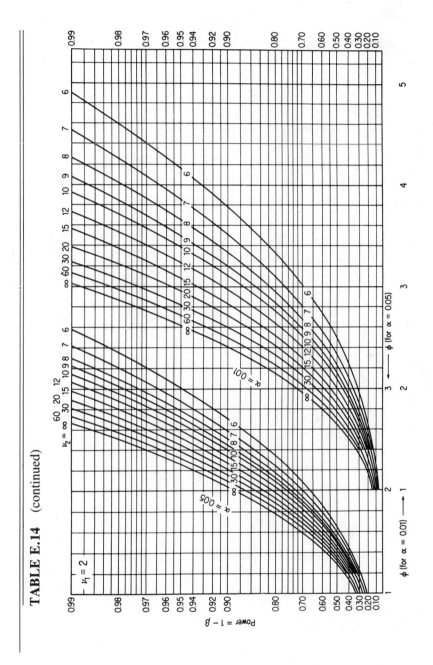

TABLE E.14   (continued)

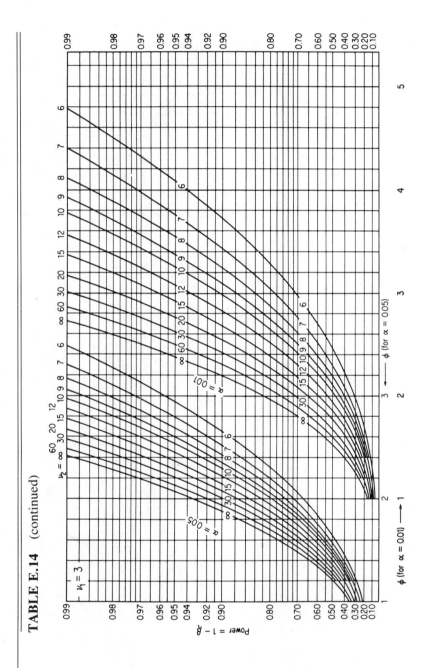

TABLE E.14 (continued)

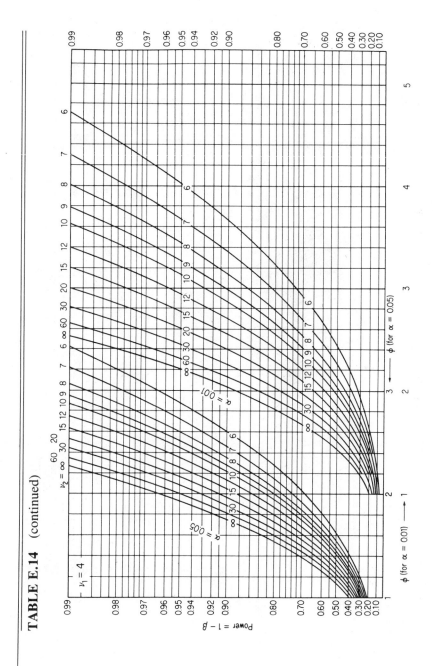

TABLE E.14 (continued)

TABLE E.14 (continued)

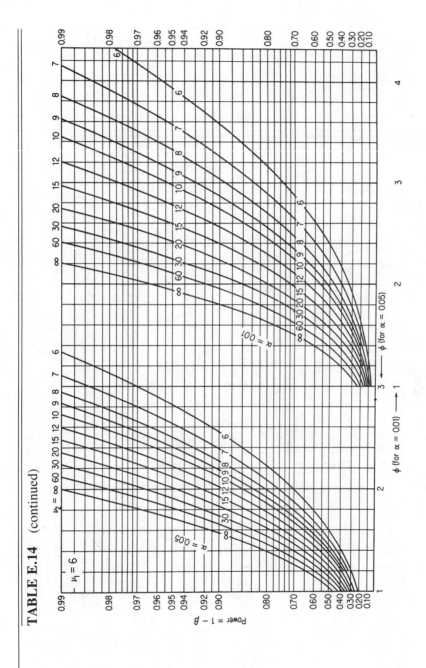

TABLE E.14   (continued)

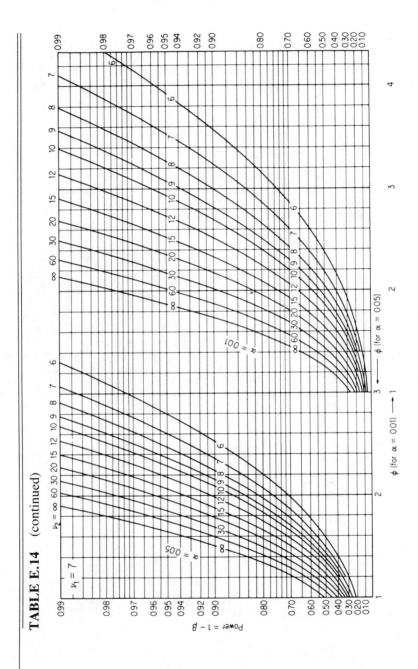

TABLE E.14 (continued)

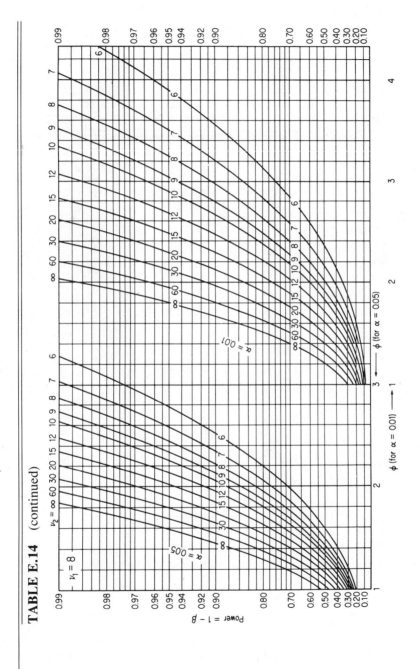

TABLE E.15   Minimum Sample Size Needed to Ensure a Given Power

p	α	$1-\beta=.70$ C 1.00	1.25	1.50	1.75	2.00	2.50	3.00	$1-\beta=.80$ C 1.00	1.25	1.50	1.75	2.00	2.50	3.00
2	.10	11	7	6	4	4	3	3	14	9	7	5	4	3	3
	.05	14	9	7	6	5	4	3	17	12	9	7	6	4	4
	.01	21	15	11	9	7	5	5	26	17	13	10	8	6	5
3	.10	13	9	7	5	4	3	3	17	11	8	6	5	4	3
	.05	17	11	8	7	5	4	3	21	14	10	8	6	5	4
	.01	25	17	12	10	8	6	5	30	20	14	11	9	7	5
4	.10	15	10	7	6	5	4	3	19	13	9	7	6	4	3
	.05	19	13	9	7	6	4	4	23	15	11	9	7	5	4
	.01	28	19	13	10	8	6	5	33	22	16	12	10	7	5
5	.10	17	11	8	6	5	4	3	21	14	10	8	6	4	4
	.05	21	14	10	8	6	5	4	25	17	12	9	7	5	4
	.01	30	20	14	11	9	6	5	35	23	17	13	10	7	6
6	.10	18	12	9	7	5	4	3	22	15	11	8	7	5	4
	.05	22	15	11	8	7	5	4	27	18	13	10	8	6	4
	.01	32	21	15	12	9	7	5	38	25	18	13	11	8	6
7	.10	19	13	9	7	6	4	3	24	16	11	9	7	5	4
	.05	24	16	11	9	7	5	4	29	19	14	10	8	6	5
	.01	34	22	16	12	10	7	5	39	26	18	14	11	8	6
8	.10	20	13	10	7	6	4	3	25	16	12	9	7	5	4
	.05	25	16	12	9	7	5	4	30	20	14	11	9	6	5
	.01	35	23	17	13	10	7	5	41	27	19	15	12	8	6
9	.10	21	14	10	8	6	4	4	26	17	12	9	7	5	4
	.05	26	17	12	9	8	5	4	31	21	15	11	9	6	5
	.01	37	24	17	13	10	7	6	43	28	20	15	12	8	6
10	.10	22	14	10	8	6	5	4	27	18	13	10	8	5	4
	.05	27	18	13	10	8	6	4	33	21	15	12	9	6	5
	.01	38	25	18	14	11	7	6	44	29	21	16	12	8	6
11	.10	23	15	11	8	7	5	4	28	18	13	10	8	6	4
	.05	28	19	13	10	8	6	4	34	22	16	12	9	7	5
	.01	39	26	18	14	11	8	6	46	30	21	16	13	9	7
13	.10	24	16	11	9	7	5	4	30	20	14	11	8	6	4
	.05	30	20	14	11	9	6	5	36	24	17	13	10	7	5
	.01	42	27	19	15	12	8	6	49	32	22	17	13	9	7

This table is abridged from Tables 1, 2, 3, and 4 in Tables of sample sizes in the analysis of variance. *Journal of Quality Technology,* 1970, 2, 156–164 with permission of the first author, T. L. Bratcher, and the editor.

TABLE E.15   (continued)

p	α	\multicolumn{7}{c}{$1 - \beta = .90$}	\multicolumn{7}{c}{$1 - \beta = .95$}
		\multicolumn{7}{c}{C}	\multicolumn{7}{c}{C}
		1.00	1.25	1.50	1.75	2.00	2.50	3.00	1.00	1.25	1.50	1.75	2.00	2.50	3.00
2	.10	18	12	9	7	6	4	3	23	15	11	8	7	5	4
	.05	23	15	11	8	7	5	4	27	18	13	10	8	6	5
	.01	32	21	15	12	10	7	6	38	25	18	14	11	8	6
3	.10	22	15	11	8	7	5	4	27	18	13	10	8	6	4
	.05	27	18	13	10	8	6	5	32	21	15	12	9	7	5
	.01	37	24	18	13	11	8	6	43	29	20	16	12	9	7
4	.10	25	16	12	9	7	5	4	30	20	14	11	9	6	5
	.05	30	20	14	11	9	6	5	36	23	17	13	10	7	5
	.01	40	27	19	15	12	8	6	47	31	22	17	13	9	7
5	.10	27	18	13	10	8	5	4	33	22	15	12	9	6	5
	.05	32	21	15	12	9	6	5	39	25	18	14	11	7	6
	.01	43	28	20	15	12	9	7	51	33	23	18	14	10	7
6	.10	29	19	14	10	8	6	4	35	23	16	12	10	7	5
	.05	34	23	16	12	10	7	5	41	27	19	14	11	8	6
	.01	46	30	21	16	13	9	7	53	35	25	19	15	10	8
7	.10	31	20	14	11	9	6	5	37	24	17	13	10	7	5
	.05	36	24	17	13	10	7	5	43	28	20	15	12	8	6
	.01	48	31	22	17	13	9	7	56	36	26	19	15	10	8
8	.10	32	21	15	11	9	6	5	39	25	18	14	11	7	5
	.05	38	25	18	13	11	7	6	45	29	21	16	12	8	6
	.01	50	33	23	17	14	9	7	58	38	27	20	16	11	8
9	.10	33	22	16	12	9	6	5	40	26	19	14	11	8	6
	.05	40	26	18	14	11	8	6	47	30	22	16	13	9	6
	.01	52	34	24	18	14	10	7	60	39	28	21	16	11	8
10	.10	35	23	16	12	10	7	5	42	27	19	15	11	8	6
	.05	41	27	19	14	11	8	6	48	31	22	17	13	9	7
	.01	54	35	25	19	15	10	7	62	40	29	21	17	11	8
11	.10	36	23	17	13	10	7	5	43	28	20	15	12	8	6
	.05	42	28	20	15	12	8	6	50	33	23	17	14	9	7
	.01	55	36	26	19	15	10	8	64	42	29	22	17	12	9
13	.10	38	25	18	13	11	7	5	46	30	21	16	12	8	6
	.05	45	29	21	16	12	8	6	53	34	24	18	14	10	7
	.01	59	38	27	20	16	11	8	68	44	31	23	18	12	9

TABLE E.16   Percentage Points of the Dunn Multiple Comparison Test

Number of Comparisons (C)	α	5	7	10	12	15	20	24	30	40	60	120	∞
							Error df						
2	.05	3.17	2.84	2.64	2.56	2.49	2.42	2.39	2.36	2.33	2.30	2.27	2.24
	.01	4.78	4.03	3.58	3.43	3.29	3.16	3.09	3.03	2.97	2.92	2.86	2.81
3	.05	3.54	3.13	2.87	2.78	2.69	2.61	2.58	2.54	2.50	2.47	2.43	2.39
	.01	5.25	4.36	3.83	3.65	3.48	3.33	3.26	3.19	3.12	3.06	2.99	2.94
4	.05	3.81	3.34	3.04	2.94	2.84	2.75	2.70	2.66	2.62	2.58	2.54	2.50
	.01	5.60	4.59	4.01	3.80	3.62	3.46	3.38	3.30	3.23	3.16	3.09	3.02
5	.05	4.04	3.50	3.17	3.06	2.95	2.85	2.80	2.75	2.71	2.66	2.62	2.58
	.01	5.89	4.78	4.15	3.93	3.74	3.55	3.47	3.39	3.31	3.24	3.16	3.09
6	.05	4.22	3.64	3.28	3.15	3.04	2.93	2.88	2.83	2.78	2.73	2.68	2.64
	.01	6.15	4.95	4.27	4.04	3.82	3.63	3.54	3.46	3.38	3.30	3.22	3.15
7	.05	4.38	3.76	3.37	3.24	3.11	3.00	2.94	2.89	2.84	2.79	2.74	2.69
	.01	6.36	5.09	4.37	4.13	3.90	3.70	3.61	3.52	3.43	3.34	3.27	3.19
8	.05	4.53	3.86	3.45	3.31	3.18	3.06	3.00	2.94	2.89	2.84	2.79	2.74
	.01	6.56	5.21	4.45	4.20	3.97	3.76	3.66	3.57	3.48	3.39	3.31	3.23
9	.05	4.66	3.95	3.52	3.37	3.24	3.11	3.05	2.99	2.93	2.88	2.83	2.77
	.01	6.70	5.31	4.53	4.26	4.02	3.80	3.70	3.61	3.51	3.42	3.34	3.26
10	.05	4.78	4.03	3.58	3.43	3.29	3.16	3.09	3.03	2.97	2.92	2.86	2.81
	.01	6.86	5.40	4.59	4.32	4.07	3.85	3.74	3.65	3.55	3.46	3.37	3.29
15	.05	5.25	4.36	3.83	3.65	3.48	3.33	3.26	3.19	3.12	3.06	2.99	2.94
	.01	7.51	5.79	4.86	4.56	4.29	4.03	3.91	3.80	3.70	3.59	3.50	3.40
20	.05	5.60	4.59	4.01	3.80	3.62	3.46	3.38	3.30	3.23	3.16	3.09	3.02
	.01	8.00	6.08	5.06	4.73	4.42	4.15	4.04	3.90	3.79	3.69	3.58	3.48
25	.05	5.89	4.78	4.15	3.93	3.74	3.55	3.47	3.39	3.31	3.24	3.16	3.09
	.01	8.37	6.30	5.20	4.86	4.53	4.25	4.1*	3.98	3.88	3.76	3.64	3.54
30	.05	6.15	4.95	4.27	4.04	3.82	3.63	3.54	3.46	3.38	3.30	3.22	3.15
	.01	8.68	6.49	5.33	4.95	4.61	4.33	4.2*	4.13	3.93	3.81	3.69	3.59
35	.05	6.36	5.09	4.37	4.13	3.90	3.70	3.61	3.52	3.43	3.34	3.27	3.19
	.01	8.95	6.67	5.44	5.04	4.71	4.39	4.3*	4.26	3.97	3.84	3.73	3.63
40	.05	6.56	5.21	4.45	4.20	3.97	3.76	3.66	3.57	3.48	3.39	3.31	3.23
	.01	9.19	6.83	5.52	5.12	4.78	4.46	4.3*	4.1*	4.01	3.89	3.77	3.66
45	.05	6.70	5.31	4.53	4.26	4.02	3.80	3.70	3.61	3.51	3.42	3.34	3.26
	.01	9.41	6.93	5.60	5.20	4.84	4.52	4.3*	4.2*	4.1*	3.93	3.80	3.69
50	.05	6.86	5.40	4.59	4.32	4.07	3.85	3.74	3.65	3.55	3.46	3.37	3.29
	.01	9.68	7.06	5.70	5.27	4.90	4.56	4.4*	4.2*	4.1*	3.97	3.83	3.72
100	.05	8.00	6.08	5.06	4.73	4.42	4.15	4.04	3.90	3.79	3.69	3.58	3.48
	.01	11.04	7.80	6.20	5.70	5.20	4.80	4.7*	4.4*	4.5*		4.00	3.89
250	.05	9.68	7.06	5.70	5.27	4.90	4.56	4.4*	4.2*	4.1*	3.97	3.83	3.72
	.01	13.26	8.83	6.9*	6.3*	5.8*	5.2*	5.0*	4.9*	4.8*			4.11

*Obtained by graphical interpolation.
Table reproduced from Multiple comparisons among means. *Journal of the American Statistical Association,* 1961, *56,* 52–64, with permission of the author, O. J. Dunn, and the editor.

TABLE E.17 Percentage Points of the Dunn-Šidák Multiple Comparison Test

Error df	α	2	3	4	5	6	7	8	9	10	15	20	25	30
							Number of Comparisons (C)							
2	.10	4.243	5.243	6.081	6.816	7.480	8.090	8.656	9.188	9.691	11.890	13.741	15.371	16.845
	.05	6.164	7.582	8.774	9.823	10.769	11.639	12.449	13.208	13.927	17.072	19.721	22.054	24.163
	.01	14.071	17.248	19.925	22.282	24.413	26.372	28.196	29.908	31.528	38.620	44.598	49.865	54.626
3	.10	3.149	3.690	4.115	4.471	4.780	5.055	5.304	5.532	5.744	6.627	7.326	7.914	8.427
	.05	4.156	4.826	5.355	5.799	6.185	6.529	6.842	7.128	7.394	8.505	9.387	10.129	10.778
	.01	7.447	8.565	9.453	10.201	10.853	11.436	11.966	12.453	12.904	14.796	16.300	17.569	18.678
4	.10	2.751	3.150	3.452	3.699	3.909	4.093	4.257	4.406	4.542	5.097	5.521	5.870	6.169
	.05	3.481	3.941	4.290	4.577	4.822	5.036	5.228	5.402	5.562	6.214	6.714	7.127	7.480
	.01	5.594	6.248	6.751	7.166	7.520	7.832	8.112	8.367	8.600	9.556	10.294	10.902	11.424
5	.10	2.549	2.882	3.129	3.327	3.493	3.638	3.765	3.880	3.985	4.403	4.718	4.972	5.187
	.05	3.152	3.518	3.791	4.012	4.197	4.358	4.501	4.630	4.747	5.219	5.573	5.861	6.105
	.01	4.771	5.243	5.599	5.888	6.133	6.346	6.535	6.706	6.862	7.491	7.968	8.355	8.684
6	.10	2.428	2.723	2.939	3.110	3.253	3.376	3.484	3.580	3.668	4.015	4.272	4.477	4.649
	.05	2.959	3.274	3.505	3.690	3.845	3.978	4.095	4.200	4.296	4.675	4.956	5.182	5.372
	.01	4.315	4.695	4.977	5.203	5.394	5.559	5.704	5.835	5.954	6.428	6.782	7.068	7.308
7	.10	2.347	2.618	2.814	2.969	3.097	3.206	3.302	3.388	3.465	3.768	3.990	4.167	4.314
	.05	2.832	3.115	3.321	3.484	3.620	3.736	3.838	3.929	4.011	4.336	4.574	4.764	4.923
	.01	4.027	4.353	4.591	4.782	4.941	5.078	5.198	5.306	5.404	5.791	6.077	6.306	6.497
8	.10	2.289	2.544	2.726	2.869	2.987	3.088	3.176	3.254	3.324	3.598	3.798	3.955	4.086
	.05	2.743	3.005	3.193	3.342	3.464	3.569	3.661	3.743	3.816	4.105	4.316	4.482	4.621
	.01	3.831	4.120	4.331	4.498	4.637	4.756	4.860	4.953	5.038	5.370	5.613	5.807	5.969
9	.10	2.246	2.488	2.661	2.796	2.907	3.001	3.083	3.155	3.221	3.474	3.658	3.802	3.921
	.05	2.677	2.923	3.099	3.237	3.351	3.448	3.532	3.607	3.675	3.938	4.129	4.280	4.405
	.01	3.688	3.952	4.143	4.294	4.419	4.526	4.619	4.703	4.778	5.072	5.287	5.457	5.598

TABLE E.17 (continued)

Error df	α	\multicolumn{13}{c}{Number of Comparisons (C)}												
		2	3	4	5	6	7	8	9	10	15	20	25	30
10	.10	2.213	2.446	2.611	2.739	2.845	2.934	3.012	3.080	3.142	3.380	3.552	3.686	3.796
	.05	2.626	2.860	3.027	3.157	3.264	3.355	3.434	3.505	3.568	3.813	3.989	4.128	4.243
	.01	3.580	3.825	4.002	4.141	4.256	4.354	4.439	4.515	4.584	4.852	5.046	5.199	5.326
11	.10	2.186	2.412	2.571	2.695	2.796	2.881	2.955	3.021	3.079	3.306	3.468	3.595	3.699
	.05	2.586	2.811	2.970	3.094	3.196	3.283	3.358	3.424	3.484	3.715	3.880	4.010	4.117
	.01	3.495	3.726	3.892	4.022	4.129	4.221	4.300	4.371	4.434	4.682	4.860	5.001	5.117
12	.10	2.164	2.384	2.539	2.658	2.756	2.838	2.910	2.973	3.029	3.247	3.402	3.522	3.621
	.05	2.553	2.770	2.924	3.044	3.141	3.224	3.296	3.359	3.416	3.636	3.793	3.916	4.017
	.01	3.427	3.647	3.804	3.927	4.029	4.114	4.189	4.256	4.315	4.547	4.714	4.845	4.953
13	.10	2.146	2.361	2.512	2.628	2.723	2.803	2.872	2.933	2.988	3.196	3.347	3.463	3.557
	.05	2.526	2.737	2.886	3.002	3.096	3.176	3.245	3.306	3.361	3.571	3.722	3.839	3.935
	.01	3.371	3.582	3.733	3.850	3.946	4.028	4.099	4.162	4.218	4.438	4.595	4.718	4.819
14	.10	2.131	2.342	2.489	2.603	2.696	2.774	2.841	2.900	2.953	3.157	3.301	3.413	3.504
	.05	2.503	2.709	2.854	2.967	3.058	3.135	3.202	3.261	3.314	3.518	3.662	3.775	3.867
	.01	3.324	3.528	3.673	3.785	3.878	3.956	4.024	4.084	4.138	4.347	4.497	4.614	4.710
15	.10	2.118	2.325	2.470	2.582	2.672	2.748	2.814	2.872	2.924	3.122	3.262	3.370	3.459
	.05	2.483	2.685	2.827	2.937	3.026	3.101	3.166	3.224	3.275	3.472	3.612	3.721	3.810
	.01	3.285	3.482	3.622	3.731	3.820	3.895	3.961	4.019	4.071	4.271	4.414	4.526	4.618
16	.10	2.106	2.311	2.453	2.563	2.652	2.726	2.791	2.848	2.898	3.092	3.228	3.334	3.420
	.05	2.467	2.665	2.804	2.911	2.998	3.072	3.135	3.191	3.241	3.433	3.569	3.675	3.761
	.01	3.251	3.443	3.579	3.684	3.771	3.844	3.907	3.963	4.013	4.206	4.344	4.451	4.540

df	α													
18	.10	2.088	2.287	2.426	2.532	2.619	2.691	2.753	2.808	2.857	3.043	3.174	3.275	3.358
	.05	2.439	2.631	2.766	2.869	2.953	3.024	3.085	3.138	3.186	3.370	3.499	3.599	3.681
	.01	3.195	3.379	3.508	3.609	3.691	3.760	3.820	3.872	3.920	4.102	4.231	4.332	4.414
20	.10	2.073	2.269	2.405	2.508	2.593	2.663	2.724	2.777	2.824	3.005	3.132	3.229	3.309
	.05	2.417	2.605	2.736	2.836	2.918	2.986	3.045	3.097	3.143	3.320	3.445	3.541	3.620
	.01	3.152	3.329	3.454	3.550	3.629	3.695	3.752	3.802	3.848	4.021	4.144	4.239	4.317
25	.10	2.047	2.236	2.367	2.466	2.547	2.614	2.672	2.722	2.767	2.938	3.058	3.149	3.224
	.05	2.379	2.558	2.683	2.779	2.856	2.921	2.976	3.025	3.069	3.235	3.351	3.440	3.513
	.01	3.077	3.243	3.359	3.449	3.521	3.583	3.635	3.682	3.723	3.882	3.995	4.081	4.152
30	.10	2.030	2.215	2.342	2.439	2.517	2.582	2.638	2.687	2.731	2.895	3.010	3.098	3.169
	.05	2.354	2.528	2.649	2.742	2.816	2.878	2.932	2.979	3.021	3.180	3.291	3.376	3.445
	.01	3.029	3.188	3.298	3.384	3.453	3.511	3.561	3.605	3.644	3.794	3.900	3.981	4.048
40	.10	2.009	2.189	2.312	2.406	2.481	2.544	2.597	2.644	2.686	2.843	2.952	3.036	3.103
	.05	2.323	2.492	2.608	2.696	2.768	2.827	2.878	2.923	2.963	3.113	3.218	3.298	3.363
	.01	2.970	3.121	3.225	3.305	3.370	3.425	3.472	3.513	3.549	3.689	3.787	3.862	3.923
60	.10	1.989	2.163	2.283	2.373	2.446	2.506	2.558	2.603	2.643	2.793	2.897	2.976	3.040
	.05	2.294	2.456	2.568	2.653	2.721	2.777	2.826	2.869	2.906	3.049	3.148	3.223	3.284
	.01	2.914	3.056	3.155	3.230	3.291	3.342	3.386	3.425	3.459	3.589	3.679	3.749	3.805
120	.10	1.968	2.138	2.254	2.342	2.411	2.469	2.519	2.562	2.600	2.744	2.843	2.918	2.978
	.05	2.265	2.422	2.529	2.610	2.675	2.729	2.776	2.816	2.852	2.987	3.081	3.152	3.209
	.01	2.859	2.994	3.087	3.158	3.215	3.263	3.304	3.340	3.372	3.493	3.577	3.641	3.693
∞	.10	1.949	2.114	2.226	2.311	2.378	2.434	2.482	2.523	2.560	2.697	2.791	2.862	2.920
	.05	2.237	2.388	2.491	2.569	2.631	2.683	2.727	2.766	2.300	2.928	3.016	3.083	3.137
	.01	2.806	2.934	3.022	3.089	3.143	3.186	3.226	3.260	3.289	3.402	3.480	3.539	3.587

This table is abridged from Table 1 in An improved *t* table for simultaneous control on *g* contrasts. *Journal of the American Statistical Association*, 1977, 72, 531–534, with permission of the author, P. A. Games, and the editor.

TABLE E.18 Percentage Points of the Studentized Augmented Range Distribution

Error df	α	Number of Means (p)						
		2	3	4	5	6	7	8
5	.10	3.060	3.772	4.282	4.671	4.982	5.239	5.458
	.05	3.832	4.654	5.236	5.680	6.036	6.331	6.583
	.01	5.903	7.030	7.823	8.429	8.916	9.322	9.669
7	.10	2.848	3.491	3.943	4.285	4.556	4.781	4.972
	.05	3.486	4.198	4.692	5.064	5.360	5.606	5.816
	.01	5.063	5.947	6.551	7.008	7.374	7.679	7.939
10	.10	2.704	3.300	3.712	4.021	4.265	4.466	4.636
	.05	3.259	3.899	4.333	4.656	4.913	5.124	5.305
	.01	4.550	5.284	5.773	6.138	6.428	6.669	6.875
12	.10	2.651	3.230	3.628	3.924	4.157	4.349	4.511
	.05	3.177	3.791	4.204	4.509	4.751	4.950	5.119
	.01	4.373	5.056	5.505	5.837	6.101	6.321	6.507
16	.10	2.588	3.146	3.526	3.806	4.027	4.207	4.360
	.05	3.080	3.663	4.050	4.334	4.557	4.741	4.897
	.01	4.169	4.792	5.194	5.489	5.722	5.915	6.079
20	.10	2.551	3.097	3.466	3.738	3.950	4.124	4.271
	.05	3.024	3.590	3.961	4.233	4.446	4.620	4.768
	.01	4.055	4.644	5.019	5.294	5.510	5.688	5.839
24	.10	2.527	3.065	3.427	3.693	3.901	4.070	4.213
	.05	2.988	3.542	3.904	4.167	4.373	4.541	4.684
	.01	3.982	4.549	4.908	5.169	5.374	5.542	5.685
30	.10	2.503	3.034	3.389	3.649	3.851	4.016	4.155
	.05	2.952	3.496	3.847	4.103	4.320	4.464	4.602
	.01	3.912	4.458	4.800	5.048	5.242	5.401	5.536
40	.10	2.480	3.003	3.352	3.605	3.803	3.963	4.099
	.05	2.918	3.450	3.792	4.040	4.232	4.389	4.521
	.01	3.844	4.370	4.696	4.931	5.115	5.265	5.392
60	.10	2.457	2.972	3.315	3.563	3.755	3.911	4.042
	.05	2.884	3.406	3.738	3.978	4.163	4.314	4.441
	.01	3.778	4.284	4.595	4.818	4.991	5.133	5.253
120	.10	2.434	2.943	3.278	3.520	3.707	3.859	3.987
	.05	2.851	3.362	3.686	3.917	4.096	4.241	4.363
	.01	3.714	4.201	4.497	4.709	4.872	5.005	5.118
∞	.10	2.412	2.913	3.243	3.479	3.661	3.808	3.931
	.05	2.819	3.320	3.634	3.858	4.030	4.170	4.286
	.01	3.653	4.121	4.403	4.603	4.757	4.882	4.987

Table reproduced from Tables of the studentized augmented range and applications to problems of multiple comparisons. *Journal of the American Statistical Association,* 1978, *73,* 656–660, with permission of the author, M. R. Stoline, and the editor.

TABLE E.19 Percentage Points of the Bryant-Paulson Generalized Studentized Range

Error df	Number of Covariates (C)	α	\multicolumn{11}{c}{Number of Means (p)}										
			2	3	4	5	6	7	8	10	12	16	20
3	1	.05	5.42	7.18	8.32	9.17	9.84	10.39	10.86	11.62	12.22	13.14	13.83
		.01	10.28	13.32	15.32	16.80	17.98	18.95	19.77	21.12	22.19	23.82	25.05
	2	.05	6.21	8.27	9.60	10.59	11.37	12.01	12.56	13.44	14.15	15.22	16.02
		.01	11.97	15.56	17.91	19.66	21.05	22.19	23.16	24.75	26.01	27.93	29.38
	3	.05	6.92	9.23	10.73	11.84	12.72	13.44	14.06	15.05	15.84	17.05	17.95
		.01	13.45	17.51	20.17	22.15	23.72	25.01	26.11	27.90	29.32	31.50	33.13
4	1	.05	4.51	5.84	6.69	7.32	7.82	8.23	8.58	9.15	9.61	10.30	10.82
		.01	7.68	9.64	10.93	11.89	12.65	13.28	13.82	14.70	15.40	16.48	17.29
	2	.05	5.04	6.54	7.51	8.23	8.80	9.26	9.66	10.31	10.83	11.61	12.21
		.01	8.69	10.95	12.43	13.54	14.41	15.14	15.76	16.77	17.58	18.81	19.74
	3	.05	5.51	7.18	8.25	9.05	9.67	10.19	10.63	11.35	11.92	12.79	13.45
		.01	9.59	12.11	13.77	15.00	15.98	16.79	17.47	18.60	19.50	20.87	21.91
5	1	.05	4.06	5.17	5.88	6.40	6.82	7.16	7.45	7.93	8.30	8.88	9.32
		.01	6.49	7.99	8.97	9.70	10.28	10.76	11.17	11.84	12.38	13.20	13.83
	2	.05	4.45	5.68	6.48	7.06	7.52	7.90	8.23	8.76	9.18	9.83	10.31
		.01	7.20	8.89	9.99	10.81	11.47	12.01	12.47	13.23	13.84	14.77	15.47
	3	.05	4.81	6.16	7.02	7.66	8.17	8.58	8.94	9.52	9.98	10.69	11.22
		.01	7.83	9.70	10.92	11.82	12.54	13.14	13.65	14.48	15.15	16.17	16.95
6	1	.05	3.79	4.78	5.40	5.86	6.23	6.53	6.78	7.20	7.53	8.04	8.43
		.01	5.83	7.08	7.88	8.48	8.96	9.36	9.70	10.25	10.70	11.38	11.90
	2	.05	4.10	5.18	5.87	6.37	6.77	7.10	7.38	7.84	8.21	8.77	9.20
		.01	6.36	7.75	8.64	9.31	9.85	10.29	10.66	11.28	11.77	12.54	13.11
	3	.05	4.38	5.55	6.30	6.84	7.28	7.64	7.94	8.44	8.83	9.44	9.90
		.01	6.85	8.36	9.34	10.07	10.65	11.13	11.54	12.22	12.75	13.59	14.21
7	1	.05	3.62	4.52	5.09	5.51	5.84	6.11	6.34	6.72	7.03	7.49	7.84
		.01	5.41	6.50	7.20	7.72	8.14	8.48	8.77	9.26	9.64	10.24	10.69
	2	.05	3.87	4.85	5.47	5.92	6.28	6.58	6.83	7.24	7.57	8.08	8.46
		.01	5.84	7.03	7.80	8.37	8.83	9.21	9.53	10.06	10.49	11.14	11.64
	3	.05	4.11	5.16	5.82	6.31	6.70	7.01	7.29	7.73	8.08	8.63	9.03
		.01	6.23	7.52	8.36	8.98	9.47	9.88	10.23	10.80	11.26	11.97	12.51
8	1	.05	3.49	4.34	4.87	5.26	5.57	5.82	6.03	6.39	6.67	7.10	7.43
		.01	5.12	6.11	6.74	7.20	7.58	7.88	8.15	8.58	8.92	9.46	9.87
	2	.05	3.70	4.61	5.19	5.61	5.94	6.21	6.44	6.82	7.12	7.59	7.94
		.01	5.48	6.54	7.23	7.74	8.14	8.48	8.76	9.23	9.61	10.19	10.63
	3	.05	3.91	4.88	5.49	5.93	6.29	6.58	6.83	7.23	7.55	8.05	8.42
		.01	5.81	6.95	7.69	8.23	8.67	9.03	9.33	9.84	10.24	10.87	11.34
10	1	.05	3.32	4.10	4.58	4.93	5.21	5.43	5.63	5.94	6.19	6.58	6.87
		.01	4.76	5.61	6.15	6.55	6.86	7.13	7.35	7.72	8.01	8.47	8.82
	2	.05	3.49	4.31	4.82	5.19	5.49	5.73	5.93	6.27	6.54	6.95	7.26
		.01	5.02	5.93	6.51	6.93	7.27	7.55	7.79	8.19	8.50	8.99	9.36
	3	.05	3.65	4.51	5.05	5.44	5.75	6.01	6.22	6.58	6.86	7.29	7.62
		.01	5.27	6.23	6.84	7.30	7.66	7.96	8.21	8.63	8.96	9.48	9.88

TABLE E.19   (continued)

Error df	Number of Covariates (C)	α	Number of Means (p) 2	3	4	5	6	7	8	10	12	16	20
12	1	.05	3.22	3.95	4.40	4.73	4.98	5.19	5.37	5.67	5.90	6.26	6.53
		.01	4.54	5.31	5.79	6.15	6.43	6.67	6.87	7.20	7.46	7.87	8.18
	2	.05	3.35	4.12	4.59	4.93	5.20	5.43	5.62	5.92	6.17	6.55	6.83
		.01	4.74	5.56	6.07	6.45	6.75	7.00	7.21	7.56	7.84	8.27	8.60
	3	.05	3.48	4.28	4.78	5.14	5.42	5.65	5.85	6.17	6.43	6.82	7.12
		.01	4.94	5.80	6.34	6.74	7.05	7.31	7.54	7.90	8.20	8.65	9.00
14	1	.05	3.15	3.85	4.28	4.59	4.83	5.03	5.20	5.48	5.70	6.03	6.29
		.01	4.39	5.11	5.56	5.89	6.15	6.36	6.55	6.85	7.09	7.47	7.75
	2	.05	3.26	3.99	4.44	4.76	5.01	5.22	5.40	5.69	5.92	6.27	6.54
		.01	4.56	5.31	5.78	6.13	6.40	6.63	6.82	7.14	7.40	7.79	8.09
	3	.05	3.37	4.13	4.59	4.93	5.19	5.41	5.59	5.89	6.13	6.50	6.78
		.01	4.72	5.51	6.00	6.36	6.65	6.89	7.09	7.42	7.69	8.10	8.41
16	1	.05	3.10	3.77	4.19	4.49	4.72	4.91	5.07	5.34	5.55	5.87	6.12
		.01	4.28	4.96	5.39	5.70	5.95	6.15	6.32	6.60	6.83	7.18	7.45
	2	.05	3.19	3.90	4.32	4.63	4.88	5.07	5.24	5.52	5.74	6.07	6.33
		.01	4.42	5.14	5.58	5.90	6.16	6.37	6.55	6.85	7.08	7.45	7.73
	3	.05	3.29	4.01	4.46	4.78	5.03	5.23	5.41	5.69	5.92	6.27	6.53
		.01	4.56	5.30	5.76	6.10	6.37	6.59	6.77	7.08	7.33	7.71	8.00
18	1	.05	3.06	3.72	4.12	4.41	4.63	4.82	4.98	5.23	5.44	5.75	5.98
		.01	4.20	4.86	5.26	5.56	5.79	5.99	6.15	6.42	6.63	6.96	7.22
	2	.05	3.14	3.82	4.24	4.54	4.77	4.96	5.13	5.39	5.60	5.92	6.17
		.01	4.32	5.00	5.43	5.73	5.98	6.18	6.35	6.63	6.85	7.19	7.46
	3	.05	3.23	3.93	4.35	4.66	4.90	5.10	5.27	5.54	5.76	6.09	6.34
		.01	4.44	5.15	5.59	5.90	6.16	6.36	6.54	6.83	7.06	7.42	7.69
20	1	.05	3.03	3.67	4.07	4.35	4.57	4.75	4.90	5.15	5.35	5.65	5.88
		.01	4.14	4.77	5.17	5.45	5.68	5.86	6.02	6.27	6.48	6.80	7.04
	2	.05	3.10	3.77	4.17	4.46	4.69	4.88	5.03	5.29	5.49	5.81	6.04
		.01	4.25	4.90	5.31	5.60	5.84	6.03	6.19	6.46	6.67	7.00	7.25
	3	.05	3.18	3.86	4.28	4.57	4.81	5.00	5.16	5.42	5.63	5.96	6.20
		.01	4.35	5.03	5.45	5.75	5.99	6.19	6.36	6.63	6.85	7.19	7.45
24	1	.05	2.98	3.61	3.99	4.26	4.47	4.65	4.79	5.03	5.22	5.51	5.73
		.01	4.05	4.65	5.02	5.29	5.50	5.68	5.83	6.07	6.26	6.56	6.78
	2	.05	3.04	3.69	4.08	4.35	4.57	4.75	4.90	5.14	5.34	5.63	5.86
		.01	4.14	4.76	5.14	5.42	5.63	5.81	5.96	6.21	6.41	6.71	6.95
	3	.05	3.11	3.76	4.16	4.44	4.67	4.85	5.00	5.25	5.45	5.75	5.98
		.01	4.22	4.86	5.25	5.54	5.76	5.94	6.10	6.35	6.55	6.87	7.11
30	1	.05	2.94	3.55	3.91	4.18	4.38	4.54	4.69	4.91	5.09	5.37	5.58
		.01	3.96	4.54	4.89	5.14	5.34	5.50	5.64	5.87	6.05	6.32	6.53
	2	.05	2.99	3.61	3.98	4.25	4.46	4.62	4.77	5.00	5.18	5.46	5.68
		.01	4.03	4.62	4.98	5.24	5.44	5.61	5.75	5.98	6.16	6.44	6.66
	3	.05	3.04	3.67	4.05	4.32	4.53	4.70	4.85	5.08	5.27	5.56	5.78
		.01	4.10	4.70	5.06	5.33	5.54	5.71	5.85	6.08	6.27	6.56	6.78

TABLE E.19   (continued)

Error df	Number of Covariates (C)	α	2	3	4	5	6	7	8	10	12	16	20
							Number of Means (p)						
40	1	.05	2.89	3.49	3.84	4.09	4.29	4.45	4.58	4.80	4.97	5.23	5.43
		.01	3.88	4.43	4.76	5.00	5.19	5.34	5.47	5.68	5.85	6.10	6.30
	2	.05	2.93	3.53	3.89	4.15	4.34	4.50	4.64	4.86	5.04	5.30	5.50
		.01	3.93	4.48	4.82	5.07	5.26	5.41	5.54	5.76	5.93	6.19	6.38
	3	.05	2.97	3.57	3.94	4.20	4.40	4.56	4.70	4.92	5.10	5.37	5.57
		.01	3.98	4.54	4.88	5.13	5.32	5.48	5.61	5.83	6.00	6.27	6.47
60	1	.05	2.85	3.43	3.77	4.01	4.20	4.35	4.48	4.69	4.85	5.10	5.29
		.01	3.79	4.32	4.64	4.86	5.04	5.18	5.30	5.50	5.65	5.89	6.07
	2	.05	2.88	3.46	3.80	4.05	4.24	4.39	4.52	4.73	4.89	5.14	5.33
		.01	3.83	4.36	4.68	4.90	5.08	5.22	5.35	5.54	5.70	5.94	6.12
	3	.05	2.90	3.49	3.83	4.08	4.27	4.43	4.56	4.77	4.93	5.19	5.38
		.01	3.86	4.39	4.72	4.95	5.12	5.27	5.39	5.59	5.75	6.00	6.18
120	1	.05	2.81	3.37	3.70	3.93	4.11	4.26	4.38	4.58	4.73	4.97	5.15
		.01	3.72	4.22	4.52	4.73	4.89	5.03	5.14	5.32	5.47	5.69	5.85
	2	.05	2.82	3.38	3.72	3.95	4.13	4.28	4.40	4.60	4.75	4.99	5.17
		.01	3.73	4.24	4.54	4.75	4.91	5.05	5.16	5.35	5.49	5.71	5.88
	3	.05	2.84	3.40	3.73	3.97	4.15	4.30	4.42	4.62	4.77	5.01	5.19
		.01	3.75	4.25	4.55	4.77	4.94	5.07	5.18	5.37	5.51	5.74	5.90

Table reproduced from An extension of Tukey's method of multiple comparisons to experimental designs with random concomitant variables. *Biometrika,* 1976, *63,* 631–638, with permission of the editor.

ANSWERS TO STARRED EXERCISES

CHAPTER 1

1.   **a)** Fraternities (EU), members (OU)   **c)** Students (EU), students (OU)

3.   **a)** Type of scene is independent variable; facial expression is dependent variable; age of subjects, sex of subjects, history of television viewing and individual differences are possible nuisance variables.

       b) Duration of morphine administration is independent variable; amount of change in EEG is dependent variable; age of rats, phase of estrous cycle, and individual differences are possible nuisance variables.

4.   Independent variables in parts (b), (c), and (d) are quantitative; independent variable in part (a) is qualitative.

8.   They do not provide a valid estimate of error variance and are not subject to powerful tools of statistical analysis, such as ANOVA.

9.   **a)** (i) type LS-3 design (ii) $\mu_{1..} = \mu_{2..} = \mu_{3..}$ (iii) $Y_{ijkl} = \mu + \alpha_j + \beta_k + \gamma_l + \epsilon_{pooled}$
         $(i = 1, \ldots, 5; j = 1, \ldots, 3; k = 1, \ldots, 3; l = 1, \ldots, 3)$

       b) (i) type CR-4 design (ii) $\mu_1 = \mu_2 = \mu_3 = \mu_4$ (iii) $Y_{ij} = \mu + \alpha_j + \epsilon_{i(j)}$ $(i = 1, \ldots, 20;$
         $j = 1, \ldots, 4)$

10.   **a)**   **b)**

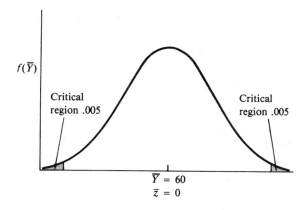

a_1	a_2	a_3
Y_{11}	Y_{12}	Y_{13}
Y_{21}	Y_{22}	Y_{23}
Y_{31}	Y_{32}	Y_{33}
$\vdots$	$\vdots$	$\vdots$
$Y_{10,1}$	$Y_{10,2}$	$Y_{10,3}$
$\overline{Y}_{.1}$	$\overline{Y}_{.2}$	$\overline{Y}_{.3}$

a_1 b_1	a_1 b_2	a_2 b_1	a_2 b_2	a_3 b_1	a_3 b_2
Y_{111}	Y_{112}	Y_{121}	Y_{122}	Y_{131}	Y_{132}
Y_{211}	Y_{212}	Y_{221}	Y_{222}	Y_{231}	Y_{232}
Y_{311}	Y_{312}	Y_{321}	Y_{322}	Y_{331}	Y_{332}
$\overline{Y}_{11}$	$\overline{Y}_{12}$	$\overline{Y}_{21}$	$\overline{Y}_{22}$	$\overline{Y}_{31}$	$\overline{Y}_{32}$

12.   History, mortality, diffusion or imitation of treatment

15.   Statistical inference must be used when the population elements are so numerous that it is impossible or impractical to observe all of them.

17. H_0: $\mu_1 - \mu_2 \leq 0$, H_1: $\mu_1 - \mu_2 > 0$

21.   **a)** yes   **c)** yes   **e)** no

22.   A sampling distribution associates a probability with every possible sample statistic that can be drawn from a population; a sample distribution associates a frequency or a relative frequency with every score in a sample.

24.   **a)** (1) State the null and alternative hypotheses—H_0: $\mu \leq 45$, H_1: $\mu > 45$. (2) Specify the test statistic—$z = (\overline{Y} - \mu_0)/(\sigma/\sqrt{n})$. (3) Specify the sample size—$n = 100$, and the sampling distribution—standard normal distribution. (4) Specify the level of significance—$\alpha = .05$. (5) Obtain a random sample of size n, compute z, and make a decision.

   b) Reject H_0 if z falls in the upper 5% of the sampling distribution of z; otherwise do not reject H_0.

28.   A knowledge of one's data might affect one's hypotheses as well as the specification of n and α. The true level of significance would not equal α if the data used in the test are also used to alter the testing procedure.

29.   **a)**

$f(\overline{Y})$

Critical
region .005

Critical
region .005

$\overline{Y} = 60$
$\overline{z} = 0$

30.   A one-tailed test is less powerful if the critical region is in the wrong tail of the sampling distribution.

32.   a) Type I error   b) Correct rejection

35.   a) The larger σ, the wider the interval.   b) The larger n, the narrower the interval.
c) The larger $1 - \alpha$, the wider the interval.

CHAPTER 2

2.   a) $E(\chi^2_{(20)}) = 20$, $V(\chi^2_{(20)}) = 40$ b) $E(\chi^2_{(16)}) = 16$, $V(\chi^2_{(16)}) = 32$

3. $\sum_{i=1}^{n} (Y_i - \overline{Y}) = \sum_{i=1}^{n} Y_i - \sum_{i=1}^{n} \overline{Y} = \sum_{i=1}^{n} Y_i - n\overline{Y} = \sum_{i=1}^{n} Y_i - \sum_{i=1}^{n} Y_i = 0$

since $n\overline{Y} = n \dfrac{\sum_{i=1}^{n} Y_i}{n} = \sum_{i=1}^{n} Y_i$

5.   a) 0.115   b) 0.103

6.   a) $E(F) = 1.07$, $V(F) = 0.50$ b) $E(F) = 1.04$, $V(F) = 0.27$

7.   a) $E(t) = 0$, $V(t) = 1.11$ b) $E(t) = 0$, $V(t) = 1.03$

9.   a) A.2   b) A.1, A.2 or A.1, A.3   c) A.3, A.6   d) A.1

10.   a) $dfTO = 29$, $dfBG = 2$, $dfWG = 27$ c) $dfTO = 15$, $dfBG = 3$, $dfWG = 12$
e) $dfTO = 47$, $dfBG = 5$, $dfWG = 42$

11.   a) Fixed-effects model   b) Random-effects model

12.   a) $n^2\mu^2$ (B.3)   b) $n^2\alpha_j^2$ (B.3)   c) $n(n\mu_\epsilon^2 + \sigma_\epsilon^2) = n\sigma_\epsilon^2$ (B.11)
f) $2n^2\alpha_j\mu_\epsilon = 0$ (B.4, B.8)   h) $np(np\mu_\epsilon^2 + \sigma_\epsilon^2) = npo_\epsilon^2$ (B.11)

13.   a) $n^2\mu^2$ (B.3)   b) $n^2(\mu_\alpha^2 + \sigma_\alpha^2) = n^2\sigma_\alpha^2$ (B.4, B.9)
c) $n(n\mu_\epsilon^2 + \sigma_\epsilon^2) = n\sigma_\epsilon^2$ (B.11)   f) $2n^2\mu_\alpha\mu_\epsilon = 0$ (B.1, B.4, B.7, B.8)
h) $np(np\mu_\epsilon^2 + \sigma_\epsilon^2) = npo_\epsilon^2$ (B.11)

15.   For a fixed-effects model, the results apply only to the p treatment levels in the experiment; for a random-effects model, the results apply to the population of P levels.

16.   If σ_α^2, the variance of the treatment effects ($\alpha_j = \mu_j - \mu$), is equal to zero, it follows that all of the effects are equal.

18.   The statement is correct provided that the populations are homogeneous in form, for example, all positively skewed and slightly leptokurtic.

21.   The statement is correct provided that the number of observations in the samples is equal.

23.   a) $F_{max} = 1.78$, $F_{max,.05} = 6.00$ b) $C = 0.46$, $C_{.05} = 0.6333$
c) $F = 0.22$, $F_{.05;2,6} = 5.14$ d) Do not reject the null hypothesis.

25.   None of the transformations discussed in the text is appropriate.

26.   None of the transformations discussed in the text is appropriate because the untransformed scores yield the smallest ratio. The ratio of the ranges for L and S is 2.25, for $\sqrt{L} + .5$ and $\sqrt{S} + .5$ is 2.68, for $\log_{10} (L + 1)$ and $\log_{10} (S + 1)$ is 3.23, and for $1/L$ and $1/S$ is 8.65.

28.   Use a square-root transformation when means and variances are proportional; use a logarithmic transformation when means and standard deviations are proportional; use a reciprocal transformation when the square of means is proportional to standard deviations; use an angular transformation when Y is a proportion and the means and variances are proportional.

30. Use a nonparametric ANOVA or choose a different criterion measure.

CHAPTER 3

2. A pairwise comparison involves only two means; a contrast involves any number of means.

3. **a)** 1 −1 **b)** 1 −½ −½
   d) ½ ½ −½ −½ **f)** 1 −⅔ −⅓

4. The following are contrasts: (a), (b), and (f).

5. Contrast (a) satisfies the requirement.

6. **a)** 3 **c)** 10

7. The contrasts in (b) and (d) are orthogonal.

9. **a)** $t = (64 − 73)/2.8284 = −3.18*$, $t = (61 − 49)/2.8284 = 4.24*$, and $t = [(64 + 73)/2 − (61 + 49)/2]/2.0000 = 6.75*$; $t_{.01/2,116} \cong 2.619$. Reject all three null hypotheses.

   b) $−16.41 \le \mu_1 − \mu_2 \le −1.59$, $4.59 \le \mu_3 − \mu_4 \le 19.41$, $8.26 \le (\mu_1 + \mu_2)/2 − (\mu_3 − \mu_4)/2 \le 18.74$

   c) $\rho = 0$ for all three correlations.

   d) $t' = (64 − 73)/2.1213 = −4.24*$; $t'_{.01/2,58} \cong 2.663$. Reject the null hypothesis. $t' = (61 − 49)/3.3912 = 3.54*$; $t'_{.01/2,45} \cong 2.689$. Reject the null hypothesis. $t' = [(64 + 73)/2 − (61 + 49)/2]/2.0000 = 6.75*$, $t'_{.01/2,78} \cong 2.640$. Reject the null hypothesis.

11. **a)** $\alpha_{PC} = .01$ **b)** $\alpha_{EW} = .0297$

13. **a)** $\alpha_{PC} = 50/6000 = .0083$ **b)** $\alpha_{PE} = 50/1000 = .05$
    c) $\alpha_{EW} = 35/1000 = .035$

15. **a)** $\alpha_{EW} = .0297$ **b)** $\alpha_{EW} \le .1855$

17. **a)**

	$\overline{Y}_{.1}$	$\overline{Y}_{.2}$	$\overline{Y}_{.3}$	$\overline{Y}_{.4}$
$\overline{Y}_{.1} = 0.28$	−	0.01	0.03	0.11*
$\overline{Y}_{.2} = 0.29$		−	0.02	0.10*
$\overline{Y}_{.3} = 0.31$			−	0.08
$\overline{Y}_{.4} = 0.39$				−

$$\hat{\psi}(D) = 3.01(0.0283) = 0.085$$

*$p < .05$

b) $−0.095 \le \mu_1 − \mu_2 \le 0.075$ $\quad −0.105 \le \mu_2 − \mu_3 \le 0.065$
$−0.115 \le \mu_1 − \mu_3 \le 0.055$ $\quad −0.185 \le \mu_2 − \mu_4 \le −0.015$
$−0.195 \le \mu_1 − \mu_4 \le −0.025$ $\quad −0.165 \le \mu_3 − \mu_4 \le 0.005$

c) $\rho_{12} = \rho_{13} = \rho_{23} = \rho_{24} = \rho_{35} = \rho_{36} = \rho_{45} = \rho_{56} = 0.5$, $\rho_{14} = \rho_{15} = \rho_{26} = \rho_{46} = −0.5$, $\rho_{16} = \rho_{25} = \rho_{34} = 0$.

d) Relative efficiency $= [(0.08480)^2/(0.08514)^2] \times 100 = 99.2\%$

e) Relative efficiency $= [(0.07326)^2/(0.07552)^2] \times 100 = 94.1\%$;
relative efficiency $= [(0.07326)^2/(0.07538)^2] \times 100 = 94.5\%$.

19.   **a)** $\hat{\psi} = \overline{Y}_{largest} - \overline{Y}_{smallest} > \hat{\psi}(HSD)$; reject the overall null hypothesis.

	$\overline{Y}_{.1}$	$\overline{Y}_{.2}$	$\overline{Y}_{.3}$	$\overline{Y}_{.4}$
$\overline{Y}_{.1} = 0.28$	–	0.01	0.03	0.11*
$\overline{Y}_{.2} = 0.29$		–	0.02	0.10*
$\overline{Y}_{.3} = 0.31$			–	0.08
$\overline{Y}_{.4} = 0.39$				–

$$\hat{\psi}(HSD) = 4.05(0.020) = 0.081$$

*$p < .05$

b) $-0.091 \leq \mu_1 - \mu_2 \leq 0.071$ $-0.101 \leq \mu_2 - \mu_3 \leq 0.061$
  $-0.111 \leq \mu_1 - \mu_3 \leq 0.051$ $-0.181 \leq \mu_2 - \mu_4 \leq -0.019$
  $-0.191 \leq \mu_1 - \mu_4 \leq -0.029$ $-0.161 \leq \mu_3 - \mu_4 \leq 0.001$

c) Relative efficiency $= [(0.0810)^2/(0.0848)^2] \times 100 = 91.2\%$.

d)

	$\overline{Y}_{.1}$	$\overline{Y}_{.2}$	$\overline{Y}_{.3}$	$\overline{Y}_{.4}$	
$\overline{Y}_{.1} = 0.28$	–	0.01	0.03	0.11*	$\hat{\psi}(W_2) = 0.060$
$\overline{Y}_{.2} = 0.29$		–	0.02	0.10*	$\hat{\psi}(W_3) = 0.073$
$\overline{Y}_{.3} = 0.31$			–	0.08*	$\hat{\psi}(W_4) = 0.081$
$\overline{Y}_{.4} = 0.39$				–	

*$p < .05$

e)

	$\overline{Y}_{.1}$	$\overline{Y}_{.2}$	$\overline{Y}_{.3}$	$\overline{Y}_{.4}$	
$\overline{Y}_{.1} = 0.28$	–	0.01	0.03	0.11*	$\hat{\psi}(W_2) = 0.060$
$\overline{Y}_{.2} = 0.29$		–	0.02	0.10*	$\hat{\psi}(W_3) = 0.063$
$\overline{Y}_{.3} = 0.31$			–	0.08*	$\hat{\psi}(W_4) = 0.065$
$\overline{Y}_{.4} = 0.39$				–	

*$p < .05$

21.   **a)** $F = (25.8 - 22.1)^2/1.5619 = 8.76^*$, $F = (26.7 - 22.1)^2/1.5619 = 13.55^*$,
    $F = [(25.8 + 26.7)/2 - 22.1]^2/1.1714 = 14.70^*$; $F' = (3 - 1)3.15 = 6.30$. Reject all three null hypotheses.

b) $\hat{\psi}_1 = 25.8 - 22.1 = 3.7^*$, $\hat{\psi}(BF) = 2.60(1.3075) = 3.40$; $\hat{\psi}_2 = 26.7 - 22.1 = 4.6^*$,
    $\hat{\psi}(BF) = 2.56(1.4655) = 3.75$; $\hat{\psi}_3 = (25.8 + 26.7)/2 - 22.1 = 4.15^*$,
    $\hat{\psi}(BF) = 2.60(1.3120) = 3.41$. Reject all three null hypotheses.

CHAPTER 4

1.   **a)**

	a_1	a_2	a_3
$\overline{Y}_{\cdot j}$	14.7	11.2	12.2
$\hat{\sigma}_j$	2.54	2.04	3.26

Y	a_1 f	a_2 f	a_3 f
19	|		
18			|
17	||		
16	|		|
15	|		
14	|	||	|
13	||	|	|
12	|	|	||
11	|	||	|
10		||	|
9		|	|
8		|	
7			|

The exploratory data analysis indicates that there are differences among the sample means. These differences would be of practical importance if they occurred in the populations. The sample variances appear to be fairly homogeneous and not related to the magnitude of the sample means. The frequency distributions do not exhibit an unusual amount of asymmetry; some asymmetry can be expected for samples of size ten due to sampling variability.

b)

	Source	SS	df	MS	F
1	Between groups (Learning approaches)	65.000	2	32.500	$[\frac{1}{2}]$ 4.59*
2	Within groups	191.300	27	7.085	
3	Total	256.300	29		

*$p < .05$

Reject the null hypothesis.

c) $\hat{\phi} = 1.302/0.842 = 1.55$; $1 - \hat{\beta} \cong .60$ **d)** $np \cong (15)(3) = 45$

e) Test statistics for $\mu_1 - \mu_2 = 0$, $\mu_1 - \mu_3 = 0$, and $\mu_2 - \mu_3 = 0$ are, respectively, $q =$

$3.500/0.842 = 4.16$, $q = 2.500/0.842 = 2.97$, and $q = -1.000/0.842 = -1.19$; $q_{.05;3,27} \cong 3.51$.
Reject the hypothesis $\mu_1 - \mu_2 = 0$.

f) $\hat{\omega}^2 = 50.830/263.385 = 0.193$.

2. **a)**

	a_1	a_2		a_1	a_2
			Y	f	f
$\overline{Y}_{.j}$	10.6	9.0	17	I	
			16		
$\hat{\sigma}_j$	4.25	3.27	15	I	I
			14	I	
			13		
			12	I	I
			11	II	I
			10		I
			9	I	II
			8	I	I
			7		I
			6	I	
			5		I
			4		I
			3	I	

The exploratory data analysis indicates that the difference between the sample means is quite small; however, this difference would be of interest if it occurred in the populations. The frequency distributions appear to be fairly symmetrical with similar dispersions.

b)

	Source	SS	df	MS	F
1	Between groups (Effect of instructions)	12.800	1	12.800	[½] 0.89
2	Within groups	258.400	18	14.356	
3	Total	271.200	19		

Do not reject the null hypothesis; instructions to use rehearsal do not aid recall 24 hours later.

c) $t = (10.6 - 9.0)/\sqrt{2(14.356)/10} = 0.944$; $t_{.05/2,18} = 2.101$. Do not reject the null hypothesis. $t^2 = F$

5. **a)** $(31)(3) = 93$ **b)** $(22)(4) = 88$

6. **a)**

	a_1	a_2	a_3	a_4
$\overline{Y}_j$	19.38	20.75	22.62	26.25
$\hat{\sigma}_j$	1.19	1.28	1.19	1.28

	a_1	a_2	a_3	a_4
Y	f	f	f	f
28				|
27				|||
26				||
25			|	|
24				|
23		|	|||	
22		|	|||	
21	|	||	|	
20	|||	|||		
19	|||	|		
18				
17	|			

The exploratory data analysis indicates that there are differences among the sample means. These differences would be of practical importance if they occurred in the populations. The sample variances are quite homogeneous and not related to the magnitude of the sample means. The frequency distributions are relatively symmetrical for such small samples.

b) $F = 71.083/1.527 = 46.55$; $F_{.05;3,28} = 2.95$. Reject the null hypothesis.

c) The F ratio for departure from linearity is $5.375/1.527 = 3.52$; $F_{.05;2,28} = 3.34$. Reject the null hypothesis.

d) (i) The F ratios for the linear, quadratic, and cubic trend components are, respectively, $F = 202.500/1.527 = 132.61$, $F = 10.125/1.527 = 6.63$, and $F = 0.625/1.527 = 0.41$; $F_{.05;1,28} = 4.20$. Reject the null hypothesis for the linear and quadratic components.
(ii) $\hat{\mu}_1 = 22.25 + 1.125(-3) + 0.562(1) = 19.437$
$\hat{\mu}_2 = 22.25 + 1.125(-1) + 0.562(-1) = 20.563$
$\hat{\mu}_3 = 22.25 + 1.125(1) + 0.562(-1) = 22.813$
$\hat{\mu}_4 = 22.25 + 1.125(3) + 0.562(1) = 26.187$

e) The linear component accounts for $(202.50/213.25) \times 100 = 94.96\%$ of the trend; the quadratic component accounts for $(10.125/213.25) \times 100 = 4.75\%$ of the trend.

f) $\hat{\omega}^2_{Y|\psi_{lin}} = 200.973/257.527 = 0.78$, $\hat{\omega}^2_{Y|\psi_{quad}} = 8.598/257.527 = 0.03$.

9. They differ in the way the levels of treatment A are selected, assumptions concerning the treatment effects, expected value of MSA, null hypothesis that is tested, and nature of the conclusions that can be drawn from the experiment.

CHAPTER 5

2.   This statement is incomplete because a linear model should always contain a model equation and associated assumptions, for example, $\sum_{j=1}^{p}\alpha_j = 0$ and $\epsilon_{i(j)}$ is $NID(0, \sigma_\epsilon^2)$.

4.   **a)**

$$\begin{bmatrix} Y_1 \\ Y_2 \\ Y_3 \end{bmatrix} = \begin{bmatrix} 1 & X_{11} & X_{12} \\ 1 & X_{21} & X_{22} \\ 1 & X_{31} & X_{32} \end{bmatrix} \begin{bmatrix} \beta_0 \\ \beta_1 \\ \beta_2 \end{bmatrix} + \begin{bmatrix} \epsilon_1 \\ \epsilon_2 \\ \epsilon_3 \end{bmatrix} \qquad \text{or} \qquad \underset{3\times1}{\mathbf{y}} = \underset{3\times3}{\mathbf{X}}\,\underset{3\times1}{\boldsymbol{\beta}} + \underset{3\times1}{\boldsymbol{\epsilon}}$$

5.   **a)**

$$\begin{bmatrix} Y_{11} \\ Y_{21} \\ Y_{12} \\ Y_{22} \end{bmatrix} = \begin{bmatrix} 1 & 1 & 0 \\ 1 & 1 & 0 \\ 1 & 0 & 1 \\ 1 & 0 & 1 \end{bmatrix} \begin{bmatrix} \mu \\ \alpha_1 \\ \alpha_2 \end{bmatrix} + \begin{bmatrix} \epsilon_{1(1)} \\ \epsilon_{2(1)} \\ \epsilon_{1(2)} \\ \epsilon_{2(2)} \end{bmatrix} \qquad \text{or} \qquad \underset{4\times1}{\mathbf{y}} = \underset{4\times3}{\mathbf{X}}\,\underset{3\times1}{\boldsymbol{\beta}} + \underset{4\times1}{\boldsymbol{\epsilon}}$$

6.   Least squares estimators are unbiased and have minimum variance among all unbiased linear estimators.

7.   **a)**

$$H_0: \underset{2\times3}{\mathbf{C}'}\,\underset{3\times1}{\boldsymbol{\beta}} = \underset{2\times1}{\mathbf{0}}, \text{ where } \underset{2\times3}{\mathbf{C}'} = \begin{bmatrix} 0 & 1 & 0 \\ 0 & 0 & 1 \end{bmatrix}, \underset{3\times1}{\boldsymbol{\beta}} = \begin{bmatrix} \beta_0 \\ \beta_1 \\ \beta_2 \end{bmatrix}, \underset{2\times1}{\mathbf{0}} = \begin{bmatrix} 0 \\ 0 \end{bmatrix}$$

c)

$$H_0: \underset{2\times3}{\mathbf{C}'}\,\underset{3\times1}{\boldsymbol{\beta}} = \underset{2\times1}{\mathbf{0}}, \text{ where } \underset{2\times3}{\mathbf{C}'} = \begin{bmatrix} 0 & 1 & 0 \\ 0 & 0 & 1 \end{bmatrix}, \underset{3\times1}{\boldsymbol{\beta}} = \begin{bmatrix} \mu \\ \alpha_1 \\ \alpha_2 \end{bmatrix}, \underset{2\times1}{\mathbf{0}} = \begin{bmatrix} 0 \\ 0 \end{bmatrix}$$

8.   **a)** (i)   (ii)

$$\underset{4\times2}{\mathbf{X}} = \begin{bmatrix} 1 & 1 \\ 1 & 1 \\ 1 & 0 \\ 1 & 0 \end{bmatrix} \qquad \underset{4\times2}{\mathbf{X}} = \begin{bmatrix} 1 & 1 \\ 1 & 1 \\ 1 & -1 \\ 1 & -1 \end{bmatrix}$$

c) (i)   (ii)

$$\underset{6\times3}{\mathbf{X}} = \begin{bmatrix} 1 & 1 & 0 \\ 1 & 1 & 0 \\ 1 & 0 & 1 \\ 1 & 0 & 1 \\ 1 & 0 & 0 \\ 1 & 0 & 0 \end{bmatrix} \qquad \underset{6\times3}{\mathbf{X}} = \begin{bmatrix} 1 & 1 & 0 \\ 1 & 1 & 0 \\ 1 & 0 & 1 \\ 1 & 0 & 1 \\ 1 & -1 & -1 \\ 1 & -1 & -1 \end{bmatrix}$$

9.   **a)** (i)   (ii)

$$\begin{aligned} \beta_0 &= \mu_3 & \beta_0 &= \mu \\ \beta_1 &= \mu_1 - \mu_3 & \beta_1 &= \alpha_1 \\ \beta_2 &= \mu_2 - \mu_3 & \beta_2 &= \alpha_2 \end{aligned}$$

b) (i)   (ii)

$$\begin{aligned} \beta_0 &= \mu_4 & \beta_0 &= \mu \\ \beta_1 &= \mu_1 - \mu_4 & \beta_1 &= \alpha_1 \\ \beta_2 &= \mu_2 - \mu_4 & \beta_2 &= \alpha_2 \\ \beta_3 &= \mu_3 - \mu_4 & \beta_3 &= \alpha_3 \end{aligned}$$

10.   **a)** $Y_i = \beta_0 + \beta_1 X_{i1} + \beta_2 X_{i2} + \epsilon_i \quad (i = 1, \ldots, N)$, where ϵ_i is $NID(0, \sigma_\epsilon^2)$.

b) (i)

$$\mathbf{X}_{6\times3} = \begin{bmatrix} 1 & 1 & 0 \\ 1 & 1 & 0 \\ 1 & 0 & 1 \\ 1 & 0 & 1 \\ 1 & 0 & 0 \\ 1 & 0 & 0 \end{bmatrix}$$

(ii)

$$\mathbf{X}_{6\times3} = \begin{bmatrix} 1 & 1 & 0 \\ 1 & 1 & 0 \\ 1 & 0 & 1 \\ 1 & 0 & 1 \\ 1 & -1 & -1 \\ 1 & -1 & -1 \end{bmatrix}$$

(iii)

$$\mathbf{X}_{6\times3} = \begin{bmatrix} 1 & 1 & 1 \\ 1 & 1 & 1 \\ 1 & -1 & 1 \\ 1 & -1 & 1 \\ 1 & 0 & -2 \\ 1 & 0 & -2 \end{bmatrix}$$

c)

$$\underset{2\times3}{\mathbf{C}'} \qquad \underset{3\times1}{\boldsymbol{\beta}} \qquad \underset{2\times1}{\mathbf{0}}$$

$$\mathrm{H_0:} \quad \begin{bmatrix} 0 & 1 & 0 \\ 0 & 0 & 1 \end{bmatrix} \begin{bmatrix} \beta_0 \\ \beta_1 \\ \beta_2 \end{bmatrix} = \begin{bmatrix} 0 \\ 0 \end{bmatrix}$$

d) $\underset{1\times1}{(\mathbf{y}'\mathbf{y})} = [123.0000]$, $\underset{3\times3}{(\mathbf{X}'\mathbf{X})} = \begin{bmatrix} 6 & 2 & 2 \\ 2 & 2 & 0 \\ 2 & 0 & 2 \end{bmatrix}$, $\underset{3\times3}{(\mathbf{X}'\mathbf{X})^{-1}} = \begin{bmatrix} .5 & -.5 & -.5 \\ -.5 & 1 & .5 \\ -.5 & .5 & 1 \end{bmatrix}$,

$$\underset{3\times1}{(\mathbf{X}'\mathbf{y})} = \begin{bmatrix} 23 \\ 14 \\ 5 \end{bmatrix}, \quad \underset{3\times1}{\hat{\boldsymbol{\beta}}} = \begin{bmatrix} 2.0 \\ 5.0 \\ 0.5 \end{bmatrix}, \quad \underset{1\times1}{(\hat{\boldsymbol{\beta}}'\mathbf{X}'\mathbf{y})} = [118.5000]$$

e)

	Source	SS	df	MS	F
1	Regression (Effect of training)	30.3333	2	15.1667	[½]10.11*
2	Error	4.5000	3	1.5000	
3	Total	34.8333	5		

*$p < .05$

f) $\beta_0 = \mu_3$, $\beta_1 = \mu_1 - \mu_3$, $\beta_2 = \mu_2 - \mu_3$

g)

$$\underset{3\times3}{\mathbf{W}} \qquad \underset{3\times1}{\hat{\boldsymbol{\beta}}} \qquad \underset{3\times1}{\bar{\mathbf{y}}}$$

$$\begin{bmatrix} 1 & 1 & 0 \\ 1 & 0 & 1 \\ 1 & 0 & 0 \end{bmatrix} \begin{bmatrix} 2.0 \\ 5.0 \\ 0.5 \end{bmatrix} = \begin{bmatrix} 7.0 \\ 2.5 \\ 2.0 \end{bmatrix}$$

h) (i) $Y_i = \beta_0 + \epsilon_i \quad (i = 1, \ldots, N)$, where ϵ_i is $NID(0, \sigma_\epsilon^2)$.

(ii)

$$\underset{6\times1}{\mathbf{X}_1} = \begin{bmatrix} 1 \\ 1 \\ 1 \\ 1 \\ 1 \\ 1 \end{bmatrix}, \quad \underset{1\times1}{(\mathbf{X}_1'\mathbf{X}_1)} = [6], \quad \underset{1\times1}{(\mathbf{X}_1'\mathbf{X}_1)^{-1}} = \begin{bmatrix} \frac{1}{6} \end{bmatrix}, \quad \underset{1\times1}{(\mathbf{X}_1'\mathbf{y})} = [23], \quad \underset{1\times1}{\hat{\boldsymbol{\beta}}_1} = [3.8333],$$

$\underset{1\times1}{(\hat{\boldsymbol{\beta}}_1'\mathbf{X}_1'\mathbf{y})} = [88.1667]$ (iii) $SSE(0) = \mathbf{y}'\mathbf{y} - \hat{\boldsymbol{\beta}}_1'\mathbf{X}_1'\mathbf{y} = 123.0000 - 88.1667 = 34.8333$,

$SSR(0|X_1 \, X_2) = 34.8333 - 4.5000 = 30.3333$.   (iv) The values are equal.

i)
$$\hat{\sigma}^2_{\psi_1} = 1.5 \left\{ [1 \quad 0 \quad -1] \begin{bmatrix} 1 & 1 & 0 \\ 1 & 0 & 1 \\ 1 & 0 & 0 \end{bmatrix} \begin{bmatrix} .5 & -.5 & -.5 \\ -.5 & 1.0 & .5 \\ -.5 & .5 & 1.0 \end{bmatrix} \begin{bmatrix} 1 & 1 & 1 \\ 1 & 0 & 0 \\ 0 & 1 & 0 \end{bmatrix} \begin{bmatrix} 1 \\ 0 \\ -1 \end{bmatrix} \right\} = 1.5$$

$$\hat{\sigma}^2_{\psi_2} = 1.5 \left\{ [0 \quad 1 \quad -1] \begin{bmatrix} 1 & 1 & 0 \\ 1 & 0 & 1 \\ 1 & 0 & 0 \end{bmatrix} \begin{bmatrix} .5 & -.5 & -.5 \\ -.5 & 1.0 & .5 \\ -.5 & .5 & 1.0 \end{bmatrix} \begin{bmatrix} 1 & 1 & 1 \\ 1 & 0 & 0 \\ 0 & 1 & 0 \end{bmatrix} \begin{bmatrix} 0 \\ 1 \\ -1 \end{bmatrix} \right\} = 1.5$$

$tD' = (7.0 - 2.0)/\sqrt{1.5} = 4.083*$, $tD' = (2.5 - 2.0)/\sqrt{1.5} = 0.408$;
$tD'_{.05/2;3,3} \cong 4$. Reject the null hypothesis for $\mu_1 - \mu_3 = 0$.

j) $\beta_0 = \mu$, $\beta_1 = \alpha_1$, $\beta_2 = \alpha_2$

k)
$$(\mathbf{X'X}) = \begin{bmatrix} 6 & 0 & 0 \\ 0 & 4 & 2 \\ 0 & 2 & 4 \end{bmatrix}_{3\times3}, \quad (\mathbf{X'X})^{-1} = \begin{bmatrix} \frac{1}{6} & 0 & 0 \\ 0 & \frac{1}{3} & -\frac{1}{6} \\ 0 & -\frac{1}{6} & \frac{1}{3} \end{bmatrix}_{3\times3}, \quad (\mathbf{X'y}) = \begin{bmatrix} 23 \\ 10 \\ 1 \end{bmatrix}_{3\times1},$$

$$\hat{\beta}_{3\times1} = \begin{bmatrix} 3.8333 \\ 3.1667 \\ -1.3333 \end{bmatrix}, \quad (\hat{\beta}'\mathbf{X'y})_{1\times1} = [118.4996]$$

l) (i) $SSR = 118.4996 - 88.1667 = 30.3329$, $SSE = 123.0000 - 118.4996 = 4.5004$,
$SSTO = 123.0000 - 88.1667 = 34.8333$; $SSR/(h-1) = 15.1667$,
$SSE/(N-h) = 1.5000$.   (ii) The results agree within rounding error.

m)

$$\underset{3\times3}{\mathbf{W}} \qquad \underset{3\times1}{\hat{\beta}} \qquad \underset{3\times1}{\bar{\mathbf{y}}}$$

$$\begin{bmatrix} 1 & 1 & 0 \\ 1 & 0 & 1 \\ 1 & -1 & -1 \end{bmatrix} \begin{bmatrix} 3.8333 \\ 3.1667 \\ -1.3333 \end{bmatrix} = \begin{bmatrix} 7.0 \\ 2.5 \\ 2.0 \end{bmatrix}$$

12. a) (i) $r = 2$ (ii) $r = 2$ (vi) $r = 4$ (vii) $r = 3$
    b) The following are of full rank: (ii), (vi), and (vii).

13. a) $Y_{ij} = \mu + \alpha_j + \epsilon_{i(j)}$ $(i = 1, 2; j = 1, 2, 3)$, where α_j is a fixed effect for all j and $\epsilon_{i(j)}$ is $NID(0, \sigma^2_\epsilon)$.

    b) (i)

$$\underset{7\times4}{\mathbf{X}} = \begin{bmatrix} 1 & 1 & 0 & 0 \\ 1 & 1 & 0 & 0 \\ 1 & 0 & 1 & 0 \\ 1 & 0 & 1 & 0 \\ 1 & 0 & 0 & 1 \\ 1 & 0 & 0 & 1 \\ 0 & 1 & 1 & 1 \end{bmatrix}$$

(ii)

$$\underset{(4-3)\times4}{\mathbf{R}} \qquad \underset{4\times1}{\beta} \qquad \underset{1\times1}{\theta}$$

$$[0 \quad 1 \quad 1 \quad 1] \begin{bmatrix} \mu \\ \alpha_1 \\ \alpha_2 \\ \alpha_3 \end{bmatrix} = [0]$$

c)

$$\underset{3\times4}{\mathbf{C'}} \qquad \underset{4\times1}{\beta} \qquad \underset{3\times1}{\mathbf{0}}$$

$$\begin{bmatrix} 0 & 1 & 0 & 0 \\ 0 & 0 & 1 & 0 \\ 0 & 0 & 0 & 1 \end{bmatrix} \begin{bmatrix} \mu \\ \alpha_1 \\ \alpha_2 \\ \alpha_3 \end{bmatrix} = \begin{bmatrix} 0 \\ 0 \\ 0 \end{bmatrix}$$

d)

$$(\mathbf{y^{*\prime}y^*})_{1\times1} = [123.0000], \quad (\mathbf{X^{*\prime}X^*})_{4\times4} = \begin{bmatrix} 6 & 2 & 2 & 2 \\ 2 & 3 & 1 & 1 \\ 2 & 1 & 3 & 1 \\ 2 & 1 & 1 & 3 \end{bmatrix},$$

$$(\mathbf{X^{\star\prime}X^{\star}})^{-1} = \begin{bmatrix} .2778 & -.1111 & -.1111 & -.1111 \\ -.1111 & .4444 & -.0556 & -.0556 \\ -.1111 & -.0556 & .4444 & -.0556 \\ -.1111 & -.0556 & -.0556 & .4444 \end{bmatrix}, (\mathbf{X^{\star\prime}y^{\star}}) = \begin{bmatrix} 23 \\ 14 \\ 5 \\ 4 \end{bmatrix},$$

$$\underset{4\times1}{\hat{\boldsymbol{\beta}}^{\star}} = \begin{bmatrix} 3.8333 \\ 3.1667 \\ -1.3333 \\ -1.8333 \end{bmatrix}, \underset{1\times1}{(\hat{\boldsymbol{\beta}}^{\star\prime}\mathbf{X^{\star\prime}y^{\star}})} = [118.5000]$$

e) (i) $SSBG = 118.5000 - 88.1667 = 30.3333$, $SSWG = 123.0000 - 118.5000 = 4.5000$, $SSTO = 123.0000 - 88.1667 = 34.8333$; $SSBG/(p-1) = 15.1667$, $SSWG/p(n-1) = 1.5000$ (ii) The results agree.

f)

$$\begin{array}{ccc} \underset{3\times4}{\mathbf{W}} & \underset{4\times1}{\hat{\boldsymbol{\beta}}^{\star}} & \underset{3\times1}{\bar{\mathbf{y}}} \end{array}$$

$$\begin{bmatrix} 1 & 1 & 0 & 0 \\ 1 & 0 & 1 & 0 \\ 1 & 0 & 0 & 1 \end{bmatrix} \begin{bmatrix} 3.8333 \\ 3.1667 \\ -1.3333 \\ -1.8333 \end{bmatrix} = \begin{bmatrix} 7.0 \\ 2.5 \\ 2.0 \end{bmatrix}$$

14. a)

$$\underset{3\times4}{\mathbf{E}} = \begin{bmatrix} 1 & \frac{1}{3} & \frac{1}{3} & \frac{1}{3} \\ 0 & 1 & 0 & -1 \\ 0 & 0 & 1 & -1 \end{bmatrix}, \underset{3\times3}{(\mathbf{EE}')^{-1}} = \begin{bmatrix} 1\frac{1}{3} & 0 & 0 \\ 0 & 2 & 1 \\ 0 & 1 & 2 \end{bmatrix},$$

$$\underset{3\times3}{(\mathbf{Z'Z})^{-1}} = \begin{bmatrix} .1667 & 0 & 0 \\ 0 & 1.0000 & .5000 \\ 0 & .5000 & 1.0000 \end{bmatrix}, \underset{3\times1}{(\mathbf{Z'y})} = \begin{bmatrix} 23.0006 \\ 6.3333 \\ -2.6667 \end{bmatrix}, \underset{3\times1}{\hat{\boldsymbol{\beta}}^{\star}} = \begin{bmatrix} 3.8332 \\ 5.0000 \\ 0.5000 \end{bmatrix}$$

b) (i) $SSBG = 118.5000 - 88.1667 = 30.3333$, $SSWG = 123.0000 - 118.5000 = 4.50000$, $SSTO = 123.0000 - 88.1667 = 34.8333$; $SSBG/(p-1) = 15.1667$, $SSWG/p(n-1) = 1.5000$ (ii) The results agree.

c)

$$\begin{array}{ccc} \underset{2\times3}{\mathbf{C}'} & \underset{3\times1}{\boldsymbol{\beta}^{\star}} & \underset{2\times1}{\mathbf{0}} \end{array}$$

$$\begin{bmatrix} 0 & 1 & 0 \\ 0 & 0 & 1 \end{bmatrix} \begin{bmatrix} \mu + \dfrac{\alpha_1 + \alpha_2 + \alpha_3}{3} \\ \alpha_1 - \alpha_3 \\ \alpha_2 - \alpha_3 \end{bmatrix} = \begin{bmatrix} 0 \\ 0 \end{bmatrix}$$

d)

$$\begin{array}{ccc} \underset{3\times4}{\mathbf{W}} & \underset{4\times1}{\hat{\boldsymbol{\beta}}^{\star}} & \underset{3\times1}{\bar{\mathbf{y}}} \end{array}$$

$$\begin{bmatrix} 1.0000 & .6667 & -.3333 \\ 1.0000 & -.3333 & .6667 \\ 1.0000 & -.3333 & -.3333 \end{bmatrix} \begin{bmatrix} 3.8332 \\ 5.0000 \\ 0.5000 \end{bmatrix} = \begin{bmatrix} 7.0 \\ 2.5 \\ 2.0 \end{bmatrix}$$

17. a) H_0: $\begin{array}{llll} \mu_1 - \mu_4 = 0 & \mu_1 - \mu_2 = 0 & \mu_1 - \mu_3 = 0 & \mu_1 - \mu_3 = 0 \\ \mu_2 - \mu_4 = 0 & \mu_2 - \mu_3 = 0 & \mu_2 - \mu_3 = 0 & \mu_2 - \mu_4 = 0 \\ \mu_3 - \mu_4 = 0 & \mu_3 - \mu_4 = 0 & \mu_4 - \mu_3 = 0 & \mu_1 - \mu_2 = 0 \end{array}$

b)

$$H_0: \begin{bmatrix} 1 & 0 & 0 & -1 \\ 0 & 1 & 0 & -1 \\ 0 & 0 & 1 & -1 \end{bmatrix} \begin{bmatrix} \mu_1 \\ \mu_2 \\ \mu_3 \\ \mu_4 \end{bmatrix} = \begin{bmatrix} 0 \\ 0 \\ 0 \end{bmatrix}, \begin{bmatrix} 1 & -1 & 0 & 0 \\ 0 & 1 & -1 & 0 \\ 0 & 0 & 1 & -1 \end{bmatrix} \begin{bmatrix} \mu_1 \\ \mu_2 \\ \mu_3 \\ \mu_4 \end{bmatrix} = \begin{bmatrix} 0 \\ 0 \\ 0 \end{bmatrix},$$

$$\begin{bmatrix} 1 & 0 & -1 & 0 \\ 0 & 1 & -1 & 0 \\ 0 & 0 & -1 & 1 \end{bmatrix} \begin{bmatrix} \mu_1 \\ \mu_2 \\ \mu_3 \\ \mu_4 \end{bmatrix} = \begin{bmatrix} 0 \\ 0 \\ 0 \end{bmatrix}, \begin{bmatrix} 1 & 0 & -1 & 0 \\ 0 & 1 & 0 & -1 \\ 1 & -1 & 0 & 0 \end{bmatrix} \begin{bmatrix} \mu_1 \\ \mu_2 \\ \mu_3 \\ \mu_4 \end{bmatrix} = \begin{bmatrix} 0 \\ 0 \\ 0 \end{bmatrix}$$

19.   **a)** $Y_{ij} = \mu_j + \epsilon_{i(j)}$ $(i = 1, 2; j = 1, 2, 3)$, where $\epsilon_{i(j)}$ is $NID(0, \sigma_\epsilon^2)$.

b) $(\mathbf{y'y}) = [123.0000], \underset{3\times3}{(\mathbf{X'X})} = \begin{bmatrix} 2 & 0 & 0 \\ 0 & 2 & 0 \\ 0 & 0 & 2 \end{bmatrix}, \underset{3\times3}{(\mathbf{X'X})^{-1}} = \begin{bmatrix} 0.5 & 0 & 0 \\ 0 & 0.5 & 0 \\ 0 & 0 & 0.5 \end{bmatrix},$
$\underset{1\times1}{}$

$\underset{3\times1}{(\mathbf{X'y})} = \begin{bmatrix} 14 \\ 5 \\ 4 \end{bmatrix}, \underset{3\times1}{\hat{\boldsymbol{\mu}}} = \begin{bmatrix} 7.0 \\ 2.5 \\ 2.0 \end{bmatrix}, \underset{1\times1}{(\hat{\boldsymbol{\mu}}'\mathbf{X'y})} = [118.5000]$

c) (i) $SSBG = 118.5000 - 88.1667 = 30.3333$, $SSWG = 123.0000 - 118.5000 = 4.5000$,
$SSTO = 123.0000 - 88.1667 = 34.8333$; $SSBG/(p - 1) = 15.1667$,
$SSWG/p(n - 1) = 1.500$ (ii) The results agree.

CHAPTER 6

2.   Use litter mates or identical twins, tests to match subjects, subjects who are matched by mutual selection, and subjects as their own controls.

4. $\sum_{j=1}^{p} \sum_{i=1}^{n} [(\overline{Y}_{.j} - \overline{Y}_{..}) + (\overline{Y}_{i.} - \overline{Y}_{..}) + (Y_{ij} - \overline{Y}_{.j} - \overline{Y}_{i.} + \overline{Y}_{..})]^2$

$= \sum_{j=1}^{p} \sum_{i=1}^{n} [(\overline{Y}_{.j} - \overline{Y}_{..})^2 + (\overline{Y}_{i.} - \overline{Y}_{..})^2 + (Y_{ij} - \overline{Y}_{.j} - \overline{Y}_{i.} + \overline{Y}_{..})^2$

$+ 2(\overline{Y}_{.j} - \overline{Y}_{..})(\overline{Y}_{i.} - \overline{Y}_{..}) + 2(\overline{Y}_{.j} - \overline{Y}_{..})(Y_{ij} - \overline{Y}_{.j} - \overline{Y}_{i.} + \overline{Y}_{..})$

$+ 2(\overline{Y}_{i.} - \overline{Y}_{..})(Y_{ij} - \overline{Y}_{.j} - \overline{Y}_{i.} + \overline{Y}_{..})]$

$= n \sum_{j=1}^{p} (\overline{Y}_{.j} - \overline{Y}_{..})^2 + p \sum_{i=1}^{n} (\overline{Y}_{i.} - \overline{Y}_{..})^2 + \sum_{j=1}^{p} \sum_{i=1}^{n} (Y_{ij} - \overline{Y}_{.j} - \overline{Y}_{i.} + \overline{Y}_{..})^2$

because

$$\sum_{j=1}^{p} (\overline{Y}_{.j} - \overline{Y}_{..}) \sum_{i=1}^{n} (\overline{Y}_{i.} - \overline{Y}_{..}) = 0$$

$$\sum_{j=1}^{p} (\overline{Y}_{.j} - \overline{Y}_{..}) \sum_{i=1}^{n} (Y_{ij} - \overline{Y}_{.j} - \overline{Y}_{i.} + \overline{Y}_{..}) = 0$$

$$\sum_{j=1}^{n} (\overline{Y}_{i.} - \overline{Y}_{..}) \sum_{j=1}^{p} (Y_{ij} - \overline{Y}_{.j} - \overline{Y}_{i.} + \overline{Y}_{..}) = 0.$$

5. $\sum_{j=1}^{p} \sum_{i=1}^{n} (Y_{ij} - \overline{Y}_{..})^2 = \sum_{j=1}^{p} \sum_{i=1}^{n} (Y_{ij}^2 - 2\overline{Y}_{..} Y_{ij} + \overline{Y}_{..}^2)$

$= \sum_{j=1}^{p} \sum_{i=1}^{n} Y_{ij}^2 - 2 \frac{\sum_{j=1}^{p} \sum_{i=1}^{n} Y_{ij}}{np} \sum_{j=1}^{p} \sum_{i=1}^{n} Y_{ij} + np \frac{\left(\sum_{j=1}^{p} \sum_{i=1}^{n} Y_{ij}\right)^2}{n^2 p^2}$

$= \sum_{j=1}^{p} \sum_{i=1}^{n} Y_{ij}^2 - \frac{\left(\sum_{j=1}^{p} \sum_{i=1}^{n} Y_{ij}\right)^2}{np}$

$$n \sum_{j=1}^{p} (\overline{Y}_{\cdot j} - \overline{Y}_{\cdot \cdot})^2 = n \sum_{j=1}^{p} (\overline{Y}_{\cdot j}^2 - 2\overline{Y}_{\cdot \cdot}\overline{Y}_{\cdot j} + \overline{Y}_{\cdot \cdot}^2)$$

$$= n \sum_{j=1}^{p} \frac{\left(\sum_{i=1}^{n} Y_{ij}\right)^2}{n^2} - 2n \frac{\sum_{j=1}^{p} \sum_{i=1}^{n} Y_{ij}}{np} \frac{\sum_{j=1}^{p} \sum_{i=1}^{n} Y_{ij}}{n} + np \frac{\left(\sum_{j=1}^{p} \sum_{i=1}^{n} Y_{ij}\right)^2}{n^2 p^2}$$

$$= \sum_{j=1}^{p} \frac{\left(\sum_{i=1}^{n} Y_{ij}\right)^2}{n} - \frac{\left(\sum_{j=1}^{p} \sum_{i=1}^{n} Y_{ij}\right)^2}{np}$$

$$p \sum_{i=1}^{n} (\overline{Y}_{i\cdot} - \overline{Y}_{\cdot \cdot})^2 = p \sum_{i=1}^{n} (\overline{Y}_{i\cdot}^2 - 2\overline{Y}_{\cdot \cdot}\overline{Y}_{i\cdot} + \overline{Y}_{\cdot \cdot}^2)$$

$$= p \sum_{i=1}^{n} \frac{\left(\sum_{j=1}^{p} Y_{ij}\right)^2}{p^2} - 2p \frac{\sum_{j=1}^{p} \sum_{i=1}^{n} Y_{ij}}{np} \frac{\sum_{j=1}^{p} \sum_{i=1}^{n} Y_{ij}}{p} + np \frac{\left(\sum_{j=1}^{p} \sum_{i=1}^{n} Y_{ij}\right)^2}{n^2 p^2}$$

$$= \sum_{i=1}^{n} \frac{\left(\sum_{j=1}^{p} Y_{ij}\right)^2}{p} - \frac{\left(\sum_{j=1}^{p} \sum_{i=1}^{n} Y_{ij}\right)^2}{np}$$

$$\sum_{j=1}^{p} \sum_{i=1}^{n} (Y_{ij} - \overline{Y}_{\cdot j} - \overline{Y}_{i\cdot} + \overline{Y}_{\cdot \cdot})^2 = \sum_{j=1}^{p} \sum_{i=1}^{n} (Y_{ij}^2 + \overline{Y}_{\cdot j}^2 + \overline{Y}_{i\cdot}^2 + \overline{Y}_{\cdot \cdot}^2 - 2\overline{Y}_{\cdot j}Y_{ij} - 2\overline{Y}_{i\cdot}Y_{ij}$$

$$+ 2\overline{Y}_{\cdot \cdot}Y_{ij} + 2\overline{Y}_{\cdot j}\overline{Y}_{i\cdot} - 2\overline{Y}_{\cdot \cdot}\overline{Y}_{\cdot j} - 2\overline{Y}_{\cdot \cdot}\overline{Y}_{i\cdot})$$

$$= \sum_{j=1}^{p} \sum_{i=1}^{n} Y_{ij}^2 + n \sum_{j=1}^{p} \frac{\left(\sum_{i=1}^{n} Y_{ij}\right)^2}{n^2} + p \sum_{i=1}^{n} \frac{\left(\sum_{j=1}^{p} Y_{ij}\right)^2}{p^2} + np \frac{\left(\sum_{j=1}^{p} \sum_{i=1}^{n} Y_{ij}\right)^2}{n^2 p^2}$$

$$- 2 \sum_{j=1}^{p} \frac{\left(\sum_{i=1}^{n} Y_{ij}\right)^2}{n} - 2\sum_{i=1}^{n} \frac{\left(\sum_{j=1}^{p} Y_{ij}\right)^2}{p} + 2 \frac{\left(\sum_{j=1}^{p} \sum_{i=1}^{n} Y_{ij}\right)^2}{np}$$

$$+ 2 \frac{\left(\sum_{j=1}^{p} \sum_{i=1}^{n} Y_{ij}\right)^2}{np} - 2n \frac{\left(\sum_{j=1}^{p} \sum_{i=1}^{n} Y_{ij}\right)^2}{n^2 p} - 2p \frac{\left(\sum_{j=1}^{p} \sum_{i=1}^{n} Y_{ij}\right)^2}{np^2}$$

$$= \sum_{j=1}^{p} \sum_{i=1}^{n} Y_{ij}^2 - \sum_{j=1}^{p} \frac{\left(\sum_{i=1}^{n} Y_{ij}\right)^2}{n} - \sum_{i=1}^{n} \frac{\left(\sum_{j=1}^{p} Y_{ij}\right)^2}{p} + \frac{\left(\sum_{j=1}^{p} \sum_{i=1}^{n} Y_{ij}\right)^2}{np}$$

6.   **a)** $F = 44.0834/1.3864 = 31.80$; reject H_0.   **b)** $\hat{\phi} = 4.53$; $1 - \hat{\beta} > .99$

c) $n = 4$ **d)** $\hat{\omega}_{Y|A}^2 = .24$, $\hat{\rho}_{IY|BL} = .63$

e) $FNADD = 2.068/1.354 = 1.53$, $F_{.10;1,21} = 2.96$; do not reject hypothesis of additivity.

f) Yes

g) $\hat{\theta} = 0.83$, $F_{.05;1.7,18.3}$ (*adjusted*) $\cong 3.69$. Since F exceeds the value required for a conservative test, we know that it would also exceed the value required for an adjusted test.

h) H_0: $\mu_{\cdot 1} \geq \mu_{\cdot 2}$, $\mu_{\cdot 1} \geq \mu_{\cdot 3}$; $tD = 1.9167/0.4807 = 3.99$, $tD = 3.8334/0.4807 = 7.97$, $tD_{.05;2,22} = 2.074$; reject both null hypotheses. The test norms were developed on different populations and are not necessarily comparable for brain-damaged patients.

i) $X_{32} = 4.5$, SSA(*corrected*) $= 88.1806$ **j)** (i) Relative efficiency $= 5.13$
(ii) $np = (62)(3) = 186$

7.   **a)** $F = 18.40/1.60 = 11.50$; reject H_0.   **b)** $\hat{\phi} = 2.96$; $1 - \hat{\beta} > .99$

c) $n = 4$ **d)** $\hat{\omega}^2_{Y|A} = .14$; $\hat{\rho}_{IY|BL} = .77$

e) $FNADD = 4.479/1.480 = 3.03$; $F_{.10;1,24} = 2.93$; reject hypothesis of additivity.

f) Yes   **g)** $\hat{\theta} = 0.48$, $F_{.01;2.39,11.95}$(*adjusted*) $\cong 6.56$; reject H_0.

h) (i) $(98 - 95)/1.095 = 2.74$, $(98 - 95)/0.730 = 4.11^*$; $(98 - 97)/0.516 = 1.94$, $(98 - 97)/0.730 = 1.37$; $(98 - 93)/0.816 = 6.12^*$, $(98 - 93)/0.730 = 6.85^*$; $(98 - 96)/0.856 = 2.34$, $(98 - 96)/0.730 = 2.74$; $(98 - 95)/0.516 = 5.81^*$, $(98 - 95)/0.730 = 4.11^*$; $tD'_{.01;6,5} = 4.60$, $tD'_{.01;6,25} = 3.11$

(ii) Yes, the null hypothesis $\mu_1 - \mu_2 = 0$ is rejected when $MSRES$ is used but not when $MSRES_i$ is used.

(iii) The value of $\hat{\theta} = 0.48$ suggests that the circularity assumption is not tenable; thus, the $MSRES_i$ error term is most appropriate.

i) (i) $F = 17.5143/1.7171 = 10.20$, $F_{.01;4,20} = 4.43$; thus, the relationship is nonlinear.

(ii) $SS\hat{\psi}_{\text{dep from lin}}/SSA \times 100 = 23.9\%$

j) $X_{52} = 97.4$, SSA(*corrected*) $= 89.4667$ **k)** (i) Relative efficiency $= 7.91$ (ii) $np = 288$

11.   **a)**

$$E\left[\sum_{j=1}^{p} \frac{\left(\sum_{i=1}^{n} Y_{ij}\right)^2}{n}\right] = E\left\{\sum_{j=1}^{p} \frac{\left[\sum_{i=1}^{n} (\mu + \alpha_j + \pi_i + \epsilon_{ij})\right]^2}{n}\right\}$$

$$= \frac{1}{n} E\left\{\sum_{j=1}^{p}\left[n^2\mu^2 + n^2\alpha_j^2 + \left(\sum_{i=1}^{n} \pi_i\right)^2 + \left(\sum_{i=1}^{n} \epsilon_{ij}\right)^2 + 2n^2\mu\alpha_j\right.\right.$$

$$\left.\left. + 2n\mu \sum_{i=1}^{n} \pi_i + 2n\mu \sum_{i=1}^{n} \epsilon_{ij} + 2n\alpha_j \sum_{i=1}^{n} \pi_i + 2n\alpha_j \sum_{i=1}^{n} \epsilon_{ij} + 2 \sum_{i=1}^{n} \pi_i \sum_{i=1}^{n} \epsilon_{ij}\right]\right\}$$

$$= \frac{1}{n}\left[pn^2\mu^2 + n^2 \sum_{j=1}^{p} \alpha_j^2 + np\sigma_\epsilon^2\right] = pn\mu^2 + n \sum_{j=1}^{p} \alpha_j^2 + p\sigma_\epsilon^2$$

$$E\left[\frac{\left(\sum_{j=1}^{p}\sum_{i=1}^{n} Y_{ij}\right)^2}{np}\right] = E\left\{\frac{\left[\sum_{j=1}^{p}\sum_{i=1}^{n} (\mu + \alpha_j + \pi_i + \epsilon_{ij})\right]^2}{np}\right\}$$

$$= \frac{1}{n}\frac{1}{p} E\left[n^2p^2\mu^2 + n^2\left(\sum_{j=1}^{p} \alpha_j\right)^2 + p^2\left(\sum_{i=1}^{n} \pi_i\right)^2 + \left(\sum_{j=1}^{p}\sum_{i=1}^{n} \epsilon_{ij}\right)^2\right.$$

$$+ 2n^2p\mu \sum_{j=1}^{p} \alpha_j + 2np^2\mu \sum_{i=1}^{n} \pi_i + 2np\mu \sum_{j=1}^{p}\sum_{i=1}^{n} \epsilon_{ij} + 2n \sum_{j=1}^{p} \alpha_j p \sum_{i=1}^{n} \pi_i$$

$$\left. + 2n \sum_{j=1}^{p} \alpha_j \sum_{j=1}^{p}\sum_{i=1}^{n} \epsilon_{ij} + 2p \sum_{i=1}^{n} \pi_i \sum_{j=1}^{p}\sum_{i=1}^{n} \epsilon_{ij}\right]$$

$$= \frac{1}{n}\frac{1}{p}[n^2p^2\mu^2 + pn\sigma_\epsilon^2] = np\mu^2 + \sigma_\epsilon^2$$

$$E(MSA) = \frac{1}{p-1}\left[\left(pn\mu^2 + n\sum_{j=1}^{p}\alpha_j^2 + p\sigma_\epsilon^2\right) - (np\mu^2 + \sigma_\epsilon^2)\right]$$

$$= \frac{1}{p-1}\left[(p-1)\epsilon^2 + n\sum_{j=1}^{p}\alpha_j^2\right] = \sigma_\epsilon^2 + n\sum_{j=1}^{p}\alpha_j^2/(p-1)$$

$$E\left[\sum_{i=1}^{n}\frac{\left(\sum_{j=1}^{p}Y_{ij}\right)^2}{p}\right] = E\left\{\sum_{i=1}^{n}\frac{\left[\sum_{j=1}^{p}(\mu + \alpha_j + \pi_i + \epsilon_{ij})\right]^2}{p}\right\}$$

$$= \frac{1}{p}E\left\{\sum_{i=1}^{n}\left[p^2\mu^2 + \left(\sum_{j=1}^{p}\alpha_j\right)^2 + p^2\pi_i^2 + \left(\sum_{j=1}^{p}\epsilon_{ij}\right)^2 + 2p\mu\sum_{j=1}^{p}\alpha_j\right.\right.$$

$$\left.\left. + 2p^2\mu\pi_i + 2p\mu\sum_{j=1}^{p}\epsilon_{ij} + 2p\sum_{j=1}^{p}\alpha_j\pi_i + 2\sum_{j=1}^{p}\alpha_j\sum_{j=1}^{p}\epsilon_{ij} + 2p\pi_i\sum_{j=1}^{p}\epsilon_{ij}\right]\right\}$$

$$= \frac{1}{p}(np^2\mu^2 + p^2\sum_{i=1}^{n}\pi_i^2 + np\sigma_\epsilon^2) = np\mu^2 + p\sum_{i=1}^{n}\pi_i^2 + n\sigma_\epsilon$$

$$E(MSBL) = \frac{1}{n-1}\left[\left(np\mu^2 + p\sum_{i=1}^{n}\pi_i^2 + n\sigma_\epsilon^2\right) - (np\mu^2 + \sigma_\epsilon^2)\right]$$

$$= \frac{1}{n-1}\left[(n-1)\sigma_\epsilon^2 + p\sum_{i=1}^{n}\pi_i^2\right] = \sigma_\epsilon^2 + p\sum_{i=1}^{n}\pi_i^2/(n-1)$$

$$E\left(\sum_{j=1}^{p}\sum_{i=1}^{n}Y_{ij}^2\right) = E\left[\sum_{j=1}^{p}\sum_{i=1}^{n}(\mu + \alpha_j + \pi_i + \epsilon_{ij})^2\right]$$

$$= E\left[\sum_{j=1}^{p}\sum_{i=1}^{n}(\mu^2 + \alpha_j^2 + \pi_i^2 + \epsilon_{ij}^2 + 2\mu\alpha_j + 2\mu\pi_i + 2\mu\epsilon_{ij}\right.$$

$$\left. + 2\alpha_j\pi_i + 2\alpha_j\epsilon_{ij} + 2\pi_i\epsilon_{ij})\right]$$

$$= np\mu^2 + n\sum_{j=1}^{p}\alpha_j^2 + p\sum_{i=1}^{n}\pi_i^2 + np\sigma_\epsilon^2$$

$$E(MSRES) = \frac{1}{(n-1)(p-1)}\left[\left(np\mu^2 + n\sum_{j=1}^{p}\alpha_j^2 + p\sum_{i=1}^{n}\pi_i^2 + np\sigma_\epsilon^2\right)\right.$$

$$\left. - \left(pn\mu^2 + n\sum_{j=1}^{p}\alpha_j^2 + p\sigma_\epsilon^2\right) - \left(np\mu^2 + p\sum_{i=1}^{n}\pi_i + n\sigma_\epsilon^2\right) + (np\mu^2 + \sigma_\epsilon^2)\right]$$

$$= \frac{1}{(n-1)(p-1)}(np\sigma_\epsilon^2 - p\sigma_\epsilon^2 - n\sigma_\epsilon^2 + \sigma_\epsilon^2) = \sigma_\epsilon^2$$

14.   **a)** The size of the average correlation coefficient for treatment levels in the randomized block design

b) The average correlation coefficient is negative.

15.   **a)** (i) Type H (ii) Type S

b) (i) $\mathbf{C}^{*\prime}\mathbf{\Sigma}\mathbf{C}^* = 5\begin{bmatrix} 1 & 0 \\ 0 & 1 \end{bmatrix}$ (ii) $\mathbf{C}^{*\prime}\mathbf{\Sigma}\mathbf{C}^* = 2.7\begin{bmatrix} 1 & 0 \\ 0 & 1 \end{bmatrix}$

17.   Estimated observations minimize the error sum of squares.

18.   **a)** $Y_i = \beta_0 X_0 + \underbrace{\beta_1 X_{i1} + \beta_2 X_{i2}} + \underbrace{\beta_3 X_{i3}} + \epsilon_i$ $(i = 1, \ldots, 6)$; ϵ_i is $NID(0, \sigma_\epsilon^2)$.

$p - 1 = 2$ treatment levels $n - 1 = 1$ block

b) (i) H_0: $\beta_1 = \beta_2 = 0$ or $\begin{bmatrix} 0 & 1 & 0 & 0 \\ 0 & 0 & 1 & 0 \end{bmatrix} \begin{bmatrix} \beta_0 \\ \beta_1 \\ \beta_2 \\ \beta_3 \end{bmatrix} = \begin{bmatrix} 0 \\ 0 \end{bmatrix}$

(ii) H_0: $\beta_3 = 0$ or $\begin{bmatrix} 0 & 0 & 0 & 1 \end{bmatrix} \begin{bmatrix} \beta_0 \\ \beta_1 \\ \beta_2 \\ \beta_3 \end{bmatrix} = [0]$

c) $\beta_0 = \mu + \alpha_3 + \pi_2$, $\beta_1 = \alpha_1 - \alpha_3$, $\beta_2 = \alpha_2 - \alpha_3$, $\beta_3 = \pi_1 - \pi_2$

d) (i) $Y_i = \beta_0 X_0 + \beta_3 X_{i3} + \epsilon_i$ $(i = 1, \ldots, n)$; ϵ_i is $NID(0, \sigma_\epsilon^2)$.

(ii) $SSR(X_1 X_2 | X_3) = SSE(X_3) - SSE(X_1 X_2 X_3)$

e)

$$(\mathbf{X_0' X_0})^{-1}_{h \times h} = \begin{bmatrix} 4/6 & -3/6 & -3/6 & -2/6 \\ -3/6 & 6/6 & 3/6 & 0 \\ -3/6 & 3/6 & 6/6 & 0 \\ -2/6 & 0 & 0 & 4/6 \end{bmatrix}, \quad (\mathbf{X_1' X_1})^{-1}_{n \times n} = \begin{bmatrix} 1/3 & -1/3 \\ -1/3 & 2/3 \end{bmatrix},$$

$$(\mathbf{X_2' X_2})^{-1}_{p \times p} = \begin{bmatrix} 1/2 & -1/2 & -1/2 \\ -1/2 & 2/2 & 1/2 \\ -1/2 & 1/2 & 2/2 \end{bmatrix};$$

$SSE(X_1 X_2 X_3) = 4.000$, $SSR(X_1 X_2 | X_3) = 9.333$, $F = (9.333/2)/(4.000/2) = 2.33$,
$F_{.05;2,2} = 19.0$; $SSR(X_3 | X_1 X_2) = 13.500$, $F = (13.500/1)/(4.000/2) = 6.75$, $F_{.05;1,2} = 18.5$

f) (i) $Y_{ij} = \mu + \alpha_j + \pi_i + \epsilon_{ij}$ $(i = 1, 2; j = 1, 2, 3)$; ϵ_{ij} is $NID(0, \sigma_\epsilon^2)$.

(ii) $Y_{ij} = \mu_{ij} + \epsilon_{ij}$ $(i = 1, 2; j = 1, 2, 3)$; ϵ_{ij} is $NID(0, \sigma_\epsilon^2)$ and μ_{ij} is subject to the restrictions that $\mu_{ij} - \mu_{i'j} - \mu_{ij'} + \mu_{i'j'} = 0$ for all, i, i', j, and j'.

g) (i) $\mathbf{Q_{1(A)}'} = \begin{bmatrix} 1 & -1 & -1 & 1 & 0 & 0 \\ 0 & 0 & 1 & -1 & -1 & 1 \\ 1/2 & 1/2 & 0 & 0 & -1/2 & -1/2 \\ 0 & 0 & 1/2 & 1/2 & -1/2 & -1/2 \end{bmatrix}$ (ii) $\mathbf{Q_{2(A)}'} = \begin{bmatrix} 1 & -1 & -1 & 1 & 0 & 0 \\ 0 & 0 & 1 & -1 & -1 & 1 \\ 1 & 0 & 0 & 0 & -1 & 0 \\ 0 & 0 & 1 & 0 & -1 & 0 \end{bmatrix}$

h) (i) $(\mathbf{R'R}) = \begin{bmatrix} 4 & -2 \\ -2 & 4 \end{bmatrix}$, $(\mathbf{R'R})^{-1} = \begin{bmatrix} 2/6 & 1/6 \\ 1/6 & 2/6 \end{bmatrix}$,

$SSRES = (\mathbf{R'\hat{\mu}})' (\mathbf{R'R})^{-1} (\mathbf{R'\hat{\mu}}) = 4.000$, $(\mathbf{Q_{2(A)}' Q_{2(A)}}) = \begin{bmatrix} 4 & -2 & 1 & -1 \\ -2 & 4 & 1 & 2 \\ 1 & 1 & 2 & 1 \\ -1 & 2 & 1 & 2 \end{bmatrix}$,

$$(\mathbf{Q_{2(A)}' Q_{2(A)}})^{-1} = \begin{bmatrix} 4/6 & 2/6 & -4/6 & 2/6 \\ 2/6 & 4/6 & -2/6 & -2/6 \\ -4/6 & -2/6 & 8/6 & -4/6 \\ 2/6 & -2/6 & -4/6 & 8/6 \end{bmatrix}$$

$(\mathbf{Q_{2(A)}' \hat{\mu}})' (\mathbf{Q_{2(A)}' Q_{2(A)}})^{-1} (\mathbf{Q_{2(A)}' \hat{\mu}}) = 13.3333$, $SSA = 13.3333 - 4.0000 = 9.3333$,
$F = (9.3333/2)/(4.0000/2) = 2.33$; $F_{.05;2,2} = 19.00$.

(ii) $(\mathbf{Q_{2(B)}' Q_{2(B)}}) = \begin{bmatrix} 4 & -2 & 2 \\ -2 & 4 & 0 \\ 2 & 0 & 2 \end{bmatrix}$, $(\mathbf{Q_{2(B)}' \hat{\mu}})' (\mathbf{Q_{2(B)}' Q_{2(B)}})^{-1} (\mathbf{Q_{2(B)}' \hat{\mu}}) = 17.5000$,

$SSBL = 17.5000 - 4.0000 = 13.5000$, $F = (13.5000/1)/(4.0000/2) = 6.75$; $F_{.05;1,2} = 18.5$.

i) $\mathbf{R}' = [0 \quad 1 \quad -1 \quad -1 \quad 1]$, $\mathbf{C}'_{3(A)} = \begin{bmatrix} 1 & 0 & 0 & -1 & 0 \\ 0 & 1 & 0 & -1 & 0 \\ 0 & 0 & 1 & 0 & -1 \end{bmatrix}$,

$\mathbf{Q}'_{3(A)} = \begin{bmatrix} 0 & 1 & -1 & -1 & 1 \\ 1 & 0 & 0 & -1 & 0 \\ 0 & 1 & 0 & -1 & 0 \end{bmatrix}$, $(\mathbf{R}'\mathbf{R}) = [4]$, $(\mathbf{R}'\mathbf{R})^{-1} = \begin{bmatrix} \frac{1}{4} \end{bmatrix}$,

$SSRES = (\mathbf{R}'\hat{\boldsymbol{\mu}})'(\mathbf{R}'\mathbf{R})^{-1}(\mathbf{R}'\hat{\boldsymbol{\mu}}) = 1.0000$, $(\mathbf{Q}'_{3(A)}\mathbf{Q}_{3(A)}) = \begin{bmatrix} 4 & 1 & 2 \\ 1 & 2 & 1 \\ 2 & 1 & 2 \end{bmatrix}$,

$(\mathbf{Q}'_{3(A)}\mathbf{Q}_{3(A)})^{-1} = \begin{bmatrix} \frac{3}{6} & 0 & -\frac{3}{6} \\ 0 & \frac{4}{6} & -\frac{2}{6} \\ -\frac{3}{6} & -\frac{2}{6} & \frac{7}{6} \end{bmatrix}$, $(\mathbf{Q}'_{3(A)}\hat{\boldsymbol{\mu}})' (\mathbf{Q}'_{3(A)}\mathbf{Q}_{3(A)})^{-1} (\mathbf{Q}'_{3(A)}\hat{\boldsymbol{\mu}}) = 13.1667$,

$SSA = 13.1667 - 1.0000 = 12.1667$, $F = (12.1667/2)/(1.0000/1) = 6.08$; $F_{.05;2,1} = 200.000$

j) (i) $X_{21} = [2(19) + 3(15) - 57]/(2 - 1)(3 - 1) = 13$

(ii) $SSA(corrected) = 12.1667$; $SSA(corrected)$ is equal to SSA for the restricted full rank experimental design model. Thus, the F statistic tests $\mu_{11} - \mu_{13} = \mu_{12} - \mu_{13} = 0$, $\mu_{22} - \mu_{23} = 0$.

19. a) $Y_i = \beta_0 X_0 + \underbrace{\beta_1 X_{i1} + \cdots + \beta_5 X_{i5}}_{p - 1 = 5 \text{ treatment levels}} + \underbrace{\beta_6 X_{i6} + \cdots + \beta_{10} X_{i10}}_{n - 1 = 5 \text{ blocks}} + \epsilon_i \quad (i = 1, \ldots, 36)$,

where ϵ_i is $NID(0, \sigma_\epsilon^2)$.

b) (i) $\beta_1 = \cdots = \beta_5 = 0$ or

$\begin{bmatrix} 0 & 1 & 0 & 0 & 0 & 0 & \cdots & 0 \\ 0 & 0 & 1 & 0 & 0 & 0 & \cdots & 0 \\ 0 & 0 & 0 & 1 & 0 & 0 & \cdots & 0 \\ 0 & 0 & 0 & 0 & 1 & 0 & \cdots & 0 \\ 0 & 0 & 0 & 0 & 0 & 1 & \cdots & 0 \end{bmatrix} \begin{bmatrix} \beta_0 \\ \beta_1 \\ \vdots \\ \beta_{10} \end{bmatrix} = \begin{bmatrix} 0 \\ 0 \\ 0 \\ 0 \\ 0 \end{bmatrix}$

(ii) $\beta_6 = \cdots = \beta_{10} = 0$ or

$\begin{bmatrix} 0 & \cdots & 1 & 0 & 0 & 0 & 0 \\ 0 & \cdots & 0 & 1 & 0 & 0 & 0 \\ 0 & \cdots & 0 & 0 & 1 & 0 & 0 \\ 0 & \cdots & 0 & 0 & 0 & 1 & 0 \\ 0 & \cdots & 0 & 0 & 0 & 0 & 1 \end{bmatrix} \begin{bmatrix} \beta_0 \\ \beta_1 \\ \vdots \\ \beta_{10} \end{bmatrix} = \begin{bmatrix} 0 \\ 0 \\ 0 \\ 0 \\ 0 \end{bmatrix}$

c) $\beta_0 = \mu + \alpha_6 + \pi_6$, $\beta_1 = \alpha_1 - \alpha_6$, ..., $\beta_5 = \alpha_5 - \alpha_6$, $\beta_6 = \pi_1 - \pi_6$, ..., $\beta_{10} = \pi_5 - \pi_6$

d) (i) $Y_i = \beta_0 X_0 + \beta_6 X_{i6} + \cdots + \beta_{10} X_{i10} + \epsilon_i \quad (i = 1, \ldots, 36)$, where ϵ_i is $NID(0, \sigma_\epsilon^2)$.

(ii) $SSR(X_1 \cdots X_5 | X_6 \cdots X_{10}) = SSE(X_6 \cdots X_{10}) - SSE(X_1 \cdots X_{10})$

e) $SSE(X_1 \cdots X_{10}) = 330008.0000 - 329967.8750 = 40.1250$, $SSE(X_6 \cdots X_{10}) = 330008.0000 - 329876.0000 = 132.0000$, $SSR(X_1 \cdots X_5 | X_6 \cdots X_{10}) = 132.0000 - 40.1250 = 91.8750$, $F = (91.8750/5)/(40.1250/25) = 11.45^*$; $SSE(X_1 \cdots X_5) = 330008.0000 - 329568.0078 = 439.9922$, $SSR(X_6 \cdots X_{10} | X_1 \cdots X_5) = 439.9922 - 40.1250 = 399.8672$, $F = (399.8672/5)/(40.1250/25) = 49.83^*$; $F_{.01;5,25} = 3.85$

f) $\bar{\mathbf{y}}' = [98 \quad 95 \quad 97 \quad 93 \quad 96 \quad 95]$

21. a) $F = MSA/MSWCELL = 4.16$, $F = MSG/MSWCELL = 1.16$, $F = MSG \times A/MSWCELL = 0.62$; $F_{.05;2,24} = 3.40$, $F_{.05;1,24} = 4.26$

b) $\Sigma\hat{\alpha}_j^2 = [(p - 1)/nw](MSA - MSWCELL) = 3.0733$, $\hat{\phi} = 1.0121/0.9866 = 1.03$; $1 - \hat{\beta} \cong .30$

c) $npw = (20)(3)(2) = 120$

d) $\hat{\omega}_{Y|A}^2 = 0.17$

e) $(11.0 - 8.7)/0.6976 = 3.30$, $(11.0 - 8.4)/0.6976 = 3.73^*$, $(8.7 - 8.4)/0.6976 = 0.43$; $q_{.05;3,24} = 3.53$

CHAPTER 7

2.   The design involves relatively restrictive requirements and assumptions: the treatment and two nuisance variables must have the same number of levels and interactions are not permitted.

5.   **a)** $F = MSA/MSWCELL = 1.54$, $F = MSB/MSWCELL = 0.51$, $F = MSC/MSWCELL = 1.83$; $F_{.05;2,18} = 3.55$. Do not reject the null hypotheses.

b) $\hat{\phi} = 0.2456/0.4108 = 0.6$; power is less than .3.   **c)** $n = 28$, $np^2 = 252$

d) $\hat{\omega}^2_{Y|A} = .04$, $\hat{\omega}^2_{Y|B} = 0$, $\omega^2_{Y|C} = .06$

6.   **a)** $F = MSA/MSRES = 7.37^*$, $F/MSB/MSRES = 5.61^*$, $F = MSC/MSRES = 1.41$; $F_{.01;4,12} = 5.41$.

b) $\hat{\phi} = 7.4544/3.3033 = 2.26$; $1 - \hat{\beta} \cong .72$.   **c)** $\hat{\omega}^2_{Y|A} = .34$, $\hat{\rho}_{IY|B} = .31$, $\hat{\rho}_{IY|C} = .03$

d) $MSNONADD/MSRES = 27.780/56.995 = 0.49$; the additive model is appropriate.

e) The values of the statistic $\hat{\psi}_i/\hat{\sigma}_{\overline{Y}}$ for $\overline{Y}_{1..} - \overline{Y}_{2..}$, $\overline{Y}_{1..} - \overline{Y}_{3..}$, et cetera are $5.800/3.303 = 1.76$, $7.000/3.303 = 2.12$, $20.200/3.303 = 6.12^*$, $19.600/3.303 = 5.93^*$, $1.200/3.303 = 0.36$, $14.400/3.303 = 4.36$, $13.800/3.303 = 4.18$, $13.200/3.303 = 4.00$, $12.600/3.303 = 3.81$, $-0.600/3.303 = -0.18$; $q_{.01;5,12} = 5.84$.

f) (i) $FNONLIN = (170.96/3)/(654.72/12) = 1.04$; do not reject the null hypothesis of linearity. (ii) The nonlinear component of the trend accounts for only 11% of the trend.

g) (i) $X_{322} = 54.4167$; $SSA(corrected) = 1568.0308 - 9.5069 = 1558.524$ using the first procedure described in Section 7.7 or $2192.4375 - 633.9077 = 1558.530$ using the second procedure; $F = (1558.530/4)/(633.9077/11) = 6.76$,* $F_{.01;4,11} = 5.67$.   (ii) $\hat{\psi}_i/\hat{\sigma}_{\psi_i} = (59.400 - 51.083)/[(633.9077/11)(2/5 + 1/12)]^{1/2} = 8.317/5.278 = 1.58$

h) Relative efficiency $= (.8667)(1.0952)(100.1433/54.5600) = 1.74$.

9.   **a)**
$$E(MSA) = \sigma^2_\epsilon + np \sum_{j=1}^{p} \alpha^2_j/(p - 1)$$

$$E(MSB) = \sigma^2_\epsilon + np \sum_{k=1}^{p} \beta^2_k/(p - 1)$$

$$E(MSC) = \sigma^2_\epsilon + np \sum_{l=1}^{p} \gamma^2_l/(p - 1)$$

$$E(MSRES) = \sigma^2_\epsilon$$
$$E(MSWCELL) = \sigma^2_\epsilon$$

b)
$$E(MSA) = \sigma^2_\epsilon + p \sum_{j=1}^{p} \alpha^2_j/(p - 1)$$

$$E(MSB) = \sigma^2_\epsilon + p\sigma^2_\beta$$
$$E(MSC) = \sigma^2_\epsilon + p\sigma^2_\gamma$$
$$E(MSRES) = \sigma^2_\epsilon$$

10.
$$SSA = np \sum_{j=1}^{p} (\overline{Y}_{.j.} - \overline{Y}_{...})^2$$

$$= np \sum_{j=1}^{p} (\overline{Y}^2_{.j.} - 2\overline{Y}_{...}\overline{Y}_{.j.} + \overline{Y}^2_{...})$$

$$= np \sum_{j=1}^{p} \overline{Y}^2_{.j.} - 2np\overline{Y}_{...} \sum_{j=1}^{p} \overline{Y}_{.j.} + np^2\overline{Y}^2_{...}$$

$$= np \sum_{j=1}^{p} \frac{\left(\sum\limits_{i=1}^{n}\sum\limits_{k=1}^{p}\sum\limits_{l=1}^{p} Y_{ijkl}\right)^2}{n^2 p^2} - 2np \frac{\left(\sum\limits_{i=1}^{n}\sum\limits_{j=1}^{p}\sum\limits_{k=1}^{p}\sum\limits_{l=1}^{p} Y_{ijkl}\right)}{np^2} \sum_{j=1}^{p} \frac{\left(\sum\limits_{i=1}^{n}\sum\limits_{k=1}^{p}\sum\limits_{l=1}^{p} Y_{ijkl}\right)}{np}$$

$$+ np^2 \frac{\left(\sum\limits_{i=1}^{n}\sum\limits_{j=1}^{p}\sum\limits_{k=1}^{p}\sum\limits_{l=1}^{p} Y_{ijkl}\right)^2}{n^2 p^4}$$

$$= \sum_{j=1}^{p} \frac{\left(\sum\limits_{i=1}^{n}\sum\limits_{k=1}^{p}\sum\limits_{l=1}^{p} Y_{ijkl}\right)^2}{np} - 2\frac{\left(\sum\limits_{i=1}^{n}\sum\limits_{j=1}^{p}\sum\limits_{k=1}^{p}\sum\limits_{l=1}^{p} Y_{ijkl}\right)^2}{np^2} + \frac{\left(\sum\limits_{i=1}^{n}\sum\limits_{j=1}^{p}\sum\limits_{k=1}^{p}\sum\limits_{l=1}^{p} Y_{ijkl}\right)^2}{np^2}$$

$$= \sum_{j=1}^{p} \frac{\left(\sum\limits_{i=1}^{n}\sum\limits_{k=1}^{p}\sum\limits_{l=1}^{p} Y_{ijkl}\right)^2}{np} - \frac{\left(\sum\limits_{i=1}^{n}\sum\limits_{j=1}^{p}\sum\limits_{k=1}^{p}\sum\limits_{l=1}^{p} Y_{ijkl}\right)^2}{np^2}$$

$$SSWCELL = \sum_{i=1}^{n}\sum_{j=1}^{p}\sum_{k=1}^{p}\sum_{l=1}^{p} (Y_{ijkl} - \overline{Y}....)^2$$

$$= \sum_{i=1}^{n}\sum_{j=1}^{p}\sum_{k=1}^{p}\sum_{l=1}^{p} (Y_{ijkl}^2 - 2\overline{Y}....Y_{ijkl} + \overline{Y}^2....)$$

$$= \sum_{i=1}^{n}\sum_{j=1}^{p}\sum_{k=1}^{p}\sum_{l=1}^{p} Y_{ijkl}^2 - 2\frac{\left(\sum\limits_{i=1}^{n}\sum\limits_{j=1}^{p}\sum\limits_{k=1}^{p}\sum\limits_{l=1}^{p} Y_{ijkl}\right)}{np^2}\left(\sum\limits_{i=1}^{n}\sum\limits_{j=1}^{p}\sum\limits_{k=1}^{p}\sum\limits_{l=1}^{p} Y_{ijkl}\right)$$

$$+ np^2 \frac{\left(\sum\limits_{i=1}^{n}\sum\limits_{j=1}^{p}\sum\limits_{k=1}^{p}\sum\limits_{l=1}^{p} Y_{ijkl}\right)^2}{n^2 p^4}$$

$$= \sum_{i=1}^{n}\sum_{j=1}^{p}\sum_{k=1}^{p}\sum_{l=1}^{p} Y_{ijkl}^2 - \frac{\left(\sum\limits_{i=1}^{n}\sum\limits_{j=1}^{p}\sum\limits_{k=1}^{p}\sum\limits_{l=1}^{p} Y_{ijkl}\right)^2}{np^2}$$

12. The value of the estimated observation is the one that minimizes the residual error sum of squares.

13. a) $e_1 = 3\tfrac{1}{3}$, $e_2 = 3$ b) $e_2 = 3$, $e_3 = 3\tfrac{1}{3}$

14. The type LS-4 design is more efficient. If the subjects had been assigned randomly to the treatment levels of a type CR-4 design, the resulting error term would have been 1.35 times as large as that of the type LS-4 design.

15. a) $Y_i = \beta_0 X_0 + \underbrace{\beta_1 X_{i1} + \beta_2 X_{i2}}_{\substack{p - 1 = 2 \text{ treat-} \\ \text{ment levels}}} + \underbrace{\beta_3 X_{i3} + \beta_4 X_{i4}}_{\substack{p - 1 = 2 \\ \text{rows}}} + \underbrace{\beta_5 X_{i5} + \beta_6 X_{i6}}_{\substack{p - 1 = 2 \\ \text{columns}}} + \epsilon_i$

$(i = 1, \ldots, 27)$, where ϵ_i is $NID(0, \sigma_\epsilon^2)$.

b) H_0: $\beta_1 = \beta_2 = 0$ or $\begin{bmatrix} 0 & 1 & 0 & 0 & 0 & 0 & 0 \\ 0 & 0 & 1 & 0 & 0 & 0 & 0 \end{bmatrix} \begin{bmatrix} \beta_0 \\ \beta_1 \\ \beta_2 \\ \vdots \\ \beta_6 \end{bmatrix} = \begin{bmatrix} 0 \\ 0 \end{bmatrix}$

c) $\beta_0 = \mu + \alpha_3 + \beta_3 + \gamma_3$, $\beta_1 = \alpha_1 - \alpha_3$, $\beta_2 = \alpha_2 - \alpha_3$, $\beta_3 = \beta_1 - \beta_3$, $\beta_4 = \beta_2 - \beta_3$, $\beta_5 = \gamma_1 - \gamma_3$, $\beta_6 = \gamma_2 - \gamma_3$

d) (i) $Y_i = \beta_0 X_0 + \beta_3 X_{i3} + \beta_4 X_{i4} + \beta_5 X_{i5} + \beta_6 X_{i6} + \epsilon_i \quad (i = 1, \ldots, 27)$, where ϵ_i is $NID(0, \sigma_\epsilon^2)$.

(ii) $SSR(X_1 X_2 | X_3 \cdots X_6) = SSE(X_3 \cdots X_6) - SSE(X_1 \cdots X_6)$

e) $Y_{ijkl} = \mu_{jkl} + \epsilon_{i(jkl)} \quad (i = 1, \ldots, 3; j, k, l = 1, \ldots, 3)$, where $\epsilon_{i(jkl)}$ is $NID(0, \sigma_\epsilon^2)$.

f)

$$\underset{2 \times 9}{\mathbf{C}_A'} = \begin{bmatrix} 1 & 1 & 1 & 0 & 0 & 0 & -1 & -1 & -1 \\ 0 & 0 & 0 & 1 & 1 & 1 & -1 & -1 & -1 \end{bmatrix}$$

$$\underset{2 \times 9}{\mathbf{C}_B'} = \begin{bmatrix} 1 & 0 & -1 & 1 & 0 & -1 & 1 & 0 & -1 \\ 0 & 1 & -1 & 0 & 1 & -1 & 0 & 1 & -1 \end{bmatrix}$$

$$\underset{2 \times 9}{\mathbf{C}_C'} = \begin{bmatrix} 1 & -1 & 0 & 0 & 1 & -1 & -1 & 0 & 1 \\ 0 & -1 & 1 & 1 & 0 & -1 & -1 & 1 & 0 \end{bmatrix}$$

$$\underset{2 \times 9}{\mathbf{R}'} = \begin{bmatrix} 1 & -1 & 0 & -1 & 0 & 1 & 0 & 1 & -1 \\ 0 & 1 & -1 & 1 & -1 & 0 & -1 & 0 & 1 \end{bmatrix}$$

g) $F = MSA/MSWCELL = 2.3334/1.5185 = 1.54$, $F = MSB/MSWCELL = 0.7778/1.5185 = 0.51$, $F = MSC/MSWCELL = 2.7778/1.5185 = 1.83$, $F = MSRES/MSWCELL = 0.7778/1.5185 = 0.51$; $F_{.05;2,18} = 3.55$. Do not reject any of the null hypotheses.

16.

a) $Y_i = \beta_0 X_0 + \underbrace{\beta_1 X_{i1} + \cdots + \beta_4 X_{i4}}_{\substack{p - 1 = 4 \text{ treatment} \\ \text{levels}}} + \underbrace{\beta_5 X_{i5} + \cdots + \beta_8 X_{i8}}_{p - 1 = 4 \text{ rows}} + \underbrace{\beta_9 X_{i9} + \cdots + \beta_{12} X_{i12}}_{p - 1 = 4 \text{ columns}} + \epsilon_i$

$(i = 1, \ldots, 25)$, where ϵ_i is $NID(0, \sigma_\epsilon^2)$.

b)

(i) $H_0: \beta_1 = \cdots = \beta_4 = 0$ or $\begin{bmatrix} 0 & 1 & 0 & 0 & 0 & 0 & 0 & 0 & 0 & 0 & 0 & 0 & 0 \\ 0 & 0 & 1 & 0 & 0 & 0 & 0 & 0 & 0 & 0 & 0 & 0 & 0 \\ 0 & 0 & 0 & 1 & 0 & 0 & 0 & 0 & 0 & 0 & 0 & 0 & 0 \\ 0 & 0 & 0 & 0 & 1 & 0 & 0 & 0 & 0 & 0 & 0 & 0 & 0 \end{bmatrix} \begin{bmatrix} \beta_0 \\ \beta_1 \\ \vdots \\ \beta_{12} \end{bmatrix} = \begin{bmatrix} 0 \\ 0 \\ 0 \\ 0 \end{bmatrix}$

(ii) $H_0: \beta_5 = \cdots = \beta_8 = 0$ or $\begin{bmatrix} 0 & 0 & 0 & 0 & 0 & 1 & 0 & 0 & 0 & 0 & 0 & 0 & 0 \\ 0 & 0 & 0 & 0 & 0 & 0 & 1 & 0 & 0 & 0 & 0 & 0 & 0 \\ 0 & 0 & 0 & 0 & 0 & 0 & 0 & 1 & 0 & 0 & 0 & 0 & 0 \\ 0 & 0 & 0 & 0 & 0 & 0 & 0 & 0 & 1 & 0 & 0 & 0 & 0 \end{bmatrix} \begin{bmatrix} \beta_0 \\ \beta_1 \\ \vdots \\ \beta_{12} \end{bmatrix} = \begin{bmatrix} 0 \\ 0 \\ 0 \\ 0 \end{bmatrix}$

(iii) $H_0: \beta_9 = \cdots = \beta_{12} = 0$ or $\begin{bmatrix} 0 & 0 & 0 & 0 & 0 & 0 & 0 & 0 & 0 & 1 & 0 & 0 & 0 \\ 0 & 0 & 0 & 0 & 0 & 0 & 0 & 0 & 0 & 0 & 1 & 0 & 0 \\ 0 & 0 & 0 & 0 & 0 & 0 & 0 & 0 & 0 & 0 & 0 & 1 & 0 \\ 0 & 0 & 0 & 0 & 0 & 0 & 0 & 0 & 0 & 0 & 0 & 0 & 1 \end{bmatrix} \begin{bmatrix} \beta_0 \\ \beta_1 \\ \vdots \\ \beta_{12} \end{bmatrix} = \begin{bmatrix} 0 \\ 0 \\ 0 \\ 0 \end{bmatrix}$

c) $\beta_0 = \mu + \alpha_5 + \beta_5 + \gamma_5$, $\beta_1 = \alpha_1 - \alpha_5$, $\ldots$, $\beta_4 = \alpha_4 - \alpha_5$, $\beta_5 = \beta_1 - \beta_5$, $\ldots$, $\beta_8 = \beta_4 - \beta_5$, $\beta_9 = \gamma_1 - \gamma_5$, $\ldots$, $\beta_{12} = \gamma_4 - \gamma_5$

d) (i) $Y_i = \beta_0 X_0 + \beta_5 X_{i5} + \cdots + \beta_{12} X_{i12} + \epsilon_i \quad (i = 1, \ldots, 25)$, where ϵ_i is $NID(0, \sigma_\epsilon^2)$.

(ii) $SSR(X_1 \ldots X_4 | X_5 \ldots X_{12}) = SSE(X_5 \ldots X_{12}) - SSE(X_1 \ldots X_{12})$

e) $Y_{jkl} = \mu_{jkl} + \epsilon_{jkl} \quad (j, k, l = 1, \ldots, 5)$ where ϵ_{jkl} is $NID(0, \sigma_\epsilon^2)$ and μ_{jkl} is subject to the restrictions that all interaction components involving A, B, and C are equal to zero.

f)

$$\mathbf{C}'_A \ (4\times25) =$$

```
1   1   1   1   1   0   0   0   0   0   0   0   0   0   0
0   0   0   0   0   1   1   1   1   1   0   0   0   0   0
0   0   0   0   0   0   0   0   0   0   1   1   1   1   1
0   0   0   0   0   0   0   0   0   0   0   0   0   0   0
                        0   0   0   0   0  -1  -1  -1  -1  -1
                        0   0   0   0   0  -1  -1  -1  -1  -1
                        0   0   0   0   0  -1  -1  -1  -1  -1
                        1   1   1   1   1  -1  -1  -1  -1  -1
```

$$\mathbf{C}'_B \ (4\times25) =$$

```
1   0   0   0  -1   1   0   0   0  -1   1   0   0   0  -1
0   1   0   0  -1   0   1   0   0  -1   0   1   0   0  -1
0   0   1   0  -1   0   0   1   0  -1   0   0   1   0  -1
0   0   0   1  -1   0   0   0   1  -1   0   0   0   1  -1
                        1   0   0   0  -1   1   0   0   0  -1
                        0   1   0   0  -1   0   1   0   0  -1
                        0   0   1   0  -1   0   0   1   0  -1
                        0   0   0   1  -1   0   0   0   1  -1
```

$$\mathbf{C}'_C \ (4\times25) =$$

```
1  -1   0   0   0   0   1  -1   0   0   0   0   1  -1   0
0  -1   0   0   1   1   0  -1   0   0   0   1   0  -1   0
0  -1   0   1   0   0   0  -1   0   1   1   0   0  -1   0
0  -1   1   0   0   0   0  -1   1   0   0   0   0  -1   1
                        0   0   0   1  -1  -1   0   0   0   1
                        0   0   1   0  -1  -1   0   0   1   0
                        0   1   0   0  -1  -1   0   1   0   0
                        1   0   0   0  -1  -1   1   0   0   0
```

$$\mathbf{R}' \ (12\times25) =$$

```
12  -3  -3  -3  -3  -3  -3   2   2   2  -3   2  -3   2   2
-3  12  -3  -3  -3   2  -3  -3   2   2   2  -3   2  -3   2
-3  -3  12  -3  -3   2   2  -3  -3   2   2   2  -3   2  -3
-3   2   2   2  -3  12  -3  -3  -3  -3  -3  -3   2   2   2
-3  -3   2   2   2  -3  12  -3  -3  -3   2  -3  -3   2   2
 2  -3  -3   2   2  -3  -3  12  -3  -3   2   2  -3  -3   2
-3   2   2  -3   2  -3   2   2   2   2  -3  12  -3  -3  -3
 2  -3   2   2  -3  -3  -3   2   2   2  -3  12  -3  -3  -3
-3   2  -3   2   2   2  -3  -3   2   2  -3  -3  12  -3  -3
-3   2  -3   2   2  -3   2   2  -3   2  -3  -3   2   2   2
 2  -3   2  -3   2   2  -3   2   2  -3  -3  -3   2   2   2
 2   2  -3   2  -3  -3   2  -3   2   2   2  -3  -3   2   2
                        -3   2   2  -3   2  -3   2   2   2  -3
                         2  -3   2   2  -3  -3  -3   2   2   2
                        -3   2  -3   2   2   2  -3  -3   2   2
                        -3   2  -3   2   2  -3   2   2  -3   2
                         2  -3   2  -3   2   2  -3   2   2  -3
                         2   2  -3   2  -3  -3   2  -3   2   2
                        -3  -3   2   2   2  -3   2  -3   2   2
                         2  -3  -3   2   2   2  -3   2  -3   2
                         2   2  -3  -3   2   2   2  -3   2  -3
                        12  -3  -3  -3  -3  -3  -3   2   2   2
                        -3  12  -3  -3  -3   2  -3  -3   2   2
                        -3  -3  12  -3  -3   2   2  -3  -3   2
```

g) $F = MSA/MSRES = 401.8600/54.5600 = 7.37^*$, $F = MSB/MSRES = 305.9600/54.5600 = 5.61^*$, $F = MSC/MSRES = 76.6600/54.5600 = 1.41$; $F_{.01;4,12} = 5.41$. Reject the null hypotheses for A and B.

h) (i) $\hat{\mu}_R' = [75.24 \quad 61.04 \quad 58.64 \quad 51.44 \quad 50.64 \quad 63.44 \quad 65.04 \quad 52.04 \quad 44.44 \quad 43.04 \quad 60.44$
$\quad 57.84 \quad 60.64 \quad 42.44 \quad 40.64 \quad 46.04 \quad 42.84 \quad 41.44 \quad 39.04 \quad 26.64 \quad 45.84 \quad 42.24 \quad 40.24$
$\quad 33.64 \quad 37.04]$

(ii) $q = (59.4 - 53.6)/3.3033 = 1.76$, $q = (59.4 - 52.4)/3.3033 = 2.12$, $q = (59.4 - 39.2)/3.3033 = 6.12^*$, $q = (59.4 - 39.8)/3.3033 = 5.93^*$, $q = (53.6 - 52.4)/3.3033 = 0.36$, $q = (53.6 - 39.2)/3.3033 = 4.35$, $q = (53.6 - 39.8)/3.3033 = 4.18$, $q = (52.4 - 39.2)/3.3033 = 4.00$, $q = (52.4 - 39.8)/3.3033 = 3.81$, $q = (39.2 - 39.8)/3.3033 = -0.18$; $q_{.01;5,12} = 5.84$. Reject the following null hypotheses: $\mu_{1..} - \mu_{4..} = 0$ and $\mu_{1..} - \mu_{5..} = 0$.

18. **a)**

$A\alpha$	$B\beta$	$C\gamma$
$B\gamma$	$C\alpha$	$A\beta$
$C\beta$	$A\gamma$	$B\alpha$

c)

$A\alpha a$	$B\beta b$	$C\gamma c$	$D\delta d$
$B\delta c$	$A\gamma d$	$D\beta a$	$C\alpha b$
$C\beta d$	$D\alpha c$	$A\delta b$	$B\gamma a$
$D\gamma b$	$C\delta a$	$B\alpha d$	$A\beta c$

19. 10×10 $(4s + 2)$, 11×11 (prime number), 12×12 (multiple of 4), 13×13 (prime number), 14×14 $(4s + 2)$, 16×16 (multiple of 4 or power of prime number), 17×17 (prime number), 18×18 $(4s + 2)$, 19×19 (prime number), 20×20 (multiple of 4).

20. **a)** $F = MSA/MSRES_2 = 3.32^*$, $F = MSB/MSRES_2 = 0.79$, $F = MSC/MSRES_2 = 1.82$, $F = MSD/MSRES_2 = 1.54$; $F_{.05;3,19} = 3.13$.

b) Since $F = MSRES_1/MSWCELL = 0.86$, it is reasonable to assume that an additive model is appropriate.

c) $\hat{\phi} = 0.4396/0.3329 = 1.32$; $1 - \hat{\beta} \cong .5$

d) The values of the statistic $\hat{\psi}_i/\hat{\sigma}_{\psi_i}$ for $\overline{Y}_{1...} - \overline{Y}_{4...}$, $\overline{Y}_{2...} - \overline{Y}_{4...}$, and $\overline{Y}_{3...} - \overline{Y}_{4...}$ are $1.2500/0.4708 = 2.66^*$, $1.1250/0.4708 = 2.39$, and $1.2500/0.4708 = 2.66^*$; $tD_{.05/2;4,19} = 2.55$. Reject the following null hypotheses: $\mu_{1...} - \mu_{4...} = 0$ and $\mu_{3...} - \mu_{4...} = 0$.

21. **a)** $Y_i = \beta_0 X_0 + \underbrace{\beta_1 X_{i1} + \beta_2 X_{i2} + \beta_3 X_{i3}}_{\substack{p - 1 = 3 \\ \text{treatment levels}}} + \underbrace{\beta_4 X_{i4} + \beta_5 X_{i5} + \beta_6 X_{i6}}_{p - 1 = 3 \text{ rows}}$
$+ \underbrace{\beta_7 X_{i7} + \beta_8 X_{i8} + \beta_9 X_{i9}}_{p - 1 = 3 \text{ columns}} + \underbrace{\beta_{10} X_{i10} + \beta_{11} X_{i11} + \beta_{12} X_{i12}}_{p - 1 = 3 \text{ Greek letters}} + \epsilon_i$
$(i = 1, \ldots, 32)$, where ϵ_i is $NID(0, \sigma_\epsilon^2)$.

b) H_0: $\beta_1 = \beta_2 = \beta_3 = 0$ or $\begin{bmatrix} 0 & 1 & 0 & 0 & 0 & \cdots & 0 \\ 0 & 0 & 1 & 0 & 0 & \cdots & 0 \\ 0 & 0 & 0 & 1 & 0 & \cdots & 0 \end{bmatrix} \begin{bmatrix} \beta_0 \\ \beta_1 \\ \vdots \\ \beta_{12} \end{bmatrix} = \begin{bmatrix} 0 \\ 0 \\ 0 \end{bmatrix}$

c) $\beta_0 = \mu + \alpha_4 + \beta_4 + \gamma_4 + \delta_4$, $\beta_1 = \alpha_1 - \alpha_4$, $\ldots$, $\beta_3 = \alpha_3 - \alpha_4$, $\beta_4 = \beta_1 - \beta_4$, $\ldots$, $\beta_6 = \beta_3 - \beta_4$, $\beta_7 = \gamma_1 - \gamma_4$, $\ldots$, $\beta_9 = \gamma_3 - \gamma_4$, $\beta_{10} = \delta_1 - \delta_4$, $\ldots$, $\beta_{12} = \delta_3 - \delta_4$

d) (i) $Y_i = \beta_0 X_0 + \beta_4 X_{i4} + \cdots + \beta_{12} X_{i12} + \epsilon_i$ $(i = 1, \ldots, 32)$, where ϵ_i is $NID(0, \sigma_\epsilon^2)$.
(ii) $SSR(X_1 X_2 X_3 | X_4 \cdots X_{12}) = SSE(X_4 \cdots X_{12}) - SSE(X_1 \cdots X_{12})$

e) $Y_{ijklm} = \mu_{jklm} + \epsilon_{i(jklm)}$ $(i = 1, 2; j, k, l, m = 1, \ldots, 4)$, where $\epsilon_{i(jkl)}$ is $NID(0, \sigma_\epsilon^2)$ and μ_{jkl} is subject to the restrictions that all interaction components involving A, B, and C are equal to zero.

f)

$$\underset{3 \times 16}{\mathbf{C}_A'} = \begin{bmatrix} 1 & 1 & 1 & 1 & 0 & 0 & 0 & 0 & 0 & 0 & 0 & 0 & -1 & -1 & -1 & -1 \\ 0 & 0 & 0 & 0 & 1 & 1 & 1 & 1 & 0 & 0 & 0 & 0 & -1 & -1 & -1 & -1 \\ 0 & 0 & 0 & 0 & 0 & 0 & 0 & 0 & 1 & 1 & 1 & 1 & -1 & -1 & -1 & -1 \end{bmatrix}$$

$$\underset{3 \times 16}{\mathbf{C}_B'} = \begin{bmatrix} 1 & 0 & 0 & -1 & 1 & 0 & 0 & -1 & 1 & 0 & 0 & -1 & 1 & 0 & 0 & -1 \\ 0 & 1 & 0 & -1 & 0 & 1 & 0 & -1 & 0 & 1 & 0 & -1 & 0 & 1 & 0 & -1 \\ 0 & 0 & 1 & -1 & 0 & 0 & 1 & -1 & 0 & 0 & 1 & -1 & 0 & 0 & 1 & -1 \end{bmatrix}$$

$$\underset{3 \times 16}{\mathbf{C}_C'} = \begin{bmatrix} 1 & 0 & 0 & -1 & 0 & 1 & -1 & 0 & 0 & -1 & 1 & 0 & -1 & 0 & 0 & 1 \\ 0 & 1 & 0 & -1 & 1 & 0 & -1 & 0 & 0 & -1 & 0 & 1 & -1 & 0 & 1 & 0 \\ 0 & 0 & 1 & -1 & 0 & 0 & -1 & 1 & 1 & -1 & 0 & 0 & -1 & 1 & 0 & 0 \end{bmatrix}$$

$$\underset{3 \times 16}{\mathbf{C}_D'} = \begin{bmatrix} 1 & 0 & -1 & 0 & 0 & -1 & 0 & 1 & 0 & 1 & 0 & -1 & -1 & 0 & 1 & 0 \\ 0 & 0 & -1 & 1 & 1 & -1 & 0 & 0 & 0 & 0 & 1 & -1 & -1 & 1 & 0 & 0 \\ 0 & 1 & -1 & 0 & 0 & -1 & 1 & 0 & 1 & 0 & 0 & -1 & -1 & 0 & 0 & 1 \end{bmatrix}$$

$$\underset{3 \times 16}{\mathbf{R}'} = \begin{bmatrix} 3 & -1 & -1 & -1 & -1 & -1 & 3 & -1 & -1 & -1 & -1 & 3 & -1 & 3 & -1 & -1 \\ -1 & -1 & 3 & -1 & 3 & -1 & -1 & -1 & -1 & 3 & -1 & -1 & -1 & -1 & -1 & 3 \\ -1 & -1 & -1 & 3 & -1 & 3 & -1 & -1 & 3 & -1 & -1 & -1 & -1 & -1 & 3 & -1 \end{bmatrix}$$

$$= \begin{bmatrix} \mathbf{r}_1' \\ \mathbf{r}_2' \\ \mathbf{r}_3' \end{bmatrix} \quad \text{or} \quad \underset{3 \times 16}{\mathbf{R}'} = \begin{bmatrix} (\mathbf{r}_1' - \mathbf{r}_2') \, \tfrac{1}{4} \\ (\mathbf{r}_1' - \mathbf{r}_3') \, \tfrac{1}{4} \\ (2\mathbf{r}_1' + \mathbf{r}_2' + \mathbf{r}_3') \, \tfrac{1}{4} \end{bmatrix}$$

$$= \begin{bmatrix} 1 & 0 & -1 & 0 & -1 & 0 & 1 & 0 & 0 & -1 & 0 & 1 & 0 & 1 & 0 & -1 \\ 1 & 0 & 0 & -1 & 0 & -1 & 1 & 0 & -1 & 0 & 0 & 1 & 0 & 1 & -1 & 0 \\ 1 & -1 & 0 & 0 & 0 & 0 & 1 & -1 & 0 & 0 & -1 & 1 & -1 & 1 & 0 & 0 \end{bmatrix}$$

g) $F = MSA/MSRES_2 = 2.9479/0.8865 = 3.32^*$, $F = MSB/MSRES_2 = 0.6979/0.8865 = 0.79$, $F = MSC/MSRES_2 = 1.6146/0.8865 = 1.82$, $F = MSD/MSRES_2 = 1.3646/0.8865 = 1.54$; $F_{.05;3,19} = 3.13$. Reject the null hypothesis for treatment A.

h) (i) $\hat{\boldsymbol{\mu}}_R' = [2.9062 \quad 3.0313 \quad 1.9062 \quad 1.1563 \quad 2.4063 \quad 2.1563 \quad 1.9062 \quad 2.0313 \quad 3.1563$ $1.9062 \quad 2.5313 \quad 1.4063 \quad 0.0312 \quad 1.4063 \quad 1.1563 \quad 1.4063]$

(ii) $tD = (2.250 - 1.000)/0.4708 = 2.66^*$, $tD = (2.125 - 1.000)/0.4708 = 2.39$, $tD = (2.250 - 1.000)/0.4708 = 2.66^*$; $tD_{.05/2;4,19} = 2.55$. Reject the null hypotheses for $\mu_{1\cdots} - \mu_{4\cdots} = 0$ and $\mu_{3\cdots} - \mu_{4\cdots} = 0$.

CHAPTER 8

2.   a) $a_1 b_1$, $a_1 b_2$, $a_2 b_1$, $a_2 b_2$

3.   a) $(2)(2)(3) = 12$

4.   a) $F = MSA/MSWCELL = 7.03^*$, $F = MSB/MSWCELL = 0.78$, $F = MSAB/MSWCELL = 9.03^*$; $F_{.05;1,16} = 4.49$. Reject the null hypotheses for treatment A and the AB interaction.

b)

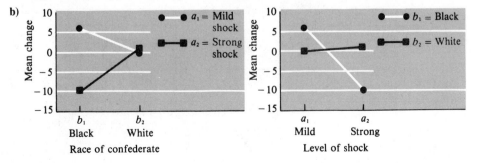

c) $1 - \hat{\beta} \cong .64$ for H_0: $\alpha_j = 0$ for all j and .76 for H_0: $(\alpha\beta)_{jk} = 0$ for all j and k.

d) $n = 8$ **e)** $\hat{\omega}^2_{Y|A} = 0.178$, $\hat{\omega}^2_{Y|AB} = .0237$

6.   **a)** $F = MSA/MSWCELL = 2.22$, $F = MSB/MSWCELL = 3.67^*$, $F = MSAB/MSWCELL = 4.82^*$; $F_{.05;1,30} = 4.17$, $F_{.05;2,30} = 3.32$. Reject the null hypothesis for treatment B and the AB interaction.

   b) (i)

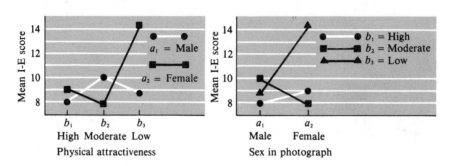

   (ii) $F = MSA\hat{\psi}_{1(B)}/MSWCELL = 1.60$, $F = MSA\hat{\psi}_{2(B)}/MSWCELL = 3.33$, $F = MSA\hat{\psi}_{3(B)}/MSWCELL = 9.54^*$; $(2F_{.05;2,30})/1 = 6.64$. Reject H_0: $\alpha\psi_{3(B)} = \delta$ for all j.

   c) $1 - \hat{\beta} \cong .45$ for H_0: $\beta_k = 0$ for all k and .65 for H_0: $(\alpha\beta)_{jk} = 0$ for all j and k.

   d) $n \cong 9$ **e)** $\hat{\omega}^2_{Y|B} = .106$, $\hat{\omega}^2_{Y|AB} = .152$

8.

$$\sum_{i=1}^{n} \sum_{j=1}^{p} \sum_{k=1}^{q} [(\overline{Y}_{\cdot j \cdot} - \overline{Y}_{\cdots}) + (\overline{Y}_{\cdot \cdot k} - \overline{Y}_{\cdots}) + (\overline{Y}_{\cdot jk} - \overline{Y}_{\cdot j \cdot} - \overline{Y}_{\cdot \cdot k} + \overline{Y}_{\cdots}) + (Y_{ijk} - \overline{Y}_{\cdot jk})]^2$$

$$= \sum_{i=1}^{n} \sum_{j=1}^{p} \sum_{k=1}^{q} [(\overline{Y}_{\cdot j \cdot} - \overline{Y}_{\cdots})^2 + (\overline{Y}_{\cdot \cdot k} - \overline{Y}_{\cdots})^2 + (\overline{Y}_{\cdot jk} - \overline{Y}_{\cdot j \cdot} - \overline{Y}_{\cdot \cdot k} + \overline{Y}_{\cdots})^2 + (Y_{ijk} - \overline{Y}_{\cdots})^2$$

$$+ \ 2(\text{all cross-product terms})].$$

It can be shown that the sum of each cross-product term is equal to zero. For example,

$$2n \sum_{j=1}^{p} (\overline{Y}_{\cdot j \cdot} - \overline{Y}_{\cdots}) \sum_{k=1}^{q} (\overline{Y}_{\cdot \cdot k} - \overline{Y}_{\cdots}) = 0, \text{ since } \sum_{k=1}^{q} (\overline{Y}_{\cdot \cdot k} - \overline{Y}_{\cdots}) = 0.$$

Summing over the remaining nonzero terms gives

$$nq \sum_{j=1}^{p} (\overline{Y}_{\cdot j \cdot} - \overline{Y}_{\cdots})^2 + np \sum_{k=1}^{q} (\overline{Y}_{\cdot \cdot k} - \overline{Y}_{\cdots})^2$$

$$+ \ n \sum_{j=1}^{p} \sum_{k=1}^{q} (\overline{Y}_{\cdot jk} - \overline{Y}_{\cdot j \cdot} - \overline{Y}_{\cdot \cdot k} + \overline{Y}_{\cdots})^2 + \sum_{i=1}^{n} \sum_{j=1}^{p} \sum_{k=1}^{q} (Y_{ijk} - \overline{Y}_{\cdot jk})^2.$$

9. $nq \displaystyle\sum_{j=1}^{p} (\overline{Y}_{\cdot j \cdot} - \overline{Y}_{\cdots})^2 = nq \sum_{j=1}^{p} (\overline{Y}^2_{\cdot j \cdot} - 2\overline{Y}_{\cdots}\overline{Y}_{\cdot j \cdot} + \overline{Y}^2_{\cdots})$

$$= nq \sum_{j=1}^{p} \frac{\left(\sum\limits_{i=1}^{n} \sum\limits_{k=1}^{q} Y_{ijk}\right)^2}{n^2 q^2} - 2nq \frac{\sum\limits_{i=1}^{n} \sum\limits_{j=1}^{p} \sum\limits_{k=1}^{q} Y_{ijk}}{npq} \sum_{j=1}^{p} \frac{\sum\limits_{i=1}^{n} \sum\limits_{k=1}^{q} Y_{ijk}}{nq} + npq \frac{\left(\sum\limits_{i=1}^{n} \sum\limits_{j=1}^{p} \sum\limits_{k=1}^{q} Y_{ijk}\right)^2}{n^2 p^2 q^2}$$

$$= \sum_{j=1}^{p} \frac{\left(\sum\limits_{i=1}^{n} \sum\limits_{k=1}^{q} Y_{ijk}\right)^2}{nq} - 2 \frac{\left(\sum\limits_{i=1}^{n} \sum\limits_{j=1}^{p} \sum\limits_{k=1}^{q} Y_{ijk}\right)^2}{npq} + \frac{\left(\sum\limits_{i=1}^{n} \sum\limits_{j=1}^{p} \sum\limits_{k=1}^{q} Y_{ijk}\right)^2}{npq}$$

$$= \sum_{j=1}^{p} \frac{\left(\sum\limits_{i=1}^{n} \sum\limits_{k=1}^{q} Y_{ijk} \right)^2}{nq} - \frac{\left(\sum\limits_{i=1}^{n} \sum\limits_{j=1}^{p} \sum\limits_{k=1}^{q} Y_{ijk} \right)^2}{npq}$$

$$\sum_{i=1}^{n} \sum_{j=1}^{p} \sum_{k=1}^{q} (Y_{ijk} - \overline{Y}_{jk})^2 = \sum_{i=1}^{n} \sum_{j=1}^{p} \sum_{k=1}^{q} (Y_{ijk}^2 - 2\overline{Y}_{jk}Y_{ijk} + \overline{Y}_{.jk}^2)$$

$$= \sum_{i=1}^{n} \sum_{j=1}^{p} \sum_{k=1}^{q} Y_{ijk}^2 - 2\sum_{j=1}^{p} \sum_{k=1}^{q} \frac{\sum\limits_{i=1}^{n} Y_{ijk}}{n} \sum_{i=1}^{n} Y_{ijk} + n\sum_{j=1}^{p} \sum_{k=1}^{q} \frac{\left(\sum\limits_{i=1}^{n} Y_{ijk} \right)^2}{n^2}$$

$$= \sum_{i=1}^{n} \sum_{j=1}^{p} \sum_{k=1}^{q} Y_{ijk}^2 - 2\sum_{j=1}^{p} \sum_{k=1}^{q} \frac{\left(\sum\limits_{i=1}^{n} Y_{ijk} \right)^2}{n} + \sum_{j=1}^{p} \sum_{k=1}^{q} \frac{\left(\sum\limits_{i=1}^{n} Y_{ijk} \right)^2}{n}$$

$$= \sum_{i=1}^{n} \sum_{j=1}^{p} \sum_{k=1}^{q} Y_{ijk}^2 - \sum_{j=1}^{p} \sum_{k=1}^{q} \frac{\left(\sum\limits_{i=1}^{n} Y_{ijk} \right)^2}{n}$$

12. The critical values are equal when the contrast has one degree of freedom.

14. When the contrasts within each set are mutually orthogonal the sum of the treatment-contrast interaction sums of squares is equal to the omnibus interaction sum of squares.

15. $E(MSA) = \sigma_\epsilon^2 + n\sigma_\beta^2 + nq \sum\limits_{j=1}^{p} \alpha_j^2/(p-1)$, $E(MSBw.A) = \sigma_\epsilon^2 + n\sigma_\beta^2$, $E(MSWCELL) = \sigma_\epsilon^2$

16. $F' = MSA/(MSBw.A + MSAC - MSBw.A \times C)$, $F'' = (MSA + MSBw.A \times C)/(MSBw.A + MSAC)$

17. **a)** and **b)**

Source	df	MS	E(MS)
A	2	21.17	$\sigma_\epsilon^2 + nq\sigma_{\alpha\gamma}^2 + nqr\sigma_\alpha^2$
B	1	8.04	$\sigma_\epsilon^2 + np\sigma_{\beta\gamma}^2 + npr\sigma_\beta^2$
C	1	12.19	$\sigma_\epsilon^2 + np\sigma_{\beta\gamma}^2 + nq\sigma_{\alpha\gamma}^2 + npq\sigma_\gamma^2$
AC	2	4.21	$\sigma_\epsilon^2 + nq\sigma_{\alpha\gamma}^2$
BC	1	7.85	$\sigma_\epsilon^2 + np\sigma_{\beta\gamma}$
RES(AB+ABC+WCELL)	16	1.57	σ_ϵ^2

c) $F' = MSC/(MSAC + MSBC - MSRES)$

19. **a)** $Y_i = \beta_0 X_0 + \beta_1 X_{i1} + \beta_2 X_{i2} + \beta_3 X_{i1}X_{i2} + \epsilon_i$ $(i = 1, \ldots, 20)$; ϵ_i is $NID(0, \sigma_\epsilon^2)$

b) H_0: $\beta_1 = 0$, H_0: $\beta_2 = 0$, H_0: $\beta_3 = 0$

c) (i) $\beta_0 = \mu$, $\beta_1 = \alpha_1$, $\beta_2 = \beta_1$, $\beta_3 = (\alpha\beta)_{11}$ (ii) $\beta_0 = \mu + \alpha_2 + \beta_2 + (\alpha\beta)_{22}$,
$\beta_1 = \alpha_1 - \alpha_2 + (\alpha\beta)_{12} - (\alpha\beta)_{22}$, $\beta_2 = \beta_1 - \beta_2 + (\alpha\beta)_{21} - (\alpha\beta)_{22}$,
$\beta_3 = (\alpha\beta)_{11} - (\alpha\beta)_{21} - (\alpha\beta)_{12} + (\alpha\beta)_{22}$

d) $SSR(X_1|X_2 \ X_1X_2) = 685.00 - 403.75 = 281.25$, $SSR(X_2|X_1 \ X_1X_2) = 685.00 - 653.75 = 31.25$,
$SSB(X_1X_2|X_1 \ X_2) = 685.00 - 323.75 = 361.25$, $SSE = 1325.00 - 685.00 = 640.00$;
$F = (281.25/1)/(640.00/16) = 7.03^*$, $F = (31.25/1)/(640.00/16) = 0.78$,
$F = (361.25/1)/(640.00/16) = 9.03^*$; $F_{.05;1,16} = 4.49$. Reject the null hypothesis for treatment A
and the AB interaction.

e) $Y_{ijk} = \mu_{ijk} + \epsilon_{i(jk)}$, where $\epsilon_{i(jk)}$ is $NID(0, \sigma_\epsilon^2)$.

f) $\mathbf{C}'_A = [1 \quad 1 -1 -1]$, $\mathbf{C}'_B = [1 -1 \quad 1 -1]$, $\mathbf{C}'_{AB} = [1 -1 -1 \quad 1]$
$\;_{(p-1)\times h}\;_{(q-1)\times h}\;_{(p-1)(q-1)\times h}$

g) $SSA = 15(1/0.80)15 = 281.25$, $SSB = -5(1/0.80)(-5) = 31.25$,
$SSAB = 17(1/0.80)17 = 361.25$, $SSWCELL = 1325 - 685 = 640.00$;
$F = (281.25/1)/(640.00/16) = 7.03^*$, $F = (31.25/1)/(640.00/16) = 0.78$,
$F = (361.25/1)/(640.00/16) = 9.03^*$; $F_{.05;1,16} = 4.49$. Reject the null hypothesis for treatment A and the AB interaction.

21.

a) $SSA = 1.6133$, $SSB = 145.3704$, $SSAB = 143.0371$, $SSWCELL = 208.6667$;
$F = (1.6133/1)/(208.6667/23) = 0.18$, $F = (145.3704/3)/(208.6667/23) = 5.34^*$,
$F = (143.0371/3)/(208.6667/23) = 5.26^*$; $F_{.05;1,23} = 4.28$, $F_{.05;3,23} = 3.03$. Reject the null hypothesis for treatment B and the AB interaction.

b) $SSA = 6.6207$, $SSB = 139.0461$, $SSAB = 142.2584$, $SSWCELL = 171.6667$;
$F = (6.6207/1)/(171.6667/20) = 0.77$, $F = (139.0461/3)/(171.6667/20) = 5.40^*$,
$F = (142.2584/2)/(171.6667/20) = 8.29^*$; $F_{.05;1,20} = 4.35$, $F_{.05;3,20} = 3.10$, $F_{.05;2,20} = 3.49$. Reject the null hypothesis for treatment B and the AB interaction.

c) Rejection of the interaction null hypothesis is interpreted to mean that at least one function of the form $\mu_{jk} - \mu_{j'k} - \mu_{jk'} + \mu_{j'k'} \neq 0$. Failure to reject the interaction null hypothesis does not mean that all functions of the preceding form equal zero because those involving a_2b_2 cannot be tested.

CHAPTER 9

1.

a) A AB BD ACD
B AC CD BCD
C AD ABC $ABCD$
D BC ABD

2. **a)** 15   **b)** 20

3.

a) $F = MSA/MSWCELL = 27.64^*$, $F = MSB/MSWCELL = 24.94^*$, $F = MSC/MSWCELL = 39.80^*$, $F = MSAB/MSWCELL = 3.39$, $F = MSAC/MSWCELL = 6.91^*$, $F = MSBC/MSWCELL = 11.68^*$, $F = MSABC/MSWCELL = 0.07$; $F_{.05;1,72} = 3.97$

b)

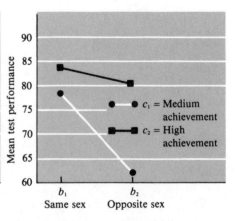

4.

a) $[ABCD] - [ABC] - [ABD] - [ACD] - [BCD] + [AB] + [AC] + [AD] + [BC] + [BD] + [CD] - [A] - [B] - [C] - [D] + [Y]$

b) $[ABCDE] - [ABCD] - [ABCE] - [ABDE] - [ACDE] - [BCDE] + [ABC] + [ABD] + [ABE] + [ACD] + [ACE] + [ADE] + [BCD] + [BCE] + [BDE] + [CDE] - [AB] - [AC] - [AD] - [AE] - [BC] - [BD] - [BE] - [CD] - [CE] - [DE] + [A] + [B] + [C] + [D] + [E] - [Y]$

5.

$$\mathbf{C}'_A \underset{1\times 12}{=} [1 \quad 1 \quad 1 \quad 1 \quad 1 \quad 1 \quad -1 \quad -1 \quad -1 \quad -1 \quad -1 \quad -1]$$

$$\mathbf{C}'_B \underset{2\times 12}{=} \begin{bmatrix} 1 & 1 & 0 & 0 & -1 & -1 & 1 & 1 & 0 & 0 & -1 & -1 \\ 0 & 0 & 1 & 1 & -1 & -1 & 0 & 0 & 1 & 1 & -1 & -1 \end{bmatrix}$$

$$\mathbf{C}'_C \underset{1\times 12}{=} [1 \quad -1 \quad 1 \quad -1 \quad 1 \quad -1 \quad 1 \quad -1 \quad 1 \quad -1 \quad 1 \quad -1]$$

$$\mathbf{C}'_{AB} \underset{2\times 12}{=} \begin{bmatrix} 1 & 1 & -1 & -1 & 0 & 0 & -1 & -1 & 1 & 1 & 0 & 0 \\ 0 & 0 & 1 & 1 & -1 & -1 & 0 & 0 & -1 & -1 & 1 & 1 \end{bmatrix}$$

$$\mathbf{C}'_{AC} \underset{1\times 12}{=} [1 \quad -1 \quad 1 \quad -1 \quad 1 \quad -1 \quad -1 \quad 1 \quad -1 \quad 1 \quad -1 \quad 1]$$

$$\mathbf{C}'_{BC} \underset{2\times 12}{=} \begin{bmatrix} 1 & -1 & -1 & 1 & 0 & 0 & 1 & -1 & -1 & 1 & 0 & 0 \\ 0 & 0 & 1 & -1 & -1 & 1 & 0 & 0 & 1 & -1 & -1 & 1 \end{bmatrix}$$

$$\mathbf{C}'_{ABC} \underset{2\times 12}{=} \begin{bmatrix} 1 & -1 & -1 & 1 & 0 & 0 & -1 & 1 & 1 & -1 & 0 & 0 \\ 0 & 0 & 1 & -1 & -1 & 1 & 0 & 0 & -1 & 1 & 1 & -1 \end{bmatrix}$$

7.

a) $F = MSBL/MSRES = 7.11^*$, $F = MSA/MSRES = 10.61^*$, $F = MSB/MSRES = 8.77^*$, $F = MSAB/MSRES = 2.46$; $F_{.05;1,5} = 6.61$, $F_{.05;2,5} = 5.79$. Reject the null hypothesis for treatments A and B.

b) $q = (2.75 - 2.75)/0.4873 = 0$, $q = (2.75 - 5.25)/0.4873 = -5.13^*$,
$q = (2.75 - 5.25)/0.4873 = -5.13^*$; $q_{.05;3,5} = 4.80$.

c) $SSBL = 11.500 - 4.7500 = 6.7500$, $SSA = 14.8333 - 4.7500 = 10.0833$,
$SSB = 21.4167 - 4.7500 = 16.6667$, $SSAB = 9.4167 - 4.7500 = 4.6667$, $SSRES = 4.7500$; F ratios are identical to those in 7(a).

10.

a) $Y_{ijk} = \mu_{ijk} + \epsilon_{ijk}$ $\quad (i = 1, \ldots, 5; j = 1, 2, 3; k = 1, 2)$, where ϵ_{ijk} is $NID(0,\sigma_\epsilon^2)$ and μ_{ijk} is subject to the restrictions that $\mu_{i(jk)} - \mu_{i'(jk)} - \mu_{i(jk)'} + \mu_{i'(jk)'} = 0$ for all i, i', (jk), and $(jk)'$, excluding μ_{211}.

b)

$$\mathbf{R}' \underset{19\times 29}{=} \begin{bmatrix} 1 & -1 & 0 & 0 & -1 & 0 & 1 & 0 & 0 & 0 & 0 & 0 & 0 & 0 & \cdots \\ 0 & 1 & -1 & 0 & 0 & 0 & -1 & 1 & 0 & 0 & 0 & 0 & 0 & 0 & \cdots \\ 0 & 0 & 1 & -1 & 0 & 0 & 0 & -1 & 1 & 0 & 0 & 0 & 0 & 0 & \cdots \\ 0 & 0 & 0 & 0 & 1 & -1 & 0 & 0 & 0 & -1 & 1 & 0 & 0 & 0 & \cdots \\ 0 & 0 & 0 & 0 & 0 & 1 & -1 & 0 & 0 & 0 & -1 & 1 & 0 & 0 & \cdots \\ 0 & 0 & 0 & 0 & 0 & 0 & 1 & -1 & 0 & 0 & 0 & -1 & 1 & 0 & \cdots \\ 0 & 0 & 0 & 0 & 0 & 0 & 0 & 1 & -1 & 0 & 0 & 0 & -1 & 1 & \cdots \\ 0 & 0 & 0 & 0 & 0 & 0 & 0 & 0 & 0 & 1 & -1 & 0 & 0 & 0 & \cdots \\ 0 & 0 & 0 & 0 & 0 & 0 & 0 & 0 & 0 & 0 & 1 & -1 & 0 & 0 & \cdots \\ 0 & 0 & 0 & 0 & 0 & 0 & 0 & 0 & 0 & 0 & 0 & 1 & -1 & 0 & \cdots \\ 0 & 0 & 0 & 0 & 0 & 0 & 0 & 0 & 0 & 0 & 0 & 0 & 1 & 1 & \cdots \\ 0 & 0 & 0 & 0 & 0 & 0 & 0 & 0 & 0 & 0 & 0 & 0 & 0 & 0 & \cdots \\ \cdot & \cdot & \cdot & \cdot & \cdot & \cdot & \cdot & \cdot & \cdot & \cdot & \cdot & \cdot & \cdot & & \\ 0 & 0 & 0 & 0 & 0 & 0 & 0 & 0 & 0 & 0 & 0 & 0 & 0 & & \end{bmatrix}$$

$$\mathbf{C}'_A_{\,2\times29} = \begin{bmatrix} 1/4 & 1/4 & 1/4 & 1/4 & 1/5 & 1/5 & 1/5 & 1/5 & 1/5 & 0 & 0 & 0 & 0 & 0 \\ 0 & 0 & 0 & 0 & 0 & 0 & 0 & 0 & 0 & 1/5 & 1/5 & 1/5 & 1/5 & 1/5 \end{bmatrix}$$

$$\begin{bmatrix} 0 & 0 & 0 & 0 & 0 & -1/5 & -1/5 & -1/5 & -1/5 & -1/5 & -1/5 & -1/5 & -1/5 & -1/5 & -1/5 \\ 1/5 & 1/5 & 1/5 & 1/5 & 1/5 & -1/5 & -1/5 & -1/5 & -1/5 & -1/5 & -1/5 & -1/5 & -1/5 & -1/5 & -1/5 \end{bmatrix}$$

$$\mathbf{C}'_B_{\,1\times29} = [\,1/4 \;\; 1/4 \;\; 1/4 \;\; 1/4 \;\; -1/5 \;\; -1/5 \;\; -1/5 \;\; -1/5 \;\; -1/5 \;\; 1/5 \;\; 1/5 \;\; 1/5$$
$$1/5 \;\; 1/5 \;\; 1/5 \;\; -1/5 \;\; -1/5 \;\; -1/5 \;\; -1/5 \;\; 1/5 \;\; 1/5 \;\; 1/5 \;\; 1/5 \;\; 1/5 \;\; -1/5 \;\; -1/5 \;\; -1/5 \;\; -1/5 \;\; -1/5\,]$$

$$\mathbf{C}'_{AB}_{\,2\times29} = \begin{bmatrix} 1/4 & 1/4 & 1/4 & 1/4 & -1/5 & -1/5 & -1/5 & -1/5 & -1/5 & -1/5 & -1/5 & -1/5 & -1/5 & -1/5 \\ 0 & 0 & 0 & 0 & 0 & 0 & 0 & 0 & 0 & 1/5 & 1/5 & 1/5 & 1/5 & 1/5 \end{bmatrix}$$

$$\begin{bmatrix} 1/5 & 1/5 & 1/5 & 1/5 & 1/5 & 0 & 0 & 0 & 0 & 0 & 0 & 0 & 0 & 0 & 0 \\ -1/5 & -1/5 & -1/5 & -1/5 & -1/5 & -1/5 & -1/5 & -1/5 & -1/5 & -1/5 & -1/5 & 1/5 & 1/5 & 1/5 & 1/5 & 1/5 \end{bmatrix}$$

$$\mathbf{C}'_{BL}_{\,4\times29} = \begin{bmatrix} 1/6 & 0 & 0 & -1/6 & 1/6 & 0 & 0 & 0 & -1/6 & 1/6 & 0 & 0 & 0 & -1/6 \\ 0 & 0 & 0 & -1/6 & 0 & 1/5 & 0 & 0 & -1/6 & 0 & 1/5 & 0 & 0 & -1/6 \\ 0 & 1/6 & 0 & -1/6 & 0 & 0 & 1/6 & 0 & -1/6 & 0 & 0 & 1/6 & 0 & -1/6 \\ 0 & 0 & 1/6 & -1/6 & 0 & 0 & 0 & 1/6 & -1/6 & 0 & 0 & 0 & 1/6 & -1/6 \end{bmatrix}$$

$$\begin{bmatrix} 1/6 & 0 & 0 & 0 & -1/6 & 1/6 & 0 & 0 & 0 & -1/6 & 1/6 & 0 & 0 & 0 & -1/6 \\ 0 & 1/5 & 0 & 0 & -1/6 & 0 & 1/5 & 0 & 0 & -1/6 & 0 & 1/5 & 0 & 0 & -1/6 \\ 0 & 0 & 1/6 & 0 & -1/6 & 0 & 0 & 1/6 & 0 & -1/6 & 0 & 0 & 1/6 & 0 & -1/6 \\ 0 & 0 & 0 & 1/6 & -1/6 & 0 & 0 & 0 & 1/6 & -1/6 & 0 & 0 & 0 & 1/6 & -1/6 \end{bmatrix}$$

c) $SSRES = 86.2917$, $SSBL = 319.7829 - 86.2917 = 233.4912$, $SSA = 336.2917 - 86.2917 = 250.0000$, $SSB = 221.4917 - 86.2917 = 135.2000$, $SSAB = 107.8302 - 86.2917 = 21.5385$, $F = (233.4912/4)/(86.2917/19) = 12.85^*$, $F = (250.0000/2)/(86.2917/19) = 27.52^*$, $F = (135.2000/1)/(86.2917/19) = 29.77^*$, $F = (21.5385/2)/(86.2917/19) = 2.37$; $F_{.05;4,19} = 2.90$, $F_{.05;2,19} = 3.52$, $F_{.05;1,19} = 4.38$. Reject the null hypotheses for blocks and treatments A and B.

CHAPTER 10

2. **a)** $F = MSA/MSB(A) = 9.82^*$, $F = MSB(A)/MSWCELL = 2.00$; $F_{.05;1,4} = 7.71$, $F_{.05;4,18} = 2.93$. Reject the hypothesis that $\alpha_j = 0$ for all j.

b) Yes, $F = MSB$ at a_1/MSB at $a_2 = 28/16 = 1.75$; $F_{.25;2,2} = 3.00$.

c) $\hat{\omega}^2_{Y|A} = 0.37$, $\hat{\rho}_{IY|B(A)} = 0.13$

3. **a)** $F = MSA/MSB(A) = 9.95^*$, $F = MSB(A)/MSWCELL = 5.87^*$; $F_{.05;2,3} = 9.55$, $F_{.05;3,24} = 3.01$. Reject both hypotheses.

b) Yes, $C = 90/220 = 0.4091$; $C_{.05,1} = 0.9669$.

c) $q = -7/2.708 = -2.58$, $q = -17/2.708 = -6.28^*$, $q = -10/2.708 = -3.69$; $q_{.05,3,3} = 5.91$. Reject the hypothesis that $\mu_{1.} - \mu_{3.} = 0$.

d) $\hat{\omega}^2_{Y|A} = 0.64$, $\hat{\rho}_{IY|B(A)} = 0.18$

e) The operators were not randomly assigned to the work spaces; thus, those at b_5 and b_6 may have been faster assemblers than those at b_1 and b_2. An examination of operator production records might shed some light on this. The experiment could have been improved by assigning the operators randomly to the six workplaces. Unfortunately, this is not always possible in industrial settings.

5. **a)** CRH-24(A)8(AB) **b)** CRPH-24(A)2   **c)** CRPH-24(A)4(A) **d)** CRF-222

j) CRPH-24(A)8(AB)2   **k)** CRPH-24(A)4(A)4(A)

7. **a)** (i) $Y_{ijklm} = \mu + \alpha_j + \beta_k + \gamma_l + \delta_{m(k)} + (\alpha\beta)_{jk} + (\alpha\gamma)_{jl} + (\alpha\delta)_{jm(k)} + (\beta\gamma)_{kl} + (\gamma\delta)_{lm(k)} + (\alpha\beta\gamma)_{jkl} + (\alpha\gamma\delta)_{jlm(k)} + \epsilon_{i(jklm)}$

(ii)

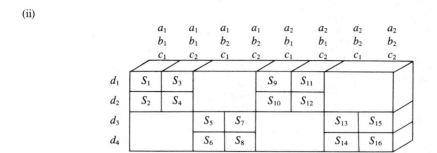

b) (i) $Y_{ijklm} = \mu + \alpha_j + \beta_k + \gamma_l + \delta_{m(jk)} + (\alpha\beta)_{jk} + (\alpha\gamma)_{jl} + (\beta\gamma)_{kl} + (\gamma\delta)_{lm(jk)} + (\alpha\beta\gamma)_{jkl} + \epsilon_{i(jklm)}$

(ii)

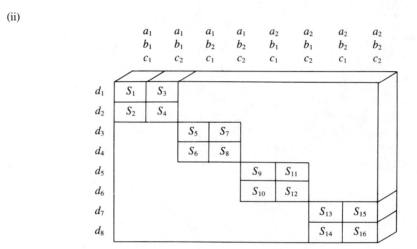

c) (i) $Y_{ijklm} = \mu + \alpha_j + \beta_{k(j)} + \gamma_{l(jk)} + \delta_{m(jkl)} + \epsilon_{i(jklm)}$

(ii)

| | b_1 | b_1 | b_1 | b_1 | b_2 | b_2 | b_2 | b_2 | b_3 | b_3 | b_3 | b_3 | b_4 | b_4 | b_4 | b_4 |
| | c_1 | c_1 | c_2 | c_2 | c_3 | c_3 | c_4 | c_4 | c_5 | c_5 | c_6 | c_6 | c_7 | c_7 | c_8 | c_8 |
	d_1	d_2	d_3	d_4	d_5	d_6	d_7	d_8	d_9	d_{10}	d_{11}	d_{12}	d_{13}	d_{14}	d_{15}	d_{16}
a_1	S_1	S_2	S_3	S_4	S_5	S_6	S_7	S_8								
a_2									S_9	S_{10}	S_{11}	S_{12}	S_{13}	S_{14}	S_{15}	S_{16}

8. (i) **a)** $SSC + SSBC$ **b)** $SSBD + SSBCD$ **d)** $SSAD + SSABD + SSACD + SSABCD$

f) $SSD + SSAD + SSBD + SSCD + SSABD + SSACD + SSBCD + SSABCD$

i) $SSBC + SSABC$ **(l)** $SSBCD + SSABCD$

9. **a)**

Sum of Squares	
CRPH-$pqrt(AB)$	CRF-$pqrt$
SSA	SSA
SSB	SSB
SSC	SSC
SSD(AB)	SSD + SSAD + SSBD + SSABD
SSAB	SSAB
SSAC	SSAC
SSBC	SSBC
SSC×D(AB)	SSCD + SSACD + SSBCD + SSABCD
SSABC	SSABC
SSWCELL	SSWCELL

Degrees of Freedom	
CRPH-$pqrt(AB)$	CRF-$pqrt$
$p - 1$	$p - 1$
$q - 1$	$q - 1$
$r - 1$	$r - 1$
$pq(t_{(jk)} - 1)$	$(t - 1) + (p - 1)(t - 1) + (q - 1)(t - 1)$ $+ (p - 1)(q - 1)(t - 1)$
$(p - 1)(q - 1)$	$(p - 1)(q - 1)$
$(p - 1)(r - 1)$	$(p - 1)(r - 1)$
$(q - 1)(r - 1)$	$(q - 1)(r - 1)$
$pq(t_{(jk)} - 1)(r - 1)$	$(r - 1)(t - 1) + (p - 1)(r - 1)(t - 1)$ $+ (q - 1)(r - 1)(t - 1)$ $+ (p - 1)(q - 1)(r - 1)(t - 1)$
$(p - 1)(q - 1)(r - 1)$	$(p - 1)(q - 1)(r - 1)$
$pqrt_{(jk)}(n - 1)$	$pqrt(n - 1)$

10.

$$\mathbf{C}'_A = [1 \quad 1 \quad 1 \quad -1 \quad -1 \quad -1] \qquad \mathbf{C}'_{B(A)} = \begin{bmatrix} 1 & 0 & -1 & 0 & 0 & 0 \\ 0 & 1 & -1 & 0 & 0 & 0 \\ 0 & 0 & 0 & 1 & 0 & -1 \\ 0 & 0 & 0 & 0 & 1 & -1 \end{bmatrix}$$

$SSA = 204.6313$, $SSB(A) = 83.6362$, $SSWCELL = 198.0004$,
$F = (204.6313/1)/(83.6362/4) = 9.79^*$, $F = (83.6362/4)/(198.0004)/17) = 1.80$; $F_{.05;1,4} = 7.71$,
$F_{.05;4,17} = 2.96$. Reject the null hypothesis for treatment A.

11. $\mathbf{C}'_A = \begin{bmatrix} 1 & 0 & 0 & -\frac{1}{2} & -\frac{1}{2} \\ 0 & \frac{1}{2} & \frac{1}{2} & -\frac{1}{2} & -\frac{1}{2} \end{bmatrix}$ $\mathbf{C}'_{B(A)} = \begin{bmatrix} 0 & 1 & -1 & 0 & 0 \\ 0 & 0 & 0 & 1 & -1 \end{bmatrix}$

$SSA = 1176.8750$, $SSB(A) = 130.0000$, $SSWCELL = 206.7500$,
$F = (1176.8750/2)/(130.0000/2) = 9.05$, $F = (130.0000/2)/(206.7500/19) = 5.97^*$;
$F_{.05;2,2} = 19.0$, $F_{.05;2,19} = 3.52$. Reject the null hypothesis for treatment B.

CHAPTER 11

1.   **a)**   **b)**

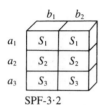

SPF-3·2

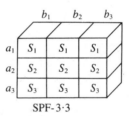

SPF-3·3

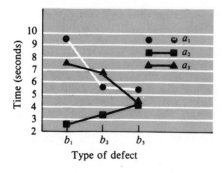

RBF-32

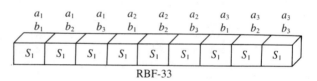

RBF-33

2.   **a)** 2(SPF), 6(RBF)   **b)** 3(SPF), 9(RBF)

3.   The precision and power of the test are reduced.

6.   **a)** $F = MSA/MSBL(A) = 6.11$, $F = MSB/MSB \times BL(A) = 11.96$, $F = MSAB/MSB \times BL(A) = 12.70$; $F_{.05;2,15} = 3.68$, $F_{05;2,30} = 3.32$, $F_{.05;4,30} = 2.69$. Reject all three null hypotheses.

      b)

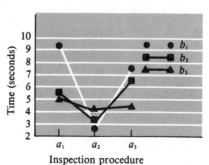

9.

$$q \sum_{i=1}^{n} \sum_{j=1}^{p} (\overline{Y}_{ij\cdot} - \overline{Y}_{\cdot j\cdot})^2 = q \sum_{i=1}^{n} \sum_{j=1}^{p} (\overline{Y}_{ij\cdot}^2 - 2\overline{Y}_{\cdot j\cdot}\overline{Y}_{ij\cdot} + \overline{Y}_{\cdot j\cdot}^2)$$

$$= q \sum_{i=1}^{n} \sum_{j=1}^{p} \overline{Y}_{ij\cdot}^2 - 2q \sum_{j=1}^{p} \overline{Y}_{\cdot j\cdot} \sum_{i=1}^{n} \overline{Y}_{ij\cdot} + nq \sum_{j=1}^{p} \overline{Y}_{\cdot j\cdot}^2$$

$$= q \sum_{i=1}^{n} \sum_{j=1}^{p} \frac{\left(\sum_{k=1}^{q} Y_{ijk}\right)^2}{q^2} - 2q \sum_{j=1}^{p} \frac{\left(\sum_{i=1}^{n} \sum_{k=1}^{q} Y_{ijk}\right)}{nq} \sum_{i=1}^{n} \left(\frac{\sum_{k=1}^{q} Y_{ijk}}{q}\right)$$

$$+ nq \sum_{j=1}^{p} \frac{\left(\sum_{i=1}^{n} \sum_{k=1}^{q} Y_{ijk}\right)^2}{n^2 q^2}$$

$$= \sum_{i=1}^{n} \sum_{j=1}^{p} \frac{\left(\sum_{k=1}^{q} Y_{ijk}\right)^2}{q} - 2 \sum_{j=1}^{p} \frac{\left(\sum_{i=1}^{n} \sum_{k=1}^{q} Y_{ijk}\right)^2}{nq} + \sum_{j=1}^{p} \frac{\left(\sum_{i=1}^{n} \sum_{k=1}^{q} Y_{ijk}\right)^2}{nq}$$

$$= \sum_{i=1}^{n} \sum_{j=1}^{p} \frac{\left(\sum_{k=1}^{q} Y_{ijk}\right)^2}{q} - \sum_{j=1}^{p} \frac{\left(\sum_{i=1}^{n} \sum_{k=1}^{q} Y_{ijk}\right)^2}{nq}$$

10.   **a)** $F_{\max} = 35.795/26.892 = 1.33$, $F_{\max,.05} = 8.38$; do not reject the homogeneity of variance null hypothesis.

11.   **a)** $\hat{\theta} = 6.5347/9.5122 = 0.687$

     b) $F_{.05;1.37,20.61} \cong 3.84$, $F_{.05;2.75,20.61} \cong 3.13$; reject the null hypothesis for treatment B and the AB interaction.

     c) The adjusted test and the conventional test lead to the same decision regarding the null hypothesis for treatment B and the AB interaction. The conservative F test would also have led to rejecting the two null hypotheses.

12.   **a)** $F = 14.8411/0.8835 = 16.80$, $(4F_{.05;4,15})/2 = 6.12$; reject H_0.

     b) $F = 17.6282/1.6729 = 10.54$, $(4F_{.05;4,15})/2 = 6.12$; reject H_0.

     c) $F = 26.0800/1.400 = 18.63$, $(4F_{.05;4,20})/2 = 5.74$; reject H_0.

     d) $F = 5.9286/1.3554 = 4.37$, $(4F_{.05;4,20})/2 = 5.74$; do not reject H_0.

16.   **a)** 0.338   **b)** 2.882

18.   **a)** $\chi^2 = -pr(n-1)d \ln(W)$, $d = 1 - [(2q^2 - 3q + 3)/6pr(n-1)(q-1)]$

19.   Between-blocks test: $MSBL(a_1), MSBL(a_2), \ldots, MSBL(a_p)$ all estimate the same population variance.

     Within blocks tests:

$$\left.\begin{array}{l} F = \dfrac{MSB}{MSB \times BL(A)} \\[2em] F = \dfrac{MSAB}{MSB \times BL(A)} \end{array}\right\} \qquad \underset{(q-1)\times qrt}{\mathbf{C}_B^{\star\prime}} \; \underset{qrt \times qrt}{\boldsymbol{\Sigma}_{aj}} \; \underset{qrt \times (q-1)}{\mathbf{C}_B^{\star}} = \lambda \mathbf{I} \qquad (j = 1, \ldots, p)$$

$$\left.\begin{array}{l} F = \dfrac{MSC}{MSC \times BL(A)} \\[2em] F = \dfrac{MSAC}{MSC \times BL(A)} \end{array}\right\} \qquad \underset{(r-1)\times qrt}{\mathbf{C}_C^{\star\prime}} \; \underset{qrt \times qrt}{\boldsymbol{\Sigma}_{aj}} \; \underset{qrt \times (r-1)}{\mathbf{C}_C^{\star}} = \lambda \mathbf{I} \qquad (j = 1, \ldots, p)$$

$$F = \frac{MSD}{MSD \times BL(A)}$$

$$F = \frac{MSAD}{MSD \times BL(A)}$$

$$\left.\right\} \quad \underset{(t-1)\times qrt}{\mathbf{C}_D^{\star\prime}} \quad \underset{qrt\times qrt}{\mathbf{\Sigma}_{a_j}} \quad \underset{qrt\times(t-1)}{\mathbf{C}_D^{\star}} = \lambda\mathbf{I} \quad (j = 1, \ldots, p)$$

$$F = \frac{MSBC}{MSB \times C \times BL(A)}$$

$$F = \frac{MSABC}{MSB \times C \times BL(A)}$$

$$\left.\right\} \quad \underset{(q-1)(r-1)\times qrt}{\mathbf{C}_{BC}^{\star\prime}} \quad \underset{qrt\times qrt}{\mathbf{\Sigma}_{a_j}} \quad \underset{qrt\times(q-1)(r-1)}{\mathbf{C}_{BC}^{\star}} = \lambda\mathbf{I} \quad (j = 1, \ldots, p)$$

. . . .

$$F = \frac{MSBCD}{MSB \times C \times D \times BL(A)}$$

$$F = \frac{MSABCD}{MSB \times C \times D \times BL(A)}$$

$$\left.\right\} \quad \underset{(q-1)(r-1)(t-1)\times qrt}{\mathbf{C}_{BCD}^{\star\prime}} \quad \underset{qrt\times qrt}{\mathbf{\Sigma}_{a_j}} \quad \underset{qrt\times(q-1)(r-1)(t-1)}{\mathbf{C}_{BCD}^{\star}} = \lambda\mathbf{I} \quad (j = 1, \ldots, p)$$

22. **a)** $Y_{ijk} = \mu_{ijk} + \epsilon_{ijk}$ $(i = 1, \ldots, n; j = 1, \ldots, p; k = 1, \ldots, q)$, where ϵ_{ijk} is $NID(0, \sigma_\epsilon^2)$ and μ_{ijk} is subject to the restrictions that for each j $\mu_{ijk} - \mu_{i'jk} - \mu_{ijk'} + \mu_{i'jk'} = 0$ for all i, i', k, and k'.

b) Treatment A H_0: $\mu_{\cdot1\cdot} - \mu_{\cdot3\cdot} = 0$
$\mu_{\cdot2\cdot} - \mu_{\cdot3\cdot} = 0$

Treatment B H_0: $\mu_{\cdot\cdot1} - \mu_{\cdot\cdot3} = 0$
$\mu_{\cdot\cdot2} - \mu_{\cdot\cdot3} = 0$

AB Interaction H_0: $\mu_{\cdot11} - \mu_{\cdot21} - \mu_{\cdot12} + \mu_{\cdot22} = 0$
$\mu_{\cdot12} - \mu_{\cdot22} - \mu_{\cdot13} + \mu_{\cdot23} = 0$
$\mu_{\cdot21} - \mu_{\cdot31} - \mu_{\cdot22} + \mu_{\cdot32} = 0$
$\mu_{\cdot22} - \mu_{\cdot32} - \mu_{\cdot23} + \mu_{\cdot33} = 0$

$BL(A)$ Error H_0: $\mu_{11\cdot} - \mu_{61\cdot} = 0$
$\mu_{21\cdot} - \mu_{61\cdot} = 0$
. . . .
$\mu_{51\cdot} - \mu_{61\cdot} = 0$
. . . .
$\mu_{13,3\cdot} - \mu_{18,3\cdot} = 0$
$\mu_{14,3\cdot} - \mu_{18,3\cdot} = 0$
. . . .
$\mu_{17,3\cdot} - \mu_{18,3\cdot} = 0$

$B \times BL(A)$ Error H_0: $\mu_{111} - \mu_{211} - \mu_{112} + \mu_{212} = 0$
$\mu_{211} - \mu_{311} - \mu_{212} + \mu_{312} = 0$
. . . .
$\mu_{512} - \mu_{612} - \mu_{513} + \mu_{613} = 0$
. . . .
$\mu_{13,31} - \mu_{14,31} - \mu_{13,32} + \mu_{14,32} = 0$
$\mu_{14,31} - \mu_{15,31} - \mu_{14,32} + \mu_{15,32} = 0$
. . . .
$\mu_{17,32} - \mu_{18,32} - \mu_{17,33} + \mu_{18,33} = 0$

c) $\mathbf{y}(54 \times 1)$, $\boldsymbol{\mu}(54 \times 1)$, $\mathbf{C}_A'(2 \times 54)$, $\mathbf{C}_B'(2 \times 54)$, $\mathbf{Q}_B'(32 \times 54)$, $\mathbf{C}_{AB}'(4 \times 54)$, $\mathbf{Q}_{AB}'(34 \times 54)$, $\mathbf{C}_{BL(A)}'(15 \times 54)$, $\mathbf{R}'(30 \times 54)$

24. **a)** $\mathbf{C}_A' = \begin{bmatrix} 1 & 1 & 1 & 1 & -1 & -1 & -1 & -1 \end{bmatrix}$ $\mathbf{C}_{BL}' = \begin{bmatrix} 1 & -1 & 1 & -1 & 0 & 0 & 0 & 0 \\ 0 & 0 & 0 & 0 & 1 & -1 & 1 & -1 \end{bmatrix}$

$\mathbf{C}_B' = \begin{bmatrix} 1 & 1 & -1 & -1 & 1 & 1 & -1 & -1 \end{bmatrix}$ $\mathbf{C}_{AB}' = \begin{bmatrix} 1 & 1 & -1 & -1 & -1 & -1 & 1 & 1 \end{bmatrix}$

$\mathbf{R}' = \begin{bmatrix} 1 & -1 & -1 & 1 & 0 & 0 & 0 & 0 \\ 0 & 0 & 0 & 0 & 1 & -1 & -1 & 1 \end{bmatrix}$

b) $F = MSA/MSBL(A) = 8.65$, $F = MSB/MSB \times BL(A) = 33.8$,
$F = MSAB/MSB \times BL(A) = 23.53$; $F_{.05;1,2} = 19.0$. Reject the null hypothesis for treatment B and the AB interaction.

26.

a) $\mathbf{C}_A' = \begin{bmatrix} \frac{2}{4} & \frac{1}{4} & \frac{1}{4} & -\frac{1}{4} & -\frac{1}{4} & -\frac{1}{4} & -\frac{1}{4} \end{bmatrix}$ $\mathbf{C}_{BL}' = \begin{bmatrix} \frac{1}{2} & \frac{1}{2} & -1 & 0 & 0 & 0 & 0 \\ 0 & 0 & 0 & \frac{1}{2} & -\frac{1}{2} & \frac{1}{2} & -\frac{1}{2} \end{bmatrix}$

$\mathbf{C}_B' = \begin{bmatrix} \frac{2}{4} & -\frac{1}{4} & -\frac{1}{4} & \frac{1}{4} & \frac{1}{4} & -\frac{1}{4} & -\frac{1}{4} \end{bmatrix}$ $\mathbf{C}_{AB}' = \begin{bmatrix} 1 & -\frac{1}{2} & -\frac{1}{2} & -\frac{1}{2} & -\frac{1}{2} & \frac{1}{2} & \frac{1}{2} \end{bmatrix}$

$\mathbf{R}' = \begin{bmatrix} 0 & 0 & 0 & 1 & -1 & -1 & 1 \end{bmatrix}$

b) $F = MSA/MSBL(A) = 1.61$, $F = MSB/MSB \times BL(A) = 18.22$, $F = MSAB/MSB \times BL(A) = 13.22$; $F_{.05;1,2} = 19.0$, $F_{.05;1,1} = 161.0$. Do not reject any of the null hypotheses.

CHAPTER 12

1.   **a)** No   **b)** Yes   **d)** Yes   **f)** Yes

2.   **a)** 1   **c)** 2   **e)** 0   **g)** 0   **i)** 4

3.   **a)** Group 0 = 00, 11; group 1 = 01, 10

c) Group 0 = 00, 14, 23, 32, 41; group 1 = 01, 10, 24, 33, 42; group 2 = 02, 11, 20, 34, 43; group 3 = 03, 12, 21, 30, 44; group 4 = 04, 13, 22, 31, 40

d) Group 0 = 000, 011, 101, 110; group 1 = 001, 010, 100, 111

4.   **a)** $F = MSA/MSAB \times BL(G) = 6.44$, $F = MSB/MSAB \times BL(G) = 1.18$; $F_{.05;1,8} = 5.32$. Reject the null hypothesis for treatment A.

b) Yes, this is a good design choice because the AB interaction is clearly not significant and treatments A and B are evaluated with maximum power given the available resources.

c) $F_{max} = 1.02$; $F_{max,.05} = 9.6$. Do not reject the homogeneity of variance null hypothesis.

d) $F_{max} = 1.21$; $F_{max,.05} = 9.6$. Do not reject the homogeneity of variance null hypothesis.

6.   **a)**

		abc	abc	abc	abc
Group 0	$(AB)_0$	000	001	110	111
	$(AB)_1$	010	011	100	101
Group 1	$(AC)_0$	000	010	101	111
	$(AC)_1$	001	011	100	110
Group 2	$(ABC)_0$	000	011	101	110
	$(ABC)_1$	001	010	100	111

c) $\frac{2}{3}(AB)$, $\frac{2}{3}(AC)$, $\frac{2}{3}(ABC)$

9.

		ab	ab	ab	ab	ab
g_0	$(AB)_0$	00	14	23	32	41
g_1	$(AB)_1$	01	10	24	33	42
g_2	$(AB)_2$	02	11	20	34	43
g_3	$(AB)_3$	03	12	21	30	44
g_4	$(AB)_4$	04	13	22	31	40

11.   **a)** (ABD) **c)** (AC^2) and (AB^2) **e)** (BC) and (AC) **f)** (BD), (DE), (ACE), $(ABCDE)$

    g) (BC^4), (AB^3C^4), (AC^2), and (AB^4C^3)

12.   **a)** (i) $Y_{ijkz} = \mu + \zeta_z + \pi_{i(z)} + \alpha_j + \beta_k + (\alpha\beta\pi)_{jki(z)} + \epsilon_{ijkz}$ $(i = 1, \ldots, n; j = 1, \ldots, p;$ $k = 1, \ldots, q; z = 1, \ldots, w)$, where ϵ_{ijkz} is $NID(0, \sigma_\epsilon^2)$.

    (ii) $Y_{ijkz} = \mu_{ijkz} + \epsilon_{ijkz}$, where ϵ_{ijkz} is $NID(0, \sigma_\epsilon^2)$ and μ_{ijkz} is subject to the restrictions that for each z $\mu_{i(jk)z} - \mu_{i'(jk)z} - \mu_{i(jk)'z} + \mu_{i'(jk)'z} = 0$ for all i, i', (jk), and $(jk)'$.

  b)

Treatment A	H_0:	$\mu_{\cdot 0 \cdot \cdot} - \mu_{\cdot 1 \cdot \cdot} = 0$
Treatment B	H_0:	$\mu_{\cdot \cdot 0 \cdot} - \mu_{\cdot \cdot 1 \cdot} = 0$
Groups	H_0:	$\mu_{\cdot \cdot \cdot 0} - \mu_{\cdot \cdot \cdot 1} = 0$
$BL(G)$ Error	H_0:	$\mu_{0 \cdot \cdot 0} - \mu_{4 \cdot \cdot 0} = 0$
		$\mu_{1 \cdot \cdot 0} - \mu_{4 \cdot \cdot 0} = 0$
		$\mu_{2 \cdot \cdot 0} - \mu_{4 \cdot \cdot 0} = 0$
		$\mu_{3 \cdot \cdot 0} - \mu_{4 \cdot \cdot 0} = 0$
		$\mu_{5 \cdot \cdot 1} - \mu_{9 \cdot \cdot 1} = 0$
		$\mu_{6 \cdot \cdot 1} - \mu_{9 \cdot \cdot 1} = 0$
		$\mu_{7 \cdot \cdot 1} - \mu_{9 \cdot \cdot 1} = 0$
		$\mu_{8 \cdot \cdot 1} - \mu_{9 \cdot \cdot 1} = 0$
$AB \times BL(G)$ Error	H_0:	$\mu_{0000} - \mu_{1000} - \mu_{0110} + \mu_{1110} = 0$
		$\mu_{1000} - \mu_{2000} - \mu_{1110} + \mu_{2110} = 0$
		$\mu_{2000} - \mu_{3000} - \mu_{2110} + \mu_{3110} = 0$
		$\mu_{3000} - \mu_{4000} - \mu_{3110} + \mu_{4110} = 0$
		$\mu_{5011} - \mu_{6011} - \mu_{5101} + \mu_{6101} = 0$
		$\mu_{6011} - \mu_{7011} - \mu_{6101} + \mu_{7101} = 0$
		$\mu_{7011} - \mu_{8011} - \mu_{7101} + \mu_{8101} = 0$
		$\mu_{8011} - \mu_{9011} - \mu_{8101} + \mu_{9101} = 0$

  c) $\mathbf{y}(20 \times 1)$, $\boldsymbol{\mu}(20 \times 1)$, $\mathbf{C}'_{BL}(9 \times 20)$, $\mathbf{C}'_G(1 \times 20)$, $\mathbf{C}'_{BL(G)}(8 \times 20)$, $\mathbf{R}'(8 \times 20)$, $\mathbf{C}'_{w.\,BL}(10 \times 20)$, $\mathbf{C}'_A(1 \times 20)$, $\mathbf{C}'_B(1 \times 20)$, $\mathbf{Q}'_{w.\,BL}(18 \times 20)$, $\mathbf{Q}'_A(9 \times 20)$, $\mathbf{Q}'_B(9 \times 20)$

  d) $F = MSG/MSBL(G) = 0.10$, $F = MSA/MSAB \times BL(G) = 6.44$, $F = MSB/MSAB \times BL(G) = 1.18$; $F_{.05;1,8} = 5.32$. Reject the null hypothesis for treatment A.

13.   **a)** $\mathbf{C}'_{BL} = \begin{bmatrix} 1 & 0 & 1 & 0 & 0 & -1 & 0 & -1 \\ 0 & 1 & 0 & 1 & 0 & -1 & 0 & -1 \\ 0 & 0 & 0 & 0 & 1 & -1 & 0 & -1 \end{bmatrix}$ $\mathbf{C}'_G = \begin{bmatrix} 1 & 1 & 1 & 1 & -1 & -1 & -1 & -1 \end{bmatrix}$

    $\mathbf{C}'_{BL(G)} = \begin{bmatrix} 1 & -1 & 1 & -1 & 0 & 0 & 0 & 0 \\ 0 & 0 & 0 & 0 & 1 & -1 & 1 & -1 \end{bmatrix}$ $\mathbf{R}' = \begin{bmatrix} 1 & -1 & -1 & 1 & 0 & 0 & 0 & 0 \\ 0 & 0 & 0 & 0 & 1 & -1 & -1 & 1 \end{bmatrix}$

    $\mathbf{C}'_{w.\,BL} = \begin{bmatrix} 1 & 0 & -1 & 0 & 0 & 0 & 0 & 0 \\ 0 & 1 & 0 & -1 & 0 & 0 & 0 & 0 \\ 0 & 0 & 0 & 0 & 1 & 0 & -1 & 0 \\ 0 & 0 & 0 & 0 & 0 & 1 & 0 & -1 \end{bmatrix}$ $\mathbf{C}'_A = \begin{bmatrix} 1 & 1 & -1 & -1 & 1 & 1 & -1 & -1 \end{bmatrix}$

    $\mathbf{C}'_B = \begin{bmatrix} 1 & 1 & -1 & -1 & -1 & -1 & 1 & 1 \end{bmatrix}$

  b) $SSG = 2.00$, $SSBL(G) = 30.50$, $SSA = 4.50$, $SSB = 2.00$, $SSAB \times BL(G) = 18.50$.

  c) $F = MSG/MSBL(G) = 0.13$, $F = MSA/MSAB \times BL(G) = 0.49$, $F = MSB/MSAB \times BL(G) = 0.22$; $F_{.05;1,2} = 18.5$. Do not reject any of the null hypotheses.

15.   **a)** For the type LSCF-3^2 design, $w = 3$ samples of $n = 2$ experimental units from a population are randomly assigned to the groups. The sequence of administration of the $v = 3$ treatment combinations within a block is randomized independently for each block. For a type RBPF-3^2 design, $w = 2$ samples of $n = 3$ units from a population are randomly assigned to the groups. The sequence of administration of the $v = 3$ treatment combinations within a block is randomized independently for each block.

b) $F = MSB/MSBL(G) = 0.27$, $F = MSA/MSAB \times BL(G) = 24.94$, $F = MSB/MSAB \times BL(G) = 41.91$, $F = MSAB'/MSAB \times BL(G) = 1.80$; $F_{.05;2,3} = 9.55$, $F_{.05;2,6} = 5.14$. Reject the null hypotheses for treatments A and B.

c) Pairwise comparisons for treatment A: $q = 9.66$, $q = 7.03$, $q = -2.63$; $q_{.05;3,6} = 4.34$. Reject H_0: $\mu_{.0..} - \mu_{.1..} = 0$ and H_0: $\mu_{.0..} - \mu_{.2..} = 0$. Pairwise comparisons for treatment B: $q = 9.66$, $q = 12.30$, $q = 2.63$; $q_{.05;3,6} = 4.34$. Reject H_0: $\mu_{..0.} - \mu_{..1.} = 0$ and H_0: $\mu_{..0.} - \mu_{..2.} = 0$.

d) The type LSCF-3^2 design provides more degrees of freedom for estimating the within-blocks error term.

CHAPTER 13

3.

Source	Alias
A	$BCDE$
B	$ACDE$
C	$ABDE$
D	$ABCE$
E	$ABCD$
AB	CDE
AC	BDE
AD	BCE

Source	Alias
AE	BCD
BC	ADE
BD	ACE
BE	ACD
CD	ABE
CE	ABD
DE	ABC

4.

Source	Aliases
A	$BC, ACDE, BDE$
B	$AC, BCDE, ADE$
C	$AB, DE, ABCDE$
D	$ABCD, CE, ABE$
E	$ABCE, CD, ABD$
AD	BCD, ACE, BE
AE	BCE, ACD, BD

6.

a_0	a_0	a_0	a_0	a_0	a_0	a_0	a_0	a_1	a_1	a_1	a_1	a_1	a_1	a_1	a_1
b_0	b_0	b_0	b_0	b_1	b_1	b_1	b_1	b_0	b_0	b_0	b_0	b_1	b_1	b_1	b_1
c_0	c_0	c_1	c_1	c_0	c_0	c_1	c_1	c_0	c_0	c_1	c_1	c_0	c_0	c_1	c_1
d_0	d_1	d_0	d_1	d_0	d_1	d_0	d_1	d_0	d_1	d_0	d_1	d_0	d_1	d_0	d_1
e_0	e_1	e_1	e_0	e_1	e_0	e_0	e_1	e_1	e_0	e_0	e_1	e_0	e_1	e_1	e_0

8.   **a)** Only two F ratios exceeded one: they are $F = MSB/MSWCELL = 40.5000/14.3125 = 2.83$ and $F = MSC/MSWCELL = 18.0000/14.3125 = 1.26$; $F_{.05;1,16} = 4.49$. Do not reject any of the null hypotheses.

b) There is no ambiguity in interpreting the results because all of the tests are insignificant.

9.   **a)** A, $(AB^2C^2D^2E^2)$, $(BCDE)$

b) A, $(BCDE^2)$, $(AB^2C^2D^2E)$

e) A, (AB^2C^2), (BC), $(ABCD)$, $(AB^2C^2D^2)$, $(ABCD^2)$, (BCD^2), (AD), D

10.   **a)** 2

c) 4

e) 8 (Note: This design has four generalized interactions: XY, XZ, YZ, and XYZ.)

f) 3

h) 27 (Note: This design has three defining contrasts and ten generalized interactions: XY, XY^2, XZ, XZ^2, YZ, YZ^2, XYZ, XYZ^2, XY^2Z, XY^2Z^2.)

i) 5

11. **a)** 8

c) 8

e) 4

f) 27

h) 9

i) 5

12. **a)**

	abcd	abcd	abcd	abcd
Group 0	0000	0110	1010	1100
Group 1	0011	0101	1001	1111

b)

Source	Alias
Groups	D
A	BCD
B	ACD
C	ABD
AB	CD
AC	BD
AD	BC

c) The ABC interaction is not a good choice to confound with groups because it also confounds treatment D with groups.

14. **a)** (i) $(ABC)_0$ **c)** (i) $(ABC)_0$ **e)** (i) $(AB^2C)_1$

     (ii) $A = BC$ (ii) $A = (BC) = (AB^2C^2)$ (ii) $A = (BC^2) = (ABC^2)$

         $B = AC$ $B = (AC) = (AB^2C)$ $B = (AC) = (ABC)$

         $C = AB$ $C = (AB) = (ABC^2)$ $C = (AB^2) = (AB^2C^2)$

                       $(AB^2) = (AC^2) = (BC^2)$ $(AB) = (AC^2) = (BC)$

16. **a)** $F = MSA/MSWCELL = 4.02$, $F = MSB/MSWCELL = 0.70$, $F = MSC/MSWCELL = 7.27$, $F = MSD/MSWCELL = 2.11$, $F = MSAD = 0.54$, $F = MSBD/MSWCELL = 0.70$, $F = MSCD/MSWCELL = 1.49$, $F = MSRES/MSWCELL = 0.21$; $F_{.05;1,18} = 4.41$, $F_{.05;2,18} = 3.55$. Reject the null hypothesis for treatment A and treatment C.

b) The assumption appears tenable because the test of the residual null hypothesis is insignificant.

c) Treatment A: $q = -0.750/0.663 = -1.13$, $q = -2.584/0.663 = -3.90$, $q = -1.834/0.663 = -2.77$; treatment C: $q = 2.917/0.663 = 4.40$, $q = 3.250/0.663 = 4.90$, $q = 0.333/0.663 = 0.50$; $q_{.05;3,18} = 3.61$. Reject the null hypothesis for $\mu_{.0..} - \mu_{.2..}$, $\mu_{...0} - \mu_{...1}$, and $\mu_{...0} - \mu_{...2}$.

17. **a)**

Treatment A	H_0:	$\mu_{0...} - \mu_{1...} = 0$
Treatment B	H_0:	$\mu_{.0..} - \mu_{.1..} = 0$
Treatment C	H_0:	$\mu_{..0.} - \mu_{..1.} = 0$
Treatment D	H_0:	$\mu_{...0} - \mu_{...1} = 0$
AB Interaction	H_0:	$\mu_{00..} - \mu_{10..} - \mu_{01..} + \mu_{11..} = 0$
AC Interaction	H_0:	$\mu_{0.0.} - \mu_{1.0.} - \mu_{0.1.} + \mu_{1.1.} = 0$
AD Interaction	H_0:	$\mu_{0..0} - \mu_{1..0} - \mu_{0..1} + \mu_{1..1} = 0$
BC Interaction	H_0:	$\mu_{.00.} - \mu_{.10.} - \mu_{.01.} + \mu_{.11.} = 0$
BD Interaction	H_0:	$\mu_{.0.0} - \mu_{.1.0} - \mu_{.0.1} + \mu_{.1.1} = 0$
CD Interaction	H_0:	$\mu_{..00} - \mu_{..10} - \mu_{..01} + \mu_{..11} = 0$
ABC Interaction	H_0:	$\mu_{000.} - \mu_{100.} - \mu_{010.} + \mu_{110.} - \mu_{001.} + \mu_{101.} + \mu_{011.} - \mu_{111.} = 0$
ABD Interaction	H_0:	$\mu_{00.0} - \mu_{10.0} - \mu_{01.0} + \mu_{11.0} - \mu_{00.1} + \mu_{10.1} + \mu_{01.1} - \mu_{11.1} = 0$
ACD Interaction	H_0:	$\mu_{0.00} - \mu_{1.00} - \mu_{0.10} + \mu_{1.10} - \mu_{0.01} + \mu_{1.01} + \mu_{0.11} - \mu_{1.11} = 0$
BCD Interaction	H_0:	$\mu_{.000} - \mu_{.100} - \mu_{.010} + \mu_{.110} - \mu_{.001} + \mu_{.101} + \mu_{.011} - \mu_{.111} = 0$
$ABCD$ Interaction	H_0:	$\mu_{0000} - \mu_{1000} - \mu_{0100} + \mu_{1100} - \mu_{0010} + \mu_{1010} + \mu_{0110} - \mu_{1110}$
		$- \mu_{0001} + \mu_{1001} + \mu_{0101} - \mu_{1101} + \mu_{0011} - \mu_{1011} - \mu_{0111} + \mu_{1111} = 0$

c) $F = $ Only two of the F ratios exceed one: they are $F = MSB/MSWCELL = 40.5000/14.3125 = 2.83$ and $F = MSC/MSWCELL = 18.0000/14.3125 = 1.26$; $F_{.05;1,16} = 4.49$. Do not reject any of the null hypotheses.

CHAPTER 14

1. **a)** (i) type of diet   (ii) weight gain   (iii) amount of each diet consumed

    b) (i) type of reading program   (ii) reading performance   (iii) achievement test scores

5.
$$\sum_{i=1}^{n} \sum_{j=1}^{p} (X_{ij} - \overline{X}_{..})(Y_{ij} - \overline{Y}_{..}) = \sum_{i=1}^{n} \sum_{j=1}^{p} \{[(X_{ij} - \overline{X}_{.j}) + (\overline{X}_{.j} - \overline{X}_{..})][(Y_{ij} - \overline{Y}_{.j}) + (\overline{Y}_{.j} - \overline{Y}_{..})]\}$$

$$= \sum_{i=1}^{n} \sum_{j=1}^{p} [(X_{ij} - \overline{X}_{.j})(Y_{ij} - \overline{Y}_{.j}) + (\overline{X}_{.j} - \overline{X}_{..})(Y_{ij} - \overline{Y}_{.j})$$

$$+ (X_{ij} - \overline{X}_{.j})(\overline{Y}_{.j} - \overline{Y}_{..}) + (\overline{X}_{.j} - \overline{X}_{..})(\overline{Y}_{.j} - \overline{Y}_{..})]$$

$$= \sum_{i=1}^{n} \sum_{j=1}^{p} (X_{ij} - \overline{X}_{.j})(Y_{ij} - \overline{Y}_{.j}) + n \sum_{j=1}^{p} (\overline{X}_{.j} - \overline{X}_{..})(\overline{Y}_{.j} - \overline{Y}_{..})$$

since $\sum_{j=1}^{p} (\overline{X}_{.j} - \overline{X}_{..}) \sum_{i=1}^{n} (Y_{ij} - \overline{Y}_{.j}) = 0$ and $\sum_{i=1}^{n} (X_{ij} - \overline{X}_{.j}) \sum_{j=1}^{p} (\overline{Y}_{.j} - \overline{Y}_{..}) = 0$.

6.

a)

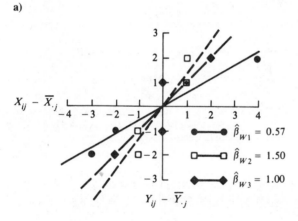

b)

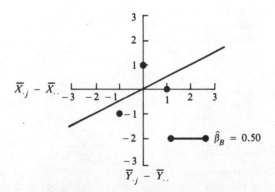

c)

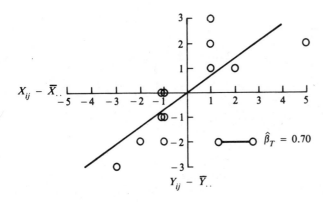

8. **a)** $F = (A_{adj}/2)/(S_{adj}/26) = 11.73^*$; $F_{.05;2,26} = 3.37$. Reject the null hypothesis.

b) (i) $F = MSBG/MSWG = 4.59$ for the ANOVA; $F = 11.73$ for the ANCOVA.
(ii) The test of associative thinking appears to be an effective concomitant variable in this experiment.

c) (i) $r_B = 0.14$, $r_W = 0.79$
(ii) When r_B is small relative to r_W the F ratio in ANCOVA will be larger than the corresponding F ratio in ANOVA. In this case the test of associative thinking appears to be an effective concomitant variable.

d) $F = (6.9629/2)/(65.2097/24) = 1.28$; $F_{.25;2,24} = 1.47$. Do not reject the null hypothesis.

e) $qBP = (14.7848 - 11.6662)/0.5327 = 5.855^*$, $qBP = (14.7848 - 11.6491)/0.5327 = 5.887^*$, $qBP = (11.6662 - 11.6491)/0.5327 = 0.032$; $q^{\circ}_{.05;1,3,26} = 3.59$. Reject H_0: $\mu_1 - \mu_2 = 0$ and H_0: $\mu_1 - \mu_3 = 0$.

10. **a)** $F = (A_{adj}/5)/(E_{adj}/24) = 11.73^*$; $F_{.01;5,24} = 3.90$. Reject the null hypothesis.

b) (i) r_B for treatment $A = -0.52$
(ii) The variables are inversely related—higher probabilities of detection are associated with shorter scanning times.
(iii) r_B for blocks $= 0.93$
(iv) Animals with higher probabilities of detection tend to have longer scanning times.

c) (i) $F = MSA/MSRES = 11.50$ for the ANOVA; $F = 11.73$ for the ANCOVA.
(ii) The concomitant variable was moderately effective. An examination of the means for X and Y at each level of treatment A and the value of r_B for treatment A reveals that the probability of correct response is inversely related to scanning time. This will affect comparisons among means.

d) (i) $\overline{Y}_{adj \cdot 1} = 98.9204$, $\overline{Y}_{adj \cdot 2} = 95.3572$, $\overline{Y}_{adj \cdot 3} = 96.9661$, $\overline{Y}_{adj \cdot 4} = 92.8409$, $\overline{Y}_{adj \cdot 5} = 95.5593$, $\overline{Y}_{adj \cdot 6} = 94.3559$
(ii) $tD' = (98.9204 - 95.3572)/0.7789 = 4.57^*$, $tD' = (98.9204 - 96.9661)/0.9020 = 2.17$, $tD' = (98.9204 - 92.8409)/0.9499 = 6.40^*$, $tD' = (98.9204 - 95.5593)/1.0686 = 3.15^*$, $tD' = (98.9204 - 94.3559)/1.1616 = 3.93^*$; $tD'_{.01;6,24} = 3.11$. Reject H_0: $\mu_1 - \mu_2 = 0$, H_0: $\mu_1 - \mu_4 = 0$, H_0: $\mu_1 - \mu_5 = 0$, and H_0: $\mu_1 - \mu_6 = 0$.

11. **a)** $F = (A_{adj}/1)/(E_{adj}/15) = 0.82$, $F = (B_{adj}/1)/(E_{adj}/15) = 2.88$, $F = (AB_{adj}/1)/(E_{adj}/15) = 4.66^*$; $F_{.05;1,15} = 4.54$. Reject the null hypothesis for the AB interaction.

b) The F ratios in ANOVA for treatment A and the AB interaction are larger than those in ANCOVA; however, the F ratio for treatment B is smaller.

c) (i) r_B for treatment $A = -1.0$, r_B for treatment $B = -1.0$, r_B for the AB interaction $= 1.0$, $r_W = 0.82$

(ii) Yes

d) $F = (8.4601/3)/(195.9096/12) = 0.17$; $F_{25;3,12} = 1.56$. Do not reject H_0: $\beta_{W11} = \beta_{W12} = \beta_{W21} = \beta_{W22}$.

e) $\overline{Y}_{\text{adj}\cdot 11} = 0.71$, $\overline{Y}_{\text{adj}\cdot 12} = -0.45$, $\overline{Y}_{\text{adj}\cdot 21} = -5.01$, $\overline{Y}_{\text{adj}\cdot 22} = 1.76$, $\overline{Y}_{11} = 6$, $\overline{Y}_{12} = 0$, $\overline{Y}_{21} = -10$, $\overline{Y}_{\cdot 22} = 1$

13.

a) $Y_i = \beta_0 X_0 + \beta_1 X_{i1} + \beta_2 X_{i2} + \beta_3 X_{i3} + \epsilon_i$ $(i = 1, \dots, 30)$

b) $\beta_0 = \mu + \alpha_3$
$\beta_1 = \alpha_1 - \alpha_3$
$\beta_2 = \alpha_2 - \alpha_3$
$\beta_3 = \beta_W$

c) $Y_i = \beta_0 X_0 + \beta_3 X_{i3} + \epsilon_i$ $(i = 1, \dots, 30)$

d) $SSE(X_1 \ X_2 \ X_3) = 72.172$, $SSE(X_3) = 137.307$, $SSR(X_1 \ X_2 | X_3) = 137.307 - 72.172 = 65.135$, $F = (65.135/2)/(72.172/26) = 11.73*$; $F_{.05;2,26} = 3.37$. Reject the null hypothesis.

e) Full model equation

$$Y_i = \beta_0 X_0 + \beta_1 X_{i1} + \beta_2 X_{i2} + \beta_3 X_{i3} + \beta_4 X_{i1} X_{i3} + \beta_5 X_{i2} X_{i3} + \epsilon_i \qquad (i = 1, \dots, 30)$$

Reduced model equation

$$Y_i = \beta_0 X_0 + \beta_1 X_{i1} + \beta_2 X_{i2} + \beta_3 X_{i3} + \epsilon_i \qquad (i = 1, \dots, 30)$$

f) $SSR(X_1 X_3 \ X_2 X_3 | X_1 \ X_2 \ X_3) = 72.172 - 65.210 = 6.962$, $F = (6.962/2)/(65.210/24) = 1.28$; $F_{.25;2,24} = 1.47$. Do not reject the null hypothesis.

References

Addelman, S. Techniques for constructing fractional replicate plans. *Journal of the American Statistical Association*, 1963, *58*, 45–71. [681]*

Addelman, S. The generalized randomized block design. *American Statistician*, 1969, *23*, 35–36. [293]

Addelman, S. Variability of treatments and experimental units in the design and analysis of experiments. *Journal of the American Statistical Association*, 1970, *65*, 1095 –1108. [293]

Addelman, S., and Kempthorne, O. *Orthogonal main-effect plans*. ASTIA Arlington Hall Station, Arlington, Va., 1961. [683]

Allan, F. E., and Wishart, J. A method of estimating yield of a missing plot in field experimental work. *Journal of Agricultural Science*, 1930, *20*, 399–406. [323]

American Anthropological Association. *Principles of professional responsibility*. Washington, D.C.: American Anthropological Association, 1971. [24]

American Psychological Association. *Ethical principles in the conduct of research with human subjects*. Washington, D.C.: Ad Hoc Committee on Ethical Standards in Psychological Research, American Psychological Association, 1973. [24]

American Psychological Association. *Publication manual of the American Psychological Association* (2nd ed.). Washington, D.C.: American Psychological Association, 1974. [141]

---

*Page on which reference is cited.

American Sociological Association. *Code of ethics.* Washington, D.C.: American Sociological Association, 1971. [24]

Appelbaum, M. I., and Cramer, E. M. Some problems in the nonorthogonal analysis of variance. *Psychological Bulletin*, 1974, *81*, 335–343. [401, 415]

Aspin, A. A. Tables for use in comparisons whose accuracy involves two variances separately estimated. *Biometrika*, 1949, *36*, 290–293. [101]

Atiqullah, M. The robustness of the covariance analysis of a one-way classification. *Biometrika*, 1964, *51*, 365–372. [732]

Barber, T. X. *Pitfalls in human research: Ten pivotal points.* New York: Pergamon Press, 1976. [21]

Barnett, V., and Lewis, T. *Outliers in statistical data.* Chichester, England: Wiley, 1978. [139]

Bartlett, M. S. A note on the analysis of covariance. *Journal of Agricultural Science*, 1936, *26*, 488–491. [741]

Bartlett, M. S. Properties of sufficiency and statistical tests. *Proceedings of the Royal Society, A901*, 1937, *160*, 268–282. [78]

Bartlett, M. S. The use of transformations. *Biometrics*, 1947, *3*, 39–52. [83]

Behrens, W. U. Ein Beitrag zur Fehlerberechnung bei wenigen Bechachtungen. *Landw. Jb.*, 1929, *68*, 807–837. [100]

Bennett, C. A., and Franklin, N. L. *Statistical analysis in chemistry and the chemical industry.* New York: Wiley, 1954. [390, 399, 672]

Betz, M. A., and Gabriel, K. R. Type IV errors and analysis of simple effects. *Journal of Educational Statistics*, 1978, *3*, 121–143. [371]

Binder, A. Further considerations on testing the null hypothesis and the strategy and tactics of investigating theoretical models. *Psychological Review*, 1963, *70*, 107–115. [26]

Bock, R. D. *Multivariate statistical methods in behavioral research.* New York: McGraw-Hill, 1975. [4, 210, 216]

Boik, R. J. *Interactions in the analysis of variance: A procedure for interpretation and a monte carlo comparison of univariate and multivariate methods for repeated measures designs.* Unpublished doctoral dissertation, Baylor University, 1975. [100, 263, 371]

Boik, R. J. Interactions, partial interactions, and interaction contrasts in the analysis of variance. *Psychological Bulletin*, 1979, *86*, 1084–1089. [371]

Boik, R. J. A priori tests in repeated measures designs: Effects of nonsphericity. *Psychometrika*, 1981, *46*, 241–255. [263]

Bose, R. C., Shrikhande, S. S., and Parker, E. T. Further results on the construction of mutually orthogonal Latin squares and the falsity of Euler's conjecture. *Canadian Journal of Mathematics*, 1960, *12*, 189–203. [341]

Box, G. E. P. Problems in the analysis of growth and wear curves. *Biometrics*, 1950, *6*, 362–389. [502]

Box, G. E. P. Nonnormality and tests on variances. *Biometrika*, 1953, *40*, 318–335. [78]

Box, G. E. P. Some theorems on quadratic forms applied in the study of analysis of variance problems, I: Effect of inequality of variance in the one-way classification. *Annals of Mathematical Statistics*, 1954a, *25*, 290–302. [77, 78]

Box, G. E. P. Some theorems on quadratic forms applied in the study of analysis of variance problems, II: Effects of inequality of variance and of correlation between errors in the two-way classification. *Annals of Mathematical Statistics*, 1954b, *25*, 484–498. [77, 259]

Box, G. E. P., and Cox, D. R. An analysis of transformations. *Journal of the Royal Statistical Society, Series B*, 1964, *26*, 211–252. [84]

Box, G. E. P., Hunter, W. G., and Hunter, J. S. *Statistics for experimenters: An introduction to design, data analysis, and model building.* New York: Wiley, 1978. [82, 104, 356, 627]

Bozivich, H., Bancroft, T. A., and Hartley, H. O. Power of analysis of variance procedures for certain incompletely specified models. *Annals of Mathematical Statistics,* 1956, *27,* 1017–1043. [396, 398, 399]

Bradu, D., and Gabriel, K. R. Simultaneous statistical inference on interactions in two-way analysis of variance. *Journal of the American Statistical Association,* 1974, *69,* 428–436. [378]

Bratcher, T. L., Moran, M. A., and Zimmer, W. J. Tables of sample sizes in the analysis of variance. *Journal of Quality Technology,* 1970, *2,* 156–164. [145, 840]

Brewer, J. K. On the power of statistical tests in the *American Educational Research Journal. American Educational Research Journal,* 1972, *9,* 391–401. [142]

Brewer, J. K. Issues of power: Clarification. *American Educational Research Journal,* 1974, *11,* 189–192. [142]

Bross, I. Fiducial intervals for variance components. *Biometrics,* 1950, *6,* 136–144. [389]

Brown, M. B., and Forsythe, A. B. The ANOVA and multiple comparisons for data with heterogeneous variances. *Biometrics,* 1974, *30,* 719–724. [120, 122]

Bryant, J. L., and Paulson, A. S. An extension of Tukey's method of multiple comparisons to experimental designs with random concomitant variables. *Biometrika,* 1976, *63,* 631–638. [736, 849]

Budescu, D. V., and Appelbaum, M. I. Variance stabilizing transformations and the power of the *F* test. *Journal of Educational Statistics,* 1981, *6,* 55–74. [79]

Campbell, D. T., and Stanley, J. C. *Experimental and quasi-experimental designs for research.* Chicago: Rand McNally Publishing Co., 1966. [20, 458]

Carlson, J. E., and Timm, N. H. Analysis of nonorthogonal fixed-effects designs. *Psychological Bulletin,* 1974, *81,* 563–570. [255, 401, 415]

Carmer, S. G., and Swanson, M. R. An evaluation of ten pairwise multiple comparison procedures by Monte Carlo methods. *Journal of the American Statistical Association,* 1973, *68,* 66–74. [115, 127]

Cochran, W. G. Recent work on the analysis of variance. *Journal of the Royal Statistical Society,* 1938, *101,* 434–449. [309]

Cochran, W. G. A survey of experimental designs. Mimeographed. U.S. Department of Agriculture, Agricultural Marketing Service, 1940. [309]

Cochran, W. G. The distribution of the largest of a set of estimated variances as a fraction of their total. *Annals of Eugenics,* 1941, *11,* 47–52. [78]

Cochran, W. G. Some consequences when the assumptions for the analysis of variance are not satisfied. *Biometrics,* 1947, *3,* 22–38. [77]

Cochran, W. G. Testing a linear relation among variances. *Biometrics,* 1951, *7,* 17–32. [395]

Cochran, W. G. Analysis of covariance: Its nature and uses. *Biometrics,* 1957, *13,* 261–281. [731, 751]

Cochran, W. G., and Cox, G. M. *Experimental designs.* New York: Wiley, 1957. [75, 100, 508, 625, 635, 636, 656, 674, 681]

Cohen, J. *Statistical power analysis for the behavioral sciences.* New York: Academic Press, 1969. [40, 144]

Cohen, J. Statistical power analysis and research results. *American Educational Research Journal,* 1973, *10,* 225–230. [142]

Collier, R. O., Baker, F. B., Mandeville, G. K., and Hayes, T. F. Estimates of test size for several test procedures based on conventional variance ratios in the repeated measures design. *Psychometrika,* 1967, *32,* 339–353. [260]

Collier, R. O., and Hummel, T. J. *Experimental design and interpretation.* Berkeley, Calif.: McCutchan, 1977. [229]

Connor, W. S., and Young, S. *Fractional factorial designs for experiments with factors at two and three levels*. National Bureau of Standards. Applied Mathematics Series 58, 1961. [681]

Connor, W. S., and Zelen, M. *Fractional factorial experimental designs for factors at three levels*. National Bureau of Standards. Applied Mathematics Series 54, 1959. [674]

Cook, T. D., and Campbell, D. T. *Quasi-experimentation, design and analysis issues for field settings*. Chicago: Rand McNally, 1979. [3,20, 21, 458, 718, 732, 752]

Cooley, W. W., and Lohnes, P. R. *Multivariate data analysis*. New York: Wiley, 1971. [4]

Coons, I. The analysis of covariance as a missing plot technique. *Biometrics*, 1957, *13*, 387–405. [720]

Cornfield, J., and Tukey, J. W. Average values of mean squares in factorials. *Annals of Mathematical Statistics*, 1956, *27*, 907–949. [390]

Cox, D. R. The use of a concomitant variable in selecting an experimental design. *Biometrika*, 1957, *44*, 150–158. [752]

Cox, D. R. *Planning of experiments*. New York: Wiley, 1958. [312]

Cox, G. M. Modernized field designs at Rothamsted. *Soil Science Society of America Proceedings*, 1943, *8*, 20–22. [9]

Cramer, E. M., and Appelbaum, M. I. Nonorthogonal analysis of variance—Once again. *Psychological Bulletin*, 1980, *87*, 51–57. [401, 415]

Cronbach, L. J., Rogosa, D. R., Floden, R. E., and Price, G. G. Analysis of covariance in nonrandomized experiments: Parameters affecting bias. Berkeley, Calif.: Stanford University, Stanford Evaluation Consortium (Occasional paper) 1977. [732]

Dalal, S. R. Simultaneous confidence procedures for univariate and multivariate Beherns-Fisher type problems. *Biometrika*, 1978, *65*, 221–224. [120]

Daniel, C. Fractional replication in industrial research. *Proceedings of the third Berkeley symposium on mathematical statistics and probability*. Berkeley: University of California Press, 1956, *5*, 87. [672]

Daniel, C. *Applications of statistics to industrial experimentation*. New York: Wiley, 1976. [672]

Das, M. N. Analysis of covariance in two-way classification with disproportionate cell frequencies. *Journal of the Indian Society of Agricultural Statistics*, 1953, *5*, 161–178. [747]

Davies, O. L. (ed.). *The design and analysis of industrial experiments*. New York: Hafner Publishing Company, 1956. [625, 672, 674]

Dayton, C. M., and Schafer, W. D. Extended tables of *t* and Chi-square for Bonferroni tests with unequal error allocation. *Journal of the American Statistical Association*, 1973, *68*, 78–83. [107]

Dayton, C. M., Schafer, W. D., and Rogers, B. G. On appropriate uses and interpretations of power analysis: A comment. *American Educational Research Journal*, 1973, *10*, 231–234. [142]

DeLury, D. B. The analysis of Latin squares when some observations are missing. *Journal of the American Statistical Association*, 1946, *41*, 370–389. [326]

Diener, D., and Crandall, R. *Ethics in social and behavioral research*. Chicago: University of Chicago Press, 1978. [24]

Dixon, W. J., and Massey, F. J., Jr. *Introduction to statistical analysis*. New York: McGraw-Hill, 1957. [100]

Dixon, W. J., and Tukey, J. W. Approximate behavior of the distribution of winsorized *t* (trimming/winsorization 2). *Technometrics*, 1968, *10*, 83–98. [139]

Doxtator, C. W., Tolman, W. B., Cormany, C. E., Bush, H. L., and Jensen, V. Standardization of experimental methods. *American Society of Sugar Beet Technology Proceedings*, 1942, *3*, 595–599. [9]

Draper, N. R., and Stoneman, D. M. Estimating missing values in unreplicated two-level factorial and fractional factorial designs. *Biometrics*, 1964, *20*, 443–458. [681]

Duncan, D. B., Multiple range and multiple *F* tests. *Biometrics*, 1955, *11*, 1–42. [105, 125, 126, 825]

Duncan, D. B. Multiple range tests for correlated and heteroscedastic means. *Biometrics*, 1957, *13*, 164–176. [126]

Dunn, O. J. Estimation of the means of dependent variables. *Annals of Mathematical Statistics*, 1958, *29*, 1095–1111. [110]

Dunn, O. J. Multiple comparisons among means. *Journal of the American Statistical Association*, 1961, *56*, 52–64. [106, 111, 842]

Dunn, O. J. On multiple tests and confidence intervals. *Communications in Statistics*, 1974, *3*, 101–103. [118]

Dunnett, C. W. A multiple comparison procedure for comparing several treatments with a control. *Journal of the American Statistical Association*, 1955, *50*, 1096–1121. [112, 826]

Dunnett, C. W. New tables for multiple comparisons with a control. *Biometrics*, 1964, *20*, 482–491. [114, 827]

Dunnett, C. W. Pairwise multiple comparisons in the homogeneous variance, unequal sample size case. *Journal of the American Statistical Association*, 1980, *75*, 789–795. [120, 127]

Dwyer, J. H. Analysis of variance and magnitude of effects: A general approach. *Psychological Bulletin*, 1974, *81*, 731–737. [163]

Edwards, W. Tactical note on the relation between scientific and statistical hypotheses. *Psychological Bulletin*, 1965, *63*, 400–402. [26]

Einot, I., and Gabriel, K. R. A study of the powers of several methods of multiple comparisons. *Journal of the American Statistical Association*, 1975, *70*, 574–583. [127]

Eisenhart, C. M., Hastay, M. W., and Wallis, W. A. *Techniques of statistical analysis*. New York: McGraw-Hill, 1947. [829]

Elashoff, J. D. Analysis of covariance: A delicate instrument. *American Educational Research Journal*, 1969, *6*, 383–401. [732]

Federer, W. T. *Experimental design: Theory and application*. New York: Macmillan, 1955. [7, 9, 312, 342, 522, 590, 600, 625, 629, 632, 636, 656, 741, 743, 747, 750]

Feldt, L. S. A comparison of the precision of three experimental designs employing a concomitant variable. *Psychometrika*, 1958, *23*, 335–354. [751, 752]

Finn, J. D. *A general model for multivariate analysis*. New York: Holt, Rinehart, and Winston, 1974. [4, 210]

Finney, D. J. The fractional replication of factorial arrangements. *Annals of Eugenics*, 1945, *12*, 291–301. [664]

Finney, D. J. Recent developments in the design of field experiments. III. Fractional replication. *Journal of Agricultural Science*. 1946a, *36*, 184–191. [664]

Finney, D. J. Standard errors of yields adjusted for regression on an independent measurement. *Biometrics Bulletin*, 1946b, *2*, 53–55. [735]

Finney, D. J. Stratification, balance, and covariance. *Biometrics*, 1957, *13*, 373–386. [752]

Finney, D. J. *An introduction to the theory of experimental design*. Chicago: The University of Chicago Press, 1960. [20, 672]

Fisher, R. A. International Mathematical Conference. Toronto, 1924. [53]

Fisher, R. A. The fiducial agreement in statistical inference. *Annals of Eugenics*, 1935, *6*, 391–398. [9, 100, 271, 327]

Fisher, R. A. *The design of experiments*. Edinburgh: Oliver and Boyd Ltd., 1949. [115]

Fisher, R. A., and MacKenzie, W. A. The correlation of weekly rainfall. *Quarterly Journal of the Royal Meterological Society*, 1922, *48*, 234–245. [9]

Fisher, R. A., and MacKenzie, W. A. Studies in crop variation. II. The manurial response of different potato varieties. *Journal of Agricultural Science*, 1923, *13*, 311–320. [9]

Fisher, R. A., and Yates, F. The six by six Latin squares. *Proceedings of the Cambridge Philosophical Society*, 1934, *30*, 492–507. [311, 341]

Fisher, R. A., and Yates, F. *Statistical tables for biological, agricultural and medical research.* Edinburgh: Oliver & Boyd Ltd., 1963. [309, 311, 312, 341, 808, 813, 821, 830]

Freeman, M. F., and Tukey, J. W. Transformations related to the angular and the square root. *Annals of Mathematical Statistics*, 1950, *21*, 607–611. [82]

Gabriel, K. R. A procedure for testing the homogeneity of all sets of means in analysis of variance. *Biometrics*, 1964, *20*, 459–477. [369, 370]

Gabriel, K. R. Simultaneous test procedure—Some theory of multiple comparisons. *Annals of Mathematical Statistics*, 1969, *40*, 224–250. [369, 370]

Gabriel, K. R. Comment. *Journal of the American Statistical Association,* 1978a, *73*, 485–487. [127]

Gabriel, K. R. A simple method of multiple comparisons of means. *Journal of the American Statistical Association*, 1978b, *73*, 724–729. [118]

Gabriel, K. R., Putter, J., and Wax, Y. Simultaneous confidence intervals for product-type interaction contrasts. *Journal of the Royal Statistical Society (B)*, 1973, *35*, 234–244. [371]

Gaito, J. Unequal intervals and unequal n in trend analyses. *Psychological Bulletin*, 1965, *63*, 125–127. [774]

Games, P. A. Multiple comparisons of means. *American Educational Research Journal*, 1971, *8*, 531–565. [97, 127]

Games, P. A. Type IV errors revisited. *Psychological Bulletin*, 1973, *80*, 304–307. [371, 396]

Games, P. A. An improved t table for simultaneous control on g contrasts. *Journal of the American Statistical Association*, 1977, *72*, 531–534. [110, 111, 845]

Games, P. A., and Howell, J. F. Pairwise multiple comparison procedures with unequal n's and/or variances: A Monte Carlo study. *Journal of Educational Statistics*, 1976, *1*, 113–125. [120, 127]

Games, P. A., Keselman, H. J., and Clinch, J. J. Tests for homogeneity of variance in factorial designs. *Psychological Bulletin*, 1979, *86*, 978–984. [79]

Games, P. A., and Lucas, P. A. Power and the analysis of variance of independent groups on nonnormal and normally transformed data. *Educational and Psychological Measurement*, 1966, *16*, 311–327. [76]

Gary, H. E., Jr. *The effects of departures from circularity on type I error rates and power for randomized block factorial experimental designs.* Unpublished doctoral dissertation, Baylor University, 1981. [261]

Gaylor, D. W., and Hopper, F. N. Estimating the degrees of freedom for linear combinations of mean squares by Satterthwaite's formula. *Technometrics*, 1969, *11*, 691–706. [395]

Geisser, S., and Greenhouse, S. W. An extension of Box's results on the use of the F distribution in multivariate analysis. *Annals of Mathematical Statistics*, 1958, *29*, 885–891. [261]

Gill, J. L. Design and analysis of experiments in the animal and medical sciences. Ames, Iowa: Iowa State University Press, 1978. [459, 483, 625]

Glass, G. V., and Hakstian, A. R. Measures of association in comparative experiments: Their development and interpretation. *American Educational Research Journal*, 1969, *6*, 403–414. [163]

Glass, G. V., Peckham, P. D., and Sanders, J. R. Consequences of failure to meet assumptions underlying the analysis of variance and covariance. *Review of Educational Research*, 1972, *42*, 237–288. [75, 732]

Gollob, H. F. Confounding of sources of variation in factor-analytic techniques. *Psychological Bulletin*, 1968a, *70*, 330–334. [378]

Gollob, H. F. A statistical model which combines features of factor analytic and analysis of variance techniques. *Psychometrika*, 1968b, *33*, 73–115. [378]

Gosset, W. S. The probable error of the mean. *Biometrika*, 1908, *6*, 1–25. [55]

Gosslee, I. J., and Lucas, H. L. Analysis of variance of disproportionate data when interaction is present. *Biometrics*, 1965, *21*, 115–133. [415, 419]

Gourlay, N. Covariance analysis and its application in psychological research. *British Journal of Statistical Psychology*, 1953, *6*, 25. [752]

Gower, J. D. Variance component estimation for unbalanced hierarchical classification. *Biometrics*, 1962, *18*, 537–542. [459]

Grandage, A. Orthogonal coefficients for unequal intervals. *Biometrics*, 1958, *14*, 287–289. [774]

Grant, D. A. Testing the null hypothesis and the strategy and tactics of investigating theoretical models. *Psychological Review*, 1962, *69*, 54–61. [26]

Graybill, F. A. *An introduction to linear models*. New York: McGraw-Hill, 1961. [216]

Graybill, F. A. *Introduction to matrices with applications in statistics*. Belmont, Calif.: Wadsworth, 1969. [173, 778]

Green, P. E., and Carroll, J. D. *Mathematical tools for applied multivariate analysis*. New York: Academic Press, 1976. [173, 778, 796, 798]

Green, B. F., and Tukey, J. Complex analysis of variance: General problems. *Psychometrika*, 1960, *25*, 127–152. [396, 398]

Grubbs, F. E. Procedures for detecting outlying observations in samples. *Technometrics*, 1969, *11*, 1–21. [139]

Guide for the care and use of laboratory animals. Washington, D.C.: U.S. Government Printing Office, 1974. [25]

Haggard, E. A. *Intraclass correlation and the analysis of variance*. New York: Dryden Press, 1958. [162]

Hahn, G. J., and Shapiro, S. S. *Catalog and computer program for the design and analysis of orthogonal symmetric and asymmetric fractional factorial experiments*. Schenectady, N.Y.: General Electric Research and Development Center, Report No. 66-C-165, 1966. [683]

Halderson, J. S., and Glasnapp, D. R. Generalized rules for calculating the magnitude of an effect in factorial and repeated measures ANOVA designs. *American Educational Research Journal*, 1972, *9*, 301–310. [163]

Harris, R. J. *A primer of multivariate statistics*. New York: Academic Press, 1975. [4]

Harter, H. L. Error rates and sample sizes for range tests in multiple comparisons. *Biometrics*, 1957, *13*, 511–536. [102]

Harter, H. L. Multiple comparison procedures for interactions. *American Statistician*, 1970, *24*(5), 30–32. [371]

Harter, H. L., and Lum, M. D. An interpretation and extension of Tukey's one degree of freedom for nonadditivity. *WADC Technical Report 62-313*. Wright-Patterson AFB, Ohio, 1962. [400]

Hartley, H. O. Testing the homogeneity of a set of variances. *Biometrika*, 1940, *31*, 249–255. [78]

Hartley, H. O. The maximum F-ratio as a short-cut test for heterogeneity of variance. *Biometrika*, 1950, *37*, 308–312. [78]

Hays, W. L. *Statistics*. New York: Holt, Rinehart, and Winston, 1981. [162]

Hazel, L. N. The covariance analysis of multiple classification tables with unequal subclass numbers. *Biometrics*, 1946, *2*, 21–25. [747]

Henderson, C. R. Estimation of variance and covariance components. *Biometrics*, 1953, *9*, 226–252. [483]

Hochberg, Y. Some generalizations of the T-method in simultaneous inference. *Journal of Multivariate Analysis*, 1974, *4*, 224–234. [118]

Hochberg, Y. A modification of the T-method of multiple comparisons for a one-way layout with unequal variances. *Journal of the American Statistical Association*, 1976, *71*, 200–203. [120]

Hocking, R. R. A discussion of the two-way mixed model. *The American Statistician*, 1973, *24* (4), 148–152. [364]

Hocking, R. R., and Speed, F. M. A full rank analysis of some linear model problems. *Journal of the American Statistical Association*, 1975, *70*, 706–712. [225]

Hohn, F. E. *Elementary matrix algebra*. New York: Macmillan, 1964. [173, 778, 798]

Hopkins, K. D., and Anderson, B. L. A guide to multiple-comparison techniques: Criteria for selecting the "method of choice." *Journal of Special Education*, 1973, *7*, 319–328. [127]

Hopkins, K. D., and Chadbourn, R. A. A schema for proper utilization of multiple comparisons in research and a case study. *American Educational Research Journal*, 1967, *4*, 407–412. [127]

Hotelling, H. The generalization of Student's ratio. *Annals of Mathematical Statistics*, 1931, *2*, 360–378. [261]

Hsu, T., and Feldt, L. S. The effect of limitations on the number of criterion score values on the significance level of the F-test. *American Education Research Journal*, 1969, *6*, 515–527. [75]

Huck, S. W., and Sandler, H. M. *Rival hypotheses*. New York: Harper and Row, 1979. [21]

Huitema, B. E. *The analysis of covariance and alternatives*. New York: Wiley, 1980. [718, 732, 733, 734]

Huynh, H. Some approximate tests for repeated measurement designs. *Psychometrika*, 1978, *43*, 161–175. [260]

Huynh, H., and Feldt, L. S. Conditions under which mean square ratios in repeated measurement designs have exact F-distributions. *Journal of the American Statistical Association*, 1970, *65*, 1582–1589. [256, 502]

Huynh, H., and Feldt, L. S. Estimation of the Box correction for degrees of freedom from sample data in randomized block and split-plot designs. *Journal of Educational Statistics*, 1976, *1*, 69–82. [260]

Huynh, H., and Mandeville, G. K. Validity conditions in repeated measures designs. *Psychological Bulletin*, 1979, *86*, 964–973. [502]

Imhof, J. P. Testing the hypothesis of fixed main effects in Scheffé's mixed model. *Annals of Mathematical Statistics*, 1962, *33*, 1086–1095. [261]

Jennings, E., and Ward, J. H., Jr. Hypothesis identification in the case of the missing cell. *American Statistician*, 1982, *36* (1), 25–27. [421]

Johnson, N. J. Modified t tests and confidence intervals for asymmetrical populations. *Journal of the American Statistical Association*, 1978, *73*, 536–544. [100]

Johnson, N. L., and Leone, F. C. *Statistics and experimental design in engineering and the physical sciences*, Vol. II. New York: Wiley, 1964. [625, 681]

Johnson, P. O., and Neyman, J. Tests of certain linear hypotheses and their application to some educational problems. *Statistical Research Memoirs*, 1936, *1*, 57–93. [733]

Kempthorne, O. A simple approach to confounding and fractional replication in factorial experiments. *Biometrika*, 1947, *34*, 255–272. [664]

Kempthorne, O. *The design and analysis of experiments*. New York: Wiley, 1952. [325, 326, 573, 608, 625, 627, 629, 636, 681]

Kendall, M. G. *The advanced theory of statistics*, Vol. II. London: Charles Griffin and Company, Ltd., 1948. [733]

Keren, G., and Lewis, C. A comment on coding in nonorthogonal designs. *Psychological Bulletin*, 1977, *84*, 346–348. [401]

Keselman, H. J., Games, P. A., and Rogan, J. C. An addendum to "A comparison of modified-Tukey and Scheffé methods of multiple comparisons for pairwise contrasts." *Journal of the American Statistical Association*, 1979, *74*, 626–627. [120, 121, 127]

Keselman, H. J., and Rogan, J. C. The Tukey multiple comparison test: 1953–1976. *Psychological Bulletin*, 1977, *84*, 1050–1056. [116, 120, 127]

Keselman, H. J., and Rogan, J. C. A comparison of the modified-Tukey and Scheffé methods of multiple comparisons for pairwise contrasts. *Journal of the American Statistical Association*, 1978, *73*, 47–51. [127]

Keuls, M. The use of studentized range in connection with an analysis of variance. *Euphytica*, 1952, *1*, 112–122. [123]

Kirk, R. E. *Statistical issues: A reader for the behavioral sciences*. Monterey, Calif.: Brooks/Cole, 1972. [9]

Kirk, R. E. *Introductory statistics*. Monterey, Calif.: Brooks/Cole, 1978. [39, 42]

Kohr, R. L., and Games, P. A. Testing complex a priori contrasts in means from independent samples. *Journal of Educational Statistics*, 1977, *1*, 207–216. [100, 101]

Kramer, C. Y. Extension of multiple range test to group means with unequal numbers of replications. *Biometrics*, 1956, *12*, 307–310. [118, 119, 126]

Kutner, M. H. Hypothesis testing in linear models (Eisenhart model I). *American Statistician*, 1974, *28*(3), 98–100. [401]

Leonard, W. H., and Clark, A. G. *Field plot techniques*. Minneapolis, Minnesota: Burgess Publishing Co., 1939. [9]

Levin, J. R. Misinterpreting the significance of "explained variation." *American Psychologist*, 1967, *22*, 675–676. [163]

Levin, J. R., and Marascuilo, L. A. Type IV errors and interactions. *Psychological Bulletin*, 1972, *78*, 368–374. [371, 396]

Levin, J. R., and Marascuilo, L. A. Type IV errors and Games. *Psychological Bulletin*, 1973, *80*, 308–309. [371, 396]

Lewis, C., and Keren, G. You can't have your cake and eat it too: Some considerations of the error term. *Psychological Bulletin*, 1977, *84*, 1150–1154. [401]

Li, J. C. R. Design and statistical analysis of some confounded factorial experiments. *Iowa Agricultural Experiment Station Research Bulletin*. Ames, Iowa, 1944, *333*, 449–492. [629, 635, 636]

Lindquist, E. F. *Design and analysis of experiments in psychology and education*. Boston: Houghton Mifflin, 1953. [9, 52, 75, 76, 77, 490]

Lord, J. M. Large-sample covariance analysis when the control variable is fallible. *Journal of the American Statistical Association*, 1960, *55*, 307–321. [732]

Lunney, G. H. Using analysis of variance with a dichotomous dependent variable: An empirical study. *Journal of Educational Measurement*, 1970, *7*, 263–269. [75]

Mandel, J. The partitioning of interaction in analysis of variance. *Journal of Research of the National Bureau of Standards*, 1969, *37B*, 309–328. [378]

Mandel, J. A new analysis of variance model for nonadditive data. *Technometrics*, 1971, *13*, 1–18. [378]

Marascuilo, L. A., and Levin, J. R. Appropriate post hoc comparisons for interactions and nested hypotheses in analysis of variance designs: The elimination of type IV errors. *American Educational Research Journal*, 1970, *7*, 397–421. [369, 371]

Marascuilo, L. A., and Levin, J. R. The simultaneous investigation of interaction and nested hypotheses in two-factor analysis of variance designs. *American Educational Research Journal*, 1976, *13*, 61–65. [371, 396]

Marascuilo, L. A., and McSweeney, M. *Nonparametric and distribution-free methods for the social sciences*. Monterey, Calif.: Brooks/Cole, 1977. [81]

Margolin, B. H. Orthogonal main-effect 2^n3^m designs and two-factor interaction aliasing. *Technometrics*, 1968, *10*, 559–573. [683]

Martin, C. G., and Games, P. A. ANOVA tests for homogeneity of variance: Nonnormality and unequal samples. *Journal of Educational Statistics*, 1977, *2*, 187–206. [79]

Mauchley, J. W. Significance test for sphericity of a normal n-variate distribution. *Annals of Mathematical Statistics*, 1940, *11*, 204–209. [259]

Maxwell, S. E. Pairwise multiple comparisons in repeated measures designs. *Journal of Educational Statistics*, 1980, *5*, 269–287. [263]

McHugh, R. B., and Ellis, D. S. The "postmortem" testing of experimental comparisons. *Psychological Bulletin*, 1955, *52*, 425–428. [105]

Mead, R., Bancroft, T. A., and Han, C. Power of analysis of variance test procedures for incompletely specified fixed models. *Annals of Statistics*, 1975, *3*, 797–808. [399]

Mendoza, G. L., Toothaker, L. E., and Crain, B. R. Necessary and sufficient conditions for F ratios in the $L \times J \times K$ factorial design with two repeated factors. *Journal of the American Statistical Association*, 1976, *71*, 992–993. [259]

Meyer, D. L. Issues of power: Rejoinder. *American Educational Research Journal*, 1974a, *11*, 193–194. [142]

Meyer, D. L. Statistical tests and surveys of power: A critique. *American Educational Research Journal*, 1974b, *11*, 179–188. [142]

Miller, R. G., Jr. *Simultaneous statistical inference*. New York: McGraw-Hill, 1966. [103, 106, 120]

Miller, R. G., Jr. Developments in multiple comparisons. *Journal of the American Statistical Association*, 1977, *72*, 779–788. [116]

Monlezun, C. J. Two-dimensional plots for interpreting interactions in the three-factor analysis of variance model. *American Statistician*, 1979, *33*(2), 63–69. [359]

Montgomery, D. C. *Design and analysis of experiments*. New York: Wiley, 1976. [625]

Mosteller, F., and Bush, R. R. Selected quantitative techniques. In G. Lindzey (ed.), *Handbook of social psychology*. Reading, Mass.: Addison-Wesley, 1954, 289–334. [82]

Nair, K. R. On a method of getting confounded arrangements in the general symmetrical type of experiments. *Sankhyā*, 1938, *4*, 121–138. [625, 629]

Nair, K. R. Balanced confounded arrangements for the 5^n type of experiment. *Sankhyā*, 1940, *5*, 57–70. [625]

National Bureau of Standards. *Fractional factorial experiment designs for factors at two levels*. National Bureau of Standards, Applied Mathematics Series 48, 1957. [674]

Neter, J., and Wasserman, W. *Applied linear statistical models*. Homewood, Ill.: Richard D. Irwin, 1974. [151]

Newman, D. The distribution of the range in samples from a normal population, expressed in terms of an independent estimate of standard deviation. *Biometrika*, 1939, *31*, 20–30. [123]

Norton, H. W. The 7×7 squares. *Annals of Eugenics*, 1939, *9*, 269–307. [311]

O'Brien, R. G. Comment on "Some problems in the nonorthogonal analysis of variance." *Psychological Bulletin*, 1976, *83*, 72–74. [401]

O'Brien, R. G. A general ANOVA method for robust tests of additive models for variances. *Journal of the American Statistical Association*, 1979, *74*, 877–880. [79]

Olds, E. G., Mattson, T. B., and Odeh, R. E. Notes on the use of transformations in the analysis of variance. *WADC Technical Report 56-308*, Wright-Patterson AFB, Ohio, 1956. [82]

Overall, J. E., and Spiegel, D. K. Concerning least squares analysis of experimental data. *Psychological Bulletin*, 1969, *72*, 311–322. [401, 412]

Overall, J. E., and Spiegel, D. K. Comment on "Regression analysis of proportional data." *Psychological Bulletin*, 1973, *80*, 28–30. [401]

Overall, J. E., Spiegel, D. K., and Cohen, J. Equivalence of orthogonal and nonorthogonal analysis of variance. *Psychological Bulletin*, 1975, *82*, 182–186. [401]

Palermo, D. S., and Jenkins, J. J. *Word association norms—Grade school through college*. Minneapolis: University of Minnesota Press, 1964. [343]

Paull, A. E. On a preliminary test for pooling mean squares in the analysis of variance. *Annals of Mathematical Statistics*, 1950, *21*, 539–556. [398, 399]

Pearson, E. S. The analysis of variance in cases of non-normal variation. *Biometrika*, 1931, *23*, 114–133. [75]

Pearson, E. S., and Hartley, H. O. The probability integral of the range in samples of *n* observations from a normal population. *Biometrika*, 1942, *32*, 301–310. [102]

Pearson, E. S., and Hartley, H. O. Tables of the probability integral of the studentized range. *Biometrika*, 1943, *33*, 89–99. [102]

Pearson, E. S., and Hartley, H. O. Charts of the power function for analysis of variance tests, derived from the noncentral *F*-distribution. *Biometrika*, 1951, *38*, 112–130. [832]

Pearson, E. S., and Hartley, H. O. *Biometrika tables for statisticians*, Vol. 1, 2nd ed. New York: Cambridge, 1958. [814, 822, 828]

Peiser, A. M. Asymptotic formulas for significance levels of certain distributions. *Annals of Mathematical Statistics*, 1943, *14*, 56–62. [108]

Ramsey, P. H. Power differences between pairwise multiple comparisons. *Journal of the American Statistical Association*, 1978, *73*, 479–487. [127]

Rawlings, R. R., Jr. Comments on the Overall and Spiegel paper. *Psychological Bulletin*, 1973, *79*, 168–169. [401]

Rawlings, R. R., Jr. Note on nonorthogonal analysis of variance. *Psychological Bulletin*, 1972, *77*, 373–374. [401]

Robson, D. S., and Atkinson, G. F. Individual degrees of freedom for testing homogeneity of regression coefficients in a one-way analysis of covariance. *Biometrics*, 1960, *16*, 593–605. [733]

Rogan, J. C., and Keselman, H. J. Is the ANOVA *F*-test robust to variance heterogeneity when sample sizes are equal?: An investigation via a coefficient of variation. *American Educational Research Journal*, 1977, *14*, 493–498. [77]

Rogan, J. C., Keselman, H. J., and Mendoza, J. L. Analysis of repeated measurements. *British Journal of Mathematical and Statistical Psychology*, 1979, *32*, 269–286. [260]

Rosenthal, R., and Rosnow, R. L. *Artifact in behavioral research*. New York: Academic Press, 1969. [21]

Rouanet, H., and Lépine, D. Comparison between treatments in a repeated-measurement design: ANOVA and multivariate methods. *British Journal of Mathematical and Statistical Psychology*, 1970, *23*, 147–163. [256, 258, 446]

Roy, S. N. On a heuristic method of test construction and its use in multivariate analysis. *Annals of Mathematical Statistics*, 1953, *24*, 220–238. [369]

Ryan, T. A. Multiple comparisons in psychological research. *Psychological Bulletin*, 1959, *56*, 26–47. [105]

Ryan, T. A. Significance tests for multiple comparisons of proportions, variances, and other statistics. *Psychological Bulletin*, 1960, *57*, 318–328. [105]

Ryan, T. A. The experiment as the unit for computing rates of error. *Psychological Bulletin*, 1962, *59*, 301–305. [105]

Sade, A. An omission in Norton's list of 7×7 squares. *Annals of Mathematical Statistics*, 1951, *22*, 306–307. [311]

Saretsky, G. The OEO P.C. experiment and the John Henry effect. *Phi Delta Kappan*, 1972, *53*, 579–581. [23]

Satterthwaite, F. E. An approximate distribution of estimates of variance components. *Biometrics Bulletin*, 1946, *2*, 110–114. [100, 394]

Scheffé, H. A method for judging all contrasts in the analysis of variance. *Biometrika*, 1953, *40*, 87–104. [111, 118, 121]

Scheffé, H. *The analysis of variance*. New York: Wiley, 1959. [78, 119, 126, 147, 163, 165, 173, 210, 263, 318, 389, 396, 399]

Scheffé, H. Practical solutions of the Behrens-Fisher problem. *Journal of the American Statistical Association*, 1970, *65*, 1501–1508. [101]

Searle, S. R. *Matrix algebra for the biological sciences*. New York: Wiley, 1966. [173, 178, 210, 778, 798]

Searle, S. R. *Linear models*. New York: Wiley, 1971a. [195, 389, 407, 414, 415, 483, 733, 796]

Searle, S. R. Topics in variance component estimation. *Biometrics*, 1971b, *27*, 1–76. [389]

Shaffer, J. P. Comparisons of means: An *F* test followed by a modified multiple range procedure. *Journal of Educational Statistics*, 1979, *4*, 14–23. [125]

Šidák, Z. Rectangular confidence regions for the means of multivariate normal distributions. *Journal of the American Statistical Association*, 1967, *62*, 626–633. [110, 118]

Smith, H. F. The problem of comparing the results of two experiments with unequal errors. *Journal of Scientific and Industrial Research*, 1936, *9*, 211–212. [100]

Smith, H. J. Interpretation of adjusted treatment means and regression in analysis of covariance. *Biometrics*, 1957, *13*, 282–308. [718]

Snedecor, G. W. *Analysis of variance and covariance*. Ames, Iowa: Iowa State University Press, 1934. [53]

Snedecor, G. W. *Statistical methods applied to experiments in agriculture and biology*. Ames, Iowa: Iowa State University Press, 1956. [459]

Snedecor, G. W., and Cochran, W. G. *Statistical methods*, 6th ed. Ames, Iowa: Iowa State University Press, 1967. [732, 743]

Snee, R. D. On the analysis of responsive curve data. *Technometrics*, 1972, *14*, 47–62. [378]

Speed, F. M. *A new approach to the analysis of linear models*. NASA Technical Memorandum, NASA TM X-58030, June 1969. [225]

Speed, F. M., and Hocking, R. R. The use of the *R* (   )-notation with unbalanced data. *American Statistician*, 1976, *30*(1), 30–33. [195]

Speed, F. M., Hocking, R. R., and Hackney, O. P. Methods of analysis of linear models with unbalanced data. *Journal of the American Statistical Association*, 1978, *73*, 105–112. [195, 415]

Spjøtvoll, E. Joint confidence intervals for all linear functions of means in ANOVA with unknown variances. *Biometrika*, 1972, *59*, 684–685. [120]

Spjøtvoll, E., and Stoline, M. R. An extension of the *T*-method of multiple comparisons to include the cases with unequal sample sizes. *Journal of the American Statistical Association*, 1973, *68*, 975–978. [118]

Steel, R. G. D., and Federer, W. T. Yield-stand analyses. *Journal of the Indian Society of Agricultural Statistics*, 1955, *7*, 27–45. [733]

Stoline, M. R. Tables of the studentized augmented range and applications to problems of multiple comparisons. *Journal of the American Statistical Association*, 1978, *73*, 656–660. [119, 846]

Stoloff, P. H. Correcting for heterogeneity of covariance for repeated measures designs of the analysis of variance. *Educational and Psychological Measurement*, 1970, *30*, 909–924. [260]

Student. Errors of routine analysis. *Biometrika*, 1927, *19*, 151–164. [123]

Tamhane, A. C. Multiple comparisons in model I one-way ANOVA with unequal variances. *Communications in Statistics*, 1977, *A6*(1), 15–32. [120]

Tamhane, A. C. A comparison of procedures for multiple comparisons of means with unequal variances. *Journal of the American Statistical Association*, 1979, *74*, 471–480. [120, 121, 127]

Tang, P. C. The power function of the analysis of variance tests with tables and illustrations of their use. *Statistics Research Memorandum*, 1938, *2*, 126–149. [142]

Tatsuoka, M. M. *Multivariate analysis: Techniques for educational and psychological research.* New York: Wiley, 1971. [4]

Tatsuoka, M. M. *The general linear model: A "new" trend in analysis of variance.* Champaign, Ill.: Institute for Personality and Ability Testing, 1975. [216]

Taylor, J. Errors of treatment comparisons when observations are missing. *Nature*, 1948, *162*, 262–263. [270]

Taylor, J. The comparison of pairs of treatments in split-plot experiments. *Biometrika*, 1950, *37*, 443–444. [509]

Thomas, D. A. H. Error rates in multiple comparisons among means—Results of a simulation exercise. *Applied Statistics*, 1974, *23*, 284–294. [127]

Timm, N. H. *Multivariate analysis with applications in education and psychology.* Monterey, Calif.: Brooks/Cole, 1975. [4, 173, 210, 218, 778]

Timm, N. H., and Carlson, J. E. Analysis of variance through full rank models. *Multivariate Behavioral Research Monographs*, 1975, No. 75-1, 1975. [225, 282, 414, 415]

Tukey, J. W. One degree of freedom for nonadditivity. *Biometrics*, 1949, *5*, 232–242. [82, 250]

Tukey, J. W. The problem of multiple comparisons. Ditto, Princeton University, 396 pp., 1953. [111, 116, 118, 119, 147, 263, 389]

Tukey, J. W. Queries. *Biometrics*, 1955, *11*, 111–113. [318]

Tukey, J. W. *Exploratory data analysis.* Reading, Mass.: Addison-Wesley, 1977. [136]

Urquhart, N. S., Weeks, D. L., and Henderson, C. R. Estimation associated with linear models, a revisitation. *Communications in Statistics*, 1973, *1*, 303–330. [225]

Ury, H. K. A comparison of four procedures for multiple comparisons among means (pairwise contrasts) for arbitrary sample sizes. *Technometrics*, 1976, *18*, 89–97. [118, 119, 127]

Ury, H. K., and Wiggins, A. D. Large sample and other multiple comparisons among means. *British Journal of Mathematical and Statistical Psychology*, 1971, *24*, 174–194. [120, 121]

Vaughan, G. M., and Corballis, M. C. Beyond tests of significance: Estimating strength of effects in selected ANOVA designs. *Psychological Bulletin*, 1969, *72*, 204–213. [163]

Wang, Y. Y. Probabilities of the type I errors of the Welch tests for the Behrens-Fisher problem. *Journal of the American Statistical Association*, 1971, *66*, 605–608. [101]

Webb, E. J., Campbell, D. T., Schwartz, R. D., and Sechrest, L. *Unobtrusive measures: nonreactive research in the social sciences.* Chicago: Rand McNally, 1966. [21]

Weinberger, C. W. Protection of human subjects. *Federal Register*, 30 May 39 (105): 18914-20 (45CFR, part 46), 1974. [24]

Weinberger, C. W. Protection of human subjects: Technical amendments. *Federal Register*, 13 March 40(50): 11854-58 (45CFR, part 46), 1975. [24]

Weisberg, H. I. Statistical adjustments and uncontrolled studies. *Psychological Bulletin*, 1979, *86*, 1149–1164. [718, 732]

Welch, B. L. The generalization of Student's problem when several different population variances are involved. *Biometrika*, 1947, *34*, 28–35. [100, 101, 122]

Welch, B. L. Further note on Mrs. Aspin's tables and on certain approximations to the tabled functions. *Biometrika*, 1949, *36*, 293–296. [100]

Wilk, M. B. The randomization analysis of a generalized randomized block design. *Biometrika*, 1955, *42*, 70–79. [293]

Wilk, M. B., and Kempthorne, O. Analysis of variance: Preliminary tests, pooling and linear models. *Wright Air Development Center Technical Report*, 1956, *2*, 55–244. [293]

Wilk, M. B., and Kempthorne, O. Nonadditivities in a Latin square design. *Journal of the American Statistical Association*, 1957, *52*, 218–236. [318]

Wilson, W. A note on the inconsistency inherent in the necessity to perform multiple comparisons. *Psychological Bulletin*, 1962, *59*, 296–300. [105]

Wilson, W. R., and Miller, H. A. A note on the inconclusiveness of accepting the null hypothesis. *Psychological Review*, 1964, *71*, 238–242. [26]

Wilson, W., Miller, H. L., and Lower, J. S. Much ado about the null hypothesis. *Psychological Bulletin*, 1967, *67*, 188–196. [26]

Winer, B. J. *Statistical principles in experimental design.* New York: McGraw-Hill, 1971. [9, 625, 627, 636, 680, 685, 748]

Yates, F. The analysis of replicated experiments when field results are incomplete. *Empire Journal of Experimental Agriculture*, 1933, *1*, 129–142. [268, 323, 325]

Yates, F. Incomplete Latin squares. *Journal of Agricultural Science*, 1936, *26*, 301–315. [326]

Yates, F. The design and analysis of factorial experiments. *Imperial Bureau of Soil Science Technical Communication No. 35*, Harpenden, England, 1937. [573, 591, 626, 629, 635, 636, 656]

Yates, F., and Hale, R. W. The analysis of Latin squares when two or more rows, columns, or treatments are missing. *Journal of the Royal Statistical Society Supplement*, 1939, *6*, 67–79. [326]

INDEX

905